SLEEP AND HEALTH

Sleep and Health

Edited by

Michael A. Grandner

Academic Press is an imprint of Elsevier
125 London Wall, London EC2Y 5AS, United Kingdom
525 B Street, Suite 1650, San Diego, CA 92101, United States
50 Hampshire Street, 5th Floor, Cambridge, MA 02139, United States
The Boulevard, Langford Lane, Kidlington, Oxford OX5 1GB, United Kingdom

© 2019 Elsevier Inc. All rights reserved.
Exception to the above, Chapter 36: 2019 Published by Elsevier Inc.

No part of this publication may be reproduced or transmitted in any form or by any means, electronic or mechanical, including photocopying, recording, or any information storage and retrieval system, without permission in writing from the publisher. Details on how to seek permission, further information about the Publisher's permissions policies and our arrangements with organizations such as the Copyright Clearance Center and the Copyright Licensing Agency, can be found at our website: www.elsevier.com/permissions.

This book and the individual contributions contained in it are protected under copyright by the Publisher (other than as may be noted herein).

Notices
Knowledge and best practice in this field are constantly changing. As new research and experience broaden our understanding, changes in research methods, professional practices, or medical treatment may become necessary.

Practitioners and researchers must always rely on their own experience and knowledge in evaluating and using any information, methods, compounds, or experiments described herein. In using such information or methods they should be mindful of their own safety and the safety of others, including parties for whom they have a professional responsibility.

To the fullest extent of the law, neither the Publisher nor the authors, contributors, or editors, assume any liability for any injury and/or damage to persons or property as a matter of products liability, negligence or otherwise, or from any use or operation of any methods, products, instructions, or ideas contained in the material herein.

Library of Congress Cataloging-in-Publication Data
A catalog record for this book is available from the Library of Congress

British Library Cataloguing-in-Publication Data
A catalogue record for this book is available from the British Library

ISBN 978-0-12-815373-4

For information on all Academic Press publications
visit our website at https://www.elsevier.com/books-and-journals

Publisher: Stacy Masucci
Acquisition Editor: Katie Chan
Editorial Project Manager: Carlos Rodriguez
Production Project Manager: Punithavathy Govindaradjane
Cover Designer: Christian Bilbow

Typeset by SPi Global, India

Dedication

For Nirav. I wish we could have done this together.

Contents

Contributors	xvii
Preface	xxi
Acknowledgments	xxiii

Part I
General concepts in sleep health

1. The basics of sleep physiology and behavior

Andrew S. Tubbs, Hannah K. Dollish, Fabian Fernandez, Michael A. Grandner

Introduction	3
The definition of sleep	3
Conceptualizing sleep as a health behavior	3
Conceptualizing sleep as a physiological process	4
Sleep and circadian rhythms	6
Basic sleep physiology	7
Quantifying sleep	8
Conclusion	10
References	10

2. Epidemiology of insufficient sleep and poor sleep quality

Michael A. Grandner

Sleep at the population level	11
Defining insufficient sleep	11
Prevalence of insufficient sleep	12
Insufficient sleep in the population	12
Insufficient sleep by age	13
Insufficient sleep by sex	14
Insufficient sleep by race/ethnicity	14
Insufficient sleep by socioeconomic status	14
Insufficient sleep by geography	15
Key limitations to population estimates of insufficient sleep	15
Prevalence of poor sleep quality	16
Prevalence of sleep disorders	16
Prevalence of sleep complaints	17
Summary and conclusions	17
References	18

3. Sex differences in sleep health

Jessica Meers, Jacqueline Stout-Aguilar, Sara Nowakowski

Introduction	21
Sex differences in infant sleep	21
Sex differences in childhood sleep	22
Sex differences in adolescent sleep	22
Sex differences in young adult sleep	24
Sex differences in middle-aged sleep	24
Sex differences in older adult sleep	25
Conclusion	26
References	26

4. Sleep and health in older adults

Junxin Li, Nalaka S. Gooneratne

Introduction	31
Sleep changes in normal aging	31
Changes in sleep parameters	31
Changes in circadian rhythm	31
Changes in sleep homeostasis	31
Common sleep disturbances in older adults	32
Insomnia	32
Sleep-disordered breathing	32
Factors associated with sleep disturbances in older adults	33
Sleep and health in older adults	33
Cognitive function	33
Cardiovascular health	35
Psychiatric illness	36
Pain	37
Conclusion	38
References	40

5. Social-ecological model of sleep health

Michael A. Grandner

Introduction	45
The social ecological model	46

Sleep as a domain of health behavior	46
Conceptualizing sleep in a social-ecological model	47
Individual level	47
Social level	48
Societal level	49
Combining upstream influences and downstream consequences	50
Applications of the model	51
References	51

Part II
Contextual factors related to sleep

6. Race, socioeconomic position and sleep

Natasha Williams, Girardin Jean-Louis, Judite Blanc, Douglas M. Wallace

Introduction	57
(Brief) history and definition of health disparities	58
Sleep characteristics	58
Self-reported sleep duration across racial/ethnic groups	58
Objective reported sleep duration across racial/ethnic groups	61
Sleep duration within racial/ethnic groups	61
Sleep duration across SES groups	61
Sleep architecture and continuity across racial/ethnic groups	62
Sleep architecture and continuity across SES groups	64
Sleep disorders	64
Sleep disordered breathing (SDB)	64
Symptoms of and risk factors for sleep disordered breathing across racial/ethnic groups	64
Diagnosis of SDB across and within racial/ethnic groups	65
SDB symptoms and diagnosis across SES groups	65
Insomnia	66
Insomnia complaints across racial/ethnic groups	66
Insomnia complaints across SES groups	69
Restless leg syndrome (RLS) and periodic limb movements during sleep (PLMS)	69
Narcolepsy	70
Circadian rhythms	71
Why do minority Americans have poor sleep?	71
Acculturation	71
Perceived discrimination	72

Worry and risk perception	72
Sleep opportunity	72
Future directions and summary	72
References	73

7. Neighborhood factors associated with sleep health

Lauren Hale, Sarah James, Qian Xiao, Martha E. Billings, Dayna A. Johnson

Neighborhoods and sleep health	77
Theoretical justification for neighborhoods and sleep health	77
Neighborhood factors associated with pediatric sleep	78
Urbanicity and population density	78
Neighborhood socioeconomic status (NSES)	78
Neighborhood access to physical activity	78
Neighborhood violence and safety concerns	78
Neighborhood factors associated with adult sleep	78
Inadequate sleep duration and delayed sleep timing	79
Insomnia	79
Obstructive sleep apnea (OSA)	79
Current limitations and future directions	79
Studying long-term trajectories of neighborhood conditions and sleep	80
Evaluating evidence from natural experiments and other causal methods	80
Using technological advances to studying neighborhoods and sleep at a larger scale	80
Are there interventions and policies to improve neighborhoods and sleep health?	80
Conclusions and public health significance	81
Acknowledgments	81
References	81

8. The impact of environmental exposures on sleep

Chandra L. Jackson, Symielle A. Gaston

The physical environment and sleep	85
The impact of light on sleep	85
The impact of temperature on sleep	89
The impact of noise on sleep	90
The impact of vibrations on sleep	91
The impact of air quality on sleep	92
The impact of seasonality and latitude/longitude on sleep	93
The social environment and sleep	94
Psychosocial stress and sleep	94

Social conditions, policies, institutions:
   The impact of socioeconomic status and racism on sleep ... 94
Community: The impact of neighborhood social and physical environments on sleep ... 95
Community: Work environment and sleep ... 96
Interpersonal relationships and sleep ... 96
Acknowledgments ... 98
References ... 98
Glossary ... 103

Part III
Addressing sleep health at the community and population level

9. Obstacles to overcome when improving sleep health at a societal level
Michael A. Grandner

Introduction ... 107
Real-world barriers to sleep health ... 107
   Lack of time ... 107
   Social norms and beliefs ... 107
   Physical environment ... 108
   Health conditions and chronic pain ... 108
   Substance use ... 108
   Distractions and on-demand culture ... 109
Conceptualizing strategies for overcoming these barriers ... 110
   The health belief model and application to sleep ... 110
   The integrated behavioral model and application to sleep ... 111
   The transtheoretical stages-of-change model ... 112
   Other health behavior models ... 112
Implementing sleep health programs ... 113
   Addressing perceived benefits ... 113
   Addressing perceived barriers ... 113
   Addressing social norms ... 113
   Addressing self-efficacy and control ... 113
   Addressing readiness ... 114
Conclusion ... 114
References ... 115

10. Screening for sleep disorders
Catherine A. McCall, Nathaniel F. Watson

Introduction ... 117
Sleep-disordered breathing ... 118
   STOP-BANG questionnaire ... 121
   Berlin questionnaire ... 121
Hypersomnolence ... 121
   Epworth Sleepiness Scale (ESS) ... 122
   Functional Outcomes of Sleep Questionnaire (FOSQ-30) ... 123
   Stanford Sleepiness Scale (SSS) ... 123
   Karolinska Sleepiness Scale (KSS) ... 123
Insomnia and sleep quality ... 124
   Insomnia Severity Index (ISI) ... 124
   Pittsburgh Sleep Quality Index (PSQI) ... 124
   Patient-Reported Outcomes Measurement Information System (PROMIS™) ... 125
Circadian rhythm disorders ... 125
   Horne-Ostberg Morningness-Eveningness Questionnaire (MEQ) ... 125
   Munich Chronotype Questionnaire (MCTQ) ... 127
Restless legs syndrome (RLS) ... 128
   International Restless Legs Syndrome Scale (IRLS) ... 128
Consumer sleep technologies ... 128
   Fitbit ... 129
   Jawbone ... 130
   SleepScore Max and SleepScore app ... 130
   Other novel technologies ... 130
Using screening data ... 131
References ... 131

11. Sleep hygiene and the prevention of chronic insomnia
Jason G. Ellis, Sarah F. Allen

Sleep hygiene ... 137
   What is sleep hygiene? ... 137
   Exercise ... 137
   Caffeine ... 138
   Alcohol ... 138
   Food and liquid intake ... 139
   Nicotine ... 139
   Bedroom environment ... 139
   Removal of electronics ... 140
   Clockwatching ... 140
   Measuring sleep hygiene ... 140
   Do people with insomnia have poorer sleep hygiene than normal sleepers? ... 140
   What is the role of sleep hygiene in the management of insomnia? ... 141
   So, is there a role for sleep hygiene in sleep medicine and practice, beyond insomnia? ... 141
The prevention of chronic insomnia ... 141
   Etiological models of insomnia ... 141
   What we know about acute insomnia? ... 142

Can we prevent acute insomnia from becoming chronic?	142
Identifying those at risk	142
Conclusions	143
Conflict of Interest	143
References	143

12. Actigraphic sleep tracking and wearables: Historical context, scientific applications and guidelines, limitations, and considerations for commercial sleep devices

Michael A. Grandner, Mary E. Rosenberger

Introduction	147
Scoring algorithms	147
Types of actigraph devices	149
Limitations of actigraphy and related considerations	150
Identifying sleep stages with actigraphy	152
Other considerations	153
Scientific guidelines	154
Evaluating commercially-available sleep trackers	155
Conclusions	156
References	156

13. Mobile technology, sleep, and circadian disruption

Cynthia K. Snyder, Anne-Marie Chang

Sleep as a biobehavioral state	160
Two-process model of sleep physiology	160
Contextual factors influencing sleep behavior and mobile technology use	161
Importance of sleep for health	161
Sleep loss impacts physical and psychological health and wellbeing	161
Negative consequences for individuals and public health concern	162
The role of mobile technology in sleep loss	163
Emergence of mobile technology	163
Impact of sleep loss	164
Conclusions	167
References	167

14. Models and theories of behavior change relevant to sleep health

Adam P. Knowlden

Foundation of theory for behavior change	171
Utility of theory for changing health behaviors	172
Causation in behavior change theories	172
Types of theories	174
Intrapersonal theories	174
Health belief model	174
Theory of reasoned action and the theory of planned behavior (continuum theory)	175
The transtheoretical model	177
Interpersonal theories	178
Social cognitive theory	178
Social network theory	179
Community level theories	179
Diffusion theory	179
Behavioral economics	180
Measurement of models and theories for behavior change interventions	181
Step 1: Define purpose of instrument	181
Step 2: Identify objects of interest	181
Step 3: Constitutively define objects of interest	181
Step 4: Operationally define objects of interest	181
Step 5: Review previously developed instruments	182
Step 6: Develop an original instrument	182
Step 7: Select appropriate scales	182
Step 8: Develop items	182
Step 9: Prepare a draft instrument	183
Step 10: Test for readability	183
Step 11: Send to panel of experts	183
Step 12: Conduct a pilot test	183
Step 13: Establish reliability and validity	183
Limitations of behavior change theories	184
Conclusion	184
References	184

Part IV
Sleep duration and cardiometabolic disease risk

15. Insufficient sleep and obesity

Andrea M. Spaeth

Sleep duration	189
Obesogenic behaviors	189
Potential physiological mechanisms	191
Group differences	192
Individual differences	193
Sleep timing	193
Sleep disorders	195
Sleep in individuals with obesity	196
The role of sleep in weight loss interventions	196
Conclusion	197
References	197

16. Insufficient sleep and cardiovascular disease risk

Sogol Javaheri, Omobimpe Omobomi, Susan Redline

Introduction	203
Defining insufficient sleep	204
Pathophysiology	205
Insufficient sleep and blood pressure	206
Insufficient sleep and coronary heart disease	207
Insufficient sleep and heart failure	208
Insufficient sleep and stroke	209
Conclusions	210
References	210

17. Sleep health and diabetes: The role of sleep duration, subjective sleep, sleep disorders, and circadian rhythms on diabetes

Azizi A. Seixas, Rebecca Robbins, Alicia Chung, Collin Popp, Tiffany Donley, Samy I. McFarlane, Jesse Moore, Girardin Jean-Louis

Sleep parameters and diabetes risk	213
Sleep duration and diabetes	213
Qualitative sleep parameters (sleep quality, excessive daytime sleepiness and social jet lag) and diabetes	216
Sleep disorders and diabetes	217
Circadian rhythm and diabetes	218
Indirect effects of sleep on diabetes	220
The exacerbating role of sleep on well-being, quality of life, health and mortality among diabetics	221
Healthy sleep and reduced diabetes risk	222
References	222

18. Social jetlag, circadian disruption, and cardiometabolic disease risk

Susan Kohl Malone, Maria A. Mendoza, Freda Patterson

Introduction	227
Definitions and epidemiology	227
Cardiometabolic syndrome	227
Circadian rhythms	228
Circadian disruption and social jetlag	229
Circadian disruption and cardiometabolic health	229
Circadian control of the cardiometabolic system	230
Cardiovascular functioning	230
Metabolism	230
Environmental rhythms and cardiometabolic health	231
Behavioral rhythms and cardiometabolic health	232
Biological rhythms and cardiometabolic health	233
Autonomic nervous system	233
Metabolically relevant hormones	234
Conclusion and future directions	236
References	236

Part V
Sleep and behavioral health

19. Sleep and food intake

Isaac Smith, Katherine Saed, Marie-Pierre St-Onge

Introduction	243
Part 1: Sleep loss and food intake	244
Part 2: Proposed mechanisms explaining the sleep-food intake relation	245
Homeostatic mechanisms	245
Nonhomeostatic mechanisms	246
Part 3: Influence of food intake on sleep duration and quality	247
Caloric consumption	247
Protein	247
Carbohydrates	248
Fat	249
Vitamins and supplements	249
Fruits	250
Alternative medicine	251
Total dietary approaches	251
Conclusion	251
References	252

20. Sleep and exercise

Christopher E. Kline

Impact of exercise on sleep	257
Observational research	257
Experimental research	258
Sedentary behavior	259
Potential mechanisms of exercise	260
Impact of exercise on sleep disorders	260
Insomnia	260
Sleep-disordered breathing	261
Restless legs syndrome/periodic limb movements during sleep	261
A bidirectional relationship: Impact of sleep on exercise	262

Combined impact of exercise and sleep on health	262
Conclusion	263
References	263

21. Sleep and alcohol use

Sean He, Brittany V. Taylor, Nina P. Thakur, Subhajit Chakravorty

Introduction	269
Neurobiology of alcohol use	269
Insomnia and alcohol use	269
Clinical findings	271
Insomnia in alcohol dependence	272
Treatments	273
Circadian rhythms and alcohol use	273
Clinical findings in shiftwork and alcohol use	274
Chronopharmacokinetic studies	274
Alcohol dependent individuals	274
Alcohol and sleep duration abnormalities	274
Breathing related sleep disorders and alcohol use	275
Summary	276
Alcohol and sleep-related movement disorders	276
Parasomnias and alcohol use	277
Other sleep-related issues associated with alcohol use	277
Discussion	277
References	278
Further reading	281

22. Improved sleep as an adjunctive treatment for smoking cessation

Freda Patterson, Rebecca Ashare

Introduction	283
Epidemiology of cigarette smoking	283
Sleep continuity and architecture in smokers versus non-smokers	284
Overview of sleep continuity and architecture	284
Sleep architecture in smokers versus non-smokers	284
Sleep continuity in smokers versus non-smokers	286
Sleep fragmentation in smokers versus non-smokers	287
Daytime sleepiness in smokers versus non-smokers	287
Summary	287
Smoking abstinence and sleep	287
Changes in sleep following abstinence	287
Relationship between sleep and cessation outcome	288
Effects of pharmacotherapy on sleep	291
Take home points: Relationship between sleep and cessation outcome	292
Possible mechanisms linking poor sleep to smoking cessation outcomes	292
Plausible adjunctive sleep therapies to promote smoking cessation	294
Overview	294
Behavioral treatments	294
Pharmacological treatments	295
Directions for future research	295
Acknowledgments	296
Conflicts of Interest	296
References	296

23. Sleep and the impact of caffeine, supplements, and other stimulants

Ninad S. Chaudhary, Priyamvada M. Pitale, Favel L. Mondesir

Introduction	303
Epidemiology	304
Epidemiology of sleep in caffeine	304
Epidemiology of sleep in energy drink supplements	306
Epidemiology of sleep in other psychostimulants	307
Relationships in specific populations	307
Physiology of caffeine in sleep-wake homeostasis	308
Role of adenosine and caffeine in sleep-wake cycle	308
Genetic factors and response to caffeine	309
Environmental factors and response to caffeine	310
Health implications of caffeine(stimulant) use—sleep disturbances model	311
Recommendations	312
Conclusion	313
References	313

24. Sleep, stress, and immunity

Aric A. Prather

Introduction	319
Overview of the immune system	319
Acquired immune system	319
Innate immune system	320
The aging immune system	320
Sleep, acquired immunity, and infectious disease risk	320
Sleep, innate immunity, and inflammatory disease risk	322

Sleep and immunological aging	323	27. Sleep and healthy decision making	
Beyond sleep: Does stress influence immunity?	324	*Kelly Glazer Baron, Elizabeth Culnan*	
Sleep and psychological stress: Reciprocal processes	325	Introduction	359
		Sleep as a health behavior	359
How does poor sleep and psychological stress affect immunity?	326	Influences on sleep and health behaviors	359
Stress-sleep connection and immunity	326	Short sleep duration is highly prevalent in the population	360
Conclusion	327	What predicts the decision to sleep or not to sleep?	360
References	327	Some individuals make time to sleep but cannot sleep	360

Part VI
Sleep loss and neurocognitive function

25. Sleep loss and impaired vigilant attention

Mathias Basner

Neurobehavioral consequences of acute and chronic sleep loss	333
Differential vulnerability to sleep loss	333
Effects of sleep loss on vigilant attention	333
The psychomotor vigilance test (PVT)	334
PVT software and hardware	335
PVT duration	335
PVT outcome metric	336
Research agenda	336
References	336

26. Sleep loss, executive function, and decision-making

Brieann C. Satterfield, William D.S. Killgore

Introduction	339
Neurobiology of sleep and fatigue	339
Alertness, sustained attention, and vigilance	340
Psychomotor vigilance	340
Wake state instability	342
Individual differences	343
Executive functions	344
Working memory	346
Inhibitory control	347
Cognitive control	347
Problem solving	350
Risk-taking, judgment, and decision-making	351
Self-rated risk propensity	351
Risky decision-making	351
Practical implications	354
Conclusions	354
References	355

Proposed pathways linking sleep to other health behaviors	361
Exposure	361
Neurocognitive factors	361
Linking sleep related changes in neurocognitive function to health behaviors	362
Neuroimaging data	362
Affective response to sleep loss	363
Effort and motivation	364
Does changing sleep make it easier to make healthy decisions?	365
Summary	365
References	366
Glossary	368

Part VII
Public health implications of sleep disorders

28. Insomnia and psychiatric disorders

Ivan Vargas, Sheila N. Garland, Jacqueline D. Kloss, Michael L. Perlis

Introduction	373
Definition, incidence, and prevalence	373
Definition	373
Incidence and prevalence	374
Theoretical perspectives on the etiology of insomnia	374
Stimulus control model	375
Behavioral model (Spielman's 3P model)	375
Neurocognitive model	375
Cognitive model	377
Psychobiological inhibition model	377
Parallel process (trans-theoretical) model	377
Insomnia and psychiatric morbidity	377
Depressive disorders	378

Suicide 379
Bipolar disorder (BPD) 379
Anxiety disorders 380
Post-traumatic stress disorder (PTSD) 380
Attention-deficit/hyperactivity disorder (ADHD) 380
Alcohol use disorder (AUD) 381
Autism spectrum disorder (ASD) 382
Schizophrenia 382
Behavioral treatment of insomnia 382
    What is CBT-I? 382
    CBT-I in the context of psychiatric disorders 382
Conclusion 383
References 383

29. Insomnia and cardiometabolic disease risk

Julio Fernandez-Mendoza

Introduction 391
Insomnia: A symptom and a chronic disorder 391
Hypertension and blood pressure 393
Type 2 diabetes and insulin resistance 398
Heart disease and stroke 399
Stress, immunity and health behaviors 400
Public health and clinical implications 402
Conclusion 403
References 403
Glossary 406

30. Sleep apnea and cardiometabolic disease risk

Andrew Kitcher, Atul Malhotra, Bernie Sunwoo

What is OSA? 409
Who gets OSA? 411
Does having OSA make you more likely to have cardiovascular disease? 411
    Hypertension 411
    Coronary artery disease 412
    Cerebrovascular disease 412
    Heart failure 412
    Arrhythmias 413
Why does OSA make you more likely to have cardiovascular disease? 413
What happens if we reduce apneic events? 414
Conclusion 414
References 415

Part VIII
Sleep health in children and adolescents

31. Sleep, obesity and cardiometabolic disease in children and adolescents

Teresa Arora, Ian Grey

Introduction 421
Defining overweight and obesity in children 421
Causes and consequences of childhood obesity 423
    Causes 423
    Consequences 423
Attempts to reduce the obesity epidemic 423
The importance of sleep in relation to health 424
Evidence for a link between sleep duration and obesity in pediatric populations 424
Other sleep parameters and childhood obesity 424
Sleep and energy homeostasis 425
Future directions 426
Metabolic disease 426
Mechanisms of diabetes 426
Sleep and type 2 diabetes mellitus 427
Sleep and children 429
Sleep, diabetes and children 429
Conclusion 431
References 431

32. Sleep and mental health in children and adolescents

Michelle A. Short, Kate Bartel, Mary A. Carskadon

Introduction 435
Sleep duration and mental health 436
Sleep quality and mental health 438
Improving sleep and mental health in children and adolescents 439
    Families 440
    Schools 440
    Clinicians 441
    Policy makers 441
Conclusion 442
    Summary 442
    Limitations and future research directions 442
    Concluding remarks 443
References 443

33. Delayed school start times and adolescent health

Aaron T. Berger, Rachel Widome, Wendy M. Troxel

Delaying high school start time improves sleep	448
Academic achievement, attention, and truancy	449
Mental health and risky behavior	450
Unintentional injury	451
Conclusions	451
References	452

Part IX
Economic and public policy implications of sleep health

34. Sleep health and the workplace

Soomi Lee, Chandra L. Jackson, Rebecca Robbins, Orfeu M. Buxton

Introduction	457
Work factors impacts nighttime sleep	457
Sleep impacts work function and productivity	457
What theories of work can tell us about modifiable work factors influencing sleep	458
Epidemiology of sleep and work	458
The relationship between sleep and work: Results from a meta-analysis (2017)	459
Theories of work and work stress that influence sleep	459
Work stress	459
Work demands and work-family conflict influence sleep	460
Micro-longitudinal (daily level) effects of work stressors on sleep	461
Sleep health and workers' future health risks	462
Workplace intervention effects on sleep	462
Business case for sleep: Considering the evidence from the employer point of view	462
Worksite wellness, and the need for more attention to sleep	462
Worksite programs targeting sleep and sleep related outcomes	463
Racial ethnic disparities in sleep health and sleep disorders	464
Future research topics and directions	466
References	467
Further reading	471

35. Sleep health equity

Judite Blanc, Jao Nunes, Natasha Williams, Rebecca Robbins, Azizi A. Seixas, Girardin Jean-Louis

Introduction: Sleep and public health	473
What is sleep health?	473
Social determinants of sleep health dimensions and associated health outcomes	474
Health differences and the historical sleep gap between blacks and whites	474
Identifying determinants of health differences	474
History behind the black–white "sleep gap"	474
Sleep health as a contributor to health disparities in modern days	476
From sleep health disparities toward sleep health equity	476
Conclusion	477
References	479

36. Obstructive sleep apnea in commercial motor vehicle operators

Indira Gurubhagavatula, Aesha M. Jobanputra, Miranda Tan

Prevalence	481
History of federally-funded research and regulatory activity	482
Screening	482
Initial evaluation	484
Diagnosis	484
Treatment	484
Monitoring PAP therapy	485
Benefits of PAP therapy	485
Education	485
Conclusion	485
References	485

37. Sleep health as an issue of public safety

Matthew D. Weaver, Laura K. Barger

Introduction	489
Demographics	489
Organizational structure	489
Individuals	490
Work hours and scheduling characteristics	490
Shift duration	490
Weekly work hours	490

The association between work schedules
  and health and safety outcomes ... 491
Implementation of schedules based on
  sleep and circadian principles ... 491
Physiological determinants of alertness ... 492
  Physiological determinants of fatigue in
    public safety ... 492
  Sleep deficiency and health ... 493

Sleep disorders ... 493
Fatigue risk management ... 495
Conclusion ... 496
References ... 496

Index ... 501

Contributors

Numbers in parentheses indicate the pages on which the authors' contributions begin.

Sarah F. Allen (137), Northumbria Sleep Research Laboratory, Northumbria University, Newcastle, United Kingdom

Teresa Arora (421), Zayed University, College of Natural and Health Sciences, Department of Psychology, Abu Dhabi, United Arab Emirates

Rebecca Ashare (283), Perelman School of Medicine at the University of Pennsylvania, Philadelphia, PA, United States

Laura K. Barger (489), Division of Sleep and Circadian Disorders, Brigham and Women's Hospital; Division of Sleep Medicine, Harvard Medical School, Boston, MA, United States

Kelly Glazer Baron (359), Division of Public Health, Department of Family and Preventive Medicine, University of Utah, Salt Lake City, UT; Department of Behavioral Sciences, Rush University Medical Center, Chicago, IL, United States

Kate Bartel (435), School of Psychology, Flinders University, Adelaide, SA, Australia

Mathias Basner (333), Unit for Experimental Psychiatry, Division of Sleep and Chronobiology, Department of Psychiatry, University of Pennsylvania Perelman School of Medicine, Philadelphia, PA, United States

Aaron T. Berger (447), Division of Epidemiology and Community Health, University of Minnesota, Minneapolis, MN, United States

Martha E. Billings (77), Division of Pulmonary, Critical Care & Sleep Medicine, University of Washington, Seattle, WA, United States

Judite Blanc (57, 473), NYU Langone Health, Department of Population Health, Center for Healthful Behavior Change, New York, NY, United States

Orfeu M. Buxton (457), Center for Healthy Aging; Department of Biobehavioral Health, Pennsylvania State University, State College, PA; Division of Sleep Medicine, Harvard Medical School; Department of Social and Behavioral Sciences, Harvard Chan School of Public Health; Department of Medicine, Brigham and Women's Hospital, Boston, MA, United States

Mary A. Carskadon (435), E.P. Bradley Hospital, Brown University, Providence, RI, United States

Subhajit Chakravorty (269), Cpl. Michael J Crescenz VA Medical Center; Perelman School of Medicine, Philadelphia, PA, United States

Anne-Marie Chang (159), College of Nursing; Department of Biobehavioral Health, Pennsylvania State University, University Park, PA, United States

Ninad S. Chaudhary (303), Department of Epidemiology, School of Public Health, University of Alabama at Birmingham; Department of Neurology, University of Alabama School of Medicine, Birmingham, AL, United States

Alicia Chung (213), NYU Langone Health, Department of Population Health, New York, NY, United States

Elizabeth Culnan (359), Department of Behavioral Sciences, Rush University Medical Center, Chicago, IL, United States

Hannah K. Dollish (3), Department of Psychology, University of Arizona, Tucson, AZ, United States

Tiffany Donley (213), NYU Langone Health, Department of Population Health, New York, NY, United States

Jason G. Ellis (137), Northumbria Sleep Research Laboratory, Northumbria University, Newcastle, United Kingdom

Fabian Fernandez (3), Department of Psychology, University of Arizona, Tucson, AZ, United States

Julio Fernandez-Mendoza (391), Sleep Research & Treatment Center, Department of Psychiatry, Penn State Health Milton S. Hershey Medical Center, Pennsylvania State University College of Medicine, Hershey, PA, United States

Sheila N. Garland (373), Department of Psychology, Faculty of Science; Division of Oncology, Faculty of Medicine, Memorial University, St. John's, NL, Canada

Symielle A. Gaston (85), Epidemiology Branch, National Institute of Environmental Health Sciences, National Institutes of Health, Department of Health and Human Services, Research Triangle Park, NC, United States

Nalaka S. Gooneratne (31), Center for Sleep and Circadian Neurobiology; Geriatrics Division, Perelman School of Medicine, University of Pennsylvania, Philadelphia, PA, United States

Michael A. Grandner (3, 11, 45, 107, 147), Sleep and Health Research Program, Department of Psychiatry, University of Arizona College of Medicine, Tucson, AZ, United States

Ian Grey (421), School of Social Sciences, Lebanese American University, Beirut, Lebanon

Indira Gurubhagavatula (481), Department of Medicine, Division of Sleep Medicine, Perelman School of Medicine at the University Hospital of Pennsylvania Medical Center; Sleep Disorders Clinic, Philadelphia VA Medical Center, Philadelphia, PA, United States

Lauren Hale (77), Program in Public Health, Department of Family, Population, and Preventive Medicine, Stony Brook University School of Medicine, Stony Brook, NY, United States

Sean He (269), Cpl. Michael J Crescenz VA Medical Center; School of Arts and Sciences, University of Pennsylvania, Philadelphia, PA, United States

Chandra L. Jackson (85, 457), Epidemiology Branch, National Institute of Environmental Health Sciences; Intramural Program, National Institute on Minority Health and Health Disparities, National Institutes of Health, Department of Health and Human Services, Research Triangle Park, NC, United States

Sarah James (77), Department of Sociology and Office of Population Research, Princeton University, Princeton, NJ, United States

Sogol Javaheri (203), Brigham and Women's Hospital, Harvard Medical School, Boston, MA, United States

Girardin Jean-Louis (57, 213, 473), NYU Langone Health, Department of Population Health; NYU Langone Health, Department of Psychiatry, New York, NY, United States

Aesha M. Jobanputra (481), Department of Medicine, Division of Pulmonary and Critical Care Medicine, Rutgers Robert Wood Johnson Medical School, New Brunswick, NJ, United States

Dayna A. Johnson (77), Division of Sleep and Circadian Disorders, Brigham and Women's Hospital and Harvard Medical School, Boston, MA, United States

William D.S. Killgore (339), Department of Psychiatry, College of Medicine, University of Arizona, Tucson, AZ, United States

Andrew Kitcher (409), Chief of Pulmonary and Critical Care Medicine, University of California San Diego; University of California, San Diego School of Medicine, La Jolla; Department of Medicine, Division of Pulmonary, Critical Care and Sleep Medicine, University of California San Diego, San Diego, CA, United States

Christopher E. Kline (257), Physical Activity and Weight Management Research Center, Department of Health and Physical Activity, University of Pittsburgh, Pittsburgh, PA, United States

Jacqueline D. Kloss (373), Behavioral Sleep Medicine Program, University of Pennsylvania, Philadelphia, PA, United States

Adam P. Knowlden (171), Department of Health Science, University of Alabama, Tuscaloosa, AL, United States

Soomi Lee (457), School of Aging Studies, University of South Florida, Tampa, FL; Center for Healthy Aging, Pennsylvania State University, State College, PA, United States

Junxin Li (31), Johns Hopkins University School of Nursing, Baltimore, MD, United States

Atul Malhotra (409), Chief of Pulmonary and Critical Care Medicine, University of California San Diego; University of California, San Diego School of Medicine, La Jolla; Department of Medicine, Division of Pulmonary, Critical Care and Sleep Medicine, University of California San Diego, San Diego, CA, United States

Susan Kohl Malone (227), Rory Meyers College of Nursing, New York University, New York, NY, United States

Catherine A. McCall (117), Department of Pulmonary, Critical Care, and Sleep Medicine, VA Puget Sound Health Care System; Department of Psychiatry, University of Washington Sleep Medicine Center, Seattle, WA, United States

Samy I. McFarlane (213), SUNY Downstate School of Medicine, New York, NY, United States

Jessica Meers (21), Department of Psychology, University of Houston, Houston, TX, United States

Maria A. Mendoza (227), Rory Meyers College of Nursing, New York University, New York, NY, United States

Favel L. Mondesir (303), Division of Cardiovascular Medicine, School of Medicine, University of Utah, Salt Lake City, UT, United States

Jesse Moore (213), NYU Langone Health, Department of Population Health, New York, NY, United States

Sara Nowakowski (21), Department of Obstetrics and Gynecology, University of Texas Medical Branch, Galveston, TX, United States

Jao Nunes (473), The City College of New York, New York, NY, United States

Omobimpe Omobomi (203), Brigham and Women's Hospital, Harvard Medical School, Boston, MA, United States

Freda Patterson (227, 283), Department of Behavioral Health and Nutrition, College of Health Sciences, University of Delaware, Newark, DE, United States

Michael L. Perlis (373), Behavioral Sleep Medicine Program; Center for Sleep and Circadian Neurobiology, University of Pennsylvania, Philadelphia, PA, United States

Priyamvada M. Pitale (303), Department of Optometry and Vision Science, School of Optometry, University of Alabama at Birmingham, Birmingham, AL, United States

Collin Popp (213), NYU Langone Health, Department of Population Health, New York, NY, United States

Aric A. Prather (319), Department of Psychiatry, University of California, San Francisco, CA, United States

Susan Redline (203), Brigham and Women's Hospital; Beth Israel Deaconess Medical Center, Harvard Medical School, Boston, MA, United States

Rebecca Robbins (213, 457, 473), NYU Langone Health, Department of Population Health, New York, NY, United States

Mary E. Rosenberger (147), Stanford Center on Longevity and Psychology Department, Stanford University, Stanford, CA, United States

Katherine Saed (243), Institute of Human Nutrition, Columbia University Irving Medical Center, New York, NY, United States

Brieann C. Satterfield (339), Department of Psychiatry, College of Medicine, University of Arizona, Tucson, AZ, United States

Azizi A. Seixas (213, 473), NYU Langone Health, Department of Population Health; NYU Langone Health, Department of Psychiatry, New York, NY, United States

Michelle A. Short (435), School of Psychology, Flinders University, Adelaide, SA, Australia

Isaac Smith (243), Institute of Human Nutrition, Columbia University Irving Medical Center, New York, NY, United States

Cynthia K. Snyder (159), College of Nursing, Pennsylvania State University, University Park, PA, United States

Andrea M. Spaeth (189), Department of Kinesiology and Health, School of Arts and Sciences, Rutgers University, New Brunswick, NJ, United States

Marie-Pierre St-Onge (243), Institute of Human Nutrition; Division of Endocrinology, Department of Medicine; Sleep Center of Excellence, Department of Medicine, Columbia University Irving Medical Center, New York, NY, United States

Jacqueline Stout-Aguilar (21), School of Nursing, University of Texas Medical Branch, Galveston, TX, United States

Bernie Sunwoo (409), Department of Medicine, Division of Pulmonary, Critical Care and Sleep Medicine, University of California San Diego, San Diego, CA, United States

Miranda Tan (481), Department of Medicine, Pulmonary Service, Section of Sleep Medicine, Memorial Sloan Kettering Cancer Center, New York, NY, United States

Brittany V. Taylor (269), Cpl. Michael J Crescenz VA Medical Center; School of Arts and Sciences, University of Pennsylvania, Philadelphia, PA, United States

Nina P. Thakur (269), Cpl. Michael J Crescenz VA Medical Center; School of Arts and Sciences, University of Pennsylvania, Philadelphia, PA, United States

Wendy M. Troxel (447), RAND Corporation, Santa Monica, CA, United States

Andrew S. Tubbs (3), Department of Psychiatry, University of Arizona, Tucson, AZ, United States

Ivan Vargas (373), Behavioral Sleep Medicine Program; Center for Sleep and Circadian Neurobiology, University of Pennsylvania, Philadelphia, PA, United States

Douglas M. Wallace (57), Department of Neurology, Sleep Medicine Division, University of Miami Miller School of Medicine; Miami VA HealthCare System, Sleep Disorders Laboratory, Miami, FL, United States

Nathaniel F. Watson (117), Department of Neurology, University of Washington Sleep Medicine Center, Seattle, WA, United States

Matthew D. Weaver (489), Division of Sleep and Circadian Disorders, Brigham and Women's Hospital; Division of Sleep Medicine, Harvard Medical School, Boston, MA, United States

Rachel Widome (447), Division of Epidemiology and Community Health, University of Minnesota, Minneapolis, MN, United States

Natasha Williams (57, 473), NYU Langone Health, Division of Health and Behavior, Department of Population Health, Center for Healthful Behavior Change, New York, NY, United States

Qian Xiao (77), Department of Health and Human Physiology and Department of Epidemiology, University of Iowa, Iowa City, IA, United States

Preface

Over the past several years, there has been an increasing interest in the topic of "Sleep and Health," which motivated the creation of this book. Although it has been known since the 1960s that population-level sleep variables were associated with health, daytime function, and even mortality—alongside diet, physical activity, smoking, alcohol use, etc.—the scientific landscape was quite sparse until more recently. In general, sleep research focused almost exclusively on basic sleep physiology and neuroscience or on clinical sleep disorders in terms of their etiology, pathophysiology, diagnosis, and treatment. The issue of sleep as a health issue in general was not widely pursued scientifically.

In the 1990s, research on sleep apnea demonstrated that not only did this condition represent a serious cardiovascular and metabolic risk factor, but also that it was much more prevalent than previously believed. This brought attention to sleep from those in the public health domain. This emerging focus on the public health impact of sleep disorders also coincided with the reemergence in the scientific literature of studies documenting the relationship between habitual sleep duration and mortality. The time was seemingly right, and this issue which had been somewhat dormant for decades was approached with renewed interest. Further, studies documenting potential mechanisms of this mortality relationship started to emerge, linking sleep duration with metabolism, brain function, obesity, cardiovascular disease, and other conditions.

In this period of time, the field of sleep research widened to address not only basic science and clinical conditions/treatments, but also issues related to sleep and health in general. Sleep duration items started to find their way into epidemiologic surveys and health surveillance efforts with increasing regularity. This led to hundreds of papers documenting relationships between habitual sleep variables at the population level and a wide range of outcomes. The emergence of this literature coincided with an increased recognition in the general public of the importance of sleep health. Popular press articles about sleep health started appearing with more frequency. Consumer technology began to address tracking and optimizing sleep. Athletes publicly bragged about how much sleep they got in order to better prepare for competition. Political and business leaders publicly lamented the fact that their schedules disallowed sufficient sleep. The conversation was changing. This book will hopefully address this interest and provide useful information about ways that sleep is related to important aspects of health.

The purpose of this book is to serve as a sort of a handbook for individuals interested in the field of sleep and health. Members of the sleep research community more familiar with basic and/or clinical science may find this volume useful for better understanding the role of these basic and clinical processes in a public health context. Especially for those unfamiliar with the role of societal factors in health, this book may be useful for understanding these topics from a scientific perspective. This is important, as a wider translational view of the sleep field is necessary for understanding the mechanisms of these phenomena and their relationships to health. This volume should also be useful for public health professionals and others who study the role of health in society. These individuals may have experience in other domains of health but may need to better understand the importance of sleep in these contexts. Those who specialize in studying obesity, behavioral health, cardiovascular disease, or other areas may find this volume useful in relating those areas to sleep. There are also sections devoted to those who have a specific interest in policy, so that better policy decisions can be made in light of current research and data. Thus, this book has a wide range of applications, including helping those with sleep expertise gain a wider view on the role of sleep in society and helping those with expertise in social factors in health who wish to get a better understand of the importance of sleep.

To accomplish these goals, this book is divided into nine sections.

- The first section is "General Concepts in Sleep Health" and includes chapters on the basics of sleep physiology and measurement, basic epidemiology of sleep health factors, basic information about sleep across age and gender, and an introduction to the social-ecological model of sleep and health. This section will orient those new to the field of sleep to some of the basic concepts in sleep physiology and ontogeny.
- The second section is "Contextual Factors Related to Sleep" and focuses on the impact of where a person is and how that affects their sleep. This section includes chapters that address issues such as race/ethnicity, socioeconomics,

neighborhood factors, and other environmental exposures. These are often important elements that serve as determinants of health and have been shown to be linked to sleep in important ways.
- The third section is about "Addressing Sleep Health at the Community and Population Level." It includes chapters on dealing with real-world obstacles in the way of healthy sleep, screening for and preventing sleep disorders, addressing sleep hygiene, understanding sleep tracking and technology, understanding the role of mobile technology and screen time in relation to sleep, and different ways of conceptualizing sleep in the context of other aspects of health. This section is somewhat broad but bridges basic understandings of sleep to real-world problems and constraints including schedule demands, technology, and access to healthcare.
- The fourth section is "Sleep Duration and Cardiometabolic Disease Risk" and includes chapters addressing connections between sleep/circadian issues and obesity, cardiovascular disease, and obesity. Since cardiometabolic diseases are still leading causes of death in society, the many connections to sleep are especially relevant for those with an interest in public health.
- The fifth section is specifically focused on "Sleep and Behavioral Health," since most of the underlying causes of chronic disease are driven by patterns in behavior. With this in mind, this section includes chapters that address many aspects of health behavior, including diet, physical activity, smoking, alcohol, caffeine and stimulants, and stress/immune function.
- The sixth section summarizes the literature on "Sleep Loss and Neurocognitive Function" and includes chapters that address vigilant attention, decision-making, learning and memory, and other brain functions. It also includes a chapter on how these effects on brain function impact not only safety and brain health but also chronic disease and healthy choices.
- The seventh section briefly discusses "Public Health Implications of Sleep Disorders" and includes chapters on the most common sleep disorders including insomnia and sleep apnea. Although many other books address sleep disorders specifically, this section focuses specifically on public health implications.
- The eighth section covers topics specifically relevant to "Sleep Health in Children and Adolescents" and includes chapters on cardiometabolic health risk, mental health, and impacts of delayed school start times. These issues have both health and policy implications.
- The final section is dedicated to "Economic and Public Policy Implications of Sleep Health" and lays out the science behind specific policy implications. These include issues such as health equity, sleep and health in the workplace, sleep and public safety, and sleep and transportation safety.

Hopefully, the next edition will address important topics that did not find their way into this one. For example, telehealth and telemedicine is a growing issue that is impacting both public health and public policy; and sleep medicine has been at the forefront of this movement. Also, even though there are chapters on insomnia and sleep apnea, these are not the only sleep disorders with public health implications—for example, Restless Legs Syndrome and other sleep-related movement disorders have cardiometabolic implications, parasomnias have important forensic implications, narcolepsy is an important issue in the context of widespread immunization as a public health strategy, etc. And although there were chapters on race/ethnicity, this volume is still limited in its treatment of cultural beliefs and practices that play important roles in sleep. In terms of societal efforts to address sleep health, a future edition of this volume should probably address the rapidly growing interest in sleep among the athletics community. The next volume could also include chapters on the functions of sleep from an evolutionary biology perspective, empirically supported interventions for promoting sleep health at the community/population level, the public health issue of sleep and dementia risk, and economic impacts of sleep loss and untreated sleep disorders. There is still a lot of work to do!

The current volume brings together many of the leaders of the field to discuss a wide range of topics that encompass the field of sleep and health. This book provides useful information for a wide range of stakeholders and will hopefully contribute to the increasing interest in the field. At the end of the day, sleep is a foundational part of human biology and physiology. And sleep health can impact many systems in the body. This book is meant to start a conversation about all of these connections. It is hoped that this conversation leads to work that helps us to lead longer, healthier, more fulfilling lives.

Acknowledgments

This book represents the combined work of a lot of people over quite a long time. I would first like to thank all of the friends, colleagues, and experts I look up to (some individuals falling into multiple categories) that took the time to write thoughtful, informative, and helpful chapters on such a wide range of topics. It is truly amazing and actually quite humbling to see the incredible names on the list of contributing authors. So, thank you.

I also want to specifically thank a number of authors specifically who shaped the thinking around the content of the book, including: Teresa Arora, Laura Barger, Kelly Baron, Mathias Basner, Orfeu Buxton, Mary Carskadon, Subhajit Chakravorty, Anne-Marie Chang, Ninad Chaudhary, Jay Ellis, Fabian Fernandez, Julio Fernandez-Mendoza, Sheila Garland, Nalaka Gooneratne, Indira Gurubhagavatula, Lauren Hale, Chandra Jackson, Girardin Jean-Louis, Dayna Johnson, Scott Killgore, Chris Kline, Adam Knowlden, Atul Malhotra, Sara Nowakowski, Freda Patterson, Michael Perlis, Aric Prather, Susan Redline, Mary Rosenberger, Azizi Seixas, Andrea Spaeth, Marie-Pierre St-Onge, Wendy Troxel, Nate Watson, and Natasha Williams.

Managing a project like this takes a lot of effort and time, and I am really fortunate to have great staff at the Sleep and Health Research Program. Jo-Ann Gehrels managed the project, kept everything moving forward, and made sure that everybody was informed throughout the process. Both Pamela Alfonso-Miller and Chloe Warlick also deserve special mention for helping to make this book a reality. And the entire team at Elsevier has been nothing short of amazing through this whole process from start to finish, especially Stacy Masucci, Carlos Rodriguez, and Punitha Govindaradjane.

I need to acknowledge my mentors, especially Michael Perlis, Allan Pack, Daniel Kripke, Sonia Ancoli-Israel, and Donna Giles. You will always be my role models for how to be a scholar, a teacher, a leader, and a communicator. I hope that this book can inspire people to work to improve others' lives as you have inspired me.

The concept of "sleep and health" is complicated and changing. Many individuals in addition to those already mentioned above have played important roles in shaping how I have approached this topic. These include: Celyne Bastien, Janet Croft, David Dinges, Reuven Ferziger, Phil Gehrman, Kristen Knutson, Sanjay Patel, Megan Petrov, Dorothy Roberts, Michael Twery, Terri Weaver, Emerson Wickwire, Shawn Youngstedt, and others. I want to specifically acknowledge the influence of Nirav Patel, who coined the term "sleep disparity" and had the original idea for what became the Social-Ecological Model of Sleep and Health. He was a great friend and colleague, and I think he would be glad to see how far the field has come in such a short amount of time. This book is dedicated to his memory.

Most importantly, I want to thank my wife Ana Liza and our boys Benjy and Charlie, who give all my work meaning and inspire me to always ask questions, push boundaries, and make a difference. I hope that this book does all those things.

Part I

General concepts in sleep health

Chapter 1

The basics of sleep physiology and behavior

Andrew S. Tubbs[a], Hannah K. Dollish[b], Fabian Fernandez[b], Michael A. Grandner[c]
[a]Department of Psychiatry, University of Arizona, Tucson, AZ, United States, [b]Department of Psychology, University of Arizona, Tucson, AZ, United States, [c]Sleep and Health Research Program, Department of Psychiatry, University of Arizona College of Medicine, Tucson, AZ, United States

INTRODUCTION

Sleep is an essential element of human health, supporting a wide range of systems including immune function, metabolism, cognition, and emotional regulation. To understand everything that sleep does, however, it is necessary to understand what sleep is. This chapter provides that foundation by discussing the conceptualization, physiology, and measurement of sleep.

The definition of sleep

Sleep is a naturally recurring and reversible biobehavioral state characterized by relative immobility, perceptual disengagement, and subdued consciousness. As a predictable and easily reversible phenomenon, sleep is distinct from states of anesthesia and coma, which typically involve the absence or suppression of neural activity. Additionally, proper sleep involves a dynamic interaction between *voluntary* decisions and *involuntary* biological activities. Turning off the lights, reducing noise, and laying down are voluntary behaviors, but the result is an involuntary increase in melatonin and a series of shifts in the activity patterns of the brain throughout the night. Sleep ultimately depends on this collaboration between behavior and biology, and a deficit in either will disrupt sleep.

Conceptualizing sleep as a health behavior

A health behavior is an action (or omission) by an individual that impacts their health. Conceptualizing sleep as a health behavior is useful because it highlights how behavior and neurobiology interact, and how individuals can modify their health through sleep. Viewed in this way, sleep can be divided into three processes: sleep need, sleep ability, and sleep opportunity. These processes are diagrammed in Fig. 1.1.

FIG. 1.1 The three process model of sleep.

Sleep need is the biological requirement for sleep, or the minimal amount of rest the body requires to prepare for the next day. This need is defined by individual genetics and physiology and does not change after losing a night of sleep or oversleeping on the weekends. Unfortunately, there is no standard method for measuring sleep need. While epidemiological studies suggest an average of 7–8 h for healthy adults, some individuals naturally need more sleep (e.g., children and adolescents), while some need less. Sleep need represents the core motivation for engaging in sleep, and consistently failing to meet this need can promote cardiometabolic disease, impair cognitive functioning, and increase risk for psychiatric disorders.

The only way to satisfy sleep need is to sleep. The amount of sleep an individual can achieve is known as **sleep ability** and is approximated by total calculated sleep time. Unlike sleep need, sleep ability can change from one night to the next depending on life circumstances. Stress, a cold, or the death of a loved one can reduce sleep ability, while one night of sleep deprivation can increase sleep ability the following evening. Thus, while sleep ability cannot be directly controlled, it can be influenced by behavior.

Whereas sleep ability is correlated to the amount of time one spends sleeping, **sleep opportunity** is the amount of time the person *makes available* for sleep. Sleep opportunity is measured by the amount of time the person stays in bed (although time spent in habitual "bedtime" activities like reading a book could theoretically be incorporated). Unlike the two previous processes, sleep opportunity is under conscious control and is the most vulnerable to environmental factors. This is illustrated by a trauma resident on a 24-h call. Fatigue accumulates over the course of the shift, slowly increasing the resident's sleep ability. However, sleep opportunity is negligible, since the resident must be ready at a moment's notice to respond to a life-threatening crisis.

While these three sleep processes can be described independently, in the real-world they work together to control sleep. Sleep need motivates the creation of sleep opportunity, which provides a context for sleep ability to produce sleep—thus satisfying sleep need and reinforcing the methods used to create sleep opportunity in the first place.

Conceptualizing sleep as a physiological process

Sleep involves a progression of neurophysiological changes in the brain. These changes are grouped (somewhat artificially) into stages based on scoring convention. To explore these changes, this section will briefly describe wakefulness and then proceed to give a description of each of the sleep stages.

Wakefulness

During wake, the brain is engaged in numerous activities, many of which are unrelated to one another. For example, someone might be watching a TV show while listening for the sound of a car in the rain and thinking about whether there is enough food in the refrigerator for lunch tomorrow. The aggregated electrical activity produced by these processes can be observed using electroencephalography (EEG), which would show a high-frequency, low-amplitude signal traveling across the surface of the brain (Fig. 1.2). To understand what this means, imagine a crowd cheering in a sports stadium. Everyone is shouting at different times. This means that while someone is always cheering (high frequency), it is impossible to discern individual words because everyone's words are drowning each other out (low amplitude). Coming back to the brain, a high frequency signal means many different processes are present in the circuitry of the brain, but the timing of these processes is scattered. Because this activity is widespread, it is hard to resolve any one particular process, resulting in a low amplitude signal. The beta frequency is the classic frequency of active wake and ranges from 12 to 30 Hz (cycles per second). When subjects lay down and close their eyes, electrical activity generally slows to an alpha frequency (8.5–12 Hz), which indicates that the person is awake but not necessarily attending to their surroundings (Fig. 1.2). Along with electrical activity, wakefulness is characterized by high levels of arousal neurotransmitters, such as dopamine, noradrenaline, and serotonin. There is also increased autonomic activity; heart rate, respiratory rate, and blood pressure are constantly responding and adapting to changes in the body and the environment throughout the day.

NREM sleep: General overview

The first half of the sleep cycle can be divided into three distinct stages of nonrapid eye movement (NREM) sleep, aptly named Stage 1, Stage 2, and Stage 3. Electrical activity throughout the brain decreases in frequency and increases in amplitude at each progressive stage. This reflects a reduction in overall neural activity, but an increasing coordination among neurons (i.e., enhanced oscillation). Recalling the example from above, imagine if the crowd slowly coordinated their cheering. At first, the sound would still be noisy and unintelligible. However, once everyone followed along, a wave of quiet alternating with cheering would emerge. The number of cheers would decrease (lower frequency), but the volume and clarity of each cheer would steadily increase (higher amplitude). By the end of NREM the electrical activity of the brain is tightly synchronized, leading to lower frequency, higher amplitude oscillating waves known as "slow waves," which cycle about

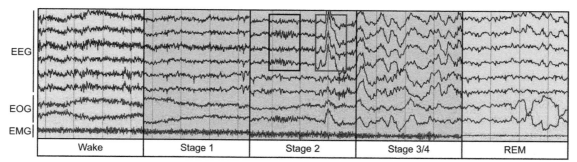

FIG. 1.2 Example traces of the stages of sleep. *Blue box* highlights a sleep spindle. *Green box* highlights a k-complex. Notice how in REM sleep the electrical activity is similar to Wake or Stage 1, but there is a complete lack of motor activity in the electromyogram (EMG). *Images taken from an anonymous human recording.*

once every second. The final stage of NREM is commonly referred to as slow wave sleep because of the dominance of these waves in the EEG record.

In addition to shifts in brain activity, NREM sleep is accompanied by a global decrease in wake-related neurotransmitters and impaired perception of external stimuli. In fact, most sensory inputs are specifically filtered out by the thalamus to protect sleep. Indices of autonomic activity such as heart rate, respiratory rate, temperature, and blood flow to the brain are reduced as one advances from one NREM stage to the next. Motor activity is markedly reduced, but not completely absent.

NREM: Stage dissection

The Rechstschaffen and Kales scoring criteria divide NREM into four stages [1], while the more current American Academy of Sleep Medicine (AASM) criteria combine the last two NREM stages [2] since the distinction is not viewed as clinically relevant [3]. Here, in our stage dissection of NREM, we observe the AASM criteria.

Stage 1 is characterized by approximately 50% *alpha* activity (waves of brain activity cycling at 8–12 Hz) and the emergence of *theta* waves (4–7 Hz) in the EEG trace. Nonelectrophysiological markers can also include slow-rolling eye movements, unusual visual sensations that take the form of clouds or flares of light (phosphenes), and hypnagogic myoclonia, which are brief jerking movements. The arousal thresholds for waking during Stage 1 are selective, as the brain determines if there is something worth attending to or whether it can commit to extended sleep. Here, for example, one's name spoken softly can awaken a person whereas a similar sounding word spoken at the same intensity might not. Stage 1 accounts for 5% of total sleep time and, in healthy sleep, is the entry point for NREM Stage 2.

Stage 2 is characterized by the absence of slow-rolling eye movements, mixed frequency neurophysiological activity, and the presence two major transient electrical phenomena: k-complexes and sleep spindles. K-complexes are large-amplitude rapidly fluctuating bursts of brain activity, while spindles are 12–15 Hz oscillating signals lasting 0.5–2 s (Fig. 1.2). While these phenomena are theorized to support memory consolidation and/or filter sensory input, their true functions remain unknown. Stage 2 sleep comprises 45%–55% of total sleep time and is viewed as a bridge between light (Stage 1) and deep (Stage 3) NREM sleep.

Stage 3 sleep is often referred to as slow wave sleep (SWS) owing to the prominence of high amplitude, low frequency *delta* oscillations recurring at ~1 Hz. The amount of time an individual spends in SWS positively correlates with lack of sleep, such that SWS is elevated during the first sleep cycle after a prolonged period of wakefulness. SWS is thought to discharge sleep pressure that has accumulated throughout the day because the amount of time spent in this stage decreases dramatically as the night progresses. SWS tends to coincide with the timing of peak growth hormone secretion, hinting at a role for this sleep stage in nightly maintenance and repair of the body. Additionally, oscillations that appear during SWS may function as a broad conduit for the repeated activation of memory centers of the brain to support memory-strengthening. The end of Stage 3 NREM sleep is usually followed by entry into REM sleep.

Description of REM sleep

Rapid eye movement (REM) sleep represents a categorical shift in sleep-related brain activity and forms the latter half of the sleep cycle. While most neurotransmitters drop to low levels, acetylcholine levels match or exceed those produced during wake. The surge in acetylcholine creates patterns of electrical activity in the sleeping brain that approximate the high frequency, low amplitude patterns usually seen in alert individuals. Despite this increase in overall excitability, there is a paradoxical loss of muscle movement (i.e., sleep paralysis). The only exceptions are eye muscles, which show the rapid, jerking movements for which the stage is named, and the diaphragm, which remains functional but contracts erratically. The loss of muscle tone leads to further narrowing of the upper airway, which can trigger snoring. Blood pressure and heart rate are also destabilized in REM sleep, in some cases leading to sympathetic "storms" of phasic arousal. At the same time, temperature regulation is impaired due to the loss of the ability to shiver.

Perhaps the most dramatic change associated with REM is in the content of dreams. In NREM, dreams are often more grounded, logical, and procedural, lacking any real visual or sensory detail. REM dreams, by contrast, are an absolute free-for-all of "sensory" experience, visual content, and emotions that can rapidly morph in content and affect with little reasoning. REM dreams may support memory consolidation processes, particularly by linking disparate concepts or connecting new ideas to old ones. They may also support emotional processing of difficult events (e.g., divorce, bereavement). This is based on functional imaging studies which show elevated activity in limbic regions during REM sleep and increases in emotional regulation inventories after subjects awake [4, 5]. In recent years, connections between REM sleep and emotion regulation have been most manifest in individuals suffering from posttraumatic stress disorder (PTSD).

Although the brain generates motor commands during dreams, the descending motor neurons are inhibited in the brainstem to prevent execution of these commands. When this process is disturbed (as in several neurological and psychiatric disorders), subjects will act out their dreams, often posing a danger to themselves or their bed partners. REM sleep accounts for approximately 25% of sleep and occurs in 4–6 episodes distributed across the night.

Moving through the sleep stages

Although the stages of sleep are presented in a particular order here, it should be noted that progression is not always linear. While all subjects start in Stage 1, they may proceed rapidly through Stage 2 to 3, or they may backtrack to earlier stages before proceeding to REM. Conversely, while it is typical to return from REM to NREM Stage 1 or 2, it is possible to return to any NREM stage following a REM episode. An example diagram of sleep stages, known as a hypnogram, is presented in Fig. 1.3.

Sleep and circadian rhythms

Across cultures, geography, seasons, and age, humans tend to sleep at night and wake up in the morning. This phenomenon is so ubiquitous that it often escapes scrutiny, but consider if it were a different biological function. What if, for example, humans only used the restroom at certain times of day? It seems banal that sleep should occur at night, but it is actually a remarkable feat of biology that humans (and many other animals) consolidate this large set of biological processes to a particular stretch of the day.

Generally speaking, there are two factors that ensure sleep occurs at night. The first is sleep propensity, or the drive for sleep. Physical and mental fatigue that accumulates during the day increases sleep propensity, and by nighttime the elevated sleep propensity drives humans to engage in sleep. The other factor is the circadian system. The molecular machinery of the circadian system is found within each cell of the body, comprised of an interlocking set of signaling proteins that produce a ~24-h rhythm of cellular functions that can be further adjusted by cues in the environment. This machinery ensures that functions such as digestion and immune system maintenance are optimized at specific points in the 24-h solar day. For example, when the clock signals that it is biological night, the body responds by shifting neurobiological activities to favor sleep.

These two factors are formally referred to as the Two Process Model of Sleep (Fig. 1.4). In the morning, sleep propensity is low, but increases over the course of wakefulness. Conversely, the circadian drive for wakefulness increases during the morning, peaks during the midday, and then drops at night. While there is a short, early evening peak that sustains wakefulness after sunset (referred to as the wake-maintenance zone), sleep propensity eventually exceeds circadian wakefulness and sleep onset occurs. This point is referred to as the sleep gate. In humans, the peak

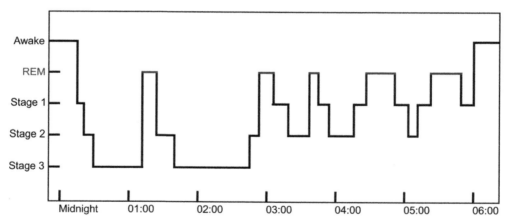

FIG. 1.3 A hypnogram, which tracks the amount of time spent in each sleep stage across the night.

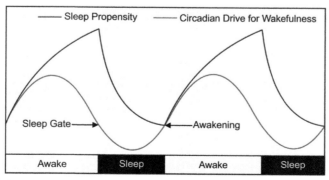

FIG. 1.4 The two process model of sleep.

of sleep propensity and the trough of wakefulness occur at night, which is why humans tend to sleep at that time.

So how exactly do these two forces work to generate the sleep gate? Sleep propensity is not well understood, but current theories focus on the buildup of certain substances (such as adenosine) that signal increasing levels of fatigue. This explains why caffeine, which opposes rising adenosine levels, is an effective stimulant.

The circadian system is largely underpinned by a part of the brain called the suprachiasmatic nucleus of the hypothalamus (SCN). Because of its circuit connections with the eye and ability to track sunrise and sunset, the SCN can operate as the master pacemaker that synchronizes all the miniature cellular clocks of the body to the light schedule set by the Earth's rotation (like a conductor of a symphony orchestra). In the absence of external photic cues, the SCN can still produce an endogenous rhythm that approximates the lengths of day and night. However, this rhythm is imprecise and follows a schedule that—depending on the person—is slightly longer or shorter than 24 h. Without any means of correction, the endogenous rhythm set by the SCN would slowly drift away from the solar day's 24-h cycle (resulting in non-24-h circadian rhythm disorder; Fig. 1.5). Fortunately, the SCN can use the light information it receives from the eye on a daily basis to adjust for the difference in timing, a process known as entrainment.

Disruptions to the circadian system can manifest as difficulties with sleep. One example is jet lag. When an individual rapidly changes time-zones, the external light/dark cues of the new destination become misaligned with the endogenous rhythm of night and day, which is still operating on the previous light schedule. Depending on the direction of the shift, an individual may awaken hours before dawn in the new location (phase advance), or take hours to fall asleep after night has fallen (phase delay). Fortunately again, after several sleep/wake intervals, the SCN will reentrain the body to the new light/dark cues, thus normalizing sleep. Examples of typical and atypical circadian rhythms are presented in Fig. 1.5.

Basic sleep physiology

There are no specific parts of the brain that act as monolithic sleep or wake centers. Rather, the neurobiological states of sleep and arousal are achieved via coordinated interactions between multiple brain regions. This section will highlight brain regions, chemical signals, and physiological processes that coordinate sleep and wake.

The brainstem

The brainstem is the most evolutionarily conserved structure within brain. As such, it is the control center for the autonomic nervous system, which regulates basic life-sustaining activities such as heart rate, blood pressure, and

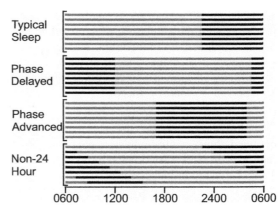

FIG. 1.5 Circadian rhythms of wake and sleep. Each bar represents a day. Wake is presented as *yellow*, while sleep is presented as *black*. Both typical and abnormal circadian rhythms are presented.

respiration. Regarding sleep and wake, the brainstem produces wake-promoting neuromodulators such as serotonin, norepinephrine, and dopamine that set the general volume of brain activity. The brainstem regions that produce these chemicals are collectively referred to as the ascending activating system because these regions project to and activate higher order brain areas located in the cerebral cortex.

The hypothalamus

The hypothalamus supports three major processes associated with sleep. First, it houses the SCN, and thus maintains circadian timekeeping. Second, the hypothalamus regulates the autonomic nervous system, particularly with regard to temperature. Third, it augments the wake-promoting neuromodulators of the brainstem with two additional chemicals: histamine and orexin (also known as hypocretin). Hypothalamic production of histamine and orexin drives wakefulness during the day, while low concentrations of histamine and orexin at night facilitate drowsiness and a tendency to sleep.

The thalamus

The thalamus is a collection of nuclei that serves as the gateway for information related to touch, taste, sight, and sound to travel to and between areas of the cerebral cortex. Although historically seen as a simple relay station, the thalamus is now understood to perform an extensive filtering function. During sleep, the thalamus blocks most sensory information from reaching the cortex. Ambient noise, whispers, and low light are all eliminated, allowing sleep to occur without having to consciously process what is going on in the environment.

Cerebrum

The cerebrum includes a variety of cortical and subcortical structures, such as the somatosensory and motor cortices, basal ganglia, and hippocampus. Sensory processing, motor commands, language, memory, and emotion all occur in or involve elements of the cerebrum, which exhibits the vast majority of neural activity in the brain. The cerebrum does not drive a specific element of sleep or wake, but the activity of billions of cortical neurons plays a large role in whether a person is awake or asleep.

Neuromodulators

As mentioned above, neuromodulators play a major role in sleep and wakefulness. Listed below are six major neuromodulators known to influence sleep.

- Dopamine: Produced by the substantia nigra and the ventral tegmental area of the brainstem, dopamine promotes wakefulness. Some other functions of dopamine include stimulation of the basal ganglia to promote voluntary movement, and stimulation of the nucleus accumbens as part of the pleasure and reward systems.
- Histamine: Produced by the tuberomammillary nucleus of the hypothalamus, histamine promotes wakefulness. This is why antihistamines such as diphenhydramine (Benadryl) cause drowsiness; they are able to enter the brain and block the wakefulness promoting effect of histamine. Second generation antihistamines, such as cetirizine (Zyrtec), do not cause drowsiness because they do not cross the blood brain barrier.
- Norepinephrine: the precursor to epinephrine (adrenalin), norepinephrine is produced by the locus coeruleus in the brainstem. Norepinephrine acts at the same receptors as epinephrine to stimulate wakefulness, although at a much reduced half-life.
- Acetylcholine: Although acetylcholine is often used as a neurotransmitter, it is also produced by the basal forebrain and multiple regions of the brainstem to act as a neuromodulator. Acetylcholine promotes wakefulness, but also supports REM sleep.
- Serotonin: Produced by the dorsal raphe nucleus of the brainstem, serotonin has more than 14 receptor subtypes, many of which differ in their activity. In general, increased serotonin levels tend to promote wakefulness and inhibit REM.
- Orexin: Produced by the lateral hypothalamus, orexin is a major wake-promoting agent in the brain. Its absence, most likely due to autoimmune destruction, is the chief cause of the sleep disorder narcolepsy. In addition, orexin enhances the activity of brainstem neuromodulators such as noradrenaline.

The autonomic nervous system

The autonomic nervous system is responsible for the subconscious regulation of heart rate, respiration, blood pressure, temperature, and other vital pieces of physiology. This system is divided into two opposing branches: the sympathetic (fight or flight) and parasympathetic (rest and digest) branches. During wakefulness, the sympathetic and parasympathetic branches are constantly adapting to environmental stimuli and emotional/mental processes. During NREM sleep, however, the sympathetic branch is largely quiescent while the parasympathetic branch remains active. This results in a progressive decrease in heart rate, temperature, and blood pressure which promotes and maintains NREM sleep moving through stages 1–3.

Quantifying sleep

Sleep incorporates a range of biological and behavioral activities which cannot be captured by a single measurement. Instead, sleep is quantified within two broad domains: sleep continuity and sleep architecture.

Sleep continuity encapsulates the timeline of how a person sleeps. Total sleep time, sleep onset latency (amount of time it takes to fall asleep), and the number and duration of awakenings in the night are all measures of sleep continuity. Sleep continuity also includes sleep efficiency, which is defined as the total sleep time divided by the time in bed. In other words, sleep efficiency is a ratio of sleep ability (sleep time) to sleep opportunity (time in bed). Sleep continuity variables are typically self-reported in a sleep diary or by using an activity monitor (discussed below). This information can help identify sleep patterns over time or diagnose specific sleep disturbances, such as insomnia or circadian rhythm disorders.

The second domain of sleep quantification is sleep architecture, which measures the electrophysiological changes throughout a sleep episode. Sleep architecture quantifies each stage of sleep, and the progression through each stage. For example, measures of sleep architecture would capture if someone enters REM very quickly after sleep onset, a key symptom in the diagnosis of narcolepsy. The gold-standard for measuring sleep architecture is polysomnography (PSG).

Capturing both the psychological and physiological elements of sleep requires subjective and objective measurements. Subjective assessments, such as questionnaires or sleep diaries, rely on self-report data. While subject perceptions may bias the results, these are the only measures that capture the subjective experience of sleep. Objective measures, such as actigraphy or PSG, replace subject perceptions with independently observable data, such as changes in brain waves or body movement. Although researchers and clinicians tend to prefer objective "hard" data to self-reports, subjective data should be seen as complementary to objective data, not subordinate. For example, suppose an objective measurement captures 8 h of sleep, but a patient only reports 30 min in between tossing and turning. It is tempting to think one of the measurements is wrong, but the reality is that something unusual is happening that neither measure adequately captures. In fact, this phenomenon is referred to as paradoxical insomnia, and is both poorly understood and difficult to treat.

Subjective measures

The simplest measure of sleep is a single question: "How much do you normally sleep?" This question varies in form, sometimes asking about weekday versus weekend sleep, sleeping alone or with a partner, and sleep before and after a child. This basic question is widely used in epidemiological studies and large datasets, such as the National Health and Nutrition Examination Survey. However, this question offers the most limited insight into a person's sleep. First, it is subject to recall bias, in that subjects can recall things differently than what actually happened. The second problem is resolution, since asking about sleep in the last month or year will lead subjects to average across many nights based only on what they can remember. This reduces temporal precision and increases recall bias. However, this question may be the only way to acquire historical information about sleep, such as when a clinician is seeking to understand the course of a sleep disorder. When patients report decreasing sleep durations and increasing sleep onset latency over the course of a month, objective measures may not be necessary to initiate treatment for early insomnia.

The next level of subjective measurements of sleep are validated questionnaires, such as the Insomnia Severity Index. Although questionnaires are both subjective and retrospective, they are usually standardized to capture specific data or screen for specific disorders. A wide variety of questionnaires exist, and a few are listed in Table 1.1.

The final subjective measurement is the sleep diary which has been used for decades in research and clinical settings. Subjects report on different sleep continuity variables shortly after waking up. Unlike questionnaires, a sleep diary is considered a prospective measurement of sleep. However, if too much time passes between the sleep episode and the recording date, recall bias affects the accuracy of the data. Sleep diaries are easy to use (the subject can complete it on paper or electronically) and can be easily collected for any length of time. For these reasons, sleep diaries are an effective tool for longitudinal assessments of sleep and sleep/wake timing.

Objective measures

Objective measures replace subjective perceptions with independent, observable phenomena. The most ubiquitous form of objective sleep measurement is actigraphy. An activity monitoring device, usually a wrist-worn device, uses an accelerometer to detect and measure bodily motion. An activity threshold is set and any activity level below the threshold is classified as either "rest" or "sleep." In addition to devices used by clinicians and researchers, there are consumer "smart" devices, such as phones and watches, that utilize actigraphy. However, the algorithms used to assess sleep efficiency and stages are proprietary and can vary between companies. It is important to choose a company and device that has been validated in many populations and against PSG data and other sleep metrics when conducting a study or for sleep assessment. An activity monitoring device can also include a light sensor. A light sensor allows the clinician to also measure changes in natural and artificial light the patient is exposed to. This is useful for capturing an individual's photoperiod (i.e., the period of light exposure), which can help determine if light exposure is related to the

TABLE 1.1 Sleep questionnaires.

Questionnaire	Description
Insomnia Severity Index [6]	Self-report measure of symptoms of insomnia, such as difficulty falling asleep or waking up too early. Answers are scored as 0–4, added together, and compared to cutoffs to determine the likelihood of clinical insomnia.
Epworth Sleepiness Scale [7] Karolinska Sleepiness Scale [8]	Sleepiness scales measure a subject's propensity to fall asleep. The Epworth measures trait sleepiness, reflected in the ability to fall asleep in a variety of environments and contexts. The Karolinska measures state sleepiness, specifically the sleepiness experienced in the last 10 min on a 1–9 scale.
Pittsburgh Sleep Quality Index [9]	The PSQI measures the frequency of sleep difficulties, particularly subjective disturbances in sleep. Commonly used in research and as an outcome measure of sleep therapies.
Sleep Disorders Symptom Checklist-25 [10]	A 25 item questionnaire that assesses the most common symptoms of several sleep disorders, such as insomnia, circadian rhythm disorders, sleep apnea, bruxism, and narcolepsy.
STOP-BANG [11]	An 8 item questionnaire that generates a risk score for obstructive sleep apnea.
Morningness-Eveningness Questionnaire [12]	A 19 item questionnaire that measures an individual's circadian preference in wakefulness. Lower scores are associated with evening preference, while higher scores are associated with morning preference.

individual's sleep. For example, blue light has a detrimental effect on sleep, and so limiting blue light late in the evening may be a treatment strategy for some patients.

Like sleep diaries, activity monitors provide day-to-day measures of sleep continuity. Additionally, multiple weeks of actigraphy data can be used to evaluate sleep/wake cycles and related circadian rhythms. Additional photopic data can show whether light exposure is affecting a person's circadian rhythm. For example, repeated blue light exposure late at night may shift sleep onset to later in the evening, resulting in a delayed circadian phase. The coupling of actigraphy to light exposure allows the comparison of sleep/wake behavior with external phototopic cues, which are helpful in assessing the synchronization of sleep/wake cycles with light/dark cycles.

The other objective measure of sleep is polysomnography (PSG), also known as a "sleep study." During a sleep study, subjects spend 1–2 nights in the sleep laboratory wearing sensors that measure brain activity, eye movements, muscle movements, heart activity, respiratory activity, and blood oxygen levels. Additional sensors may be placed on the legs to measure periodic limb movements, which can occur naturally or as part of sleep movement disorders.

PSG measures the electrical and physiological changes that occur during sleep and is currently the only way to determine sleep stages. The primary clinical utility of PSG, however, is for the diagnosis and treatment of sleep apnea. Sleep apnea is a condition where patients cease breathing during sleep, often due to upper airway collapse. The PSG captures these events as a decrease in both nasal airflow and blood oxygenation, and the number of events per hour is used as a measure of the severity of the sleep apnea. In some cases, physicians will order a "split-night" study, in which sleep apnea is measured in the first part of the night, and then positive airway pressure therapy is initiated to control the apneas in the later half.

CONCLUSION

Without sleep, a wide variety of systems such as cell division, metabolism, neurological functions, and mental and emotional health would all be greatly impaired. Diseases that affect sleep are life-altering, and if left untreated, can decrease quality of life and increase risk of death. It is also important to synchronize sleep to our external world, a job done exceedingly well by the biological clock. The rhythmicity and predictability of sleep highlights where and how disruptions are occurring in various conditions and disorders. Understanding sleep at the fundamental level is critical in understanding the clinical significance sleep has on overall health.

REFERENCES

[1] Kales A, Rechtschaffen A. A manual of standardized terminology, techniques and scoring system for sleep stages of human subjects. vol. 57. Bethesda, MD: U.S. National Institute of Neurological Diseases and Blindness, Neurological Information Network; 1968.

[2] Berry RB, et al. AASM scoring manual updates for 2017 (version 2.4). J Clin Sleep Med 2017;13(5):665–6.

[3] Moser D, et al. Sleep classification according to AASM and Rechtschaffen & Kales: effects on sleep scoring parameters. Sleep 2009;32(2):139–49.

[4] Rothbaum BO, Mellman TA. Dreams and exposure therapy in PTSD. J Trauma Stress 2001;14(3):481–90.

[5] Desseilles M, et al. Cognitive and emotional processes during dreaming: a neuroimaging view. Conscious Cogn 2011;20(4):998–1008.

[6] Bastien CH, Vallieres A, Morin CM. Validation of the insomnia severity index as an outcome measure for insomnia research. Sleep Med 2001;2(4):297–307.

[7] Johns MW. Reliability and factor analysis of the Epworth sleepiness scale. Sleep 1992;15(4):376–81.

[8] Kaida K, et al. Validation of the Karolinska sleepiness scale against performance and EEG variables. Clin Neurophysiol 2006;117(7):1574–81.

[9] Buysse DJ, et al. The Pittsburgh sleep quality index: a new instrument for psychiatric practice and research. Psychiatry Res 1989;28(2):193–213.

[10] Klingman KJ, Jungquist CR, Perlis ML. Questionnaires that screen for multiple sleep disorders. Sleep Med Rev 2017;32:37–44.

[11] Chung F, et al. STOP questionnaire: a tool to screen patients for obstructive sleep apnea. Anesthesiology 2008;108(5):812–21.

[12] Horne JA, Ostberg O. A self-assessment questionnaire to determine morningness-eveningness in human circadian rhythms. Int J Chronobiol 1976;4(2):97–110.

Chapter 2

Epidemiology of insufficient sleep and poor sleep quality

Michael A. Grandner
Sleep and Health Research Program, Department of Psychiatry, University of Arizona College of Medicine, Tucson, AZ, United States

SLEEP AT THE POPULATION LEVEL

Sleep is a universal human phenomenon and impacts every person, every day (whether or not they actually get to sleep). For this reason, population-level estimates of sleep are important. However, they may be difficult to obtain. Since an individual is unconscious while they are sleeping (and for the time surrounding sleep onset and awakening), accurate assessment of the population burden of sleep disturbance may be difficult. Methods typically exist on a continuum whereby increased generalizability is compromised by reduced precision. For example, most population-level estimates are based on a retrospective self-report, which lacks precision. More precise measures, such as polysomnography and even actigraphy, have been thus far impractical for truly large and population-level assessments. Still, several tentative conclusions about the population can be drawn regarding sleep health.

DEFINING INSUFFICIENT SLEEP

There has been a general lack of consensus on the definition of what constitutes "insufficient sleep" in the general population since at least 1964 [1], when Hammond published the finding that habitual short and long sleep duration were associated with increased mortality rates. Since that time, there has been considerable debate regarding how sleep insufficiency should be defined. Laboratory studies where sleep is manipulated in an experimental protocol are preferred by some (because of their precision) and population-based studies where individuals are observed relative to habitual sleep behaviors are preferred by others (because of their generalizability).

Regarding the former, information about the physiologic and health consequences of sleep duration often come from studies that employ *total sleep deprivation* (defined as an experimental manipulation where an individual is kept awake for at least an entire sleep period) and *partial sleep deprivation* (defined as an experimental manipulation where an individual's sleep period is restricted over a period of days). This is also sometimes called *sleep restriction*. Sometimes, partial sleep deprivation can be characterized as *chronic partial sleep deprivation* (defined as partial sleep deprivation over a period of weeks). All of these experimental manipulations can be useful to discern physiologic effects of changes in sleep duration, but they are generally poor approximations of real-world sleep. As such, *total sleep deprivation, partial sleep deprivation/sleep restriction,* and *chronic partial sleep deprivation* sacrifice generalizability for precision [2–4].

Other studies use population-based studies of sleep. These studies can characterize *habitual sleep duration* (defined as typical perceived sleep duration experienced in real-world settings), often categorized as *short sleep duration, normal/normative sleep duration,* and *long sleep duration* based on cutoffs that often vary by study. These studies may also model *sleep loss* (reduction in sleep duration over time). They may also capture aspects of sleep continuity, including *total sleep time* (calculated sleep duration based on time in bed, subtracting sleep latency and wake time after sleep onset). These parameters may be assessed retrospectively (e.g., through surveys and questionnaires) or prospectively; prospective assessments can be subjective (e.g., sleep diary) or objective (e.g., actigraphy). These studies often sacrifice precision for generalizability [2–4].

But what is "insufficient sleep?" Often, terms such as *sleep deprivation, sleep loss, short sleep,* and others are used interchangeably. Also, "insufficient sleep" is sometimes used interchangeably with concepts such as *sleep deficiency* (insufficient sleep duration or inadequate sleep quality), *poor sleep quality,* and even *insomnia* despite these concepts being misapplied to insufficient sleep [2–4].

With this in mind, defining insufficient sleep has been problematic, since all of these concepts have appropriated the label of "insufficient sleep." For the purposes of this chapter, "insufficient sleep" will refer to sleep duration that is likely too brief to meet physiologic needs. Also, this chapter focuses on habitual sleep duration in the population and thus experimental terms such as *sleep deprivation* are not appropriate. Even at the population level, there is disagreement regarding how much sleep is "insufficient." Various studies use cutoffs of 4, 5, 6, or 7 h as representing insufficient sleep.

Recently, a consensus panel was convened by the American Academy of Sleep Medicine and Sleep Research Society to determine the recommended amount of sleep for a healthy adult. This panel recommended that 7 or more hours was recommended [5, 6]. In a follow-up manuscript, the panel members discussed in detail how this was reached, pointing out that the consensus was most clear that 6 h or less was likely insufficient and less clear for sleep durations between 6 and 7 h [5, 6]. This finding was echoed in similar consensus statements issues by the National Sleep Foundation [7, 8], the American Thoracic Society [9], and the American Heart Association [10]. Therefore, for the purposes of this chapter, "insufficient sleep" will generally refer to habitual sleep duration of 6 h or less.

PREVALENCE OF INSUFFICIENT SLEEP

In order to estimate the prevalence of insufficient sleep in the population, data sources that assess habitual sleep duration in large samples that are representative of the general population. Existing work in this area is limited, as most studies that investigate sleep in such samples do so without using well-validated assessments of sleep. It is important to note that most population estimates of habitual sleep duration are based on subjective, retrospective self-report, which presents biases in assessing sleep [11, 12]. These estimates may better reflect time in bed than actual physiologic sleep and should be interpreted with appropriate caution.

Insufficient sleep in the population

Estimates of the prevalence of insufficient sleep have used the Behavioral Risk Factor Surveillance System (BRFSS) in the United States. The BRFSS is an annual telephone survey of hundreds of thousands of US adults, conducted by the Centers for Disease Control and Prevention (CDC) (http://www.cdc.gov/brfss). It is state-based, with population-weighted samples representing each strata of age, sex, race/ethnicity, and geographic region. Sleep duration in the BRFSS is assessed with the item, "On average, how many hours of sleep do you get in a 24-h period?" Responses are coded in whole numbers. Liu and colleagues reported population-weighted prevalence estimates for sleep duration around a cutoff of 7 h (based on the consensus statement [13] from the 2014 BRFSS ($N=444,306$). Overall, the age-adjusted estimated prevalence of insufficient sleep ($\leq$6 h) was reported to be 35.1% of the US population. Grandner and colleagues [14] reported prevalence estimates also using the 2014 BRFSS. Estimated prevalence by hour was calculated, such that the estimated prevalence by hour of sleep duration was 1.12% for $\leq$3 h, 3.19% for 4 h, 7.75% for 5 h, 23.55% for 6 h, 28.72% for 7 h, 27.64% for 8 h, 4.42% for 9 h, 2.35% for 10 h, and 1.27% for $\geq$11 h. See Fig. 2.1 for a graphical representation of these data.

Other prevalence estimates have also been calculated using the National Health and Nutrition Examination Survey (NHANES). The NHANES is a survey that is also conducted by the CDC that includes a nationally-representative

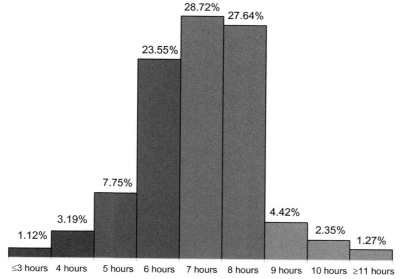

FIG. 2.1 Distribution of sleep duration in the US Population using 2014 BRFSS. *Data from Grandner MA, Seixas A, Shetty S, Shenoy S. Sleep duration and diabetes risk: population trends and potential mechanisms. Curr Diab Rep 2016;16(11):106. PubMed PMID: 27664039.*

sample (http://www.cdc.gov/nchs/nhanes). The sample size is much smaller than the BRFSS, though reliability of data may be improved since surveys were administered in person rather than over the phone. Similar to the BRFSS, NHANES assesses sleep duration by whole number hour (no partial hours). Unlike the BRFSS, though, NHANES assesses sleep duration with the item, "How much sleep do you usually get at night on weekdays or workdays?" Thus, this item may capture modal nighttime sleep, rather than 24-h sleep, which may include naps. Using the 2007–2008 wave of NHANES, Grandner and colleagues calculated prevalence estimates for sleep duration by category, with 4.96% reporting ≤4h, 32.16% reporting 5–6h, 55.68% reporting 7–8h, and 7.20% reporting ≥9h [15]. Thus, insufficient sleep (≤6h) was reported by 37.12% of the US population. The higher estimate relative to BRFSS may be explained by the wording of the item, which does not include naps or weekends. See Fig. 2.1 for an illustration of these values.

Lower estimates of short sleep duration are reported by Basner and colleagues using data from the American Time Use Survey (ATUS) [16]. The ATUS is conducted annually by the US Bureau of Labor Statistics and assigns activity codes to each 15-min increment of the 24-h day in a representative sample of US adults (http://www.bls.gov/tus). Because ATUS does not distinguish time in bed from time asleep, values will generally overestimate sleep and understate insufficient sleep [16]. Using ATUS from 2003 to 2011 (N=124,517), the estimated prevalence of insufficient sleep (≤6h) was 10.6%, compared to 78.4% for 6–11h and 11.0% for ≥11h.

Thus, estimates for insufficient sleep (≤6h) from relatively recent, nationally representative surveys, are 10.6% from ATUS, 35.1% from BRFSS, and 37.12% from NHANES. These may vary as a result of the survey item asked, as well as other factors including the years included and sampling methodologies. Although other studies have examined large samples using more well-validated measures, none of these studies are nationally-representative and thus cannot be used to develop population prevalence estimates.

Rather than assess insufficient sleep relative to a benchmark (sleep hours), an alternative approach would be to ask individuals how often they perceive their sleep to be insufficient. The 2008 BRFSS asked, "During the past 30 days, for about how many days have you felt that you did not get enough rest or sleep?" Based on this variable, Mcknight-Eily and colleagues [17] reported prevalence estimates based on responses to this variable. They estimate that 30.7% of the population reports 0/30 days of insufficient sleep, with 1–13 days reported by 41.3% of the population, 14–29 days reported by 16.8% of the population, and 30/30 days reported by 11.1% of the population. Based on these estimates, 27.9% of the US population reports perceived sleep insufficiency at least 2 weeks out of the month. Interestingly, this estimate is similar to the ~1/3 of the population who experience insufficient sleep based on sleep duration, though the overlap between these groups is only moderate [18].

Insufficient sleep by age

Based on BRFSS data, Liu and colleagues [19] provided age-based prevalence estimates for insufficient sleep (≤6h). They reported estimated of 32.2% for those age 18–24, 37.9% for 25–34, 38.3% for 35–44, 37.3% for 45–64, and 26.3% for those 65 or older (see Fig. 2.2). Of note, the lowest rate of insufficient sleep was seen among the oldest adults. This is consistent with other studies that showed that perceived insufficient sleep declines with age [20], as does self-reported sleep disturbance [21–23]. This is in contrast to more objective sleep disturbances, which are well-characterized to increase in older adults [24–26]. There are a number of potential reasons for this, including retirement offering greater sleep opportunity and differing expectations regarding sleep [27].

Similar prevalence estimates of sleep duration by age in NHANES were reported by Grandner and colleagues [15]. Among teenagers aged 16–17, prevalence of sleep duration was 0.63% for ≤4h, 19.38% for 5–6h, 62.47% for 7–8h, and 17.52% for ≥9h. For younger adults aged 18–30, prevalence was 4.83% for ≤4h, 31.02% for 5–6h, 54.44% for 7–8h, and 9.81% for ≥9h. For adults aged 30–50, prevalence was 5.86% for ≤4h, 33.61% for 5–6h, 55.49% for 7–8h, and 5.03% for ≥9h. For adults aged 50–65, prevalence was 4.95% for ≤4h, 35.41% for 5–6h, 56.04% for 7–8h, and 3.61% for ≥9h. For older adults 65 and older, prevalence was 4.17% for ≤4h, 28.31% for 5–6h, 55.58% for 7–8h, and 11.94% for ≥9h. Thus, prevalence of insufficient sleep (≤6h) was reported to be 20.01% for those aged 16–17, 35.85% for those aged 18–30, 39.47% for adults 30–50, 40.36% for adults age 50–65, and 32.48% for older adults over 65. Again, prevalence of insufficient sleep is highest in working age adults.

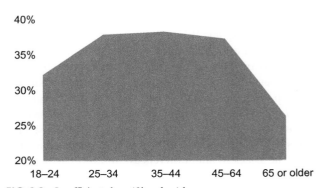

FIG. 2.2 Insufficient sleep (6h or less) by age.

Using the ATUS data, Basner and colleagues [16] found that, compared to 15–24 year olds, increased likelihood of insufficient sleep (≤6h) was seen in those aged 25–34 (OR=1.38; 95% CI=1.18;1.61), 35–44 (OR=1.40; 95% CI=1.22;1.62), 45–54 (OR=1.68; 95% CI=1.44;1.94), and 55–64 (OR=1.41; 95% CI=1.18;1.68), but not those 65 or older. Similarly, shortest sleep durations were seen in working age adults.

Using self-reported insufficiency from the BRFSS, Mcknight-Eily and colleagues [17] report that the prevalence of self-reported insufficient sleep at least 14 of the past 30 days was reported by 31.3% of 18–24 year olds. Estimated prevalence was 34.2% for 35–34 year olds, 32.1% for 35–44 year olds, 27.2% for 45–64 year olds, and 15.0% for those 65 or older.

Insufficient sleep by sex

Several studies have examined sex relative to insufficient sleep. Liu and colleagues reports that based on the 2014 BRFSS data, insufficient sleep (≤6h) is reported by 35.4% of men and 34.8% of women [19]. Using data from the 2007–2008 NHANES, Whinnery and colleagues report no sex differences in likelihood of insufficient sleep (though they report that women are 35% less likely to report long sleep duration after adjusting for covariates) [28]. Using NHIS data, Krueger and Friedman report that men are 7% less likely to report ≤5 vs 7h of sleep [29]. Basner and colleagues report that men are more likely to report insufficient sleep (OR=1.27; 95% CI=1.20; 1.35) [16]. McKnight-Eily reports that self-reported insufficient sleep at least 14 out of the past 30 days was reported by 25.5% of men and 30.4% of women [17]. Taken together, sex differences in insufficient sleep are likely small and difficult to observe. This is in contrast to self-reported sleep disturbances, which are much more prevalent in women [30–32].

Insufficient sleep by race/ethnicity

Many studies have documented differences in sleep duration by race/ethnicity. In general, racial/ethnic minorities are more likely to experience insufficient sleep duration. Actigraphic studies have shown that racial/ethnic minorities demonstrate a sleep duration between 40 and 60 min less than non-Hispanic White counterparts [33–35].

More data are available from survey studies that included larger numbers of people but lack the precision of objective measurements. For example, data from the NHIS has shown that sleep duration of 6h or less was more prevalent among Blacks/African-Americans, non-Mexican Hispanics/Latinos, and Asians/Others, compared to non-Hispanic Whites [36, 37]. Longitudinal analysis of NHIS data suggests that Black-White differences in insufficient sleep have persisted, relatively unchanged since 1977 [38, 39]. See Fig. 2.3 for an illustration of this.

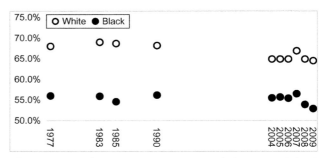

FIG. 2.3 Black-white differences in 7–8h sleep in the US Population in NHIS. *Data from Jean-Louis G, Grandner MA, Youngstedt SD, Williams NJ, Zizi F, Sarpong DF, Ogedegbe GG. Differential increase in prevalence estimates of inadequate sleep among black and white Americans. BMC Public Health 2015;15:1185. PubMed PMID: 26611643; PMCID: PMC4661980; Jean-Louis G, Youngstedt S, Grandner M, Williams NJ, Sarpong D, Zizi F, Ogedegbe G. Unequal burden of sleep-related obesity among black and white Americans. Sleep Health. 2015;1(3):169–176. PubMed PMID: 26937487; PMCID: PMC4770938.*

Other population-level studies have found similar patterns. For example, Stamatakis showed in the Alameda County study that African-Americans were about twice as likely to report short sleep duration [40]. Using NHANES data, Whinnery and colleagues showed that Blacks/African-Americans are about 2.5 times as likely to sleep <5h and about twice as likely to sleep 5–6h, compared to non-Hispanic Whites. Non-Mexican Hispanics/Latinos were about 2.7 times as likely to sleep <5h and Asians/Others were about four times as likely to sleep <5h and about twice as likely to sleep 5–6h. Mexican-Americans were the only minority group not more likely to report insufficient sleep [28].

Insufficient sleep by socioeconomic status

Perhaps due to environmental stressors, those of lower socioeconomic status are more likely to experience insufficient sleep. Kruger and Friedman used NHIS data to compute mean family income according to sleep duration [29]. They found that the highest mean income was reported among 7-h sleepers ($48,065), with the lowest income levels in those sleeping 5h or less ($36,819) or 9h or more ($34,883). Stamatakis evaluated likelihood of insufficient sleep relative to income quintile [40]. This study reported that compared to the highest income quintile, short sleep duration (6h or less) was increasingly reported in the fourth (3% more likely), third (11% more likely), second (29% more likely), and first quintile (54% more likely). Using BRFSS data, days of perceived insufficient sleep decreased at higher levels of household income [20].

Using NHANES data, Whinnery and colleagues examined several socioeconomic indices relative to sleep duration [28]. Compared to those with family income over $75,000, increased likelihood of <5h of sleep ($P<0.05$) was observed for all categories, including <$20,000 (OR=5.5), $20,000–$25,000 (OR=2.9), $25,000–$35,000 (OR=4.1),

$35,000–$45,000 (OR=2.4), $45,000–$55,000 (OR=2.8), $55,000–$65,000 (OR=2.4), and even $65,000–$75,000 (OR=3.8). Increased likelihood of 5–6h sleep relative to those earning over $75,000 was only seen in the lowest income group earning <$20,000 (OR=1.3). Education level was another socioeconomic indicator that was associated with sleep duration in this sample. Those with less than a high school education were approximately four times as likely to report <5h of sleep, compared to college graduates. Similarly, those who completed some high school were more likely than college graduates to report <5 (OR=5.3) and 5–6 (OR=1.7) hours of sleep, those who completed high school were more likely than college graduates to report <5 (OR=4.3) or 5–6 (OR=1.6) hours, and those with some college were also more likely than college graduates to report <5 (OR=3.6) or 5–6 (OR=1.6) hours of sleep [28]. Another socioeconomic indicator evaluated in this study was lack of access to healthcare, which was more common among those reporting <5h of sleep. Food insecurity—a measure of inability to financially provide healthy access to enough food—was also more common among those reporting <5 and 5–6h of sleep [28].

Insufficient sleep by geography

Insufficient sleep in the United States is differentially experienced across varying regions of the country. An analysis of self-reported perceived insufficient sleep using BRFSS data was reported [41]. Using a geospatial hotspot analysis, several key "hotspots" of insufficient sleep were identified in the United States, including parts of the southeast, parts of the Texas/Louisiana border, areas in the Midwest, and the largest hotspot in central Appalachia. "Coldspots" with abnormally low levels of insufficient sleep were seen in the northern Midwest (Wisconsin/Minnesota/Iowa), central Texas, central Virginia, and areas in along the West Coast. See Fig. 2.4 for a map of US counties relative to their proportion of insufficient sleep and Fig. 2.5 for a map of hotspots and coldspots.

Rather than examine statistical hotspots of perceived insufficient sleep, researchers at the CDC used BRFSS data to map prevalence of ≤6h of sleep across the United States[19]. The US states with the highest prevalence were (in order) Hawaii (43.9%), Kentucky (39.7%), Maryland (38.9%), Alabama (38.8%), Georgia (38.7%) and Michigan (38.7%). The US states with the lowest prevalence were (in order) South Dakota (28.4%), Colorado (28.5%), Minnesota (29.2%), Nebraska (30.4%), and Idaho (30.6%).

KEY LIMITATIONS TO POPULATION ESTIMATES OF INSUFFICIENT SLEEP

There are several key limitations in the existing literature on insufficient sleep epidemiology. First, there is a lack of clarity of gold-standard methods for estimating population levels of sleep duration. Most of these studies used

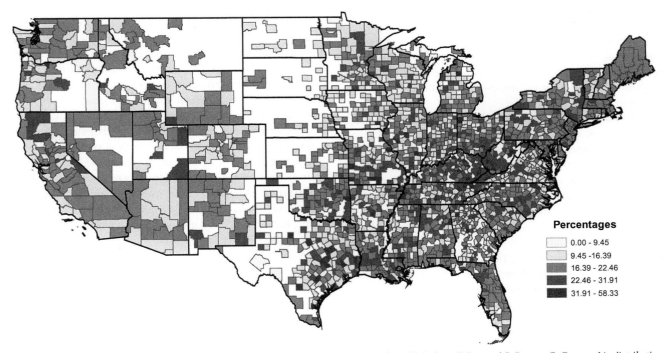

FIG. 2.4 County-level insufficient sleep in the US. *From Grandner MA, Smith TE, Jackson N, Jackson T, Burgard S, Branas C. Geographic distribution of insufficient sleep across the United States: a county-level hotspot analysis. Sleep Health 2015;1(3):158–165. PubMed PMID: 26989761; PMCID: 4790125.*

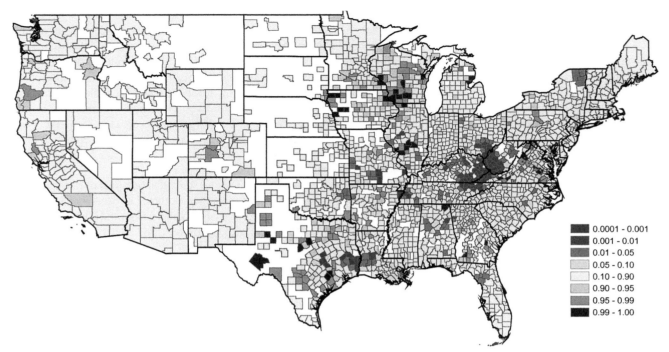

FIG. 2.5 Hotspots and coldspots of insufficient sleep in the US. *From Grandner MA, Smith TE, Jackson N, Jackson T, Burgard S, Branas C. Geographic distribution of insufficient sleep across the United States: a county-level hotspot analysis. Sleep Health 2015;1(3):158–165. PubMed PMID: 26989761; PMCID: 4790125.*

single-item self-report measures from surveys, which are fraught with psychometric problems [2, 11, 42]. Not only do self-report measures tend to over-report sleep relative to physiologic recordings and likely better approximate time in bed than physiologic sleep, they may be subject to a number of other biases, demand characteristics, and social desirability. There still exists no nationally-representative dataset that estimates sleep duration based on gold-standard approaches, especially those that record physiologic sleep.

Second, the definition of insufficient sleep varies widely across studies, and most studies do not allow enough resolution to examine different cutoffs. Given recent consensus statements [5–7, 9, 10], a cutoff of 7h seems reasonable, but there is yet no clear consensus on the range between 6 and 7h, where many Americans fall regarding their typical sleep habits. Also, it is not clear whether a determination of insufficient sleep should be made on the basis of physiologic sleep or perceived sleep.

Third, definitions of insufficient sleep are based on nomothetic, population-level recommendations which don't take into account individual differences in sleep need, sleep ability, and resilience to sleep loss. Also, these do not necessarily take into account sleep sufficiency relative to any particular outcome. Future work should consider these issues in order to take a more personalized/precision medicine view of sleep duration, as it relates to an individual and impacts on specific outcome measures, in a specific set of contexts.

PREVALENCE OF POOR SLEEP QUALITY

Poor sleep quality, like insufficient sleep, has been variably defined. The National Sleep Foundation has recently attempted to develop a coherent conceptualization of sleep quality [43, 44]. In a consensus document, elements of sleep quality included sleep latency (amount of time to fall asleep), wake time after sleep onset (amount of time awake at night), and sleep efficiency (proportion of the time in bed spent sleeping). Thus, sleep quality was generally defined as good sleep continuity. Recognizing the limitations of this, the National Sleep Foundation has begun work on a tool to measure sleep satisfaction with is presented as another key element of overall sleep quality [45]. In addition to sleep-focused elements as indicators of sleep quality, perhaps daytime indicators can be useful as well. For example, daytime sleepiness is often an indicator of poor nighttime sleep [46, 47] and may also serve as an indicator of poor sleep quality.

Prevalence of sleep disorders

Poor sleep quality can refer to a relatively wide range of problems, including sleep disorders as well as sleep symptoms. The most common types of sleep disorders in the population are insomnia and sleep apnea. Although other chapters in this volume focus specifically on these issues at the population level, it is important to note that the population prevalence of acute insomnia is high (about 4% per month) [48, 49] and that although most of these resolve, approximately 10% of the population likely meets criteria for an insomnia disorder [50, 51].

Regarding sleep apnea, prevalence estimates need to account for sex and body mass index. Relatively recent estimates of the prevalence of sleep apnea estimate that among men age 30–49, rates are 7.0%, 18.3%, 44.6%, and 79.5% for those with BMI of <25, 25–29.9, 30–39.9, and 40 or above, respectively. For men 50–70, the rates increase to 18.9%, 36.6%, 61.4%, and 82.8%, respectively. For women age 30–49, the rates of sleep apnea are lower, at 1.4$, 4.2%, 13.5%, and 43.0% for women with a BMI of <25, 25–29.9, 30–39.9, and 40 or higher, respectively. As with men, these numbers are higher in women age 50–70, with 9.3%, 20.2%, 41.1%, and 67.9% with sleep apnea among those with BMI of <25, 25–29.9, 30–39.9, and 40 or greater, respectively. This high prevalence of sleep apnea (Fig. 2.6) is particularly notable [52], especially since recent estimates suggest that approximately 85% of sleep apnea cases are never diagnosed, and up to half of diagnosed cases remain insufficiently treated [53].

Regarding circadian rhythm sleep disorders, the prevalence of delayed sleep phase disorder is estimated to be about 0.2% of the general population but 7–16% of adolescents [54]. Prevalence of other circadian rhythm sleep disorders is largely unknown, though the prevalence of shift work disorder is estimated to be about 5–10% of the population, based on prevalence estimates of night shift work and the prevalence of the disorder among shift workers [55].

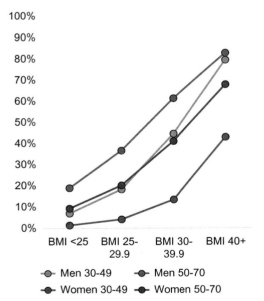

FIG. 2.6 Estimated prevalence of sleep apnea by age group, sex, and BMI. *From Peppard PE, Young T, Barnet JH, Palta M, Hagen EW, Hla KM. Increased prevalence of sleep-disordered breathing in adults. Am J Epidemiol. 2013. PubMed PMID: 23589584; PMCID: 3639722.*

Prevalence of sleep complaints

Several studies have examined prevalence of sleep complaints in the general population. For example, Grandner and colleagues [21] found that the rate of general sleep disturbance in the US population was about 16% in men and 21% in women, and general daytime fatigue was 18% in men and 26% in women. However, this depended on age. Fig. 2.7 depicts the rates of these across age groups, illustrating a general decline in reports with age. Of note, in women, increased sleep duration and tiredness are evident around the age typical of menopause and in both men and women, fatigue increases starting at age 70. When odds ratios for these outcomes were computed after adjusting for covariates that included sociodemographics, health and depression, the decrease in symptoms with age was even more pronounced. This has been replicated by several others, using other databases and addressing the issue of subjective sleep complaint in different ways [20, 22]. In general, self-reported sleep complaints generally decrease with age. This is in contrast to objective sleep disturbances, which generally increase with age [24].

This general sleep complaint may be differentially experienced across demographic groups. Grandner and colleagues [56] showed that in addition to age and sex, general sleep disturbance was reported more frequently among non-Hispanic Whites, compared to other groups. It was also more frequently reported by those with less education, less income, and lack of employment. In addition, it is reported more frequently among those in worse health overall and less healthcare access [21].

Regarding specific sleep complaints, Grandner and colleagues [57] examined data from the NHANES. In a nationally-representative sample, the prevalence of self-reported sleep latency >30 min was 18.8%. Regarding other insomnia symptoms, the prevalence of difficulty at least once per week was 19.4% for falling asleep, 20.9% for resuming sleep during the night, and 16.5% for early morning awakenings; regarding problems at least three nights per week, these rates were reduced to 7.7%, 7.7%, and 5.8%, respectively. Regarding daytime symptoms, daytime sleepiness and nonrestorative sleep at least once per week were reported by 18.8% and 28.7% of the population, respectively; when the criterion was increased to three nights per week, this was reduced to 5.8% and 10.9%, respectively. In this sample, 70.6% of adults reported snoring at least once per week and 51% reported snoring at least three nights per week.

SUMMARY AND CONCLUSIONS

Although accurately measuring sleeping individuals in large numbers is difficult, prevalence estimates for insufficient and poor quality sleep can be obtained from large-scale studies of health. Despite limitations of these estimates,

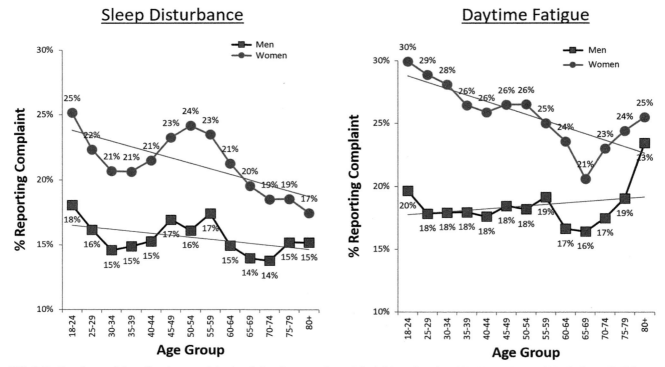

FIG. 2.7 Prevalence of sleep disturbance and daytime fatigue by age and sex. *Adapted from Grandner MA, Martin JL, Patel NP, Jackson NJ, Gehrman PR, Pien G, Perlis ML, Xie D, Sha D, Weaver T, Gooneratne NS. Age and sleep disturbances among American men and women: data from the U.S. Behavioral risk factor surveillance system. Sleep. 2012;35(3):395–406. Epub 2012/03/02. PubMed PMID: 22379246; PMCID: 3274341.*

it is clear that many adults are achieving insufficient sleep duration and/or inadequate sleep quality. This is concerning, since sleep is associated with so many important outcomes including health, daytime functioning, and mental well-being. Public health surveillance efforts should aim to improve measurement of sleep health at the population level across multiple domains. In addition, public health intervention efforts should address healthy sleep as an important population health goal.

REFERENCES

[1] Hammond EC. Some preliminary findings on physical complaints from a prospective study of 1,064,004 men and women. Am J Public Health Nations Health 1964;54:11–23. Epub 1964/01/01. PubMed PMID: 14117648.

[2] Grandner MA, Patel NP, Gehrman PR, Perlis ML, Pack AI. Problems associated with short sleep: bridging the gap between laboratory and epidemiological studies. Sleep Med Rev 2010;14:239–47. Epub 2009/11/10. PubMed PMID: 19896872.

[3] Grandner MA. Sleep deprivation: societal impact and long-term consequences. In: Chokroverty S, Billiard M, editors. Sleep medicine: a comprehensive guide to its development, clinical milestones, and advances in treatment. New York: Spinger; 2016. p. 495–509.

[4] Watson NF, Badr MS, Belenky G, Bliwise DL, Buxton OM, Buysse D, Dinges DF, Gangwisch J, Grandner MA, Kushida C, Malhotra RK, Martin JL, Patel SR, Quan SF, Tasali E. Joint Consensus statement of the American Academy of Sleep Medicine and Sleep Research Society on the recommended amount of sleep for a healthy adult: methodology and discussion. J Clin Sleep Med 2015;11(8):931–52. PubMed PMID: 26235159.

[5] Watson NF, Badr MS, Belenky G, Bliwise DL, Buxton OM, Buysse D, Dinges DF, Gangwisch J, Grandner MA, Kushida C, Malhotra RK, Martin JL, Patel SR, Quan SF, Tasali E. Recommended amount of sleep for a healthy adult: a joint consensus statement of the American Academy of Sleep Medicine and Sleep Research Society. Sleep 2015;38(6):843–4. PubMed PMID: 26039963.

[6] Watson NF, Badr MS, Belenky G, Bliwise DL, Buxton OM, Buysse D, Dinges DF, Gangwisch J, Grandner MA, Kushida C, Malhotra RK, Martin JL, Patel SR, Quan SF, Tasali E, Non-Participating O, Twery M, Croft JB, Maher E, American Academy of Sleep Medicine Society, Barrett JA, Thomas SM, Heald JL. Recommended amount of sleep for a healthy adult: a joint consensus statement of the American Academy of sleep medicine and Sleep Research Society. J Clin Sleep Med 2015;11(6):591–2. PubMed PMID: 25979105.

[7] Hirshkowitz M, Whiton K, Alpert SM, Alessi C, Bruni O, DonCarlos L, Hazen N, Herman J, Hillard PJA, Katz ES, Kheirandish-Gozal L, Neubauer DN, O'Donnell AE, Ohayon M, Peever J, Rawding R, Sachdeva RC, Setters B, Vitiello MV, Ware JC. National Sleep Foundation's updated sleep duration recommendations: final report. Sleep Health 2015;1:233–43.

[8] Hirshkowitz M, Whiton K, Alpert SM, Alessi C, Bruni O, DonCarlos L, Hazen N, Herman J, Katz ES, Kheirandish-Gozal L, Neubauer DN, O'Donnell AE, Ohayon M, Peever J, Rawding R, Sachdeva RC, Setters B, Vitiello MV, Ware JC, Hillard PJA. National Sleep Foundation's sleep time duration recommendations: methodology and results summary. Sleep Health 2015;1:40–3.

[9] Mukherjee S, Patel SR, Kales SN, Ayas NT, Strohl KP, Gozal D, Malhotra A, American thoracic society ad hoc committee on healthy S. An official American thoracic society statement: the importance of healthy sleep. Recommendations and future priorities. Am J Respir Crit Care Med 2015;191(12):1450–8. PubMed PMID: 26075423.

[10] St-Onge MP, Grandner MA, Brown D, Conroy MB, Jean-Louis G, Coons M, Bhatt DL, American Heart Association obesity committee, American Heart Association behavior change committee, American Heart Association diabetes committee, American Heart Association nutrition committee, American Heart Association Council on lifestyle and cardiometabolic health, American Heart Association Council on cardiovascular disease in the Young, American Heart Association Council on clinical cardiology, American Heart Association stroke council. Sleep duration and quality: impact on lifestyle behaviors and cardiometabolic health: a scientific statement from the American Heart Association. Circulation 2016;134(18):e367–86. PubMed PMID: [27647451].

[11] Kurina LM, McClintock MK, Chen JH, Waite LJ, Thisted RA, Lauderdale DS. Sleep duration and all-cause mortality: a critical review of measurement and associations. Ann Epidemiol 2013;23(6):361–70. PubMed PMID: 23622956.

[12] Lauderdale DS, Knutson KL, Yan LL, Liu K, Rathouz PJ. Self-reported and measured sleep duration: how similar are they? Epidemiology 2008;19(6):838–45. Epub 2008/10/16. PubMed PMID: 18854708.

[13] Centers for Disease Control and Prevention. Behavioral risk factor surveillance system 2014 codebook report. Atlanta, GA: CDC; 2015.

[14] Grandner MA, Seixas A, Shetty S, Shenoy S. Sleep duration and diabetes risk: population trends and potential mechanisms. Curr Diab Rep 2016;16(11):106. PubMed PMID: 27664039.

[15] Grandner MA, Schopfer EA, Sands-Lincoln M, Jackson N, Malhotra A. Relationship between sleep duration and body mass index depends on age. Obesity (Silver Spring) 2015;23(12):2491–8. PubMed PMID: 26727118.

[16] Basner M, Spaeth AM, Dinges DF. Sociodemographic characteristics and waking activities and their role in the timing and duration of sleep. Sleep 2014;37(12):1889–906. PubMed PMID: 25325472.

[17] McKnight-Eily LR, Liu Y, Perry GS, Presley-Cantrell LR, Strine TW, Lu H, Croft JB. Perceived insufficient rest or sleep among adults—United States, 2008. MMWR Morb Mortal Wkly Rep 2009;58(42):1175–9.

[18] Altman NG, Izci-Balserak B, Schopfer E, Jackson N, Rattanaumpawan P, Gehrman PR, Patel NP, Grandner MA. Sleep duration versus sleep insufficiency as predictors of cardiometabolic health outcomes. Sleep Med 2012;13(10):1261–70. PubMed PMID: 23141932.

[19] Liu Y, Wheaton AG, Chapman DP, Cunningham TJ, Lu H, Croft JB. Prevalence of healthy sleep duration among adults—United States, 2014. MMWR Morb Mortal Wkly Rep 2016;65(6):137–41. PubMed PMID: 26890214.

[20] Grandner MA, Jackson NJ, Izci-Balserak B, Gallagher RA, Murray-Bachmann R, Williams NJ, Patel NP, Jean-Louis G. Social and Behavioral determinants of perceived insufficient sleep. Front Neurol 2015;6:112. PubMed PMID: 26097464.

[21] Grandner MA, Martin JL, Patel NP, Jackson NJ, Gehrman PR, Pien G, Perlis ML, Xie D, Sha D, Weaver T, Gooneratne NS. Age and sleep disturbances among American men and women: data from the U.S. behavioral risk factor surveillance system. Sleep 2012;35(3):395–406. Epub 2012/03/02. PubMed PMID: 22379246.

[22] Soldatos CR, Allaert FA, Ohta T, Dikeos DG. How do individuals sleep around the world? Results from a single-day survey in ten countries. Sleep Med 2005;6(1):5–13. Epub 2005/02/01. PubMed PMID: 15680289.

[23] Zilli I, Ficca G, Salzarulo P. Factors involved in sleep satisfaction in the elderly. Sleep Med 2009;10(2):233–9. Epub 2008/04/05. PubMed PMID: 18387848.

[24] Ohayon MM, Carskadon MA, Guilleminault C, Vitiello MV. Meta-analysis of quantitative sleep parameters from childhood to old age in healthy individuals: developing normative sleep values across the human lifespan. Sleep 2004;27(7):1255–73. Epub 2004/12/14. PubMed PMID: 15586779.

[25] Lindstrom V, Andersson K, Lintrup M, Holst G, Berglund J. Prevalence of sleep problems and pain among the elderly in Sweden. J Nutr Health Aging 2012;16(2):180–3. Epub 2012/02/11. PubMed PMID: 22323355.

[26] Cooke JR, Ancoli-Israel S. Normal and abnormal sleep in the elderly. In: Montagna P, Chokroverty S, editors. Sleep disorders. Edinburgh: Elsevier; 2011. p. 653–65.

[27] Grandner MA, Patel NP, Gooneratne NS. Difficulties sleeping: a natural part of growing older? Aging Health 2012;8(3):219–21.

[28] Whinnery J, Jackson N, Rattanaumpawan P, Grandner MA. Short and long sleep duration associated with race/ethnicity, sociodemographics, and socioeconomic position. Sleep 2014;37(3):601–11. PubMed PMID: 24587584.

[29] Krueger PM, Friedman EM. Sleep duration in the United States: a cross-sectional population-based study. Am J Epidemiol 2009;169(9):1052–63. Epub 2009/03/21. PubMed PMID: 19299406.

[30] Schredl M, Reinhard I. Gender differences in nightmare frequency: a meta-analysis. Sleep Med Rev 2011;15(2):115–21. Epub 2010/09/08. PubMed PMID: 20817509.

[31] Subramanian S, Guntupalli B, Murugan T, Bopparaju S, Chanamolu S, Casturi L, Surani S. Gender and ethnic differences in prevalence of self-reported insomnia among patients with obstructive sleep apnea. Sleep Breath 2011;15(4):711–5. Epub 2010/10/19. PubMed PMID: 20953842.

[32] Zhang B, Wing YK. Sex differences in insomnia: a meta-analysis. Sleep 2006;29(1):85–93. Epub 2006/02/04. PubMed PMID: 16453985.

[33] Jean-Louis G, Kripke DF, Ancoli-Israel S, Klauber MR, Sepulveda RS. Sleep duration, illumination, and activity patterns in a population sample: effects of gender and ethnicity. Biol Psychiatry 2000;47(10):921–7. PubMed PMID: 10807965.

[34] Lauderdale DS, Knutson KL, Yan LL, Rathouz PJ, Hulley SB, Sidney S, Liu K. Objectively measured sleep characteristics among early-middle-aged adults: the CARDIA study. Am J Epidemiol 2006;164(1):5–16. Epub 2006/06/03. PubMed PMID: 16740591.

[35] Ertel KA, Berkman LF, Buxton OM. Socioeconomic status, occupational characteristics, and sleep duration in African/Caribbean immigrants and US white health care workers. Sleep 2011;34(4):509–18. PubMed PMID: 21461330.

[36] Hale L, Do DP. Racial differences in self-reports of sleep duration in a population-based study. Sleep 2007;30(9):1096–103. Epub 2007/10/04. PubMed PMID: 17910381.

[37] Nunes J, Jean-Louis G, Zizi F, Casimir GJ, von Gizycki H, Brown CD, McFarlane SI. Sleep duration among black and white Americans: results of the National Health Interview Survey. J Natl Med Assoc 2008;100(3):317–22. Epub 2008/04/09. PubMed PMID: 18390025.

[38] Jean-Louis G, Grandner MA, Youngstedt SD, Williams NJ, Zizi F, Sarpong DF, Ogedegbe GG. Differential increase in prevalence estimates of inadequate sleep among black and white Americans. BMC Public Health 2015;15:1185. PubMed PMID: 26611643.

[39] Jean-Louis G, Youngstedt S, Grandner M, Williams NJ, Sarpong D, Zizi F, Ogedegbe G. Unequal burden of sleep-related obesity among black and white Americans. Sleep Health 2015;1(3):169–76. PubMed PMID: 26937487.

[40] Stamatakis KA, Kaplan GA, Roberts RE. Short sleep duration across income, education, and race/ethnic groups: population prevalence and growing disparities during 34 years of follow-up. Ann Epidemiol 2007;17(12):948–55. Epub 2007/09/15. PubMed PMID: 17855122.

[41] Grandner MA, Smith TE, Jackson N, Jackson T, Burgard S, Branas C. Geographic distribution of insufficient sleep across the United States: a county-level hotspot analysis. Sleep Health 2015;1(3):158–65. PubMed PMID: 26989761.

[42] Watson NF, Badr MS, Belenky G, Bliwise DL, Buxton OM, Buysse D, Dinges DF, Gangwisch J, Grandner MA, Kushida C, Malhotra RK, Martin JL, Patel SR, Quan SF, Tasali E. Joint Consensus statement of the American Academy of sleep medicine and Sleep Research Society on the recommended amount of sleep for a healthy adult: methodology and discussion. Sleep 2015;38(8):1161–83. PubMed PMID: 26194576.

[43] Knutson KL, Phelan J, Paskow MJ, Roach A, Whiton K, Langer G, Hillygus DS, Mokrzycki M, Broughton WA, Chokroverty S, Lichstein KL, Weaver TE, Hirshkowitz M. The National Sleep Foundation's Sleep Health Index. Sleep Health 2017;3(4):234–40. Epub 2017/07/16. PubMed PMID: 28709508.

[44] Ohayon M, Wickwire EM, Hirshkowitz M, Albert SM, Avidan A, Daly FJ, Dauvilliers Y, Ferri R, Fung C, Gozal D, Hazen N, Krystal A, Lichstein K, Mallampalli M, Plazzi G, Rawding R, Scheer FA, Somers V, Vitiello MV. National Sleep Foundation's sleep quality recommendations: first report. Sleep Health 2017;3(1):6–19. Epub 2017/03/28. PubMed PMID: 28346153.

[45] Ohayon MM, Chen MC, Bixler E, Dauvilliers Y, Gozal D, Plazzi G, Vitiello MV, Paskow M, Roach A, Hirshkowitz M. A provisional tool for the measurement of sleep satisfaction. Sleep Health 2018;4(1):6–12. Epub 2018/01/16. PubMed PMID: 29332682.

[46] Ferini-Strambi L, Sforza M, Poletti M, Giarrusso F, Galbiati A. Daytime sleepiness: more than just obstructive sleep apnea (OSA). Med Lav 2017;108(4):260–6. PubMed PMID: 28853423.

[47] Malhotra RK. Sleepy or sleepless: clinical approach to the sleep patient. Heidelberg: Springer; 2015.

[48] Ellis JG, Gehrman P, Espie CA, Riemann D, Perlis ML. Acute insomnia: current conceptualizations and future directions. Sleep Med Rev 2012;16(1):5–14. Epub 2011/05/21. PubMed PMID: 21596596.

[49] Ellis JG, Perlis ML, Neale LF, Espie CA, Bastien CH. The natural history of insomnia: focus on prevalence and incidence of acute insomnia. J Psychiatr Res 2012;46(10):1278–85. PubMed PMID: 22800714.

[50] Ohayon MM, Guilleminault C. Epidemiology of sleep disorders. In: Lee-Chiong TL, editor. Sleep: a comprehensive handbook. Hoboken, N.J: Wiley; 2006. p. 73–82.

[51] Ohayon MM. Epidemiology of insomnia: what we know and what we still need to learn. Sleep Med Rev 2002;6(2):97–111. Epub 2003/01/18. PubMed PMID: 12531146.

[52] Peppard PE, Young T, Barnet JH, Palta M, Hagen EW, Hla KM. Increased prevalence of sleep-disordered breathing in adults. Am J Epidemio 2013;. 23589584.

[53] American Academy of Sleep Medicine. Underdiagnosing and undertreating obstructive sleep apnea draining healthcare system. Mountain View, CA: Frost & Sullivan; 2016.

[54] Zhu L, Zee PC. Circadian rhythm sleep disorders. Neurol Clin 2012;30(4):1167–91. PubMed PMID: 23099133.

[55] American Academy of Sleep Medicine. International classification of sleep disorders. 3rd ed. Darien, IL: American Academy of Sleep Medicine; 2014.

[56] Grandner MA, Patel NP, Gehrman PR, Xie D, Sha D, Weaver T, Gooneratne N. Who gets the best sleep? Ethnic and socioeconomic factors related to sleep disturbance. Sleep Med 2010;11:470–9.

[57] Grandner MA, Petrov MER, Rattanaumpawan P, Jackson N, Platt A, Patel NP. Sleep symptoms, race/ethnicity, and socioeconomic position. J Clin Sleep Med 2013;9(9):897–905. PubMed PMID: WOS:000324375100009.

Chapter 3

Sex differences in sleep health

Jessica Meers[a], Jacqueline Stout-Aguilar[b], Sara Nowakowski[c]

[a]Department of Psychology, University of Houston, Houston, TX, United States, [b]School of Nursing, University of Texas Medical Branch, Galveston, TX, United States, [c]Department of Obstetrics and Gynecology, University of Texas Medical Branch, Galveston, TX, United States

INTRODUCTION

Biologically based sex differences contribute to sleep-related differences in men and women and may help explain the differential risk for sleep disorders [1–5]. Sex differences refer to biological and physiological differences between men and women, with the sex chromosomes and the gonadal hormones primarily contributing to these differences at the cellular, organ, and system levels. Emerging clinical evidence suggests that sleep dysregulation may have more severe health consequences for women than men. Several studies have demonstrated that women report more sleep difficulties [6, 7] and are at greater risk for a diagnosis of insomnia compared to men [8, 9]. In general, there is a higher prevalence of insomnia and dissatisfaction with sleep in women across a wide age range. In subjective studies, women report poorer sleep quality, difficulties falling asleep, frequent night awakenings and longer periods of time awake throughout the night [3]. In contrast, objective measures of sleep, measured by actigraphy and polysomnography (PSG), have demonstrated shorter sleep onset latency, increased sleep efficiency and total sleep time in women compared to men [10–12], whereas a metaanalysis of sex differences of sleep behaviors in older adults (aged 58+) revealed no sex differences in total sleep time [13]. In addition to sex differences found in complaint of sleep disturbances, sex differences also exist for treatment of sleep disorders. For example, in 2013 the US Food and Drug Administration (FDA) required the manufacturers of Ambien to lower the recommended dose of zolpidem for women from 10 to 5 mg for immediate-release products and from 12.5 to 6.25 mg for extended-release products due to the risk of next-morning impairment and motor vehicle accidents. Women appear to be more susceptible to this risk because they eliminate zolpidem from their bodies more slowly than men. Zolpidem is the first drug in the United States to have different recommended doses for women versus men, but it seems likely pharmacokinetic sex differences would lead to differences in rates of absorption, metabolism, and excretion of other medications as well. This review will focus on sex differences in sleep across the lifespan.

SEX DIFFERENCES IN INFANT SLEEP

Newborns spend 60%–70% of the 24 h day in sleep, compared to just 20%–25% in adults [14, 15]. The brain structures and systems that regulate sleep and the circadian process are still relatively immature in newborns, and neonates are not synchronized to the 24 h day. By the end of first month, though, infants begin to spend more time awake during the day and more time sleeping at night. Infants possess a very strong homeostatic drive for sleep and, combined with the relative lack of a circadian rhythm, infants frequently cycle between sleep and wakefulness throughout the 24 h day, requiring frequent naps to dissipate the quickly building sleep pressure [14]. Sleep in early infancy has not yet developed the cyclic patterns found in later life, and is more appropriately characterized as active sleep, quiet sleep, or indeterminate sleep [16]. It is not until roughly 3 months of age that infant sleep begins to consolidate to resemble the organization and rhythmicity that it will maintain throughout the lifespan [17].

The limited research into infant sleep has indicated small and inconsistent sex differences [18]. Even so, there are some important distinctions that could play a vital role in infant health. Males tend to lag behind females in the development of the central nervous system (CNS), leading females to have more mature respiration during the first 6 months [19] and more organized sleep patterns [20]. These immaturities have been linked with Sudden Infant Death Syndrome (SIDS), which affects males at rates about 1.5 times those of females [21]. Intuitively, the earlier development of the CNS system in infant girls suggests that circadian entrainment may happen at an earlier chronological age in girls compared to boys. In fact, there is some evidence that suggests that girls do, in fact, demonstrate earlier rhythm development in terms of core body temperature [22]. Few studies, however, have examined sex differences of early circadian development, so conclusions cannot yet be drawn.

In terms of sleep continuity, although findings have been inconsistent, infant boys tend to have shorter sleep duration (by 5–10 min, on average) and wake earlier than girls [23–25]. Boys spend less time in motionless or quiet sleep compared to girls [25]. Thus, in the first year of life, boys are more active sleepers [26], wake more frequently, and tend to have lower sleep efficiency compared to girls [27]. By the end of the first 18 months, however, sex differences in nocturnal awakenings have been shown to diminish [18].

Sleep duration in infants has been linked to rates of physical growth and weight gain. For all infants, shorter sleep duration is correlated with a higher weight to length ratio. When examining the sleep duration—growth relationship by sex, poorer sleep quality (more fragmented sleep) is associated with a greater weight to length ratio in male infants at 6 months of age compared to female infants [28]. Thus, the link between poor sleep and weight gain likely begins in infancy and is particularly relevant for males.

SEX DIFFERENCES IN CHILDHOOD SLEEP

As development progresses into early childhood, homeostatic sleep drive remains higher compared to adolescents and adults [29]. Over the course of childhood homeostatic sleep pressure trends toward a slower daytime build up, allowing for children to maintain longer periods of wakefulness during the day. By 12 months of age, children are napping on average twice daily for several hours [30]. By the third and fourth year, children begin to need shorter and less frequent naps as they move toward a more consolidated sleep-wake pattern [31]. In accordance with these changes, total sleep need decreases and total sleep duration (including naps) moves from around 13 h at age two, 11 h at age six, to 10 h by age nine [32]. In childhood, daily sleep-wake patterns stay fairly consistent; wake typically occurs spontaneously and at consistent times, even on weekends [33].

Few studies have explored sex differences in childhood sleep and the findings remain generally inconsistent and relatively small. Although some studies have found no sex differences at all [34], there is mounting evidence for general trends that mirror those found in infancy. Across the length of childhood, girls tend to sleep longer than boys [24, 35, 36], potentially attributed to later morning wake times in girls, as opposed to differing bed times [24]. It should be noted, however, that other studies have either found no such differences [37], or have found earlier wake times in girls [38], and later bed times in boys [39]. Regardless, differences in sleep efficiency become more apparent during childhood, with girls showing higher sleep efficiency and smaller proportions of wakefulness compared to boys [36, 40].

Despite the appearance of better sleep suggested by higher sleep efficiency and shorter awakenings, girls are more vulnerable to sleep complaints, even in childhood. These include problems with bedtime resistance, sleep anxiety, and daytime sleepiness [41]. In a study assessing insomnia across 5–12-year-olds, whereas rates in males remained steady, rates of insomnia for females reached their peak in late childhood (aged 11–12), at the cusp of pubertal onset [42].

Sleep problems such as these are correlated with emotional and affective problems throughout the lifespan. Sleep disturbance is both a risk factor for and a response to many emotional and behavioral problems [43–45]. In children, insomnia symptoms have been linked to behavioral problems, including both internalizing and externalizing behaviors, mood variability, and school problems [46]. Further, both prepubertal and pubertal girls with depression show a lower amplitude of circadian rest-activity cycles, indicating a lack of circadian entrainment to the 24 h clock, resulting in irregular sleep-wake schedules and sleep complaints [47]. Thus, early sleep problems may be one initiating factor related to emotional problems that may persist into adolescence, and girls may be especially vulnerable to the negative effects of poor sleep.

Health-related outcomes of poor sleep in childhood are highlighted by the associations that exist between short sleep duration and obesity in childhood through young adulthood. In children as young as 3 years of age, short sleep duration is associated with a greater body mass index (BMI) for boys, but not for girls [39]. In a longitudinal study of 313 children/adolescents ages 8–19 years, Storfer-Isser et al. found sleep duration to have a negative linear association with BMI for boys but not girls, and the magnitude of the association decreased with age [48]. Thus, short sleep duration is associated with BMI and weight gain in boys during their middle childhood through young adulthood but less so through their middle-to-older adult years. Findings are contradictory, however, owing in part to differences in study design (cross-sectional vs longitudinal), measurement (self-report vs objective measures of sleep), race/ethnicity, and country of origin.

SEX DIFFERENCES IN ADOLESCENT SLEEP

Puberty brings with it significant physical and psychological change and marks the emergence of most sex related differences in sleep health. Sleep itself is closely related to the regulation of endocrine functioning closely tied to puberty initiation and reproductive functioning, including growth hormone, melatonin, and sex steroids, such as testosterone in males and estrogen and progestin in females [49, 50].

Because girls typically begin puberty at an earlier chronological age, they begin to show circadian phase advancement before boys [51–54] and the sex differences found in infancy and childhood become more pronounced. Homeostatic sleep pressure slows significantly at the onset of puberty, allowing adolescents to be able to maintain wakefulness for even longer periods of time [31]. As adolescents mature, the circadian phase shifts later, causing more wakefulness in the evenings and later bedtimes [55, 56]. Some studies have found as much as an hour difference in bedtimes once puberty is initiated [54]. The puberty-related changes in sleep and circadian timing can be attributed to the complex relationship between sleep and endocrine functioning. Melatonin is a hormone secreted by the pineal gland that is closely tied to the circadian rhythm. In humans, melatonin levels are lowest during the day and gradually increase in the hours just prior to sleep onset [57]. During childhood, melatonin peaks in the early evening; but during puberty, the timing of the melatonin peak begins to occur later [58].

Despite the biological shift toward a later bedtime, total amount of sleep need is not significantly lower than that of children. When this phase shift is combined with the increased demands that adolescents face in terms of school start times and other obligations, the result is often a sleep duration that does not meet recommendations [33, 54, 59]. As a result, daytime sleepiness increases dramatically during adolescence along with the physical, cognitive, and emotional consequences of inadequate sleep [60]. In boys, inadequate sleep is related to risk behaviors such as smoking, alcohol use, and excessive caffeine usage. For girls, however, poor sleep is associated with more emotional and relationship difficulties [61].

Adolescent girls continue exhibit a longer sleep period than boys [52, 62, 63], with later weekend wake times [54]. Sleep efficiency also remains higher for girls compared to boys [62]. Conversely, boys are, on average, more active sleepers and spend less time in motionless sleep compared to girls. In fact, as males mature and begin puberty, wake time after sleep onset increases [62, 64]. Males also report later bedtimes and more weekday to weekend discrepancy in bed and wake times [65]. Conversely, girls tend to be more extreme sleepers, either sleeping much shorter (less than 6h), or much longer (greater than 10h) than average [66].

Prior to adolescence, no gender differences have been consistently detected in sleep architecture. Concurrent with the phase delay that occurs with puberty, slow wave sleep also begins to decline and is indicative of brain maturation. In prepubertal children, delta power remains the same across both sexes. Girls, however, begin to experience this decline in slow wave activity before boys. By the age of 12, boys have greater delta power per minute comparatively, likely attributed to girls' earlier entry into puberty [67]. Sex differences remain to a degree, however, when pubertal status is controlled, suggesting that gender differences exist beyond pubertal timing [68].

Pubertal onset is a time when sex differences in sleep complaints tend to arise and are maintained into adulthood. Although the degree to which these hormonal changes are mechanistic of these differences remain unclear, adolescent girls report greater sleep difficulties than boys. Girls report longer sleep onset latency and exhibit greater rates of insomnia compared to boys [65, 69]. The increased insomnia risk in females persists across the lifespan [69, 70]. Increased vulnerability to restless legs symptoms also co-occurs with pubertal onset, and affects adolescent girls at higher rates than males contributing to poorer sleep quality and difficulty in sleep initiation [71, 72]. Furthermore, because puberty is a time of increased vulnerability to affective disorders [73], the relationship between poor sleep and general emotional problems becomes more pronounced. Insufficient sleep is associated with higher ratings of anxiety, depression, and negative perceptions of health [74]. This is particularly problematic given that even when psychiatric disturbances are treated, sleep problems often persist [75].

Beyond emotional difficulties, insufficient sleep in adolescence continues to be linked with metabolic and cardiovascular consequences. In the National Longitudinal Study of Adolescent Health, Suglia et al. [76] found that short sleep duration was associated with obesity in adolescent males (mean age 16 years) but not females. However in longitudinal analyses of the same study, the investigators found short sleep in adolescents to be predictive of obesity in young adulthood (mean age 21 years) in both males and females. Also, during puberty, physiological changes that occur lead to differences in fat mass, body composition, and hormonal changes between adolescent boys and girls such that boys tend to decrease fat mass due to increases in growth hormone and testosterone and girls tend to increase fat mass due to estradiol. Thus, the relationship between sleep duration and metabolism, postpuberty into younger adulthood, may start favoring a greater obesogenic propensity among women compared to men. Further, sleep and obesity are both risk factors for hypertension, a particularly dangerous and growing health concern [77–79]. A study by Peach et al. found that sleep quality and daytime sleepiness directly serve as direct risk factors for greater BMI, which indirectly predicts hypertension. For males, however, shorter sleep duration was found to differentially drive these risks whereas daytime sleepiness (a proxy for poor quality sleep) was more likely to be a risk factor for females. Therefore, when considering the link between sleep and physiological health complaints, risk factors are differentially tied to biological sex, and thus findings of studies that do not take into sex into account may be misleading [80].

Sex differences in sleep disordered breathing also emerge during adolescence and are related to obesity. Whereas rates of obstructive sleep apnea are roughly equivalent in childhood, beginning in adolescence, males are affected in rates that exceed those of females [81]. These rates do not emerge until pubertal onset, indicating the potential role of sex steroids in muscle and fat distribution affecting airway structure, and further perpetuated by higher rates of obesity [82]. Poor sleep may therefore be both a risk factor for and a consequence of obesity, indicating that a focus on sleep should be at the center of the obesity crisis and that biologically based sex differences cannot be ignored.

SEX DIFFERENCES IN YOUNG ADULT SLEEP

The sleep disturbances that initially emerge in adolescence frequently carry forward into adulthood. The multitude of biological, psychosocial, and environmental factors that contribute to insufficient sleep and sleep disturbance among adolescents similarly exist for young adults. This includes continued biological changes in the accumulation of homeostatic sleep pressure, increasing academic and vocational demands, and use of substances such as alcohol and caffeine [83, 84]. Additionally, the form and function of sleep continues to change as we age [85]. General recommendations for sleep duration in young and midlife adults suggest 7–9 h of sleep per night [86]. Forty percent of American adults, however, report obtaining less than 7 h of sleep per night on weeknights [87]. Moreover, 38% of young and midlife adults report waking up feeling unrefreshed with 21% have difficulty falling asleep several nights per week. Among young adults ages 19–29, 67% reported not getting enough sleep to function properly [88]. In general, sleep duration decreases with age across the reproductive years. In a study by Campbell and Murphy evaluating the spontaneous sleep across the 24-h day among young, middle-aged, and older adults. Findings indicated that compared with young adults (10.5 h), middle-aged (9.1 h) and older adults (8.1 h) had significantly shorter average nighttime sleep duration [89]. Data from 160 healthy adults (without sleep complaints) aged between 20 and 90 years from the SIESTA database showed that sleep duration decreased about 8 min per decade in males and 10 min per decade in females. Further, they found an age-related increase in lighter sleep (NREM stage 1) associated with concomitant decreases in slow wave sleep for males and NREM stage 2 sleep for females [90]. Three separate metaanalyses found that age was linearly correlated with decreased sleep duration. It was further concluded that there was an approximately 10–12 min reduction per decade of age in the adult population (with a stronger association in women and young adults when compared to men and middle-aged or older adults). Findings also indicated that sleep duration plateaued after the age of 60.

The sex differences in sleep that exist in adolescence through adulthood are attributed in part to the complex interaction of neuroendocrine changes associated with reproductive functioning [91]. In healthy males, this includes production of testosterone that remains fairly stable through most of young adulthood. In females, however, puberty initiates monthly cyclic changes in the production of various hormones, including follicle-stimulating hormone, luteinizing hormone, estrogen, and progesterone. During the first half of the cycle, estrogen predominates, whereas progesterone dominates the latter portion of the cycle prior to the sharp decline in both hormones if fertilization does not occur (the premenstrual period) [92]. At least one third of women report sleep disruptions related to their menstrual cycle [87]. Sleep quality decreases and sleep disturbances increase toward the end of each menstrual cycle [93, 94]. These subjective sleep complaints mirror objective findings showing decreased sleep efficiency and total sleep time just prior to menses [95], although objective findings are mixed [95, 96]. Regardless, these sleep disruptions have been linked to the increase in progesterone and a related rise in core body temperature [97].

It is well known that sleep plays a multitude of roles (including promoting growth, learning and cognitive development), has a role in immunoprotective functionalities. In addition to obesity-related risk, studies have reported a significant association between poor sleep and heart disease in adults [98]. Furthermore, both short and long sleep durations have been associated with increased risk of coronary heart disease and type 2 diabetes as well as with daytime sleepiness and waking unrefreshed [99, 100]. Short sleep in men in particular has been shown to decrease insulin sensitivity, increasing metabolic risk [101]. Similarly, short sleep duration has long been identified as a risk factor for hypertension, but this relationship is shown more consistently in females [102]. The precise mechanisms behind these findings, however, are poorly understood.

SEX DIFFERENCES IN MIDDLE-AGED SLEEP

The middle years of life are a significant time of change in sleep health. As much as 35% of the population aged 40–60 years old report sleep difficulties [103]. The sex differences in subjective sleep complaints found across adolescence and early adulthood tend to be amplified with aging, with women demonstrating increased risk of insomnia, poorer sleep quality, and more frequent awakenings, despite reporting earlier bedtimes and longer sleep duration compared to men [3, 104]. Conversely, the effects of aging on objectively measured sleep have been found to be more pronounced in men, including reductions in the deeper, more restorative sleep characterized by slow wave sleep and increases in lighter sleep stages, such as NREM stage 1 (Fig. 3.1) [105, 106].

The modulating role of endocrine functioning on sleep continues to change through adulthood. The slowing of growth hormone secretion since adolescence is accompanied by age-related decreases of slow wave sleep throughout the lifespan [50, 107]. Rates of melatonin secretion decrease

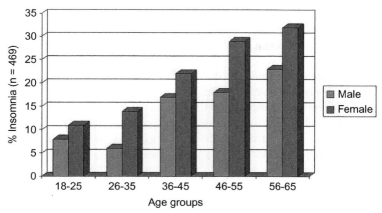

FIG. 3.1 Age and sex distribution of insomnia. *(Adapted from Hohagen, Eur Arch Psychiatry Clin Neurosci 1993; with permission.)*

with age, as well. There is evidence that suggests that the typical nightly increase in nocturnal melatonin is reduced in older compared to younger adults [108]. The changes in gonadotropins and sex steroids that naturally occur with aging are arguably more strongly associated with sleep changes in mid-life. In men, testosterone levels decrease progressively with age after year 30, and the diurnal patterning of testosterone may be reduced in older men [109]. Poorer sleep quality is associated with lower testosterone concentrations in men [110, 111]. These naturally occurring decreases in testosterone may relate to the increased sleep fragmentation commonly seen in older adults [112]. In women, sleep may be even more sensitive to changes in the sex steroidal milieu. Menopause, the permanent cessation of menstrual periods, occurs between 50 and 52 years of age for most women (although variations related to race/ethnicity, health, and lifestyle factors do exist) [113]. During and after menopause, estradiol levels decrease and follicle-stimulating hormone levels increase significantly. These changes in reproductive hormones have been linked to increased complaints of sleep disturbance. Insomnia prevalence rates increase from 33%–36% in premenopausal women to 44%–61% in postmenopausal women [114–116]. Hot flashes occur in 60%–80% of women during the menopausal transition [117], and when they occur during the night, they are often associated with subjectively reported arousals and sleep disturbance, although when objective measures are used, findings are mixed [118–125].

Rates of sleep apnea increase during the middle age, related to increased incidence of obesity and aging [126]. Rates of obstructive sleep apnea in men far exceed those of premenopausal women [127]. In men, the relationship between weight and sleep apnea is stronger than that of women [128, 129]. Following the menopausal transition, however, risk for obstructive sleep apnea increase by 3.5 times for women [130]. Although the relationship between steroidal hormones and sleep apnea are not yet clear, evidence suggests that estrogen and progesterone are protective factors, enhancing respiratory functioning in women [131, 132]. Further, the changes in endocrine milieu associated with the menopause transition, i.e., the decrease in estrogen has been found to be a prominent risk factor for weight gain [133], further contributing to postmenopausal sleep apnea risk in women. The increased risk for weight gain and sleep apnea further increase risk to cardiovascular and metabolic health in midlife. Poor sleep further compounds these risks for both men and women.

SEX DIFFERENCES IN OLDER ADULT SLEEP

Normative aging is often accompanied by greater difficulties in sleep initiation and maintenance. Older adults report longer sleep-onset latency, shorter sleep duration, and generally more fragmented sleep, as they spend more time awake during the night [134, 135]. The timing of sleep changes as well, as older adults show a general advancing of circadian timing to an earlier nocturnal sleep period [136]. Napping frequency increases, with as many as 25% of adults aged 75–84 reporting daytime naps [137]. In fact, with the advancement of age, a greater vulnerability to circadian rhythm disorders such has advanced sleep phase and the deleterious effects of jet lag emerge [136]. Emerging evidence also suggests that sleep disruptions and complaints precede the emergence of dementia and related disorders, and may increase risk [138].

Compared to women, men generally exhibit greater age-related changes objectively measured sleep. In a study of 2500 older adults, age was associated with continued decreases in slow wave sleep and increases in NREM stages 1 and 2, particularly for men. In fact, slow wave sleep (with a concomitant increase in NREM stages 1 and 2) was reduced by as much as 50% when comparing those over 70 to those aged 55—growing indeed, reductions in rates of slow wave sleep have been consistently shown to increase across the lifespan in males relative to females [105, 106, 134]. Relatedly, following sleep deprivation, men experience reduced rebound effects on slow wave sleep compared to women [139]. It should be noted, however, that several studies have found no such sex difference in sleep architecture in older adults [10, 140].

Unsurprisingly, subjective sleep quality appears to decrease in older adulthood. Interestingly, however, despite the greater objectively measured sleep disturbances found in men, older women exhibit greater subjective sleep complaints. Women have been shown to have greater self-reported sleep onset latency and report poorer sleep quality [3, 104, 106]. While the mechanisms behind these sex differences are unclear, it may be that variables impact women's perception of their sleep, such as health-related conditions like fibromyalgia, chronic pain, and overactive bladder.

CONCLUSION

There is an ever-growing body of work that demonstrates sex differences in sleep health across the lifespan. Research studies, in general, should be conscious of these differences and consider effects of sex when examining sleep. Further, investigators should take into account the timing of menstrual phase cycle when studying premenopausal women. The field has come a long way in recognizing and measuring sex differences in sleep health. Much work remains to be completed in examining sex differences related the differential effects of subjective complaint versus objective measure of sleep and the varying impact of sleep duration and sleep disturbance as it relates to health outcomes for men versus women.

REFERENCES

[1] Manber R, Armitage R. Sex, steroids, and sleep: a review. Sleep 1999;22(5):540–55.

[2] Sowers MF, Zheng H, Kravitz HM, Matthews K, Bromberger JT, Gold EB, Ownes J, Consens F, Hall M. Sex steroid hormone profiles are related to sleep measures from polysomnography and the Pittsburgh Sleep Quality Index. Sleep 2008;31(10):1339–49.

[3] Zhang B, Wing YK. Sex differences in insomnia: a meta-analysis. Sleep 2006;29(1):85–93.

[4] Krishnan V, Collop NA. Gender differences in sleep disorders. Curr Opin Pulm Med 2006;12(6):383–9.

[5] Mallampalli MP, Carter CL. Exploring sex and gender differences in sleep health: a Society for Women's Health Research Report. J Women's Health (Larchmt) 2014;23(7):553–62.

[6] Akerstedt T, Knutsson A, Westerholm P, Theorell T, Alfredsson L, Kecklund G. Sleep disturbances, work stress and work hours: a cross-sectional study. J Psychosom Res 2002;53(3):741–8.

[7] Lindberg E, Janson C, Gislason T, Bjornsson E, Hetta J, Boman G. Sleep disturbances in a young adult population: can gender differences be explained by differences in psychological status? Sleep 1997;20(6):381–7.

[8] Jaussent I, Dauvilliers Y, Ancelin ML, Dartigues JF, Tavernier B, Touchon J, et al. Insomnia symptoms in older adults: associated factors and gender differences. Am J Geriatr Psychiatry 2011;19(1):88–97.

[9] Singareddy R, Vgontzas AN, Fernandez-Mendoza J, Liao D, Calhoun S, Shaffer ML, et al. Risk factors for incident chronic insomnia: a general population prospective study. Sleep Med 2012;13(4):346–53.

[10] Carrier J, Land S, Buysse DJ, Kupfer DJ, Monk TH. The effects of age and gender on sleep EEG power spectral density in the middle years of life (ages 20–60 years old). Psychophysiology 2001;38(2):232–42.

[11] Bixler EO, Papaliaga MN, Vgontzas AN, Lin HM, Pejovic S, Karataraki M, et al. Women sleep objectively better than men and the sleep of young women is more resilient to external stressors: effects of age and menopause. J Sleep Res 2009;18(2):221–8.

[12] Jean-Louis G, Mendlowicz MV, Von Gizycki H, Zizi F, Nunes J. Assessment of physical activity and sleep by actigraphy: examination of gender differences. J Womens Health Gend Based Med 1999;8(8):1113–7.

[13] Rediehs MH, Reis JS, Creason NS. Sleep in old age: focus on gender differences. Sleep 1990;13(5):410–24.

[14] Hilliard T. Principles and practice of pediatric sleep medicine. Arch Dis Child 2006;91:546–7.

[15] Galland BC, Taylor BJ, Elder DE, Herbison P. Normal sleep patterns in infants and children: a systematic review of observational studies. Sleep Med Rev 2012;16(3):213–22.

[16] Peirano P, Algarin C, Uauy R. Sleep-wake states and their regulatory mechanisms throughout early human development. J Pediatr 2003;143(4 Suppl):S70–9.

[17] Lushington K, Pamula Y, Martin J, Kennedy J. Developmental changes in sleep: infancy and preschool years. In: Wolfson A, Montgomery-Downs H, editors. The Oxford handbook of infant, child, and adolescent sleep and behavior. New York, NY: Oxford University Press; 2013.

[18] Weinraub M, Bender RH, Friedman SL, Susman EJ, Knoke B, Bradley R, et al. Patterns of developmental change in infants' nighttime sleep awakenings from 6 through 36 months of age. Dev Psychol 2012;48(6):1511–28.

[19] Hoppenbrouwers T, Hodgman JE, Harper RM, Sterman MB. Respiration during the first six months of life in normal infants: IV. Gender differences. Early Hum Dev 1980;4(2):167–77.

[20] Mirmiran M, Maas YG, Ariagno RL. Development of fetal and neonatal sleep and circadian rhythms. Sleep Med Rev 2003;7(4):321–34.

[21] Mage DT, Donner M. A unifying theory for SIDS. Int J Pediatr 2009;2009:368270.

[22] Lodemore MR, Petersen SA, Wailoo MP. Factors affecting the development of night time temperature rhythms. Arch Dis Child 1992;67(10):1259–61.

[23] McDonald L, Wardle J, Llewellyn CH, van Jaarsveld CH, Fisher A. Predictors of shorter sleep in early childhood. Sleep Med 2014;15(5):536–40.

[24] Blair PS, Humphreys JS, Gringras P, Taheri S, Scott N, Emond A, et al. Childhood sleep duration and associated demographic characteristics in an English cohort. Sleep 2012;35(3):353–60.

[25] Goodlin-Jones BL, Burnham MM, Gaylor EE, Anders TF. Night waking, sleep-wake organization, and self-soothing in the first year of life. J Dev Behav Pediatr 2001;22(4):226–33.

[26] Almli CR, Ball RH, Wheeler ME. Human fetal and neonatal movement patterns: gender differences and fetal-to-neonatal continuity. Dev Psychobiol 2001;38(4):252–73.

[27] Saenz J, Yaugher A, Alexander GM. Sleep in infancy predicts gender specific social-emotional problems in toddlers. Front Pediatr 2015;3:42.

[28] Tikotzky L, De Marcas G, Har-Toov J, Dollberg S, Bar-Haim Y, Sadeh A. Sleep and physical growth in infants during the first 6 months. J Sleep Res 2010;19(1 Pt 1):103–10.

[29] Jenni OG, O'Connor BB. Children's sleep: an interplay between culture and biology. Pediatrics 2005;115(1 Suppl):204–16.

[30] Weissbluth M. Naps in children: 6 months-7 years. Sleep 1995;18(2):82–7.

[31] Jenni OG, LeBourgeois MK. Understanding sleep-wake behavior and sleep disorders in children: the value of a model. Curr Opin Psychiatry 2006;19(3):282–7.

[32] Iglowstein I, Jenni OG, Molinari L, Largo RH. Sleep duration from infancy to adolescence: reference values and generational trends. Pediatrics 2003;111(2):302–7.

[33] Acebo C, Sadeh A, Tzischinsky O, Hafer A, Carskadon MA. Sleep/wake patterns derived from activity monitoring and maternal report for health 1- to 5-year old children. Sleep 2005;28(12):1568–77. https://www.ncbi.nlm.nih.gov/pubmed/16408417.

[34] Wolfson AR. Sleeping patterns of children and adolescents. Child Adolesc Psychiatr Clin N Am 1996;5:549–68.

[35] Meijer AM, Habekothe HT, Van Den Wittenboer GL. Time in bed, quality of sleep and school functioning of children. J Sleep Res 2000;9(2):145–53.

[36] Lemola S, Raikkonen K, Scheier MF, Matthews KA, Pesonen AK, Heinonen K, et al. Sleep quantity, quality and optimism in children. J Sleep Res 2011;20(1 Pt 1):12–20.

[37] Seo WS, Sung HM, Lee JH, Koo BH, Kim MJ, Kim SY, et al. Sleep patterns and their age-related changes in elementary-school children. Sleep Med 2010;11(6):569–75.

[38] Iwata S, Iwata O, Iemura A, Iwasaki M, Matsuishi T. Determinants of sleep patterns in healthy Japanese 5-year-old children. Int J Dev Neurosci 2011;29(1):57–62.

[39] Plancoulaine S, Lioret S, Regnault N, Heude B, Charles MA. Gender-specific factors associated with shorter sleep duration at age 3 years. J Sleep Res 2015;24(6):610–20.

[40] Montgomery-Downs HE, O'Brien LM, Gulliver TE, Gozal D. Polysomnographic characteristics in normal preschool and early school-aged children. Pediatrics 2006;117(3):741–53.

[41] Wang G, Xu G, Liu Z, Lu N, Ma R, Zhang E. Sleep patterns and sleep disturbances among Chinese school-aged children: prevalence and associated factors. Sleep Med 2013;14(1):45–52.

[42] Calhoun SL, Fernandez-Mendoza J, Vgontzas AN, Liao D, Bixler EO. Prevalence of insomnia symptoms in a general population sample of young children and preadolescents: gender effects. Sleep Med 2014;15(1):91–5.

[43] Gregory AM, O'Connor TG. Sleep problems in childhood: a longitudinal study of developmental change and association with behavioral problems. J Am Acad Child Adolesc Psychiatry 2002;41(8):964–71.

[44] Gregory AM, Caspi A, Eley TC, Moffitt TE, Oconnor TG, Poulton R. Prospective longitudinal associations between persistent sleep problems in childhood and anxiety and depression disorders in adulthood. J Abnorm Child Psychol 2005;33(2):157–63.

[45] Johnson EO, Chilcoat HD, Breslau N. Trouble sleeping and anxiety/depression in childhood. Psychiatry Res 2000;94(2):93–102.

[46] Calhoun SL, Fernandez-Mendoza J, Vgontzas AN, Mayes SD, Liao D, Bixler EO. Behavioral profiles associated with objective sleep duration in young children with insomnia symptoms. J Abnorm Child Psychol 2017;45:337–44.

[47] Armitage R, Hoffmann R, Emslie G, Rintelman J, Moore J, Lewis K. Rest-activity cycles in childhood and adolescent depression. J Am Acad Child Adolesc Psychiatry 2004;43(6):761–9.

[48] Storfer-Isser A, Patel SR, Babineau DC, Redline S. Relation between sleep duration and BMI varies by age and sex in youth age 8–19. Pediatr Obes 2012;7(1):53–64.

[49] Lord C, Sekerovic Z, Carrier J. Sleep regulation and sex hormones exposure in men and women across adulthood. Pathol Biol 2014;62(5):302–10.

[50] Van Cauter E, Leproult R, Plat L. Age-related changes in slow wave sleep and REM sleep and relationship with growth hormone and cortisol levels in healthy men. JAMA 2000;284(7):861–8.

[51] Roenneberg T, Kuehnle T, Pramstaller PP, Ricken J, Havel M, Guth A, et al. A marker for the end of adolescence. Curr Biol 2004;14:R1038–9.

[52] Tonetti L, Fabbri M, Natale V. Sex difference in sleep-time preference and sleep need: a cross-sectional survey among Italian pre-adolescents, adolescents, and adults. Chronobiol Int 2008;25(5):745–59.

[53] Petersen AC, Crockett L, Richards M, Boxer A. A self-report measure of pubertal status: reliability, validity, and initial norms. J Youth Adolesc 1988;17(2):117–33.

[54] Laberge L, Petit D, Simard C, Vitaro F, Tremblay RE, Montplaisir J. Development of sleep patterns in early adolescence. J Sleep Res 2001;10(1):59–67.

[55] Carskadon MA, Acebo C, Richardson GS, Tate BA, Seifer R. An approach to studying circadian rhythms of adolescent humans. J Biol Rhythm 1997;12(3):278–89.

[56] Carskadon MA, Vieira C, Acebo C. Association between puberty and delayed phase preference. Sleep 1993;16(3):258–62.

[57] Lewy AJ, Wehr TA, Goodwin FK, Newsome DA, Markey SP. Light suppresses melatonin secretion in humans. Science 1980;210(4475):1267–9.

[58] Crowley SJ, Van Reen E, LeBourgeois MK, Acebo C, Tarokh L, Seifer R, et al. A longitudinal assessment of sleep timing, circadian phase, and phase angle of entrainment across human adolescence. PLoS ONE 2014;9(11):e112199.

[59] Anders TF, Carskadon MA, Dement WC, Harvey K. Sleep habits of children and the identification of pathologically sleepy children. Child Psychiatry Hum Dev 1978;9(1):56–63.

[60] Carskadon MA, Acebo C. Regulation of sleepiness in adolescents: update, insights, and speculation. Sleep 2002;25(6):606–14.

[61] Zhang J, Chan NY, Lam SP, Li SX, Liu Y, Chan JW, et al. Emergence of sex differences in insomnia symptoms in adolescents: a large-scale school-based study. Sleep 2016;39(8):1563–70.

[62] Gaina A, Sekine M, Hamanishi S, Chen X, Kagamimori S. Gender and temporal differences in sleep-wake patterns in Japanese schoolchildren. Sleep 2005;28(3):337–44.

[63] Olds T, Maher C, Blunden S, Matricciani L. Normative data on the sleep habits of Australian children and adolescents. Sleep 2010;33(10):1381–8.

[64] Short MA, Gradisar M, Lack LC, Wright H, Carskadon MA. The discrepancy between actigraphic and sleep diary measures of sleep in adolescents. Sleep Med 2012;13(4):378–84.

[65] Hysing M, Pallesen S, Stormark KM, Lundervold AJ, Sivertsen B. Sleep patterns and insomnia among adolescents: a population-based study. J Sleep Res 2013;22(5):549–56.

[66] Maslowsky J, Ozer EJ. Developmental trends in sleep duration in adolescence and young adulthood: evidence from a national United States sample. J Adolesc Health 2014;54(6):691–7.
[67] Campbell IG, Darchia N, Khaw WY, Higgins LM, Feinberg I. Sleep EEG evidence of sex differences in adolescent brain maturation. Sleep 2005;28(5):637–43.
[68] Campbell IG, Grimm KJ, de Bie E, Feinberg I. Sex, puberty, and the timing of sleep EEG measured adolescent brain maturation. Proc Natl Acad Sci U S A 2012;109(15):5740–3.
[69] Knutson KL. The association between pubertal status and sleep duration and quality among a nationally representative sample of U. S. adolescents. Am J Hum Biol 2005;17(4):418–24.
[70] Johnson EO, Roth T, Schultz L, Breslau N. Epidemiology of DSM-IV insomnia in adolescence: lifetime prevalence, chronicity, and an emergent gender difference. Pediatrics 2006;117(2):e247–56.
[71] Zhang J, Lam SP, Li SX, Li AM, Kong AP, Wing YK. Restless legs symptoms in adolescents: epidemiology, heritability, and pubertal effects. J Psychosom Res 2014;76(2):158–64.
[72] Silva GE, Goodwin JL, Vana KD, Vasquez MM, Wilcox PG, Quan SF. Restless legs syndrome, sleep, and quality of life among adolescents and young adults. J Clin Sleep Med 2014;10(7):779–86.
[73] Angold A, Costello EJ, Worthman CM. Puberty and depression: the roles of age, pubertal status and pubertal timing. Psychol Med 1998;28(1):51–61.
[74] Moore M, Kirchner HL, Drotar D, Johnson N, Rosen C, Ancoli-Israel S, et al. Relationships among sleepiness, sleep time, and psychological functioning in adolescents. J Pediatr Psychol 2009;34(10):1175–83.
[75] Puig-Antich J, Goetz R, Hanlon C, Tabrizi MA, Davies M, Weitzman ED. Sleep architecture and REM sleep measures in prepubertal major depressives. Studies during recovery from the depressive episode in a drug-free state. Arch Gen Psychiatry 1983;40(2):187–92.
[76] Suglia SF, Kara S, Robinson WR. Sleep duration and obesity among adolescents transitioning to adulthood: do results differ by sex? J Pediatr 2014;165(4):750–4.
[77] Babinska K, Kovacs L, Janko V, Dallos T, Feber J. Association between obesity and the severity of ambulatory hypertension in children and adolescents. J Am Soc Hypertens 2012;6(5):356–63.
[78] Javaheri S, Storfer-Isser A, Rosen CL, Redline S. Sleep quality and elevated blood pressure in adolescents. Circulation 2008;118(10):1034–40.
[79] Assadi F. The growing epidemic of hypertension among children and adolescents: a challenging road ahead. Pediatr Cardiol 2012;33(7):1013–20.
[80] Peach H, Gaultney JF, Reeve CL. Sleep characteristics, body mass index, and risk for hypertension in young adolescents. J Youth Adolesc 2015;44(2):271–84.
[81] Spilsbury JC, Storfer-Isser A, Rosen CL, Redline S. Remission and incidence of obstructive sleep apnea from middle childhood to late adolescence. Sleep 2015;38(1):23–9.
[82] Fuentes-Pradera MA, Sanchez-Armengol A, Capote-Gil F, Quintana-Gallego E, Carmona-Bernal C, Polo J, et al. Effects of sex on sleep-disordered breathing in adolescents. Eur Respir J 2004;23(2):250–4.
[83] Moore M, Meltzer LJ. The sleepy adolescent: causes and consequences of sleepiness in teens. Paediatr Respir Rev 2008;9(2):114–20 [quiz 20–1].
[84] Hershner SD, Chervin RD. Causes and consequences of sleepiness among college students. Nat Sci Sleep 2014;6:73–84.
[85] Espiritu JR. Aging-related sleep changes. Clin Geriatr Med 2008;24(1):1–14 [v].
[86] Hirshkowitz M, Whiton K, Albert SM, Alessi C, Bruni O, DonCarlos L, et al. National Sleep Foundation's updated sleep duration recommendations: final report. Sleep Health 2015;1(4):233–43.
[87] Gillin JC. The sleep therapies of depression. Prog Neuro-Psychopharmacol Biol Psychiatry 1983;7(2–3):351–64.
[88] Gradisar M, Wolfson AR, Harvey AG, Hale L, Rosenberg R, Czeisler CA. The sleep and technology use of Americans: findings from the National Sleep Foundation's 2011 Sleep in America poll. J Clin Sleep Med 2013;9(12):1291–9.
[89] Campbell SS, Murphy PJ. The nature of spontaneous sleep across adulthood. J Sleep Res 2007;16(1):24–32.
[90] Dorffner G, Vitr M, Anderer P. The effects of aging on sleep architecture in healthy subjects. Adv Exp Med Biol 2015;821:93–100.
[91] Conley CS, Rudolph KD, Bryant FB. Explaining the longitudinal association between puberty and depression: sex differences in the mediating effects of peer stress. Dev Psychopathol 2012;24(2):691–701.
[92] Knudtson J, McLaughlin J. Menstrual cycle: Merck manuals. Available from: http://www.merckmanuals.com/home/women-s-health-issues/biology-of-the-female-reproductive-system/menstrual-cycle#v801614; 2017.
[93] Romans SE, Kreindler D, Einstein G, Laredo S, Petrovic MJ, Stanley J. Sleep quality and the menstrual cycle. Sleep Med 2015;16(4):489–95.
[94] Baker FC, Driver HS. Circadian rhythms, sleep, and the menstrual cycle. Sleep Med 2007;8(6):613–22.
[95] Zheng H, Harlow SD, Kravitz HM, Bromberger J, Buysse DJ, Matthews KA, et al. Actigraphy-defined measures of sleep and movement across the menstrual cycle in midlife menstruating women: study of Women's Health Across the Nation Sleep Study. Menopause 2015;22(1):66–74.
[96] Baker FC, Sassoon SA, Kahan T, Palaniappan L, Nicholas CL, Trinder J, et al. Perceived poor sleep quality in the absence of polysomnographic sleep disturbance in women with severe premenstrual syndrome. J Sleep Res 2012;21(5):535–45.
[97] Sharkey KM, Crawford SL, Kim S, Joffe H. Objective sleep interruption and reproductive hormone dynamics in the menstrual cycle. Sleep Med 2014;15(6):688–93.
[98] Khan MS, Aouad R. The effects of insomnia and sleep loss on cardiovascular disease. Sleep Med Clin 2017;12(2):167–77.
[99] Wang D, Li W, Cui X, Meng Y, Zhou M, Xiao L, et al. Sleep duration and risk of coronary heart disease: a systematic review and meta-analysis of prospective cohort studies. Int J Cardiol 2016;219:231–9.
[100] Shan Z, Ma H, Xie M, Yan P, Guo Y, Bao W, et al. Sleep duration and risk of type 2 diabetes: a meta-analysis of prospective studies. Diabetes Care 2015;38(3):529–37.
[101] Wong PM, Manuck SB, DiNardo MM, Korytkowski M, Muldoon MF. Shorter sleep duration is associated with decreased insulin sensitivity in healthy white men. Sleep 2015;38(2):223–31.
[102] Pepin JL, Borel AL, Tamisier R, Baguet JP, Levy P, Dauvilliers Y. Hypertension and sleep: overview of a tight relationship. Sleep Med Rev 2014;18(6):509–19.
[103] Phillips B, Mannino D. Correlates of sleep complaints in adults: the ARIC study. J Clin Sleep Med 2005;1(3):277–83.

[104] Reyner LA, Horne JA, Reyner A. Gender- and age-related differences in sleep determined by home-recorded sleep logs and actimetry from 400 adults. Sleep 1995;18(2):127–34.

[105] Hume KI, Van F, Watson A. A field study of age and gender differences in habitual adult sleep. J Sleep Res 1998;7(2):85–94.

[106] Luca G, Haba Rubio J, Andries D, Tobback N, Vollenweider P, Waeber G, et al. Age and gender variations of sleep in subjects without sleep disorders. Ann Med 2015;47(6):482–91.

[107] Copinschi G, Caufriez A. Sleep and hormonal changes in aging. Endocrinol Metab Clin N Am 2013;42(2):371–89.

[108] Zeitzer JM, Duffy JF, Lockley SW, Dijk DJ, Czeisler CA. Plasma melatonin rhythms in young and older humans during sleep, sleep deprivation, and wake. Sleep 2007;30(11):1437–43.

[109] Bremner WJ, Vitiello MV, Prinz PN. Loss of circadian rhythmicity in blood testosterone levels with aging in normal men. J Clin Endocrinol Metab 1983;56(6):1278–81.

[110] Penev PD. Association between sleep and morning testosterone levels in older men. Sleep 2007;30(4):427–32.

[111] Schiavi RC, White D, Mandeli J. Pituitary-gonadal function during sleep in healthy aging men. Psychoneuroendocrinology 1992;17(6):599–609.

[112] Pandi-Perumal SR, Monti JM, Monjan AA. Principles and practice of geriatric sleep medicine: treatment of sleep disorders in the elderly. 2018.

[113] Gold EB. The timing of the age at which natural menopause occurs. Obstet Gynecol Clin N Am 2011;38(3):425–40.

[114] Kravitz HM, Ganz PA, Bromberger J, Powell LH, Sutton-Tyrrell K, Meyer PM. Sleep difficulty in women at midlife: a community survey of sleep and the menopausal transition. Menopause 2003;10(1):19–28.

[115] Kravitz HM, Zhao X, Bromberger JT, Gold EB, Hall MH, Matthews KA, et al. Sleep disturbance during the menopausal transition in a multi-ethnic community sample of women. Sleep 2008;31(7):979–90.

[116] National Institutes of Health. National Institutes of Health State-of-the Science Conference statement. Management of menopause-related symptoms. Ann Intern Med 2005;142(12):1003–13.

[117] Gold EB, Sternfeld B, Kelsey JL, Brown C, Mouton C, Reame N, et al. Relation of demographic and lifestyle factors to symptoms in a multi-racial/ethnic population of women 40–55 years of age. Am J Epidemiol 2000;152(5):463–73.

[118] Young T, Rabago D, Zgierska A, Austin D, Laurel F. Objective and subjective sleep quality in premenopausal, perimenopausal, and postmenopausal women in the Wisconsin Sleep Cohort Study. Sleep 2003;26(6):667–72.

[119] Shaver J, Giblin E, Lentz M, Lee K. Sleep patterns and stability in perimenopausal women. Sleep 1988;11(6):556–61.

[120] Ensrud KE, Stone KL, Blackwell TL, Sawaya GF, Tagliaferri M, Diem SJ, et al. Frequency and severity of hot flashes and sleep disturbance in postmenopausal women with hot flashes. Menopause 2009;16(2):286–92.

[121] Savard J, Davidson JR, Ivers H, Quesnel C, Rioux D, Dupere V, et al. The association between nocturnal hot flashes and sleep in breast cancer survivors. J Pain Symptom Manag 2004;27(6):513–22.

[122] Freedman RR, Roehrs TA. Lack of sleep disturbance from menopausal hot flashes. Fertil Steril 2004;82(1):138–44.

[123] Erlik Y, Tataryn IV, Meldrum DR, Lomax P, Bajorek JG, Judd HL. Association of waking episodes with menopausal hot flushes. JAMA 1981;245(17):1741–4.

[124] Freedman RR, Benton MD, Genik 2nd RJ, Graydon FX. Cortical activation during menopausal hot flashes. Fertil Steril 2006;85(3):674–8.

[125] Woodward S, Freedman RR. The thermoregulatory effects of menopausal hot flashes on sleep. Sleep 1994;17(6):497–501.

[126] Hall MH, Kline CE, Nowakowski S. Insomnia and sleep apnea in midlife women: prevalence and consequences to health and functioning. F1000Prime Rep 2015;7:63.

[127] Bixler EO, Vgontzas AN, Lin HM, Ten Have T, Rein J, Vela-Bueno A, et al. Prevalence of sleep-disordered breathing in women: effects of gender. Am J Respir Crit Care Med 2001;163(3 Pt 1):608–13.

[128] Newman AB, Foster G, Givelber R, Nieto FJ, Redline S, Young T. Progression and regression of sleep-disordered breathing with changes in weight: the Sleep Heart Health Study. Arch Intern Med 2005;165(20):2408–13.

[129] Redline S, Schluchter MD, Larkin EK, Tishler PV. Predictors of longitudinal change in sleep-disordered breathing in a nonclinic population. Sleep 2003;26(6):703–9.

[130] Young T, Finn L, Austin D, Peterson A. Menopausal status and sleep-disordered breathing in the Wisconsin Sleep Cohort Study. Am J Respir Crit Care Med 2003;167(9):1181–5.

[131] Driver HS, McLean H, Kumar DV, Farr N, Day AG, Fitzpatrick MF. The influence of the menstrual cycle on upper airway resistance and breathing during sleep. Sleep 2005;28(4):449–56.

[132] Popovic RM, White DP. Upper airway muscle activity in normal women: influence of hormonal status. J Appl Physiol (1985) 1998;84(3):1055–62.

[133] Davis SR, Castelo-Branco C, Chedraui P, Lumsden MA, Nappi RE, Shah D, et al. Understanding weight gain at menopause. Climacteric 2012;15(5):419–29.

[134] Ohayon MM, Carskadon MA, Guilleminault C, Vitiello MV. Meta-analysis of quantitative sleep parameters from childhood to old age in healthy individuals: developing normative sleep values across the human lifespan. Sleep 2004;27(7):1255–73.

[135] Kales A, Wilson T, Kales JD, Jacobson A, Paulson MJ, Kollar E, et al. Measurements of all-night sleep in normal elderly persons: effects of aging. J Am Geriatr Soc 1967;15(5):405–14.

[136] Duffy JF, Zitting KM, Chinoy ED. Aging and circadian rhythms. Sleep Med Clin 2015;10(4):423–34.

[137] Foley DJ, Vitiello MV, Bliwise DL, Ancoli-Israel S, Monjan AA, Walsh JK. Frequent napping is associated with excessive daytime sleepiness, depression, pain, and nocturia in older adults: findings from the National Sleep Foundation '2003 Sleep in America' Poll. Am J Geriatr Psychiatry 2007;15(4):344–50.

[138] Pase MP, Himali JJ, Grima NA, Beiser AS, Satizabal CL, Aparicio HJ, et al. Sleep architecture and the risk of incident dementia in the community. Neurology 2017;89(12):1244–50.

[139] Reynolds 3rd CF, Kupfer DJ, Hoch CC, Stack JA, Houck PR, Berman SR. Sleep deprivation in healthy elderly men and women: effects on mood and on sleep during recovery. Sleep 1986;9(4):492–501.

[140] Svetnik V, Snyder ES, Ma J, Tao P, Lines C, Herring WJ. EEG spectral analysis of NREM sleep in a large sample of patients with insomnia and good sleepers: effects of age, sex and part of the night. J Sleep Res 2017;26(1):92–104.

Chapter 4

Sleep and health in older adults

Junxin Li[a], Nalaka S. Gooneratne[b,c]
[a]Johns Hopkins University School of Nursing, Baltimore, MD, United States, [b]Center for Sleep and Circadian Neurobiology, Perelman School of Medicine, University of Pennsylvania, Philadelphia, PA, United States, [c]Geriatrics Division, Perelman School of Medicine, University of Pennsylvania, Philadelphia, PA, United States

INTRODUCTION

The impact of sleep on health outcomes in older adults has become a topic of great interest in the geriatric research world. Current evidence supports a mostly bidirectional relationship between sleep and health. Characteristics of sleep often change with age; for example, sleep complaints and certain sleep disorders, such as sleep-disordered breathing (SDB) and insomnia, are more common in older adults than in young or middle-aged adults. Additionally, homeostatic sleep drive and circadian rhythm are less robust in older adults than in young adults [1]. Age-related changes in sleep make older adults more prone to develop sleep problems; however, disturbed sleep is not a part of normal aging. Disturbed sleep or sleep problems are largely due to existing comorbidities and polypharmacy use in older adults. At the same time, research suggests that disturbed sleep can result in a range of adverse health outcomes in older adults. In this chapter, we will review sleep changes in normal aging, primary sleep disorders in older adults, and the relationship between sleep and health outcomes concerning cognitive function, cardiovascular health, psychiatric illness, and pain in older adults.

SLEEP CHANGES IN NORMAL AGING

Changes in sleep parameters

Sleep clearly changes with aging in multiple ways. For example, total sleep time (TST), the ability to maintain sleep, and the proportion of slow wave sleep decrease as people age from pediatric to older adulthood [2]. However, sleep does not change much in aging older adults with excellent health. For example, further age-associated decreases in these sleep parameters have not been observed after 60 years of age in relatively healthy older adults. Evidence suggests that older adults maintain their ability to initiate sleep (i.e., sleep latency) and fall back to sleep after awakenings as rapidly as younger adults after the age of 60 years. In contrast, sleep efficiency continues to decline slowly with increasing age [1].

Changes in circadian rhythm

Circadian timing and circadian amplitudes change with increasing age [3]. Older adults commonly experience an age-related advance (1 h) in circadian phases and some degree of reduction in circadian amplitudes compared to young adults [1]. For instance, older adults often feel sleepy early in the evening and wake up early in the morning [1]. Body temperature, rhythm, and timing of secretion of melatonin and cortisol can also advance with aging [4]. These age-related changes in circadian systems may contribute to sleep disruption and daytime napping which further decreases the circadian amplitude in older adults [3]. In addition, the ability to adjust to and recover from phase shifting (e.g., shift work or jet lag) decreases with aging, which may contribute to longer periods of sleep disruption and daytime dysfunction [1].

Changes in sleep homeostasis

Sleep homeostasis regulates wakefulness and sleep. Sleep pressure is generated by sleep homeostasis and builds up with the amount of time being awake. It increases during wakefulness and decreases during sleep [5]. Sleep homeostasis is reduced with aging [1, 6], which may cause decreased TST and sleep efficiency as well as increased nocturnal awakenings, early morning awakening, and daytime sleepiness in older adults [7]. The age-related changes in sleep homeostasis co-occur with changes in circadian rhythm and contribute to earlier sleep times and less consolidated sleep in older adults [8].

COMMON SLEEP DISTURBANCES IN OLDER ADULTS

Sleep disturbances such as difficulty initiating sleep, frequent nocturnal awakenings, early morning awakenings, nonrestorative sleep, and daytime sleepiness are prevalent in older adults. In addition, compared to young adults, the prevalence of primary sleep disorders (e.g., insomnia, sleep-disordered breathing, periodic limb movements in sleep, restless legs syndrome, and REM sleep behavior disorder) is considerably higher in older adults. These primary sleep disorders contribute to poor sleep in older adults.

Insomnia

The incidence of insomnia increases continuously with age and is more frequent in women than in men [9]. Although up to 50% of older adults in epidemiological studies reported at least one insomnia symptom, such as difficulty falling asleep, maintaining asleep, or early morning awakenings [9], a smaller percentage of older adults (5%–20%) met clinical criteria for the diagnosis of insomnia depending on the definition [6]. In addition, since many older adults perceive these symptoms as expected sleep changes with normal aging rather than disturbed sleep, they do not seek help from healthcare professionals [1]. Furthermore, older adults are more likely to consider difficulty falling asleep as a sleep problem than difficulty staying asleep, even though the latter is more common [10]. These nocturnal sleep symptoms may cause impairments in daytime functioning, such as fatigue, daytime sleepiness, and mood/behavioral disturbances and lead to insomnia diagnosis.

According to the Diagnostic Criteria for Chronic Insomnia (ICSD-3), insomnia is defined as dissatisfactory nocturnal sleep (i.e., difficulty falling, staying asleep, or early morning awakening) at least 3 nights per week for at least 3 months with clinically significant distress or impairment in social, occupational, educational, academic, behavioral, or other important areas of functioning [9, 11]. Insomnia may be primary or secondary to medical comorbidities and psychiatric illness, medication use, and lifestyle or environmental changes associated with aging [6].

Both pharmacological and nonpharmacological treatments are used to treat insomnia in older adults. The goal of treatment is to improve quantity and/or quality of sleep and reduce insomina-realted daytime impairments [9]. The treatment plan is usually selected based on the severity and duration of the insomnia symptoms, the patient's existing comorbid conditions, the patient's preference of treatment options, and vulnerability of patients to the adverse effects of medications [12]. Pharmacologic treatments, such as benzodiazepines, nonbenzodiazepine hypnotics, melatonin receptor agonists, antidepressant, orexin-receptor agents and antihistamine, should start with low doses and in most cases only be used for the short-term management of insomnia due to the side effects and long-term safety concerns [9]. In addition, the treatment effects usually do not sustain once the patient stops using the medications [13]. Behavioral treatments (e.g., cognitive behavioral therapy for insomnia (CBTi), physical activity, social engagement, and sleep hygiene), bright light therapy, and acupuncture are commonly used nonpharmacological treatments for insomnia. CBTi is a multicomponent cognitive and behavioral intervention which involves cognitive restructuring, sleep hygiene education, sleep restriction, relaxation, and stimulus control. Evidence suggests that CBTi is the most effective nonpharmacological treatment for insomnia and has more long-lasting effects than medications [13, 14].

Sleep-disordered breathing

Sleep disordered breathing (SDB) is an umbrella term for chronic conditions in which repeated episodes of hypopnea or apnea (not breathing) during sleep occur throughout the night. SDB is a more prevalent condition in older adults than younger adults and in men than women. Obstructive sleep apnea (OSA) is the most common form of SDB. Several epidemiologic studies have showed that the prevalence of SDB ranged from 27% to 80% in people aged 60 years or older [15]. Approximately 80% of older adults aged 71 and older had obstructive sleep apnea (OSA) (AHI > 5) and the incidence increased 2.2 times for each 10-year increase of age [16].

Snoring and excessive daytime sleepiness (EDS) are main symptoms of SDB in older adults [15, 17]. In an epidemiological study of older men, people with SDB were 50% more likely to have daytime sleepiness than those didn't have sleep apnea [18]. In addition, studies also found that male gender, Asian race, advancing age, higher body mass index/obesity, neck girth, habitual snoring, hypertension, cardiovascular (CV) disease, and heart failure were independently associated with SDB in older adults and the magnitude of some of these associations decreased with advancing age [18, 19]. Older adults with SDB may also experience insomnia and those with both sleep disorders have poorer sleep quality. In addition, the incidence of comorbid insomnia may be higher in women with SDB than in men with SDB [20].

The gold standard for diagnosing SDB in older adults is overnight polysomnography (PSG) [15]. Obtaining in-laboratory PSG recording from older adults, especially those with cognitive impairment can be challenging due to difficulties with transformation, understanding complicated instructions, and tolerance of equipment. Alternative diagnosing tests need to be explored in older adults. For instance, using in-home unattended home sleep apnea testing that records airflow, symptoms of sleep apnea (e.g., snoring and sleepiness), BMI, neck circumference, age, and sex might be an effective and reliable way to diagnose obstructive sleep apnea (OSA) in older adults [21].

The first-line therapy for SDB is the continuous positive airway pressure therapy (CPAP). The compliance/noncompliance to CPAP treatment has been associated with gender, BMI, severity of OSA and symptoms, early experience of treatment, side effects, level of education and support received, and behavioral (e.g., cigarette smoking) and cost factors [22, 23]. Cognitive impairment may also be associated with poor CPAP adherence. Studies found that patients with Alzheimer's disease (AD) and depressive symptoms had worse adherence than people with mild to moderate AD and Parkinson's disease. People with mild cognitive impairment and Parkinson's disease could wear CPAP around 5 h per night [24, 25]. In addition, oral appliances are used to treat OSA, but its effectiveness in managing OSA varies by patient anatomy.

FACTORS ASSOCIATED WITH SLEEP DISTURBANCES IN OLDER ADULTS

Sleep disturbances reported in older adults are usually multifactorial and can not simply be explained by age alone. In addition to age and primary sleep disorders, medical conditions and changes in social engagement, lifestyle, and living environment associated with aging can contribute to sleep problems in older adults [15]. For example, older adults may be more sedentary and less engaged in daytime physical and social activities, which impacts sleep homeostasis and circadian regulation and leads to sleep disturbances [26]. In addition, life events such as loss of loved ones and transitioning to long term care settings can create physical and emotional stressors that can cause acute sleep problems or worsen the sleep quality in older adults. These acute sleep problems can develop into long lasting chronic sleep disturbances if they are not treated properly. Additionally, environmental temperature, noise, and bright light exposure affect sleep quality in older adults [1].

SLEEP AND HEALTH IN OLDER ADULTS

Approximately 67%–75% of older adults aged 65 and over have two or more concurrent chronic medical conditions, the most common being osteoarthritis, cardiovascular disease, lung disease, cancer, diabetes, Alzheimer's disease/dementia, anxiety and depression [27, 28]. The majority of these older adults take prescription medications to treat these chronic conditions [29, 30], which can impact their sleep. Epidemiological studies found that older adults with chronic medical conditions report more difficulty initiating sleep, difficulty maintaining sleep, daytime sleepiness, and fatigue than healthy older adults [28]. Growing evidence shows that sleep problems seen in older adults are more commonly related to comorbidities rather than normal aging [1]. On the other hand, in some cases sleep disturbances can also trigger or worsen medical conditions in older adults.

Older adults with sleep disturbance and comorbid medical conditions have increased risks of mortality and hospitalization, and may also receive inappropriate polypharmacy [1, 28]. Therefore, coexisting sleep disturbances and chronic medical conditions must be managed together to promote health and quality of life in older adults. We will review the relationship between sleep and multiple health outcomes in older adults, including cognitive function, cardiovascular health, mental health, and pain.

Cognitive function

Cognitive abilities change across the life span, generally declining with advancing age during midlife and older adulthood [31]. A number of health and behavioral modifiable factors contribute to cognitive decline in older adults. Growing evidence suggests a potential connection between sleep and cognitive function. However, the mechanisms underlying the association are not fully understood. In this section, we present evidence that examines the association between cognitive function/cognitive impairment and sleep disturbances in older adults, including studies of short/long sleep duration, poor sleep quality, self-reported sleep complaints, objectively measured sleep disturbances, daytime napping, excessive daytime sleepiness, insomnia, and sleep-disordered breathing.

Sleep duration

Several large-scale epidemiologic studies have investigated the association between sleep duration and cognitive functioning in older adults and demonstrate inconsistent results. In general, current evidence suggests a potential U-shaped association between sleep duration and cognitive outcomes in older adults. Many studies showed that short and long sleep duration negatively impact older adults' cognitive function [32–34]. However, other studies have found no significant influence of sleep on cognitive function [35, 36]. For instance, a recent systematic review and metaanalysis synthesized findings from 18 studies (11 cross-sectional and 7 prospective cohort studies; total $N=97,264$) and found that self-reported short and long sleep durations increased the likelihood for poor cognitive function by 1.40 and 1.58 times, respectively [37]. These associations were not moderated by gender or age. In terms of specific cognitive domains, the analysis of the cross-sectional studies revealed both short and long sleep durations were significantly associated with poor performance in multiple-domain tasks, executive function, verbal memory, and working memory capacity but not associated with processing speed. Short and long sleep durations were only associated with subsequent poor multiple-domain performance in the analysis of prospective studies. Another systematic review of 32 observational studies in older adults demonstrated similar findings: negative associations were found between short and long sleep durations and cognitive performance [38]. Additionally, this review found that greater changes in sleep duration from earlier life were associated with cognitive decline in later life [38]. Evidence from the prospective studies found that both long and short sleep durations were associated with an increased risk of cognitive impairment or dementia [39]. Most existing studies use self-reported sleep duration to examine the association between sleep duration and cognitive outcomes; however, there is an increasing need for prospective studies to use objective sleep measures (e.g., actigraphy).

Self-reported sleep complaints

Self-reported sleep complaints such as insomnia symptoms and poor sleep quality are more prevalent in older adults than young and middle-aged adults. Research has provided mixed findings, but generally supports an association between these sleep complaints and worse cognitive performance or increased likelihood of cognitive decline/impairment [39]. Three cross-sectional [40, 41] and four prospective analyses [42–45] showed that self-reported poor sleep quality or sleep complaints were independently associated with worse performance in global cognition and measures of specific cognitive domains, or increased risk of cognitive impairment. The association was nonsignificant in two cross-sectional [46, 47] and one prospective [48] studies. Schmutte et al. found that a longer sleep onset latency was cross-sectionally associated with worse performance in tests of verbal knowledge, long-term memory, and visuospatial ability in cognitive intact older adults [40]. The prospective studies suggest that sleep complaints were associated with an increased risk of developing cognitive impairment or dementia in older adults [39].

Objectively measured sleep disturbances

Actigraphy and polysomnography (PSG) are the two most commonly used objective measures of sleep that provide more specific and detailed information on sleep than self-reported data. For example, actigraphy and PSG can reveal information about sleep fragmentation, sleep efficiency, night to night sleep variability, and sleep architecture, which are believed to be key sleep characteristics that impact changes in cognitive function in older adults [39]. Although researchers have shown growing interests in this field in the last decade, the quantity of research is inadequate to draw clear associations. So far, not enough evidence suggests a specific association between cognitive function and certain PSG measured sleep parameters in older adults population, even though a link between less slow-wave sleep and worse cognitive function was established in young and middle aged adults [49]. There are more actigraphy based sleep studies that mostly support the association between disturbed sleep and adverse cognitive outcomes. For example, in studies of older men and women in the United States, disturbed sleep as measured by actigraphy (lower sleep efficiency, longer sleep onset latency, longer wake after sleep onset, and prolonged napping) was associated with an increased risk of cognitive impairment [47, 50, 51] and executive function [50]. In addition, actigraphically measured sleep fragmentation was associated with higher severity of cognitive impairment [52] or increased risk of Alzheimer's disease/cognitive decline [53]. Current evidence suggests a possible link between poor sleep and cognitive outcomes in older adults; however, more research using objectively measured sleep are needed to demonstrate explicit associations.

Daytime napping

Napping is more prevalent in older adults than young and middle-aged adults across the world [54, 55]. Older adults nap for numerous reasons such as to compensate for disturbed nighttime sleep, restore energy, reduce excessive daytime sleepiness or fatigue from comorbidities or medications, or simply as a habit [1, 56]. The impact of daytime napping and cognitive function in older adults largely depends on features of the nap, the population, and napping measures used in the study. Current evidence yields mixed findings due to the heterogeneity of these aspects among studies. The association of long nap (e.g., ≥90 min or 2 h) with adverse cognitive outcomes is consistent in most epidemiological studies [34, 57, 58]. One recent epidemiological study among over 3000 Chinese older adults found that those who reported napping for <90 min at both baseline and follow up had better cognitive trajectories over 2 years than those who did not nap or napped longer [58]. In addition, positive cognitive effects were found in patients who engaged in nap intervention studies (nap intervention lasted from single session to 4-week); these patients demonstrated improvements in various domains of cognitive function, such as attention, alertness, and visual detection [59–61]. One study with 4-week nap intervention found that both 45 min and 2-h napping opportunities in the afternoon were associated with improved logical reasoning, mathematical processing, and memory [59]. This evidence suggests that intentional naps within a certain duration may provide cognitive benefits in later life; however, this finding would benefit from future research to determine optimal nap duration.

Excessive daytime sleepiness

Approximately 20%–30% of older adults report excessive daytime sleepiness (EDS) [1, 39]. EDS can be a symptom of various disorders (e.g., SDB, cardiovascular disease, and depression) or a result of sedative medications. These disorders and sedative medications have proven to increase the risk of cognitive impairment [39]. Both cross-sectional and prospective studies have found that EDS independently predicted cognitive impairment [44, 62, 63] and was associated with higher incidence of dementia [64]. EDS was measured subjectively in these studies. Future studies will need to explore the association between EDS and cognitive function using both subjective and objective measures of sleepiness to elucidate the relationship and underlying mechanisms.

Insomnia

Insomnia has been linked with worse daytime cognitive performance (e.g., working memory and episodic memory) in young and middle-aged adults [65]. Only a few studies have assessed this association in older adults and the findings are inconsistent. Thus, no firm conclusion can be drawn based on the current state of the science. For example, in a large scale epidemiological study, self-reported insomnia was associated with increased risk of cognitive decline after 3 years in older men, but not in older women [66]. The association was not significant in population based samples of older Japanese-American men [62] and Italian older adults [64]. Some research examined the impact of insomnia intervention on cognitive outcomes and also yielded inconsistent findings. One study showed that a multiple component insomnia intervention improved sleep (sleep onset latency and sleep efficiency) and performance on simple vigilance tasks [67]. However, another insomnia intervention that used brief behavioral therapy failed to achieved any cognitive benefits in community dwelling older adults with insomnia [68].

Sleep disordered breathing

Numerous studies support a link between SDB and worse cognitive outcomes in older adults [39]. A recent systematic review and metaanalysis on the association between SDB and cognitive function and risk of cognitive impairment reviewed 14 population-based study (total $N=4,288,419$). Analysis of the cross-sectional studies suggested SDB was associated with worse executive function but revealed no significant association of SDB with memory or global cognition. Analyses of six prospective studies showed that people with SDB were more likely to develop cognitive impairment or were at risk of dementia. These findings suggest SDB is an essential modifiable risk factor of cognitive impairment in older adults and highlight the importance of early identification and treatment of SDB.

In addition, studies found that the severity of SDB may be associated with cognitive outcomes in older adults [69]. According to a recent narrative review, studies have found that higher apnea-hypopnea index and respiratory disturbance index were associated with worse global cognition and domain specific cognitive outcomes, including vigilance, executive function, attention, and memory [69].

There are promising findings on positive airway pressure (PAP) treatments' improvements on cognitive function in older adults with SDB. Studies had shown both short-term and long-term PAP therapy could benefit cognitive functions in older adults [69]. Importantly, long-term PAP treatment may delay the onset of cognitive impairment in older adults with SDB and slow down the speed of cognitive decline in Alzheimer's disease patients with SDB [69, 70]. Compliance with PAP therapy may be even more challenging in older adults with cognitive impairment than in the general population. More research that aims to improve adherence of PAP therapy in older adults with cognitive impairment are needed to treat SDB and potentially promote cognitive outcomes in this population.

Both sleep disturbances and cognitive decline are prevalent in older adults. In general, current evidence suggests a link between disturbed sleep and adverse cognitive outcomes in later life, which directs us toward a potential approach to improve cognitive health through modifying sleep health. However, most of the reviewed research included observational studies, which do not lead to concrete conclusion based on causality. Future research investigating the effects of sleep promoting interventions on cognitive function in older adults is clearly warranted, especially given the growing number of older adults at risk for cognitive impairment.

Cardiovascular health

Cardiovascular diseases, including hypertension, coronary heart disease (CHD), peripheral arterial disease (PAD), heart failure (HF), valvular heart disease, and strokes are prevalent in older adults. In 2016, it was estimated that two-thirds of older adults between the ages of 60 and 70 and around 85% of older adults aged 80 and above have cardiovascular disease (CVD). Heart disease is the leading cause of death in both men and women aged 65 and above [71]. Sleep characteristics and sleep disorders have been linked to cardiovascular health in older adults. For example, epidemiological studies found that the prevalence of hypertension and coronary artery disease is higher in people with obstructive sleep apnea (OSA) than those without OSA [72, 73]. Studies also suggest an association between OSA and increased risk of cardiovascular diseases, including hypertension, arrhythmias and heart failure [74].

Sleep duration

Literature suggests that both short and long sleep durations are predictors of poor cardiovascular outcomes [75]. However, most research has been focused on the general adult population rather than older adults. Although short sleep duration (e.g., <5 h) is a risk factor for hypertension in young and middle-aged adults [76, 77], research on older adults has suggested that only long sleep duration was associated with poor cardiovascular health and has found no clear association between short sleep duration and cardiovascular health. A large epidemiological study examined the association between sleep duration and strokes in over 150,000 US adults and found that only long sleep duration (and not short sleep duration) was associated with higher prevalence of history of stroke [78]. One study that examined the association in both middle aged and older adults found that long sleep duration (>9 h) is a risk factor for hypertension in older adults, but not in middle aged adults [76]. A population based study of Japanese older adults found that long sleep duration was associated with higher risk of cardiovascular mortality among those with poor sleep quality [79]. In addition, findings from a recent systematic review with metaanalysis suggested long sleep duration was associated 43% increased risk of cardiovascular mortality in older adults [80].

Sleep disordered breathing

Adverse impacts of SDB on cardiovascular outcomes has been established in middle-aged adults, but is not clear in older adults [15]. Research on whether SDB was associated with increased risk of hypertension are inconsistent in older adults. A study in 372 French older adults found that OSA was associated with an increased risk of new onset of hypertension. Severe OSA (AHI ≥ 30 per hour) was independently associated with 1.8-fold of increased risk of incident hypertension after 3 years [81]. However, this association was not found in the Sleep Heart Health Study [82] and the Osteoporotic Fractures in Men Study (MrOS) [83]. In addition, observational studies suggest that SDB independently increased the risk for stroke in older adults [84–86].

The association between SDB and heart failure (HF) might be bidirectional. There is evidence that suggests that chronic heart failure (CHF) may contribute to the pathogenesis of SDB [28]. For instance, the nocturnal fluid that shifts from the legs to the neck in patient with chronic heart failure could increase the pharyngeal wall edema, which can contribute to SDB. In addition, sedentary lifestyle in CHF patients may lead to weight gain [87]. It is not clear whether higher prevalence of SDB increases the risk of heart failure (HF) in older adults. One analysis from an MrOS study found that central apnea and the Cheyne-Stokes respiration (CSR), but not obstructive apnea, significantly predicted incident heart failure over 7 years in older men [88]. Another prospective, longitudinal study found that older men with an apnea-hypopnea index (AHI) of 30 or greater had increased risk for HF compared to older men with an AHI < 5. However, no significant association was found in women [89]. The possible bidirectional association between HF and SDB suggests that treating either condition could potentially benefit both conditions in patients with both SDB and HF. This potential benefit needs to be examined in future prospective, randomized, controlled trials.

Insomnia and other sleep disturbances

Systematic reviews suggest that insomnia is significantly associated with increased risk of cardiovascular outcomes (e.g., myocardial infarction, stroke, and CHD) and mortality after adjusting for established cardiovascular risk factors in the general adult population [90, 91]. This may remain true in the older adults since older adults were included in most of the reviewed studies and age was controlled for as a confounding factor. There are only a few studies that examine the association specifically in the older adult population and the results support the association. For example, a cross-sectional study of approximate 3000 Chinese older adults who self-reported occasional insomnia and frequent insomnia were more likely to have CHD than those reported no insomnia [92].

The association between other sleep complaints and cardiovascular disease has also been evaluated. Overall, there is a paucity of evidence to support the association between self-reported sleep quality and cardiovascular health in older adults. One population-based cohort study of Japanese older adults found no association between self-reported sleep quality and cardiovascular mortality [79]. In regards to daytime sleepiness, one prospective cohort study of around 6000 older adults found that daytime sleepiness at baseline predicted incident CHF, MI and cardiovascular morbidity and mortality over 5 years [93].

Psychiatric illness

Psychiatric illnesses in older adults are common, but less prevalent than in young adults. It has been estimated that up to 15% and 13% of the population have anxiety and depression, respectively [94]. Anxiety and depression are serious concerns in older adults due to their association with poor health outcomes, including decreased functional status and an increased risk of morbidity [94]. Sleep disturbances, including daytime sleepiness, poor sleep quality, prolonged sleep latency, and long wake after sleep onset, are common in older adults with psychiatric illness, such as anxiety [95]. Epidemiological and metaanalytic studies have linked sleep disturbances with increased risks of developing depression and anxiety among older adults [96]. We will review current evidence on the relationship between sleep characteristics and psychiatric diseases, with a focus on depression and anxiety in older adults.

Sleep duration

A metaanalysis of seven prospective studies involving approximately 49,000 adults revealed that both short and long sleep durations were associated with increased risk of depression in the general adult population [97]. However, the association may not be significant in older adults. Two large prospective cohort studies found no significant association between sleep duration and later onset of depression in older men [98] and older women [99].

Insomnia

Evidence suggests that insomnia is a significant risk factor for both depression and anxiety in the general adult population. We did not find any studies on anxiety and diagnosis of insomnia specifically in older adults. Studies focused on older adults support this association between depression and insomnia [100, 101]. Additionally, a cross-sectional study showed that depressive symptoms were more likely to be sustained in depressed older adults with insomnia than depressed older adults without insomnia [102]. Patients who had persistent insomnia had more residual depressive symptoms, suicidal ideation remission, and higher incidences of relapse than those who did not have insomnia [103, 104].

When considering specific insomnia symptoms, as opposed to an insomnia diagnosis, certain insomnia symptoms may prospectively predict depression in older adults. A longitudinal study of approximately 5000 older men revealed a strong relationship between difficulty initiating sleep and incidence of depression after 3 years, whereas there was no association between early morning awakening or difficulty maintaining sleep and depression [101]. In addition, sleep disturbances may also be associated with anxiety symptoms. A study of 2759 older adults found that older adults with prolonged sleep latency (>30 min) were more likely to have diagnosis of anxiety or other mood disorders [105]. A cross-sectional study found that after adjusting for depressive symptoms, medical conditions, and use of antianxiety medications, anxiety symptoms were independently associated with poor sleep efficiency and higher sleep fragmentation in older women [106].

Sleep-disordered breathing

One review of the literature noted that current evidence is not sufficient to illustrate the relationship between SDB and anxiety in the adult population [107]. A limited number of studies in the adult population yielded inconsistent findings, with 70% individuals with OSA having anxiety in some studies, to no association between the severity of apnea symptoms and anxiety symptoms in other studies [94]. One study showed that the PAP treatment may be effective in reducing anxiety symptoms among adult patients [108]. However, no study examined the relationship in older adults. Generally, current evidence suggests a relationship between OSA and depressive symptoms in the adult population [94].

Sleep quality and other sleep disturbances

A recent metaanalysis of nine studies explored the relationship between sleep quality and depression. The findings suggest that poor sleep quality is significantly associated with depression in older adults [109]. A population-based study of 2393 older adults found that short sleep duration, daytime sleepiness, and sleep disturbance are independently associated with anxiety, and sleep medication is associated with depression [110].

The number of studies in older adults is limited compared to the body of research in the broader adult population. In general, sleep disturbances and sleep disorders are associated with increased risks of developing or maintaining depression and anxiety among the adult population. However, the strength of association in older adults may differ from that in young and middle-aged adults. Future perspective studies should further examine the association in older adults and explore whether adding strategies that address treatment for sleep problems can improve the efficacy of treatment for psychiatric disorders.

Pain

The prevalence of pain increases with advancing age. Over 50% of community-dwelling older adults and 80% of nursing home residents are affected by pain [28]. Pain makes older adults more prone to falls and is associated with decreased quality of life and increased risks of all-cause mortality. In older adults, pain is usually a symptom of one or more existing medical conditions [111]. The relationship between sleep and pain has been well examined in many epidemiological and experimental studies and suggests a likely bidirectional association. Cross-sectional studies also found this/a relation between pain and insomnia or insomnia symptoms in older adults [28]. Older adults with chronic pain are more likely to report clinically significant insomnia symptoms than those without pain. Up to 80% older adults with pain reported at least one sleep compliant [111]. Acute or chronic pain contributes to sleep disturbances and these changes in sleep can subsequently/cyclically impact pain perception and tolerance [28]. One study found day-to-day associations between objectively measured sleep and self-reported morning pain in community-dwelling older adults with insomnia [112]. In addition, some studies suggest that sleep disturbances are stronger predictors of pain than using pain to predict sleep disturbances [113], though this needs to be further explored through additional research.

Psychiatric conditions may play an important role in the relation between sleep disturbances and pain. A number of studies that explored this topic have suggested that psychiatric illnesses may mediate the relationship between sleep and pain. Osteoarthritis (OA) has been frequently associated with pain in older adults [28]. One study of 367 people with OA (mean age of 68 years) found that sleep was associated with pain and depression and depressive symptoms mediated the sleep-pain relationship in the cross-sectional analysis. However, baseline sleep disturbance pain was not associated with pain at follow-up [114]. Similarly, a longitudinal analysis of a sample of 1860 participants examined whether insomnia and sleep duration predict the onset of chronic multisite musculoskeletal pain over 6 years [113]. The findings suggest that insomnia and short sleep duration are risk factors for developing chronic pain, and depressive symptoms partially mediate the effect for insomnia and short sleep with developing chronic pain. Another population-based prospective study found that insomnia symptoms were associated with an increased risk for new onset of pain after 2 years [111]. The relationship was not mediated by depression in this study. Anxiety symptoms accounted for 17% of the total effect of difficulty in initiating sleep and 15% of the total effect of difficulty in sleep maintenance on the new onset of pain, respectively.

There might be an association between OSA and pain in older adults. A randomized, double blind crossover study found significant improvement in electrical pain tolerance when OSA patients were treated with CPAP [115]. In addition, improved sleep in older adults may benefit pain management. A recent study found that older adults with clinically significant improvements in insomnia symptoms within 2 months sustained the improvements in sleep measures over an 18 months period, and showed improvements in measures of pain (Tables 4.1 and 4.2) [116].

CONCLUSION

Sleep patterns change with normal aging across the lifespan from pediatric to middle age to older age, but generally have minimal further progressive change in healthy older adulthood. Prevalent disturbed sleep or sleep problems in older adults are largely attributable to existing medical conditions and polypharmacy used to treat these conditions, rather than aging. Older adults with chronic medical conditions commonly experience insomnia symptoms, daytime sleepiness, and fatigue. Those with sleep disturbance have an increased risk of multiple adverse health outcomes, such as impaired cognition, cardiovascular morbidity, depression, pain, etc. Likewise, these chronic conditions can contribute to sleep disturbances or sleep disorders in older adults, causing a cyclical effect. Sleep interventions need to be incorporated in the management of chronic conditions as they often occur concurrent with sleep problems and vice versa. Future studies, in addition to testing mechanistic pathways and associations, should also test whether treating sleep and chronic medical condition concurrently could add treatment effects to both conditions.

TABLE 4.1 Risk factors of poor sleep in older adults.

Risk factor	Description
Advancing age	• Sleep changes from pediatric to older adulthood. Advancing age is associated with advanced sleep timing, shortened nocturnal sleep duration, increased frequency of daytime naps, increased nocturnal awakenings and time spent awake, and decreased slow wave sleep. • Age related changes in sleep may reduce after 60 years of age among older adults with good health.
Chronic medical conditions	Cardiovascular diseases; pulmonary disease, cancer, Parkinson's disease, dementia, depression, anxiety and pain related illness, such as arthritis.
Medication	Diuretics, antidepressant, hypnotics, inappropriate use of OCT medications, etc.
Primary sleep disorders	Insomnia, sleep disordered breathing, REM behavior disorder, restless legs syndrome
Lifestyle factors	Sedentary lifestyle, lack of social engagement, irregular sleep schedules, caffeine use later in the day, excessive daytime napping.
Stressful events	Transition to live in a nursing home; death of loved ones
Environmental factors	Lack of daytime bright light exposure, excessive nighttime light exposure, too cold or hot room temperature, excessive noise, uncomfortable bedding, etc.

TABLE 4.2 Summary of reviewed evidence on sleep and reviewed health outcomes in older adults.

Health outcomes	Sleep duration		Insomnia symptoms/ sleep complaints	Poor sleep quality	Prolonged daytime napping	Excessive daytime sleepiness	Insomnia	Sleep disordered breathing
	Short	Long						
Cognitive function								
Global cognition	+	+	+	+	+	N/A	+/−	+
Cognitive decline/ impairment/dementia	+	+	+	+	+	+	+/−	+
Cardiovascular health								
Hypertension	−	+	N/A	N/A	N/A	N/A	+	+/−
Stroke	−	+	N/A	N/A	N/A	N/A	+	N/A
Cardiovascular morbidity	−	+	N/A	+	N/A	N/A	+	N/A
Heart failure	N/A	N/A	N/A	N/A	N/A	+	N/A	+
Psychiatric illness								
Depression	−	−	+	+	N/A	+	+	N/A
Anxiety	+	−	+	N/A	N/A	+	N/A	+/−
Pain	+	−	+	+	N/A	N/A	+	+

− denotes current evidence shows no clear association between the sleep character and the poor health outcome in general.
+ denotes current evidence supports association between the sleep character and the poor health outcome in general.
+/− denotes current evidence yielded conflict findings.
N/A denotes the association either not reviewed or no related literature was found.

REFERENCES

[1] Li J, Vitiello MV, Gooneratne NS. Sleep in normal aging. Sleep Med Clin 2018;13(1):1–11.
[2] Espiritu JR. Aging-related sleep changes. Clin Geriatr Med 2008;24(1):1–14. [v].
[3] Mattis J, Sehgal A. Circadian rhythms, sleep, and disorders of aging. Trends Endocrinol Metab 2016;27(4):192–203.
[4] Duffy JF, Zitting KM, Chinoy ED. Aging and circadian rhythms. Sleep Med Clin 2015;10(4):423–34.
[5] Taillard J, Philip P, Coste O, Sagaspe P, Bioulac B. The circadian and homeostatic modulation of sleep pressure during wakefulness differs between morning and evening chronotypes. J Sleep Res 2003;12(4):275–82.
[6] Wennberg AM, Canham SL, Smith MT, Spira AP. Optimizing sleep in older adults: treating insomnia. Maturitas 2013;76(3):247–52.
[7] Dijk DJ, Groeger JA, Stanley N, Deacon S. Age-related reduction in daytime sleep propensity and nocturnal slow wave sleep. Sleep 2010;33(2):211–23.
[8] Dorffner G, Vitr M, Anderer P. The effects of aging on sleep architecture in healthy subjects. Adv Exp Med Biol 2015;821:93–100.
[9] Brewster GS, Riegel B, Gehrman PR. Insomnia in the older adult. Sleep Med Clin 2018;13(1):13–9.
[10] Cochen V, Arbus C, Soto ME, Villars H, Tiberge M, Montemayor T, Hein C, Veccherini MF, Onen SH, Ghorayeb I, et al. Sleep disorders and their impacts on healthy, dependent, and frail older adults. J Nutr Health Aging 2009;13(4):322–9.
[11] Medicine AAoS. International classification of sleep disorders–third edition (ICSD-3). American Academy of Sleep Medicine: Darien, IL; 2014.
[12] Schutte-Rodin S, Broch L, Buysse D, Dorsey C, Sateia M. Clinical guideline for the evaluation and management of chronic insomnia in adults. J Clin Sleep Med 2008;4(5):487–504.
[13] Mitchell MD, Gehrman P, Perlis M, Umscheid CA. Comparative effectiveness of cognitive behavioral therapy for insomnia: a systematic review. BMC Fam Pract 2012;13:40.
[14] Morin CM, Bootzin RR, Buysse DJ, Edinger JD, Espie CA, Lichstein KL. Psychological and behavioral treatment of insomnia: update of the recent evidence (1998–2004). Sleep 2006;29(11):1398–414.
[15] Chowdhuri S, Patel P, Badr MS. Apnea in older adults. Sleep Med Clin 2018;13(1):21–37.
[16] Duran J, Esnaola S, Rubio R, Iztueta A. Obstructive sleep apnea-hypopnea and related clinical features in a population-based sample of subjects aged 30 to 70 yr. Am J Respir Crit Care Med 2001;163(3 Pt 1):685–9.
[17] Kleisiaris CF, Kritsotakis EI, Daniil Z, Tzanakis N, Papaioannou A, Gourgoulianis KI. The prevalence of obstructive sleep apnea-hypopnea syndrome-related symptoms and their relation to airflow limitation in an elderly population receiving home care. Int J Chron Obstruct Pulmon Dis 2014;9:1111–7.
[18] Mehra R, Stone KL, Blackwell T, Ancoli Israel S, Dam TT, Stefanick ML, Redline S, Osteoporotic Fractures in Men Study. Prevalence and correlates of sleep-disordered breathing in older men: osteoporotic fractures in men sleep study. J Am Geriatr Soc 2007;55(9):1356–64.
[19] Young T, Shahar E, Nieto FJ, Redline S, Newman AB, Gottlieb DJ, Walsleben JA, Finn L, Enright P, Samet JM, et al. Predictors of sleep-disordered breathing in community-dwelling adults: the Sleep Heart Health Study. Arch Intern Med 2002;162(8):893–900.
[20] Lavie P. Insomnia and sleep-disordered breathing. Sleep Med 2007;8(Suppl. 4):S21–5.
[21] Morales CR, Hurley S, Wick LC, Staley B, Pack FM, Gooneratne NS, Maislin G, Pack A, Gurubhagavatula I. In-home, self-assembled sleep studies are useful in diagnosing sleep apnea in the elderly. Sleep 2012;35(11):1491–501.
[22] Catcheside PG. Predictors of continuous positive airway pressure adherence. F1000 Med Rep 2010;2:1–6.
[23] Russo-Magno P, O'Brien A, Panciera T, Rounds S. Compliance with CPAP therapy in older men with obstructive sleep apnea. J Am Geriatr Soc 2001;49(9):1205–11.
[24] Ayalon L, Ancoli-Israel S, Stepnowsky C, Marler M, Palmer BW, Liu L, Loredo JS, Corey-Bloom J, Greenfield D, Cooke J. Adherence to continuous positive airway pressure treatment in patients with Alzheimer's disease and obstructive sleep apnea. Am J Geriatr Psychiatry 2006;14(2):176–80.
[25] Chong MS, Ayalon L, Marler M, Loredo JS, Corey-Bloom J, Palmer BW, Liu L, Ancoli-Israel S. Continuous positive airway pressure reduces subjective daytime sleepiness in patients with mild to moderate Alzheimer's disease with sleep disordered breathing. J Am Geriatr Soc 2006;54(5):777–81.
[26] Li J, Yang B, Varrasse M, Li K. Sleep among long-term care residents in China: a narrative review of literature. Clin Nurs Res 2018;27(1):35–60.
[27] Fillenbaum GG, Pieper CF, Cohen HJ, Cornoni-Huntley JC, Guralnik JM. Comorbidity of five chronic health conditions in elderly community residents: determinants and impact on mortality. J Gerontol A Biol Sci Med Sci 2000;55(2):M84–9.
[28] Onen SH, Onen F. Chronic medical conditions and sleep in the older adult. Sleep Med Clin 2018;13(1):71–9.
[29] Marengoni A, Angleman S, Melis R, Mangialasche F, Karp A, Garmen A, Meinow B, Fratiglioni L. Aging with multimorbidity: a systematic review of the literature. Ageing Res Rev 2011;10(4):430–9.
[30] Salive ME. Multimorbidity in older adults. Epidemiol Rev 2013;35:75–83.
[31] Lehert P, Villaseca P, Hogervorst E, Maki PM, Henderson VW. Individually modifiable risk factors to ameliorate cognitive aging: a systematic review and meta-analysis. Climacteric 2015;18(5):678–89.
[32] Lo JC, Loh KK, Zheng H, Sim SK, Chee MW. Sleep duration and age-related changes in brain structure and cognitive performance. Sleep 2014;37(7):1171–8.
[33] Xu L, Jiang CQ, Lam TH, Liu B, Jin YL, Zhu T, Zhang WS, Cheng KK, Thomas GN. Short or long sleep duration is associated with memory impairment in older Chinese: the Guangzhou Biobank Cohort Study. Sleep 2011;34(5):575–80.
[34] Blackwell T, Yaffe K, Laffan A, Ancoli-Israel S, Redline S, Ensrud KE, Song Y, Stone KL, Osteoporotic Fractures in Men (MrOS) Study Group. Associations of objectively and subjectively measured sleep quality with subsequent cognitive decline in older community-dwelling men: the MrOS sleep study. Sleep 2014;37(4):655–63.
[35] Saint Martin M, Sforza E, Barthelemy JC, Thomas-Anterion C, Roche F. Does subjective sleep affect cognitive function in healthy elderly subjects? The Proof cohort. Sleep Med 2012;13(9):1146–52.
[36] Loerbroks A, Debling D, Amelang M, Sturmer T. Nocturnal sleep duration and cognitive impairment in a population-based study of older adults. Int J Geriatr Psychiatry 2010;25(1):100–9.

[37] Lo JC, Groeger JA, Cheng GH, Dijk DJ, Chee MW. Self-reported sleep duration and cognitive performance in older adults: a systematic review and meta-analysis. Sleep Med 2016;17:87–98.

[38] Devore EE, Grodstein F, Schernhammer ES. Sleep duration in relation to cognitive function among older adults: a systematic review of observational studies. Neuroepidemiology 2016;46(1):57–78.

[39] Yaffe K, Falvey CM, Hoang T. Connections between sleep and cognition in older adults. Lancet Neurol 2014;13(10):1017–28.

[40] Schmutte T, Harris S, Levin R, Zweig R, Katz M, Lipton R. The relation between cognitive functioning and self-reported sleep complaints in nondemented older adults: results from the Bronx aging study. Behav Sleep Med 2007;5(1):39–56.

[41] Tworoger SS, Lee S, Schernhammer ES, Grodstein F. The association of self-reported sleep duration, difficulty sleeping, and snoring with cognitive function in older women. Alzheimer Dis Assoc Disord 2006;20(1):41–8.

[42] Jelicic M, Bosma H, Ponds RW, Van Boxtel MP, Houx PJ, Jolles J. Subjective sleep problems in later life as predictors of cognitive decline. Report from the Maastricht Ageing Study (MAAS). Int J Geriatr Psychiatry 2002;17(1):73–7.

[43] Potvin O, Lorrain D, Forget H, Dube M, Grenier S, Preville M, Hudon C. Sleep quality and 1-year incident cognitive impairment in community-dwelling older adults. Sleep 2012;35(4):491–9.

[44] Elwood PC, Bayer AJ, Fish M, Pickering J, Mitchell C, Gallacher JE. Sleep disturbance and daytime sleepiness predict vascular dementia. J Epidemiol Community Health 2011;65(9):820–4.

[45] Sterniczuk R, Theou O, Rusak B, Rockwood K. Sleep disturbance is associated with incident dementia and mortality. Curr Alzheimer Res 2013;10(7):767–75.

[46] Hsieh S, Li TH, Tsai LL. Impact of monetary incentives on cognitive performance and error monitoring following sleep deprivation. Sleep 2010;33(4):499–507.

[47] Blackwell T, Yaffe K, Ancoli-Israel S, Redline S, Ensrud KE, Stefanick ML, Laffan A, Stone KL, Osteoporotic Fractures in Men (MrOS) Study Group. Association of sleep characteristics and cognition in older community-dwelling men: the MrOS sleep study. Sleep 2011;34(10):1347–56.

[48] Jaussent I, Bouyer J, Ancelin ML, Berr C, Foubert-Samier A, Ritchie K, Ohayon MM, Besset A, Dauvilliers Y. Excessive sleepiness is predictive of cognitive decline in the elderly. Sleep 2012;35(9):1201–7.

[49] Scullin MK, Bliwise DL. Sleep, cognition, and normal aging: integrating a half century of multidisciplinary research. Perspect Psychol Sci 2015;10(1):97–137.

[50] Blackwell T, Yaffe K, Ancoli-Israel S, Schneider JL, Cauley JA, Hillier TA, Fink HA, Stone KL, Study of Osteoporotic Fractures Group. Poor sleep is associated with impaired cognitive function in older women: the study of osteoporotic fractures. J Gerontol A Biol Sci Med Sci 2006;61(4):405–10.

[51] Lambiase MJ, Gabriel KP, Kuller LH, Matthews KA. Sleep and executive function in older women: the moderating effect of physical activity. J Gerontol A Biol Sci Med Sci 2014;69(9):1170–6.

[52] Naismith SL, Rogers NL, Hickie IB, Mackenzie J, Norrie LM, Lewis SJ. Sleep well, think well: sleep-wake disturbance in mild cognitive impairment. J Geriatr Psychiatry Neurol 2010;23(2):123–30.

[53] Lim AS, Kowgier M, Yu L, Buchman AS, Bennett DA. Sleep fragmentation and the risk of incident Alzheimer's disease and cognitive decline in older persons. Sleep 2013;36(7):1027–32.

[54] Campbell SS, Murphy PJ. The nature of spontaneous sleep across adulthood. J Sleep Res 2007;16(1):24–32.

[55] Furihata R, Kaneita Y, Jike M, Ohida T, Uchiyama M. Napping and associated factors: a Japanese nationwide general population survey. Sleep Med 2016;20:72–9.

[56] Spira AP. Sleep and health in older adulthood: recent advances and the path forward. J Gerontol A Biol Sci Med Sci 2018;73(3):357–9.

[57] Li J, Cacchione PZ, Hodgson N, Riegel B, Keenan BT, Scharf MT, Richards KC, Gooneratne NS. Afternoon napping and cognition in Chinese older adults: findings from the China health and retirement longitudinal study baseline assessment. J Am Geriatr Soc 2017;65(2):373–80.

[58] Li J, Chang YP, Riegel B, Keenan BT, Varrasse M, Pack AI, Gooneratne NS. Intermediate, but not extended, afternoon naps may preserve cognition in Chinese older adults. J Gerontol A Biol Sci Med Sci 2018;73(3):360–6.

[59] Campbell SS, Stanchina MD, Schlang JR, Murphy PJ. Effects of a month-long napping regimen in older individuals. J Am Geriatr Soc 2011;59(2):224–32.

[60] Milner CE, Cote KA. A dose-response investigation of the benefits of napping in healthy young, middle-aged and older adults. Sleep Biol Rhythms 2008;6(1):2–15.

[61] Korman M, Dagan Y, Karni A. Nap it or leave it in the elderly: a nap after practice relaxes age-related limitations in procedural memory consolidation. Neurosci Lett 2015;606:173–6.

[62] Foley D, Monjan A, Masaki K, Ross W, Havlik R, White L, Launer L. Daytime sleepiness is associated with 3-year incident dementia and cognitive decline in older Japanese-American men. J Am Geriatr Soc 2001;49(12):1628–32.

[63] Ohayon MM, Vecchierini MF. Daytime sleepiness and cognitive impairment in the elderly population. Arch Intern Med 2002;162(2):201–8.

[64] Merlino G, Piani A, Gigli GL, Cancelli I, Rinaldi A, Baroselli A, Serafini A, Zanchettin B, Valente M. Daytime sleepiness is associated with dementia and cognitive decline in older Italian adults: a population-based study. Sleep Med 2010;11(4):372–7.

[65] Fortier-Brochu E, Beaulieu-Bonneau S, Ivers H, Morin CM. Insomnia and daytime cognitive performance: a meta-analysis. Sleep Med Rev 2012;16(1):83–94.

[66] Cricco M, Simonsick EM, Foley DJ. The impact of insomnia on cognitive functioning in older adults. J Am Geriatr Soc 2001;49(9):1185–9.

[67] Altena E, Van Der Werf YD, Strijers RL, Van Someren EJ. Sleep loss affects vigilance: effects of chronic insomnia and sleep therapy. J Sleep Res 2008;17(3):335–43.

[68] Wilckens KA, Hall MH, Nebes RD, Monk TH, Buysse DJ. Changes in cognitive performance are associated with changes in sleep in older adults with insomnia. Behav Sleep Med 2016;14(3):295–310.

[69] Dzierzewski JM, Dautovich N, Ravyts S. Sleep and cognition in older adults. Sleep Med Clin 2018;13(1):93–106.

[70] Osorio RS, Gumb T, Pirraglia E, Varga AW, Lu SE, Lim J, Wohlleber ME, Ducca EL, Koushyk V, Glodzik L, et al. Sleep-disordered breathing advances cognitive decline in the elderly. Neurology 2015;84(19):1964–71.

[71] American Heart Association, American Stroke Association. Older Americans & cardiovascular diseases—statistical fact sheet 2016 update. 2016.

[72] Nieto FJ, Young TB, Lind BK, Shahar E, Samet JM, Redline S, D'Agostino RB, Newman AB, Lebowitz MD, Pickering TG. Association of sleep-disordered breathing, sleep apnea, and hypertension in a large community-based study. Sleep Heart Health Study. JAMA 2000;283(14):1829–36.

[73] Shahar E, Whitney CW, Redline S, Lee ET, Newman AB, Nieto FJ, O'Connor GT, Boland LL, Schwartz JE, Samet JM. Sleep-disordered breathing and cardiovascular disease: cross-sectional results of the Sleep Heart Health Study. Am J Respir Crit Care Med 2001;163(1):19–25.

[74] Thomas JJ, Ren J. Obstructive sleep apnoea and cardiovascular complications: perception versus knowledge. Clin Exp Pharmacol Physiol 2012;39(12):995–1003.

[75] Cappuccio FP, Cooper D, D'Elia L, Strazzullo P, Miller MA. Sleep duration predicts cardiovascular outcomes: a systematic review and meta-analysis of prospective studies. Eur Heart J 2011;32(12):1484–92.

[76] Gangwisch JE, Heymsfield SB, Boden-Albala B, Buijs RM, Kreier F, Pickering TG, Rundle AG, Zammit GK, Malaspina D. Short sleep duration as a risk factor for hypertension: analyses of the first National Health and Nutrition Examination Survey. Hypertension 2006;47(5):833–9.

[77] Gangwisch JE. A review of evidence for the link between sleep duration and hypertension. Am J Hypertens 2014;27(10):1235–42.

[78] Fang J, Wheaton AG, Ayala C. Sleep duration and history of stroke among adults from the USA. J Sleep Res 2014;23(5):531–7.

[79] Suzuki E, Yorifuji T, Ueshima K, Takao S, Sugiyama M, Ohta T, Ishikawa-Takata K, Doi H. Sleep duration, sleep quality and cardiovascular disease mortality among the elderly: a population-based cohort study. Prev Med 2009;49(2–3):135–41.

[80] da Silva AA, de Mello RG, Schaan CW, Fuchs FD, Redline S, Fuchs SC. Sleep duration and mortality in the elderly: a systematic review with meta-analysis. BMJ Open 2016;6(2):e008119.

[81] Guillot M, Sforza E, Achour-Crawford E, Maudoux D, Saint-Martin M, Barthelemy JC, Roche F. Association between severe obstructive sleep apnea and incident arterial hypertension in the older people population. Sleep Med 2013;14(9):838–42.

[82] Haas DC, Foster GL, Nieto FJ, Redline S, Resnick HE, Robbins JA, Young T, Pickering TG. Age-dependent associations between sleep-disordered breathing and hypertension: importance of discriminating between systolic/diastolic hypertension and isolated systolic hypertension in the Sleep Heart Health Study. Circulation 2005;111(5):614–21.

[83] Fung MM, Peters K, Redline S, Ziegler MG, Ancoli-Israel S, Barrett-Connor E, Stone KL, Osteoporotic Fractures in Men Research Group. Decreased slow wave sleep increases risk of developing hypertension in elderly men. Hypertension 2011;58(4):596–603.

[84] Stone KL, Blackwell TL, Ancoli-Israel S, Barrett-Connor E, Bauer DC, Cauley JA, Ensrud KE, Hoffman AR, Mehra R, Stefanick ML, et al. Sleep disordered breathing and risk of stroke in older community-dwelling men. Sleep 2016;39(3):531–40.

[85] Munoz R, Duran-Cantolla J, Martinez-Vila E, Gallego J, Rubio R, Aizpuru F, De La Torre G. Severe sleep apnea and risk of ischemic stroke in the elderly. Stroke 2006;37(9):2317–21.

[86] Yaggi HK, Concato J, Kernan WN, Lichtman JH, Brass LM, Mohsenin V. Obstructive sleep apnea as a risk factor for stroke and death. N Engl J Med 2005;353(19):2034–41.

[87] Yumino D, Redolfi S, Ruttanaumpawan P, Su MC, Smith S, Newton GE, Mak S, Bradley TD. Nocturnal rostral fluid shift: a unifying concept for the pathogenesis of obstructive and central sleep apnea in men with heart failure. Circulation 2010;121(14):1598–605.

[88] Javaheri S, Blackwell T, Ancoli-Israel S, Ensrud KE, Stone KL, Redline S, Osteoporotic Fractures in Men Study Research Group. Sleep-disordered breathing and incident heart failure in older men. Am J Respir Crit Care Med 2016;193(5):561–8.

[89] Gottlieb DJ, Yenokyan G, Newman AB, O'Connor GT, Punjabi NM, Quan SF, Redline S, Resnick HE, Tong EK, Diener-West M, et al. Prospective study of obstructive sleep apnea and incident coronary heart disease and heart failure: the sleep heart health study. Circulation 2010;122(4):352–60.

[90] Li M, Zhang XW, Hou WS, Tang ZY. Insomnia and risk of cardiovascular disease: a meta-analysis of cohort studies. Int J Cardiol 2014;176(3):1044–7.

[91] Sofi F, Cesari F, Casini A, Macchi C, Abbate R, Gensini GF. Insomnia and risk of cardiovascular disease: a meta-analysis. Eur J Prev Cardiol 2014;21(1):57–64.

[92] Zhuang J, Zhan Y, Zhang F, Tang Z, Wang J, Sun Y, Ding R, Hu D, Yu J. Self-reported insomnia and coronary heart disease in the elderly. Clin Exp Hypertens 2016;38(1):51–5.

[93] Newman AB, Spiekerman CF, Lefkowitz D, Manolio T, Reynolds CF, Robbins J. Daytime sleepiness predicts mortality and cardiovascular disease in older adults. J Am Geriatr Soc 2000;48(2):115–23.

[94] Nadorff MR, Drapeau CW, Pigeon WR. Psychiatric illness and sleep in older adults: comorbidity and opportunities for intervention. Sleep Med Clin 2018;13(1):81–91.

[95] Wetherell JL, Le Roux H, Gatz M. DSM-IV criteria for generalized anxiety disorder in older adults: distinguishing the worried from the well. Psychol Aging 2003;18(3):622–7.

[96] Cole MG, Dendukuri N. Risk factors for depression among elderly community subjects: a systematic review and meta-analysis. Am J Psychiatry 2003;160(6):1147–56.

[97] Zhai L, Zhang H, Zhang D. Sleep duration and depression among adults: a meta-analysis of prospective studies. Depress Anxiety 2015;32(9):664–70.

[98] Paudel M, Taylor BC, Ancoli-Israel S, Blackwell T, Maglione JE, Stone K, Redline S, Ensrud KE. Sleep disturbances and risk of depression in older men. Sleep 2013;36(7):1033–40.

[99] Maglione JE, Ancoli-Israel S, Peters KW, Paudel ML, Yaffe K, Ensrud KE, Stone KL, Study of Osteoporotic Fractures Research Group. Subjective and objective sleep disturbance and longitudinal risk of depression in a cohort of older women. Sleep 2014;37(7):1179–87.

[100] Perlis ML, Smith LJ, Lyness JM, Matteson SR, Pigeon WR, Jungquist CR, Tu X. Insomnia as a risk factor for onset of depression in the elderly. Behav Sleep Med 2006;4(2):104–13.

[101] Yokoyama E, Kaneita Y, Saito Y, Uchiyama M, Matsuzaki Y, Tamaki T, Munezawa T, Ohida T. Association between depression and insomnia subtypes: a longitudinal study on the elderly in Japan. Sleep 2010;33(12):1693–702.

[102] Pigeon WR, Hegel M, Unutzer J, Fan MY, Sateia MJ, Lyness JM, Phillips C, Perlis ML. Is insomnia a perpetuating factor for late-life depression in the IMPACT cohort? Sleep 2008;31(4):481–8.

[103] Taylor DJ, Walters HM, Vittengl JR, Krebaum S, Jarrett RB. Which depressive symptoms remain after response to cognitive therapy of depression and predict relapse and recurrence? J Affect Disord 2010;123(1–3):181–7.

[104] Nadorff MR, Ellis TE, Allen JG, Winer ES, Herrera S. Presence and persistence of sleep-related symptoms and suicidal ideation in psychiatric inpatients. Crisis 2014;35(6):398–405.

[105] Leblanc M-F, Desjardins S, Desgagné A. Sleep problems in anxious and depressive older adults. Psychol Res Behav Manag 2015;8:161.

[106] Spira AP, Stone K, Beaudreau SA, Ancoli-Israel S, Yaffe K. Anxiety symptoms and objectively measured sleep quality in older women. Am J Geriatr Psychiatry 2009;17(2):136–43.

[107] Diaz SV, Brown LK. Relationships between obstructive sleep apnea and anxiety. Curr Opin Pulm Med 2016;22(6):563–9.

[108] Lee E, Cho HJ, Olmstead R, Levin MJ, Oxman MN, Irwin MR. Persistent sleep disturbance: a risk factor for recurrent depression in community-dwelling older adults. Sleep 2013;36(11):1685–91.

[109] Becker NB, Jesus SN, João KA, Viseu JN, Martins RI. Depression and sleep quality in older adults: a meta-analysis. Psychol Health Med 2017;22(8):889–95.

[110] Potvin O, Lorrain D, Belleville G, Grenier S, Preville M. Subjective sleep characteristics associated with anxiety and depression in older adults: a population-based study. Int J Geriatr Psychiatry 2014;29(12):1262–70.

[111] Dunietz GL, Swanson LM, Jansen EC, Chervin RD. Key insomnia symptoms and incident pain in older adults: direct and mediated pathways through depression and anxiety. Sleep 2018;41(9):zsy125.

[112] Dzierzewski JM, Williams JM, Roditi D, Marsiske M, McCoy K, McNamara J, Dautovich N, Robinson ME, McCrae CS. Daily variations in objective nighttime sleep and subjective morning pain in older adults with insomnia: evidence of covariation over time. J Am Geriatr Soc 2010;58(5):925–30.

[113] Generaal E, Vogelzangs N, Penninx BW, Dekker J. Insomnia, sleep duration, depressive symptoms, and the onset of chronic multisite musculoskeletal pain. Sleep 2017;40(1):zsw030.

[114] Parmelee PA, Tighe CA, Dautovich ND. Sleep disturbance in osteoarthritis: linkages with pain, disability, and depressive symptoms. Arthritis Care Res 2015;67(3):358–65.

[115] Onen SH, Onen F, Albrand G, Decullier E, Chapuis F, Dubray C. Pain tolerance and obstructive sleep apnea in the elderly. J Am Med Dir Assoc 2010;11(9):612–6.

[116] Vitiello MV, McCurry SM, Shortreed SM, Baker LD, Rybarczyk BD, Keefe FJ, Von Korff M. Short-term improvement in insomnia symptoms predicts long-term improvements in sleep, pain, and fatigue in older adults with comorbid osteoarthritis and insomnia. Pain 2014;155(8):1547–54.

Chapter 5

Social-ecological model of sleep health

Michael A. Grandner
Sleep and Health Research Program, Department of Psychiatry, University of Arizona College of Medicine, Tucson, AZ, United States

INTRODUCTION

Insufficient sleep duration and/or poor sleep quality represents a significant, unmet public health problem [1]. In epidemiological studies, spanning over 40 years and multiple continents, short and/or long sleep duration, as well as poor sleep quality, is associated with increased mortality risk [2, 3]. Additionally, insufficient and/or poor quality sleep is associated with (and thought to play a causal role in) 4 of the 7 leading causes of death, including heart disease, stroke, accidents, and diabetes, as well as other important health outcomes, such as weight gain and obesity, depression, and cognitive deficits. Current research is exploring mechanistic aspects of these relationships, such as isolating genetic vulnerabilities, identifying biomarkers for daytime sleepiness, and determining ways in which sleep plays a role in protective signaling pathways.

What these studies have in common is that they are exploring the "downstream" effects of sleep-related problems, clarifying how sleep plays a role in cardiometabolic disease, in addition to other adverse health outcomes, and how these pathways may explain the well-documented relationships with mortality. Increasingly, attention has focused on determinants of sleep—"upstream" influences that play a role in the development of the problematic sleep patterns that are predictive of worse health. A better understanding of the determinants of sleep will aid in the identification of modifiable factors and intervention targets that can be manipulated. For example, as poor diet is known to be associated with a number of negative health states, understanding the determinants of obesity (e.g., advertising, access to healthy food, socioeconomic status, sedentary lifestyle) has identified useful targets for change.

Accordingly, this chapter proposes a theoretical model for considering insufficient and/or poor quality sleep in the context of its associated negative health outcomes (e.g., obesity, diabetes, cardiovascular disease, depression), as well as its likely determinants. Since there is sparse literature on determinants of sleep, we constructed our model with input from existing models for other health behaviors (e.g., diet, exercise). These models are typically based on a Social-Ecological framework, which conceives of a behavior of interest (e.g., a person's diet) in the context of individual-level factors, which are embedded within social networks (e.g., family, work), which are themselves interrelated and embedded within larger networks (e.g., community, religion), which exist in a context of society that influences these networks in a number of ways (e.g., laws, technology, economics). Using a traditional Social-Ecological approach as a starting point, we constructed our model as a series of embedded systems (Individual Level, Social Level, and Societal Level), identifying key components of those systems believed to be determinants of sleep.

In summary, this model presents a conceptual framework for the "downstream" negative effects of insufficient and/or poor quality sleep, as well as the "upstream" determinants. The upper part of the model (determinants) is constructed based on existing theoretical models for other health behaviors, focusing on aspects thought to be particularly germane to sleep. The second part of the model (outcomes) is constructed as a synthesis of available data from epidemiological and experimental studies. Together, the model considers a global view of sleep and health, establishing a framework for future research to explore the determinants of sleep from a societal standpoint. Future studies will add clarity to the model, discerning unique and combined influences from the Individual, Community, and Society levels. Finally, interventions developed based on this model can address problematic sleep at the Individual (e.g., improving an individual's sleep), Social (e.g., promoting workplace initiatives that minimize sleep-related impairments or increasing healthy sleep habits in families), and Societal (e.g., public policy initiatives and educational campaigns) levels.

THE SOCIAL ECOLOGICAL MODEL

The Social-Ecological Model was originally proposed by Bronfenbrenner [4]. This model was intended to conceptualize the role of the individual in their environment. A key feature of this model is that the individual exists at the center of a nested set of constructs that describe levels of the environment in relation to the individual. The idea is that each level is nested within the next, and so on. A schematic of the main components of the model is displayed in Fig. 5.1. First, the model starts with the individual. Each individual person is believed to exist within their own specific social-ecological framework of nested systems.

The first layer beyond the individual is the "microsystem." This is the system in which the individual is embedded. According to this model, the microsystem refers to the set of interactions between the individual and elements of their environment at home, at school or work, etc. These are specific environments that the individual interacts with. In each of these environments, the individual takes on a specific role (e.g., mother or father, worker, son or daughter, teacher, friend) for specific periods of time. A key element of the microsystem is that this is where the individual specifically acts in relation to those around them.

The next layer around the microsystem is the "mesosystem." Just as the individual is embedded within the microsystem of people, places, and roles, the microsystem is itself embedded within the mesosystem. This system describes the interrelations among elements of the microsystem that are outside of the individual. For example, the interactions of a child at home with their parents and at school with their teachers exist within the microsystem, but the parent-teacher meeting would, for example, exist within the mesosystem. The mesosystem represents the collected microsystems of the other people that the individual interacts with in their own microsystem.

The next layer around the mesosystem is the "exosystem." This system of interactions encompasses elements of the mesosystem that do not specifically interact with the microsystem of the individual. For example, the neighborhood, a person's industry, the media, the consumer landscape, and the communication system of mobile phones represent discrete and conceptual members of the exosystem. The exosystem is the milieu within which the mesosystem exists. It represents the interactions and human processes that facilitate, hinder, control, or modify elements of the mesosystem (e.g., a company or a school) that then influences the microsystem (e.g., a workplace or a classroom), which influences the individual.

Beyond the exosystem, which exists outside of the mesosystem, is the "macrosystem." The macrosystem includes the constructs within which the exosystem exists. The mesosystem rarely includes specific entities (as these would likely exist in the exosystem). Rather, the macrosystem reflects the common ideas, expectations, prototypes, and stereotypes that guide the exosystem. For example, there are sets of ideas in a culture about how a workplace should be, which influences the exosystem of an industry, which influences a specific company, which influences a specific workplace, which influences a specific individual. Thus, the macrosystem is the most abstract of the layers of the social-ecological model.

This concept of embedded systems reflecting layers of abstraction (the social-ecological framework) has remained a useful construct in understanding health behavior. For example, it has guided the development of interventions for diet [5], physical activity [6, 7], substance abuse [8], stress management [9], vaccination [10], suicide prevention [11], environmental change [12], and other domains. It is possible that this model can also be applied to sleep, which itself is an important domain of health behavior.

SLEEP AS A DOMAIN OF HEALTH BEHAVIOR

The concept of a "health behavior" refers to a behavioral domain that can have broad impact on health, functioning, and longevity. This idea gained strength when it was found that behavioral factors were the leading "actual" causes of death in the United States in 2000 [13]. The leading "actual" causes of death were smoking, poor diet, lack of physical activity, and alcohol consumption, which accounted for >38% of all deaths combined. Insufficient and/or poor quality sleep was not considered in these analyses, but assessment of the mortality data for sleep duration and sleep apnea suggests that many deaths may be at least partially sleep-related. Further, insufficient and/or poor quality sleep has wide-ranging physiologic and psychological outcomes (described below). For these reasons, we propose that sleep represents a domain of health behavior.

Many studies have documented adverse physiologic, medical and psychological outcomes associated with habitual short sleep duration and experimental sleep deprivation (i.e., insufficient sleep duration), as well as poor sleep quality in general and sleep disorders such as insomnia,

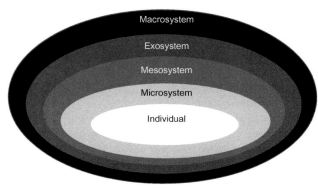

FIG. 5.1 Social-ecological model.

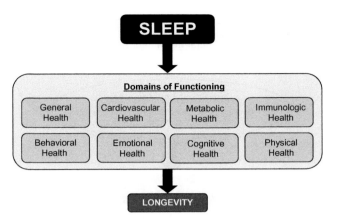

FIG. 5.2 Sleep as factor in health and mortality.

sleep apnea, and others (see other chapters in this volume). Accordingly, in building the model, we began with a proposed causal pathway linking sleep to adverse outcomes (Fig. 5.2). As depicted, the main domains impacted by sleep include general health, cardiovascular health, metabolic health, immunologic health, behavioral health, emotional health, cognitive health, and physical health.

Several studies have shown that habitual short sleep duration is associated with increased risk of hypertension, heart attack, and stroke [14]. In addition, short sleep duration has been associated with elevated cholesterol and inflammatory markers [15]. Habitual poor quality sleep has also been associated with elevated risk of cardiovascular disease [16]. Insomnia—especially in the context of short sleep duration—has been shown to be associated with elevated cardiovascular disease risk [17]. Regarding sleep apnea, a large body of literature describes very strong relationships between sleep apnea and cardiovascular function [18]. Although the primary mechanism proposed for this link is through intermittent hypoxia, there is evidence that the sleep fragmentation that occurs as part of sleep apnea is independently associated with negative effects (see chapter in this volume).

Many studies have also found that short sleep duration is associated with increased body mass index and/or risk of obesity [19]. These findings are supported by a growing literature of epidemiologic studies that demonstrate prospective weight gain associated with short sleep duration [20–22] and laboratory studies that show that sleep deprivation is associated with increased calorie intake (despite no difference in energy use) [23]. Although fewer studies have explored sleep quality in this regard, a number of studies have shown that sleep disturbance is associated with increased risk of obesity (see chapter in this volume).

The role of sleep in mental health is also well-established. Poor sleep is a well-characterized risk factor for stress and is related to dysregulation of neuroendocrine stress systems. Insomnia is a major risk factor for depression [24, 25] and is a core feature of nearly all psychiatric conditions. Poor sleep has been linked to suicide [26], as has insufficient sleep duration [27] and nocturnal wakefulness [28, 29]. There is a growing literature demonstrating the role of sleep in neuroaffective regulation [30, 31]. And sleep plays important roles in substance abuse disorders involving alcohol, nicotine, opiates, cannabis, and other substance use disorders.

The impact of sleep loss on cognitive functioning is well-characterized as well. Extensive work has demonstrated that sleep loss impairs attention and ability to remain vigilant [32]. Sleep difficulties have also been associated with processes such as working memory, executive function, and function in other neurocognitive domains [33]. This may underlie other findings showing that sleep loss impairs strategic decision-making [34], risk-based decision-making [35, 36], and work productivity [37].

Taken together, the findings described above describe associations between sleep and a number of outcomes. It should be noted that in Fig. 5.2, these outcomes are in a common box; this represents the many intercorrelations among these domains. It should be noted that the model also takes into account the many studies that have shown that poor health states cause disruptions in sleep—either by shortening sleep duration or worsening sleep quality.

CONCEPTUALIZING SLEEP IN A SOCIAL-ECOLOGICAL MODEL

The social-ecological model, outlined above, describes the individual as embedded within the microsystem within which they interact, which is embedded within the mesosystem in which elements of the microsystem interact, which is embedded within the exosystem which includes the structures that surround the mesosystem, and the macrosystem which includes the broader constructs that guide the social environment. For the purposes of the proposed model, sleep is conceptualized along a similar nested system. The proposed model conceptualizes the determinants of sleep to exist on three levels: the individual level, the social level, and the societal level.

INDIVIDUAL LEVEL

Fig. 5.3 shows the causal relationship between individual-level factors and sleep. These individual-level factors are conceptualized to reflect aspects of the individual that proximally and/or directly relate to that individual's sleep. These factors represent aspects of the individual that impact sleep, as well as cognitive and behavioral phenomena that impact sleep. This level contains all factors that could fit into this category. For example, this level could include individual genetics, beliefs about sleep, attitudes about sleep, sleep-related behaviors, aspects of individual physiology, health status, and sleep-related choices.

FIG. 5.3 Individual-level factors, sleep, and health.

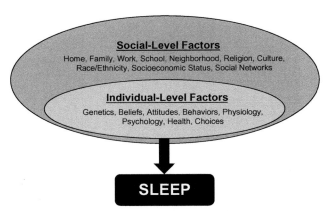

FIG. 5.4 Role of social factors in sleep and health.

An individual's genotype may exert influence over sleep in many ways. For example, genetics may influence circadian preference, sleep need, resilience against sleep loss, risk for sleep disorders, or even risk for other disorders that may impact sleep. Similarly, some aspects of physiology may impact sleep. For example, individuals who have lowered arousal thresholds may have more difficulty sleeping, as might those with disrupted neuroendocrine stress systems or even raised body temperature. Along these lines, health status is well known to influence sleep. Individuals may experience problems sleeping if they are ill, are in pain, have a sleep disorder, or another medical condition that interferes with sleep. Beliefs about sleep are also important factors at this level. Individuals who believe that they need less sleep, for example, may engage in more sleep-promoting behaviors. Previous studies have shown associations between sleep-related beliefs and sleep health [38], and changing sleep beliefs may impact on sleep-related behaviors. In addition to beliefs, individuals maintain attitudes about sleep. Those who express generally positive attitudes about sleep may be more likely to experience better quality sleep in general [39]. Further, these attitudes may be additional important contributors to behavior. An individual's sleep-related behaviors can encompass a wide range of possibilities, including sleep hygiene practices, bedtime routines, schedules maintained, etc.

Individual-level factors represent the proximal determinants of an individual's sleep, in a specific place, at a specific time. Some of these factors will represent aspects of the individual, some of these factors will represent aspects of the individual's current state, and some will represent aspects of what the individual is believing, thinking, and doing. These factors are conceptualized as generally time-dependent in that their influence on sleep depends on their characteristics at the time in which they exert influence. Genetics, for example, may not change, but beliefs may change, as might health conditions.

SOCIAL LEVEL

The proposed model conceptualizes the individual level factors as being embedded within social-level factors. These exist outside of the individual but include the individual.

A construct exists at the social level if the individual is part of that construct but that construct would exist whether or not the individual exists. For example, work represents a social level construct, as the workplace represents a construct that contains the individual but would theoretically exist without them. Fig. 5.4 depicts the embedded relationship between the individual and social levels. Some factors that could exist on the social level are also depicted and include Home, Family, Work, School, Neighborhood, Religion, Culture, Race/Ethnicity, Socioeconomic Status, and Social Networks.

The home represents the most proximal social-level factor since it includes many elements, including a sleeping environment, social dynamics, access to other behaviors such as eating, socializing, working, relating to family members, etc. It is (often) the place where people generally sleep, which gives it a prominent place. Many previous studies have described ways in which elements of the home might impact sleep. Related to the home is the family. The family can play an important role in an individual's sleep. The family may represent the source or model for sleep-related beliefs, attitudes, and behaviors. The family may also represent a source for aspects of the individual such as health and even genetics. The family—which represents both the nuclear or extended family—plays many roles in an individual's sleep.

Another key factor at the social level is work or school. This is another complex factor that subsumes schedules, social circles, stresses, expectations, hierarchies, physical environments, logistics, and other factors. For those in school, school start times have been explored as key determinants of sleep [40–43] and modifying them has been shown to be helpful for promoting sleep. School-related stresses (in many forms) can impact sleep of young children and adolescents alike. For working adults, the workplace similarly can impact sleep—especially by dictating work demands, schedule demands, and sometimes even out-of-work activities (including evening activities). Problems at work can impact sleep and poor sleep can impact performance at work [37].

Aspects of the built environment are also relevant at this level. The built environment represents the buildings, streets, and other man-made elements of the environment. Ways in which these elements may impact sleep can be direct (for example via light or noise) or indirect (for example via crime and stress). The social environment is also relevant at this level. This includes an individual's social network of friends, acquaintances, and other people that may contribute to norming, behaviors, beliefs, and other individual-level factors. It can also represent aspects such as race/ethnicity and socioeconomics, which are also related to sleep at an individual and population level.

The model suggests that it is factors at the social level that are largely responsible for the factors at the individual level. In some cases, the social-level factors represent an explanation for the origin of some of the individual-level factors (e.g., predisposition for a physical or mental health condition). In other cases, individual-level factors are facilitated or impinged on by the social-level factors (e.g., work schedules). Alternatively, social-level factors may more subtly shape individual-level factors (e.g., beliefs, practices and attitudes). In some cases, a sleep intervention at the individual level will be inert unless social-level factors are considered (e.g., individuals working multiple jobs).

SOCIETAL LEVEL

Just as the individual is embedded within a social context, these factors, too, are embedded within an even broader societal context. Just as the social level describe the forces that converge to act upon the individual, the societal level represents the forces that impact the social context and, in turn, the individual. Fig. 5.5 adds the societal layer to the model, listing societal factors such as globalization, 24/7 Society, geography, public policy, technology and progress, racism and discrimination, economics, and the natural environment as factors that directly and indirectly affect sleep.

The social level represented constructs that exist outside of the individual but of which the individual is a part (with a key element that if the individual ceased to exist, those structures would still exist). Similarly, the societal level consists of constructs of which the social-level factors are a part, and would persist even if those factors would cease to exist. For example, globalization exists outside of any one workplace or organization, but arguably most workplaces and organizations participate in globalization in ways ranging from outsourcing to calling tech support in another country. Similar constructs that exist at a regional, national, and/or global level are included in this level.

Although individual-level factors are proximal and more direct causes of an individual's sleep experience, societal factors are also critically important. Although they may

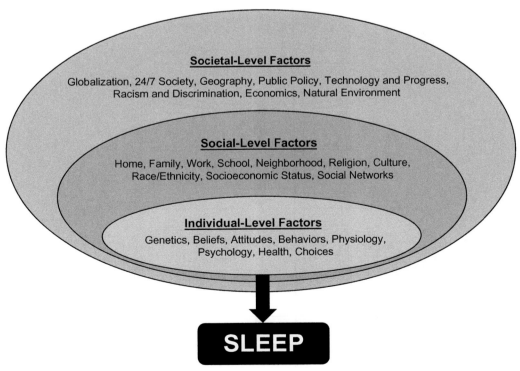

FIG. 5.5 Individual, social, and societal influences on sleep.

be less readily modifiable, conceptualizing an individual's sleep without considering the role of these factors omits important contextual information. For example, the increased drive toward globalization has interacted with the 24/7 society brought on by automation and the industrial revolution. This has led to products and services available at all hours of the day, and, consequently, the awake consumers taking advantage of this opportunity and the shiftworkers employed to meet these demands. Public policy and economics may also impact sleep, whether it be reflected in school start times or regulations about sleep disorders testing, or workplace rules, or light/noise ordinances, or other policies. Factors such as economic stress can impact an individual's sleep whether that stress is felt at the individual level [44], at the level of the neighborhood [45–47], or even at the national level [48, 49]. The physical environment and geography may dictate elements that impact sleep such as sunlight duration, weather patterns, disease risk, local culture, green space, pollution, traffic, or other aspects of the environment. Technology represents a particularly salient factor that indirectly influences sleep. There has been much discussion about the sleep-related effects of using technology in the bedroom [50], which has quickly become ubiquitous in our society.

Taken together, the combined social-ecological model of sleep, represented in Figs. 5.5 and 5.6, attempts to conceptualize the upstream influences on sleep as being primarily driven at the individual level (representing aspects of the individual as well as what the individual thinks/does). But the model recognizes that these individual-level factors exist in a social context and it is these social-level factors that may dictate many aspects of the individual-level factors. These social factors, though, exist in the context of societal factors that operate at a macro level, influencing communities, workplaces, schools, and families, which then subsequently influence individuals. In this way, a societal-level factor (e.g., development and adoption of mobile technology) influences social-level factors (e.g., workplace emails at all hours of the day), which subsequently influences individual behavior (e.g., sleeping with a smartphone by the bed).

COMBINING UPSTREAM INFLUENCES AND DOWNSTREAM CONSEQUENCES

The role of sleep as a mediator or moderator on aspects of health is still relatively unclear. Sleep could mediate or moderate the relationships between individual-level factors and adverse health outcomes. Namely, these relationships could

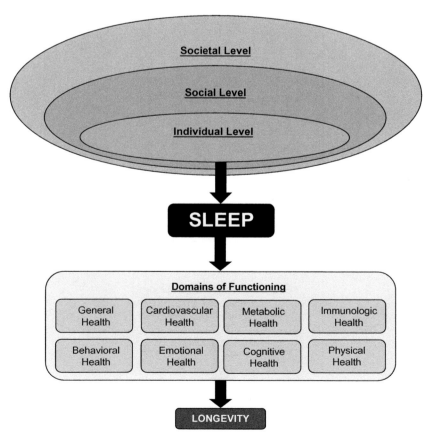

FIG. 5.6 Full social-ecological model of sleep and health.

be partly explained by the effects of the individual-level factors on sleep, which, in turn produces adverse outcomes (mediator), or process involved in sleep could change the strength of relationships between individual-level factors and adverse outcomes (moderator). Likely, these relationships are complex. Future research will need to better clarify this. For parsimony, figures only display the mediation relationship, though this is not the only possibility.

APPLICATIONS OF THE MODEL

The first version of this model for sleep health and mortality risk was originally proposed by Grandner et al. [2] and has since appeared in various forms in several other publications (e.g., [16, 51–53]). This model has been modified to fit various publications because the concepts underlying it are flexible. For example, the model could be used to develop interventions by examining modifiable targets that account for the social environmental context. The model can also be used to conceptualize the role of sleep in health, with sleep being more than just a set of physiologic processes.

This is somewhat more evident in the simplified version of the model presented in Fig. 5.7. This version highlights the core of what the model is describing—sleep exists at the intersection of upstream influences (individual, embedded within social, embedded within societal) and downstream influences that encompass the combination of health outcomes and functioning/performance. Although this version is simplified, the core elements remain, including the nested determinants and the explicit implication that health and functioning are inseparably overlapping as outcomes of sleep. Future studies could use this model to better understand the health context of factors related to sleep (such as health disparities, public policy, etc.), to better understand the role of upstream factors in sleep-related outcomes (such as sleep-related cardiometabolic risk), and to better develop and contextualize healthy sleep interventions that aim to improve downstream factors while addressing the contextualized upstream determinants.

REFERENCES

[1] Colten HR, Altevogt BM, Institute of Medicine Committee on Sleep Medicine and Research. Sleep disorders and sleep deprivation: an unmet public health problem. Washington, DC: Institute of Medicine: National Academies Press; 2006. xviii, 404 p..

[2] Grandner MA, Patel NP, Hale L, Moore M. Mortality associated with sleep duration: the evidence, the possible mechanisms, and the future. Sleep Med Rev 2010;14:191–203.

[3] Gallicchio L, Kalesan B. Sleep duration and mortality: a systematic review and meta-analysis. J Sleep Res 2009;18(2):148–58.

[4] Bronfenbrenner U. Toward an experimental ecology of human development. Am Psychol 1977;32:513–31.

[5] Chang J, Guy MC, Rosales C, de Zapien JG, Staten LK, Fernandez ML, Carvajal SC. Investigating social ecological contributors to diabetes within Hispanics in an underserved U.S.-Mexico border community. Int J Environ Res Public Health 2013;10(8):3217–32. PubMed PMID: 23912202.

[6] Sisson SB, Broyles ST. Social-ecological correlates of excessive TV viewing: difference by race and sex. J Phys Act Health 2012;9(3):449–55. Epub 2011/09/22. PubMed PMID: 21934164.

[7] Hinkley T, Salmon J, Okely AD, Hesketh K, Crawford D. Correlates of preschool children's physical activity. Am J Prev Med 2012;43(2):159–67. PubMed PMID: 22813680.

[8] Stellefson M, Barry AE, Stewart M, Paige SR, Apperson A, Garris E, Russell A. Resources to reduce underage drinking risks and associated harms: social ecological perspectives. Health Promot Pract 2018:1524839918814736. PubMed PMID: 30466329.

[9] Loewenstein K. Parent psychological distress in the neonatal intensive care unit within the context of the social ecological model: a scoping review. J Am Psychiatr Nurses Assoc 2018;24(6):495–509. PubMed PMID: 29577790.

[10] Nyambe A, Van Hal G, Kampen JK. Screening and vaccination as determined by the social ecological model and the theory of triadic influence: a systematic review. BMC Public Health 2016;16(1):1166. PubMed PMID: 27855680.

[11] Cramer RJ, Kapusta ND. A Social-ecological framework of theory, assessment and prevention of suicide, Front Psychol 2017;8:1756. PubMed PMID: 29062296.

[12] Golden SD, McLeroy KR, Green LW, Earp JA, Lieberman LD. Upending the social ecological model to guide health promotion efforts toward policy and environmental change. Health Educ Behav 2015;42(1 Suppl):8S–14S. PubMed PMID: 25829123.

[13] Mokdad AH, Marks JS, Stroup DF, Gerberding JL. Actual causes of death in the United States, 2000. JAMA 2004;291(10):1238–45. Epub 2004/03/11. PubMed PMID: 15010446.

[14] Grandner MA, Alfonso-Miller P, Fernandez-Mendoza J, Shetty S, Shenoy S, Combs D. Sleep: important considerations for the prevention of cardiovascular disease. Curr Opin Cardiol 2016;31(5):551–65. PubMed PMID: 27467177.

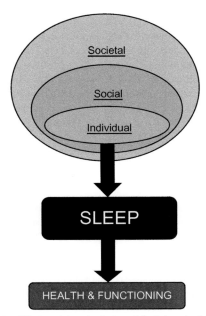

FIG. 5.7 Simplified social-ecological model of sleep health.

[15] Grandner MA, Sands-Lincoln MR, Pak VM, Garland SN. Sleep duration, cardiovascular disease and proinflammatory biomarkers, Nat Sci Sleep 2013;5:93–107. PubMed PMID: 23901303.

[16] Grandner MA. Addressing sleep disturbances: an opportunity to prevent cardiometabolic disease? Int Rev Psychiatry 2014;26(2):155–76. PubMed PMID: 24892892.

[17] Grandner MA. Sleep, health, and society. Sleep Med Clin 2017;12(1):1–22. PubMed PMID: 28159089].

[18] Ge X, Han F, Huang Y, Zhang Y, Yang T, Bai C, Guo X. Is obstructive sleep apnea associated with cardiovascular and all-cause mortality? PLoS One 2013;8(7):e69432. PubMed PMID: 23936014.

[19] Koren D, Taveras EM. Association of sleep disturbances with obesity, insulin resistance and the metabolic syndrome. Metabolism 2018. PubMed PMID: 29630921.

[20] Markwald RR, Melanson EL, Smith MR, Higgins J, Perreault L, Eckel RH, Wright KP, Jr. Impact of insufficient sleep on total daily energy expenditure, food intake and weight gain, Proc Natl Acad Sci USA 2013;110(14):5695–700. PubMed PMID: 23479616.

[21] Nagai M, Tomata Y, Watanabe T, Kakizaki M, Tsuji I. Association between sleep duration, weight gain, and obesity for long period. Sleep Med 2013;14(2):206–10. PubMed PMID: 23218534.

[22] St-Onge MP. The role of sleep duration in the regulation of energy balance: effects on energy intakes and expenditure. J Clin Sleep Med 2013;9(1):73–80. PubMed PMID: 23319909.

[23] St-Onge MP, Roberts AL, Chen J, Kelleman M, O'Keeffe M, RoyChoudhury A, Jones PJ. Short sleep duration increases energy intakes but does not change energy expenditure in normal-weight individuals. Am J Clin Nutr. 2011;94(2):410–6. Epub 2011/07/01. PubMed PMID: 21715510.

[24] Baglioni C, Riemann D. Is chronic insomnia a precursor to major depression? Epidemiological and biological findings. Curr Psychiatry Rep 2012;14(5):511–8. PubMed PMID: 22865155.

[25] Baglioni C, Spiegelhalder K, Lombardo C, Riemann D. Sleep and emotions: a focus on insomnia. Sleep Med Rev 2010;14(4):227–38. Epub 2010/02/09. PubMed PMID: [20137989].

[26] Pigeon WR, Pinquart M, Conner K. Meta-analysis of sleep disturbance and suicidal thoughts and behaviors. J Clin Psychiatry 2012;73(9):e1160–7. PubMed PMID: 23059158.

[27] Chakravorty S, Siu HY, Lalley-Chareczko L, Brown GK, Findley JC, Perlis ML, Grandner MA. Sleep duration and insomnia symptoms as risk factors for suicidal ideation in a nationally representative sample. The primary care companion to CNS disorders. 2015;17(6). PubMed PMID: 27057399.

[28] Perlis ML, Grandner MA, Brown GK, Basner M, Chakravorty S, Morales KH, Gehrman PR, Chaudhary NS, Thase ME, Dinges DF. Nocturnal wakefulness as a previously unrecognized risk factor for suicide. J Clin Psychiatry 2016;77(6):e726–33. PubMed PMID: 27337421.

[29] Perlis ML, Grandner MA, Chakravorty S, Bernert RA, Brown GK, Thase ME. Suicide and sleep: is it a bad thing to be awake when reason sleeps? Sleep Med Rev 2016;29:101–7. PubMed PMID: 26706755.

[30] Perogamvros L, Schwartz S. The roles of the reward system in sleep and dreaming. Neurosci Biobehav Rev 2012;36(8):1934–51. PubMed PMID: 22669078.

[31] van der Helm E, Walker MP. Sleep and emotional memory processing. Sleep Med Clin 2011;6:31–43.

[32] Banks S, Dinges DF. Behavioral and physiological consequences of sleep restriction. J Clin Sleep Med 2007;3(5):519–28. Epub 2007/09/07. PubMed PMID: 17803017.

[33] Jackson ML, Gunzelmann G, Whitney P, Hinson JM, Belenky G, Rabat A, Van Dongen HP. Deconstructing and reconstructing cognitive performance in sleep deprivation. Sleep Med Rev 2013;17(3):215–25. PubMed PMID: 22884948.

[34] Killgore WD, Grugle NL, Balkin TJ. Gambling when sleep deprived: don't bet on stimulants. Chronobiol Int 2012;29(1):43–54. Epub 2012/01/06. PubMed PMID: [22217100].

[35] Hisler G, Krizan Z. Sleepiness and behavioral risk-taking: do sleepy people take more or less risk? Behav Sleep Med 2017:1–14. PubMed PMID: 28745529.

[36] Fraser M, Conduit R, Phillips JG. Effects of sleep deprivation on decisional support utilisation. Ergonomics 2013;56(2):235–45. PubMed PMID: 23419086.

[37] Hui SK, Grandner MA. Trouble sleeping associated with lower work performance and greater health care costs: longitudinal data from Kansas state employee wellness program. J Occup Environ Med 2015;57(10):1031–8. PubMed PMID: 26461857.

[38] Meridew CM, Jaszewski A, Newman-Smith K, Killgore WDS, Gallagher RA, Carrazco N, Alfonso-Miller P, Gehrels J, Grandner MA. Sleep practices, beliefs, and attitudes associated with overall health. Sleep 2016;39(Abstract Supplement):A268–9.

[39] Grandner MA, Patel NP, Jean-Louis G, Jackson N, Gehrman PR, Perlis ML, Gooneratne NS. Sleep-related behaviors and beliefs associated with race/ethnicity in women. J Natl Med Assoc 2013;105(1):4–15. PubMed PMID: 23862291.

[40] Barnes M, Davis K, Mancini M, Ruffin J, Simpson T, Casazza K. Setting adolescents up for success: promoting a policy to delay high school start times. J Sch Health 2016;86(7):552–7. PubMed PMID: 27246680.

[41] Millman RP, Boergers J, Owens J. Healthy school start times: can we do a better job in reaching our goals? Sleep 2016;39(2):267–8. PubMed PMID: 26943474.

[42] Minges KE, Redeker NS. Delayed school start times and adolescent sleep: a systematic review of the experimental evidence. Sleep Med Rev 2016;28:86–95. PubMed PMID: 26545246.

[43] Thacher PV, Onyper SV. Longitudinal outcomes of start time delay on sleep, behavior and achievement in high school, Sleep 2016;39(2):271–81. PubMed PMID: 26446106.

[44] Grandner MA, Patel NP, Gehrman PR, Xie D, Sha D, Weaver T, Gooneratne N. Who gets the best sleep? Ethnic and socioeconomic factors related to sleep disturbance. Sleep Med 2010;11:470–9.

[45] Hale L, Hill TD, Friedman E, Nieto FJ, Galvao LW, Engelman CD, Malecki KM, Peppard PE. Perceived neighborhood quality, sleep quality, and health status: evidence from the survey of the health of Wisconsin. Soc Sci Med 2013;79:16–22. PubMed PMID: 22901794.

[46] Hale L, Hill TD, Burdette AM. Does sleep quality mediate the association between neighborhood disorder and self-rated physical health? Prev Med 2010;51(3–4):275–8. Epub 2010/07/06. PubMed PMID: [20600254].

[47] Hill TD, Burdette AM, Hale L. Neighborhood disorder, sleep quality, and psychological distress: testing a model of structural amplification. Health Place 2009;15(4):1006–13. Epub 2009/05/19. PubMed PMID: [19447667].

[48] Whinnery J, Jackson N, Rattanaumpawan P, Grandner MA. Short and long sleep duration associated with race/ethnicity, sociodemographics, and socioeconomic position. Sleep 2014;37(3):601–11. PubMed PMID: 24587584.

[49] Grandner MA, Petrov MER, Rattanaumpawan P, Jackson N, Platt A, Patel NP. Sleep symptoms, race/ethnicity, and socioeconomic position. J Clin Sleep Med 2013;9(9):897–905. PubMed PMID: WOS:000324375100009.

[50] Ko PR, Kientz JA, Choe EK, Kay M, Landis CA, Watson NF. Consumer sleep technologies: a review of the landscape. J Clin Sleep Med 2015;11(12):1455–61. PubMed PMID: 26156958.

[51] Grandner MA, Williams NJ, Knutson KL, Roberts D, Jean-Louis G. Sleep disparity, race/ethnicity, and socioeconomic position. Sleep Med 2016;18:7–18. PubMed PMID: 26431755.

[52] Watson NF, Badr MS, Belenky G, Bliwise DL, Buxton OM, Buysse D, Dinges DF, Gangwisch J, Grandner MA, Kushida C, Malhotra RK, Martin JL, Patel SR, Quan SF, Tasali E. Joint consensus statement of the American academy of sleep medicine and sleep research society on the recommended amount of sleep for a healthy adult: methodology and discussion. J Clin Sleep Med 2015;11(8):931–52. PubMed PMID: 26235159.

[53] Watson NF, Badr MS, Belenky G, Bliwise DL, Buxton OM, Buysse D, Dinges DF, Gangwisch J, Grandner MA, Kushida C, Malhotra RK, Martin JL, Patel SR, Quan SF, Tasali E. Joint consensus statement of the American academy of sleep medicine and sleep research society on the recommended amount of sleep for a healthy adult: methodology and discussion. Sleep 2015;38(8):1161–83. PubMed PMID: 26194576.

Part II

Contextual factors related to sleep

Chapter 6

Race, socioeconomic position and sleep

Natasha Williams[a,*], Girardin Jean-Louis[b,c], Judite Blanc[d], Douglas M. Wallace[e,f,*]

[a]NYU Langone Health, Division of Health and Behavior, Department of Population Health, Center for Healthful Behavior Change, New York, NY, United States, [b]NYU Langone Health, Department of Population Health, New York, NY, United States, [c]NYU Langone Health, Department of Psychiatry, New York, NY, United States, [d]NYU Langone Health, Department of Population Health, Center for Healthful Behavior Change, New York, NY, United States, [e]Department of Neurology, Sleep Medicine Division, University of Miami Miller School of Medicine, Miami, FL, United States, [f]Miami VA HealthCare System, Sleep Disorders Laboratory, Miami, FL, United States

Much of the history of thinking about inequality in the United States, including health inequality, has usually been framed in terms of race or class, but seldom both

[1, 347, pp.]

ABBREVIATIONS

ACT	actigraphy
DFA	difficulty falling asleep
DMS	difficulty maintaining sleep
EDS	excessive daytime sleepiness
EMA	early morning awakening
LS	long sleep
NHANES	National Health and Nutrition Examination Survey
NHIS	National Health Interview Survey
NRS	nonrestorative sleep
OSA	obstructive sleep apnea
PLMS	periodic limb movements during sleep
PSG	polysomnography
RLS	restless leg syndrome
SC	sleep complaints
SDB	sleep disordered breathing
SES	socioeconomic status
SL	sleep latency
SS	short sleep
TST	total sleep time
WASO	wake after sleep onset

*Contributed equally to this work.

INTRODUCTION

There is no doubt that the US economy has experienced strong growth. Broadly defined as a measure of economic wellbeing [2], the gross domestic product (GDP) is approximately 4.2%, an 18% increase from the 2009 financial crisis [2]. The unemployment rate is 4%, also approaching the same rate as in 2009 [3]. Lamentably, only a small segment of the population has benefited from this recent economic growth. For example, Blacks with a college degree earn less than their white counterparts without a college degree [4]. Only 70% of Native American/Alaska Natives earned a high school diploma in 4 years compared to 87% of whites [5]. The total household income of Blacks has remained relatively unchanged over the past 30 years [6]. In 2015, the total household income of Hispanics was less than it was in 1972 and 1 in 4 Hispanics were living at the US federal poverty level [6]. Black and white Americans have significant differences in life expectancy. Life expectancy at birth for white men is 76.1, approximately 4 years longer than Black men and in 2016 white women were expected to live until age 81 compared to Black women at 77.9 [7]. Hispanics make up the highest percentage of Americans without health insurance coverage (19.5%) whereas only 6% of non-Hispanic whites do not have health insurance coverage. Given that there are disparities at nearly every level of socioeconomic status (SES) and racial/ethnic disparities by nearly every indicator of health status, addressing these persistent health disparities has become one of the most pressing public health challenges.

Beginning with Healthy People 2010 (officially launched in January 2000), the nation's public health agenda has included the broad goal of eliminating disparities [8]. Healthy People 2020 includes sleep health objectives, specifically to increase sleep duration among adults and adolescents and to increase the proportion of adults seeking a medical evaluation for obstructive sleep apnea [5]. Aligned with the overarching goals, the inclusion of the sleep health objectives provides an important opportunity for the sleep community to: (1) address health disparities, (2) identify the complex conditions that contribute to and/or exacerbate these disparities, and (3) develop, evaluate, and implement evidenced-based interventions for achieving these objectives among socioeconomically disadvantaged Americans and racial/ethnic minorities.

The purpose of this chapter is not to present an exhaustive and comprehensive review of the literature, because several reviews about sleep among racial/ethnic minorities have been conducted [9–11]. Rather, we describe the current landscape in health disparities science in the United States. First, we provide a brief overview of health disparities. Second, we summarize the available influential published studies on inadequate sleep duration, poor sleep quality, and a selection of the sleep disorders. Finally, we identify gaps in the literature and suggestions for future inquiry.

(BRIEF) HISTORY AND DEFINITION OF HEALTH DISPARITIES

Presumably, the earliest remark of a "health disparity" in the United States was noted in the Secretary's Task Force on Black and Minority Human Services. Congress convened the report in 1984 in order to comprehend the health of minority populations [5]. In the report, former Secretary of the Department of Health and Human Services, Margaret Heckler, described "…the disparity has existed ever since accurate federal record keeping began." The definition of disparity was best defined as 'excess deaths,' or "the difference between the number of deaths observed in minority populations and the number of deaths which would have been expected in the minority population and the same age- and sex-specific death rate as the non-minority population [5]." While the Secretary's remarks stated that the report should serve as a "generating force" for a "national assault on the persisting health disparity, (p. 1)," the disparities widened.

In 1999, Congress requested that the National Academy of Medicine (formerly known as the Institute of Medicine), to convene its seminal study that synthesized the evidence of disparities in health care [12]. In this work titled, "Unequal Treatment: What Healthcare Providers need to Know about Racial and Ethnic Disparities in Health Care" the committee defined disparities in healthcare as "racial or ethnic differences in the quality of healthcare that are not due to access-related factors or clinical needs, preferences, and appropriateness of interventions." The Committee reviewed over 100 published studies and concluded that even among the insured, there were differences in health care utilization and treatment and these differences occurred beyond the traditional individual level factors such as smoking and patients' attitudes about treatment. Rather, these differences could be attributed to factors within the healthcare system, prejudice and discrimination and clinical uncertainty [12]. Moreover, eliminating these disparities is possible. In contrast to efforts in monitoring disparities in the United States, European countries consistently report health status by SES indicators [13]. For example, the Whitehall studies of British civilian workers were most striking as it related to SES and health inequities. In those studies, a homogeneous sample, with access to national health insurance, demonstrated that SES was related to age-adjusted mortality in 10 years and that there was an SES-gradient effect for nearly every cause of death. That is, higher-ranking employees had a lower relative risk of mortality compared to lower ranking employees [14]. These studies have demonstrated the importance of exploring disparities not only by educational level and income but also occupation status and within various economic strata to assess whether relationships are graded.

SLEEP CHARACTERISTICS

Self-reported sleep duration across racial/ethnic groups

Population-based studies have shown that individuals belonging to minority racial/ethnic groups are more likely to experience self-reported extremes in sleep duration than white individuals. In the Alameda County Health and Ways of Living Study, Stamatakis et al. found that disparities among minorities for habitual short sleep (SS; ≤6 h daily) have existed for several decades and have widened over time [15]. In 1965, Blacks had nearly a twofold higher odds for SS while non-Hispanic individuals of other race-ethnicity had a 40% increased odds for SS than whites (Table 6.1) [15]. Over the 34 years of the study (1965–99), the age-adjusted mean probability of SS among black (26%–54%) and Hispanic (12%–37%) individuals grew disproportionately relative to whites (15%–25%). In a nationally representative sample (National Health Interview Survey 1990 data) adjusting for individual SES factors, health behaviors, and urban living environments, Hale and Do found that individuals of black, non-Mexican Hispanic, and other non-Hispanic race-ethnicity had increased odds of SS relative to whites (Table 6.1) [16]. Blacks also had increased odds of long sleep (LS; ≥9 h; OR 1.62 [1.40–1.88]) relative to whites. Examining more recent data from the National Health and Interview Survey (NHIS) (2004–07), Krueger et al. reported similar racial/ethnic findings for SS. However, they also found that Blacks (OR 1.46 [1.33–1.59]) and Mexican-Americans (1.42 [1.27–1.60]) were more likely to report LS [18, 25]. In the National Health and Nutrition Examination Survey (NHANES), Whinnery et al. examined

TABLE 6.1 Representative studies examining sleep duration in adult racial/ethnic minorities and lower SES populations.

Study	Design/source	Sample	Racial/ethnic/SES group (% of sample)	Sleep duration assessment	Sleep duration definitions (h/night)	Covariates	Findings (racial/ethnic reference group is white unless specified)
Stamatakis et al. [15]	Longitudinal; Alameda County Health and Ways of Living Study (1965–99)	N=6928 (1965); N=2123 (1999); age: 16–94	Black: 12.4 Hispanic: 3.9 Other: 4.7 White: 78.9 — Lowest income quintile: 20	Self-report at each study wave	Short: ≤6 Reference: 7–8 Excluded: ≥9	Age, BMI, living conditions, health behaviors, health status, depression, insomnia	*Short sleep (<6h):* Black OR 1.97 (1.7–2.3) Other OR 1.4 (1.1–1.9) — Lowest income quintile OR 1.6 (1.3–1.9) relative to highest quintile
Hale and Do [16]	Cross-sectional; National Health Interview Survey 1990	N=32,749; 52% women; age 43	Black: 10.1 Mex-Amer: 4.1 Other Hispanic: 3.5 Other non-Hisp: 3.3 White: 79.0	Self-report: "total hours slept over 24-h day"	Short: ≤6 Reference: 7–8 Long sleep: ≥9	Age, sex, individual SES, health conditions/behaviors, residence type, urban environment characteristics	*Short sleep (<6h):* Black OR 1.4 (1.3–1.6) Other Hispanic OR 1.3 (1.1–1.5) Other non-Hispanic OR 1.4 (1.1–1.6) *Long sleep (>9h):* Black OR 1.6 (1.4–1.9)
Whinnery et al. [17]	Cross-sectional; National Health and Nutrition Examination Survey 2007–08	N=4850; 52% women; age 25–64: 71%	Black: 11.4 Mex-Amer: 8.5 Other Hispanic: 5.0 Asian/other: 6.0 White: 69.2 Income <20k:16.4 >75K:33.4	Self-report: "sleep on weekday or workday nights"	Very short: <5 Short: 5–6 Normal: 7–8 Long: ≥9	Demographics, self-rated health, country of origin, language, income, education, health insurance, food security	*Very short sleep (<5h):* Black OR 2.3 (1.6–3.4) Other Hispanic OR 2.7 (1.1–6.3) Asians/others OR 3.99 (1.7–9.5) *Short sleep (5–6h):* Black OR 1.85 (1.5–2.2) Asians/others OR 2.1 (1.3–3.3) *Long sleep (≥9h):* Mex-Amer OR 0.4 (0.1–0.9)
Krueger and Friedman [18]	Cross-sectional; National Health Interview Survey 2004–07	N=110,441	NR	Self-report: "total hours slept over 24-h day"	5, 6, 8, 9 7 (reference)	Demographics, foreign born status, family structure, SES, health behaviors, health status	*6 vs 7h:* Blacks OR 1.5 (1.4–1.6) Other Hispanics OR 1.15 (1.03–1.3) Other non-Hispanic OR 1.4 (1.2–1.5) *9 vs 7h* Blacks OR 1.5 (1.3–1.6) Mex-Amer OR 1.4 (1.3–1.6)
Lauderdale et al. [19]	Cross-sectional; CARDIA 2003–04	N=669; 58% women; age 43±4	Black: 44 White: 56	3 days of ACT	Continuous	Age, BMI, individual SES, employment, household, health behaviors, SDB-risk, shift work	White women **6.5 (6.0–6.9)** h White men 5.8 (5.3–6.3) h Black women 5.9 (5.5–6.3) h Black men **5.1 (4.6–5.6)** h

Continued

TABLE 6.1 Representative studies examining sleep duration in adult racial/ethnic minorities and lower SES populations.—cont'd

Study	Design/source	Sample	Racial/ethnic/SES group (% of sample)	Sleep duration assessment	Sleep duration definitions (h/night)	Covariates	Findings (racial/ethnic reference group is white unless specified)
Mezick et al. [20]	Cross-sectional; Pittsburgh Sleep SCORE	N=187; 47% women; age 60±7	Black: 41 White/Asians: 59 (Asians; n=3)	9 days of ACT, 2-nights of home PSG	Continuous	Age, sex, medication use	ACT: black **5.4±0.8** vs white/Asian 6.0±0.9 h PSG: black **5.8±1.1** vs white/Asian 6.2±0.9 h
Hall et al. [21]	Cross-sectional; Study of Women's Health Across the Nation (SWAN) Sleep ancillary	N=368 adults; 100% women; age 51±2	Black: 37.5 Chinese: 16.0 White: 46.5 — Difficulty paying for basics: 27.4	3-nights of home PSG (1st night excluded)	Continuous	Age, menopausal status, vasomotor symptoms, BMI, depression, perceived health, sleep meds	Beta for blacks: **−0.27** Beta for Chinese: −0.04; P=NS Beta for difficulty paying for basics: 0.02; P=NS
Song et al. [27]	Cross-sectional; Outcomes of Sleep Disorders in Older Men Study (MrOs Sleep)	N=2862 adults; 100% men; age 76±6	Black: 3.4 Hispanic: 1.9 Asian-Amer: 2.9 Other: 1.1 White: 90.8	5 days of ACT	Continuous	Age, social status, BMI, education, marital status, health behaviors/conditions, study site	Black: 6.1 (5.8–6.3) Hispanic: 6.7 (6.4–7.0) Asian-Amer: 6.1 (5.8–6.4) Other: 6.5 (6.1–7.0) White: 6.4 (6.4–6.5)
Chen et al. [22]	Cross-sectional; Multi-Ethnic Study of Atherosclerosis (MESA) Sleep cohort	N=2230 adults; 54% women; age 68±9	Black: 28.0 Hispanic: 23.9 Chinese: 12.1 White: 36.1	7 days of ACT	Short: <6,6–7 Reference: 7–8 Long: >8	Age, sex, and site	Short sleep (<6h): Blacks OR 4.95 (3.6–6.9) Hispanics OR 1.80 (1.3–2.6) Chinese OR 2.3 (1.5–3.6) Long sleep (>8h): No racial differences
Carnethon et al. [23]	Cross-sectional; Chicago Area Sleep Study (CASS)	N=496 adults; 60% women; age 48±8	Black: 31.3 Hispanic: 20.8 Asian: 22.0 White: 26.0	7 days of ACT	Continuous	Age, sex, BMI, education, shift work, health behaviors, depression	Black: **6.7 (6.5–6.8)** h Hispanic: **6.9 (6.6–7.1)** h Asian: **6.8 (6.6–7.1)** h White: 7.5 (7.3–7.7) h
Dudley et al. [24]	Cross-sectional; Hispanic Community Health Study/Study of Latinos (HCHS/SOL)	N=2087 adults; 64.6% women; age 47±12	Mexico: 26.9 Central Am: 13.6 Cuban: 18.1 Dominican: 12.5 Puerto Rico: 20.7 South Am: 8.2	5–7 days of ACT	Continuous	Age, sex, education, employment status, shift work status, income, BMI, SDB, depression, health behaviors, sleep medication, season	Regression coeff beta [SE] min Mexico (reference) Central Am: **−13.2 [5.2]** Cuban −5.2 [4.9] Dominican: **−10.5 [5.1]** Puerto Rico: **−12.4 [6.7]** South Am: **−19.3 [6.7]**

Data presented as mean±SD, mean [SE], or OR (95% CI). Bolded values represent P<.05; ACT, actigraphy; BMI, body mass index; CARDIA, Coronary Artery Risk Development in Young Adults; LS, long sleep; OR, odds ratio; SDB, sleep disordered breathing; SES, socioeconomic; SOL, sleep onset latency; SS, short sleep; VSS, very short sleep.

the relationship between racial/ethnic groups and extremes of sleep duration while adjusting not only for individual SES characteristics, but also including homeownership, health insurance, food insecurity, and acculturation measures [17]. Blacks and non-Mexican Hispanics were more than two times the odds to report very SS (<5h nightly) compared to whites, while Asians/individuals of other race/ethnicity were nearly four times more likely to report very SS relative to whites (Table 6.1). Blacks and Asians/others also reported about two times the odds of SS (5–6h) compared to whites. In contrast, Mexican Americans were about one-third as likely to report LS (>9h) relative to whites. Finally, Jean-Louis et al. examined three decades of data from NHIS (1977–2007), among whites, the prevalence of VSS increased from 1.5% in 1977 to 2.3% in 2009, and the prevalence of SS increased from 19.3% to 25.4%, whereas prevalence among blacks, the prevalence of VSS increased from 3.3% to 4.0%, and the prevalence of SS increased from 24.6% to 33.7% [26].

These nationally representative self-reported data suggest that specific racial/ethnic groups may be at greater risk for very SS, SS and LS than whites and that these disparities in sleep duration may have widened over time.

Objective reported sleep duration across racial/ethnic groups

In addition to self-reported measures, some population-based studies have employed objective measures of total sleep time (TST). Objective measures of TST is important because (1) the disparity between subjectively and objectively measured TST can exceed 1h and (2) comorbid sleep disorders (e.g., undiagnosed sleep disordered breathing can influence an individual's perception of sleep duration). Using three nights of actigraphy (ACT) in black and white participants aged 39–50 years, Lauderdale et al. examined race–sex interactions for sleep measures in the Coronary Artery Risk Development in Young Adults (CARDIA) study [19]. In models adjusting for individual SES factors, employment, health behaviors, and sleep disordered breathing (SDB) risk, white women slept 6.5h while black men slept 5.1h with white men and black women having similar intermediate sleep durations (Table 6.1). Mezick et al. used two nights of home polysomnography (PSG) and nine nights of ACT finding that Blacks, on average, had significantly shorter TST (about 30 min less) than white/Asian participants by both methods [20] (Table 6.1). In the Study of Women's Health Across the Nation (SWAN) sleep cohort, participants completed three nights of home PSG in a diverse sample of middle-aged women from seven US cities [21]. Adjusting for financial strain and education attainment, black, but not Chinese, women had significantly shorter PSG-measured TST than white women. Similarly, Song et al. assessed the sleep characteristics of older men (>65 years of age) enrolled in the Outcomes of Sleep Disorders in Older Men study (MrOs Sleep) using 5 days of ACT and one night of PSG to measure TST and SDB, respectively [27]. In adjusted models accounting for social status and a number of health factors, black men slept less than white men (6.0 vs 6.4h) while Hispanic men [28] slept more than black (6.0h) and Asian American (6.1h) participants. Similar results were observed in further analyses adjusting for SDB. More recently, Chen et al. used 7 days of ACT to measure TST and home PSG to assess for SDB in the Multi-Ethnic Study of Atherosclerosis (MESA) in participants from six US cities [22]. After adjusting for age, sex, and study site, Blacks were nearly five times as likely to have objectively SS (<6h) compared to whites while Hispanics and Chinese had nearly twice the odds of SS relative to white participants. These racial differences in odds for SS persisted despite additional adjustments for SDB and insomnia. As in CARDIA, the shortest sleep duration was found among black men who on average slept 75 min less than white women. Black women on average slept 43 min less than white women. Finally, Carnethon et al. expanded the results of CARDIA by examining additional racial/ethnic groups (Hispanics and Asians), and screening for SDB with PSG in the Chicago Area Sleep Study [23, 29]. Using 7 days of ACT, Carnethon et al. found that, in adjusted analyses, Blacks, Asians, and Hispanics each had significantly shorter TST than whites (Table 6.1). Most, but not all, studies using objective TST and accounting for comorbid sleep disorders suggest individuals of minority background have greater risk for shorter sleep duration than whites. Consistently across these studies with varying adjustments (e.g., SES), black men and women had the shortest objective TST relative to their white counterparts.

Sleep duration within racial/ethnic groups

There are fewer data concerning heterogeneity of sleep duration within racial/ethnic groups. Recently, the Hispanic Community Health Study/Study of Latinos (HCHS/SOL) has provided detailed sleep characteristics concerning intra-ethnic variability within US Hispanics from the Bronx, Chicago, San Diego, and Miami [24]. Using 7 days of ACT in participants evaluated for SDB with PSG, Dudley et al. found that mean TST for US Hispanics was >30 min longer than for non-Hispanic Blacks and whites who participated in CARDIA. In age- and sex-adjusted analyses, Mexican Americans had the longest TST (6.82h) while individuals of Puerto Rican (6.57h) and South American (6.44h) heritage had the shortest TST (Table 6.1).

Sleep duration across SES groups

Above and beyond the effects of race-ethnicity, there is evidence for individuals belonging to lower SES groups having greater risk for extremes of TST than individuals from higher SES groups. For example, in the Alameda County Health and Ways of Living Study, individuals in the lowest income quintile had over a 60% increased odds of SS relative to those in the highest quintile [15] (Table 6.1). In the Nurses Health Study II, Patel et al. observed that the odds of subjective LS (>9h) relative to normal sleep duration (7–8h) was increased for never having been married (OR 1.4 [1.2,1.5]), unemployment (OR 2.4 [2.3,2.6]), and having an annual income less than $30,000 [30]. In NHANES, Whinnery et al. reported that lower educational attainment and very low food security was associated with SS, while having public health insurance relative to being uninsured was associated with LS [17]. Within US Hispanics ($n=11,860$), Patel found that having full-time employment and less than a high school education predicted self-reported SS (<7h); whereas unemployment, less than a high school education, and household income (<$10,000 annually) predicted LS (>9h) [31]. A similar association between lower SES position and LS (>9h) has been reported within US Blacks in the Jackson Heart Study (JHS) [32]. In contrast, among middle aged women in the SWAN study, financial strain was not associated with PSG-measured TST in adjusted analyses [21]. In CARDIA, ACT-defined TST was not associated to traditional SES measures (income, education) but was associated with other employment factors (i.e., unemployment, shift work) [19]. In a sample of healthcare workers, Ertel et al. used 7 days of ACT to compare the sleep duration of black Caribbean/African immigrants to those of whites [33]. In age- and gender-adjusted analyses, the sleep duration disparity between black Caribbean/African immigrants was >1h relative to whites, which was attenuated by 41% with adjustment for individual economic indicators and occupational characteristics [34]. Finally, in a nationally representative sample (NHIS), a race by occupation/industry interaction has been observed for SS. With increasing professional/managerial roles, SS increases among Blacks but decreases among whites [35]. These data suggest that psychosocial measures of SES (e.g., workplace, social, and environmental features) may be more closely linked to sleep duration for racial/ethnic groups than traditional measures of SES. Of note, these studies used objective measures of TST, which may be more accurate than self-reported sleep duration [19].

Sleep architecture and continuity across racial/ethnic groups

Most studies examining sleep architecture and continuity across racial/ethnic groups have shown that minority individuals tend to have "lighter" and more fragmented sleep than whites [36, 37]. For example, Redline et al. used home PSG to characterize sleep architecture in the Sleep Heart Health Study (SHHS) [38] (Table 6.2). In analyses adjusting for demographics, health comorbidities and SDB, American Indians had higher percentage N1 sleep than Blacks (6.7% vs 5.3%, $P=.01$) or whites (6.7% vs 5.4%, $P<.001$). Additionally, American Indians had higher percentage N2 sleep than Hispanics (65.1% vs 58.4%, $P<.001$) or whites (65.1% vs 58.4%, $P<.001$). The same group also had significantly less percentage of N3 (deep) sleep than any other racial and ethnic group. Blacks had higher percentage of N2 sleep than whites (62.2% vs 58.4%, $P<.001$) or Hispanics (62.2% vs 58.4%, $P=.02$) and lower N3 sleep than whites (11.0% vs 14.8%) or Hispanics (11.0% vs 15.1%). In a study examining sleep architecture in the sleep laboratory, Tomfohr et al. found that Blacks spent approximately 4.5% more TST in N2 sleep and 4.7% less in N3 sleep than whites [46]. Similar sleep architecture findings were reported in the MrOs Sleep study (one night of home PSG) where black men had less N3 sleep than white men (4.9% vs 8.8%, $P<.001$) [27]. Actigraphic measures also found that black men had longer sleep latencies (28.7 vs 21.9 min, $P=.02$) and lower sleep efficiencies (80.6% vs 83.4%, $P=.02$). Similarly, in CASS, Carnethon et al. reported that Blacks significantly increased sleep fragmentation index and greater wake after sleep onset time (50.2 vs 41.2 min, $P<.01$) than whites. However, these sleep parameters were similar in Asians and Hispanics relative to whites [23]. In the SWAN sleep study, using home PSG, black women had longer sleep latency, poorer sleep efficiency, lower N3 stage sleep than white women [21]. Chinese women also had significantly lower N3 sleep than white women. Additionally, this study used EEG spectral analysis to quantify sleep microstructural differences and found that beta power, a marker of hyperarousal during sleep, was significantly higher in black women than white women [21]. Using ACT in CARDIA, white women had shorter sleep latency (34 vs 48 min, $P<.01$) and greater sleep efficiency (77% vs 69%, $P<.01$) than black men [19]. White men and black women had similar intermediate sleep latency and sleep efficiency. In a metaanalysis of studies comparing subjective and objective sleep parameters [47] and among Black ($n=1010$) and white ($n=3156$) healthy sleepers, Ruiter et al. reported that Blacks had significantly shorter TST (objective and subjective), longer sleep latency (SL) greater N2 and lower N3 sleep percentage than whites [48]. However, racial/ethnic differences in many of these variables disappeared when studies examined participants in sleep laboratories and excluded participants with (1) undiagnosed mental illness or (2) using prescription medications. Finally, the Hispanic Community Health Study/Study of Latinos (HCHS/SOL) has shown that there is significant heterogeneity in ACT sleep patterns among US Hispanics [24]. In SES-adjusted models, Dudley et al. showed that individuals of Mexican heritage had the most consolidated sleep and healthiest sleep schedule while individuals of Puerto Rican descent had the most fragmented sleep and irregular sleep. To our knowledge, there are no other objective sleep data examining sleep heterogeneity within other US racial/ethnic minority groups.

TABLE 6.2 Representative studies examining sleep disordered breathing in racial/ethnic minorities and low SES populations.

Study	Design/source	Sample	Racial/ethnic/SES (% of sample)	SDB assessment	SDB definition	Covariates	Findings; OR (95% CI) (reference racial/ethnic group is white)
Young et al. [39, 40]	Cross-sectional; Sleep Heart Health Study (SHHS)	N=5615; 53% women; age 64±11	Black: 7.4 American Indian: 10.4 Other: 5.0 White: 77.1	1-night home PSG	AHI ≥15	Age, sex	Black 1.23 (0.97–1.6) American Indian 1.70 (1.4–2.1) Other 0.94 (0.65–1.37)
Fulop et al. [41]	Cross-sectional; Jackson Heart Study (JHS)	N=5301; 63% women; age 55±13	Black: 100 Annual income <25K: 27.5	5 items: sex, snoring, witnessed apneas, BMI, age	SDB risk score	None	*High SDB risk:* women 16.8% men 3.5% *Habitual loud snoring:* women 58.1% men 66.3% *Excessive daytime sleepiness:* women 61.4% men 68.6%
Johnson et al. [42]	Cross-sectional; Jackson Heart Study (JHS)	N=825; 66% women; age 63±11	Black: 100	1-night home PSG	AHI 5–15; AHI 15–30; AHI ≥30	None	Mild SDB 38.4% Moderate SDB 21.3% Severe SDB 15.8%
Chen et al. [22]	Cross-sectional; Multi-Ethnic Study of Atherosclerosis (MESA) Sleep cohort	N=2230; 54% women; age 68±9	Black: 28.0 Hispanic: 23.9 Chinese: 12.1 White: 36.1	1-night home PSG	SAS: AHI ≥5 and ESS ≥10 Severe SDB: AHI ≥30	Age, sex, and site	SAS / AHI ≥30 Black 1.8 (1.2–2.6) / 1.4 (0.9–2.0) Hispanic 1.5 (0.9–2.4) / 2.1 (1.4–3.3) Chinese 1.3 (0.7–2.2) / 1.4 (0.8–2.3)
Mihaere et al. [43]	Cross-sectional; New Zealand	N=358; 27% women; age 30–59	Maori: 46.4 White: 53.6	1-night of home PSG	RDI ≥15	Age, sex	Maori 4.26 (1.31–13.9)
Kripke et al. [44]	Cross-sectional; San Diego, CA	N=355; 54% women; age 40–64	Black: 3.4 Hispanic: 12.4 Other: 3.9 White: 80.3	4-nights of oximetry and actigraphy	ODI 4 ≥20	None	Black 16.7 Hispanic 15.9 Other 21.5 White 5.6
Redline et al. [45]	Cross-sectional; Hispanic Community Health Study/Study of Latinos (HCHS/SOL)	N=14,440; 60% women; age 46±12	Central Am:10.3 Cuba: 13.2 Dom Rep: 8.9 Puerto Rico:15.9 South Am: 6.5 Mexico: 42.3 Mixed: 2.8 Annual income <30K: 64	1-night of home PSG	AHI ≥15	Age, BMI	Men / Women Central Am 15.2 / 4.9 Cuba 17.8 / 6.1 Dom Rep 16.0 / 4.9 Puerto Rico 11.7 / 7.0 South Am 13.0 / 5.5 Mexico 14.4 / 5.8 mixed 16.4 / 4.6

Data presented as means±SD, frequency (95% CI) or OR (95% CI). Bolded values represent $P<.05$ for Chi-square comparisons. *AHI*, apnea-hypopnea index (events/h of sleep); *BMI*, body mass index; *ESS*, Epworth sleepiness scale; *PSG*, polysomnography; *RDI*, respiratory disturbance index; *SAS*, sleep apnea syndrome; *SDB*, sleep disordered breathing; *SES*, socioeconomic status.

Sleep architecture and continuity across SES groups

Currently, sparse data concerning the influence of SES factors on sleep architecture and continuity independent of race/ethnicity exists in the literature. In ACT-measured sleep in CARDIA, analyses adjusting for education and employment status showed that lower income was associated with longer SL and lower sleep efficiency [19]. In Pittsburgh, lower SES (assessed by a composite of education and income) was associated with longer ACT SL and greater PSG wake after sleep onset (WASO) time [20]. Similarly, in the SWAN study, financial strain was associated with poorer PSG-assessed sleep continuity and efficiency and this relationship was equivalent across race-ethnicity [21]. Thus, using multiple proxies for SES, most existing data suggest more disruptive and less efficient sleep in lower SES individuals.

SLEEP DISORDERS

Sleep disordered breathing (SDB)

SDB encompasses a number of chronic breathing abnormalities occurring during sleep, of which OSA is the most common [49] Diagnosis of SDB requires PSG with detection of breathing abnormalities (>5 per hour of sleep) associated with sleep irregularities/daytime consequences [49]. Estimates of the prevalence of moderate to severe SDB (apnea–hypopnea index [AHI] ≥ 15) in the US middle-aged (50–70 years of age) population are 17% for men and 9% for women [50].

Symptoms of and risk factors for sleep disordered breathing across racial/ethnic groups

Population-based studies have shown that SDB symptoms may vary across racial/ethnic groups. In the Sleep Heart Health Study, habitual loud snoring was significantly more common among Black (OR 1.6 [1.1–2.1]) and Hispanic (OR 2.3 [1.5–3.4]) women than among white women [51]. Hispanic men had more than twice the odds (OR 2.3 [1.4–3.7]) of reporting loud snoring than white men; However, the prevalence of loud snoring was similar among Black, Asian/Pacific islander, American Indian and white men [51]. Subjective excessive daytime sleepiness (EDS), one of the hallmark consequences of SDB, varies by racial/ethnic groups and assessment method. In analyses adjusted for demographics, medical comorbidities, sleep and psychosocial variables in the MESA study, Baron et al. found that Blacks had significantly higher odds (OR 1.5 [1.2–1.9]) for EDS (Epworth sleepiness scale >12) than whites [52]. The odds of reporting EDS (ESS>12) among Hispanics and Chinese participants was similar to that of whites. However, when EDS was measured as frequency reporting feeling excessively sleepy on 5 days or more over the previous month, Black (OR 0.57 [0.5–0.7]) and Hispanic (OR 0.62 [0.5–0.8]) participants were less likely to report EDS than whites. These contradictory data may suggest cultural differences in the interpretations of questions assessing sleepiness or normative levels of daytime sleepiness [52, 53]. Normalization of the consequences of the nocturnal symptoms may lead to lower screening for SDB among racial/ethnic individuals.

Some recognized risk factors for SDB may vary among racial/ethnic groups compared to whites. For example, one of the strongest risk factors for SDB is obesity, which is prevalent among US Blacks, Hispanics, and American Indians [32, 54–56]. In age- and gender-adjusted analysis of SHHS, Blacks and American Indians had a significantly higher odds of moderate to severe SDB (AHI ≥ 15) [39]. However, after adjusting for body habitus, racial/ethnic background was no longer associated with increased risk for SDB relative to whites. Thus, much of the increased risk of SDB among racial/ethnic individuals may be attributable to increased prevalence of obesity. Also, racial/ethnic individuals may distribute excess body fat differently. Therefore, typical body mass index (BMI) definitions of obesity (i.e., 30kg/m^2) may not adequately measure SDB risk across racial/ethnic groups. As Asians store greater amounts of body fat at the same level of BMI, the World Health Organizations has recommended using a lower BMI threshold (i.e., 25kg/m^2) to define obesity in Asian populations [57]. Data from MESA shows that each unit BMI increase may have a greater effect on AHI in Asian Americans that among other US racial/ethnic groups [58]. Thus, alternative measurements of obesity among racial/ethnic individuals should be considered in SDB risk-stratification.

Racial/ethnic differences in craniofacial anatomy have also been implicated in differential risk for SDB among racial/ethnic groups. Studies have suggested that soft tissue factors (e.g., tongue size, soft palate, tonsils) may be more relevant in predicting SDB risk among Blacks while skeletal components of the upper airways (e.g., maxillary-mandibular shape, inferior hyoid position) may be more important among Asians [59]. Studies comparing the anatomical differences between Blacks and whites have noted that the volume of the tongue is significantly larger among Blacks with SDB [60]. Relative to whites, Asian individuals have been found to have greater skeletal restrictions assessed by shorter cranial base as well as decreased thyromental distance and larger thyromental angles, suggesting the presence of these skeletal features contributed to their upper airway obstruction [61]. Among the Maori of New Zealand, reductions in mandibular prognathism and wider bony nasal aperture were associated with SDB, while reduced retropalatal airway size was more important among whites participants [62]. Less is known about Hispanic craniofacial structure and SDB risk. One cephalometric study found that Hispanics had greater bimaxillary retroposition relative to whites, but another failed to find any anatomical differences between Hispanics and whites [63, 64]. However, these anatomical data are limited by the relatively small sample size ($n < 100$) of these studies.

Diagnosis of SDB across and within racial/ethnic groups

The prevalence of SDB among some racial/ethnic groups has been reported to be higher than among whites. For example, in the SHHS, the prevalence of moderate to severe SDB (AHI $\geq$ 15) was 23%, 20%, and 17% among American Indians, Blacks, and whites, respectively [40]. A metaanalysis examining studies of SBD between Blacks and whites ($n = 10$ studies) found a small but significant effect size for Blacks having more prevalent and severe SDB relative to whites [65]. A recent study ($n = 512$; 48% women; 66% black) has specifically identified that this Black–white difference in SBD severity may be primarily driven by younger [66] and middle-aged (50–59 years of age) Black men [67]. In the Jackson Heart Study ($n = 5301$), Fulop et al. reported that 16.8% of Black women and 3.5% of Black men were at high risk for SDB and that individual symptoms of SDB were extremely common [41] (Table 6.2). In a subset of this cohort completing PSG ($n = 825$), 37.1% of this sample had moderate to severe SDB.

Early studies that explored diverse samples suggested that Hispanics may carry an excessive burden of SDB relative to their white counterparts. In a population-based study in San Diego, Kripke et al. showed that Hispanic participants ($n = 44$, mainly of Mexican-descent) had higher prevalence of moderate to severe SDB than whites (15.9% vs 5.6%) [44]. In the MESA study, US Hispanics had more than twice the odds of severe SDB than whites [22]. Recently, data from the HCHS/SOL ($n = 14,440$) showed that the large heterogeneity within US Hispanics regarding SDB [45]. Overall, 19% of Hispanic women and 33% of Hispanic men had at least mild SDB (AHI $\geq$ 5) with 6% of women and 14% of men having moderate to severe SBD (AHI $\geq$ 15). However, among US Hispanic men, moderate to severe SDB was most common among Cubans and least common among those of South American and Puerto Rican background (Table 6.2). Among Hispanic women, SDB was most prevalent among individuals of Puerto Rican heritage and least common among women of South American background [45]. Thus, although the overall SDB estimates among US Hispanics may be similar to that of the US general population, these data suggest that specific Hispanic subgroups may be particularly high risk for SDB.

There is also evidence of heterogeneity in the prevalence of SDB within individuals of Asian descent. Early studies comparing SDB among white and Asian individuals (ethnicity not reported) presenting to a sleep center reported that Asians had significantly greater prevalence of severe SDB (25% vs 11%, $P = .03$) [68]. In the MESA study, adjusted analyses revealed that Chinese individuals were 37% more likely to have severe SDB (AHI $\geq$ 30) than whites [22]. However, a cross-cultural SDB comparison study between individuals from Japan ($n = 978$) and the US (Hispanics $n = 211$; whites $n = 246$) found that the prevalence of SDB (RDI $\geq$ 15) among the Japanese (18.4%) was significantly lower than among US Hispanics (36.5%) or whites (33.3%) [69]. The SDB prevalence differences in these groups were largely explained by differences in their BMIs. There are few data for direct SDB comparison studies of other Asian ethnic subgroups living in the United States. Finally, in population-based studies from New Zealand, the Maori people have been reported to have four times the risk of moderate to severe SDB than white individuals (6% vs 1.5%) [43]. However, in analyses that adjusted for BMI, the association between ethnicity and SDB disappears. This suggests that increased body habitus among the Maori, as in US racial/ethnic groups, imparts greater SDB risk. Overall, these studies propose that a great deal of heterogeneity for SDB may exist in individuals of Asian descent.

SDB symptoms and diagnosis across SES groups

Beyond the risk factors of race/ethnicity, low SES individuals may have higher risk for SDB mediated by increased obesity rates, unhealthy behaviors (e.g., smoking, alcohol use), toxic environmental exposure, or other factors [20, 32, 42, 70]. However, as seen in Table 6.2, most epidemiological studies examining PSG-verified SDB among racial/ethnic groups have not adjusted for SES measures [22, 40, 44, 68]. Indirect evidence for SDB comparisons among socioeconomically diverse samples may be gained from randomized clinical trials that eliminate some of the financial barriers associated with sleep testing. For example, in the Home PAP study (seven US cities; $n = 183$) [71], individuals who lived in the poorest neighborhoods had similar SDB severity to those who lived in more affluent neighborhoods (AHI 43 ± 28 vs 44 ± 26, $P = .81$). In other smaller studies (SWAN, Sleep Score Table 6.1), different SES measures have not been found to be associated with SBD severity [20, 21]. In NHANES analyses ($N = 4081$; household income <$20,000 annually; 15.5% of the sample) adjusting for demographics, income, and other SES factors, Grandner et al. described an inverse relationship between habitual snoring and educational attainment with those achieving higher educational degrees reporting lower odds of snoring relative to highly educated individuals [72]. In addition, low household food security (versus high) was associated to greater odds of symptoms suggestive of SDB [73]. Similarly, analyses from the JHS adjusting for demographics, health behaviors, and comorbidities, Fulop et al. reported that lower SES individuals and physical inactivity were associated with high SDB risk among Blacks [41]. To date, the prevalence of PSG-diagnosed SDB between different SES levels has not been comprehensively studied.

Insomnia

Approximately 33% of the US general population reports insomnia complaints, including difficulty falling asleep, maintaining sleep, and/or premature awakening [74]. Even when more stringent diagnostic criteria are used, the prevalence of chronic insomnia disorder ranges from 6% to 15% [75]. Most population-based studies have focused on insomnia complaints (nocturnal symptoms alone) as opposed to insomnia disorder (complaints associated with daytime impairments) [49]. We have included studies examining sleep quality in this section in accordance with one presentation of insomnia: nonrestorative sleep.

Insomnia complaints across racial/ethnic groups

Studies assessing insomnia complaints in racially/ethnically diverse samples have used a varying assessment of insomnia symptoms. One of the first studies was conducted among Blacks and whites living in a small county in Florida ($n=1645$) [76]. In that study, 58% reported on survey questionnaire insomnia complaints, and this finding was greater among lower SES, but unrelated to race. An early population-based study concerning racial/ethnic differences in sleep complaints was the Established Populations for Epidemiologic Studies of the Elderly Study [77]. Using a racially diverse sample representing older individuals from North Carolina ($n=3976$, 54% Black), Blazer et al. evaluated sleep complaints using questions: difficulty falling asleep (DFA), difficulty maintaining sleep (DMS), early morning awakenings (EMA), and nonrestorative sleep (NRS). Whites were significantly more likely to endorse all insomnia complaints than Blacks (Table 6.3). In addition, among those with high levels of insomnia complaints, whites were more likely to be prescribed sedative-hypnotic medications than Blacks (83% vs 17%, $P<.001$) [77]. Significant Black–white disparity in insomnia complaints were also found in the Atherosclerosis Communities at Risk (ARIC) study, a population-based study of cardiovascular risk in four US cities [78]. Assessing symptoms of DFA, DMS, or NRS, Phillips and Mannino found that Blacks had significantly lower odds of reporting DFA or DMS (OR 0.8 [0.7–0.9] for both sleep complaints) [78]. In an analysis of the Women's Health Initiative ($n=98,705$; 8% Black, 3% Hispanic, 4% other), Kripke et al. assessed the frequency of sleep complaints [81] in the previous month in postmenopausal women [82]. Black, Hispanic, and women of other racial/ethnic minorities reported more DFA but less DMS and use of sleep aids. Similarly, in sample of older women in Brooklyn, Jean Louis et al. ($n=1274$, 72% Black) found that white women reported insomnia complaints more often than Black women (74% vs 46%, $P<.001$). Specifically, white women complained of greater rates of DFA (42% vs 16%, $P<.01$), DMS (64% vs 40%, $P<.01$), EMA (53% vs 27%, $P<.01$), and regular use of sleep medicine (19% vs 4%, $P<.01$) than Black women [83]. In a metaanalysis comparing insomnia studies among Black and whites ($n=13$ studies; 21,685 Black, 108,964 white participants), Ruiter et al. found small negative effect sizes for sleep complaints and subjectively measured wake after sleep onset (WASO) and terminal wakefulness but not for subjective SL [65]. These data show that Blacks report significantly less sleep complaints and report less subjective time awake after falling asleep than whites.

Other community-based and nationally representative samples have been able to provide data on insomnia complaints of other racial/ethnic groups. Analyzing the 2006 Behavioral Risk Factor Surveillance System (BRFSS), Grandner et al. used responses to the question "Over the last 2 weeks, how many days have you had trouble falling asleep or staying asleep or sleeping too much?" to categorize sleep complaints (SC) and their association with social determinants [79]. Participants categorized with SCs were those reporting problems most days of the week. In adjusted models for age, education, income, marital status, and unemployment, Asian/other men had significantly lower odds (OR 0.43 [0.24–0.76]) of reporting SCs than white men. The odds for SC in men of other racial/ethnic groups (Black, Asian/other, Hispanic) were equivalent to those of white men [79]. For women, Black, Asian/other, and Hispanic women all had significantly lower odds of reporting SCs than white women. In contrast, multiracial women had increased odds of SCs (OR 1.67 [1.09–2.55]) compared to white women. Similar results were found in the MESA study, where Chen et al. used the Women's Health Initiative Insomnia Rating Scale (WHIIRS) to diagnose insomnia complaints over the previous 4 weeks [22]. Black and Hispanic participants had similar odds of insomnia than whites while Chinese participants had lower odds of insomnia (OR 0.66 [0.44–1.00]) than whites. In NHANES, Grandner et al. assessed insomnia complaints via two methods: (1) asking how long it took an individual to fall asleep and (2) inquiring about frequency of *difficulty* with falling asleep, maintaining sleep, premature awakenings, or nonrestorative sleep [73]. Mexican Americans and other Hispanic individuals had a 40% and 30%, respectively, lower odds of having DFA than whites. In addition, Mexican Americans had 20% lower odds of reporting DMS. Relative to whites, Blacks had greater odds of reporting SL > 30 min

TABLE 6.3 Representative studies examining insomnia in adult racial/ethnic minority and lower SES populations.

Study	Design/source	Sample	Racial/ethnic/SES (% of sample)	Insomnia assessment	Insomnia definition	Covariates	Findings; OR (95% CI) (reference racial/ethnic group is white)
Blazer et al. [77]	Cross-sectional; Established Populations for Epidemiologic Studies of the Elderly study (EPESES)	N=3976; 65% women; age 73 ± 7	Black: 54.2 White: 45.8	4 items; no data on symptom duration	DFA, DMS, EMA, NRS	None	<u>DFA</u> White **16.3%** vs black 13.4% <u>DMS</u> White **33.8%** vs black 19.9% <u>EMA</u> White **16.0%** vs black 12.9% <u>NRS</u> White **13.3%** vs black 10.4%
Phillips and Mannino [78]	Cross-sectional; Atherosclerosis Risk in Community Study (ARIC) 1990–92 wave	N=13,563; 55% women; 50–59: 53%	Black: 23.7 White: 76.3 Annual income <16K: 19.1	3 items; no data on symptom duration or frequency	DFA, DMS, or NRS	Age, sex, BMI, individual SES factors, health behaviors, comorbidities, depression, hypnotic use, menopausal status	<u>DFA</u> Black 0.8 (0.7–0.9) <u>DMS</u> Black 0.8 (0.7–0.9) <u>NRS</u> Black 1.0 (0.9–1.1)
Grandner et al. [79]	Cross-sectional; Behavioral Risk Factor Surveillance System (BRFSS)	N=159,856; 60% women; age 52 ± 16	Black: 8.9 Hispanic: 17.4 Asian/other:4.8 Multirace: 1.8 White: 67.1	Telephone survey items	1 item: "trouble falling asleep, staying asleep, or sleeping too much" over 2 weeks	Age, education, income, marital status, employment, interactions	OR (95% CI) <u>Women</u> <u>Men</u> Black 0.7 (0.6–1.0) 0.8 (0.5–1.3) Hisp 0.7 (0.6–1.0) 0.9 (0.5–1.5) Asian 0.4 (0.3–0.6) 0.4 (0.2–0.8) Multi 1.7 (1.1–2.6) 0.7 (0.3–1.6)
Chen et al. [22]	Cross-sectional; Multi-Ethnic Study of Atherosclerosis (MESA) Sleep cohort	N=2230 adults; 54% women; age 68 ± 9	Black: 27.4 Hispanic: 23.7 Chinese: 11.7 White: 37.1	5 items: Women's Health Initiative Insomnia Rating Scale	WHIIRS >10	Age, sex, and site	Black 1.11 (0.86–1.45) Hispanic 1.28 (0.95–1.72) Chinese 0.66 (0.44–1.00)
Patel et al. [30]	Cross-sectional; Philadelphia Health Management Corporation	N=9553; 67% women; age 40–64:51%	Black: 21.1 Hispanic: 10.6 Other: 2.6 White: 65.7 Poor: 26.8	Telephone survey item	1 item: "How would you rate quality of sleep in past week?"	Age, sex, BMI, education, employment, marital status, general and mental health, health behaviors	OR (95% CI) Black, not poor 1.45 (1.1–1.9) Black, poor 1.2 (0.9–1.5) Hispanic, not poor 1.3 (0.8–2.1) Hispanic, poor 1.05 (0.7–1.5) Other, not poor 1.01 (0.5–2.1) Other, poor 0.67 (0.2–2.3) White, poor 4.20 (3.3–5.4)

Continued

TABLE 6.3 Representative studies examining insomnia in adult racial/ethnic minority and lower SES populations.—cont'd

Study	Design/source	Sample	Racial/ethnic/SES (% of sample)	Insomnia assessment	Insomnia definition	Covariates	Findings; OR (95% CI) (reference racial/ethnic group is white)
Grandner et al. [72, 73]	Cross-sectional; National Health and Nutrition Examination Survey 2007–08	N=4081; 48% women; age 47±17	Black: 10.3 Mex-Amer: 7.6 Other Hisp: 4.6 Asian/other: 4.8 White: 72.8 Income <20K: 13.5	5 items; "How long does it take to fall asleep at bedtime?"; "In the past month, how often had you had difficulty with…."	SOL >30min, DFA, DMS, EMA, or NR sleep	Age, sex, marital status, individual SES factors, marital status, immigrant status, physical and mental health	<u>SOL ≥30 min</u> Black 1.6 (1.3–2.0) <u>DFA</u> Black 0.6 (0.5–0.7) Mex-Amer 0.6 (0.5–0.8) Other Hispanic 0.7 (0.5–0.9) <u>DMS</u> Black 0.8 (0.7–0.98) Mex-Amer 0.8 (0.6–0.98) <u>EMA</u> Black 0.8 (0.7–0.96) <u>NRS</u> Black 1.6 (1.3–2.0)
Paine et al. [80]	Cross-sectional; New Zealand	N=2670; 56% women; age 20–59	Maori: 45.8 White: 54.2	Mailed questionnaire	DFA, DMS, EMA, NRS	Age, sex, shift work, employment status, SES deprivation score	<u>DMS</u> Maori 1.2 (1.0–1.5) <u>EMA</u> Maori 1.4 (1.2–1.7) Unemployed vs employed <u>DFA</u> 1.4 (1.1–1.7) <u>DMS</u> 1.4 (1.2–1.9)

Data presented as means±SD or OR (95% CI). Bolded values represent $P<.05$ for Chi-square comparisons. *AHI*, apnea-hypopnea index; *BMI*, body mass index; *DFA*, difficulty falling asleep; *DMS*, difficulty maintaining sleep; *EMA*, early morning awakening; *NRS*, nonrestorative sleep; *SES*, socioeconomic status; *SOL*, sleep onset latency; *WASO*, wake after sleep onset; *WHIIRS*, women's health initiative insomnia rating scale.

but lower odds of reporting DFA, DMS, EMA or NRS (Table 6.3) [73]. These data among minority individuals suggest that the wording of insomnia assessment may influence results among racially/ethnically diverse individuals.

In a community-based sample of three indigenous North American groups (56% women, age 43 ± 1 years) [56], Froese et al. found that 17% of this sample reported insomnia symptoms (DFA, DMS, EMA, or NRS) "every night or almost every night." Using the Pittsburgh Sleep Quality Index in participants from eight rural Indian reservations ($n = 386$, 54% women, age 31 ± 14 years), Ehlers et al. reported that slightly >50% of the sample had SL > 30 min and reported overall poor sleep quality (PSQI score > 5) [84]. In age-, gender-, and SES-adjusted analyses in New Zealand, Maori ethnicity was associated with greater odds of DMS and EMA complaints than among whites [80].

Insomnia complaints across SES groups

Many of the same studies evaluating insomnia complaints or poor sleep among racial/ethnic groups have also explored their associations with SES variables. Among elderly individuals in EPSES, lower education was associated with greater insomnia complaints after adjustment for demographics, health status, depression, cognitive impairment, and use of sedative-hypnotics [77]. In a study using sleep diaries and daytime impairments to define insomnia disorder ($n = 575$, 50% women, 27% Black), Gellis et al. found that individuals with lower educational attainment had greater odds of insomnia (Table 6.3) [85]. Compared to college graduates, high school drop outs were 3.9 times more likely to have insomnia and high school graduates were 2.3 times more likely to have insomnia. Analyzing data from the BRFSS, Grandner et al. found that having lower educational attainment, living in poverty (annual income < $10,000), or unemployment were all associated with significantly greater odds of SS than college graduates, earning more than $75,000 annually, or having full-time employment [79]. An inverse linear relationship was noted between education, income, and SS with increasing odds for SS as income and education categories each decreased. Similarly, in NHANES, Grandner et al. reported that long SL was associated with lower educational attainment, lacking private insurance, and food insecurity [73]. In nationally representative sample of New Zealand, after adjusting for ethnicity and demographic factors, living in areas of socioeconomic deprivation and being unemployed were also associated with increased odds of DFA and DMS complaints [80].

Studies that have utilized sleep quality as a proxy for insomnia have produced similar findings. In the SWAN Sleep study cohort, analyses adjusted for demographics, health status, depression, and sleep medications showed that greater financial strain was related to more sleep quality complaints [21]. Similarly, in Pittsburgh, Mezick et al. found that a lower SES composite score (income and education) was associated with poorer sleep quality [20]. In another study focusing on sleep quality, Patel et al. examined the relationship between sociodemographic factors and perceived poor sleep [30]. In models adjusting for education, employment, and health covariates, impoverished whites (OR 4.20 [3.3–5.4]) and nonpoor Blacks (OR 1.45 [1.1–1.9]) had significantly increased odds of poor sleep quality than nonpoor whites. These covariates mediated the association of poor sleep quality among poor Blacks observed in unadjusted analyses. These data suggest a complex relationship between SES measures and perceived sleep health that may vary by racial/ethnic background. Not all studies have found an association between insomnia and SES. For example, in a diverse sample of low-income individuals in Brooklyn ($n = 1118$; mean income < $20,000 annually), adjusting for health status, social support, and demographics, neither education nor income predicted a composite sleep disturbance index score [86]. These contradictory data suggest that the relationship with SES and insomnia may vary in racial/ethnic samples. They may also indicate a need for comprehensive measures of SES and/or measurement of protective factors (e.g., coping style, social networks) that may explain the moderation of race on the relationship between SES and insomnia symptoms.

Restless leg syndrome (RLS) and periodic limb movements during sleep (PLMS)

The prevalence of RLS and PLMS is estimated to be between over 1% and 15% in the general population, depending on the diagnostic criteria used [87]. Almost all patients with RLS exhibit PLMS [88]. In a comprehensive sleep study including 24-h PSG, 592 participants from Detroit ($n = 186$ Blacks) completed subjective and objective sleep measurements to determine racial/ethnic differences in the prevalence of PLMS. The authors found a lower prevalence of PLMS in Blacks compared to whites (4.3% vs 9.3%, $P < .05$) [89]. While the other racial/ethnic groups were too small to analyze, they included American Indian or Alaskan Native, Asian and "other racial" groups. The "other" racial category had similar prevalence of PLMS to whites. Given that lower stores in the blood is associated with PLMS/RLS [90], one study examined serum ferritin—a protein that stores iron—and complaints of RLS in dialysis patients comparing Blacks with whites. Using a combination of in-person interviews and medical chart review, 210 chronic kidney disease patients (48% Black) were asked: "During the past 4 weeks, to what extent were you bothered by restless legs?" Blacks had a lower odds of complaint or "bothered by restless legs symptoms" compared with whites. Blacks also had a lower odds of reporting RLS than white patients (OR, 0.44; $P = .03$) [91]. The results were similar to an earlier investigation conducted by the same group.

In that study, 48% of older Blacks reported a complaint of RLS compared to 68% of whites ($P=.0006$). The findings remained significant after adjusting for gender, education, BMI, cardiovascular disease (CVD) morbidity, hours receiving hemodialysis treatment, and months receiving hemodialysis treatment [92].

Most recently, in a large representative cohort of veterans, Molnar et al. conducted a 1:1 propensity score matched analysis in a cohort consisting of 3696 patients (17% Black) in each group. In adjusted analysis, Black, Hispanic, and "other race" was associated with a decreased risk of incident RLS than whites. In an assessment of clinical outcomes, Blacks reported a similar risk to whites for CVD and stroke, but reduced risk for mortality. No other racial/ethnic group was included in the analysis. Finally, incident RLS was associated with an 88% increased hazard ratio of mortality across all racial groups, except for Blacks [87].

Using data from the Baltimore Health and Mental Health Study, which included a seven-item RLS validated questionnaire, unadjusted results revealed similar prevalence rates of RLS (4.7%) in Blacks and (3.8%) whites. After adjusting for age, gender, SES, and medical comorbidities, there were no differences in rates of RLS found between Blacks and whites [93]. The authors speculated that a phenotypically difference in RLS symptoms could be different in Blacks than other racial groups. Given the lower prevalence of RLS and/or PLMS among Blacks, one group of investigators speculated that Blacks are less likely to seek treatment for RLS/PLMS. This may partly explain the lower prevalence [93], but to our knowledge, this hypothesis has not been tested.

The Sleep Health and Knowledge in US Hispanics Project conducted the first population-based study to assess RLS in Hispanics [94]. 1754 Hispanics of Mexican descent and 1913 non-Hispanic whites in San Diego, were queried about RLS, based on the four diagnostic criteria from the International Restless Legs Syndrome Study Group [95]. Hispanics had a lower prevalence of RLS than whites (14.4% vs 18.3%, $P=.002$). Koo et al. analyzed data from the MESA sleep study, and found that Blacks had less PLMS (10.5%) than other ethnic groups. A higher prevalence of PLMS was found in whites (18.8%), Hispanics (20.1%), and Chinese Americans (19.1%). When analyzing race/ethnicity and prevalence of hypertension among those with PLMS, middle aged to older Blacks had 20% increase in their odds of having hypertension and Chinese Americans had a 10% increase in their odds of having hypertension relative to whites [96]. In sensitivity analysis, PLMS was associated with 2.47 mmHg systolic and as high as 3.71 mmHg systolic, depending on the unit of PLMS (periodic leg movement index [97] 10-unit versus PLMAI 1-unit) ($P<.0001$). Similar findings were reported for Chinese Americans, 10-unit increase in PLMI was associated with 1.31 mmHg higher SBP ($P=.03$). For diastolic blood pressure, there were modest significant findings with PLMI among Blacks and Chinese Americans ($P=.09$ and $P=.08$) but not for Hispanics.

To summarize, in all previous studies Blacks have reported lower prevalence of RLS and PLMS, other racial/ethnic groups have similar prevalence of PLMS compared to whites, and all studies compared one racial/ethnic groups (e.g., Blacks to whites) with the exception of the MESA study. The clinical outcomes suggest that despite lower risk, when the disorder is present it confers greater morbidity among racial/ethnic minorities. Given that there are few studies on these conditions, and because not all studies have adjusted for the same set of covariates, and population differences (e.g., older men only, hemodialysis patients), it is too soon to definitively make an assessment about the lower prevalence of RLS/PLMS in Blacks compared to whites. Notably, none of the aforementioned studies investigated the role of SES.

Narcolepsy

The first study to assess the prevalence of narcolepsy in the United States was conducted among young (16–34 years of age) naval recruit men at the US Marine Corps in North Carolina [98]. In this study, the prevalence of narcolepsy was estimated to be 19 cases out of 10,000 recruits [98]. The author described Black men as "constantly in a state of readiness for sleep" and referenced an impoverished environment as potential contributions to narcolepsy among Black men. Later in 2002, Okun et al. conducted a retrospective study of patient data at the Stanford Sleep Clinic ($n=64$ Blacks, $n=353$ whites, $n=32$ Asians, $n=26$ Latinos, $n=9$ mixed ethnicity). There were no differences across ethnic groups in symptomatology and severity [99]. In 2009, a population-based study was conducted in King County, Washington to determine the prevalence of narcolepsy among a diverse patient population. Patients were recruited from local sleep centers, neurologists, or self-referred. Patients were eligible if they reported a diagnosis by a physician of narcolepsy. Blacks (42.8%) had the highest prevalence of narcolepsy followed by whites (32.2%), Asians (15.0%), and other races (27.9%) [100]. At Stanford University, the largest study to date that compared ethnic differences ($n=839$ whites, $n=182$ Blacks, $n=35$ Asians, $n=41$ Latinos) in the clinical presentation of narcolepsy type 1 [100] and type 2 [101] was conducted. Adjusting for age, sex and BMI, Black patients with narcolepsy type 1 and patients with narcolepsy type 2 were diagnosed at an earlier age and scored higher on the Epworth Sleepiness Scale ($P<.001$) than whites, Asians, and Hispanics [101]. In a subgroup analysis, Hispanics with type 2 also presented at an earlier age ($P<.003$). Blacks with narcolepsy type 2 exhibited cerebrospinal fluid (CSF) hypocretin-1 deficiency and lower mean CSF hypocretin-1 levels than whites (45.7 ± 22.1 pg/mL vs 262.6 ± 16.4 pg/mL). The sample size was too small among other groups to analyze CSF hypocretin-1 levels. Importantly, from this data, genetic typing of human leukocyte antigen (HLA) DQB1*0602 allele, was analyzed. HLA DQB1*06:02 positivity was significantly higher in blacks than other ethnicities (91.6% vs 77.4%, 80.4%, and 71.7%, respectively; $P<.001$). Blacks were also obese and of younger age. The early age finding is likely related to a strong genetic component and a difference in clinical presentation of the disorder in Blacks versus whites. Another interesting finding is that Blacks were less likely to report complaints of cataplexy. It is plausible that there is overlap between cataplexy and other psychological problems (e.g., nightmares, trauma) leading to an underestimation of cataplexy among Blacks.

Circadian rhythms

The results on the circadian processes of racial/ethnic minorities compared with whites is mixed. Among a multiethnic cohort of postmenopausal women, circadian rhythm was measured by 24-h collection of urine specimens for melatonin and an Actillume wrist monitor worn for up to 1 week. Daily sleep logs and 1 week of actigraphic recording measured subjective sleep duration. Findings revealed that ethnicity was significantly associated with illumination and European Americans received greater illumination than Blacks, Hispanics and American Indian/Alaska Native ($P<.001$). Melatonin secretion was not significantly related to ethnicity and declined with increasing age ($P<.001$) [82]. Jean-Louis et al. analyzed data from a group of men and women from San Diego and found no differences in activity patterns, illumination, and timing of sleep between non-Hispanic whites and minorities [102]. More recent studies suggest that Blacks have a shorter free-running (tau) period. One study conducted by Eastman et al. included 94 healthy adults. Participants provided saliva every 30 min over the course of 5 days and slept in windowless rooms. Data collection was conducted in dim light (<5 lx). Findings revealed that Black women had a tau that was approximately 14 min shorter than white women [103]. The authors findings were confirmed based on a previous analysis conducted in an earlier study which found that Blacks have larger phase shifts than whites

[104]. In another study conducted by Eastman's group, researchers observed Blacks and whites in a laboratory and conducted phase sleep shift (i.e., changing their sleep time as if on eastern time zone). The authors explored cognition and found that Blacks performed less on cognitive assessments than whites [105]. But these findings were in contrast to a later study in which Blacks and whites performed equally on cognitive tasks [106]. While these studies have been conducted in the laboratory environment and with small samples, these early findings raise important questions as circadian sleep–wake processes can have a profound effect on the trajectory of psychiatric disorders [107] and metabolic health [108, 109]. Additional research is needed to understand the mechanisms of how these processes influence the development and outcomes of psychiatric and medical morbidity and how these vary by race/ethnicity and sex.

WHY DO MINORITY AMERICANS HAVE POOR SLEEP?

Thomas et al. [110] described health disparities research occurring across generations including first, second, third, and fourth (see Fig. 6.1). First generation was characterized as measuring the gap in a health outcome based on epidemiological, observational studies using cohort studies and large population data sets. Second generation focused on identifying the mechanisms, third generation focused on designing or identifying solutions to address challenges, and fourth and current generation focuses on taking action and addressing health equity. In the field of sleep medicine, significant work has been done as part of first and second generation, which we have outlined in this chapter. Far less has been done with respect to third and fourth generation research. In the next section, we highlight a select set of potential mediators that could explain disparities: acculturation, discrimination, worry and risk perception, and sleep opportunity, as well as potential solutions.

Acculturation

Acculturation is a multidimensional process in which migrants maintain aspects of their culture of origin while

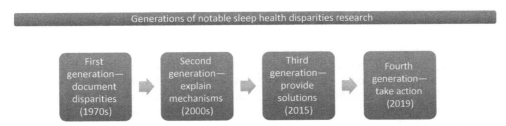

FIG. 6.1 Historical timeline of notable studies examining sleep and race/ethnicity and health disparities research. *(Adapted from Thomas SB, Quinn SC, Butler J, Fryer CS, Garza MA. Toward a fourth generation of disparities research to achieve health equity. Annu Rev Public Health 2011;32:399–416. WOS:000290776200022.)*

adopting elements of their new cultural group. Acculturation has been measured via multiple proxies (e.g., nativity, language) in studies examining its relationship to sleep [55]. In the NHIS, foreign-born individuals had lower risk for SS (6h OR 0.92 [0.85–0.99]) and LS (9h OR 0.85 [0.76–0.95]) than US-born individuals [18]. Similarly, in NHANES, individuals who were born in Mexico reported less SS and those who spoke Spanish only at home reported less very SS (<5h) compared to whites [17]. In the SWAN study cohort, Hale et al. reported that US-born Latinas, Japanese, and Chinese women were more likely to report SC than their first generation ethnic counterparts [54]. In addition, language acculturation mediated 40% of the association between immigrant status and SC. Among US Hispanics in the HCHS/SOL, greater levels of acculturation stress have been linked to lower ACT-measured TST, greater sleep fragmentation, and more variable sleep timing [111]. Although individuals in the Sleep Health and Knowledge in US Hispanics Project [94] reported a lower prevalence of RLS than whites, additional analyses revealed that among Hispanics with high acculturation, prevalence of RLS was greater than Hispanics with a lower acculturation score (17.4% vs 12.8%, $P=.008$). Highly acculturated individuals had similar prevalence to whites (17.4% vs 18.3%, $P=.637$). There was also a difference in acculturation and gender with high acculturated younger men reporting higher prevalence of RLS than low acculturated younger men (15.5% vs 7.4%, $P=.003$) [94]. In general, greater acculturation is associated with worsened sleep quantity and quality possibly via adoption of unhealthy lifestyle factors of the new culture (e.g., electronics use, lower levels of physical activity) or weakening of protective factors associated with the culture of origin (e.g., social cohesion).

Perceived discrimination

Racial/ethnic minorities experience perceived discrimination and several studies have documented that perceived discrimination is associated with sleep duration and sleep quality [112]. In a review by Slopen et al. [112], absent from these studies were sufficient objective measurements of sleep; 13 out of 17 studies used self-report to measure sleep. Of those studies using an objective measure, the results were inconsistent. Additionally, these studies lacked a richness of measures of discrimination and none explored the role of coping including racial/ethnic identity development, which could serve as protective factors [113]. One peer reviewed abstract reported that experiences of racial identity may moderate the relationship between sleep duration and discrimination [114]. Additional evidence on potential mediators to explain this relationship as well as more novel race-discrimination factors including internalized racism and structural racism should be explored. Studying this further may be particularly important as many investigations have described the physiological and psychological consequences of discrimination [115–117], which could have significant implications for sleep health.

Worry and risk perception

Worry and risk perception in the health psychology literature are often seen as contradictory but overlapping constructs [118, 119]. One potential reason is that cognitive risk perception (the degree of perceived susceptibility) and worry (affective perceptions) may influence health behavior differently and may also interact in influencing health behavior. These psychosocial variables have been shown to influence cancer screening [120] and uptake of flu vaccination [121]. Little is known about risk perception and worry in sleep medicine. Understanding the role of these factors in sleep is important because these could influence the uptake of screening for PSG, the ability to engage in healthy sleep and adherence to treatment for sleep disorders.

Sleep opportunity

Optimizing sleep entails a balance in sleep opportunity (how much time you spend in bed) and sleep ability (how long you are able to sleep) [122]. If there is a mismatch between sleep ability (low) and opportunity by habitual napping and sleeping in (high) it is likely that the individual will experience too little sleep, too much sleep, NRS, or low sleep efficiency [123]. For example, if you are in bed for 10h and you can only sleep 5h then there is a mismatch. This hypothesis has been tested in Drosophila flies where extending a dark period (sleep) from 12 to 14h and 16h resulted in impaired sleep [123]. Indeed, when sleep opportunity is aligned with sleep ability, as is done through sleep restriction, individuals achieve improve sleep efficiency and quality sleep. This research is a promising beginning and raises the importance to test this model in human populations.

FUTURE DIRECTIONS AND SUMMARY

Converging evidence from several studies over the past two decades demonstrate that racial/ethnic minorities report inadequate sleep, experience poor sleep quality, and have a greater risk for certain sleep disorders, than their white counterparts. Four broad questions remain: First, most of the sleep disparities literature has focused on Blacks, but some studies have described the experience of other racial/ethnic groups. There is a marked paucity in good data on other racial/ethnic groups, rural populations and sexual minorities. How will researchers engage these populations in order to recruit and retain a rich and diverse sample for further inquiry? Second, few studies provide sufficient data to understand the mechanisms of sleep symptoms and sleep disorders by race/ethnicity and SES limiting the ability to infer causality. How will researchers argue for large-scale multiethnic epidemiological studies with long-term follow-up to fully elucidate potential mechanisms that may contribute to disparities in sleep? Simultaneously, intensive naturalistic studies with a smaller sample size could also be undertaken. Third, as there is increasing national, local and global recognition of social determinants of health (SDH), the question for the sleep community is: how do we act on the SDH and implement important public health policy as it relates to closing the disparities gap? Fourth, as we work to determine a strategy to assess these important challenges, at the same time, there is an urgent need to test the efficacy of evidence-based sleep interventions (e.g., PAP adherence, sleep extension, cognitive behavioral therapy for insomnia) to improve health and well-being (e.g., obesity, impaired glucose intolerance, depression, quality of life). A few investigators have reported on evidence-based sleep interventions and results are promising [124, 125]. But, the question still holds which is how does the community ensure that established efficacious treatments (e.g., cognitive behavioral therapy for insomnia) for sleep disorders are generalizable and also personalized to the needs of vulnerable populations? Overall, more could be done to explore variability in sleep disorders by race/ethnicity, SES, and the inclusion of sexual minorities. Importantly, studies must be designed with the explicit purpose of addressing these questions.

In our observation, the field of sleep medicine could advance the discussion by examining both racial/ethnic disparities *and* SES. In the pursuit of effective policy that could close the disparities gap, we hope that the materials outlined in this chapter will move the field forward.

REFERENCES

[1] Kawachi I, Daniels N, Robinson DE. Health disparities by race and class: why both matter—we must link efforts to address the injuries of race and class simultaneously if we are to reduce health disparities. Health Aff 2005;24(2):343–52. WOS:000227835700008.

[2] Analysis BoE. Gross domestic product: second quarter 2018; 2018.

[3] Analysis BoE. Available from: https://www.bea.gov/news/2018/gross-domestic-product-second-quarter-2018-second-estimate-corporate-profits-second; 2018.

[4] Musu-Gilette L, de Brey C, McFarland J, Hussar W, Sonnenberg W. Status and trends in the education of racial and ethnic groups. Education Do. Washington, DC: National Center for Education Statistics; 2017.

[5] USDHHS. Healthy People 2020. Available from: http://www.healthypeople.gov/2020/topicsobjectives2020/objectiveslist.aspx?topicid=38; 2014.

[6] Bureau USC. QuickFacts. Income and Poverty in the United States: 2017. Available from: https://www.census.gov/content/dam/Census/library/publications/2018/demo/p60-263.pdf; 2018.

[7] Services UDoHaH. Health, United States, 2012: with special feature on emergency care. Hyattsville, MD: National Center for Health Statistics; 2013.

[8] MMWR. Health objectives for the Nation Healthy People 2000: National Health Promotion and Disease Prevention Objectives for the year 2000. Washington, DC: U.S. Department of Health and Human Services; 1990. p. 695–7.

[9] Grandner MA, Alfonso-Miller P, Fernandez-Mendoza J, Shetty S, Shenoy S, Combs D. Sleep: important considerations for the prevention of cardiovascular disease. Curr Opin Cardiol 2016;31(5):551–65, WOS:000382559100011. [English].

[10] Grandner MA, Williams NJ, Knutson KL, Roberts D, Jean-Louis G. Sleep disparity, race/ethnicity, and socioeconomic position. Sleep Med 2016;18:7–18, WOS:000371837500003.

[11] Petrov ME, Lichstein KL. Differences in sleep between black and white adults: an update and future directions. Sleep Med 2016;18:74–81, WOS:000371837500010.

[12] Smedley BD, Stith AY, Nelson AR. In: Smedley BD, Stith AY, Nelson AR, editors. Unequal treatment: confronting racial and ethnic disparities in health care (with CD). The National Academies Press; 2003.

[13] Adler N. Socioeconomic status and health. The challenge of the gradient effect. Am Psychol 1994;49:15–24.

[14] Marmot MG, Shipley MJ, Rose G. Inequalities in death—specific explanations of a general pattern? Lancet 1984;1(8384):1003–6. 6143919, [eng].

[15] Stamatakis KA, Kaplan GA, Roberts RE. Short sleep duration across income, education, and race/ethnic groups: population prevalence and growing disparities during 34 years of follow-up. Ann Epidemiol 2007;17(12):948–55. 17855122. [Epub 2007/09/15. eng.].

[16] Hale L, Do P. Racial differences in self-reports of sleep duration in a population-based study. Sleep 2007;30(9):1096–103. WOS:000249293000006 [English].

[17] Whinnery J, Jackson N, Rattanaumpawan P, Grandner MA. Short and long sleep duration associated with race/ethnicity, sociodemographics, and socioeconomic position. Sleep 2014;37(3):601. WOS:000332520400021.

[18] Krueger PM, Friedman EM. Sleep duration in the United States: a cross-sectional population-based study. Am J Epidemiol 2009;169(9):1052–63, 19299406. [Epub 2009/03/21. eng].

[19] Lauderdale DS, Knutson KL, Yan LJL, Rathouz PJ, Hulley SB, Sidney S, et al. Objectively measured sleep characteristics among early-middle-aged adults—the CARDIA study. Am J Epidemiol 2006;164(1):5–16, WOS:000238536900002. [English].

[20] Mezick EJ, Matthews KA, Hall M, Strollo PJ, Buysse DJ, Kamarck TW, et al. Influence of race and socioeconomic status on sleep: Pittsburgh SleepSCORE project. Psychosom Med 2008;70(4):410–6, WOS:000255922400004. [English].

[21] Hall MH, Matthews KA, Kravitz HM, Gold EB, Buysse DJ, Bromberger JT, et al. Race and financial strain are independent correlates of sleep in midlife women: the SWAN Sleep Study. Sleep 2009;32(1):73–82, WOS:000262075600013. [English].

[22] Chen XL, Wang R, Zee P, Lutsey PL, Javaheri S, Alcantara C, et al. Racial/ethnic differences in sleep disturbances: the multi-ethnic study of atherosclerosis (MESA). Sleep 2015;38(6):877–88, WOS:000355617000009. [English].

[23] Carnethon MR, De Chavez PJ, Zee PC, Kim KYA, Liu K, Goldberger JJ, et al. Disparities in sleep characteristics by race/ethnicity in a population-based sample: Chicago Area Sleep Study. Sleep Med 2016;18:50–5, WOS:000371837500006. [English].

[24] Dudley KA, Weng J, Sotres-Alvarez D, Simonelli G, Feliciano EC, Ramirez M, et al. Actigraphic Sleep Patterns of US Hispanics: the Hispanic Community Health Study/Study of Latinos. Sleep 2017;40(2):8, WOS:000394129900012. [English].

[25] Altman NG, Izci-Balserak B, Schopfer E, Jackson N, Rattanaumpawan P, Gehrman PR, et al. Sleep duration versus sleep insufficiency as predictors of cardiometabolic health outcomes. Sleep Med 2012;13(10):1261–70, WOS:000311940400010.

[26] Jean-Louis G, Grandner MA, Youngstedt SD, Williams NJ, Zizi F, Sarpong DF, et al. Differential increase in prevalence estimates of inadequate sleep among black and white Americans. BMC Public Health 2015;15, WOS:000365476300002.

[27] Song YS, Ancoli-Israel S, Lewis CE, Redline S, Harrison SL, Stone KL. The association of race/ethnicity with objectively measured sleep characteristics in older men. Behav Sleep Med 2012;10(1):54–69, WOS:000300169500005. [English].

[28] Chen X, Gelaye B, Williams MA. Sleep characteristics and health-related quality of life among a national sample of American young adults: assessment of possible health disparities. Qual Life Res Int J Qual Life Asp Treat Care Rehab 2014;23(2):613–25. MEDLINE:23860850.

[29] Edwards C, Mukherjee S, Simpson L, Palmer LJ, Almeida OP, Hillman DR. Depressive symptoms before and after treatment of obstructive sleep apnea in men and women. J Clin Sleep Med 2015;11(9):1029–38, WOS:000366292600010.

[30] Patel NP, Grandner MA, Xie DW, Branas CC, Gooneratne N. "Sleep disparity" in the population: poor sleep quality is strongly associated with poverty and ethnicity. BMC Public Health 2010;10:11, WOS:000282236700002. [English].

[31] Patel SR, Malhotra A, Gottlieb DJ, White DP, Hu FB. Correlates of long sleep duration. Sleep 2006;29(7):881–9, WOS:000238961500004. [English].

[32] Johnson DA, Lisabeth L, Hickson D, Johnson-Lawrence V, Samdarshi T, Taylor H, et al. The social patterning of sleep in African Americans: associations of socioeconomic position and neighborhood characteristics with sleep in the Jackson Heart Study. Sleep 2016;39(9):1749–59, WOS:000384333400015. [English].

[33] Ertel KA, Berkman LF, Buxton OM. Socioeconomic status, occupational characteristics, and sleep duration in African/Caribbean immigrants and US White health care workers. Sleep 2011;34(4):509–18, WOS:000289061800016.

[34] Akerstedt T, Fredlund P, Gillberg M, Jansson B. Work load and work hours in relation to disturbed sleep and fatigue in a large representative sample. J Psychosom Res 2002;53(1):585–8, WOS:000177230200009.

[35] Jackson CL, Redline S, Kawachi I, Williams MA, Hu FB. Racial disparities in short sleep duration by occupation and industry. Am J Epidemiol 2014;178(9):1442–51, WOS:000326642300012.

[36] Giles DE, Kupfer DJ. Effects of race on EEG sleep in major depression. Biol Psychiatry 1995;37(9):624, WOS:A1995QX03700114.

[37] Rao U, Poland RE, Lutchmansingh P, Ott GE, McCracken JT, Lim KM. Relationship between ethnicity and sleep patterns in normal controls: implications for psychopathology and treatment. J Psychiatr Res 1999;33(5):419–26, WOS:000082296000007.

[38] Redline S, Kirchner L, Quan SF, Gottlieb DJ, Kapur V, Newman A. The effects of age, sex, ethnicity, and sleep-disordered breathing on sleep architecture. Arch Intern Med 2004;164(4):406–18, WOS:000189148600008.

[39] Young T, Peppard PE, Gottlieb DJ. Epidemiology of obstructive sleep apnea—a population health perspective. Am J Respir Crit Care Med 2002;165(9):1217–39, WOS:000175314900009.

[40] Young T, Shahar E, Nieto FJ, Redline S, Newman AB, Gottlieb DJ, et al. Predictors of sleep-disordered breathing in community-dwelling adults—the sleep heart health study. Arch Intern Med 2002;162(8):893–900, WOS:000175039400005. [English].

[41] Fulop T, Hickson DA, Wyatt SB, Bhagat R, Rack M, Gowdy O, et al. Sleep-disordered breathing symptoms among African-Americans in the Jackson Heart Study. Sleep Med 2012;13(8):1039–49, WOS:000309038300011. [English].

[42] Johnson DA, Simonelli G, Moore K, Billings M, Mujahid MS, Rueschman M, et al. The neighborhood social environment and objective measures of sleep in the Multi-Ethnic Study of Atherosclerosis. Sleep 2017;40(1):8, WOS:000394125700016. [English].

[43] Mihaere KM, Harris R, Gander PH, Reid PM, Purdie G, Robson B, et al. Obstructive sleep apnea in New Zealand adults: prevalence and risk factors among Maori and non-Maori. Sleep 2009;32(7):949–56, WOS:000268126000015.

[44] Kripke DF, AncoliIsrael S, Klauber MR, Wingard DL, Mason WJ, Mullaney DJ. Prevalence of sleep-disordered breathing in ages 40–64 years: a population-based survey. Sleep 1997;20(1):65–76, WOS:A1997WV60300011. [English].

[45] Redline S, Sotres-Alvarez D, Loredo J, Hall M, Patel SR, Ramos A, et al. Sleep-disordered breathing in Hispanic/Latino individuals of diverse backgrounds the Hispanic Community Health Study/Study of Latinos. Am J Respir Crit Care Med 2014;189(3):335–44, WOS:000331793400016. [English].

[46] Tomfohr L, Pung MA, Edwards KM, Dimsdale JE. Racial differences in sleep architecture: the role of ethnic discrimination. Biol Psychol 2012;89(1):34–8, WOS:000299714500004. [English].

[47] Yon A, Scogin F, DiNapoli EA, McPherron J, Arean PA, Bowman D, et al. Do manualized treatments for depression reduce insomnia symptoms? J Clin Psychol 2014;70(7):616–30, WOS:000337623800002.

[48] Ruiter ME, Decoster J, Jacobs L, Lichstein KL. Normal sleep in African-Americans and Caucasian-Americans: a meta-analysis. Sleep Med 2011;12(3):209–14. 21317037. [eng].

[49] Medicine AAoS. International classification of sleep disorders. 3rd ed. Darien, IL: American Academy of Sleep Medicine; 2014.

[50] Peppard PE, Young T, Barnet JH, Palta M, Hagen E, Hla KM. Increased prevalence of sleep-disordered breathing in adults. Am J Epidemiol 2013;177(9):1006–14, WOS:000318576300019.

[51] O'Connor GT, Lind B, Eea L. Variation in symptoms of sleep-disordered breathing with race and ethnicity: the Sleep Heart Health Study. Sleep 2003;26:74–9.

[52] Baron KG, Liu KA, Chan CL, Shahar E, Hasnain-Wynia R, Zee P. Race and ethnic variation in excessive daytime sleepiness: the Multi-Ethnic Study of Atherosclerosis. Behav Sleep Med 2010;8(4):231–45, WOS:000282578800005. [English].

[53] Hayes AL, Spilsbury JC, Patel SR. The Epworth score in African American populations. J Clin Sleep Med 2009;5(4):344–8, WOS:000270263300009.

[54] Hale L, Troxel WM, Kravitz HM, Hall MH, Matthews KA. Acculturation and sleep among a multiethnic sample of women: the Study of Women's Health Across the Nation (SWAN). Sleep 2014;37(2):309–17, WOS:000332519100011. [English].

[55] Loredo JS, Soler X, Bardwell W, Ancoli-Israel S, Dimsdale JE, Palinkas LA. Sleep health in US hispanic population. Sleep 2010;33(7):962–7, WOS:000279365600016. [English].

[56] Froese CL, Butt A, Mulgrew A, Cheema R, Speirs MA, Gosnell C, et al. Depression and sleep-related symptoms in an adult, Indigenous, North American population. J Clin Sleep Med 2008;4(4):356–61, WOS:000209777000010. [English].

[57] WHO Expert Consultation. Appropriate body-mass index for Asian populations and its implications for policy and intervention strategies. Lancet 2004;363(9403):157–63. MEDLINE:14726171. [English].

[58] Chen XL, Wang R, Lutsey PL, Zee PC, Javaheri S, Alcantara C, et al. Racial/ethnic differences in the associations between obesity measures and severity of sleep-disordered breathing: the Multi-Ethnic Study of Atherosclerosis. Sleep Med 2016;26:46–53, WOS:000390720900009. [English].

[59] Sutherland K, Lee RWW, Cistulli PA. Obesity and craniofacial structure as risk factors for obstructive sleep apnoea: impact of ethnicity. Respirology 2012;17(2):213–22, WOS:000299416100004. [English].

[60] Cakirer B, Hans MG, Graham G, Aylor J, Tishler PV, Redline S. The relationship between craniofacial morphology and obstructive sleep apnea in whites and in African-Americans. Am J Respir Crit Care Med 2001;163(4):947–50, WOS:000168057700032. [English].

[61] Lam B, Ip MSM, Tench E, Ryan CF. Craniofacial profile in Asian and white subjects with obstructive sleep apnoea. Thorax 2005;60(6):504–10, WOS:000229433900013. [English].

[62] Coltman R, Taylor DR, Whyte K, Harkness M. Craniofacial form and obstructive sleep apnea in Polynesian and Caucasian men. Sleep 2000;23(7):943–50, WOS:000165175500011. [English].

[63] Will MJ, Ester MS, Ramirez SG, Tiner BD, McAnear JT, Epstein L. Comparison of cephalometric analysis with ethnicity in obstructive sleep apnea syndrome. Sleep 1995;18(10):873–5, WOS:A1995TQ23600008. [English].

[64] Lee JJ, Ramirez SG, Will MJ. Gender and racial variations in cephalometric analysis. Otolaryngol Head Neck Surg 1997;117(4):326–9, WOS:A1997YA33900006. [English].

[65] Ruiter ME, DeCoster J, Jacobs L, Lichstein KL. Sleep disorders in African Americans and Caucasian Americans: a meta-analysis. Behav Sleep Med 2010;8(4):246–59, 20924837. [eng].

[66] Zanobetti A, Redline S, Schwartz J, Rosen D, Patel S, O'Connor GT, et al. Associations of PM10 with sleep and sleep-disordered breathing in adults from seven US Urban areas. Am J Respir Crit Care Med 2010;182(6):819–25, WOS:000282162100015.

[67] Pranathiageswaran S, Badr MS, Severson R, Rowley JA. The influence of race on the severity of sleep disordered breathing. J Clin Sleep Med 2013;9(4):303–9, WOS:000318604100002.

[68] Ong KC, Clerk AA. Comparison of the severity of sleep-disordered breathing in Asian and Caucasian patients seen at a sleep disorders center. Respir Med 1998;92(6):843–8, WOS:000075263000007. [English].

[69] Yamagishi K, Ohira T, Nakano H, Bielinski SJ, Sakurai S, Imano H, et al. Cross-cultural comparison of the sleep-disordered breathing prevalence among Americans and Japanese. Eur Respir J 2010;36(2):379–84, WOS:000281601800023. [English].

[70] Billings ME, Rosen CL, Wang R, Auckley D, Benca R, Foldvary-Schaefer N, et al. Is the relationship between race and continuous positive airway pressure adherence mediated by sleep duration? Sleep 2013;36(2):221–7, WOS:000314393700011.

[71] Billings ME, Auckley D, Benca R, Foldvary-Schaefer N, Iber C, Redline S, et al. Race and residential socioeconomics as predictors of CPAP adherence. Sleep 2011;34(12):1653–8, 22131602. [eng].

[72] Grandner MA, Jackson N, Gerstner JR, Knutson KL. Dietary nutrients associated with short and long sleep duration. Data from a nationally representative sample. Appetite 2013;64:71–80, WOS:000317325500010.

[73] Grandner MA, Petrov MER, Rattanaumpawan P, Jackson N, Platt A, Patel NP. Sleep symptoms, race/ethnicity, and socioeconomic position. J Clin Sleep Med 2013;9(9):897–905, WOS:000324375100009. [English].

[74] Ohayon MM. Epidemiology of insomnia: what we know and what we still need to learn. Sleep Med Rev 2002;6(2):97–111, WOS:000176231600003.

[75] Roth T, Drake C. Evolution of insomnia: current status and future direction. Sleep Med 2004;5:S23–30, WOS:000222352700005. [English].

[76] Karacan I, Thornby JI, Anch M, Holzer CE, Warheit GJ, Schwab JJ, et al. Prevalence of sleep disturbance in a primarily Urban Florida County. Soc Sci Med 1976;10(5):239–44, WOS:A1976CB00600006.

[77] Blazer DG, Hays JC, Foley DJ. Sleep complaints in older adults—a racial comparison. J Gerontol A Biol Sci Med Sci 1995;50(5):M280–4, WOS:A1995RY60500019.

[78] Phillips B, Mannino D. Correlates of sleep complaints in adults: the ARIC study. J Clin Sleep Med 2005;1(3):277–83, MEDLINE:17566189.

[79] Grandner MA, Patel NP, Gehrman PR, Xie DW, Sha DH, Weaver T, et al. Who gets the best sleep? Ethnic and socioeconomic factors related to sleep complaints. Sleep Med 2010;11(5):470–8, WOS:000277878500008. [English].

[80] Paine SJ, Gander PH, Harris R, Reid P. Who reports insomnia? Relationships with age, sex, ethnicity, and socioeconomic deprivation. Sleep 2004;27(6):1163–9, WOS:000224445600017.

[81] Barile JP, Reeve BB, Smith AW, Zack MM, Mitchell SA, Kobau R, et al. Monitoring population health for Healthy People 2020: evaluation of the NIH PROMISA (R) Global Health, CDC Healthy Days, and satisfaction with life instruments. Qual Life Res 2013;22(6):1201–11, WOS:000322735700004.

[82] Kripke DF, Jean-Louis G, Elliott JA, Klauber MR, Rex KM, Tuunainen A, et al. Ethnicity, sleep, mood, and illumination in postmenopausal women. BMC Psychiatry 2004;4:8, 15070419. [eng].

[83] Jean-Louis G, Magai C, Consedine NS, Pierre-Louis J, Zizi F, Casimir GJ, et al. Insomnia symptoms and repressive coping in a sample of older Black and White women. BMC Womens Health 2007;7(1), MEDLINE:17261187.

[84] Ehlers CL, Wills DN, Lau P, Gilder DA. Sleep quality in an adult American Indian community sample. J Clin Sleep Med 2017;13(3):385–91, WOS:000397051100006. [English].

[85] Gellis LA, Lichstein KL, Scarinci IC, Durrence HH, Taylor DJ, Bush AJ, et al. Socioeconomic status and insomnia. J Abnorm Psychol 2005;114(1):111–8, WOS:000227146600011.

[86] Jean-Louis G, Magai C, Cohen C, Zizi F, von Gizycki H, DiPalma J, et al. Ethnic differencs in self reported sleep problems in older adults. Sleep 2001;926–33.

[87] Molnar M. Association of incident restless legs syndrome with outcomes in a large cohort of US veterans. J Sleep Res 2016;25(1):47–56.

[88] Chesson AL, Wise M, Davila D, Johnson S, Littner M, Anderson WM, et al. Practice parameters for the treatment of Restless Legs Syndrome and periodic limb movement disorder. Sleep 1999;22(7):961–8, WOS:000083566100015.

[89] Scofield H, Roth T, Drake C. Periodic limb movements during sleep: population prevalence, clinical correlates, and racial differences. Sleep 2008;31(9):1221–7, WOS:000258891100005.

[90] O'Brien LM, Koo J, Fan L, Owusu JT, Chotinaiwattarakul W, Felt BT, et al. Iron stores, periodic leg movements, and sleepiness in obstructive sleep apnea. J Clin Sleep Med 2009;5(6):525–31, WOS:000272780400006.

[91] Kutner NG, Zhang R, Huang YJ, Bliwise DL. Racial differences in restless legs symptoms and serum ferritin in an incident

[91] dialysis patient cohort. Int Urol Nephrol 2012;44(6):1825–31, WOS:000313523000030.
[92] Kutner NG, Bliwise DL. Restless legs complaint in African-American and Caucasian hemodialysis patients. Sleep Med 2002;3:497–500.
[93] Lee HB, Hening WA, Allen RP, Earley CJ, Eaton WW, Lyketsos CG. Race and restless legs syndrome symptoms in an adult community sample in east Baltimore. Sleep Med 2006;7(8):642–5, WOS:000243272500008.
[94] Sawanyawisuth K, Palinkas LA, Ancoli-Israel S, Dimsdale JE, Loredo JS. Ethnic differences in the prevalence and predictors of restless legs syndrome between Hispanics of Mexican descent and non-Hispanic Whites in San Diego county: a population-based study. J Clin Sleep Med 2013;9(1):47–53, 23319904. [eng].
[95] Allen RP. Race, iron status and restless legs syndrome. Sleep Med 2002;3(6):467–8, MEDLINE:14592139. [English].
[96] Koo BB. Restless leg syndrome across the globe: epidemiology of the restless legs syndrome/Willis-Ekbom disease. Sleep Med Clin 2015;10(3):189, WOS:000218432500003.
[97] Koo BB, Blackwell T, Ancoli-Israel S, Stone KL, Stefanick ML, Redline S, et al. Association of incident cardiovascular disease with periodic limb movements during sleep in older men outcomes of sleep disorders in older men (MrOS) study. Circulation 2011;124(11):1223–31, WOS:000294779000016.
[98] Solomon P. Narcolepsy in negroes. Dis Nerv Syst 1945;6:179–83.
[99] Okun ML, Lin L, Pelin Z, Hong S, Mignot E. Clinical aspects of narcolepsy-cataplexy across ethnic groups. Sleep 2002;25(1):27–35, 11833858. [eng].
[100] Longstreth WT, Ton TG, Koepsell T, Gersuk VH, Hendrickson A, Velde S. Prevalence of narcolepsy in King County, Washington, USA. Sleep Med 2009;10(4):422–6, 19013100. [Epub 2008/11/13. eng].
[101] Kawai M, O'Hara R, Einen M, Lin L, Mignot E. Narcolepsy in African Americans. Sleep 2015;38(11):1673–81, 26158891. [Epub 2015/11/01. eng].
[102] Jean-Louis G, Kripke DF, Ancoli-Israel S, Klauber MR, Sepulveda RS, Mowen MA, et al. Circadian sleep, illumination, and activity patterns in women: influences of aging and time reference. Physiol Behav 2000;68(3):347–52, 10716544. [eng].
[103] Eastman CI, Molina TA, Dziepak ME, Smith MR. Blacks (African Americans) have shorter free-running circadian periods than whites (Caucasian Americans). Chronobiol Int 2012;29(8):1072–7, WOS:000308654000010.
[104] Smith MR, Burgess HJ, Fogg LF, Eastman CI. Racial differences in the human endogenous circadian period. PLoS ONE 2009;4(6), WOS:000267515700001.
[105] Eastman CI, Suh C, Tomaka VA, Crowley SJ. Circadian rhythm phase shifts and endogenous free-running circadian period differ between African-Americans and European-Americans. Sci Rep 2015;5, WOS:000349240000005.
[106] Emens JS, Eastman CI. Diagnosis and treatment of non-24-h sleep-wake disorder in the blind. Drugs 2017;77(6):637–50, WOS:000398036100004.
[107] Boivin DB. Influence of sleep-wake and circadian rhythm disturbances in psychiatric disorders. J Psychiatry Neurosci 2000;25(5):446–58, WOS:000165383200004.
[108] Gu D, Sautter J, Pipkin R, Zeng Y. Sociodemographic and health correlates of sleep quality and duration among very old Chinese. Sleep 2010;33(5):601–10, 20469802. [Epub 2010/05/18. eng].
[109] Dinges DF, Douglas SD, Hamarman S, Zaugg L, Kapoor S. Sleep-deprivation and human immune function. Adv Neuroimmunol 1995;5(2):97–110, WOS:A1995RR29300002.
[110] Thomas SB, Quinn SC, Butler J, Fryer CS, Garza MA. Toward a fourth generation of disparities research to achieve health equity. Annu Rev Public Health 2011;32:399–416, WOS:000290776200022.
[111] Alcantara C, Patel SR, Carnethon M, Castaneda S, Isasi CR, Davis S, et al. Stress and sleep: results from the Hispanic Community Health Study/Study of Latinos Sociocultural Ancillary Study. SSM Popul Health 2017;3:713–21, MEDLINE:29104908. [English].
[112] Slopen N, Williams DR. Discrimination, other psychosocial stressors, and self-reported sleep duration and difficulties. Sleep 2014;37(1):147–56, 24381373. [eng].
[113] Brondolo E, ver Halen NB, Pencille M, Beatty D, Contrada RJ. Coping with racism: a selective review of the literature and a theoretical and methodological critique. J Behav Med 2009;32(1):64–88, WOS:000262434000005.
[114] Williams NJ, Nuru-Jeter A. Does raical identity moderate the association between racial discrimination and sleep quality? Psychosom Med 2017;79(4):A81, WOS:000401250500216.
[115] Krieger N. Embodying inequality: a review of concepts, measures, and methods for studying health consequences of discrimination. Int J Health Serv 1999;29(2):295–352, WOS:000080804700004.
[116] Williams DR, Williams-Morris R. Racism and mental health: the African American experience. Ethn Health 2000;5(3–4):243–68, WOS:000168231000006.
[117] Harrell JP, Hall S, Taliaferro J. Physiological responses to racism and discrimination: an assessment of the evidence. Am J Public Health 2003;93(2):243–8, WOS:000180721000017.
[118] Acheson LS, Wang C, Zyzanski SJ, Lynn A, Ruffin MT, Gramling R, et al. Family history and perceptions about risk and prevention for chronic diseases in primary care: a report from the family healthware impact trial. Genet Med 2010;12(4):212–8, 20216073. [eng].
[119] Lipkus IM, Skinner CS, Dement J, Pompeii L, Moser B, Samsa GP, et al. Increasing colorectal cancer screening among individuals in the carpentry trade: test of risk communication interventions. Prev Med 2005;40(5):489–501, 15749130. [eng].
[120] Leventhal H, Kelly K, Leventhal EA. Population risk, actual risk, perceived risk, and cancer control: a discussion. J Natl Cancer Inst Monogr 1999;(25)81–5, 10854461. [eng].
[121] Quinn SC, Jamison A, Freimuth VS, An J, Hancock GR, Musa D. Exploring racial influences on flu vaccine attitudes and behavior: results of a national survey of White and African American adults. Vaccine 2017;35(8):1167–74, 28126202. [Epub 2017/01/17. eng].
[122] Smith MT, Perlis ML. Who is a candidate for cognitive-behavioral therapy for insomnia? Health Psychol 2006;25(1):15–9, WOS:000235123300003.
[123] Belfer S, Perlis M, Kayser M. A neurobiological basis for behavioral therapy using Drosophila. Biol Psychiatry 2017;81(10):S69, WOS:000400348700168.
[124] Jean-Louis G, Newsome V, Williams NJ, Zizi F, Ravenell J, Ogedegbe G. Tailored behavioral intervention among Blacks with metabolic syndrome and sleep apnea: results of the MetSO trial. Sleep 2017;40(1), WOS:000394125700008.
[125] Cukor D, Pencille M, Ver Halen N, Primus N, Gordon-Peters V, Fraser M, et al. An RCT comparing remotely delivered adherence promotion for sleep apnea assessment against an information control in a black community sample. Sleep Health 2018;4(4):369–76. WOS:000439075900011.

Chapter 7

Neighborhood factors associated with sleep health

Lauren Hale[a], Sarah James[b], Qian Xiao[c], Martha E. Billings[d], Dayna A. Johnson[e]

[a]Program in Public Health, Department of Family, Population, and Preventive Medicine, Stony Brook University School of Medicine, Stony Brook, NY, United States, [b]Department of Sociology and Office of Population Research, Princeton University, Princeton, NJ, United States, [c]Department of Health and Human Physiology and Department of Epidemiology, University of Iowa, Iowa City, IA, United States, [d]Division of Pulmonary, Critical Care & Sleep Medicine, University of Washington, Seattle, WA, United States, [e]Division of Sleep and Circadian Disorders, Brigham and Women's Hospital and Harvard Medical School, Boston, MA, United States

NEIGHBORHOODS AND SLEEP HEALTH

Sleep, a modifiable health behavior, is increasingly recognized as integral for optimal health and well-being [1,2]. One-third of Americans report obtaining <7h of sleep: an insufficient amount according to expert consensus panels [3,4]. Highly prevalent sleep disorders such as sleep apnea and insomnia [5,6] are underdiagnosed and a pressing public health burden [7]. Additionally, there are substantial disparities in sleep health; insufficient sleep and unrecognized and undertreated sleep disorders are highly prevalent among racial/ethnic minorities and lower socioeconomic status populations who disproportionately reside in under-resourced neighborhoods [8,9]. In this chapter, we first present a theoretical justification for the link between neighborhoods and sleep health, review current literature on the neighborhood determinants of sleep among children and adolescents, followed by a separate section on neighborhoods and sleep among adults. We conclude with opportunities and challenges for advancing the research on neighborhoods and sleep health and their implications for developing interventions and reducing health disparities.

THEORETICAL JUSTIFICATION FOR NEIGHBORHOODS AND SLEEP HEALTH

As part of the emerging literature on the social determinants of sleep health, one active line of research investigates the neighborhood factors that are associated with sleep health across the life course [10]. The high prevalence of poor sleep health, particularly among vulnerable populations, is associated with neighborhood features such as noise disturbances, crime, crowding, excess light, and social isolation; these same factors are also associated with poor health outcomes [11,12]. The theoretical rationale for such an association is rooted in an evolutionary understanding of sleep as being highly contextually dependent [13]. When an external threat puts a sleeping individual at risk, we can expect sleep to be affected through a reduction in sleep quality or duration to minimize the time vulnerable to threats [14,15]. Thus, the study of sleep health must embrace a socioecological model in which neighborhood factors are a key component [2].

A growing number of studies have evaluated the association between neighborhood factors and sleep outcomes [10–12,16–22]. As with other neighborhood research, associations between neighborhoods and an outcome of interest may be due to causal processes or due simply to compositional differences across the neighborhoods. Insufficient sleep may be a cause of poor health outcomes observed among residents of disadvantaged neighborhoods [18,23,24]. Previous research has identified key environmental factors that may be causally linked to sleep quality and duration, including (1) safety concerns, (2) neighborhood socioeconomic status (NSES), (3) noise, (4) temperature, (5) neighborhood disorder, (6) pollution, and (7) cultural factors. For example, the regional clustering of insufficient sleep, such as in American Appalachia, may be related to poor health behaviors, reduced access to health care, and/or economic disparity [25]. In contrast, a compositional explanation for the association between neighborhood characteristics and sleep means that individuals with less or poorer quality sleep are more likely to live in the same neighborhoods, and it is not the features of the neighborhoods themselves that contribute to the sleep outcomes. That is, there may be collinearity of disadvantage with other causal environmental factors contributing to sleep health (such as pollution, noise, and disorder, or other items listed before) [26,27]. Therefore, in this chapter, we limit our discussion to articles that adjust for individual risk factors to minimize the role of compositional explanations for the associations between neighborhood factors and sleep outcomes.

NEIGHBORHOOD FACTORS ASSOCIATED WITH PEDIATRIC SLEEP

Research on the association between neighborhood factors and pediatric sleep health has shown that a variety of neighborhood factors are associated with child and adolescent sleep outcomes. These factors include urbanicity and neighborhood density, neighborhood disadvantage, low walkability, and neighborhood violence.

Urbanicity and population density

Infants, children and adolescents living in more urban areas and/or areas with higher population density have shorter sleep durations [28,29], higher odds of inadequate sleep [29,30], and higher rates of obstructive sleep apnea [31] than do children living in less urban or dense areas.

Neighborhood socioeconomic status (NSES)

Low NSES and related measures of neighborhood disadvantage are consistently associated with worse sleep health among both children and adolescents. Children and adolescents living in disadvantaged neighborhoods have shorter nightly sleep durations [32,33] and greater odds of inadequate sleep [30] than do children and adolescents living in more advantaged neighborhoods. Notably, one recent study of children ages 5–10 years living in urban California counties found that children living in neighborhoods with 40-year histories of consistently high poverty have higher odds of inadequate sleep than children living in neighborhoods with consistently low- or moderate-poverty trajectories [34]; the same study finds that current neighborhood poverty is not associated with sleep adequacy [34]. Additionally, adolescents living in more disadvantaged neighborhoods have more variable sleep times [35] and more sleep problems [36,37] than those living in more advantaged neighborhoods. Finally, obstructive sleep apnea is also substantially more common [31,38] and the sleep apnea severity is greater [39] among children living in disadvantaged neighborhoods compared to more advantaged areas.

Neighborhood access to physical activity

The association between neighborhoods and sleep may partially operate through the promotion of physical activity, which is necessary for good sleep. Using data from the National Survey of Children's Health, including waves from 2003, 2007, and 2011–2012, Singh and Kenney [30] found that children ages 6–17 years residing in neighborhoods with fewer amenities—such as a lack of parks/playgrounds, recreation/community center, or access to a library/bookmobile—had higher odds of inadequate sleep than their peers living in neighborhoods with these amenities. A smaller study of adolescents living in the Southeast found that recreation facilities located closer to the adolescent's home lead to higher physical activity, which in turn predicted more daily sleep minutes, better-quality sleep, and less variability in sleep schedules [40]. Relatedly, sleep is more variable for adolescents living on busier streets [35].

Neighborhood violence and safety concerns

Exposure to violence within the neighborhood may also impact sleep. Concerns about violence and crime or exposure to community violence are associated with a range of sleep problems and inadequate sleep among children and adolescents [36,37,41–47]. Similarly, concerns about school and community violence are associated with poorer sleep quality in a sample of adolescents living in the Southeastern United States, with stronger associations for girls than boys. In less violent settings, girls slept longer each night than boys but in violent contexts there was no sex difference in nightly sleep duration [45]. Finally, acute exposure to violent events is associated with delayed sleep timing and shorter duration [48]. Using a rigorous within-person study design in a small sample of adolescents, Heissel et al. [48] found that adolescents go to sleep 30 min later and sleep for 39 min less on the night after a violent crime occurred within half a mile of their home.

NEIGHBORHOOD FACTORS ASSOCIATED WITH ADULT SLEEP

Social characteristics of the neighborhood environment (e.g., social cohesion, safety, violence, disorder) are associated with sleep duration, daytime sleepiness, sleep difficulties and a sleep quality among adults [12,17,18,20,49,50]. Adverse neighborhood social environments—those low in social cohesion and high in violence and disorder—are associated with sleeping between 7 and 11 min less per night on average, after adjustment for age and sex [12,50,51]. Perceived neighborhood safety and social cohesion are associated with both self-reported and objectively measured sleep duration, with longer sleep in safer and more cohesive neighborhoods [12,50]. Neighborhood features are also associated with common adult sleep problems as detailed as follows: inadequate sleep duration and sleep timing, insomnia, and obstructive sleep apnea.

Inadequate sleep duration and delayed sleep timing

Physical neighborhood features such as artificial light, vehicular traffic and noise related to crowding all impact sleep. Traffic (including air, road, and rail), and other urban noise (such as that of alarms, construction, sirens, etc.) can lead to sleep fragmentation, delay sleep onset or contribute to early awakenings. Bright light exposure from street lights, houses, business and commercial space can similarly impact sleep timing, typically delaying sleep onset (circadian phase delay) [52–59]. Excess artificial light may depress melatonin secretion, which impacts the initiation of sleep by causing circadian phase delay and prolonging sleep latency [60]. In a US study, those with greater nighttime exposure to outdoor lights had a 28% greater odds of a circadian phase delay. Similarly, those living in areas that are brighter at night (typically dense cities) have a later bedtime [61]. Thus, city dwellers often sleep less than their rural counterparts as a result of these physical features of urban neighborhoods.

Other features of the built neighborhood environment, such as walkability, green space, density, street connectivity, and mixed land use, may also impact sleep. Observational studies show that adults living in neighborhoods with more green space or natural water features have a lower likelihood of insufficient sleep [62,63]. However, data from the Multi-Ethnic Study of Atherosclerosis showed that living in neighborhoods with higher street smart walk scores, more social engagement destinations, street intersections, and population density are associated with 17%–23% higher odds of short sleep duration (≤6h) [64]. This finding demonstrates the complex relation of the built environment with health—while built environment features may be favorable for physical activity [65,66] simultaneously there may be a cost to sleep opportunity.

Insomnia

An estimated 10% of adults suffer from chronic insomnia with 35% of the adult population experiencing insomnia symptoms annually. Insomnia, a clinical diagnosis, is characterized as difficulty initiating and/or maintaining sleep, awakening too early, with a resulting daytime impairment [67]. Contextual features of the neighborhood likely contribute to insomnia. Living in disadvantaged neighborhoods is associated with insomnia symptoms [9,12,19,68]. Objective measures of insomnia such as a greater period of wake after sleep onset are also associated with neighborhood disadvantage [69]. One possible mechanism underlying the association of neighborhood features with insomnia symptoms may be that crime, noise, disorder promote hypervigilance and lead to increased trouble falling asleep, staying asleep, among other sleep disturbances [11,23]. Neighborhood physical disorder and low social cohesion are associated with greater odds of difficulty falling asleep among older adults [70]. Neighborhood disadvantage may also have an indirect association with sleep through increased psychosocial distress, which is associated with insufficient sleep [71,72]. Residing in an adverse neighborhood environment, fear of crime and violence, discrimination, and/or social disorganization may increase anxiety or depression, which may lead to dysregulation of the hypothalamic-pituitary-adrenal (HPA) axis, impacting biological rhythms and mood [73].

Neighborhood light and noise pollution can also foster insomnia in susceptible urbanites. A study in Oslo, Norway found a 5% greater odds of difficulty falling asleep and too early awakenings per 5dB increase in traffic noise. Loud noises from trucks, trains, planes, sirens, and highways—all sounds related to a high population density—also disrupt sleep and may lead to insomnia symptoms [53,54,74,75]. For example, noisy neighborhoods were associated with a 4% greater prevalence of insomnia symptoms in a US national epidemiological study of Hispanics and Latinos [19]. In the elderly, artificial light exposure at night can decrease melatonin secretion and lead to increased objective sleep disturbance and subjective insomnia [76].

Obstructive sleep apnea (OSA)

Neighborhood features which promote obesity, sedentary behaviors and metabolic disease [77,78] likely increase the risk of OSA. Sleep apnea is highly correlated with obesity, with greater prevalence and severity among the morbidly obese [6]. Neighborhood built characteristics associated with body mass index (BMI) and physical activity levels include walkability, access to healthy food, recreation, street connectivity and green spaces [65,66,79–81]. Living in neighborhoods with lower-rated walking environments is associated with a greater severity of OSA, with stronger associations in persons with obesity [16]. Neighborhood crowding is also associated with OSA, and BMI partially mediates the association [82].

Neighborhood physical features such as traffic and ambient air pollution are associated with OSA. Individuals exposed to higher levels of ozone and particulate matter have a greater severity of OSA, particularly in the summer [83–85]. In a recent study in Taiwan, higher exposure to traffic pollution is associated with a 4%–5% greater OSA severity. Similarly, proximity to traffic is associated with greater OSA symptoms such as snoring and daytime sleepiness, possibly through noise and air pollution mechanisms [86,87]. An adverse physical environment with greater inflammatory irritants may also increase OSA propensity [88,89].

CURRENT LIMITATIONS AND FUTURE DIRECTIONS

Though recent research has generated a growing body of evidence supporting an important role of neighborhood environment in sleep health, we have identified several gaps in the current literature. Specifically, the field would benefit from more studies that (1) characterize long-term neighborhood conditions, (2) evaluate evidence from quasi- or natural experiments that use statistical methods to strengthen the argument for causality, and (3) use technological advances to objectively measure neighborhood characteristics at a larger spatial scale.

Studying long-term trajectories of neighborhood conditions and sleep

The vast majority of the literature on neighborhood factors associated with pediatric and adult sleep health examined neighborhood conditions at a single time point. Yet neighborhoods are not static and people are mobile, both of which may lead to changes in neighborhood environment that may impact sleep behaviors. Similarly, sleep health may be differently influenced by exposure to neighborhood factors at different points in the lifespan (such as during childhood). However, only a few studies have sought to understand the impact of long-term exposure patterns to neighborhood conditions on sleep. In one study, compared to neighborhoods with historically high poverty, neighborhoods that showed upward mobility over one decade were associated with lower odds of insufficient sleep [90]. In another study of middle-to-older aged adults, a decrease in NSES was associated with very short sleep (<5h) in women; while an improvement in NSES was associated with long sleep (≥9h) in men [91]. Additional studies with this longitudinal approach could help better characterize neighborhood-related sleep disparities in the population and identify vulnerable groups that are at a high risk of adverse health outcomes related to sleep deficiency.

Evaluating evidence from natural experiments and other causal methods

Most of the empirical studies of neighborhood factors and sleep health have relied on observational data. However, observational studies are plagued by residual confounding from individual backgrounds, which limit their ability to make causal inferences [92]. Due to the scarcity of interventional studies that change neighborhood conditions, quasi-experimental studies take advantage of public policy, funding and physical environmental changes to examine the impact of neighborhood factors on health behaviors and related outcomes. Many such studies have provided valuable insight about designing interventions that aim at improving environmental conditions to reduce health disparities related to physical activity and nutrition [93]. Using a similar approach, the effect of neighborhood factors on sleep could be more accurately assessed. In addition, some studies use methods designed to approximate causal estimates, such as propensity score analysis, marginal structural models, and fixed effect models, to examine neighborhood conditions

and health behaviors and outcomes longitudinally [94–97]. The application of such methods in the research of neighborhood and sleep is limited, with the notable exception of a recent paper that examined neighborhood disorders and sleep problems in older adults using fixed-effect models to control for confounding of unmeasured personal traits [98]. Future research on the neighborhood determinants of sleep health would be strengthened by evaluating evidence from experimentally designed studies and the use of more rigorous statistical methods.

Using technological advances to studying neighborhoods and sleep at a larger scale

Though recent studies have taken advantage of satellite imagery, national exposure maps, and other technologies to investigate environmental exposures associated with health outcomes [99–101], these data have been underutilized to understand sleep health. Because many population-based studies have collected both sleep data and participants' addresses, more large-scale epidemiological studies could link information such as street connectivity, land use, vegetation, environmental pollutants, and outdoor light and noise exposures to participants' neighborhoods [102]. For example, two recent studies examined satellite measurements of nighttime outdoor artificial light as predictors of sleep health variables and found that a higher level of outdoor light at night was associated with insufficient sleep in Korean adults [103], and a stronger evening-type orientation in adolescents in Germany [104]. In addition, widespread use of commercial sleep tracking devices may allow objective assessments of sleep on a large-scale population level [105,106]. More studies are needed to link sleep data with large-scale exposure databases, and such linkages will provide novel means of assessing the environmental determinants of sleep in the population.

ARE THERE INTERVENTIONS AND POLICIES TO IMPROVE NEIGHBORHOODS AND SLEEP HEALTH?

Substantial evidence demonstrates that adverse physical and social neighborhood environments negatively impact sleep health and likely contribute to sleep health disparities. Thus, there is a clear need for community interventions and policies to improve neighborhood conditions and promote healthy sleep. Governmental agencies (e.g. World Health Organization, US Centers for Disease Control and Prevention) have highlighted the need to improve housing and neighborhood conditions as a strategy to improve health and address health disparities [107–111]. Although limited, there are examples in the sleep research literature that have demonstrated that interventions which consider the neighborhood environment or the neighborhood environment in combination with household factors may improve sleep outcomes. A randomized controlled trial of households in five communities in New Zealand found that children in the intervention group (installation of nonpolluting, more effective home heater before winter) had less sleep disturbed by wheezing than children in the control group [112]. Another in the Peruvian Andes showed reduced sleep apnea symptoms with less indoor air pollution by modifying biomass exposure [113]. Cross-sectional data have shown that access to neighborhood green space is associated with a lower risk of short sleep [62,114]; therefore policies that promote green spaces may also promote healthy sleep. As a result of home and neighborhood structures, exposure to daylight may be limited, thus interventions that include light boxes in the home can be efficacious in community settings for improving sleep [115,116]. Also, interventions that target walking outside can also have a positive effect on sleep. For example, results of a small randomized controlled trial among individuals with Alzheimer's disease ($n=132$) showed that a combination of walking and light exposure were effective treatments for improving sleep in terms of reduced actigraphic total wake time at night [115]. However, for neighborhood interventions to be successful, walkable and safe neighborhoods are necessary. Racial/ethnic minorities and lower SES populations disproportionately reside in adverse neighborhoods [117], which may impact the uptake of neighborhood-sleep interventions. Based on evidence that neighborhood factors are more adverse for the sleep health of minority populations [30,33,50,64,69,118], there is a clear need for policy that will increase safety and development in the neighborhood that promotes physical activity, social cohesion, and improved esthetics. Such policy changes could potentially address sleep health disparities. Further, interventions that target changes in the environment to promote healthy sleep should be developed, tested, and evaluated as possible pathways for ameliorating sleep health disparities and subsequently health disparities. Interventions to improve sleep should target the home sleeping environment (e.g., shades, thermal comfort, smart home lighting), the physical neighborhood (e.g., traffic, light, noise, and pollution reduction; access to parks), and the social atmosphere neighborhood (e.g. improved neighborhood safety and social cohesion). Targeting these salient neighborhood factors will help to identify priorities for public health intervention and policies. Improving the neighborhood environment has the potential to improve sleep health as well as population health.

CONCLUSIONS AND PUBLIC HEALTH SIGNIFICANCE

Identifying the mechanisms underlying the association between neighborhood characteristics and sleep health may provide opportunities to reduce population health disparities, as adequate and high-quality sleep are fundamental to physical and mental health. To further advance this field, rigorous experimental studies with urban planning and policy interventions are necessary to confirm how neighborhood context might be modified to improve sleep health and reduce overall health disparities.

ACKNOWLEDGMENTS

Funding: Research reported in this publication was supported by the Eunice Kennedy Shriver National Institute for Child Health and Human Development (NICHD) R01 HD073352 (to LH), The Eunice Kennedy Shriver National Institute of Child Health & Human Development of the National Institutes of Health under Award Number P2CHD047879 (for SJ) and by the National Heart, Lung, and Blood Institute, (NHLBI) K01HL138211 (to DAJ).

REFERENCES

[1] Czeisler CA. Duration, timing and quality of sleep are each vital for health, performance and safety. Sleep Health 2015;1(1):5–8.

[2] Grandner MA, Hale L, Moore M, Patel NP. Mortality associated with short sleep duration: the evidence, the possible mechanisms, and the future. Sleep Med Rev 2010;14(3):191–203.

[3] Watson NF, Badr MS, Belenky G, et al. Recommended amount of sleep for a healthy adult: a joint consensus statement of the American Academy of Sleep Medicine and Sleep Research Society. Sleep 2015;38(6):843–4.

[4] Hirshkowitz M, Whiton K, Albert SM, et al. National Sleep Foundation's updated sleep duration recommendations: final report. Sleep Health 2015;1(4):233–43.

[5] Roth T. Insomnia: definition, prevalence, etiology, and consequences. J Clin Sleep Med 2007;3(5 Suppl):S7–10.

[6] Peppard PE, Young T, Barnet JH, Palta M, Hagen EW, Hla KM. Increased prevalence of sleep-disordered breathing in adults. Am J Epidemiol 2013;177:1006–14.

[7] Rosen RC, Zozula R, Jahn EG, Carson JL. Low rates of recognition of sleep disorders in primary care: comparison of a community-based versus clinical academic setting. Sleep Med 2001;2(1):47–55.

[8] Grandner MA, Williams NJ, Knutson KL, Roberts D, Jean-Louis G. Sleep disparity, race/ethnicity, and socioeconomic position. Sleep Med 2016;18:7–18.

[9] Patel NP, Grandner MA, Xie D, Branas CC, Gooneratne N. "Sleep disparity" in the population: poor sleep quality is strongly associated with poverty and ethnicity. BMC Public Health 2010;10:475.

[10] Hale L, Emanuele E, James S. Recent updates in the social and environmental determinants of sleep health. Curr Sleep Med Rep 2015;1(4):212–7.

[11] Hill TD, Trinh HN, Wen M, Hale L. Perceived neighborhood safety and sleep quality: a global analysis of six countries. Sleep Med 2016;18:56–60.

[12] Desantis AS, Diez Roux AV, Moore K, Baron KG, Mujahid MS, Nieto FJ. Associations of neighborhood characteristics with sleep timing and quality: The multi-ethnic study of atherosclerosis. Sleep 2013;36(10):1543–51.

[13] Nunn CL, Samson DR, Krystal AD. Shining evolutionary light on human sleep and sleep disorders. Evol Med Public Health 2016;2016(1):227–43.

[14] Samson DR, Crittenden AN, Mabulla IA, Mabulla AZP, Nunn CL. Chronotype variation drives night-time sentinel-like behaviour in hunter-gatherers. Proc Biol Sci 2017;284(1858).
[15] Dahl RE. The regulation of sleep and arousal: Development and psychopathology. Dev Psychopathol 1996;8(1):3–27.
[16] Billings ME, Johnson DA, Simonelli G, et al. Neighborhood walking environment and activity level are associated with OSA: the multi-ethnic study of atherosclerosis. Chest 2016;150(5):1042–9.
[17] Johnson DA, Brown DL, Morgenstern LB, Meurer WJ, Lisabeth LD. The association of neighborhood characteristics with sleep duration and daytime sleepiness. Sleep Health 2015;1:148–55.
[18] Hale L, Hill TD, Friedman E, et al. Perceived neighborhood quality, sleep quality, and health status: evidence from the Survey of the Health of Wisconsin. Soc Sci Med 2013;79:16–22.
[19] Simonelli G, Dudley KA, Weng J, et al. Neighborhood factors as predictors of poor sleep in the Sueno Ancillary Study of the Hispanic Community Health Study/Study of Latinos. Sleep 2017;40(1).
[20] Bassett E, Moore S. Neighbourhood disadvantage, network capital and restless sleep: is the association moderated by gender in urban-dwelling adults? Soc Sci Med 2014;108:185–93.
[21] Chambers EC, Pichardo MS, Rosenbaum E. Sleep and the housing and neighborhood environment of urban Latino adults living in low-income housing: The AHOME study. Behav Sleep Med 2016;14(2):169–84.
[22] Simonelli G, Patel SR, Rodriguez-Espinola S, et al. The impact of home safety on sleep in a Latin American country. Sleep Health 2015;1(2):98–103.
[23] Hale L, Hill TD, Burdette AM. Does sleep quality mediate the association between neighborhood disorder and self-rated physical health? Prev Med 2010;51(3–4):275–8.
[24] Curtis DS, Fuller-Rowell TE, El-Sheikh M, Carnethon MR, Ryff CD. Habitual sleep as a contributor to racial differences in cardiometabolic risk. Proc Natl Acad Sci U S A 2017;114(33):8889–94.
[25] Grandner MA, Smith TE, Jackson N, Jackson T, Burgard S, Branas C. Geographic distribution of insufficient sleep across the United States: a county-level hotspot analysis☆. Sleep Health 2015;1(3):158–65.
[26] Hajat A, Diez-Roux AV, Adar SD, et al. Air pollution and individual and neighborhood socioeconomic status: evidence from the multi-ethnic study of atherosclerosis (MESA). Environ Health Perspect 2013;121(11–12):1325–33.
[27] Ross CE, Mirowsky J. Neighborhood disadvantage, disorder, and health. J Health Soc Behav 2001;42(3):258–76.
[28] Bottino CJ, Rifas-Shiman SL, Kleinman KP, et al. The association of urbanicity with infant sleep duration. Health Place 2012;18(5):1000–5.
[29] Patte KA, Qian W, Leatherdale ST. Sleep duration trends and trajectories among youth in the COMPASS study. Sleep Health 2017;3(5):309–16.
[30] Singh GK, Kenney MK. Rising prevalence and neighborhood, social, and behavioral determinants of sleep problems in US children and adolescents, 2003–2012. Sleep Disord 2013;2013:394320.
[31] Brouillette RT, Horwood L, Constantin E, Brown K, Ross NA. Childhood sleep apnea and neighborhood disadvantage. J Pediatr 2011;158(5):789–95. [e781].
[32] McLaughlin Crabtree V, Beal Korhonen J, Montgomery-Downs HE, Faye Jones V, O'Brien LM, Gozal D. Cultural influences on the bedtime behaviors of young children. Sleep Med 2005;6(4):319–24.
[33] Bagley EJ, Fuller-Rowell TE, Saini EK, Philbrook LE, El-Sheikh M. Neighborhood economic deprivation and social fragmentation: Associations with Children's sleep. Behav Sleep Med 2016;1–13.
[34] Sheehan C, Powers D, Margerison-Zilko C, McDevitt T, Cubbin C. Historical neighborhood poverty trajectories and child sleep. Sleep Health 2018;4(2):127–34.
[35] Marco CA, Wolfson AR, Sparling M, Azuaje A. Family socioeconomic status and sleep patterns of young adolescents. Behav Sleep Med 2011;10(1):70–80.
[36] Rubens SL, Gudino OG, Fite PJ, Grande JM. Individual and neighborhood stressors, sleep problems, and symptoms of anxiety and depression among Latino youth. Am J Orthopsychiatry 2018;88(2):161–8.
[37] Rubens SL, Fite PJ, Cooley JL, Canter KS. The role of sleep in the relation between community violence exposure and delinquency among Latino adolescence. J Community Psychol 2014;42(6):723–34.
[38] Spilsbury JC, Storfer-Isser A, Kirchner HL, et al. Neighborhood disadvantage as a risk factor for pediatric obstructive sleep apnea. J Pediatr 2006;149(3):342–7.
[39] Wang R, Dong Y, Weng J, et al. Associations among neighborhood, race, and sleep apnea severity in children. A Six-City analysis. Ann Am Thorac Soc 2017;14(1):76–84.
[40] Philbrook LE, El-Sheikh M. Associations between neighborhood context, physical activity, and sleep in adolescnets. Sleep Health 2016;2(3):205–10.
[41] Kliewer W, Lepore SJ. Exposure to violence, social cognitive processing, and sleep problems in urban adolescents. J Youth Adolesc 2015;44(2):507–17.
[42] Umlauf MG, Bolland JM, Lian BE. Sleep disturbance and risk behaviors among inner-city African-American adolescents. J Urban Health 2011;88(6):1130–42.
[43] Umlauf MG, Bolland AC, Bolland KA, Tomek S, Bolland JM. The effects of age, gender, hopelessness, and exposure to violence on sleep disorder symptoms and daytime sleepiness among adolescents in impoverished neighborhoods. J Youth Adolesc 2015;44(2):518–42.
[44] McHale SM, Kim JY, Kan M, Updegraff KA. Sleep in Mexican-American adolescents: social ecological and well-being correlates. J Youth Adolesc 2011;40(6):666–79.
[45] Bagley EJ, Tu KM, Buckhalt JA, El-Sheikh M. Community violence concerns and adolescent sleep. Sleep Health 2016;2(1):57–62.
[46] Smaldone A, Honig JC, Byrne MW. Sleepless in America: inadequate sleep and relationships to health and well-being of our nation's children. Pediatrics 2007;119(Suppl 1):S29–37.
[47] Cooley-Quille M, Lorion R. Adolescents exposure to community violence: Sleep and psychophysiological functioning. J Community Psychol 1999;27(4):367–75.
[48] Heissel JA, Sharkey PT, Torrats-Espinosa G, Grant K, Adam EK. Violence and vigilance: the acute effects of community violent crime on sleep and cortisol. Child Dev 2017;89(4):e323–31.
[49] Hill TD, Burdette AM, Hale L. Neighborhood disorder, sleep quality, and psychological distress: testing a model of structural amplification. Health Place 2009;15(4):1006–13.
[50] Johnson DA, Simonelli G, Moore K, et al. The neighborhood social environment and objective measures of sleep in the multi-ethnic study of atherosclerosis. Sleep 2017;40(1).
[51] Johnson DA, Lisabeth L, Hickson D, et al. The social patterning of sleep in African Americans: Associations of socioeconomic position and neighborhood characteristics with sleep in the Jackson Heart Study. Sleep 2016;39(9):1749–59.

[52] Basner M, Brink M, Elmenhorst EM. Critical appraisal of methods for the assessment of noise effects on sleep. Noise Health 2012;14(61):321–9.

[53] Basner M, Muller U, Elmenhorst EM. Single and combined effects of air, road, and rail traffic noise on sleep and recuperation. Sleep 2011;34(1):11–23.

[54] Halonen JI, Vahtera J, Stansfeld S, et al. Associations between Nighttime Traffic Noise and Sleep: The Finnish Public Sector Study. Environ Health Perspect 2012;120:1391–6.

[55] Hume KI, Brink M, Basner M. Effects of environmental noise on sleep. Noise Health 2012;14(61):297–302.

[56] Muzet A. Environmental noise, sleep and health. Sleep Med Rev 2007;11(2):135–42.

[57] Pirrera S, De Valck E, Cluydts R. Nocturnal road traffic noise: a review on its assessment and consequences on sleep and health. Environ Int 2010;36(5):492–8.

[58] Pirrera S, De Valck E, Cluydts R. Field study on the impact of nocturnal road traffic noise on sleep: the importance of in- and outdoor noise assessment, the bedroom location and nighttime noise disturbances. Sci Total Environ 2014;500:84–90.

[59] Perron S, Plante C, Ragettli MS, Kaiser DJ, Goudreau S, Smargiassi A. Sleep disturbance from road traffic, railways, airplanes and from total environmental noise levels in montreal. Int J Environ Res Public Health 2016;13(8).

[60] Cho Y, Ryu SH, Lee BR, Kim KH, Lee E, Choi J. Effects of artificial light at night on human health: a literature review of observational and experimental studies applied to exposure assessment. Chronobiol Int 2015;32(9):1294–310.

[61] Ohayon MM, Milesi C. Artificial outdoor nighttime lights associate with altered sleep behavior in the American general population. Sleep 2016;39(6):1311–20.

[62] Grigsby-Toussaint DS, Turi KN, Krupa M, Williams NJ, Pandi-Perumal SR, Jean-Louis G. Sleep insufficiency and the natural environment: Results from the US behavioral risk factor surveillance system survey. Prev Med 2015;78:78–84.

[63] Bodin T, Bjork J, Ardo J, Albin M. Annoyance, sleep and concentration problems due to combined traffic noise and the benefit of quiet side. Int J Environ Res Public Health 2015;12(2):1612–28.

[64] Johnson DA, Hirsh JA, Moore K, Redline S, Diez Roux AV. Associations between the built environment and objective measures of sleep: The multi-ethnic study of atherosclerosis (MESA). Am J Epidemiol 2017;.

[65] Lovasi GS, Hutson MA, Guerra M, Neckerman KM. Built environments and obesity in disadvantaged populations. Epidemiol Rev 2009;31:7–20.

[66] Papas MA, Alberg AJ, Ewing R, Helzlsouer KJ, Gary TL, Klassen AC. The built environment and obesity. Epidemiol Rev 2007;29:129–43.

[67] Sateia MJ. International classification of sleep disorders-third edition: Highlights and modifications. Chest 2014;146(5):1387–94.

[68] Riedel N, Fuks K, Hoffmann B, et al. Insomnia and urban neighbourhood contexts—Are associations modified by individual social characteristics and change of residence? Results from a population-based study using residential histories. BMC Public Health 2012;12:810.

[69] Fuller-Rowell TE, Curtis DS, El-Sheikh M, Chae DH, Boylan JM, Ryff CD. Racial disparities in sleep: the role of neighborhood disadvantage. Sleep Med 2016;27–28:1–8.

[70] Chen-Edinboro LP, Kaufmann CN, Augustinavicius JL, et al. Neighborhood physical disorder, social cohesion, and insomnia: Results from participants over age 50 in the health and retirement study. Int Psychogeriatr 2014;1–8.

[71] Akerstedt T. Psychosocial stress and impaired sleep. Scand J Work Environ Health 2006;32(6):493–501.

[72] Johnson DA, Lisabeth L, Lewis TT, et al. The contribution of psychosocial stressors to sleep among African Americans in the Jackson Heart Study. Sleep 2016;39(7):1411–9.

[73] Hirotsu C, Tufik S, Andersen ML. Interactions between sleep, stress, and metabolism: from physiological to pathological conditions. Sleep Sci 2015;8(3):143–52.

[74] Evandt J, Oftedal B, Hjertager Krog N, Nafstad P, Schwarze PE, Marit AG. A population-based study on nighttime road traffic noise and insomnia. Sleep 2017;40(2).

[75] Kim M, Chang SI, Seong JC, et al. Road traffic noise: annoyance, sleep disturbance, and public health implications. Am J Prev Med 2012;43(4):353–60.

[76] Obayashi K, Saeki K, Kurumatani N. Association between light exposure at night and insomnia in the general elderly population: the HEIJO-KYO cohort. Chronobiol Int 2014;31(9):976–82.

[77] Diez Roux AV, Mair C. Neighborhoods and health. Ann N Y Acad Sci 2010;1186:125–45.

[78] Christine PJ, Auchincloss AH, Bertoni AG, et al. Longitudinal associations between neighborhood physical and social environments and incident type 2 diabetes mellitus: the multi-ethnic study of atherosclerosis (MESA). JAMA Intern Med 2015;.

[79] Auchincloss AH, Mujahid MS, Shen M, Michos ED, Whitt-Glover MC, Diez Roux AV. Neighborhood health-promoting resources and obesity risk (the multi-ethnic study of atherosclerosis). Obesity (Silver Spring) 2013;21(3):621–8.

[80] Fish JS, Ettner S, Ang A, Brown AF. Association of perceived neighborhood safety with [corrected] body mass index. Am J Public Health 2010;100(11):2296–303.

[81] Brownson RC, Hoehner CM, Day K, Forsyth A, Sallis JF. Measuring the built environment for physical activity: state of the science. Am J Prev Med 2009;36(4 Suppl). S99–123.e112.

[82] Johnson DA, Drake C, Joseph CL, Krajenta R, Hudgel DW, Cassidy-Bushrow AE. Influence of neighbourhood-level crowding on sleep-disordered breathing severity: mediation by body size. J Sleep Res 2015;24(5):559–65.

[83] Zanobetti A, Redline S, Schwartz J, et al. Associations of PM10 with sleep and sleep-disordered breathing in adults from seven U.S. urban areas. Am J Respir Crit Care Med 2010;182(6):819–25.

[84] DeMeo DL, Zanobetti A, Litonjua AA, Coull BA, Schwartz J, Gold DR. Ambient air pollution and oxygen saturation. Am J Respir Crit Care Med 2004;170(4):383–7.

[85] Shen YL, Liu WT, Lee KY, Chuang HC, Chen HW, Chuang KJ. Association of PM2.5 with sleep-disordered breathing from a population-based study in Northern Taiwan urban areas. Environ Poll (Barking, Essex : 1987) 2017;233:109–13.

[86] Gerbase MW, Dratva J, Germond M, et al. Sleep fragmentation and sleep-disordered breathing in individuals living close to main roads: results from a population-based study. Sleep Med 2014;15(3):322–8.

[87] Gislason T, Bertelsen RJ, Real FG, et al. Self-reported exposure to traffic pollution in relation to daytime sleepiness and habitual snoring: a questionnaire study in seven north-European cities. Sleep Med 2016;24:93–9.

[88] Mehra R, Redline S. Sleep apnea: a proinflammatory disorder that coaggregates with obesity. J Allergy Clin Immunol 2008;121(5):1096–102.

[89] Lopez-Jimenez F, Sert Kuniyoshi FH, Gami A, Somers VK. Obstructive sleep apnea: implications for cardiac and vascular disease. Chest 2008;133(3):793–804.

[90] Sheehan CM, Cantu PA, Powers DA, Margerison-Zilko CE, Cubbin C. Long-term neighborhood poverty trajectories and obesity in a sample of California mothers. Health Place 2017;46:49–57.

[91] Xiao Q, Hale L. Neighborhood socioeconomic status, sleep duration and napping in middle-to-old aged US men and women. Sleep 2018;.

[92] Oakes JM, Andrade KE, Biyoow IM, Cowan LT. Twenty years of neighborhood effect research: an assessment. Curr Epidemiol Rep 2015;2(1):80–7.

[93] Mayne SL, Auchincloss AH, Michael YL. Impact of policy and built environment changes on obesity-related outcomes: a systematic review of naturally occurring experiments. Obes Rev 2015;16(5):362–75.

[94] Hearst MO, Oakes JM, Johnson PJ. The effect of racial residential segregation on black infant mortality. Am J Epidemiol 2008;168(11):1247–54.

[95] Oakes JM, Forsyth A, Schmitz KH. The effects of neighborhood density and street connectivity on walking behavior: The twin cities walking study. Epidemiol Perspect Innov 2007;4:16.

[96] Glymour MM, Mujahid M, Wu Q, White K, Tchetgen Tchetgen EJ. Neighborhood disadvantage and self-assessed health, disability, and depressive symptoms: longitudinal results from the health and retirement study. Ann Epidemiol 2010;20(11):856–61.

[97] Jokela M. Are neighborhood health associations causal? A 10-year prospective cohort study with repeated measurements. Am J Epidemiol 2014;180(8):776–84.

[98] Bierman A, Lee Y, Schieman S. Neighborhood disorder and sleep problems in older adults: subjective social power as mediator and moderator. Gerontologist 2018;58(1):170–80.

[99] James P, Bertrand KA, Hart JE, Schernhammer ES, Tamimi RM, Laden F. Outdoor light at night and breast Cancer incidence in the Nurses' health study II. Environ Health Perspect 2017;125(8):087010.

[100] Sarkar C. Residential greenness and adiposity: Findings from the UK biobank. Environ Int 2017;106:1–10.

[101] Casey JA, Morello-Frosch R, Mennitt DJ, Fristrup K, Ogburn EL, James P. Race/ethnicity, socioeconomic status, residential segregation, and spatial variation in noise exposure in the contiguous United States. Environ Health Perspect 2017;125(7):077017.

[102] Johnson DA, Hirsch JA, Moore KA, Redline S, Diez Roux AV. Associations between the built environment and objective measures of sleep: the multi-ethnic study of atherosclerosis. Am J Epidemiol 2018;187(5):941–50.

[103] Koo YS, Song JY, Joo EY, et al. Outdoor artificial light at night, obesity, and sleep health: Cross-sectional analysis in the KoGES study. Chronobiol Int 2016;33(3):301–14.

[104] Vollmer C, Michel U, Randler C. Outdoor light at night (LAN) is correlated with eveningness in adolescents. Chronobiol Int 2012;29(4):502–8.

[105] Baron KG, Duffecy J, Berendsen MA, Cheung Mason I, Lattie EG, Manalo NC. Feeling validated yet? A scoping review of the use of consumer-targeted wearable and mobile technology to measure and improve sleep. Sleep Med Rev 2017;40:151–9.

[106] Ko PR, Kientz JA, Choe EK, Kay M, Landis CA, Watson NF. Consumer sleep technologies: A review of the landscape. J Clin Sleep Med 2015;11(12):1455–61.

[107] Kjellstrom T, Mercado S, Sami M, Havemann K, Iwao S. Achieving health equity in urban settings. J Urban Health 2007;84(3 Suppl):i1–6.

[108] Services UTFoCP. Recommendations to promote healthy social environments. Am J Prev Med 2003;24(3):4.

[109] Shaw M. Housing and public health. Annu Rev Public Health 2004;25:397–418.

[110] Howden-Chapman P. Housing and inequalities in health. J Epidemiol Community Health 2002;56(9):645–6.

[111] Services TFoCP. Recommmendations to promote healthy social environments. Am J Prev Med 2003;24(3S):21–4.

[112] Howden-Chapman P, Pierse N, Nicholls S, et al. Effects of improved home heating on asthma in community dwelling children: randomised controlled trial. BMJ 2008;337:a1411.

[113] Castaneda JL, Kheirandish-Gozal L, Gozal D, Accinelli RA. Pampa Cangallo Instituto de Investigaciones de la Altura research G. Effect of reductions in biomass fuel exposure on symptoms of sleep apnea in children living in the peruvian Andes: A preliminary field study. Pediatr Pulmonol 2013;48(10):996–9.

[114] Astell-Burt T, Feng X, Kolt GS. Does access to neighbourhood green space promote a healthy duration of sleep? Novel findings from a cross-sectional study of 259 319 Australians. BMJ Open 2013;3(8).

[115] McCurry SM, Pike KC, Vitiello MV, Logsdon RG, Larson EB, Teri L. Increasing walking and bright light exposure to improve sleep in community-dwelling persons with Alzheimer's disease: results of a randomized, controlled trial. J Am Geriatr Soc 2011;59(8):1393–402.

[116] McCurry SM, Gibbons LE, Logsdon RG, Vitiello MV, Teri L. Nighttime insomnia treatment and education for Alzheimer's disease: a randomized, controlled trial. J Am Geriatr Soc 2005;53(5):793–802.

[117] Reardon SF, Fox L, Townsend J. Neighborhood income composition by household race and income, 1990–2009. Ann Am Acad Pol Soc Sci 2015;660(1):78–97.

[118] Hale L, Do DP. Racial differences in self-reports of sleep duration in a population-based study. Sleep 2007;30(9):1096–103.

Chapter 8

The impact of environmental exposures on sleep

Chandra L. Jackson[a,b], Symielle A. Gaston[a]

[a]Epidemiology Branch, National Institute of Environmental Health Sciences, National Institutes of Health, Department of Health and Human Services, Research Triangle Park, NC, United States, [b]Intramural Program, National Institute on Minority Health and Health Disparities, National Institutes of Health, Department of Health and Human Services, Research Triangle Park, NC, United States

ABBREVIATIONS

ALAN	Artificial light at night
C	Celsius
CVD	Cardiovascular disease
dB	Decibel
F	Fahrenheit
HHCM	High heat capacity mattress
JHS	Jackson Heart Study
LED	Light emitting diode
LHCM	Low heat capacity mattress
mm/s	Millimeter per second
MESA	Multi-Ethnic Study of Atherosclerosis
nm	Nanometer
NO$_2$	Nitrogen dioxide
PM	Particulate matter
PSG	Polysomnography
REM	Rapid eye movement
SWS	Slow wave sleep
SDB	Sleep disordered breathing
SEP	Socioeconomic position
SES	Socioeconomic status
W	Watt
WASO	Wake after sleep onset

THE PHYSICAL ENVIRONMENT AND SLEEP

Sleep is an essential human need for maintaining biological homeostasis. While many internal biological mechanisms act in concert to regulate sleep-wake cycles, sleep is not entirely endogenous. In fact, many naturally-occurring and artificial external factors can have either a negative or positive effect on sleep. Fig. 8.1. displays our conceptual framework for how physical and social environments across the lifecourse may (1) influence a person's health behaviors (e.g., sleep) that are related to risk, protection, and/or resiliency and (2) "get under their skin" to subsequently influence health conditions (e.g., cardiovascular disease). Of note, individuals spend the 24-h period in either their residential, work and/or school, or recreational environments. These environments have structural factors (e.g., policies), community stressors (e.g., low social cohesion), both social and material resources to promote or maintain health and mitigate individual- and community-level stressors, and potential environmental hazards/pollutants (e.g., light, noise, and/or air pollution) that could either directly or indirectly impact the individual's health through various pathways.

In this section of the chapter, we describe the most salient known—albeit understudied—exposures in the physical environment (i.e., light, temperature, noise, vibrations, air quality, seasonality, and latitude/longitude) that influence sleep health. We also summarize prior observational and experimental studies (including interventions) that have investigated the impact of these external exposures on sleep before providing future directions based on gaps in our current understanding.

The impact of light on sleep

Greatly influencing human health and well-being, environmental light is the strongest synchronizing agent between the external physical environment and a person's circadian clock, which helps regulate internal biological systems [1, 2]. As one of the most pervasive and fastest growing forms of environmental pollution with an annual increase of approximately 2%, excessive light has become a substantial public health concern [3]. Light pollution is generally considered artificial lighting that is either (1) unnecessary or inefficient (e.g., not targeted for a specific task and that can trespass into homes and bedrooms), (2) brighter than natural light like lighting often used for advertising commercial goods and services such as gas stations and shopping centers, (3) uncomfortable/annoying to the human eye, and/or (4) unsafe (e.g., causes glare among drivers

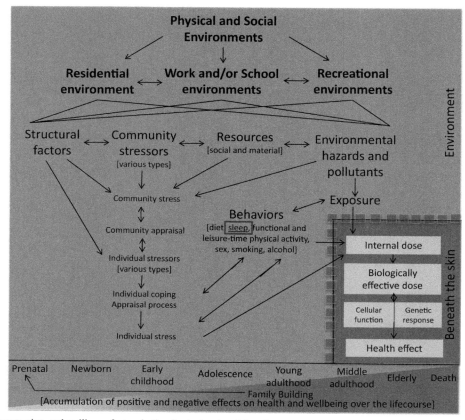

FIG. 8.1 How risk, protective, and resiliency factors in the physical and social environments may "get under the skin" to influence health behaviors and subsequently impact health outcomes.

and pedestrians) [4–6]. Various forms include overillumination, light trespass, glare, and sky glow [5]. In addition to wasting energy by shining non-targeted light upward into space and creating sky glow above cities that obscures views of the stars and other planets, light pollution can even harm the biological integrity of ecosystems and has been shown to affect both flora and fauna [6].

Although light pollution thresholds have not been officially established, the International Astronomical Union recommends considering light pollution as artificial sky brightness >10% of the natural sky brightness above 45° of elevation (see Table 8.1) [7]. It is also recommended that the visible light spectrum wavelength range (i.e., 440–540 nm) that corresponds to the maximum scotopic vision sensitivity (responsible for night vision) serve as the established protected range. Therefore, one should not install outdoor lamps with a wavelength that exceeds 15% of the energy flux emitted in the photopic (responsible for daytime vision) response pass band (measured in watts [W]) and where emissions in the scotopic response pass band exceeds two-thirds of that emitted in the photopic response pass band (measured in lumens) [8].

Regarding potential biological mechanisms linking light pollution to poor sleep, humans are genetically adapted to a natural environment consisting of sunlight during the day and darkness at night; thus, making biological processes in rhythm with the 24-h light/dark cycle [9]. However, artificial light has replaced natural sunlight during the day and artificial light at night (ALAN) has also replaced darkness at night. Exposure to ALAN now begins in early life, continues throughout one's lifespan, and presumably affects biological processes in ways that raise the risk of poor sleep and health conditions like obesity, type 2 diabetes, certain cancers (e.g., breast), depression, and cardiovascular disease (CVD) [10, 11]. Inopportune light exposure could lead to misalignment between one's external environment and their internal biological circadian clock. Circadian rhythm mechanisms that may underlie or contribute to the aforementioned health conditions include: melatonin synthesis suppression, circadian disruption/misalignment, 24-h sleep-wake cycle perturbations, sleep deprivation, or a combination of all of these factors [23].

ALAN (based on both light intensity and wavelength) has been shown to suppress melatonin production, which is a hormone secreted by the pineal gland in the epithalamus of the brain that helps regulate sleeping patterns by, for instance, enhancing sleep onset. Human health can be negatively affected due to inhibited melatonin production because of exposure to bright light at night, especially blue light (described in greater detail later in the chapter) since it, in particular, disrupts normal melatonin rhythms [6].

TABLE 8.1 Potential pollution definitions and thresholds.

	Definition of pollution	Threshold	Organization defining threshold
Light			
Indoor	No definition identified	No threshold identified	No threshold identified
Outdoor	Light that is not targeted for a specific task, is bright and uncomfortable to the human eye, causes unsafe glare, trespasses in homes and bedrooms, or creates sky glow above cities [4, 6]	Artificial brightness >10% of the natural sky brightness above 45° of elevation [160]	International Astronomical Union
Noise			
Indoor	No definition identified	Indoor A-weighted sound equivalent level >35 dB [161]	World Health Organization
Outdoor	No definition identified	Outdoor noise at night >40 dB [48]	World Health Organization Occupational and Environmental Health Team
Temperature			
Indoor	No definition identified	(a) minimum: 18°C [162] (b) 17–19°C in winter and 23–25°C in summer [41] (c) minimum for heating: 20°C and maximum for cooling: 26°C [163]	(a) World Health Organization Regional Office for Europe (b) Chartered Institution of Building Services Engineers (c) Comitè Europèen de Normalization
Outdoor	No definition identified	No threshold identified	No threshold identified
Vibration			
Outdoor	Indoor and outdoor—ground-borne vibration from railway movements created by the interaction between the wheels and the rail, where regularities from either surface can generate considerable vibration energy [63]	No threshold identified	No threshold identified
Air			
Indoor	Indoor air quality defined as the air quality within and around buildings and structures, especially as it relates to the health and comfort of building occupants [164]	State-specific (for the United States) indoor air quality information is available at: https://www.epa.gov/indoor-air-quality-iaq/epa-regional-office-and-state-indoor-air-quality-information	US Environmental Protection Agency
Outdoor	Air pollution represents ambient concentrations of particulate matter, ozone, nitrogen dioxide, and sulfur dioxide	Air quality guidelines for particulate matter, ozone, nitrogen dioxide, and sulfur dioxide are found online: (a) World Health Organization: http://apps.who.int/iris/bitstream/handle/10665/69477/WHO_SDE_PHE_OEH_06.02_eng.pdf;jsessionid=1F2335180E99DF2100546E2FA14FBF8B?sequence=1 (b) US Environmental Protection Agency: https://www.epa.gov/indoor-air-quality-iaq/introduction-indoor-air-quality	(a) World Health Organization (b) US Environmental Protection Agency

Of note, bedroom illumination that is typical for most homes in the US is sufficient to reduce and delay melatonin production.

Schulmeister et al. found that monochromatic red light at 100 lux would take 403 h to suppress melatonin exposure by 50% and other sources take a much shorter time period: candle (66 min), 60 W incandescent bulb (39 min), 58 W deluxe daylight fluorescent light (15 min), and pure white high-output light emitting diode (LED) (13 min) [12]. LED light is an important source of light pollution that emits a large amount of blue light, which actually appears white to the naked human eye. LED light can cause glare and reduces visibility or the ability to resolve spatial detail. This glare is considered worse than conventional lighting, and can create a safety hazard when, for example, drivers are affected. The main advantages of popular LED lighting are energy efficiency (in terms of reduced energy consumption and a decrease in fossil fuel use) along with cost savings [12a]. It, however, takes years to accrue cost savings.

Previous studies have found that bright residential lighting from LED, for example, is associated with reduced sleep time, dissatisfaction with sleep quality, night-time awakenings, excessive sleepiness, impaired daytime functioning, and obesity [13, 14]. Therefore, outdoor LED lighting may contribute to chronic disease risk in cities where they are installed. In studies of outdoor ALAN, individual or group residential areas have been found to be key contributors. Outdoor ALAN has been positively associated with an increased risk of breast cancer among women and prostate cancer in men [15–17]. However, there have been no relevant links made between outdoor ALAN and melatonin levels (based on a urinary biomarker) in cross-sectional studies [10]. Also, the threshold of light intensity that triggers a response in human health effects is currently unknown, which is an important topic for future research.

In terms of indoor illumination levels, melatonin levels have been found to be significantly lower in nurses who work night versus day shifts [18–20]. Nurses working night shifts were exposed to more light during sleep. Rotating shift workers with erratic light exposure also had abnormal melatonin levels [10]. Intense ALAN was also associated with a higher prevalence of self-reported insomnia symptoms [14]. It has been proposed that interventions among night-shift workers could seek to reset circadian rhythms to match work schedules with dark/light cycles by using bright light exposure during work and avoiding light exposure before sleep. Additionally, entrainment to scheduled time cues for food and exercise may further reduce health effects despite disrupted photoperiods. Night workers could also curtail non-essential shift work and establish or maintain good sleep hygiene practices (e.g., regular sleep-wake schedule, avoid stimulants before bed) [5, 6, 21], but more research is needed.

Higher ALAN exposure and reduced nocturnal melatonin concentrations have also been observed among adolescents with delayed sleep phase disorder compared to controls [10, 22, 23]. Increased evening and nighttime light exposure also significantly raised sleep onset latency [24]. Bright light 1 m away from eyes during sleep was associated with less deep sleep as demonstrated by periodic arousal and altered brain activity [25]. It was also associated with delayed sleep initiation and reduced overall sleep quality and fatigue [26, 27]. Of note, an increase in duration of exposure to bright light at night may be more important than intensity [10], and melatonin levels were negatively affected by blue light exposure, in particular. Blue light exposure was associated with decreased sleepiness and raised alertness [25].

Although more research is warranted, chronic sleep disruption or shift work associated with exposure to brighter light in the evening and at night appears to be associated with a long-term increased risk of obesity, diabetes, certain cancers (e.g., breast and prostate), and cardiovascular disease [28]. One study found low bedroom brightness was associated with reduced obesity rates in children [29]. Another study that examined the association between ALAN and obesity among 113,343 women in the United Kingdom found that body mass index, waist-to-hip ratio, waist-to-height ratio, and waist circumference increased with increasing lightness of the room that the person sleeps in at night, even after adjusting for factors like sleep duration [30]. The worldwide rise of obesity and use of electric lighting have paralleled each other, and both epidemiological evidence and animal experimental models support the belief that electric lighting is not merely a marker for a developed society in which behavioral factors (e.g., increased consumption, less energy expenditure) increase likelihood of obesity [30a]. Chronic circadian disruption from a 24-h light dark cycle could mediate increased susceptibility to obesity [30a], and more studies of, for instance, gene-by-environment interactions that take lighting and other environmental factors into account are needed since obesity and internal circadian clocks both have a high genetic component.

To protect sleep and avoid over lighting (greater than the minimum required for the task at hand), it is recommended by Falchi et al. that indoor light be dim and limited to the area that needs illumination before being turned off when deemed unnecessary. Since vision provided by rods and cones has sensitivity to certain wavelengths, the light also needs to have wavelengths towards the red, yellow, and orange rather than the blue end of the spectrum [5]. For instance, incandescent lights are preferred over fluorescent lights although they are considered less energy efficient [5]. International Dark Sky Association recommends use of low-pressure sodium lights generally and high-pressure sodium lights when color perception is important [5]. Ultimately, there are various types of lighting recommended in the following order from the most to least recommended: low pressure sodium, high pressure sodium, incandescent, metal halide, and LED are the least recommended because of their scotopic-to-photopic ratios and melatonin suppression effects [5]. A randomized controlled trial found that wearing amber versus clear lenses 2 h before bed was associated with reduced severity in insomnia symptoms and significant improvements in sleep, which may represent one effective intervention if behaviors prove difficult to change [31]. Practically, there should be complete darkness during sleep; therefore, televisions and other devices should not remain on during the sleep period [6]. Blinds should be closed to keep illumination from street lights out. To minimize melatonin suppression, elderly patients should be exposed to natural light via skylights and other daytime lighting ranging from 2500 to 3000 lux. Red lights should light the way to bathrooms at night [6]. Furthermore, recommendations for limiting outdoor artificial light at night includes avoiding the over lighting of outdoor spaces by turning lights off when not in use and banning outdoor emission of light at wavelengths shorter than 540 nm to reduce the adverse health effects of decreased melatonin production and circadian rhythm disruption in both humans and animals [5]. Also, luminaires should not send any light directly at and above the horizontal plane [5]. While it is important to note that some suggested interventions are counter to "public safety" (e.g., mall parking lots, etc. lighting to deter crime-even though not proven effective), these recommendations should be evaluated for effectiveness in improving sleep. A previous study sought to identify protected areas that could be a refuge from light pollution and daily noise in European countries using spatial mapping and regression modeling to define areas without light pollution and that are quiet [31a]. However, the authors have recommended that future studies go beyond finding refuge from light and noise pollution towards finding ways to reduce these forms of pollution. Future studies should also use more advanced mapping tools to create finer map scales for more local areas (e.g., regional) versus country, and define classes of areas needed beyond just dichotomous levels of illuminated versus unilluminated.

There are several gaps in the literature that can serve as important directions for future research. For instance, more longitudinal studies assessing the impact of ALAN on melatonin and subsequent poor sleep and disease risk are needed. More research is needed to determine how varying levels of bedroom illumination affect melatonin production in different types of sleepers, which may be especially important for people with sleep disorders (e.g., insomnia) and disparities in sleep architecture [6]. We need more research on illumination levels, duration, and colors of the light spectrum required to suppress human melatonin production. Furthermore, more research should be conducted on indoor light pollution; sleep studies should ascertain outdoor and indoor ALAN along with their independent and combined effects, and the combined effects of insufficient exposure to light during the day (or wakefulness periods) and too much exposure to light at night (or during desired sleeping periods) should be increasingly investigated. Future research should also study communities and cities that have adopted outdoor LED lighting, and investigate its impact on sleep and health conditions affected by light-dark cycles. Additional recommended strategies are outlined in a National Heart, Lung, and Blood Institute workshop report on circadian health and light [32].

The impact of temperature on sleep

In addition to light, outdoor, indoor, and an individual's core body temperature can also affect sleep health, which has noteworthy implications for disease risk [33, 34]. The core body temperature of humans is regulated by the thermoregulatory control center (known as the preoptic-anterior-hypothalamus), and is typically 37.0°C (or 98.6°F) [35]. There is an overlap in neurons that are sensitive to heat and neurons that change their firing pattern before and during sleep [36]. Control of both body and brain temperature is closely tied to sleep regulation [37]. During normal sleep, a 2°F decrease in core body temperature occurs from the person's peak in body temperature in the early evening to their lowest point before waking up [36]. External factors can influence core body temperature, and distal skin temperature appears to be more important than the proximal skin temperature for sleep regulation [37a]. For instance, air conditioners are usually unnecessarily low at night (when compared to thermal comfort temperatures) [38]. Although cycling temperatures within the individual's personal thermoneutral temperature range do not appear to significantly affect sleep stages, the thermal environment is one of the primary causes of sleep disturbance, and rapid-eye movement (REM) sleep is more vulnerable to thermal discomfort than other sleep stages [39]. Cold temperatures can increase the number and duration of wakefulness periods and the length of REM cycles. Furthermore, a high temperature can reduce total sleep time, duration of REM and slow wave sleep (SWS), and increase sleep onset latency and wakefulness. Higher ambient air temperatures have been associated with afternoon rest periods, but there appears to be no relationship with afternoon resting periods when temperatures were lower than an 18–25°C threshold [40].

Of note, an agreed upon thermal neutral temperature for sleeping people has not been established. Bed microenvironments vary by person, season, location, indoor air temperature, bedding, clothing, etc. Females have also been shown to be more sensitive to ambient air temperature than males [37a]. Furthermore, humidity at night/during sleep increases heart rate, sweat rate, and thermal load, which suppresses SWS and increases wakefulness [39]. Effects of humidity vary by time of sleep with stronger effects observed during the initial versus later sleeping period [39]. Negative effects of heat exposure are aggravated by high humidity as compromised air flow can reduce the heat load and wakefulness in warm, humid climates. Even exposure to bright light at night maintains high skin as well as rectal temperature and increases both heart rate and systolic blood pressure [39]. Therefore, temperature measurements are insufficient to thoroughly describe the thermal environment. While there is no threshold for outdoor temperature, the minimum recommended indoor threshold for European homes is approximately 18°C (17–19°C in winter and 23–25° in summer) (see Table 8.1) [41]. Limitations of current thresholds include thermal standards being defined for waking people, but the thermal comfort/neutral temperature is often warmer for sleeping compared to awake individuals [37a]. There is also important natural/endogenously driven and unnatural/exogenously driven variation in thermal comfort/neutral temperature that is difficult to capture.

A previous study sought to examine the impact of a high heat capacity mattress (HHCM) compared to a low heat capacity mattress (LHCM) on sleep and to determine whether core body temperature decline is enhanced and SWS is increased under gentle core body cooling on a HHCM [42]. The investigators found that participants on a HHCM had significantly reduced core body temperature, proximal skin temperatures on the back, and mattress surface temperature compared to participants on a LHCM. They also had significantly increased N3 (or slow wave) sleep (27% of total sleep time for HHCM vs 23% for LHCM, $P=.03$) [42]. Since Japanese preschool children often share beds with mothers and distal skin temperature has been shown to have a strong modulating effect on sleep onset and sleep depth, a study among Japanese mother-child dyads investigated the relationship between sleep and distal skin temperature in preschool children and compared the temperature to their mother's temperature [43]. The study found that proximal and distal skin temperature were lower in children than in their mothers in the early segment of sleep or the sleep onset period. The authors concluded that behavioral thermoregulation (e.g., removal of bed covers) may be important for maintaining sleep in preschoolers. Future interventions could focus on changing the indoor environment in ways that meet the varied thermal comfort requirements throughout the sleep period. Although more research is needed, bedside personalized ventilation systems could help with thermal comfort maintenance in addition to improving indoor air quality [37a].

Important gaps in the literature on temperature and sleep currently exist. For instance, objective measures of both the environment and sleep are necessary. It is important to measure heart rate variability and skin temperature concurrently with sleep to monitor human thermal comfort state as comfort reported during subjective sleep can be influenced by emotional or psychological stress. These influences may affect associations of interest if sleep is subjectively measured. More studies of the micro-environment/climate of bed and its interaction with, for example, ventilation systems are needed to define a comfortable thermal neutral zone or sleep environment.

The impact of noise on sleep

Excessive natural and anthropogenic noise is an environmental pollutant and an increasingly recognized public health issue that can negatively impact sleep health, which is considered the most serious non-auditory effect of excessive noise [46]. Noise is perceived by the auditory system in humans, and important studied sources of outdoor noise include traffic on the road, railways, and aircraft near homes [47]. Indoor noise/sound pollution is considered >35 decibels (dB) and outdoor noise at night is >40 dB (see Table 8.1) [48], but even ambient noise from external stimuli can be processed by the sleeper's sensory functions despite a non-conscious perception of their presence [47]. Environmental noise (and other external environmental events) can also activate processes during sleep that lead to eventual consciousness [49].

Previous studies have found that noise can negatively affect sleep architecture [47]. For instance, one study concluded that railway noise was associated with a higher percentage of people self-reporting sleep disturbance compared to those exposed to the same amount of road noise [50]. In addition to railway noise leading to sleep disturbance [51], studies have also found delayed sleep onset, induced awakenings, earlier final awakening or nocturnal awakenings, arousals along with hormonal responses as well as acute autonomic and cardiac activations that may explain associations between noise and compromised cardiovascular health [47, 52, 53]. In addition to reductions in REM sleep [54], there are also reports of a greater likelihood of taking sleep medications and experiencing objectively-measured body movements during sleep [51, 55], increased wakefulness [51], performance decrements [56], and impaired daytime functioning in terms of increased tiredness, daytime sleepiness, and need for compensatory resting periods [52]. Railway noise exposure has even been linked with a higher risk for certain types of breast cancer compared to non-exposure to railway noise [50]. Moreover, freight train noise may be particularly deleterious as it has been shown to cause more frequent awakenings [57, 58], a stronger cardiac response [53], and greater night time annoyance compared to passenger trains [59]. Wind turbine noise also has the potential to adversely affect sleep through frequent physiological activation in response to disturbance; however, prior studies have had inconsistent results and additional studies are needed [60].

In a meta-analysis of pooled data from 28 data sources, Miedema and Vos investigated the association between night-time noise and sleep disturbance to both establish functions that specify self-reported sleep disturbance in relation to nighttime transportation noise exposure (ranging from 45 to 65 dB) and to quantify the impact of transportation types (i.e., air, road, rail) [61]. The authors found that exposure to nighttime transportation noise was related to self-reported sleep in a dose-response manner [61]. Also, aircraft noise was associated with more self-reported sleep disturbance than road traffic, and road traffic noise had a stronger association with sleep disturbance than railways. Furthermore, the association between noise-induced sleep disturbance and age had an inverse U-shape, with the strongest reaction found between 50 and 56 years of age. The investigators concluded that this counterintuitive finding could be due to hearing decline in older age. Another study investigating the independent and combined effects of air, road, and rail traffic noise on sleep and recuperation, found small changes in SWS latency (+8.3 min), stage 1 sleep (+4 min), and SWS (−6 min) associated with traffic noise [52]. Changes in sleep continuity were also significant in that small noise-induced changes affected subjective assessments of sleep quality but did not affect daytime performance. Road traffic noise led to the most prominent changes in sleep structure and continuity even though nights with air and rail traffic noise were reported as being more disturbing than road traffic noise [52]. Relatively moderate cumulative effects of noise on sleep structure were only observed for REM latency, SWS latency, and time spent in REM. Awakening probability decreased in the order of rail, road, and air traffic noise, which is the reverse order for annoyance rankings reported during the day. A different study among hospitalized patients found a new-onset insomnia prevalence of 36% and that noise/brightness were contributors after frequent blood draws and vital sign checks due to the their illnesses [62].

Challenges associated with research on noise and sleep include the association being modified by personal/non-auditory factors including sensitivity to noise [62a]. Also, sound levels in communities constantly change. Intermittent noise is more likely to affect sleep (than constant noise) [62a], which can vary by source like rail versus road versus aircrafts. Moreover, bedroom noise depends on a variety of factors (e.g., insulation of house, sounds of household appliances, if windows are open) and is difficult to measure [62a]. Overall, for this type of research to advance, it is necessary for some fundamental questions to be answered regarding whether environmental noise has long term detrimental effects on health and quality of life and, if so, what these effects are for night-time, noise-exposed populations. Studies of individual differences in sensitivity of noise and effects of daytime noise on shift/night workers are also needed. Furthermore, more longitudinal studies need to investigate how sleep is affected by noise and its subsequent impact on health outcomes like CVD. As previously recommended by researchers who study the health effects of noise, future research should also focus on long-term effects of night-time noise exposure among different populations, specific subgroups that are "at risk" (e.g., children, elderly people, self-estimated noise-sensitive people, individuals with a severe disease that may be disturbed by noise, insomniacs, sleep disorder patients, night and shift workers), and the combined effects of noise exposure and other physical agents or stressors during sleep.

The impact of vibrations on sleep

Noise is often accompanied by vibrations. Ground-born vibration from railway movements, as an example, are created by the interaction between the wheels and the rail, where irregularities from either surface can generate considerable vibrational energy [63]. Ground vibrations near railway lines are often in the range of 0.4–1.5 mm/s [64]. Previous studies have found that increasing vibration results in increased reporting of waking during the night and waking too early in the morning [65]. Annoyance due to railway vibration is higher during the evening than the day, and higher during night than evening [66]. Both lower sleep quality and alertness have been reported following the night after vibration exposure of 1.4 mm/s compared to nights with vibration of 0.4 mm/s. This finding corresponded with more restlessness, greater difficulty falling asleep, and more awakenings once asleep [66]. Although stronger vibrations from railway freight were associated with reduced subjective sleep quality and increased self-reported sleep disturbance [67], self-reported sleep parameters do not necessarily correlate strongly with actual physiological responses [68]. Like with noise, even minor sleep disturbances due to vibrations can cause degradation of executive functions despite a lack of subjective response [69]. There is also some evidence that the number of trains can influence human response as annoyance was found to be higher in areas with a greater number of trains [64]. Pass-by frequency during the night has also been linked to self-reported noise-induced disturbances [51]. However, the number of nocturnal pass-bys was not found to be statistically significantly associated with wakefulness and light sleep time as objectively determined by polysomnography [51]. The vibration amplitude contributed to arousal as well as awakening and sleep stage change probabilities in that the number of trains during the night led to self-reports of sleep disturbance at moderate vibration amplitudes. Furthermore, both vibration amplitude and the number of trains contributed to effects on sleep architecture, whereby the number of sleep depth changes, SWS continuity and nocturnal wakefulness were negatively affected during the high vibration condition with a high number of trains. Some studies also suggest that low-frequency vibration (like from a car or inside a train) can, in turn, induce sleep [70]. Nonetheless, future studies need to investigate the long-term effect of nocturnal vibrations, examine the combined impact of noise and vibration, and elucidate mechanisms involved in potential vibration habituation and/or sensitization.

The impact of air quality on sleep

It is biologically plausible that elevations in ambient pollution may affect sleep through pollutant-associated effects on central or peripheral neurotransmitters that influence sleep state stability, upper airway patency, and/or ventilatory control. Pollutants may directly contribute to nasal or pharyngeal inflammatory and oxidative stress responses that increase upper and lower airway resistance and reduce airway patency, which could lead to, for instance, oxyhemoglobin desaturation [71]. Fine and ultrafine particles may alter ventilation-perfusion relations, exacerbating the hypoxia of sleep disordered breathing (SDB) [71, 72], and pollution may increase SDB through influencing central ventilatory control centers. In patients with asthma and SDB, elevated air pollution has been demonstrated to worsen lower airway inflammation and airflow obstruction through allergic and nonallergic mechanisms [73]; this may also contribute to the propensity for desaturation with sleep-associated reductions in ventilation. In patients with hay fever, upper airflow obstruction may worsen on an allergic basis when air pollution particles also contain allergen fragments [71, 72]. Pollutants entering the blood could have an effect on the brain and hence the regulation of breathing.

Of the few studies that have investigated the relationship between outdoor air pollution and sleep apnea or SDB, results have been mixed among adults [74–77]. Among US adults aged 39 years and older, Zanobetti et al. found that an interquartile increase in short-term exposure to fine particulate matter $\leq 10\,\mu m$ (PM_{10}) in diameter was associated with an approximately 12% increase in the percentage of sleep time at <90% oxygen saturation as well as a 1% decrease in sleep efficiency (affected by sleep state stability) [74]. However, population-based studies among German and Taiwanese adults observed no associations between PM_{10} and sleep apnea severity [75]. Furthermore, Weinreich et al. suggest that short-term exposure to ozone was associated with sleep apnea severity [76]. Shen et al., found no relationship between sleep apnea severity and ozone, but observed positive associations between sleep apnea severity and both short-term and annual exposure to particulate matter with aerodynamic diameters ≤ 2.5 ($PM_{2.5}$) as well as nitrogen dioxide (NO_2) [75]. These associations were stronger in the winter and spring seasons. There also may be associations between $PM_{2.5}$ and average sleep duration. For instance, a prospective study of 12,291 college freshman in China found that students reported longer sleep duration (in hours) for each standard deviation (36.5 mg/m^3) increase in $PM_{2.5}$ exposure ($\beta = 1.07(95\%\ CI: 1.04–1.11)$) [78].

Tenero et al. conducted a review highlighting studies of air pollution and SDB among children [77]. They found that, compared to children with low exposure, Australian and Italian children with high exposure to NO_2 and PM_{10} (Italian children only for PM_{10}) were more likely to report SDB or have objectively-measured SDB [77]. Similarly, a study of elementary-aged children in Tehran, Iran suggested that odds of subjectively-measured habitual snoring were higher among children exposed to greater air pollution compared to children exposed to less pollution [77]. Conversely, Abou-Khadra et al. did not observe a relationship between PM_{10} exposure and SDB among Egyptian elementary school-aged children [77].

While the relationship between outdoor air pollution and sleep may prove important, humans spend approximately 80%–90% of their time indoors with approximately 34% of their day in the sleep microenvironment [79–81], which — as delineated by Boor et al.—is characterized as the physical space generally involving a mattress, pillow, bedding materials, bed frame, breathing zone, and a buoyant thermal plume [82, 83]. This physical environment (especially beds as reservoirs for a complex mixture of chemical- and non-chemical-laden dust) can harbor diverse indoor pollutants like house dust mites comprising a wide range of organisms and their associated allergens, fungi, bacteria (mainly from human origins [e.g. skin, intestinal, genital]), volatile organic compounds, plasticizers like phthalates, flame retardants such as polybrominated diphenyl ethers in and on mattresses, pillows, and bed sheets (as accumulation zones) [82]. Inhalation and dermal absorption of these potentially detrimental chemical and non-chemical environmental exposures while sleeping could be hazardous to health. In fact, data suggests that the sleep microenvironment may even play a critical role in characterizing exposures of very young children to indoor pollutants. Of note, if a child and an adult are exposed to the same breathing zone concentration of a pollutant released from, for example, a mattress, then the child-normalized dose will be much greater than the adult.

There are several key factors in the sleep microenvironment that contribute to a source-proximity effect also described by Boor et al., including the spatial proximity of the breathing zone to the source of the exposure, incomplete mixing and common poor ventilation of bedroom air that leads to concentration gradients in the space near an actively emitting source, the personal cloud due to human body movement-induced particle resuspension, the buoyant human thermal plume, heat transfer from the human body to the source (which may elevate the emissions of gaseous pollutants), and direct dermal contact with the source [83]. Sleep microenvironments are important, but they have not been extensively researched in contrast to exposures in other types of indoor environments, such as classrooms, kitchens, and occupational workplaces. To investigate subsequent health effects and to develop strategies to promote healthy bedrooms, more research is needed to understand the pollutants commonly found in sleep microenvironments, the mechanisms by which pollutants are transported around the human body to an individual's breathing zone, pollutant concentrations and exposure levels that individuals experience while sleeping, and the total amount of pollutants that are inhaled or absorbed via dermal exposure. As summarized by Boor et al., future research should focus on personal exposure monitoring of particles and gases, which will permit more reliable associations with health outcomes beyond basic mattress dust sampling/analysis [83]. Studies should also investigate the contribution of early-life sleep exposures to microbes, allergens, and semi-volatile organic compounds as well as total exposures to these agents. In fact, a systematic review found an overall positive association between allergic rhinitis and sleep-disordered breathing among children [84]. It is also important to understand variation in exposures during sleep among various demographic groups (e.g., ethnically-diverse populations and age groups) as well as to elucidate the impact of human exposure to the full range (indoor and outdoor) of particulate and gaseous air pollutants. Investigators should also identify the role of bed partners since 48%–72% of Americans sleep with a significant other and 2%–16% with their pets [83]. Studies should also evaluate the impact of bedside personalized ventilation systems helping to filter and deliver fresh air in addition to providing thermal comfort maintenance (as described earlier in the chapter). Ventilation systems are also considered energy efficient. In addition to fan air ventilation, opening windows is associated with lower carbon monoxide levels and better subjective sleep quality although there have been weak or nonsignificant associations with objective sleep as measured by wrist actigraphy [44]. Furthermore, stoves with external versus internal exhaust also appear to be an effective intervention to help address parent reported child sleep problems [45]. Lastly, indoor air quality and its effects on sleep quality as well as special settings for sleep microenvironments (e.g., hospital beds, camping tents, dormitories) should be studied.

The impact of seasonality and latitude/longitude on sleep

Although the health effects are not well understood, seasonal variation in sleep/wake cycles likely exist due to differences in photoperiod, which refers to variability in hours of light throughout the four weather seasons [85]. Since light entrains circadian rhythms that depend on when the light exposure occurs, light exposure prior to the normal temperature nadir will phase-delay the circadian rhythm and exposure to light after the normal temperature nadir will phase-advance the circadian rhythm [86]. During summer sleep times, core body temperature and melatonin secretion tend to be slightly advanced due to the longer photoperiods. Users of a smartphone ENTRAIN app found that later sunset times were associated with longer sleep duration among >20 different countries located around the globe [87]. They also found that those reporting being typically exposed to outdoor light go to sleep earlier and sleep more than those who reported more indoor light exposure [87].

Furthermore, seasonal changes during the year are more marked in extreme Arctic and Antarctic latitudes than in equatorial regions. Thus, people living at higher latitudes have a higher variability of sunlight duration, depending on the season. Even in pre-industrial societies, the specific photopic period related to latitude seems to markedly influence sleep duration [85]. For instance, geographic latitude showed significant associations with self-reported sleep duration in a large population-based sample of people living between 18°29′S and 53°18′S in Chile [88]. Those individuals who lived closer to the Antarctic Circle had significantly longer sleep durations compared to their counterparts who resided closer to the equator. This association was stronger in men, especially for sleep during weekends. People living in latitudes more proximal to the equator had a greater than threefold odds of being short sleepers during weekends than those living near the Antarctic Circle [88]. Another study investigated the impact of seasonality on sleep by comparing Norway and Ghana where Norway experiences large seasonal variation while Ghana does not (due to its position near the equator). While not found in Ghana, the bed and wake times in Norway were earlier in the summer while insomnia, fatigue, and poor mood were more common in the winter [89]. Furthermore, Stothard et al. aimed to [1] quantify the impact of a week-long exposure to the natural winter light-dark cycle compared to exposure to modern electrical lighting on the timing of the human circadian clock and quantify the circadian response to a weekend of exposure to the natural summer light-dark cycle [90]. Investigators found that the beginning of the biological night and sleep occurred earlier after a week's exposure to a natural winter light-dark cycle as compared to the modern electrical lighting environment [90]. They also found that the human circadian clock was sensitive to seasonal changes in the natural light-dark cycle in that it was longer in the winter than the summer [90]. Circadian and sleep timing occurred earlier after spending a week camping in a summer natural light-dark cycle compared to a typical weekend in the modern environment; thus, the circadian clock appears timed later in the modern environment in both the winter and the summer. Furthermore, participants in cities along the same latitude had similar resting periods, and resting periods over the year reflected the length of the night throughout the year [40]. Since better characterizing the impact of seasonality and latitude/longitude on sleep health could lead to more effective interventions, more research in this area is warranted.

THE SOCIAL ENVIRONMENT AND SLEEP

In addition to naturally-occurring and man-made exposures in the physical environment (e.g., light, temperature, noise, vibrations, air quality in the microenvironment, seasonality, latitude/longitude) affecting sleep health, many factors in one's social environment also appear important. The social environment includes social conditions (e.g., culture), policies (e.g., workplace flexibility), and institutions (e.g., the economic system), the community (e.g., neighborhood) immediately surrounding an individual, interpersonal factors (e.g., intimate relationships), and intrapersonal activities (e.g., actions like coffee consumption directed toward curtailing sleep) (see Fig. 8.1) [91]. With psychosocial stress as a main driver connecting the social environment to poor sleep, this section of the chapter summarizes prior observational studies that have investigated how sleep is impacted by psychosocial stressors one may experience across multiple social domains and in various settings within the social environment (e.g., neighborhood, home, workplace, social/recreational). For example, we discuss relationships between sleep and low socioeconomic status; suboptimal neighborhood characteristics like low safety and high disorder; racism and other forms of discrimination at multiple levels (e.g., institutional and interpersonal); suboptimal work-related conditions like shift work and job strain; as well as interpersonal factors such as family structure and trauma.

Psychosocial stress and sleep

An adverse social environment whether at the policy, community, or interpersonal level can lead to psychosocial stress in individuals. Defined as stress that occurs when a person perceives that the external demands of a stressor exceed their ability to cope with the stressor [92], there are two types of psychosocial stress: acute stress (e.g., major event like a flood) and chronic stress (e.g., the aftermath of such disasters like homelessness or financial strain). Chronic stress is associated with increased vulnerability to disease [93] and poor sleep may be on the causal pathway from stress to disease.

Each form of psychosocial stress has the potential to affect sleep as shown in the results of a prior review [94]. Although most of the investigated stressors have been chronic, acute stress has been associated with worse sleep efficiency, frequent awakenings, decreased REM sleep, and less SWS [94]. Particularly, major stressful life events can affect sleep through decreases in REM latency, increased REM sleep percentage, and reduced SWS [94]. It has been difficult to observe consistent patterns between various chronic stressors and sleep, likely due to differences in study populations and the daily life stressors measured; yet, everyday emotional stress was found to be consistently associated with unfavorable changes in sleep architecture, namely less SWS [94]. Since a variety of acute and chronic, everyday psychosocial stressors within an individual's social environment can affect sleep, we summarize the most salient factors that have been investigated in the sleep literature.

Social conditions, policies, institutions: The impact of socioeconomic status and racism on sleep

Socioeconomic status (SES) is a fundamental cause and strong driver of differences in population health, and it likely affects sleep by determining an individual's physical as well as social environmental exposures [95]. Within several studies highlighted in a literature review by Grandner et al., lower socioeconomic position (SEP) has been associated with higher rates of sleep disturbance, such as difficulty falling asleep, difficulty staying asleep, waking too early, and worse sleep quality [95]. SES characteristics associated with worse sleep include not graduating from college, lower household income, unemployment, a lack of access to private health insurance, and food insecurity [96, 97]. Studies also suggest a negative association between educational attainment and insomnia [98] and that lower childhood SEP is associated with decreased SWS and increased stage 2 sleep in adulthood [99].

Discussed in greater detail in Chapter 7, many racial/ethnic groups are disproportionately represented in lower socioeconomic status groups, which is likely a manifestation of discriminatory policies and practices on the societal level. Discrimination is generally defined as differential or unfair treatment based on actual or perceived membership in a group [100] and can occur based on race/ethnicity, national origin, religion, sex/gender, sexual orientation, SES, or other social factors related to an identity [101]. In fact, the pervasive, widespread observation of racial/ethnic disparities across a range of health outcomes likely results from a race-based or race-conscious society involving systematic discrimination across institutions and cultural practices on the basis of race [102]. In terms of sleep, racial/ethnic minorities are even more likely to experience poor sleep, especially African Americans [95]. For instance, non-whites like African Americans and Hispanics/Latinos in the US are generally 2–4 times more likely than whites to report shorter and longer sleep duration [95], both of which are associated with suboptimal health. Certain racial/ethnic minorities are also more likely to experience suboptimal sleep architecture and quality [95]. One manifestation of an individual living in a stressful/oppressive environment is known as John Henryism in the epidemiological literature. It is the hypothesis that members of marginalized groups are routinely exposed to psychosocial stressors (e.g., discrimination, job strain) that require a consistently high level of active coping or use of considerable energy to manage the psychosocial stress [103]. When continual engagement in effortful active coping and ambitions or interests are not met with support or resources, strain ensues. This strain or stress can lead to disrupted sleep, which may (as an example) lead to less SWS and non-dipping of blood pressure during normal sleep [104]. Discriminatory practices related to, for example, residential segregation, un- or subpar-employment, and/or low education quality based on race likely play an important role in determining differential exposure to factors that promote or harm health.

Community: The impact of neighborhood social and physical environments on sleep

Described in greater detail in Chapter 7, the neighborhood environment has been linked to individual health outcomes and sleep may be on the causal pathway through which factors in the environment—both physical and social—affect health. In short, psychosocial stress in the neighborhood (like relative social disadvantage and a lack of social cohesion, social capital, and/or safety) can affect sleep health and subsequent risk of disease. Aspects of the neighborhood environment are measured differently across studies, but most suggest that suboptimal neighborhood conditions can contribute to poor sleep. However, these studies are often cross-sectional and rely on self-reported data. Therefore, more research with longitudinal designs and objective measures are needed.

Nonetheless, neighborhood physical and social disorder were associated with self-reported trouble falling asleep and waking too early [105] as well as shorter sleep duration [106]. In one study, no social or physical neighborhood factors were associated with insomnia symptoms [106], but insomnia symptoms were found to be more prevalent in noisy compared to non-noisy neighborhoods in another study [107]. Johnson et al. found that neither sleep fragmentation nor sleep efficiency varied by social environment, but participants residing in increasingly more favorable social environments had approximately 6 min longer sleep duration and an earlier sleep midpoint [108]. Lower neighborhood social cohesion has been found to be associated with self-reported trouble falling asleep and feeling unrested [105] as well as worse sleep quality [109]. There was a suggestion that a cohesive, positive social environment was associated with less daytime sleepiness [106].

Studies about neighborhood socioeconomic factors and sleep have had mixed results. Neighborhood disadvantage, measured by census tract-level data (e.g., percentage of residents with a Bachelor's degree, median home value, median household income), was not associated with sleep duration in the Jackson Heart Study (JHS) after adjustment for individual-level SES factors, demographic factors, and health characteristics [109]. However, Fang et al. observed that participants with low- and middle-neighborhood SES had higher odds of very short and short sleep compared to individuals in high SES neighborhoods in Boston communities [110]. Furthermore, DeSantis et al. showed higher neighborhood SES was associated with less daytime sleepiness within US adults enrolled in the Multi-Ethnic Study of Atherosclerosis (MESA) [106]. Similarly, Xiao and Hale found that adults residing in lower-SES neighborhoods were more likely to report very short sleep, long sleep, and long (≥ 1 h) napping [111].

Relationships between neighborhood characteristics and various sleep dimensions may vary by sex and an individual's perception of their neighborhood. For instance, neighborhood disadvantage has been associated with higher odds of restless sleep among women and not men, but census tract density was associated with higher odds of restless sleep among men and not women [112]. An individual's perception as well as outsiders' perceptions of their neighborhood may also affect sleep, but such relationships vary by the neighborhood perceptions measured. Perceived low neighborhood quality was associated with fair/poor sleep quality [113, 114] and prolonged sleep latency [114]. Studies have shown that perceived neighborhood safety was not associated with sleep disruption [115] nor insomnia symptoms [107], but neighborhood safety was associated with longer sleep duration in a different study [106]. High levels of neighborhood violence were associated with insomnia symptoms among Hispanics/Latinos [107] as well as shorter sleep duration and poor sleep quality among African Americans [109]. In the JHS, participants with perceptions of neighborhood problems like traffic, trash/litter, and lack of parks had lower sleep quality than participants with more positive neighborhood perceptions, which remained after adjustment for individual-level sociodemographic and socioeconomic factors as well as health characteristics [109]. Neighborhood stigma in the form of negative media perception was negatively associated with self-reported sleep duration and quality among 120 Hispanic/Latino adults enrolled in the New York City Low Income Housing, Neighborhoods, and Health Study [116].

Physical features of the neighborhood environment may also be associated with sleep. For instance, low neighborhood walkability was associated with greater severity of sleep apnea (especially among male individuals who were obese) [117], greater built environment was associated with shorter average sleep duration, and larger intersection density was associated with lower sleep efficiency [118]. Neighborhood-level crowding has also been suggested as a factor related to increased severity of SDB [119].

Community: Work environment and sleep

The workplace is an important setting for potential exposure to suboptimal physical and social environmental factors that can negatively affect sleep, which are described in greater detail in Chapter 36. Briefly, shift work has been identified as a probable human carcinogen [120], and shift workers have been shown to have long sleep latency, decreased SWS, increased REM, and shorter REM latency, which is attributed largely to emotional stress or worry regarding the next morning [94]. A meta-analysis of three longitudinal studies suggested that steady shift work was a risk factor for future sleep disturbances [121]. Shift work can also lead to circadian misalignment, resulting in deleterious outcomes like chronic sleep loss, shift work disorder (a clinically diagnosed circadian rhythm sleep disorder), and morbidity [21]. Shift workers are more likely to have low SES and be non-white (particularly, African American), which has several implications regarding racial/ethnic and socioeconomic sleep disparities discussed previously.

Meta-analyses of longitudinal and cross-sectional studies provide evidence that stressful work environments may lead to poor sleep [121–123]. Higher job demands are positively associated with sleep disturbances and lower sleep quality in longitudinal studies [121, 122]. In terms of job control, there is evidence that higher job control is positively associated with higher sleep quality in longitudinal studies [122]. Positive associations have also been observed between job control and lower levels of sleep disturbances in fixed effects models, although the relationship was no longer significant in the random effects model [121]. Job strain may be a risk factor for sleep disturbances [121]. A meta-analysis found moderately strong evidence that social support within the work environment may induce better sleep quality [122] and lower frequency of sleep disturbances [121]. Effort-reward imbalance (imbalance between the effort to perform a job and the reward it provides), although negatively associated with sleep quality [122], was not found to influence future sleep disturbances [121]. However, self-reported and actual workplace bullying was positively associated with future sleep disturbances [121]. Lastly, organizational justice/influence over decisions was positively related to better sleep quality and lack thereof was related to future sleep disturbances [121, 122]. Despite these findings among longitudinal studies, no significant effect of changes to psychosocial work characteristics on sleep quality have been established [122].

Interpersonal relationships and sleep

In addition to the aforementioned institutional discrimination, interpersonal discrimination occurs in ways that have been shown to affect sleep. For instance, there are consistent positive associations between various forms of interpersonal discrimination and self-reported sleep difficulties as well as insomnia [101]. However, one study among older adult participants aged 50 years and older in the Health and Retirement Study reported no longitudinal association between discrimination and sleep quality [124]. Findings related to sleep duration are inconsistent, and vary by type of discrimination (major vs everyday discrimination). Among four studies that used PSG or actigraphy, one study found that everyday discrimination was associated with shorter sleep duration, but no significant association was observed in the other studies [101].

In terms of sleep architecture, across four studies of objectively-measured sleep, discrimination was positively associated with wake after sleep onset (WASO) in two of the studies as well as a smaller proportion of REM sleep in one study, and experiencing discrimination was inversely related to the proportion of SWS (stages 3 and 4) in two studies [101]. One study examined and found a positive relationship between discrimination and light (stage 2) sleep [101]. Of the three studies that examined discrimination and sleep efficiency, two reported no association with discrimination, but one suggested an inverse relationship [101]. Associations between discrimination and sleep were generally attenuated after adjustment for depressive symptoms [101].

An additionally important social stressor that may affect sleep is acculturation, which is the process by which immigrants adopt, internalize, and exhibit the behaviors of the host society [125]. Among Hispanics/Latinos, one's level of acculturation has been associated with changes in health status when living in the United States [126]. Most studies regarding acculturation and sleep among the United States population have focused on Hispanics/Latinos, were cross-sectional, and investigated associations with self-reported sleep [97, 125–132]. Although the measure of acculturation varied across studies, less acculturation was consistently associated with better sleep. Among adolescents, middle-aged adults, and older adults, less acculturated Hispanic/Latinos were more likely to report longer sleep duration and better sleep quality and were also less likely to report very short sleep, short sleep, and insomnia symptoms—although relationships with insomnia symptoms were attenuated after adjustment for sociodemographic and behavioral characteristics [97, 125, 132, 133]. Furthermore, greater acculturation and acculturation stress were marginally associated with long sleep and positively associated with poor sleep quality including daytime sleepiness, short sleep (prior to adjustment for sociodemographic and behavioral factors), and more sleep complaints [127, 129–131]. One repeated measures study among Mexican-American pregnant women found that women with higher acculturation reported more nighttime disruptions in sleep compared to less acculturated women throughout pregnancy [128]. Further research regarding acculturation inclusive of non-Mexicans and other immigrant populations (e.g., Caribbean-born Blacks, non-US-born Whites) with longitudinal study designs and objective measures of sleep throughout the lifecourse is warranted.

A few studies with a focus on personal stigma and sleep have been conducted, and stigma may be associated with worse sleep. For instance, in a study of HIV-infected individuals, internalized (integration of others attitudes and opinions into one's identity/sense of self) HIV stigma was indirectly associated with poorer sleep quality, as measured by the Pittsburgh Sleep Quality Index, through depression and loneliness [134]. Similarly, among 64 adults aged 18–64 years that reported substance abuse, internalized stigma mediated the positive association between perceived stigma and self-reported poorer sleep [135].

In terms of trauma as a stressor that appears to affect sleep, a systematic review conducted by Kajeepeta et al. showed significant associations between history of childhood adversity and a variety of sleep disorders including sleep apnea, narcolepsy, nightmare distress, sleep paralysis, and psychiatric sleep disorders in adulthood among women [136]. The strengths of the associations increased with the number and severity of experiences [136]. These findings, particularly among adult women of childbearing age, have potential implications for child health because poor maternal mental health may be associated with WASO among infants [137]. Another early-life trauma, bullying victimization at 8 and 10 years, has also been associated with nightmares, night terrors, sleepwalking and any type of parasomnia at age 12 years [137]. Furthermore, there is evidence that peer victimization is associated with insomnia symptoms over time among both preschool-aged children and adolescents [138, 139]. In terms of violence, a review of literature investigating relationships between community violence and physical health outcomes among youth aged 0–18 years found that exposure to community violence was associated with sleep problems in all of the included studies [140]. Bailey et al. reported that children who were victims of community violence had a 94% increased risk of self-reported sleep difficulty [141]. Youth who reported witnessing a homicide had a twofold increase in WASO at baseline in one of the studies; but, associations were no longer statistically significant at study follow-up [142]. One study suggests that relationships between witnessing violence and sleep problems may be stronger among females compared to males [143]. Lastly, high exposure to violence was also associated with self-reports of short sleep and sleep interruption among the mothers of young children [115].

Family structure or dynamics are important for sleep throughout the lifecourse as noted in a review by Meltzer and Montgomery-Downs [144]. During pregnancy and in the postpartum period, mothers are likely to experience less sleep, lower sleep efficiency, less deep sleep, more daytime sleepiness, and have more frequent awakenings [145–147]. Newborn sleep is often fragmented; thus, the parents' sleep is likely to be fragmented. Conversely, parents are considered the main agents affecting infant sleep [148]. Therefore, there may be a bidirectional relationship between parent and infant sleep [144]. Toddler, preschool-aged, and school-aged children's sleep problems can disrupt parents' sleep and family functioning [144]. Among adolescents, there may be bidirectional relationships between adolescent sleep quantity, quality, and problems and family factors including parenting style, family problems, and the home environment [144]. Adolescent behaviors and activities like being involved in extracurricular activities/events that result in time constraints can affect parents' ability to initiate and maintain sleep [144]. Conversely, adolescent sleep can also be affected by parental behaviors. While parents are not typically strongly involved in establishing adolescent sleep routines, total sleep time has been increased with parental rules like setting bedtimes [149]. Additionally, when parents are more involved with monitoring adolescents, adolescents have been shown to experience less psychosocial distress and greater sleep efficiency [150]. Single parent family structure has also been found to be associated with worse sleep efficiency and shorter sleep duration among both black and white adolescents [151]. Relationships between family conflict and sleep may remain when children reach adulthood: family conflict at ages 7–15 years has been associated with higher odds of insomnia in adulthood [136].

Many adults share a bed with a partner or spouse, and there is evidence that couple-level factors affect sleep (such as through relational conflicts, abuse, differential preferences for morningness-eveningness, and timing of sleep-wake cycles) [152]. Sleep can be affected by other relationships such as those with caregivers, including parents of infants and children, healthcare professionals, and family or other informal caregivers of individuals with physical or cognitive disabilities, for instance. In a study of caregivers who live with adult care-receivers, in addition to predisposing risk factors (e.g., female sex, chronic health conditions, and precipitating events like menopause), factors like reduced physical activity and irregular sleep/wake schedules due to caregiving could lead to sleep problems such as insomnia in caregivers [153].

Early attachment style, defined as the interpersonal style that develops during childhood based on the type of bond and amount of trust with primary caregivers, may be associated with sleep throughout the lifecourse [154]. In all studies except for one among 2–18 year-olds, lower attachment was associated with poorer subjective and objective sleep measures, and studies also suggest bidirectional relationships among parental attachment, emotional security, and sleep in children [154]. Of 14 studies among individuals 18–64 years old, only one had a longitudinal design, and the majority identified a relationship between insecure attachment and poorer sleep as well as differences in sleep architecture [154]. Among studies of individuals >65 years of age, two found that those with high attachment reported better sleep, but less securely attached individuals were more likely to report daytime napping, use of sleep-inducing medication, and tended to sleep less at night [154].

Lastly, a variety of other social factors may positively or negatively affect sleep, but additional studies are warranted. Social support is considered a protective/resiliency factor against a variety of poor health outcomes, which is also plausible for sleep given that studies show lack of social support as being associated with shorter subjectively measured sleep [155–157]. A study using actigraphy among US adults aged 24–81 years enrolled in the MacArthur Study on Aging: Midlife in the United States found that neither social support nor social strain were associated with total sleep time, but social strain, defined as strained aspects of the individual's social network, was associated with lower sleep efficiency and night-to-night sleep variability [158]. Other psychosocial factors like financial strain, social isolation, low emotional support, and negative social interactions were related to self-reported sleep problems in a cross-sectional study of 736 men and women aged 58–72 years in the Whitehall II civil servants cohort [159].

In conclusion, this chapter has discussed many of the most salient physical and social environmental factors that can either positively or negatively impact sleep, which is important for maintaining and restoring health across the lifecourse. Some organizations or researchers have provided threshold recommendations for certain exposures in the physical environment like light or temperature that are based on scientific evidence or expert opinion. We have complied a non-exhaustive list of potential pollution definitions and safety thresholds (see Table 8.1), but recommendations for defining and determining the maximum acceptable level across the range of factors that impact sleep health and subsequent risk of disease are desperately needed. There are many gaps in our current understanding of how environmental exposures affect sleep, and we have provided directions for future research; but, a wide range of experimental studies and interventions across multiple domains from the individual- to societal-level and across the lifecourse should be designed and evaluated to optimize the impact of the physical and social environments on sleep health.

ACKNOWLEDGMENTS

This work was funded by the Intramural Program at the NIH, National Institute of Environmental Health Sciences (Z1AES103325-01). We are grateful to Erin Knight, Eleanor Weston, and Stacey Mantooth for their help with preparing this draft.

REFERENCES

[1] Emens JS, Burgess HJ. Effect of light and melatonin and other melatonin receptor agonists on human circadian physiology. Sleep Med Clin 2015;10(4):435–53.

[2] Pendergast JS, Yamazaki S. Effects of light, food, and methamphetamine on the circadian activity rhythm in mice. Physiol Behav 2014;128:92–8.

[3] Kyba CCM, Kuester T, Sanchez de Miguel A, Baugh K, Jechow A, Holker F, et al. Artificially lit surface of Earth at night increasing in radiance and extent. Sci Adv 2017;3(11):e1701528.

[4] Chepesiuk R. Missing the dark: health effects of light pollution. Environ Health Perspect 2009;117(1):A20–7.

[5] Falchi F, Cinzano P, Elvidge CD, Keith DM, Haim A. Limiting the impact of light pollution on human health, environment and stellar visibility. J Environ Manage 2011;92(10):2714–22.

[6] Pauley SM. Lighting for the human circadian clock: recent research indicates that lighting has become a public health issue. Med Hypotheses 2004;63(4):588–96.

[7] Smith FG. Report and recommendations of IAU Commission Reports on astronomy. IAU Trans XVIIA 1979;50:218–22.

[8] Falchi F, Cinzano P, Duriscoe D, Kyba CC, Elvidge CD, Baugh K, et al. The new world atlas of artificial night sky brightness. Sci Adv 2016;2(6):e1600377.

[9] Turner PL, Van Someren EJ, Mainster MA. The role of environmental light in sleep and health: effects of ocular aging and cataract surgery. Sleep Med Rev 2010;14(4):269–80.

[10] Cho Y, Ryu SH, Lee BR, Kim KH, Lee E, Choi J. Effects of artificial light at night on human health: A literature review of observational and experimental studies applied to exposure assessment. Chronobiol Int 2015;32(9):1294–310.

[11] Haus E, Smolensky M. Biological clocks and shift work: circadian dysregulation and potential long-term effects. Cancer Causes Control 2006;17(4):489–500.

[12] Schulmeister K, Weber M, Bogner W, Schernhammer E. Application of melatonin action spectra on practical lighting issues. Palo Alto, CA: The Electric Power Research Institute; 2004. Contract No.: 1009370.

[12a] Kraus LJ. Human and environmental effects of light emitting diode (LED) community lighting. 2016. Report No. 2-A-16.

[13] Koo YS, Song JY, Joo EY, Lee HJ, Lee E, Lee SK, et al. Outdoor artificial light at night, obesity, and sleep health: cross-sectional analysis in the KoGES study. Chronobiol Int 2016;33(3):301–14.

[14] Obayashi K, Saeki K, Kurumatani N. Association between light exposure at night and insomnia in the general elderly population: the HEIJO-KYO cohort. Chronobiol Int 2014;31(9):976–82.

[15] James P, Bertrand KA, Hart JE, Schernhammer ES, Tamimi RM, Laden F. Outdoor light at night and breast cancer incidence in the nurses' health study II. Environ Health Perspect 2017;125(8):087010.

[16] Al-Naggar RA, Anil S. Artificial light at night and cancer: global study. Asian Pac J Cancer Prev 2016;17(10):4661–4.

[17] Spivey A. Light at night and breast cancer risk worldwide. Environ Health Perspect 2010;118(12):a525.

[18] Lewy AJ, Wehr TA, Goodwin FK, Newsome DA, Markey SP. Light suppresses melatonin secretion in humans. Science 1980;210(4475):1267–9.

[19] Bojkowski CJ, Aldhous ME, English J, Franey C, Poulton AL, Skene DJ, et al. Suppression of nocturnal plasma melatonin and 6-sulphatoxymelatonin by bright and dim light in man. Horm Metab Res 1987;19(9):437–40.

[20] Trinder J, Armstrong SM, O'Brien C, Luke D, Martin MJ. Inhibition of melatonin secretion onset by low levels of illumination. J Sleep Res 1996;5(2):77–82.

[21] Schaefer EW, Williams MV, Zee PC. Sleep and circadian misalignment for the hospitalist: a review. J Hosp Med 2012;7(6):489–96.

[22] Bartel KA, Gradisar M, Williamson P. Protective and risk factors for adolescent sleep: a meta-analytic review. Sleep Med Rev 2015;21:72–85.

[23] Smolensky MH, Sackett-Lundeen LL, Portaluppi F. Nocturnal light pollution and underexposure to daytime sunlight: complementary mechanisms of circadian disruption and related diseases. Chronobiol Int 2015;32(8):1029–48.

[24] Gamble AL, D'Rozario AL, Bartlett DJ, Williams S, Bin YS, Grunstein RR, et al. Adolescent sleep patterns and night-time technology use: results of the Australian Broadcasting Corporation's Big Sleep Survey. PLoS One 2014;9(11):e111700.

[25] Cho JR, Joo EY, Koo DL, Hong SB. Let there be no light: the effect of bedside light on sleep quality and background electroencephalographic rhythms. Sleep Med 2013;14(12):1422–5.

[26] Martin JS, Hebert M, Ledoux E, Gaudreault M, Laberge L. Relationship of chronotype to sleep, light exposure, and work-related fatigue in student workers. Chronobiol Int 2012;29(3):295–304.

[27] Auger RR, Burgess HJ, Dierkhising RA, Sharma RG, Slocumb NL. Light exposure among adolescents with delayed sleep phase disorder: a prospective cohort study. Chronobiol Int 2011;28(10):911–20.

[28] Touitou Y, Reinberg A, Touitou D. Association between light at night, melatonin secretion, sleep deprivation, and the internal clock: health impacts and mechanisms of circadian disruption. Life Sci 2017;173:94–106.

[29] Pattinson CL, Allan AC, Staton SL, Thorpe KJ, Smith SS. Environmental light exposure is associated with increased body mass in children. PLoS One 2016;11(1):e0143578.

[30] McFadden E, Jones ME, Schoemaker MJ, Ashworth A, Swerdlow AJ. The relationship between obesity and exposure to light at night: cross-sectional analyses of over 100,000 women in the breakthrough generations study. Am J Epidemiol 2014;180(3):245–50.

[30a] Wyse CA, Selman C, Page MM, Coogan AN, Hazlerigg DG. Circadian desynchrony and metabolic dysfunction; did light pollution make us fat? Med Hypotheses 2011;77(6):1139–44.

[31] Shechter A, Kim EW, St-Onge MP, Westwood AJ. Blocking nocturnal blue light for insomnia: a randomized controlled trial. J Psychiatr Res 2018;96:196–202.

[31a] Votsi N-EP, Kallimanis AS, Pantis ID. An environmental index of noise and light pollution at EU by spatial correlation of quiet and unlit areas. Environ Pollut 2017;221:459–69.

[32] Mason IC, Boubekri M, Figueiro MG, Hasler BP, Hattar S, Hill SM, et al. Circadian health and light: a report on the national heart, lung, and Blood Institute"s workshop. J Biol Rhythms 2018;33(5):451–7.

[33] Bach V, Telliez F, Chardon K, Tourneux P, Cardot V, Libert JP. Thermoregulation in wakefulness and sleep in humans. Handb Clin Neurol 2011;98:215–27.

[34] Krauchi K, Deboer T. The interrelationship between sleep regulation and thermoregulation. Front Biosci 2010;15:604–25.

[35] Boulant JA. Role of the preoptic-anterior hypothalamus in thermoregulation and fever. Clin Infect Dis 2000;31(Suppl 5):S157–61.

[36] Van Someren EJ. More than a marker: interaction between the circadian regulation of temperature and sleep, age-related changes, and treatment possibilities. Chronobiol Int 2000;17(3):313–54.

[37] Kharakoz DP. Brain temperature and sleep. Zh Vyssh Nerv Deiat Im I P Pavlova 2013;63(1):113–24.

[37a] Lan L, Lian Z. Ten questions concerning thermal environment and sleep quality. Build Environ 2016;99:252–9.

[38] Song C, Liu Y, Zhou X, Liu J. Investigation of human thermal comfort in sleeping environments based on the effects of bed climate. Procedia Eng 2015;121:1126–32.

[39] Okamoto-Mizuno K, Mizuno K. Effects of thermal environment on sleep and circadian rhythm. J Physiol Anthropol 2012;31:14.

[40] Monsivais D, Bhattacharya K, Ghosh A, Dunbar RIM, Kaski K. Seasonal and geographical impact on human resting periods. Sci Rep 2017;7(1):10717.

[41] CISBE. Guide A: environmental design. London: Chartered Institution of Building Services Engineers (CISBE); 2006.

[42] Krauchi K, Fattori E, Giordano A, Falbo M, Iadarola A, Agli F, et al. Sleep on a high heat capacity mattress increases conductive body heat loss and slow wave sleep. Physiol Behav 2018;185:23–30.

[43] Okamoto-Mizuno K, Mizuno K, Shirakawa S. Sleep and skin temperature in preschool children and their mothers. Behav Sleep Med 2018;16(1):64–78.

[44] Strom-Tejsen P, Zukowska D, Wargocki P, Wyon DP. The effects of bedroom air quality on sleep and next-day performance. Indoor Air 2016;26(5):679–86.

[45] Accinelli RA, Llanos O, Lopez LM, Pino MI, Bravo YA, Salinas V, et al. Adherence to reduced-polluting biomass fuel stoves improves respiratory and sleep symptoms in children. BMC Pediatr 2014;14:12.

[46] Basner M, Babisch W, Davis A, Brink M, Clark C, Janssen S, et al. Auditory and non-auditory effects of noise on health. Lancet 2014;383(9925):1325–32.

[47] Muzet A. Environmental noise, sleep and health. Sleep Med Rev 2007;11(2):135–42.

[48] Berglund B, Lindvall T, Schwela DH. Team WHOOaEH. Guidelines for Community Noise. Geneva: World Health Organization; 1999. Available from: http://www.who.int/iris/handle/10665/66217.

[49] Tavakoli P, Varma S, Campbell K. Highly relevant stimuli may passively elicit processes associated with consciousness during the sleep onset period. Conscious Cogn 2018;58:60–74.

[50] Sorensen M, Ketzel M, Overvad K, Tjonneland A, Raaschou-Nielsen O. Exposure to road traffic and railway noise and postmenopausal breast cancer: a cohort study. Int J Cancer 2014;134(11):2691–8.

[51] Aasvang GM, Overland B, Ursin R, Moum T. A field study of effects of road traffic and railway noise on polysomnographic sleep parameters. J Acoust Soc Am 2011;129(6):3716–26.

[52] Basner M, Muller U, Elmenhorst EM. Single and combined effects of air, road, and rail traffic noise on sleep and recuperation. Sleep 2011;34(1):11–23.

[53] Munzel T, Gori T, Babisch W, Basner M. Cardiovascular effects of environmental noise exposure. Eur Heart J 2014;35(13):829–36.

[54] Halperin D. Environmental noise and sleep disturbances: a threat to health? Sleep Sci 2014;7(4):209–12.

[55] Lercher P, Brink M, Rudisser J, Van Renterghem T, Botteldooren D, Baulac M, et al. The effects of railway noise on sleep medication intake: results from the ALPNAP-study. Noise Health 2010;12(47):110–9.

[56] Stansfeld S, Hygge S, Clark C, Alfred T. Night time aircraft noise exposure and children's cognitive performance. Noise Health 2010;12(49):255–62.

[57] Saremi M, Greneche J, Bonnefond A, Rohmer O, Eschenlauer A, Tassi P. Effects of nocturnal railway noise on sleep fragmentation in young and middle-aged subjects as a function of type of train and sound level. Int J Psychophysiol 2008;70(3):184–91.

[58] Elmenhorst EM, Pennig S, Rolny V, Quehl J, Mueller U, Maass H, et al. Examining nocturnal railway noise and aircraft noise in the field: sleep, psychomotor performance, and annoyance. Sci Total Environ 2012;424:48–56.

[59] Pennig S, Quehl J, Mueller U, Rolny V, Maass H, Basner M, et al. Annoyance and self-reported sleep disturbance due to night-time railway noise examined in the field. J Acoust Soc Am 2012;132(5):3109–17.

[60] Micic G, Zajamsek B, Lack L, Hansen K, Doolan C, Hansen C, et al. A review of the potential impacts of wind farm noise on sleep. Acoust Australia 2018;46(1):87–97.

[61] Miedema HM, Vos H. Associations between self-reported sleep disturbance and environmental noise based on reanalyses of pooled data from 24 studies. Behav Sleep Med 2007;5(1):1–20.

[62] Ho A, Raja B, Waldhorn R, Baez V, Mohammed I. New onset of insomnia in hospitalized patients in general medical wards: incidence, causes, and resolution rate. J Community Hosp Intern Med Perspect 2017;7(5):309–13.

[62a] Kageyama T. Adverse effects of community noise as a public health issue. Sleep Biol Rhythms 2016;14(3):223–9.

[63] EPA. Guidelines for the assessment of noise from rail infrastructure. Adelaide, SA: EPA; 2013.

[64] Gidlof-Gunnarsson A, Ogren M, Jerson T, Ohrstrom E. Railway noise annoyance and the importance of number of trains, ground vibration, and building situational factors. Noise Health 2012;14(59):190–201.

[65] Klaeboea R, Turunen-Riseb IH, Hårvikc L, Madshusc C. Vibration in dwellings from road and rail traffic—part II: exposure–effect relationships based on ordinal logit and logistic regression models. Appl Acoust 2003;64(1):89–109.

[66] Peris E, Woodcock J, Sica G, Moorhouse AT, Waddington DC. Annoyance due to railway vibration at different times of the day. J Acoust Soc Am 2012;131(2):EL191–6.

[67] Smith MG, Croy I, Ogren M, Persson WK. On the influence of freight trains on humans: a laboratory investigation of the impact of nocturnal low frequency vibration and noise on sleep and heart rate. PLoS One 2013;8(2):e55829.

[68] Baker FC, Maloney S, Driver HS. A comparison of subjective estimates of sleep with objective polysomnographic data in healthy men and women. J Psychosom Res 1999;47(4):335–41.

[69] Marks A, Griefahn B. Associations between noise sensitivity and sleep, subjectively evaluated sleep quality, annoyance, and performance after exposure to nocturnal traffic noise. Noise Health 2007;9(34):1–7.

[70] Kimura H, Kuramoto A, Inui Y, Inou N. Mechanical bed for investigating sleep-inducing vibration. J Healthc Eng 2017;2017:2364659.

[71] Mehra R, Redline S. Sleep apnea: a proinflammatory disorder that coaggregates with obesity. J Allergy Clin Immunol 2008;121(5):1096–102.

[72] DeMeo DL, Zanobetti A, Litonjua AA, Coull BA, Schwartz J, Gold DR. Ambient air pollution and oxygen saturation. Am J Respir Crit Care Med 2004;170(4):383–7.

[73] Jerrett M, Shankardass K, Berhane K, Gauderman WJ, Kunzli N, Avol E, et al. Traffic-related air pollution and asthma onset in children: a prospective cohort study with individual exposure measurement. Environ Health Perspect 2008;116(10):1433–8.

[74] Zanobetti A, Redline S, Schwartz J, Rosen D, Patel S, O'Connor GT, et al. Associations of PM10 with sleep and sleep-disordered breathing in adults from seven US urban areas. Am J Respir Crit Care Med 2010;182(6):819–25.

[75] Shen YL, Liu WT, Lee KY, Chuang HC, Chen HW, Chuang KJ. Association of PM2.5 with sleep-disordered breathing from a population-based study in Northern Taiwan urban areas. Environmental pollution (Barking, Essex: 1987). 233:2018;109–13.

[76] Weinreich G, Wessendorf TE, Pundt N, Weinmayr G, Hennig F, Moebus S, et al. Association of short-term ozone and temperature with sleep disordered breathing. Eur Respir J 2015;46(5):1361–9.

[77] Tenero L, Piacentini G, Nosetti L, Gasperi E, Piazza M, Zaffanello M. Indoor/outdoor not-voluptuary-habit pollution and sleep-disordered breathing in children: a systematic review. Transl Pediatr 2017;6(2):104–10.

[78] An R, Yu H. Impact of ambient fine particulate matter air pollution on health behaviors: a longitudinal study of university students in Beijing, China. Public Health 2018;159:107–15.

[79] Klepeis NE, Nelson WC, Ott WR, Robinson JP, Tsang AM, Switzer P, Behar JV, Hern SC, Engelmann WH. The National Human Activity Pattern Survey (NHAPS): a resource for assessing exposure to environmental pollutants. J Expo Anal Environ Epidemiol 2001;11:231–52.

[80] Odeh I, Hussein T. Activity pattern of urban adult students in an eastern mediterranean society. Int J Environ Res Public Health 2016;13(10):960e5.

[81] Schweizer C, Edwards RD, Bayer-Oglesby L, Gauderman WJ, Ilacqua V, Jantunen MJ, et al. Indoor time-microenvironment-activity patterns in seven regions of Europe. J Expo Sci Environ Epidemiol 2007;17(2):170–81.

[82] Boor BE, Järnström H, Novoselac A, Xu Y. Infant exposure to emissions of volatile organic compounds from crib mattresses. Environ Sci Technol 2014;48(6):3541–9.

[83] Boor BE, Spilak MP, Corsi RL, Novoselac A. Characterizing particle resuspension from mattresses: chamber study. Indoor Air 2015;25(4):441–56.

[84] Lin SY, Melvin TA, Boss EF, Ishman SL. The association between allergic rhinitis and sleep-disordered breathing in children: a systematic review. Int Forum Allergy Rhinol 2013;3(6):504–9.

[85] Yetish G, Kaplan H, Gurven M, Wood B, Pontzer H, Manger PR, et al. Natural sleep and its seasonal variations in three pre-industrial societies. Curr Biol 2015;25(21):2862–8.

[86] Khalsa SBS, Jewett ME, Cajochen C, Czeisler CA. A phase response curve to single bright light pulses in human subjects. J Physiol 2003;549:945–52.

[87] Walch OJ, Cochran A, Forger DB. A global quantification of "normal" sleep schedules using smartphone data. Sci Adv 2016;2(5):e1501705.

[88] Brockmann PE, Gozal D, Villarroel L, Damiani F, Nuñez F, Cajochen C. Geographic latitude and sleep duration: a population-based survey from the Tropic of Capricorn to the Antarctic Circle. Chronobiol Int 2017;34(3):373–81.

[89] Friborg O, Bjorvatn B, Amponsah B, Pallesen S. Associations between seasonal variations in day length (photoperiod), sleep timing, sleep quality and mood: a comparison between Ghana (5 degrees) and Norway (69 degrees). J Sleep Res 2012;21(2):176–84.

[90] Stothard ER, McHill AW, Depner CM, Birks BR, Moehlman TM, Ritchie HK, et al. Circadian entrainment to the natural light-dark cycle across seasons and the weekend. Curr Biol 2017;27(4):508–13.

[91] Warnecke RB, Oh A, Breen N, Gehlert S, Paskett E, Tucker KL, et al. Approaching health disparities from a population perspective: the National Institutes of Health Centers for Population Health and Health Disparities. Am J Public Health 2008;98(9):1608–15.

[92] Lazarus RS. Theory-based stress measurement. Psychol Inq 1990;1(1):3–13.

[93] Pitsavos C, Panagiotakos DB, Papageorgiou C, Tsetsekou E, Soldatos C, Stefanadis C. Anxiety in relation to inflammation and

[94] Kim EJ, Dimsdale JE. The effect of psychosocial stress on sleep: a review of polysomnographic evidence. Behav Sleep Med 2007;5(4):256–78.
[95] Grandner MA, Williams NJ, Knutson KL, Roberts D, Jean-Louis G. Sleep disparity, race/ethnicity, and socioeconomic position. Sleep Med 2016;18:7–18.
[96] Grandner MA, Patel NP, Gehrman PR, Xie D, Sha D, Weaver T, et al. Who gets the best sleep? Ethnic and socioeconomic factors related to sleep complaints. Sleep Med 2010;11(5):470–8.
[97] Whinnery J, Jackson N, Rattanaumpawan P, Grandner MA. Short and long sleep duration associated with race/ethnicity, sociodemographics, and socioeconomic position. Sleep 2014;37(3):601–11.
[98] Gellis LA, Lichstein KL, Scarinci IC, Durrence HH, Taylor DJ, Bush AJ, et al. Socioeconomic status and insomnia. J Abnorm Psychol 2005;114(1):111–8.
[99] Tomfohr LM, Ancoli-Israel S, Dimsdale JE. Childhood socioeconomic status and race are associated with adult sleep. Behav Sleep Med 2010;8(4):219–30.
[100] Williams DR, Lavizzo-Mourey R, Warren RC. The concept of race and health status in America. Public Health Rep 1994;109(1):26–41. (Washington, DC: 1974).
[101] Slopen N, Lewis TT, Williams DR. Discrimination and sleep: a systematic review. Sleep Med 2016;18:88–95.
[102] Reskin B. The race discrimination system. Annu Rev Sociol 2012;38(1):17–35.
[103] James SA. John Henryism and the health of African-Americans. Cult Med Psychiatry 1994;18(2):163–82.
[104] Barksdale DJ, Woods-Giscombe C, Logan JG. Stress, cortisol, and nighttime blood pressure dipping in nonhypertensive Black American women. Biol Res Nurs 2013;15(3):330–7.
[105] Chen-Edinboro LP, Kaufmann CN, Augustinavicius JL, Mojtabai R, Parisi JM, Wennberg AM, et al. Neighborhood physical disorder, social cohesion, and insomnia: results from participants over age 50 in the Health and Retirement Study. Int Psychogeriatr 2014;27(2):289–96.
[106] Desantis AS, Diez Roux AV, Moore K, Baron KG, Mujahid MS, Nieto FJ. Associations of neighborhood characteristics with sleep timing and quality: the multi-ethnic study of atherosclerosis. Sleep 2013;36(10):1543–51.
[107] Simonelli G, Dudley KA, Weng J, Gallo LC, Perreira K, Shah NA, et al. Neighborhood factors as predictors of poor sleep in the sueno ancillary study of the hispanic community health study/study of latinos. Sleep 2017;40(1):zsw025.
[108] Johnson DA, Simonelli G, Moore K, Billings M, Mujahid MS, Rueschman M, et al. The neighborhood social environment and objective measures of sleep in the multi-ethnic study of atherosclerosis. Sleep 2017;40(1).
[109] Johnson DA, Lisabeth L, Hickson D, Johnson-Lawrence V, Samdarshi T, Taylor H, et al. The social patterning of sleep in African Americans: associations of socioeconomic position and neighborhood characteristics with sleep in the Jackson heart study. Sleep 2016;39(9):1749–59.
[110] Fang SC, Subramanian SV, Piccolo R, Yang M, Yaggi HK, Bliwise DL, et al. Geographic variations in sleep duration: a multilevel analysis from the Boston Area Community Health (BACH) Survey. J Epidemiol Community Health 2015;69(1):63–9.
[111] Xiao Q, Hale L. Neighborhood socioeconomic status, sleep duration, and napping in middle-to-old aged US men and women. Sleep 2018;41(7).
[112] Bassett E, Moore S. Neighbourhood disadvantage, network capital and restless sleep: is the association moderated by gender in urban-dwelling adults? Soc Sci Med 2014;108:185–93.
[113] Hale L, Hill TD, Friedman E, Nieto FJ, Galvao LW, Engelman CD, et al. Perceived neighborhood quality, sleep quality, and health status: evidence from the Survey of the Health of Wisconsin. Soc Sci Med 2013;79:16–22.
[114] Chambers EC, Pichardo MS, Rosenbaum E. Sleep and the housing and neighborhood environment of urban latino adults living in low-income housing: The AHOME study. Behav Sleep Med 2016;14(2):169–84.
[115] Johnson SL, Solomon BS, Shields WC, McDonald EM, McKenzie LB, Gielen AC. Neighborhood violence and its association with mothers' health: assessing the relative importance of perceived safety and exposure to violence. J Urban Health 2009;86(4):538–50.
[116] Ruff RR, Ng J, Jean-Louis G, Elbel B, Chaix B, Duncan DT. Neighborhood stigma and sleep: findings from a pilot study of low-income housing residents in New York city. Behav Med 2016;44:1–6.
[117] Billings ME, Johnson DA, Simonelli G, Moore K, Patel SR, Diez Roux AV, et al. Neighborhood walking environment and activity level are associated with OSA: the multi-ethnic study of atherosclerosis. Chest 2016;150(5):1042–9.
[118] Johnson DA, Hirsch JA, Moore KA, Redline S, Diez Roux AV. Associations between the built environment and objective measures of sleep: the multi-ethnic study of atherosclerosis. Am J Epidemiol 2018;187(5):941–50.
[119] Johnson DA, Drake C, Joseph CL, Krajenta R, Hudgel DW, Cassidy-Bushrow AE. Influence of neighbourhood-level crowding on sleep-disordered breathing severity: mediation by body size. J Sleep Res 2015;24(5):559–65.
[120] IARC. Shift-work, painting, and fire-fighting. International Agency for Research on Cancer; 2010.
[121] Linton SJ, Kecklund G, Franklin KA, Leissner LC, Sivertsen B, Lindberg E, et al. The effect of the work environment on future sleep disturbances: a systematic review. Sleep Med Rev 2015;23:10–9.
[122] Van Laethem M, Beckers DG, Kompier MA, Dijksterhuis A, Geurts SA. Psychosocial work characteristics and sleep quality: a systematic review of longitudinal and intervention research. Scand J Work Environ Health 2013;39(6):535–49.
[123] Yang B, Wang Y, Cui F, Huang T, Sheng P, Shi T, et al. Association between insomnia and job stress: a meta-analysis. Sleep Breath 2018;22(4):1221–31.
[124] Bierman A, Lee Y, Schieman S. Chronic discrimination and sleep problems in late life: religious involvement as buffer. Res Aging 2018;40(10):933–55.
[125] Ebin VJ, Sneed CD, Morisky DE, Rotheram-Borus MJ, Magnusson AM, Malotte CK. Acculturation and interrelationships between problem and health-promoting behaviors among Latino adolescents. J Adolesc Health 2001;28(1):62–72.
[126] Loredo JS, Soler X, Bardwell W, Ancoli-Israel S, Dimsdale JE, Palinkas LA. Sleep health in U.S. Hispanic population. Sleep 2010;33(7):962–7.
[127] Alcantara C, Patel SR, Carnethon M, Castaneda S, Isasi CR, Davis S, et al. Stress and sleep: results from the Hispanic Community Health Study/Study of Latinos Sociocultural Ancillary Study. SSM Popul Health 2017;3:713–21.

[128] D'Anna-Hernandez KL, Garcia E, Coussons-Read M, Laudenslager ML, Ross RG. Sleep moderates and mediates the relationship between acculturation and depressive symptoms in pregnant Mexican-American women. Matern Child Health J 2016;20(2):422–33.

[129] Ehlers CL, Gilder DA, Criado JR, Caetano R. Sleep quality and alcohol-use disorders in a select population of young-adult Mexican Americans. J Stud Alcohol Drugs 2010;71(6):879–84.

[130] Hale L, Rivero-Fuentes E. Negative acculturation in sleep duration among Mexican immigrants and Mexican Americans. J Immigr Minor Health 2011;13(2):402–7.

[131] Hale L, Troxel WM, Kravitz HM, Hall MH, Matthews KA. Acculturation and sleep among a multiethnic sample of women: the Study of Women's Health Across the Nation (SWAN). Sleep 2014;37(2):309–17.

[132] Roberts RE, Lee ES, Hemandez M, Solari AC. Symptoms of insomnia among adolescents in the lower Rio Grande Valley of Texas. Sleep 2004;27(4):751–60.

[133] Kachikis AB, Breitkopf CR. Predictors of sleep characteristics among women in southeast Texas. Womens Health Issues 2012;22(1):e99–109.

[134] Fekete EM, Williams SL, Skinta MD. Internalised HIV-stigma, loneliness, depressive symptoms and sleep quality in people living with HIV. Psychol Health 2017;1–18.

[135] Birtel MD, Wood L, Kempa NJ. Stigma and social support in substance abuse: implications for mental health and well-being. Psychiatry Res 2017;252:1–8.

[136] Kajeepeta S, Gelaye B, Jackson CL, Williams MA. Adverse childhood experiences are associated with adult sleep disorders: a systematic review. Sleep Med 2015;16(3):320–30.

[137] Oh DL, Jerman P, Silverio Marques S, Koita K, Purewal Boparai SK, Burke Harris N, et al. Systematic review of pediatric health outcomes associated with childhood adversity. BMC Pediatr 2018;18(1):83.

[138] Bilodeau F, Brendgen M, Vitaro F, et al. Longitudinal association between peer victimization and sleep problems in preschoolers: the moderating role of parenting. J Clin Child Adolesc Psychol 2018;47(suppl 1):S555–68.

[139] Chang LY, Chang HY, Lin LN, Wu CC, Yen LL. Transitions in sleep problems from late adolescence to young adulthood: a longitudinal analysis of the effects of peer victimization. Aggress Behav 2018;44(1):69–82.

[140] Wright AW, Austin M, Booth C, Kliewer W. Systematic review: exposure to community violence and physical health outcomes in youth. J Pediatr Psychol 2017;42(4):364–78.

[141] Bailey BN, Delaney-Black V, Hannigan JH, Ager J, Sokol RJ, Covington CY. Somatic complaints in children and community violence exposure. J Dev Behav Pediatr 2005;26(5):341–8.

[142] Spilsbury JC, Babineau DC, Frame J, Juhas K, Rork K. Association between children's exposure to a violent event and objectively and subjectively measured sleep characteristics: a pilot longitudinal study. J Sleep Res 2014;23(5):585–94.

[143] Umlauf MG, Bolland AC, Bolland KA, Tomek S, Bolland JM. The effects of age, gender, hopelessness, and exposure to violence on sleep disorder symptoms and daytime sleepiness among adolescents in impoverished neighborhoods. J Youth Adolesc 2015;44(2):518–42.

[144] Meltzer LJ, Montgomery-Downs HE. Sleep in the family. Pediatr Clin North Am 2011;58(3):765–74.

[145] Schulmeister K, Weber M, Bogner W, Schernhammer E. Application of melatonin action spectra on practical lighting issues. Palo Alto, CA: Electric Power Research Institute; 2004. Medium: X; Size: vp p.

[146] Lee KA, Zaffke ME, McEnany G. Parity and sleep patterns during and after pregnancy. Obstet Gynecol 2000;95(1):14–8.

[147] Hunter LP, Rychnovsky JD, Yount SM. A selective review of maternal sleep characteristics in the postpartum period. J Obstet Gynecol Neonatal Nurs 2009;38(1):60–8.

[148] Tikotzky L. Parenting and sleep in early childhood. Curr Opin Psychol 2017;15:118–24.

[149] Adam EK, Snell EK, Pendry P. Sleep timing and quantity in ecological and family context: a nationally representative time-diary study. J Fam Psychol 2007;21(1):4–19.

[150] Cousins JC, Bootzin RR, Stevens SJ, Ruiz BS, Haynes PL. Parental involvement, psychological distress, and sleep: a preliminary examination in sleep-disturbed adolescents with a history of substance abuse. J Fam Psychol 2007;21(1):104–13.

[151] Troxel WM, Lee L, Hall M, Matthews KA. Single-parent family structure and sleep problems in black and white adolescents. Sleep Med 2014;15(2):255–61.

[152] Rogojanski J, Carney CE, Monson CM. Interpersonal factors in insomnia: a model for integrating bed partners into cognitive behavioral therapy for insomnia. Sleep Med Rev 2013;17(1):55–64.

[153] McCurry SM, Song Y, Martin JL. Sleep in caregivers: what we know and what we need to learn. Curr Opin Psychiatry 2015;28(6):497–503.

[154] Adams GC, Stoops MA, Skomro RP. Sleep tight: exploring the relationship between sleep and attachment style across the life span. Sleep Med Rev 2014;18(6):495–507.

[155] Glenn C, Enwerem N, Odeyemi Y, Mehari A, Gillum RF. Social support and sleep symptoms in US adults. J Clin Sleep Med 2015;11(8):957.

[156] Grandner MA, Jackson NJ, Izci-Balserak B, Gallagher RA, Murray-Bachmann R, Williams NJ, et al. Social and behavioral determinants of perceived insufficient sleep. Front Neurol 2015;6:112.

[157] Williams NJ, Grandner MA, Wallace DM, Cuffee Y, Airhihenbuwa C, Okuyemi K, et al. Social and behavioral predictors of insufficient sleep among African Americans and Caucasians. Sleep Med 2016;18:103–7.

[158] Chung J. Social support, social strain, sleep quality, and actigraphic sleep characteristics: evidence from a national survey of US adults. Sleep Health 2017;3(1):22–7.

[159] Steptoe A, O'Donnell K, Marmot M, Wardle J. Positive affect, psychological well-being, and good sleep. J Psychosom Res 2008;64(4):409–15.

[160] Cinzano P, Falchi F, Elvidge CD. The first World Atlas of the artificial night sky brightness. Mon Not R Astron Soc 2001;328(3):689–707.

[161] WHO. Night Noise Guidelines for Europe Copenhagen: World Health Organization Regional Office for Europe. Available from: http://www.eruo.who.int/__data/assets/pdf_file/0017/43316/E92845.pdf; 2009.

[162] Ranson RP. Guidelines for healthy housing. Copenhagen: WHO Regional Office for Europe; 1988. 244 p.. Available from: http://www.who.int/iris/handle/10665/191555.

[163] CEN. Indoor environmental input parameters for design and assessment of energy performance of buildings. In: Normalisation CE, editors. Addressing Indoor Air Quality, Thermal Environment, Lighting, and Acoustics. 2007. Brussels.

[164] EPA. Indoor air pollution and health. Available from: https://www.epa.gov/indoor-air-quality-iaq/introduction-indoor-air-quality; 2018.

GLOSSARY

Air pollution   Term representing ambient concentrations of particulate matter, ozone, nitrogen dioxide, and sulfur dioxide.

Buoyant thermal plume   Upwelling and downwelling features in an element (e.g., fluid) that are maintained by thermal buoyancy, which is the ability or tendency to float in water, air, or some other fluid.

Circadian misalignment   The incorrect arrangement or position of the sleep-wake cycle to the biological night; incorrect arrangement or position of feeding rhythms to the sleep-wake cycle; internal misalignment of central and peripheral rhythms.

Circadian rhythm   A daily rhythmic activity cycle, based on 24-h intervals, that is exhibited by many organisms.

Circadian time structure disruption or circadian disruption   Disturbance or problems that interrupt naturally recurring 24-h cycles of biological processes.

Effort-reward imbalance   "High cost/low gain" situation at work, in which individuals spend high effort while receiving low rewards (in terms of monetary gratification, career opportunities, esteem, respect, and job security).

Environment   The complex physical, chemical, and biotic factors that act upon an organism or an ecological community and ultimately determine its form and survival; the aggregate of social and cultural conditions that influence the life of an individual or community.

Glare   A visual sensation caused by excessive and uncontrolled brightness.

Health disparity   A type of difference in health that is closely linked with social or economic disadvantage and negatively affects groups of people who have systematically experienced greater social or economic obstacles to health. These obstacles stem from characteristics historically linked to discrimination or exclusion such as race or ethnicity, religion, socioeconomic status, gender, mental health, sexual orientation, or geographic location.

Health inequity   A difference or disparity in health outcomes between groups that is systematic, avoidable, and unjust.

Job control   Two theoretically distinct subdimensions of decision latitude, namely skill discretion and decision authority.

Job demands   Stressors present in the work environment.

Job strain   Work condition that occurs when job demand is greater than job control.

Light trespass   Light that is cast where it is not wanted.

Light pollution   Light that is not targeted for a specific task, is bright and uncomfortable to the human eye, causes unsafe glare, trespasses in homes and bedrooms, or creates sky glow above cities.

Microenvironment   A small or relatively small usually distinctly specialized and effectively isolated habitat or environment.

Noise pollution   Annoying or harmful noise (as of automobiles or jet airplanes) in an environment.

Organizational justice   A multidimensional concept, which refers to the fairness of decision-making processes, how equally supervisors treat employees and share information, and whether employees themselves perceive that their viewpoints are taken into account.

Overillumination   Excessive light supplied beyond the amount required for a given task that can produce glare, annoyance, and adverse health effects.

Photopic   Relating to vision in bright light with light-adapted eyes, which involves the cones of the retina.

Sky glow   A glow in the night sky deriving from an artificial source.

Sleep onset latency   The length of time that it takes to accomplish the transition from full wakefulness to sleep, which is normally to the lightest of the non-REM sleep stages.

Scotopic   Relating to vision in dim or low light with dark-adapted eyes, which involves the retinal rods as light receptors.

Thermal comfort/neutral temperature   Satisfaction with thermal environment assessed by subjective evaluation.

Thermal neutral zone   An endotherm's temperature tolerance range where the basal rate of heat production is in equilibrium with the rate of heat loss to the external environment.

Part III

Addressing sleep health at the community and population level

Chapter 9

Obstacles to overcome when improving sleep health at a societal level

Michael A. Grandner

Sleep and Health Research Program, Department of Psychiatry, University of Arizona College of Medicine, Tucson, AZ, United States

INTRODUCTION

Sleep is important for overall health. Yet, achieving health sleep is difficult for a large number of people. This is not likely because people dislike sleep. Rather, there are obstacles to overcome that are imposed by individuals themselves and their contexts. This chapter will address some of the common real-world barriers to sleep health, conceptualizing how understanding these real-world barriers can be thought of in context, and recommendations for using this information to design and implement sleep health interventions in real-world settings.

REAL-WORLD BARRIERS TO SLEEP HEALTH

Lack of time

One of the most common explanations given for insufficient sleep is a lack of time. To examine this phenomenon, Basner and colleagues examined data from the American Time Use Survey (ATUS). The first study to explore these data [1] found that there was a relationship between sleep time and time spent socializing/leisure activities and traveling, such that less sleep associated with more of these activities. But the largest relationship was seen with work—clearly it is work that Americans are trading for time to sleep. This relationship is displayed in Fig. 9.1. In a follow-up analysis [2], the authors examined the activities in the 2 h prior to bed and following getting up. Fig. 9.2 shows the results of this analysis. In examining the 2 h before bed, the activity that seems to be most likely to interfere with getting into bed is not work, though, but watching television. And in examining what gets people up, that seems to be where work comes into play—the likelihood of working in after awakening increases steadily in the first 2 h, as does travel (e.g., commute). Still, many people find it difficult to make time for sleep, and this is a key barrier for interventions.

Social norms and beliefs

There is a general perception that sleep might be important, but it is difficult to achieve for people who are busy; this is a sentiment repeatedly echoed in the popular press but has received relatively little scientific attention. In a study by Henry and colleagues, the cultural impact of work being paramount and more important than sleep was reported by US adults [3]. In a study of a general population sample, Grandner and colleagues found that perceived social norms about the importance of sleep varied based on the outcome [4]. As depicted in Fig. 9.3, respondents generally agreed that lack of sleep could lead to cognitive/functional problems including tiredness, sleepiness, lack of energy, and memory/concentration problems. There was also general agreement that friends and family believed that lack of sleep could lead to moodiness and difficulties at work or school. The only outcomes where >40% of respondents reported that their friends and family strongly agreed that lack of sleep could lead to that outcome were tiredness, lower energy, and daytime sleepiness. The minority of respondents agreed that their friends and family believed that lack of sleep could lead to missed days, lower sex drive, weight gain, heart disease, hypertension, diabetes, and high cholesterol. Of note, most respondents who did not agree that their friends and family believed that lack of sleep could lead to outcomes indicated that they were unsure of their friends' and family's beliefs, rather than indicating that their friends and family do not believe those things.

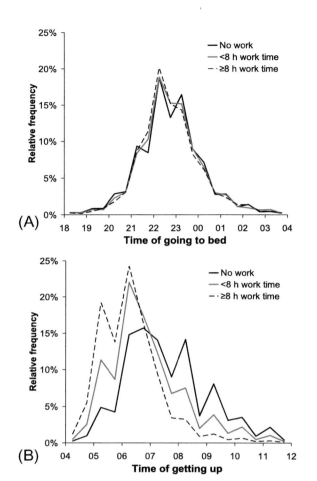

FIG. 9.1 Panels A and B show distributions of time of going to bed (A) and time of getting up (B) for respondents who did not work (*black lines*, N = 9770), who worked less than 8 h (*gray lines*, N = 5589), and for those who worked 8 h or more (*broken black lines*, N = 6116). *From Basner M, Fomberstein KM, Razavi FM, Banks S, William JH, Rosa RR, Dinges DF. American time use survey: sleep time and its relationship to waking activities. Sleep 2007;30(9):1085–95. Epub 2007/10/04. PubMed PMID: 17910380.*

Physical environment

Aspects of the physical environment propose unique barriers to sleep. Since sleep requires an environment conducive to a lack of sensory input, environments with an uncomfortable temperature level, too much or too little light, too much or too little noise, or some other aspect of the physical environment that is uncomfortable can interfere with sleep [5]. In the bedroom, this can present as environmental temperature that is too hot or too cold, or contains too much light or noise. It can also be represented in an uncomfortable sleeping surface or pillows or blankets. Outside of the bedroom, barriers to sleep become apparent when aspects of the physical environment become out of the individual's control. For example, regions that experience excessive light pollution have been shown to present a unique risk for medically-treated insomnia [6]. Other studies have shown that environmental noise from airplanes, traffic, and other sources can systematically interfere with sleep [7].

Addressing these aspects of the physical environment may be important ingredients of a sleep health intervention.

Health conditions and chronic pain

Chronic pain impacts approximately 1/5 of US adults [8], though this depends on age—rates ranged from 7.0% among 18–24 year olds to 33.6% among those 85 years old or older. High-impact chronic pain is estimated to impact 8% of US adults, and this also is strongly related to age, such that the least prevalence was among 18–24 year olds (1.5%) and the highest was among those 85 or older (15.8%). Chronic pain can directly interfere with sleep, causing reduced sleep duration and impaired sleep quality [9]. Other relatively common health problems can impact on sleep health as well. Sleep disorders such as sleep apnea, restless legs syndrome, and insomnia disorder are often undertreated in the population. Autoimmune and rheumatological diseases, as well as musculoskeletal and other orthopedic conditions can also be common and impact sleep. Cardiometabolic diseases such as diabetes, obesity, and hypertension can be reflected in poor sleep [10]. Cancer and remitted cancer are both associated with sleep difficulties [11]. Of note, nearly every domain of medical diagnosis has sleep disturbance as a common symptom either via pain/discomfort or through some other mechanism.

Substance use

Many substances that are commonly consumed in society have adverse impacts on sleep. Among these, alcohol, nicotine, caffeine, and cannabis have undergone the most study. Alcohol is consumed by the majority of US adults. It is frequently consumed as a sleep aid. Yet, alcohol can directly interfere with sleep by reducing sleep duration, increasing sleep fragmentation, and reducing sleep quality [12]. Although rates of smoking are declining, may US adults smoke. Although many people smoke to relax, and/or smoke in bed, nicotine is a stimulant. Smokers are more likely to have sleeping difficulties, especially in the beginning of the night [13, 14]. Further, smoking cessation treatments also tend to be insomnogenic [15]. As cannabis use becomes legitimized in society, its effects on sleep require much more study—especially since many people are using it as a sleep aid. Existing evidence suggests that it may provide some short-term benefits in terms of sleep continuity, but this may come at a cost of impaired sleep architecture [16, 17]. Further, those who regularly use cannabis are not more likely to have better sleep outcomes; rather, they frequently report worse sleep. This may indicate that although people use this substance to improve sleep, it is ineffective. Caffeine is possibly the most frequently consumed psychoactive drug in the world. Caffeine is frequently used to improve daytime functioning, and it is quite effective as a relatively safe stimulant. Yet, the effects of caffeine may last for several hours after consumption and this may interfere with sleep [18]. These and other commonly-used

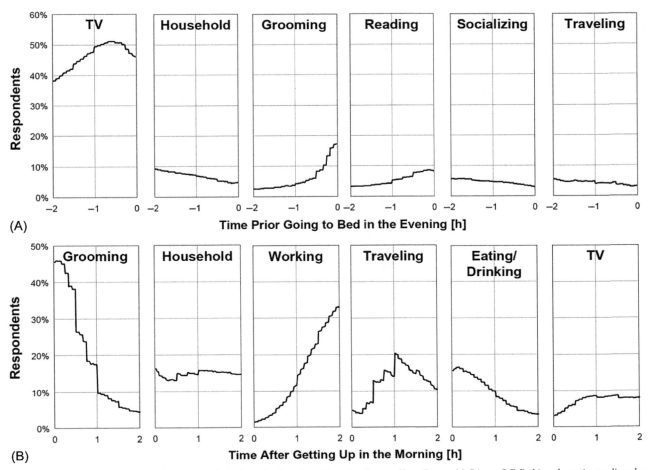

FIG. 9.2 Activities just before and after sleep periods, from the American Time Use Survey. *From Basner M, Dinges DF. Dubious bargain: trading sleep for Leno and Letterman. Sleep 2009;32(6):747–752. Epub 2009/06/24. PubMed PMID: 19544750; PMCID: 2690561.*

substances—which for many adults have become a regular part of daily life—may represent important barriers to sleep.

Distractions and on-demand culture

As the dominant culture has become one that discourages downtime and relaxation in favor of work and productivity, this has led to several widespread phenomena that are interfering with sleep. As highlighted earlier, television watching is probably the most universally common activity in the 2 h before bedtime, often leading right up until sleep [1]. Television is just one of many possible electronic distractions that people engage in the evening. For example, using mobile phones (which are actually more like handheld computers) to browse the internet, engage with social media, communicate via messaging system, play video games, and watch videos has become an indispensable part of the evening routines of many people. There is a growing body of evidence that the mental engagement (and distraction) that electronic devices provide may interfere with sleep just as they do so by emitting light [19]. This may reflect the cultural attitude toward constantly remaining mentally active—when not actively working or fulfilling other obligations, we keep our minds busy with these distractions. Yet, distraction is not relaxation, which is an active process of reducing physical and mental strain and tension. And this may result in extended time spent engaged in distracting activities in the hopes that they will aid in relaxation when they do not. This is evident in the practice of "binge-watching" television shows for hours rather than sleeping. This is related to the concept of "on-demand" culture, where things and experiences are available at all times, immediately. For example, whole seasons of television shows are available for bingeing at any time, day or night. Stores and especially websites are open 24 h a day, every day. In the middle of the night, in a relatively quiet town, a person can—if they so choose—watch nearly any movie or television show every recorded, shop online for everything from housewares to clothing to real estate to a new car, access nearly any book that has ever been written, immediately download and play nearly every video game ever developed, send a message to anyone that is instantly delivered, etc. It is possible that this cultural shift has provided additional opportunities for people to engage in distracting activities rather than sleeping.

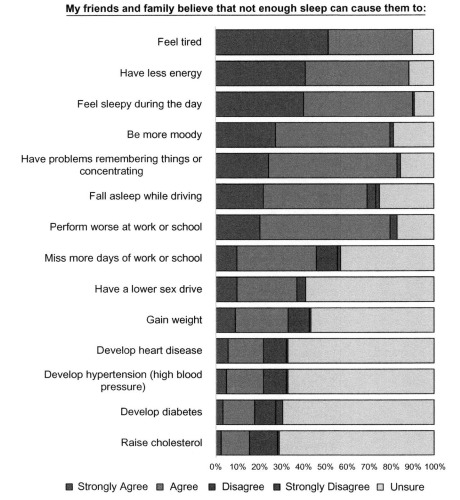

FIG. 9.3 Perceived social norms about sleep. *From Grandner MA, Jackson N, Gooneratne NS, Patel NP. The development of a questionnaire to assess sleep-related practices, beliefs, and attitudes. Behav Sleep Med 2014;12(2):123–142. PubMed PMID: 23514261; PMCID: 3795978.*

CONCEPTUALIZING STRATEGIES FOR OVERCOMING THESE BARRIERS

Although sleep is a biological imperative, it is driven by volitional behaviors and, thus, many of the same factors that drive other health-related behaviors. Several theories of health behavior change have been developed to understand how individuals decide to engage in healthful behaviors, and these might be applicable to sleep.

The health belief model and application to sleep

The Health Belief Model has been utilized for a wide range of possible health behaviors. Briefly, the Health Belief Model conceptualizes that a person will engage in a behavior based on the following conditions:

1. Perceived Susceptibility: "How vulnerable am I?"
2. Perceived Severity: "Are the consequences severe?"
3. Perceived Benefits: "Will taking action remove consequences?"
4. Perceived Barriers: "What is preventing me from taking action?"
5. Cues to Action: "Will I remember to take action?"
6. Self-Efficacy: "Is taking action in my control?"

See Fig. 9.4 for a schematic of this model (reprinted from Ref. [20]). There are many ways in which this model can be applied to sleep. For example, interventions should aim to address all six components of the model. Interventions should provide education about the overall relationship of sleep to the outcomes that people care about and also how they, personally, may be at risk for those adverse outcomes. These interventions should also focus on the importance of the outcomes and place them in the context that would be relevant for daily life. For example, an intervention aimed at getting an individual to prioritize sleep should not only provide education about how sleep is relevant to outcomes that the individual is specifically interested in (e.g., weight gain,

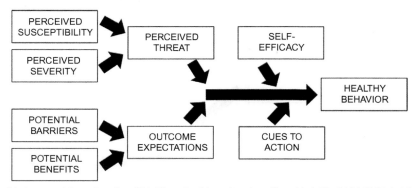

FIG. 9.4 Health belief model. *Reprinted from Grandner MA. Sleep, health, and society. Sleep Med Clin 2017;12(1):1–22. PubMed PMID: 28159089.*

work performance, relationship quality), but that the effects are large enough to warrant action. Interventions should also address consequences of performing the action—communicate that improving sleep can produce measurable changes in the outcomes that the individual cares about. Addressing barriers is absolutely critical. As outlined above, barriers to achieving healthy sleep are often cultural, physical, and (seemingly) external. If individuals do not consider these barriers to be surmountable, the intervention will be ineffective. A thorough understanding of the barriers, and the proposed solutions to those barriers, is needed. Self-efficacy is also very important. If a person feels that they have decreased ability to get sleep, even if those barriers are removed, then the intervention will be unlikely to be effective. Finally, the health belief model notes the importance of cues to action—reminders and encouragement needed in order for the intervention to be successful. Thus, in order for change to sleep to be possible, then a broad approach needs to be undertaken that connects the individual to the outcomes that they care about, convinces them that sleep will improve those outcomes, and facilitating that change by removing barriers, increasing self-efficacy, and providing cues to action.

The integrated behavioral model and application to sleep

The Integrated Behavioral Model arose from the Theory of Planned Behavior and Theory of Reasoned Action [21] to describe why people engage in behaviors. These are described in more detail in Chapter 14. According to the integrated behavior model, an individual needs to make a formal motivated decision (intention) to perform that behavior. That intention is thought to be the product of three separate influences:

1. Attitudes (overall cognitive and emotional perception of performing the behavior), which consist of
   a. Experiential Attitudes (emotional responses to the idea of performing the action)
   b. Instrumental Attitudes (beliefs about what would happen if the outcome is performed)
2. Subjective Norms (perception of the social influences over performing the action), which consist of
   a. Injunctive Norms (beliefs about what others think an individual should do and motivation to comply with those pressures)
   b. Descriptive Norms (beliefs about what other people in an individual's social group are actually doing)
3. Personal Agency (capability to actually perform the behavior, influenced by knowledge, sills, environmental constraints, etc.), which consists of
   a. Perceived Control (beliefs about whether or not an individual can actually perform the behavior in the context of constraints)
   b. Self-Efficacy (beliefs about whether or not an individual can perform a behavior well)

A schematic for this model is presented in Fig. 9.5. Regarding the understanding of sleep health in society, this model can help toward understanding how to build a stronger motivation for individuals to prioritize healthy sleep. According to this model, attitudes, norms, and agency all need to be addressed.

Interventions will need to focus on helping individuals to endorse helpful beliefs and attitudes about sleep. It will also necessitate that individuals generally have positive feelings about sleep—which may be difficult given the dominant culture. Norms can be addressed by better examining how the sleep of a person's (perceived) peers impacts behavior, as well as the (perceived) sleep of those to which that individual wishes to emulate. This would involve gaining a better understanding of what an individual believes their social group believes and does regarding sleep and determining the value of this information. For example, a person in a workplace where working late is prioritized over sufficient sleep will have a difficult time maintaining healthy habits no matter what positive sleep attitudes they hold. The social pressure to work late is a powerful force that will need to be addressed and not ignored. Regarding agency, this involves helping individuals feel empowered in engaging in positive sleep behaviors. When an individual has positive attitudes about sleep, perceives healthy sleep to be consistent with their social group, and perceives

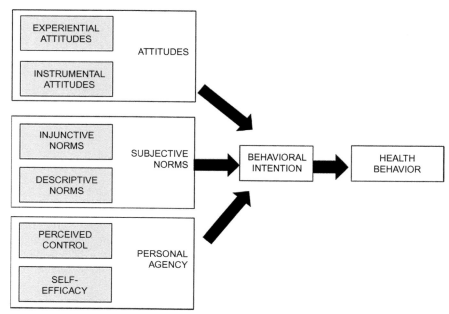

FIG. 9.5 Integrated behavior model. *Reprinted from Grandner MA. Sleep, health, and society. Sleep Med Clin 2017;12(1):1–22. PubMed PMID: 28159089.*

themselves to have the power to influence their sleep, positive changes in society are possible.

The transtheoretical stages-of-change model

The transtheoretical model of behavior change is sometimes called the "Stages of Change" model because it addresses the issue of readiness for change. According to this model, behavior change depends on an individual's readiness to engage in that healthy behavior. This is operationalized along five levels (depicted in Fig. 9.6):

1. Precontemplation (not even begun to consider change)
2. Contemplation (considered change, have not decided to act)
3. Preparation (decided to act, but have not begun)
4. Action (started to act, in early stages)
5. Maintenance (maintaining action over time)

Based on this model, an individual's readiness to engage in a behavior exists on a spectrum, and any attempt to promote healthy behavior should be tailored to the specific stage that the individual is currently inhabiting. For example, demanding action of someone in precontemplation is premature. And for those in the action stage, relapse prevention rather than motivation to begin an action is more appropriate.

This model can help in the understanding of the role of healthy sleep in society. Many individuals are likely at the precontemplation and contemplation stages, whereas others are at other stages. Understanding this concept of readiness for change can aid in the education and provision of sleep interventions. Many interventions aim for action, but if the individual is not yet at the action stage, then the intervention will be less likely to be successful. This implies that additional efforts are needed to simply increase "contemplation" and awareness, and separate efforts need to motivate action. It also implies that efforts should also be focused on maintenance once people engage in healthy sleep behaviors, so that they are maintained.

Other health behavior models

Other health behavior models exist, and many of these are reviewed in another chapter. Taken together, they may

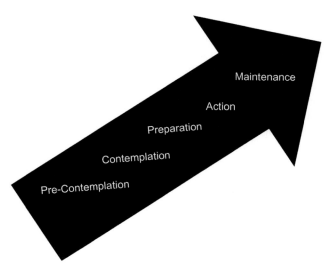

FIG. 9.6 Transtheoretical model.

provide insights on how sleep health interventions should be implemented in order to motivate sustained change and, eventually, better population sleep health.

IMPLEMENTING SLEEP HEALTH PROGRAMS

Addressing perceived benefits

In order for a sleep health intervention to be successful, it needs to appropriately aid the individual in perceiving benefits from improved sleep. Some suggestions include:

- Providing education on the importance of sleep for several domains of health and functioning, including tiredness/sleepiness/fatigue, behavioral and emotional health and well-being, cardiovascular and metabolic health, immune system functioning, cognitive performance, and overall physical and mental performance. In particular, interventions should focus on the outcomes that are the most relevant to the individual. These can be specific, such as absenteeism, financial stability, etc.
- Highlighting how sleep can impact both short-term and long-term outcomes. That the effects may be measurable in some outcomes very quickly and others over time. This will help reinforce that there are aspects of the person's daily life that can be impacted quickly so that they may see benefits soon, but that longer term changes may need more sustained efforts. Still, these benefits can be expected, eve if they take time.
- Demonstrating which areas of the individual's life might be amenable to better sleep. This may provide opportunities for helping the individual see the ways that improved sleep can benefit them.
- Pointing out how their functioning is improved on days when they sleep better. This will help individuals see that benefits of better sleep are possible and that they have, in fact, already been experienced.

Addressing perceived barriers

In addition to supporting perceived benefits, interventions should address perceived barriers as well. Some suggestions include:

- Comprehensively addressing the individual's perceived barriers to better sleep. Work with the individual to identify what all of those barriers are and develop strategies for overcoming them. These could be relatively simple or complicated, depending on the presenting problem. And some barriers may not be readily overcome (e.g., shiftwork).
- Review a chronology of a typical day or week, so that all of the barriers to healthy sleep could be identified. This can be accomplished as an interview, which would allow for follow-up questions regarding clarifying specific barriers that may be less obvious.
- Inventory which barriers are the most relevant and most likely to impede progress. Attention should be focused on those barriers that impede sleep the most.
- Develop strategies for mitigating the impact of barriers that are less amenable to change. For example, if a person's work shift is precluding healthy sleep and this cannot be changed, perhaps fatigue countermeasures could be employed or other approaches can be taken to support healthy sleep.

Addressing social norms

Even if the individual perceives benefits to healthy sleep behaviors and perceives that the barriers are addressable, the presence of social norms about sleep is important. Some suggestions for incorporating this concept into interventions include:

- Inventory perceived social norms regarding sleep, sleep health, taking time for sleep, winding down, and other relevant concepts. This inventory should address the norms perceived across various social factors such as work, family, friends, relevant social groups and demographic groups with which the person identified, "society," and other groups deemed important.
- Question the degree to which unhelpful attitudes about sleep are actually held by these groups, and/or question the degree to which disagreeing with those attitudes will result in social or other adverse outcomes.
- Focus on the benefits of better sleep, and identify ways in which those benefits would be valuable according to the norms of the identified social groups. For example, if an individual believes that their workplace does not value sleep, perhaps focus on the fact that the workplace values productivity, which may be enhanced with better sleep. Thus, even in the presence of social norms that are counter to healthy sleep, other social norms may be used to support a sleep health intervention.
- Include multiple individuals in group-wide sleep health interventions, and reinforce the importance of sleep from those in leadership positions. For example, if a workplace wellness program includes sleep health, then the program should reinforce the idea that all of the employees in the program are valuing sleep together, and that the company leadership also values sleep because they are supporting this program.

Addressing self-efficacy and control

There are a number of strategies that can be appropriated in order to increase an individual's sense of control over their sleep and their sense of self-efficacy regarding taking steps to improve their sleep. Some suggestions include:

- Inventory perceived control over sleep. This can be accomplished with the Brief Index of Sleep Control

(BrISC) or some other tool. Generally, any inventory should assess perceived control over sleep timing, sleep duration, and sleep quality. This may identify which aspects of sleep are seen as most problematic and should be the focus of a more intensive approach.

- Provide education about the basic functions of sleep and how basic sleep health interventions work. The goal of this approach is to help take some of the "mystery" out of sleep and help individuals see sleep as a set of processes that they can understand (at some level) and, therefore, control.
- Interventions should focus on measurable and attainable goals. Setting goals that are too difficult to achieve (e.g., sleep 8 h a night and feel refreshed every morning) may be too difficult. And repeated, daily failure to meet those goals may lead to frustration, burnout, and giving up. Rather, interventions should focus on incremental, achievable goals (e.g., go to bed 10 min earlier tonight) that can result in a track record of successes that accumulate over time. This can reinforce a sense of mastery rather than failure.

Addressing readiness

Interventions focused on immediate action may be premature. And those struggling to maintain behaviors despite already attempting to change behavior may be left without options. Some suggestions for promoting readiness and adapting to stages of change are:

- Assess the individual's stage of change. This would ascertain whether the person is in a stage of precontemplation, contemplation, preparation, action, or maintenance. Depending on the stage of change for sleep, interventions could be targeted.
- Individuals in the precontemplation stage should be funneled to interventions focused simply on increasing awareness. These interventions should aim to bring those individuals to the contemplation stage, whether or not they engage in behaviors yet.
- Individuals in the contemplation stage should be met with interventions aimed are achieving a behavioral intention to improve sleep and progress to the preparation stage. These interventions should aim to understand what the individual will need in order to make the decision to change their sleep.
- Individuals in the preparation stage, who are intending or planning to make a change, do not need to be convinced about the importance of sleep. Rather, interventions at this stage should focus on removing barriers and facilitating the first steps of actual action.
- Individuals at the action stage should be given all of the tools possible in order to remain successful for as long as possible. This includes a focus on the barriers

FIG. 9.7 Recommended elements of a sleep health intervention.

and benefits, as well as maintaining focus on the issues that are most relevant to them. The goal at this stage is to turn the intervention into a habit and facilitate its maintenance.

- Individuals who have been working on their sleep for some time should be given the tools needed in order to identify problem areas, address those, and periodically reinforce their healthy practices. Dealing with lapses is important here as well. The goal at this stage is to keep individuals performing healthy sleep behaviors and preventing relapse.

Taken together, these recommendations focus on six main areas (depicted in Fig. 9.7). Interventions should include education about sleep, about the importance of sleep, and about how to make improvements to sleep. Assessment is another key component; this entails not just assessment of sleep, but also assessment of the contextual factors that play a role in sleep, including social-level factors. Interventions should also include an element of facilitation—identifying and removing perceived barriers. Related to this issue is the one of control; interventions should explicitly aim to increase individuals' sense of control over their sleep and their self-efficacy regarding sleep-related behaviors. Another key element is addressing the social norms—this includes assessment but also may include strategies for mitigating unhelpful sleep-related norms. Finally, interventions should be adaptive regarding an individual's level of readiness and stage of change.

CONCLUSION

Real-world barriers to sleep health include intra- and interpersonal factors, as well as social and contextual factors.

These can encompass a wide range of possibilities, most prominently perceived lack of time for sleep, social norms, physical environment, health issues, substances used, and the ubiquity of distractions. These barriers and others can be conceptualized in any number of models, including the Health Belief Model, Integrated Behavior Model, and Transtheoretical Model. Using these frameworks, some recommendations for implementing sleep behavior change interventions include (1) maximizing perceived benefits, (2) minimizing perceived barriers, (3) mitigating unhelpful social norms, (4) increasing self-efficacy and feelings of control, and (5) leveraging the concept of readiness. Future research needs to study the implementation of these approaches into existing sleep health interventions in order to evaluate which elements of sleep health interventions are most helpful, and in what contexts.

REFERENCES

[1] Basner M, Fomberstein KM, Razavi FM, Banks S, William JH, Rosa RR, Dinges DF. American time use survey: sleep time and its relationship to waking activities. Sleep 2007;30(9):1085–95. Epub 2007/10/04. PubMed PMID: 17910380.

[2] Basner M, Dinges DF. Dubious bargain: trading sleep for leno and letterman. Sleep 2009;32(6):747–52. Epub 2009/06/24. PubMed PMID: 19544750.

[3] Henry D, McClellen D, Rosenthal L, Dedrick D, Gosdin M. Is sleep really for sissies? Understanding the role of work in insomnia in the US. Soc Sci Med 2008;66(3):715–26. Epub 2007/11/17. PubMed PMID: [18006129].

[4] Grandner MA, Jackson N, Gooneratne NS, Patel NP. The development of a questionnaire to assess sleep-related practices, beliefs, and attitudes. Behav Sleep Med 2014;12(2):123–42. PubMed PMID: 23514261.

[5] Pigeon WR, Grandner MA. Creating an optimal sleep environment. In: Kushida CA, editor. Encyclopedia of sleep. Oxford: Elsevier; 2013. p. 72–6.

[6] Min JY, Min KB. Outdoor artificial nighttime light and use of hypnotic medications in older adults: a population-based cohort study. J Clin Sleep Med 2018;14(11):1903–10. PubMed PMID: 30373695.

[7] Omlin S, Bauer GF, Brink M. Effects of noise from non-traffic-related ambient sources on sleep: review of the literature of 1990-2010. Noise Health 2011;13(53):299–309. Epub 2011/07/20. PubMed PMID: [21768734].

[8] Dahlhamer J, Lucas J, Zelaya C, Nahin R, Mackey S, DeBar L, Kerns R, Von Korff M, Porter L, Helmick C. Prevalence of chronic pain and high-impact chronic pain among adults—United States, 2016. MMWR Morb Mortal Wkly Rep 2018;67(36):1001–6. PubMed PMID: 30212442.

[9] Vitiello MV. A step toward solving the sleep/pain puzzle. Sleep 2012;35(5):593–4. PubMed PMID: 22547883.

[10] Grandner MA, Alfonso-Miller P, Fernandez-Mendoza J, Shetty S, Shenoy S, Combs D. Sleep: important considerations for the prevention of cardiovascular disease. Curr Opin Cardiol 2016;31(5):551–65. PubMed PMID: 27467177.

[11] Phillips KM, Jim HS, Donovan KA, Pinder-Schenck MC, Jacobsen PB. Characteristics and correlates of sleep disturbances in cancer patients. Support Care Cancer 2012;20(2):357–65. Epub 2011/02/12. PubMed PMID: [21311913].

[12] Ebrahim IO, Shapiro CM, Williams AJ, Fenwick PB. Alcohol and sleep I: effects on normal sleep. Alcohol Clin Exp Res 2013;37(4):539–49. PubMed PMID: [23347102].

[13] Patterson F, Grandner MA, Lozano A, Satti A, Ma G. Transitioning from adequate to inadequate sleep duration associated with higher smoking rate and greater nicotine dependence in a population sample. Addict Behav 2018;77:47–50. PubMed PMID: 28950118.

[14] de Leeuw R, Eisenlohr-Moul T, Bertrand P. The association of smoking status with sleep disturbance, psychological functioning, and pain severity in patients with temporomandibular disorders. J Orofac Pain 2013;27(1):32–41. PubMed PMID: 23424718.

[15] Patterson F, Grandner MA, Malone SK, Rizzo A, Davey A, Edwards DG. Sleep as a target for optimized response to smoking cessation treatment. Nicotine Tob Res 2019. 21, 139–148. PubMed PMID: 29069464.

[16] Babson KA, Boden MT, Bonn-Miller MO. The impact of perceived sleep quality and sleep efficiency/duration on cannabis use during a self-guided quit attempt. Addict Behav 2013;38(11):2707–13. PubMed PMID: 23906725.

[17] Bolla KI, Lesage SR, Gamaldo CE, Neubauer DN, Wang NY, Funderburk FR, Allen RP, David PM, Cadet JL. Polysomnogram changes in marijuana users who report sleep disturbances during prior abstinence. Sleep Med 2010;11(9):882–9. Epub 2010/08/06. PubMed PMID: 20685163.

[18] Roehrs T, Roth T. Caffeine: sleep and daytime sleepiness. Sleep Med Rev 2008;12(2):153–62. Epub 2007/10/24. PubMed PMID: [17950009].

[19] Weaver E, Gradisar M, Dohnt H, Lovato N, Douglas P. The effect of presleep video-game playing on adolescent sleep. J Clin Sleep Med 2010;6(2):184–9. Epub 2010/04/24. PubMed PMID: 20411697.

[20] Grandner MA. Sleep, health, and society. Sleep Med Clin 2017;12(1):1–22. PubMed PMID: [28159089].

[21] Montano DE, Kasprzyk D. Theory of reasoned action, theory of planned behavior, and the integrated behavioral model. In: Glanz K, Rimer BK, Viswanath K, editors. Health behavior and health education: theory, research, and practice. San Francisco: Jossey-Bass; 2008. p. 68–96.

Chapter 10

Screening for sleep disorders

Catherine A. McCall[a,b], Nathaniel F. Watson[c]
[a]Department of Pulmonary, Critical Care, and Sleep Medicine, VA Puget Sound Health Care System, Seattle, WA, United States, [b]Department of Psychiatry, University of Washington Sleep Medicine Center, Seattle, WA, United States, [c]Department of Neurology, University of Washington Sleep Medicine Center, Seattle, WA, United States

ABBREVIATIONS

AASM	American Academy of Sleep Medicine
AHI	Apnea-hypopnea index
AIS	Athens Insomnia Scale
BMI	Body mass index
CAT	Computer-adaptive test
CPAP	Continuous positive airway pressure
DLMO	Dim light melatonin onset
EEG	Electroencephalography
ESS	Epworth Sleepiness Scale
FOSQ-30	Functional Outcomes of Sleep Questionnaire
HSAT	Home sleep apnea test
HSROC	Hierarchical summary receiver operating characteristic
ICSD	International Classification of Sleep Disorders
IRLS	International Restless Legs Syndrome Scale
IRT	Item response theory
ISI	Insomnia Severity Index
KSS	Karolinska Sleepiness Scale
MCTQ	Munich Chronotype Questionnaire
MEQ	Horne-Ostberg Morningness-Eveningness Questionnaire
MSLT	Multiple sleep latency test
MWT	Maintenance of wakefulness test
NIH	National Institutes of Health
NPV	Negative predictive value
ODI	Oxygen desaturation index
OSA	Obstructive sleep apnea
PPV	Positive predictive value
PROMIS	Patient-Reported Outcomes Measurement Information System
PSG	Polysomnography
PSQI	Pittsburgh Sleep Quality Index
REM	Rapid eye movement
RLS	Restless legs syndrome
SD	Sleep disturbance
SRI	Sleep-related impairment
SSS	Stanford Sleepiness Scale

INTRODUCTION

An estimated 50–70 million people worldwide have chronic disorders of sleep and wakefulness [1]. The most common sleep disorders in the general population include insomnia, obstructive sleep apnea (OSA), circadian rhythm disorders, restless legs syndrome (RLS), and chronically insufficient sleep. Up to 33% of adults report difficulty with sleep onset or maintenance, with chronic insomnia affecting approximately 30 million Americans [2]. As obesity incidence increases and the population ages, the prevalence of OSA has increased to ~26%, of whom 10% have moderate-to-severe illness [3]. Despite the high prevalence in the general population, routine screening for these disorders does not commonly occur in the primary care setting [4]. For OSA alone, between 80% and 85% of individuals with the disorder remain undiagnosed [5].

In addition to primary sleep disorders, industrialized countries face an epidemic of chronic sleep deprivation perpetuated by long work hours and commute times, early school start times, and increasing exposure to electronic devices during normal sleep hours. These societal pressures and technologies have led to increasingly later bedtimes and earlier rise times for both children and adults. In health surveys conducted by the Centers for Disease Control and Prevention, 35%–40% of U.S. adults report sleeping <7–8h per night, with 15% sleeping less than 6h [6]. In 2018, the makers of a popular consumer sleep technology (Fitbit Inc., San Francisco, CA, United States) reported an analysis of their 2017 worldwide user database indicating an average nightly sleep duration for women of 6h and 50min, with men averaging 6h and 26min [7], both substantially less than the minimum sleep duration of 7h or more per night recommended by the American Academy of Sleep Medicine and Sleep Research Society [8].

Poor sleep is associated with numerous adverse health outcomes. Chronic insufficient sleep (<7h per 24) has been linked with increased risk of all-cause mortality in longitudinal studies, even after controlling for age and health comorbidities [9]. In addition, chronic sleep deprivation has been associated with risk for hypertension [10], major coronary events [11], diabetes [12], poor memory, decreased concentration, slowed reaction times, motor vehicle accidents including fatal fall asleep crashes [13], and impaired judgment [14]. Adverse health outcomes have also been well-studied in most primary sleep disorders. Untreated OSA is associated with increased risk for hypertension [15], myocardial infarction [16], stroke [17], depression [18], congestive heart failure [19], COPD [20], arrhythmias [21], type 2 diabetes [22], motor vehicle and industrial accidents [23], and overall mortality [24]. Untreated insomnia is linked to the development of depression, substance use disorders, poor quality of life, and suicide [25, 26].

The costs of sleep loss and sleep disorders are estimated to be billions of dollars and are a huge economic drain at both the individual and societal levels. Recent reports indicate sleep deprivation costs the US economy $411 billion dollars per year [27], sleep apnea $100 billion, and insomnia $30–100 billion [28–30]. In all three cases, most of these economic losses are due to workplace absenteeism and presenteeism, and to a lesser extent health care utilization costs. Excessive sleepiness is associated with industrial accidents, motor vehicle crashes, medical errors, work-related injuries, reduced academic performance, and impaired work productivity [1].

Over the past several years, increased recognition of the morbidity and mortality associated with sleep disorders has led to a rising demand for sleep medicine services. However, access to trained sleep clinicians and diagnostic testing resources is limited. Laboratory polysomnography (PSG) is considered the gold standard for the diagnostic assessment of most sleep disorders, but requires extensive resources for equipment, facilities, staffing, and interpretation of sleep data. The cost and limited availability of PSG increases wait times for diagnosis and delays in treatment. Home sleep apnea testing (HSAT) is also available as a more limited tool for detecting OSA; however still typically requires an evaluation by a sleep medicine practitioner prior to undergoing testing. An increasing need to appropriately triage high-risk individuals for further sleep evaluation underscores the need for screening tools with high predictive validity for sleep disorders.

Various screening instruments have been developed to evaluate patients who would benefit the most from further evaluation and diagnostic testing. Screening questionnaires have been found to be an efficient and relatively simple way to identify patients at risk for sleep disorders in the primary care setting [4]. Some of these instruments assess nonspecific symptoms such as subjective sleepiness, while others target symptoms of specific sleep disorders such as OSA. An effective screening tool is cost-effective, minimally burdensome to complete and score, and has a high sensitivity for the targeted symptoms or disorder. With the advent of mobile technology, patients are also increasingly turning to consumer sleep technologies that track sleep parameters such as timing, duration, and even sleep architecture. In the following sections, we discuss the more common screening tools that identify patients at elevated risk for sleep-related illnesses. See Table 10.1 for more information regarding each instrument, including format, scoring, and accessibility.

SLEEP-DISORDERED BREATHING

OSA is a sleep disorder characterized by partial or complete closure of the upper airway during sleep. Symptoms of OSA include snoring, choking during sleep, and breathing pauses associated with oxygen desaturation and/or cortical arousal. Daytime symptoms associated with this disorder include excessive sleepiness, lack of energy, morning headaches, nocturnal gastroesophageal reflux, and erectile dysfunction. Approximately 13% of men and 6% of women aged 30–70 years have an apnea-hypopnea index (AHI, a measure on polysomnography of breathing pauses or irregularity) >15 events/h, whereas 14% of men and 5% of women have an AHI of >5 events/h with symptoms of daytime sleepiness [3]. Both of these scenarios meet criteria for obstructive sleep apnea as defined by the International Classification of Sleep Disorders, 3rd Edition [41]. The prevalence of OSA is higher in individuals with elevated body mass index (BMI), male gender, advancing age, increased neck circumference, and post-menopausal status [42, 43].

Despite the increased risk conferred by signs and symptoms of OSA, subjective clinical impressions of this disorder appear to have low diagnostic sensitivity and specificity [44]. In one study of patients referred to a sleep center, a report of snoring was found to have high sensitivity (80%–90%) and low specificity (20%–50%) for the diagnosis of OSA, whereas a report of nocturnal choking or gasping was less sensitive (52%) and more specific (84%). The positive predictive value (PPV) of nocturnal choking or gasping for the diagnosis of OSA was 35%, which is greater than the PPV of morning headache, witnessed apneas, snoring, or excessive daytime sleepiness [44]. Optimal screening of patients with OSA may include a combination of signs and symptoms of OSA, including habitual snoring, witnessed apneas, large neck circumference, elevated BMI, and systemic hypertension [45]. Various screening instruments have been designed to evaluate risk for OSA by assessing these factors. Other screening tools commonly used for OSA include those assessing nonspecific symptoms such as excessive daytime sleepiness, reduced daytime function,

TABLE 10.1 Common screening questionnaires.

Questionnaire	Symptoms measured	Timing of symptoms	Number of questions	Response options	Scoring	How to obtain
STOP-BANG [31]	Sleep-disordered breathing	No specified timing	8	Four yes/no questions plus four clinical attributes	Item responses are summed to obtain a total score. Scores range from 0 to 8. Suggested guidelines: 0–2 = low risk; 3–4 = intermediate risk; 5–8 = high risk	http://www.stopbang.ca
Berlin questionnaire [32]	Sleep-disordered breathing	No specified timing	9	Multiple choice questions in three categories: OSA symptoms, daytime sleepiness, and the presence of hypertension	Each item is given a unique score. Categories are considered "positive" with a specific number of total points. 2–3 positive categories = high risk for OSA. 0–1 positive categories = low risk for OSA	Freely available online
Epworth Sleepiness Scale [33]	Hypersomnolence	"Recently"; no specific timing	8	4-point Likert scale response format regarding the likelihood of falling asleep in various situations, from 0 (would never doze) to 3 (high chance of dozing)	Item responses are summed to obtain a total score. Scores range from 0 to 28. Scores of ≥10 indicate clinically significant hypersomnia	http://epworthsleepinessscale.com
Functional Outcomes of Sleep Questionnaire [34]	Hypersomnolence	No specified timing	FOSQ-30: 30 FOSQ-10: 10	4-point Likert scale response format, divided into five subscales rating the current difficulty of performing various tasks due to sleepiness. Answers range from 1 (extreme difficulty) to 4 (no difficulty)	An average score is calculated for each sub-scale. The five sub-scales are totaled to produce a total score ranging from 5 to 20, with higher scores indicating better functional status. A total score <18 indicates a clinically significant impact of sleepiness on quality of life. A change of two or more points indicates a significant change in sleep-related daily functioning	Available from the authors. Permission for use is required. Contact Terri E. Weaver, PhD, RN, University of Illinois at Chicago, teweaver@uic.edu
Stanford Sleepiness Scale [35]	Hypersomnolence	Current point in time	1	One item in which respondents select one of seven statements best representing their current level of sleepiness	Score is 1–7, with 1 indicating no sleepiness, and 7 indicating excessive sleepiness	Freely available online
Karolinska Sleepiness Scale [64]	Hypersomnolence	Current point in time	1	One item in which respondents select one of nine statements representing their level of alertness or sleepiness	Score is 1–9, with 7 or above indicating excessive sleepiness	A copy can be obtained from the author: Torbjörn Åkerstedt, IPM & Karolinska Institutet, Torbjorn.Akerstedt@ki.se
Insomnia Severity Index [36]	Insomnia	Last 2 weeks	7	5-point Likert response format ranging from 0 (None) to 4 (very severe) in two sections: severity of insomnia, and impact of symptoms	Responses are added to determine the total score. Score range is 0–28. Suggested guidelines: 0–7 = no clinically significantly insomnia; 8–14 = subthreshold insomnia; 15–21 = moderate clinical insomnia; 22–28 = severe clinical insomnia	Permission for usage can be obtained from the author: Charles M. Morin, PhD, Université Laval and Centre de recherche Université, cmorin@psy.ulaval.ca

Continued

TABLE 10.1 Common screening questionnaires.—cont'd

Questionnaire	Symptoms measured	Timing of symptoms	Number of questions	Response options	Scoring	How to obtain
Pittsburgh Sleep Quality Index [37]	Subjective sleep quality and insomnia	1 month	19, plus 5 additional items completed by a bed partner	Seven components: subjective sleep quality, sleep latency, sleep duration, habitual sleep efficiency, sleep disturbance, use of sleep medications, and daytime dysfunction. Items 1–4 are free entry fields for sleep timing. Remaining items have 4-point Likert scale response format with 0 indicating lowest severity/frequency, and 4 indicating greatest severity/frequency	A scoring algorithm is used to calculate each component score from 0 to 3, which are then summed to obtain a total score. Scores range from 0 to 21, with higher scores indicating worse sleep quality. A score ≥5 indicates poor sleep quality	Permission for usage can be obtained from the author: Daniel J. Buysse, MD, University of Pittsburgh, buyssedj@upmc.edu
Patient-Reported Outcomes Measurement Information System [38]	Sleep disturbance, sleep-related impairment	7 days	SD Scale: 27; SRI Scale: 16; CAT and short versions with 4, 6, and 8 items also available	5-point Likert scale response format from 1 to 5, with 5 indicating more severe sleep disturbance or impairment	A scoring algorithm converts the raw score to a standardized T-score with a mean of 50 and standard deviation of 10, with higher scores indicating greater sleep disturbance or impairment	Freely available online at: http://www.healthmeasures.net/explore-measurement-systems/promis
Horne-Ostberg Morningness-Eveningness Questionnaire [39]	Chronotype	"Recent weeks"	19	Multiple choice answers ranging from 0 to 6	Responses are added to determine the total score. Score range is 16–86, with lower scores indicating an evening chronotype and higher scores indicating a morning chronotype	Freely available online
Munich Chronotype Questionnaire [40]	Chronotype, circadian rhythm disorders	Last 4 weeks	Full version: 32 core version: 13	Combination of multiple choice and free entry in six sections: personal data, sleep schedule on work days and free days, work and commute times, time spent outdoors, and frequency of stimulants		Permission for usage can be obtained from the author: Till Roenneberg, till.roenneberg@med.uni-muenchen.de
International Restless Legs Syndrome Scale [87]	Restless legs syndrome	Most recent 2 weeks	10	5-point Likert scale response format with 0 indicating no symptoms and 4 indicating greater severity or frequency of symptoms	Responses are added to determine the total score. Score range is 0–40, with higher numbers indicating more severe RLS symptoms	Permission for usage can be obtained from: Caroline Anfray, Information Resources Centre, MAPI Research Institute, canfray@mapi.fr or instdoc@mapi.fr

OSA, obstructive sleep apnea; SD, sleep disturbance; SRI, sleep-related impairment; CAT, computer-adaptive test.

and poor sleep quality. Screening for OSA may improve the sensitivity and specificity of laboratory and home sleep apnea tests, and reduce overall cost by limiting expensive diagnostic studies to those who have higher pre-test probability of having the disease. The following sections describe the most commonly used screening tools designed to assess sleep-disordered breathing.

STOP-BANG questionnaire

The STOP-BANG is an eight-item screening tool for OSA that consists of four yes/no questions and four clinical attributes, with a total possible score of 0–8 [31]. The tool is a simple mnemonic in which "S" stands for snoring, "T" for tiredness/fatigue, "O" for observed apneas, "P" for high blood pressure, "B" for BMI >35, "A" for age >50, "N" for neck circumference >40cm, and "G" for male gender. This tool was originally developed as the four-item STOP questionnaire to screen for OSA in pre-surgical patients, and was validated against the AHI recorded during overnight PSG [31]. The authors found that, of those patients with an AHI of >5 events/h, there was a higher percentage of males (57%) versus females (43%). This group was also about 10 years older, had greater BMI, and larger neck size than patients with AHI ≤5. The STOP questionnaire was then revised to include these additional measures.

Since then, a number of studies have validated the effectiveness of the STOP-BANG for OSA screening in various populations. A systematic review and meta-analysis including 17 studies with 9206 subjects validating STOP-BANG scores by polysomnographic testing concluded that in the sleep clinic population, sensitivity was 90% for detecting any OSA (defined as AHI ≥5), 94% for detecting moderate-severe OSA (defined as AHI ≥15), and 96% for detecting severe OSA (defined as AHI ≥30). However, specificity was relatively low at 49%, 34% and 25% respectively. The positive predictive value (PPV) was 91% for any OSA, 72% for moderate-severe OSA, and 48% for severe OSA, and the negative predictive values (NPV) were 46%, 75%, and 90%, respectively. In the sleep clinic population, the probability of severe OSA with a STOP-Bang score of 3 was 25%. With a stepwise increase of the STOP-BANG score to 4, 5, 6 and 7/8, the probability rose proportionally to 35%, 45%, 55% and 75%, respectively [46].

The American Academy of Sleep Medicine (AASM) clinical practice guideline for the diagnostic testing of OSA notes the STOP-BANG tool demonstrates overall high sensitivity, but low specificity, for the detection of OSA in studies validating the instrument against PSG. The findings were similar for home sleep apnea tests [47].

Berlin questionnaire

The Berlin questionnaire is an 11-item self-report measurement, divided into three categories: five questions describing snoring and witnessed apneas, four questions rating daytime sleepiness, and a yes/no question regarding the presence or absence of hypertension, scored in conjunction with the patient's BMI. (Fig. 10.1) Each category is graded as "high risk" or "low risk" for OSA based on separate criteria. The patient is considered at overall "high risk" if the patient reports having persistent (>3–4 times/week) symptoms in at least two symptom categories [32].

A meta-analysis comparing the summary sensitivity and specificity of the Berlin Questionnaire, STOP-BANG, STOP, and Epworth Sleepiness Scale (ESS) found pooled sensitivity levels of 76% for AHI ≥5, 77% for AHI ≥15, and 84% for AHI ≥30. Pooled specificity was 59% for AHI ≥5, 44% for AHI ≥15, and 38% for AHI ≥30. This was less sensitive and more specific than the STOP-BANG for all severities of OSA, but less specific than the ESS [48].

The AASM clinical practice guideline found similar results in a meta-analysis of 19 studies that evaluated the performance of the Berlin Questionnaire against PSG in various patient populations and geographic studies, finding a pooled sensitivity of 76% and pooled specificity of 45% for AHI ≥5. The authors additionally noted suboptimal accuracy, derived by hierarchical summary receiver operating characteristic (HSROC) curves, ranging from 56% to 70% that was progressively reduced with higher AHI thresholds [47].

HYPERSOMNOLENCE

A large proportion of adults presenting to sleep clinics report symptoms of excessive daytime sleepiness. The prevalence of chronic excessive sleepiness is as high as 20% in the general population [49]. Sleepiness is associated with a wide array of sleep disorders, medical illnesses, and medication side effects. Assessing the severity of sleepiness, as opposed to fatigue, is an important aspect of diagnosis and management of sleep disorders such as OSA, idiopathic hypersomnia, and narcolepsy.

The multiple sleep latency test (MSLT) is a widely used assessment of objective sleepiness. It involves four to five sequential daytime nap opportunities, each separated by 2h. This test is based on the premise that, given an adequate opportunity to sleep the night prior, a patient with hypersomnia will have a shorter mean sleep latency than a patient without hypersomnia. In addition, observation of two or more naps containing rapid eye movement (REM) sleep within 15 min of falling asleep is indicative of narcolepsy, a central nervous system hypersomnic disorder. The MSLT has a high test-retest reliability in normal subjects and is the standard method for objectively measuring daytime sleepiness despite being time-consuming and expensive. Another laboratory test that indirectly assesses sleepiness is the maintenance of wakefulness test (MWT), which measures the individual's ability to stay awake in a quiet, nonstimulating environment. Other sleepiness assessments are subjective and reviewed below.

Category 1:
1. Do you snore?
    a. Yes
    b. No
    c. Don't know

If you answered 'yes':
2. Your snoring is:
    a. Slightly louder than breathing
    b. As loud as talking
    c. Louder than talking
3. How often do you snore?
    a. Almost every day
    b. 3-4 times per week
    c. 1-2 times per week
    d. 1-2 times per month
    e. Rarely or never
4. Has your snoring ever bothered other people?
    a. Yes
    b. No
    c. Don't know
5. Has anyone noticed that you stop breathing during your sleep?
    a. Almost every day
    b. 3-4 times per week
    c. 1-2 times per week
    d. 1-2 times per month
    e. Rarely or never

Category 2
6. How often do you feel tired or fatigued after your sleep?
    a. Almost every day
    b. 3-4 times per week
    c. 1-2 times per week
    d. 1-2 times per month
    e. Rarely or never
7. During your waking time, do you feel tired, fatigued or not up to par?
    a. Almost every day
    b. 3-4 times per week
    c. 1-2 times per week
    d. 1-2 times per month
    e. Rarely or never
8. Have you ever nodded off or fallen asleep while driving a vehicle?
    a. Yes
    b. No

If you answered 'yes':
9. How often does this occur?
    a. Almost every day
    b. 3-4 times per week
    c. 1-2 times per week
    d. 1-2 times per month
    e. Rarely or never

Category 3
10. Do you have high blood pressure?
    a. Yes
    b. No
    c. Don't know

FIG. 10.1 Berlin questionnaire.

Epworth Sleepiness Scale (ESS)

The ESS is an eight-item self-report measure assessing sleep propensity in a variety of common situations [33]. Respondents are asked, "How likely are you to doze off or fall asleep in the following situations, in contrast to feeling just tired?" Answer choices for each scenario range from 0 ("would never doze") to 3 ("high chance of dozing"). Responses are summed to yield a total score of 0–24, with higher scores indicating greater sleep propensity. The ESS is the most widely used tool for assessing subjective sleepiness in both research and clinical practice. Unlike other sleepiness scales such as the Stanford Sleepiness Scale, the ESS measures sleep propensity based on recall of situational tendencies to fall asleep, rather than subjective sleepiness itself.

The ESS reliability and efficacy were initially assessed in OSA patients. Fifty-four OSA patients completed the ESS before and after starting treatment with continuous positive airway pressure (CPAP). When compared to 104 medical student controls, a significant difference in baseline ESS score was observed in OSA patients before, but not after, CPAP treatment [50]. Additional evidence of validity includes a study suggesting ESS predicts narcolepsy with greater accuracy than the MSLT [51]. Since then, a number of studies have shown an association between ESS score and OSA severity using PSG or objective sleepiness using MSLT [10, 52]; however other studies show no significant association [53–55].

The AASM clinical practice guideline task force reviewed studies evaluating the performance of the ESS against PSG for identifying OSA, and found a sensitivity of 27%–72%, with a specificity of 50%–76% for identifying patients with AHI ≥ 5. When compared to home sleep apnea tests, the ESS showed low sensitivity (36%) and higher specificity (77%) for identifying OSA [47].

One challenge when evaluating ESS validity for OSA diagnosis is that sleepiness is commonly reported in OSA as fatigue and lack of energy [56, 57]. Patients with moderate-to-severe OSA may report feeling unrested or tired but deny sleepiness and have a normal ESS score [57]. Additionally, some patients underestimate their own sleep propensity. One study in which both the patient and their bed partner completed the ESS, the scores completed by the partner were significantly higher than those completed by the patient. Additionally, the partner and patient-partner consensus scores were both significantly correlated with the presence of OSA, whereas patient scores alone were not [58]. Physician-administered ESS scores may also be greater than patient self-reported scores. Physician-administered ESS scores also correlate better with PSG determined AHI and oxygen desaturation indices (ODI) than those self-administered by the patient [59].

Functional Outcomes of Sleep Questionnaire (FOSQ-30)

The FOSQ-30 is a 30-item questionnaire assessing the respondent's current difficulty performing tasks due to sleepiness, based on a Likert-type scale from 1 to 4, with a rating of 1 indicating "extreme difficulty," and 4 indicating "no difficulty." [34] The difference between "sleepy" and "tired" is clarified in the instructions. All 30 items are categorized into one of five sub-scales: activity level (9 items), vigilance (7 items), intimacy and sexual relationships (4 items), general productivity (8 items), and social outcomes (2 items). There is additionally a shorter version of the FOSQ with only 10 items.

In clinical practice and research, the FOSQ assesses sleep disorders impact on daytime function and evaluates therapeutic response. The FOSQ has good sensitivity in measuring CPAP adherence differences [60]. A score >17.9 represents the lower limit for normal sleep-related quality of life, whereas a change of two or more points indicates a clinically meaningful improvement in daily functioning [61].

Stanford Sleepiness Scale (SSS)

The Stanford Sleepiness Scale (SSS) is a simple test where patients select one of seven statements best representing their present level of perceived sleepiness [35]. (Fig. 10.2) The SSS is a momentary sleepiness scale often administered repeatedly to detect sleepiness variation over the course of a day. The SSS was initially validated to predict performance deficits during acute total or short-term partial sleep deprivation [35, 62]; however, it is less valid for assessing cumulative partial sleep deprivation [62] or identifying narcolepsy [63]. The SSS has not been validated, to our knowledge, as a predictive measure for OSA.

Karolinska Sleepiness Scale (KSS)

Similar to the SSS, the Karolinska Sleepiness Scale (KSS) is a tool measuring momentary sleepiness using a nine-point Likert-type scale with possible ratings ranging from a score of 1 signifying "very alert" to 9, signifying "very sleepy, great effort to stay awake, fighting sleep." Scores of ≥ 7 are pathologic (Fig. 10.3) [64]. The original version provided labels at every other number of the scale, however studies showed a bias where subjects selected labeled numbers over unlabeled numbers. This resulted in labels being added to all numbers of the scale [65].

The KSS focuses on sleep propensity, whereas the SSS assesses levels of fatigue or boredom [66]. The KSS is validated against electroencephalography (EEG) and various performance measurements, and is most commonly used to evaluate sleepiness in drug trial participants, professional drivers, flight crews, train engineers, and oil rig workers [64, 67]. The KSS varies over the diurnal cycle, and is impacted by physical activity, social interaction, and light exposure [66].

Data is limited regarding efficacy of sleepiness measures in clinical populations, including predictive value for OSA diagnosis. One study showed positive correlations between subjective sleepiness on the KSS, EEG changes indicative of sleepiness, and likelihood of errors on a 2-h monotonous

Degree of Sleepiness	Scale Rating
Feeling active and vital, alert, wide awake	1
Functioning at a high level, but not at peak, able to concentrate	2
Relaxed, awake, not at full alertness, responsive	3
A little foggy, not at peak, let down	4
Foggy, beginning to lose interest in remaining awake, slowed down	5
Sleepy, prefer to be lying down, fighting sleep, woozy	6
Almost in reverie, sleep onset soon, lost struggle to remain awake	7

FIG. 10.2 Stanford Sleepiness Scale.

Select the one statement that best describes your sleepiness during the previous 5 minutes.

_____ 1. Extremely alert
_____ 2. Very alert
_____ 3. Alert
_____ 4. Rather alert
_____ 5. Neither alert nor sleepy
_____ 6. Some signs of sleepiness
_____ 7. Sleepy, but no effort to keep awake
_____ 8. Sleepy, some effort to keep awake
_____ 9. Very sleepy, great effort to keep awake, fighting sleep

FIG. 10.3 Karolinska Sleepiness Scale.

driving simulator task in OSA patients treated with CPAP. After sleep restriction, OSA patients had significantly shorter safe driving times relative to controls, and underestimation of KSS reported sleepiness level [68]. Another study with OSA patients found significantly higher baseline KSS scores in OSA patients compared to controls, but no difference in sleepiness or performance on the psychomotor vigilance task after 40 h of sustained wakefulness [69].

INSOMNIA AND SLEEP QUALITY

Insomnia is a disorder characterized by persistent difficulty with sleep initiation, sleep maintenance, sleep consolidation, and/or sleep quality. Diagnosing insomnia requires a complete history identifying medical, neurologic, psychiatric, medication and/or substance-related causes of the sleep impairment. Unlike with other sleep disorders, PSG is not indicated for insomnia diagnosis, although it is necessary to rule out other sleep disorders such as OSA that may cause persistent difficulty sleeping. Screening instruments are highly predictive for insomnia because symptoms define the disorder [70]. Some instruments such as the Insomnia Severity Index (ISI) identify core symptoms of insomnia, whereas others such as the Pittsburgh Sleep Quality Index (PSQI) include additional questions on sleep quality and daytime dysfunction.

Insomnia Severity Index (ISI)

The ISI is a self-report instrument with seven items that characterize insomnia symptoms and the degree of concern or distress caused by those symptoms [36]. Initially developed to measure insomnia research outcomes, including clinical trials and morbidity studies, the ISI is now a commonly used insomnia screening instrument. The first three questions rate respondents' difficulty falling asleep, staying asleep, or waking up too early over the past 2 weeks, with 0 indicating "None" and four indicating "Very Severe." The last four questions rate how satisfied/dissatisfied they are with their current sleep pattern, the noticeability of their problem, how worried/distressed they are, and to what extent their sleep problem interferes with daily functioning. Each of these questions is also rated from 0 to 4, with 4 indicating greater problem severity. The total scores range from 0 to 28, with higher numbers indicating more severe insomnia.

Suggested guidelines for interpreting scores state 0–7 indicates no clinically significant insomnia, 8–14 indicates subthreshold (mild) insomnia, 15–21 indicates moderate clinical insomnia, and 22–28 indicates severe clinical insomnia. Although few studies validate these cutoffs, one community-based study found a score of ≥10 identified insomnia with a sensitivity of 86% and a specificity of 88%. In a clinical sample in the same study, a score reduction of 8.4 points revealed moderate insomnia improvement following treatment as rated by an independent assessor [71, 72]. A study with 1670 cancer patients showed significant correlations between the ISI rating of sleep onset insomnia and PSG sleep onset latency, and ISI rating of sleep maintenance insomnia and PSG number of nocturnal awakenings. Receiver operating characteristic analysis showed a score of 8 represents an optimal cutoff score for clinically significant insomnia, with a sensitivity of 95% and specificity of 47% [73].

A meta-analysis of 19 studies comprising 4693 participants evaluating the diagnostic accuracy of the ISI, Athens Insomnia Scale (AIS), and Pittsburg Sleep Quality Index (PSQI) found a pooled sensitivity of 88% and specificity of 85% when compared to a diagnostic reference standard such as the International Classification of Sleep Disorders (ICSD). This was less sensitive but more specific than the PSQI for identifying individuals with insomnia [74].

Pittsburgh Sleep Quality Index (PSQI)

The PSQI is a self-report measure of sleep quality consisting of 19 items, plus a five-item rating completed by a bed partner not included in scoring [37]. Respondents are instructed to record their typical sleep schedule and frequency of specific sleep problems over the past month on a four-point Likert-type scale, with 0 indicating lower frequency or severity, and 3 indicating greater frequency or severity. Unlike sleep instruments targeted to screen only for the presence of insomnia, the PSQI also includes questions on symptoms causing sleep disturbance such as snoring, nightmares, discomfort, and pain. These questions yield scores on seven subscales: subjective sleep efficiency, sleep latency, sleep duration, sleep quality, sleep disturbance, sleep medication use, and daytime dysfunction due to sleepiness. The subscale scores are added to obtain a total score ranging from 0 to 21, with higher total scores indicating poorer sleep quality.

The PSQI has high sensitivity in studies assessing insomnia screening accuracy. A meta-analysis comparing the PSQI to a diagnostic reference standard such as the International Classification of Sleep Disorders (ICSD) found a sensitivity of 94% and specificity of 76%. This was more sensitive but less specific in identifying insomnia than the ISI [74]. Studies examining other clinical populations including individuals with obstructive sleep apnea, periodic limb movement disorder, rapid eye movement sleep behavior disorder, and narcolepsy found low criterion validity for these diagnoses when compared to PSG and MSLT. However significant correlations were observed with symptoms of depression and anxiety in these populations [75, 76]. The PSQI is useful for identifying factors contributing to insomnia, including psychiatric and medical symptoms, but requires more time to complete than shorter insomnia questionnaires.

Patient-Reported Outcomes Measurement Information System (PROMIS™)

The PROMIS is a set of scales created as an NIH Roadmap initiative for improving patient-reported outcomes for a variety of symptoms, including sleep and wakefulness [38]. The scales were developed based on rigorous psychometric testing methods to provide continuous, relative values for individuals rather than categorizing disorders. There are two sleep-related item banks in this System: the Sleep Disturbance (SD) scale with 27 items, and the Sleep-Related Impairment (SRI) scale with 16 items. Questions are posed as statements with potential responses ranging from 1 to 5 for symptoms occurring within the last 7 days. Individual items can be selected from these item banks to create short forms. All of the scales are free and available in multiple "short forms" of 4, 6, and 8-item questionnaires, as well as computer-adaptive tests (CAT) that display only relevant questions based on the respondent's prior answers. The short forms allow providers to combine questions from other domains of the PROMIS database, for example fatigue, depression, and pain. Scoring is performed using Item Response Theory (IRT), a family of statistical models linking individual questions to a theoretical underlying syndrome of sleep disturbance. Using the PROMIS scoring manual or scoring service, the respondent's score is converted into a standardized T-score, which can be viewed relative to a mean of 50 and standard deviation of 10. Higher scores indicate greater sleep disturbance or impairment. (For more information, see www.nihpromis.org.)

One study investigating the validity of customized eight-item short forms for the SD and SRI item banks developed from post-hoc CAT simulations found these short forms demonstrated greater measurement precision than the PSQI and ESS, although the full item banks provided greater test information [77]. A study in ambulatory cancer patients with PROMIS CAT item banks found high correlation with ISI scores, and 98% of the patients reported the screening was not burdensome [78]. We were unable to find any studies comparing the PROMIS instrument to an objective reference standard, such as PSG, for the diagnosis of a sleep disorder.

Some advantages of using the PROMIS instruments include scale-specific psychometric validity, customizability in combination with other symptom domains, and measurement power regarding an individual's fit along a spectrum of symptom severity. The PROMIS scales help providers follow trends in symptoms or treatment outcomes over time. The National Institutes of Health (NIH) integrates PROMIS measures into electronic health records, including Epic and Cerner. However, this instrument does not query respondents on quantitative data such as sleep timing, sleep duration, or symptoms of specific sleep disorders. Although it would help identify and track sleep-related symptoms, it does not provide the necessary screening data to quantify risk for a specific diagnosis or support PSG testing. Scoring the instrument requires time and knowledge if a scoring service is not utilized.

CIRCADIAN RHYTHM DISORDERS

Although most humans sleep at night and are awake during the day, considerable inter-individual variation exists in chronotype, with extreme morning types often described as "larks" and extreme evening types described as "owls."

Circadian rhythm sleep disorders are typically due to discrepancies between the internal circadian cycle and the external light-dark cycle. These disorders can be caused by alterations in the external cycle (e.g., shift work or jet lag), internal cycle (e.g., delayed sleep-wake phase disorder or advanced sleep-wake phase disorder), or dysfunction in the clock circuitry of the brain (e.g., irregular sleep-wake rhythm disorder). Delayed sleep-wake phase disorder typically presents as nighttime insomnia with morning hypersomnolence, whereas advanced sleep-wake phase disorder presents as evening hypersomnolence with early morning awakenings. Approximately 7%–16% of patients presenting to sleep disorders clinics with insomnia complaints are diagnosed with delayed sleep-wake phase disorder [79]. Questionnaires such as the Munich Chronotype Questionnaire (MCTQ) and the Horne-Ostberg Morningness-Eveningness Questionnaire (MEQ) may be useful to distinguish a circadian rhythm disorder from chronic insomnia by identifying the individual's chronotype.

Horne-Ostberg Morningness-Eveningness Questionnaire (MEQ)

The MEQ is a 19-item self-report questionnaire in which respondents report their optimal sleep and wake schedule, how they feel at various points in their current schedule, and what they would do faced with a variety of sleep scheduling scenarios [39]. (Fig. 10.4) Each item has a range of possible scores from 0 to 6. The sum of the individual items ranges from 16 to 86, with lower scores corresponding to evening types or "owls," and higher scores indicating morning types or "larks." The MEQ has become a widely used instrument to classify circadian type in research on normal subjects and patients.

A review of 14 studies using the MEQ in normal subjects in comparison to an objective circadian phase marker such as core body temperature or dim light melatonin onset (DLMO) found a strong negative correlation between the MEQ score and the objective marker, indicating congruence between these measures. Four of the studies used additional measures such as actigraphy or sleep diaries and found the MEQ score correlated with the ability to adapt to night shift

For each question, please select the answer that best describes you by circling the point value that best indicates how you have felt in recent weeks.

1. Approximately what time would you get up if you were entirely free to plan your day?
    - [5] 5:00 AM–6:30 AM (05:00–06:30 h)
    - [4] 6:30 AM–7:45 AM (06:30–07:45 h)
    - [3] 7:45 AM–9:45 AM (07:45–09:45 h)
    - [2] 9:45 AM–11:00 AM (09:45–11:00 h)
    - [1] 11:00 AM–12 noon (11:00–12:00 h)
2. Approximately what time would you go to bed if you were entirely free to plan your evening?
    - [5] 8:00 PM–9:00 PM (20:00–21:00 h)
    - [4] 9:00 PM–10:15 PM (21:00–22:15 h)
    - [3] 10:15 PM–12:30 AM (22:15–00:30 h)
    - [2] 12:30 AM–1:45 AM (00:30–01:45 h)
    - [1] 1:45 AM–3:00 AM (01:45–03:00 h)
3. If you usually have to get up at a specific time in the morning, how much do you depend on an alarm clock?
    - [4] Not at all
    - [3] Slightly
    - [2] Somewhat
    - [1] Very much
4. How easy do you find it to get up in the morning (when you are not awakened unexpectedly)?
    - [1] Very difficult
    - [2] Somewhat difficult
    - [3] Fairly easy
    - [4] Very easy
5. How alert do you feel during the first half hour after you wake up in the morning?
    - [1] Not at all alert
    - [2] Slightly alert
    - [3] Fairly alert
    - [4] Very alert
6. How hungry do you feel during the first half hour after you wake up?
    - [1] Not at all hungry
    - [2] Slightly hungry
    - [3] Fairly hungry
    - [4] Very hungry
7. During the first half hour after you wake up in the morning, how do you feel?
    - [1] Very tired
    - [2] Fairly tired
    - [3] Fairly refreshed
    - [4] Very refreshed
8. If you had no commitments the next day, what time would you go to bed compared to your usual bedtime?
    - [4] Seldom or never later
    - [3] Less than 1 hour later
    - [2] 1–2 hours later
    - [1] More than 2 hours later
9. You have decided to do physical exercise. A friend suggests that you do this for one hour twice a week, and the best time for him is between 7-8 AM (07-08 h). Bearing in mind nothing but your own internal "clock," how do you think you would perform?
    - [4] Would be in good form
    - [3] Would be in reasonable form
    - [2] Would find it difficult
    - [1] Would find it very difficult
10. At approximately what time in the evening do you feel tired, and, as a result, in need of sleep?
    - [5] 8:00 PM–9:00 PM (20:00–21:00 h)
    - [4] 9:00 PM–10:15 PM (21:00–22:15 h)
    - [3] 10:15 PM–12:45 AM (22:15–00:45 h)
    - [2] 12:45 AM–2:00 AM (00:45–02:00 h)
    - [1] 2:00 AM–3:00 AM (02:00–03:00 h)
11. You want to be at your peak performance for a test that you know is going to be mentally exhausting and will last two hours. You are entirely free to plan your day. Considering only your "internal clock," which one of the four testing times would you choose?
    - [6] 8 AM–10 AM (08–10 h)
    - [4] 11 AM–1 PM (11–13 h)
    - [2] 3 PM–5 PM (15–17 h)
    - [0] 7 PM–9 PM (19–21 h)

FIG. 10.4 Morningness-Eveningness Questionnaire. *English version prepared by Terman M, Rifkin JB, Jacobs J, White TM (2001), New York State Psychiatric Institute, 1051 Riverside Drive, Unit 50, New York, NY, 10032.*

12. If you got into bed at 11 PM (23 h), how tired would you be?
    [0] Not at all tired
    [2] A little tired
    [3] Fairly tired
    [5] Very tired
13. For some reason you have gone to bed several hours later than usual, but there is no need to get up at any particular time the next morning. Which one of the following are you most likely to do?
    [4] Will wake up at usual time, but will not fall back asleep
    [3] Will wake up at usual time and will doze thereafter
    [2] Will wake up at usual time, but will fall asleep again
    [1] Will not wake up until later than usual
14. One night you have to remain awake between 4-6 AM (04-06 h) in order to carry out a night watch. You have no time commitments the next day. Which one of the alternatives would suit you best?
    [1] Would not go to bed until the watch is over
    [2] Would take a nap before and sleep after
    [3] Would take a good sleep before and nap after
    [4] Would sleep only before the watch
15. You have two hours of hard physical work. You are entirely free to plan your day. Considering only your internal "clock," which of the following times would you choose?
    [4] 8 AM–10 AM (08–10 h)
    [3] 11 AM–1 PM (11–13 h)
    [2] 3 PM–5 PM (15–17 h)
    [1] 7 PM–9 PM (19–21 h)
16. You have decided to do physical exercise. A friend suggests that you do this for one hour twice a week. The best time for her is between 10-11 PM (22-23 h). Bearing in mind only your internal "clock," how well do you think you would perform?
    [1] Would be in good form
    [2] Would be in reasonable form
    [3] Would find it difficult
    [4] Would find it very difficult
17. Suppose you can choose your own work hours. Assume that you work a five-hour day (including breaks), your job is interesting, and you are paid based on your performance. At approximately what time would you choose to begin?
    [5] 5 hours starting between 4–8 AM (05–08 h)
    [4] 5 hours starting between 8–9 AM (08–09 h)
    [3] 5 hours starting between 9 AM–2 PM (09–14 h)
    [2] 5 hours starting between 2–5 PM (14–17 h)
    [1] 5 hours starting between 5 PM–4 AM (17–04 h)
18. At approximately what time of day do you usually feel your best?
    [5] 5–8 AM (05–08 h)
    [4] 8–10 AM (08–10 h)
    [3] 10 AM–5 PM (10–17 h)
    [2] 5–10 PM (17–22 h)
    [1] 10 PM–5 AM (22–05 h)
19. One hears about "morning types" and "evening types." Which one of these types do you consider yourself to be?
    [6] Definitely a morning type
    [4] Rather more a morning type than an evening type
    [2] Rather more an evening type than a morning type
    [1] Definitely an evening type
Total points for all 19 questions:

FIG. 10.4, cont'd

work, preferred exercise time, age, and characteristic circadian sleep difficulties relative to diurnal preference [80].

In clinical populations, two studies of families with familial advanced sleep-wake phase disorder showed affected family members scored significantly higher on the MEQ than unaffected members, and unaffected first-degree relatives scored higher than controls [81, 82].

One criticism of the MEQ is that respondents do not report actual sleep times, nor differences between work days and free days, or circadian cues such as exposure to outdoor light [40]. The Munich Chronotype Questionnaire (MCTQ) was developed to obtain this information in addition to diurnal preferences.

Munich Chronotype Questionnaire (MCTQ)

On the MCTQ, respondents report their current sleep times, how they feel at various times of the day, and the amount of daylight exposure. They also rate themselves as one of seven chronotypes: Extreme Early, Moderate Early, Slightly Early, Normal, Slightly Late, Moderate Late, and Extreme Late [40]. Respondents are additionally asked to rate their chronotype at different life stages, and the chronotypes of family members. The authors of the MCTC initially developed the questionnaire to record the discrepancy between self-assessed chronotype and current sleep schedule. In their study of 500 subjects in Germany and Switzerland, they found that late chronotypes tended to accumulate considerable sleep debt during the work week, with subsequent compensation of several additional hours sleep on free days [40].

A study using both the MEQ and the MCTQ in 2481 respondents completing both questionnaires online found the midpoint sleep time on non-work days and the self-rated chronotype on the MCTQ both correlated strongly with the chronotype based on the MEQ score. Because late chronotypes tend to have more variability in sleep times between free days and work days, the timing of mid-sleep was a better predictor of chronotype than the timing of sleep onset or wake time [83]. The midpoint sleep time on non-work days on the MCTQ additionally correlates with DLMO [84].

RESTLESS LEGS SYNDROME (RLS)

RLS, also known as Willis-Ekbom disease, is a common disorder characterized by an unpleasant sensation in the legs temporarily alleviated with movement, typically occurring at night and during rest. Up to 88% of those with RLS report at least one sleep-related symptom, with the majority reporting chronically impaired sleep consistent with insomnia [85]. RLS symptoms are exacerbated or precipitated by low ferritin levels, pregnancy, peripheral neuropathy, end-stage renal disease, and antidepressant medications [86]. Although RLS is common, the majority of sufferers go undiagnosed and untreated even after consulting a physician about their symptoms [85]. Most RLS screening instruments are brief and helpful for identifying patients at risk of disease and monitoring treatment outcomes.

International Restless Legs Syndrome Scale (IRLS)

The IRLS is a 10-item self-report questionnaire designed by the International Restless Legs Syndrome Study Group [87]. This questionnaire evaluates the severity of RLS symptoms and their impact on sleep, mood, and daily life over a 1-week period. Ratings describe RLS symptoms and timing with each question ranging from 0 for "None" to 4 for "Very severe" or "Very often". Total scores range from 0 to 40, with higher scores indicating greater severity and impact of RLS symptoms.

The IRLS is the most extensively used RLS severity scale for research studies. It is validated for outcome evaluation in clinical trials, and strongly correlated with measures of disease severity at baseline and after treatment [88]. Criticisms of the IRLS include the absence of questions targeting symptom timing, severity of symptoms at rest versus during activity, or the presence of symptoms in other body parts. A systematic review of RLS ratings scales conclude the IRLS is validated under baseline conditions and responsive to symptom change, although validation as a self-administered scale without clinician intervention is yet to be done [89].

CONSUMER SLEEP TECHNOLOGIES

Consumer technologies designed to monitor sleep and other health-related data have become increasingly widespread and track sleep timing, duration, sleep stage, and even sleep pathology. These technologies include wearable devices, stand-alone bedside devices, and apps installed on smart phones leveraging the device's intrinsic accelerometer function to track sleep activity. Major consumer advantages include the ability to measure sleep longitudinally in the subject's typical sleep environment. Some technologies also measure aspects of the sleep environment itself such as temperature, light and noise levels. The most widely used commercial sleep trackers are designed to be user-friendly, colorful, and provide immediate information about sleep and activity. Some leaders in the healthcare and technology consumables industry such as Fitbit, Jawbone, Beddit, and SleepScore Labs provide an array of sleep tracking functions. In more recent years, companies such as Apple, Philips, ResMed, Nokia, and Microsoft have entered the consumer health-tracking wearable technology space. Presently, validation research is limited on many of these devices, obviating users' ability to substantiate their claims in both general and clinical populations.

Most consumer sleep technologies provide some type of movement detection, often based on wrist actigraphy, although some devices, such as the SleepScore Max, monitor sleep in a contactless manner using radiofrequency biomotion sensor technology, similar to echolocation. This device is not worn, but rather placed at the bedside. SleepScore also offers a stand-alone app converting the mobile phone into an active sonar to objectify aspects of sleep in a contactless manner. Individual device technologies use proprietary algorithms to calculate wake and sleep, in some cases categorizing sleep stage as "light sleep" or "deep sleep," and in other cases as rapid eye movement (REM) or non-REM sleep. In recent years, devices have also begun to use more sophisticated sensors of physiologic data to calculate sleep parameters, including aspects of heart rate and respiration.

Advantages to using consumer sleep technologies include the ability for personal empowerment in tracking, viewing, and changing sleep patterns. Consumers can assess objectively in real time the impact that behavior change has on their sleep health, increasing the probability that these healthy behaviors will become habits. Consumers are able to choose devices with appealing features and appearance, increasing the likelihood of using and gaining benefit from the device. Another potential advantage of these technologies is the ability to analyze data from all users. SleepScore Labs analyzed data from SleepScore Max, and found that on average, people who sleep in a room with a temperature of 65 degrees or lower sleep 30 min more than those sleeping in a room that is 77 degrees or higher. Additionally, they found that SleepScore Max users of any gender with obesity (BMI >30) sleep 18 min less per night than those with a normal BMI. Fitbit analyzed data from its worldwide user base (6 billion nights of data) and reported mean sleep durations by country and gender. They also found that bedtime variation between free days and workdays was inversely correlated with total mean sleep duration. Variation of 120 min was associated with approximately 30 min lower sleep duration compared to users whose bedtime varied by only 30 min [7]. In the future, data from consumer sleep technologies may be used to identify causes of short sleep duration in individual users, including sleep variation, similarly to the MCTQ.

For sleep disorder screening, the most immediate potential utility for consumer sleep technologies lie in their ability to measure sleep continuity, duration, quality, and regularity. Accurate, ecologically valid, longitudinal sleep monitoring can help in the assessment and potential treatment of insomnia, hypersomnia, circadian rhythm sleep disorders, and insufficient sleep. For insomnia sufferers, evidence exists that symptoms improve when both objective and subjective data are assessed together [90]. Of less clear benefit are device claims of accurate sleep-disordered breathing detection, although some technologies are developing this function. A device with the capability of detecting nocturnal respiratory events may be helpful for risk stratification purposes when used in conjunction with clinical data.

Downsides of consumer sleep technologies include a lack of validation and proprietary algorithms preventing assessment of how sleep stages or wake are calculated. Validation of devices against PSG exists for some devices, most notably SleepScore Max, however most studies validate with normal subjects devoid of sleep disorders. Fewer studies investigate validity in clinical populations. This paradigm is further limited by the rapid development and release of new versions of these technologies, which typically outpace the research timeline. As a result, available studies on devices such as Fitbit may be applicable only to the actual device studied, and not on newer devices with revised or enhanced features. Additionally, research on consumer sleep technologies is limited by variation of analytic methods. Other challenges facing consumer sleep technologies include inaccessibility to large segments of the population due to cost, and privacy concerns regarding the use of personal data by developers. In the following sections, we focus on the most studied consumer sleep monitoring devices to date, with a focus on those devices with validation studies in a clinical population.

Fitbit

Fitbit now produces a number of wearable devices that track movements and position via a triaxial accelerometer. In devices not using heart rate variability, two settings exist: a "normal" setting in which epochs with significant movements during sleep are counted as wake periods, and a "sensitive" setting in which epochs with any movement are counted as wake periods. Fitbit states the sensitive mode is helpful for users with sleep disorders, or users who wear the device on a body location other than the wrist. More recent devices additionally monitor heart rate. From this data, Fitbit uses a proprietary algorithm to estimate sleep stages of "light sleep," "deep sleep," and REM sleep using a combination of movement and heart rate variability. The device requires at least 3 h of sleep data to estimate sleep stages, and thus does not calculate stages for naps. (For more information, see help.fitbit.com.)

Several Fitbit devices have been validated against PSG in subjects without sleep disorders. Validation studies of the Fitbit Classic [91], the Fitbit Flex [92], and the Fitbit Ultra [93] against PSG have varied with age and the type of analysis. Both Fitbit Classic and actigraphy were found to overestimate total sleep time and sleep efficiency, with high sensitivity for sleep and low specificity for detecting wake when compared to PSG [91]. A similar pattern of findings was demonstrated in adolescents between Fitbit Ultra, actigraphy, and PSG [93]. However, another study found comparable total sleep time between Fitbit Flex and PSG in young adults. Time in "deep sleep" did not differ from PSG-measured slow-wave sleep plus REM sleep, but time in "light sleep" did differ from PSG-measured N1 and N2 sleep [92]. In depressed patients, one study evaluating the Fitbit Flex found that the "normal" setting significantly overestimated sleep time and sleep efficiency, and had low specificity relative to PSG. In the "sensitive" setting, the Fitbit significantly underestimated sleep time and sleep efficiency relative to PSG [94].

Few studies have been performed in clinical sleep populations. A study comparing the Fitbit Flex with actigraphy and overnight PSG in patients with insomnia disorder and in good sleepers found a strong correlation for total sleep time between the Fitbit and PSG in the insomnia group (ICC=0.886) and healthy sleepers (ICC=0.974). However, the Fitbit overestimated the total sleep time and sleep efficiency in both groups with respect to PSG. The sensitivity (97%) and accuracy (87%) of the Fitbit in an epoch-by-epoch comparison with PSG was similar to that of actigraphy for insomniac patients, but specificity was low at 36% [95].

Jawbone

Jawbone UP (Jawbone, San Francisco, CA) is a wrist device that uses actigraphy to monitor sleep. The data are downloaded from the device into the MotionX (Fullpower Technologies, Inc.) smart phone application. A small button on the band allows the user to switch between "active mode" and "sleep mode." (For more information, see https://jawbone.com/up.)

Few studies have been performed with normal subjects or sleep-disordered patients. In a study comparing Jawbone UP MOVE to actigraphy and home PSG, significant correlations were found for total sleep time and time in bed, but not for wake, deep sleep, light sleep, or sleep efficiency [96]. A study of midlife women found that the Jawbone UP showed 97% sensitivity for detecting sleep and 37% specificity for detecting wake. It also overestimated total sleep time (26.6 ± 35.3 min) and sleep onset latency (5.2 ± 9.6 min), and underestimated wake after sleep onset (31.2 ± 32.3 min) relative to PSG [97]. A study of adolescents found that Jawbone UP overestimated total sleep duration and sleep efficiency, and underestimated wake time after sleep onset, but the differences were small. Jawbone "light sleep" was predicted by PSG time in N2 sleep, N3 sleep, and arousal index. "Sound sleep" was predicted by PSG time in N2 sleep, REM sleep, and arousal index [98].

A study of 78 children and adolescents being evaluated for sleep-disordered breathing tested two commercial sleep devices (Jawbone UP and a smartphone application called MotionX 24/7) with actigraphy and PSG. They found no difference in mean total sleep time, wake time after sleep onset, or sleep efficiency between PSG and Jawbone UP. Actigraphy significantly underestimated sleep onset latency compared to PSG. Jawbone showed sensitivity of 92% and accuracy of 86%, but poor specificity of 66% for identifying sleep when compared to PSG [99].

SleepScore Max and SleepScore app

Another novel device is the SleepScore Max by SleepScore Labs, formerly called the ResMed S+ (https://www.sleepscore.com/sleepscore-max-sleep-tracker), a small non-contact device that is placed by the bedside. It measures sleep by using radiofrequency signals to sense breathing and body movements, as well as light, noise, and temperature levels in the bedroom. A study comparing algorithm versions 1 and 2 of the S+ to PSG found full sleep staging agreement of V1 was 61%, while V2 was 62%. Sleep sensitivity of V1 and V2 were 93% and 94%, while wake specificity of V1 and V2 were 70% and 73%. Specificity of V1 and V2 for wake after sleep onset were 51% and 53%, respectively. Specificity of V1 and V2 for wake before sleep onset were 88% and 90%. The authors conclude that this device better identifies wake before sleep onset relative to published evaluations of other wearable sleep-tracking devices [100]. The SleepScore Max is based on technology acquired from BiancaMed, which developed SleepMinder, another non-contact device that measured both sleep and respiration. SleepMinder monitored movement by transmitting low-power radio waves and recording phase changes caused by movement, including respiratory changes. Validation studies comparing the SleepMinder device to PSG found high sensitivity for detecting sleep [101] and OSA, with one study showing sensitivity of 90%, specificity of 92%, and accuracy of 91% for detecting AHI ≥15 [102]. In multiple studies, however, the SleepMinder was found to overestimate AHI in patients with periodic limb movements [102, 103].

SleepScore Labs recently launched the SleepScore app, distinguished from the SleepScore Max device and app by utilizing 18–20 kHz sound waves (as opposed to radiofrequency waves) to turn the phone into an active sonar that monitors fine and gross body movements in a contactless manner without need of an accessory appliance. The ability to monitor breathing related movements allows the device to assess sleep latency, duration, architecture, and quality. Regarding sleep architecture, validation against PSG showed the following sensitivity and specificity: wake (66%, 87%), light sleep (58%, 70%), deep sleep (60%, 89%), and REM sleep (59%, 93%) [104].

Other novel technologies

A variety of novel devices have been developed in recent years that claim to track sleep, wake, and even sleep-disordered breathing using technology other than wrist actigraphy. For the most part, these devices are either still in development, or have not yet published validation studies.

Multiple consumer sleep trackers based on electroencephalographic (EEG) monitoring have been developed and released. These devices include Neuroon (https://neuroonopen.com), Kokoon (https://kokoon.io), Sleep Shepherd (https://sleepshepherd.com), and Philips SmartSleep (https://www.usa.philips.com/c-e/smartsleep-ces.html), which embed EEG channels in a sleep mask, headphones, and headbands, respectively. These devices claim to offer the capability of monitoring sleep architecture and providing biofeedback cues. The Kickstarter-funded Neuroon Intelligent Sleep Mask additionally claims to provide bright light therapy, "smart" napping cues, and monitoring of heart rate, skin temperature, and oxygen saturation. The Philips SmartSleep headband claims to use sound tones that enhance slow wave sleep, specifically for people between the ages of 18 and 40 years, who sleep 5–7 h per night for at least four nights per week, and have no issues falling or staying asleep.

USING SCREENING DATA

A large array of screening instruments are available for identifying individuals at risk for sleep-disordered breathing. In addition to traditional paper-and-pencil questionnaires, avenues for screening have increasingly included consumer-driven technologies that provide more long-term quantitative measures of sleep duration, timing, and quality. The challenge for the primary care provider is how to utilize this potential array of information to refer high-risk patients appropriately for further evaluation and diagnostic testing.

Screening tools offer the opportunity to gather information from the patient on a variety of symptoms, including symptoms of sleep-disordered breathing, insomnia, hypersomnia, and circadian rhythm sleep disorders. They also provide the opportunity to recognize insufficient sleep, the most common cause of excessive sleepiness [49]. One strategy for using this data is to combine individual screening tools with high predictive power for detecting specific sleep disorders such as OSA and insomnia, along with tools to detect nonspecific symptoms of excessive sleepiness. Some of the higher-sensitivity instruments, such as the STOP-BANG and the ISI, require little time for the patient to take and the clinician to score. Alternately, more comprehensive instruments such as the PSQI, that cover multiple domains of sleep, may be used as stand-alone tools for gathering basic information from which to proceed with further evaluation.

Additionally, questionnaires may be used to help guide primary care providers in the types of questions to ask patients suspected of having sleep disorders. For example, use of a circadian rhythm sleep disorder screening instrument such as the MEQ or MCTQ may be given selectively to patients with unusual patterns of insomnia and hypersomnia, or for patients with sleep schedules suggesting a phase shift.

As consumer technologies continue to develop, additional screening modalities may become available to both clinicians and patients. Sleep tracking devices now boast the capability to monitor physiologic data previously unavailable outside the clinic, including movement, respiration, body position, heart rate variability, and even EEG. Devices that monitor respiratory patterns have been validated with high sensitivity. Unfortunately, technology companies do not release their algorithms to allow verification of clinical validity, and new versions of consumer technologies are released far more quickly than research studies performed to validate them. This leaves healthcare providers without the ability to evaluate the screening validity of individual sleep devices. Ultimately, the best usage of consumer sleep technologies may be in their intra-user variability, allowing the patient and their provider to track changes in sleep over time, in conjunction with more traditional measures of sleep problems.

Over time, further collaboration between technology companies and researchers may increase the usability of technology in the clinical environment. One example is the PROMIS scales, with NIH-funded validation and integration of clinical scales into electronic medical records. These scales are completed by the patient on a tablet in clinic, using computer-adaptive testing to limit the total number of questions. Integration of sleep screening tools into consumer devices would also improve the clinical application of these devices. In the meantime, most clinical providers would benefit from using routine tools currently available today to screen for the most common sleep problems.

REFERENCES

[1] Colten HR, Altevogt BM, Institute of Medicine Committee on Sleep Medicine and Research. Sleep disorders and sleep deprivation: an unmet public health problem. Washington, DC: National Academies Press; 2006. Available from: http://www.ncbi.nlm.nih.gov/pubmed/20669438. [Cited 26 February 2018].

[2] Ohayon MM. Epidemiology of insomnia: what we know and what we still need to learn. Sleep Med Rev 2002;6(2):97–111. Available from: http://www.ncbi.nlm.nih.gov/pubmed/12531146. [Cited 26 February 2018].

[3] Peppard PE, Young T, Barnet JH, Palta M, Hagen EW, Hla KM. Increased prevalence of sleep-disordered breathing in adults. Am J Epidemiol 2013;177(9):1006–14. Available from: https://academic.oup.com/aje/article-lookup/doi/10.1093/aje/kws342. [Cited 26 February 2018].

[4] Senthilvel E, Auckley D, Dasarathy J. Evaluation of sleep disorders in the primary care setting: history taking compared to questionnaires. J Clin Sleep Med 2011;7(1):41–8. Available from: http://www.ncbi.nlm.nih.gov/pubmed/21344054. [Cited 2018 Feb 26].

[5] Lee W, Nagubadi S, Kryger MH, Mokhlesi B. Epidemiology of obstructive sleep apnea: a population-based perspective. Expert Rev Respir Med 2008;2(3):349–64. Available from: http://www.ncbi.nlm.nih.gov/pubmed/19690624. [Cited 7 January 2019].

[6] Centers for Disease Control and Prevention (CDC). Effect of short sleep duration on daily activities—United States, 2005-2008. MMWR Morb Mortal Wkly Rep 2011;60(8):239–42. Available from: http://www.ncbi.nlm.nih.gov/pubmed/21368739. [Cited 26 February 2018].

[7] Pogue D. What Fitbit discovered from 6 billion nights of sleep data, 2018. Accessed at: https://finance.yahoo.com/news/exclusive-fitbits-6-billion-nights-sleep-data-reveals-us-110058417.html. Available from: https://finance.yahoo.com/news/exclusive-fitbits-6-billion-nights-sleep-data-reveals-us-110058417.html [Cited 26 February 2018]

[8] Watson NF, Badr MS, Belenky G, Bliwise DL, Buxton OM, Buysse D, et al. Joint consensus statement of the American Academy of sleep medicine and sleep research society on the recommended amount of sleep for a healthy adult: methodology and discussion. Sleep 2015;38(8):1161–83. Available from: http://www.ncbi.nlm.nih.gov/pubmed/26194576. [Cited 26 February 2018].

[9] Grandner MA, Hale L, Moore M, Patel NP. Mortality associated with short sleep duration: the evidence, the possible mechanisms, and the future. Sleep Med Rev 2010;14(3):191–203. Available from http://linkinghub.elsevier.com/retrieve/pii/S1087079209000720. [Cited 26 February 2018].

[10] Gottlieb DJ, Whitney CW, Bonekat WH, Iber C, James GD, Lebowitz M, et al. Relation of sleepiness to respiratory disturbance index. Am J Respir Crit Care Med 1999;159(2):502–7. Available from http://www.ncbi.nlm.nih.gov/pubmed/9927364. [Cited 26 February 2018].

[11] Barger LK, Rajaratnam SMW, Cannon CP, Lukas MA, Im K, Goodrich EL, et al. Short sleep duration, obstructive sleep apnea, shiftwork, and the risk of adverse cardiovascular events in patients after an acute coronary syndrome. J Am Heart Assoc 2017;6(10):e006959. Available from: http://jaha.ahajournals.org/lookup/doi/10.1161/JAHA.117.006959. [Cited 26 February 2018].

[12] Yaggi HK, Araujo AB, McKinlay JB. Sleep duration as a risk factor for the development of type 2 diabetes. Diabetes Care 2006;29(3):657–61. Available from http://www.ncbi.nlm.nih.gov/pubmed/16505522. [Cited 26 February 2018].

[13] Czeisler CA, Wickwire EM, Barger LK, Dement WC, Gamble K, Hartenbaum N, et al. Sleep-deprived motor vehicle operators are unfit to drive: a multidisciplinary expert consensus statement on drowsy driving. Sleep Health 2016;2(2):94–9. Available from: http://www.ncbi.nlm.nih.gov/pubmed/28923267. [Cited 26 March 2018].

[14] Belenky G, Wesensten NJ, Thorne DR, Thomas ML, Sing HC, Redmond DP, et al. Patterns of performance degradation and restoration during sleep restriction and subsequent recovery: a sleep dose-response study. J Sleep Res 2003;12(1):1–12. Available from: http://www.ncbi.nlm.nih.gov/pubmed/12603781. [Cited 26 February 2018].

[15] Nieto FJ, Young TB, Lind BK, Shahar E, Samet JM, Redline S, et al. Association of sleep-disordered breathing, sleep apnea, and hypertension in a large community-based study. Sleep Heart Health Study. JAMA 2000;283(14):1829–36. Available from: http://www.ncbi.nlm.nih.gov/pubmed/10770144. [Cited 26 February 2018].

[16] Hung J, Whitford EG, Parsons RW, Hillman DR. Association of sleep apnoea with myocardial infarction in men. Lancet 1990;336(8710):261–4. Available from: http://www.ncbi.nlm.nih.gov/pubmed/1973968. [Cited 26 February 2018].

[17] Dyken ME, Bin IK. Obstructive sleep apnea and stroke. Chest 2009;136(6):1668–77. Available from http://www.ncbi.nlm.nih.gov/pubmed/19995768. [Cited 26 February 2018].

[18] Gagnadoux F, Le Vaillant M, Goupil F, Pigeanne T, Chollet S, Masson P, et al. Depressive symptoms before and after long-term CPAP therapy in patients with sleep apnea. Chest 2014;145(5):1025–31. Available from http://www.ncbi.nlm.nih.gov/pubmed/24435294. [Cited 26 February 2018].

[19] Oldenburg O, Lamp B, Faber L, Teschler H, Horstkotte D, Töpfer V. Sleep-disordered breathing in patients with symptomatic heart failure: a contemporary study of prevalence in and characteristics of 700 patients. Eur J Heart Fail 2007;9(3):251–7. Available from http://doi.wiley.com/10.1016/j.ejheart.2006.08.003. [Cited 26 February 2018].

[20] Romem A, Iacono A, McIlmoyle E, Patel KP, Reed RM, Verceles AC, et al. Obstructive sleep apnea in patients with end-stage lung disease. J Clin Sleep Med 2013;9(7):687–93. Available from: http://www.ncbi.nlm.nih.gov/pubmed/23853563. [Cited 26 February 2018].

[21] Digby GC, Baranchuk A. Sleep apnea and atrial fibrillation; 2012 update. Curr Cardiol Rev 2012;8(4):265–72. Available from: http://www.ncbi.nlm.nih.gov/pubmed/23003203. [Cited 26 February 2018].

[22] Foster GD, Sanders MH, Millman R, Zammit G, Borradaile KE, Newman AB, et al. Obstructive sleep apnea among obese patients with type 2 diabetes. Diabetes Care 2009;32(6):1017–9. Available from: http://www.ncbi.nlm.nih.gov/pubmed/19279303. [Cited 2018 Feb 26].

[23] Young T, Blustein J, Finn L, Palta M. Sleep-disordered breathing and motor vehicle accidents in a population-based sample of employed adults. Sleep 1997;20(8):608–13. Available from: http://www.ncbi.nlm.nih.gov/pubmed/9351127. [Cited 26 February 2018].

[24] Punjabi NM, Caffo BS, Goodwin JL, Gottlieb DJ, Newman AB, O'Connor GT, et al. Sleep-disordered breathing and mortality: a prospective cohort study. Patel A, editor PLoS Med 2009;6(8):e1000132. Available from: http://www.ncbi.nlm.nih.gov/pubmed/19688045. [Cited 26 February 2018].

[25] Breslau N, Roth T, Rosenthal L, Andreski P. Sleep disturbance and psychiatric disorders: a longitudinal epidemiological study of young adults. Biol Psychiatry 1996;39(6):411–8. Available from: http://www.ncbi.nlm.nih.gov/pubmed/8679786. [Cited 26 February 2018].

[26] Fawcett J, Scheftner WA, Fogg L, Clark DC, Young MA, Hedeker D, et al. Time-related predictors of suicide in major affective disorder. Am J Psychiatry 1990;147(9):1189–94. Available from: http://psychiatryonline.org/doi/abs/10.1176/ajp.147.9.1189. [Cited 26 February 2018].

[27] Hafner M, Stepanek M, Taylor J, Troxel W, Stolk C. Why sleep matters—the economic costs of insufficient sleep: a cross-country comparative analysis. Rand Health Q 2016;6(4):11. Available from: http://www.rand.org/pubs/research_reports/RR1791.html. [Cited 31 March 2018].

[28] Chilcott LA, Shapiro CM. The socioeconomic impact of insomnia. An overview. Pharmacoeconomics 1996;10(Suppl 1):1–14. Available from: http://www.ncbi.nlm.nih.gov/pubmed/10163422. [Cited 31 March 2018].

[29] Wickwire EM, Shaya FT, Scharf SM. Health economics of insomnia treatments: the return on investment for a good night's sleep. Sleep Med Rev 2016;30:72–82. Available from: http://linkinghub.elsevier.com/retrieve/pii/S1087079215001550. [Cited 31 March 2018].

[30] Frost & Sullivan. Hidden health crisis costing America billions: underdiagnosing and undertreating obstructive sleep apnea draining healthcare system. Darien, IL: American Academy of Sleep Medicine; 2016. Available from: http://www.aasmnet.org/sleep-apnea-economic-impact.aspx.

[31] Chung F, Yegneswaran B, Liao P, Chung SA, Vairavanathan S, Islam S, et al. STOP Questionnaire: a tool to screen patients for obstructive sleep apnea. Anesthesiology 2008;108(5):812–21. Available from: http://www.ncbi.nlm.nih.gov/pubmed/18431116. [Cited 26 February 2018].

[32] Netzer NC, Stoohs RA, Netzer CM, Clark K, Strohl KP. Using the Berlin Questionnaire to identify patients at risk for the sleep apnea syndrome. Ann Intern Med 1999;131(7):485–91. Available from: http://www.ncbi.nlm.nih.gov/pubmed/10507956. [Cited 26 February 2018].

[33] Johns MW. A new method for measuring daytime sleepiness: the Epworth Sleepiness Scale. Sleep 1991;14(6):540–5. Available from: http://www.ncbi.nlm.nih.gov/pubmed/1798888. [Cited 26 February 2018].

[34] Weaver TE, Laizner AM, Evans LK, Maislin G, Chugh DK, Lyon K, et al. An instrument to measure functional status outcomes for disorders of excessive sleepiness. Sleep 1997;20(10):835–43. Available from: http://www.ncbi.nlm.nih.gov/pubmed/9415942. [Cited 26 February 2018].

[35] Hoddes E, Dement W, Zarcone V. The development and use of the Stanford Sleepiness Scale (SSS). Psychophysiology 1972;9:150.

[36] Bastien CH, Vallières A, Morin CM. Validation of the Insomnia Severity Index as an outcome measure for insomnia research. Sleep Med 2001;2(4):297–307. Available from: http://www.ncbi.nlm.nih.gov/pubmed/11438246. [Cited 26 February 2018].

[37] Buysse DJ, Reynolds CF, Monk TH, Berman SR, Kupfer DJ. The Pittsburgh Sleep Quality Index: a new instrument for psychiatric practice and research. Psychiatry Res 1989;28(2):193–213. Available from: http://www.ncbi.nlm.nih.gov/pubmed/2748771. [Cited 26 February 2018].

[38] Buysse DJ, Yu L, Moul DE, Germain A, Stover A, Dodds NE, et al. Development and validation of patient-reported outcome measures for sleep disturbance and sleep-related impairments. Sleep 2010;33(6):781–92. Available from: http://www.ncbi.nlm.nih.gov/pubmed/20550019. [Cited 26 February 2018].

[39] Horne JA, Ostberg O. A self-assessment questionnaire to determine morningness-eveningness in human circadian rhythms. Int J Chronobiol 1976;4(2):97–110. Available from: http://www.ncbi.nlm.nih.gov/pubmed/1027738. [Cited 26 February 2018].

[40] Roenneberg T, Wirz-Justice A, Merrow M. Life between clocks: daily temporal patterns of human chronotypes. J Biol Rhythm 2003;18(1):80–90. Available from: http://www.ncbi.nlm.nih.gov/pubmed/12568247. [Cited 26 February 2018].

[41] American Academy of Sleep Medicine. International Classification of Sleep Disorders. 3rd ed. Darien, IL: AASM; 2014.

[42] Young T, Peppard PE, Gottlieb DJ. Epidemiology of obstructive sleep apnea: a population health perspective. Am J Respir Crit Care Med 2002;165(9):1217–39. Available from: http://www.ncbi.nlm.nih.gov/pubmed/11991871. [Cited 26 February 2018].

[43] Young T, Finn L, Austin D, Peterson A. Menopausal status and sleep-disordered breathing in the Wisconsin sleep cohort study. Am J Respir Crit Care Med 2003;167(9):1181–5. Available from: http://www.ncbi.nlm.nih.gov/pubmed/12615621. [Cited 26 February 2018].

[44] Myers KA, Mrkobrada M, Simel DL. Does this patient have obstructive sleep apnea? JAMA 2013;310(7):731. Available from; :http://www.ncbi.nlm.nih.gov/pubmed/23989984. [Cited 26 February 2018].

[45] Flemons WW, Whitelaw WA, Brant R, Remmers JE. Likelihood ratios for a sleep apnea clinical prediction rule. Am J Respir Crit Care Med 1994;150(5):1279–85. Available from; :http://www.ncbi.nlm.nih.gov/pubmed/7952553. Cited 26 February 2018.

[46] Nagappa M, Liao P, Wong J, Auckley D, Ramachandran SK, Memtsoudis S, et al. Validation of the STOP-bang questionnaire as a screening tool for obstructive sleep apnea among different populations: a systematic review and meta-analysis. Arias-Carrion O PLoS One 2015;10(12):e0143697. Available from: http://www.ncbi.nlm.nih.gov/pubmed/26658438. [Cited 2018 February 10].

[47] Kapur VK, Auckley DH, Chowdhuri S, Kuhlmann DC, Mehra R, Ramar K, et al. Clinical practice guideline for diagnostic testing for adult obstructive sleep apnea: an American academy of sleep medicine clinical practice guideline. J Clin Sleep Med 2017;13(03):479–504. Available from: http://www.ncbi.nlm.nih.gov/pubmed/28162150. [Cited 2018 Feb 26].

[48] Chiu H-Y, Chen P-Y, Chuang L-P, Chen N-H, Tu Y-K, Hsieh Y-J, et al. Diagnostic accuracy of the Berlin questionnaire, STOP-BANG, STOP, and Epworth Sleepiness Scale in detecting obstructive sleep apnea: a bivariate meta-analysis. Sleep Med Rev 2017;36:57–70. Available from: https://www.ncbi.nlm.nih.gov/pubmed/27919588. [Cited 10 February 2018].

[49] Guilleminault C, Brooks SN. Excessive daytime sleepiness: a challenge for the practising neurologist. Brain 2001;124(Pt 8):1482–91. Available from: http://www.ncbi.nlm.nih.gov/pubmed/11459741. [Cited 26 February 2018].

[50] Johns MW. Reliability and factor analysis of the Epworth Sleepiness Scale. Sleep 1992;15(4):376–81. Available from: http://www.ncbi.nlm.nih.gov/pubmed/1519015. [Cited 26 February 2018].

[51] Johns MW. Sensitivity and specificity of the multiple sleep latency test (MSLT), the maintenance of wakefulness test and the Epworth Sleepiness Scale: failure of the MSLT as a gold standard. J Sleep Res 2000;9(1):5–11. Available from: http://www.ncbi.nlm.nih.gov/pubmed/10733683. [Cited 26 February 2018].

[52] Chervin RD, Aldrich MS, Pickett R, Guilleminault C. Comparison of the results of the Epworth Sleepiness Scale and the multiple sleep latency test. J Psychosom Res 1997;42(2):145–55. Available from: http://www.ncbi.nlm.nih.gov/pubmed/9076642. [Cited 26 February 2018].

[53] Pouliot Z, Peters M, Neufeld H, Kryger MH. Using self-reported questionnaire data to prioritize OSA patients for polysomnography. Sleep 1997;20(3):232–6. Available from: http://www.ncbi.nlm.nih.gov/pubmed/9178919. [Cited 2018 February 26].

[54] Chervin RD, Aldrich MS. The Epworth Sleepiness Scale may not reflect objective measures of sleepiness or sleep apnea. Neurology 1999;52(1):125–31. Available from: http://www.ncbi.nlm.nih.gov/pubmed/9921859. [Cited 26 February 2018].

[55] Fong SYY, Ho CKW, Wing YK. Comparing MSLT and ESS in the measurement of excessive daytime sleepiness in obstructive sleep apnoea syndrome. J Psychosom Res 2005;58(1):55–60. Available from: http://linkinghub.elsevier.com/retrieve/pii/S0022399904004945. [Cited 2018 February 27].

[56] Chervin RD. Sleepiness, fatigue, tiredness, and lack of energy in obstructive sleep apnea. Chest 2000;118(2):372–9. Available from: http://www.ncbi.nlm.nih.gov/pubmed/10936127. [Cited 2018 February 27].

[57] He K, Kapur VK. Sleep-disordered breathing and excessive daytime sleepiness. Sleep Med Clin 2017;12(3):369–82. Available from: http://linkinghub.elsevier.com/retrieve/pii/S1556407X17300218. [Cited 26 February 2018].

[58] Bonzelaar LB, Salapatas AM, Yang J, Friedman M. Validity of the Epworth Sleepiness Scale as a screening tool for obstructive sleep apnea. Laryngoscope 2017;127(2):525–31. Available from: http://doi.wiley.com/10.1002/lary.26206. [Cited 2018 February 11].

[59] Damiani MF, Quaranta VN, Falcone VA, Gadaleta F, Maiellari M, Ranieri T, et al. The Epworth Sleepiness Scale: conventional self vs physician administration. Chest 2013;143(6):1569–75. Available from: http://linkinghub.elsevier.com/retrieve/pii/S0012369213603840. [Cited 26 February 2018].

[60] Walia HK, Thompson NR, Katzan I, Foldvary-Schaefer N, Moul DE, Mehra R. Impact of sleep-disordered breathing treatment on quality of life measures in a large clinic-based cohort. J Clin Sleep Med 2017;13(11):1255–63. Available from: http://jcsm.aasm.org/ViewAbstract.aspx?pid=31115. [Cited 26 February 2018].

[61] Chasens ER, Ratcliffe SJ, Weaver TE. Development of the FOSQ-10: a short version of the Functional Outcomes of Sleep Questionnaire. Sleep 2009;32(7):915–9. Available from: http://www.ncbi.nlm.nih.gov/pubmed/19639754. [Cited 26 February 2018].

[62] Herscovitch J, Broughton R. Sensitivity of the Stanford Sleepiness Scale to the effects of cumulative partial sleep deprivation and recovery oversleeping. Sleep 1981;4(1):83–91. Available from: http://www.ncbi.nlm.nih.gov/pubmed/7232973. [Cited 26 February 2018].

[63] Valley V, Broughton R. Daytime performance deficits and physiological vigilance in untreated patients with narcolepsy-cataplexy compared to controls. Rev Electroencephalogr Neurophysiol Clin 1981;11(1):133–9. Available from: http://www.ncbi.nlm.nih.gov/pubmed/7313247. [Cited 26 February 2018].

[64] Akerstedt T, Gillberg M. Subjective and objective sleepiness in the active individual. Int J Neurosci 1990;52(1–2):29–37. Available from: http://www.ncbi.nlm.nih.gov/pubmed/2265922. [Cited 2018 Feb 26].

[65] Baulk SD, Reyner LA, Horne JA. Driver sleepiness—Evaluation of reaction time measurement as a secondary task. Sleep 2001;24(6):695–8. Available from: http://www.ncbi.nlm.nih.gov/pubmed/11560183. [Cited 26 February 2018].

[66] Akerstedt T, Anund A, Axelsson J, Kecklund G. Subjective sleepiness is a sensitive indicator of insufficient sleep and impaired waking function. J Sleep Res 2014;23(3):240–52. Available from: http://doi.wiley.com/10.1111/jsr.12158. [Cited 26 February 2018].

[67] Kaida K, Takahashi M, Akerstedt T, Nakata A, Otsuka Y, Haratani T, et al. Validation of the Karolinska Sleepiness Scale against performance and EEG variables. Clin Neurophysiol 2006;117(7):1574–81. Available from: http://linkinghub.elsevier.com/retrieve/pii/S1388245706001428. [Cited 26 February 2018].

[68] Filtness AJ, Reyner LA, Horne JA. Moderate sleep restriction in treated older male OSA participants: greater impairment during monotonous driving compared with controls. Sleep Med 2011;12(9):838–43. Available from: http://linkinghub.elsevier.com/retrieve/pii/S138994571100222X. [Cited 2018 February 13].

[69] KKH W, Marshall NS, Grunstein RR, Dodd MJ, Rogers NL. Comparing the neurocognitive effects of 40 h sustained wakefulness in patients with untreated OSA and healthy controls. J Sleep Res 2008;17(3):322–30. Available from: http://doi.wiley.com/10.1111/j.1365-2869.2008.00665.x. [Cited 2018 February 13].

[70] Littner M, Hirshkowitz M, Kramer M, Kapen S, Anderson WM, Bailey D, et al. Practice parameters for using polysomnography to evaluate insomnia: an update. Sleep 2003;26(6):754–60. Available from: http://www.ncbi.nlm.nih.gov/pubmed/14572131. [Cited 26 February 2018].

[71] Morin CM, Belleville G, Bélanger L, Ivers H. The Insomnia Severity Index: psychometric indicators to detect insomnia cases and evaluate treatment response. Sleep 2011;34(5):601–8. Available from:. http://www.ncbi.nlm.nih.gov/pubmed/21532953; [Cited 26 February 2018].

[72] Gagnon C, Belanger L, Ivers H, Morin CM. Validation of the Insomnia Severity Index in primary care. J Am Board Fam Med 2013;26(6):701–10. Available from: http://www.ncbi.nlm.nih.gov/pubmed/24204066. [Cited 2019 January 7].

[73] Savard M-H, Savard J, Simard S, Ivers H. Empirical validation of the Insomnia Severity Index in cancer patients. Psychooncology 2005;14(6):429–41. Available from: http://doi.wiley.com/10.1002/pon.860. [Cited 26 February 2018].

[74] Chiu H-Y, Chang L-Y, Hsieh Y-J, Tsai P-S. A meta-analysis of diagnostic accuracy of three screening tools for insomnia. J Psychosom Res 2016;87:85–92. Available from: http://linkinghub.elsevier.com/retrieve/pii/S0022399916303324. [Cited 26 February 2018].

[75] Wells RD, Day RC, Carney RM, Freedland KE, Duntley SP. Depression predicts self-reported sleep quality in patients with obstructive sleep apnea. Psychosom Med 2004;66(5):692–7. Available from: https://insights.ovid.com/crossref?an=00006842-200409000-00011. [Cited 26 February 2018].

[76] Nishiyama T, Mizuno T, Kojima M, Suzuki S, Kitajima T, Ando KB, et al. Criterion validity of the Pittsburgh Sleep Quality Index and Epworth Sleepiness Scale for the diagnosis of sleep disorders. Sleep Med 2014;15(4):422–9. Available from: http://www.ncbi.nlm.nih.gov/pubmed/24657203. [Cited 2018 February 11].

[77] Yu L, Buysse DJ, Germain A, Moul DE, Stover A, Dodds NE, et al. Development of short forms from the PROMIS™ sleep disturbance and sleep-related impairment item banks. Behav Sleep Med 2012;10(1):6–24. Available from: http://www.ncbi.nlm.nih.gov/pubmed/22250775. [Cited 26 February 2018].

[78] Leung YW, Brown C, Cosio AP, Dobriyal A, Malik N, Pat V, et al. Feasibility and diagnostic accuracy of the Patient-Reported Outcomes Measurement Information System (PROMIS) item banks for routine surveillance of sleep and fatigue problems in ambulatory cancer care. Cancer 2016;122(18):2906–17. Available from: http://doi.wiley.com/10.1002/cncr.30134. [Cited 26 February 2018].

[79] Weitzman ED, Czeisler CA, Coleman RM, Spielman AJ, Zimmerman JC, Dement W, et al. Delayed sleep phase syndrome. A chronobiological disorder with sleep-onset insomnia. Arch Gen Psychiatry 1981;38(7):737–46. Available from: http://www.ncbi.nlm.nih.gov/pubmed/7247637. [Cited 26 February 2018].

[80] Sack RL, Auckley D, Auger RR, Carskadon MA, Wright KP, Vitiello MV, et al. Circadian rhythm sleep disorders: part I, basic principles, shift work and jet lag disorders. An American Academy of Sleep Medicine review. Sleep 2007;30(11):1460–83. Available from; http://www.ncbi.nlm.nih.gov/pubmed/18041480. [Cited 26 February 2018].

[81] Jones CR, Campbell SS, Zone SE, Cooper F, DeSano A, Murphy PJ, et al. Familial advanced sleep-phase syndrome: a short-period circadian rhythm variant in humans. Nat Med 1999;5(9):1062–5. Available from: http://www.nature.com/articles/nm0999_1062. [Cited 26 February 2018].

[82] Satoh K, Mishima K, Inoue Y, Ebisawa T, Shimizu T. Two pedigrees of familial advanced sleep phase syndrome in Japan. Sleep 2003;26(4):416–7. Available from: http://www.ncbi.nlm.nih.gov/pubmed/12841366. [Cited 26 February 2018].

[83] Zavada A, Gordijn MCM, Beersma DGM, Daan S, Roenneberg T. Comparison of the Munich Chronotype Questionnaire with the Horne-Ostberg's morningness-eveningness score. Chronobiol Int 2005;22(2):267–78. Available from: http://www.ncbi.nlm.nih.gov/pubmed/16021843. [Cited 26 February 2018].

[84] Kitamura S, Hida A, Aritake S, Higuchi S, Enomoto M, Kato M, et al. Validity of the Japanese version of the Munich Chronotype Questionnaire. Chronobiol Int 2014;31(7):845–50. Available from http://www.tandfonline.com/doi/full/10.3109/07420528.2014.914035. [Cited 26 February 2018].

[85] Hening W, Walters AS, Allen RP, Montplaisir J, Myers A, Ferini-Strambi L. Impact, diagnosis and treatment of restless legs syndrome (RLS) in a primary care population: the REST (RLS epidemiology, symptoms, and treatment) primary care study. Sleep Med

[86] Becker PM, Novak M. Diagnosis, comorbidities, and management of restless legs syndrome. Curr Med Res Opin 2014;30(8):1441–60. Available from: http://www.ncbi.nlm.nih.gov/pubmed/24805265. [Cited 26 February 2018].

[87] Walters AS, LeBrocq C, Dhar A, Hening W, Rosen R, Allen RP, et al. Validation of the International restless legs syndrome study group rating scale for restless legs syndrome. Sleep Med 2003;4(2):121–32. Available from: http://www.ncbi.nlm.nih.gov/pubmed/14592342. [Cited 26 February 2018].

[88] Allen R, Oertel W, Walters A, Benes H, Schollmayer E, Grieger F, et al. Relation of the International restless legs syndrome study group rating scale with the Clinical global impression severity scale, the restless legs syndrome 6-item questionnaire, and the restless legs syndrome-quality of life questionnaire. Sleep Med 2013;14(12):1375–80. Available from: http://linkinghub.elsevier.com/retrieve/pii/S1389945713011374. [Cited 26 February 2018].

[89] Walters AS, Frauscher B, Allen R, Benes H, Chaudhuri KR, Garcia-Borreguero D, et al. Review of severity rating scales for restless legs syndrome: critique and recommendations. Mov Disord Clin Pract 2014;1(4):317–24. Available from: http://doi.wiley.com/10.1002/mdc3.12088. [Cited 26 February 2018].

[90] Tang NKY, Harvey AG. Altering misperception of sleep in insomnia: behavioral experiment versus verbal feedback. J Consult Clin Psychol 2006;74(4):767–76. Available from: http://doi.apa.org/getdoi.cfm?doi=10.1037/0022-006X.74.4.767. [Cited 26 February 2018].

[91] Montgomery-Downs HE, Insana SP, Bond JA. Movement toward a novel activity monitoring device. Sleep Breath 2012;16(3):913–7. Available from: http://link.springer.com/10.1007/s11325-011-0585-y. [Cited 26 February 2018].

[92] Mantua J, Gravel N, Spencer RMC. Reliability of sleep measures from four personal health monitoring devices compared to research-based actigraphy and polysomnography. Sensors (Basel) 2016;16(5):646. Available from: http://www.mdpi.com/1424-8220/16/5/646. [Cited 26 February 2018].

[93] Meltzer LJ, Hiruma LS, Avis K, Montgomery-Downs H, Valentin J. Comparison of a commercial accelerometer with polysomnography and actigraphy in children and adolescents. Sleep 2015;38(8):1323–30. Available from: http://www.ncbi.nlm.nih.gov/pubmed/26118555. [Cited 26 February 2018].

[94] Cook JD, Prairie ML, Plante DT. Utility of the Fitbit flex to evaluate sleep in major depressive disorder: a comparison against polysomnography and wrist-worn actigraphy. J Affect Disord 2017;217:299–305. Available from: http://linkinghub.elsevier.com/retrieve/pii/S0165032716317700. [Cited 26 February 2018].

[95] Kang S-G, Kang JM, Ko K-P, Park S-C, Mariani S, Weng J. Validity of a commercial wearable sleep tracker in adult insomnia disorder patients and good sleepers. J Psychosom Res 2017;97:38–44. Available from: http://linkinghub.elsevier.com/retrieve/pii/S002239991630561X. [Cited 26 February 2018].

[96] Gruwez A, Libert W, Ameye L, Bruyneel M. Reliability of commercially available sleep and activity trackers with manual switch-to-sleep mode activation in free-living healthy individuals. Int J Med Inform 2017;102:87–92. Available from: http://linkinghub.elsevier.com/retrieve/pii/S1386505617300692. [Cited 26 February 2018].

[97] de Zambotti M, Claudatos S, Inkelis S, Colrain IM, Baker FC. Evaluation of a consumer fitness-tracking device to assess sleep in adults. Chronobiol Int 2015;32(7):1024–8. Available from http://www.tandfonline.com/doi/full/10.3109/07420528.2015.1054395. [Cited 26 February 2018].

[98] de Zambotti M, Baker FC, Colrain IM. Validation of sleep-tracking technology compared with polysomnography in adolescents. Sleep 2015;38(9):1461–8. Available from: https://academic.oup.com/sleep/article-lookup/doi/10.5665/sleep.4990. [Cited 26 February 2018].

[99] Toon E, Davey MJ, Hollis SL, Nixon GM, Horne RSC, Biggs SN. Comparison of commercial wrist-based and smartphone accelerometers, actigraphy, and PSG in a clinical cohort of children and adolescents. J Clin Sleep Med 2016;12(3):343–50. Available from: http://jcsm.aasm.org/ViewAbstract.aspx?pid=30509. [Cited 12 February 2018].

[100] Schade M, Bauer C, Murray B, Gahan L, Doheny E, Kilroy H, et al. 0784 Sleep validity of a non-contact bedside movement and respiration-sensing device. Sleep 2017;40(suppl_1):A290–1. Oxford University Press. Available from: https://academic.oup.com/sleep/article-lookup/doi/10.1093/sleepj/zsx050.783. [Cited 26 March 2018].

[101] De Chazal P, Fox N, O'Hare E, Heneghan C, Zaffaroni A, Boyle P, et al. Sleep/wake measurement using a non-contact biomotion sensor. J Sleep Res 2011;20(2):356–66. Available from: http://www.ncbi.nlm.nih.gov/pubmed/20704645. [Cited 26 February 2018].

[102] Zaffaroni A, Kent B, O'Hare E, Heneghan C, Boyle P, O'Connell G, et al. Assessment of sleep-disordered breathing using a non-contact bio-motion sensor. J Sleep Res 2013;22(2):231–6. Available from: http://doi.wiley.com/10.1111/j.1365-2869.2012.01056.x. [Cited 26 February 2018].

[103] Weinreich G, Terjung S, Wang Y, Werther S, Zaffaroni A, Teschler H. Validation of a non-contact screening device for the combination of sleep-disordered breathing and periodic limb movements in sleep. Sleep Breath 2018;22(1):131–8. Available from: http://link.springer.com/10.1007/s11325-017-1546-x. [Cited 26 February 2018].

[104] SleepScore Labs. Evaluation of the SleepScore Labs sonar and radiofrequency sleep sensor technologies by ResMed(TM). Calsbad, CA: SleepScore Labs; 2018. Available from: https://gkng5olag22mpz1r551iq1ddwpengine.netdna-ssl.com/wp-content/uploads/2018/06/Sonar-and-RF-Validation-info-sheet-1.pdf.

Chapter 11

Sleep hygiene and the prevention of chronic insomnia

Jason G. Ellis, Sarah F. Allen
Northumbria Sleep Research Laboratory, Northumbria University, Newcastle, United Kingdom

SLEEP HYGIENE

What is sleep hygiene?

One of the main difficulties when several things are lumped together under one overarching term is standardization. As such, asking one person what they consider sleep hygiene to entail is likely to result in a different set of rules or techniques compared to another individuals' list [1]. As you will come to see, this can create challenges and confusion for the sleep researcher, the clinician and indeed the public. Hauri developed the first set of "sleep hygiene" recommendations in the late 1970s with the specific aim of helping patient's manage their insomnia [2]. The original list contained 10 items ranging from not going to bed hungry, regularizing wake times and avoiding caffeine in the evening, to sound attenuating the bedroom and making sure the bedroom is not too warm (see Table 11.1). While based on the available evidence at the time, in addition to Hauri's own clinical observations, the original list has been modified extensively over the years, including by Hauri himself, as more information became available [3–8]. Where some recommendations appear to have been removed and new ones added (see Table 11.1), there are, however, a few recommendations which appear in the majority of definitions, namely, exercise, limiting caffeine, avoiding alcohol, and having a snack before bedtime.

There also appears to be some overlap between some of the recommendations outlined as "sleep hygiene" by some authors and what others might consider elements from other components of traditional Cognitive Behavioral Therapy for Insomnia (CBT-I). For example, leaving the bed if not awake, only using the bed for sleep and avoiding napping during the day are commonly contained within some definitions of sleep hygiene while also being aspects of stimulus control instructions [9]. Similarly, decreasing time in bed is generally associated with sleep restriction [10], as is, albeit more implicitly, keeping a regular sleep–wake schedule. Not actively trying to sleep and keeping a worry list are also commonly included in the cognitive components of CBT-I [11, 12]. The challenge is, however, as packaged in sleep hygiene instructions, these latter items do not represent the full instructions for stimulus control, sleep restriction, or cognitive therapy, respectively. Are they, therefore, likely to be as effective as the full instructions? Moreover, an individual who has been exposed to these brief instructions is likely to be more resistant to full CBT-I because they have "done that before." So, the question remains; are these items aspects of sleep hygiene, or not? For the purpose of this chapter, we will take a psycho-educational approach and explore only those recommendations that do not overlap with other components of traditional CBT-I, namely, exercise, caffeine, alcohol, food and liquid intake, nicotine, the bedroom environment, and clockwatching. So, what are the specific recommendations for each and how do these elements impact on sleep? In terms of the first part of this question, for most aspects there are no standardized specific recommendations, but more general guidelines, which again, vary from definition to definition [13].

Exercise

Broadly, there are two recommendations associated with exercise in the context of sleep hygiene: (i) exercise is good for sleep and should be encouraged but (ii) exercise too close to bedtime is detrimental to sleep and should be discouraged. The benefits of both acute and regular exercise on sleep have been documented in numerous studies including children, adolescents, and older adults, with and without sleep [14, 15]. These benefits appear to include increases in Slow Wave Sleep and total sleep time; reductions in sleep latencies and time awake after sleep onset and a slight delay, and minimal reductions, in REM Sleep [16]. That said, a recent review points to several moderators underpinning these relationships, including sex, age, type of exercise, baseline physical activity levels, and the timing of the exercise [17]. In terms of exercising in the evening, the general consensus

TABLE 11.1 OVERVIEW OF SLEEP HYGIENE RECOMMENDATIONS OVER TIME.

Recommendations:	Hauri (1977)	Schoicket et al. (1988)	Hauri (1992)	Hauri (1993)	Guilleminault et al. (1995)	Friedman et al. (2000)	Perlis et al. (2005)
Environmental							
Eliminate bedroom noise	●	●				●	●
Regulate temperature of bedroom	●		●	●	●	●	●
Eliminate clocks from the bedroom		●	●	●			●
Make bedroom comfortable			●			●	
Sleep/wake schedule							
Eliminate napping	●	●	●		●	●	●
Decrease time in bed	●		●	●		●	●
Regular bed/wake times	●	●	●	●	●	●	●
Positive things to do in the evening							
Undertake relaxing activities	●	●	●			●	●
Exercise	●	●	●	●	●	●	●
Make worry list	●		●	●		●	●
Host bath	●	●	●			●	
Eat a light snack	●	●	●	●		●	●
Limit liquid intake		●				●	
Decrease or avoid smoking			●	●	●	●	●
Limit or avoid caffeine	●	●	●	●	●	●	●
Avoid Alcohol	●	●	●	●		●	●
Sleep/wake behaviours							
Don't try to sleep	●	●	●	●		●	●
Leave bed if awake	●	●	●	●	●	●	●
Use the bedroom only for sleep	●		●		●	●	●
Avoid use of sleeping pills	●		●				

is that exercising too close to bedtime could disrupt the circadian system, elevate body core temperature, and/or create a form of physiological arousal—each of which could be detrimental to sleep [18]. A recent study, however, suggests otherwise, finding that evening exercise, at least within 4 h of bedtime, was not associated with poorer sleep [19]. While it appears that exercise, in general, is good for sleep, the detrimental impact of evening exercise on sleep is less clear.

Caffeine

It is widely believed that caffeine can have a negative impact on sleep [20]. In fact, acute caffeine administration has been used as an analogous model for sleep disruption and disturbance in several studies [21, 22]. In these cases a dose of caffeine before bedtime results in a prolonged sleep latency, decreased total sleep time, and reductions in slow wave sleep, usually in a dose–response manner [23]. The general rule relating to caffeine, with respect to sleep hygiene, is that it should be avoided after midday, although it is unclear where the timing for this rule came from. There is experimental evidence that caffeine administration 6 h before bed can negatively impact on sleep [24]. Outside the laboratory, however, the impact of caffeine on sleep is less clear with one large survey finding that caffeine consumption of up to seven–eight cups per day was unrelated to self-reported levels of sleep duration and daytime somnolence, although more than eight cups per day was associated with a reduction in total sleep time [25]. Further, another study showed caffeine was unrelated to insomnia severity once anxiety and race/ethnicity were controlled for [26]. Individual differences may play a part in explaining the inconsistencies as it has been shown that the adenosine A_{2A} receptor gene may influence the relationship between caffeine and sleep [27]. That said, public awareness of the effects of caffeine on sleep might also explain why experimental studies using caffeine show an impact on sleep but naturalistic studies tend to find only a limited relationship [28].

Alcohol

Like caffeine, alcohol is widely regarded to impact on sleep. The main issue in this context is that alcohol is a sedative and as such can be an appealing hypnotic [29]. The review by Ebrahim and colleagues [30] suggests that although alcohol results in a reduction in sleep onset latency and a more consolidated first half sleep, it is also associated with an increase in disrupted and fragmented sleep in the second half of sleep. Moreover, where low and moderate doses show little effect on REM sleep, high doses can significantly reduce REM, especially in the first half of the night. The recommendation regarding alcohol is that it should be avoided in the evening and certainly not to be used as a sleep aid. With regard to the first point, however, as the effects of alcohol on sleep appear to be dose dependent, should alcohol be avoided altogether? Will that dissuade an individual from complying? As for the latter recommendation, considering that insomnia has been shown to be a significant risk factor for the development of alcohol problems [31] due to increasing tolerance to its sleep-promoting effects, it makes sense to suggest that alcohol must not be used as a sleep aid.

Food and liquid intake

While there is a considerable literature on the impact of sleep deprivation on hunger and food intake, there is surprisingly little information on the impact of hunger or being over sated on how an individual sleeps. The story appears to be similar for liquid intake. The general recommendation is that an individual should not go to sleep hungry or eat a heavy meal before bed and should minimize liquid intake in the evening (2–3h before bedtime). Certainly these recommendations make physiological sense as trying to sleep is going to be more challenging if the individual is digesting a heavy meal, especially if the meal contained high levels of spice or fat. Additionally, going to bed hungry is likely to increase the chances of a blood sugar drop in the night, waking the individual. Excessive liquid during the evening is likely to increase nocturnal awakenings with the need to use the bathroom at night. The impact of which is likely to be more so for older adults [32]. These rules aside, it is important to account for individual and circumstantial (e.g., illnesses characterized by dehydration) differences in the timing and amount of food and liquid intake.

Nicotine

As with alcohol, the general recommendation is that nicotine should be avoided close to bedtime. Nicotine is a stimulant and as such has the capacity to disrupt sleep [33]. Jaehne and colleagues [33] reviewed the literature regarding nicotine and sleep and from the nine human studies that met criteria they found that smokers had double the risk of developing sleep disturbances, compared to nonsmokers. Additionally, from their review of polysomnography studies, smokers tend to demonstrate longer sleep onset latencies, reduced SWS, and a longer REM latency compared to nonsmokers [33]. Where this would suggest curtailment or even cessation of nicotine would be good for sleep, a challenge here is the negative impact of withdrawal on sleep, which can begin 6h following abstinence and last for 3 weeks or longer [34]. Nicotine abstinence has been consistently shown to relate to subjectively and objectively defined poor sleep [33]. As such it appears that a balance between abstinence and cessation needs to be struck. Passive smoking is also negatively related to sleep [35]. Sabanayagam and Shankar [36] found that while tobacco users had a twofold increase in risk of insufficient sleep compared to nontobacco users, second-hand smoke exposure was associated with insufficient sleep among nonsmokers.

Bedroom environment

The general rule with regard to the sleep environment is ensuring that it should be cool, dark, quiet, and comfortable, and more recently, free from electronics. Having a hot bedroom or an uncomfortable bed was independently associated with reports of nonrestorative sleep in one large pan-European study [37]. Environmental modifications such as blackout blinds, eye masks, earplugs, new mattresses, and the use of suitable bedding and sleepwear are encouraged under this recommendation.

Bedroom temperature—both excessively hot and cold environments can negatively influence sleep. Above 71°F/21.6°C has been shown to disrupt sleep, as has below 41°F/5°C [38, 39]. The National Sleep Foundation [40] suggests between 60°F/15.6°C and 67°F/19.4°C is ideal.

Bedroom light—light is likely to wake the individual earlier than desired or prevent the individual from sleeping due to its influence on the circadian system and the suppression of melatonin. Indoor lighting as low as <500 Lux has been shown to suppress melatonin [41]. In one study participants exposed to <200 Lux for 5 consecutive days showed a later melatonin onset and a shortened melatonin duration (90 min) compared to those who were exposed to dim light <3 Lux [42].

Bedroom noise—of all the bedroom environmental factors, noise has been the most studied [43]. The findings are clear in that excessive noise disrupts sleep, and subsequent daytime performance, even if the individual does not recall awakening during the night [43]. The World Health Organization (WHO) suggests nighttime noise should be below 40 dB [44]. That said, there are individual differences in noise tolerance with one study demonstrating that 15 dB was sufficient to wake one individual whereas for another it took 100 dB [45].

Bedroom Comfort—a comfortable sleep environment could mean very different things to different people. For some this may include specific elements regarding the bed itself—good bedding, a supportive mattress, the number and density of pillows. There is little evidence for these factors making a specific impact on sleep in general, however, as personal preference is likely to be key [46]. That said, one study has demonstrated poorer sleep outcomes when sleeping on a hard surface compared to a softer one [47]. Irrespective of the definition of bedroom comfort it stands to reason that if the bedroom environment is not perceived as comfortable, achieving sleep can be problematic.

Removal of electronics

Although not included explicitly in sleep hygiene recommendations, presumably due to historical reasons, recently the impact of electronics in the bedroom, and use close to bedtime, in relation to sleep has become a focus of attention [48]. The review by Cain and Gradisar [49] showed that the most consistent aspects of sleep disturbed were a delayed bedtime and reduced total sleep time. Similarly, Hale and Guan's review of 67 studies showed similar findings for screen use with 90% of those studies showing adverse sleep outcomes [50]. The question remains as to the relative contribution of the blue light emitted from these devices and the cognitive arousal that using these devices can have on sleep [51]. As such it would appear that removing electronics would be the advisable thing to do but this may be resisted, especially in younger populations.

Clockwatching

The general recommendation is that clocks are not good in the bedroom and should be removed or, at the least, not be visible. The rationale is that if an individual is awake in bed (either at bedtime or during the night) they are likely to check the clock and calculate (i) how long they have been asleep/awake and (ii) how long they have left before they need to get up. This is likely to provoke an anxious response, prolonging sleep initiation/reinitiation, resulting in further clockwatching and a vicious cycle of checking anxiety. Surprisingly, there is very little empirical data on the impact of clockwatching on sleep. From the data that does exist, however, it appears that monitoring the clock increases presleep worry whether an individual has insomnia or not [52]. That said, people with insomnia do tend to demonstrate an attentional bias toward clocks [53] so would be more likely to gravitate toward them if awake in bed.

Measuring sleep hygiene

In the majority of cases sleep hygiene information is gathered as part of a routine clinical interview (see Morin [54] for a good example). There are, however, three scales that specifically measure sleep hygiene—The Sleep Hygiene Awareness and Practice Scale [55], the Sleep Hygiene Self Test [56], and the Sleep Hygiene Index [57]. Additionally, there are at least two that are more developmentally focused—The Children's Sleep Hygiene Scale [58] and the Adolescent Sleep Hygiene Scale [59]. Again, as with the recommendations on sleep hygiene, there is no consensus between the scales as to what does and what does not constitute sleep hygiene. As such, comparing the results from studies is challenging.

Do people with insomnia have poorer sleep hygiene than normal sleepers?

Several studies have examined levels of sleep hygiene in individuals with insomnia, comparative to controls. The results are mixed, both in terms of overall findings and the influence of individual recommendations. People with insomnia were more likely to smoke and drink alcohol before bed, compared to their normally sleeping counterparts, in one study [60]. Interestingly, in the same study caffeine consumption did not differ between the groups [60]. This latter finding contradicts an earlier study whereby levels of caffeine consumption were significantly higher in those with insomnia compared to normal sleepers [55]. Further, in an internet-based study in the United States, Gellis and Lichstein [61] found that poor sleepers engaged in poorer sleep hygiene practices; specifically they reported sleeping in environments perceived to be uncomfortable, in terms of temperature and noise, compared to good sleepers. Another study found people with insomnia were more likely to smoke within 5 min of bedtime, drink more alcohol in general, use alcohol to sleep, and consume alcohol within 30 min of bedtime, compared to controls. Despite this evidence, others have found no significant differences between people with insomnia and normal sleepers on a range of sleep hygiene behaviors [63]. For example, in Harveys' study on individuals with sleep onset insomnia, she observed no differences on several sleep hygiene measures including bedroom noise; mattress comfort; general alcohol use; and caffeine, alcohol, and nicotine use close to bedtime [63].

As such, it appears there is no real consistent pattern of sleep hygiene behaviors that can be linked to insomnia. Whether this inconsistency speaks to different behaviors being measured or to actual inconsistencies within different populations is difficult to ascertain. Another issue is causality—is it the case that any differences observed between those with insomnia and those without are a consequence of having insomnia? Alternatively, if poor sleep hygiene causes insomnia then it could reasonably be assumed that knowledge of sleep hygiene would relate to behavior.

With respect to the latter question, Brown and colleagues [1] found that knowledge of sleep hygiene rules was associated with sleep hygiene practices in college students, which in turn was related to sleep quality. Conversely, however, Voinescu and Szentagotal-Tarar [64] found that moderate to low levels of sleep hygiene awareness were unrelated to sleep quality. As such it is unclear as to whether insomnia is related to poor sleep hygiene practices.

What is the role of sleep hygiene in the management of insomnia?

The premise here is that if poor sleep hygiene can cause insomnia then the reverse would also be applicable, that is, correcting poor sleep hygiene can fix insomnia. This supposition is implied in the International Classification of Sleep Disorders (ICSD), which had, in its first and second iterations (ICSD [65]; ICSD-2 [66]) a specific subtype of insomnia labeled "inadequate sleep hygiene" (it is still included in the ICSD-3 but as a variant form of insomnia). Under the old ICSD framework it was suggested that there were 11 contributors to inadequate sleep hygiene. However, as we saw in the last section the findings of an association between poor sleep hygiene and insomnia are quite mixed. One study, which examined diagnoses according to the ICSD, DSM, and ICD, found Inadequate Sleep Hygiene was the primary diagnosis in only 6.2% of 257 patients although it was a very commonly ascribed as a secondary diagnosis—34.2% [67]. Further, a review of the literature by Reynolds and Kupfer [68] concluded that "inadequate sleep hygiene" was unlikely to be a primary cause of insomnia but rather should be seen as an exacerbating factor. The American Academy of Sleep Medicine (AASM) summarized the data from sleep hygiene studies and concluded that there is insufficient evidence to suggest that sleep hygiene is an effective stand-alone therapy for insomnia (AASM [69, 70]). A more recent meta-analytic comparison of 15 studies, which employed sleep hygiene interventions, concurs with the AASM [62]. They found that the effect sizes from the sleep hygiene interventions were similar to those observed for psychological placebo interventions for insomnia. Interestingly, although comparable treatment improvements were observed for a sleep hygiene intervention compared to meditation and stimulus control in one study, participants rated the sleep hygiene intervention less favorably [3].

Due to this accumulated knowledge and lack of consistency in findings, over time, sleep hygiene, in its broadest form, has three main functions today: (i) serving as an "active control" in insomnia intervention trials, (ii) as the psycho-educational component in CBT-I, and (iii) as an alternative to pharmacotherapy in primary care [71]. In the latter case, despite the belief that sleep hygiene has limited value by those providing it (i.e., General Practitioners).

So, is there a role for sleep hygiene in sleep medicine and practice, beyond insomnia?

If sleep hygiene is not recommended as a stand-alone therapy for chronic insomnia, does it have a part to play elsewhere in the arena of sleep? One suggestion is that it be used as part of a broader public health campaign with the aim of increasing overall sleep health [72]. Only recently has the concept of Sleep Health been defined [73, 74] with the premise that improvements can be made, in terms of sleep for those without sleep problems. Further that improvements in sleep health will have an impact on overall health and wellbeing. In line with that suggestion, an interesting study by Barber, Grawitch and Munz [75] found that those with poorer sleep hygiene were at risk of poorer work performance due to its presumed impact on the after-work recovery process. Herein lies a further consideration, however, if someone has poor sleep hygiene but considers himself or herself an average or even good sleeper are they going to change their sleep habits and rituals? As yet, however, sleep hygiene has not been systematically employed in the arena of sleep health or a preventative public health campaign [72].

THE PREVENTION OF CHRONIC INSOMNIA

In order to outline a preventative agenda for any disorder, let alone insomnia, we must first understand its etiology. Only by knowing how and why it occurs, as well as how it evolves or changes over time, can we start to determine how and when to intervene. Intervening too early may be detrimental in that we may be altering/affecting what may be a normal biological process whereas intervening too late is also likely to be as detrimental, if not more so, and prolong suffering. This level of understanding of a disorder can be achieved using several methods. For example, both descriptive and analytic epidemiological studies are likely to give us an insight into which individuals are most likely to be affected by the disorder and when in their lifespan they are most likely to be at risk. Unfortunately there is scarce empirical data on the development of insomnia, with sufficient sensitivity, to understand its early progression [76]. That said, there is one suggestion, outlined in several of the models of insomnia, as to what initiates insomnia.

Etiological models of insomnia

The most widely regarded model of the etiology of insomnia comes from Spielman [77, 78]. Spielman proposed that insomnia occurs as a combination of two factors—predisposing factors and precipitating events, respectively. Where predisposing characteristics, such as personality factors, demographic factors, or biologic traits are not sufficient to give someone insomnia on their own, a triggering event occurs (precipitant) which then pushes the individual over a threshold, into insomnia. This phase is termed "Acute" or "Short Term" Insomnia. Further, Spielman suggests that perpetuating factors (behaviors the individual with insomnia engages in to manage their initial insomnia—such as going to bed early, napping, staying in bed in the morning) start during this acute phase and gain momentum as the insomnia progresses (Fig. 11.1). Although the model gives a starting point of what factors may be involved in

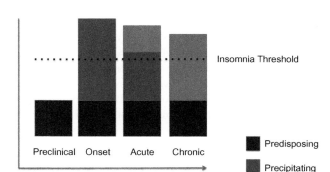

FIG. 11.1 Spielman's 3 P model of insomnia.

the etiology of insomnia, and their relative contributions over time, what the model does not tell us is when insomnia becomes insomnia. Without that knowledge it is difficult to determine when best to intervene. Several other models followed Spielman's (e.g., Perlis and colleagues [79] Espie [11]; Harvey [12]; Buysse and colleagues [80]) and central to all is the concept of a precipitating event and the beginning of perpetuating factors during the acute phase (although later models emphasized cognitive as well as behavioral perpetuating factors) but none as yet have provided a duration element to their respective model.

What we know about acute insomnia?

Up to one-third (31%–36%) of the population will suffer from Acute Insomnia (i.e., up to 3 months) in a given year and between 7.9% and 9.5% of adults report having acute insomnia at any given time, with 51.2% of them reporting it as a first episode [81]. The duration criterion used in this study was based upon the Diagnostic and Statistical Manual of Mental Disorders (DSM-5 [82]) and International Classification of Sleep Disorders (ICSD-3 [83]) definitions for Insomnia Disorder. That said, the genesis of that timeline in both classification systems has never been explicated. Those with Acute Insomnia differ from normal sleepers in both subjective and objective sleep characteristics. Those with Acute Insomnia report longer Sleep Latencies, more awakenings during the night, longer periods of nocturnal wakefulness, and lower sleep efficiencies compared to normal sleepers [84]. Further, actigraphically defined sleep fragmentation is higher in those with Acute Insomnia, relative to normal sleepers, as is polysomnographically defined light sleep (N2) with lower amounts of Slow Wave Sleep [84, 85]. Further, in support of Spielman, it appears that those with Acute Insomnia do engage in both cognitive and behavioral activities that can perpetuate insomnia in a similar manner to those with Chronic Insomnia (i.e., worrying about sleep, becoming preoccupied with sleep, spending more time in bed) [86, 87].

Can we prevent acute insomnia from becoming chronic?

There is currently only one intervention that has attempted to circumvent the transition from acute to chronic Insomnia [88]. Based on the assumption that acute insomnia should be easier to manage during the acute phase due to perpetuating factors being in their infancy [89] and the relative success of brief variants of CBT-I (e.g., [90, 91]), the intervention—termed the "one shot" involves a single 60–70 min treatment session and an accompanying pamphlet. Three relatively small studies have been undertaken on the "one shot" [88, 92, 93]. The first, a Randomized Controlled Trial in a community sample of adults demonstrated a significant improvement in insomnia symptoms 1 month following the intervention. Furthermore, the 1-month remission rate was 60% compared to 15% in the control group [88]. Interestingly, at 3 months postintervention the remission rate in this group had increased to 73%. The second study aimed to determine whether the intervention could be delivered in groups, which it could although group treatment impacts negatively on adherence, and in that instance the overall remission rate at 1 month was 72%. The final study aimed to use the intervention in a male prison setting due to the high levels of insomnia in this population [94]. The remission rate in this final study was similar to that observed in the second trial (73%). In the final two studies, significant reductions in anxiety and depression symptomology were also observed [92, 93].

Identifying those at risk

While there have been, albeit small, advances in the arena of acute insomnia, there is even less work aimed at primary prevention. One avenue that has not, until recently, been systematically explored is determining individuals' who may be vulnerable to insomnia (i.e., Spielman's predisposing factors). Where previous research has tended to characterize personality characteristics and other presumed predisposing factors in those who already have insomnia, Drake and colleagues created the Ford Insomnia Response to Stress Test (FIRST [95]). The FIRST is a brief self-report scale that asks the likelihood that an individual would lose sleep over several stressful situations. Several studies have been undertaken using the FIRST and it appears to be a good indicator of a first episode of insomnia [96] in addition to having a strong genetic component [97]. Importantly, recent research using the FIRST has demonstrated that the first episode of insomnia sensitizes the sleep system making the individual vulnerable to insomnia again in the future as well as depression and anxiety [98]. What would be interesting is to determine whether a sleep hygiene intervention or the "one-shot" would act as a buffer to getting insomnia, in the face of a stressor, in those who score highly on the FIRST.

CONCLUSIONS

Sleep hygiene is clearly a complex area in terms of defining and evidencing the specific recommendations. While there have been significant levels of investigation for some aspects (e.g., alcohol, caffeine, bedroom noise) there has been very little in other areas (e.g., clockwatching, food and liquid intake). While each aspect of sleep hygiene has the potential to influence sleep, this data has not translated well in terms of both differentiating those with insomnia from normal sleepers and in the management of insomnia. It is plausible to assume that lay knowledge of sleep hygiene, especially in those with insomnia who are more likely to seek out this information, and self-motivated behavioral change may be responsible for this lack of association between sleep hygiene and poor sleep. Although the existing evidence base would suggest a limited role for sleep hygiene in the management of chronic insomnia, beyond a psycho-education component within Cognitive Behavior Therapy for Insomnia (CBT-I), an alternative perspective is proposed. That sleep hygiene should still be routinely assessed in all sleep disturbed patients but any corresponding advice should be tailored specifically to what the patient has not yet tried or adopted successfully. This recommendation is to account for individual differences in tolerability and vulnerability to insomnia based upon specific aspects of sleep hygiene. Furthermore, with the advent of Sleep Health as a concept and recent moves toward a more preventative approach to sleep medicine, sleep hygiene may still have a role to play alongside briefer forms of CBT-I. When considering the costs and consequences of Chronic Insomnia, prevention is clearly the way forward, preferably at a public health level.

CONFLICT OF INTEREST

The Sleep Council, in part, funded the production of this chapter but had no role in its content.

REFERENCES

[1] Brown FC, Buboltz Jr. WC, Soper B. Relationship of sleep hygiene awareness, sleep hygiene practices, and sleep quality in university students. Behav Med 2002;28(1):33–8.
[2] Hauri P. The sleep disorders. Upjohn Company: Kalamazoo; 1977.
[3] Schoicket SL, Bertelson AD, Lacks P. Is sleep hygiene a sufficient treatment for sleep-maintenance insomnia? Behav Ther 1988;19(2):183–90.
[4] Hauri P. Sleep hygiene, relaxation therapy, and cognitive interventions. In: Hauri PJ, editor. Case Studies in Insomnia. New York: Plenum Press; 1992.
[5] Hauri PJ. Consulting about insomnia: A method and some preliminary data. Sleep 1993;16(4):344–50.
[6] Guilleminault C, Clark A, Black J, Labanowski M, Pelayo R, Claman D. Nondrug treatment trials in psychophysiologic insomnia. Arch Intern Med 1995;155:838–44.
[7] Friedman L, Benson K, Noda A, Zarcone V, Wicks DA, O'Connell K, ... Yesavage JA. An actigraphic comparison of sleep restriction and sleep hygiene treatments for insomnia in older adults. J Geriatr Psychiatry Neurol 2000;13(1):17–27.
[8] Perlis ML, Jungquist C, Smith MT, Posner D. Cognitive behavioral treatment of insomnia: A session-by-session guide. vol. 1. Springer Science & Business Media; 2006.
[9] Bootzin RR. Stimulus control treatment for insomnia. Proc Am Psychol Asso 1972;7:395–6.
[10] Spielman AJ, Saskin P, Thorpy MJ. Treatment of chronic insomnia by restriction of time in bed. Sleep 1987;10(1):45–56.
[11] Espie CA. Insomnia: conceptual issues in the development, persistence, and treatment of sleep disorder in adults. Annu Rev Psychol 2002;53(1):215–43.
[12] Harvey AG. A cognitive model of insomnia. Behav Res Ther 2002;40(8):869–93.
[13] Stepanski EJ, Wyatt JK. Use of sleep hygiene in the treatment of insomnia. Sleep Med Rev 2003;7(3):215–25.
[14] Lang C, Kalak N, Brand S, Holsboer-Trachsler E, Pühse U, Gerber M. The relationship between physical activity and sleep from mid adolescence to early adulthood. A systematic review of methodological approaches and meta-analysis. Sleep Med Rev 2016;28:32–45.
[15] Hartescu I, Morgan K, Stevinson CD. Increased physical activity improves sleep and mood outcomes in inactive people with insomnia: a randomized controlled trial. J Sleep Res 2015;24(5):526–34.
[16] Driver HS, Taylor SR. Exercise and sleep. Sleep Med Rev 2000;4(4):387–402.
[17] Kredlow MA, Capozzoli MC, Hearon BA, Calkins AW, Otto MW. The effects of physical activity on sleep: a meta-analytic review. J Behav Med 2015;38(3):427–49.
[18] Morin CM, Hauri PJ, Espie CA, Spielman AJ, Buysse DJ, Bootzin RR. Nonpharmacologic treatment of chronic insomnia. Sleep 1999;22(8):1134–56.
[19] Buman MP, Phillips BA, Youngstedt SD, Kline CE, Hirshkowitz M. Does nighttime exercise really disturb sleep? Results from the 2013 National Sleep Foundation Sleep in America Poll. Sleep Med 2014;15(7):755–61.
[20] Landolt HP, Werth E, Borbély AA, Dijk DJ. Caffeine intake (200 mg) in the morning affects human sleep and EEG power spectra at night. Brain Res 1995;675(1–2):67–74.
[21] Bonnet MH, Arand DL. Caffeine use as a model of acute and chronic insomnia. Sleep 1992;15(6):526–36.
[22] Paterson LM, Wilson SJ, Nutt DJ, Hutson PH, Ivarsson M. A translational, caffeine-induced model of onset insomnia in rats and healthy volunteers. Psychopharmacology (Berl) 2007;191(4):943–50.
[23] Roehrs T, Roth T. Caffeine: sleep and daytime sleepiness. Sleep Med Rev 2008;12(2):153–62.
[24] Drake C, Roehrs T, Shambroom J, Roth T. Caffeine effects on sleep taken 0, 3, or 6 hours before going to bed. J Clin Sleep Med 2013;9(11):1195–200.
[25] Sanchez-Ortuno M, Moore N, Taillard J, Valtat C, Leger D, Bioulac B, Philip P. Sleep duration and caffeine consumption in a French middle-aged working population. Sleep Med 2005;6(3):247–51.

[26] Chaudhary NS, Grandner MA, Jackson NJ, Chakravorty S. Caffeine consumption, insomnia, and sleep duration: Results from a nationally representative sample. Nutrition 2016;32(11):1193–9.

[27] Retey JV, Adam M, Khatami R, Luhmann UFO, Jung HH, Berger W, Landolt HP. A genetic variation in the adenosine A2A receptor gene (ADORA2A) contributes to individual sensitivity to caffeine effects on sleep. Clin Pharmacol Ther 2007;81(5):692–8.

[28] Kerpershoek ML, Antypa N, Van den Berg JF. Evening use of caffeine moderates the relationship between caffeine consumption and subjective sleep quality in students. J Sleep Res 2018;e12670.

[29] Thakkar MM, Sharma R, Sahota P. Alcohol disrupts sleep homeostasis. Alcohol 2015;49(4):299–310.

[30] Ebrahim IO, Shapiro CM, Williams AJ, Fenwick PB. Alcohol and sleep I: effects on normal sleep. Alcohol Clin Exp Res 2013;37(4):539–49.

[31] Roehrs T, Roth T. Insomnia as a path to alcoholism: tolerance development and dose escalation. Sleep 2018;41(8):zsy091.

[32] Morin CM, Mimeault V, Gagné A. Nonpharmacological treatment of late-life insomnia. J Psychosom Res 1999;46(2):103–16.

[33] Jaehne A, Loessl B, Bárkai Z, Riemann D, Hornyak M. Effects of nicotine on sleep during consumption, withdrawal and replacement therapy. Sleep Med Rev 2009;13(5):363–77.

[34] Hughes JR, Higgins ST, Bickel WK. Nicotine withdrawal versus other drug withdrawal syndromes: similarities and dissimilarities. Addiction 1994;89(11):1461–70.

[35] Nakata A, Takahashi M, Haratani T, Ikeda T, Hojou M, Fujioka Y, Araki S. Association of active and passive smoking with sleep disturbances and short sleep duration among Japanese working population. Int J Behav Med 2008;15(2):81.

[36] Sabanayagam C, Shankar A. The association between active smoking, smokeless tobacco, second-hand smoke exposure and insufficient sleep. Sleep Med 2011;12(1):7–11.

[37] Ohayon MM. Prevalence and correlates of nonrestorative sleep complaints. Arch Intern Med 2005;165(1):35–41.

[38] Otto E, Kramer H, Bräuer D. Effects of increasing air temperature on heart rate, body movements, rectal temperature and electroencephalogram of sleeping subjects. Int Arch Arbeitsmed 1971;28(3):189–202.

[39] Angus RG, Pearce DG, Buguet AG, Olsen L. Vigilance performance of men sleeping under arctic conditions. Aviat Space Environ Med 1979;50:692–6.

[40] National Sleep Foundation. https://www.sleep.org/articles/temperature-for-sleep/; 2018.

[41] Boivin DB, Duffy JF, Kronauer RE, Czeisler CA. Dose-response relationships for resetting of human circadian clock by light. Nature 1996;379:540–2.

[42] Gooley JJ, Chamberlain K, Smith KA, Khalsa SBS, Rajaratnam SMW, Van Reen E, ... Lockley SW. Exposure to room light before bedtime suppresses melatonin onset and shortens melatonin duration in humans. J Clin Endocrinol Metab 2011;96(3):E463–72.

[43] Muzet A. Environmental noise, sleep and health. Sleep Med Rev 2007;11(2):135–42.

[44] World Health Organization (2018). http: //www.euro.who.int/en/health-topics/environment-and health/noise/policy/who-night-noise-guidelines-for-europe.

[45] Rechtschaffen A, Hauri P, Zeitlin M. Auditory awakening thresholds in REM and NREM sleep stages. Percept Mot Skills 1966;22(3):927–42.

[46] Bader GG, Engdal S. The influence of bed firmness on sleep quality. Appl Ergon 2000;31(5):487–97.

[47] Kinkel HJ, Maxion H. Physiological sleep studies for the evaluation of different mattresses. Internationale Zeitschrift fur angewandte Physiologie, einschliesslich Arbeitsphysiologie 1970;28(3):247–62.

[48] Villani S. Impact of media on children and adolescents: a 10-year review of the research. J Am Acad Child Adolesc Psychiatry 2001;40(4):392–401.

[49] Cain N, Gradisar M. Electronic media use and sleep in school-aged children and adolescents: a review. Sleep Med 2010;11(8):735–42.

[50] Hale L, Guan S. Screen time and sleep among school-aged children and adolescents: a systematic literature review. Sleep Med Rev 2015;21:50–8.

[51] Gradisar M, Wolfson AR, Harvey AG, Hale L, Rosenberg R, Czeisler CA. The sleep and technology use of Americans: findings from the National Sleep Foundation's 2011 Sleep in America poll. J Clin Sleep Med 2013;9(12):1291–9.

[52] Tang NK, Schmidt DA, Harvey AG. Sleeping with the enemy: clock monitoring in the maintenance of insomnia. J Behav Ther Exp Psychiatry 2007;38(1):40–55.

[53] Woods H, Marchetti LM, Biello SM, Espie CA. The clock as a focus of selective attention in those with primary insomnia: an experimental study using a modified Posner paradigm. Behav Res Ther 2009;47(3):231–6.

[54] Morin CM. Insomnia: psychological assessment and management. Guilford Press; 1993.

[55] Lacks P, Rotert M. Knowledge and practice of sleep hygiene techniques in insomniacs and good sleepers. Behav Res Ther 1986;24(3):365–8.

[56] Blake DD, Gomez MH. A scale for assessing sleep hygiene: preliminary data. Psychol Rep 1998;83(3_suppl):1175–8.

[57] Mastin DF, Bryson J, Corwyn R. Assessment of sleep hygiene using the Sleep Hygiene Index. J Behav Med 2006;29(3):223–7.

[58] Harsh JR, Easley A, LeBourgeois MK. A measure of children's sleep hygiene. Sleep 2002;25:A316.

[59] Storfer-Isser A, Lebourgeois MK, Harsh J, Tompsett CJ, Redline S. Psychometric properties of the Adolescent Sleep Hygiene Scale. J Sleep Res 2013;22(6):707–16.

[60] Jefferson CD, Drake CL, Scofield HM, Myers E, McClure T, Roehrs T, Roth T. Sleep hygiene practices in a population-based sample of insomniacs. Sleep 2005;28(5):611–5.

[61] Gellis LA, Lichstein KL. Sleep hygiene practices of good and poor sleepers in the United States: an internet-based study. Behav Ther 2009;40(1):1–9.

[62] Chung KF, Lee CT, Yeung WF, Chan MS, Chung EWY, Lin WL. Sleep hygiene education as a treatment of insomnia: a systematic review and meta-analysis. Family Practice 2017;35(4):365–75.

[63] Harvey AG. Sleep hygiene and sleep-onset insomnia. J Nerv Ment Dis 2000;188(1):53–5.

[64] Voinescu BI, Szentagotai-Tatar A. Sleep hygiene awareness: its relation to sleep quality and diurnal preference. J Mol Psychiatry 2015;3(1):1.

[65] American Sleep Disorders Association, Diagnostic Classification Steering Committee. The international classification of sleep disorders: diagnostic and coding manual. American sleep disorders association; 1990.

[66] American Academy of Sleep Medicine. International classification of sleep disorders. Diagnostic and coding manual. Westchester IL: American Academy of Sleep Medicine; 200551–5.

[67] Buysse DJ, Reynolds III CF, Kupfer DJ, Thorpy MJ, Bixler E, Manfredi R, ... Hauri P. Clinical diagnoses in 216 insomnia patients using the International Classification of Sleep Disorders (ICSD), DSM-IV and ICD-10 categories: a report from the APA/NIMH DSM-IV Field Trial. Sleep 1994;17(7):630–7.

[68] Reynolds III CF, Kupfer DJ. Subtyping DSM-III-R primary insomnia: a literature review by the DSM-IV Work Group on Sleep Disorders. Am J Psychiatry 1991;148(4):432.

[69] Chesson Jr. AL, Anderson WM, Littner M, Davila D, Hartse K, Johnson S, ... Rafecas J. Practice parameters for the nonpharmacologic treatment of chronic insomnia. Sleep 1999;22(8):1128–33.

[70] Morin CM, Bootzin RR, Buysse DJ, Edinger JD, Espie CA, Lichstein KL. Psychological and behavioral treatment of insomnia: update of the recent evidence (1998–2004). Sleep 2006;29(11):1398–414.

[71] Everitt H, McDermott L, Leydon G, Yules H, Baldwin D, Little P. GPs' management strategies for patients with insomnia: a survey and qualitative interview study. Br J Gen Pract 2014;64(619):e112–9.

[72] Irish LA, Kline CE, Gunn HE, Buysse DJ, Hall MH. The role of sleep hygiene in promoting public health: a review of empirical evidence. Sleep Med Rev 2015;22:23–36.

[73] Buysse DJ. Sleep health: can we define it? Does it matter? Sleep 2014;37(1):9–17.

[74] Knutson KL, Phelan J, Paskow MJ, Roach A, Whiton K, Langer G, ... Lichstein KL. The National Sleep Foundation's sleep health index. Sleep Health 2017;3(4):234–40.

[75] Barber L, Grawitch MJ, Munz DC. Are better sleepers more engaged workers? A self-regulatory approach to sleep hygiene and work engagement. Stress Health 2013;29(4):307–16.

[76] Perlis ML, Gehrman P, Ellis JG. The natural history of insomnia: what we know, don't know, and need to know. Sleep Med Res 2011;2:79–88.

[77] Spielman AJ. Assessment of insomnia. Clin Psychol Rev 1986;6(1):11–25.

[78] Spielman AJ, Caruso LS, Glovinsky PB. A behavioral perspective on insomnia treatment. Psychiatr Clin North Am 1987;10(4):541–53.

[79] Perlis ML, Giles DE, Mendelson WB, Bootzin RR, Wyatt JK. Psychophysiological insomnia: the behavioural model and a neurocognitive perspective. J Sleep Res 1997;6(3):179–88.

[80] Buysse DJ, Germain A, Hall M, Monk TH, Nofzinger EA. A neurobiological model of insomnia. Drug Discov Today Dis Model 2011;8(4):129–37.

[81] Ellis JG, Perlis ML, Neale LF, Espie CA, Bastien CH. The natural history of insomnia: focus on prevalence and incidence of acute insomnia. J Psychiatr Res 2012;46(10):1278–85.

[82] American Psychiatric Association. Diagnostic and statistical manual of mental disorders (DSM-5®). American Psychiatric Pub; 2013.

[83] American Academy of Sleep Medicine. International classification of sleep disorders–third edition (ICSD-3). Darien, IL: American Academy of Sleep Medicine; 2014.

[84] Ellis JG, Perlis ML, Bastien CH, Gardani M, Espie CA. The natural history of insomnia: acute insomnia and first-onset depression. Sleep 2014;37(1):97–106.

[85] Fossion R, Rivera AL, Toledo-Roy JC, Ellis J, Angelova M. Multiscale adaptive analysis of circadian rhythms and intradaily variability: Application to actigraphy time series in acute insomnia subjects. PloS one 2017;12(7):e0181762.

[86] Ellis J, Cropley M. An examination of thought control strategies employed by acute and chronic insomniacs. Sleep Med 2002;3(5):393–400.

[87] Man S, Freeston M, Ellis JG, Lee DR. A pilot study investigating differences in sleep and life preoccupations in chronic and acute insomnia. Sleep Medicine Research (SMR) 2013;4(2):43–50.

[88] Ellis JG, Cushing T, Germain A. Treating acute insomnia: a randomized controlled trial of a "single-shot" of cognitive behavioral therapy for insomnia. Sleep 2015;38(6):971–8.

[89] Ellis JG, Gehrman P, Espie CA, Riemann D, Perlis ML. Acute insomnia: current conceptualizations and future directions. Sleep Med Rev 2012;16(1):5–14.

[90] Germain A, Moul DE, Franzen PL, Miewald JM, Reynolds CF, Monk TH, Buysse DJ. Effects of a brief behavioral treatment for late-life insomnia: preliminary findings. J Clin Sleep Med 2006;2(04):407–8.

[91] Edinger JD, Sampson WS. A primary care "friendly" cognitive behavioral insomnia therapy. Sleep 2003;26(2):177–82.

[92] Boullin P, Ellwood C, Ellis JG. Group vs. individual treatment for acute insomnia: A pilot study evaluating a "one-shot" treatment strategy. Brain Sci 2016;7(1):1.

[93] Randall C, Nowakowski S, Ellis JG. Managing acute insomnia in prison: evaluation of a "one-shot" cognitive behavioral therapy for insomnia (CBT-I) intervention. Behav Sleep Med 2018;1–10.

[94] Dewa LH, Kyle SD, Hassan L, Shaw J, Senior J. Prevalence, associated factors and management of insomnia in prison populations: an integrative review. Sleep Med Rev 2015;24:13–27.

[95] Drake C, Richardson G, Roehrs T, Scofield H, Roth T. Vulnerability to stress-related sleep disturbance and hyperarousal. Sleep 2004;27(2):285–91.

[96] Jarrin DC, Chen IY, Ivers H, Morin CM. The role of vulnerability in stress-related insomnia, social support and coping styles on incidence and persistence of insomnia. J Sleep Res 2014;23(6):681–8.

[97] Drake CL, Friedman NP, Wright Jr. KP, Roth T. Sleep reactivity and insomnia: genetic and environmental influences. Sleep 2011;34(9):1179–88.

[98] Kalmbach DA, Pillai V, Arnedt JT, Anderson JR, Drake CL. Sleep system sensitization: evidence for changing roles of etiological factors in insomnia. Sleep Med 2016;21:63–9.

Chapter 12

Actigraphic sleep tracking and wearables: Historical context, scientific applications and guidelines, limitations, and considerations for commercial sleep devices

Michael A. Grandner[a], Mary E. Rosenberger[b]

[a]*Sleep and Health Research Program, Department of Psychiatry, University of Arizona College of Medicine, Tucson, AZ, United States,* [b]*Stanford Center on Longevity and Psychology Department, Stanford University, Stanford, CA, United States*

INTRODUCTION

Human sleep is naturally-recurring and easily reversible state that is characterized by reduced or absent consciousness, perceptual disengagement, immobility, and adoption of a characteristic sleeping posture. Regulation of the sleep-wake system includes both homeostatic and circadian components and is modified by genetic, physiologic, environmental, and behavioral factors (See Chapter 1 in this volume). Importantly, determining whether an individual is "asleep" can only be accomplished through indirect methods—it is typically impossible to make a direct measurement of sleep or wake, as this represents a complex output of neurologic systems that are primarily contained in the midbrain. Instead, we rely in indirect measures of sleep.

The accepted "gold-standard" measure of sleep is polysomnography, which is a combination of physiologic recording channels including electroencephalography, electromyography, electrooculography, electrocardiography, oximetry, and measures of respiration [1]. Polysomnography assesses cortical synchronization activity and can be used to discern "sleep stages" which are distinct states that occur across the sleep period and reflect characteristically distinct brainwave patterns that themselves represent and/or correlate with other physiologic processes. Polysomnography, while considered the gold-standard, is an indirect measure of sleep, and has some important limitations. Most notably, it is usually very expensive and burdensome. The expense precludes repeated assessment in large samples and the burden on patients is such that it can itself cause changes to sleep. Because of this, polysomnography is known to interfere with sleep, is rarely recorded over several nights, and polysomnographic recordings are not well-suited to reflect habitual sleep.

Field-based measurement of sleep is different from polysomnography in that it utilizes a movement-detection apparatus to assess patterns of mobility and immobility in order to estimate whether an individual is asleep or awake [2]. Newer wearable devices also incorporate heart-rate based measurements of sleep [3]. Although actigraphy cannot assess sleep stages, it can estimate whether an individual is awake or asleep with (typically) 1-min resolution. With this information over a whole recording period, determinations can be made regarding sleep duration and time awake (e.g., sleep latency and wake time after initial sleep onset); this can be used to calculate variables such as sleep efficiency (the ratio of time asleep to total time in bed). See Table 12.1 for a broad comparison between Actigraphy and Polysomnography.

SCORING ALGORITHMS

The first use of movement-based recordings to determine sleep and wake were published in 1972 [4,5] on a set of psychiatric patients. Even in this first study, with little temporal resolution or precision in movement estimation, it was

TABLE 12.1 Comparison of actigraphy to polysomnography.

	Actigraphy	Polysomnography
Standard epoch length	1 min	30 s
Cost	Low	High
Nights typically evaluated	7–14	1–2
Captures daytime sleep	Yes	No
Channels recorded	1 or few	Many
Measures immobility	Yes	Yes
Measures reduced or absent consciousness	No	Indirectly
Measures sensory inhibition	No	Indirectly
Measures sleep-related breathing	No	Yes
Measures sleep continuity	Yes	Yes
Measures sleep timing	Yes	No
Measures sleep regularity	Yes	No
Measures sleep satisfaction	No	No

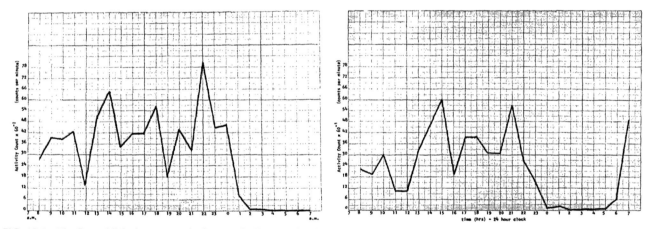

FIG. 12.1 The first published actograms, in the paper by Foster and colleagues [4].

clear that this method could be useful for determining sleep schedules. See Fig. 12.1 for a reproduction of the first images of 24-h actigraphy recordings.

In 1978, Kripke and colleagues would publish results of a similar study using a different device, referring to the method as "actigraphy" [6]. These initial devices used crude movement-transducers to quantify movement counts. The first studies examined the raw data outputs to score sleep versus wake [6,7], but it was soon hypothesized that this scoring could be automated. The first sleep-wake scoring algorithms were developed with the assumption that simply quantifying movement was insufficient—other factors, including the context of that movement and time of day, are also essential.

Several actigraphy scoring algorithms have been developed over the course of the past ~40 years. As actigraphic technology changed, algorithms changed as well. The first validated scoring algorithm was described by Webster and colleagues [8] and derived their algorithm against polysomnography as a gold standard. Their algorithm, which included weights of the 1-min epoch of interest, as well as the previous four epochs and subsequent two epochs for context, set the standard for future actigraphy scoring algorithms. Of note, the Webster algorithm was meant to be followed up by additional scoring rules that represented areas where the scoring algorithm often failed (such as brief apparent sleep episodes that were likely actually time awake). Overall, rate of agreement between actigraphy and polysomnography was >90%.

Subsequent attempts at developing scoring algorithms typically adopted a similar approach (weighting the epoch of interest against several previous and subsequent epochs), though the adoption of follow-up rules were often dropped in favor of less structured hand-scoring for clear anomalies and errors. These included algorithms developed for analog motion transducers [9], analog accelerometers [10–12], and newer accelerometers that included multiaxial assessment and digital recording [13,14]. Overall, as the technology has progressed and the use of actigraphy has expanded, validation studies often included larger and more diverse samples. Still rates of agreement with polysomnography are typically around 85%–90% [2,15–17].

Of note, each of these algorithms were validated in a specific context. For example, these validation studies typically are restricted to a particular device or type of device (e.g., linear analog transducer) and it is not clear how well these algorithms will fare in other devices. There is evidence that algorithms are somewhat transferrable across devices and even movement recording modalities [8,14], but this limitation should be acknowledged. Also, these validation studies typically only compared actigraphy to polysomnography within a defined in-bed interval. Methods for using actigraphy to detect sleep during the daytime or methods to automatically determine an in-bed interval in real-world settings are often not well-validated and therefore caution should be taken with these approaches.

Most newer actigraphic devices use microelectromechanical systems [14,18] to record movement. These approaches, which apply nanotechnology to accelerometry and produce accelerometers that fit on a tiny microchip, often apply similar principles to traditional accelerometers, just on a smaller scale [18]. Several devices that use MEMS chips to record accelerometry for sleep detection have been validated [14], though many of these chips are optimized for physical activity measurement more than sleep.

TYPES OF ACTIGRAPH DEVICES

The most common devices in the scientific literature include those developed by Ambulatory Monitoring, Inc. (AMI) and Mini-Mitter Inc. (later acquired by Respironics, which was acquired by Philips) [12,13,19–24]. These devices are similar and have changed relatively little over the past several years. Several AMI devices have been validated in the scientific literature for sleep, though the ones most frequently used are the Actillume (older device) and Motionlogger (newer device). These devices, and others, are depicted in Fig. 12.2. The Actillume was an analog device with a basic accelerometer array that could capture movement in 3 dimensions. The Motionlogger replaced the Actillume as a digital device that was smaller and actually had a watch face. Both the Actillume and Motionlogger record environmental light with a single photometer channel, but only the Motionlogger device has a mode that can estimate off-wrist time, which is important to distinguish from on-wrist lack of movement and is also water-resistant. The device made by Mini-Mitter was the Actiwatch (also depicted in Fig. 12.2, in several iterations). The Actiwatch was similar to the Motionlogger in that it digitally recorded both movement and light and stored data in 1-min epochs. Since the manufacturer was acquired by Respironics and then Philips, new versions of the device include the Actiwatch-2 (pictured) and the Actiwatch Spectrum (pictured). These devices

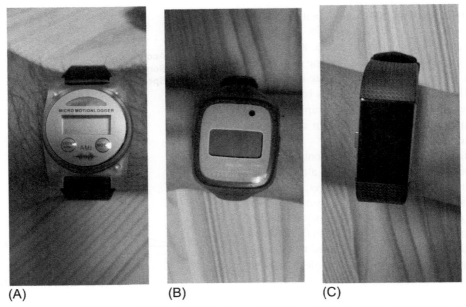

FIG. 12.2 Images of common actigraphic devices, including the Motionlogger (A), Actiwatch Spectrum (B), and Fitbit Charge 2 (C).

similarly record movement and light; the Actiwatch-2 has a rechargeable battery and is water-resistant. The current models of the Actiwatch Spectrum are also water-resistant and also have rechargeable batteries; the added features of the Spectrum include off-wrist detection and light channels for red, green, and blue light spectra.

Several other devices have also been used with relative frequency in scientific settings, though validation of these devices is less robust. For example, the GT3X and related devices from Actigraph, Inc. (pictured in Fig. 12.2) are frequently less expensive than AMI or Philips devices, though they have been less rigorously validated [25,26]. Of note, no scoring algorithm has been developed for this device and the scoring software simply co-opts algorithms from other devices, assuming relative accuracy. Also, these devices do not contain a light channel. Other devices that have shown some degree of scientific utility include the GENEActiv [14] (made by ActivInsights), the Motion Watch [27] (made by Cambridge Neurotechnology), Fitbit devices [3,19,26,28–31] (made by Fitbit), and the Readiband (made by Fatigue Science). These devices may be useful but are less well validated.

Other actigraphy-like devices have also emerged on the market. For example, some products are to be worn as an armband, such as the SenseWear [32] band (made by Body Media). Other devices, such as the Oura [33] (made by Oura) are to be worn as a ring on the finger. Some devices use movement recorded by the mattress using pressure-sensing technology, such as the EarlySense [34] (made by EarlySense Ltd.) device. Another device, called the SleepScore MAX (made by SleepScore, formerly the device called S+ made by ResMed) uses contactless infrared technology to assess sleep-related parameters. Although it does not measure movement in a similar way, it has been used as an alternative to actigraphy for individuals who cannot tolerate a wearable and validation data for the device are relatively strong for sleep-wake detection. A portable, FDA approved, one-lead EEG sensor called the Z-machine has been validated [26,35], and used to validate other field-based measures. The validity of these devices is much less established than that of more traditional actigraphy, but this may change as more literature is published on specific devices and modalities.

It should be noted that many smartphones contain MEMS chips and many software programs (i.e., apps) attempt to leverage these for sleep-wake detection. However, these apps typically do not estimate sleep based on movement at the wrist or anywhere else on the person; rather they frequently attempt to assess movement by being placed on the mattress or otherwise near the sleeper. It should be noted that despite the apparent popularity of these apps, none has demonstrated good validity relative to validated sleep measures. For example, one very popular app was shown to be unrelated to any observed values obtained from a validated source [36].

Newer commercial wearable devices use MEMS chips in addition to heart rate sensors to improve sleep algorithmic scoring, such as the Fitbit Charge 3, the Motiv ring, and others. Heart rate measurement is relatively simple compared to other measurements during sleep, and a study of heart rate during sleep is well established [37]. Specifically heart rate variability, or the changes in time between heart beats measured as an R-R interval (referring to the EKG recording of heart rate), is specifically studied as a method for determining sleep stages [38,39]. The combination of actigraphy and heart rate measures during sleep hold promise for more accurate sleep scoring algorithms.

LIMITATIONS OF ACTIGRAPHY AND RELATED CONSIDERATIONS

All measures of sleep in humans are indirect. And, thus, all measures of sleep in humans are imperfect. There is no way to accurately estimate a person's sleep duration and sleep architecture over a period of days or weeks in such a way that habitual parameters can be estimated. Actigraphy may be the best and most accepted solution, but it is not without limitations.

For example, compared to polysomnography, actigraphy has relatively good sensitivity (it correctly identifies sleep most of the time), but it has relatively poor specificity (it incorrectly identifies wake much of the time). See Fig. 12.3 for an image from a paper by Marino and colleagues [15] that illustrates this based on data from a large sample of real-world adults. What this means is that, relative to polysomnography, it will identify nearly all of the sleep epochs as sleep, but it will mis-identify many of the wake epochs also as sleep, thus over-estimating sleep relative to polysomnography.

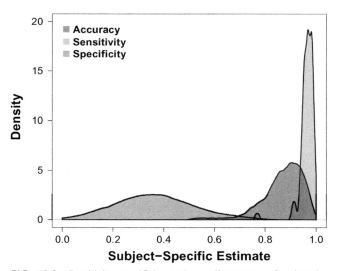

FIG. 12.3 Sensitivity, specificity, and overall accuracy of actigraphy devices in the study by Marino and colleagues [15].

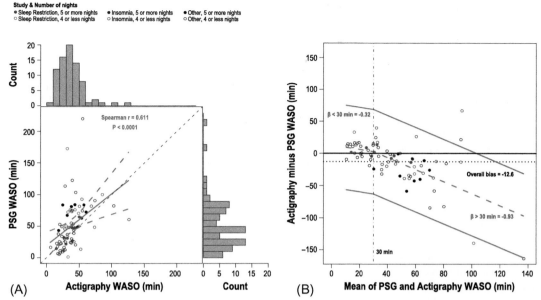

FIG. 12.4 Underestimation of Wake after Sleep Onset (WASO) by actigraphy, as a function of increasing polysomnographic WASO. *(From Marino M, Li Y, Rueschman MN, et al. Measuring sleep: accuracy, sensitivity, and specificity of wrist actigraphy compared to polysomnography. Sleep 2013;36(11):1747–1755.)*

This pattern of findings is frequently reported in the actigraphy validation literature and is probably explained by individuals lying in bed awake immobile when they are trying to sleep. Fortunately, most people spend relatively little time immobile but awake in bed, and this measurement error is minimized. But as the previously-referenced paper by Marino and colleagues showed, the more time individuals spent awake in bed not sleeping, the greater the proportion of those epochs were misidentified as sleep by actigraphy (see Fig. 12.4). Fig. 12.4 shows that although there is a strong relationship between actigraph and polysomnographic wake time after sleep onset, there is a systematic increase in underestimation as wake time after sleep onset increases beyond about 30 min. Therefore, actigraphy may be relatively accurate overall, but it is typically better at detecting sleep than wake and thus may overestimate sleep, especially when individuals spend excessive time in bed awake.

It should be noted that these studies, by design, have clearly identified in-bed intervals (defined by the time hooked up to polysomnography). Real-world implementation of actigraphy, which records over days, typically lacks this defined period in bed. Although this issue applies to nighttime sleep, it is unclear how it applies to daytime sleep. Since actigraphy algorithms typically identify many periods of brief sleep during the day, it may over-estimate sleep out of bed, compared to what polysomnography would presumably record. Also of note, sleep diary is known to over-estimate sleep compared to polysomnography, especially in people without insomnia. Actigraphy, though, does not approximate sleep diary in this way. It is typically a better approximation of objective sleep recorded with polysomnography than it is an approximation of self-reported sleep recorded by sleep diary. Therefore, it will appear to underestimate sleep, compared to sleep diary.

Another limitation of actigraphy is that its utility in sleep disorders is different than in individuals without sleep disorders. Of note, actigraphy can be quite useful for measuring sleep in insomnia patients, being relatively accurate at estimating sleep duration, sleep timing, sleep efficiency, wake after sleep onset, and other parameters. Yet, it has been shown to be relatively inaccurate regarding sleep latency, as this is likely when individuals are most likely to be awake but immobile [22]. For sleep apnea, actigraphy can often assess sleep characteristics, but the frequent movements associated with arousals may interfere with the device's accuracy [16,40–42].

Children move more during the night, especially boys. Therefore, scoring rules for actigraphy in children may need to be different and may need vary by developmental stage. Many previous studies have validated actigraphy for pediatric populations including infants [43–47], children [19,43,48], and adolescents [19,32,40,49,50]. Especially since parent reports are unreliable and child reports of sleep are also unreliable, a method for accurately representing sleep is imperative. In this case, actigraphy is a useful option, but caution should be used in scoring in order to ensure accuracy. Additionally, sleep changes throughout the lifespan, and therefore scoring algorithms in adult populations may need modifications for specific physiological conditions, such as pregnancy [51] and in older adults [11].

A critical issue with actigraphic sleep assessment is that many sleep parameters require the estimation of values in the context of being in bed and trying to sleep. For example, sleep efficiency is based on total time in bed, and sleep latency represents the amount of time elapsed between when the person goes to bed and when they fall asleep. Therefore, knowing when a person is in bed is important for actigraphic data to be useful for characterizing sleep. Yet, actigraphic devices cannot sense when an individual is in bed. When scoring actigraphic sleep, most software packages have proprietary methods for estimating the time a person enters and leaves the bed. These intervals are critical for determining sleep parameters in actigraphy, as illustrated in Fig. 12.5. In Fig. 12.5, a typical night is depicted, where the software estimated an in-bed interval based on movement that is clearly incorrect. The software estimate that the person was in bed where indicated in blue (the darker blue reflects the estimated interval where the individual was mostly asleep). The black lines represent movement, and the broken red line at the bottom indicates estimated sleep versus wake. The software estimated an in-bed interval from approximately 5:30 PM until approximately 8:00 AM. Yet, based on the movement patterns, it is likely that the individual did not get into bed until about 11:30 PM. Thus, rather than 14.5 h in bed, the individual likely spent about 8.5 h in bed. There is currently no validated algorithm for choosing these in-bed intervals in the software. Hand-scoring approaches have been the gold-standard [7,52] and many scorers use a combination of experience, sleep diary data, and data from the light channel (on/off) to determine the interval in which to look for sleep [2]. Along these lines, detecting time in bed for naps is similarly (and more) difficult.

Another key issue with actigraphy is that many of the most well-established devices (e.g., Actiwatch and Motionlogger) download data directly into a computer. Because of this, data loss and/or device failure is not known until the device is downloaded at the end of the assessment period. This is in contrast to devices that use Bluetooth connectivity to a mobile phone device, which can provide updates into the cloud, allowing for more prospective and/or active monitoring of data to better identify problems.

IDENTIFYING SLEEP STAGES WITH ACTIGRAPHY

Movement-based sleep estimation is unable to distinguish between sleep stages. Scientific-grade devices that assess sleep using actigraphy should not (and typically do not) claim to be able to estimate sleep architecture or depth in any way. Yet, several commercially-available devices have claimed this ability, despite available data.

There is one exception to this, in that a group led by Roenneberg [53] has characterized what they describe as "Locomotor Inactivity During Sleep" (LIDS), which is a way to characterize actigraphic recordings in more detail in such a way that may discern sleep stages. Fig. 12.6 shows some of the data from the first study of LIDS, showing a rough approximation of sleep cycles with cycling LIDS. Of note, this is still controversial and the methods for using LIDS have not been well-characterized.

One way in which actigraphic devices have been used to relatively accurately predict sleep stages is through the use of an additional physiologic channel (typically heart rate). The WatchPAT (Itamar Medical) is a wrist-worn device that is used to screen for sleep apnea and includes an actigraphy channel, as well as a channel for peripheral arterial tone. Using this additional data, the device has been shown to relatively accurately distinguish between sleep stages [54,55]. Another device that uses a second channel to approximate sleep stages is Fitbit, who validated their devices which use a combination of accelerometry and optical plethysmography to get a measure of continuous heart rate [3]. This validation study showed that by combining channels, the devices were able to achieve 94.6% sensitivity (comparable to other devices) and specificity of 69.3% (which is higher than that typically seen for devices that use movement alone). Less accuracy was seen

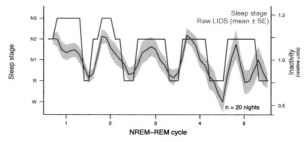

FIG. 12.6 Approximation of locomotor activity during sleep (LIDS) values relative to sleep stages. *(From the paper by Winnebeck EC, Fischer D, Leise T, Roenneberg T. Dynamics and ultradian structure of human sleep in real life. Curr Biol 2018;28(1):49–59 [e45].)*

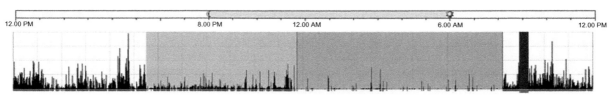

FIG. 12.5 A typical 24-h actigraphic recording.

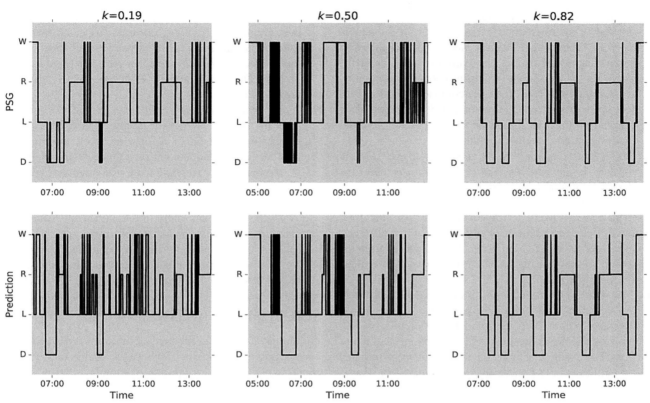

FIG. 12.7 A comparison of hypnograms for which the Fitbit device accurately represented sleep, compared to hypnograms for which it was less accurate. *(From Beattie Z, Pantelopoulos A, Ghoreyshi A, Oyang Y, Statan A, Heneghan C. Estimation of sleep stages using cardiac and accelerometer data from a wrist-worn device. Sleep 2017;40(Abstract Supplement):A26.)*

for "light" sleep (stage N1 and N2 combined, 69.2%), "deep" sleep (stage N3, 62.4%), and REM sleep (71.6%) individually, though these values should be taken in context relative to the benefits of a repeatable and inexpensive recording. Of note, these values are similar to those seen for WatchPAT.

The paper by Beattie and colleagues further notes that the device was more accurate for some sleep patterns than others [3]. Fig. 12.7 depicts examples of where the device's native scoring algorithm was most accurate and least accurate, and a typical record. This suggests that these types of devices may be more accurate for some people than others. Further work is needed to better model these discrepancies and improve accuracy.

The OURA ring has also undergone validation for sleep stages [33]. Similar to the other devices, the combination of movement and heart rate data yields sensitivity (95.5%) that is generally comparable to scientific-grade devices and specificity (48.1%) that is generally better than devices that use movement alone. The relative accuracy for "light" sleep (64.6%), "deep" sleep (50.9%), and REM sleep (61.4%) are comparable to Fitbit and other similar devices.

The SleepScore MAX device, mentioned above as a contactless infrared device that assesses sleep, has also been validated for sleep staging based on the signals it collects that relate to respiration and heart rate data.

Despite the existence of validation data for these devices, sleep staging using these methods is still not endorsed by any organization or guidelines as accurate. Future guidelines may wish to take these into account, though.

OTHER CONSIDERATIONS

There are a number of additional factors that should be considered when using actigraphy to record sleep:

1. *Type of movement recording.* Devices use single-axis or multi-axis recording for omnidirectional recording. In addition, some devices may use MEMS chips for accelerometry. These considerations may play a role in deciding which device is best for a situation.
2. *Recording modes.* Most actigraphic devices record in one (or more) of three recording modes. Zero-crossing mode (ZCM) is akin to a movement count approach, where each time a movement is recorded above a threshold is registered as a count, with more counts representing greater movement. Time above threshold (TAT) refers to the sum of time above a set threshold representing movement versus nonmovement. This approach

is similar to ZCM but includes a time element so it can not only capture frequency but duration of movement as well. Proportional integral mode (PIM) is the most sophisticated and uses a calculation of area under the curve for all movement recordings to most accurately represent movement. In studies comparing modes, PIM has been shown to be superior [11,56], though many older algorithms are based on ZCM because older devices recorded movement counts only.

3. *Ability to assess whether the device is off wrist.* Many devices do not have the ability to discern whether the device is off wrist; this can lead to difficulties in scoring. Off-wrist time will often be recorded as no movement and will likely be mis-scored as sleep. Therefore, if a device cannot automatically tell when it is off-wrist, hand scoring is needed to delete these instances from the record so that scoring does not incorporate these missing values as sleep.

4. *Assessment of brief changes in activity level.* Different devices quantify activity in different ways. This applies to epoch length (which should be 30 or 60s, in line with validation studies), as well as sampling within epochs. For example, a large, quick movement surrounded by stillness may be more likely to reflect sleep than a smaller but more sustained movement. As technology changes and greater resolution of sampling within epochs becomes available, approaches to handling these sorts of movements may refine scoring.

5. *Dynamic range.* Actigraphic devices use piezoelectric materials, which convert pressure changes due to acceleration into changes in voltage. These materials are often ceramic or crystal in nature, but piezoelectric materials have varying properties. This may lead to different dynamic ranges across devices—ranges that dictate the precision of the types of pressure changes to elicit accurately recorded changes in voltage. For example, some materials may be more accurate for larger types of movements but not as accurate for smaller movements. For this reason, devices optimized for sleep are often not optimized for physical activity and vice versa.

6. *Frequency response and output deviation.* This refers to the reliability of the output voltage change recorded in the presence of movement. Some devices have very narrow ranges, such that voltage changes consistently reflect acceleration changes. Other devices may allow for more deviations, such that a single movement can produce a range of outputs that may or may not accurately reflect that movement.

7. *High and low frequency limit.* All devices have high and low limits regarding what types of activities are accurately recorded. Measurements above and below these thresholds are clipped. For sleep recording, lower thresholds are preferred, since sleep assessment requires resolution in the context of little or no movement.

8. *Noise.* All electrical systems contain noise, as do all measurements. This source of error, especially when it reflects background noise that can lead an epoch to be mis-scored as sleep or wake, can create difficulties with scoring. Devices should be evaluated relative to their noise levels.

9. *Temperature sensitivity and range.* Accelerometers measure changes in voltage output from piezoelectric materials as a result of a force of acceleration acting on a mass that interacts with the piezoelectric element. As temperatures change, density of matter changes and this may impact voltage measurements as the density of the space around the mass expands and contracts, for example. This is especially important for MEMS chips, where a very small change in density and pressure can cause a proportionately large change in recordings. Devices should be evaluated relative to their ability to record across temperature ranges.

SCIENTIFIC GUIDELINES

Guidelines for the conduct of actigraphy have been recently published by the Society of Behavioral Sleep Medicine [2]. This guideline document outlines key considerations for those interested in using actigraphy for sleep research. In particular, this document outlines suggestions for actigraphs regarding their accelerometers (omnidirectional and/or triaxial measurements are required, recording modes other than ZCM are recommended, and epoch lengths should be 30 or 60s, though some newer devices can make use of the raw recordings), appearance (appropriate to the population and intended use), event markers (which are recommended to increase precision of hand scoring), light sensors (recommended to aid in hand scoring), battery type (life of 3 days is required and 2 weeks is recommended), data storage (nonvolatile memory is recommended), and customer support (required). In addition, guidelines are provided regarding data collection, recording of ambient light, integration of rating scales, inclusion of patient instructions, device placement, and use of event markers and other techniques for gaining peripheral information to assist in scoring. Further, this document discusses three instances where data should be considered invalid: device removal (mentioned earlier), artifacts or abnormal data (i.e., suspected device malfunction), or artifacts of parenting (in pediatrics; e.g., rocking a sleeping infant). For use in research, the SBSM guidelines and other documents outline the importance of considering several key issues, including making sure scoring algorithms are appropriate and defining the major sleep period, which can be difficult and particularly error-prone using automatic methods.

When evaluating devices for use in research, all of the above issues should be considered. This includes specifications of the device and software as well as plans for implementation. For example, some external indicator of time in bed should be applied, including a sleep diary or event marker. Automated scoring of sleep periods and sleep/wake determinations should be supplemented by hand scoring to eliminate anomalies and/or device malfunctions. In evaluating specific devices, in addition to specifications, validation data should be critically evaluated. Validation by an independent research group is preferred over internal efforts, so that any appearances of conflict of interest are addressed. Whether internal or independent, characteristics of the validation study should be evaluated as well. For example, sample size should be large enough and representativeness of the sample should be adequate to the use case being considered. Some validation studies compare to overnight polysomnography as a gold standard, while others compare against other previously-validated actigraphic devices. Either or both of these approaches may be appropriate, depending on the intended use of the device. For example, when using sleep measurement as a diagnostic tool, polysomnography is still the accepted method and all devices intended for diagnostic use will need to be similarly accurate, but devices used as population surveillance and public health could be validated in the field.

Validation studies should not rely on correlation coefficients. Ideally kappa scores, Bland-Altman plots, and other statistics should be employed. In particular, values should examine epoch-by-epoch agreement across modalities, reporting values for sensitivity (number of sleep epochs identified by the gold standard that were also identified as sleep epochs by the study device), specificity (number of wake epochs identified by the gold standard that were also identified as wake epochs by the study device), and overall rate of agreement (number of epochs for which the gold standard and study device agreed. Regarding sleep, since most evaluation periods will primarily consist of sleep, a high rate of agreement can be maintained even in the case of poor specificity. For example, if a sample achieves 90% sleep efficiency, and a device scores every epoch as sleep without even employing an algorithm, they would obtain a 90% rate of agreement, 100% sensitivity, and 0% specificity. Thus, a high rate of agreement does not alone suggest an accurate device.

EVALUATING COMMERCIALLY-AVAILABLE SLEEP TRACKERS

Commercially-available sleep trackers are widely used with relatively little validation in independent laboratories, but their popularity and effect on user behavior makes these devices extremely important in a public health context. With all new technology, it is important to keep previous work in this area in mind when evaluating new devices. For example, if the Society for Behavioral Sleep Medicine does not recommend the use of actigraphy for sleep-staging, then claims of sleep-staging from commercial motion-based devices should not be trusted until validation is provided in a peer-reviewed scientific journal. Unfortunately, scientific guidelines are rarely taken into account with commercial device development and independent validation is very late if available at all.

There are several ways to demonstrate the accuracy of commercial sleep devices. An optimal validation cycle would include: *First*, a comparison to polysomnography in a laboratory setting; *Second*, a comparison to validated field-based sleep measures; and *Third*, validation for specific populations (such as older adults or teens). Devices should undergo this process in order to be considered validated. Unfortunately, objective scientific validation is well behind evaluating technology at the pace in which new technology is introduced to the public. In general, the quality of the data and the consumer usability of the devices are in opposition.

It should be noted that insufficient validation does not imply that a device is inaccurate; rather, it implies that the accuracy is undetermined. There are several ways in which devices are insufficiently validated. First, device manufacturers may not sufficiently determine that the electrical output of the device in question is consistent with what would be expected (e.g., that the apparatus is functioning properly). Second, manufacturers may not sufficiently determine that the electrical output is encoded, recorded, and processed in a way that is consistent with the intended function of the device. Third, device manufacturers may not sufficiently determine that the scoring is consistent with what would be expected (e.g., what is typically referred to as validation). Fourth, devices may not be tested in the field in addition to a laboratory setting, in order to characterize their performance under real-world conditions. Completing all of these steps would be ideal. It is possible that a product could be deemed minimally sufficient if it meets some if not all of these criteria.

According to market reports [57], smart sleep tracking is expected to grow over the next 5 years, with companies such as Samsung, Philips, Nokia, Fitbit, Emfit, Garmin, ResMed, Sleepace, and Apple, Inc. all interested in sleep measurement. There have even been sleep sensors which are completely housed under the bed's leg which can measure heart rate, respiratory rate and body movements [58]. In bed sensors differ from the traditional wearable devices, but rely on the same estimation of sleep from body movement and physiological measures. Validation of any of these devices usually relies on one or no studies published on their ability to measure sleep. Despite this limitation, sleep measurement using devices is probably better than the large-scale questionnaires that measure sleep in sleep epidemiology.

CONCLUSIONS

Sleep tracking is rooted in a long history of actigraphic sleep recording across days and weeks. Actigraphy is a technique for measuring sleep using movement at the wrist (and sometimes hip), and mathematically predicting the likelihood that an individual is awake or asleep based on the pattern of movement. This approach is well-validated and there are a number of devices that have been shown to reliably and validly assess sleep, with some limitations and caveats that should be noted. Newer devices that include other channels such as heart rate can improve the detection of sleep and even offer a limited but useful estimation of sleep architecture. Devices, including commercial devices, should be used in the context of the limitations of the measurement approach and data should be interpreted in that context. Commercially available devices are often changing and rarely provide validation data and thus should not be used for diagnosis and treatment of sleep-related disorders, but their use is limited to simple measurement in a healthy population.

REFERENCES

[1] Collop NA. Polysomnography. In: Lee-Chiong TL, editor. Sleep: a comprehensive handbook. Hoboken, N.J: Wiley; 2006, p. 973–6.

[2] Ancoli-Israel S, Martin JL, Blackwell T, et al. The SBSM guide to actigraphy monitoring: clinical and research applications. Behav Sleep Med 2015;13(Suppl 1):S4–38.

[3] Beattie Z, Pantelopoulos A, Ghoreyshi A, Oyang Y, Statan A, Heneghan C. Estimation of sleep stages using cardiac and accelerometer data from a wrist-worn device. Sleep 2017;40:. Abstract Supplement):A26.

[4] Foster FG, Kupfer D, Weiss G, Lipponen V, McPartland RJ, Delgado J. Mobility recording and cycle research in neuropsychiatry. J Interdis Cycle Res 1972;3(1):61–72.

[5] Kupfer DJ, Detre TP, Foster G, Tucker GJ, Delgado J. The application of Delgado's telemetric mobility recorder for human studies. Behav Biol 1972;7(4):585–90.

[6] Kripke DF, Mullaney DJ, Messin S, Wyborney VG. Wrist actigraphic measures of sleep and rhythms. Electroencephalogr Clin Neurophysiol 1978;44(5):674–6.

[7] Mullaney DJ, Kripke DF, Messin S. Wrist-actigraphic estimation of sleep time. Sleep 1980;3(1):83–92.

[8] Webster JB, Kripke DF, Messin S, Mullaney DJ, Wyborney G. An activity-based sleep monitor system for ambulatory use. Sleep 1982;5(4):389–99.

[9] Cole RJ, Kripke DF, Gruen W, Mullaney DJ, Gillin JC. Automatic sleep/wake identification from wrist activity. Sleep 1992;15(5):461–9.

[10] Jean-Louis G, Kripke DF, Ancoli-Israel S, Klauber MR, Sepulveda RS. Sleep duration, illumination, and activity patterns in a population sample: effects of gender and ethnicity. Biol Psychiatry 2000;47(10):921–7.

[11] Jean-Louis G, Kripke DF, Cole RJ, Assmus JD, Langer RD. Sleep detection with an accelerometer actigraph: comparisons with polysomnography. Physiol Behav 2001;72(1–2):21–8.

[12] Jean-Louis G, Kripke DF, Mason WJ, Elliott JA, Youngstedt SD. Sleep estimation from wrist movement quantified by different actigraphic modalities. J Neurosci Methods 2001;105(2):185–91.

[13] Kripke DF, Hahn EK, Grizas AP, et al. Wrist actigraphic scoring for sleep laboratory patients: algorithm development. J Sleep Res 2010;19(4):612–9.

[14] te Lindert BH, Van Someren EJ. Sleep estimates using microelectromechanical systems (MEMS). Sleep 2013;36(5):781–9.

[15] Marino M, Li Y, Rueschman MN, et al. Measuring sleep: accuracy, sensitivity, and specificity of wrist actigraphy compared to polysomnography. Sleep 2013;36(11):1747–55.

[16] Ancoli-Israel S, Cole R, Alessi C, Chambers M, Moorcroft W, Pollak CP. The role of actigraphy in the study of sleep and circadian rhythms. Sleep 2003;26(3):342–92.

[17] Blackwell T, Ancoli-Israel S, Gehrman PR, Schneider JL, Pedula KL, Stone KL. Actigraphy scoring reliability in the study of osteoporotic fractures. Sleep 2005;28(12):1599–605.

[18] Stein GJ. Some recent developments in acceleration sensors. Measurement Sci Rev 2001;1(1):183–6.

[19] Meltzer LJ, Hiruma LS, Avis K, Montgomery-Downs H, Valentin J. Comparison of a commercial accelerometer with polysomnography and actigraphy in children and adolescents. Sleep 2015;38(8):1323–30.

[20] Rupp TL, Balkin TJ. Comparison of Motionlogger Watch and Actiwatch actigraphs to polysomnography for sleep/wake estimation in healthy young adults. Behav Res Methods 2011;43(4):1152–60.

[21] Terrill PI, Mason DG, Wilson SJ. Development of a continuous multisite accelerometry system for studying movements during sleep. Conf Proc IEEE Eng Med Biol Soc 2010;1:6150–3.

[22] Lichstein KL, Stone KC, Donaldson J, et al. Actigraphy validation with insomnia. Sleep 2006;29(2):232–9.

[23] Paquet J, Kawinska A, Carrier J. Wake detection capacity of actigraphy during sleep. Sleep 2007;30(10):1362–9.

[24] Weiss AR, Johnson NL, Berger NA, Redline S. Validity of activity-based devices to estimate sleep. J Clin Sleep Med 2010;6(4):336–42.

[25] Full KM, Kerr J, Grandner MA, et al. Validation of a physical activity accelerometer device worn on the hip and wrist against polysomnography. Sleep Health 2018;4(2):209–16.

[26] Rosenberger ME, Buman MP, Haskell WL, McConnell MV, Carstensen LL. Twenty-four hours of sleep, sedentary behavior, and physical activity with nine wearable devices. Med Sci Sports Exerc 2016;48(3):457–65.

[27] Landry GJ, Falck RS, Beets MW, Liu-Ambrose T. Measuring physical activity in older adults: calibrating cut-points for the MotionWatch 8(c). Front Aging Neurosci 2015;7:165.

[28] Baroni A, Bruzzese JM, Di Bartolo CA, Shatkin JP. Fitbit Flex: an unreliable device for longitudinal sleep measures in a non-clinical population. Sleep Breath 2016;20(2):853–4.

[29] de Zambotti M, Baker FC, Willoughby AR, et al. Measures of sleep and cardiac functioning during sleep using a multi-sensory commercially-available wristband in adolescents. Physiol Behav 2016;158:143–9.

[30] Mantua J, Gravel N, Spencer RM. Reliability of sleep measures from four personal health monitoring devices compared to research-based actigraphy and polysomnography. Sensors (Basel) 2016;16(5).

[31] Montgomery-Downs HE, Insana SP, Bond JA. Movement toward a novel activity monitoring device. Sleep Breath 2012;16(3):913–7.

[32] Roane BM, Van Reen E, Hart CN, Wing R, Carskadon MA. Estimating sleep from multisensory armband measurements: validity and reliability in teens. J Sleep Res 2015;24(6):714–21.

[33] de Zambotti M, Rosas L, Colrain IM, Baker FC. The sleep of the ring: comparison of the OURA sleep tracker against polysomnography. Behav Sleep Med 2017;1–15.

[34] Tal A, Shinar Z, Shaki D, Codish S, Goldbart A. Validation of contact-free sleep monitoring device with comparison to polysomnography. J Clin Sleep Med 2017;13(3):517–22.

[35] Kaplan RF, Wang Y, Loparo KA, Kelly MR, Bootzin RR. Performance evaluation of an automated single-channel sleep-wake detection algorithm. Nat Sci Sleep 2014;6:113–22.

[36] Patel P, Kim JY, Brooks LJ. Accuracy of a smartphone application in estimating sleep in children. Sleep Breath 2017;21(2):505–11.

[37] Snyder F, Hobson JA, Morrison DF, Goldfrank F. Changes in respiration, heart rate, and systolic blood pressure in human sleep. J Appl Physiol 1964;19:417–22.

[38] Otzenberger H, Gronfier C, Simon C, et al. Dynamic heart rate variability: a tool for exploring sympathovagal balance continuously during sleep in men. Am J Physiol 1998;275(3 Pt 2):H946–50.

[39] Busek P, Vankova J, Opavsky J, Salinger J, Nevsimalova S. Spectral analysis of the heart rate variability in sleep. Physiol Res 2005;54(4):369–76.

[40] Johnson NL, Kirchner HL, Rosen CL, et al. Sleep estimation using wrist actigraphy in adolescents with and without sleep disordered breathing: a comparison of three data modes. Sleep 2007;30(7):899–905.

[41] Morgenthaler T, Alessi C, Friedman L, et al. Practice parameters for the use of actigraphy in the assessment of sleep and sleep disorders: an update for 2007. Sleep 2007;30(4):519–29.

[42] Littner M, Kushida CA, Anderson WM, et al. Practice parameters for the role of actigraphy in the study of sleep and circadian rhythms: an update for 2002. Sleep 2003;26(3):337–41.

[43] Galland BC, Kennedy GJ, Mitchell EA, Taylor BJ. Algorithms for using an activity-based accelerometer for identification of infant sleep-wake states during nap studies. Sleep Med 2012;13(6):743–51.

[44] Scher A. Continuity and change in infants' sleep from 8 to 14 months: a longitudinal actigraphy study. Infant Behav Dev 2012;35(4):870–5.

[45] Shinohara H, Kodama H. Relationship between duration of crying/fussy behavior and actigraphic sleep measures in early infancy. Early Hum Dev 2012;88(11):847–52.

[46] Tikotzky L, DEM G, Har-Toov J, Dollberg S, Bar-Haim Y, Sadeh A. Sleep and physical growth in infants during the first 6 months. J Sleep Res 2010;19(1 Pt 1):103–10.

[47] Tsai SY, Thomas KA. Actigraphy as a measure of activity and sleep for infants: a methodologic study. Arch Pediatr Adolesc Med 2010;164(11):1071–2.

[48] Meltzer LJ, Westin AM. A comparison of actigraphy scoring rules used in pediatric research. Sleep Med 2011;12(8):793–6.

[49] Arora T, Broglia E, Pushpakumar D, Lodhi T, Taheri S. An investigation into the strength of the association and agreement levels between subjective and objective sleep duration in adolescents. PLoS One 2013;8(8):e72406.

[50] Short MA, Gradisar M, Lack LC, Wright H, Carskadon MA. The discrepancy between actigraphic and sleep diary measures of sleep in adolescents. Sleep Med 2012;13(4):378–84.

[51] Zhu B, Calvo RS, Wu L, et al. Objective sleep in pregnant women: a comparison of actigraphy and polysomnography. Sleep Health 2018;4(5):390–6.

[52] Sadeh A, Hauri PJ, Kripke DF, Lavie P. The role of actigraphy in the evaluation of sleep disorders. Sleep 1995;18(4):288–302.

[53] Winnebeck EC, Fischer D, Leise T, Roenneberg T. Dynamics and ultradian structure of human sleep in real life. Curr Biol 2018;28(1). 49–59 e45.

[54] Hedner J, White DP, Malhotra A, et al. Sleep staging based on autonomic signals: a multi-center validation study. J Clin Sleep Med 2011;7(3):301–6.

[55] Choi JH, Kim EJ, Kim YS, et al. Validation study of portable device for the diagnosis of obstructive sleep apnea according to the new AASM scoring criteria: Watch-PAT 100. Acta Otolaryngol 2010;130(7):838–43.

[56] Blackwell T, Ancoli-Israel S, Redline S, Stone KL. Factors that may influence the classification of sleep-wake by wrist actigraphy: the MrOS Sleep Study. J Clin Sleep Med 2011;7(4): 357–67.

[57] Research Nester. Smart sleep tracking device market : global demand analysis & opportunity outlook 2024. New York: Research Nester; 2018.

[58] Brink M, Muller CH, Schierz C. Contact-free measurement of heart rate, respiration rate, and body movements during sleep. Behav Res Methods 2006;38(3):511–21.

Chapter 13

Mobile technology, sleep, and circadian disruption

Cynthia K. Snyder[a], Anne-Marie Chang[a,b]
[a]College of Nursing, Pennsylvania State University, University Park, PA, United States, [b]Department of Biobehavioral Health, Pennsylvania State University, University Park, PA, United States

ABBREVIATIONS

CRP	c-reactive protein
DVD	digital video disc
IBM	International Business Machines Corporation
IL-6	interleukin 6
ipRGC	intrinsically photosensitive retinal ganglion cell
LED	light-emitting diode
SCN	suprachiasmatic nucleus
TNF-α	tumor necrosis factor alpha

Mobile technology use is ubiquitous in modern society. It is common to observe people of all ages actively engaged with mobile phone devices while walking, within vehicles, and during daily life. The convenience of instant communication provides many benefits in modern life; however, engagement with mobile devices within the bedroom setting negatively impacts sleep patterns and the normal circadian rhythm, ultimately causing sleep loss [1]. Insufficient sleep is a public health concern [2]. One of the Healthy People 2020 goals for disease prevention and health promotion of all groups of people includes increasing the proportion of the population who obtain the minimum amount of sleep in a 24-h period [2]. The National Sleep Foundation consensus panel issued recommendations for minimum sleep duration for 8–10 h for teens (14–17 years), 7–9 h for adults (18–64 years), and 7–8 h for older adults (65 years and older) [3]. Nearly three quarters of high school students do not obtain the minimum amount of sleep (8 h) according to the Center for Disease Control's 2017 Youth Risk Behavior Surveillance Survey [4], and about 25% of adults obtain less than the recommended amount of sleep 50% of the time [5].

Using electronic/technological devices in the bedroom setting before and after lights out is prevalent among Americans of all ages in the 2014 and 2011 Sleep in America polls [1, 6]. More than 90% of adolescents report using mobile phones after bedtime and during the night [7]. Among Americans, 95% of people 18 years and older own some type of mobile phone, and 77% of those own a smart phone. Mobile technology use is more robust among individuals under the age of 30 years. In the young adult age group, 18–29 years, 100% own some type of mobile phone, 94% of those own a smart phone, 53% own a tablet computer, and 22% own an eReader [8]. Among teens, 13–17 years, 88% either own a mobile phone or have regular access to a mobile phone, and 58% have a tablet computer [9a]. The use of smartphones without concurrent broadband services within the home is more common among younger adults, minorities, and lower income Americans, including 31% of those earning less than $30,000 per year, 35% of Hispanics, and 24% of Black/African Americans [8].

There is a pervasive presence of mobile phones among residents of emerging countries, with a higher incidence of conventional mobile phones rather than smart phones [9b]. The lack of availability of landlines contributes to the ubiquitous presence of mobile phone use in these countries, and some emerging countries have moved directly to mobile phones rather than adding infrastructure for landlines [9b]. Those residents who have a college degree or those who are under the age of 30 are more likely to own a smartphone than a conventional phone. Text messaging is the most predominantly used mobile phone feature. Other mobile phone features include photographs and videos, as well as internet and social media access, with Facebook® and Twitter® being utilized as the most common social media platforms [9b]. The coincidence of sleep deficiency reaching its highest levels and ubiquitous mobile technology use throughout the world's population are likely related, as a growing body of evidence would suggest.

SLEEP AS A BIOBEHAVIORAL STATE

Sleep is a vital biopsychosocial state, a naturally occurring process necessary for brain activity and body functions, psychological health and behavior, and individual, family and sociocultural norms. In addition to the physiologic functions, sleep is also necessary for mental health, such as internalizing and externalizing behaviors.

Two-process model of sleep physiology

Sleep is regulated by a two-process mechanism: the sleep-dependent sleep-wake (process S) and the sleep-independent circadian oscillator (process C) [10]. These processes work in tandem to regulate sleep. The sleep-wake homeostatic process is a compensatory brain mechanism, and is a function of sleep and wake homeostasis. Borbély proposed that the greater period of time one is awake, the longer the period of sleep that follows. The state of wakefulness produces a pressure to fall asleep, which becomes stronger with the length of time awake, and dissipates with sleep.

The circadian process is a cyclic rhythmic process independent of the sleep cycle occurring approximately every 24 h, and the propensity for sleep corresponds with the lowest body temperature [10]. Process C is regulated by two suprachiasmatic nuclei (SCN) oscillators located within the anterior hypothalamus [11]. Light enters the eyes through the retina and is transmitted by intrinsically photosensitive retinal ganglion cells (ipRGC), then relayed via a nonvisual retinal hypothalamic tract to the SCN [12]. The SCN regulates circadian body processes, such as sleep, to maintain a 24-h cycle. The ipRGC neurons are highly sensitive to light cues, especially short wavelength light that is enriched in light-emitting electronic devices, and mediate the nonimage forming circadian responses to these cues that serve to entrain the body's biological rhythms [13, 14].

One such rhythm is the release of melatonin, a hormone secreted by the pineal gland. Melatonin release is inhibited by natural or artificial light and has a circadian pattern that reflects this: normally at a low level during the day, rises prior to sleep and increases sleep propensity, and is inhibited by morning light cues [11, 15a]. Melatonin suppression and phase resetting, or shifting, of the melatonin rhythm are two examples of nonimage forming circadian responses to light. Full-spectrum light, even of moderately low intensity, can suppress melatonin levels and shift the timing of its cycle; however, short wavelength light, in the blue-green range, induces greater suppression and phase shifts [15b]. The human eye is sensitive to wavelengths in the range of 390–720 nm of full spectrum light but the peak sensitivity for the visual system, which is mediated by the rod and cone photoreceptors, occurs at 555 nm [15c]; whereas the peak sensitivity for circadian responses to light is at shorter wavelengths (~460–500 nm) and is mediated by the ipRGC photoreceptors [15d] (see Fig. 13.1).

Another factor modulating the effects of light on the circadian system is the timing of the exposure. Light in the early part of the biological day (early to mid-morning) shifts circadian rhythms (e.g., melatonin release, sleep propensity, etc.) to an earlier phase, and light exposure at the end of the day (evening to early night) delays circadian phase [15e]. Furthermore, given that (1) portable electronic devices (e.g., cell phones, tablets) emit blue light; (2) evening light exposure suppresses melatonin, which is released prior to bedtime and increases sleep propensity at this time of day; and (3) evening/nighttime exposure to short wavelength light, compared to full-spectrum light, induces greater phase delays of circadian rhythms suggests there are likely multiple and interacting ways in which mobile technology use close to bedtime or during the night negatively impact sleep.

The regulation of sleep by processes S and C is balanced between the often-opposing pressure for the sleep or wake

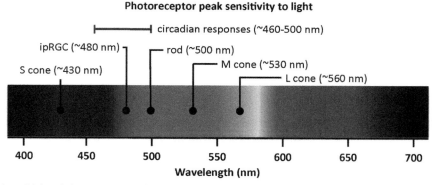

FIG. 13.1 Peak spectral sensitivity of photoreceptors and circadian responses to light. Photoreceptors of the visual system exhibit peak sensitivities to specific wavelengths: rods (~500 nm), S cones (short wavelength; ~430 nm), M cones (medium wavelength; ~530 nm), and L cones (long wavelength; ~560 nm). Intrinsically photosensitive retinal ganglion cells (ipRGC) have been more recently identified photoreceptors that mediate the nonimage forming circadian responses (e.g., melatonin suppression and phase resetting). IpRGCs have a peak spectral sensitivity of ~480 nm, which is in the peak range for circadian responses (~460–500 nm) [15c].

state. For example, sleep drive accumulates with longer duration of wake, increasing throughout the waking day and reaches a peak at sleep onset. It is just prior to this point, however, that the circadian drive for wakefulness is at high levels. In contrast to the sleep homeostatic process, the circadian system is increasing its drive for alertness throughout the day until reaching maximal levels in the late evening (an hour or two before bedtime). After this peak, the circadian process decreases its alertness drive and circadian-regulated levels of melatonin, a sleep-promoting hormone, begin to rise. Perturbations to either process can result in altered sleep propensity and/or timing.

Contextual factors influencing sleep behavior and mobile technology use

Other factors can influence sleep behavior, such as sociodemographic factors, socioeconomic factors, racial and cultural factors, and neighborhood and community factors. Age and gender influence sleep patterns in epidemiologic studies of adults [16], and children and adolescents [17]. Among adults surveyed in the Behavioral Risk Factor Surveillance System, young adults reported the highest levels of insufficient sleep and older adults reported the lowest levels [16]. In children and adolescents in the National Survey of Children's Health, parents reported increased sleep insufficiency in their children with increasing age. Some children as young as 6 years of age obtained insufficient sleep, and the highest levels of sleep insufficiency were among the older adolescents [17]. Those children and adolescents in single parent homes were more likely to report sleep insufficiency; however, increasing income level and educational level were associated with more sleep insufficiency among all age groups [16, 17]. Black/African American and Hispanic children and adolescents were more likely to obtain insufficient sleep [17]; however, Black/African American, Hispanic and Asian adults were less likely to report sleep insufficiency [16]. Perceptions of environmental safety within neighborhoods affected sleep sufficiency, with increasing sleep insufficiency observed in communities where neighbors did not watch out for other children within the neighborhood [17].

Contextual factors have also been shown to affect behaviors related to digital media use. Engagement with mobile technology near, at, and after bedtime is prevalent, including homework on computers or laptops, internet usage, emailing, social networking, text messaging, phone calling, video viewing, music, and reading an e-book [1]. Among adolescents, 98% engage in text messaging in the prebedtime period, and 70% engage in sending text messages after lights out [18]. Young adults are heavy users of multiple social media platforms, including YouTube, Facebook, Snapchat, Instagram, and Twitter, with 94% using at least one of these platforms, with declining usage with increasing age [19].

High school and college students, especially females, are more likely to use mobile phones near and after bedtime [20–24], while males spend more time video gaming [22, 25]. Females are more likely to take a mobile phone to bed than males [21], and more likely to use text messaging at night [26].

Importance of sleep for health

Sleep is important for physical, psychological, emotional, and developmental functions. Although sleep may be considered an unconscious state, whereby the body is not responsive to the external environment, sleep is a time of rest and renewal for the body, and enhanced brain activity [27]. The glymphatic system of the brain circulates cerebral spinal fluid throughout to clear the brain of accumulated toxins and cellular debris, which is necessary for tissue homeostasis [28, 29]. Sleep is necessary for growth and development in children and adolescents with hormonal regulation and secretion [30]. Circadian rhythm of hormones, such as growth hormone, parathyroid hormone, prolactin, and cortisol [31], and hormonal control of appetite regulation for satiety and hunger, with secretion of leptin and suppression of ghrelin occur [30] occur during sleep. The immune system is strengthened by peak fluctuations of circulating cytokines and naïve T-cells during the sleep cycle [31]. Other physiological processes occurring during sleep include muscle and tissue repair [32], and circadian variability of body temperature [33] and blood pressure [34]. Sleep is necessary for short-term memory consolidation, protection of memories, and re-establishment of chronology of daily life [35]. Cognitive function [36], learning [37], and decision-making ability [38] are enhanced during the sleep cycle.

Sleep loss impacts physical and psychological health and wellbeing

Several factors play a role in the impact of sleep loss and physical and psychological health and wellbeing. Sleep is a vital life process, and is essential for survival [39]. The prevalence of short sleep among adults and adolescents predisposes individuals to ill-being and ill health, and chronic sleep loss has become a societal norm [40]. Less priority is placed on the importance of sleep health, with an increasingly busy lifestyle, shiftwork among various workers, and the belief that optimal sleep is optional [39].

Sleep loss and sleep deficiency

Short sleep leads to an increasing sleep debt, which is difficult to replace. Sleep deficiency, described as inadequate or mistimed sleep, leads to an increased mortality risk, with predisposition to cardiovascular disease, diabetes, and obesity [39]. Sleep deficiency is associated with increases in blood pressure, which may lead to cardiac events or stroke

[39]. Obesity results in increased body mass index, risk of developing type 2 diabetes, impaired glucose tolerance, and altered hormonal regulation of insulin, cortisol, leptin (regulation of satiety), and ghrelin (regulation of hunger) [39]. Inflammatory markers, c-reactive protein (CRP), tumor necrosis factor alpha (TNF-α), and interleukin 6 (IL-6), are elevated with sleep deficiency [39]. Risk of cancer (breast, colorectal, prostate, and endometrial) is increased in persons with sleep deficiency [39]. Additionally, sleep deficiency results in increased daytime sleepiness, with an increased risk of accidents, reduced cognitive function, and reduced motor performance, similar to the effects produced by alcohol intoxication [39].

Sleep deprivation

Sleep deprivation has been related to negative effects on cognition and performance. Psychomotor performance that is affected by sleep deprivation may include delays in reaction time, lapses in attention, omission errors, and poor short term memory. Mood symptoms related to sleep deprivation may include an increase in fatigue, confusion, stress, and irritability. Lack of motivation and increased distractibility are observed with sleep deprivation. Performance-related impairments are most apparent with sedentary, long or difficult tasks, and in conditions of decreased light. Tasks related to driving a vehicle are markedly affected by sleep deprivation [40].

Sleep restriction

Sleep restriction, routinely obtaining less sleep than is needed, is a common phenomenon among today's society. Even mild sleep restriction may cause significant metabolic effects. Impaired glucose tolerance and decreased insulin sensitivity affect energy balance and body weight. Alterations in hormone secretion, such as cortisol, leptin, and ghrelin, affect caloric intake and energy balance. Individuals who experience sleep restriction are more likely to feel hunger, eat more, and gain weight, with predisposition to obesity, and type 2 diabetes [11].

Sleep fragmentation

Sleep fragmentation occurs with night awakenings, such as those caused by sleep apnea or periodic limb movements. Physiologic effects of sleep fragmentation include elevated blood pressure. Cognition and psychomotor performance is affected with increased daytime sleepiness, negative mood, and poorer psychomotor performance comparable to the responses of alcohol intoxication [40].

Circadian dysfunction

The circadian timing system is dependent upon the time of day. Altered circadian timing and function is reflected in physiologic and psychologic changes in body patterns [11]. Circadian dysfunction can be caused by several factors: changes in weekday compared to weekend habitual sleep patterns, light exposure, and shift work. Daily sleep and work patterns may result in sleep loss and sleep deficiency, especially among persons who exhibit an evening chronotype. Short sleep duration during weekday or work days, with longer sleep duration on weekend days or nonwork days contribute to sleep loss and sleep deficiency. Metabolic consequences from sleep insufficiency may occur even with mild sleep deprivation [11]. Impaired glucose tolerance, decreased insulin sensitivity, insulin resistance, abnormal cortisol secretion, and altered lipid metabolism are observed with sleep loss [11].

Modern society provides a preponderance of exposure to light, with sunlight and natural light, artificial light within homes and other buildings, and artificial light from electronic devices. Light exposure affects circadian function with later melatonin onset, promoting sleep loss with fixed work or school schedules, and increased daytime sleepiness symptoms [11]. Light exposure combined with altered energy balance contributes to an increased prevalence of obesity [11]. Exposure to shift work, with entrainment to the night schedule, affects circadian function with sleep deprivation and poorer metabolic health, performance impairment, and impaired vigilance [11].

Sleep disorder

Insomnia and sleep apnea are the most common sleep disorders resulting in sleep loss. Insomnia or sleep difficulty may affect up to 15% of the US population [41]. Insomnia results in sleep loss from difficulty falling asleep, difficulty maintaining sleep, or awakening too early. Sleep loss related to insomnia can negatively affect psychological health with altered mood and cognition, diminished vigilance and performance, and lack of attention [41]. Physical manifestations of sleep loss from insomnia may include immune and metabolic activity, increased risk of glucose intolerance and developing diabetes, elevated blood pressure, and increased risk for developing cardiovascular disease [41]. Sleep loss in sleep apnea is caused by frequent episodes of apnea or hypopnea with sleep arousals [42]. Sleep apnea is associated with physical and neurobehavioral outcomes. Physical outcomes related to sleep apnea and sleep loss include obesity, hypertension, cardiovascular, and cerebrovascular health conditions [42]. Neurobehavioral outcomes related to sleep loss and sleep apnea include increased daytime sleepiness, cognitive function impairment, and risk for accidental injury [42].

Negative consequences for individuals and public health concern

Today's modern lifestyle can promote sleep insufficiency by many causes: cultural, social, psychological, behavioral, and environmental factors [43]. Societal demands of long working hours, including around-the-clock working hours, coupled with constant availability of merchandise and services, and global availability of communication can influence the opportunity for sleep. The association between chronic disease development and sleep insufficiency among all ages is concerning with respect to population health.

Chronic disease

Sleep insufficiency is associated with development of chronic disease, such as obesity, and hypertension, with resultant cardiovascular disease, diabetes, metabolic syndrome, and cancer in adults [34, 43–49], as well as children and adolescents [50–55]. Short sleep duration leads to increased levels of energy intake, lower levels of leptin, and higher levels of ghrelin, resulting in hunger, increased appetite and fat storage, leading to weight gain in adults and children [43]. Sleeping less than 5h [43], 6h [45] or 7h [47] per night was associated with an increased risk of obesity in adults. Among Puerto Rican youth, 10–19 years of age, sleeping less than 7–9h was associated with a threefold increase in the risk of obesity, compared with those sleeping more than this amount nightly [52]. Increased amounts of carbohydrate consumption and lowered fat consumption was associated with shorter sleep duration among a population of Mexican-American children 9–11 years of age at a high risk for obesity [54]. A systematic review [51] assessed the association of childhood obesity and sleep duration, measured by wrist or waist accelerometry, in children from birth to 19 years of age. A strong relationship was observed between short sleep duration, insulin resistance, and obesity. Insufficient sleep duration was observed to have a 45% increased risk of developing obesity in children in a systematic review and meta-analysis of cohort studies [53]. Cumulative sleep loss was associated with an increased risk of diabetes, elevated levels of fasting glucose, glucose intolerance, and insulin resistance, leading to metabolic syndrome in adults [43–45]. Videogame addiction among a sample of adolescents was associated with reduced sleep time, and negative health outcomes, including obesity and cardiovascular outcomes [56]. Sleeping less than 5h increased the risk of hypertension, and elevated triglycerides, leading to cardiovascular disease in cohort studies in women [44, 49]. Increased prevalence of coronary heart disease associated with short sleep was observed in the Nurses' Health Study [44] and the Women's Health Initiative [49]. In the Behavioral Risk Factor Surveillance System, a cross-sectional national study, shorter sleep duration increased the odds of obesity and cardiovascular events among adults [34].

Cognitive function

Sleep is vital for cognition, concentration, memory, and learning throughout the life cycle [27, 37, 57, 58]. Memory and cognition are affected by sleep loss, including tasks such as working memory, attention, speed and accuracy, and long-term memory consolidation [57]. Executive function for these cognitive tasks is impaired in individuals with chronic sleep restriction, i.e., chronically insufficient sleep duration, with decreases in accuracy on cognitive tasks and impaired decision-making [57]. Meta-analytic review of the literature revealed an association between school performance and sleep quantity, sleep quality, and daytime sleepiness among children and adolescents [27]. The strongest relationship among these variables was the association between daytime sleepiness and school performance, although insufficient sleep and poor sleep quality were also associated with poorer academic performance. Younger children fared worse on cognitive performance than adolescents, suggesting an influence of pubertal maturation. Meta-analysis on adolescent, i.e., middle school and high school students, and early college students, school performance and sleep revealed a relationship between these factors [37]. Shortened sleep duration, physiologic sleep timing, and sleep quality were negatively related to academic performance among adolescents. Epidemiologic studies demonstrate an association between sleep and cognitive decline in the older adult population. Poorer sleep quality and insomnia in adults over the age of 55 was associated with worse cognitive performance [58]. Altered cognitive performance among all age groups related to sleep loss negatively impacts normal daily functioning, such as school or work performance, driving, decision-making, and creativity.

THE ROLE OF MOBILE TECHNOLOGY IN SLEEP LOSS

Emergence of mobile technology

The origin of smart phones dates to the early 1990s with the advent of the IBM Simon® device, which was a large, cumbersome and costly mobile device (nearly $900 in 1993); Simon® provided mobile phone, personal data assistant, and fax services within the confines of the device [59]. From there, mobile technology evolved through technology devices such as Palm Pilot® for organizing data and Blackberry® to access e-mail and internet into more sophisticated devices with more features. However, the advent of the iPhone® in 2007 revolutionized mobile phone technology with release of a touchscreen device with easy accessibility of web browsing, gaming, music technology, photography, and applications, in addition to phone services and text messaging. Combined with ready availability of wireless services within the home setting, mobile phones became widely available to most households [59]. Mobile technology was further enhanced by tablet computing with the release of the iPad® in 2010 [60]. Tablets provided similar features of mobile phones, such as internet access, music technology, photographs, and applications, except for telephone and text message capability.

Portable gaming devices emerged in the 1980s with Nintendo's Game & Watch® and later Game Boy® as handheld portable devices that could be carried in a pocket [61]. Like the evolution of mobile phone technology, handheld gaming systems developed more advanced graphics, color, sleek design, touchpad, light-emitting diode (LED) screen, and ability to play with others via wireless communication among other features [61]. Nintendo Switch®, one of the latest in a stream of portable gaming systems, combines a game console with a handheld portable device. This system can be attached to a television screen or played as a handheld portable device, and is able to connect remotely for play with online gamers [62].

Although the idea for an electronic book reader dates back to the 1930s, it took 40 years until the first eBook emerged. Electronic book readers, eReaders, emerged in the late 1990s with release of Gemstar's Rocket eBook Reader [63]. Similar to other technologies, eReaders morphed from a basic device into a number of devices with touchscreen interface, color graphics, and enhanced features. Amazon's Kindle eBook reader debuted in 2007 [63], and currently, electronic books are available for download on a variety of electronic devices in addition to eReaders, such as laptop computers, tablet computers and smart phones.

Impact of sleep loss

Use of technology devices at night is common among all age groups, with 90% of Americans using technology within the hour prior to bedtime [1]. It is common practice for mobile phones to be taken to bed, especially among the adolescent and young adult populations [1]. Mobile media devices are used within the bedroom setting during the time just prior to bedtime, at bedtime, and after bedtime, including during the night. Adults may access e-mail or other technology features related to their employment prior to turning in for the night. Young adults and adolescents frequently access social media sites, send and/or receive text messaging, engage in video game play, or phone calls during this time. Mobile phones are often used as an alarm for morning awakening. Proposed mechanisms of the impact on sleep loss with mobile technology exposure include time displacement, artificial light exposure, and psychological stimulation or stress by media content (Fig. 13.2) [64].

Time displacement

A conventional day and night have a finite number of hours. Activities such as employment, academics, leisure, personal hygiene, and sleep occupy much of these hours. Time spent on one activity displaces time available for another activity. Current lifestyles can fill the available hours with many opportunities, often at the expense of sleep. Use of mobile technology devices in bed is convenient due to the handheld capability, illuminated screen, and multifunctional amenities. Engagement with technology devices within the bedroom setting may cause sleep deprivation, sleep restriction, and/or sleep fragmentation with both passive (sleeping near a mobile device which is powered on) or active use prior to and after bedtime. Mobile media devices left on in the bedroom allow visual and/or auditory alerts signaling incoming text messages, phone calls, and social media activity at any time of the day or night.

Sleep deprivation

Time spent on media activities on a mobile device limits the amount of time available for sleep, causing sleep deprivation. Engagement with mobile media devices in the prebedtime period delays bedtime in adolescents and adults. These activities may include late night use of a mobile phone [65–67], use of nighttime text messaging [68], engagement with social media [69], and excessive video gaming in adults [70] and adolescents [71].

Lanaj and colleagues [65] evaluated the use of smartphones for work purposes in the late night period, and the effects on workday outcomes. Increased engagement with smartphone devices late at night resulted in short sleep duration, increased morning depletion, and impaired work

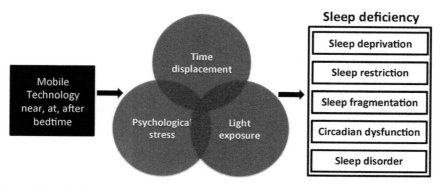

FIG. 13.2 Schematic of potential mechanisms by which mobile technology use near or during the time of sleep may lead to sleep deficiency. Time displacement, psychological stress or arousal, and exposure to light (particularly short-wavelength or *blue light*) may act separately or in combination to alter multiple types of sleep loss.

engagement the next day. A study of adults using mobile phones at night resulted in delayed sleep onset and sleep loss [67]. Younger adults were more likely to use mobile phones at night than older adults, and were more likely to have shorter sleep duration, and increased tiredness. Symptoms of depression, anxiety, and stress were associated with use of mobile phones at night [67]. Less ability to control use of a mobile phone at night was associated with later bedtime, poorer academic performance, and poorer sleep quality in a sample of undergraduate college students [66].

Among a sample of emerging young adults, the use of text messaging at night was predictive of sleep disturbance [68]. In this population of college students, increased frequency of text message use, awareness of nighttime smartphone notifications, and compulsivity to check the smartphone at night was predictive of reduced sleep quality by self-reported sleep diary and questionnaires. Use of social media among a nationally representative sample of young adults (19–32 years of age), observed an association between social media use and sleep disturbance [69]. Median social media usage was 61 min per day. The frequency of social media use had a greater effect on sleep disturbance than volume of social media use. In a study of videogaming in a sample of adults, each hour of videogame play was associated with later bedtime and wake time, longer time to fall asleep, and poorer sleep quality [70]. Perceptions of risk-taking (playing videogames at night) and consequences (bedtime) were studied in a sample of older adolescent high school and college students [71]. Those adolescents who perceived a lower risk of videogaming stayed up later playing videogames than those who perceived a higher risk of consequences with and without availability of a clock to monitor time [71].

Sleep restriction

Sleeping near a mobile technology device (passive use) is associated with delayed bedtime [21], and night awakenings [1, 72] among adolescents. Use of mobile technology prior to bedtime has been associated with shortened sleep duration, delayed bedtimes, and earlier wake up times among adolescents [24–26, 72–77]. University students [78] and adults 18–94 years of age [79] who engaged in mobile phone use after lights out demonstrated an association with insomnia, fatigue, poor sleep quality, and daytime tiredness, and had more symptoms with increasing phone use after lights out. Meta-analysis of studies exploring use of mobile media device use among children and adolescents, reported significant associations between inadequate sleep duration, sleep quality, and excessive daytime sleepiness [80].

Sleep fragmentation

Sleep fragmentation may be caused by mobile technology use at night by an individual's device or that of another household member, such as spouse, roommate, or child. Studies among adolescents [1, 21, 72, 74, 75, 81–83] and adults [67] using mobile phones at night report frequent night awakenings from mobile phone activity related to text messaging or incoming phone calls resulting from taking the phone to bed or accessing the phone during the night. In the National Sleep Foundation's 2011 Sleep in America poll, 18% of adolescents and 20% of young adults reported being awoken by phone ringers during the night several times per week [1]. Increased prevalence of insomnia complaints was associated with use of mobile technology at bedtime among German adolescents [84]. In an online survey of adults, up to 75% of adults reported being awakened by a mobile phone, either their own phone or that of another person, at least once a month [67]. Most participants reported text message rather than phone calling as the source of the disturbance. The severity of next day tiredness were dependent upon the time of awakening from mobile phones. Text messages sent or received immediately after lights out had less effect on sleep disturbance and next day tiredness [67].

Exposure to artificial light

Light is a potent biological and behavioral stimulator of alertness. In addition to the brightness of light, the wavelength in the short wavelength blue light range has more effect on alertness than longer wavelength light exposure [85]. Mobile technology devices frequently have an LED screen display which provides bright short wavelength blue light to the eyes [86]. Artificial light exposure can affect sleep by circadian disruption and sleep restriction. Considering the importance of exposure to artificial light at night, with the resultant impact on sleep health, the American Medical Association has issued recommendations for further research and interventions aimed at minimizing circadian disruption [87].

Circadian disruption

Circadian disruption by exposure to artificial light at night is reported with using mobile technology devices near bedtime in multiple studies of both adults [86, 88–91] and adolescents [92–95]. Melatonin hormone suppression and delayed circadian timing were observed in a group of healthy young adults after exposure to a 4-h period of reading from an eReader at maximum brightness prior to bedtime, compared to reading a print copy of a book over the same period in a crossover study [88]. Similarly, later melatonin release was observed by Heo and colleagues [86] in a group of adult men following 150 min of gaming on a conventional smartphone compared with a blue light-suppressed LED device. Green [89] observed an increase in melatonin suppression among young adults following exposure to a computer screen for 2 h.

Exposure to varying light conditions during the day and early evening produced effects on melatonin suppression. Use of blue-blocker glasses during the evening for a 1-week period while using LED screen devices followed by exposure to an unaltered LED screen prior to bedtime decreased melatonin suppression compared with clear lenses, but did not change sleep architecture among a group of male adolescents [94]. Rangtell and colleagues [91] observed that exposure to bright light for a period of 6.5 h prior to a 2-h exposure to a self-luminous tablet or print book prior to bedtime attenuated the melatonin suppression in adults.

Differences in melatonin suppression among studies suggest there may be a dose response to artificial light exposure and melatonin levels in the prebedtime period. Gronli [90] evaluated a 30-min exposure to reading from an iPad in bed in a group of young adults, compared to reading a print book. Among this group of participants, no changes in melatonin levels were noted between the groups. Among adolescent participants comparing exposure to bright unfiltered screenlight versus short wavelength filtered light via the *f.lux* application for reducing blue light emissions from a technology device for a 1-h period prior to bedtime demonstrated a small effect on prebedtime cognitive alertness, but did not show significant effects on sleep onset latency or sleep architecture on polysomnography [93]. A period of less than 2 h of screen exposure in the prebedtime period did not have an effect on circadian dysfunction.

Psychological stimulation and stress by media content

Video gaming is a popular pastime among children, adolescents, and young adults, and can be accessed freely from a mobile device. Playing videogames [96], engagement with social media [69, 97], as well as viewing internet content such as videos [69], in the prebedtime and after lights out periods can provide exposure to stress-producing media content in adolescents, young adults and adults, causing emotional arousal or psychological stress. While social media can provide a way to communicate with others from both near and far, constant exposure to the nuances of the personal details of the social network of friends and family life events can cause stress from awareness of the stressful events of others, comparison of life details, and fears of missing out on the activities of those within the social network [98]. These effects can have unintended consequences related to healthy sleep patterns, including sleep deprivation, sleep restriction, sleep fragmentation, and sleep disorders.

Video game playing in the prebedtime period has been studied primarily in adolescents and young adults [96, 99, 100]; however, more studies are now using adult videogaming participants [70]. Comparisons of violent and nonviolent video game play in the prebedtime period have been conducted in adolescents. King [100] compared duration of violent videogame play in experienced gamers between 50-min and 150-min sessions in adolescents prior to bedtime. Players in the 50-min session were not satisfied with the game time and desired to continue play. The 150-min session was associated with shorter sleep time, lower sleep efficiency, and mildly increased sleep latency. Violent and nonviolent videogaming was compared in either a short session of less than 1 h and a longer session of greater than 3 h in adolescents [99]. Violent videogaming produced increased stress at bedtime and poorer sleep quality in both time sessions, and poorer sleep quality was observed in the nonviolent session.

There is some evidence related to differences in type of media exposure and the effects on sleep outcomes. Weaver [96] compared a 50-min video game exposure period with a control period of watching a documentary on DVD to evaluate sleep outcomes in adolescent participants. Participants either played a violent first-person shooter video game, or viewed a leisurely paced nature documentary film to assess differences in sleep outcomes between type of media content. The video game exposure group displayed cognitive alertness without physiological alertness and longer sleep onset latency, whereas the control group (documentary film) did not experience either of these outcomes, and some participants fell asleep during the documentary. Smith [101] compared video game play with easy and hard playing conditions on two consecutive nights in the sleep lab. Participants played video games and self-selected bedtime after completion of play. Physiological arousal with increased heart rate was observed in the easy playing condition after 150 min of play. Later bedtimes were selected by all game players in both playing conditions. Exelmans and Van den Bulck [70] studied adult videogame participants to evaluate sleep outcomes of prebedtime gaming. A higher volume of video game play predicted sleep quality (longer sleep onset, poorer sleep efficiency, and increase use of sleep medication), fatigue, insomnia, and later bedtime and rise times. A 31% increase in risk of poor sleep was observed with a 1-h increase in video game time.

Mobile media devices are commonly used to access social media. Engagement with social media in the prebedtime period may cause emotional arousal in studies of adolescents and young adults [69, 97, 102]. In adolescents, nighttime use of social media led to poorer sleep quality, increased anxiety and depression, and lower self-esteem [102]. The concept of social media stress was examined in adolescents accessing social media in the prebedtime period [97]. Social media stress was caused by interactions with social media in females and was predictive of daytime sleepiness. In a study of young adults and bedtime social media usage, emotional, psychological, and physiological stimulation was observed [69]. The frequency of access to social media was more predictive of sleep disturbances than actual time spent on social media activities, suggesting there may be a directionality to the social media use and sleep disturbances [69].

Insomnia is the most commonly reported sleep disorder associated with excessive use of media at night in adults [78, 79, 103]. Case studies of young adult military personnel reported videogaming of 30–60 h per week, in addition to work responsibilities of 40 h per week, was associated with significant insomnia and daytime tiredness. Reduction in the hours of videogaming mitigated negative sleep symptoms [103]. Shortened sleep duration and increased daytime sleepiness were associated with more than 2 h of videogaming prior to bedtime in adolescents [104]. Viewing emotionally charged or disturbing content on social media platforms, such as Facebook and YouTube, in the bedroom setting resulted in emotional and psychological arousal, insomnia and short sleep duration in adults [69]. Social media stress related to engagement with social media near the bedtime period was associated with increased sleep onset and daytime sleepiness among female adolescents [97].

Mobile technology use as an epiphenomenon of insomnia

The reverse hypothesis of the use of nighttime mobile technology as an epiphenomenon of insomnia needs to be considered [24, 105]. The question remains as to whether individuals who experience insomnia use mobile technology during the night when not sleeping, or whether use of these devices result in the symptoms of insomnia. There is currently a lack of empirical studies to evaluate this phenomenon. Future research should address this potential issue.

CONCLUSIONS

Mobile technology is a modern convenience that provides instant communication with others, as well as access to the world, social network, and multiple applications, in a compact handheld portable device. Use of mobile devices is ubiquitous among individuals throughout the life spectrum. Use of mobile technology devices surrounding the sleep period can have negative health consequences, creating a public health crisis regarding healthy sleep among individuals of all ages. Correlational studies have identified negative relationships between sleep outcomes and mobile media consumption; however, most studies are cross-sectional and do not establish causality. More research is needed with longitudinal studies to assess causality, to identify predictors, moderators, and mediators of sleep and health outcomes related to mobile media exposure. Health education, especially in the most vulnerable age groups of children and adolescents, is needed to apprise the public of the detrimental effects of unchecked bedtime media exposure on sleep and health outcomes. Intervention studies to evaluate strategies to reduce effects of mobile media exposure are needed to expand the state of the science. Finally, advocacy for healthy mobile media use, especially among children and adolescents, is needed to guard the health of future generations.

REFERENCES

[1] Gradisar M, Wolfson AR, Harvey AG, Hale L, Rosenberg R, Czeisler CA. The sleep and technology use of Americans: findings from the National Sleep Foundation's 2011 Sleep in America poll. J Clin Sleep Med 2013;9(12):1291.

[2] Office of Disease Prevention and Health Promotion, Healthy People 2020. Sleep health objectives. [homepage on the internet]. Available from: https://www.healthypeople.gov/2020/topics-objectives/topic/sleep-health/objectives; 2018. [Updated 20 March 2018; Cited 1 April 2018].

[3] Hirshkowitz M, Whiton K, Albert SM, Alessi C, Bruni O, DonCarlos L, Hazen N, Herman J, Hillard PJ, Katz ES, Kheirandish-Gozal L. National Sleep Foundation's updated sleep duration recommendations. Sleep Health 2015;1(4):233–43.

[4] Kann L, McManus T, Harris WA, Shanklin SL, Flint KH, Queen B, Lowry R, Chyen D, Whittle L, Thornton J, Lim C, Bradford D, Yamakawa Y, Leon M, Brener N, Ethier KA. Youth risk behavior surveillance-United States, 2017. Morb Mortal Wkly Rep 2018;67(8):1–479.

[5] Centers for Disease Control and Prevention (CDC). Perceived insufficient rest or sleep among adults-United States, 2008. Morb Mortal Wkly Rep 2009;58(42):1175.

[6] Buxton OM, Chang AM, Spilsbury JC, Bos T, Emsellem H, Knutson KL. Sleep in the modern family: protective family routines for child and adolescent sleep. Sleep Health 2015;1(1):15–27.

[7] Calamaro CJ, Mason TB, Ratcliffe SJ. Adolescents living the 24/7 lifestyle: effects of caffeine and technology on sleep duration and daytime functioning. Pediatrics 2009;123(6):e1005–10.

[8] Pew Research Center. Mobile fact sheet. [homepage on the internet]. Available from: http://www.pewinternet.org/fact-sheet/mobile/; 2018. [Updated 5 February 2018; Cited 1 April 2018].

[9] (a) Lenhart A. Teen, social media and technology overview 2015. [homepage on the internet]. Available from Pew Research Center website: http://assets.pewresearch.org/wp-content/uploads/sites/14/2015/04/PI_TeensandTech_Update2015_0409151.pdf; 2015. [Updated April 2015; Cited 1 April 2018].(b) Poushter J. Smartphone ownership and internet usage continues to climb in emerging countries. [homepage on the internet]. Available from Pew Research Center website: http://assets.pewresearch.org/wp-content/uploads/sites/2/2016/02/pew_research_center_global_technology_report_final_february_22__2016.pdf; 2016. [Updated 22 February 2016; Cited 15 July 2018].

[10] Borbély AA. A two process model of sleep regulation. Hum Neurobiol 1982;1(3):195–204.

[11] Potter GD, Skene DJ, Arendt J, Cade JE, Grant PJ, Hardie LJ. Circadian rhythm and sleep disruption: causes, metabolic consequences, and countermeasures. Endocr Rev 2016;37(6):584–608.

[12] Moore RY, Speh JC, Patrick Card J. The retinohypothalamic tract originates from a distinct subset of retinal ganglion cells. J Comp Neurol 1995;352(3):351–66.

[13] Berson DM, Dunn FA, Takao M. Phototransduction by retinal ganglion cells that set the circadian clock. Science 2002;295(5557):1070–3.

[14] Hattar S, Liao HW, Takao M, Berson DM, Yau KW. Melanopsin-containing retinal ganglion cells: architecture, projections, and intrinsic photosensitivity. Science 2002;295(5557):1065–70.

[15] (a) Krauchi K, Cajochen C, Wirz-Justice A. A relationship between heat loss and sleepiness: effects of postural change and melatonin administration. J Appl Physiol 1997;83. (1):134–9. (b) Lockley SW, Brainard GC, Czeisler CA. High sensitivity of the human circadian melatonin rhythm to resetting by short wavelength light. J Clin Endocrinol Metab 2003;88(9):4502–5. (c) Gross H, Blechinger F, Achtner B. Human eye. Handbook of optical systems: volume 4: survey of optical instruments. vol. 420081–87. .(d)Hatori M, Panda S. The emerging roles of melanopsin in behavioral adaptation to light. Trends Mol Med. 2010;16(10):435–46.(e)Khalsa SB, Jewett ME, Cajochen C, Czeisler CA. A phase response curve to single bright light pulses in human subjects. J Physiol 2003;549(3):945–52.

[16] Grandner MA, Jackson NJ, Izci-Balserak B, Gallagher RA, Murray-Bachmann R, Williams NJ, Patel NP, Jean-Louis G. Social and behavioral determinants of perceived insufficient sleep. Front Neurol 2015;6:112.

[17] Hawkins SS, Takeuchi DT. Social determinants of inadequate sleep in US children and adolescents. Public Health 2016;138:119–26.

[18] Troxel WM, Hunter G, Scharf D. Say "GDNT": frequency of adolescent texting at night. Sleep Health 2015;1(4):300–3.

[19] Smith A, Anderson M. Social media use in 2018. [homepage on the internet]. Available from Pew Research Center website: http://www.pewinternet.org/2018/03/01/social-media-use-in-2018/; 2018. [Updated 1 March 2018; Cited 12 April 2018].

[20] Bruni O, Sette S, Fontanesi L, Baiocco R, Laghi F, Baumgartner E. Technology use and sleep quality in preadolescence and adolescence. J Clin Sleep Med 2015;11(12):1433.

[21] Falbe J, Davison KK, Franckle RL, Ganter C, Gortmaker SL, Smith L, Land T, Taveras EM. Sleep duration, restfulness, and screens in the sleep environment. Pediatrics 2015;135(2):e367–75.

[22] King DL, Delfabbro PH, Zwaans T, Kaptsis D. Sleep interference effects of pathological electronic media use during adolescence. Int J Ment Heal Addict 2014;12(1):21–35.

[23] Kubiszewski V, Fontaine R, Rusch E, Hazouard E. Association between electronic media use and sleep habits: an eight-day follow-up study. Int J Adolesc Youth 2014;19(3):395–407.

[24] Munezawa T, Kaneita Y, Osaki Y, Kanda H, Minowa M, Suzuki K, Higuchi S, Mori J, Yamamoto R, Ohida T. The association between use of mobile phones after lights out and sleep disturbances among Japanese adolescents: a nationwide cross-sectional survey. Sleep 2011;34(8):1013–20.

[25] Hysing M, Pallesen S, Stormark KM, Jakobsen R, Lundervold AJ, Sivertsen B. Sleep and use of electronic devices in adolescence: results from a large population-based study. BMJ Open 2015;5(1):e006748.

[26] Grover K, Pecor K, Malkowski M, Kang L, Machado S, Lulla R, Heisey D, Ming X. Effects of instant messaging on school performance in adolescents. J Child Neurol 2016;31(7):850–7.

[27] Dewald JF, Meijer AM, Oort FJ, Kerkhof GA, Bögels SM. The influence of sleep quality, sleep duration and sleepiness on school performance in children and adolescents: a meta-analytic review. Sleep Med Rev 2010;14(3):179–89.

[28] Eugene AR, Masiak J. The neuroprotective aspects of sleep. MEDtube Sci 2015;3(1):35–40.

[29] Jessen NA, Munk AS, Lundgaard I, Nedergaard M. The glymphatic system: a beginner's guide. Neurochem Res 2015;40(12):2583–99.

[30] Leproult R, Van Cauter E. Role of sleep and sleep loss in hormonal release and metabolism. In: Pediatric neuroendocrinology. vol. 17. Karger Publishers; 2010. p. 11–21.

[31] Lange T, Dimitrov S, Born J. Effects of sleep and circadian rhythm on the human immune system. Ann N Y Acad Sci 2010;1193(1):48–59.

[32] Oswald I. Sleep as a restorative process: human clues. In: McConnell PS, editor. Progress in brain research: adaptive capabilities of the nervous system. vol. 53. Elsevier; 1980. p. 279–88.

[33] Charles AC, Janet CZ, Joseph MR, Martin CM, Elliot DW. Timing of REM sleep is coupled to the circadian rhythm of body temperature in man. Sleep 1980;2(3):329–46.

[34] Grandner MA, Jackson NJ, Pak VM, Gehrman PR. Sleep disturbance is associated with cardiovascular and metabolic disorders. J Sleep Res 2012;21(4):427–33.

[35] Stickgold R, Walker MP. Sleep-dependent memory consolidation and reconsolidation. Sleep Med 2007;8(4):331–43.

[36] Dahl RE. The impact of inadequate sleep on children's daytime cognitive function. Semin Pediatr Neurol 1996;3(1):44–50.

[37] Wolfson AR, Carskadon MA. Understanding adolescent's sleep patterns and school performance: a critical appraisal. Sleep Med Rev 2003;7(6):491–506.

[38] Harrison Y, Horne JA. The impact of sleep deprivation on decision making: a review. J Exp Psychol Appl 2000;6(3):236.

[39] Luyster FS, Strollo PJ, Zee PC, Walsh JK. Sleep: a health imperative. Sleep 2012;35(6):727–34.

[40] Bonnet MH, Arand DL. Clinical effects of sleep fragmentation versus sleep deprivation. Sleep Med Rev 2003;7(4):297–310.

[41] Khan MS, Aouad R. The effects of insomnia and sleep loss on cardiovascular disease. Sleep Med Clin 2017;12(2):167–77.

[42] Young T, Peppard PE, Gottlieb DJ. Epidemiology of obstructive sleep apnea: a population health perspective. Am J Respir Crit Care Med 2002;165(9):1217–39.

[43] Cappuccio FP, Miller MA. Sleep and cardio-metabolic disease. Curr Cardiol Rep 2017;19(11):110.

[44] Ayas NT, White DP, Manson JE, Stampfer MJ, Speizer FE, Malhotra A, Hu FB. A prospective study of sleep duration and coronary heart disease in women. Arch Intern Med 2003;163(2):205–9.

[45] Deng HB, Tam T, Zee BC, Chung RY, Su X, Jin L, Chan TC, Chang LY, Yeoh EK, Lao XQ. Short sleep duration increases metabolic impact in healthy adults: a population-based cohort study. Sleep 2017;40(10):1–11.

[46] Im HJ, Baek SH, Chu MK, Yang KI, Kim WJ, Park SH, Thomas RJ, Yun CH. Association between weekend catch-up sleep and lower body mass: population-based study. Sleep 2017;40(7):1–8.

[47] Kim K, Shin D, Jung GU, Lee D, Park SM. Association between sleep duration, fat mass, lean mass and obesity in Korean adults: the fourth and fifth Korea National Health and Nutrition Examination Surveys. J Sleep Res 2017;26(4):453–60.

[48] McHill AW, Wright KP. Role of sleep and circadian disruption on energy expenditure and in metabolic predisposition to human obesity and metabolic disease. Obes Rev 2017;18(S1):15–24.

[49] Sands-Lincoln M, Loucks EB, Lu B, Carskadon MA, Sharkey K, Stefanick ML, Ockene J, Shah N, Hairston KG, Robinson JG, Limacher M. Sleep duration, insomnia, and coronary heart disease among postmenopausal women in the Women's Health Initiative. J Women's Health 2013;22(6):477–86.

[50] Dutil C, Chaput JP. Inadequate sleep as a contributor to type 2 diabetes in children and adolescents. Nutr Diabetes 2017;7(5):e266.

[51] Felső R, Lohner S, Hollódy K, Erhardt É, Molnár D. Relationship between sleep duration and childhood obesity: systematic review including the potential underlying mechanisms. Nutr Metab Cardiovasc Dis 2017;27(9):751–61.

[52] Koinis-Mitchell D, Rosario-Matos N, Ramírez RR, García P, Canino GJ, Ortega AN. Sleep, depressive/anxiety disorders, and obesity in Puerto Rican youth. J Clin Psychol Med Settings 2017;24(1):59–73.

[53] Li L, Zhang S, Huang Y, Chen K. Sleep duration and obesity in children: a systematic review and meta-analysis of prospective cohort studies. J Paediatr Child Health 2017;53(4):378–85.

[54] Martinez SM, Tschann JM, Butte NF, Gregorich SE, Penilla C, Flores E, Greenspan LC, Pasch LA, Deardorff J. Short sleep duration is associated with eating more carbohydrates and less dietary fat in Mexican American children. Sleep 2017;40(2):1–7.

[55] Navarro-Solera M, Carrasco-Luna J, Pin-Arboledas G, González-Carrascosa R, Soriano JM, Codoñer-Franch P. Short sleep duration is related to emerging cardiovascular risk factors in obese children. J Pediatr Gastroenterol Nutr 2015;61(5):571–6.

[56] Turel O, Romashkin A, Morrison KM. Health outcomes of information system use lifestyles among adolescents: videogame addiction, sleep curtailment and cardio-metabolic deficiencies. PLoS ONE 2016;11(5):e0154764.

[57] Alhola P, Polo-Kantola P. Sleep deprivation: impact on cognitive performance. Neuropsychiatr Dis Treat 2007;3(5):553–67.

[58] Yaffe K, Falvey CM, Hoang T. Connections between sleep and cognition in older adults. Lancet Neurol 2014;13(10):1017–28.

[59] Reed B. A brief history of smartphones. [homepage on the internet]. Available from: https://www.networkworld.com/article/2869645/network-security/a-brief-history-of-smartphones.html; 2010. [Updated 5 June 2010; Cited 1 April 2018].

[60] Arthur C. The history of smartphones: timeline. In: The guardian. 2012. Available from: https://www.theguardian.com/technology/2012/jan/24/smartphones-timeline. [Updated 24 January 2012; Cited 1 April 2018].

[61] Gamble R. The evolution of handheld video gaming in 17 consoles. [homepage on the internet]. Available from: http://www.denofgeek.com/us/games/video-games/269035/the-evolution-of-handheld-video-gaming-in-17-consoles; 2017. [Updated 16 November 2017; Cited 1 April 2018].

[62] Nintendo switch. n.d. [homepage on the internet]. Available from: https://www.nintendo.com/switch/features/ [Cited 1 April 2018].

[63] Bartram M. The history of eBooks from 1930s "Readies" to todays GPO eBook services. [homepage on the internet]. Available from: https://govbooktalk.gpo.gov/2014/03/10/the-history-of-ebooks-from-1930s-readies-to-todays-gpo-ebook-services/; 2014. [Updated 10 March 2014; Cited 1 April 2018].

[64] LeBourgeois MK, Hale L, Chang AM, Akacem LD, Montgomery-Downs HE, Buxton OM. Digital media and sleep in childhood and adolescence. Pediatrics 2017;140(Suppl. 2):S92–6.

[65] Lanaj K, Johnson RE, Barnes CM. Beginning the workday yet already depleted? Consequences of late-night smartphone use and sleep. Organ Behav Hum Decis Process 2014;124(1):11–23.

[66] Li J, Lepp A, Barkley JE. Locus of control and cell phone use: implications for sleep quality, academic performance, and subjective well-being. Comput Hum Behav 2015;52:450–7.

[67] Saling LL, Haire M. Are you awake? Mobile phone use after lights out. Comput Hum Behav 2016;64:932–7.

[68] Murdock KK, Horissian M, Crichlow-Ball C. Emerging adults' text message use and sleep characteristics: a multimethod, naturalistic study. Behav Sleep Med 2017;15(3):228–41.

[69] Levenson JC, Shensa A, Sidani JE, Colditz JB, Primack BA. The association between social media use and sleep disturbance among young adults. Prev Med 2016;85:36–41.

[70] Exelmans L, Van den Bulck J. Sleep quality is negatively related to video gaming volume in adults. J Sleep Res 2015;24(2):189–96.

[71] Reynolds CM, Gradisar M, Kar K, Perry A, Wolfe J, Short MA. Adolescents who perceive fewer consequences of risk-taking choose to switch off games later at night. Acta Paediatr 2015;104(5):e222–7.

[72] Arora T, Broglia E, Thomas GN, Taheri S. Associations between specific technologies and adolescent sleep quantity, sleep quality, and parasomnias. Sleep Med 2014;15(2):240–7.

[73] Gamble AL, D'Rozario AL, Bartlett DJ, Williams S, Bin YS, Grunstein RR, Marshall NS. Adolescent sleep patterns and nighttime technology use: results of the Australian Broadcasting Corporation's Big Sleep Survey. PLoS ONE 2014;9(11):e111700.

[74] Garmy P, Ward TM. Sleep habits and nighttime texting among adolescents. J Sch Nurs 2017.https://doi.org/10.1177/1059840517704964.

[75] Johansson AE, Petrisko MA, Chasens ER. Adolescent sleep and the impact of technology use before sleep on daytime function. J Pediatr Nurs 2016;31(5):498–504.

[76] Oshima N, Nishida A, Shimodera S, Tochigi M, Ando S, Yamasaki S, Okazaki Y, Sasaki T. The suicidal feelings, self-injury, and mobile phone use after lights out in adolescents. J Pediatr Psychol 2012;37(9):1023–30.

[77] Pieters D, De Valck E, Vandekerckhove M, Pirrera S, Wuyts J, Exadaktylos V, Haex B, Michiels N, Verbraecken J, Cluydts R. Effects of pre-sleep media use on sleep/wake patterns and daytime functioning among adolescents: the moderating role of parental control. Behav Sleep Med 2014;12(6):427–43.

[78] Zarghami M, Khalilian A, Setareh J, Salehpour G. The impact of using cell phones after light-out on sleep quality, headache, tiredness, and distractibility among students of a university in north of Iran. Iran J Psychiatry Behav Sci 2015;9(4).

[79] Exelmans L, Van den Bulck J. Bedtime mobile phone use and sleep in adults. Soc Sci Med 2016;148:93–101.

[80] Carter B, Rees P, Hale L, Bhattacharjee D, Paradkar MS. Association between portable screen-based media device access or use and sleep outcomes: a systematic review and meta-analysis. JAMA Pediatr 2016;170(12):1202–8.

[81] Adachi-Mejia AM, Edwards PM, Gilbert-Diamond D, Greenough GP, Olson AL. TXT me I'm only sleeping: adolescents with mobile phones in their bedroom. Fam Community Health 2014;37(4):252–7.

[82] Fobian AD, Avis K, Schwebel DC. The impact of media use on adolescent sleep efficiency. J Dev Behav Pediatr 2016;37(1):9.

[83] Schoeni A, Roser K, Röösli M. Symptoms and cognitive functions in Adolescents in relation to mobile phone use during night. PLoS ONE 2015;10(7):e0133528.

[84] Lange K, Cohrs S, Skarupke C, Görke M, Szagun B, Schlack R. Electronic media use and insomnia complaints in German adolescents: gender differences in use patterns and sleep problems. J Neural Transm 2017;124(1):79–87.

[85] Cajochen C. Alerting effects of light. Sleep Med Rev 2007;11(6):453–64.

[86] Heo JY, Kim K, Fava M, Mischoulon D, Papakostas GI, Kim MJ, Kim DJ, Chang KA, Oh Y, Yu BH, Jeon HJ. Effects of smartphone use with and without blue light at night in healthy adults: a randomized, double-blind, cross-over, placebo-controlled comparison. J Psychiatr Res 2017;87:61–70.

[87] Stevens RG, Brainard GC, Blask DE, Lockley SW, Motta ME. Adverse health effects of nighttime lighting: comments on American Medical Association policy statement. Am J Prev Med 2013;45(3):343–6.

[88] Chang AM, Aeschbach D, Duffy JF, Czeisler CA. Evening use of light-emitting eReaders negatively affects sleep, circadian timing, and next-morning alertness. Proc Natl Acad Sci 2015;112(4):1232–7.

[89] Green A, Cohen-Zion M, Haim A, Dagan Y. Comparing the response to acute and chronic exposure to short wavelength lighting emitted from computer screens. Chronobiol Int 2018;35(1):90–100.

[90] Grønli J, Byrkjedal IK, Bjorvatn B, Nødtvedt Ø, Hamre B, Pallesen S. Reading from an iPad or from a book in bed: the impact on human sleep. A randomized controlled crossover trial. Sleep Med 2016;21:86–92.

[91] Rångtell FH, Ekstrand E, Rapp L, Lagermalm A, Liethof L, Búcaro MO, Lingfors D, Broman JE, Schiöth HB, Benedict C. Two hours of evening reading on a self-luminous tablet vs. reading a physical book does not alter sleep after daytime bright light exposure. Sleep Med 2016;23:111–8.

[92] Figueiro M, Overington D. Self-luminous devices and melatonin suppression in adolescents. Light Res Technol 2016;48(8):966–75.

[93] Heath M, Sutherland C, Bartel K, Gradisar M, Williamson P, Lovato N, Micic G. Does one hour of bright or short-wavelength filtered tablet screenlight have a meaningful effect on adolescents' pre-bedtime alertness, sleep, and daytime functioning? Chronobiol Int 2014;31(4):496–505.

[94] van der Lely S, Frey S, Garbazza C, Wirz-Justice A, Jenni OG, Steiner R, Wolf S, Cajochen C, Bromundt V, Schmidt C. Blue blocker glasses as a countermeasure for alerting effects of evening light-emitting diode screen exposure in male teenagers. J Adolesc Health 2015;56(1):113–9.

[95] Wood B, Rea MS, Plitnick B, Figueiro MG. Light level and duration of exposure determine the impact of self-luminous tablets on melatonin suppression. Appl Ergon 2013;44(2):237–40.

[96] Weaver E, Gradisar M, Dohnt H, Lovato N, Douglas P. The effect of presleep video-game playing on adolescent sleep. J Clin Sleep Med 2010;6(2):184.

[97] van der Schuur WA, Baumgartner SE, Sumter SR. Social media use, social media stress, and sleep: examining cross-sectional and longitudinal relationships in adolescents. Health Commun 2018;1–8.

[98] Hampton K, Rainie L, Lu W, Shin I, Purcell K. Social media and the cost of caring. [homepage on the internet]. Available from Pew Research Center website: http://www.pewinternet.org/2015/01/15/psychological-stress-and-social-media-use-2/; 2015. [Updated 15 January 2015; Cited 12 April 2018].

[99] Ivarsson M, Anderson M, Åkerstedt T, Lindblad F. The effect of violent and nonviolent video games on heart rate variability, sleep, and emotions in adolescents with different violent gaming habits. Psychosom Med 2013;75(4):390–6.

[100] King DL, Gradisar M, Drummond A, Lovato N, Wessel J, Micic G, Douglas P, Delfabbro P. The impact of prolonged violent videogaming on adolescent sleep: an experimental study. J Sleep Res 2013;22(2):137–43.

[101] Smith LJ, King DL, Richardson C, Roane BM, Gradisar M. Mechanisms influencing older adolescents' bedtimes during videogaming: the roles of game difficulty and flow. Sleep Med 2017;39:70–6.

[102] Woods HC, Scott H. #Sleepyteens: social media use in adolescence is associated with poor sleep quality, anxiety, depression and low self-esteem. J Adolesc 2016;51:41–9.

[103] Eickhoff E, Yung K, Davis DL, Bishop F, Klam WP, Doan AP. Excessive video game use, sleep deprivation, and poor work performance among US Marines treated in a military mental health clinic: a case series. Mil Med 2015;180(7):e839–43.

[104] Brunetti VC, O'Loughlin EK, O'Loughlin J, Constantin E, Pigeon É. Screen and nonscreen sedentary behavior and sleep in adolescents. Sleep Health 2016;2(4):335–40.

[105] Fossum IN, Nordnes LT, Storemark SS, Bjorvatn B, Pallesen S. The association between use of electronic media in bed before going to sleep and insomnia symptoms, daytime sleepiness, morningness, and chronotype. Behav Sleep Med 2014;12(5):343–57.

Chapter 14

Models and theories of behavior change relevant to sleep health

Adam P. Knowlden
Department of Health Science, University of Alabama, Tuscaloosa, AL, United States

ABBREVIATIONS

HBM	health belief model
MGDB	model of goal-directed behavior
SCT	social cognitive theory
SNT	social network theory
TACT	target, action, context, and time
TPB	theory of planned behavior
TRA	theory of reasoned action
TTM	transtheoretical model

FOUNDATION OF THEORY FOR BEHAVIOR CHANGE

Conceptually, theory is grounded in the discipline of *epistemology*: the study of knowledge [1]. Theories can be conceptualized as *structural*, *functional*, and *dynamic* in nature [1]. From a *structural* perspective, theories comprise a set of abstract concepts, systematically organized into a logical and causal progression. *Functionally*, theories need to demonstrate that a specific cause, or set of causes, produces a specific effect(s). From the perspective of behavior change, a theory must demonstrate that a hypothesized set of constructs are linked to a specified behavioral outcome. Such a theory would elucidate the relationship between the set of variables, and the conditions under which these relationships do or do not occur [2]. The advancement of a theory is tied closely to its ability to demonstrate causation. Causation is the foundation of empiricism and in this regard, theories are *dynamic* entities which can be subjected to scientific scrutiny [1].

Given their dynamic nature, theories serve a functional role for predicting, interpreting, and explaining phenomena of interest. *Concepts* are the building blocks of a theory [3]. *Constructs* are concepts which are explicitly defined and causally ordered into a theoretical framework. To subject a theory to empirical testing, it's constructs must be operationalized into quantifiable, measurable variables. *Operationalization* is the process of tailoring and quantifying constructs from a theory or model to the purpose of the intervention or program [3]. Operationalized constructs with a range of possible quantitative outcomes are classified as *variables* [3]. Closely related to the concept of a theory, is that of a *model*. Although the terms "theory" and "model" are often used interchangeably, there are subtle differences between the two. *Models* are scaled representations of reality [4]. Depending on a model's comprehensiveness, it may contain multiple theories. For example, the social-ecological approach (covered later in this chapter), is comprised of multiple layers; each layer potentially including a variety of behavioral and/or environmental theories. Typically, models are conceptual and comprehensive in nature. While models may not be at the level of specificity of a theory, they must be rooted in rationality. In the field of public health, *logic models* are often used during the program planning stage to illustrate the processes applied during intervention implementation [5]. Like the term model and theory, the terms intervention and program, while often used interchangeably, have subtle distinctions between them. Behavior change *interventions* are systematically planned and organized activities carried out over time to accomplish specific behavior-related goals and objectives [6]. Alternatively, behavior change *programs* may consist of a set or series of behavior change interventions [6].

For example, a community-based, public health program designed to improve sleep behaviors may include a mass media campaign, a mobile phone sleep tracking application, and interactive telephone health counseling sessions. Combined, this set of interventions would be considered a behavior change program. During the planning stages of the program, a logic model may be conceptualized to show how available community resources will be utilized to maximize the practicality, robustness, and feasibility of the program. Furthermore, the specific intervention activities may be rooted in an evidence-based behavior change theory. To evaluate the program scientifically, the constructs of the behavior change theory could then be measured using a valid and reliable instrument at pretest and posttest. Assuming significant behavior change occurred, the intervention team could link the improvements in sleep behavior to the specific theoretical constructs targeted by the program. During this process, the team could determine which constructs were responsible for the most behavior change. If the program were replicated, those constructs responsible for the greatest change could be emphasized more, potentially leading to more robust programs in the future.

UTILITY OF THEORY FOR CHANGING HEALTH BEHAVIORS

Historically, health was contextualized as the relative absence or presence of physical ailment. However, in 1946, the World Health Organization (WHO) published a new, holistic definition of health which stated that "health is a state of complete physical, mental, and social well-being and not merely the absence of disease or infirmity" [7]. In 1986, WHO expanded upon this definition, stating that health is "a resource for everyday life, not the objective of living" [8]. From these definitions, several proprieties of health emerge [6]. *First*, health extends beyond the mere presence or absence of physical disease and includes quality of life factors such as an individual's social, emotional, spiritual, environmental, occupational, intellectual, and physical well-being. As such, sleep health can include both the increased risk of chronic disease associated with long-term sleep deprivation, and acute effects of sleep deprivation such as stymied emotional regulation. *Second*, the various dimensions of health overlap and influence one another to various degrees. Sleep for example, influences both physical and mental health outcomes [9]. *Third*, health cannot not be categorized as a static state of being, but instead is a dynamic condition of the human organism. For example, an individual may acquire consistent sleep quality on a regular basis; yet, they may occasionally experience periods of acute insomnia. *Fourth*, health should be considered a resource for living, rather than an end unto itself. In Western culture, sleep is often considered an expendable behavior. Individuals may believe they will experience more opportunities in their lives if they restrict their sleep, when in fact, research indicates quality of life is significantly improved when a consistent, healthy sleep pattern is adopted and maintained.

To improve health, it is necessary to change behaviors. *Behaviors* are overt actions which consist of a measurable frequency, intensity, and duration. In the context of sleep, sleep behavior may be defined as obtaining 7–9h of sleep (duration), every night (frequency), while experiencing all sleep stages (intensity). *Health behaviors* are actions undertaken by individuals to maintain, restore, or improve health. To apply theory for measurement purposes, the targeted health behavior must be operationally defined. The theory of planned behavior (described later in this chapter) for example, requires that any given behavior must be defined in terms of its relevant target, action, context, and time (TACT).

There are many potential benefits of incorporating theory into sleep health programs and interventions. *First*, theory can assist in specifying measurable program outcomes. Primarily, this occurs through the operationalization of theory, which will be detailed later in this chapter. Once operationalized, theory can be used to test scientific hypotheses or to evaluate program objectives. *Second*, theory can guide best practices for actualizing behavior change. For example, the theoretical construct of self-efficacy can be increased by: (1) breaking a complex behavior down into small steps; (2) reducing stress associated with behavior change; (3) applying verbal reinforcement; and (4) including role models into the intervention activities. *Third*, theory can assist in optimizing the timing of an intervention. For example, the transtheoretical model can categorize individuals into stages of readiness for behavior change. An individual in the contemplation stage of behavior change requires a different type of intervention relative to an individual in the action stage of behavior change. *Fourth*, theory can assist in selecting the right mix of strategies. Once operationalized, a theory can be modeled and the relative importance of the various constructs of a theory can be ascertained (e.g., beta weights in a regression model). Modeling of theory can inform which constructs should receive the most emphasis when implementing an intervention. This approach lends itself to reduced intervention dosage, efficient allocation of resources, and greater likelihood of behavior change. *Fifth*, theory can aid in replication. If the constructs of a theory are measured in valid and reliable ways, researchers can conduct systematic reviews and metaanalyses to help advance theory-based research. *Sixth*, theory can enhance communication among professionals. For example, a consensus on a definition of self-efficacy among sleep researchers can streamline research presentations and publications.

CAUSATION IN BEHAVIOR CHANGE THEORIES

Perhaps the foremost advantage of applying theory to behavior change interventions is theory's capacity to empirically examine causal relationships between constructs. Incorporation of theory into an intervention should be set up such that causation can be evaluated as part of the intervention. For example, a theory-based intervention for changing sleep behaviors, must be able to demonstrate the causal connection between the constructs of theory and any observed changes in the targeted sleep behavior. The capacity to ascertain causation primarily lies in the design of the intervention (e.g., randomized control trial design), random selection and assignment, as well as operationalization and measurement of the targeted theoretical constructs.

In 1965, Sir Austin Bradford Hill published, "The Environment and Disease: Association or Causation?" [10]. In his publication, Hill outlined nine criteria for evaluating the casual connection between variables. These criteria, known as *Hill's Criteria of Causation*, form the basis of modern epidemiological research. Hill's criteria were a consequence of his work in the British Doctor's Study; often considered the first definitive publication demonstrating a causal link between smoking and lung cancer [11]. Hill's criteria originate in the axioms of causation delineated by the philosopher, David Hume (1711–76) [1]. Hume argued that causality can never be proven. Rather, causation is ultimately a judgment-call based on the interpretation of the available evidence. In applying this principle, for example, the American Academy of Sleep Medicine may review all available evidence about a treatment option, and make judgment calls about causation. While Hill's criteria cannot prove causation, they can assist in ruling out explanations outside of a causal relationship. Incorporation of Hill's criteria into intervention design (e.g., randomized controlled trial design) maximizes the utility of theory for advancing evidence-based treatment approaches.

The first criterion of causation which Hill described is the *strength of association*. The strength of association between two or more variables describes the magnitude of their relationship. For example, when evaluating a data set, a researcher may find a strong association between short sleep and high levels of stress. This may provide a clue that a causal connection exists between these variables; however, when considering strength of association, it is important to keep in mind the often-cited mantra that "correlation does not equal causation." It is entirely possible for two variables to strongly associated yet have no causal connection. It is also possible that a confounding, mediating, or moderating variable is influencing the strength of the association. All three of these types of variables have the potential to produce biased results. *Confounding variables* are external influences that change the relationship between two variables. The greatest risk of confounding variables is failure to collect data on the confounder and thus not being able to explain the anomalous results. A *mediating variable* creates an indirect effect between the independent and dependent variables. When present, the path between the two hypothesized causally connected variables is partially explained by a third variable. A *moderating variable* creates an interaction effect. When present, two variables may share no relationship, unless in the presence of the moderating variable. Measurement of theory as part of a behavior change intervention requires careful consideration of these variable types. To avoid these issues, it is best to conduct a thorough operationalization process (discussed later in this chapter). Even if the behavior change program is limited in such a way that collecting data related to all possible confounding/mediating/moderating variables is not feasible, thorough formative research may assist in explaining those variables most likely to be the most important to measure, as well as provide a context for any anomalous results.

Hill's second criterion of causation is *consistency*. Consistency is built upon the scientific axiom of repeatability. In other words, if a causal relationship exists between variables, evidence of the relationship must be demonstrated consistently, even if tested by different researchers at different locations, and at different times. Importantly, and use of theory for changing sleep outcomes must have a thorough methods section so that other researchers can independently test the theory for gauging consistency. The third criterion of causation is *temporality*. Often considered the only necessary criterion of causation, demonstrating a causal requires the cause precede the effect. Typically, temporality is assessed using a pretest-posttest design whereby researchers can demonstrate changes in a theoretical construct leads to changes in the targeted behavior. The fourth criterion is the *dose-response relationship*. This criterion dictates that an increase in dose will lead to an increase in the response. Often, the dose-response relationship between a theorized cause and effect is illustrated using a dose-response curve. The fifth criterion is *coherence*. The hypothesized causal connection should be consistent with existing theory and knowledge. Similarity, the association should be plausible with currently accepted pathological processes (the sixth criterion). *Specificity*, the seventh criterion, is confirmed when a single cause produces a specific effect. In terms of behavior change, this is considered the weakest criterion as most behaviors are the consequence of multiple causes. The eighth criterion is *experimental evidence*. If it can be shown that a condition was prevented or ameliorated by an appropriate experimental treatment/intervention it lends evidence to a causal connection. In judging whether a reported association is causal, it is necessary to determine the extent to which researchers have taken *alternative explanations* (ninth criterion) into account and have effectively ruled out such alternate explanations (e.g., controlled for bias).

Hill's criteria serve to provide the foundation for the composition of a quality theory. First, theory must be causally *predictive*. Predictability entails that a given theory be testable, and therefore amenable to hypothesis testing. Second, theory must be *measurable*. Measurability allows researchers to gauge consistency and to determine if a given health outcome can be changed through the manipulation of a theory's constructs (experimental evidence). Furthermore, it can assist in determine the amount of change to a construct required to elicit meaningful change (dose-response). Third, a quality theory must be *generalizable*. Although health behavior theories can be tailored for specific uses and demographics, the core constructs of a theory should hold true for all populations (consistency). Fourth, a theory must be *practical*. From a health behavior prospective, practicality is important in two ways. The first is parsimony. The principle of parsimony states that an optimal theory will explain the greatest amount of variability, using the smallest number of constructs. Interventions seeking to change health behavior are inevitably constrained by time and resources. Although a theory comprised of a large ensemble of constructs may improve the theory's predictive capacity, such a theory may not be practical to implement from an intervention perspective. Often, a balance between what is ideal and what is practical must be identified. Secondly, practically requires that a theory's constructs be modifiable. For a theory to have utility, there must established methods of changing a theory's constructs (e.g., self-efficacy, discussed later in this chapter).

TYPES OF THEORIES

While there are numerous ways to categorize behavior change theories, perhaps the two most common are whether the theory is targeting a level of influence (socio-ecological approach) or whether the theory attempts to explain a behavior (continuum or stage theories) [5]. From the socio-ecological perspective, the health behavior of individuals is shaped in part by the social context in which they exist. Social-ecological approaches apply a systems perspective to model the upstream and downstream environmental and behavioral factors that influence a given behavior [12]. McLeroy, Bibeau, Steckler, and Glanz [13] have defined five socio-ecological levels of influence on behavior: (1) intrapersonal factors; (2) interpersonal factors; (3) institutional or organizational factors; (4) community factors; and (5) public policy factors.

In this context, the intrapersonal level of the socio-ecological approach primarily focuses on factors such as knowledge, emotions, beliefs, experiences, motivation, skills, and behavior. The interpersonal level examines a person's immediate social context and seeks to determine how social interactions influence health outcomes. Behavior change theories at the interpersonal level explain the influence of social environments an individual's behaviors. People's attitudes about health and their health behaviors do not exist in isolation; they effect, and are affected by, the attitudes and behaviors of others. The more proximal any two individuals are, the greater role this social context will have on their health behaviors. Furthermore, this influence will have a reciprocal effect on those individuals that make up their social environment. The organizational level considers how one's workplace environment impacts a given health outcome. For example, an individual may desire to change a health behavior, but if their employer does not provide access to health insurance it may prohibit their ability to actualize behavior change. The community level explores how settings, such as schools, workplaces, and neighborhoods, in which social relationships occur influence health outcomes. Finally, the policy level seeks to understand how policies and social climates influence health outcomes. Simons-Morton et al. [14] included two additional levels of socio-ecological influence on behavior: (1) the physical environment; and (2) culture [14]. The socio-ecological model elicits that behavior is shaped by multiple levels of influence. Therefore, a central tenant of the socio-ecological approach is that interventions must target multiple levels of influence to create sustainable changes in health behaviors. Fig. 14.1 illustrates a social-ecological model of sleep and health.

Theories can also be categorized by the approach they use to explain behavior. *Continuum theories* seek to identify variables that influence a behavior and quantify them to predict the likelihood of behavior change. *Stage theories* comprise an ordered set of categories by which individuals can be classified according to their progress in behavior change. Stage theories also identify factors that will induce movement from one category of behavior change to the next. Weinstein, Rothman, and Sutton have identified four characteristics of stage theories: (1) a categorical system to define the stages of change; (2) an ordering of the stages of change; (3) identification of the common barriers to change which may prevent people from transition to the next stage; and (4) identification of the different barriers to change facing people in different stages [15]. For the purposes of this chapter, the behavior change theories presented will be organized by the level of influence they are designed to target, and whether they can constitute a continuum or stage theory. While there are numerous models and theories of behavior change, the ones presented in this chapter are some the most frequently applied theories cited in the literature.

INTRAPERSONAL THEORIES

Health belief model

Continuum theory

The *health belief model* (HBM) is a value-expectancy framework. Value-expectancy theories attempt to explain how an individual's behaviors are influenced by the expectancy that an action will be followed by certain consequences. In this context, the HBM seeks to predict and explain the cognitions individuals employ when engaging in a specific health behavior. Developed in the 1950s to understand why individuals were not utilizing free, mobile, tuberculosis screening services [16], the HBM is considered one of the first theories of health behavior [17]. It hypothesizes that health-related behaviors depend on the simultaneous occurrence of three classes of factors: (1) sufficient motivation; (2) the belief that one is susceptible to the deleterious outcomes of a health problem; and (3) the belief that a certain behavior would be beneficial in reducing the perceived threat and that the action can be initiated at an acceptable cost.

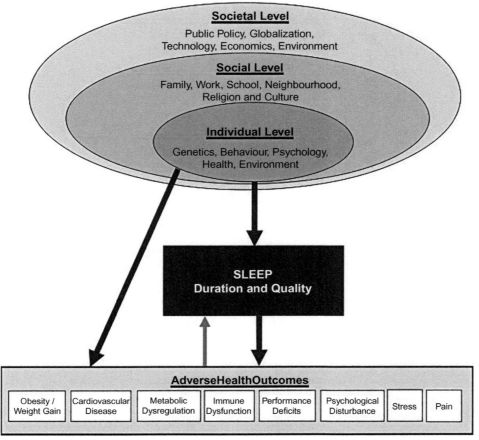

FIG. 14.1 Socio-ecological model of sleep and health. *(Adapted from Grandner MA. Addressing sleep disturbances: an opportunity to prevent cardiometabolic disease? Int Rev Psychiatry 2014;26(2):155–76. Reprinted by permission of Taylor and Francis.)*

The HBM was originally comprised of five theoretical constructs. *Perceived susceptibility* describes an individual's subjective risk to procurement of a negative health outcome. *Perceived severity* describes the subjective extent of harm incurred through engagement of a behavior. Often, perceived susceptibility and perceived severity and combined into the construct of *perceived threat*. *Perceived benefits* encompass the personal advantages of engaging in a given behavior. *Perceived barriers* include the subjective and objective costs of adopting a behavior. *Cues to action* include the cognitive triggers that motivate a given behavior. In recent years, the construct of self-efficacy has become central to the HBM. *Self-efficacy* is an individual's confidence to engage in a specific behavior and is the sixth construct of the HBM.

Application

The HBM may be a useful model for researchers seeking to predict the negative health outcomes from poor sleep. Knowlden et al. specified a HBM model to measure and predict the sleep behavior of employed college students [18].

This cross-sectional study found the HBM explained 34% of the variance in sleep behavior, with perceived severity, perceived barriers, cues to action, and self-efficacy identified as significant predictors. Fig. 14.2 illustrates the final HBM with standardized regression weights. In this study, self-efficacy was identified as the strongest predictor of adequate sleep. Fig. 14.2 illustrates the final model from this study.

Theory of reasoned action and the theory of planned behavior (continuum theory)

The *theory of reasoned action* (TRA) and the *theory of planned behavior* (TPB) [19] assist in predicting the complex nature of health behaviors [20]. As value-expectancy theories, the TRA and TPB assume that altering domain-specific beliefs can assist in modifying unhealthy behaviors. Both models posit that behavior is goal-oriented and results from defined cognitive processes that arise as individuals assess their environment [21]. The TRA and the TPB define *behavior* as an observable action delineated in terms

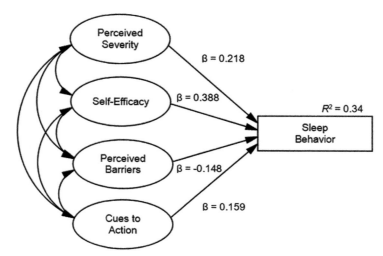

FIG. 14.2 Health belief structural model illustrating standardized regression weights for the sample of employed undergraduate college students (n = 188). *(Reprinted by permission of Wolters Kluwer Health, Inc.)*

of its target, action, context, and time [22]. According to these theories, *behavioral intention* is the most immediate antecedent of behavior. *Behavioral intention* is described as an individuals' readiness to perform a specific behavior [23]. Conceptually, the constructs of the TRA and the TPB are considered independent predictors of intention.

Theory of reasoned action

The constructs of the TRA predict volitional behaviors; that is, behaviors that are intentional and explicitly under the control of the individual. The TRA postulates that behavioral intention is the function of two factors: attitude towards the behavior and subjective norm. The construct of *attitude towards the behavior* refers to the overall feeling of like or dislike towards a given behavior. Attitude is shaped by *behavioral beliefs*, which reflect an individual's disposition toward performing a specific behavior, and *outcome evaluations*, which refers to the value an individual associates with engaging in a behavior. In addition to attitudes, the TRA hypothesizes that *subjective norms* are a salient determinant of behavioral intentions. The subjective norm construct is comprised of *normative beliefs* and *motivation to comply*. Normative beliefs are an individual's perception about how referent others would like them to act regarding an explicit behavior. Motivation to comply, in turn, is the degree to which an individual is willing to act in accordance with the referent group.

Theory of planned behavior

Although the TRA is a strong predictor of volitional behavior [24], it is unable to account for behaviors that are beyond the complete control of the individual. To compensate, Azjen [23] proposed the addition of the *perceived behavioral control* construct. Perceived behavioral control is divided into two components: *control beliefs* and *perceived power*. Control beliefs describe beliefs related to external and internal factors that can impede or promote the performance of a behavior. Perceived power is the perception of ease or difficulty an individual ascribes to a behavior. Additionally, Azjen [23] posited that to the extent to which perceived behavioral control was accurate, it could serve as a proximal measure of actual control and predict behavior. The addition of this construct led to the evolution of the TRA into its modern form of the TPB.

Application

The TRA and TPB may be useful models for researchers seeking to understand the behavioral intentions which underlie sleep behavior. Knowlden et al. [18] operationalized the TPB to predict the sleep intentions and behaviors of undergraduate college students. Their results found each of the three predictors regressed on behavioral intention were deemed significant: perceived behavioral control ($\beta = 0.457$, $P < .001$), subjective norm ($\beta = 0.179$, $P = .003$), and attitude towards the behavior ($\beta = 0.231$, $P < .001$). Collectively, the predictors produced an $R^2_{adjusted}$ value of 0.362 ($P < .001$), suggesting the model accounted for 36.2% of the variance in the behavioral intention to obtain adequate sleep in the sample of participants. Behavioral intention was identified as a significant predictor of sleep behavior ($P < .001$), explaining 18.5% of the variance of the participants' sleep behavior. To the best of the author's knowledge, this was the first study to operationalize a behavior change model for predicting sleep behaviors.

The model of goal-directed behavior

The model of goal-directed behavior (MGDB) is an extension of Ajzen's TPB [23, 25, 26]. Hypothetically, additional constructs can be included along with the primary TPB constructs to improve the TPB's predictive and behavior change capacity. The MGDB is one such theory which attempts to incorporate this feature of the TPB. The MGDB posits behavioral desires, a motivational state of mind in which reasons to act are formed, are a direct determinant of behavioral intentions. These behavioral desires mediate the effects of attitudes toward a behavior (evaluation of the behavior), perceived behavioral control (self-efficacy assessments), subjective norms (perceptions of social pressures), and anticipated emotions (prefactuals posited to influence desires to perform a behavior) on behavioral intentions. According to the MGDB, the standard predictors of TPB (attitudes, perceived behavioral control, and subjective norms) are not directly related to behavioral intentions, but indirectly through behavioral desires. The MGDB posits that anticipated emotions, both positive and negative, predict behavioral desires along with the standard TPB variables.

Application

The MGDB may be useful for researchers which seek to understand the role affect plays in explaining sleep behavior. Knowlden et al. [27] tested a MGDB-based theoretical framework for its capacity to measure and predict the sleep desires, intentions, and behavior of employed college students. Significant paths were identified between attitude toward the behavior, positive emotions, and negative emotions for behavioral desires ($R^2 = 0.654$). Direct paths were identified between perceived behavioral control and behavioral desires for behavioral intentions ($R^2 = 0.513$). Finally, direct paths were identified between perceived behavioral control and behavioral intentions for sleep behavior ($R^2 = 0.464$).

The transtheoretical model

Stage theory

The *transtheoretical model* (TTM) focuses on explaining how individuals and populations progress toward adoption and maintenance of health behavior change [28]. The TTM, sometimes called the *stages of change model*, was developed in the 1977 by James Prochaska and Carlo DiClemente after completing a comparative analysis of therapy systems as well as a critical review of therapy outcome studies. The core constructs of the TTM include the stages of change, the process of change, decisional balance, self-efficacy, and temptation. The stage construct is comprised of categories of change along a continuum of readiness to change a problematic behavior [29]. The first stage of change is *precontemplation*. During this stage the person is not considering changing their behavior [30]. The second stage of change is *contemplation* in which the person is considering change within the next 6 months. Following contemplation, a person enters the *preparation* stage in which the person is actively planning for change in the near future. During the action stage the person has made meaningful change leading into *maintenance* in which a person has maintained this change for an extended period. During the *termination* stage, the person has no temptation to revert to their prior behavior and has reached maximum self-efficacy. Aside from the maintenance stage, each stage is susceptible to relapse; that is, reverting to a previous stage of change.

The second primary construct of the TTM is *processes of change* [31]. These processes are the activities people utilize to progress through the stages of change. The 10 processes of change include: (1) consciousness raising which involves raising awareness about the causes, consequences, and cures for a particular problem; (2) dramatic relief which focuses on enhancing emotional arousal about the behavior and the relief that can come from changing it; (3) environmental re-evaluation which explains how the behavior impacts a persona's proximal social and physical environment and how changing it would benefit the environment; (4) self-evaluation which includes assessment of one's self-image if they employed the new behavior; (5) self-liberation which is a person's belief that they can change and making a commitment to act on the change; (6) counter-conditioning which entails learning a new healthy behavior to replace the old, unhealthy behaviors; (7) reinforcement management which includes applying reinforcements and punishments for taking steps toward behavior change; (8) stimulus control which requires modification of one's environment to increase cues for a healthy behavior and decrease cues for unhealthy behavior; (9) helping relationships which includes seeking relationships that will reinforce adherence to the new behavior; and (10) social liberation which includes realizing social norms are changing in the direction of supporting the healthy behavior change.

Research studies have found certain processes are more useful at specific stages of change. The processes of consciousness raising, dramatic relief, environmental re-evaluation are typically used to help individuals progress from the precontemplation to the contemplation stage. The process of self-evaluation can assist in progression from contemplation to preparation while the process of self-liberation can assist in progression from preparation to action. The final four processes—counter-conditioning, stimulus control, helping relationships, and reinforcement management—can help individuals progress from action to maintenance.

The construct of *decisional balance* includes the pros and cons associated with changing of the behavioral change [32]. Characteristically, the pros of healthy behavior are low during the initial stages of change and increase as the individual progresses through the stages of change. If the pros of the behavior change outweigh the cons, it is more likely the change in behavior will transpire. The fourth construct of the TTM is *self-efficacy* [33]. That is, the confidence to perform a specific behavior. Increase in self-efficacy is incremental until 100% self-efficacy is reached. The final construct is *temptation*, which is the converse of self-efficacy. Temptation reflects the urges to revert back to an unhealthy behavior when in a difficult situation such as being under emotional distress, social situations, and cravings [34]. Temptation decreases as the individual moves through the stages of change and progresses toward termination.

The TTM not only includes five core constructs but it is based on five critical assumptions. (1) No single theory can account for all the complexities of behavior change; (2) Behavior change is a process that progresses over time through a sequence of stages of change; (3) stages of change are both stable and open to change in the same way behavioral risk factors are stable and open to change; (4) Most at-risk populations are not prepared for action; therefore, traditional action-oriented behavior change programs. Determining a population's current stage of change is helpful for identifying the optimal intervention for the population; and (5) Specific processes should be emphasized at specific stages to optimize behavior change efficacy.

Application

The TTM can be useful to understand the current stage of change toward sleep health in which intervention participants currently reside. Based on such results, different intervention strategies can be applied. Using data from an online health risk assessment (HRA) survey completed by participants of the Kansas State employee wellness program, Hui and Grandner found poor sleep quality was associated with an increased likelihood of contemplation, preparation, and action [35]. However, the likelihood of maintenance of the healthy behavior was generally lower.

INTERPERSONAL THEORIES
Social cognitive theory
Continuum theory

Social cognitive theory (SCT) [33] operates on the premise of reciprocal determinism; a causal model that assumes behaviors are influenced by triadic reciprocal causation between personal (e.g., cognition, affect), behavioral (e.g., behavioral patterns, biological traits), and environmental factors (e.g., social dynamics) [36]. Several constructs influence these factors. The *reinforcement* construct can be direct, such as a physician that provides verbal feedback to a patient. Reinforcement can also occur vicariously, by observing the actions of others. Reinforcement can also be self-induced. Self-reinforcement would pertain to an individual that keeps records of their own behaviors and set up a system of rewards based on whether they accomplished their self-set behavioral objectives.

For an individual to undergo a behavior, they must understand the behavior and have the *behavioral capacity* to perform it. Closely aligned to behavioral capacity is the construct of *expectations*. *Outcome expectations* refer to an individual's perception of the likely outcomes that would ensue because of engaging in the behavior. *Outcome expectancies*, in turn, refer to the value a person places on the probable outcomes resulting from performing that behavior. The construct of *self-control* or *self-regulation* includes setting goals and developing plans to accomplish a behavior. Bandura identified six methods for achieving self-regulation: (1) self-monitoring of one's behavior; (2) setting both short and long-term goals; (3) obtaining feedback on the behavior and how it can be improved; (4) self-reinforcement or rewarding one's self; (5) self-instructing; and (6) and obtaining social support [37].

Likely, the most popular construct in all of behavior change theories is the staple of SCT; namely, the construct of self-efficacy. *Self-efficacy* is the confidence to perform a specific behavior. Self-efficacy includes the capacity to overcome barriers one may face in executing a specific behavior. Individuals become self-efficacious toward a behavior in four main ways: (1) personal mastery of a task, which can be achieved by breaking a complex task down into smaller steps; (2) vivacious learning, such as observing a credible individual carry out the behavior; (3) persuasion or re-assurance, even in the face of behavioral relapse; and (4) emotional arousal, such as reducing emotional stress that comes from adopting a new behavior.

Application

Due to its core construct of self-efficacy, SCT will perhaps have the most utility for researchers seeking to include skill-acquisition as a component of an intervention. Baron, Berg, Czajkowski, Smith, Gunn, and Jones [38] investigated individual differences in the daily associations between CPAP use and improvements in affect and sleepiness patients beginning CPAP. They found those with greater treatment self-efficacy and moderate outcome expectancies reported stronger daily benefits from CPAP. In terms of general sleep behavior, Knowlden, Robbins, and Grandner [39] tested the capacity of Bandura's social cognitive model of health behavior to account for variance in fruit and vegetable consumption, moderate physical activity, and sleep behavior in overweight and obese men. In this study, self-efficacy had the greatest total effect on sleep behavior ($\beta_{total}=0.406$). Self-efficacy also had a significant indirect ($\beta_{indirect}=0.194$) effect on sleep behavior through its influence on outcome expectations ($\beta_{direct}=0.265$), socio-structural factors ($\beta_{direct}=0.679$), and goals ($\beta_{direct}=0.700$). Fig. 14.3 illustrates the final SCT structural model of sleep behavior with standardized regression weights.

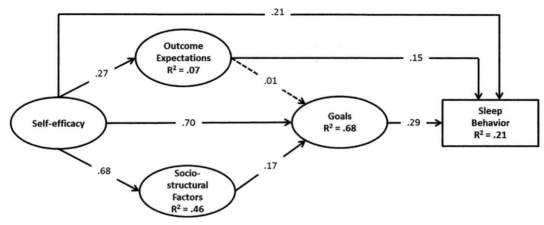

FIG. 14.3 Social cognitive theory model of sleep behavior ($N=303$). *Notes: Filled-in lines* represent significant direct pathways; *dotted lines* represent nonsignificant pathways ($P > .05$). Direct model pathways reported as standardized beta coefficients while squared multiple correlations are presented in bold text.

Social network theory

Continuum theory

Social network theory (SNT) refers to the web of social relationships that surround people and includes the structural characteristics of that web [40]. Social epidemiological observational studies have documented the positive effects that support networks have on health outcomes. However, due to the lack of intervention studies using SNT, some experts caution that social networks do not constitute a theory, but rather a set of concepts that describe the processes, functions, and structure(s) of social relationships [41]. In considering the influence of a network, the following must be assessed to understand the role of the network on a given health behavior: (1) centrality versus marginality (how much involvement does the person have on the network?); (2) reciprocity of relationships (are relationships one way or two way?); (3) complexity and intensity of relationships; (4) homogeneity or heterogeneity of those existing within the network; (5) subgroups within the network; and (6) communication patterns in the network (how is information transmitted?). While it is known that social networks impact health status, specifics of who in the network is impacted, and to what extent the impact influences health behavior outcomes, is unknown.

Application

While there are no known studies applying SNT to sleep, in the field of health promotion, social networks play an integral role in interpersonal-level interventions. Robbins et al. [42] implemented a sleep health, education intervention tailored to African American populations. The recruitment strategy for their intervention relied heavily upon social networks. For example, the team developed relationships with local community-and faith-based organizations as well as barber shops and churches. They also assembled a community steering committee, comprised of community stakeholders, patients, and health advocates to provide feedback on the content of the intervention. Researchers seeking to increase awareness about sleep health, recruit participants, or design interventions may wish to incorporate SNT.

COMMUNITY LEVEL THEORIES

Diffusion theory

Stage theory

Diffusion theory provides a framework to understand how new innovations, such as new ideas or behaviors, spread across various channels in a population or community [43]. In this context, there are three types of innovations. Incremental innovations represent a relatively small improvement over previous ideas or products. Distinctive innovations reflect significant improvement, but do not rely on any new technology or approach. Breakthrough innovations are based on a new approach or new technology. Similar to the transtheoretical model, diffusion theory hypothesizes that populations go through stages of change when assimilating innovations. Rogers has outlined the following five stages: (1) acquiring knowledge of the innovation; (2) persuasion or attitudes about the innovation; (3) decision about whether to adopt or reject the innovation; (4) implementation of the innovation; and (5) confirmation or ongoing use of the innovation.

Diffusion of an innovation in a population can be represented by the standard bell-curve. *Innovators*, represent approximately 2%–3% of a given population, are the individuals which are the first to adopt an innovation. Innovators are characterized as risk takers and adventurous. *Early adopters*, which represent approximately 14% of a given population, are those that accept an innovation early on. Early adopters are characterized as respected opinion leaders. *Early majority* are characterized as being ahead of the average and comprise approximately 34% of a given population. Late majority (34%) are skeptical and will wait until most people in a social system have adopted the innovation. *Laggards*, which represent about 16% of the population, are not interested in the innovation and will be the last to adopt the innovation, if they do at all. Because diffusions transpire over time, the rate of adoption can be plotted against time. In doing so the rate of adoption can be modeled as an S-shaped curve.

In terms of promoting a diffusion into a system, it is important to identify population characteristics. Homophily, or similarity among groups (e.g., culture), can speed the diffusion of an innovation. Additionally, the identification of physical and virtual social networks used by the population can be leveraged to hasten the speed in which an innovation is adopted. When working in communities, identifying change agents and opinion leaders is crucial. These individuals can be considered gatekeepers to innovations and their favorable perspective of an innovation can assist in its permeation through a social system. Diffusions can also be promoted using such strategies as *social marketing*. Social marketing is the application of commercial marketing techniques to promote beneficial behaviors. Social marketing methods such as audience segmentation can be used to develop messages targeted to specific groups.

Application

Although no current sleep health interventions have directly applied diffusion theory, sleep researchers are increasingly incorporating consumer sleep technology into their interventions. Mobile phones and wearable devices are have diffused into the marketplace at a rapid pace. In 2018, the American Academy of Sleep Medicine released a position statement regarding consumer sleep technologies [44]. In brief, they indicated that consumer sleep technology can be useful as a tool to bolster patient-clinician interaction when used as part of a clinical evaluation. They did note, however, that the general lack of validation, access to algorithms, and Food and Drug Administration oversight limits their current applicability.

Behavioral economics

Continuum theory

In recent years, the potential application of economic theory to behavior change has received greater attention. *Economics* is a social science concerned with description and analysis of the production, distribution, and consumption of goods and services. Standard economic theory describe how people should behave when acting in their own best interests. It considers individuals to be acting as purely rational agents. Under this model, if an individual fails to acquire sufficient duration of sleep, it is assumed decision is best for them and they have thought through all the costs and benefits of this action. However, behaviors are not completely rational actions. When engaging in behaviors, emotions, heuristics, present-tense biases, etc. factor into the decision. *Behavioral economics* operates under the assumption that individuals often act irrationally, but that they do so in predictable ways. Experts have suggested that behavioral economics can be applied to move individuals toward environments that can motivate healthier choices [45].

Behavioral economic approaches seek to overcome three key biases which often factor into the behavioral decision-making process [46]. The first is *present-tense bias*. Often, individuals overemphasize immediate benefits relative to long-term costs. For example, an individual may drive all night to arrive at a destination in a more timelier manner. While this may provide a more immediate convenience to the individual, their decision does not factor in the long-term costs of a potential accident due to drowsy driving. *Visceral factors* also tend to bias behavioral decision making. Cues or stimuli in the social or physical environment may override rational decision making. Visceral factors tend to be characterized by a direct hedonic impact (usually deleterious) and relative desirability of other related actions. A final bias which behavioral economics attempts to address is *status quo bias*. People tend to adhere to the current or default option. If a person has always voluntarily restricted their sleep throughout the week and engaged in oversleeping on the weekends, they are more likely to maintain the status quo, even if a healthier behavior, such as consistent a sleep/wake schedule is available. In considering these biases, behavioral economics suggests both unconscious and nonrational forces influence people's behaviors and may partly explain why simply informing individuals about, for example, their need for adequate sleep, does not always result in behavior change.

Application

Behavioral economics may assist in researchers seeking to influence the commonly held believe in Western cultures that sleep is an expendable resource. Although no current behavioral economic interventions have been implemented in the realm of sleep health, Malone, Ziporyn, and Buttenheim [47] suggested four behavioral economics strategies for individuals and communities seeking to address school start time policies. Their first recommendation is to offer later start times as the default option. They also recommend promoting social norms using success stories. To counter omission bias, the researchers recommend visually depicting deleterious outcomes related to early school start times such as poorer test scores and automobile accidents related to drowsy driving. Finally, they recommend increasing messaging salience. One method they recommend is to color code those districts with later start times in school publications and Web sites to better inform community stakeholders.

MEASUREMENT OF MODELS AND THEORIES FOR BEHAVIOR CHANGE INTERVENTIONS

Empirical evaluation of theory-based, behavior change programs requires measurement of the theoretical constructs targeted by the intervention. Often, researchers will state that an intervention is theory-based, yet they do not conduct measurement of the theoretical constructs used in the intervention. Failure to measure theoretical constructs makes it impossible to scientifically assess hypothesized causal connections between the applied theory and the intervention results. If, for example, operationalization, does not occur other researchers cannot apply the same methods to gauge consistency, an essential criterion for assessment of causation (see Hill's Criteria).

Application of theoretical frameworks for behavior change interventions involves three steps: (a) explicit operationalization of the theoretical constructs, (b) actualization of the theoretical constructs for intervention application, and (c) measurement of the theoretical constructs at pre- and postintervention to ascertain whether the theoretical constructs indeed improved. For example, if self-efficacy for obtaining 8h of sleep is targeted by an intervention. Self-efficacy should be measured both before and after the intervention to determine if self-efficacy increased. Instruments for measuring for theory-based interventions are developed through psychometric modeling. Validated theoretical models can provide detailed insight into the dynamics that underlie behavior change. The following section will provide a framework for incorporation of instruments for measuring theoretical constructs.

Step 1: Define purpose of instrument

The first step in developing an instrument (instrumentation) is to *define its purpose*. If the purpose is not explicitly defined there is a risk the instrument will not be relevant for the study. The purpose of the instrument should align with the study research questions and hypotheses.

Step 2: Identify objects of interest

An object of interest is any factor or determinant that is theoretically connected to the behavior. Objects of interest should be grounded in the literature (e.g., correlates, predictors, descriptive epidemiological studies). Sometimes, objects of interest are not well known or understood. In such a case, formative qualitative research may be required (e.g., grounded theory). The socio-ecological approach can be used to identify objects of interest at all levels of influence.

Step 3: Constitutively define objects of interest

For this step, the standardized, universal definitions of the object of interest should be considered. At this point, the research team may opt to use a theoretical framework to guide the objects of interest they will measure. For example, if the research team decides the TPB will best address their research questions, they will need to define each TPB construct. For instance, the universal definition of attitude toward the behavior from the TPB is "the degree to which performance of the behavior is positively or negatively valued." Once each object of interest is defined, the objects of interest must be prioritized in terms of their ability to evaluate the research questions and hypotheses. It is unlikely that each object of interest will be become part of the instrument. In the field of health education and promotion, for example, Golden et al. found fewer than 10% of interventions incorporated the socio-ecological approach as the foundation of their interventions [48]. Even if an intrapersonal level theoretical framework is used for the study, it is possible that not all constructs from the theory will be measured. For example, the SCT includes approximately 10 constructs. This is a relatively large number of construct. Based on time and resources, it may not be feasible to include the full set of constructs. As an aside, one reason the TPB is popular, is its parsimonious composition. However, even if the instrument will only seek to measure a select number of objects of interest, it still important to identify all objects. Doing so may help account unexplained variance in the theoretical model. For example, if the TPB is used to model sleep behavior and it is found the model only accounts for 30% variance in the outcome variable, the remaining 70% variance could be attributed to objects of interest which were not measured.

Step 4: Operationally define objects of interest

Operationalization requires two main features: (1) tailoring of objects for the need of the research and (2) assigning numerical values to the objects for measurement purposes. The following example demonstrates the translation of the universal definitions of sleep behavior and attitude toward the behavior into operational definitions of these objects.

- Universal Definition: A behavior is an overt human action, conscious or unconscious, with measurable frequency, duration, and intensity. Behavior from the vantage point of the TPB is defined in terms of the target, action, context, and time (TACT) principle.
- Operational Definition: For the purposes of this program, adequate sleep behavior is defined as adults receiving 7–9h of continuous sleep in a 24-h period. Applying the TACT principle, adequate sleep behavior is defined as

adults (target) achieving 7–9 h (time) of sleep (action) every 24 h (context). The sleep behavior variable will be assessed through self-report.
- Universal Definition: Attitude towards a behavior is the degree to which performance of the behavior is positively or negatively valued.
- Operational Definition: For the purpose of this study, attitude towards the behavior of obtaining adequate sleep is operationally defined as the individual's overall feeling of like or dislike toward obtaining adequate sleep behavior. A total of six items will be used to measure this construct. Seven-point semantic differential scale will be used to measure each item. A score of 6–42 is possible for this construct. The mean of the item scores will be calculated to provide an overall attitude score. A higher score is indicative of a more positive attitude towards the behavior.

Step 5: Review previously developed instruments

For example, if the researcher conducts a TPB-based sleep program for college students, they may want to use the instrument developed by Knowlden et al. [49]. In such a case, this previously developed instrument may be an optimal fit. If the research team opts to use a previously validated instrument for their intervention, they must carefully consider the original purpose of the instrument and the demographics that were sampled. Instruments developed with different demographics, literacy levels, online delivery versus paper-and-pencil administration, etc., may impact measurement accuracy. When adopting previously developed instruments, it also important to consider the psychometrics and limitations of the methods applied during the original instrumentation process. For example, if the instrument has weak internal consistency it may not introduce bias into the measurement process. If the data collected for instrumentation were from a nonprobability samples, the validity of the findings may not be generalizable (external validity).

Researchers can search research databases instruments that may fit the needs of their interventions. Popular research databases include: Medical Literature Analysis and Retrieval System Online (MEDLINE) & PubMed, Education Resources Information Center (ERIC), Cumulative Index to Nursing and Allied Health CINAHL, and Google Scholar. Health and Psychological Instruments (HaPI) is a specialized database that focuses on instruments in health-related, behavioral/social sciences, and organizational behavior disciplines. Inter-Nomological Network (INN) is a beta project designed to catalog variables and scales and may also be helpful. Some questions researchers may wish to address when considering whether to use a previously validated instrument include: (1) Was the instrument developed with similar demographics? (2) How did the researchers go about conducting the process of instrumentation? (3) Does the instrument have adequate psychometric properties (reliability/validity)?

Step 6: Develop an original instrument

If there is no previously developed instrument that meets the needs of the study, the researchers may need to develop an original instrument. It is important to note that if the research team opts to develop an original instrument, it is likely to add considerable time and resources to the research study. Therefore, it may be beneficial to use a previously developed instrument, even if it will not provide the highest degree of measurement precision. This will be largely determined by the resources available to the research team and the need for measurement precision.

Step 7: Select appropriate scales

For this step, consider the level of measurement (e.g., nominal, ordinal, interval, and/or ratio) used to evaluate the study hypotheses. Ratio level variables provide the most precision and allow for a wide-range of statistical tests. However, ratio levels variables are not always possible. Furthermore, higher levels of measurement are limited by the way in which the data are collected. For example, the precision of the ratio-level data can vary if self-report (indirect monitoring) is used as opposed to accelerometers (direct monitoring) collecting ratio level data on sleep; however. It is often good to remember that there is no perfect method of measurement. Researches operate within a range of confidence.

In terms of selecting scales, some theories provide guidance on the type of scale that should be applied. For example, the developer of the TPB, Icek Ajzen, recommends use of semantic differential scales for measure TPB constructs [23]. If no recommendation exists, the literature can also help inform which scale the researchers should use. For example, the research team may look at five different research papers describing development of questionnaires using SCT. Based on the psychometric properties of the papers or some other feature, such as similarity in the type of behavior, the researchers may opt for one type of scale over another. In considering the type of scale for questionnaires, it is often advisable to maintain, as much as possible, the same type of scale throughout the questionnaire. For example, if Likert-type scales are selected, strive to use Likert-type scales throughout, as opposed to integrating multiple types of scales. Multiple scales can increase questionnaire complexity and lead to participant bias.

Step 8: Develop items

One item should be developed corresponding to each property of a construct. Items must be clear and only tap one attribute of a construct. When a concept has been reduced to a variable, the scale of measurement chosen, and the items developed the instrument is also called a questionnaire. Paper-and-pencil questionnaires is the conventional method of delivery. Web-based questionnaires are becoming more popular. Each medium of questionnaire delivery possesses inherent strengths and limitations. Developing items is a tedious, iterative process. It is skill that requires time and experience to develop. A good first step is to examine items that have already been developed and tested. It might be possible to slightly alter such items. A common paradox for researchers is developing items that fully tap into a construct, while simultaneously seeking to minimize participant burden. If too many items are present, it can increase participant burden and introduce bias (e.g., acquiescence bias). One method for accomplishing this is separating "need to know" information from "nice to know" information. Every item on a questionnaire should answer something the researcher needs to know to test their hypotheses. If an item is not directly linked to the study hypotheses, it is likely "want to know" information.

Step 9: Prepare a draft instrument

Directions are important in any instrument. They should be optimum in length and easy to understand. The instrument should have clear guidance about scoring: how each item will be scored, how different subscales will be scored, whether there will be one scorer or more, what will be the range of scores, and what high and low scores mean (develop a code book). The instrument should have a clean, organized, professional layout. Developing an appealing hard copy or online questionnaire layout can take considerable time. A feature of online questionnaires is many interfaces are smart phone accessible, allowing participants to complete the questionnaire when it is convenient for them. Participants can also save their answers and return later.

Step 10: Test for readability

Health literacy is a topic of growing importance in public health. Health literacy is conceptualized as an individual's ability to obtain, process, and comprehend health information in order to make informed health decisions [50]. Readability metrics can serve as a starting point for gauging health literacy. Readability scores are based on the number of syllables in words and should be a beginning point for gauging health literacy. There are numerous readability metrics available. Two such examples are the Flesch Reading Ease and Flesch-Kincaid Grading tests. Reading ease between 60 and 70 is generally good. Grade level scores coincide with US Education Grade levels; e.g., a level of 5 indicates general compatibility with fifth-grade US education standards. If a questionnaire indicates a high level of literacy is required, a good first step is to simplify words with three or more syllables. Further refinement should occur through pilot testing.

Step 11: Send to panel of experts

A panel of experts is a group of individuals with sufficient expertise to gauge the properties of an instrument. A panel of experts is typically comprised of six jurors: two subject experts (including theory, if theory has been used), two experts in measurement and instrument development, and two experts of the priority population. Primarily, they will be involved in evaluating readability as well as face and content validity. Face validity contrasts operational definitions against universal definitions of the construct. Content validity subjectively assesses whether the items fully captures all the dimensions of the intended construct as operationally defined. Panel members should also evaluate the directions, layout, and readability of the questionnaire. Experts can be identified by emailing corresponding authors from research articles. The letter to panel members should be formal, professional, and ask for their assistance. Typically, a form for completing the evaluation of the instrument is supplied along with the questionnaire to ease juror burden. After receiving input from the first round with the panel of experts, changes are made and then in the second round once again the panelists are approached to check if suggestions have been sufficiently incorporated; add more rounds of review if necessary.

Step 12: Conduct a pilot test

This step can be conducted before or after forming the panel of experts. In this pilot test, the target population members are instructed to encircle any words they do not understand or any statements that are unclear. They are also timed to determine approximately how long it takes to complete the instrument in a practical setting. The pilot sample is also asked to provide any suggestions regarding improving readability of the instrument. If possible, the pilot test should be followed with a debriefing session in which participants provide verbal feedback.

Step 13: Establish reliability and validity

Establishing reliability and validity of an instrument requires proficiency in statistics and statistical software. The following section will provide a brief overview of reliability and validity but is by no means exhaustive. Let us consider the following example to help conceptualize reliability and validity. First, let us consider that we are interested in measuring body weight for a weight loss intervention. A couple of factors may go into to our methods for measuring body weight in a consistent way. First, we may want participants to weigh themselves on an empty stomach, first thing in the morning.

Second, we may want participants to place the scale on a flat surface. Third, we may want participants to use an electronic scale that is appropriately calibrated. We could say that these three properties of measuring body weight must be internally consistent to reliably capture bodyweight. This same idea is applied when using a collection of cohesive instrument items, or scales, to tap into a psychological construct. Internal consistency gauges how well the items gel together to tap into a construct. A common statistic to gauge internal consistency is Cronbach's alpha. Typically, values of 0.70 or higher are recommended [51].

Now, let us say that we are interested in testing the stability of our construct. We request that participants step onto the scale at the beginning of the trial. Let us assume, the pretest mean body weight of the sample is 135 lbs. At the end of the program, participants' weight is once again measured. The posttest mean bodyweight of the sample is 125 lbs. It appears the intervention worked! But we think back to many times in which we stepped onto a weighing scale. We stepped on and recorded our weight. We waited for it to re-calibrate and we stepped on it again, only to find the scale produced a weight that differed by a few pounds compared to the first time we stepped on. If this is the case with objective instrument like a scale, it is likely that intangible concepts like attitude are more prone to measurement error. Even if consistent procedures are applied it is possible there will be fluctuations, or measurement error, inherent to the instrument. The less objective the measurement tool, the more likely error will be present. For example, with body weighing scales, a cheaper bathroom scale will have less accuracy than a sophisticated electronic body weighing scale. If we do not attempt to determine if the fluctuation is reasonable, it is possible we could find changes from pretest to posttest that are due to random fluctuation as opposed to actual change attributed to the intervention. Stability reliability, then, is the degree of association between a measurement taken at one point in time against the same measurement taken at a second point in time. The method used to assess stability reliability is called the test-retest method. The statistic used to gauge the degree of association between the two measurements is called a correlation coefficient. Often, Pearson's *r* correlation coefficient is applied. Although, there are much more sophisticated metrics, such as intraclass correlations that can be applied. Typically, Pearson's *r* values of .70 or higher are considered adequate for gauging stability.

Validity would be, for example, the ability of the scale to measure fat loss as, opposed to some other feature such as water weight. There are several types of validity [51]. Construct validity is gauging whether a construct confirms to an a priori theoretical framework. Methods for construct validity are complex and require advanced training (typically doctoral statistics classes). Two types other types of validity include: concurrent validity and predictive validity.

Concurrent validity attempts to gauge the ability of a field-based instrument such as an accelerometer or self-report questionnaire to correlate to the current gold standard. For example, how valid are accelerometers for measuring sleep when compared to the gold standard of polysomnography. Predictive validity attempts to determine how well a set of constructs predicts an outcome of interest. For example, how accurate is the TPB for predicting sleep behavior intentions.

LIMITATIONS OF BEHAVIOR CHANGE THEORIES

Although models and theories can be a helpful tool in explaining and predicting health behavior change, they are not without their limitations. No single theory has been shown to be useful in all situations. Additionally, each theory presented in this chapter has its own inherent limitations. Continuum, value-expectancy theories tend to focus on cognitive factors but do not consider that behavior change occurs over time. Meanwhile, many theorists disagree with the way stage theories classify the process of behavior change. In addition to these limitations, behavior change theories are rarely operationalized and measured in intervention research. To improve the utility of behavior change models and theories, more interventions which can demonstrate a causal connection between theoretical constructs and behavior change are needed.

CONCLUSION

A range of behavior change models and theories are available to assist in the process of modifying sleep behaviors. Theories can assist in explaining why individuals do, or do not, engage in healthful sleep actions. Effective implementation of behavior change models and theories can reduce intervention costs, make more efficient use of time, and elicit greater behavioral outcomes. Models and theories can be classified in two general domains: (1) continuum theories, which seek to quantify theoretical constructs correlated to a behavior and (2) stage theories, which attempt to classify individuals according to their progress toward permanent behavior change. To advance the efficaciousness of behavior change models and theories for improving sleep health, researchers must operationalize the theoretical constructs they apply. Poor sleep impacts a large segment of the United States population. As such, an emphasis on theory-based research at the community-level is greatly needed.

REFERENCES

[1] Gopnik A, Meltzoff AN, Bryant P. Words, thoughts, and theories. Cambridge, MA: Mit Press; 1997.

[2] Nutbeam D, Harris E. Theory in a nutshell: a guide practitioner's guide to health promotion theory. Sydney: McGraw Hill; 1999.

[3] Glanz K, Rimer BK, Viswanath K. Health behavior and health education: theory, research, and practice. 4th ed. John Wiley & Sons; 2008.

[4] Chaplin J, Krawlec T. Systems and theories of psychology. 4th ed. New York: Holt, Rinehart & Winston; 1979.

[5] McKenzie J, Neiger B, Thackeray R. Planning, implementing & evaluating health promotion programs: a primer. 7th ed. San Francisco: Pearson Benjamin Cummings; 2017.

[6] Randall R, Girvan J, McKenzie J, Seabert D. Principles and foundations of health promotion and education. 6th ed. San Francisco: Pearson; 2015.

[7] World Health Organization. Constitution of the world health organization. Geneva: World Health Organization; 1948.

[8] World Health Organization. The Ottawa charter for health promotion. Geneva: World Health Organization; 1986. Available from: http://www.who.int/healthpromotion/conferences/previous/ottawa/en/.

[9] Bixler E. Sleep and society: an epidemiological perspective. Sleep Med 2009;10:S3–6.

[10] Hill AB. The environment and disease: association or causation? SAGE Publications; 1965.

[11] Doll R, Hill AB. The mortality of doctors in relation to their smoking habits. Br Med J 1954;1(4877):1451.

[12] Rimer BK, Glanz K. Theory at a glance: a guide for health promotion practice. 2nd ed. Washington, DC: National Cancer Institute; 2005. NIH Pub. No. 05-3896.

[13] McLeroy KR, Bibeau D, Steckler A, Glanz K. An ecological perspective on health promotion programs. Health Educ Q 1988;15(4):351–77.

[14] Simons-Morton B, McLeroy KR, Wendel ML. Behavior theory in health promotion practice and research. Jones & Bartlett Publishers; 2012.

[15] Weinstein ND, Rothman AJ, Sutton SR. Stage theories of health behavior: conceptual and methodological issues. Health Psychol 1998;17(3):290.

[16] Hochbaum GM. Public participation in medical screening programs: a socio-psychological study. (Public health service publication no. 572) Washington, DC: Government Printing Office; 1958.

[17] Rosenstock IM. Historical origins of the health belief model. In: Becker MH, editor. The health belief model and personal health behavior. Thorofare, NJ: Charles B. Slack; 1947. p. 1–8.

[18] Knowlden AP, Sharma M. Health belief structural equation model predicting sleep behavior of employed college students. Fam Community Health 2014;37(4):271–8.

[19] Ajzen I. From intentions to action: a theory of planned behavior. In: Huhl J, Beckman J, editors. Will; performance; control (psychology); motivation (psychology). Berlin and New York: Springer-Verlag; 1985. p. 11–39.

[20] Casper ES. The theory of planned behavior applied to continuing education for mental health professionals. Psychiatr Serv 2007;58(10):1324–9.

[21] Fishbein M, Ajzen I. Predicting and changing behavior: the reasoned action approach. New York: Psychology Press (Taylor & Francis); 2010.

[22] Sharma M, Romas JA. Theoretical foundations of health education and health promotion. Jones & Bartlett Publishers; 2008.

[23] Ajzen I. The theory of planned behavior. Organ Behav Hum Decis Process 1991;50(2):179–211.

[24] Madden TJ, Ellen PS, Ajzen I. A comparison of the theory of planned behavior and the theory of reasoned action. Personal Soc Psychol Bull 1992;18(1):3–9.

[25] Perugini M, Bagozzi RP. The role of desires and anticipated emotions in goal-directed behaviours: broadening and deepening the theory of planned behaviour. Br J Soc Psychol 2001;40(1):79–98.

[26] Perugini M, Conner M. Predicting and understanding behavioral volitions: the interplay between goals and behaviors. Eur J Soc Psychol 2000;30(5):705–31.

[27] Knowlden AP, Sharma M, Shewmake ME. Testing the model of goal-directed behavior for predicting sleep behaviors. Health Behav Policy Rev 2016;3(3):238–47.

[28] Prochaska JO. Systems of psychotherapy: a transtheoretical analysis. Homewood, IL: Dorsey Press; 1979.

[29] Prochaska JO. Changing at differing stages. In: Snyder CR, Ingram RE, editors. Handbook of psychological change: psychotherapy processes and practices for the 21st century. New York: Wiley; 2000. p. 109–27.

[30] Prochaska JO, DiClemente CC, Norcross JC. In search of how people change: application to addictive behaviors. Am Psychol 1992;47(9):1102–14.

[31] Prochaska JO, DiClemente CC. Transtheoretical therapy: toward a more integrative model of change. Psychother Theory Res Pract 1992;20:161–73.

[32] Janis IL, Mann L. Decision making: a psychological analysis of conflict, choice, and commitment. New York: Free Press; 1977.

[33] Bandura A. Social foundations of thought and action. Englewood Cliffs, NJ: Prentice-Hall; 1986.

[34] Prochaska JO. An eclectic and integrative approach: transtheoretical therapy. In: Gurman AS, Messer SB, editors. Essential psychotherapies: theory and practice. New York: Guilford Press; 1995. p. 403–40.

[35] Hui S-K, Grandner MA. Associations between poor sleep quality and stages of change of multiple health behaviors among participants of employee wellness program. Prev Med Rep 2015;2:292–9.

[36] Crosby RA, Salazar LF, DiClemente RJ. How theory informs health promotion and public health practice. In: DiClemente RJ, Salazar LF, Crosby RA, editors. Health behavior theory for public health: principles, foundations, and application. Burlington, MA: Jones & Bartlett; 2013. p. 163–85.

[37] Bandura A. Self-efficacy: toward a unifying theory of behavioral change. Psychol Rev 1977;84(2):191.

[38] Baron KG, Berg CA, Czajkowski LA, Smith TW, Gunn HE, Jones CR. Self-efficacy contributes to individual differences in subjective improvements using CPAP. Sleep Breath 2011;15(3):599–606.

[39] Knowlden AP, Robbins R, Grandner M. Social cognitive models of fruit and vegetable consumption, moderate physical activity, and sleep behavior in overweight and obese men. Health Behav Res 2018;1(2):5.

[40] Edberg M. Essentials of health behavior: social and behavioral theory in public health. Sudbury, MA: Jones & Bartlett; 2007.

[41] Heaney CA, Israel BA. Social networks and social support. In: Glanz K, Rimer BK, Viswanath K, editors. Health behavior and health education: theory, research, and practice. San Francisco, CA: Jossey-Bass; 2008. p. 189–210.

[42] Robbins R, Allegrante J, Rapoport D, Senathirajah Y, Rogers A, Williams N, Cohalll A, Butler M, Ogedegbe O, Jean-Louis G. Tailored approach to sleep health education (TASHE): preliminary results for a randomized controlled trial of a web-based educational tool to promote self-efficacy for OSA diagnosis and treatment among blacks. Sleep 2018;41(suppl_1):A212.

[43] Rogers EM. Diffusion of innovations. 5th ed. New York: Free Press; 2003.

[44] Khosla S, Deak MC, Gault D, Goldstein CA, Hwang D, Kwon Y, et al. Consumer sleep technology: an American Academy of Sleep Medicine position statement. J Clin Sleep Med 2018;14(05):877–80.

[45] Bickel W, Vuchinich R. Reframing health behavior change with behavioral economics. New York: Psychology Press; 2000.

[46] Loewenstein G. Out of control: visceral influences on behavior. Organ Behav Hum Decis Process 1996;65(3):272–92.

[47] Malone SK, Ziporyn T, Buttenheim AM. Applying behavioral insights to delay school start times. Sleep Health 2017;3(6):483–5.

[48] Golden SD, Earp JAL. Social ecological approaches to individuals and their contexts: twenty years of health education & behavior health promotion interventions. Health Educ Behav 2012;39(3):364–72.

[49] Knowlden AP, Sharma M, Bernard AL. A theory of planned behavior research model for predicting the sleep intentions and behaviors of undergraduate college students. J Prim Prev 2012;33(1):19–31.

[50] Parker R, Ratzan SC. Health literacy: a second decade of distinction for Americans. J Health Commun 2010;15(S2):20–33.

[51] Polit DF, Beck CT. Nursing research: principles and methods. Lippincott Williams & Wilkins; 2004.

Part IV

Sleep duration and cardiometabolic disease risk

Chapter 15

Insufficient sleep and obesity

Andrea M. Spaeth
Department of Kinesiology and Health, School of Arts and Sciences, Rutgers University, New Brunswick, NJ, United States

More than two-thirds of adults and one-third of children are considered to have overweight/obesity in the United States and the prevalence rate of obesity among adults has increased from 33.7% to 39.6% in just the past decade [1]. The high prevalence of obesity and its associated diseases, such as type 2 diabetes and cardiovascular disease, have prompted calls for the evaluation of innovative approaches to decrease obesity risk and promote healthy weight management. Emerging evidence suggests that sleep plays an important role in energy balance and metabolism and that enhancing sleep may represent a novel, modifiable behavior for weight regulation in children, adolescents and adults. Recent research has also begun to examine how sleep and circadian rhythm disorders affect energy balance regulation as well as how changes in weight status, diet and exercise impact sleep physiology.

SLEEP DURATION

Epidemiologic studies consistently demonstrate that habitual short sleep duration is a risk factor for obesity. Cross-sectional and longitudinal studies show that adults who report sleeping ≤5–6 h/day have more adiposity, larger waist circumferences, gain more weight over time and are more likely to have obesity [2–4]. In one study of nearly 14,000 adults, short sleepers (≤6 h/day) were 1.0 kg/m² heavier and had a 2.2 cm larger waistline than sufficient sleepers (7–9 h/day) [5]. Similarly, adults who slept <5 h a night exhibited a 16% higher prevalence of general obesity and a 9% higher prevalence of abdominal obesity compared to those who slept 7–8 h [6]. Consistent with these correlational findings, two in-laboratory experimental protocols demonstrated that healthy adults randomized to a sleep restriction condition gained more weight than those randomized to a sufficient sleep condition [7, 8].

The relationship between sleep duration and obesity has also been studied in pediatric populations. Short sleep duration during infancy predicts increased risk for obesity in preschool-aged children [9, 10] and cross-sectional studies have revealed an association between short sleep duration and obesity in preschoolers, school-aged children and teenagers [10–13]. A recent metaanalysis examining the longitudinal impact of sleep duration on weight status in children and adolescents found that short sleepers had twice the risk for becoming overweight/obese compared to sufficient sleepers [14] and, in a large cohort followed longitudinally, self-reported sleep problems in adolescence (mean age 13 years) predicted general obesity in young adulthood (mean age 20 years) [15]. Furthermore, in a counterbalanced crossover experiment, school-aged children exhibited greater weight gain when sleep was restricted by 1.5 h/night for 1 week compared to when sleep was prolonged by 1.5 h/night for 1 week [16]. Collectively, these studies provide strong support for the identification of habitual short sleep duration as a risk factor for obesity in children and adults.

Obesogenic behaviors

Modifiable behaviors that contribute to uncontrolled weight gain include (1) being sedentary/having low levels of physical activity, (2) overeating and consuming a poor diet, and (3) exhibiting a delayed meal timing pattern. Several epidemiological studies have demonstrated that sleep duration associates with physical activity levels, diet quality and caloric intake (Fig. 15.1). Children and adolescents who habitually obtain sufficient sleep exhibit more physical activity, greater fruit/vegetable intake, lower total caloric intake (kcal/day), reduced diet energy density, less added sugar intake and soda consumption [17–22]. Data from the UCLA Energetics Study using self-report measures of sleep duration and objective measures of energy intake (estimated based on doubly labeled water measurement of total energy expenditure) revealed that adults sleeping adults ≤6 h/night consumed approximately 50 more calories than those sleeping 7 h/night, 160 more calories than those sleeping 8 h/night and 440 more calories than those sleeping 9 h/night [23]. These differences in energy intake were also reflected

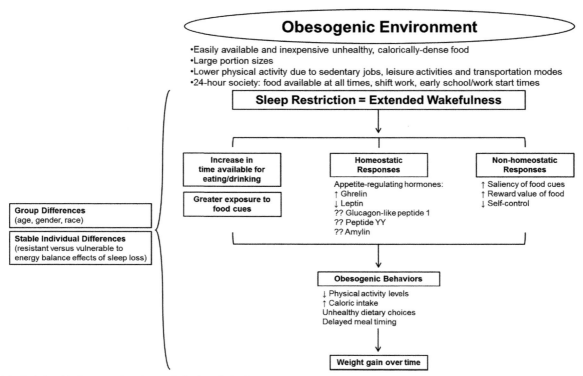

FIG. 15.1 Relationship between sleep loss and obesity within the context of our current environment.

in BMI and the prevalence of obesity (≤6h: BMI 28.0, 34% obese; 7h: BMI 25.0, 23%; 8h: BMI 23.7, 14% obese, 9h: BMI 24.6, 10% obese) [23].

In children [24] and adults [25] participating in experimental protocols, physical activity was lower during days following sleep restriction compared to days following sufficient sleep and this overall decrease is due to less time spent in moderate-vigorous activity. Laboratory studies in adults have also demonstrated that sleep restriction leads to increased caloric intake, more food purchases, greater consumption of snacks, increased portion sizes, and increased impulsivity in response to food cues [7, 8, 26–31]. For example, when given ad libitum access to food in a laboratory setting, participants consumed nearly 500 additional calories in one study [7] and nearly 300 additional calories in another study [28] during days following sleep restriction compared to days following sufficient sleep. In both studies, participants also consumed calories more frequently when sleep restricted suggesting increased snacking behavior.

Regarding macronutrient intake, some studies have shown that sleep restriction leads to increased craving and consumption of carbohydrates [27, 32], others have observed greater consumption of fats [28, 33, 34]. Normal-to-overweight adolescents assigned 6.5h TIB/night for 5 nights consumed foods with a higher glycemic index and more desserts/sweets than when assigned 10h TIB/night for 5 nights [35]. Similarly, during an in-laboratory study, during days following sleep restriction (compared to days following sufficient sleep), adults consumed significantly more calories from these categories: (1) grains and pasta; (2) condiments; (3) desserts; (4) salty snacks and (5) caffeine-free soda and juice but did not consume more calories from healthier food categories (meat, eggs, and fish; fruit, vegetables, and salad; milk) [33]. Therefore, although macronutrient intake findings are mixed, in general it seems that sleep restriction is associated with greater intake of unhealthy foods [36].

Delayed meal timing has also been identified as important for weight regulation. Women who are were dieting and consumed the majority of calories earlier in the day lost more weight than women who consumed the majority of calories later in the day, even when total daily caloric intake was held constant [37, 38]. In a cross-sectional cohort of healthy young adults (18–22 years), delayed circadian timing of food intake (calculated relative to each participant's melatonin onset) was significantly associated with body fat percentage and body mass [39].

Recent findings suggest that sleep duration may impact meal timing. When bedtime was delayed during an in-laboratory sleep restriction protocol (5 nights of 4h TIB/night), adults consumed approximately 500 additional calories during the late-night hours when they were kept awake instead of going to bed (10:00 p.m.–04:00 a.m.) and then consumed approximately 100 fewer calories the following morning (08:00 a.m.–03:00 p.m.) [7, 33]. This led to a shift such that participants consumed the majority of daily calories before 03:00 p.m. during sufficient sleep and consumed the majority of daily calories after 03:00 p.m. during sleep restriction. Furthermore, the percentage of daily calories consumed during the late-night period (10:00 p.m.–04:00 a.m.) positively associated with weight gained during the study [40]. Calories consumed during the late-night period were also higher in fat compared to calories consumed during the two earlier time periods [7]. Evening fat intake may be particularly contributory to weight gain; Baron and colleagues found that the percentage of fat consumed after 10:00 p.m. associated with greater total caloric intake and a higher BMI among adults [41].

Thus, habitual short sleep duration may lead to uncontrolled weight gain by decreasing physical activity, increasing energy intake and promoting a low-quality diet, and by shifting the timing of caloric intake. Findings suggest that behavioral interventions that simultaneously target sleep and these obesogenic behaviors may be particularly effective in promoting healthy weight management.

Potential physiological mechanisms

Laboratory studies have begun to examine possible physiological mechanisms—such as appetite-regulating hormones and changes in brain activation—underlying the relationship between sleep duration and weight gain (Fig. 15.1). Leptin, an anorexigenic hormone released from adipocytes, and ghrelin, an orexigenic hormone released from the stomach, have been the most studied appetite-regulating hormones in regards to sleep duration. Two large cross-sectional studies reported that short sleepers (5 h/night) exhibit lower leptin and higher ghrelin levels than normal sleepers (8 h/night) [42, 43]. Spiegel and colleagues found that men, undergoing 2 nights of sleep restriction (4 h TIB/night) with controlled energy intake via an intravenous glucose infusion, exhibited increased levels of ghrelin (+28%) and decreased levels of leptin (−18%) and these neuroendocrine changes were accompanied by significant increases in self-reported ratings of hunger and appetite [32]. A recent study [44] collected blood samples from healthy men at 15 to 30 min intervals for 24 h on the third consecutive night of either sufficient sleep (8.5 h TIB/night) or sleep restriction (4.5 h TIB/night). Caloric intake was strictly controlled during blood sampling but was then ad libitum on the day following the fourth consecutive night of each sleep condition. Sleep restriction led to significantly higher mean 24 h ghrelin levels (ghrelin was elevated during the nocturnal period) and postprandial ghrelin levels (meal-related peaks were higher and postmeal nadirs were attenuated) compared to the sufficient sleep condition. Furthermore, peak ghrelin levels at the dinner meal predicted calories consumed from sweet snacks [44]. However, leptin and pancreatic polypeptide levels were similar across conditions [44].

Other laboratories have either not observed changes in ghrelin or leptin, or have observed increases in leptin levels [45]. Glucagon-like peptide 1 and peptide YY have also been studied in relation to changes in sleep duration in a few studies; however results have been mixed [46]. One reason for the heterogeneity in results may be gender differences—St-Onge and colleagues [46, 47] found that sleep restriction led to increased ghrelin levels in men but not in women. Another reason may be differences in protocol procedures, with some studies allowing participants ad libitum food/drink intake and others implementing dietary control. Given that most studies demonstrate an increase in the frequency of eating during sleep restriction, future studies should examine postprandial signals, such as amylin, which influence the interval length to the next meal.

In addition to these homeostatic, appetite-regulating hormones, it may be particularly important to also focus on nonhomeostatic processes (reward and limbic systems) given that laboratory studies have observed overeating during sleep loss. The amount of additional calories that adults consume when sleep restricted (~300–500 kcal) exceeds the amount of additional energy they need to maintain prolonged wakefulness (~100–150 kcal) and these calories are derived from unhealthy, palatable, more rewarding foods [7, 8, 28, 48]. In today's obesogenic environment, there is an abundance of food cues and energy dense foods are easily available at any time of day. Palatability and pleasantness are as equally or more powerful determinants of food intake than hunger or homeostatic drive.

Neuroimaging studies examining the effect of sleep loss on brain activation have demonstrated that participants undergoing sleep restriction display greater overall neuronal activity in response to food stimuli, particularly in areas related to reward including the putamen, nucleus accumbens, thalamus, insula and prefrontal cortex compared to sufficient sleep [49, 50]. Participants also display greater activation in the insular cortex, orbitofrontal cortex and dorsolateral prefrontal cortex in response to images of healthy versus unhealthy foods during sleep restriction [50, 51]. Increased activation has also been observed in regions associated with sensory/motor signaling (right paracentral lobule), inhibitory control (right inferior frontal gyrus) and reward coding and decision-making (ventral medial prefrontal cortex) [49].

Compared with normal sleep, participants undergoing total sleep deprivation exhibited increased activation in the right anterior cingulate cortex (ACC) in response to food images, and the change in activation correlated with postscan appetite ratings for pictures of high calorie foods [52]. Conversely, in another study involving total sleep deprivation, participants displayed decreased activation in higher-order cortical evaluation regions (i.e., ACC, left lateral orbitofrontal cortex, and anterior insula) and enhanced activation in the amygdala while rating their desirability for various foods displayed in pictures [53]. Thus, in parallel to the homeostatic metabolic pathways, sleep restriction impacts both reward and control processing in response to food cues.

Maintaining a healthy weight in our obesogenic environment requires self-control and the maintenance of healthy habits. For example, eating egg whites instead of pastry for breakfast, packing a lunch instead of eating fast-food, or taking a walk instead of drinking a glass of wine to destress after work. Sleep loss impairs decision-making and self-control and is associated with decreases in activity within the prefrontal cortex and thalamus—two areas that play critical roles in exerting self-control [54]. Therefore, habitually restricting sleep likely contributes to the accumulation of unhealthy choices that can lead to obesity. More research is needed to directly examine the relationship between sleep, self-control, decision-making and weight management.

Two recent studies have used resting state functional connectivity to assess changes in network activity in response to sleep. The first focused on salience network connectivity following a night of total sleep deprivation relative to sufficient sleep [34]. In this protocol caloric intake was ad libitum and participants were healthy adults. During total sleep deprivation, participants consumed nearly 1000 cal during the overnight period of wakefulness. Despite consuming these additional calories overnight, participants consumed a similar number of calories during the day following total sleep deprivation compared to the day following sufficient sleep. In addition, participants consumed a greater percentage of calories from fat and a smaller percentage of calories from carbohydrates during the day following total sleep deprivation compared to the day following sufficient sleep. At the neural level, one night of total sleep deprivation enhanced dorsal ACC functional connectivity with bilateral putamen and bilateral insula, which are core regions of the salience network. Moreover, dorsal ACC connectivity with these two regions positively correlated with the percentage of calories consumed from fat and negatively correlated with the percentage of calories consumed from carbohydrates after total sleep deprivation. Thus, total sleep deprivation altered salience network functional connectivity and the increased co-activation of dorsal ACC and insula as well as dorsal ACC and putamen predicted subsequent macronutrient intake [34]. At any given moment, our sensory systems receive multiple sources of stimuli; the salience network selects which of these stimuli are relevant and deserving of our attention. Thus, sleep deprivation may alter food choices by affecting how much attention is focused on food stimuli.

The second resting-state study, a pilot investigation, focused on reward and interoception-related brain circuitry following three nights of normal (Midnight to 08:00 a.m.) or late (03:30–11:30 a.m.) sleep timing with either normal or late meal timing [55]. Resting state functional connectivity between insula and somatosensory cortex, postcentral gyrus and precuneus as well as between central opercular cortex and somatosensory cortex and pre-/postcentral gyrus was stronger during the late sleep schedule compared to the normal sleep schedule. As neuroimaging technology continues to advance, these studies will be critical in our understanding of the relationship between habitual short sleep duration and increased risk for obesity.

Group differences

Demographic characteristics, such as age, gender and race, may influence the relationship between sleep duration and obesity (Fig. 15.1). Although the significant association between sleep duration and risk for obesity has been observed in nearly every age range (see summary above), there is evidence that the association between short sleep duration and body mass index (BMI) may be stronger among young (18–29 years) and middle-aged (30–64 years) adults than among older (>65 years) adults [56, 57]. A longitudinal study examined 5-year change in computed tomography (CT)-derived visceral adipose tissue and subcutaneous adipose tissue in African Americans and Hispanic Americans (IRAS Family Study) [58]. Sleep duration (assessed by questionnaire and categorized as ≤5h, 6–7h, and ≥8h) at baseline interacted with age to predict change in fat measures. Individuals 18–40 years who slept ≤5h exhibited greater accumulation of visceral adipose tissue ($13\,cm^2$) and subcutaneous adipose tissue ($42\,cm^2$) compared to those sleeping 6–7h. However, there was no significant association between sleep duration and fat depot change in participants older than 40 years old.

Findings related to gender differences have been mixed, with some population studies observing a stronger association between sleep duration and BMI in women [59], and others demonstrating a stronger association in men [60, 61]. In a prospective cohort study, short sleep duration was associated with weight gain and the development of obesity at 1-year follow-up in men but not in women [62]. During an in-laboratory protocol, sleep restricted men exhibited a greater increase in daily caloric intake, consumed a greater percentage of daily calories during late-night hours (10:00 p.m.–04:00 a.m.) and gained more weight than women [7, 33]. Several studies in adolescents and children have also observed stronger effects in males compared to females [63–65]; however, it is not the case in all studies.

Finally, race may also play an important role in the relationship between sleep duration and weight. African American children and adults are more likely to be short sleepers than Caucasian children and adults [66–68] and two epidemiological studies found that the association between short sleep duration and increased odds for obesity was stronger in African Americans than Caucasians [69, 70]. During an in-laboratory experimental protocol, sleep restricted African Americans gained more weight and exhibited lower resting metabolic rates compared to sleep-restricted Caucasians [7, 71]. In addition, although the two groups did not differ in daily caloric intake or meal timing patterns, there were differences in macronutrient intake. African Americans consumed a lower percentage of calories from protein and a higher percentage of calories from carbohydrates [33]. Racial differences in the relationship between sleep duration and risk for obesity requires more research, particularly in Hispanic and Asian populations. Although the relationship between sleep duration and BMI has also been demonstrated in these groups, whether or not the relationship is stronger compared to African Americans or Caucasians remains unknown.

In conclusion, epidemiologic and laboratory studies have indicated that younger adults, men and African Americans may be more vulnerable to the weight-gain effect of sleep loss than older adults, women and Caucasians, respectively. More research is needed to understand why these groups may be more susceptible and to determine if short-sleeping Hispanics or Asians are also at greater risk. This is particularly important given that African American and Hispanic minority groups exhibit higher prevalence rates of obesity as well as diseases associated with obesity (e.g., cardiovascular disease, type 2 diabetes).

Individual differences

In addition to group differences, emerging evidence suggests that there are individual differences in the way people respond to sleep loss (Fig. 15.1). Two recent in-laboratory studies examined these individual differences in detail [40, 72]. The first study examined data collected during two randomized crossover trials from 43 healthy adults and calculated the difference in objectively measured caloric intake during a sufficient sleep condition and a sleep restriction condition [72]. Large interindividual variability was observed—the change in caloric intake during sleep restriction ranged from −813 kcal (participants consumed fewer calories when sleep restricted) to 1437 kcal/day (participants consumed significantly more calories when sleep restricted).

The second study assessed caloric intake, meal timing and weight change in a group of healthy adults who participated in two separate sleep restriction experiments in the same laboratory [40]. Participation in each experiment was separated by at least 60 days and both protocols involved two nights of sufficient sleep followed by five nights of sleep restriction. Large interindividual differences were observed for all three variables (when averaging across both experiments for each participant). Some participants experienced weight loss whereas others experienced substantial weight gain, change in caloric intake during sleep restriction ranged from −500.7 to 1178.2 kcal, and late-night caloric intake ranged from 11.9 to 1434.1 kcal. In addition to examining these interindividual differences, Spaeth and colleagues also assessed intra-individual consistency to determine if the same individuals respond the same way during two separate exposures to sleep restriction [40]. Change in weight during the protocol and change in caloric intake during sleep restriction were very consistent for men but not women. Men who gained a substantial amount of weight and increased their caloric intake to a significant degree during sleep restriction did so consistently during both exposures, suggesting they may be particularly vulnerable to the energy balance effects of sleep restriction. Conversely, men who lost or maintained weight during the protocol and did not show a substantial increase in caloric intake during sleep restriction also did so consistently during both exposures, suggesting they may be resistant to the energy balance effects of sleep restriction. Late-night caloric intake was very consistent among men and women; suggesting that some individuals are more prone to late-night eating than others. Therefore, adults who are particularly vulnerable to late-night eating and who are awake during late-night hours due to shift work or other lifestyle circumstances may be at heightened risk for weight gain.

Collectively, these findings suggest that although there are consistent group-average increases in weight, caloric intake and late-night eating during sleep restriction, there is also considerable variability between individuals. Furthermore, obtaining sufficient sleep may be particularly important for weight maintenance in individuals who are the most vulnerable. Future research is needed to identify biomarkers for predicting this vulnerability and to establish countermeasures for helping vulnerable individuals who experience sleep loss due to shiftwork or other lifestyle factors.

SLEEP TIMING

In addition to short sleep duration, disruptions in the timing of sleep have also been associated with increased risk for obesity; however, it can be difficult to separate the two constructs. Humans are less efficient sleepers when attempting to sleep outside of their endogenous circadian rhythm; therefore, altering the timing of sleep can also lead to shortened sleep. Four areas of research examining the relationship between sleep timing and obesity risk are chronotype, social jetlag, shift work, and delayed bedtime.

Chronotype refers to an individual's optimal timing of wakefulness and sleep. Humans exhibit a diurnal circadian rhythm (active during the day, sleep during the night) but some individuals prefer activity in the morning (larks) and exhibit an advanced (earlier) sleep period whereas others prefer activity in the evening (owls) and exhibit a delayed (later) sleep period. Age, gender, and genetic factors influence morningness versus eveningness preference. The interaction between Earth's light/dark cycle and current school/work schedules complement individuals who function best in the morning rather than in the evening. Because owls experience heightened alertness in the late evening, they often delay bedtime but still have to wake up early in the morning to accommodate school/work schedules which produces sleep restriction during the work week. There is evidence that adolescents and adults with an evening preference are at increased risk for weight gain/obesity [73–76], consume a less healthy diet [77–79] and exhibit delayed meal timing [80–82]. For example, a prospective study in college freshmen found that students characterized as evening-types gained significantly more weight over an 8-week period compared to morning-types [74]. Baron and colleagues found that late sleepers (sleep midpoint >0530h) exhibited a shorter sleep duration, consumed more calories at dinner and after 2000h, more fast food and full-calorie soda, and had a higher BMI compared to normal sleepers (sleep midpoint <0530h) [41]. Finally, in a large sample of severely obese adults undergoing bariatric surgery, evening-type individuals weighed more before surgery, lost less weight after surgery, and regained more weight at follow-up [83].

Social jetlag describes a misalignment between biological and social time. For example, a person's internal clock may prefer a wake time of 07:30 a.m. but a 5 a.m. wake time is needed in order to fulfill personal and household tasks before attending work. Individuals with social jetlag sleep longer on "free" nights (such as weekends when work obligations do not dictate sleep timing) and exhibit large differences in sleep duration and/or the midpoint of sleep between free and nonfree days. Many students experience social jetlag due to early school-start times and many adults experience chronic social jetlag for the duration of their working career. When quantifying social jetlag as the difference in mid-sleep time between free days and workdays, Roenneberg and colleagues observed that 69% of participants experienced at least 1 h of social jetlag and that social jetlag significantly increased the probability of participants being overweight/obese [84]. Among those who were overweight/obese, social jetlag positively correlated with weight. Social jetlag has also been associated with cardiovascular risk factors, fat mass and incidence of metabolic syndrome [85–87]. In a cross-sectional study of children aged 8–10 years, social jetlag was associated with adiposity (body fat %, fat mass, fat mass index (FMI, kg/m^2), waist to hip ratio, and body mass index (kg/m^2); with body fat increasing by 3% per 1 h of social jetlag [88].

It is unclear if short sleep during nonfree days drives these energy balance responses or the inconsistency in sleep timing between free and nonfree days. Ideally, individuals would be able to obtain adequate sleep during free and nonfree nights and thus maintain sufficient sleep duration *and* a consistent sleep schedule. However, given that school/work schedules and household/personal/social obligations influence sleep opportunity, the effects of "catching up" on sleep during free days warrants more study. One recent study found that adults who engaged in catch-up sleep on weekends exhibited a lower BMI than those who did not [89] and there is evidence that "banking sleep" (providing an extended sleep opportunity, ~10h in bed) before engaging in sleep restriction attenuates deficits in neurobehavioral function caused by sleep loss [90].

Shift work, an essential component of our 24 h economy that requires work performed outside of the traditional 9 a.m.–5 p.m. business day, represents more extreme circadian misalignment than social jetlag. Shift work schedules often require an individual to work during the night when the circadian system is promoting sleep, and require an individual to sleep during the day when the circadian system is promoting wakefulness. Night shift workers sleep 2–4h less per day than day shift workers and are more likely to experience excessive sleepiness. Shift workers are at increased risk for weight gain and obesity as well as diseases associated with obesity including type 2 diabetes, cardiovascular disease, and gastrointestinal disorders [87, 91, 92]. Interestingly, daily caloric intake does not seem to differ between shift workers and traditional day workers; however, poor diet quality, lower physical activity levels, delayed meal timing, and altered nutritional metabolism have all been observed as possible mechanisms underlying the relationship between shift work and obesity [87, 91, 92]. A recent metaanalysis found that the overall odds ratio of night shift work was 1.23 (95% confidence interval = 1.17 − 1.29) for risk of overweight/obesity, shift workers had a higher frequency of developing abdominal obesity (odds ratio = 1.35) than other obesity types, and permanent night workers had a 29% higher risk of developing abdominal obesity than rotating shift workers (odds ratio 1.43 vs 1.14) [93].

Experimental studies mimicking shift work schedules have shown that this type of circadian misalignment leads to decreased energy expenditure and impaired glucose metabolism, which are also likely contributors to the propensity for weight gain [94]. Research is ongoing to examine how an individual's morning/evening preference affects his/her response to shift work. For example, the relationship between night shift work and obesity may be stronger for morning-types than for evening-types.

Finally, when examining the relationship between bedtime and weight gain/obesity (independent from sleep duration) studies have shown that increased variability in bedtime and going to bed later associate with unhealthy diet and higher BMI in children, adolescents and adults [95–100]. These associations are consistent with findings from an experimental study [7]. During this protocol, adult participants experienced two nights of sufficient sleep (10h TIB/night) followed by five nights of sleep restriction (4h TIB/night) and sleep restriction was implemented by delaying bedtime until 04:00 a.m. Participants exhibited increased caloric intake on all days when bedtime was delayed to 04:00 a.m., including the first day of the sleep restriction phase, when participants woke up after sufficient sleep but were kept awake for 20 h (until 04:00 a.m.). Caloric intake was not increased on the last day of the sleep restriction phase, when subjects woke up after 5 consecutive nights of insufficient sleep but were only kept awake for 14 h and went to bed at 10:00 p.m. This pattern suggested that bedtime and/or hours of wakefulness are better predictors of daily intake than how much sleep was achieved during the preceding night [7]. The timing and variability of bedtime may be targets for improving sleep and promoting weight maintenance and this may be particularly important for low-income families where sleep habits are less stable and the risk for obesity is greater [100, 101].

SLEEP DISORDERS

Sleep disorders such as insomnia, obstructive sleep apnea and narcolepsy, can also lead to changes in sleep duration or timing and energy balance. Insomnia occurs when an individual complains of having difficulty initiating or maintaining sleep, waking up earlier than desired and impaired daytime functioning despite having sufficient opportunities for sleep. Insomnia is highly comorbid with psychiatric disorders and is not always associated with objectively measured short sleep duration. There is a paucity of research examining the relationship between insomnia and obesity and results have been mixed. A recent metaanalysis showed that the odds of having obesity among individuals with an insomnia diagnosis was not significantly greater than among those who did not have an insomnia diagnosis but there was a small, significant cross-sectional correlation between insomnia symptoms and BMI [102]. Longitudinal data on the association between insomnia symptoms and future incidence of obesity were inconclusive [102]. Insomnia with objectively measured short sleep duration has been associated other markers of metabolic dysregulation (i.e., hypertension and type 2 diabetes) and insomnia symptoms have been associated with development of the metabolic syndrome [103–105]. More work is needed in this area, particularly since the primary treatment for insomnia involves decreasing sleep opportunity which may impact weight regulation via the behaviors and mechanisms described above.

Obstructive sleep apnea (OSA) occurs when an individual exhibits shallow breathing or ceases to breathe during sleep; symptoms include loud snoring, gasping for air and reduced airflow during sleep and impaired daytime functioning. The most common cause of OSA is obesity. It is estimated that 50% of children and adults with obesity have OSA and studies have consistently demonstrated that weight loss improves OSA symptoms (see Refs. [106–108] for reviews on this topic). OSA is also associated with metabolic dysregulation, independent of obesity, and treatment of OSA with continuous positive airway pressure (CPAP) leads to improvements in daytime functioning and metabolic health [109]. Given the serious adverse cognitive and health consequences of untreated OSA, it is critical for physicians to screen for and treat OSA in patients with obesity.

Narcolepsy occurs when an individual experiences excessive sleepiness with uncontrolled need for sleep or lapses into sleep during the day and be accompanied with (Type 1) or without (Type 2) cataplexy and cerebral spinal fluid hypocretin-1 deficiency. Hypocretin (also referred to as orexin) also plays an important role in appetite, reward and motivation. Patients with narcolepsy exhibit a higher BMI than those without narcolepsy and this association has been observed in children, adolescents and adults [110–112]. Orexin deficiency and decreased energy expenditure have been identified as mechanisms that may underlie the relationship between narcolepsy and obesity [113–116]. Narcoleptics are also more likely to experience persistent food cravings, binge eat, and have an eating disorder [117–119].

A recent study used a behavioral paradigm to explored the effects of satiation on food choice and caloric intake in patients with narcolepsy type 1 compared with healthy matched controls [120]. First, participants were trained on a choice task to earn their preferred salty or sweet snacks. One of the snack outcomes was devalued by having participants actually consume it until they were sated. Participants then completed the choice task again. Control participants decreased choosing the devalued snack by 14% whereas participants with narcolepsy only decreased choosing the devalued snack by 4%. Finally, when participants were given access to snacks at the end of the task, participants with narcolepsy consumed nearly four times as many calories as control participants (~400 kcal vs ~120 kcal). Findings suggest that individuals with narcolepsy may not experience sensory-specific satiety to the same degree as healthy individuals. In addition, as expected, healthy controls were less hungry and wanted the devalued snack less after consuming it until sated, this decrease in hunger and wanting associated with choosing the devalued snack less in the choice task. By contrast, there was no association between decreased hunger/wanting and performance on the choice task in participants with narcolepsy, suggesting that these individuals may not experience the connection between subjective experiences to the same degree as healthy individuals [120]. Thus, changes in satiety and subjective experiences may increase risk of overeating in individuals with narcolepsy.

Interestingly, recent studies have also demonstrated differences in the energy balance response to two common treatments for narcolepsy. After beginning treatment, patients using sodium oxybate (a GABA receptor agonist) lost weight (women: $-2.56\,kg/m^2$, men: $-0.84\,kg/m^2$) between the first and last measurement whereas patients using modafinil (an atypical, selective dopamine transporter inhibitor) gained weight (women: $+0.57\,kg/m^2$, men: $+0.67\,kg/m^2$); the effect of drug treatment on BMI was most pronounced in those with a higher baseline BMI [121].

Future research examining weight regulation in patients with narcolepsy will not only inform clinical practice but also help elucidate the physiological role of hypocretin/orexin in both sleep and energy balance regulation.

SLEEP IN INDIVIDUALS WITH OBESITY

Poor sleep quality and excessive daytime sleepiness are frequent complaints among individuals with obesity [122, 123]. Weight loss after either diet/exercise programs or bariatric surgery leads to improvements in sleep and daytime functioning [124–126]. Psychological distress may play an important role in the relationships between obesity, sleep and daytime functioning [127]. Obese youth are more likely to report difficulties with sleep, symptoms of depression, and lower quality of life; in a cross-sectional study that was conducted in a specialized obesity clinic, degree of obesity predicted increased sleep difficulties and decreased quality-of-life scores [128]. In addition, obese children with more symptoms of depression had more sleep problems.

Few studies have objectively measured sleep in obese individuals; however, evidence suggests that excessive adiposity relates to changes in sleep architecture in children, adolescents and adults. Sleep is comprised of rapid eye movement (REM) sleep and non-REM sleep, and non-REM sleep is further categorized as stage 1, stage 2 and slow-wave sleep (SWS). SWS duration has been negatively correlated with BMI, waist circumference, ghrelin levels, intake during an ad libitum meal, saturated fat intake, and hunger ratings [129–131] and has been positively related with fiber intake, lean body mass and growth hormone release [132–134]. REM sleep duration has been correlated with increased hunger ratings, body fat percentage, higher BMI, and positive energy balance due to overeating [130, 132, 135, 136]. However, other studies have shown that REM sleep duration is inversely correlated with waist circumference and BMI [129, 131].

In 12 healthy normal weight men participating in an in-laboratory study involving 2 days of caloric restriction to 10% of energy requirements followed by 2 days of ad libitum/free feeding, sleep architecture was measured by polysomnography [137]. Two days of caloric restriction significantly increased the duration of SWS and this effect was entirely reversed after ad libitum feeding. Interestingly, caloric restriction also decreased orexin levels and the change in orexin levels positively correlated with duration of SWS during caloric restriction.

Normal weight adults exhibit higher sleep efficiency than overweight or obese individuals in young, middle-aged and older adults [138–140]; however results have been more mixed in children and adolescent populations [73, 135, 136, 141, 142]. More research is needed to examine differences in sleep architecture between normal, overweight and obese individuals and assess how changes in weight and/or body composition affect subjective and objective measures of sleep.

Two eating disorders associated with increased risk for obesity are Binge Eating Disorder and Night Eating Syndrome. Although it is beyond the scope of this chapter, interested readers are referred to Ref. 143 for a review of changes in sleep (e.g., subjective measures of sleep quality as well as objective measures of sleep duration, latency, efficiency and architecture) in these and other eating disorders.

THE ROLE OF SLEEP IN WEIGHT LOSS INTERVENTIONS

Recent research has highlighted the importance of sleep during weight-loss interventions and assessed the efficacy of sleep extension as a behavioral modification to promoting health. In children (2–5 years) enrolled in a randomized trial to improve household routines (6-month intervention, promoted family meals, adequate sleep, limiting TV time, and removing the TV from the child's bedroom), intervention participants exhibited increased sleep duration and decreased BMI [144]. In a sample of obese preschool-aged children enrolled in a weight management program, longer sleep duration associated with lower BMI and caloric intake posttreatment [145]. In obese adolescents attending a clinical multidisciplinary weight management program, longer weekly sleep duration at baseline predicted greater weight loss after 3 months of treatment; those who reduced their BMI by $\geq 1\,kg/m^2$ reported approximately 4 more hours of sleep/week compared to those who reduced their BMI by $<1\,kg/m^2$ [146]. Similarly, in obese adolescents participating in a summer camp-based immersion treatment program, shorter sleep duration, lower sleep quality and more sleep debt associated with larger waist circumference and higher BMI preintervention and smaller weight reduction during the intervention [147].

Compared to adults currently enrolled in weight loss interventions, those who successfully maintained weight loss for at least a year (National Weight Control Registry) were more likely to be a morning-type, less likely to be a short sleeper (<6 h/night) and reported better sleep quality [148]. Among women randomized to a weight-loss program, better subjective sleep quality and sleeping >7 h/night at baseline significantly increased the likelihood of weight-loss success [149]. Similarly, baseline sleep duration and sleep quality predicted greater fat mass loss (assessed by dual-energy X-ray absorptiometry) during a 15–24 weeks weight loss intervention consisting of a targeted 600–700 kcal/day decrease in energy intake in overweight and obese men and women; an increase by 1 h in sleep duration at baseline was associated with a decrease of 0.7 kg in fat mass (after adjustment for covariates) [150]. Collectively these findings suggest that sleep plays an important role in weight-loss across many age groups and types of weight loss programs, future studies are needed to examine how sleep can be used to increase weight-loss success and what mechanisms underlie this relationship.

During an in-laboratory experimental study [151], overweight/obese women were placed on a hypocaloric diet for 14 days with either 8.5 or 5.5 h sleep opportunity each night. During the 5.5 h sleep condition, woman lost a similar amount of weight as during the 8.5 h condition; however, they lost less fat, reported greater hunger and exhibited a higher respiratory quotient (RQ). Recently, a similar experiment was conducted outside of the laboratory; overweight or obese adults were randomized to 8-week of caloric restriction with either sufficient sleep or sleep restriction [152]. Participants were instructed to restrict daily calorie intake to 95% of their measured resting metabolic rate and participants in the sleep restriction group were instructed to reduce time in bed on 5 nights by 1 h per night and to sleep ad libitum on the other 2 nights during each week. Although both groups lost similar amounts of weight, lean mass, and fat mass, the proportion of total mass lost as fat was significantly less in the sleep restricted group. Participants in the sufficient sleep condition exhibited a significant reduction in body fat percentage and RQ over the 8 weeks period whereas participants in the sleep restriction condition did not. Although more work is needed in this area, these experiments demonstrate that sleep influences changes in physiology that occur during weight loss. Therefore addressing sleep hygiene may help individuals achieve success in losing weight and maintaining a healthy weight.

Sleep extension interventions in children [153], adolescents [154] and adults [155] have been proposed for weight management and metabolic health. These interventions focus on using behavioral modification strategies (self-monitoring, goal-setting, positive reinforcement, etc.) to increase the amount of time participants set aside for sleep (time-in-bed). Preliminary data demonstrates that sleep extension decreases desire for high calorie foods and sugar intake [156, 157], improves blood pressure [158] and associates with increased insulin sensitivity in adults [159]. Thus, sleep extension interventions prove to be a promising new behavioral approach to weight management and metabolic health.

CONCLUSION

Habitual short sleep, due to lifestyle factors, evening chronotype or sleep disorders, associates with an increased risk for obesity in children, adolescents and adults. Experimental studies demonstrate that sleep restriction leads to increased daily caloric intake, greater consumption of unhealthy food and drink, and delayed meal timing as well as alterations in appetite-regulating hormones and brain activity that promote positive energy balance and weight gain over time. Addressing sleep issues with individuals who are at risk for uncontrolled weight gain or are obese will improve daytime functioning and may increase the likelihood of weight-loss success.

REFERENCES

[1] Hales CM, Fryar CD, Carroll MD, Freedman DS, Ogden CL. Trends in obesity and severe obesity prevalence in US youth and adults by sex and age, 2007–2008 to 2015–2016. JAMA 2018;319(16):1723–5.

[2] Vezina-Im LA, Nicklas TA, Baranowski T. Associations among sleep, body mass index, waist circumference, and risk of type 2 diabetes among US childbearing-age women: National Health and Nutrition Examination Survey. J Women's Health 2018;.

[3] Deng HB, Tam T, Zee BC, Chung RY, Su X, Jin L, et al. Short sleep duration increases metabolic impact in healthy adults: a population-based cohort study. Sleep 2017;40(10).

[4] Potter GDM, Cade JE, Hardie LJ. Longer sleep is associated with lower BMI and favorable metabolic profiles in UK adults: findings from the National Diet and Nutrition Survey. PLoS ONE 2017;12(7):e0182195.

[5] Ford ES, Li CY, Wheaton AG, Chapman DP, Perry GS, Croft JB. Sleep duration and body mass index and waist circumference among US adults. Obesity 2014;22(2):598–607.

[6] Ogilvie RP, Redline S, Bertoni AG, Chen XL, Ouyang P, Szklo M, et al. Actigraphy measured sleep indices and adiposity: the multi-ethnic study of atherosclerosis (MESA). Sleep 2016;39(9):1701–8.

[7] Spaeth AM, Dinges DF, Goel N. Effects of experimental sleep restriction on weight gain, caloric intake, and meal timing in healthy adults. Sleep 2013;36(7):981–90.

[8] Markwald RR, Melanson EL, Smith MR, Higgins J, Perreault L, Eckel RH, et al. Impact of insufficient sleep on total daily energy expenditure, food intake, and weight gain. Proc Natl Acad Sci U S A 2013;110(14):5695–700.

[9] Halal CSE, Matijasevich A, Howe LD, Santos IS, Barros FC, Nunes ML. Short sleep duration in the first years of life and obesity/overweight at age 4 years: a birth cohort study. J Pediatr 2016;168:99–103.

[10] Chaput JP, Gray CE, Poitras VJ, Carson V, Gruber R, Birken CS, et al. Systematic review of the relationships between sleep duration and health indicators in the early years (0–4 years). BMC Public Health 2017;17:91–107.

[11] Katzmarzyk PT, Barreira TV, Broyles ST, Champagne CM, Chaput JP, Fogelholm M, et al. Relationship between lifestyle behaviors and obesity in children ages 9–11: results from a 12-country study. Obesity 2015;23(8):1696–702.

[12] Mitchell JA, Rodriguez D, Schmitz KH, Audrain-McGovern J. Sleep duration and adolescent obesity. Pediatrics 2013;131(5):E1428–34.

[13] Wu YH, Gong QH, Zou ZQ, Li H, Zhang XH. Short sleep duration and obesity among children: a systematic review and meta-analysis of prospective studies. Obes Res Clin Pract 2017;11(2):140–50.

[14] Fatima Y, Doi SA, Mamun AA. Longitudinal impact of sleep on overweight and obesity in children and adolescents: a systematic review and bias-adjusted meta-analysis. Obes Rev 2015;16(2):137–49.

[15] Fatima Y, Doi SAR, Al MA. Sleep problems in adolescence and overweight/obesity in young adults: is there a causal link? Sleep Health 2018;4(2):154–9.

[16] Hart CN, Carskadon MA, Considine RV, Fava JL, Lawton J, Raynor HA, et al. Changes in children's sleep duration on food intake, weight, and leptin. Pediatrics 2013;132(6):e1473–80.

[17] Bel S, Michels N, De Vriendt T, Patterson E, Cuenca-Garcia M, Diethelm K, et al. Association between self-reported sleep

[18] duration and dietary quality in European adolescents. Br J Nutr 2013;110(5):949–59.
[18] Bornhorst C, Wijnhoven TMA, Kunesova M, Yngve A, Rito AI, Lissner L, et al. WHO European Childhood Obesity Surveillance Initiative: associations between sleep duration, screen time and food consumption frequencies. BMC Public Health 2015;15.
[19] Fisher A, McDonald L, van Jaarsveld CHM, Llewellyn C, Fildes A, Schrempft S, et al. Sleep and energy intake in early childhood. Int J Obes 2014;38(7):926–9.
[20] Franckle RL, Falbe J, Gortmaker S, Ganter C, Taveras EM, Land T, et al. Insufficient sleep among elementary and middle school students is linked with elevated soda consumption and other unhealthy dietary behaviors. Prev Med 2015;74:36–41.
[21] Hjorth MF, Quist JS, Andersen R, Michaelsen KF, Tetens I, Astrup A, et al. Change in sleep duration and proposed dietary risk factors for obesity in Danish school children. Pediatr Obes 2014;9(6):e156–9.
[22] Weiss A, Xu F, Storfer-Isser A, Thomas A, Ievers-Landis CE, Redline S. The association of sleep duration with adolescents' fat and carbohydrate consumption. Sleep 2010;33(9):1201–9.
[23] Kjeldsen JS, Hjorth MF, Andersen R, Michaelsen KF, Tetens I, Astrup A, et al. Short sleep duration and large variability in sleep duration are independently associated with dietary risk factors for obesity in Danish school children. Int J Obes 2014;38(1):32–9.
[24] Hart CN, Hawley N, Davey A, Carskadon M, Raynor H, Jelalian E, et al. Effect of experimental change in children's sleep duration on television viewing and physical activity. Pediatr Obes 2017;12(6):462–7.
[25] Bromley LE, Booth 3rd JN, Kilkus JM, Imperial JG, Penev PD. Sleep restriction decreases the physical activity of adults at risk for type 2 diabetes. Sleep 2012;35(7):977–84.
[26] Chapman CD, Nilsson EK, Nilsson VC, Cedernaes J, Rangtell FH, Vogel H, et al. Acute sleep deprivation increases food purchasing in men. Obesity (Silver Spring) 2013;21(12):E555–60.
[27] Nedeltcheva AV, Kessler L, Imperial J, Penev PD. Exposure to recurrent sleep restriction in the setting of high caloric intake and physical inactivity results in increased insulin resistance and reduced glucose tolerance. J Clin Endocrinol Metab 2009;94(9):3242–50.
[28] St-Onge MP, Roberts AL, Chen J, Kelleman M, O'Keeffe M, RoyChoudhury A, et al. Short sleep duration increases energy intakes but does not change energy expenditure in normal-weight individuals. Am J Clin Nutr 2011;94(2):410–6.
[29] Cedernaes J, Brandell J, Ros O, Broman JE, Hogenkamp PS, Schioth HB, et al. Increased impulsivity in response to food cues after sleep loss in healthy young men. Obesity 2014;22(8):1786–91.
[30] Hogenkamp PS, Nilsson E, Nilsson VC, Chapman CD, Vogel H, Lundberg LS, et al. Acute sleep deprivation increases portion size and affects food choice in young men. Psychoneuroendocrinology 2013;38(9):1668–74.
[31] Calvin AD, Carter RE, Adachi T, Macedo PG, Albuquerque FN, van der Walt C, et al. Effects of experimental sleep restriction on caloric intake and activity energy expenditure. Chest 2013;144(1):79–86.
[32] Spiegel K, Tasali E, Penev P, Van Cauter E. Brief communication: sleep curtailment in healthy young men is associated with decreased leptin levels, elevated ghrelin levels, and increased hunger and appetite. Ann Intern Med 2004;141(11):846–50.
[33] Spaeth AM, Dinges DF, Goel N. Sex and race differences in caloric intake during sleep restriction in healthy adults. Am J Clin Nutr 2014;100(2):559–66.
[34] Fang Z, Spaeth AM, Ma N, Zhu S, Hu S, Goel N, et al. Altered salience network connectivity predicts macronutrient intake after sleep deprivation. Sci Rep 2015;5:8215.
[35] Beebe DW, Simon S, Summer S, Hemmer S, Strotman D, Dolan LM. Dietary intake following experimentally restricted sleep in adolescents. Sleep 2013;36(6):827–34.
[36] Capers PL, Fobian AD, Kaiser KA, Borah R, Allison DB. A systematic review and meta-analysis of randomized controlled trials of the impact of sleep duration on adiposity and components of energy balance. Obes Rev 2015;16(9):771–82.
[37] Garaulet M, Gomez-Abellan P, Alburquerque-Bejar JJ, Lee YC, Ordovas JM, Scheer FAJL. Timing of food intake predicts weight loss effectiveness. Int J Obes 2013;37(4):604–11.
[38] Jakubowicz D, Barnea M, Wainstein J, Froy O. High Caloric intake at breakfast vs. dinner differentially influences weight loss of overweight and obese women. Obesity 2013;21(12):2504–12.
[39] McHill AW, Phillips AJ, Czeisler CA, Keating L, Yee K, Barger LK, et al. Later circadian timing of food intake is associated with increased body fat. Am J Clin Nutr 2017;106(5):1213–9.
[40] Spaeth AM, Dinges DF, Goel N. Phenotypic vulnerability of energy balance responses to sleep loss in healthy adults. Sci Rep 2015;5:14920.
[41] Baron KG, Reid KJ, Kim T, Van Horn L, Attarian H, Wolfe L, et al. Circadian timing and alignment in healthy adults: associations with BMI, body fat, caloric intake and physical activity. Int J Obes (Lond) 2017;41(2):203–9.
[42] Taheri S, Lin L, Austin D, Young T, Mignot E. Short sleep duration is associated with reduced leptin, elevated ghrelin, and increased body mass index. PLoS Med 2004;1(3):210–7.
[43] Chaput JP, Despres JP, Bouchard C, Tremblay A. Short sleep duration is associated with reduced leptin levels and increased adiposity: results from the Quebec Family Study. Obesity 2007;15(1):253–61.
[44] Broussard JL, Kilkus JM, Delebecque F, Abraham V, Day A, Whitmore HR, et al. Elevated ghrelin predicts food intake during experimental sleep restriction. Obesity 2016;24(1):132–8.
[45] St-Onge MP. The role of sleep duration in the regulation of energy balance: effects on energy intakes and expenditure. J Clin Sleep Med 2013;9(1):73–80.
[46] St-Onge MP. Sleep-obesity relation: underlying mechanisms and consequences for treatment. Obes Rev 2017;18:34–9.
[47] St-Onge MP, O'Keeffe M, Roberts AL, RoyChoudhury A, Laferrere B. Short sleep duration, glucose dysregulation and hormonal regulation of appetite in men and women. Sleep 2012;35(11):1503–10.
[48] Shechter A, Rising R, Albu JB, St-Onge MP. Experimental sleep curtailment causes wake-dependent increases in 24-h energy expenditure as measured by whole-room indirect calorimetry. Am J Clin Nutr 2013;98(6):1433–9.
[49] Demos KE, Sweet LH, Hart CN, McCaffery JM, Williams SE, Mailloux KA, et al. The effects of experimental manipulation of sleep duration on neural response to food cues. Sleep 2017;40(11).
[50] St-Onge MP, McReynolds A, Trivedi ZB, Roberts AL, Sy M, Hirsch J. Sleep restriction leads to increased activation of brain regions sensitive to food stimuli. Am J Clin Nutr 2012;95(4):818–24.
[51] St-Onge MP, Wolfe S, Sy M, Shechter A, Hirsch J. Sleep restriction increases the neuronal response to unhealthy food in normal-weight individuals. Int J Obes (Lond) 2014;38(3):411–6.
[52] Benedict C, Brooks SJ, O'Daly OG, Almen MS, Morell A, Aberg K, et al. Acute sleep deprivation enhances the brain's response to

hedonic food stimuli: an fMRI study. J Clin Endocrinol Metab 2012;97(3):E443–7.

[53] Greer SM, Goldstein AN, Walker MP. The impact of sleep deprivation on food desire in the human brain. Nat Commun 2013;4:2259.

[54] Pilcher JJ, Morris DM, Donnelly J, Feigl HB. Interactions between sleep habits and self-control. Front Hum Neurosci 2015;9:284.

[55] Yoncheva YN, Castellanos FX, Pizinger T, Kovtun K, St-Onge MP. Sleep and meal-time misalignment alters functional connectivity: a pilot resting-state study. Int J Obes (Lond) 2016;40(11):1813–6.

[56] Canning KL, Brown RE, Jamnik VK, Kuk JL. Relationship between obesity and obesity-related morbidities weakens with aging. J Gerontol A Biol Sci Med Sci 2014;69(1):87–92.

[57] Grandner MA, Schopfer EA, Sands-Lincoln M, Jackson N, Malhotra A. Relationship between sleep duration and body mass index depends on age. Obesity 2015;23(12):2491–8.

[58] Hairston KG, Bryer-Ash M, Norris JM, Haffner S, Bowden DW, Wagenknecht LE. Sleep duration and five-year abdominal fat accumulation in a minority cohort: the IRAS Family Study. Sleep 2010;33(3):289–95.

[59] St-Onge MP, Perumean-Chaney S, Desmond R, Lewis CE, Yan LL, Person SD, et al. Gender differences in the association between sleep duration and body composition: the Cardia study. Int J Endocrinol 2010;2010:726071.

[60] Meyer KA, Wall MM, Larson NI, Laska MN, Neumark-Sztainer D. Sleep duration and BMI in a sample of young adults. Obesity 2012;20(6):1279–87.

[61] Yang TC, Matthews SA, Chen VYJ. Stochastic variability in stress, sleep duration, and sleep quality across the distribution of body mass index: insights from quantile regression. Int J Behav Med 2014;21(2):282–91.

[62] Watanabe M, Kikuchi H, Tanaka K, Takahashi M. Association of short sleep duration with weight gain and obesity at 1-year follow-up: a large-scale prospective study. Sleep 2010;33(2):161–7.

[63] Araujo J, Severo M, Ramos E. Sleep duration and adiposity during adolescence. Pediatrics 2012;130(5):E1146–54.

[64] Suglia SF, Kara S, Robinson WR. Sleep duration and obesity among adolescents transitioning to adulthood: do results differ by sex? J Pediatr 2014;165(4):750–4.

[65] Tatone-Tokuda F, Dubois L, Ramsay T, Girard M, Touchette E, Petit D, et al. Sex differences in the association between sleep duration, diet and body mass index: a birth cohort study. J Sleep Res 2012;21(4):448–60.

[66] Pena MM, Rifas-Shiman SL, Gillman MW, Redline S, Taveras EM. Racial/ethnic and socio-contextual correlates of chronic sleep curtailment in childhood. Sleep 2016;39(9):1653–61.

[67] Adenekan B, Pandey A, McKenzie S, Zili F, Casimir GJ, Jean-Louis G. Sleep in America: role of racial/ethnic differences. Sleep Med Rev 2013;17(4):255–62.

[68] Whinnery J, Jackson N, Rattanaumpawan P, Grandner MA. Short and long sleep duration associated with race/ethnicity, sociodemographics, and socioeconomic position. Sleep 2014;37(3):601.

[69] Donat M, Brown C, Williams N, Pandey A, Racine C, McFarlane SI, et al. Linking sleep duration and obesity among black and white US adults. Clin Pract (Lond) 2013;10(5).

[70] Grandner MA, Chakravorty S, Perlis ML, Oliver L, Gurubhagavatula I. Habitual sleep duration associated with self-reported and objectively determined cardiometabolic risk factors. Sleep Med 2014;15(1):42–50.

[71] Spaeth AM, Dinges DF, Goel N. Resting metabolic rate varies by race and by sleep duration. Obesity (Silver Spring) 2015;23(12):2349–56.

[72] McNeil J, St-Onge MP. Increased energy intake following sleep restriction in men and women: a one-size-fits-all conclusion? Obesity (Silver Spring) 2017;25(6):989–92.

[73] Arora T, Taheri S. Associations among late chronotype, body mass index and dietary behaviors in young adolescents. Int J Obes (Lond) 2015;39(1):39–44.

[74] Culnan E, Kloss JD, Grandner M. A prospective study of weight gain associated with chronotype among college freshmen. Chronobiol Int 2013;30(5):682–90.

[75] Maukonen M, Kanerva N, Partonen T, Kronholm E, Konttinen H, Wennman H, et al. The associations between chronotype, a healthy diet and obesity. Chronobiol Int 2016;33(8):972–81.

[76] Yu JH, Yun CH, Ahn JH, Suh S, Cho HJ, Lee SK, et al. Evening chronotype is associated with metabolic disorders and body composition in middle-aged adults. J Clin Endocrinol Metab 2015;100(4):1494–502.

[77] Maukonen M, Kanerva N, Partonen T, Kronholm E, Tapanainen H, Kontto J, et al. Chronotype differences in timing of energy and macronutrient intakes: a population-based study in adults. Obesity (Silver Spring) 2017;25(3):608–15.

[78] Mota MC, Waterhouse J, De-Souza DA, Rossato LT, Silva CM, Araujo MBJ, et al. Association between chronotype, food intake and physical activity in medical residents. Chronobiol Int 2016;33(6):730–9.

[79] Patterson F, Malone SK, Lozano A, Grandner MA, Hanlon AL. Smoking, screen-based sedentary behavior, and diet associated with habitual sleep duration and chronotype: data from the UK biobank. Ann Behav Med 2016;50(5):715–26.

[80] Lucassen EA, Zhao XC, Rother KI, Mattingly MS, Courville AB, de Jonge L, et al. Evening chronotype is associated with changes in eating behavior, more sleep apnea, and increased stress hormones in short sleeping obese individuals. PLoS ONE 2013;8(3).

[81] Reutrakul S, Hood MM, Crowley SJ, Morgan MK, Teodori M, Knutson KL. The relationship between breakfast skipping, chronotype, and glycemic control in type 2 diabetes. Chronobiol Int 2014;31(1):64–71.

[82] Munoz JSG, Canavate R, Hernandez CM, Cara-Salmeron V, Morante JJH. The association among chronotype, timing of food intake and food preferences depends on body mass status. Eur J Clin Nutr 2017;71(6):736–42.

[83] Ruiz-Lozano T, Vidal J, de Hollanda A, Scheer FAJL, Garaulet M, Izquierdo-Pulido M. Timing of food intake is associated with weight loss evolution in severe obese patients after bariatric surgery. Clin Nutr 2016;35(6):1308–14.

[84] Roenneberg T, Allebrandt KV, Merrow M, Vetter C. Social jetlag and obesity. Curr Biol 2012;22(10):939–43.

[85] Parsons MJ, Moffitt TE, Gregory AM, Goldman-Mellor S, Nolan PM, Poulton R, et al. Social jetlag, obesity and metabolic disorder: investigation in a cohort study. Int J Obes (Lond) 2015;39(5):842–8.

[86] Rutters F, Lemmens SG, Adam TC, Bremmer MA, Elders PJ, Nijpels G, et al. Is social jetlag associated with an adverse endocrine, behavioral, and cardiovascular risk profile? J Biol Rhythm 2014;29(5):377–83.

[87] Reutrakul S, Knutson KL. Consequences of circadian disruption on cardiometabolic health. Sleep Med Clin 2015;10(4):455–68.

[88] Stoner L, Castro N, Signal L, Skidmore P, Faulkner J, Lark S, et al. Sleep and adiposity in preadolescent children: the importance of social jetlag. Child Obes 2018.

[89] Im HJ, Baek SH, Chu MK, Yang KI, Kim WJ, Park SH, et al. Association between weekend catch-up sleep and lower body mass: population-based mass population-based study. Sleep 2017;40(10).

[90] Rupp TL, Wesensten NJ, Bliese PD, Balkin TJ. Banking sleep: realization of benefits during subsequent sleep restriction and recovery. Sleep 2009;32(3):311–21.

[91] Depner CM, Stothard ER, Wright Jr. KP. Metabolic consequences of sleep and circadian disorders. Curr Diab Rep 2014;14(7):507.

[92] Laermans J, Depoortere I. Chronobesity: role of the circadian system in the obesity epidemic. Obes Rev 2016;17(2):108–25.

[93] Sun M, Feng W, Wang F, Li P, Li Z, Li M, et al. Meta-analysis on shift work and risks of specific obesity types. Obes Rev 2018;19(1):28–40.

[94] Buxton OM, Cain SW, O'Connor SP, McLaren D, Czeisler CA, Shea SA. Metabolic consequences of chronic sleep restriction combined with circadian misalignment. Sleep 2010;33:A86.

[95] Asarnow LD, McGlinchey E, Harvey AG. Evidence for a possible link between bedtime and change in body mass index. Sleep 2015;38(10):1523–7.

[96] Golley RK, Maher CA, Matricciani L, Olds TS. Sleep duration or bedtime? Exploring the association between sleep timing behaviour, diet and BMI in children and adolescents. Int J Obes (Lond) 2013;37(4):546–51.

[97] Taylor BJ, Matthews KA, Hasler BP, Roecklein KA, Kline CE, Buysse DJ, et al. Bedtime variability and metabolic health in midlife women: the SWAN sleep study. Sleep 2016;39(2):457–65.

[98] Scharf RJ, DeBoer MD. Sleep timing and longitudinal weight gain in 4-and 5-year-old children. Pediatr Obes 2015;10(2):141–8.

[99] Thivel D, Isacco L, Aucouturier J, Pereira B, Lazaar N, Ratel S, et al. Bedtime and sleep timing but not sleep duration are associated with eating habits in primary school children. J Dev Behav Pediatr 2015;36(3):158–65.

[100] Miller AL, Kaciroti N, Lebourgeois MK, Chen YP, Sturza J, Lumeng JC. Sleep timing moderates the concurrent sleep duration-body mass index association in low-income preschool-age children. Acad Pediatr 2014;14(2):207–13.

[101] Appelhans BM, Fitzpatrick SL, Li H, Cail V, Waring ME, Schneider KL, et al. The home environment and childhood obesity in low-income households: indirect effects via sleep duration and screen time. BMC Public Health 2014;14:1160.

[102] Chan WS, Levsen MP, McCrae CS. A meta-analysis of associations between obesity and insomnia diagnosis and symptoms. Sleep Med Rev 2018;40:170–82.

[103] Vgontzas AN, Calhoun S, Liao DP, Karataraki M, Pejovic S, Bixler EO. Insomnia with objective short sleep duration is associated with type 2 diabetes a population-based study. Diabetes Care 2009;32(11):1980–5.

[104] Vgontzas AN, Liao DP, Bixler EO, Chrousos GP, Vela-Bueno A. Insomnia with objective short sleep duration is associated with a high risk for hypertension. Sleep 2009;32(4):491–7.

[105] Troxel WM, Buysse DJ, Matthews KA, Kip KE, Strollo PJ, Hall M, et al. Sleep symptoms predict the development of the metabolic syndrome. Sleep 2010;33(12):1633–40.

[106] Jordan AS, McSharry DG, Malhotra A. Adult obstructive sleep apnoea. Lancet 2014;383(9918):736–47.

[107] Drager LF, Togeiro SM, Polotsky VY, Lorenzi G. Obstructive sleep apnea a cardiometabolic risk in obesity and the metabolic syndrome. J Am Coll Cardiol 2013;62(7):569–76.

[108] Araghi MH, Chen YF, Jagielski A, Choudhury S, Banerjee D, Hussain S, et al. Effectiveness of lifestyle interventions on obstructive sleep apnea (OSA): systematic review and meta-analysis. Sleep 2013;36(10):1553–62. [62A-62E].

[109] Chirinos JA, Gurubhagavatula I, Teff K, Rader DJ, Wadden TA, Townsend R, et al. CPAP, weight loss, or both for obstructive sleep apnea. N Engl J Med 2014;370(24):2265–75.

[110] Ponziani V, Gennari M, Pizza F, Balsamo A, Bernardi F, Plazzi G. Growing up with type 1 narcolepsy: its anthropometric and endocrine features. J Clin Sleep Med 2016;12(12):1649–57.

[111] Dahmen N, Bierbrauer J, Kasten M. Increased prevalence of obesity in narcoleptic patients and relatives. Eur Arch Psychiatry Clin Neurosci 2001;251(2):85–9.

[112] Poli F, Plazzi G, Di Dalmazi G, Ribichini D, Vicennati V, Pizza F, et al. Body mass index-independent metabolic alterations in narcolepsy with cataplexy. Sleep 2009;32(11):1491–7.

[113] Sonka K, Kemlink D, Buskova J, Pretl M, Srutkova Z, Maurovich Horvat E, et al. Obesity accompanies narcolepsy with cataplexy but not narcolepsy without cataplexy. Neuro Endocrinol Lett 2010;31(5):631–4.

[114] Wang ZW, Wu HJ, Stone WS, Zhuang JH, Qiu LL, Xu X, et al. Body weight and basal metabolic rate in childhood narcolepsy: a longitudinal study. Sleep Med 2016;25:139–44.

[115] Dahmen N, Tonn P, Messroghli L, Ghezel-Ahmadi D, Engel A. Basal metabolic rate in narcoleptic patients. Sleep 2009;32(7):962–4.

[116] Overeem S, Scammell TE, Lammers GJ. Hypocretin/orexin and sleep: implications for the pathophysiology and diagnosis of narcolepsy. Curr Opin Neurol 2002;15(6):739–45.

[117] Chabas D, Foulon C, Gonzalez J, Nasr M, Lyon-Caen O, Willer JC, et al. Eating disorder and metabolism in narcoleptic patients. Sleep 2007;30(10):1267–73.

[118] Fortuyn HA, Swinkels S, Buitelaar J, Renier WO, Furer JW, Rijnders CA, et al. High prevalence of eating disorders in narcolepsy with cataplexy: a case-control study. Sleep 2008;31(3):335–41.

[119] Dimitrova A, Fronczek R, Van der Ploeg J, Scammell T, Gautam S, Pascual-Leone A, et al. Reward-seeking behavior in human narcolepsy. J Clin Sleep Med 2011;7(3):293–300.

[120] van Holst RJ, van der Cruijsen L, van Mierlo P, Lammers GJ, Cools R, Overeem S, et al. Aberrant food choices after satiation in human orexin-deficient narcolepsy type 1. Sleep 2016;39(11):1951–9.

[121] Schinkelshoek MS, Smolders IM, Donjacour CE, van der Meijden WP, van Zwet EW, Fronczek R, et al. Decreased body mass index during treatment with sodium oxybate in narcolepsy type 1. J Sleep Res 2018;e12684.

[122] Fatima Y, Doi SA, Mamun AA. Sleep quality and obesity in young subjects: a meta-analysis. Obes Rev 2016;17(11):1154–66.

[123] Rahe C, Czira ME, Teismann H, Berger K. Associations between poor sleep quality and different measures of obesity. Sleep Med 2015;16(10):1225–8.

[124] Toor P, Kim K, Buffington CK. Sleep quality and duration before and after bariatric surgery. Obes Surg 2012;22(6):890–5.

[125] Thomson CA, Morrow KL, Flatt SW, Wertheim BC, Perfect MM, Ravia JJ, et al. Relationship between sleep quality and quantity and weight loss in women participating in a weight-loss intervention trial. Obesity 2012;20(7):1419–25.

[126] Fernandez-Mendoza J, Vgontzas AN, Kritikou I, Calhoun SL, Liao DP, Bixler EO. Natural history of excessive daytime sleepi-

[127] Vgontzas AN, Lin HM, Papaliaga M, Calhoun S, Vela-Bueno A, Chrousos GP, et al. Short sleep duration and obesity: the role of emotional stress and sleep disturbances. Int J Obes (Lond) 2008;32(5):801–9.

[128] Whitaker BN, Fisher PL, Jambhekar S, Com G, Razzaq S, Thompson JE, et al. Impact of degree of obesity on sleep, quality of life, and depression in youth. J Pediatr Health Care 2018;32(2):e37–44.

[129] Rao MN, Blackwell T, Susan R, Stefanick ML, Ancoli-Israel S, Stone KL, et al. Association between sleep architecture and measures of body composition. Sleep 2009;32(4):483–90.

[130] Rutters F, Gonnissen HK, Hursel R, Lemmens SG, Martens EA, Westerterp-Plantenga MS. Distinct associations between energy balance and the sleep characteristics slow wave sleep and rapid eye movement sleep. Int J Obes 2012;36(10):1346–52.

[131] Theorell-Haglow J, Berne C, Janson C, Sahlin C, Lindberg E. Associations between short sleep duration and central obesity in women. Sleep 2010;33(5):593–8.

[132] Spaeth AM, Dinges DF, Goel N. Objective measurements of energy balance are associated with sleep architecture in healthy adults. Sleep 2017;40(1).

[133] St-Onge MP, Roberts A, Shechter A, Choudhury AR. Fiber and saturated fat are associated with sleep arousals and slow wave sleep. J Clin Sleep Med 2016;12(1):19–24.

[134] VanCauter E, Plat L, Scharf MB, Leproult R, Cespedes S, LHermiteBaleriaux M. Simultaneous stimulation of slow-wave sleep and growth hormone secretion by gamma-hydroxybutyrate in normal young men. J Clin Invest 1997;100(3):745–53.

[135] Arun R, Pina P, Rubin D, Erichsen D. Association between sleep stages and hunger scores in 36 children. Pediatr Obes 2016;11(5):E9–11.

[136] Wojnar J, Brower KJ, Dopp R, Wojnar M, Emslie G, Rintelmann J, et al. Sleep and body mass index in depressed children and healthy controls. Sleep Med 2010;11(3):295–301.

[137] Collet TH, van der Klaauw AA, Henning E, Keogh JM, Suddaby D, Dachi SV, et al. The sleep/wake cycle is directly modulated by changes in energy balance. Sleep 2016;39(9):1691–700.

[138] Kahlhofer J, Karschin J, Breusing N, Bosy-Westphal A. Relationship between actigraphy-assessed sleep quality and fat mass in college students. Obesity 2016;24(2):335–41.

[139] Moraes W, Poyares D, Zalcman I, de Mello MT, Bittencourt LR, Santos-Silva R, et al. Association between body mass index and sleep duration assessed by objective methods in a representative sample of the adult population. Sleep Med 2013;14(4):312–8.

[140] Wirth MD, Hebert JR, Hand GA, Youngstedt SD, Hurley TG, Shook RP, et al. Association between actigraphic sleep metrics and body composition. Ann Epidemiol 2015;25(10):773–8.

[141] Chamorro R, Algarin C, Garrido M, Causa L, Held C, Lozoff B, et al. Night time sleep macrostructure is altered in otherwise healthy 10-year-old overweight children. Int J Obes 2014;38(8):1120–5.

[142] Mcneil J, Tremblay MS, Leduc G, Boyer C, Belanger P, Leblanc AG, et al. Objectively-measured sleep and its association with adiposity and physical activity in a sample of Canadian children. J Sleep Res 2015;24(2):131–9.

[143] Allison KC, Spaeth A, Hopkins CM. Sleep and eating disorders. Curr Psychiatry Rep 2016;18(10):92.

[144] Haines J, McDonald J, O'Brien A, Sherry B, Bottino CJ, Schmidt ME, Taveras EM. Healthy Habits, Happy Homes: randomized trial to improve household routines for obesity prevention among preschool-aged children. JAMA Pediatr 2013;167(11):1072–9.

[145] Clifford LM, Beebe DW, Simon SL, Kuhl ES, Filigno SS, Rausch JR, Stark LJ. The association between sleep duration and weight in treatment-seeking preschoolers with obesity. Sleep Med 2012;13(8):1102–5.

[146] Sallinen BJ, Hassan F, Olszewski A, Maupin A, Hoban TF, Chervin RD, Woolford SJ. Longer weekly sleep duration predicts greater 3-month BMI reduction among obese adolescents attending a clinical multidisciplinary weight management program. Obes Facts 2013;6(3):239–46.

[147] Valrie CR, Bond K, Lutes LD, Carraway M, Collier DN. Relationship of sleep quality, baseline weight status, and weight-loss responsiveness in obese adolescents in an immersion treatment program. Sleep Med 2015;16(3):432–4.

[148] Ross KM, Graham Thomas J, Wing RR. Successful weight loss maintenance associated with morning chronotype and better sleep quality. J Behav Med 2016;39(3):465–71.

[149] Thomson CA, Morrow KL, Flatt SW, Wertheim BC, Perfect MM, Ravia JJ, Sherwood NE, Karanja N, Rock CL. Relationship between sleep quality and quantity and weight loss in women participating in a weight-loss intervention trial. Obesity (Silver Spring) 2012;20(7):1419–25.

[150] Chaput JP, Tremblay A. Sleeping habits predict the magnitude of fat loss in adults exposed to moderate caloric restriction. Obes Facts 2012;5(4):561–6.

[151] Nedeltcheva AV, Kilkus JM, Imperial J, Schoeller DA, Penev PD. Insufficient sleep undermines dietary efforts to reduce adiposity. Ann Intern Med 2010;153(7):435–41.

[152] Wang X, Sparks JR, Bowyer KP, Youngstedt SD. Influence of sleep restriction on weight loss outcomes associated with caloric restriction Sleep 2018;41(5):zsy027.

[153] Hart CN, Hawley NL, Wing RR. Development of a behavioral sleep intervention as a novel approach for pediatric obesity in school-aged children. Pediatr Clin N Am 2016;63(3):511–23.

[154] Van Dyk TR, Zhang N, Catlin PA, Cornist K, McAlister S, Whitacre C, Beebe DW. Feasibility and emotional impact of experimentally extending sleep in short-sleeping adolescents. Sleep 2017;40(9).

[155] Cizza G, Marincola P, Mattingly M, Williams L, Mitler M, Skarulis M, Csako G. Treatment of obesity with extension of sleep duration: a randomized, prospective, controlled trial. Clin Trials 2010;7(3):274–85.

[156] Tasali E, Chapotot F, Wroblewski K, Schoeller D. The effects of extended bedtimes on sleep duration and food desire in overweight young adults: a home-based intervention. Appetite 2014;80:220–4.

[157] Al Khatib HK, Hall WL, Creedon A, Ooi E, Masri T, McGowan L, Harding SV, Darzi J, Pot GK. Sleep extension is a feasible lifestyle intervention in free-living adults who are habitually short sleepers: a potential strategy for decreasing intake of free sugars? A randomized controlled pilot study. Am J Clin Nutr 2018;107(1):43–53.

[158] Haack M, Serrador J, Cohen D, Simpson N, Meier-Ewert H, Mullington JM. Increasing sleep duration to lower beat-to-beat blood pressure: a pilot study. J Sleep Res 2013;22(3):295–304.

[159] Leproult R, Deliens G, Gilson M, Peigneux P. Beneficial impact of sleep extension on fasting insulin sensitivity in adults with habitual sleep restriction. Sleep 2015;38(5):707–15.

Chapter 16

Insufficient sleep and cardiovascular disease risk

Sogol Javaheri[a], Omobimpe Omobomi[a], Susan Redline[a,b]
[a]Brigham and Women's Hospital, Harvard Medical School, Boston, MA, United States, [b]Beth Israel Deaconess Medical Center, Harvard Medical School, Boston, MA, United States

ABBREVIATIONS

ACTH	adrenocorticotropic hormone
CARDIA	Coronary Artery Risk Development in Young Adults
CHD	coronary heart disease
CRP	C-reactive protein
IL-6	interleukin-6
MONICA	monitoring trends and determinants on cardiovascular disease
NHANES	National Health and Nutrition Examination Survey
PSG	polysomnography
REGARDS	Reasons for Geographic And Racial Differences in Stroke
TNF-α	tumor necrosis factor-alpha

INTRODUCTION

Sleep deficiency, commonly used to represent habitual short sleep duration or reduced sleep quality, is a distinct entity yet overlaps with the clinical phenotype, insomnia. According to the National Sleep Foundation, healthy sleep duration in adults is 7–9 h each night [1]. However, elements of sleep quality may play a role in whether sleep is sufficient (see Fig. 16.1), including the continuity, depth, night to night variability, and perceived quality of sleep. Objective measurements of these aspects of sleep, made using actigraphy or polysomnography, include sleep efficiency (time spent sleeping divided by time spent in bed), wake after sleep onset, sleep latency (time to sleep onset), time and progression of sleep stages (particular time in deep, slow wave sleep and rapid eye movement sleep), and various summary measurements of numbers and patterns of sleep-wake bouts. Subjective sleep quality is assessed using standardized questionnaires or sleep diaries assessing an individual's satisfaction with his or her sleep and includes perceived concerns falling asleep, maintaining sleep, early awakenings and feeling refreshed after sleep. Most large epidemiologic studies examining insufficient sleep and cardiovascular disease have used the more readily available self-reported measurements of sleep duration and quality rather than objective recordings, resulting in a relative limitation of data on the association of sleep fragmentation or depth on health outcomes.

Chronic insomnia is defined as difficulty initiating or maintaining sleep, or early morning awakening, coupled with daytime impairment, for at least 3 months' duration. An estimated 10%–20% of Americans suffer from chronic insomnia [2], the most prevalent sleep disorder in the United States, while an estimated 1/3 of Americans are considered short or insufficient sleepers. Both insomnia and insufficient sleep impair attention and can contribute to impaired cognition and increase in risk of motor vehicle and industrial accidents, reduced work productivity, and increased healthcare costs [2]. Based on results of primarily observational data, both insufficient sleep and insomnia confer increased risk of cardiovascular disease, including hypertension, coronary heart disease, heart failure, stroke, arrhythmia, and cardiovascular mortality. Further, emerging data suggest that insomnia coupled with short sleep duration is associated with even stronger cardiovascular disease risk than either of these entities alone [3, 4]. While there is some conflicting data on the associations between cardiovascular disease and sleep, this may in part be due to varying definitions of insomnia and insufficient sleep, an active area of interest in the sleep research community. The mechanisms by which insufficient sleep and insomnia may contribute to cardiovascular disease risk are overlapping, and future trials and experimental research are needed to further delineate the pathways linking insufficient sleep with cardiovascular disease, and better define sleep-related risk factors. Throughout this chapter, we will attempt to distinguish associations with cardiovascular risk that are related

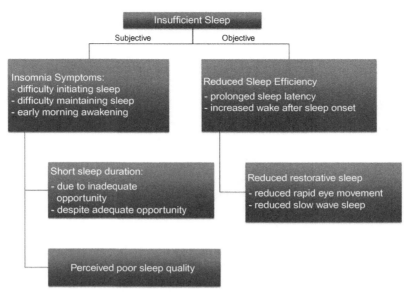

FIG. 16.1 Subjective and objective components of sleep quality and duration that comprise the umbrella term "insufficient sleep."

to insufficient sleep as well as with insomnia specifically. The high prevalence of these conditions and the associated risks and adverse health outcomes highlight the need for ongoing research as well as public education for improved identification and treatment in the population.

Defining insufficient sleep

Insufficient sleep may be defined as voluntary or involuntary sleep restriction. With behaviorally induced sleep deprivation, the individual has inadequate sleep due to failure to allow adequate time for sleep. Involuntary sleep restriction, however, may occur as a result of insomnia, circadian rhythm or other sleep disorders, another co-morbidity, or for a variety of other reasons. Varying cut-offs are used to define insufficient sleep, most commonly ranging from <5 to <7 h. The CDC defines healthy sleep in adults at least 7 h each night based on analysis of data showing consistency of findings associating sleeping <7 h with increased risk for adverse cardio-metabolic disorders including obesity, diabetes, heart disease, and all-cause mortality. However, epidemiologic studies have found that short sleep durations, particularly <5 and <6 h, predict higher rates of obesity or adverse cardiovascular outcomes as compared to 6 and 7 h of sleep [5–7].

Self-reported sleep duration can vary significantly compared to sleep duration measured by actigraphy or polysomnography (PSG), with correlations often <0.40 [8] and kappa values estimating the degree of agreement across categories of short, intermediate and long sleep duration weak or modest. Therefore, extrapolating thresholds for defining sleep duration from studies using one set of measurements to other settings needs to be done very cautiously.

Population-based data on other aspects of sleep deficiency and cardiovascular risk are sparse, limiting the ability to use standardized definitions for defining thresholds for prolonged sleep onset, low sleep efficiency, and poor sleep quality.

A variety of tools may be used to characterize insufficient sleep, including PSG, actigraphy, and subjective self-reporting tools such as questionnaires and sleep diaries. These tools may capture different aspects of sleep quality and provide different estimates of duration that reflect different pathophysiologic process. While polysomnography is not recommended for diagnosis of insomnia, it provides information on secondary contributors to insufficient sleep (e.g., sleep apnea, periodic limb movements) and quantifies sleep stage distributions. The latter may be particularly relevant to understanding cardiometabolic disease given that selected reduction of slow wave sleep is associated with incident hypertension [9], obesity [10], and metabolic dysfunction [11]. However, polysomnography is more burdensome than other methods and its general use for only one night in laboratory settings may reduce the representativeness of its estimates of sleep. Actigraphy, a watch-like device with an accelerometer to detect movement, is typically worn on the wrist for sleep estimation, and can be used to depict rest-activity patterns for days to weeks in the individual's typical environments. Various algorithms are applied to the measured activity counts that provide estimates of 24 h sleep patterns, including average and night to night variability in sleep duration, wake after sleep onset, and sleep efficiency. However, sleep onset latency, a key feature used to gauge insomnia severity, can be difficult to estimate due to vagaries in knowing the time of "lights off" or "in-bed" timing. Compared to polysomnography, sleep may be systematically overestimated for individuals who move little, or underestimated in individuals with sleep disorders or who are very active during sleep [12]. Questionnaires, including the Pittsburgh Sleep Quality Index, the Insomnia Severity Index, the Women's Health Initiative Insomnia Rating Scale, among others, are commonly used for evaluating insomnia in the research setting. Single-item questions on sleep duration and/or quality are also frequently used in large epidemiologic studies. These tools are generally administered at a single point in time and therefore may not reflect habitual sleep duration over time. Sleep diaries record subjective daily estimation of sleep and wake times and are considered a standard assessment of insomnia and sleep duration in clinical settings. The growing availability of electronic communication devices that can deliver research surveys provides the ability to deliver sleep diaries electronically, potentially improving compliance. Finally, questions on sleep can be asked at random and repeated intervals over time to gain information on changing sleep behaviors in real time using an approach known as ecological momentary assessment. Each of these tools ultimately capture different aspects of insomnia and insufficient sleep and pose their own challenges regarding accuracy and what specific aspect of sleep health is being measured.

PATHOPHYSIOLOGY

The pathophysiologic mechanisms underlying the associations between insufficient sleep and insomnia with cardiovascular disease are multifactorial and, though not fully understood, are also overlapping and will be discussed together here. General mechanisms include increased sympathetic activity resulting in elevated heart rate and blood pressure [13], dysregulation of the hypothalamic-pituitary-adrenal axis [14–16], increased inflammation [17–19], impaired glucose metabolism, vascular dysfunction [20], increased atherogenesis [17, 21] and obesity [22]. A prior review article on insomnia and risk of cardiovascular disease details the pathogenesis, and Fig. 16.2, modified from this review, summarizes the relationship between insufficient sleep, insomnia, and cardiovascular disease.

Compared to normal controls, human studies show that individuals with either insomnia or short sleep have elevated plasma and urine norepinephrine levels, increased heart rate and blood pressure, as well as blunted heart rate variability [23–25]. Increased sympathetic nervous system activity, one of the primary mechanisms thought to underlie the relationship between short sleep, insomnia, and cardiovascular disease, will be further addressed in the "Insufficient sleep and blood pressure" section.

Evidence from human studies also demonstrates increased ACTH and cortisol secretion, mechanisms that may lead to cardiovascular disease directly or indirectly through mediating pathways such as impaired glucose tolerance and diabetes. Elevations in C-reactive protein (CRP), tumor necrosis factor-alpha (TNF-α), and interleukin-6 (IL-6) have been demonstrated in short sleepers and individuals with insomnia [17, 19], markers that are implicated in multiple disease processes including obesity, diabetes, and atherosclerosis, among others. A causal association between inflammation and insomnia is supported by the results of a randomized control trial assessing the effect of cognitive behavioral therapy in older adults with insomnia, finding that effective insomnia treatment and remission resulted in

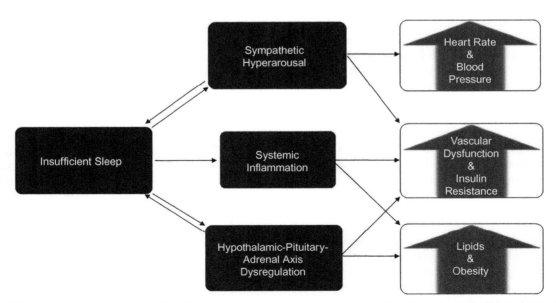

FIG. 16.2 Flow diagram demonstrating possible pathophysiologic mechanisms underlying the relationship between insufficient sleep and cardiovascular disease.

reductions in CRP levels [26]. Short sleep also results in diminished nitrous oxide availability, one potential pathway by which short sleep contributes to impaired endothelium-dependent vasodilation [20]. Finally, sleep loss potentiates weight gain and obesity by tilting the balance between energy intake and expenditure via increased appetite, altered eating behaviors, and reduced physical activity [27, 28]. While early studies suggested that increased caloric intake resulting from effects of sleep deprivation on increasing the appetite stimulating hormone ghrelin and reducing the appetite suppressing hormone leptin [29], recent research suggests that the mechanisms may be through effects on central brain reward centers [28].

Insufficient sleep and blood pressure

Arterial blood pressure is modulated by the autonomic nervous system and baroreceptors that are in part regulated by sleep and circadian rhythm. In normotensive individuals, the typical circadian rhythm of blood pressure is characterized by a 10%–20% nocturnal decrease, a phenomenon referred to as "dipping." This is followed by a morning surge in blood pressure. Nondipping, the absence of this nocturnal blood pressure decrease, has been observed in cardio-metabolic disorders including hypertension [30, 31], and is associated with increased mortality independent of daytime blood pressure [32]. It has been hypothesized that increased sympathetic outflow, decreased baroreflex sensitivity, and a higher baroreflex set-point [13] are some mechanisms by which sleep deprivation contributes to increased blood pressure [31]. Increased inflammation [33] and vascular dysfunction [34, 35] are other potential mechanisms for hypertension. Experimental studies of acute sleep deprivation in adult volunteers have generally demonstrated elevations in blood pressure and heart rate [35], particularly during the night and the following morning [36]. A number of epidemiologic studies throughout the world, both cross-sectional and prospective, have also demonstrated an association between reduced sleep duration and consolidation (measured both subjectively and objectively) and elevated blood pressure or hypertension. The Coronary Artery Risk Development in Young Adults (CARDIA) sleep study, an ancillary study to the CARDIA cohort study, showed that reduced sleep duration and lower sleep efficiency maintenance measured by actigraphy predicted increased higher systolic and diastolic blood pressures cross-sectionally as well as increased odds of incident hypertension [37]. A prospective study of 1715 Korean adults aged 40–70 who were free of hypertension were followed for a median of 2.6 years and participants with <6 h self-reported sleep duration demonstrated increased odds of incident hypertension (odds ratio 1.71, 95% confidence interval 1.01–2.89) compared to normal sleepers (6–7.9 h of sleep) [38]. In the Penn State Cohort, a random, general population sample of 1741 men and women, objective short sleep duration measured by polysomnography associated with elevated blood pressure [39], and objective short sleep was estimated to partially mediate the relationship between hypertension and all-cause mortality [40].

Two prospective studies have found interactions between age and sleep duration that influence the association of short sleep with incident hypertension. A prospective study of 3086 English men and women aged 50 and over demonstrated that self-reported short sleep was predictive of incident hypertension in men and women <60 years old but not in older people [33]. The National Health and Nutrition Examination Survey (NHANES) also found a significant interaction between age and sleep duration. They followed 4810 participants for 8–10 years for incident HTN and found that ≤5 h of self-reported short sleep was significantly associated with increased risk of incident hypertension (hazard ratio 1.51; 95% confidence interval 1.17–1.95) in younger (age 32–59 years) participants after adjusting for confounders but not in older individuals. They reported a higher percentage of patients developing hypertension who reported <7 h of sleep in their cohort compared to those with 7–8 h per night in the younger group only. They also found that obesity and diabetes were partial mediators of this association, as shown by the attenuation of associations after adding these variables to the models relating short sleep and incident HTN [41]. One of the most recent large epidemiologic studies on short sleep duration (sleep <6 h) showed an increase in the risk for elevated blood pressure by 8% (adjusted hazards ratio 1.08, confidence interval 1.04–1.13) in a population of 162,121 adult men and women free from major diseases including obesity [42]. This supports NHANES in that although obesity may be a partial mediator on the pathway between short sleep and high blood pressure, there are other underlying mechanisms independent of obesity at play.

The above findings have also been confirmed with meta-analyses that demonstrate increased risk of hypertension or elevated blood pressure in short sleepers [43], including metaanalysis restricted to prospective cohort studies [44]. Finally, short sleep and reduced sleep efficiency have also been associated with elevated blood pressure in adolescents [45, 46]. The latter is of particular importance given that confounding factors are likely to be fewer in younger generally healthy individuals compared to older individuals with multiple chronic diseases.

The insomnia literature also generally supports an association between insomnia and incident hypertension, though the data are more conflicting than that for short sleep, which may at least in part be due to the wider variation in how insomnia is defined [24]. There is also evidence that chronic insomnia coupled with objective short sleep duration is an even stronger predictor of incident hypertension than either sleep parameter alone. In the Penn State Cohort, chronic insomnia (present for at least 1 year) coupled with objective short sleep (<6h by polysomnography) was associated with an almost fourfold increased odds of incident hypertension (odds ratio 3.75, 95% confidence interval 1.58–8.95) as compared to chronic insomnia alone (odds ratio 2.24, 95% confidence interval 1.19–4.19) [47]. A cross-sectional observational study of 255 adults with clinically diagnosed insomnia found that individuals with insomnia and <6h sleep by polysomnography had increased prevalence of reported hypertension compared to those with ≥6h sleep, but found no significant differences in hypertension in those with insomnia and subjectively reported short sleep [48].

The associations between poorer sleep quality (reduced sleep efficiency and adverse changes in sleep architecture) and blood pressure are not as well defined as with sleep duration and insomnia. Multiple small studies demonstrate that reduced time in deeper stages of sleep (slow wave and REM sleep) as well as lower sleep efficiency are associated with blunted blood pressure dipping in healthy adults [49–51]. Poorer sleep quality may also contribute to the reduction in blood pressure dipping African-Americans experience compared to Caucasians [52]. Reduced sleep efficiency has been associated with prehypertension in adolescents in one study [45]. Ultimately, larger prospective studies are needed to further understand these associations, particularly since blunted dipping is associated with increased risk of cardiovascular mortality [53].

The evidence overwhelmingly suggests that insufficient sleep, particularly short sleep duration, increases risk of elevated blood pressure and incident hypertension, and there is also strong evidence supporting an association between insomnia and incident hypertension. To date however there are few published randomized control trials and no large studies assessing whether therapies targeted at extending sleep or improving sleep may improve blood pressure. A small randomized control study provides encouragement for future investigation. In this pilot study 22 participants with prehypertension or stage 1 hypertension and habitual sleep duration of <7h were randomized to receive 6-week of sleep extension or to a sleep-maintenance control group. 24h beat to beat systolic and diastolic blood pressure was collected over a 24h period pre and postintervention. 24h recordings in the sleep maintenance group demonstrated large decreases in systolic (14 ± 3 mmHg) and diastolic (8 ± 3 mmHg) blood pressure from pre to postintervention. While the decrease in the sleep-maintenance group was not significant (7 ± 5 and 5 ± 4 mmHg reduction in systolic and diastolic blood pressure, respectively), the difference between groups was not significant, which may reflect a lack of power in this small sample. Larger investigations are needed to test the effect of behavioral interventions on blood pressure reduction, and this is an emerging area of interest [54]. Currently there are trials underway evaluating cognitive behavioral therapy for insomnia and brief behavioral therapy for insomnia as a therapeutic target for lowering blood pressure.

Insufficient sleep and coronary heart disease

The effect of insufficient sleep on coronary heart disease (CHD) is more conflicting than that of insufficient sleep and blood pressure, and the relative effects of short sleep versus reduced sleep quality on incident CHD are also debated. There are no randomized control trials assessing sleep as a therapeutic target for CHD beyond treatment of obstructive sleep apnea. General mechanisms underlying a relationship between poor sleep and increased risk of CHD include increased blood pressure [55], diabetes [55], increased body mass index, and adverse alterations in markers of inflammation [56] and lipid levels [57] as already described. Mechanisms more specific to coronary heart disease include endothelial vascular dysfunction that could contribute to development and progression of atherosclerosis and subsequent acute coronary events. To this end, a study of 30 adult men demonstrated lower forearm blood flow response to intra-arterial infusion of acetylcholine in short sleepers (total sleep time approximately 6h) compared to normal sleeping controls (sleep duration of 7.7h) as well as diminished nitrous oxide availability. Their data suggest that impaired nitrous-oxide mediated endothelium-dependent vasodilation may be an important mechanism contributing to increased coronary heart disease risk in insufficient sleepers [20].

Overall, the largest prospective epidemiologic studies support the trend that short or insufficient sleep is associated with increased risk of CHD in both men and women [58, 59], despite some smaller studies finding an association in only one sex or the other [60, 61]. Additionally, some data suggest that the combination of short sleep and disturbed sleep better predicts incident adverse CHD events compared to short sleep duration or insomnia alone [62]. We will present data from some of the largest prospective studies below, and discuss potential reasons for discrepancies in the literature.

The monitoring trends and determinants on cardiovascular disease (MONICA) Augsburg cohort study, a population sample of 3508 and 3388 middle aged German men and women, reported a significant association between self-reported short sleep (≤5h) with incident myocardial infarction in women (hazard ratio 2.98, 95% confidence interval 1.48–6.03) but not men. They found no associations between self-reported difficulty initiating sleep or difficulty maintaining sleep and incident myocardial infarction. Participants were followed for a mean of 10 years, and health outcomes were ascertained by medical registries and death certificates. These results need to be cautiously interpreted due to the low number of incident cases (295 cases in men and 85 cases among women), particularly in those with shortest sleep duration, and while stratified analyses suggested sex differences, the test for statistical interaction between sex and sleep duration categories was not significant [61]. However, these findings suggest that there may be sex-specific variation in the effects of short sleep on cardiovascular outcomes due to different pathophysiologic manifestations of sleep deficit or responses to sleep disturbance in men compared to women and different expression of chronic inflammation in women compared to men [63].

In contrast, a smaller population-based Swedish cohort of 1870 men and women aged 45–65 found that poor sleep quality was associated with adverse cardiovascular outcomes in men rather than women and reported no relationship with short sleep in either sex. Specifically, self-reported difficulty initiating sleep, a key symptom of insomnia, was associated with increased coronary artery disease mortality in men but not women (relative risk 3.1, 95% confidence interval 1.5–6.3) [60]. Because there were fewer subjects in this study, the negative findings for short sleep and adverse cardiovascular outcomes may be related to lack of power. Alternatively, this study highlights that considering sleep duration without other information on sleep quality, particularly in small studies, may inadequately characterize sleep-related stressors relevant to cardiovascular disease.

When evaluating cohort studies with larger numbers of participants, short sleep duration has a positive association with risk of CHD. For example, a large prospective study of 71,617 women aged 45–65 from the Nurse's Health Study found that ≤5h self-reported sleep duration was predictive of incident CHD compared to 6 and >7h (relative risk 1.45, 95% confidence interval 1.1–1.92) [59].

The Whitehall II study, a large prospective study in London, England, sought to address the discrepancy of effect of insufficient sleep in men versus women as well as the relative effects of sleep duration versus sleep quality. 10,308 men and women were evaluated for sleep duration and sleep quality using questions from the General Health Quesitonnaire-30 and then were followed for a mean of 15 years for development of CHD events including fatal CHD deaths, incident nonfatal myocardial infarction, or incident angina. They found no differences by sex in sleep complaints and risk of CHD, and in multivariate models combining men and women found an increased risk of CHD events in those reporting disturbed sleep (hazard ratio 1.23, 95% confidence interval 1.07–1.43) but not self-reported short sleep (≤5h as well as between 5 and 6h). There was also some evidence of an interaction effect between short sleep and disturbed sleep, with participants with <6h sleep and report of low quality sleep showing the highest hazard rate for incident CHD events (hazard ratio 1.45, 95% confidence interval 1.24–1.7) after adjusting for confounders [62]. The authors suggest that disturbances of the physiologic processes occurring during sleep, rather than the duration of sleep itself, may be driving the increased CHD risk.

A large prospective study to date examining sleep quality, sleep duration, and risk of incident coronary heart disease in 60,586 Asian adults found that <6h self-reported sleep duration or difficulty falling asleep/use of sleeping pills were associated with increased risk of CHD (hazard ratio 1.13, 95% confidence interval 1.04–1.23 and 1.31, 95% confidence interval 1.16–1.47, respectively) [58]. The primary weaknesses in this study were that incident CHD events were also self-reported and sleep apnea was not accounted included as a covariate. The combined effects of sleep duration and sleep quality were not assessed.

A number of prospective observational studies have also shown an association between insomnia and risk of CHD and recurrent acute coronary syndrome [24]. The largest study collected information on subjective insomnia symptoms that impair work performance in 52,610 men and women followed for 11.4 years for first acute myocardial infarction. They found that difficulty initiating sleep, difficulty maintaining sleep, and complaint of nonrestorative sleep are all associated with increased risk of acute myocardial infarction in men and women, and that difficulty initiating sleep was the sleep symptom most strongly associated with incident myocardial infarction [64]. The second largest study used data from the Taiwan National Health Insurance Research Database and matched individuals with and without insomnia ($n=44,080$) by age, sex and comorbidity. Individuals were followed for 10 years for acute myocardial infarction, and those with insomnia had a 68% increased risk of developing an incident myocardial infarction (95% confidence interval 1.31–2.16) [65].

In summary, the largest prospective cohort studies report associations between reduced sleep duration and quality and risk of incident coronary heart disease [58, 59], with some evidence that combining information on short sleep and poor sleep quality identify individuals at high risk for coronary heart disease [62]. The insomnia literature also supports associations between insomnia and incident CHD. Generally <5h of sleep is more likely to be associated with risk of incident CHD events than sleep of higher durations, particularly if individuals also have reduced sleep quality. This is consistent with the insomnia literature that demonstrates insomnia coupled with objective short sleep may be a stronger predictor of adverse cardiovascular outcomes such as hypertension [47].

Insufficient sleep and heart failure

Inasmuch as insufficient sleep has been associated with hypertension and cardiovascular disease, it would seem reasonable to presume that a similar association exists with heart failure given the common underlying mechanisms for these conditions. However, data on the association between insufficient sleep and heart failure is limited and the relationship between heart failure and insufficient sleep is more difficult to understand given potential for bidirectionality. Heart failure may cause insufficient sleep due to symptoms such as paroxysmal nocturnal dyspnea or Cheyne Stoke Respiration, treatment with medications resulting in sleep fragmentation such as diuretics, or disease-related anxiety and depression. Subjective sleep assessments with use of standardized questionnaires in chronic heart failure patients have demonstrated difficulty initiating sleep, maintaining sleep, and shorter total sleep duration, and that heart failure patients with sleep complaints have reduced health-care quality of life as compared to those without sleep complaints [66].

One of the few prospective studies to examine the association between self-reported sleep duration and incident heart failure, the British Regional Study, followed 3723 older men with and without preexisting cardiovascular disease but without prevalent heart failure for approximately 9 years. Self-reported nighttime sleep duration of <6h was associated with heart failure risk in men with preexisting cardiovascular disease. Regardless of duration of nighttime sleep, men without preexisting cardiovascular disease did not have increased risk of heart failure, although daytime napping appeared to associate with increased heart failure risk in this group [67]. Notably, obstructive sleep apnea was not assessed in this cohort, and given that daytime napping may reflect an underlying diagnosis of obstructive sleep apnea, a common sleep disorder in older men, and that obstructive sleep apnea may be an independent risk factor for heart failure, it cannot be assumed that daytime napping was solely related to insufficient sleep in this cohort. As such, the relationship between insufficient sleep and heart failure in patients without preexisting cardiovascular disease is not as clear.

Currently, prospective studies examining the relationship between sleep duration and incident heart failure are very limited. Some of these studies have demonstrated a lack of association between self-reported nighttime sleep duration and heart failure in the overall population [68, 69]. However it is of interest that when symptoms of sleep disturbance coexisted with reported short sleep duration, there was a demonstrable association with a composite cardiovascular disease outcome that included heart failure, though not specific to heart failure [69].

Large prospective cohort studies have demonstrated similar findings with regards to an association between self-reported insomnia symptoms (difficulty initiating, difficulty maintaining sleep and nonrestorative sleep) and increased risk of incident heart failure [70, 71]. One such study was limited to a cohort of overweight middle-aged men and demonstrated that the observed relationship was independent of their established risk factors for heart failure [71]. Based on these data, insufficient sleep may potentially play a role in incident heart failure but it appears that overall sleep-related cardiovascular consequences are more pronounced in the presence of coexisting sleep symptoms. Short sleep duration appears to have a more evident impact on heart failure risk in patients with preexisting cardiovascular disease.

Insufficient sleep and stroke

Several of the changes observed with insufficient sleep are linked to mechanisms that potentiate risk factors for stroke discussed earlier such as hypertension, vascular dysfunction and inflammation.

The majority of these studies are limited by use of subjective reports of sleep duration. Short sleep (<6h) has been associated with increased risk for stroke in several studies. The European Prospective Investigation into Cancer and Nutrition (EPIC)-Potsdam Study followed 23,620 middle-aged men and women over 8 years and short sleepers were found to have a significantly increased risk for ischemic and hemorrhagic stroke [72]. A study of 93,175 postmenopausal women aged 50–79 years followed over 7.5 years in the Women's Health Initiative study found an association between short sleep (≤6h) and ischemic stroke in subjects without preexisting cardiovascular disease or diabetes at baseline [73]. The MONICA/KORA Augsburg cohort study followed subjects aged 25–74 years over a 14-year period and found short sleep (≤5h) to be significantly associated with a 2.3-fold increase in stroke in men [74]. Another study followed 5666 employed participants from the Reasons for Geographic And Racial Differences in Stroke (REGARDS) study aged ≥45 years, over a 3-year period, collecting data on self reported sleep duration and self reported stroke symptoms. In the overall sample short sleep duration was not associated with incident stroke symptoms in fully adjusted models, but normal weight individuals were at increased risk for stroke symptoms after stratification by BMI. The study concluded that short sleep duration (<6h) is prospectively associated with a fourfold increased risk of self-reported stroke symptoms in normal weight individuals after adjusting for demographics, stroke risk factors, health behaviors and diet (hazard ratio 4.2, 95% confidence interval 1.62–10.84) [75].

On the other hand, there have also been prospective studies that have not supported these associations [76–78]. A prospective study of 9692 stroke-free participants aged 42–81 from the European Prospective Investigation into Cancer-Norfolk cohort found that long sleep but not short sleep was associated with a higher risk of stroke [78].

There are also studies that have found that short sleep is independently associated with an increased risk of stroke [79, 80]. The Singapore Chinese Health Study, a cohort of 63,257 adults aged 45–74 showed that short sleep (≤5h), when compared to 7h of sleep duration was significantly associated with increased risk for ischemic and nonspecified stroke mortality (but not hemorrhagic stroke) in subjects with hypertension [80].

Insomnia is also implicated as a risk factor for stroke [81]. When insomnia symptoms coexist with objective short sleep duration (defined as <5 h on polysomnography), it represents a more severe phenotype and carries a higher risk for increased heart rate variability, hypertension, diabetes, neurocognitive impairment, and mortality when compared with insomnia with longer objective sleep [4, 47, 82].

A retrospective cohort study of 21,438 individuals with clinically diagnosed insomnia (with an ICD 9 diagnosis code of insomnia) and 64,314 age and sex matched noninsomniac controls taken from the Taiwan National Health Insurance Research database and tracked for 4 years, showed that individuals with insomnia have a 54% higher risk of developing stroke; this risk was highest in young adults aged 18–34 years. In addition, those with persistent insomnia had a higher 3-year cumulative incidence of stroke compared to the remission group. Patients with sleep apnea were excluded [81].

In conclusion, several large studies have established an association between short sleep and incident stroke. Mechanisms are likely similar to those relating sleep insufficiency with CHD. The coexistence of insomnia symptoms may confer a higher risk than short sleep alone [69].

CONCLUSIONS

The overall literature supports an association between insufficient sleep (whether defined as short sleep duration, reduced sleep quality, or insomnia symptoms) and incident cardiovascular disease, particularly hypertension, stroke, and coronary heart disease. Additionally, insomnia coupled with objective short sleep duration appears to be a high-risk phenotype with increased risk of developing incident cardiovascular disease. Mechanisms underlying the pathophysiology of this relationship including increased sympathetic activity, cortisol dysregulation, increased inflammation, and vascular dysfunction resulting from insufficient sleep. Larger epidemiologic studies evaluating insufficient sleep and heart failure are needed to have sufficient statistical power to address this potential association. The majority of studies do not account for obstructive sleep apnea or other underlying sleep disorders that may reduce sleep duration and/or quality, and therefore it is unclear the extent to which sleep disordered breathing may be confounding or mediating this relationship. Also, since most studies are based on self-report data, there is little known about how sleep fragmentation and sleep architecture influences cardiovascular risk. Additionally, future studies are needed assessing sleep quality/duration as a therapeutic target in modifying cardiovascular disease risk, particularly for hypertension, coronary heart disease, and stroke.

REFERENCES

[1] Hirshkowitz M, Whiton K, Albert SM, Alessi C, Bruni O, DonCarlos L, et al. National Sleep Foundation's sleep time duration recommendations: methodology and results summary. Sleep Health 2015;1(1):40–3.

[2] Buysse DJ. Insomnia. JAMA 2013;309(7):706–16.

[3] Vgontzas AN, Liao D, Pejovic S, Calhoun S, Karataraki M, Basta M, et al. Insomnia with short sleep duration and mortality: the Penn State cohort. Sleep 2010;33(9):1159–64.

[4] Vgontzas AN, Fernandez-Mendoza J, Liao D, Bixler EO. Insomnia with objective short sleep duration: the most biologically severe phenotype of the disorder. Sleep Med Rev 2013;17(4):241–54.

[5] Xiao Q, Arem H, Moore SC, Hollenbeck AR, Matthews CE. A large prospective investigation of sleep duration, weight change, and obesity in the NIH-AARP Diet and Health Study cohort. Am J Epidemiol 2013;178(11):1600–10.

[6] Ogilvie RP, Redline S, Bertoni AG, Chen X, Ouyang P, Szklo M, et al. Actigraphy measured sleep indices and adiposity: the multiethnic study of atherosclerosis (MESA). Sleep 2016;39(9):1701–8.

[7] Sepahvand E, Jalali R, Mirzaei M, Kargar Jahromi M. Association between short sleep and body mass index, hypertension among acute coronary syndrome patients in coronary care unit. Global J Health Sci 2014;7(3):134–9.

[8] Jackson CL, Patel SR, Jackson WB, Lutsey PL, Redline S. Agreement between self-reported and objectively measured sleep duration among white, black, Hispanic, and Chinese adults in the United States: multi-ethnic study of atherosclerosis. Sleep 2018;41(6). https://doi.org/10.1093/sleep/zsy057.

[9] Javaheri S, Zhao YY, Punjabi NM, Quan SF, Gottlieb DJ, Redline S. Slow-wave sleep is associated with incident hypertension: the Sleep Heart Health Study. Sleep 2018;41(1). https://doi.org/10.1093/sleep/zsx179.

[10] Patel SR, Hayes AL, Blackwell T, Evans DS, Ancoli-Israel S, Wing YK, et al. The association between sleep patterns and obesity in older adults. Int J Obes 2014;38(9):1159–64.

[11] Tasali E, Leproult R, Ehrmann DA, Van Cauter E. Slow-wave sleep and the risk of type 2 diabetes in humans. Proc Natl Acad Sci U S A 2008;105(3):1044–9.

[12] Blackwell T, Paudel M, Redline S, Ancoli-Israel S, Stone KL. Group OFiMMS. A novel approach using actigraphy to quantify the level of disruption of sleep by in-home polysomnography: the MrOS Sleep Study: sleep disruption by polysomnography. Sleep Med 2017;32:97–104.

[13] Ogawa Y, Kanbayashi T, Saito Y, Takahashi Y, Kitajima T, Takahashi K, et al. Total sleep deprivation elevates blood pressure through arterial baroreflex resetting: a study with microneurographic technique. Sleep 2003;26(8):986–9.

[14] Castro-Diehl C, Diez Roux AV, Redline S, Seeman T, Shrager SE, Shea S. Association of sleep duration and quality with alterations in the hypothalamic-pituitary adrenocortical axis: the multiethnic study of atherosclerosis (MESA). J Clin Endocrinol Metab 2015;100(8):3149–58.

[15] Floam S, Simpson N, Nemeth E, Scott-Sutherland J, Gautam S, Haack M. Sleep characteristics as predictor variables of stress systems markers in insomnia disorder. J Sleep Res 2015;24(3):296–304.

[16] Vgontzas AN, Bixler EO, Lin HM, Prolo P, Mastorakos G, Vela-Bueno A, et al. Chronic insomnia is associated with nyctohemeral activation of the hypothalamic-pituitary-adrenal axis: clinical implications. J Clin Endocrinol Metab 2001;86(8):3787–94.

[17] Meier-Ewert HK, Ridker PM, Rifai N, Regan MM, Price NJ, Dinges DF, et al. Effect of sleep loss on C-reactive protein, an inflammatory marker of cardiovascular risk. J Am Coll Cardiol 2004;43(4):678–83.

[18] Parthasarathy S, Vasquez MM, Halonen M, Bootzin R, Quan SF, Martinez FD, et al. Persistent insomnia is associated with mortality risk. Am J Med 2015;128(3):268–75. e2.

[19] Shearer WT, Reuben JM, Mullington JM, Price NJ, Lee BN, Smith EO, et al. Soluble TNF-alpha receptor 1 and IL-6 plasma levels in humans subjected to the sleep deprivation model of spaceflight. J Allergy Clin Immunol 2001;107(1):165–70.

[20] Bain AR, Weil BR, Diehl KJ, Greiner JJ, Stauffer BL, DeSouza CA. Insufficient sleep is associated with impaired nitric oxide-mediated endothelium-dependent vasodilation. Atherosclerosis 2017;265:41–6.

[21] Alter-Wolf S, Blomberg BB, Riley RL. Deviation of the B cell pathway in senescent mice is associated with reduced surrogate light chain expression and altered immature B cell generation, phenotype, and light chain expression. J Immunol 2009;182(1):138–47.

[22] Wu Y, Zhai L, Zhang D. Sleep duration and obesity among adults: a meta-analysis of prospective studies. Sleep Med 2014;15(12):1456–62.

[23] Khan MS, Aouad R. The effects of insomnia and sleep loss on cardiovascular disease. Sleep Med Clin 2017;12(2):167–77.

[24] Javaheri S, Redline S. Insomnia and risk of cardiovascular disease. Chest 2017;152(2):435–44.

[25] Spiegelhalder K, Fuchs L, Ladwig J, Kyle SD, Nissen C, Voderholzer U, et al. Heart rate and heart rate variability in subjectively reported insomnia. J Sleep Res 2011;20(1 Pt 2):137–45.

[26] Irwin MR, Olmstead R, Carrillo C, Sadeghi N, Breen EC, Witarama T, et al. Cognitive behavioral therapy vs. Tai Chi for late life insomnia and inflammatory risk: a randomized controlled comparative efficacy trial. Sleep 2014;37(9):1543–52.

[27] Cassidy S, Chau JY, Catt M, Bauman A, Trenell MI. Low physical activity, high television viewing and poor sleep duration cluster in overweight and obese adults; a cross-sectional study of 398,984 participants from the UK Biobank. Int J Behav Nutr Phys Act 2017;14(1):57.

[28] St-Onge MP, Wolfe S, Sy M, Shechter A, Hirsch J. Sleep restriction increases the neuronal response to unhealthy food in normal-weight individuals. Int J Obes 2014;38(3):411–6.

[29] Koo DL, Nam H, Thomas RJ, Yun CH. Sleep disturbances as a risk factor for stroke. J Stroke 2018;20(1):12–32.

[30] Rahman A, Hasan AU, Nishiyama A, Kobori H. Altered circadian timing system-mediated non-dipping pattern of blood pressure and associated cardiovascular disorders in metabolic and kidney diseases. Int J Mol Sci 2018;19(2). https://doi.org/10.3390/ijms19020400.

[31] Carrillo-Larco RM, Bernabe-Ortiz A, Sacksteder KA, Diez-Canseco F, Cárdenas MK, Gilman RH, et al. Association between sleep difficulties as well as duration and hypertension: is BMI a mediator? Glob Health Epidemiol Genom 2017;2:e12.

[32] Brotman DJ, Davidson MB, Boumitri M, Vidt DG. Impaired diurnal blood pressure variation and all-cause mortality. Am J Hypertens 2008;21(1):92–7.

[33] Jackowska M, Steptoe A. Sleep and future cardiovascular risk: prospective analysis from the English Longitudinal Study of Ageing. Sleep Med 2015;16(6):768–74.

[34] Sauvet F, Leftheriotis G, Gomez-Merino D, Langrume C, Drogou C, Van Beers P, et al. Effect of acute sleep deprivation on vascular function in healthy subjects. J Appl Physiol (1985) 2010;108(1):68–75.

[35] Sauvet F, Drogou C, Bougard C, Arnal PJ, Dispersyn G, Bourrilhon C, et al. Vascular response to 1 week of sleep restriction in healthy subjects. A metabolic response? Int J Cardiol 2015;190:246–55.

[36] Lusardi P, Zoppi A, Preti P, Pesce RM, Piazza E, Fogari R. Effects of insufficient sleep on blood pressure in hypertensive patients: a 24-h study. Am J Hypertens 1999;12(1 Pt 1):63–8.

[37] Knutson KL, Van Cauter E, Rathouz PJ, Yan LL, Hulley SB, Liu K, et al. Association between sleep and blood pressure in midlife: the CARDIA sleep study. Arch Intern Med 2009;169(11):1055–61.

[38] Yadav D, Hyun DS, Ahn SV, Koh SB, Kim JY. A prospective study of the association between total sleep duration and incident hypertension. J Clin Hypertens (Greenwich) 2017;19(5):550–7.

[39] Fernandez-Mendoza J, He F, LaGrotte C, Vgontzas AN, Liao D, Bixler EO. Impact of the metabolic syndrome on mortality is modified by objective short sleep duration. J Am Heart Assoc 2017;6(5). https://doi.org/10.1161/JAHA.117.005479.

[40] Fernandez-Mendoza J, He F, Vgontzas AN, Liao D, Bixler EO. Objective short sleep duration modifies the relationship between hypertension and all-cause mortality. J Hypertens 2017;35(4):830–6.

[41] Gangwisch JE, Heymsfield SB, Boden-Albala B, Buijs RM, Kreier F, Pickering TG, et al. Short sleep duration as a risk factor for hypertension: analyses of the first National Health and Nutrition Examination Survey. Hypertension 2006;47(5):833–9.

[42] Deng HB, Tam T, Zee BC, Chung RY, Su X, Jin L, et al. Short sleep duration increases metabolic impact in healthy adults: a population-based cohort study. Sleep 2017;40(10). https://doi.org/10.1093/sleep/zsx130.

[43] Wang Y, Mei H, Jiang YR, Sun WQ, Song YJ, Liu SJ, et al. Relationship between duration of sleep and hypertension in adults: a meta-analysis. J Clin Sleep Med 2015;11(9):1047–56.

[44] Meng L, Zheng Y, Hui R. The relationship of sleep duration and insomnia to risk of hypertension incidence: a meta-analysis of prospective cohort studies. Hypertens Res 2013;36(11):985–95.

[45] Javaheri S, Storfer-Isser A, Rosen CL, Redline S. Sleep quality and elevated blood pressure in adolescents. Circulation 2008;118(10):1034–40.

[46] Bal C, Öztürk A, Çiçek B, Özdemir A, Zararsız G, Ünalan D, et al. The relationship between blood pressure and sleep duration in Turkish children: a cross-sectional study. J Clin Res Pediatr Endocrinol 2018;10(1):51–8.

[47] Fernandez-Mendoza J, Vgontzas AN, Liao D, Shaffer ML, Vela-Bueno A, Basta M, et al. Insomnia with objective short sleep duration and incident hypertension: the Penn State Cohort. Hypertension 2012;60(4):929–35.

[48] Bathgate CJ, Edinger JD, Wyatt JK, Krystal AD. Objective but not subjective short sleep duration associated with increased risk for hypertension in individuals with insomnia. Sleep 2016;39(5):1037–45.

[49] Hinderliter AL, Routledge FS, Blumenthal JA, Koch G, Hussey MA, Wohlgemuth WK, et al. Reproducibility of blood pressure dipping: relation to day-to-day variability in sleep quality. J Am Soc Hypertens 2013;7(6):432–9.

[50] Loredo JS, Nelesen R, Ancoli-Israel S, Dimsdale JE. Sleep quality and blood pressure dipping in normal adults. Sleep 2004;27(6):1097–103.

[51] Silva AP, Moreira C, Bicho M, Paiva T, Clara JG. Nocturnal sleep quality and circadian blood pressure variation. Rev Port Cardiol 2000;19(10):991–1005.

[52] Sherwood A, Routledge FS, Wohlgemuth WK, Hinderliter AL, Kuhn CM, Blumenthal JA. Blood pressure dipping: ethnicity, sleep quality, and sympathetic nervous system activity. Am J Hypertens 2011;24(9):982–8.

[53] Dolan E, Stanton A, Thijs L, Hinedi K, Atkins N, McClory S, et al. Superiority of ambulatory over clinic blood pressure measurement in predicting mortality: the Dublin outcome study. Hypertension 2005;46(1):156–61.

[54] Haack M, Serrador J, Cohen D, Simpson N, Meier-Ewert H, Mullington JM. Increasing sleep duration to lower beat-to-beat blood pressure: a pilot study. J Sleep Res 2013;22(3):295–304.

[55] Yaggi HK, Araujo AB, McKinlay JB. Sleep duration as a risk factor for the development of type 2 diabetes. Diabetes Care 2006;29(3):657–61.

[56] Vgontzas AN, Zoumakis E, Bixler EO, Lin HM, Follett H, Kales A, et al. Adverse effects of modest sleep restriction on sleepiness, performance, and inflammatory cytokines. J Clin Endocrinol Metab 2004;89(5):2119–26.

[57] Kinuhata S, Hayashi T, Sato KK, Uehara S, Oue K, Endo G, et al. Sleep duration and the risk of future lipid profile abnormalities in middle-aged men: the Kansai Healthcare Study. Sleep Med 2014;15(11):1379–85.

[58] Lao XQ, Liu X, Deng HB, Chan TC, Ho KF, Wang F, et al. Sleep quality, sleep duration, and the risk of coronary heart disease: a prospective cohort study with 60,586 adults. J Clin Sleep Med 2018;14(1):109–17.

[59] Ayas NT, White DP, Manson JE, Stampfer MJ, Speizer FE, Malhotra A, et al. A prospective study of sleep duration and coronary heart disease in women. Arch Intern Med 2003;163(2):205–9.

[60] Mallon L, Broman JE, Hetta J. Sleep complaints predict coronary artery disease mortality in males: a 12-year follow-up study of a middle-aged Swedish population. J Intern Med 2002;251(3):207–16.

[61] Meisinger C, Heier M, Löwel H, Schneider A, Döring A. Sleep duration and sleep complaints and risk of myocardial infarction in middle-aged men and women from the general population: the MONICA/KORA Augsburg cohort study. Sleep 2007;30(9):1121–7.

[62] Chandola T, Ferrie JE, Perski A, Akbaraly T, Marmot MG. The effect of short sleep duration on coronary heart disease risk is greatest among those with sleep disturbance: a prospective study from the Whitehall II cohort. Sleep 2010;33(6):739–44.

[63] Ishii S, Karlamangla AS, Bote M, Irwin MR, Jacobs DR, Cho HJ, et al. Gender, obesity and repeated elevation of C-reactive protein: data from the CARDIA cohort. PLoS ONE 2012;7(4):e36062.

[64] Laugsand LE, Vatten LJ, Platou C, Janszky I. Insomnia and the risk of acute myocardial infarction: a population study. Circulation 2011;124(19):2073–81.

[65] Hsu CY, Chen YT, Chen MH, Huang CC, Chiang CH, Huang PH, et al. The association between insomnia and increased future cardiovascular events: a nationwide population-based study. Psychosom Med 2015;77(7):743–51.

[66] Broström A, Strömberg A, Dahlström U, Fridlund B. Sleep difficulties, daytime sleepiness, and health-related quality of life in patients with chronic heart failure. J Cardiovasc Nurs 2004;19(4):234–42.

[67] Wannamethee SG, Papacosta O, Lennon L, Whincup PH. Self-reported sleep duration, napping, and incident heart failure: prospective associations in the British regional heart study. J Am Geriatr Soc 2016;64(9):1845–50.

[68] Kim Y, Wilkens LR, Schembre SM, Henderson BE, Kolonel LN, Goodman MT. Insufficient and excessive amounts of sleep increase the risk of premature death from cardiovascular and other diseases: the Multiethnic Cohort Study. Prev Med 2013;57(4):377–85.

[69] Westerlund A, Bellocco R, Sundström J, Adami HO, Åkerstedt T, Trolle Lagerros Y. Sleep characteristics and cardiovascular events in a large Swedish cohort. Eur J Epidemiol 2013;28(6):463–73.

[70] Laugsand LE, Strand LB, Platou C, Vatten LJ, Janszky I. Insomnia and the risk of incident heart failure: a population study. Eur Heart J 2014;35(21):1382–93.

[71] Ingelsson E, Lind L, Arnlöv J, Sundström J. Sleep disturbances independently predict heart failure in overweight middle-aged men. Eur J Heart Fail 2007;9(2):184–90.

[72] von Ruesten A, Weikert C, Fietze I, Boeing H. Association of sleep duration with chronic diseases in the European Prospective Investigation into Cancer and Nutrition (EPIC)-Potsdam study. PLoS ONE 2012;7(1):e30972.

[73] Chen JC, Brunner RL, Ren H, Wassertheil-Smoller S, Larson JC, Levine DW, et al. Sleep duration and risk of ischemic stroke in postmenopausal women. Stroke 2008;39(12):3185–92.

[74] Helbig AK, Stöckl D, Heier M, Ladwig KH, Meisinger C. Symptoms of insomnia and sleep duration and their association with incident strokes: findings from the population-based MONICA/KORA augsburg cohort study. PLoS ONE 2015;10(7):e0134480.

[75] Ruiter Petrov ME, Letter AJ, Howard VJ, Kleindorfer D. Self-reported sleep duration in relation to incident stroke symptoms: nuances by body mass and race from the REGARDS study. J Stroke Cerebrovasc Dis 2014;23(2):e123–32.

[76] Leng Y, Cappuccio FP, Wainwright NW, Surtees PG, Luben R, Brayne C, et al. Sleep duration and risk of fatal and nonfatal stroke: a prospective study and meta-analysis. Neurology 2015;84(11):1072–9.

[77] Kawachi T, Wada K, Nakamura K, Tsuji M, Tamura T, Konishi K, et al. Sleep duration and the risk of mortality from stroke in Japan: the takayama cohort study. J Epidemiol 2016;26(3):123–30.

[78] He Q, Sun H, Wu X, Zhang P, Dai H, Ai C, et al. Sleep duration and risk of stroke: a dose-response meta-analysis of prospective cohort studies. Sleep Med 2017;32:66–74.

[79] Li W, Wang D, Cao S, Yin X, Gong Y, Gan Y, et al. Sleep duration and risk of stroke events and stroke mortality: a systematic review and meta-analysis of prospective cohort studies. Int J Cardiol 2016;223:870–6.

[80] Pan A, De Silva DA, Yuan JM, Koh WP. Sleep duration and risk of stroke mortality among Chinese adults: Singapore Chinese health study. Stroke 2014;45(6):1620–5.

[81] Wu MP, Lin HJ, Weng SF, Ho CH, Wang JJ, Hsu YW. Insomnia subtypes and the subsequent risks of stroke: report from a nationally representative cohort. Stroke 2014;45(5):1349–54.

[82] Fernandez-Mendoza J, Calhoun S, Bixler EO, Pejovic S, Karataraki M, Liao D, et al. Insomnia with objective short sleep duration is associated with deficits in neuropsychological performance: a general population study. Sleep 2010;33(4):459–65.

Chapter 17

Sleep health and diabetes: The role of sleep duration, subjective sleep, sleep disorders, and circadian rhythms on diabetes

Azizi A. Seixas[a,b], Rebecca Robbins[a], Alicia Chung[a], Collin Popp[a], Tiffany Donley[a], Samy I. McFarlane[c], Jesse Moore[a], Girardin Jean-Louis[a,b]

[a]NYU Langone Health, Department of Population Health, New York, NY, United States, [b]NYU Langone Health, Department of Psychiatry, New York, NY, United States, [c]SUNY Downstate School of Medicine, New York, NY, United States

SLEEP PARAMETERS AND DIABETES RISK

Sleep health is a multifactorial and heterogeneous phenomenon that describes an organism's circadian 24-h sleep-wake cycle biological and behavioral profile. Framing sleep within a circadian biology perspective re-conceptualizes how we understand the sleep-wake cycle and how behaviors and biological processes during day and night influence each other. From this perspective, sleep consists of several: (1) subjective parameters, such as sleep quality/satisfaction, alertness, sleepiness; and (2) objective parameters, such as sleep duration, efficiency (sleep latency, wake after sleep onset), and timing. Abnormal levels of these parameters have been linked with adverse functional and health outcomes, such as accidents, cognitive impairment, impaired performance, cardiovascular-cardiometabolic health conditions (obesity, hypertension, and diabetes), poor brain health (dementia and accelerated aging), mental health, and mortality. The focus of the current chapter is to describe extant evidence linking sleep parameters and diabetes risk/diabetes outcomes. Our epistemological stance throughout the chapter is that associations between sleep disturbance and diabetes are bidirectional, where poor sleep is considered an antecedent and a consequence of diabetes, which has been buttressed by seminal systematic reviews and metaanalyses. Relative to other behavioral risk factors, such as diet and exercise, sleep is as strong of a risk factor for cardiometabolic conditions, such as diabetes [1, 2].

Sleep duration and diabetes

The unequivocal links between sleep duration and diabetes risk, outcomes, and surrogate biomarkers, such as unhealthy glucose levels (indicative of poor glycemic control), insulin resistance, advanced glycated end products, and obesity, are supported by overwhelming epidemiological and experimental evidence. Epidemiological studies in the United States and globally have shown a high prevalence of short (≤6h/24h period) and long sleep (≥9h/24h period) durations being strongly associated with diabetes risk and diabetes outcomes. Additionally, experimental and biological studies have substantiated the link between sleep duration and diabetes by providing mechanistic explanations and pathways as to how habitual short or long sleep durations engender diabetes.

Insufficient/short sleep duration's influence on type 2 diabetes, insulin resistance, and obesity: Mechanisms

Insufficient/short sleep duration (≤6h in a 24h period) is associated with type II diabetes, diabetes outcomes (such as insulin resistance and unhealthy glucose), and preclinical and surrogate biomarkers such as obesity and prediabetes, in cross-sectional, experimental, and prospective studies. Some of the most convincing cross-sectional and epidemiological evidence indicates that chronic sleep deprivation is associated with diabetes and related outcomes such as unhealthy glucose levels, insulin resistance, and insulin sensitivity. Gangwisch et al. [3], in a metaanalysis of seven prospective studies, reported that individuals with insufficient/short sleep were 28% more likely to be at risk for type 2 diabetes. Engeda et al. [4] found that individuals who reported sleeping ≤5 h/night were two times more likely to report prediabetes (fasting glucose 5.6–6.9 mmol/L) than those who slept 7–9 h. In a nationally representative study with a sample of 11, 815 individuals from the National Health and Nutrition Examination Survey 2005–10 dataset, Ford et al. [5] found that insufficient/short sleepers compared to average sleepers (7–9 h in a 24 h period) were more likely to have greater levels of adjusted mean levels of insulin, 2-h glucose and hemoglobin A1c. They also found that levels of insulin and 2-h glucose varied by sex and race/ethnicity. Specifically, adjusted mean levels of insulin varied by sex between insufficient/short sleepers and average sleepers (insufficient/short sleepers: male = 60.7 ± 1.5 pmol/L and female = 51.2 ± 1.4 pmol/L; average sleepers: male = 56.0 ± 1.1 and female = 52.3 ± 1.1 pmol/L). While, adjusted mean levels of 2-h glucose varied by race/ethnicity between insufficient/short sleepers and average sleepers (insufficient/short sleepers: Whites = 6.6 ± 0.1 mmol/L, Blacks = 6.1 ± 0.1 pmol/L, Mexican American = 6.8 ± 0.2 mmol/L; average sleepers: Whites = 6.4 ± 0.1, Blacks = 6.3 ± 0.1 mmol/L, and Mexican American = 6.5 ± 0.2 mmol/L).

Several mechanisms have been proffered to better understand the sleep-diabetes association. In one widely accepted mechanism, chronic sleep deprivation resulting in sleep debt induces metabolic dysfunction (such as excess adiposity and insulin resistance) as well as the development and progression of diabetes. Arora et al. [6] demonstrated this in their 12-month prospective investigation of an intensive lifestyle intervention (usual care, diet, and physical activity) on diabetes outcomes across five sites in the United Kingdom. They found that individuals with sleep debt (chronic sleep deprivation and insufficient sleep) compared to those without were 72% more likely to be obese. Also, individuals with sleep debt became progressively worse over time, where at 6 and 12 months they were two-three times more likely to be obese (body mass index [BMI]; ≥30 kg/m^2) or insulin resistant. Similarly, Kim et al. [7] found in a cohort study of 17, 983 Korean adults that prediabetics who reported insufficient/short sleep duration were 44%–68% more likely to develop diabetes and that these associations were mediated by adiposity, fatty liver, and insulin resistance.

It appears that the adverse effects of chronic sleep deprivation and habitual short sleep on diabetes risk occurs across all age groups. Dutil and Chaput [8] in a narrative review of 23 studies and Hancox and Landhuis [8a] in a cohort study found that children, adolescents, and young adults who are habitual short sleepers are at increased risk of developing prediabetes and diabetes [8, 9]. Leng et al. [10] found in an 8-year prospective study that napping, a compensatory method often used to mitigate adverse effects of sleep deprivation and sleepiness, was not associated with reduced odds of diabetes. In fact, they found that individuals who: (1) napped had a 30%–58% greater likelihood of developing diabetes over time adjusting for age, sex, BMI and waist circumference across different regression models; (2) were short sleepers were 46% more likely to develop diabetes compared to average sleepers; and (3) were habitual short sleepers and napped during the day had double the odds of diabetes, as compared to average sleepers who did not nap. These findings suggest that napping and insufficient/short sleep may have an adverse synergistic effect on diabetes.

A second mechanism that has garnered significant traction posits that significant sleep deprivation may lead to lower glucose tolerance and insulin sensitivity. Spiegel et al. [11] and Buxton et al. [12] found that individuals who slept 4–5 h per night for a week had lower glucose tolerance and insulin sensitivity. Chronic sleep deprivation (4–5 h for >4 days) and the reduction of insulin sensitivity has significant physiological and health consequences as they may modify cellular integrity in subcutaneous adipocytes [13]. However, the adverse effects of chronic sleep deprivation on abnormal insulin production and metabolism may only be ephemeral as data shows that the body compensates over a period of time. Robertson et al. [14] found that after restricting individuals' sleep by 1.5 h for 3 weeks, their insulin sensitivity either rebounded to normal levels or did not worsen after the first week of unstable insulin levels and insulin insensitivity.

A third mechanism suggests that chronic sleep loss, as a result of insufficient/short sleep, modifies cellular and tissue activity and integrity (i.e., Beta cell functioning responsible for the production of insulin) which in turn compromises homeostatic metabolic pathways and engenders insulin resistance. Rao and colleagues found that sleep restriction (4 h in bed) induced a reduction in whole-body and tissue-specific insulin sensitivity [15]. In another study, early morning nonesterified/free fatty acid levels were greater in young healthy men after sleep restriction compared to normal sleep which may partially contribute to insulin resistance [16]. At the cellular level, adipose tissue insulin sensitivity, determined by the insulin concentration for the half-maximal stimulation of the pAkt/tAkt ratio, decreased during sleep restriction [13]. In summary, repeated bouts of restricted sleep may induce chronic hyperinsulinemia, stimulating downstream pathways like pancreatic beta cell failure and lipogenesis, driving the development of diabetes and obesity, respectively.

Indirect relationship between insufficient/short sleep and diabetes

A fierce debate ensues about whether the relationship between sleep and diabetes is direct or indirect. A growing body of research indicates that the association between insufficient/short sleep and diabetes may be moderated or mediated by sleep disorders (obstructive sleep apnea and insomnia) as well as biological, demographic, and psychosocial factors such as obesity, sex, and glycemic control. Overwhelming evidence from observational and experimental studies indicate that the insufficient/short sleep and diabetes relationship is moderated or mediated by excess adiposity (central or peripheral), either in the form of high body mass index or waist circumference. According to Gohil and Hannon [17], insufficient/short sleep is associated with an unhealthy diet and increased cardiovascular and cardiometabolic risk factors including insulin resistance and hyperglycemia, which in turn leads to elevated diabetes risk. In spite of compelling evidence that obesity and elevated waist circumference may mediate the relationship between insufficient/short sleep and diabetes, there is contradictory evidence that there may be a genetic/hereditary component to this link.

A study found that lean young adults with a history of diabetes and habitual short sleep duration were susceptible to diabetes compared to those who slept >6h each day [18]. The authors suggest that chronic insufficient/short sleep can alter an individual's circadian biology (a topic we will address in a later section and is described in greater detail in another chapter in this book), thus altering the endogenous timing of key biological and cardiometabolic processes. It is believed that such endogenous alterations and dysregulations adversely affect the quality of sleep and may lead to negative health outcomes, specifically maladaptive and nonhomeostatic glycemic control. There is further debate as to whether the biological and cardiometabolic processes disrupted by habitual insufficient/short sleep are reversible, as some believe that chronic sleep deprivation epigenetically induces an unhealthy metabolic memory whereby an individual is unable to process insulin regardless of the quantity and quality of food ingested [18, 19]. Though the literature is unsettled, new findings indicating that extending sleep among short sleepers may have positive effects on cardiometabolic health may further buttress the argument that insufficient/short sleep is inextricably linked to diabetes (Fig. 17.1) [20].

Long sleep duration's influence on type 2 diabetes, insulin resistance, and obesity

Based on the overwhelming evidence that short sleep is a significant risk factor in the development and maintenance of diabetes, the perfunctory assumption is that increasing sleep duration will mitigate any adverse consequences of habitual insufficient/short sleep. In fact, studies show that extending sleep was associated with improvements in insulin sensitivity [20]. However, the evidence appears to be more nuanced where longer sleep duration is generally associated with decreased diabetes risk. However, studies have indicated that long sleep duration (>8h/24h period) and increasing sleep duration drastically may have adverse health effects [21, 22].

Protective effects of longer sleep and diabetes

Byberg et al. [23] in a study of 771 participants in Denmark investigated the effect short or long sleep durations have on glucose homeostasis and tolerance. The researchers found that an additional hour of sleep was associated with 0.3mmol/mol (0.3%) and 25% reduction in HbA(1c) and impaired glucose regulation, respectively. These findings support a conventional view that longer sleep may protect or stave off diabetes. However, recent findings are beginning to show contradictory evidence.

Negative associations between long sleep on diabetes

Although longer sleep duration may protect an individual from diabetes, there is growing evidence that too much sleep may have an adverse effect. Cespedes et al. [21] found that an extreme increase in sleep duration by 2 or more hours among female habitual short sleepers increased their risk for diabetes. Ferrie et al. [23a] found similar findings as Cespedes

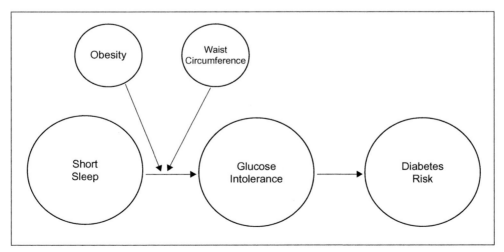

FIG. 17.1 The main and interactive effects of short sleep on diabetes risk, via glucose intolerance, obesity and waist circumference.

and colleagues in the UK Whitehall Study, when they investigated the association between change in sleep duration and incident diabetes, over a 20-year span [24]. They found that compared to individuals who slept 7h/24h period, those who reported a ≥2h sleep increase had a 65% greater likelihood of diabetes (OR=1.65, 95% CI=1.15–2.37) after adjusting for age, employment status ethnicity, and sex. Interestingly, BMI and weight change attenuated the association between long sleep and diabetes (OR=1.50, 95% CI=1.04–2.16) suggesting that the relationship between long sleep and diabetes may be indirect or mediated by obesity.

Indirect relationship between long sleep and diabetes

Similar to Gohil and Hannon [17] (in the short sleep duration section above) and Ferrie et al. [23a], Brady et al. [24a] found that diabetes and long sleep were indirectly associated via obesity. In a sample of 2848 participants, Brady and colleagues found that both short sleep and long sleep durations were independently associated with elevated body mass index and waist circumference. They also found that fasting insulin was positively associated with sleep duration, while plasma adiponectin levels (a marker inversely associated with insulin resistance) were negatively associated with sleep duration.

A second factor that might explain the indirect relationship between long sleep and diabetes is frequent napping. Han et al. [25], in a prospective study with middle-aged Chinese adults, found that long sleep and habitual napping were associated with an increased risk of diabetes. Specifically, individuals who slept 10 or more hours were 42% more likely to develop diabetes, compared to those who slept 7–8h. Additionally, individuals who frequently napped for >90min were 28% more likely to develop diabetes compared to those who did not nap. These associations were further strengthened when long sleep and napping were combined. Individuals who slept for 10h and more and who napped for >60min were 72% more likely to have diabetes compared to those who slept 7–8h and did not nap.

Summary

Based on the evidence above, there appears to be a U-shaped relationship between sleep duration (short and long sleep) and diabetes and diabetes outcomes [26]. Epidemiological (cross-sectional and longitudinal) and experimental studies provide compelling evidence that short sleep and long sleep durations are directly and indirectly associated with diabetes. However, within this large body of evidence, several studies have highlighted that other sleep and circadian factors might also be responsible for the development and maintenance of diabetes, such as subjective parameters of sleep (sleep quality, social jetlag, and excessive daytime sleepiness), sleep disorders, and circadian misalignment.

Qualitative sleep parameters (sleep quality, excessive daytime sleepiness and social jet lag) and diabetes

Outside of sleep quantity (e.g., duration), qualitative parameters of sleep such as sleep quality, excessive daytime sleepiness, and social jet lag, are considered some of the strongest correlates and predictors of diabetes and its outcomes. Epidemiological and experimental studies provide overwhelming evidence that qualitative parameters gathered through subjective (via self-report) or objective (via devices) means are reliably and consistently associated with diabetes and diabetes outcomes.

Sleep quality

Sleep quality has been linked to the development of diabetes and the worsening of symptoms. In a metaanalysis that analyzed the relationship between sleep quality and diabetes, Cappuccio et al. [27] found that quantity and quality of sleep predicted diabetes risk. Specifically, difficulty initiating sleep (RR=1.57, 95% CI=1.25–1.97) and difficulty maintaining sleep (RR=1.84, CI: 1.39–2.43) were both strongly associated with diabetes risk at follow-up. In addition to cross-sectional evidence, Martyn-Nemeth et al. [28] investigated whether sleep quality was associated with glycemic control and variability (GV), or fear of hypoglycemia (FOH) over a 3-year period with continuous glucose monitoring among 48 men and women with type I diabetes. Poor sleep quality was positively associated with higher levels of nocturnal GV and FOH. These findings highlight a potential bi-directional relationship between sleep quality and diabetes outcomes, with the implication that improving diabetes outcomes may improve sleep quality and vice versa.

Sleep quality may also be associated with diabetes management. In a metaanalysis comparing the relative risks of sleep disturbances (e.g., insufficient sleep duration, insomnia, obstructive sleep apnea, and abnormal sleep timing) and traditional risk factors (overweight, family history of diabetes, and physical inactivity), individuals who reported poor sleep quality had a 40% increased likelihood of having diabetes which was comparable with family history of diabetes and overweight and higher than physical inactivity [2]. In another study, Byberg et al. [23] in a study of 771 participants in Denmark found that a 1-point increase in sleep quality was associated with a 2% increase in insulin sensitivity and a 1% decrease in the homeostatic function of β-cell function, cells responsible for metabolizing glucose. Taken together, sleep disturbances and poor sleep quality play a key role in the development of diabetes among healthy populations and the management of diabetes among diabetics, as healthy sleep is responsible for maintaining the homeostasis of glucose metabolism.

Excessive daytime sleepiness and social jetlag

Raj et al. [29] investigated the prevalence of excessive daytime sleepiness, among individuals with Type II diabetes and whether excessive daytime sleepiness was associated with glycemic control, in a sample of 102 individuals in India. Though the prevalence of excessive daytime sleepiness was only 17.5%, the researchers found in regression analyses that a 1 unit increase in Epworth Sleepiness Scale (a subjective measure of daytime sleepiness) was significantly associated with a 0.143.1dL HbA1c increase. In addition to daytime sleepiness, social jetlag is another qualitative sleep parameter that is associated with diabetes. Koopman et al. [30], in a cross-sectional study from the New Hoorn Study cohort ($n=1585$), investigated the association between social jetlag (defined as the difference between the midpoint of weekday sleep hours and the midpoint of weekends sleep hours) and metabolic syndrome and diabetes. The researchers found that individuals <61 years of age who reported social jetlag (1–2h) had approximately a twofold greater risk of metabolic syndrome and prediabetes/diabetes compared to their counterparts who reported less than 1 h of social jetlag.

Physiological and biological mechanisms

Despite overwhelming evidence that sleep quality and diabetes are related, less is known about biological mechanisms that engender this relationship. Several mechanisms are argued to be at play and for this book chapter we only discuss a few, those that have the most robust evidence. They are insulin resistance, leptin and ghrelin hormones, and inflammation.

Insulin resistance

Evidence suggests an association between diabetes risk and sleep deprivation, even among healthy subjects [31]. For instance, in a seminal study, healthy subjects restricted to 4h in bed demonstrated increased markers of insulin resistance [11]. Thus, sleep deprivation and poor sleep quality are potential mechanisms through which insulin resistance manifests.

Leptin and ghrelin hormones

Research has implicated metabolic hormones leptin and ghrelin in the relationship between sleep and diabetes risk. Leptin, a molecule released from adipose tissue, signals feelings of satiety in the brain and increases energy expenditure. Leptin release from adipose tissue binds to leptin receptors in the hypothalamus to reduce appetite, and thus decrease energy intake. Research demonstrates sleep deprivation is associated with decreased leptin and associated perceptions of hunger [32–34]. While, ghrelin, a hormone secreted by the stomach that triggers hunger, is associated with sleep deprivation [35–37].

Inflammation

Evidence also points to inflammatory cytokines, tumor necrosis factor (a cell signaling protein indicative of inflammation), and interleukin-6, as potential risk factors at play in the relationship between diabetes and sleep disturbances, specifically chronic sleep deprivation. Although there is some dispute about exact etiology, research has linked tumor necrosis factor with sleep restriction and inflammation [38, 39]. Also, interleukin-6, a pro-inflammatory molecule secreted by T-cells and macrophages like tumor necrosis factor, is associated with chronic inflammation and also plays a role in stimulating energy utilization in adipose tissue [31]. Interleukin-6 operates within the human circadian rhythm, with its peak expression at night around sleep onset, but is suppressed by slow wave sleep linked with homeostatic restorative biological processes such as processing glucose [40]. Therefore, the presence of inflammation will compromise homeostatic processes in metabolizing glucose and thus may lead to insulin insensitivity.

Sleep disorders and diabetes

Sleep and circadian disorders, such as obstructive sleep apnea, insomnia, and circadian misalignment disorders represent some of the most robust evidence linking sleep and diabetes. The evidence can be compartmentalized in two: (1) evidence showing that individuals with sleep and circadian disorders are at greater risk of developing diabetes compared to those without a sleep or circadian disorder history and (2) evidence indicating that treatment of sleep and circadian disorders lowers an individual's risk of developing diabetes and improves their ability to better manage their diabetes.

Obstructive sleep apnea

Obstructive sleep apnea (OSA) is a sleep disordered breathing condition characterized by partial or full blockage of the upper airway during sleep thus causing repetitive bouts of reflexive awakenings to receive sufficient air and oxygen (apneas and hypopneas). Apnea and hypopnea episodes often result in oxygen desaturation and physiological stress, which compromises important homeostatic physiological processes. Repetitive cycles of apnea and hypopnea compromise the cardiometabolic system where untreated OSA induces metabolic abnormalities. Studies have demonstrated that intermittent hypoxia (loss of oxygen due to apnea and hypopnea), reduced sleep duration, and sleep fragmentation, exert adverse effects on glucose metabolism. Based on the foregoing evidence, it is imperative that clinicians address the high likelihood of comorbid OSA risk and type 2 diabetes among their patients.

Insomnia

Insomnia, like OSA, is a prevalent sleep disorder and is characterized by three chief complaints: difficulty falling and staying asleep as well as involuntary early awakenings. Secondary complaints may include excessive sleep deprivation, daytime sleepiness, not feeling restored after sleep, and fatigue, which can all affect an individual's mental and physical functioning and health, such as mood problems and chronic health conditions like diabetes. Insomnia increases diabetes risk through several mechanisms such as constant elevation of cardiometabolic homeostatic markers due to chronic sleeplessness [2, 23, 41].

Epidemiological and population level studies

Epidemiological evidence on the relationship between sleep disorders and diabetes is quite compelling [42–45] citing direct and indirect associations. Bakker et al. [46] found, in a sample of 2151 participants in the Multi-Ethnic Study of Atherosclerosis, that individuals with an apnea-hypopnea index (AHI) ≥15/h (moderate to severe range) had a twofold greater likelihood of having abnormal fasting glucose levels, compared to individuals with a low AHI score—low sleep apnea risk (0–4.9/h). Interestingly, sleep duration was not associated with abnormal fasting glucose after adjusting for the effects of apnea-hypopnea index, a proxy for OSA. The putative relationship between sleep apnea and diabetes might vary by race/ethnicity and gender, as men compared to women and whites compared to blacks have a greater odds of having abnormal glucose levels and beta cell functioning if they have sleep apnea [47]. The relationship between sleep apnea and diabetes are further buttressed by evidence indicating that treatment of sleep apnea, via continuous positive airway pressure, is associated with diabetes risk reduction. Labarca et al. [48] found that continuous positive airway pressure treatment significantly lowered diabetes risk and reduced abnormal glucose levels.

Similarly, the epidemiological evidence linking insomnia and diabetes is robust. Hein et al. [41] conducted a study among insomnia patients to better understand the prevalence of and risk factors associated with type 2 diabetes. They found that 21.3% of 1311 individuals with insomnia had type 2 diabetes and alcohol consumption of ≥4 units/day, BMI ≥ 25 kg/m^2, age 50 ≥ 50, being male, C-reactive protein ≥4.5 mg/L, early morning awakenings, high blood pressure, total sleep hours <6.5 h, apnea-hypopnea index ≥15/h, elevated triglyceride, and periodic limb movements index ≥26/h. were risk factors in the insomnia-diabetes association [41]. In a sample of 5078 diabetic individuals from China, Tan et al. [22] and Zhang et al. [49] found that the prevalence of insomnia was 20.2% and that individuals with insomnia compared to those without insomnia were 31% more likely to have type 2 diabetes, after controlling age, alcohol, body mass index, chronic health conditions, depression and smoking. Men and individuals between the ages 40–59 years with insomnia were more at risk for diabetes compared to their counterparts. Further proof of the insomnia-diabetes association is provided in evidence showing that successful management of insomnia may lead to improved glycemic control, glucose tolerance, and reduced risk of diabetes. Carroll et al. [50] found that treatment of insomnia symptoms improved sleep quality and reduced the risk of chronic disease, like diabetes, in older adults with sleep disturbances.

Mechanistic studies

The relationship between sleep disorders (i.e., sleep apnea and insomnia) and diabetes is multifactorial and thus difficult to simplify. Sleep loss, a common resultant of sleep disorders, has a negative impact on cardiometabolic health, inducing glucose intolerance and insulin resistance, which if not treated over time will lead to diabetes. Sleep deprivation also has an effect on leptin and ghrelin, hormones responsible for physiological drives of appetite and satiety. Specifically, sleep loss decreases leptin levels and increases ghrelin levels [51]. These hormonal changes promote overeating and increase the risk of obesity and in turn diabetes [52]. Deng and colleagues found that compared to average sleep duration, short sleep increased the risk of obesity by 12% [53] but these effects might be reversed or muted through adequate sleep (7–8 h) [54]. Chirinos et al. [54a] found that continuous positive airway pressure (CPAP), an effective treatment for OSA, significantly reduced cardiometabolic risk clinical risk markers (such as insulin sensitivity, triglycerides and blood pressure) greater than weight loss alone at week 8 of a 24-week. At 24 weeks, both CPAP and weight loss had the strongest effect on reducing metabolic risk highlighting synergistic effect sleep has on other health risk behaviors. Other studies found that increasing sleep duration can reverse this process, especially among those who have abnormal glucose levels and prediabetes (Fig. 17.2) [20].

Circadian rhythm and diabetes

The circadian rhythm of living organisms relates to the 24 h wake-sleep clock that regulates all biological, chemical, and physiological processes in the central and peripheral nervous systems. Human circadian rhythm is influenced by internal (endogenous) and exogenous (external) cues. Endogenous cues include genetic markers such as the MTNR1B genotype, chronotype, and intrinsic circadian rhythm. While, exogenous cues include environmental factors such as light, dark, noise, and temperature that influence wake-sleep timing and behaviors, such as activity, food consumption, and sleep. Circadian rhythm regulates gluconeogenesis, such as insulin sensitivity, insulin secretion and energy expenditure over a 24-h period [55–57].

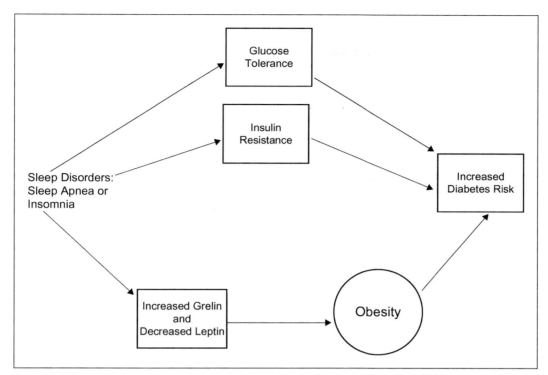

FIG. 17.2 Pathophysiological mechanisms linking sleep loss and diabetes risk via glucose tolerance, insulin resistance, appetite and obesity.

Misalignment and de-synchrony: (1) of the wake-sleep cycle and clock and (2) between central and peripheral nervous systems will disrupt the innate biological timing of the suprachiasmatic nucleus (master clock) and peripheral clocks of cells, tissues and organs which can disrupt normal homeostatic metabolic processes, such as glucose metabolism. These disruptions may lead to mistimed or irregular eating schedules and thus compromise healthy metabolism of foods leading to spikes in glucose and insulin resistance. Circadian misalignment may also lead to elevated plasma cortisol levels or activation of the sympathetic nervous system thus increasing Type 2 diabetes mellitus (T2DM) and obesity risk via insulin sensitivity in adipose tissue [58–60].

Independent and interactive associations between endogenous circadian rhythm and diabetes

Evidence for the association between circadian rhythm and diabetes can be categorized into two categories: independent and interactive effects. In a cross-sectional study of nearly 10,000 European individuals enrolled in five studies as part of the Candidate Gene Association Resource (CARe), Tare et al. [61] examined 16 fasting glucose variants and aggregate genetic risk scores. The researchers found that short sleep duration was associated with T2DM [61]. However, sleep duration did not mediate or modify the association between circadian rhythm genes and diabetes, thus suggesting that circadian rhythm is independently associated with diabetes risk.

However, other studies have found an interactive effect between circadian rhythm and sleep duration. Nisa et al. [62] found significant changes in oral glucose tolerance tests in 1025 Chinese women carrying the circadian rhythm-related melatonin receptor 1B genotype MTNR1B. Study findings indicate that women with different MTNR1B genotypes and the short sleep duration-related G allele had significant long-term postpartum changes in their 2 h oral glucose tolerance tests regardless of inadequate, adequate, or excessive gestational weight gain, albeit women with greater gestational weight gain were more at risk for gestational diabetes. These findings highlight that both circadian rhythm and short sleep combined increases an individual's risk for diabetes.

Other circadian-related studies have found that people with evening chronotype, those who prefer to conduct activities of daily living later in the day, have higher metabolic syndrome and diabetes risk. Anothaisintawee et al. [62a] found that evening chronotype in individuals living with diabetes reported poor glycemic control, independent of sleep health compared to individuals who are not evening chronotypes. In their cross-sectional study, Anothaisintawee et al. [62a] explored the relationship between chronotype, social rhythms, and hemoglobin A1c levels in 1014 adults living with prediabetes. The researchers found that later mid-sleep time on a nonworkday adjusted for sleep debt was significantly associated with HbA1c levels ($P=.049$), after adjusting for confounding factors [62a]. Similarly, Reutrakul and Van Cauter [44] found that evening chronotype and larger dinner size were associated with poor glycemic control in patients living with T2DM, independent of poor sleep health.

Endogenous misalignment between fatty acids that regulate glucose metabolism and the circadian clock could adversely affect the primary systems it regulates such as appetite, energy expenditure, dyslipidemia, and insulin resistance [63]. Insulin regulation may be affected by time of day. Circadian rhythm balance which maintained at the cellular level in adipose tissue may affect insulin levels. At the cellular level, there appears to be a circadian component in regulating glucose levels in adipose tissue. Carrasco-Benso et al. [64] investigated whether human adipose tissue expressed intrinsic circadian rhythms related to insulin sensitivity mechanisms in obese participants. The researchers found significant differences in subcutaneous adipose tissue and circadian rhythm insulin signaling ($P < .00001$). Insulin sensitivities reached their highest at noon, climbing 54% higher than at midnight. Thus, an underlying endogenous circadian rhythm may serve as a mechanism for insulin sensitivity.

Exogenous

Sleep timing, a component of circadian rhythm, and diet have been reported as significant factors that affect insulin sensitivity. Delayed sleep times may increase food intake during times of peak appetite. This may result in higher energy intake and contribute to positive energy balance [58]. Consuming the majority of calories earlier in the day rather than in the evening is one approach to overcoming these adverse effects as it regulates metabolism and prevents weight gain [65]. Similarly, Shapiro et al. [66] reported glucose levels reached their peak in the early morning, hovering nearly 30% above the lowest daytime glucose level. Insulin secretion rates and glucose levels were nearly in sync in half of the participants. Overlapping morning glucose levels with circadian rhythm alignment affects insulin sensitivity.

Endogenous and exogenous

In some instances, both endogenous and exogenous circadian processes may be associated with diabetes risk. Sleep timing and/or duration combined with later chronotype have been associated with high insulin resistance in Hispanic adults. Egan et al. [67] examined cross-sectional data from over 13,000 individuals enrolled in a community-based study. Researchers evaluated sleep timing and chronotype against fasting glucose levels, insulin resistance, glucose levels 2 h post oral glucose ingestion and hemoglobin A1c. Chronotype (+1.2%/h later, $P < .05$) and midpoint of sleep duration (+1.5%/h later, $P < .05$) were positively associated with insulin resistance.

Circadian misalignment and diabetes

Circadian misalignment can occur if: (1) sleep and wake are inappropriately timed, (2) sleep-wake cycle is not aligned with feeding rhythm, and (3) central and peripheral rhythms are misaligned. Potential adverse health outcomes include hormonal changes that affect appetite, unregulated eating patterns, poor glucose metabolism, and mood problems. Circadian misalignment is linked with cardiovascular disease, diabetes, obesity, cancer, and mental illness [68].

Circadian misalignment combined with short sleep duration has been associated with reduced glucose tolerance. Eckel et al. [69] investigated the influence of morning circadian misalignment due to short nighttime sleep on insulin sensitivity. High sustained melatonin levels after wake resulted in poor insulin sensitivity. Additionally, sleeping 5-h/night during the work week and irregular food intake times resulted in approximately 20% reduced insulin sensitivity. Reduced insulin sensitivity triggers a higher glucose response, which can lead to diabetes mellitus. Insulin sensitivity was increased over time as a result habitual short sleep duration. Buxton et al. [70] examined the prolonged effect of sleep restriction combined with circadian disruption on metabolic health. The researchers found that sleep restriction for 3 weeks (average total sleep time of 5.6 h) and inducing circadian disruption across 28 h days decreased metabolic rate and increased postprandial plasma, thus increasing obesity and diabetes risk. Circadian balance was restored within 9 days of sleep recovery highlighting restorative effects of healthy.

Additionally, Rao et al. [15] found that circadian misalignment and the central circadian pacemaker influence glucose tolerance differently. Decreased pancreatic β-cell function in the evening hours and decreased insulin sensitivity, resulting in reduced glucose tolerance in the evening compared to the morning was directly related to circadian misalignment. The above-mentioned evidence provides compelling evidence for the association between synchrony of the peripheral circadian clock and glucose rhythm alignment.

Circadian misalignment due to short sleep not only affects glucose tolerance but increases carbohydrate cravings. Al Khatib et al. [71] assessed the effect of sleep extension on dietary intake among adult short sleepers (5–<7 h). Individuals whose sleep was extended had reduced fat (percentage), carbohydrate (grams), and free sugar (grams) intake compared to those whose sleep was not extended. No significant differences were found in energy balance or markers of cardiometabolic health (Fig. 17.3).

Indirect effects of sleep on diabetes

Although the proximal effects of sleep on diabetes is undeniable, evidence for the distal and indirect effects are as compelling, albeit less causal. Both epidemiological and experimental studies have linked sleep and diabetes via obesity/visceral adiposity.

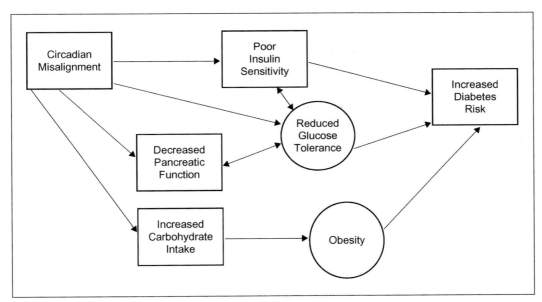

FIG. 17.3 Pathophysiological effects of circadian misalignment on diabetes risk.

Obesity moderates or mediates the relationship between sleep and diabetes

Sleep curtailment is also associated with a dysregulation of the neuroendocrine control of appetite, as well as reduction of the satiety hormone leptin and an increase in the hunger-promoting hormone ghrelin. Thus, sleep loss may alter the ability of leptin and ghrelin to accurately signal caloric need, acting in concert to produce an internal misperception of insufficient energy availability [72].

The relationship between obesity and short sleep duration may be mediated by changes in energy intake. The energy balance equation states that under conditions in which energy intake equates energy expenditure, no change in body weight will occur. Excess energy intake (i.e., positive energy balance) above energy expenditure results in weight gain leading to overweight and obesity. Pooled epidemiological evidence suggests a consistent association between short sleep duration and higher total energy intake and higher total fat intake [73]. Healthy men increased caloric intake by 22% the day after one night of sleep restriction [74]. Similar results on sleep restriction and increase energy intake were also found in healthy women resulting in net weight gain after 4 nights [75]. The changes in energy intake that accompany sleep loss may be explained by changes in the hormonal environment, previously described by leptin and ghrelin. The hormonal imbalance of leptin and ghrelin stimulate the desire to consume energy-dense, nutrient poor foods.

Along with energy intake, physical activity levels are affected by sleep duration. Individuals who reported shorter sleep (<6h/nigh) spend more time in sedentary behaviors, have fewer daily activity counts and spent less time in moderate to vigorous physical activity (MVPA) compared to individuals with >6h/night [76]. Adults at risk for type 2 diabetes restricted to sleep of 5.5h per night compared to 8.5h per night reduced moderate-and vigorous physical activity MVPA by 24% and increased sedentary behavior [76]. Short sleep duration may confer conditions of increased energy intake and reductions in physical activity, thus perpetuating obesity status.

Visceral adiposity and not obesity is responsible for the relationship between sleep disturbances and diabetes

Diabetes is well known to be associated with obesity. However, the prevalence of newly diagnosed diabetics who are not overweight has increased in recent years, thus prompting more exploration of the association of visceral adiposity and diabetes. Visceral adipose tissue or visceral fat is tissue located deep in the abdomen and around internal organs. Studies have shown that visceral adiposity is strongly linked to type 2 diabetes [77]. Similar to sleep apnea, it also causes a decrease in insulin sensitivity deterioration [77].

The exacerbating role of sleep on well-being, quality of life, health and mortality among diabetics

The co-occurrence of sleep disturbances and diabetes risk is high and if not addressed can have deleterious effects on an individual's health. For example, diabetic retinopathy (DR) a common microvascular complication of diabetes is associated with sleep disturbances. DR affects 40%–50% of diabetic patients and can cause blindness if not treated [78]. Sleep apnea has been well defined as a direct risk of DR. Altaf et al. [78] found that in as short as 4 years, sleep apnea advances the development of retinopathy for people with diabetes.

Sleep apnea has also been associated with endothelial dysfunction for diabetic patients. Endothelial cells release substances that control vascular relaxation and contraction as well as enzymes that control blood clotting, immune function, and platelet adhesion. Endothelial dysfunction is caused by diabetes, and sleep disturbances decrease endothelial production [79, 80]. A third consequence of the sleep and diabetes association is poor quality of life and well-being. Recognizing that sleep duration and sleep quality are related to diabetes, insulin resistance, and poor glycemic control, researchers have found that poor sleep in diabetes patients was inversely associated with quality of life [81, 82].

Healthy sleep and reduced diabetes risk

The National Sleep Foundation, the American Academy of Sleep Medicine and the Sleep Research Society all recommend that adults should sleep approximately 7–9 h daily to reduce health risk and optimize wellness [83]. Adults who sleep between the recommended hours are less likely to be at risk for diabetes compared to those who sleep fewer hours [84–86]. In today's society, habitual sleep deprivation and disturbances and poor sleep health have become increasingly common. Most people self-report sleeping an average of <7 h each day and as evidenced above may increase the risk of cardiometabolic diseases such as diabetes [87]. However, there is hope to arrest this public health issue as there is a growing body of research indicating that improving sleep health parameters such as duration, sleepiness/alertness, timing, efficiency, satisfaction/quality, sleep disorders and architecture may reduce the risk of cardiometabolic diseases [20, 87].

REFERENCES

[1] Arora T, Taheri S. Sleep optimization and diabetes control: a review of the literature. Diabetes Ther 2015;6(4):425–68. https://doi.org/10.1007/s13300-015-0141-z.

[2] Anothaisintawee T, Reutrakul S, Van Cauter E, Thakkinstian A. Sleep disturbances compared to traditional risk factors for diabetes development: systematic review and meta-analysis. Sleep Med Rev 2016; https://doi.org/10.1016/j.smrv.2015.10.002.

[3] Gangwisch JE, Malaspina D, Boden-Albala B, Heymsfield SB. Inadequate sleep as a risk factor for obesity: analyses of the NHANES I. Sleep 2005;28(10):1289–96. https://doi.org/10.1093/sleep/28.10.1289.

[4] Engeda J, Mezuk B, Ratliff S, Ning Y. Association between duration and quality of sleep and the risk of pre-diabetes: evidence from NHANES. Diabet Med 2013;30(6):676–80. https://doi.org/10.1111/dme.12165.

[5] Ford ES, Wheaton AG, Chapman DP, Li C, Perry GS, Croft JB. Associations between self-reported sleep duration and sleeping disorder with concentrations of fasting and 2-h glucose, insulin, and glycosylated hemoglobin among adults without diagnosed diabetes. J Diabetes 2014;6(4):338–50. https://doi.org/10.1111/1753-0407.12101.

[6] Arora T, Chen MZ, Cooper AR, Andrews RC, Taheri S. The impact of sleep debt on excess adiposity and insulin sensitivity in patients with early type 2 diabetes mellitus. J Clin Sleep Med 2016;12(5):673–80. https://doi.org/10.5664/jcsm.5792.

[7] Kim A, Yu HY, Lim J, Ryu C-M, Kim YH, Heo J, ... Choo M-S. Improved efficacy and in vivo cellular properties of human embryonic stem cell derivative in a preclinical model of bladder pain syndrome. Sci Rep 2017;7(1):8872. https://doi.org/10.1038/s41598-017-09330-x.

[8] Dutil C, Chaput J-P. Inadequate sleep as a contributor to type 2 diabetes in children and adolescents. Nutr Diabetes 2017;7(5): https://doi.org/10.1038/nutd.2017.19.

[8a] Hancox RJ, Landhuis CE. Association between sleep duration and haemoglobin A1c in young adults. J Epidemiol Community Health 2012;66(10):957–61.

[9] Landhuis CE, Perry DK, Hancox RJ. Association between childhood and adolescent television viewing and unemployment in adulthood. Prev Med 2012;54(2):168–73. https://doi.org/10.1016/j.ypmed.2011.11.007.

[10] Leng Y, Cappuccio FP, Surtees PG, Luben R, Brayne C, Khaw KT. Daytime napping, sleep duration and increased 8-year risk of type 2 diabetes in a British population. Nutr Metab Cardiovasc Dis 2016;26(11):996–1003. https://doi.org/10.1016/j.numecd.2016.06.006.

[11] Spiegel K, Leproult R, Van Cauter E. Impact of sleep debt on metabolic and endocrine function. Lancet 1999;354(9188):1435–9. https://doi.org/10.1016/S0140-6736(99)01376-8.

[12] Buxton OM, Pavlova M, Reid EW, Wang W, Simonson DC, Adler GK. Sleep restriction for 1 week reduces insulin sensitivity in healthy men. Diabetes 2010; https://doi.org/10.2337/db09-0699.

[13] Broussard JL, Ehrmann DA, Van Cauter E, Tasali E, Brady MJ. Impaired insulin signaling in human adipocytes after experimental sleep restriction. Ann Intern Med 2012;157(8):549. https://doi.org/10.7326/0003-4819-157-8-201210160-00005.

[14] Robertson MD, Russell-Jones D, Umpleby AM, Dijk DJ. Effects of three weeks of mild sleep restriction implemented in the home environment on multiple metabolic and endocrine markers in healthy young men. Metab Clin Exp 2013;62(2):204–11. https://doi.org/10.1016/j.metabol.2012.07.016.

[15] Rao MN, Neylan TC, Grunfeld C, Mulligan K, Schambelan M, Schwarz J-M. Subchronic sleep restriction causes tissue-specific insulin resistance. J Clin Endocrinol Metab 2015;100(4):1664–71. https://doi.org/10.1210/jc.2014-3911.

[16] Broussard JL, Chapotot F, Abraham V, Day A, Delebecque F, Whitmore HR, Tasali E. Sleep restriction increases free fatty acids in healthy men. Diabetologia 2015;58(4):791–8. https://doi.org/10.1007/s00125-015-3500-4.

[17] Gohil A, Hannon TS. Poor sleep and obesity: concurrent epidemics in adolescent youth. Front Endocrinol 2018; https://doi.org/10.3389/fendo.2018.00364.

[18] Darukhanavala A, Booth JN, Bromley L, Whitmore H, Imperial J, Penev PD. Changes in insulin secretion and action in adults with familial risk for type 2 diabetes who curtail their sleep. Diabetes Care 2011;34(10):2259–64. https://doi.org/10.2337/dc11-0777.

[19] Trento M, Broglio F, Riganti F, Basile M, Borgo E, Kucich C, ... Porta M. Sleep abnormalities in type 2 diabetes may be associated with glycemic control. Acta Diabetol 2008;45(4):225–9. https://doi.org/10.1007/s00592-008-0047-6.

[20] Leproult R, Deliens G, Gilson M, Peigneux P. Beneficial impact of sleep extension on fasting insulin sensitivity in adults with habitual sleep restriction. Sleep 2015;38(5):707–15. https://doi.org/10.5665/sleep.4660.

[21] Cespedes EM, Bhupathiraju SN, Li Y, Rosner B, Redline S, Hu FB. Long-term changes in sleep duration, energy balance and risk of type 2 diabetes. Diabetologia 2016;59(1):101–9. https://doi.org/10.1007/s00125-015-3775-5.

[22] Tan X, Chapman CD, Cedernaes J, Benedict C. Association between long sleep duration and increased risk of obesity and type 2 diabetes: a review of possible mechanisms. Sleep Med Rev 2017;4–11. https://doi.org/10.1016/j.smrv.2017.11.001.

[23] Byberg S, Hansen ALS, Christensen DL, Vistisen D, Aadahl M, Linneberg A, Witte DR. Sleep duration and sleep quality are associated differently with alterations of glucose homeostasis. Diabet Med 2012;29(9): https://doi.org/10.1111/j.1464-5491.2012.03711.x.

[23a] Ferrie JE, Shipley MJ, Stansfeld SA, Marmot MG. Effects of chronic job insecurity and change in job security on self reported health, minor psychiatric morbidity, physiological measures, and health related behaviours in British civil servants: the Whitehall II study. J Epidemiol Community Health 2002; https://doi.org/10.1136/jech.56.6.450.

[24] Ferrie JE, Kivimäki M, Akbaraly TN, Tabak A, Abell J, Davey Smith G, ... Shipley MJ. Change in sleep duration and type 2 diabetes: the Whitehall II Study. Diabetes Care 2015;38(8):1467–72. https://doi.org/10.2337/dc15-0186.

[24a] Brady EM, Bodicoat DH, Hall AP, Khunti K, Yates T, Edwardson C, Davies MJ. Sleep duration, obesity and insulin resistance in a multiethnic UK population at high risk of diabetes. Diabetes Res Clin Pract 2018;139:195–202.

[25] Han X, Liu B, Wang J, Pan A, Li Y, Hu H, ... He M. Long sleep duration and afternoon napping are associated with higher risk of incident diabetes in middle-aged and older Chinese: the Dongfeng-Tongji cohort study. Ann Med 2016; https://doi.org/10.3109/07853890.2016.1155229.

[26] Bliwise DL, Young TB. The parable of parabola: what the U-shaped curve can and cannot tell us about sleep. Sleep 2007;30(12):1614–5.

[27] Cappuccio FP, D'Elia L, Strazzullo P, Miller MA. Quantity and quality of sleep and incidence of type 2 diabetes: a systematic review and meta-analysis. Diabetes Care 2010;33(2):414–20. https://doi.org/10.2337/dc09-1124.

[28] Martyn-Nemeth P, Schwarz Farabi S, Mihailescu D, Nemeth J, Quinn L. Fear of hypoglycemia in adults with type 1 diabetes: impact of therapeutic advances and strategies for prevention—a review. J Diabetes Complicat 2016; https://doi.org/10.1016/j.jdiacomp.2015.09.003.

[29] Raj JP, Hansdak SG, Naik D, Mahendri NV, Thomas N. SLEep among diabetic patients and their GlycaEmic control (SLEDGE)—a pilot observational study. J Diabetes 2018; https://doi.org/10.1111/1753-0407.12825.

[30] Koopman ADM, Rauh SP, Van'T Riet E, Groeneveld L, Van Der Heijden AA, Elders PJ, ... Rutters F. The association between social jetlag, the metabolic syndrome, and type 2 diabetes mellitus in the general population: the New Hoorn Study. J Biol Rhythm 2017;32(4):359–68. https://doi.org/10.1177/0748730417713572.

[31] Grandner MA, Seixas A, Shetty S, Shenoy S. Sleep duration and diabetes risk: population trends and potential mechanisms. Curr Diab Rep 2016; https://doi.org/10.1007/s11892-016-0805-8.

[32] Omisade A, Buxton OM, Rusak B. Impact of acute sleep restriction on cortisol and leptin levels in young women. Physiol Behav 2010;99(5):651–6. https://doi.org/10.1016/j.physbeh.2010.01.028.

[33] Spiegel K, Tasali E, Penev P, Van Cauter E. Brief communication: sleep curtailment in healthy young men is associated with decreased leptin levels, elevated ghrelin levels, and increased hunger and appetite. Ann Intern Med 2004;141(11):846–50.

[34] St-Onge M-P. The role of sleep duration in the regulation of energy balance: effects on energy intakes and expenditure. J Clin Sleep Med 2013; https://doi.org/10.5664/jcsm.2348.

[35] Knutson KL, Spiegel K, Penev P, Van Cauter E. The metabolic consequences of sleep deprivation. Sleep Med Rev 2007; https://doi.org/10.1016/j.smrv.2007.01.002.

[36] Spiegel K, Leproult R, Van Cauter E. Rythmes et sommeil Impact d'une dette de sommeil sur les rythmes physiologiques. Rev Neurol (Paris) 2003;159:6–11.

[37] Van Cauter E. Sleep disturbances and insulin resistance. Diabet Med 2011; https://doi.org/10.1111/j.1464-5491.2011.03459.x.

[38] Pfeffer K. Biological functions of tumor necrosis factor cytokines and their receptors. Cytokine Growth Factor Rev 2003;14(3–4):185–91. https://doi.org/10.1016/S1359-6101(03)00022-4.

[39] Vgontzas AN, Zoumakis E, Bixler EO, Lin H-M, Follett H, Kales A, Chrousos GP. Adverse effects of modest sleep restriction on sleepiness, performance, and inflammatory cytokines. J Clin Endocrinol Metab 2004;89(5):2119–26. https://doi.org/10.1210/jc.2003-031562.

[40] Kapsimalis F, Basta M, Varouchakis G, Gourgoulianis K, Vgontzas A, Kryger M. Cytokines and pathological sleep. Sleep Med 2008;9(6):603–14. https://doi.org/10.1016/j.sleep.2007.08.019.

[41] Hein M, Lanquart J-P, Loas G, Hubain P, Linkowski P. Prevalence and risk factors of type 2 diabetes in insomnia sufferers: a study on 1311 individuals referred for sleep examinations. Sleep Med 2018;46:37–45. https://doi.org/10.1016/J.SLEEP.2018.02.006.

[42] Buxton OM, Marcelli E. Short and long sleep are positively associated with obesity, diabetes, hypertension, and cardiovascular disease among adults in the United States. Soc Sci Med 2010;71(5):1027–36. https://doi.org/10.1016/J.SOCSCIMED.2010.05.041.

[43] Lin C-L, Chien W-C, Chung C-H, Wu F-L. Risk of type 2 diabetes in patients with insomnia: a population-based historical cohort study. Diabetes Metab Res Rev 2017;e2930 https://doi.org/10.1002/dmrr.2930.

[44] Reutrakul S, Van Cauter E. Interactions between sleep, circadian function, and glucose metabolism: implications for risk and severity of diabetes. Ann N Y Acad Sci 2014;1311(1):151–73. https://doi.org/10.1111/nyas.12355.

[45] Vgontzas AN, Liao D, Pejovic S, Calhoun S, Karataraki M, Bixler EO. Insomnia with objective short sleep duration is associated with type 2 diabetes: a population-based study. Diabetes Care 2009;32(11):1980–5. https://doi.org/10.2337/dc09-0284.

[46] Bakker JP, Weng J, Wang R, Redline S, Punjabi NM, Patel SR. Associations between obstructive sleep apnea, sleep duration, and abnormal fasting glucose the multi-ethnic study of atherosclerosis. Am J Respir Crit Care Med 2015;192(6):745–53. https://doi.org/10.1164/rccm.201502-0366OC.

[47] Temple KA, Leproult R, Morselli L, Ehrmann DA, Van Cauter V, Mokhlesi E. Sex differences in the impact of obstructive sleep apnea on glucose metabolism. Front Endocrinol (Lausanne) 2018;9:376. https://doi.org/10.3389/fendo.2018.00376.

[48] Labarca G, Ortega F, Arenas A, Reyes T, Rada G, Jorquera J. Extrapulmonary effects of continuous airway pressure on patients with obstructive sleep apnoea: protocol for an overview of systematic reviews. BMJ Open 2017;7(6): https://doi.org/10.1136/bmjopen-2016-015315.

[49] Zhang Y, Lin Y, Zhang J, Li L, Liu X, Wang T, Gao Z. Association between insomnia and type 2 diabetes mellitus in Han Chinese individuals in Shandong Province, China. Sleep Breath 2018;1–6. https://doi.org/10.1007/s11325-018-1687-6.

[50] Carroll JE, Seeman TE, Olmstead R, Melendez G, Sadakane R, Bootzin R, … Irwin MR. Improved sleep quality in older adults with insomnia reduces biomarkers of disease risk: pilot results from a randomized controlled comparative efficacy trial. Psychoneuroendocrinology 2015;55:184–92. https://doi.org/10.1016/j.psyneuen.2015.02.010.

[51] McHill AW, Hull JT, McMullan CJ, Klerman EB. Chronic insufficient sleep has a limited impact on circadian rhythmicity of subjective hunger and awakening fasted metabolic hormones. Front Endocrinol 2018;9:https://doi.org/10.3389/fendo.2018.00319.

[52] Nedeltcheva AV, Scheer FAJL. Metabolic effects of sleep disruption, links to obesity and diabetes. Curr Opin Endocrinol Diabetes Obes 2014;21(4):293–8. https://doi.org/10.1097/MED.0000000000000082.

[53] Deng H-B, Tam T, Zee BC-Y, Chung RY-N, Su X, Jin L, … Lao XQ. Short sleep duration increases metabolic impact in healthy adults: a population-based cohort study. Sleep 2017;https://doi.org/10.1093/sleep/zsx130.

[54] Kim CE, Shin S, Lee H-W, Lim J, Lee J, Shin A, Kang D. Association between sleep duration and metabolic syndrome: a cross-sectional study. BMC Public Health 2018;18(1):720. https://doi.org/10.1186/s12889-018-5557-8.

[54a] Chirinos JA, Gurubhagavatula I, Teff K, Rader DJ, Wadden TA, Townsend R, Foster GD, Maislin G, Saif H, Broderick P, Chittams J, Hanlon AL, et al. CPAP, weight loss, or both for obstructive sleep apnea. New Engl J Med 2014;370(24):2265–75.

[55] Gerhart-Hines Z, Lazar MA. Circadian metabolism in the light of evolution. Endocr Rev 2015;https://doi.org/10.1210/er.2015-1007.

[56] Marcheva B, Ramsey KM, Peek CB, Affinati A, Maury E, Bass J. Circadian clocks and metabolism. Handb Exp Pharmacol 2013;217:127–55. https://doi.org/10.1007/978-3-642-25950-0-6.

[57] Wang Y-H, Wu H-H, Ding H, Li Y, Wang Z-H, Li F, Zhang J-P. Changes of insulin resistance and β-cell function in women with gestational diabetes mellitus and normal pregnant women during mid- and late pregnant period: a case-control study. J Obstet Gynaecol Res 2013;39(3):647–52. https://doi.org/10.1111/j.1447-0756.2012.02009.x.

[58] Potter GDM, Skene DJ, Arendt J, Cade JE, Grant PJ, Hardie LJ. Circadian rhythm and sleep disruption: causes, metabolic consequences, and countermeasures. Endocr Rev 2016; https://doi.org/10.1210/er.2016-1083.

[59] Qian J, Scheer FAJL. Circadian system and glucose metabolism: implications for physiology and disease. Trends Endocrinol Metab 2016;27(5):282–93. https://doi.org/10.1016/j.tem.2016.03.005.

[60] Shimizu I, Yoshida Y, Minamino T. A role for circadian clock in metabolic disease. Hypertens Res 2016; https://doi.org/10.1038/hr.2016.12.

[61] Tare A, Lane JM, Cade BE, Grant SF, Chen TH, Punjabi NM, … Saxena R. Sleep duration does not mediate or modify association of common genetic variants with type 2 diabetes. Diabetologia 2014; https://doi.org/10.1007/s00125-013-3110-y.

[62] Nisa H, Qi KHT, Leng J, Zhou T, Liu H, Li W, … Qi L. The circadian rhythm-related MTNR1B genotype, gestational weight gain, and postpartum glycemic changes. J Clin Endocrinol Metab 2018;103(6):2284–90. https://doi.org/10.1210/jc.2018-00071.

[62a] Anothaisintawee T, Lertrattananon D, Thamakaison S, Knutson KL, Thakkinstian A, Reutrakul S. Later chronotype is associated with higher hemoglobin A1c in prediabetes patients. Chronobiol Int 2017;34(3):393–402. https://doi.org/10.1080/07420528.2017.1279624.

[63] Poggiogalle E, Jamshed H, Peterson CM. Circadian regulation of glucose, lipid, and energy metabolism in humans. Metabolism 2017; https://doi.org/10.1016/j.metabol.2017.11.017.

[64] Carrasco-Benso MP, Rivero-Gutierrez B, Lopez-Minguez J, Anzola A, Diez-Noguera A, Madrid JA, … Garaulet M. Human adipose tissue expresses intrinsic circadian rhythm in insulin sensitivity. FASEB J 2016;30(9):3117–23. https://doi.org/10.1096/fj.201600269RR.

[65] Jakubowicz D, Barnea M, Wainstein J, Froy O. High Caloric intake at breakfast vs. dinner differentially influences weight loss of overweight and obese women. Obesity 2013;21(12):2504–12. https://doi.org/10.1002/oby.20460.

[66] Shapiro ET, Tillil H, Polonsky KS, Fang VS, Rubenstein AH, Van Cauter E. Oscillations in insulin secretion during constant glucose infusion in normal man: relationship to changes in plasma glucose. J Clin Endocrinol Metab 1988;67(2):307–14.

[67] Egan KJ, Knutson KL, Pereira AC, von Schantz M. The role of race and ethnicity in sleep, circadian rhythms and cardiovascular health. Sleep Med Rev 2017; https://doi.org/10.1016/j.smrv.2016.05.004.

[68] Baron KG, Reid KJ. Circadian misalignment and health. Int Rev Psychiatry 2014;26(2):139–54. https://doi.org/10.3109/09540261.2014.911149.

[69] Eckel RH, Depner CM, Perreault L, Markwald RR, Smith MR, McHill AW, … Wright KP. Morning circadian misalignment during short sleep duration impacts insulin sensitivity. Curr Biol 2015;25(22):3004–10. https://doi.org/10.1016/j.cub.2015.10.011.

[70] Buxton OM, Cain SW, O'Connor SP, Porter JH, Duffy JF, Wang W, … Shea SA. Adverse metabolic consequences in humans of prolonged sleep restriction combined with circadian disruption. Sci Transl Med 2012; https://doi.org/10.1126/scitranslmed.3003200.

[71] Al Khatib HK, Hall WL, Creedon A, Ooi E, Masri T, McGowan L, … Pot GK. Sleep extension is a feasible lifestyle intervention in free-living adults who are habitually short sleepers: a potential strategy for decreasing intake of free sugars? A randomized controlled pilot study. Am J Clin Nutr 2018;107(1):43–53. https://doi.org/10.1093/ajcn/nqx030.

[72] Knutson KL, Van Cauter E. Associations between sleep loss and increased risk of obesity and diabetes. Ann N Y Acad Sci 2008;1129:287–304. https://doi.org/10.1196/annals.1417.033.

[73] Dashti HS, Scheer FA, Jacques PF, Lamon-Fava S, Ordovás JM. Short sleep duration and dietary intake: epidemiologic evidence, mechanisms, and health implications. Adv Nutr 2015;6(6):648–59. https://doi.org/10.3945/an.115.008623.

[74] Brondel L, Romer MA, Nougues PM, Touyarou P, Davenne D. Acute partial sleep deprivation increases food intake in healthy men. Am J Clin Nutr 2010;91(6):1550–9. https://doi.org/10.3945/ajcn.2009.28523.

[75] Bosy-Westphal A, Hinrichs S, Jauch-Chara K, Hitze B, Later W, Wilms B, … Müller MJ. Influence of partial sleep deprivation on energy balance and insulin sensitivity in healthy women. Obes Facts 2008;1(5):266–73. https://doi.org/10.1159/000158874.

[76] Bromley LE, Booth JN, Kilkus JM, Imperial JG, Penev PD. Sleep restriction decreases the physical activity of adults at risk for type 2 diabetes. Sleep 2012;35(7):977–84. https://doi.org/10.5665/sleep.1964.

[77] Bozorgmanesh M, Hadaegh F, Azizi F. Predictive performance of the visceral adiposity index for a visceral adiposity-related risk: type 2 diabetes. Lipids Health Dis 2011;10(1):88. https://doi.org/10.1186/1476-511X-10-88.

[78] Altaf QA, Dodson P, Ali A, Raymond NT, Wharton H, Fellows H, … Tahrani AA. Obstructive sleep apnea and retinopathy in patients with type 2 diabetes: a longitudinal study. Am J Respir Crit Care Med 2017;196(7):892–900. https://doi.org/10.1164/rccm.201701-0175OC.

[79] Bironneau V, Goupil F, Ducluzeau PH, Le Vaillant M, Abraham P, Henni S, … Gagnadoux F. Association between obstructive sleep apnea severity and endothelial dysfunction in patients with type 2 diabetes. Cardiovasc Diabetol 2017;16(1):39. https://doi.org/10.1186/s12933-017-0521-y.

[80] Rajendran P, Rengarajan T, Thangavel J, Nishigaki Y, Sakthisekaran D, Sethi G, Nishigaki I. The vascular endothelium and human diseases. Int J Biol Sci 2013;9(10):1057–69. https://doi.org/10.7150/ijbs.7502.

[81] Gabric K, Matetic A, Vilovic M, Kurir TT, Rusic D, Galic T, … Bozic J. Health-related quality of life in type 2 diabetes mellitus patients with different risk for obstructive sleep apnea. Patient Prefer Adherence 2018;12:765–73. https://doi.org/10.2147/PPA.S165203.

[82] Lou P, Qin Y, Zhang P, Chen P, Zhang L, Chang G, … Zhang N. Association of sleep quality and quality of life in type 2 diabetes mellitus: a cross-sectional study in China. Diabetes Res Clin Pract 2015;107(1):69–76. https://doi.org/10.1016/J.DIABRES.2014.09.060.

[83] Hirshkowitz M, Whiton K, Albert SM, Alessi C, Bruni O, Doncarlos L, … Ware JC. National Sleep Foundation's sleep time duration recommendations: methodology and results summary. Sleep Health 2015;1(1):40–3. https://doi.org/10.1016/j.sleh.2014.12.010.

[84] Chao C-Y, Wu J-S, Yang Y-C, Shih C-C, Wang R-H, Lu F-H, Chang C-J. Sleep duration is a potential risk factor for newly diagnosed type 2 diabetes mellitus. Metabolism 2011;60(6):799–804. https://doi.org/10.1016/J.METABOL.2010.07.031.

[85] Chaput J-P, Després J-P, Bouchard C, Tremblay a. Association of sleep duration with type 2 diabetes and impaired glucose tolerance. Diabetologia 2007;50(11):2298–304. https://doi.org/10.1007/s00125-007-0786-x.

[86] Maskarinec G, Jacobs S, Amshoff Y, Setiawan VW, Shvetsov YB, Franke AA, … Le Marchand L. Sleep duration and incidence of type 2 diabetes: the Multiethnic Cohort. Sleep Health 2018;4(1):27–32. https://doi.org/10.1016/J.SLEH.2017.08.008.

[87] Tasali E, Chapotot F, Wroblewski K, Schoeller D. The effects of extended bedtimes on sleep duration and food desire in overweight young adults: a home-based intervention. Appetite 2014;80:220–4. https://doi.org/10.1016/J.APPET.2014.05.021.

Chapter 18

Social jetlag, circadian disruption, and cardiometabolic disease risk

Susan Kohl Malone[a], Maria A. Mendoza[a], Freda Patterson[b]
[a]Rory Meyers College of Nursing, New York University, New York, NY, United States, [b]Department of Behavioral Health and Nutrition, College of Health Sciences, University of Delaware, Newark, DE, United States

INTRODUCTION

The sun rises and sets over the earth in a predictable pattern. This pattern has existed for billions of years and has influenced the behavior of all living things. Behavioral rhythms have aligned with these light-dark rhythms and conferred an evolutionary advantage. Humans have adapted to the light-dark cycle so that activity occurs during the day and rest occurs during the night. Increased visibility afforded by daylight optimizes foraging and safety while being active. Reduced visibility during the night optimizes sleeping and fasting. Daily rhythms, such as light-dark, are known as circadian rhythms from the Latin words "circa," for about, and "dias," for a day. Physiological processes rely on predictable circadian rhythms. These processes include sleeping and waking, cardiac function, such as heart rate and blood pressure, and metabolic processes, such as glucose, lipid, and energy metabolism. Disrupting circadian rhythms can profoundly impact cardiometabolic health and well-being. Poor cardiometabolic health can also disrupt the circadian system. This chapter will briefly introduce the cardiometabolic syndrome, the circadian system, circadian disruption, and social jetlag as a form of circadian disruption.

Circadian regulation of cardiac and metabolic functioning will be reviewed and the influence of environmental, behavioral, and biological rhythms on cardiometabolic health will be discussed (Fig. 18.1).

DEFINITIONS AND EPIDEMIOLOGY
Cardiometabolic syndrome

The cardiometabolic syndrome also known as metabolic syndrome (MetS) or syndrome X is a cluster of metabolic dysfunctions that include insulin resistance, impaired glucose tolerance, dyslipidemia, hypertension, and central adiposity. The presence of three or more of these cardiometabolic risk factors are predictive of diabetes and cardiovascular disease. Individuals with MetS are more likely to die from coronary heart disease and stroke than individuals without MetS.

The pathophysiologic mechanism in MetS originates from insulin resistance. The pancreatic beta cells produce insulin which binds with cells in the liver, fat, muscles, and blood vessels. Normally in the liver, insulin suppresses glucose production. In the setting of insulin resistance, there is impaired glucose production or hepatic gluconeogenesis. It is not clearly understood why insulin also plays a role in hepatic lipogenesis causing increased free fatty acids and triglycerides in the blood stream causing dyslipidemia. This leads to an atherogenic state characterized by increased triglycerides, low high-density lipoprotein cholesterol (HDL-c) and increased small, dense low-density lipoproteins.

Other key pathologic mechanisms in MetS include obesity, specifically abnormal ectopic fat distribution and inflammation. The obesity is described as increased visceral rather than subcutaneous fat. There is also accumulation of fat in the liver causing nonalcoholic fatty liver disease. This adipose tissue hypertrophy leads to insulin resistance. Moreover, the hypertrophic adipocytes result in a low inflammatory state from the release of pro-inflammatory factors such as adipocytokines, plasminogen activator inhibitor-1, tumor necrosis factor alpha, interleukin 6, C-reactive protein and fibrinogen.

Data from the National Health and Nutrition Examination Survey (NHANES) indicate that in 2011–12, approximately one third of adults in the United States (34.7%) had MetS [1], representing a 35% prevalence increase as compared to 1998–94 [2]. The distribution of MetS prevalence varies according to sex, race, age, and education levels. Specifically, a greater proportion of women meet the diagnostic criteria for MetS as compared to men (36.6% versus 32.8%, respectively) [1]. In terms of racial differences, 38.6% of Hispanic, 37.4% of nonHispanic White, and 35.5% of Black adults

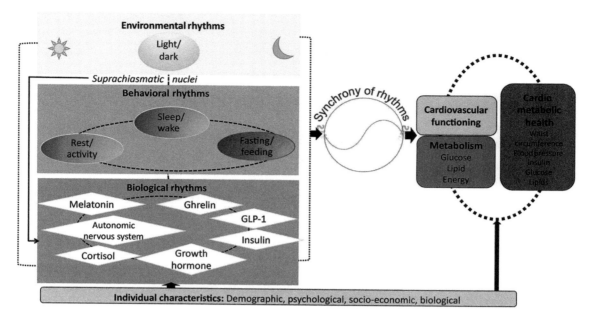

FIG. 18.1 The influence of environmental, behavioral, and biological rhythms on cardiometabolic health.

reported having METs in 2012 [1]. Between 1988 and 1994 and 2007–12, the largest increase in the prevalence of MetS was reported in nonHispanic black men (55%), then nonHispanic white women (44%), and nonHispanic black women (41%) [2]. Advanced age and lower educational attainment are associated with increased rates of MetS. Approximately one in five (18.3%) 20–39 year olds, as compared to one in two (46.7%) adults aged 60 and older have MetS [1]. As compared to college graduates, those with a less than high school education had a 56% increased odds of having METs (OR = 1.56, 95% CI = 1.32–1.84) [2].

Circadian rhythms

Circadian rhythms have been increasingly linked to cardiometabolic health. Circadian rhythms provide a temporal structure for events. Knowing when predictable events will occur allows for preparation and optimal functioning. Being prepared is so important that living things have translated the light-dark rhythm of the solar clock into their very being by creating internal biological clocks. These biological clocks predict daily changes in the environment and prepare the body for anticipated behaviors. As a result, the right hormones are in the right place at the right time. For example, cortisol, the alerting and arousal hormone, increases prior to and just after waking to prepare the body for the upcoming demands of the day [3, 4]. Optimal performance follows and survival advantage is conferred.

Life would be chaos without the temporal structure of circadian rhythms. Biological clocks must be set to the same time as the solar clock to effectively anticipate predictable 24-h patterns. However, biological clocks tick slightly slower than the solar clock, a little over 24 h each day. It is essential to synchronize biological and solar clocks to prevent chaos and to optimize function. Light is the primary synchronizer. The light/dark cycle resets the biological clock to tick in sync with the solar clock every day.

The eye is key to this synchronization. Specialized cells in the eye's retina, the retinal ganglion cells, perceive light and dark. This light/dark message is carried along a neural pathway, the retino-hypothalamic tract, and delivered to the master clock, a specialized cluster of cells in the hypothalamus known as the suprachiasmatic nuclei (SCN). The SCN transmits stimulatory and inhibitory messages to about 35 brain regions, particularly hypothalamic regions controlling hormone release and autonomic nervous system control [5]. The SCN also transmits these messages to the autonomic nervous system which regulates tissue specific sensitivity to these hormones. These hormonal and neuronal messages convey a sense of time to the body.

But the SCN is not the only clock. Every cell of the body possesses a clock, liver cells, pancreatic cells, muscle tissue cells, etc. These specific body cell clocks synchronize behavioral and physiological processes to the light-dark cycle. The SCN plays an important role in this synchronization, however, the synchronization process is complex and still not completely understood. What is known is that disrupting the synchrony between the clocks impacts cognitive, behavioral, and psychological functioning.

Sleep-wake rhythms are one of the most pronounced 24-h behavioral rhythms in humans. Sleeping and waking partition essential cognitive and physiological tasks to certain times of day. Sleep opens the door for information processing and memory consolidation because sensory input is minimal compared to the wake period. Energy consumed during the day can be diverted to the brain rather than used to meet the energy needs of daytime activities. Sleep also opens the door for the body to "clean up" toxins that have built up along the metabolic pathways from daytime energy consumption. Despite a universal 24-h cycle for sleeping and waking in humans, sleep-wake times vary between individuals. Some people prefer early bedtimes and wake times; others prefer late bedtimes and wake times.

Sleep-wake times, or the clock times for sleeping and waking, are influenced by multiple factors. These factors include light-dark cycles within a time zone, developmental stage, social obligations, and genetic factors. For example, sunrise and sunset are 4 min later for every one-degree of longitude from east to west. Sleep-wake times replicate this by delaying 4-min for every one-degree longitude. Developmentally, sleep-wake times are early in childhood, delay throughout adolescents, reach a peak in lateness around 20 years of age, and gradually advance thereafter until approximately 60 years of age. Social obligations such as late-night gatherings and early morning work commitments also impact sleep wake times. Individual differences in the intrinsic pace of the biological clock also affects sleep-wake times, such that faster paced clock (e.g., 24.2 h) contribute to earlier sleep periods than slower paced clocks (e.g., 24.4 h) [6]. Racial differences in clock pace have been reported with Blacks of African descent exhibiting faster internal clocks than their White counterparts [7, 8]. Genetic factors also contribute to sleep-wake times as several single nucleotide polymorphisms have been linked to genes with well-known circadian roles, such as PER2 [9, 10].

Circadian disruption and social jetlag

Circadian disruption is defined as an altered tau, amplitude, and/or mean level of a rhythm [11]. These disrupted rhythms can be environmental, exemplified by reduced daytime and increased nighttime light exposure; behavioral, exemplified by reduced nocturnal and increased daytime sleep; as well as biological. Suboptimal functioning follows when the synchrony between environmental, behavioral, and biological rhythms is disrupted. Staying awake during the night when your body is secreting sleep-promoting hormones, such as melatonin results in suboptimal wakefulness. Conversely, sleeping during the day when hormones promoting sleep are missing lead to curtailed sleep duration and altered sleep architecture. Specifically, REM and SWS sleep are reduced and REM sleep shifts from the later to the earlier part of the sleep period [12].

Irregular sleep-wake times contribute to a loss of synchrony between environmental, behavioral, and biological rhythms. Irregular sleep-wake times can be chronic, as in shift work, or transient, as in trans-meridian travel. Chronic, irregular sleep-wake times also characterize social jetlag. Social jetlag is defined as the difference in sleep-wake times between work and free days [13]. Different work and free day sleep-wake times result from the different schedules people adopt. For example, people tend to go to bed late but rise early for work on work days, contributing to short work-day sleep. To compensate for sleep debt incurred during the work week, people often sleep longer on free days by waking later. The result is late-bed/early-rise times on work days and late-bed/late-rise times on free days. Social jetlag can also occur when early bedtimes and early rise times are adopted during the work week, but social obligations on free-days lead to late bedtimes. Seventy percent of a European population sample has reported social jetlag of at least 1 h or more. The widespread and chronic nature of social jetlag have raised concern about potential adverse cardiometabolic health effects like those found in shift work [14].

Circadian disruption and cardiometabolic health

Data clearly show that the odds of MetS is significantly higher in shift workers as compared to nonshift workers [15]. Specifically, across 13 pooled observational studies, the association between 'ever exposed to night shift work' and MetS risk was 1.57 (95% CI=1.24–1.98, $p=0.001$), while a higher risk was indicated in workers with longer exposure to night shifts (RR=1.77, 95% CI=1.32–2.36, $p=0.936$) [16]. Consistent with these pooled data, prospective data over a median 6.6 year period showed that the risk of METs development increased with accumulated years of shift work, independent of possible confounders [17]. Not yet clear from this body of work is the cut-point of time spent doing shiftwork that confers exponentially greater risk for MetS.

In addition to shift work being independently and prospectively associated with an increased risk for MetS, there is evidence that the individual risk factors for MetS (abdominal obesity, hypertension, atherogenic dyslipidemia and impaired glucose tolerance) [18] are associated with shift work. For example, shift workers were shown to have a 50% increased odds of hypertension [19], while an elevated waist circumference was shown to be more common in night and rotating shift workers: specifically, waist circumference increased by 1.089 cm per 1000 night duties and by 0.99 cm per 10,000 night shift hours in a sample of nurses [20]. Prospective data also show that among persistent shift workers, the presence of 1–2 MetS factors conferred a more than 12-fold risk of developing MetS 5 years later [21]. Thus, the trajectory toward a MetS diagnosis is shown to accelerate with each accumulated risk factor diagnosis in populations whose circadian function is disrupted, such as shift workers.

Data linking social jetlag and MetS are limited and findings are inconsistent. Social jetlag has been associated with a greater likelihood of having MetS, or individual MetS risk factors in some [22–24], but not all [25, 26], studies. Adults with type 2 diabetes reporting >30 min of social jetlag did not have poorer glucose regulation than adults with ≤30 min of social jetlag [25, 26]. Adults younger than 61 years of age were more likely to have MetS if they reported 1–2 h of social jetlag (1.29, 95% CI=0.9–1.9) or ≥2 h of social jetlag (2.13, 95% CI=1.3–3.4) than adults with <1 h of social jetlag [27]. However, there was no association between social jetlag and MetS in adults ≥61 years of age [27]. Social jetlag has been associated with poorer metabolic, but not cardiac outcomes as evidenced by higher triglyceride, fasting insulin, and insulin resistance levels as well as greater waist circumferences and lower HDL-c levels, but not heart rate or blood pressure [23].

Reasons for these inconsistent findings are uncertain. Chronological age may be a contributing factor. Younger adults may be more likely to suffer the deleterious effects of social jetlaf, but they are also more likely to have greater social jetlag than older adults. Differences may also stem from variability in vulnerability to the adverse cardiometabolic effects from social jetlag between individuals. Support for interindividual variability is garnered from evidence of significant differences in weight gain and neurobehavioral deficits following sleep loss among individuals, but stability of weight gain and neurobehavioral deficits within individuals [28, 29].

CIRCADIAN CONTROL OF THE CARDIOMETABOLIC SYSTEM

Links between MetS and circadian disruption are plausible because the cardiometabolic system is influenced by multiple biological, behavioral, and environmental factors that have been shown to have endogenous circadian rhythms. Identifying endogenous biological rhythms requires separating these biological factors from the behavioral and environmental factors affecting them, such as sleeping and waking or light and dark. This is accomplished by keeping individuals constantly awake with low activity levels in very dim light and providing evenly spaced meals of equal caloric content, or in other words, keeping individuals in constant routine conditions [30]. Endogenous rhythms can also be separated from behavioral and environmental factors by scheduling sleeping and waking every 28 h rather than every 24 h, or forced de-synchrony conditions. Both constant routine and forced desynchrony conditions have been used in experimental studies to elucidate the circadian regulation of the cardiometabolic system. Simulated shift work protocols have also been used to elucidate the effects of circadian disruption on individual cardiometabolic outcomes in controlled laboratory environments.

Cardiovascular functioning

Cardiovascular demands change drastically in response to activity levels. Faster heart rates and higher blood pressures are needed to deliver sufficient nutrients throughout the body during activity as opposed to rest. The circadian system can anticipate and prepare for predictable changes in cardiovascular demands. Several constant routine and forced desynchrony studies have revealed that cardiovascular parameters, including heart rate and blood pressure, have an endogenous rhythm [31–34]. Heart rates peak between 11:00 and 12:00 and dip by about seven beats per minute between 2:00 and 3:00 [31, 32]. Blood pressures peak between 18:00 and 21:00 with a 3–6 mmHg greater systolic and a 2–3 mmHg greater diastolic blood pressure than the trough blood pressure [35, 36]. Chronotype, or an individual's preference for morning or evening activities [37], may advance or delay this endogenous rhythm. For example, peak heart rates are 6 h later for evening compared to morning types [33].

Metabolism

The metabolic system is tightly coupled to the circadian system [38–40] for molecular and cellular reviews. The circadian system informs the metabolic system of anticipated periods for activity/feeding and rest/fasting, thereby optimizing energy utilization and storage across the 24-h day. The SCN conveys day/night messages to the body through neuronal and hormonal messages that orchestrate biological and behavioral rhythms for glucose and lipid metabolism, energy expenditure, and appetite. The SCN ensures that the right hormones and right behaviors are in the right place at the right time of day. These biological rhythms collectively favor efficient metabolism to coincide with daytime feeding and activity in humans. But the SCN is also influenced by the metabolic system through nutrient messages that convey the energy status of the body. This bidirectional regulation between the circadian and the metabolic systems ensures that glucose homeostasis is maintained.

Glucose metabolism

A robust daily rhythm for glucose metabolism has been well-established. 24/7 glucose availability is critical to support the central nervous system which uses 20% of glucose consumed but cannot store or synthesize glucose. Morning energy needs for activity and feeding are anticipated by the circadian system and several reports of higher morning-versus-evening fasting glucose levels that are offset by greater insulin secretion support this premise [41–44]. Morris et al. reported a 5% higher morning, compared to evening, fasting glucose level in healthy young adults [42]. Yet not all studies report higher morning-versus-evening fasting glucose levels [45, 46]. Age has been conjectured as an underlying reason for these disparate findings. Higher morning-versus-evening fasting glucose has been reported in younger, but not older adults (<35 and >65 years of age, respectively) [47]. But others have reported that healthy youth do not have a morning-evening differences in fasting glucose [46]. This mixed evidence raises speculation that age is an underlying factor in morning versus evening differences in fasting glucose.

On the other hand, glucose responses to meals consistently demonstrate an endogenous rhythm. Evening glucose responses to meals, oral glucose challenges, and intravenous glucose are higher and remain elevated longer than morning responses in healthy adults [42, 48–54]. Postmeal glucose excursions to identical meals in the morning and the evening are 2.3 mmol/L and 3.3 mmol/L respectively, in healthy young adults [48]. Glucose tolerance is 17% lower in the evening versus the morning [42]. These morning-to-evening declines in glucose tolerance are partially driven by circadian modulation of target tissue sensitivity. Insulin-producing beta cells of the pancreas become less sensitive to glucose across the day; other peripheral tissues become less sensitive to insulin across the day. SCN excised rats lack daily glucose and insulin sensitivity rhythms exemplifying the critical role of the circadian system in glucose regulation [55].

Lipid metabolism

Endogenous rhythms may optimize lipid absorption, storage, and transport by synchronizing these metabolic activities with anticipated sleep-wake and feeding-fasting behaviors. Fifteen percent to twenty percent of lipid metabolites have demonstrated endogenous rhythms that are characterized by morning peaks and afternoon/evening declines during constant routine conditions [56, 57]. However, large interindividual differences in the specific lipid metabolites exhibiting rhythmicity exist and up to12-h differences in the timing of the rhythms have been demonstrated [57]. This evidence suggests that several distinct circadian phenotypes for lipid metabolism exist [57].

Several studies using different study protocols have demonstrated that low density lipoprotein cholesterol (LDL-c) and total cholesterol levels are higher in the morning compared to the evening [58–60]. Miida reported that LDL-c was 0.09–0.13 mmol/L higher at waking than at midnight [60]. Declining LDL-c and total cholesterol levels from morning to evening may be explained by enhanced daytime lipid clearance from the circulation. This premise is supported by evidence that apolipoproteins which are used to transport lipids are lower following night versus day time meals [61]. Higher triglyceride [61, 62] and very low density lipoproteins [61] levels have also been reported following night versus day time meals in rotating shift workers and healthy young adults.

Energy metabolism

It is uncertain whether there is an endogenous rhythm for energy metabolism. One method for estimating energy metabolism is through indirect calorimetry. Indirect calorimetry calculates metabolic rate by measuring the heat produced by oxidizing macronutrients, such as carbohydrates and fats. Some [63], but not [31] all studies report an endogenous rhythm for postprandial energy metabolism based on diet-induced thermogenesis [63, 64]. Disparate findings may stem from methodological differences, such as responses to meals versus snacks. Morris et al. found that diet-induced thermogenesis was 44% lower in the evening than in the morning in healthy young adults [63]. These same authors reported that postprandial energy expenditure was 4% lower in the evening than the morning [63]. This suggests that eating larger meals in the morning and smaller meals in the evening would be beneficial for weight homeostasis and/or weight loss. Yet, large breakfast and small dinner eating patterns are not the norm in Western cultures. Dinner is the largest meal of the day [65].

Some contend that changes in appetite across the day may explain why dinner is the largest meal [66]. Healthy adults rate their hunger as lowest after waking in the morning and highest in the evening at approximately 20:00 [67, 68]. These rhythms are exacerbated during sleep restriction [68] and are unrelated to meal times [69]. However, time pressure and social cues significantly influence meal sizes [70, 71].

ENVIRONMENTAL RHYTHMS AND CARDIOMETABOLIC HEALTH

The 24-h transitions from light to darkness (LD) provide a strong zeitgeber for metabolic system regulation in both animals and humans [72]. In free-living conditions characterized by bright daylight and the absence of artificial nightlight, behavioral rhythms such as feeding and activity shift earlier [73]. However, in human-constructed environments where daylight can be reduced and exposure to artificial nightlight increased, circadian rhythms and the regulation of the metabolic system can be negatively impacted.

The widespread use of electric lighting in the early-to-mid twentieth century enabled the artificial extension of daytime to the point where today virtually all adults living in developed countries experience light pollution [74]. Sources of LAN include urban street lighting (5–15 lx) and electronic tablets (~40 lx depending on size of screen) [75]. Data suggest that approximately one third of adults and youth leave an electronic device (e.g., television or tablet) on while sleeping [76]. Constant exposure to light desynchronizes circadian activity in rodents [77], while experimental studies show that such exposure is associated with poor metabolic indices including increased body mass and reduced glucose tolerance [78].

In terms of human studies into the association between LAN, circadian rhythms, and metabolic health, light exposure >180 lx have been shown to phase shift circadian rhythms and suppress melatonin concentration [79]. Shift workers represent a population who are chronically exposed to LAN and experience disproportionately higher rates of poor metabolic health. A large body of work consistently reports the positive association between shift work with increased hypertension, cholesterol, and obesity [80, 81]. Meanwhile, observational studies in the general population have also shown higher levels of LAN to be significantly associated with obesity even after adjustment for confounders such as sleep duration, tobacco use, and physical activity [82, 83]. Larger waist circumferences, higher LDL-c and lower HDL-c levels have been associated with bedroom LAN ≥ 3 lx versus <3 lx in older community dwelling adults [84]. Morning and evening blue light exposure increases insulin resistance in healthy adults [85].

One mechanism that may go toward explaining the association between LAN, circadian disruption, and poor metabolic health is the suppression of melatonin and glucagon like peptide-1 (GLP-1). Melatonin is an endogenously synthesized molecule that is produced by the pineal gland and serves to help regulate sleep and wakefulness. Melatonin production may be inhibited by exposure to nighttime light of sufficient intensity and duration [86]—even 1 h of exposure to 45 lx was shown to decrease melatonin levels by 60% [87]. Reduced melatonin secretion has been associated with an increased risk for obesity [88] and type-2 diabetes [89]. Relatedly, suppressed melatonin production delays sleep onset, and is associated with shorter sleep duration. Short sleep duration is a reliable determinant of metabolic syndrome, and as such represents a viable mechanism through which LAN is a risk factor for metabolic syndrome. GLP-1 limits postmeal glucose excursions by stimulating insulin secretion. Constant light exposure reduces the amplitude of GLP-1 rhythms in adults with obesity and type 2 diabetes [90].

LAN exposure also impacts heart rate and heart rhythm. Brighter lights increase heart rates and the effect of light on heart rates is strongest during the middle of the night and early morning [91]. Disrupting light-dark cycles has also been shown to disrupt heart rate rhythms and lengthen QT intervals in mice [92]. These lines of evidence suggest an association also exists between LAN, circadian rhythms, and cardiovascular health.

BEHAVIORAL RHYTHMS AND CARDIOMETABOLIC HEALTH

Behavioral rhythms also influence cardiometabolic regulation. Synchronizing behaviors such as sleep and fasting with darkness, as well as activity and feeding with daylight lead to appropriately timed levels of glucose, insulin, glucocorticoids, and metabolically relevant hormones. Behavioral circadian disruption can occur when the SCN, entrained by light, sends sleep-promoting signals and lipid mobilization messages but food intake sends wake-promoting signals and fat storage messages. This mixed circadian signaling between the SCN and peripheral clocks in other tissues, such as the pancreas and adipose tissue has been shown to negatively impact the cardiometabolic regulation in several controlled laboratory studies [12, 70, 93, 94].

Acutely disrupting sleep-wake patterns may impact cardiometabolic regulation through changes in metabolically relevant hormones and diet induced thermogenesis [70]. Cortisol, a hormone that raises glucose, peaks at sleep onset rather than at waking and has a blunted rhythm during forced desynchrony protocols [70, 95]. The morning rise in cortisol is also significantly lower and the evening nadir is significantly higher when adults are deprived of night time sleep [96]. Leptin, a satiety hormone, has been shown to decrease [70] and glucose has been shown to rise [95], even to prediabetic levels in some participants [70] under forced desynchrony protocols. Adverse metabolic effects have been demonstrated regardless of whether sleep is advanced or delayed [95]. Daytime sleep has also been associated with a decrease in total daily energy expenditure, diet induced thermogenesis, and leptin levels compared to night time sleep [97].

Several studies suggest that humans' metabolic systems do not adapt to disrupted sleep-wake patterns. Rather, adverse metabolic effects persist even after exposure to longer disrupted sleep-wake patterns [93, 94, 98]. Fasting insulin levels increased by 18%, postmeal glucose levels rose 14%, and resting metabolic rate decreased by 8% during a 3-week protocol in healthy younger and older adults [93]. Chronic shift workers still demonstrated higher postmeal glucose levels despite a 10% increase in late phase insulin secretion during daytime versus nighttime sleep in a controlled laboratory environment [94]. Smaller and more prolonged disruptions in sleep-wake patterns have also been shown to significantly reduce leptin levels [98]. Blunted cortisol rhythms have been associated with irregular sleep patterns in community dwelling adults [99]. All told, chronic disruption in sleep-wake patterns as might occur during shift work or social jetlag may be a contributing factor for adverse cardiometabolic health outcomes.

Short sleep duration has also been associated with poor cardiometabolic health and is often linked with irregular sleep-wake patterns [12]. Several laboratory studies have demonstrated that the effects of disrupted sleep-wake patterns on the metabolic system are independent of and in addition to that of short sleep duration. Leproult et al. demonstrated that insulin resistance was doubled during circadian disruption plus sleep restriction versus sleep restriction alone in men but not women [100]. Leptin levels were significantly reduced following irregular sleep patterns despite sleeping longer than 6.5 h [98]. Eckel et al. reported that insulin resistance increased 20% during a simulated social jetlag study [101]. These studies provide evidence into the additive deleterious metabolic effects of circadian disruption.

Meal timing has been linked to cardiometabolic health, particularly weight regulation. Several weight loss studies support the benefits of earlier eating habits for optimal weight and glucose regulation. Shifting caloric intake earlier in the day optimizes weight loss efforts. A 10% decrease in BMI was achieved by eating 50% of total daily calories at breakfast in women with MetS undergoing weight loss treatment [102]. Only a 5% decrease in BMI was achieved by eating 50% of total daily calories at dinner [102]. Improvements in waist circumference were also greater in the early eaters. Waist circumference decreased 8.5 cm versus 3.9 cm in the earlier versus later eaters respectively ($p < 0.0001$) [102]. Similarly, overweight and obese adults who ate their main meal earlier in the day lost 25% more weight than overweight and obese individuals who ate their main meal earlier in the day despite similar caloric intake and physical activity [103].

Potential mechanisms linking meal timing with weight regulation include morning-evening differences in diet induced thermogenesis and changes in the gut microbiome. Diet induced thermogenesis is greater following morning, versus evening meals [63]. Morning-evening difference in diet induced thermogenesis have been shown to be driven by endogenous circadian rhythms rather than behavioral rhythms [63]. Therefore, eating early in the day is required to take advantage of the increased metabolic rates for eating. The gut microbiome is primed to support energy metabolism during the day and detoxification during the night [104]. Irregular sleeping and eating patterns are associated with changes in the gut microbiome that are characterized by increased firmicutes, a condition associated with obesity [104].

Other metabolic benefits are reaped by eating earlier versus later meals. Early eaters with MetS had greater reductions in total cholesterol, fasting insulin, insulin resistance, and fasting glucose, as well as a greater increase in HDL-cholesterol compared to later eaters [102]. In a randomized cross over study, earlier meal times were associated with greater nighttime peaks in leptin and higher melatonin levels overall in healthy young adults during the earlier versus later sleep time (1:30 bedtimes-8:30 wake times versus 22:30 bedtimes-6:30 wake times, respectively) [105]. Later meals are associated with greater and longer lasting triglyceride elevations compared to earlier meals [106]. Regular eating habits have also been linked to improved fasting lipid and postprandial insulin profiles and thermogenesis [107].

All told, the relationship between behavioral rhythms and cardiometabolic health suggest that increasing regularity and promoting earlier timing for sleeping and eating habits may improve cardiometabolic health for many people.

BIOLOGICAL RHYTHMS AND CARDIOMETABOLIC HEALTH

Autonomic nervous system

An important determinant of endogenous cardiac rhythms is the circadian nature of the autonomic nervous system [108]. The SCN projects directly to the paraventricular nucleus, the hypothalamic area regulating autonomic nervous system control [109]. Through this pathway, the SCN conveys day/night messages to the autonomic nervous system and facilitates cardiovascular changes needed to support anticipated day and night time behaviors. During the day, the sympathetic system dominates and contributes to faster heart rates [110]. During the night, the parasympathetic system dominates and contributes to slower heart rates which open the door for greater nocturnal heart rate variability [111]. Hence, heart rate variability is used to noninvasively estimate cardiac autonomic regulation [112, 113]. Greater night time, compared to daytime, heart rate variability exemplifies nocturnal parasympathetic dominance [114]. The critical role of the circadian system in cardiac autonomic regulation is illustrated by the loss of heart rate variability and nocturnal blood pressure dips in SCN-excised rats [109, 115].

The placement of sleep (or the timing of sleep and activity) within the 24-h day matters, in part because the endogenous circadian system modulates the parasympathetic-sympathetic activity that occurs during sleep. When sleep occurs during the day, parasympathetic activity is diminished compared to the same amount of sleep during the night [116]. Enhanced parasympathetic activity of night time sleep provides optimal cardiovascular protection underscoring the importance of syncing sleep-wake patterns with day-night. Circadian disruption may diminish the cardiovascular protection afforded by night time sleep.

Parasympathetic-sympathetic activity also varies across sleep stages [117, 118]. Heart rates are slower and blood pressures are lower due to greater parasympathetic activity during nonrapid eye movement sleep (N-REM) compared to rapid eye movement sleep (REM) [119]. And parasympathetic activity increases from stages one to three of N-REM. Heart rates and blood pressures approach waking levels during REM sleep due to sympathetic activity [116, 119]. This suggests that interrupted sleep or daytime sleep, as needed in shift work, result in overall greater sympathetic activity.

The circadian nature of the autonomic nervous system is also a determinant of endogenous metabolic rhythms. Through SCN projections to the paraventricular nucleus [109], the SCN conveys day/night messages to the autonomic nervous system. The autonomic nervous system then signals peripheral organs through its extensive network of parasympathetic and sympathetic nerve fibers. Key metabolic organs including the liver, pancreas, and visceral adipose tissue share a common neuronal connection with the paraventricular nucleus, dorsal medial nucleus, and SCN [120]. Paraventricular neuronal stimulation activates sympathetic input to the liver and adipose tissue stimulating gluconeogenesis and lipolysis respectively. Paraventricular autonomic fibers also project through other pathways to the pineal gland and regulate the release of melatonin. These neuronal messages contribute to the metabolic changes needed to support anticipated day and night time behaviors. Glucose metabolism rhythms are lost when connections between the liver and autonomic nervous system are disrupted. Impaired glucose tolerance and the loss of glucose stimulated insulin secretion rhythms and adipose tissue insulin sensitivity are caused by removing the pineal gland in animals.

Metabolically relevant hormones

Important determinants of endogenous metabolic rhythms, such as glucose and lipid metabolism, energy expenditure, and appetite, are the circadian nature of the several metabolically relevant hormones and target tissue sensitivity to these hormones. These hormones include insulin and glucagon-like peptide 1 (GLP-1), as well as counter-regulatory hormones, such as cortisol, growth hormone, melatonin, and appetite regulatory hormones, such as ghrelin and leptin.

Insulin

Insulin is central to glucose homeostasis. Insulin promotes glucose entry into metabolically active cells such as muscles for energy use. Insulin also promotes energy storage by suppressing fat breakdown in adipose tissue and by stimulating the liver to store glucose as glycogen. Evidence for an endogenous rhythm for insulin is mixed [70, 93, 121, 122]. One study reported 21% lower fasting insulin levels in the morning versus evening in healthy adults [42]. Moreover, insulin secretion rates are lowest between midnight and 6 a.m. and highest between noon and 6 p.m. [122] corresponding with daytime feeding and nocturnal fasting rhythms.

Optimal glucose regulation in the early day is conferred by greater beta cell and peripheral tissue sensitivity to glucose and insulin respectively. Early phase insulin release is 27% lower in the evening versus the morning exemplifying pancreatic beta cell sensitivity decline [42, 54]. Postmeal insulin secretion increases 25%–50% in the evening versus the morning exemplifying peripheral tissue sensitivity decline [96, 123]. Greater declines in insulin sensitivity are estimated for subcutaneous adipose tissue [124]. Carrasco-Benso et al. reported a 54% decline in subcutaneous adipose tissue insulin sensitivity from noon to evening in obese adults [124]. These findings may have important implications for the insulin resistant state in many obese individuals. Moreover, subcutaneous adipose tissue insulin resistance is greater in people reporting shorter sleep and later bedtimes suggesting a complex interplay between biological and behavioral rhythms for adipose tissue insulin sensitivity [124].

Glucagon-like peptide 1 (GLP-1)

GLP-1 is a nutrient-sensing hormone secreted by the L-cells of the intestine. GLP-1 influences the magnitude of the insulin response from the beta cells. Basal GLP-1 secretion and L-cell sensitivity have demonstrated endogenous rhythms. GLP-1 secretion peaks at 6 a.m. coinciding with increased beta cell sensitivity to GLP-1 [90, 125]. Early phase GLP-1 responses to meals is also greatest following morning meals suggesting enhanced L-cell sensitivity to glucose in the morning [90, 125]. Collectively, greater basal and early phase postmeal secretion contributes to enhanced insulin secretion and optimal glucose regulation upon waking. Evidence is conflicted for variations in afternoon and evening postmeal L-cell sensitivity. GLP-1 secretion was greater following morning/afternoon versus evening meals in one study (08:00 versus 17:00) [126] but not another (11:00 versus 23:00) [90]. In the latter study, GLP-1 secretion was 23% higher at 23:00 versus 11:00 [90]. Reasons for these disparate findings are uncertain and suggest that GLP-1's role in postmeal glucose regulation is minor and/or that beta cell sensitivity to GLP-1 varies. Short sleep as well as delaying sleep reduces GLP-1 secretion [95, 127], although these results may differ by sex [127]. The amplitude of GLP-1 rhythms are flattened in adults with obesity and T2D [90].

Cortisol

Cortisol secretions are strongly regulated by the circadian system and, to a lesser extent, behavioral rhythms [70]. Cortisol begins rising 1–2 h after sleep onset, peaks within 1 h of morning waking, and declines thereafter across the 24-h day [70]. Cortisol rhythms parallel those of insulin sensitivity, such that high cortisol levels coincide with increased insulin sensitivity and vice versa [53]. Similarly, postmeal cortisol secretions decline by 33% from morning to evening [48]. These parallel declines in cortisol secretion and insulin sensitivity counter what is known about cortisol's insulin antagonizing effects [128]. Plat et al. reconciled this seeming incongruence by demonstrating that cortisol profoundly decreases insulin sensitivity, however insulin's response lags [129]. Insulin sensitivity begins declining 4–6 h after cortisol's peak and continues declining for >16 h [129]. This lag between cortisol and insulin sensitivity partially explains morning-to-evening increases in insulin resistance and declines in glucose tolerance. Cortisol rhythms persist with aging, albeit cortisol peaks earlier in the morning and the amplitude of cortisol rhythms are diminished [130]. Cortisol rhythm amplitudes are diminished by irregular and daytime, as opposed to night time sleep [95]. These diminished cortisol amplitudes may be associated with poorer health outcomes [131].

Melatonin

Melatonin secretion is strongly regulated by the circadian system [132]. Norepinephrine stimulates melatonin release from the pineal gland with the onset of darkness or approximately 14 h after wake times [133, 134]. Melatonin secretion varies with the length of the dark period, is suppressed by light, and directly opposes insulin secretion [122].

Melatonin directly inhibits nocturnal insulin secretion at melatonin levels well within the physiological range (1:00 p.m.) [135, 136]. Exogenous melatonin administration also reduces insulin secretion and worsens glucose levels [137–139]. Early morning waking in sleep-restricted, young healthy adults has been linked to increased insulin resistance [101]. Melatonin was implicated in these findings because insulin resistance increased the longer melatonin levels remained high after waking [101].

Melatonin receptors exist throughout the body [140]. When melatonin binds to melatonin receptor 1B (MTNR1B) on the insulin-producing beta cells of the pancreas [141, 142], insulin secretion is suppressed [136, 143], nocturnal hypoglycemia is prevented, and beta cell oxidative stress is reduced [144, 145]. Common genetic variants in MTNR1B [143, 146–148] overexpress MTNR1B receptors and lead to reduced insulin secretion [149]. Although melatonin suppresses insulin secretion in all persons, it may do so more in certain genotypes [150].

Contrasting melatonin's initial insulin suppressing effects, prolonged melatonin exposure has been shown to sensitize the beta cells to the action of the intestinal incretin hormone, GLP-1, on stimulating insulin secretion [135]. This evidence suggests that following nocturnal sleep, melatonin plays a role in preparing the body for morning feeding [135]. Melatonin may also influence other metabolically relevant hormones, such as cortisol and leptin. For example, cortisol and leptin rhythmicity is lost when the pineal gland is removed in rats or hamsters [142].

Reduced melatonin levels and a lower amplitude of melatonin rhythms have been reported in adults with type 2 diabetes and MetS [151, 152], older versus younger adults, and late versus early chronotypes. Autonomic neuropathy and/or diabetic retinopathy may underlie reduced melatonin levels and flattened melatonin amplitudes in adults with type 2 diabetes because of impaired transmission of light-dark information to the pineal gland [153]. Age associated changes in sleep-wake patterns relative to melatonin rhythms [154] suggest that some older adults may be waking during the biological night when melatonin is still being secreted. It is unknown whether these age-associated changes in sleep-wake patterns relative to melatonin secretion.

Growth hormone

Growth hormone lacks a diurnal rhythm but is linked to sleep-wake behaviors [96, 155]. Growth hormone is secreted almost immediately following the onset of nonREM sleep [156] and typically occurs between 22:00 and 02:00 in nocturnal sleepers [96]. Glucose rises and insulin resistance increases by 50% following the growth hormone surge [157] marking the prominent shift in energy substrates from glucose to lipids during the night. Growth hormone secretion decreases with aging [158].

Ghrelin and leptin

Ghrelin and leptin are appetite regulatory hormones associated with hunger and satiety respectively. Ghrelin is produced by the parietal cell of the stomach. Leptin is produced by adipocytes and the stomach. Leptin and between-meal ghrelin exhibit diurnal rhythms. Leptin rises across the day and peaks at approximately 3:00 in healthy young adults [105]. Loss of this diurnal rhythm in SCN excised rats exemplifies that leptin is regulated by the circadian system and is not solely regulated by feeding behaviors [159, 160]. Ghrelin levels are strongly regulated by meals with levels rising before eating and falling sharply after eating. However, between-meal ghrelin levels have been shown to also follow a diurnal rhythm that rises rising progressively across the day, peaks at 01:00 and falls to its lowest level in the morning 06:00 [161].

Bi-directional regulation

Information from the body is also relayed back to the brain. Many SCN neural fibers terminate near the arcuate nucleus (ARC) of the hypothalamus providing a plausible anatomical way that circulating metabolic hormones and nutrients may interact with the central circadian pacemaker [162]. It is here at the ARC-SCN connection where hunger and satiety messages (ghrelin and leptin) intersect with glucose homeostasis messages (insulin) and convey the status of the body to the brain [162]. Nutrient signals may modulate the circadian system itself. Switching from high carbohydrate/low fat diet to a low carbohydrate/high fat diet alters peripheral and central circadian clocks [163]. The tau, or period, of the central clock becomes longer and may be uncoupled from the clocks of peripheral organs that are important for metabolic regulation [163].

Other behavioral patterns can also impact the circadian clock and alter the endogenous rhythms of metabolically relevant hormones. For example, melatonin onset is delayed by 1.1 h following periods of no exercise in healthy young adults [164, 165]. Heart rate variability increases following morning exercise indicating greater parasympathetic stimulation [164]. Heart rates increase following evening exercise indicating greater sympathetic activity [164]. Irregular sleep-wake patterns, irregular or late eating patterns, daytime inactivity, and prolonged nighttime activity (short sleep) may lead to the hypothalamus getting confused biological signals across the 24 h day that then contribute to flattened rhythms for metabolically relevant hormones and metabolic processes [162].

Metabolic disorders, such as obesity and type 2 diabetes may also disrupt the circadian system. Obese adults lack diurnal rhythms in glucose tolerance [123]. Beta cell sensitivity to glucose and peripheral tissue sensitivity to insulin fails to decline from morning to evening [123] leading to disruptions between cortisol and insulin sensitivity rhythms [166]. Evening-versus-morning glucose responses to meals, oral glucose challenges, and intravenous glucose are consistently higher and remain elevated longer in healthy [42, 48–54], but not obese [123], adults. Reduced amplitudes for heart rate, blood pressure, and body temperature rhythms have been reported in adults with overweight/obesity, prediabetes and type 2 diabetes [167]. It is uncertain whether disrupted circadian rhythms contribute to the onset of these metabolic disorders. It may be that the onset of these metabolic disorders underlies the development of circadian disruption and ongoing circadian disruption may further negatively impact metabolic health.

CONCLUSION AND FUTURE DIRECTIONS

The circadian system regulates cardiac and metabolic functioning. Altering the synchrony between environmental, behavioral, and biological rhythms disrupts the circadian system and adversely impacts cardiometabolic health. Understanding these rhythms will be important for advancing, accelerating, and personalizing future cardiometabolic health promotion and treatment interventions. Behavioral rhythms such as sleep-wake and feeding-fasting are potentially modifiable. Meal timing, or chrono-nutrition, is a promising area of research for optimizing weight loss intervention strategies. Medication timing to maximize effectiveness and mitigate side effects is an important area of research for treating cardiometabolic disease [168]. The wide-spread prevalence of social jetlag and potential implications of social jetlag for cardiometabolic health warrant further investigation in community dwelling persons. Open questions persist. How much circadian disruption is too much? Who is most vulnerable? Who is resistant? Answering these questions may open a new horizon for personalizing cardiometabolic health promotion and treatment interventions.

REFERENCES

[1] Aguilar M, Bhuket T, Torres S, Liu B, Wong RJ. Prevalence of the metabolic syndrome in the United States, 2003-2012. JAMA 2015;313(19):1973–4.

[2] Moore JX, Chaudhary N, Akinyemiju T. Metabolic syndrome prevalence by race/ethnicity and sex in the United States, National Health and Nutrition Examination Survey, 1988-2012. Prev Chronic Dis 2017;14:E24.

[3] Fries E, Dettenborn L, Kirschbaum C. The cortisol awakening response (CAR): facts and future directions. Int J Psychophysiol 2009;72(1):67–73.

[4] Clow A, Hucklebridge F, Thorn L. The cortisol awakening response in context. Int Rev Neurobiol 2010;153–75.

[5] Buijs RM, Escobar C, Swaab DF. The circadian system and the balance of the autonomic nervous system. Handb Clin Neurol 2013;117:173–91.

[6] Duffy JF, Rimmer DW, Czeisler CA. Association of intrinsic circadian period with morningness-eveningness, usual wake time, and circadian phase. Behav Neurosci 2001;115(4):895–9.

[7] Eastman CI, Molina TA, Dziepak ME, Smith MR. Blacks (African Americans) have shorter free-running circadian periods than whites (Caucasian Americans). Chronobiol Int 2012;29(8):1072–7.

[8] Eastman CI, Tomaka VA, Crowley SJ. Sex and ancestry determine the free-running circadian period. J Sleep Res 2017;.

[9] Lane JM, Vlasac I, Anderson SG, et al. Genome-wide association analysis identifies novel loci for chronotype in 100,420 individuals from the UK Biobank. Nat Commun 2016;7:10889.

[10] Hu Y, Shmygelska A, Tran D, Eriksson N, Tung JY, Hinds DA. GWAS of 89,283 individuals identifies genetic variants associated with self-reporting of being a morning person. Nat Commun 2016;7:10448.

[11] Smolensky MH, Hermida RC, Reinberg A, Sackett-Lundeen L, Portaluppi F. Circadian disruption: new clinical perspective of disease pathology and basis for chronotherapeutic intervention. Chronobiol Int 2016;33(8):1101–19.

[12] Gonnissen HK, Mazuy C, Rutters F, Martens EA, Adam TC, Westerterp-Plantenga MS. Sleep architecture when sleeping at an unusual circadian time and associations with insulin sensitivity. PLoS ONE 2013;8(8):e72877.

[13] Wittmann M, Dinich J, Merrow M, Roenneberg T. Social jetlag: misalignment of biological and social time. Chronobiol Int 2006;23(1-2):497–509.

[14] Torquati L, Mielke GI, Brown WJ, Kolbe-Alexander T. Shift work and the risk of cardiovascular disease. A systematic review and meta-analysis including dose-response relationship. Scand J Work Environ Health 2017;.

[15] Lu YC, Wang CP, Yu TH, et al. Shift work is associated with metabolic syndrome in male steel workers-the role of resistin and WBC count-related metabolic derangements. Diabetol Metab Syndr 2017;9:83.

[16] Wang F, Zhang L, Zhang Y, et al. Meta-analysis on night shift work and risk of metabolic syndrome. Obes Rev 2014;15(9):709–20.

[17] De Bacquer D, Van Risseghem M, Clays E, Kittel F, De Backer G, Braeckman L. Rotating shift work and the metabolic syndrome: a prospective study. Int J Epidemiol 2009;38(3):848–54.

[18] Sookoian S, Gemma C, Fernandez Gianotti T, et al. Effects of rotating shift work on biomarkers of metabolic syndrome and inflammation. J Intern Med 2007;261(3):285–92.

[19] Yeom JH, Sim CS, Lee J, et al. Effect of shift work on hypertension: cross sectional study. Ann Occup Environ Med 2017;29:11.

[20] Peplonska B, Bukowska A, Sobala W. Association of rotating night shift work with BMI and abdominal obesity among nurses and midwives. PLoS ONE 2015;10(7):e0133761.

[21] Lin YC, Hsiao TJ, Chen PC. Persistent rotating shift-work exposure accelerates development of metabolic syndrome among middle-aged female employees: a five-year follow-up. Chronobiol Int 2009;26(4):740–55.

[22] Parsons MJ, Moffitt TE, Gregory AM, et al. Social jetlag, obesity and metabolic disorder: investigation in a cohort study. Int J Obes 2015;39(5):842–8.

[23] Wong PM, Hasler BP, Kamarck TW, Muldoon MF, Manuck SB. Social jetlag, chronotype, and cardiometabolic risk. J Clin Endocrinol Metab 2015;100(12):4612–20.

[24] Roenneberg T, Allebrandt KV, Merrow M, Vetter C. Social jetlag and obesity. Curr Biol 2012;22(10):939–43.

[25] Reutrakul S, Siwasaranond N, Nimitphong H, et al. Relationships among sleep timing, sleep duration and glycemic control in Type 2 diabetes in Thailand. Chronobiol Int 2015;32(10):1469–76.

[26] Reutrakul S, Hood MM, Crowley SJ, et al. Chronotype is independently associated with glycemic control in type 2 diabetes. Diabetes Care 2013;36(9):2523–9.

[27] Koopman A, Rauh S, van't Riet E, et al. The association between social jetlag, the metabolic syndrome, and type 2 diabetes mellitus in the general population: the New Hoorn Study. J Biol Rhythm 2017;32(4):359–68.

[28] Van Dongen HP, Baynard MD, Maislin G, Dinges DF. Systematic interindividual differences in neurobehavioral impairment from sleep loss: evidence of trait-like differential vulnerability. Sleep 2004;27(3):423–33.

[29] Spaeth AM, Dinges DF, Goel N. Phenotypic vulnerability of energy balance responses to sleep loss in healthy adults. Sci Rep 2015;5:14920.

[30] Duffy J, Dijk DJ. Getting through to circadian oscillators: why use constant routines? J Biol Rhythm 2002;17(1):4–13.

[31] Krauchi K, Wirz-Justice A. Circadian rhythm of heat production, heart rate, and skin and core temperature under unmasking conditions in men. Am J Physiol Heart Circ Physiol 1994;267:R819–29.

[32] Van Dongen HPA, Maislin G, Kerkhof GA. Repeated assessment of the endogenous 24-hour profile of blood pressure under constant routine*. Chronobiol Int 2009;18(1):85–98.

[33] Kerkhof G, Van Dongen H, Bobbert A. Absence of endogenous circadian rhythmicity in blood pressure? Am J Hypertens 1998;11:373–7.

[34] Hu K, Ivanov P, Hilton M, et al. Endogenous circadian rhythm in an index of cardiac vulnerability independent of changes in behavior. Proc Natl Acad Sci 2004;101:18223–7.

[35] Scheer F, Hu K, Evoniuk H, et al. Impact of the human circadian system, exercise, and their interaction on cardiovascular function. PNAS 2010;107(47):20541–6.

[36] Shea SA, Hilton MF, Hu K, Scheer FA. Existence of an endogenous circadian blood pressure rhythm in humans that peaks in the evening. Circ Res 2011;108(8):980–4.

[37] Adan A, Archer SN, Hidalgo MP, Di Milia L, Natale V, Randler C. Circadian typology: a comprehensive review. Chronobiol Int 2012;29(9):1153–75.

[38] Bechtold DA. Energy-responsive timekeeping. J Genet 2008;87(5):447–58.

[39] Perelis M, Bass J. The molecular clock as a metabolic rheostat. Diabetes Obes Metab 2015;17(Suppl. 1):99–105.

[40] Green C, Takahashi JS, Bass J. The meter of metabolism. Cell 2008;134:728–42.

[41] Troisi R, Cowie CC, Harris M. Diurnal variation in fasting plasma glucose: implications for diagnosis of diabetes in patients examined in the afternoon. JAMA 2000;284:3157–9.

[42] Morris CJ, Yang JN, Garcia JI, et al. Endogenous circadian system and circadian misalignment impact glucose tolerance via separate mechanisms in humans. Proc Natl Acad Sci U S A 2015;112(17):E2225–34.

[43] Hulman A, Faerch K, Vistisen D, et al. Effect of time of day and fasting duration on measures of glycaemia: analysis from the Whitehall II Study. Diabetologia 2013;56:294–7.

[44] Saad A, Dalla Man C, Nandy D, et al. Diurnal pattern to insulin secretion and insulin action in healthy individuals. Diabetes 2012;61:2691–700.

[45] Meneilly G, Elahi D, Minaker K, Rowes J. The dawn phenomenon does not occur in normal elderly subjects. J Clin Endocrinol Metab 1986;63:292.

[46] Marin G, Rose S, Kibarian M, Barnes K, Cassorla F. Absence of dawn phenomenon in normal children and adolescents. Diabetes Care 1988;11:393–6.

[47] Rosenthal M, Argoud G. Absence of the dawn glucose rise in nondiabetic men compared by age. J Gerontol 1989;44(2):M57–61.

[48] Van Cauter E, Shapiro T, Tillil H, Polonsky K. Circadian modulation of glucose and insulin responses to meals: relationship to cortisol rhythm. Am J Physiologyendocrinol Metab 1992;262:E467–75.

[49] Aparicio N, Puchulu F, Gagliardino J, et al. Circadian variation of the blood glucose, plasma insulin and human growth hormone levles in response to an oral glucose load in normal subjects. Diabetes 1974;23:132–7.

[50] Zimmet P, Wall J, Rome R, Stimmler L, Jarrett R. Diurnal variation in glucose tolerance: associated changes in plasma insulin, growth hormone, and non-esterified fatty acids. Br Med J 1974;16:485–91.

[51] Pinkhasov B, Selyatinskaya V, Astrakhantseva E, Anufrienko E. Circadian rhythms of carbohydrate metabolism in women with different types of obesity. Bull Exp Biol Med 2016;161(3):323–6.

[52] Jarrett R, Baker I, Keen H, Oakley N. Diurnal variation in oral glucose tolerance: blood sugar and plasma insulin levels morning, afternoon, and evening. Br Med J 1972;1:199–201.

[53] Van Cauter E, Polonsky K, Scheen A. Roles of circadian rhythmicity and sleep in human glucose regulation. Endocr Rev 1997;18(5):716–38.

[54] Carroll K, Nestel P. Diurnal variation in glucose tolerance and in insulin secretion in man. Diabetes 1973;22:333–48.

[55] Yamamoto H, Nagai K, KNakagawa H. Role of the SCN in daily rhythms of plasma glucose, FFA, insulin and glucagon. Chronobiol Int 1987;4:483–91.

[56] Dallmann R, Viola A, Tarokh L, Cajochen C, Brown S. The human circadian metabolome. PNAS 2012;109(7):2625–9.

[57] Chua E, Shui G, Lee IM, et al. Extensive diversity in circadian regulation of plasma lipids and evidence for different circadian metabolic phenotypes in humans. PNAS 2013;110(35):14468–73.

[58] Sennels HP, Jorgensen HL, Fahrenkrug J. Diurnal changes of biochemical metabolic markers in healthy young males—the Bispebjerg study of diurnal variations. Scand J Clin Lab Invest 2015;75(8):686–92.

[59] van Kerkhof LW, Van Dycke KC, Jansen EH, et al. Diurnal variation of hormonal and lipid biomarkers in a molecular epidemiology-like setting. PLoS ONE 2015;10(8):e0135652.

[60] Miida T, Nakamura Y, Mezaki T, et al. LDL-cholesterol and HDL-cholesterol concentrations decrease during the day. Ann Clin Biochem 2002;39:241–9.

[61] Romon M, Le Fur C, Lebel P, Edme J, Fruchart J, Dalloneville J. Circadian variation of postprandial lipemia. Am J Clin Nutr 1997;65:934–40.

[62] Lund J, Arendt J, Hampton S, Morgan L. Postprandial hormone and metabolic responses amongst shift workers in Antarctica. J Endocrinol 2001;171:557–64.

[63] Morris CJ, Garcia JI, Myers S, Yang JN, Trienekens N, Scheer FA. The human circadian system has a dominating role in causing the morning/evening difference in diet-induced thermogenesis. Obesity (Silver Spring) 2015;23(10):2053–8.

[64] Romon M, Edme J, Boulenguez C, Lescroart L, Frimat P. Circadian variation of diet-induced thermogenesis. Am J Clin Nutr 1993;57:476–80.

[65] US Department of Agriculture ARS, Beltsville Human Nutrition Research Center, Food Surveys Research Group & US Department of Health and Human Services, Centers for Disease Control and Prevention. Statistics NCfH. What We Eat in America, NHANES 2009–2010, http://www.ars.usda.gov/ba/bhnrc/fsrg; 2012. (Accessed June 4, 2018).

[66] Waterhouse J, Jones K, Edwards B, Harrison Y, Nevill A, Reilly T. Lack of evidence for a marked endogenous component determining food intake in humans during forced desynchrony. Chronobiol Int 2009;21(3):445–68.

[67] Scheer FA, Morris CJ, Shea SA. The internal circadian clock increases hunger and appetite in the evening independent of food intake and other behaviors. Obesity (Silver Spring) 2013;21(3):421–3.

[68] Sargent C, Zhou X, Matthews RW, Darwent D, Roach GD. Daily rhythms of hunger and satiety in healthy men during one week of sleep restriction and circadian misalignment. Int J Environ Res Public Health 2016;13(2).

[69] Wehrens SMT, Christou S, Isherwood C, et al. Meal timing regulates the human circadian system. Curr Biol 2017;27(12):1768–75. e1763.

[70] Scheer FA, Hilton MF, Mantzoros CS, Shea SA. Adverse metabolic and cardiovascular consequences of circadian misalignment. Proc Natl Acad Sci U S A 2009;106(11):4453–8.

[71] Schoeller D, Cella L, Sinha M, Caro J. Entrainment of the diurnal rhythm of plasma leptin to meal timing. J Clin Investig 1997;100(7):1882–7.

[72] Plano SA, Casiraghi LP, Garcia Moro P, Paladino N, Golombek DA, Chiesa JJ. Circadian and metabolic effects of light: implications in weight homeostasis and health. Front Neurol 2017;8:558.

[73] Wright Jr. KP, McHill AW, Birks BR, Griffin BR, Rusterholz T, Chinoy ED. Entrainment of the human circadian clock to the natural light-dark cycle. Curr Biol 2013.

[74] Falchi F, Cinzano P, Duriscoe D, et al. The new world atlas of artificial night sky brightness. Sci Adv 2016;2(6):e1600377.

[75] Gaston KJ, Bennie J, Davies TW, Hopkins J. The ecological impacts of nighttime light pollution: a mechanistic appraisal. Biol Rev Camb Philos Soc 2013;88(4):912–27.

[76] National Sleep Foundation. 2014 Sleep in America poll, sleep in the modern family. 2014. 2014. https://sleepfoundation.org/sites/default/files/2014-NSF-Sleep-in-America-poll-summary-of-findings---FINAL-Updated-3-26-14-.pdf; (Accessed June 14, 2018).

[77] Coomans CP, van den Berg SA, Houben T, et al. Detrimental effects of constant light exposure and high-fat diet on circadian energy metabolism and insulin sensitivity. FASEB J 2013;27(4):1721–32.

[78] Fonken L, Neslon R. The effects of light at night on circadian clocks and metabolism. Endocr Rev 2014;35(4):648–70.

[79] Boivin DB, Duffy J, Kronauer R, Czeisler C. Dose-response relationships for resetting of human circadian clock by light. Nature 1996;379:540–2.

[80] Wang XS, Armstrong ME, Cairns BJ, Key TJ, Travis RC. Shift work and chronic disease: the epidemiological evidence. Occup Med (Lond) 2011;61(2):78–89.

[81] Pietroiusti A, Neri A, Somma G, et al. Incidence of metabolic syndrome among night-shift healthcare workers. Occup Environ Med 2010;67(1):54–7.

[82] McFadden E, Jones ME, Schoemaker MJ, Ashworth A, Swerdlow AJ. The relationship between obesity and exposure to light at night: cross-sectional analyses of over 100,000 women in the breakthrough generations study. Am J Epidemiol 2014;.

[83] Koo YS, Song JY, Joo EY, et al. Outdoor artificial light at night, obesity, and sleep health: cross-sectional analysis in the KoGES study. Chronobiol Int 2016;33(3):301–14.

[84] Obayashi K, Saeki K, Iwamoto J, et al. Exposure to light at night, nocturnal urinary melatonin excretion, and obesity/dyslipidemia in the elderly: a cross-sectional analysis of the HEIJO-KYO study. J Clin Endocrinol Metab 2013;98(1):337–44.

[85] Cheung I, Zee P, Shalman D, Malkani R, Kang J, Reid KJ. Morning and evening blue-enriched light exposure alters metabolic function in normal weight adults. PLoS ONE 2016.

[86] Brainard G, Rollag M, Hanifin J. Photic regulation of melatonin in humans: ocular and neural signal transduction. J Biol Rhythm 1997;12(6):537–46.

[87] Brainard G, Lewy A, Menaker M, et al. Dose-response relationship between light irradiance and the suppression of plasma melatonic in human volunteers. Brain Res 1988;454(1–2):212–8.

[88] Tan DX, Manchester LC, Fuentes-Broto L, Paredes SD, Reiter RJ. Significance and application of melatonin in the regulation of brown adipose tissue metabolism: relation to human obesity. Obes Rev 2011;12(3):167–88.

[89] McMullan CJ, Schernhammer ES, Rimm EB, Hu FB, Forman JP. Melatonin secretion and the incidence of type 2 diabetes. JAMA 2013;309(13):1388.

[90] Gil-Lozano M, Hunter P, Behan L, Gladanac B, Casper R, Brubaker PL. Short-term sleep deprivation with nocturnal light exposure alters time- dependent glucagon-like peptide-1 and insulin secretion in male volunteers. Am J Physiol Endocrinol Metab 2016;310:E41–50.

[91] Scheer F, van Doornen L, Buijs RM. Light and diurnal cycle affect human heart rate: possible role for the circadian. J Biol Rhythm 1999;14:202–12.

[92] West AC, Smith L, Ray DW, Loudon ASI, Brown TM, Bechtold DA. Misalignment with the external light environment drives metabolic and cardiac dysfunction. Nat Commun 2017;8(1):417.

[93] Buxton OM, Cain SW, O'Connor SP, et al. Adverse metabolic consequences in humans of prolonged sleep restriction combined with circadian disruption. Sci Transl Med 2012;4(129):129–43.

[94] Morris CJ, Purvis TE, Mistretta J, Scheer FA. Effects of the internal circadian system and circadian misalignment on glucose tolerance in chronic shift workers. J Clin Endocrinol Metab 2016;101(3):1066–74.

[95] Gonnissen H, Rutters F, Mazuy C, Martens E, Adam TC, Westerterp-Plantenga M. Effect of a phase advance and phase delay of the 24-h cycle on energy metabolism, appetite, and related hormones. Am J Clin Nutr 2012;96:689–97.

[96] Van Cauter E, Blackman J, Roland D, Spire J, Refetoff S, Polonsky K. Modulation of glucose regulation and insulin secretion by circadian rhythmicity and sleep. J Clin Investig 1991;88:934–42.

[97] McHill AW, Melanson EL, Higgins J, et al. Impact of circadian misalignment on energy metabolism during simulated nightshift work. Proc Natl Acad Sci 2014;111(48):17302–7.

[98] Nguyen J, Wright K. Influence of weeks of circadian misalignment on leptin levels. Nat Sci Sleep 2010;2:9–18.

[99] Bei B, Seeman TE, Carroll JE, Wiley JF. Sleep and physiological dysregulation: a closer look at sleep intraindividual variability. Sleep 2017.

[100] Leproult R, Holmback U, Van Cauter E. Circadian misalignment augments markers of insulin resistance and inflammation, independently of sleep loss. Diabetes 2014;63:1860–9.

[101] Eckel RH, Depner CM, Perreault L, et al. Morning circadian misalignment during short sleep duration impacts insulin sensitivity. Curr Biol 2015;25(22):3004–10.

[102] Jakubowicz D, Barnea M, Wainstein J, Froy O. High caloric intake at breakfast vs. dinner differentially influences weight loss of overweight and obese women. Obesity (Silver Spring) 2013;21(12):2504–12.

[103] Garaulet M, Gomez-Abellan P, Alburquerque-Bejar JJ, Lee YC, Ordovas JM, Scheer FA. Timing of food intake predicts weight loss effectiveness. Int J Obes 2013;37(4):604–11.

[104] Thaiss CA, Zeevi D, Levy M, et al. Transkingdom control of microbiota diurnal oscillations promotes metabolic homeostasis. Cell 2014;159(3):514–29.

[105] Qin L-Q, Li J, Wang Y, Wang J, Xu J-Y, Kaneko T. The effects of nocturnal life on endocrine circadian patterns in healthy adults. Life Sci 2003;73(19):2467–75.

[106] Morgan L, Arendt J, Owens D, et al. Effects of the endogenous clock and sleep time on melatonin, insulin, glucose and lipid metabolism. J Endocrinol 1998;157:443–51.

[107] Farshchi HR, Taylor MA, Macdonald IA. Beneficial metabolic effects of regular meal frequency on dietary, insulin sensitivity, and fasting lipid profiles in healthy obese women. Am J Clin Nutr 2005;81:16–24.

[108] Guo Y-F, Stein PK. Circadian rhythm in the cardiovascular system: chronocardiology. Am Heart J 2003;145(5):779–86.

[109] Scheer F, Horst G, Van der Vliet J, Buijs RM. Physiological and anatomic evidence for regulation of the heart by suprachiasmatic nucleus in rats. Am J Physiol Heart Circ Physiol 2001;280:H1392–9.

[110] White W. Ambulatory blood pressure monitoring: dippers compared to non-dippers. Blood Press Monit 2000;5(suppl 1):S17–23.

[111] Vandewalle G, Middleton B, Rajaratnam S, et al. Robust circadian rhythm in heart rate and its variability: influence of exogenous melatonin and photoperiod. Sleep 2007;16:148–55.

[112] Akselrod S, Gordon D, Ubel F, Shannon D, Berger A, Cohen R. Power spectrum analysis of heart rate fluctuation: a quantitative probe of beat-to-beat cardiovascular control. Science 1981;213:220–2.

[113] Saul J, Rea R, Eckberg D, Berger R, Cohen R. Heart rate and muscle sympathetic nerve variability during reflex changes of autonomic activity. Am J Physiol Heart Circ Physiol 1990;258(3):H713–21.

[114] Hilton M, Umali M, Czeisler C, Wyatt J, Shea S. Endogenous circadian control of the human autonomic nervous system. Comput Cardiol 2000;27:197–200.

[115] Witte K, Schnecko A, Buijs RM, et al. Effects of Scn lesions on orcadian blood pressure rhythm in normotensive and transgenic hypertensive rats. Chronobiol Int 2009;15(2):135–45.

[116] Boudreau P, Yeh WH, Dumont GA, Boivin DB. Circadian variation of heart rate variability across sleep stages. Sleep 2013;36(12):1919–28.

[117] Ahnve S, Theorell T, Akerstedt T, Froberg J, Halberg F. Circadian variations in cardiovascular parameters during sleep deprivation a noninvasive study of young healthy men*. Eur J Appl Physiol 1981;46:9–19.

[118] Bernardi L, Valle F, Coco M, Calciati A, Sleight P. Physical activity influences heart rate variability and very-low-frequency components in Holter electrocardiograms. Cardiovasc Res 1996;32:234–7.

[119] Mancini G. Autonomic modulation of the cardiovascular system during sleep. N Engl J Med 1993;328(5):347–9.

[120] Bartness T, Song C, Demas G. SCN efferents to peripheral tissues: implications for biological rhythms. J Biol Rhythm 2001;16(2):196–204.

[121] Van Cauter E, Desir D, DeCoster C, Fery F, Balasse E. Nocturnal decrease in glucose tolerance during constant glucose infusion. J Clin Endocrinol Metab 1989;69:604–11.

[122] Boden G, Ruiz J, Urbain J, Chen X. Evidence for a circadian rhythm of insulin secretion. Am J Phys 1996;271:E246–52.

[123] Lee A, Ader M, Bray G, Bergman R. Diurnal variation in glucose tolerance cyclic suppression of insulin action and insulin secretion in normal-weight, but not obese, subjects. Diabetes 1992;41:750–9.

[124] Carrasco-Benso M, Rivero-Gutierrez B, Lopez-Minguez J, Anzola A, Diez-Noguera A. Human adipose tissue expresses intrinsic circadian rhythm in insulin sensitivity. J Fed Am Soc Exp Biol 2016;30(9):3117–23.

[125] Lindgren O, Mari A, Deacon CF, et al. Differential islet and incretin hormone responses in morning versus afternoon after standardized meal in healthy men. J Clin Endocrinol Metab 2009;94(8):2887–92.

[126] Lindgren CM, Heid IM, Randall JC, et al. Genome-wide association scan meta-analysis identifies three Loci influencing adiposity and fat distribution. PLoS Genet 2009;5(6):e1000508.

[127] St-Onge MP, O'Keeffe M, Roberts AL, RoyChoudhury A, Laferrere B. Short sleep duration, glucose dysregulation and hormonal regulation of appetite in men and women. Sleep 2012;35(11):1503–10.

[128] Dinneen S, Alzaid A, Miles J, Rizza R. Metabolic effects of the nocturnal rise in cortisol on carbohydrate metabolism in normal humans. J Clin Investig 1993;91:2283–90.

[129] Plat L, Byrne M, Sturis J, et al. Effects of morning cortisol elevation on insulin secretion and glucose regulation in humans. Am J Physiol 1996;270:E36–42.

[130] Van Cauter E, Leproult R, Kupfer D. Effects of gender and age on the levels and circadian rhythmicity of plasma cortisol. J Clin Endocrinol Metab 1996;81(7):2468–73.

[131] Sephton SE, Lush E, Dedert EA, et al. Diurnal cortisol rhythm as a predictor of lung cancer survival. Brain Behav Immun 2013;30 (Suppl):S163–70.

[132] Arendt J. Melatonin and human rhythms. Chronobiol Int 2006;23(1–2):21–37.

[133] Peschke E, Muhlbauer E. New evidence for a role of melatonin in glucose regulation. Best Pract Res Clin Endocrinol Metab 2010;24(5):829–41.

[134] Burgess HJ, Savic N, Sletten T, Roach G, Gilbert SS, Dawson D. The relationship between the dim light melatonin onset and sleep on a regular schedule in young healthy adults. Behav Sleep Med 2003;1(2):102–14.

[135] Kemp D, Ubeda M, Habener J. Identification and functional characterization of melatonin Mel 1a receptors in pancreatic b cells: potential role in incretin-mediated cell function by sensitization of cAMP signaling. Mol Cell Endocrinol 2002;191:157–66.

[136] Muhlbauer E, Albrecht E, Hofmann K, Bazwinsky-Wutschke I, Peschke E. Melatonin inhibits insulin secretion in rat insulinoma b-cells (INS-1) heterologously expressing the human melatonin receptor isoform MT2. J Pineal Res 2011;51:361–72.

[137] Cagnacci A, Arangino S, Renzi A, et al. Influence of melatonin administration on glucose tolerance and insulin sensitivity of postmenopausal women. Clin Endocrinol 2001;54:339–46.

[138] Rubio-Sastre P, Scheer FA, Gomez-Abellan P, Madrid JA, Garaulet M. Acute melatonin administration in humans impairs glucose tolerance in both the morning and evening. Sleep 2014;37(10):1715–9.

[139] Garaulet M, Gomez-Abellan P, Rubio-Sastre P, Madrid JA, Saxena R, Scheer FA. Common type 2 diabetes risk variant in MTNR1B worsens the deleterious effect of melatonin on glucose tolerance in humans. Metabolism 2015;64(12):1650–7.

[140] Lardone PJ, Álvarez-Sánchez N, Guerrero JM, Carrillo-Vico A. Melatonin and glucose metabolism: clinical relevance. Curr Pharm Des 2014;20:4841–53.

[141] Vanecek J. Cellular mechanisms of melatonin action. Physiol Rev 1998;78:687–721.

[142] Peschke E, Peschke D. Evidence for a circadian rhythm of insulin release from perifused rat pancreatic islets. Diabetologia 1998;41:1085–92.

[143] Lyssenko V, Nagorny CL, Erdos MR, et al. Common variant in MTNR1B associated with increased risk of type 2 diabetes and impaired early insulin secretion. Nat Genet 2009;41(1):82–8.

[144] Costes S, Boss M, Thomas AP, Matveyenko AV. Activation of melatonin signaling promotes beta-cell survival and function. Mol Endocrinol 2015;29(5):682–92.

[145] Park J, Shim H, Na A, et al. Melatonin prevents pancreatic b-cell loss due to glucotoxicity: the relationship between oxidative stress and endoplasmic reticulum stress. J Pineal Res 2014;2014(56):143–53.

[146] Prokopenko I, Langenberg C, Florez JC, et al. Variants in MTNR1B influence fasting glucose levels. Nat Genet 2009;41(1):77–81.

[147] Bouatia-Naji N, Bonnefond A, Cavalcanti-Proenca C, et al. A variant near MTNR1B is associated with increased fasting plasma glucose levels and type 2 diabetes risk. Nat Genet 2009;41(1):89–94.

[148] Zheng C, Dalla Man C, Cobelli C, et al. A common variant in the MTNR1b gene is associated with increased risk of impaired fasting glucose (IFG) in youth with obesity. Obesity (Silver Spring) 2015;23(5):1022–9.

[149] Jonsson A, Ladenvall C, Ahluwalia T, et al. Effects of common genetic variants associated with type 2 diabetes and glycemic traits on a- and b-cell function and insulin action in humans. Diabetes 2013;62:2978–83.

[150] Tuomi C, Nagorny TLF, Singh P, Wierup N, Groop l, Mulder H. Increased melatonin signaling is a risk factor for type 2 diabetes. Cell Metab 2016;23:1067–77.

[151] Corbalan-Tutau D, Madrid JA, Nicolas F, Garaulet M. Daily profile in two circadian markers "melatonin and cortisol" and associations with metabolic syndrome components. Physiol Behav 2014;123:231–5.

[152] Robeva R, Kirilov G, Tomova A, Kumanov P. Melatonin–insulin interactions in patients with metabolic syndrome. J Pineal Res 2008;44:52–6.

[153] Hikichi T, Tateda N, Miura T. Alteration of melatonin secretion in patients with type 2 diabetes and proliferative diabetic retinopathy. Clin Ophthalmol 2011;5:655–60.

[154] Scholtens RM, van Munster BC, van Kempen MF, de Rooij SE. Physiological melatonin levels in healthy older people: a systematic review. J Psychosom Res 2016;86:20–7.

[155] Pietrowsky R, Meyrer R, Kern W, Born J, Fehm HL. Effects of diurnal sleep on secretion of cortisol, luteinizing hormone, and growth hormone in man. J Clin Endocrinol Metab 1994;78:683–7.

[156] Holl R, Hartman M, Veldhaus J, Taylor W, Thorner M. Thirty-second sampling of plasma growth hormone in man: correlation with sleep stages. J Clin Endocrinol Metab 1991;72:854–61.

[157] Scheen A, Van Cauter E. The roles of time of day and sleep quality in modulating glucose regulation: clinical implications. Horm Res 1998;49:191–201.

[158] Iranmanesh A, Lizarralde G, Veldhaus J. Age and relative adiposity are specific negative determinants of the frequency and amplitude of growth hormone (GH) secretory bursts and the half-life of endogenous GH in healthy men*. J Clin Endocrinol Metab 1991;73(5):1081–8.

[159] Kalsbeek A, Fliers E, Romijn JA, et al. The suprachiasmatic nucleus generates the diurnal changes in plasma leptin levels. Endocrinology 2001;142:2677–85.

[160] Otway DT, Frost G, Johnston JD. Circadian rhythmicity in murine pre-adipocyte and adipocyte cells. Chronobiol Int 2009;26(7):1340–54.

[161] Cummings DE, Purnell JQ, Frayo RS, Schmidova K, Wisse BE, Weigle DS. A preprandial rise in plasma ghrelin levels suggests a role in meal initiation in humans. Diabetes 2001;50(8):1714–9.

[162] Buijs RM, Scheer FA, Kreier F, et al. Organization of circadian functions: interaction with the body. Prog Brain Res 2006;153:341–60.

[163] Pivovarova O, Jurchott K, Rudovich N, et al. Changes of dietary fat and carbohydrate content alter central and peripheral clock in humans. J Clin Endocrinol Metab 2015;100(6):2291–302.

[164] Yamanaka Y, Hashimoto S, Takasu NN, et al. Morning and evening physical exercise differentially regulate the autonomic nervous system during nocturnal sleep in humans. Am J Phys Regul Integr Comp Phys 2015;309(9):R1112–21.

[165] Buxton OM, Frank S, L'Hermite-Baleriaux M, Leproult R, Turek FW, Van Cauter E. Roles of intensity and duration of nocturnal exercise in causing phase delays of human circadian rhythms. Am J Physiol 1997;273(3):E536–42.

[166] Van Cauter E, Polonsky K, Blackman J, et al. Abnormal temporal patterns of glucose tolerance in obesity: relationship to sleep-related growth hormone secretion and circadian cortisol rhythmicity. J Clin Endocrinol Metab 1994;1994(79).

[167] Gubin D, Nelaeva A, Uzhakova A, Hasanova Y, Cornelissen G, Weinert D. Disrupted circadian rhythms of body temperature, heart rate and fasting blood glucose in prediabetes and type 2 diabetes mellitus. Chronobiol Int 2017;.

[168] Roush GC, Fapohunda J, Kostis JB. Evening dosing of antihypertensive therapy to reduce cardiovascular events: a third type of evidence based on a systematic review and meta-analysis of randomized trials. J Clin Hypertens (Greenwich) 2014;16(8):561–8.

Part V

Sleep and behavioral health

Chapter 19

Sleep and food intake

Isaac Smith[a], Katherine Saed[a], Marie-Pierre St-Onge[a,b,c]
[a]Institute of Human Nutrition, Columbia University Irving Medical Center, New York, NY, United States,
[b]Division of Endocrinology, Department of Medicine, Columbia University Irving Medical Center, New York, NY, United States, [c]Sleep Center of Excellence, Department of Medicine, Columbia University Irving Medical Center, New York, NY, United States

ABBREVIATIONS

2-AG	2-arachidonoylglycerol
GABA	gamma aminobutyric acid
GLP-1	glucagon-like peptide 1
HPA	hypothalamic pituitary axis
LNAA	large neutral amino acids
NHANES	National Health and Nutrition Examination Survey
REM	rapid-eye movement
TIB	time in bed
TSD	total sleep deprivation

INTRODUCTION

As explored in previous chapters, it is clear that getting a good night's sleep is critical to one's overall health and well-being. But average sleep duration in adults deviates from the joint recommendation of 7–9 h/night by the American Academy of Sleep Medicine and the Sleep Research Society [1]. In fact, in 2007–10, 37.3% of U.S. adults older than 20 years of age reported sleeping 6 h/night or less [2]. Yet, insufficient sleep hasn't always been such a pervasive issue in the United States: between 1985 and 2004, the percentage of short sleepers rose by up to 31% while average sleep duration declined slightly [3].

Interestingly, the trend toward shorter sleep duration in the United States developed alongside the dramatic rise in obesity. Between 1980 and 2000, the prevalence of obesity increased substantially amongst adult men and women [4]. Partly because it presents with numerous health consequences including heart disease, diabetes, and certain forms of cancer, the obesity epidemic continues to be targeted as one of the most important public health concerns of the 21st century [5]. The co-occurrence of the obesity epidemic and the increasing prevalence of short sleep duration, however, may be more deeply rooted. As detailed in Chapter 17, there is substantial epidemiological evidence to support the association between short sleep duration and obesity [6–8]. There is also accumulating evidence of a causal impact of short sleep on obesity risk. But, as described previously by Capers et al. [9] and St-Onge [10], the majority of clinical interventional studies have not been long enough to truly determine if causality exists between habitual short sleep duration and obesity. Obesity is the result of prolonged positive energy balance, whereby intake exceeds expenditures. If short sleep duration is a causal factor in this pathway, it must influence either one of the two main drivers of energy balance [11]—a process that may require a threshold "sleep debt" to be attained in order for significant body weight change to occur [10]. While the influence of sleep restriction on total daily energy expenditure remains tentative, the causal influence of short sleep on food intake has been well-established and will be discussed in detail in Part 1 of this chapter.

In Part 2 of this chapter, a number of factors that may play a role in the relation between sleep and food intake will be discussed. For instance, some have proposed that decreased time in bed leads to greater wake time available to consume and thus increased food intake [8]. However, more than merely a simple time availability issue, some have illustrated a physiological mechanism implicating hormonal variation, largely ghrelin and leptin, as driving increased hunger and appetite, supporting the effect of short sleep in increasing food intake [12]. In addition, since appetite is regulated by both physiological and psychological processes, nonhormonal mechanisms have also been described implicating reward pathways of the brain [13] and impaired inhibitory executive functioning/impulse control as influencers of food intake due to short sleep [14]. A number of novel hypotheses have been proposed as well, including stress [15] and alterations in the endocannabinoid system [16] but those have not been sufficiently studied to warrant in-depth description in this chapter.

While there has been a wealth of research that strongly links short sleep duration to increased food intake, there has also been a suggestion of bi-directionality in this relationship, implicating dietary factors in modulating sleep duration and/or quality. Part 3 will focus on how food intake can influence sleep duration and quality.

PART 1: SLEEP LOSS AND FOOD INTAKE

Between 1969 and 2001, per capita caloric consumption in developed nations rose by 400 kcal/day, and, given a continuation of this trend, is expected to increase past 3500 kcal/day by the year 2050 [17]. Interestingly, this trend toward increased food intake parallels both trends toward greater obesity rates [4] and shorter sleep duration [3]. Indeed, numerous observational studies of habitual sleep and dietary intake have found an inverse association between sleep duration and energy intake [18–20] or diet quality [21–25]. Data from the National Health and Nutrition Examination Survey (NHANES), for instance, revealed that short sleepers (5–6 h) had greater total energy intake than average sleepers [26]. Sleep duration is not only associated with total energy intake but may be particularly related to intakes of specific macronutrients. In a large epidemiological study of Korean adults, sleep duration was positively correlated with protein intake and negatively correlated with carbohydrate intake [27]. In a separate analysis of the same population, women with short sleep duration consumed more carbohydrates, men with short sleep duration consumed more fat, and both men and women with short sleep duration consumed less protein [28]. In the United States, sleep duration was negatively correlated with fat intake [25]. However, despite the large number of cross-sectional population studies that demonstrate this association, causality cannot be assumed from epidemiological studies alone. Prospective studies and randomized controlled trials are necessary to ascertain the causal link between sleep and food intake.

Many short-term, highly restrictive, experimentally controlled sleep studies have been conducted to determine whether sleep restriction increases food intake in adults. In the longest of these studies to date, Nedeltcheva et al. demonstrated that restricting time in bed (TIB) to 5.5 h/night as compared to 8.5 h/night for 14 nights increased food intake by 297 kcal amongst 11 healthy overweight volunteers [29]. Although the difference in total energy intake or intake from meals were not significant between sleep duration periods, energy intake from snacks, in particular, was increased in the short sleep condition. The lack of statistical significance on total energy intake may have been due to inadequate sample size since St-Onge et al. noted a similar increase in total energy intakes of 296 kcal/day when 26 men and women were restricted to 4 h TIB for 4 nights relative to a control period of 9 h TIB [30]. The authors further noted that women tended to have greater increase in energy intakes relative to men (unpublished data) but others have reported the converse [31]. Many other short-term studies have presented similar findings on the link between sleep restriction and food intake [32–34]. In fact, a recent meta-analysis concluded that sleep restriction, in normal sleepers, leads to an increase in energy intakes of ~385 kcal/day [35], an amount that, if sustained over time, could lead to substantial weight gain. These conclusions have been corroborated in multiple opinion and systematic reviews [2, 9, 10, 14].

Experimentally induced sleep restriction studies have also provided evidence of macronutrient-specific increases in intakes as a result of short sleep. Two previously mentioned intervention studies found that acute sleep restriction increased carbohydrate intakes in healthy adults [29, 33]. In a separate study, fat and saturated fat intakes were specifically increased after 5 days of restricted sleep, compared to adequate sleep, in 15 healthy men and women [36]. Other intervention studies have found similar effects of acute sleep restriction on fat intake in healthy adults [34, 37]. Yet whether the combined effects of short sleep duration on increased energy, carbohydrate, and fat intakes contribute to positive energy balance and weight gain long-term is relatively understudied given that most studies have been of relatively short duration (mostly 4–5 days, and < 14 days).

Ultimately, body weight is the outcome of the balance between energy intake and energy expenditure. Although the effect of short sleep duration on energy balance regulation is outside the scope of this chapter, it is interesting to note that reviews and meta-analyses have found no significant effect of sleep restriction on energy expenditure [9, 10]. This is mostly due to high between-study heterogeneity and the diverse components of energy expenditure that are difficult to measure under controlled conditions of acute sleep restriction studies. Nonetheless, the evidence suggests that a positive energy balance as a result of increased food intake, given no net compensation in energy expenditure, may be achieved as a result of sleep restriction. In fact, Markwald et al. found that, compared to 5 nights of 9 h TIB, 5 nights of sleep restriction to 5 h TIB increased food intake and promoted positive energy balance and weight gain [33]. Few others have attempted to replicate the same finding. Unfortunately, given the paucity of evidence that supports this association, it would be premature to claim that short sleep duration causing increased food intake, in short-term studies, leads to positive energy balance and long-term weight gain. However, one thing seems clear: how we sleep affects how much we eat and the types of food that we choose.

PART 2: PROPOSED MECHANISMS EXPLAINING THE SLEEP-FOOD INTAKE RELATION

A number of mechanisms have been proposed to explain the relation between short sleep duration and increased food intake. These are categorized as homeostatic (i.e., hormonal) and nonhomeostatic (i.e., hedonic). The differences between these concepts were highlighted in one study that compared chosen portion sizes after sleep deprivation in either a fasted or sated state [38]. Participants were asked to rate desired portion sizes, using computer-generated pictures of foods of different energy content. Interestingly, after one night of total sleep deprivation (TSD), average portion sizes chosen were higher both before and after consumption of a 600-kcal breakfast, as compared to those chosen after a night of adequate sleep (8 h TIB). Greater portion sizes chosen in the fasted state represent an increased homeostatic drive to eat after TSD relative to adequate sleep, whereas greater portions sizes chosen after breakfast consumption, in a sated state, represent an increased hedonic drive to eat. The most studied of these homeostatic and nonhomeostatic mechanisms are graphically displayed in Fig. 19.1. It is important to keep in mind that each pathway may contribute to food intake dependently and with varying magnitude. Additionally, many of these pathways lack adequate evidence to fully support causality. For this reason, more research into each of these fields should be completed in order to fully understand the relationship between short sleep duration and increased food intake.

Homeostatic mechanisms

Changes in hormonal modulation of food intake by sleep duration was first demonstrated by Spiegel et al., whereby ghrelin levels increased by 28% and leptin levels decreased by 18% after 3 nights of 4h TIB compared to 10h TIB in young healthy men [39]. These findings are consistent with the notion that restricting sleep leads to positive energy balance since leptin is an adipose-derived cytokine that sends satiety signals to the brain [40] and ghrelin is an orexigenic hormone that sends appetitive signals from the stomach to the brain [41]. But limitations of this study included unnatural feeding conditions and low external generalizability. Since that landmark study, a plethora of studies assessing changes in leptin and ghrelin in response to sleep manipulations have produced mixed results [13, 42–46]. Such heterogeneity was highlighted in a meta-analysis that noted no

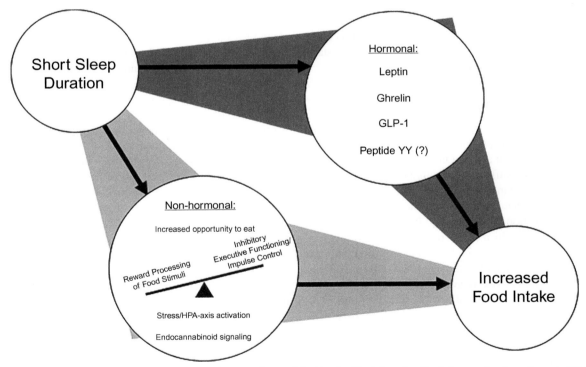

FIG. 19.1 Proposed mechanisms to explain the relation between sleep and food intake. Short sleep duration in humans has been shown to (A) decrease leptin levels, (B) increase ghrelin levels, and (C) decrease GLP-1 levels. The effect of short sleep duration on (D) peptide YY is not as well-known. Short sleep duration (E) increases opportunity to eat when conditions favor abundance, (F, G) shift the balance of reward and impulse control toward enhanced reward and impaired inhibitory executive functioning and impulse control, (H) increase stress and activate the HPA-axis, and (I) delay and enhance the endocannabinoid signaling pathway. Each of these pathways has been shown to increase food intake in human participants.

net effect of sleep restriction on leptin and ghrelin levels [9]. It is clear that additional studies are needed to draw definitive conclusions about the impact of sleep restriction on leptin and ghrelin concentrations and how changes in these hormones as a result of sleep disturbances further impact food intake regulation.

While the majority of research on hormonal variation has focused on leptin and ghrelin, other hormones, such as glucagon-like peptide 1 (GLP-1) and peptide YY, satiety factors released by the small and large intestines in response to food [47, 48], have also been implicated in the relationship between short sleep duration and increased food intake. St-Onge et al., for instance, demonstrated that GLP-1 was downregulated following meals consumed after sleep restriction (4h TIB) compared to habitual sleep (9h TIB) in young healthy adult women [30]. This finding was not observed in men. In fact, in 8 healthy young men, postprandial GLP-1 was actually upregulated after one night of TSD [49]. In contrast, the postprandial GLP-1 response was delayed, but not diminished, following TSD in 12 healthy young men [50]. Thus, considering these dramatic sex differences and the scarcity and inconsistency of evidence linking GLP-1 variations to sleep restriction, more research considering the role of GLP-1 in modulating food intake after sleep restriction is imperative. Interestingly, the same can be said for peptide YY, of which there is even less experimental research related to its relationship with sleep. Although systemic reviews have considered the role of short sleep duration on peptide YY concentration [10], no experimental studies have established any substantial relationship between the two.

Nonhomeostatic mechanisms

Although food intake is under hormonal regulation, food consumption occurs despite physiological hunger (hedonic control). In fact, Chaput and St-Onge have argued that sleep may have a stronger influence on nonhormonal controls of food intake, rather than hormonal controls [13]. To that effect, Sivak noted that hormonal variance accounted for a very small proportion of the average body mass gained by participants enrolled in a sleep restriction study: 3% for leptin and 1% for ghrelin [51]. Some have posited that in a society with consistent and abundant access to food, the less one sleeps, the longer one is awake, the more opportunity is available to fall prey to the allure of palatable food, leading to overconsumption [8, 10, 51]. Indeed, the explanation may be more complex than a simple time-availability condition. It has been hypothesized that eating and food choice are likely at the center of a balancing act between reward-driven motivations to eat and inhibitory executive functioning and impulse control [52]. It seems likely that if either of these processes are impaired, overeating—or poor food choice—may prevail. This relationship is graphically displayed in Fig. 19.1.

One of these pathways that has been quite well-established and is continuing to accumulate evidence is the action of hedonic stimulus processing in the brain in response to food stimuli. Functional magnetic resonance imaging of normal-weight men after one night of TSD showed an increase in activity in response to food stimuli in the right anterior cingulate cortex, a brain region often associated with reward processing, when compared to one night of sleep [53]. This result was found after participants were given a caloric load, further suggesting the influence of nonhomeostatic mechanisms. These findings have been corroborated in another functional magnetic resonance imaging study that showed increased activity in response to food stimuli in other brain regions associated with pleasure and reward processing after sleep was restricted to 4h/night compared to 9h/night for 5 nights [54]. Greater activation in reward pathways or greater sensitivity to reward has been correlated with obesity and a greater risk for overeating [55, 56].

On the other end of this balancing act is impaired executive functioning and impulse control. The first experimental study analyzing the effect of sleep deprivation on cognitive performance took place in 1896 [57]. Since then, hundreds of studies have been published on this relationship [58]. Nilsson et al., for instance, found that executive functioning was impaired after one night of TSD in 22 healthy adults [59]. Various aspects of cognitive performance, executive function, and decision-making have been linked with both sleep restriction and deprivation [60–62]. Relatedly, studies have shown that TSD leads to greater impulsivity compared to adequate sleep [63, 64]. And, after sleep restriction, brain regions associated with cognitive processing, decision-making, and self-control were activated to a lesser extent in response to food stimuli as compared to habitual sleep [54]. Moreover, impaired executive functioning is associated with uncontrolled eating [65] and impulsivity and impaired executive functioning have been shown to increase food intake in ad libitum eating tasks [52]. Therefore, short sleep duration seems to deactivate brain regions associated with executive functioning and impulse control and activate those associated with reward processing in response to food stimuli. Both of these effects are associated with a greater risk for overeating. Thus, taken together, evidence suggests that the imbalance between hedonic stimulus processing and inhibitory executive functioning and impulse control may partly mediate the effect of short sleep duration on food intake.

Due to the insufficiency of evidence linking any one particular mechanism to the relationship as graphically displayed in Fig. 19.1, however, more novel mechanisms have been proposed. Of these, the modulating effect of stress on the relationship between sleep and food intake is relatively well-established. In a recent review analyzing the effect of sleep on stress, Wolkow et al. concluded that sleep may act as a potent stressor to elicit both physiological and psychological responses [66]. Additionally, stress has a causal relationship with hyperactivity in the hypothalamic-pituitary-adrenal (HPA) axis, although this relationship is complex and bi-directional [67]. Nonetheless, hyperactivity in the HPA-axis has been observed in response to both TSD and sleep restriction [68–70]. Activation of the HPA-axis, self-reported levels of stress, and objective stressful conditions have all been correlated with increased food intake [68]. Acute exposure to stressful conditions increased food intake, when compared to a stress-free condition, in normal weight women [71]. Others have replicated these results using various stimuli to promote a stressful environment, many in the absence of physiological hunger [72–75]. Although not specifically conducted in the context of sleep restriction, these illustrate the potential role of sleep restriction, as a stressor, in eliciting increased food intake and suggest that the effect of short sleep duration on food intake may be partly mediated by the stress-inducing/HPA-axis activating effects of inadequate sleep.

A more novel mechanism that has been implicated in the relationship between short sleep duration and increased food intake is the effect of circulating levels of 2-arachidonoylglycerol (2-AG), the most abundant endocannabinoid [76]. The endocannabinoid system mediates hedonic food intake, reaches a nadir during mid-sleep, and exhibits a large diurnal peak during lunch hours [16]. Sleep restriction results in an increase and delay of the diurnal peak in 2-AG along with reports of increased hunger and appetite, and inability to inhibit intake of palatable snacks [77]. A recent review of novel mechanisms in the development of obesity suggests that, due to its ability to control appetite, the endocannabinoid system may play an important role in increased food intake due to short sleep duration [16].

Many other mechanisms have been proposed. Such mechanisms, including circadian disruptions in the gut microbiota [16] and emotional dysregulation [14], have been studied to a much lesser extent, have a paucity of evidence to explain any association, and will not be discussed herein. Nonetheless, these mechanisms are included in this discussion as they require additional study.

PART 3: INFLUENCE OF FOOD INTAKE ON SLEEP DURATION AND QUALITY

Although the previous sections covered the evidence surrounding the impact of sleep duration on food intake and food intake regulation, there is evidence that food choice may also influence sleep duration and quality. As described below, intakes of specific macronutrients and foods may have an impact on the quality of overnight sleep. In addition, specific foods and various dietary supplements have been studied for their potential beneficial impact on sleep. A holistic dietary approach that incorporates several of these components may be the ideal solution for long-term healthy sleep. As discussed earlier in this chapter, studies have shown that short sleepers have higher energy intake, most commonly from fat [22] and snacks [25], than normal sleepers. On the other hand, numerous epidemiological studies have opined the reverse: a negative association between carbohydrate intakes and sleep quality [78–80]. In Japanese women, high intake of noodles and confectionary was shown to be associated with poor sleep quality while high intake of fish and vegetables was associated with good sleep quality [81]. There was also a trend toward declining sleep quality with increased carbohydrate intake. As was highlighted earlier, the directionality between these variables cannot be established from cross-sectional studies. Just as poor sleep could result in unhealthy dietary intakes, it may very well be that unhealthful diets lead to poor sleep. If such is the case, then improving dietary intakes could be a target for improving sleep quality.

Caloric consumption

Energy restriction has been shown to reduce melatonin secretion [82], which would be expected to lead to difficulties initiating sleep. Indeed, short-term, voluntary fasting by total rejection of food or intake of <300 kcal per day for 2–7 days significantly reduced circulating melatonin concentrations. Glucose supplementation during these short-term fasts returned melatonin concentrations to normal [82]. However, the study did not report on sleep duration. Nonetheless, patients with anorexia have been shown to have poorer sleep quality and greater sleep disturbances than healthy control participants [83], pointing to a negative influence of energy restriction on sleep.

Protein

Neurotransmitters regulating sleep are influenced by what we eat, including serotonin [84], gamma-amino butyric acid (GABA), orexin, melanin-concentrating hormone, galanin, noradrenaline, and histamine [85]. The synthesis of serotonin is dependent on the precursor tryptophan, an amino acid found in high concentration in poultry and soy. Serotonin is the precursor to melatonin, which is released by the pineal gland and a known regulator of sleep. Tryptophan consumption could therefore contribute to increased serotonin production, melatonin production, and sleep onset [84]. In fact, studies have shown that tryptophan doses as low as 1 g improve sleep latency and subjective sleep quality [84]. However, tryptophan uptake in the brain is mediated through a competitive mechanism, and consumption of other macronutrients and competing amino acids can influence serotonin production from tryptophan. High plasma concentration of tryptophan relative to other amino acids promotes entry of tryptophan through the blood-brain barrier promoting production of serotonin and triggering release of melatonin from the pineal gland (Fig. 19.2).

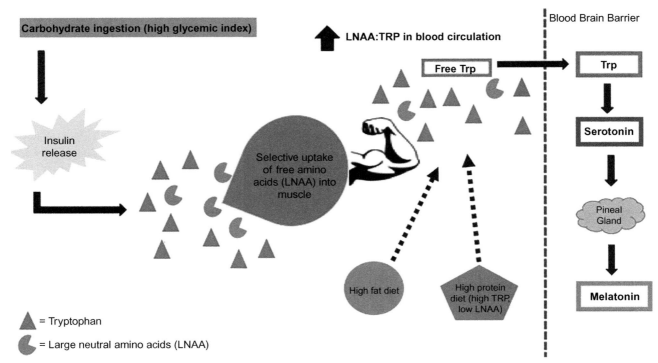

FIG. 19.2 Dietary tryptophan triggers biosynthesis of melatonin. The synthesis of serotonin is dependent on the availability of tryptophan. Serotonin is necessary for increased melatonin release from the pineal gland. Macronutrient intake influences this process and plays a role in sleep induction. Carbohydrate ingestion, particularly high glycemic index food choices, increases circulating glucose concentration which triggers insulin release resulting in selective uptake of large neutral amino acids into the muscle. Tryptophan is taken up with lower efficiency further increasing the ratio of tryptophan to large neutral amino acids (LNAA) in the blood, sending large amounts of tryptophan across the blood brain barrier leading to sleep induction. High fat diets and diets high in plant proteins rich in tryptophan and low in LNAA, may also favorably influence the Trp:LNAA ratio in blood, leading to increased serotonin and melatonin production in the brain.ART: The font here should be white or background changed to a lighter color to make the writing more legible

Increasing tryptophan access to the brain can be accomplished through multiple mechanisms: pure tryptophan supplementation, increased carbohydrate intake [86], or consumption of tryptophan-rich α-lactalbumin protein [87]. Tryptophan loading has been shown to dose-dependently increase circulating melatonin in humans throughout the day but this effect was more pronounced at night [87]. The nocturnal administration of tryptophan has been shown to increase physiological concentrations of serotonin and melatonin [88], therefore facilitating sleep [89].

Evidence for a role of tryptophan on sleep quality can be extrapolated from studies in aging adults. There is a reduction in the ability of tryptophan to cross the blood-brain barrier with increasing age, leading to a reduction in biosynthesis of serotonin and melatonin, which may explain the reduction in sleep quality and duration that occurs with aging [90]. It can be hypothesized that low intakes of tryptophan may lead to increased occurrence of sleep problems and supplementation with this amino acid may improve sleep. In one study, normal weight elderly volunteers suffering from sleep difficulties were provided with 30 g of tryptophan-enriched ready-to-eat cereal, containing 60 mg tryptophan, at breakfast and dinner. Most sleep parameters improved following this treatment compared to the control period consisting of a diet with standard ready-to-eat cereal containing 22.5 mg of tryptophan per dose and compared to participants' habitual diet. Of particular note, there was also an increase in total sleep time indicating that tryptophan consumption may be a potential treatment for those suffering from sleep disorders and disturbances [90].

Carbohydrates

Increasing carbohydrate intake promotes tryptophan entry in the brain. This is accomplished by triggering a high insulinemic response, which results in the uptake of amino acids into the muscle. High-glycemic index, carbohydrate-rich meals are particularly efficient in increasing circulating glucose concentrations and triggering this response (Fig. 19.2). This then leads to reduced tryptophan uptake up by muscle, therefore increasing its ratio relative to other amino acids in the blood, the net effect being greater amounts of tryptophan reaching the brain to trigger serotonin synthesis and ultimately promoting sleep induction [91, 92]. This mechanism provides a possible explanation as to how high carbohydrate meals increase drowsiness and promote sleep. Indeed, a study by Porter and Horne found that a high-carbohydrate meal resulted in increased rapid-eye movement (REM) sleep, decreased light sleep, and decreased wakefulness throughout the night [93]. Similarly, clinical trials have reported that participants tended to feel sleepier and less awake after being given a high-glycemic index meal compared to a low-glycemic index meal [94] and sleep latency was shorter when participants were provided a high carbohydrate diet (56% carbohydrate, 22% protein, and 22% fat) compared to control diets (50% carbohydrate, 35% fat, and 15% protein) [95]. However, in contrast to the prevailing theoretical mechanism as presented in Fig. 19.2, research into the effects of carbohydrate intake on sleep quality show mixed results.

In an intervention study, providing six healthy women with a low-carbohydrate diet (50 g/day) for 7 days increased REM latency compared to baseline [96]. In a secondary analysis, St-Onge et al. noted that higher intakes of sugar and nonsugar carbohydrates were also associated with more nocturnal arousals during sleep [97]. Notably, lower intake of fiber and greater intake of saturated fat were associated with a lighter, less deep sleep profile. That both a positive and negative association has been demonstrated between intake of carbohydrates and sleep indicates that the study design, namely observational versus interventional and the quality of the carbohydrate may be important factors when evaluating the impact of carbohydrate intake on sleep.

Evidence is also accumulating to suggest a relationship between timing of carbohydrate intake and sleep quality [98, 99]. For instance, high-glycemic index meals ingested 4h before bedtime reduced sleep onset latency to a greater extent than the same meal ingested 1h before bedtime [94]. Thus, a small adjustment in the composition of the evening meal may be a potential therapeutic strategy to improve sleep initiation. Yet the evidence is insufficient to support this recommendation.

Three mechanisms have been proposed to explain the apparent relationship between timing of carbohydrate ingestion and sleep quality. First, a longer time lapse between meal and bedtimes may be necessary to allow for the digestive and absorptive processes to occur and impact serotonin synthesis. Second, late eating may contribute to a phase delay, which exacerbates circadian disruption, leading to further reduced sleep duration [100]. Third, late eating may influence food choice due to the availability of foods at that time.

Fat

Consumption of a fatty meal has been shown to increase self-reported sleepiness in an intervention study [101]. This confirms data from a longitudinal study of Chinese adults showing that those with high dinner fat intake at baseline had persistent short sleep over 5 years in a prospective cohort study [99]. This study showed that the association between macronutrient intake and sleep differs by meal timing, because similar associations were not observed when the authors considered total daily fat intake or fat intakes earlier in the day. High-fat dinner intake was positively associated with short sleep duration in Chinese adults while high-fat breakfast intake was inversely associated with falling asleep during the day [99]. The authors postulated that fat intake from dinner could alter circadian regulation in humans leading to shorter sleep and thus recommended limiting fat intake particularly at dinner to maintain adequate sleep. However, this study was not an intervention and causality cannot be implied from these results.

One potential mechanism to explain the link between increased fat intake and improved sleep quality is via circulating tryptophan levels. Greater free fatty acid levels increase free tryptophan in circulation, resulting in increased uptake of tryptophan into the central nervous system and biosynthesis of serotonin and melatonin [84]. Another potential hypothesis that explains the relationship between increased fat intake and poor sleep quality is based on the function of sleep in macromolecular biosynthesis of proteins, lipids, cholesterol, and heme [102]. It is hypothesized that an increased consumption of high fat foods results in a decreased need for biosynthesis of these macromolecules, thus decreasing the signal for sleep [25]. An alternative explanation is based on the function of sleep in energy conservation. Relative to protein, fat ingestion has a lower thermogenic effect [102]. Thus, the need for energy conservation is not as great following consumption of high fat foods. The overall effect of decreased energy expenditure after increased fat intake is a reduction in the need for energy conservation and, thus, reduced sleep duration and/or quality [25]. As with carbohydrate intake, because both positive and negative associations have been found between fat intake and sleep, it is likely that the quantity, quality, and timing of fat intake may play a role in this relationship.

To that effect, omega-3 fatty acid deficiencies have been associated with poor sleep [103], while supplementation of omega-3 or consumption of high omega-3 foods are potentially beneficial in promoting sleep [104]. In a randomized controlled trial, daily dietary supplementation with 600 mg/day of docosahexaenoic acid increased sleep duration by 58 min in children relative to placebo [104]. This was accompanied with fewer night-waking episodes, although no subjective improvements were evident.

Vitamins and supplements

Melatonin

Melatonin has been found in many foods; eggs and fish are the best animal sources, and nuts, cereals, legumes, seeds, and mushrooms are rich plant-based food sources. Melatonin is also a natural compound found in cow's milk. This might explain why milk has traditionally been used for its sleep inducing effects [105]. However, a double-blinded study in elderly participants did not find any effect of ingesting 500 g/day of normal commercial milk on sleep at night or alertness the following morning [106]. On the other hand, consumption of a melatonin-rich milk led to better sleep at night in 80 elderly participants [105]. This discrepancy is likely explained by differences in quantity of melatonin contained in the milk.

Sleep disorders and difficultly sleeping have been shown to increase circulating reactive oxygen species [107, 108]. Melatonin can directly scavenge free radicals and stimulate antioxidant enzymes [109]. Thus, in addition to its role in the regulation of circadian rhythms [110, 111], intake of melatonin-rich foods may provide benefits in improving sleep and weight management due to its antioxidant, antiinflammatory, and antiobesity effects [112]. There is some evidence that melatonin may help treat delayed sleep phase syndrome with short-term use and help those struggling to fall asleep by initiating sleep onset [113]. That said, the short-term use of melatonin for the treatment of primary sleep disorders is not recommended [114].

B vitamins

The secretion of endogenous melatonin is also influenced by external factors such as artificial light exposure and vitamin B12 intake [105]. Yet supplementation with B12 to treat circadian rhythms disorders is not recommended [113]. Other B vitamins, such as niacin, have been shown to improve sleep quality [115]. Administration of niacin to 6 participants with normal sleep increased rapid-eye movement sleep and improved sleep efficiency in individuals with insomnia. Given that niacin is biosynthesized from dietary tryptophan, researchers postulated that the administration of niacin may lead to decreased biosynthesis, and thus, increased tryptophan available for the production of serotonin and melatonin [105].

Isoflavones

Isoflavones are soy-based phytoestrogens known to offer potential therapeutic effects for many hormone-dependent conditions including certain forms of cancer, menopausal symptoms, and cardiovascular disease [116]. Additionally, phytoestrogens may influence serotonin levels and the sleep-wake cycle. Isoflavones are purported to have similar effects on sleep quality as estrogen replacement therapy, which has been shown to alleviate symptoms of insomnia and increase sleep efficiency in postmenopausal women [117]. In a cross-sectional study conducted in Japanese men and women, it was reported that daily isoflavone intake was associated with an optimal sleep duration of 7–8 h/night and better sleep quality [118]. However, clinical studies are warranted to confirm the effects of isoflavone intakes on insomnia symptoms and overall sleep quality.

Magnesium

Magnesium is an essential macro-mineral in the body and is involved in over 300 biochemical reactions [119]. Magnesium requirements remain relatively unchanged through adulthood yet aging increases the risk of magnesium deficiency [120]. Coincidently, sleep quality has also been shown to decrease with age. For this reason, some have proposed that the increased risk for magnesium deficiency in the elderly mediates the decline in sleep quality [120]. Indeed, in the elderly, magnesium supplementation has been shown to increase sleep duration and efficiency, decrease sleep onset latency, and improve insomnia severity index scores. Daily dietary or supplemental magnesium has been shown to improve subjective and objective measures of insomnia and may become useful in managing these symptoms [120]. Almonds, a good source of both magnesium and melatonin, may be a natural method to improve sleep [121]. Studies are needed to evaluate the influence of food sources of magnesium on sleep quality and whether magnesium supplementation, in the absence of inadequate intakes, can improve sleep quality.

Fruits

Tart cherries contain phenolic compounds that have antiinflammatory and antioxidant properties [122]. As a result of these properties, tart cherries have been shown to improve sleep quality and reduce insomnia symptoms. Additionally, this may be explained, in part, by the rise in circulating melatonin concentrations that occurs after daily ingestion of tart cherry juice. In healthy adults with no reported sleep disturbances, consumption of tart cherry juice decreased napping throughout the day and resulted in improved sleep quality throughout the night.

Similarly, consumption of kiwifruit has been shown to improve sleep quality [107]. Currently, three mechanisms have been proposed to explain this effect. First, kiwifruit has the highest proportion of vitamin C, an antioxidant vitamin, relative to other fruits. Given the relation between reactive oxidant species and sleep, a high concentration of vitamin C might partly explain the improvement in sleep quality. Second, kiwifruit contain relatively high amounts of serotonin, which is involved in sleep-wake regulation. Third, kiwifruit is a rich source of folate. Folate deficiency has been linked with poor sleep and insomnia [26].

Walnuts have also been associated with improved sleep [123]. Walnuts are known to be rich in many nutrients including phosphorus, copper, and magnesium [124]. Likewise, they are a good source of melatonin and healthy fats, such as omega-3 fatty acids [123]. In addition, walnut consumption increases blood melatonin concentration which has been shown to correlate with increased antioxidant capacity and may lead to improved sleep.

Alternative medicine

Valerian

Valerian is a flowering plant that has been shown to improve symptoms associated with insomnia and anxiety [125]. Valerian is thought to induce a calming effect by binding to GABA type A receptors [126]. Yet, the evidence for a role of valerian on sleep remains inconclusive due to contradictory findings. Some studies have found no effect on sleep of a single dose of valerian [127–129] while others have reported improved sleep without the negative side effects normally associated with insomnia medication or sleep aids [125, 130, 131]. Results of a recent meta-analysis showed significant improvement in subjective sleep quality, while objective parameters remained unchanged [132]. Unfortunately, significant drowsiness and dizziness may follow supplementation with valerian [133]. As a result, valerian supplementation has not been advised for management of insomnia symptoms [134].

Kava

Kava, a plant extracted from Piper tnethysticum Forst, has been used for its sedative effect for millennia [135]. Research has shown positive effects on sleep, but supplementation may have lingering effects that hinder normal day-to-day activities. In the past, kava has been used to treat symptoms of anxiety and depression [136]. The suspected receptors of kavalactones, the primary psychoactive constituents of kava, include GABA, serotonin, and dopamine receptors [137]. No randomized controlled trials have been conducted to explore the efficacy of kava in the treatment of sleep disorders and therefore, its effect on sleep remains unknown.

Total dietary approaches

The Mediterranean Diet has a healthy profile of fat, protein, and fiber obtained mainly from the consumption of fish, olive oil, fruits and vegetables, and nuts [138]. It has been postulated that the antiinflammatory and antioxidant pathways link the Mediterranean Diet to improved sleep, largely because it consists of foods high in antioxidants, healthy fats, resveratrol, and polyphenols, including phytoestrogens [139, 140]. One study concluded that a lower risk of having poor sleep quality was associated with a higher adherence to a Mediterranean Diet pattern [141]. Additionally, plant-based proteins tend to be relatively high in the amino acid tryptophan [97], which, as mentioned previously, is a precursor to melatonin and serotonin, the two neurotransmitters involved in sleep regulation. In addition, diets higher in complex carbohydrates and lower in saturated fats, such as plant-based diets and the Mediterranean Diet, may have benefits on sleep quality [97].

While information on the relation of total dietary approaches, such as the Mediterranean Diet, and sleep quality is emerging, knowledge on the effects of their individual components on sleep quality support a potential causal role. For instance, polyunsaturated fats, omega-3 fatty acids, B vitamins, melatonin, magnesium, and high-quality plant proteins are all highly concentrated in the Mediterranean Diet. As described previously, each of these individual components has been associated with improvements in sleep quality and/or duration. For that reason, it may be worthwhile to mention that following a total dietary approach might be a more effective and feasible method to improve sleep quality than addressing individual dietary components.

CONCLUSION

As detailed in Part 1 of this chapter, acute short sleep duration has a marked effect on food intake. Further, a resulting positive energy balance and weight gain is likely implicated with increased food intake, given no net compensation in energy expenditure. That said, the long-term effects of chronic short sleep duration as a causal factor for obesity and chronic diseases are relatively unknown and understudied. Likewise, it is important to keep in mind that the majority of the mechanisms that explain the relationship between sleep and food intake presented in Part 2 have mixed results or are not yet fully developed. Nonetheless, the presented hypotheses represent areas of active research that will propel our understanding of the relationship between sleep and food intake and lead to developments of potential targets for intervention.

Obesity is one of the leading causes of preventable death in the United States [142]. Any effort to reduce the impact and proliferation of this epidemic is a worthwhile pursuit. As more evidence to support the relation between sleep and obesity accumulates, it becomes clear that short sleep duration has wide-spread public health implications. This brings about important questions as to the potential for sleep extension as a potential strategy for decreasing food intake. Studies to assess sleep extension strategies as adjunct therapy to standard weight management protocols should also be undertaken. The answer to these questions necessitates more research into the chronic effects of sleep on food intake, the mechanisms underlying the sleep-food intake relation, the effects of short sleep duration on energy balance and weight gain, and the feasibility and efficacy of sleep extension.

While the evidence strongly links short sleep duration to increased food intake, there has also been a suggestion of bi-directionality in this relationship. As explored in Part 3 of this chapter, food choice appears to have an impact on sleep. Most documented studies have been epidemiological and have not provided objective measures of sleep and diet. As a result, the link between food choice and sleep quality may not be as well established as the link between short sleep duration and increased food intake, as discussed in Part 1 and Part 2 of this chapter. That said, emerging research into these fields promises insight into the role that food choice may play in the management of sleep disorders, particularly short sleep, insomnia, and sleep apnea. Impacting both mental and physical quality of life, these sleep disorders have become important public health concerns that warrant further study into preventive strategies. Moreover, sleep disorders may increase the risk of morbidity. Because current medications come with numerous side-effects and limited efficacy, lifestyle strategies to improve sleep quality and duration may be preferable. Based on the best evidence, a balanced and varied diet rich in fresh fruits, vegetables, whole grains and high-quality protein sources, which contain plenty of tryptophan as well as adequate micronutrients, including several vitamins and minerals, may be a step in the right direction. Yet, in order to recommend these changes with confidence, further interventional research is needed to bring us closer to a defined conclusion.

REFERENCES

[1] Watson NF, et al. Recommended amount of sleep for a healthy adult: a joint consensus statement of the American Academy of sleep medicine and Sleep Research Society. Sleep 2015;38(6):843–4.

[2] Dashti HS, et al. Short sleep duration and dietary intake: epidemiologic evidence, mechanisms, and health implications. Adv Nutr 2015;6(6):648–59.

[3] Ford ES, Cunningham TJ, Croft JB. Trends in self-reported sleep duration among US adults from 1985 to 2012. Sleep 2015;38(5):829–32.

[4] Flegal KM, et al. Trends in obesity among adults in the United States, 2005 to 2014. JAMA 2016;315(21):2284–91.

[5] Bray GA. Medical consequences of obesity. J Clin Endocrinol Metab 2004;89(6):2583–9.

[6] Cappuccio FP, et al. Meta-analysis of short sleep duration and obesity in children and adults. Sleep 2008;31(5):619–26.

[7] Chen X, Beydoun MA, Wang Y. Is sleep duration associated with childhood obesity? A systematic review and meta-analysis. Obesity (Silver Spring) 2008;16(2):265–74.

[8] Patel SR, Hu FB. Short sleep duration and weight gain: a systematic review. Obesity (Silver Spring) 2008;16(3):643–53.

[9] Capers PL, et al. A systematic review and meta-analysis of randomized controlled trials of the impact of sleep duration on adiposity and components of energy balance. Obes Rev 2015;16(9):771–82.

[10] St-Onge MP. Sleep-obesity relation: underlying mechanisms and consequences for treatment. Obes Rev 2017;18(Suppl 1):34–9.

[11] Romieu I, et al. Energy balance and obesity: what are the main drivers? Cancer Causes Control 2017;28(3):247–58.

[12] Knutson KL, et al. The metabolic consequences of sleep deprivation. Sleep Med Rev 2007;11(3):163–78.

[13] Chaput JP, St-Onge MP. Increased food intake by insufficient sleep in humans: are we jumping the gun on the hormonal explanation? Front Endocrinol 2014;5:116.

[14] Lundahl A, Nelson TD. Sleep and food intake: a multisystem review of mechanisms in children and adults. J Health Psychol 2015;20(6):794–805.

[15] Geiker NRW, et al. Does stress influence sleep patterns, food intake, weight gain, abdominal obesity and weight loss interventions and vice versa? Obes Rev 2018;19(1):81–97.

[16] Broussard JL, Van Cauter E. Disturbances of sleep and circadian rhythms: novel risk factors for obesity. Curr Opin Endocrinol Diabetes Obes 2016;23(5):353–9.

[17] Kearney J. Food consumption trends and drivers. Philos Trans R Soc B 2010;365(1554):2793–807.

[18] Haghighatdoost F, et al. Sleep deprivation is associated with lower diet quality indices and higher rate of general and central obesity among young female students in Iran. Nutrition 2012;28(11–12):1146–50.

[19] Patterson RE, et al. Short sleep duration is associated with higher energy intake and expenditure among African-American and non-Hispanic white adults. J Nutr 2014;144(4):461–6.

[20] Stern JH, et al. Short sleep duration is associated with decreased serum leptin, increased energy intake and decreased diet quality in postmenopausal women. Obesity (Silver Spring) 2014;22(5):E55–61.

[21] Al-Disi D, et al. Subjective sleep duration and quality influence diet composition and circulating adipocytokines and ghrelin levels in teen-age girls. Endocr J 2010;57(10):915–23.

[22] Weiss A, et al. The association of sleep duration with adolescents' fat and carbohydrate consumption. Sleep 2010;33(9):1201–9.

[23] Westerlund L, Ray C, Roos E. Associations between sleeping habits and food consumption patterns among 10-11-year-old children in Finland. Br J Nutr 2009;102(10):1531–7.

[24] Moreira P, et al. Food patterns according to sociodemographics, physical activity, sleeping and obesity in Portuguese children. Int J Environ Res Public Health 2010;7(3):1121–38.

[25] Grandner MA, et al. Relationships among dietary nutrients and subjective sleep, objective sleep, and napping in women. Sleep Med 2010;11(2):180–4.

[26] Grandner MA, et al. Dietary nutrients associated with short and long sleep duration. Data from a nationally representative sample. Appetite 2013;64:71–80.

[27] Doo H, Chun H, Doo M. Associations of daily sleep duration and dietary macronutrient consumption with obesity and dyslipidemia in Koreans: a cross-sectional study. Medicine (Baltimore) 2016;95(45).

[28] Doo M, Kim Y. Association between sleep duration and obesity is modified by dietary macronutrients intake in Korean. Obes Res Clin Pract 2016;10(4):424–31.

[29] Nedeltcheva AV, et al. Sleep curtailment is accompanied by increased intake of calories from snacks. Am J Clin Nutr 2009;89(1):126–33.

[30] St-Onge MP, et al. Short sleep duration, glucose dysregulation and hormonal regulation of appetite in men and women. Sleep 2012;35(11):1503–10.

[31] Spaeth AM, Dinges DF, Goel N. Sex and race differences in caloric intake during sleep restriction in healthy adults. Am J Clin Nutr 2014;100(2):559–66.

[32] Calvin AD, et al. Effects of experimental sleep restriction on caloric intake and activity energy expenditure. Chest 2013;144(1):79–86.

[33] Markwald RR, et al. Impact of insufficient sleep on total daily energy expenditure, food intake, and weight gain. Proc Natl Acad Sci U S A 2013;110(14):5695–700.

[34] Brondel L, et al. Acute partial sleep deprivation increases food intake in healthy men. Am J Clin Nutr 2010;91(6):1550–9.

[35] Al Khatib HK, et al. The effects of partial sleep deprivation on energy balance: a systematic review and meta-analysis. Eur J Clin Nutr 2017;71(5):614–24.

[36] St-Onge MP, et al. Short sleep duration increases energy intakes but does not change energy expenditure in normal-weight individuals. Am J Clin Nutr 2011;94(2):410–6.

[37] Schmid SM, et al. Short-term sleep loss decreases physical activity under free-living conditions but does not increase food intake under time-deprived laboratory conditions in healthy men. Am J Clin Nutr 2009;90(6):1476–82.

[38] Hogenkamp PS, et al. Acute sleep deprivation increases portion size and affects food choice in young men. Psychoneuroendocrinology 2013;38(9):1668–74.

[39] Spiegel K, et al. Brief communication: sleep curtailment in healthy young men is associated with decreased leptin levels, elevated ghrelin levels, and increased hunger and appetite. Ann Intern Med 2004;141(11):846–50.

[40] Ahima RS, et al. Leptin regulation of neuroendocrine systems. Front Neuroendocrinol 2000;21(3):263–307.

[41] van der Lely AJ, et al. Biological, physiological, pathophysiological, and pharmacological aspects of ghrelin. Endocr Rev 2004;25(3):426–57.

[42] Bosy-Westphal A, et al. Influence of partial sleep deprivation on energy balance and insulin sensitivity in healthy women. Obes Facts 2008;1(5):266–73.

[43] Omisade A, Buxton OM, Rusak B. Impact of acute sleep restriction on cortisol and leptin levels in young women. Physiol Behav 2010;99(5):651–6.

[44] Pejovic S, et al. Leptin and hunger levels in young healthy adults after one night of sleep loss. J Sleep Res 2010;19(4):552–8.

[45] Simpson NS, Banks S, Dinges DF. Sleep restriction is associated with increased morning plasma leptin concentrations, especially in women. Biol Res Nurs 2010;12(1):47–53.

[46] St-Onge MP. The role of sleep duration in the regulation of energy balance: effects on energy intakes and expenditure. J Clin Sleep Med 2013;9(1):73–80.

[47] Shah M, Vella A. Effects of GLP-1 on appetite and weight. Rev Endocr Metab Disord 2014;15(3):181–7.

[48] Vincent RP, le Roux CW. The satiety hormone peptide YY as a regulator of appetite. J Clin Pathol 2008;61(5):548–52.

[49] Gil-Lozano M, et al. Short-term sleep deprivation with nocturnal light exposure alters time-dependent glucagon-like peptide-1 and insulin secretion in male volunteers. Am J Physiol Endocrinol Metab 2016;310(1):E41–50.

[50] Benedict C, et al. Acute sleep deprivation delays the glucagon-like peptide 1 peak response to breakfast in healthy men. Nutr Diabetes 2013;3:e78.

[51] Sivak M. Sleeping more as a way to lose weight. Obes Rev 2006;7(3):295–6.

[52] Rollins BY, Dearing KK, Epstein LH. Delay discounting moderates the effect of food reinforcement on energy intake among non-obese women. Appetite 2010;55(3):420–5.

[53] Benedict C, et al. Acute sleep deprivation enhances the brain's response to hedonic food stimuli: an fMRI study. J Clin Endocrinol Metab 2012;97(3):E443–7.

[54] St-Onge MP, et al. Sleep restriction increases the neuronal response to unhealthy food in normal-weight individuals. Int J Obes 2014;38(3):411–6.

[55] Davis C, Strachan S, Berkson M. Sensitivity to reward: implications for overeating and overweight. Appetite 2004;42(2):131–8.

[56] Stice E, et al. Relation of reward from food intake and anticipated food intake to obesity: a functional magnetic resonance imaging study. J Abnorm Psychol 2008;117(4):924–35.

[57] Patrick GT, Gilbert J. On the effects of loss of sleep. Psychol Rev 1896;3:469–83.

[58] Goel N, et al. Neurocognitive consequences of sleep deprivation. Semin Neurol 2009;29(4):320–39.

[59] Nilsson JP, et al. Less effective executive functioning after one night's sleep deprivation. J Sleep Res 2005;14(1):1–6.

[60] Beebe DW, et al. Feasibility and behavioral effects of an at-home multi-night sleep restriction protocol for adolescents. J Child Psychol Psychiatry 2008;49(9):915–23.

[61] Belenky G, et al. Patterns of performance degradation and restoration during sleep restriction and subsequent recovery: a sleep dose-response study. J Sleep Res 2003;12(1):1–12.

[62] Harrison Y, Horne JA. The impact of sleep deprivation on decision making: a review. J Exp Psychol Appl 2000;6(3):236–49.

[63] Anderson C, Platten CR. Sleep deprivation lowers inhibition and enhances impulsivity to negative stimuli. Behav Brain Res 2011;217(2):463–6.

[64] Drummond SP, Paulus MP, Tapert SF. Effects of two nights sleep deprivation and two nights recovery sleep on response inhibition. J Sleep Res 2006;15(3):261–5.

[65] Calvo D, et al. Uncontrolled eating is associated with reduced executive functioning. Clin Obes 2014;4(3):172–9.

[66] Wolkow A, et al. Effects of work-related sleep restriction on acute physiological and psychological stress responses and their interactions: a review among emergency service personnel. Int J Occup Med Environ Health 2015;28(2):183–208.

[67] Zhu LJ, et al. The different roles of glucocorticoids in the hippocampus and hypothalamus in chronic stress-induced HPA axis hyperactivity. PLoS ONE 2014;9(5):e97689.

[68] Hirotsu C, Tufik S, Andersen ML. Interactions between sleep, stress, and metabolism: from physiological to pathological conditions. Sleep Sci 2015;8(3):143–52.

[69] Meerlo P, et al. Sleep restriction alters the hypothalamic-pituitary-adrenal response to stress. J Neuroendocrinol 2002;14(5):397–402.

[70] Meerlo P, Sgoifo A, Suchecki D. Restricted and disrupted sleep: effects on autonomic function, neuroendocrine stress systems and stress responsivity. Sleep Med Rev 2008;12(3):197–210.

[71] Born JM, et al. Acute stress and food-related reward activation in the brain during food choice during eating in the absence of hunger. Int J Obes 2010;34(1):172–81.

[72] Lemmens SG, et al. Stress augments food 'wanting' and energy intake in visceral overweight subjects in the absence of hunger. Physiol Behav 2011;103(2):157–63.

[73] Chaput JP, et al. Glycemic instability and spontaneous energy intake: association with knowledge-based work. Psychosom Med 2008;70(7):797–804.

[74] Rutters F, et al. Acute stress-related changes in eating in the absence of hunger. Obesity (Silver Spring) 2009;17(1):72–7.

[75] Epel E, et al. Stress may add bite to appetite in women: a laboratory study of stress-induced cortisol and eating behavior. Psychoneuroendocrinology 2001;26(1):37–49.

[76] Hanlon EC, et al. Circadian rhythm of circulating levels of the endocannabinoid 2-arachidonoylglycerol. J Clin Endocrinol Metab 2015;100(1):220–6.

[77] Hanlon EC, et al. Sleep restriction enhances the daily rhythm of circulating levels of endocannabinoid 2-arachidonoylglycerol. Sleep 2016;39(3):653–64.

[78] Tanaka E, et al. Associations of protein, fat, and carbohydrate intakes with insomnia symptoms among middle-aged Japanese workers. J Epidemiol 2013;23(2):132–8.

[79] Tan X, et al. Associations of disordered sleep with body fat distribution, physical activity and diet among overweight middle-aged men. J Sleep Res 2015;24(4):414–24.

[80] Jaussent I, et al. Insomnia symptoms in older adults: associated factors and gender differences. Am J Geriatr Psychiatry 2011;19(1):88–97.

[81] Katagiri R, et al. Low intake of vegetables, high intake of confectionary, and unhealthy eating habits are associated with poor sleep quality among middle-aged female Japanese workers. J Occup Health 2014;56(5):359–68.

[82] Peuhkuri K, Sihvola N, Korpela R. Dietary factors and fluctuating levels of melatonin. Food Nutr Res 2012;56.

[83] Sauchelli S, et al. Orexin and sleep quality in anorexia nervosa: clinical relevance and influence on treatment outcome. Psychoneuroendocrinology 2016;65:102–8.

[84] Halson SL. Sleep in elite athletes and nutritional interventions to enhance sleep. Sports Med 2014;44(Suppl 1):S13–23.

[85] Saper CB, Scammell TE, Lu J. Hypothalamic regulation of sleep and circadian rhythms. Nature 2005;437(7063):1257–63.

[86] Markus CR, et al. Effect of different tryptophan sources on amino acids availability to the brain and mood in healthy volunteers. Psychopharmacology 2008;201(1):107–14.

[87] Silber BY, Schmitt JA. Effects of tryptophan loading on human cognition, mood, and sleep. Neurosci Biobehav Rev 2010;34(3):387–407.

[88] Esteban S, et al. Effect of orally administered L-tryptophan on serotonin, melatonin, and the innate immune response in the rat. Mol Cell Biochem 2004;267(1-2):39–46.

[89] Andreas S, et al. ST segmental changes and arrhythmias in obstructive sleep apnea. Pneumologie 1991;45(9):720–4.

[90] Bravo R, et al. Tryptophan-enriched cereal intake improves nocturnal sleep, melatonin, serotonin, and total antioxidant capacity levels and mood in elderly humans. Age (Dordr) 2013;35(4):1277–85.

[91] Behall KM, Howe JC. Effect of long-term consumption of amylose vs amylopectin starch on metabolic variables in human subjects. Am J Clin Nutr 1995;61(2):334–40.

[92] Lyons PM, Truswell AS. Serotonin precursor influenced by type of carbohydrate meal in healthy adults. Am J Clin Nutr 1988;47(3):433–9.

[93] Porter JM, Horne JA. Bed-time food supplements and sleep: effects of different carbohydrate levels. Electroencephalogr Clin Neurophysiol 1981;51(4):426–33.

[94] Afaghi A, O'Connor H, Chow CM. High-glycemic-index carbohydrate meals shorten sleep onset. Am J Clin Nutr 2007;85(2):426–30.

[95] Lindseth G, Lindseth P, Thompson M. Nutritional effects on sleep. West J Nurs Res 2013;35(4):497–513.

[96] Kwan RM, Thomas S, Mir MA. Effects of a low carbohydrate isoenergetic diet on sleep behavior and pulmonary functions in healthy female adult humans. J Nutr 1986;116(12):2393–402.

[97] St-Onge M-P, Crawford A, Aggarwal B. Plant-based diets: reducing cardiovascular risk by improving sleep quality? Curr Sleep Med Rep 2018;4(1).

[98] Roky R, et al. Sleep during Ramadan intermittent fasting. J Sleep Res 2001;10(4):319–27.

[99] Cao Y, et al. Dinner fat intake and sleep duration and self-reported sleep parameters over five years: findings from the Jiangsu nutrition study of Chinese adults. Nutrition 2016;32(9):970–4.

[100] Baron KG, et al. Contribution of evening macronutrient intake to total caloric intake and body mass index. Appetite 2013;60(1):246–51.

[101] Wells AS, et al. Effects of meals on objective and subjective measures of daytime sleepiness. J Appl Physiol (1985) 1998;84(2):507–15.

[102] Mackiewicz M, et al. Macromolecule biosynthesis: a key function of sleep. Physiol Genomics 2007;31(3):441–57.

[103] Kidd PM. Omega-3 DHA and EPA for cognition, behavior, and mood: clinical findings and structural-functional synergies with cell membrane phospholipids. Altern Med Rev 2007;12(3):207–27.

[104] Montgomery P, et al. Fatty acids and sleep in UK children: subjective and pilot objective sleep results from the DOLAB study—a randomized controlled trial. J Sleep Res 2014;23(4):364–88.

[105] Peuhkuri K, Sihvola N, Korpela R. Diet promotes sleep duration and quality. Nutr Res 2012;32(5):309–19.

[106] Valtonen M, et al. Effect of melatonin-rich night-time milk on sleep and activity in elderly institutionalized subjects. Nord J Psychiatry 2005;59(3):217–21.

[107] Lin HH, et al. Effect of kiwifruit consumption on sleep quality in adults with sleep problems. Asia Pac J Clin Nutr 2011;20(2):169–74.

[108] Tsaluchidu S, et al. Fatty acids and oxidative stress in psychiatric disorders. BMC Psychiatry 2008;8(Suppl 1):S5.

[109] Reiter RJ, et al. Reducing oxidative/nitrosative stress: a newly-discovered genre for melatonin. Crit Rev Biochem Mol Biol 2009;44(4):175–200.

[110] Srinivasan V, et al. Melatonin and melatonergic drugs on sleep: possible mechanisms of action. Int J Neurosci 2009;119(6):821–46.

[111] Wehr TA, et al. A circadian signal of change of season in patients with seasonal affective disorder. Arch Gen Psychiatry 2001;58(12):1108–14.

[112] Meng X, et al. Dietary sources and bioactivities of melatonin. Nutrients 2017;9(4).

[113] Auger RR, et al. Clinical practice guideline for the treatment of intrinsic circadian rhythm sleep-wake disorders: advanced sleep-wake phase disorder (ASWPD), delayed sleep-wake phase disorder (DSWPD), non-24-hour sleep-wake rhythm disorder (N24SWD), and irregular sleep-wake rhythm disorder (ISWRD). An update for 2015: an American Academy of sleep medicine clinical practice guideline. J Clin Sleep Med 2015;11(10):1199–236.

[114] Schutte-Rodin S, et al. Clinical guideline for the evaluation and management of chronic insomnia in adults. J Clin Sleep Med 2008;4(5):487–504.

[115] Robinson CR, et al. The effects of nicotinamide upon sleep in humans. Biol Psychiatry 1977;12(1):139–43.

[116] Setchell KD, Cassidy A. Dietary isoflavones: biological effects and relevance to human health. J Nutr 1999;129(3):758S–67S.

[117] Hachul H, et al. Isoflavones decrease insomnia in postmenopause. Menopause 2011;18(2):178–84.

[118] Cui Y, et al. Relationship between daily isoflavone intake and sleep in Japanese adults: a cross-sectional study. Nutr J 2015;14:127.

[119] Altura BM. Basic biochemistry and physiology of magnesium: a brief review. Magnes Trace Elem 1991;10(2-4):167–71.

[120] Abbasi B, et al. The effect of magnesium supplementation on primary insomnia in elderly: a double-blind placebo-controlled clinical trial. J Res Med Sci 2012;17(12):1161–9.

[121] Zeng Y, et al. Strategies of functional foods promote sleep in human being. Curr Signal Transduct Ther 2014;9(3):148–55.

[122] Howatson G, et al. Effect of tart cherry juice (Prunus cerasus) on melatonin levels and enhanced sleep quality. Eur J Nutr 2012;51(8):909–16.

[123] Reiter RJ, Manchester LC, Tan DX. Melatonin in walnuts: influence on levels of melatonin and total antioxidant capacity of blood. Nutrition 2005;21(9):920–4.

[124] Brennan AM, et al. Walnut consumption increases satiation but has no effect on insulin resistance or the metabolic profile over a 4-day period. Obesity (Silver Spring) 2010;18(6):1176–82.

[125] Bent S, et al. Valerian for sleep: a systematic review and meta-analysis. Am J Med 2006;119(12):1005–12.

[126] Wheatley D. Medicinal plants for insomnia: a review of their pharmacology, efficacy and tolerability. J Psychopharmacol 2005;19(4):414–21.

[127] Donath F, et al. Critical evaluation of the effect of valerian extract on sleep structure and sleep quality. Pharmacopsychiatry 2000;33(2):47–53.

[128] Diaper A, Hindmarch I. A double-blind, placebo-controlled investigation of the effects of two doses of a valerian preparation on the sleep, cognitive and psychomotor function of sleep-disturbed older adults. Phytother Res 2004;18(10):831–6.

[129] Hallam KT, et al. Comparative cognitive and psychomotor effects of single doses of *Valeriana officianalis* and triazolam in healthy volunteers. Hum Psychopharmacol 2003;18(8):619–25.

[130] Leathwood PD, et al. Aqueous extract of valerian root (*Valeriana officinalis* L.) improves sleep quality in man. Pharmacol Biochem Behav 1982;17(1):65–71.

[131] Jacobs BP, et al. An internet-based randomized, placebo-controlled trial of kava and valerian for anxiety and insomnia. Medicine (Baltimore) 2005;84(4):197–207.

[132] Fernandez-San-Martin MI, et al. Effectiveness of valerian on insomnia: a meta-analysis of randomized placebo-controlled trials. Sleep Med 2010;11(6):505–11.

[133] Morin CM, Benca R. Chronic insomnia. Lancet 2012;379(9821):1129–41.

[134] Sateia MJ, et al. Clinical practice guideline for the pharmacologic treatment of chronic insomnia in adults: an American Academy of sleep medicine clinical practice guideline. J Clin Sleep Med 2017;13(2):307–49.

[135] Wheatley D. Kava and valerian in the treatment of stress-induced insomnia. Phytother Res 2001;15(6):549–51.

[136] Yurcheshen M, Seehuus M, Pigeon W. Updates on nutraceutical sleep therapeutics and investigational research. Evid Based Complement Alternat Med 2015;2015:105256.

[137] Rowe A, et al. Kavalactone pharmacophores for major cellular drug targets. Mini-Rev Med Chem 2011;11(1):79–83.

[138] Castro-Quezada I, Roman-Vinas B, Serra-Majem L. The Mediterranean diet and nutritional adequacy: a review. Nutrients 2014;6(1):231–48.

[139] Perez-Jimenez J, Diaz-Rubio ME, Saura-Calixto F. Contribution of macromolecular antioxidants to dietary antioxidant capacity: a study in the Spanish Mediterranean diet. Plant Foods Hum Nutr 2015;70(4):365–70.

[140] Estruch R. Anti-inflammatory effects of the Mediterranean diet: the experience of the PREDIMED study. Proc Nutr Soc 2010;69(3):333–40.

[141] Campanini MZ, et al. Mediterranean diet and changes in sleep duration and indicators of sleep quality in older adults. Sleep 2017;40(3).

[142] Hurt RT, et al. The obesity epidemic: challenges, health initiatives, and implications for gastroenterologists. Gastroenterol Hepatol (NY) 2010;6(12):780–92.

Chapter 20

Sleep and exercise

Christopher E. Kline
Physical Activity and Weight Management Research Center, Department of Health and Physical Activity, University of Pittsburgh, Pittsburgh, PA, United States

Exercise and sleep are interrelated behaviors that are each vital to optimal health and functioning. There has long been a consensus that exercise improves sleep, and accumulating evidence highlights the potential for exercise as a nonpharmacologic treatment option for disturbed sleep. Recent research has also emphasized that disturbed or insufficient sleep may impact physical activity levels. Together, these studies suggest a robust bidirectional relationship between exercise and sleep.

This chapter will provide an overview regarding the available research on the relationship between sleep and exercise. The majority of the chapter will review the impact of exercise on sleep, focusing on experimental research in adults with and without sleep disorders and discussing potential mechanisms of effect. The chapter will also discuss recent research on the bidirectional relationship between exercise and sleep and will conclude with a brief discussion regarding the potential synergistic impact of improving sleep and exercise on health.

IMPACT OF EXERCISE ON SLEEP

Exercise has long been associated with better sleep [1] and is one of the most common sleep hygiene recommendations [2]. There are a number of reasons why exercise is an attractive nonpharmacologic alternative for treating disturbed sleep. Exercise is a relatively inexpensive, simple, and easily accessible treatment modality. Exercise does not possess the side effects (e.g., hypnotic medications), low acceptability (e.g., positive airway pressure), and/or barriers to access (e.g., cognitive-behavioral therapy for insomnia) that plague many of the standard sleep medicine treatments. Finally, exercise confers a number of mental and physical health benefits [3] that are independent of its effects on sleep, which is notable given the daytime impairment and comorbid health conditions that commonly plague those with disturbed sleep.

Observational research

Observational research involving cross-sectional samples and prospective epidemiologic cohorts has frequently reported a robust association between exercise and better sleep [1]. In general, higher levels of physical activity have been consistently associated with a lower likelihood of reporting insufficient sleep, general sleep disturbance, or insomnia symptoms [4, 5]. These studies have traditionally relied upon self-reported (and often unvalidated) measures of leisure-time physical activity and sleep [1]. However, recent research involving objective assessment of exercise, fitness, and/or sleep has supported these earlier findings. For example, Dishman and colleagues found that the preservation of objectively measured cardiorespiratory fitness, a physiological measure that is strongly influenced by physical activity, was protective against the onset of self-reported sleep complaints in a sample of middle-aged adults [6]. Kline and colleagues reported that greater amounts of self-reported leisure-time physical activity were associated with better self-reported and polysomnographic (PSG) measures of sleep quality, continuity, and depth in a sample of middle-aged women [7]. More recently, studies have linked greater amounts of accelerometer-measured physical activity with higher accelerometer-measured sleep efficiency [8] and a moderate level of self-reported activity with lower incidence of PSG-measured short sleep duration and low sleep efficiency [9]. Overall, the finding that higher levels of physical activity are associated with better sleep has been consistent regardless of age, sex, and race, and across a wide range of sample sizes [1].

Observational research offers a number of advantages when examining the association between exercise and sleep, including the ability to explore relationships in potentially large samples with wide variability in sleep and activity habits. In addition, with observational research it is possible to examine whether and how potential confounding factors (e.g., age, health status) influence the association between exercise and sleep. However, because of the common expectation that exercise improves sleep, observational research that is limited to self-report measures may simply reflect this assumption [10]. Moreover, since exercise is often discontinued during times of elevated psychosocial stress [11], better sleep on days in which exercise occurs may be more indicative of reduced stress rather than exercise per se. Finally, the link between exercise and sleep could be explained by a variety of factors that are often poorly accounted for in epidemiologic analyses, including other daytime health behaviors [12], light exposure [13], and mental health [14].

Experimental research

Despite the intuitive appeal of exercise and supportive observational evidence of an association between greater amounts of physical activity and better sleep, the majority of experimental research has evaluated the efficacy of exercise for improving sleep in samples without consideration of baseline sleep disturbance. Nevertheless, the available research suggests that both acute exercise and chronic exercise training have mild to modest benefits on sleep among adults with no to mild sleep impairment and more robust effects on those with significant sleep disturbance.

Acute exercise

A large number of studies have examined the effect of an acute bout of exercise on the subsequent night's sleep, often in comparison with an inactive control day. However, the majority of studies have focused on adults without sleep complaints, thereby limiting any room for sleep improvement with exercise [15]. Nevertheless, an acute bout of exercise appears to result in small-to-moderate improvements in sleep, at least among relatively normal sleepers, though the effects may differ according to individual and/or exercise characteristics.

A 2015 meta-analysis of 41 studies found that an acute bout of exercise significantly increases total sleep time (TST; effect size $d=0.22$), sleep efficiency (SE; $d=0.25$), and slow-wave sleep (SWS; $d=0.19$) and decreases sleep onset latency (SOL; $d=-0.17$), wake after sleep onset (WASO; $d=-0.38$), stage 1 nonrapid eye movement (NREM) sleep ($d=-0.35$), and rapid eye movement (REM) sleep ($d=-0.27$) in comparison with a day without exercise [16]. Interestingly, the meta-analysis also found several significant moderators of the effect of acute exercise on specific sleep parameters. Sample characteristics seemed to matter, as acute exercise increased SWS to a greater extent among those with a high baseline activity level compared to those with a low baseline level of activity, and greater reductions in stage 1 NREM sleep and WASO were observed for men compared to women. Moreover, characteristics of the exercise bout significantly impacted the association between acute exercise and sleep; studies involving cycling resulted in a greater increase in SWS compared to studies involving running, and longer exercise duration was linked to greater increases in TST and SWS and greater reductions in SOL, REM sleep, and REM sleep latency. Participant age and exercise intensity did not moderate any associations, though. Caution is urged regarding the robustness of these potential moderators, since very few studies directly compared these factors for their impact on sleep. Nevertheless, the findings of this meta-analysis suggest that the effects of acute exercise on sleep may vary according to the characteristics of the individual and exercise bout [16].

In contrast to the potential moderators noted above, time of day has often been examined as a moderator of the effect of acute exercise on sleep. In the 2015 meta-analysis by Kredlow and colleagues, exercising <3h prior to bedtime was associated with greater reductions of WASO and stage 1 NREM sleep, whereas exercising 3–8h prior to bedtime did not lead to reductions in these sleep parameters [16]. This finding contradicts the common warning provided in sleep hygiene recommendations that exercise should be avoided close to bedtime due to the possibility of sleep impairment [17]. However, epidemiologic and experimental studies have found, with few exceptions [18], that late-night exercise does not impair sleep and, in some cases, improves sleep. For instance, a survey of 1000 adults found that evening exercisers (<4h before bed) did not differ in any self-reported sleep parameters compared to nonexercisers, with over 90% of individuals who performed vigorous exercise in the evening actually reporting that their sleep was of equal or better quality and duration on days they exercised compared to days without exercise [19]. Likewise, moderate- to vigorous-intensity exercise ending as close as 30min before bedtime has not impaired objective or subjective sleep despite causing increases in heart rate and core body temperature that persisted into the early hours of sleep [20]. Other studies have shown better subjective and/or objective sleep following late-night exercise compared to a nonexercise day [21, 22]. However, the focus on trained adults without sleep complaints in these experimental studies is a prominent limitation of this literature, as adults who are physically inactive and/or with significant sleep complaints may be more reactive to and/or recover less quickly from late-night exercise than those with high aerobic fitness and healthy sleep. Thus, consistent with the preliminary nature of other potential moderators, the optimal time of day of exercise to improve sleep remains unclear, though active adults without sleep complaints should not avoid late-night exercise for fear of sleep disruption.

Overall, while the small- to moderate-sized effects of acute exercise on sleep may prompt skepticism regarding their practical significance, it is important to reiterate that the vast majority of experimental studies examining acute exercise have focused on adults without sleep complaints. Thus, it is assumed that acute exercise would be even more efficacious among adults with disturbed sleep. Although this does appear to be the case for adults with insomnia (discussed later), future research should focus on the effect of acute exercise in adults with subclinical sleep complaints and/or sleep disorders and probe whether participant characteristics (e.g., age, sex, activity level) and exercise bout characteristics (e.g., duration, mode, timing) moderate these effects.

Chronic exercise training

Multiple studies have evaluated whether maintaining an exercise training regimen for a sustained duration of time (e.g., 12 weeks) improves sleep. Unfortunately, similar to studies focused on acute exercise and sleep, until recently most exercise training studies had focused on individuals with relatively normal sleep patterns at baseline; as one might expect, these studies demonstrated only mild, if any, sleep improvements following exercise training [23]. Recent work, noted below, has provided more robust evidence that exercise training improves sleep among those with significant sleep disturbance.

Meta-analyses of exercise training studies have reported significant effects of exercise training on multiple sleep parameters. In addition to reviewing the literature on the effect of acute exercise on sleep, Kredlow and colleagues also evaluated 25 exercise training studies in their 2015 meta-analysis. Exercise training significantly increased objective measures of TST ($d=0.25$) and SE ($d=0.30$) and self-reported sleep quality ($d=0.74$) while also decreasing objective SOL ($d=-0.35$) [16]. In moderator analyses, the only participant characteristic found to moderate the effect of exercise training on sleep was age; exercise training was less effective at reducing SOL among older adults. Among exercise characteristics, longer durations of individual exercise bouts resulted in a greater SOL reduction and greater adherence to the exercise training regimen was associated with greater improvement in sleep quality, whereas a longer duration of the exercise regimen was associated with a smaller TST improvement [16]. Other meta-analyses, focused on middle-aged women [24] or older adults with sleep problems [25], have also noted significant improvements in self-reported sleep quality following exercise training.

Although Kredlow and colleagues suggested several potential moderators of the effect of exercise training on sleep [16], minimal research has directly examined which participant and/or exercise characteristics result in the greatest improvement in sleep following exercise training. Recent research has focused on midlife women and older adults, two specific populations in which physical inactivity and sleep disturbance are both prevalent. As noted above, Kredlow found exercise to be less effective at reducing SOL among older adults; however, the overall body of evidence suggests that exercise is efficacious at improving sleep in these populations [24–26]. It is also unknown whether the type of exercise is important for sleep benefits. Although studies have shown that aerobic exercise and resistance exercise each can improve sleep [16, 27], they have not been directly compared [28]. Moreover, significant improvements in sleep have been observed with meditative mind-body exercise (e.g., yoga, tai chi, qigong) [29]; however, these studies have often been limited to self-reported sleep and have not been directly compared against aerobic or resistance exercise. Although tangentially addressed by Kredlow and colleagues, additional questions remain regarding whether a dose-response relationship exists between exercise and sleep and whether exercise intensity differentially impacts sleep; minimal research has directly compared different doses or intensities of exercise on sleep [30, 31]. Finally, it remains unclear whether the effects of exercise training on sleep are dependent upon the extent to which fitness is improved [30, 32]. Future research needs to elucidate the effects of different modes, intensities, and doses of exercise in various samples of adults with sleep disturbances and evaluate sleep using both subjective and objective measures to better understand the impact of exercise training on sleep.

Sedentary behavior

In addition to the health benefits of exercise, the public health impact of the entire physical activity continuum is increasingly recognized, with particular focus on the potentially deleterious effects of excessive sedentary behavior. As a behavior that is distinct from a lack of exercise, sedentary behavior occurs at the lowest end of the physical activity continuum and is formally defined as any waking behavior characterized by an energy expenditure ≤ 1.5 metabolic equivalents that is performed in a sitting, reclining, or lying posture [33].

Research on the potential relationship between sedentary behavior and sleep is in its infancy. A recent meta-analysis found that greater sedentary behavior was associated with 18% greater odds for insomnia and 38% greater odds for sleep disturbance [34]. Multiple cross-sectional studies have recently reported that greater sedentary behavior is associated with lower sleep efficiency [8], higher daytime sleepiness [35], and greater odds of having short sleep duration [36], poor sleep quality [36, 37], and sleep problems [36]. In addition, cross-sectional studies have found higher amounts of sedentary behavior to be associated with greater OSA severity, including higher objectively measured AHI [38, 39] and greater odds of OSA [38] or OSA 'high risk' classification [8, 37]. However, the vast majority of the current evidence is limited to cross-sectional and observational study designs, with many relying on self-reported sedentary behavior [36–39] and/or self-reported sleep [35–37]. While sparse, experimental data have been consistent with observational findings. In a crossover trial of 6 adults whose sleep was experimentally restricted to 5 h/night for 3 nights, breaking up prolonged sedentary behavior with periodic walking breaks led to a 26% reduction in WASO and a 9% increase in SWS compared to days with uninterrupted sedentary behavior [40]. Finally, in a crossover trial of 25 adults with elevated

BP, Kline and colleagues recently found that alternating sitting with standing every 30 min during a simulated workday led to lower WASO on that night compared to uninterrupted sedentary behavior [41]. Together, these data provide preliminary support that excessive sedentary behavior is associated with worse sleep and that reducing sedentary behavior may improve sleep; however, much remains unknown about how best to reduce sedentary behavior for sleep benefits and which sleep parameters are most impacted.

Potential mechanisms of exercise

The mechanisms underlying how exercise improves sleep remain poorly understood. However, a number of potential mechanisms have been advanced.

Anxiolytic and antidepressant effects

Anxiety and depression are key consequences of disturbed sleep but also significant risk factors in the development and perpetuation of sleep disturbance [14]. Acute exercise reduces both subjective and physiologic markers of anxiety [42], and exercise training significantly reduces anxiety symptoms [43]. Similarly, acute exercise and exercise training have antidepressant effects that are similar in efficacy to pharmacotherapy [44]. Underlying the anxiolytic and antidepressant effects of exercise on sleep may be improvements in autonomic function, increases in circulating endocannabinoids, reductions in inflammation, and/or normalization of hypothalamic-pituitary-adrenal axis dysfunction [45, 46].

Circadian phase-shifting effects

Exercise may improve sleep via its effects on the circadian system. Sleep/wake patterns are heavily influenced by the circadian system, and disturbances in circadian timing and blunted amplitude of the circadian system are primary causes of circadian rhythm sleep disorders such as advanced sleep phase disorder, delayed sleep phase disorder, and irregular sleep/wake disorder [47]. In a laboratory setting, single and multiple bouts of aerobic exercise result in time-dependent alterations in the phase of the central circadian system [48], while muscular activity may serve as a synchronizing cue for the peripheral circadian system and provide input to the central clock [49]. Because the current phase-response curve for exercise is preliminary [48], the optimal timing of exercise for phase-shifting effects is unknown. Regardless of alterations in phase, though, there is evidence that greater amounts of physical activity are associated with greater amplitude of circadian clock gene expression [50] and circadian output markers [51].

Body temperature effects

Impaired heat loss mechanisms and blunted temperature downregulation at the time of sleep initiation have been observed among adults with insomnia [52]. Acute exercise performed in the late afternoon or evening may improve sleep by activating heat loss mechanisms to dissipate the exercise-induced increase in core temperature [53], as distal vasodilation has previously been shown to hasten sleep onset [54]. In addition, exercise training produces thermoregulatory adaptations (e.g., improved distal vasodilatory capacity, increased distal blood flow) [55] that may facilitate a more rapid nocturnal dissipation of core body temperature.

Adenosine

Adenosine has been strongly linked to the homeostatic regulation of sleep [56]. Because acute exercise significantly increases extracellular adenosine levels, this may be a mechanism by which exercise promotes sleep. In rats, high-intensity exercise increased levels of adenosine in the brain [57]. Following acute exercise in humans, homeostatic sleep need (as measured by SWS) was significantly higher under a placebo condition compared to high caffeine intake, which blocks adenosine receptors [58].

IMPACT OF EXERCISE ON SLEEP DISORDERS

Perhaps the most compelling evidence regarding the effect of exercise on sleep has been observed among those with insomnia, sleep-disordered breathing, or restless legs syndrome. Although limited by a small evidence base, these data suggest that there is a role for exercise in the prevention and/or management of each of these sleep disorders.

Insomnia

Insomnia is characterized by difficulty initiating and/or maintaining sleep, awakening for the final time earlier than desired, or unrefreshing or nonrestorative sleep. Diagnostic criteria for insomnia disorder, however, require that these sleep complaints occur with a minimum frequency and chronicity (e.g., ≥3 nights/week for ≥3 months) despite an adequate opportunity to sleep and are accompanied by dissatisfaction with sleep and significant daytime impairment [59]. Although hypnotic medication is the most common treatment, most medications are intended for short-term use and numerous studies have demonstrated an unfavorable risk/benefit ratio for hypnotics [60]. In contrast, cognitive behavioral therapy for insomnia is the recommended first-line treatment option for insomnia, but it is not readily available [61, 62].

There is limited, yet reliable, evidence that exercise reduces the likelihood and severity of insomnia across both observational and experimental research [63]. In epidemiologic cohorts, greater amounts of physical activity have been consistently associated with a lower prevalence and incidence of insomnia [7, 64, 65]. Likewise, the few experimental studies have found that both acute exercise and chronic exercise training improve the sleep of adults with insomnia. In one of only two trials to evaluate acute exercise, moderate-intensity aerobic exercise performed in the late afternoon, but not moderate-intensity resistance exercise or vigorous aerobic exercise performed at the same time of day, improved subjective reports of SOL and TST and PSG-based measures of SOL, TST, and SE, compared with a nonexercise control night [66]. In contrast, another trial reported that an acute bout of aerobic exercise performed in the morning, but not in the afternoon, improved sleep in a sample of older adults with insomnia [67]. Meanwhile, randomized trials involving 4–6 months of exercise training produced significant improvements in sleep and indices of daytime function among samples of middle-aged and older adults with chronic insomnia [68, 69]. More recently, a pilot study found that integrating light-intensity exercise into one's lifestyle for 8 weeks reduced insomnia severity but not self-reported or actigraphy-assessed sleep [70]. However, with sample sizes of <50 adults in each of these experimental trials, larger trials are needed to substantiate these results and identify the timing, mode, and dose of exercise for optimal improvement in sleep. Moreover, whether exercise reduces insomnia via separate mechanisms than those already discussed remains unclear.

Sleep-disordered breathing

Sleep-disordered breathing (SDB) serves as an umbrella term for a variety of sleep-related breathing disorders. As the most common form of SDB, obstructive sleep apnea (OSA) is characterized by repetitive episodes of airflow reduction or cessation due to upper airway collapse despite continued attempts at respiration, and is strongly linked with excess weight. As another form of SDB that is much less common than OSA, central sleep apnea (CSA) is characterized by recurring bouts of airflow reduction or cessation that are due to a reduced central drive for respiration; CSA is most commonly observed in those with neurological conditions or congestive heart failure. Although positive airway pressure (PAP) is highly efficacious for adults with SDB, its effectiveness is greatly limited by low patient acceptance and adherence [71]. While oral appliances are a treatment option for adults with OSA who cannot tolerate PAP, they have only modest efficacy at reducing OSA severity and there is risk for dental side effects with long-term use [72].

In epidemiologic research, habitual exercise has consistently been associated with lower OSA risk [73, 74], with some studies observing a dose-response relationship between weekly exercise duration and OSA risk [75, 76]. Even among those with OSA, greater levels of exercise are associated with a more favorable cardiometabolic profile [77]. Although no study has focused on the effect of acute exercise on OSA, multiple studies have examined exercise training. Meta-analyses have found that 4–24 weeks of exercise training results in an approximately 30% reduction in OSA severity despite nonsignificant reductions in body weight [78, 79]. In addition to reduced OSA severity, exercise training also improves multiple dimensions of daytime functioning (e.g., reduced sleepiness, improved quality of life) in samples of adults with OSA [79, 80]. Moreover, preliminary evidence suggests that combining exercise with PAP may provide greater improvements in daytime functioning than can be achieved with PAP alone [81, 82]. A variety of mechanisms have speculated how exercise reduces OSA severity. In addition to its modest impact on weight loss [83], exercise may strengthen and/or increase the fatigue resistance of upper airway dilator muscles [84] and/or minimize the respiratory instability that accompanies lighter sleep stages and initial sleep onset by inducing more SWS and greater sleep continuity [85]. The most widely explored mechanism, though, involves whether exercise reduces OSA by attenuating overnight fluid redistribution from the lower extremities to the upper body [86]. Multiple studies have observed concurrent reductions in the magnitude of the overnight rostral fluid shift with OSA reduction following 1–4 weeks of exercise training [87, 88].

The effects of exercise training on CSA have been sparsely evaluated, with two experimental trials conducted involving patients with chronic heart failure. Exercise training reduced the overall apnea-hypopnea index (AHI) by 36%–65% in these studies; however, divergent effects were observed on the efficacy of specific SDB events. In one trial, exercise reduced central but not obstructive events [89], while exercise reduced the AHI of those with predominantly OSA but not those with CSA in another trial [90]. Exercise training is hypothesized to reduce CSA severity by augmenting cardiac function and/or normalizing chemoreflex sensitivity, resulting in decreased circulation time and stable respiration.

Restless legs syndrome/periodic limb movements during sleep

Restless legs syndrome (RLS), also known as Willis Ekbom disease, is characterized by an irresistible urge to move one's limbs, most often the legs. The overwhelming sensation to move, commonly described as a burning, prickly feeling, is usually most severe in the evening. Most adults with RLS also have periodic limb movements during sleep (PLMS), which involves recurring involuntary movements of the limbs during sleep. Unfortunately, the most common pharmacologic treatments for RLS and PLMS (e.g., dopaminergic agents, opioids, benzodiazepines, anticonvulsants) are accompanied by significant side effects [91].

Exercise is a common recommendation for RLS management, since symptoms are often exacerbated by prolonged inactivity yet relieved with physical movement [92].

Though sparse, epidemiological research indicates that low levels of physical activity are associated with greater risk for RLS [93]. Likewise, there is limited experimental evidence regarding the effect of acute exercise or chronic exercise training on RLS or PLMS. Both acute exercise and exercise training have been shown to significantly reduce RLS symptoms [94, 95]. In the only randomized trial on the topic, 12 weeks of exercise training (i.e., lower body resistance exercise and treadmill walking) significantly reduced RLS symptoms [96]. Finally, both acute vigorous exercise and 6 months of moderate-intensity exercise training have been shown to improve sleep and reduce PLMS severity [97]. Exercise is hypothesized to reduce RLS and PLMS symptoms via activation of the dopaminergic and opiate systems [97].

A BIDIRECTIONAL RELATIONSHIP: IMPACT OF SLEEP ON EXERCISE

Although research on the relationship between exercise and sleep has generally focused on the impact of exercise on sleep, disturbed sleep may also precipitate reduced exercise [98]. Observational studies have found that disturbed sleep predicts lower levels of physical activity 2–7 years later [99, 100]. Moreover, a number of recently published studies have concurrently assessed physical activity and sleep on a daily basis and explored whether sleep on a given night predicts the next day's physical activity and whether physical activity on a given day predicts that night's sleep. Findings from these studies have been inconsistent, as they have observed that physical activity improves [101, 102] or has no impact [103–105] on the subsequent night's sleep and that sleep predicts [103, 106, 107] or does not influence [101, 104, 105, 108] the next day's activity levels. Thus, while these studies provide some evidence of a bidirectional relationship between sleep and exercise, it appears that small daily fluctuations in these behaviors may have negligible effects on each other and/or that the association may differ according to how these behaviors are measured and across different samples.

Cross-sectional studies comparing adults with poor sleep to those without sleep complaints provide additional evidence that poor sleep may impede physical activity. For instance, adults with short sleep duration or later sleep timing have less physical activity and greater sedentary behavior [109, 110]. Moreover, adults with OSA have low levels of daytime activity [78] and are less active than adults without OSA [111, 112]. In crossover experimental trials, restricting sleep to 4–5.5 h/night for 1–7 nights led to reduced daytime activity and/or increased sedentary behavior [113, 114], potentially due to decreased vigor and alertness following sleep restriction [113].

By demonstrating reductions in physical activity with disturbed or restricted sleep, these studies raise the possibility that improving sleep may lead to greater daytime activity. Unfortunately, the limited experimental evidence that is currently available provides only weak support for this possibility. To date, research has been limited to adults whose OSA was treated with PAP; in only one of these studies has treatment increased physical activity levels relative to baseline [115]; in contrast, most studies have failed to observe any change in physical activity with treatment [78, 116, 117]. Overall, these studies suggest that improving sleep may be insufficient to change physical activity and that low activity levels may be linked to other factors (e.g., poor lifestyle habits) in addition to disturbed sleep [78].

COMBINED IMPACT OF EXERCISE AND SLEEP ON HEALTH

Given that exercise and sleep are each independently linked with health outcomes yet are interrelated behaviors (Fig. 20.1), research has recently begun to explore the health impact when one behavior is promoted at the expense of another [118, 119]. By recognizing that physical activity, sedentary, and sleep behaviors occur within the time constraint of a 24-h day, these analyses have examined how reallocating a specific duration of these movement behaviors is associated with various health outcomes [120]. While some studies have found that reallocating a portion of time spent in sedentary behavior to sleep was associated with lower mortality risk [121] and greater insulin sensitivity [122], most studies have observed the greatest health benefits when moderate-vigorous physical activity was increased at the expense of other movement behaviors [123, 124]. Of note, though, many studies have neglected to even include sleep in these analyses [120]. Moreover, isotemporal substitution and compositional data analysis studies are limited in their examination of sleep since sleep duration is the only sleep-related predictor or outcome that can be utilized in the models. While duration is an important characteristic of sleep, sleep health encompasses a variety of other dimensions (e.g., timing, quality, depth) [125] that cannot be easily probed with these statistical approaches.

There is also preliminary research exploring whether health improvements resulting from modification of sleep or exercise are dependent upon the other behavior. For example, there is increasing recognition that the effect of exercise on cognition may be at least partially mediated by the effect of exercise on sleep [126] and that the association between sleep and cardiometabolic health is impacted by physical activity [127, 128]. Conversely, recent studies have documented that the potential health benefits of exercise or healthy sleep may be limited by the impact of disturbed sleep or physical inactivity, respectively. Among adults with OSA, reduced arterial stiffness following a physical activity and dietary intervention was dependent upon the extent of OSA severity improvement [129]. As an additional

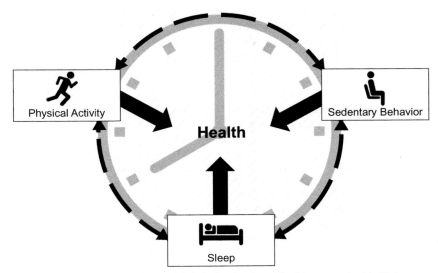

FIG. 20.1 Interrelationships between movement behaviors and their isolated and combined impact on health. Various movement behaviors comprise each 24-h cycle. Each of these behaviors—sedentary behavior, various intensities of physical activity (light, moderate, vigorous), and sleep—significantly impact health. However, each of these behaviors are related to each other in a bidirectional manner. Understanding the impact of different allocations of these behaviors is a current research emphasis. *Adapted with permission from: Smith K, Rosenberger ME. Keeping seniors active—a 24 hour approach. Stanford, CA: Stanford Center on Longevity; 2017. Available from: http://longevity.stanford.edu/2017/08/11/keeping-seniors-active-a-24-hour-approach/.*

example, the typical improvement in postprandial glucose metabolism resulting from breaking up prolonged sedentary time was not observed when the participants had their sleep restricted in a recent study [40].

Finally, research is now beginning to examine whether simultaneously intervening on sleep and exercise provides additive or even synergistic benefits on health compared to isolated intervention on these behaviors. Observational studies suggest that clusters of poor sleep and physical inactivity are linked to worse cardiovascular health [130, 131] and greater mortality [132] than either behavior in isolation. However, experimental evidence is currently limited to one trial of adults with insomnia. In this study, adults who received physical activity counseling and sleep restriction therapy had greater improvements in sleep than those who received only sleep restriction therapy [133]. While the current evidence is scarce, multiple studies are currently underway [134, 135] that will add insight into the health benefits related to simultaneous intervention on sleep and exercise.

CONCLUSION

Exercise is commonly assumed to lead to improved sleep. This assumption is consistently supported by observational research, and experimental research involving acute exercise and exercise training in samples with disturbed sleep have largely corroborated this claim. Nevertheless, much remains unknown regarding the efficacy of exercise for alleviating sleep disturbances and the underlying mechanisms through which exercise improves sleep. With increasing recognition that the relationship between exercise and sleep is bidirectional, research is now seeking to identify whether combining sleep and exercise interventions are efficacious and whether the health benefits of exercise or sleep are dependent upon its impact on the other behavior. As we continue to refine our understanding of the relationship between exercise and sleep, it is clear that there is great potential for both exercise and sleep to be optimized for the purpose of improving health.

REFERENCES

[1] Youngstedt SD, Kline CE. Epidemiology of exercise and sleep. Sleep Biol Rhythms 2006;4(3):215–21.

[2] Irish LA, Kline CE, Gunn HE, Buysse DJ, Hall MH. The role of sleep hygiene in promoting public health: a review of empirical evidence. Sleep Med Rev 2015;22:23–36.

[3] Blair SN, Morris JN. Healthy hearts—and the universal benefits of being physically active: physical activity and health. Ann Epidemiol 2009;19(4):253–6.

[4] Sherrill DL, Kotchou K, Quan SF. Association of physical activity and human sleep disorders. Arch Intern Med 1998;158(17):1894–8.

[5] Zheng B, Yu C, Lin L, Du H, Lv J, Guo Y, Bian Z, Chen Y, Yu M, Li J, Chen J, Chen Z, Li L. Associations of domain-specific physical activities with insomnia symptoms among 0.5 million Chinese adults. J Sleep Res 2017;26(3):330–7.

[6] Dishman RK, Sui X, Church TS, Kline CE, Youngstedt SD, Blair SN. Decline in cardiorespiratory fitness and odds of incident sleep complaints. Med Sci Sports Exerc 2015;47(5):960–6.

[7] Kline CE, Irish LA, Krafty RT, Sternfeld B, Kravitz HM, Buysse DJ, Bromberger JT, Dugan SA, Hall MH. Consistently high sports/exercise activity is associated with better sleep quality, continuity and depth in midlife women: the SWAN Sleep Study. Sleep 2013;36(9):1279–88.

[8] Gubelmann C, Heinzer R, Haba-Rubio J, Vollenweider P, Marques-Vidal P. Physical activity is associated with higher sleep efficiency in the general population: the CoLaus study. Sleep 2018;41(7). zsy070.

[9] Mesas AE, Hagen EW, Peppard PE. The bidirectional association between physical activity and sleep in middle-aged and older adults: a prospective study based on polysomnography. Sleep 2018;41(9). zsy114.

[10] Gerber M, Brand S, Holsboer-Trachsler E, Puhse U. Fitness and exercise as correlates of sleep complaints: is it all in our minds? Med Sci Sports Exerc 2010;42(5):893–901.

[11] Burg MM, Schwartz JE, Kronish IM, Diaz KM, Alcantara C, Duer-Hefele J, Davidson KW. Does stress result in you exercising less? Or does exercising result in you being less stressed? Or is it both? Testing the bi-directional stress-exercise association at the group and person (N of 1) level. Ann Behav Med 2017;51(6):799–809.

[12] Strine TW, Chapman DP. Associations of frequent sleep insufficiency with health-related quality of life and health behaviors. Sleep Med 2005;6(1):23–7.

[13] Chesson Jr. AL, Littner M, Davila D, Anderson WM, Grigg-Damberger M, Hartse K, Johnson S, Wise M. Practice parameters for the use of light therapy in the treatment of sleep disorders. Standards of Practice Committee, American Academy of Sleep Medicine. Sleep 1999;22(5):641–60.

[14] Alvaro PK, Roberts RM, Harris JK. A systematic review assessing bidirectionality between sleep disturbances, anxiety, and depression. Sleep 2013;36(7):1059–68.

[15] Youngstedt SD. Ceiling and floor effects in sleep research. Sleep Med Rev 2003;7(4):351–65.

[16] Kredlow MA, Capozzoli MC, Hearon BA, Calkins AW, Otto MW. The effects of physical activity on sleep: a meta-analytic review. J Behav Med 2015;38(3):427–49.

[17] National Sleep Foundation. What is sleep hygiene? Available from https://www.sleepfoundation.org/sleep-topics/sleep-hygiene; (Accessed August 20, 2018).

[18] Oda S, Shirakawa K. Sleep onset is disrupted following pre-sleep exercise that causes large physiological excitement at bedtime. Eur J Appl Physiol 2014;114(9):1789–99.

[19] Buman MP, Phillips BA, Youngstedt SD, Kline CE, Hirshkowitz M. Does nighttime exercise really disturb sleep? Results from the 2013 National Sleep Foundation Sleep in America Poll. Sleep Med 2014;15(7):755–61.

[20] Myllymaki T, Kyrolainen H, Savolainen K, Hokka L, Jakonen R, Juuti T, Martinmaki K, Kaartinen J, Kinnunen ML, Rusko H. Effects of vigorous late-night exercise on sleep quality and cardiac autonomic activity. J Sleep Res 2011;20(1 Pt 2):146–53.

[21] Flausino NH, Da Silva Prado JM, de Queiroz SS, Tufik S, de Mello MT. Physical exercise performed before bedtime improves the sleep pattern of healthy young good sleepers. Psychophysiology 2012;49(2):186–92.

[22] Brand S, Kalak N, Gerber M, Kirov R, Puhse U, Holsboer-Trachsler E. High self-perceived exercise exertion before bedtime is associated with greater objectively assessed sleep efficiency. Sleep Med 2014;15(9):1031–6.

[23] Youngstedt SD. Effects of exercise on sleep. Clin Sports Med 2005;24(2):355–65.

[24] Rubio-Arias JA, Marin-Cascales E, Ramos-Campo DJ, Hernandez AV, Perez-Lopez FR. Effect of exercise on sleep quality and insomnia in middle-aged women: a systematic review and meta-analysis of randomized controlled trials. Maturitas 2017;100:49–56.

[25] Yang PY, Ho KH, Chen HC, Chien MY. Exercise training improves sleep quality in middle-aged and older adults with sleep problems: a systematic review. J Phys 2012;58(3):157–63.

[26] Varrasse M, Li J, Gooneratne N. Exercise and sleep in community-dwelling older adults. Curr Sleep Med Rep 2015;1(4):232–40.

[27] Kovacevic A, Mavros Y, Heisz JJ, Fiatarone Singh MA. The effect of resistance exercise on sleep: a systematic review of randomized controlled trials. Sleep Med Rev 2018;39:52–68.

[28] Bertani RF, Campos GO, Perseguin DM, Bonardi J, Ferriolli E, Moriguti JC, Lima NKC. Resistance exercise training is more effective than interval aerobic training in reducing blood pressure during sleep in hypertensive elderly patients. J Strength Cond Res 2017;32(7):2085–90.

[29] Wang F, Eun-Kyoung Lee O, Feng F, Vitiello MV, Wang W, Benson H, Fricchione GL, Denninger JW. The effect of meditative movement on sleep quality: a systematic review. Sleep Med Rev 2016;30:43–52.

[30] Kline CE, Sui X, Hall MH, Youngstedt SD, Blair SN, Earnest CP, Church TS. Dose-response effects of exercise training on the subjective sleep quality of postmenopausal women: exploratory analyses of a randomised controlled trial. BMJ Open 2012;2(4):e001044.

[31] Singh NA, Stavrinos TM, Scarbek Y, Galambos G, Liber C, Fiatarone Singh MA. A randomized controlled trial of high versus low intensity weight training versus general practitioner care for clinical depression in older adults. J Gerontol A Biol Sci Med Sci 2005;60(6):768–76.

[32] Tworoger SS, Yasui Y, Vitiello MV, Schwartz RS, Ulrich CM, Aiello EJ, Irwin ML, Bowen D, Potter JD, McTiernan A. Effects of a yearlong moderate-intensity exercise and a stretching intervention on sleep quality in postmenopausal women. Sleep 2003;26(7):830–6.

[33] Tremblay MS, Aubert S, Barnes JD, Saunders TJ, Carson V, Latimer-Cheung AE, Chastin SFM, Altenburg TM, Chinapaw MJM. Sedentary Behavior Research Network (SBRN)—Terminology Consensus Project process and outcome. Int J Behav Nutr Phys Act 2017;14(1):75.

[34] Yang Y, Shin JC, Li D, An R. Sedentary behavior and sleep problems: a systematic review and meta-analysis. Int J Behav Med 2017;24(4):481–92.

[35] Loprinzi PD, Nalley C, Selk A. Objectively-measured sedentary behavior with sleep duration and daytime sleepiness among US Adults. J Behav Health 2014;3(2):141–4.

[36] Vancampfort D, Stubbs B, Firth J, Hagemann N, Myin-Germeys I, Rintala A, Probst M, Veronese N, Koyanagi A. Sedentary behaviour and sleep problems among 42,489 community-dwelling adults in six low- and middle-income countries. J Sleep Res 2018:e12714.

[37] Buman MP, Kline CE, Youngstedt SD, Phillips B, Tulio de Mello M, Hirshkowitz M. Sitting and television viewing: novel risk factors for sleep disturbance and apnea risk? results from the 2013 National Sleep Foundation Sleep in America Poll. Chest 2015;147(3):728–34.

[38] Kline CE, Krafty RT, Mulukutla S, Hall MH. Associations of sedentary time and moderate-vigorous physical activity with sleep-disordered breathing and polysomnographic sleep in community-dwelling adults. Sleep Breath 2017;21(2):427–34.

[39] Redolfi S, Yumino D, Ruttanaumpawan P, Yau B, Su MC, Lam J, Bradley TD. Relationship between overnight rostral fluid shift and obstructive sleep apnea in nonobese men. Am J Respir Crit Care Med 2009;179(3):241–6.

[40] Vincent GE, Jay SM, Sargent C, Kovac K, Lastella M, Vandelanotte C, Ridgers ND, Ferguson SA. Does breaking up prolonged sitting when sleep restricted affect postprandial glucose responses and subsequent sleep architecture?—a pilot study. Chronobiol Int 2018;35(6):821–6.

[41] Kline CE, Kowalsky RJ, Perdomo SJ, Gibbs BB. Use of a sit-stand desk reduces wake time during the subsequent night's sleep. Med Sci Sports Exerc 2017;49(5S):S640.

[42] Ensari I, Greenlee TA, Motl RW, Petruzzello SJ. Meta-analysis of acute exercise effects on state anxiety: an update of randomized controlled trials over the past 25 years. Depress Anxiety 2015;32(8):624–34.

[43] Stubbs B, Vancampfort D, Rosenbaum S, Firth J, Cosco T, Veronese N, Salum GA, Schuch FB. An examination of the anxiolytic effects of exercise for people with anxiety and stress-related disorders: a meta-analysis. Psychiatry Res 2017;249:102–8.

[44] Blumenthal JA, Babyak MA, Doraiswamy PM, Watkins L, Hoffman BM, Barbour KA, Herman S, Craighead WE, Brosse AL, Waugh R, Hinderliter A, Sherwood A. Exercise and pharmacotherapy in the treatment of major depressive disorder. Psychosom Med 2007;69(7):587–96.

[45] Kandola A, Vancampfort D, Herring M, Rebar A, Hallgren M, Firth J, Stubbs B. Moving to beat anxiety: epidemiology and therapeutic issues with physical activity for anxiety. Curr Psychiatry Rep 2018;20(8):63.

[46] Hallgren M, Herring MP, Owen N, Dunstan D, Ekblom O, Helgadottir B, Nakitanda OA, Forsell Y. Exercise, physical activity, and sedentary behavior in the treatment of depression: broadening the scientific perspectives and clinical opportunities. Front Psychiatry 2016;7:36.

[47] Morgenthaler TI, Lee-Chiong T, Alessi C, Friedman L, Aurora RN, Boehlecke B, Brown T, Chesson Jr. AL, Kapur V, Maganti R, Owens J, Pancer J, Swick TJ, Zak R. Practice parameters for the clinical evaluation and treatment of circadian rhythm sleep disorders: an American Academy of Sleep Medicine report. Sleep 2007;30(11):1445–59.

[48] Atkinson G, Edwards B, Reilly T, Waterhouse J. Exercise as a synchroniser of human circadian rhythms: an update and discussion of the methodological problems. Eur J Appl Physiol 2007;99(4):331–41.

[49] Schroder EA, Esser KA. Circadian rhythms, skeletal muscle molecular clocks, and exercise. Exerc Sport Sci Rev 2013;41(4):224–9.

[50] Takahashi M, Haraguchi A, Tahara Y, Aoki N, Fukazawa M, Tanisawa K, Ito T, Nakaoka T, Higuchi M, Shibata S. Positive association between physical activity and PER3 expression in older adults. Sci Rep 2017;7:39771.

[51] Tranel HR, Schroder EA, England J, Black WS, Bush H, Hughes ME, Esser KA, Clasey JL. Physical activity, and not fat mass is a primary predictor of circadian parameters in young men. Chronobiol Int 2015;32(6):832–41.

[52] Raymann RJ, Swaab DF, Van Someren EJ. Skin temperature and sleep-onset latency: changes with age and insomnia. Physiol Behav 2007;90(2–3):257–66.

[53] Horne JA, Staff LH. Exercise and sleep: body-heating effects. Sleep 1983;6(1):36–46.

[54] Krauchi K, Cajochen C, Werth E, Wirz-Justice A. Functional link between distal vasodilation and sleep-onset latency? Am J Phys Regul Integr Comp Phys 2000;278(3):R741–8.

[55] Simmons GH, Wong BJ, Holowatz LA, Kenney WL. Changes in the control of skin blood flow with exercise training: where do cutaneous vascular adaptations fit in? Exp Physiol 2011;96(9):822–8.

[56] Landolt HP. Sleep homeostasis: a role for adenosine in humans? Biochem Pharmacol 2008;75(11):2070–9.

[57] Dworak M, Diel P, Voss S, Hollmann W, Struder HK. Intense exercise increases adenosine concentrations in rat brain: implications for a homeostatic sleep drive. Neuroscience 2007;150(4):789–95.

[58] Youngstedt SD, O'Connor PJ, Crabbe JB, Dishman RK. The influence of acute exercise on sleep following high caffeine intake. Physiol Behav 2000;68(4):563–70.

[59] Chung KF, Yeung WF, Ho FY, Yung KP, Yu YM, Kwok CW. Cross-cultural and comparative epidemiology of insomnia: the Diagnostic and Statistical Manual (DSM), International Classification of Diseases (ICD) and International Classification of Sleep Disorders (ICSD). Sleep Med 2015;16(4):477–82.

[60] Wilt TJ, MacDonald R, Brasure M, Olson CM, Carlyle M, Fuchs E, Khawaja IS, Diem S, Koffel E, Ouellette J, Butler M, Kane RL. Pharmacologic treatment of insomnia disorder: an evidence report for a clinical practice guideline by the American College of Physicians. Ann Intern Med 2016;165(2):103–12.

[61] Thomas A, Grandner M, Nowakowski S, Nesom G, Corbitt C, Perlis ML. Where are the behavioral sleep medicine providers and where are they needed? A geographic assessment. Behav Sleep Med 2016;14(6):687–98.

[62] Qaseem A, Kansagara D, Forciea MA, Cooke M, Denberg TD. Management of chronic insomnia disorder in adults: a clinical practice guideline from the American College of Physicians. Ann Intern Med 2016;165(2):125–33.

[63] Passos GS, Poyares DL, Santana MG, Tufik S, Mello MT. Is exercise an alternative treatment for chronic insomnia? Clinics 2012;67(6):653–60.

[64] Sporndly-Nees S, Asenlof P, Lindberg E. High or increasing levels of physical activity protect women from future insomnia. Sleep Med 2017;32:22–7.

[65] Inoue S, Yorifuji T, Sugiyama M, Ohta T, Ishikawa-Takata K, Doi H. Does habitual physical activity prevent insomnia? A cross-sectional and longitudinal study of elderly Japanese. J Aging Phys Act 2013;21(2):119–39.

[66] Passos GS, Poyares D, Santana MG, Garbuio SA, Tufik S, Mello MT. Effect of acute physical exercise on patients with chronic primary insomnia. J Clin Sleep Med 2010;6(3):270–5.

[67] Morita Y, Sasai-Sakuma T, Inoue Y. Effects of acute morning and evening exercise on subjective and objective sleep quality in older individuals with insomnia. Sleep Med 2017;34:200–8.

[68] Reid KJ, Baron KG, Lu B, Naylor E, Wolfe L, Zee PC. Aerobic exercise improves self-reported sleep and quality of life in older adults with insomnia. Sleep Med 2010;11(9):934–40.

[69] Hartescu I, Morgan K, Stevinson CD. Increased physical activity improves sleep and mood outcomes in inactive people with insomnia: a randomized controlled trial. J Sleep Res 2015;24(5):526–34.

[70] Yeung WF, Lai AY, Ho FY, Suen LK, Chung KF, Ho JY, Ho LM, Yu BY, Chan LY, Lam TH. Effects of zero-time exercise on inactive adults with insomnia disorder: a pilot randomized controlled trial. Sleep Med 2018;52:118–27.

[71] Sawyer AM, Gooneratne NS, Marcus CL, Ofer D, Richards KC, Weaver TE. A systematic review of CPAP adherence across age groups: clinical and empiric insights for developing CPAP adherence interventions. Sleep Med Rev 2011;15(6):343–56.

[72] Ramar K, Dort LC, Katz SG, Lettieri CJ, Harrod CG, Thomas SM, Chervin RD. Clinical practice guideline for the treatment of obstructive sleep apnea and snoring with oral appliance therapy: an update for 2015. J Clin Sleep Med 2015;11(7):773–827.

[73] da Silva RP, Martinez D, Pedroso MM, Righi CG, Martins EF, Silva LM, Lenz MD, Fiori CZ. Exercise, occupational activity, and risk of sleep apnea: a cross-sectional study. J Clin Sleep Med 2017;13(2):197–204.

[74] Murillo R, Reid KJ, Arredondo EM, Cai J, Gellman MD, Gotman NM, Marquez DX, Penedo FJ, Ramos AR, Zee PC, Daviglus ML. Association of self-reported physical activity with obstructive sleep apnea: results from the Hispanic Community Health Study/Study of Latinos (HCHS/SOL). Prev Med 2016;93:183–8.

[75] Simpson L, McArdle N, Eastwood PR, Ward KL, Cooper MN, Wilson AC, Hillman DR, Palmer LJ, Mukherjee S. Physical inactivity is associated with moderate-severe obstructive sleep apnea. J Clin Sleep Med 2015;11(10):1091–9.

[76] Peppard PE, Young T. Exercise and sleep-disordered breathing: an association independent of body habitus. Sleep 2004;27(3):480–4.

[77] Monico-Neto M, Moreira Antunes HK, RVT DS, V D'A, Alves Lino de Souza A, Azeredo Bittencourt LR, Tufik S. Physical activity as a moderator for obstructive sleep apnoea and cardiometabolic risk in the EPISONO study. Eur Respir J 2018;52(4):1701972.

[78] Mendelson M, Bailly S, Marillier M, Flore P, Borel JC, Vivodtzev I, Doutreleau S, Verges S, Tamisier R, Pepin JL. Obstructive sleep apnea syndrome, objectively measured physical activity and exercise training interventions: a systematic review and meta-analysis. Front Neurol 2018;9:73.

[79] Iftikhar IH, Kline CE, Youngstedt SD. Effects of exercise training on sleep apnea: a meta-analysis. Lung 2014;192(1):175–84.

[80] Kline CE, Ewing GB, Burch JB, Blair SN, Durstine JL, Davis JM, Youngstedt SD. Exercise training improves selected aspects of daytime functioning in adults with obstructive sleep apnea. J Clin Sleep Med 2012;8(4):357–65.

[81] Servantes DM, Javaheri S, Kravchychyn ACP, Storti LJ, Almeida DR, de Mello MT, Cintra FD, Tufik S, Bittencourt L. Effects of exercise training and CPAP in patients with heart failure and OSA: a preliminary study. Chest 2018;154(4):808–17.

[82] Ackel-D'Elia C, da Silva AC, Silva RS, Truksinas E, Sousa BS, Tufik S, de Mello MT, Bittencourt LR. Effects of exercise training associated with continuous positive airway pressure treatment in patients with obstructive sleep apnea syndrome. Sleep Breath 2012;16(3):723–35.

[83] Jakicic JM, Otto AD. Treatment and prevention of obesity: what is the role of exercise? Nutr Rev 2006;64(2 Pt 2):S57–61.

[84] Vincent HK, Shanely RA, Stewart DJ, Demirel HA, Hamilton KL, Ray AD, Michlin C, Farkas GA, Powers SK. Adaptation of upper airway muscles to chronic endurance exercise. Am J Respir Crit Care Med 2002;166(3):287–93.

[85] Series F, Roy N, Marc I. Effects of sleep deprivation and sleep fragmentation on upper airway collapsibility in normal subjects. Am J Respir Crit Care Med 1994;150(2):481–5.

[86] Kline CE. Exercise: shifting fluid and sleep apnoea away. Eur Respir J 2016;48(1):23–5.

[87] Redolfi S, Bettinzoli M, Venturoli N, Ravanelli M, Pedroni L, Taranto-Montemurro L, Arnulf I, Similowski T, Tantucci C. Attenuation of obstructive sleep apnea and overnight rostral fluid shift by physical activity. Am J Respir Crit Care Med 2015;191(7):856–8.

[88] Mendelson M, Lyons OD, Yadollahi A, Inami T, Oh P, Bradley TD. Effects of exercise training on sleep apnoea in patients with coronary artery disease: a randomised trial. Eur Respir J 2016;48(1):142–50.

[89] Yamamoto U, Mohri M, Shimada K, Origuchi H, Miyata K, Ito K, Abe K, Yamamoto H. Six-month aerobic exercise training ameliorates central sleep apnea in patients with chronic heart failure. J Card Fail 2007;13(10):825–9.

[90] Ueno LM, Drager LF, Rodrigues AC, Rondon MU, Braga AM, Mathias Jr. W, Krieger EM, Barretto AC, Middlekauff HR, Lorenzi-Filho G, Negrao CE. Effects of exercise training in patients with chronic heart failure and sleep apnea. Sleep 2009;32(5):637–47.

[91] Stiasny K, Oertel WH, Trenkwalder C. Clinical symptomatology and treatment of restless legs syndrome and periodic limb movement disorder. Sleep Med Rev 2002;6(4):253–65.

[92] Pigeon WR, Yurcheshen M. Behavioral sleep medicine interventions for restless legs syndrome and periodic limb movement disorder. Sleep Med Clin 2009;4(4):487–94.

[93] Batool-Anwar S, Li Y, De Vito K, Malhotra A, Winkelman J, Gao X. Lifestyle factors and risk of restless legs syndrome: prospective cohort study. J Clin Sleep Med 2016;12(2):187–94.

[94] Cederberg KL, Motl RW, Burnham TR. Magnitude and duration of acute-exercise intensity effects on symptoms of restless legs syndrome: a pilot study. Sleep Biol Rhythms 2018;16(3):337–44.

[95] Esteves AM, de Mello MT, Benedito-Silva AA, Tufik S. Impact of aerobic physical exercise on restless legs syndrome. Sleep Sci 2011;4(2):45–8.

[96] Aukerman MM, Aukerman D, Bayard M, Tudiver F, Thorp L, Bailey B. Exercise and restless legs syndrome: a randomized controlled trial. J Am Board Fam Med 2006;19(5):487–93.

[97] Esteves AM, de Mello MT, Pradella-Hallinan M, Tufik S. Effect of acute and chronic physical exercise on patients with periodic leg movements. Med Sci Sports Exerc 2009;41(1):237–42.

[98] Kline CE. The bidirectional relationship between exercise and sleep: implications for exercise adherence and sleep improvement. Am J Lifestyle Med 2014;8(6):375–9.

[99] Haario P, Rahkonen O, Laaksonen M, Lahelma E, Lallukka T. Bidirectional associations between insomnia symptoms and unhealthy behaviours. J Sleep Res 2013;22(1):89–95.

[100] Holfeld B, Ruthig JC. A longitudinal examination of sleep quality and physical activity in older adults. J Appl Gerontol 2014;33(7):791–807.

[101] Best JR, Falck RS, Landry GJ, Liu-Ambrose T. Analysis of dynamic, bidirectional associations in older adult physical activity and sleep quality. J Sleep Res 2018;e12769.

[102] Dzierzewski JM, Buman MP, Giacobbi Jr. PR, Roberts BL, Aiken-Morgan AT, Marsiske M, McCrae CS. Exercise and sleep in community-dwelling older adults: evidence for a reciprocal relationship. J Sleep Res 2014;23(1):61–8.

[103] Baron KG, Reid KJ, Zee PC. Exercise to improve sleep in insomnia: exploration of the bidirectional effects. J Clin Sleep Med 2013;9(8):819–24.

[104] Mitchell JA, Godbole S, Moran K, Murray K, James P, Laden F, Hipp JA, Kerr J, Glanz K. No evidence of reciprocal associations between daily sleep and physical activity. Med Sci Sports Exerc 2016;48(10):1950–6.

[105] Irish LA, Kline CE, Rothenberger SD, Krafty RT, Buysse DJ, Kravitz HM, Bromberger JT, Zheng H, Hall MH. A 24-hour approach to the study of health behaviors: temporal relationships between waking health behaviors and sleep. Ann Behav Med 2014;47(2):189–97.

[106] Lambiase MJ, Gabriel KP, Kuller LH, Matthews KA. Temporal relationships between physical activity and sleep in older women. Med Sci Sports Exerc 2013;45(12):2362–8.

[107] Tang NK, Sanborn AN. Better quality sleep promotes daytime physical activity in patients with chronic pain? A multilevel analysis of the within-person relationship. PLoS ONE 2014;9(3):e92158.

[108] Fortier MS, Guerin E, Williams T, Strachan S. Should I exercise or sleep to feel better? A daily analysis with physically active working mothers. Ment Health Phys Act 2015;8:56–61.

[109] Booth JN, Bromley LE, Darukhanavala AP, Whitmore HR, Imperial JG, Penev PD. Reduced physical activity in adults at risk for type 2 diabetes who curtail their sleep. Obesity 2012;20(2):278–84.

[110] Shechter A, St-Onge MP. Delayed sleep timing is associated with low levels of free-living physical activity in normal sleeping adults. Sleep Med 2014;15(12):1586–9.

[111] Hargens T.A., Martin R.A., Strosnider C.L., Giersch GEW, Womack C.J. Obstructive sleep apnea negatively impacts objectively measured physical activity. Sleep Breath [in press]. https://doi.org/10.1007/s11325-018-1700-0.

[112] Kline CE, Irish LA, Buysse DJ, Kravitz HM, Okun ML, Owens JF, Hall MH. Sleep hygiene behaviors among midlife women with insomnia or sleep-disordered breathing: the SWAN Sleep Study. J Women's Health 2014;23(11):894–903.

[113] Bromley LE, Booth III JN, Kilkus JM, Imperial JG, Penev PD. Sleep restriction decreases the physical activity of adults at risk for type 2 diabetes. Sleep 2012;35(7):977–84.

[114] Tajiri E, Yoshimura E, Hatamoto Y, Tanaka H, Shimoda S. Effect of sleep curtailment on dietary behavior and physical activity: a randomized crossover trial. Physiol Behav 2018;184:60–7.

[115] Jean RE, Duttuluri M, Gibson CD, Mir S, Fuhrmann K, Eden E, Supariwala A. Improvement in physical activity in persons with obstructive sleep apnea treated with continuous positive airway pressure. J Phys Act Health 2017;14(3):176–82.

[116] West SD, Kohler M, Nicoll DJ, Stradling JR. The effect of continuous positive airway pressure treatment on physical activity in patients with obstructive sleep apnoea: a randomised controlled trial. Sleep Med 2009;10(9):1056–8.

[117] Batool-Anwar S, Goodwin JL, Drescher AA, Baldwin CM, Simon RD, Smith TW, Quan SF. Impact of CPAP on activity patterns and diet in patients with obstructive sleep apnea (OSA). J Clin Sleep Med 2014;10(5):465–72.

[118] Rosenberger ME, Fulton JE, Buman MP, Troiano RP, Grandner MA, Buchner DM, Haskell WL. The 24-hour activity cycle: a new paradigm for physical activity. Med Sci Sports Exerc 2019;51(3):454–64.

[119] Smith K, Rosenberger ME. Keeping seniors active—a 24 hour approach. Stanford, CA: Stanford Center on Longevity; 2017. Available from http://longevity.stanford.edu/2017/08/11/keeping-seniors-active-a-24-hour-approach/.

[120] Grgic J, Dumuid D, Bengoechea EG, Shrestha N, Bauman A, Olds T, Pedisic Z. Health outcomes associated with reallocations of time between sleep, sedentary behaviour, and physical activity: a systematic scoping review of isotemporal substitution studies. Int J Behav Nutr Phys Act 2018;15(1):69.

[121] Stamatakis E, Rogers K, Ding D, Berrigan D, Chau J, Hamer M, Bauman A. All-cause mortality effects of replacing sedentary time with physical activity and sleeping using an isotemporal substitution model: a prospective study of 201,129 mid-aged and older adults. Int J Behav Nutr Phys Act 2015;12:121.

[122] Buman MP, Winkler EA, Kurka JM, Hekler EB, Baldwin CM, Owen N, Ainsworth BE, Healy GN, Gardiner PA. Reallocating time to sleep, sedentary behaviors, or active behaviors: associations with cardiovascular disease risk biomarkers, NHANES 2005-2006. Am J Epidemiol 2014;179(3):323–34.

[123] Biddle GJH, Edwardson CL, Henson J, Davies MJ, Khunti K, Rowlands AV, Yates T. Associations of physical behaviours and behavioural reallocations with markers of metabolic health: a compositional data analysis. Int J Environ Res Public Health 2018;15(10):E2280.

[124] Dumuid D, Lewis LK, Olds TS, Maher C, Bondarenko C, Norton L. Relationships between older adults' use of time and cardiorespiratory fitness, obesity and cardio-metabolic risk: a compositional isotemporal substitution analysis. Maturitas 2018;110:104–10.

[125] Buysse DJ. Sleep health: can we define it? Does it matter? Sleep 2014;37(1):9–17.

[126] Wilckens KA, Ferrarelli F, Walker MP, Buysse DJ. Slow-wave activity enhancement to improve cognition. Trends Neurosci 2018;41(7):470–82.

[127] Kanagasabai T, Riddell MC, Ardern CI. Physical activity contributes to several sleep-cardiometabolic health relationships. Metab Syndr Relat Disord 2017;15(1):44–51.

[128] Seixas AA, Vallon J, Barnes-Grant A, Butler M, Langford AT, Grandner MA, Schneeberger AR, Huthchinson J, Zizi F, Jean-Louis G. Mediating effects of body mass index, physical activity, and emotional distress on the relationship between short sleep and cardiovascular disease. Medicine 2018;97(37):e11939.

[129] Dobrosielski DA, Patil S, Schwartz AR, Bandeen-Roche K, Stewart KJ. Effects of exercise and weight loss in older adults with obstructive sleep apnea. Med Sci Sports Exerc 2015;47(1):20–6.

[130] Seixas AA, Henclewood DA, Williams SK, Jagannathan R, Ramos A, Zizi F, Jean-Louis G. Sleep duration and physical activity profiles associated with self-reported stroke in the United States: application of Bayesian belief network modeling techniques. Front Neurol 2018;9:534.

[131] Wennman H, Kronholm E, Partonen T, Tolvanen A, Peltonen M, Vasankari T, Borodulin K. Interrelationships of physical activity and sleep with cardiovascular risk factors: a person-oriented approach. Int J Behav Med 2015;22(6):735–47.

[132] Wennman H, Kronholm E, Heinonen OJ, Kujala UM, Kaprio J, Partonen T, Bäckmand H, Sarna S, Borodulin K. Leisure time physical activity and sleep predict mortality in men irrespective of background in competitive sports. Prog Prev Med 2017;2(6):e0009.

[133] Wang J, Yin G, Li G, Liang W, Wei Q. Efficacy of physical activity counseling plus sleep restriction therapy on the patients with chronic insomnia. Neuropsychiatr Dis Treat 2015;11:2771–8.

[134] Buman MP, Epstein DR, Gutierrez M, Herb C, Hollingshead K, Huberty JL, Hekler EB, Vega-Lopez S, Ohri-Vachaspati P, Hekler AC, Baldwin CM. BeWell24: development and process evaluation of a smartphone "app" to improve sleep, sedentary, and active behaviors in US Veterans with increased metabolic risk. Transl Behav Med 2016;6(3):438–48.

[135] Rayward AT, Murawski B, Plotnikoff RC, Vandelanotte C, Brown WJ, Holliday EG, Duncan MJ. A randomised controlled trial to test the efficacy of an m-health delivered physical activity and sleep intervention to improve sleep quality in middle-aged adults: the Refresh Study protocol. Contemp Clin Trials 2018;73:36–50.

Chapter 21

Sleep and alcohol use

Sean He[a,b], Brittany V. Taylor[a,b], Nina P. Thakur[a,b], Subhajit Chakravorty[a,c]

[a]Cpl. Michael J Crescenz VA Medical Center, Philadelphia, PA, United States, [b]School of Arts and Sciences, University of Pennsylvania, Philadelphia, PA, United States, [c]Perelman School of Medicine, Philadelphia, PA, United States

INTRODUCTION

Alcohol is one of the most commonly used psychoactive substances in the community, with 56% of individuals 18 years or older reporting alcohol use in the past month [1]. Its use has been linked to a wide range of sleep-related disorders and their associated daytime consequences such as daytime sedation and tiredness. These daytime ramifications may affect the individual's ability to function satisfactorily.

Our understanding of alcohol's effect on sleep has improved dramatically over the last few decades because of advances in translational and clinical research at this interface. Alcohol may help an individual to fall asleep relatively quickly, but its deleterious effects involve disruption of the underlying sleep mechanisms, interference with the normal circadian mechanism, aggravating snoring and breathing-related sleep abnormalities and triggering movements of the limbs during sleep. In this chapter, we examine the link between alcohol and a range of sleep-related disorders.

NEUROBIOLOGY OF ALCOHOL USE

A vast majority of metabolism of alcohol occurs in the liver via the alcohol dehydrogenase pathway [2]. A fraction of the alcohol consumed is metabolized in the brain through alternate mechanisms such as, the catalase (in cellular peroxisomes) and cytochrome P450 pathways [3]. In the liver, alcohol (ethanol) is metabolized to acetaldehyde by an enzyme, alcohol dehydrogenase, Fig. 21.1. This acetaldehyde is then converted to acetate by the enzyme, aldehyde dehydrogenase. Acetate is finally converted to acetyl-CoA, which then enters the tricarboxylic acid (TCA) cycle for its final metabolism. This metabolism of alcohol decreases the blood alcohol level in a time dependent fashion, Fig. 21.2. In other words, the blood alcohol level is directly related to the number of drinks consumed and the time elapsed since the last drink was ingested. Acetaldehyde, the direct metabolite of alcohol, is also the most toxic byproduct of this pathway. In animal studies, acetaldehyde has been shown to reduce REM sleep and CNS serotonin levels, which may contribute to the sleep disturbance observed in alcohol use [4].

INSOMNIA AND ALCOHOL USE

Introduction. Insomnia is probably the most investigated sleep disorder, although some studies in the literature have evaluated insomnia symptoms only, instead of it as a disorder. Insomnia disorder, as defined by the ICSD-3, necessitates the presence of one or more of the following complaints: difficulty initiating sleep, difficulty maintaining sleep, or waking up earlier than desired. These symptoms are associated with at least one of the following impairments: fatigue or malaise, attention or memory problems, impairment of psychosocial functioning, mood disturbance, daytime sleepiness, behavioral problems, reduced motivation or energy, proneness for errors, and concern or dissatisfaction with sleep. These complaints must occur despite an adequate opportunity and circumstance for sleep which is present for most nights of the week for 3 months or more [5].

The above definition of insomnia is different from the DSM-IV-TR and from that of ICSD-2 for several reasons. First, primary insomnia and secondary insomnia terminologies have been replaced by insomnia, which is now viewed as a disorder, with symptoms and burden related to these symptoms, independent of primary or comorbid status. Second, a new insomnia symptom frequency criterion has been inserted. Third, the duration criterion for chronicity has been extended from 1 to 3 months. Finally, non-restorative sleep (or poor sleep quality) has been removed as a diagnostic criterion for insomnia.

Epidemiology of alcohol use or misuse and sleep-related problems. While many studies show that insomnia leads to alcohol consumption some studies have the contrary, that is, alcohol consumption leading to insomnia. But, there is no consensus on the causality of this relationship. Data from studies such as the Virginia Adult Twin Studies of

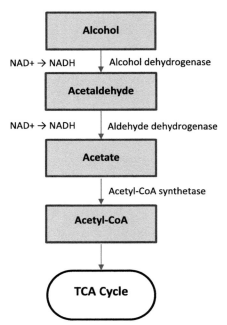

FIG. 21.1 The metabolism of alcohol *Adapted from Cederbaum AI. Alcohol metabolism. Clin Liver Dis 2012;16:667–685.*

Psychiatric and Substance Use Disorders and the Helsinki Health Study demonstrate the presence of a bi-directional effect. We will systematically approach this concept in the following sections. We will start by describing whether early life experiences or factors influence the relationship between insomnia and alcohol use. We will then briefly review the temporal trend in adults, whereby insomnia predicts drinking, followed by a brief explanation of this relationship in adults. We will conclude by elaborating on the risk factors for insomnia or sleep disturbance in the context of alcohol use.

Sleep problems and future alcohol use in children. A study of young boys from Western Pennsylvania ($N=145$) showed that poor sleep quality was associated with subsequent alcohol use as a young adult (hazard ratio = 1.09, 1.10, and 1.12) [6]. In adolescents, complaints of trouble sleeping as well as overtiredness have been associated with the onset of alcohol use in cross-sectional and longitudinal epidemiologic investigations [7–10]. But, one study failed to demonstrate a relationship between a history of alcohol use in the last 6 months of high school and adverse sleep patterns among first-year college students [11].

Sleep problems and alcohol use in adults. Data on the relationship between insomnia and sleep is relatively more robust in adults, with studies having been conducted across many countries. Among male Japanese industrial workers ($N=271$), insomnia symptoms were associated with alcohol consumption more than half the days of the week (OR = 2.6, 95% CI: 1.1–5.7) [12]. Data from the Epidemiologic Catchment Area study (ECA, $N=7954$) have shown that adults complaining of insomnia symptoms (10.2%) were at an increased risk for subsequently developing alcohol abuse disorder when evaluated a year later (odds ratio [OR] = 2.4, 95% CI: 1.0–6.1) [13]. In a subsequent analysis of this dataset, uncomplicated insomnia symptoms (and without a lifetime history of psychiatric disorder) was associated with an increased risk of developing alcohol abuse a year later, as

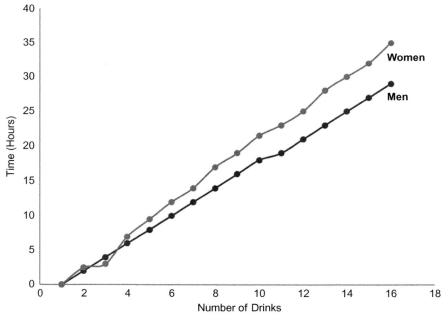

FIG. 21.2 The time until blood alcohol level returns to zero versus number of drinks. Legend: Based on a body weight of 120 lbs. *Adapted from data reported at http://www.selfcounseling.com/help/alcohol/hourstozerobac.html).*

compared to those without baseline insomnia or any psychiatric disorder (OR=2.3, 95% CI: 1.2–4.3) [14]. Similarly, in a longitudinal survey in Michigan, a lifetime history of insomnia symptoms was associated with an increased risk for incident diagnosis of alcohol dependence, 3.5 years later (OR=1.72, 95% CI: 0.85–3.52) [15]. In a clinical sample of subjects with comorbid alcohol dependence and insomnia (N=63), about half of the subjects (52%) reported having insomnia prior to their onset of alcohol dependence [16]. In summary, the above data demonstrate that sleep problems at baseline may be linked to subsequent alcohol use or alcohol use disorder.

Do persons with sleep problems gravitate towards alcohol? Some studies have demonstrated that individuals with insomnia may have a preference for alcoholic drinks over non-alcoholic ones. In a laboratory-based study, subjects with insomnia (N=11) compared to healthy sleepers (N=9) chose more doses of alcohol and on more nights [17]. Furthermore, those with insomnia may prefer alcohol to hypnotic medications. An epidemiologic investigation evaluated the use of alcohol (N=9226) and hypnotic medications (N=9021) as a sleep aid [18]. The results showed that the use of alcohol as a sleep aid was increased in a stepwise fashion up to the age groups of 55–59 years and 40–44 years, in men and women, respectively. Beyond these age categories, alcohol use decreased in a stepwise manner. Among the insomnia symptoms, use of alcohol as a sleep aid (more than once a week) was associated with a higher risk of having difficulty maintaining sleep in both genders and with difficulty initiating sleep only in women. Similar associations between insomnia and the use of alcohol as a hypnotic have been reported in other epidemiologic studies [19, 20]. A survey of primary care patients demonstrated that use of alcohol for sleep was positively associated with hazardous alcohol consumption (OR=4.53, 95% CI: 2.9–6.9) [21]. In sum, individuals with insomnia may prefer alcohol use, possibly to self-medicate their sleep problems.

Polysomnographic studies. High-frequency EEG activity in the beta and gamma range is increased in those with primary insomnia [22, 23]. The main polysomnographic sleep disturbances in active drinkers appear to be an increased sleep onset latency and increased time spent awake after initially falling asleep. These changes may decrease the time spent asleep [24].

Genetic studies. There is an emerging interest in evaluating associations between alcohol consumption and circadian clock genes as well as other genes. Neuropeptide Y-deficient mice had increased use of alcohol and are less sensitive to the sedative or hypnotic effects of ethanol and show rapid recovery from ethanol-induced sleep [25]. The AA genotype of the single nucleotide polymorphism 10,870 on the PER 2 gene has been associated with increased alcohol consumption in the presence of sleep problems only in adolescent males [26]. In a case-control study of subjects with alcohol abuse or dependence (N=512) versus controls (N=511), polymorphism of specific genes including some clock genes such as RNTL, ARNTL2, VIP, and ADCYAP1 was associated with alcohol consumption and alcohol abuse. No association of PER2 with alcohol consumption or alcohol abuse/dependence was seen in either gender [27]. In a study of subjects drinking alcohol (0.5 g/kg) daily for 7 days, PER2 levels were not affected. In a sample of non-obese Japanese adult males (N=29), the following associations with the clock genes were seen: BMAL1 with alcohol consumption; PER2 and serum GGT levels (GGT); CLOCK and GGT [28]. Thus, genetic polymorphisns linked to alcohol use may also be linked to sleep problems.

Clinical findings

Insomnia and alcohol use among different populations. Heavy drinking or binge drinking is defined as the consumption of ≥5 standard alcoholic beverages. This behavior has been associated with insomnia symptoms in multiple populations, including adolescents, young adults, college students, older adults, Veterans, and firefighters.

(A) *Adolescents.* The KiGGS cross-sectional survey in Germany involving youth between the ages 11–17 showed that use of alcohol >5 drinks a week was associated with a twofold increased risk of insomnia complaints in both males and females. Moreover, in females only, the risk of insomnia complaints was exacerbated with high alcohol use (OR: 5.14, 95% CI: 2.89–9.13). In women, even low alcohol use (0–1 drinks/week) showed an increased risk for insomnia complaints (OR: 1.95, 95% CI: 1.42–2.66). But, after adjusting for caffeine use, the predictive validity of alcohol on insomnia complaints was attenuated, suggesting an over-estimation of alcohol's effect size on insomnia [29]. In other studies among adolescents aged 14–20 in Michigan presenting to the ED, unhealthy alcohol use and consuming alcohol to fall sleep was associated with sleep problems (bivariate correlation of 0.11 and 0.16, respectively) [30]. Finally, among adolescents aged 12–21.9, alcohol use was associated with worse sleep quality [31].

(B) *College students.* Among heavy drinking college students in New England. a reduction in weekly alcohol drinking was associated with an improvement in sleep quality [32].

(C) *Adults.* In an epidemiologic investigation, young adults, a positive relationship between binge drinking and insomnia symptoms was seen. This effect of alcohol on sleep increased in magnitude with increased frequency of binge drinking [33]. In a nationally representative sample of Australian women between the ages of 25–30 years, binge drinking was associated with difficulty sleeping, in a model adjusted for covariates [34]. In the Virginia adult twin studies of psychiatric and substance use disorders ($N=7500$), there was an 18% genetic overlap between insomnia and AD (alcohol abuse and dependence) [35]. This genetic overlap further suggests that insomnia and alcohol misuse are genetically related and share similar etiological contributors, supporting the bi-directional nature of insomnia and alcohol use.

Although binge drinking has been associated with insomnia symptoms in multiple populations, the causality is unclear from a multitude of these cross-sectional studies. Haario and colleagues recently evaluated the relationship between insomnia and alcohol prospectively over 6 years, using longitudinal data from the Helsinki Health Study, a cohort of 40–60-year-old employees of the City of Helsinki ($N=8960$) [36]. Their results showed that in adjusted models, frequent insomnia symptoms at baseline were associated with subsequent heavy drinking (OR=1.34). Conversely, heavy drinking at baseline was associated with future insomnia symptoms (OR=1.48). This study added to the evidence of a bi-directional effect of alcohol use and insomnia.

(D) *Older Adults.* Canham and colleagues evaluated relationship between insomnia and binge drinking (≥4 drinks per session) in adults 50 years or older in age. Their results demonstrated that older adults who binged on two or more days per week had 64% greater odds of complaining about insomnia than non-binge drinkers. The odds of insomnia decreased to 35% in those with two or more binge drinking days per week as compared to non-binge drinkers. However, once smoking was added to the model, both these relationships fell just below the level of significance [37]. In another study involving elderly Chinese respondents ($N=3176$), frequent drinking (two or more times per week) was associated with increased insomnia in men ($P=0.04$). Conversely, occasional drinking in men was associated with lower odds of insomnia (OR=0.59, 95% CI=0.40–0.86) [38].

(E) *Veterans and firefighters.* Veterans are at a higher risk for insomnia. Among active US Army members ($N=4101$) before military deployment, alcohol use disorder was linked to one and a half fold increased risk of insomnia [39]. Cucciare and colleagues also demonstrated that binge drinking was linked to sleep-related complaints involving nightmares and sleep continuity disturbances [40]. Among firefighters ($N=112$), Casey and colleagues demonstrated that sleep disturbance was prevalent in 59%, whereas binge drinking was prevalent in 59% of respondents [41]. It is unclear how many of these firefighters had post-traumatic stress disorder, a condition that may be linked to a higher prevalence of insomnia in those responding to traumatic events in the community.

(F) *Pregnancy*—In a study involving Japanese women, those drinking during pregnancy were demonstrated to have higher odds of difficulty initiating and maintaining sleep, as well as early morning awakening [42].

(G) *Perimenopausal state.* Although women in the perimenopausal phase of life have been shown to have a higher prevalence of insomnia and psychiatric disorders, very little is known about the association of insomnia with alcohol consumption in this population. In a recent study, Blumel and colleagues evaluated sleep in middle-aged women across 11 Latin American countries. They found that 41.7% of perimenopausal women reported insomnia on the Athens Insomnia Scale. The most prevalent insomnia symptom was an "awakening during the night." [43].

(H) *Alcohol-dependent individuals.* A growing body of literature has demonstrated a higher prevalence of insomnia or sleep disturbance in alcohol dependent individuals as compared to those in the community [44].

Insomnia in alcohol dependence

Insomnia in alcohol dependence is estimated to be between 36% and 91% [24, 45]. Consequences of insomnia in the alcohol-dependent population include continued relapse to drinking, lower quality of life, and lower sleep duration, which if sleep is less than 6 h, is known to have a variety of physical and psychosocial health effects. Below are findings related to insomnia from all stages of alcohol dependence—active drinking, withdrawal, early recovery, and sustained recovery.

1. *Active alcohol dependence.* This population is estimated to have insomnia approximately 74% using the Insomnia Severity Index and 76% using the Pittsburgh Sleep Quality Index (PSQI). The main sleep continuity disturbances in active drinkers include increased SOL, shorter REM sleep duration, and increased WASO, which consequently decreased the TST.
2. *Alcohol dependence in acute withdrawal.* It is estimated that as many as 92% of patients in withdrawal from alcohol dependence have sleep disturbance. As alcohol withdrawal ends, insomnia may improve in some but persists in others [24]. The cause of this heterogeneity is currently unclear.
3. *Alcohol dependent individuals during early recovery.* About 65% of alcohol dependent patients report insomnia in early recovery, 4–8 weeks after cessation from alcohol use. The symptoms consist of greater Sleep Onset Latency (SOL), greater Wake After Sleep Onset time (WASO), and lower Total Sleep Time (TST) are found in alcoholics in early recovery [24].

4. *Alcohol dependence individuals during sustained recovery.* We define sustained recovery as abstinence from alcohol lasting longer than 3 months. The sleep related abnormalities include greater SOL, lower TST, and lower Rapid Eye Movement sleep (REM) irregularities. SOL is the first to reach normality by around 9 months in recovery, whereas sleep fragmentation may persist up to 21 months. Napping during the day in alcoholics may contribute to this effect, as napping increases WASO and decreases TST and SE [24].

Treatments

Does treatment of risky drinking improve insomnia? Treatment of drinking does appear to improve sleep in several studies. Berman and colleagues evaluated a modified version of screening and online personalized feedback in 633 individuals with problematic drinking. Their results demonstrated that 36% of the participants decreased their alcohol use during the 12-month period with the intervention. Furthermore, those who decreased their alcohol use reported an improvement of their sleep quality, when compared to those who did not decrease their drinking. The magnitude of improvement in their sleep demonstrated a medium effect size, Hedge's $g=0.39$ [46]. In another study of heavy drinking college students in New England ($N=42$), for both the sleep intervention and control interventions which involved online self-monitoring treatment, there was reduced weekly drinking, reduced alcohol-related consequences, and improved sleep quality [32].

Pharmacologic treatments for insomnia in alcohol dependence. The results of medication trials have demonstrated conflicting results for treatment of insomnia. Some of the medications that have been evaluated include, acamprosate, agomelatine, ramelteon, and triazolam. Trazodone has been shown to improve sleep quality, but it may decrease abstinence from alcohol, when compared to placebo. Quetiapine demonstrated a short term improvement in insomnia when compared to placebo [44].

Behavioral treatments for insomnia and alcohol dependence. Progressive muscle relaxation (PMR) and cognitive behavioral therapy for insomnia (CBT-I) are the main behavioral interventions evaluated for the treatment of insomnia in alcohol dependence. PMR improved sleep in patients although it's effects on alcohol consumption is unknown. CBT-I, a treatment modality that primarily involves sleep restriction and stimulus control has demonstrated preliminary efficacy in treating insomnia in this population. However, the available data shows that it appears that an improvement in insomnia may not have any effect in improving abstinence from alcohol use [44].

In summary, the above body of knowledge shows us that it appears likely that insomnia has a bidirectional association with heavy drinking, such that one disorder may lead to the other. Insomnia is also highly prevalent across all the stages of alcohol dependence and preliminary data has shown that behavioral treatment may be efficacious for treating insomnia in the alcohol-dependent population.

CIRCADIAN RHYTHMS AND ALCOHOL USE

Recent studies have demonstrated a relationship between alcohol consumption and disrupted circadian sleep-wake rhythms. Circadian rhythms are a manifestation of the activity of the primary endogenous pacemaker, the suprachiasmatic nucleus (SCN) in the hypothalamus. Melatonin serves as a link between the circadian clock and the sleep-wake rhythms. The onset of melatonin secretion under dim light conditions (Dim Light Melatonin Onset (DLMO)) is a commonly used marker for evaluating the activity of the circadian pacemaker, and for assessing the changes in circadian phase, i.e., advanced or delayed [47]. This melatonin can be measured in blood or saliva samples. The peak of the salivary melatonin curve occurs around 2 AM in middle-aged males [48]. Another marker of sleep-wake rhythms is the body temperature. Core body temperature varies across the circadian period (Tb). The nadir of Tb is around 5 AM. Circadian rhythms may be "advanced" or "delayed" based on these shifts from normal variation, adjusting for an individual's age.

Chronotype. The term chronotype denotes the propensity of a person to sleep during the time of the day. There are three main categories of chronotypes, which are the "evening," "morning," and the "indeterminate" type. The "evening" type (Eveningness) person prefers to go to bed later and wake up later and has a greater need for sleep. The "evening" chronotype has been associated with psychiatric disorders and is commonly seen during adolescence. The "morning" type (Morningness) is also referred to as advanced sleep phase and is diametrically opposite to delayed sleep phase on the chronotype spectrum. An individual with the "morning" type prefers to go to bed earlier and wakes up earlier during the 24-h circadian day. Those in the "indeterminate" type do not meet criteria for either the "morning" or the "evening" type and constitute about 50% of the population. The "morning" and the "evening" types constitute about 25% of the population each [49, 50].

Clinical findings on eveningness and alcohol use. A growing body of literature has demonstrated that eveningness is associated with higher alcohol consumption in adolescents [51–53], young adults, college students [54, 55], and adults [56]. A genetic predisposition to eveningness may exist, as demonstrated in a study of 1127 twin pairs. Among these twins, the "evening" type twins, as compared to the "morning" type twins, were more likely to consume higher amounts of alcohol and binge drink [57]. However, a study involving high school students in Canada failed to demonstrate an association between chronotypes and alcohol consumption [58].

Under dim light conditions (that typically occurs at night), melatonin is released into the bloodstream. This preliminary beginning of melatonin secretion is called DLMO (Dim Light Melatonin Onset). A later start of DLMO has been associated with greater severity of substance use problems in adolescent subjects [59] as well as in adolescents, college students, and young adults [11, 31, 60]. However, a recent study of adults who were moderate drinkers did not demonstrate a circadian phase delay [61].

A preference for a later sleep-wake schedule in adolescents and the societal requirements of an earlier schedule may lead to sleep-related complaints and increase the risk of substance abuse. A recent study in adolescents demonstrated that circadian shifts (in the form of greater advances in sleep between weekends and weekdays) were associated with the decreased reactivity of the medial prefrontal cortex and striatum to reward [62]. The reduced responsiveness of these areas may play a role in the development of substance abuse problems for adolescents.

Clinical findings in shiftwork and alcohol use

The body of knowledge on alcohol's effect on shift work disorder is currently unclear due to the variability of findings. Individuals who work evening shift jobs may have a higher risk of phase delay with alcohol use. In a study of 22 healthy adult volunteers, consumption of 0.5 g/kg of alcohol for 7 days delayed the circadian phase in those working on the night shift. In contrast, workers on the daytime shift failed to demonstrate a phase delay [63]. In another study evaluating shift workers in four industries (printing, postal, nursing, oil), long, rotating shifts were associated with an increased risk of binge drinking, but not an overall increase in alcohol consumption [64]. Thus, emerging trends suggest that workers on the night shift are uniquely vulnerable to circadian phase delay as well as risky drinking pattern [64].

Chronopharmacokinetic studies

The bioavailability and metabolism of alcohol may be higher in the morning as compared to later in the day [65]. Alcohol consumption may also dampen circadian rhythms in healthy young adults. In one study, alcohol consumption led to an initial decrease in core body temperature (Tb) followed by an increase during the circadian nadir and an overall blunting of the diurnal temperature rhythm [66]. Similarly, in another study, moderate alcohol consumption in young adults (with a breath alcohol level of 0.05 g%) decreased salivary melatonin levels, a measure of circadian activity [67].

Alcohol dependent individuals

Circadian disruption has been seen in alcohol-dependent individuals. Individuals recovering from alcohol dependence showed a delayed evening rise of melatonin, and lower melatonin levels during the earlier part of the night, in addition to an increased sleep latency [68]. In another study, individuals with alcohol dependence had a slower rate of rising melatonin and lower maximum amplitude of endogenous melatonin secretion [69]. This blunting of the circadian rhythm with alcohol use may be one reason why patients with pathological alcohol use may complain of difficulty falling asleep.

In conclusion, there is a growing body of literature demonstrating a bi-directional link between alcohol use and circadian rhythms. Among the different work shifts, night shifts may uniquely increase the risk of sleep and alcohol use-related problems. Alcohol use blunt circadian rhythms and may be linked to complaints of insomnia in alcohol-dependent individuals.

ALCOHOL AND SLEEP DURATION ABNORMALITIES

Sleep duration has been defined as "the total amount of sleep obtained, either during the nocturnal sleep episode or across the 24-hour period" [70]. The recommended range of sleep duration to support optimal health in adults is 7–9 h [71]. Abnormalities of sleep duration have generally been categorized into short sleep duration (<6 h a night) and long sleep duration (≥9 h a night).

Short and long sleep duration. The estimated prevalence of short sleep duration in the general community varies from 9.3% to 40% [72, 73]. In adults, it has been linked to an increased risk of mortality, physical injuries, cardio-metabolic and psychiatric problems and suicide [74]. In contrast to short sleep duration, very little is known about long sleep duration, although some studies have linked it to cardiovascular diseases, anxiety, and depressive disorder [75–77].

Short sleep duration and alcohol consumption. Some prior epidemiologic studies have demonstrated an association between drinking and short sleep duration. In a survey of respondents in the Oxfordshire area in the United Kingdom [78], an inverse relationship was demonstrated between alcohol consumption and sleep duration, especially in the male respondents. In another survey involving 21–25-year-old males ($N=955$), those who complained of short sleep duration (<6 h a day) had the highest number of alcohol-related problems (missed school/work and blackouts) in contrast to those with long sleep duration [79]. Moreover, they also had a higher proportion of parents who sought psychiatric help, as compared to those with normal sleep duration. A recent study evaluated the drinking patterns and sleep duration of healthy adults in the Quebec Family Study Quebec metropolitan area of Canada [80]. The results demonstrated that short sleepers (<6 h a night) consumed more alcohol than those with normal and long sleep duration. In models adjusted for covariates, men and women reporting short sleep duration had a higher risk of risky drinking through the week. Respondents of either sex reporting short sleep duration were more likely to consume alcohol on a daily basis.

Long sleep duration and alcohol. In the abovementioned study of young adults, long sleepers (>9h per day) reported a later mean age of onset of drinking and fewer number of drinking days per month, when compared to those with short sleep duration and normal sleep duration (7–8h a night). They were concerned about the effects of alcohol and were considering abstinence. A large epidemiologic study evaluated the relationship between sleep duration and alcohol consumption in 110,441 Americans [81]. Their results demonstrated that increased alcohol consumption through the week was associated with increased risk for abnormal sleep duration. For both short (≤6h a night) and long sleep duration (≥9h), the lowest risk was for individuals who consumed 6–12 drinks a week.

Clinical findings in adolescents and young adults. Abnormalities in sleep duration have also been reported in actively drinking adolescents and college students. In one study involving preadolescent boys, insufficient sleep duration was linked to a higher risk of subsequent alcohol use as adults [6]. In another study of adolescent students, longer duration of sleep was linked to a lower risk of heavy drinking [81a]. A similar trend was seen in college students where inadequate sleep was correlated with stronger alcohol-related consequences [82]. In contrast to the above findings, two studies failed to demonstrate an association between alcohol use and sleep duration, one in adolescents and young adults [31] and the other in college students [11].

In summary, increased alcohol consumption, especially heavy drinking and alcohol use disorder (or alcohol dependence) is associated with abnormalities of sleep duration, especially short sleep duration. Future studies should evaluate the relationship between alcohol consumption and long sleep duration, and how insomnia interacts with short sleep duration in the context of heavier alcohol use.

BREATHING RELATED SLEEP DISORDERS AND ALCOHOL USE

Breathing related sleep events. Breathing related sleep events consist of snoring, apneas, and hypopneas. Snoring results from a mismatch between an increased airflow and a reduced upper airway tone. Apneas are defined as a complete cessation of airflow for 10 or more seconds. Hypopneas are considered as a partial airflow obstruction that is 30% or greater (for 10 or more seconds), and are associated with a decrease in arterial oxygen desaturation that is at least 4% in magnitude or an arousal (for 10 or more seconds) [83].

Breathing related sleep disorders. Breathing related sleep disorders are broadly classified into obstructive sleep apnea syndrome and central sleep apnea syndromes, with the main difference between them being the presence of respiratory effort in the former and absence in the latter. Both these conditions are diagnosed with either an in-laboratory polysomnography (PSG) or home sleep testing (HST) using a portable monitor.

The ICSD-3 recommends a diagnosis of obstructive sleep apnea syndrome (OSA) based solely on overnight sleep study measures or a combination of symptoms and findings on a sleep study. Subjective information suggestive of OSA includes, a history of habitual snoring, waking up with breath holding/gasping/choking, complains of sleepiness/fatigue/insomnia, and a prior diagnosis of cardiovascular disease or diabetes or mood disorder or cognitive dysfunction. To diagnose OSA, one or more of these abovementioned criteria is required along with PSG/HST testing demonstrating at least five respiratory events/hour of sleep that are predominantly obstructive in nature. Alternately, OSA may be diagnosed with the presence of ≥15 obstructive respiratory events per hour of sleep on PSG/HST testing [5]. A similar criterion is used for central sleep apnea syndrome, although the respiratory events here are predominantly central sleep events and consist of episodes consisting of complete cessation of airflow and breathing effort for 10 or more seconds.

Effect of alcohol use on breathing during sleep. Alcohol can impair normal breathing during sleep by one of two possible mechanisms: first, it can impair the normal arousal response to airway occlusion [84]; second, it can relax a muscle at the base of the tongue (the genioglossus muscle). This relaxation leads to an increased resistance in the upper airway [85]. This increased resistance in the upper airway may lead to an induction of snoring in some healthy adults and aggravation of snoring in habitual snorers [86–88]. In addition to snoring, moderate alcohol consumption has been demonstrated to aggravate respiratory events with or without a drop in the arterial oxygen saturation levels [86, 87, 89, 90], especially within the first 2–3h of sleep [86, 87]. However, one study did not demonstrate this association between alcohol use and increased sleep related breathing events in the laboratory [91]. A recent meta-analysis of 14 studies examined the effect of alcohol on breathing parameters in sleep replicated some of these findings. The authors demonstrated that when compared to placebo, alcohol increased the apnea hypopnea index (breathing related sleep events) and led to a reduction in the oxyhemoglobin saturation in the blood. Furthermore, these differences were greater in those with a history of habitual snoring and those who were already diagnosed with obstructive sleep apnea [92].

Obstructive sleep apnea has also been linked to alcohol dependence in prior studies. Alcohol dependent individuals have been demonstrated to have more breathing related sleep events, as compared to healthy controls, during early withdrawal from alcohol [93]. Treatment-seeking alcohol dependent patients may have a higher prevalence of obstructive sleep apnea as compared to control subjects [94]. Whereas airway obstruction has a normal arousal response, alcohol use disrupts this response. The consequences include exacerbation of snoring and sleep fragmentation. In one study, there was an 18% difference in sleep disordered breathing between the alcohol dependent group and control group [24]. Thus, the prevalence of breathing-related sleep events may be higher in alcohol dependent individuals.

SUMMARY

Alcohol consumption is linked to breathing related sleep disorders. It may aggravate snoring and increase the respiratory events during sleep, especially those with pre-existing snoring and obstructive sleep apnea. These relationships may be important in middle-aged subjects with alcohol dependence, a population with a higher prevalence of breathing-related sleep events.

Finally, to the best of our knowledge, there are no studies that have evaluated the effects of alcohol on central sleep apnea or mixed apnea (in those with or without AD) or have evaluated the longitudinal trends in breathing-related sleep indices when subjects transition from heavy drinking to sustained recovery. In the absence of data on treatment for OSA in AD, current treatments include recommendations to avoid or minimize alcohol use, weight loss in overweight patients, treatment with a mandibular device (for mild OSA) or with a positive airway pressure device, or upper airway surgery.

ALCOHOL AND SLEEP-RELATED MOVEMENT DISORDERS

These disorders primarily involve abnormal movements of the limbs and consists of restless leg syndrome (RLS) and periodic leg movement disorder (PLMD). The abnormal limb movements interfere with falling asleep or staying asleep through the night. It is for this reason that some individuals with these disorders may complain of sleep initiation or sleep maintenance insomnia and may use alcohol to self-medicate their problems, although very little is known about the relationship between these sleep problems and alcohol consumption. Other conditions in this category of sleep disorders include sleep-related bruxism, sleep-related leg cramps, and sleep-related rhythmic movement disorder.

Restless leg syndrome. RLS is a condition that predominantly occurs in the evening or at night-time. In this condition, the individual complains of an uncomfortable and unpleasant sensation in the legs that begin or worsens during rest or inactivity. The person attempts activities such as walking or stretching to suppress these unpleasant sensations, and in trying to do so, they are unable to fall asleep within a reasonable time frame.

There is limited data linking alcohol consumption to RLS. In one study, alcohol consumption increased the risk for symptoms of RLS by 1.5 times [42] and another study demonstrated a lower risk of RLS symptoms with alcohol consumption of one drink a month or less, as compared to higher drinking levels [95]. However, two other studies have reported a protective effect of alcohol consumption against RLS symptoms [96, 97]. To further complicate this relationship, two studies failed to demonstrate a link between alcohol use and RLS symptoms [98, 99]. These findings show that the association between RLS and alcohol consumption requires more study.

Periodic limb movement disorders. PLMD is diagnosed using polysomnography employing a criterion of >15 limb movements per hour of sleep in adults, that are mostly in the lower extremities. In this condition, the patient may present to the clinic with sleep disturbance and resultant impairment of functioning, which are not explained by another sleep, medical, neurologic or psychiatric disorder [5].

Alcohol consumption has been associated with an increased risk of having periodic limb movements (PLMs) during sleep. Aldrich and colleagues evaluated the effect of alcohol consumption on periodic limb movements during sleep using overnight sleep studies [100]. They demonstrated that in women, consumption of two or more alcoholic beverages was linked to a higher index of limb movements during sleep, as compared to the use of fewer than two drinks per day. They observed a similar trend in men, but the results in men did not reach statistical significance. Among those with alcohol dependence, alcohol use has been linked to periodic limb movement disorder. Alcohol-dependent individuals during early recovery were seen to have a higher index of periodic limb movements as compared to healthy control subjects, especially those who were older in age [101]. In another study, recovering alcohol-dependent individuals with a higher PLM index have an increased risk of relapse to drinking than those in sustained recovery from alcohol use [102].

Treatments for RLS and PLMD have been traditionally included dopaminergic medications such as pramipexole and ropinirole, as well as gabapentin [103]. Interestingly, magnesium may be beneficial for some alcohol-dependent patients with PLMD during early recovery [104].

Sleep-related bruxism consists of frequent or regular tooth grinding sounds during sleep that may lead to abnormal tooth wear, jaw pain, and headaches. Alcohol consumption has been associated with self-reported complaints of bruxism [105, 106] and with objective activity of the jaw muscles during sleep [107]. The relationship between alcohol use and Sleep-related leg cramps and Sleep-related rhythmic movement disorder is unknown due to the lack of research at this interface.

In summary, alcohol consumption has been shown to be associated with a higher risk of having periodic limb movement disorder and possibly sleep-related bruxism, but more research is needed to demonstrate how alcohol consumption and restless leg syndrome are linked.

PARASOMNIAS AND ALCOHOL USE

Parasomnias are defined as "undesirable physical events or experiences that occur during entry into sleep, within sleep, or during arousal from sleep" [5] resulting in disturbed sleep and include, sleepwalking, sleep terrors, and sleep-disordered breathing that occurs during sleep. These disorders can result in psychosocial problems and place the person at risk of injury. Parasomnias may occur during non-REM (non-rapid eye movement) or REM (rapid eye movement) sleep, or during transitions to and from sleep.

NREM-related parasomnias in adults primarily include sleepwalking and sleep-related eating disorder. Sleepwalking episodes involve getting out of bed and involvement in movements that are usually non-goal directed and may be complex in nature. Sleep-related eating disorder involves recurrent episodes of dysfunctional eating which may include eating inappropriate items such as inedible and/or toxic items. In the process of pursuing these activities or its consequences, the person may place them at risk of harm. The association of alcohol with these NREM-related parasomnias is unclear at this current time. REM-related parasomnias mainly include REM behavioral disorder. REM sleep behavior disorder (RBD) involves repeated vocalization and/or complex motor behaviors in sleep and occur during REM sleep. RBD may include the person acting out his/her dreams. This condition is usually chronic in nature. However, an acute form of RBD has been reported during acute withdrawal from alcohol and sedative-hypnotic agents and may occur during times of aggravated REM sleep rebound states, which is defined as a longer and more pronounced REM sleep due to previous REM deprivation possibly linked to alcohol use [5].

Specific studies evaluating the association of alcohol use with parasomnias is lacking in the current body of knowledge. In a study conducted by Aldrich and colleagues, parasomnias were reported by 3% of men and 5% of women who consumed <2 alcoholic drinks per day. This contrasts with 2% of women who consumed ≥2 alcoholic drinks per day [100], which suggests that greater alcohol use may be linked to a lower risk for parasomnia. It is unclear which specific sub-types of parasomnias were reported. We acknowledge the ongoing debate about slow wave sleep, alcohol use, and parasomnias [108, 109] and the need for future studies, including spectral analysis scoring of PSGs to discriminate sleepwalkers and controls [110].

No specific guidelines exist for the treatment of parasomnias, except for REM behavior disorder, where low dose clonazepam or high dose of melatonin has been seen to be beneficial. A general recommendation is to avoid alcohol consumption in those at risk.

OTHER SLEEP-RELATED ISSUES ASSOCIATED WITH ALCOHOL USE

It is well known that alcohol use during pregnancy has many deleterious effects on the infant, and its impact on sleep is not surprising. Habitual use of alcohol by mothers may have implications in the sleep of their children. Infants born to mothers with a history of binge-drinking or alcoholism, as compared to abstinent mothers, demonstrated an increased power of their EEG in REM sleep and NREM sleep. Mothers who indulged in binge drinking during the first 6 weeks of pregnancy rather than just heavy drinking had a five-times higher risk of having infants with sleep problems [111]. An increased frequency of binge drinking during the first 6 weeks of pregnancy was linked to a sixfold higher risk of the infant subsequently developing sleep-related problems [112].

In adults, some studies have evaluated the effect of alcohol on napping and narcolepsy. Napping is defined as a bout of sleep lasting from several minutes to several hours during an individual's waking period [113]. Social drinking may be correlated with fewer daytime naps [114]. In addition to napping, a study demonstrated that alcohol consumption might be related to a diagnosis of narcolepsy. In this study, moderate alcohol use was linked to a higher probability of having narcolepsy, in contrast to heavy drinking with a lower likelihood of having narcolepsy [115].

In summary, risky alcohol use in the mother may increase the risk of the infants subsequently developing sleep-related problems. Patients with narcolepsy may be protected against risky alcohol use, and this effect may be due to the destruction of the orexinergic neurons in the hypothalamus that mediate appetite and sleep drives.

DISCUSSION

As seen above, a growing body of literature is showing us that alcohol use, especially use that may be considered heavy, has a deleterious effect on the sleep of the individual. In this chapter, we have shown that insomnia is the most studied sleep disorder in the context of alcohol use. This relationship between alcohol use and insomnia is bidirectional as data from multiple sources has demonstrated that one persisting disorder (insomnia or pathological drinking) may lead to the other in the future. This bidirectional relationship has direct implications related to the clinical care of patients in clinics which are proving the treatment of patients with sleep disorders and addictive disorders, as well as for patients in primary care clinics.

In addition to sleep continuity disorder, alcohol consumption has been linked to circadian rhythm sleep disorders, sleep duration irregularities, breathing-related sleep disorders, sleep-related movement disorders, parasomnias, and other sleep-related issues. A lesser-known health issue is maternal use during pregnancy and its correlation with an increased risk of the infant developing sleep disturbance. Future studies should rigorously evaluate the effect of exposure to alcohol on parasomnia, sleep-related movement disorders, and the impact on the sleep-wake rhythms in neonates and infants.

REFERENCES

[1] NIAAA. Alcohol facts and statistics. [online]. Available from: National Institute on Alcohol Abuse and Alcoholism, 2017.

[2] Heit C, Dong H, Chen Y, Thompson DC, Deitrich RA, Vasiliou VK. The role of CYP2E1 in alcohol metabolism and sensitivity in the central nervous system. Subcell Biochem 2013;67:235–47. https://doi.org/10.1007/978-94-007-5881-0_8.

[3] Cederbaum AI. Alcohol metabolism. Clin Liver Dis 2012;16(4):667–85. https://doi.org/10.1016/j.cld.2012.08.002.

[4] Franco-Perez J, Padilla M, Paz C. Sleep and brain monoamine changes produced by acute and chronic acetaldehyde administration in rats. Behav Brain Res 2006;174(1):86–92. https://doi.org/10.1016/j.bbr.2006.07.008.

[5] AASM. ICSD-3: International Classification of Sleep Disorders. 3rd ed. Darien, IL: AASM; 2014.

[6] Mike TB, Shaw DS, Forbes EE, Sitnick SL, Hasler BP. The hazards of bad sleep—Sleep duration and quality as predictors of adolescent alcohol and cannabis use. Drug Alcohol Depend 2016;168:335–9. https://doi.org/10.1016/j.drugalcdep.2016.08.009.

[7] Roane BM, Taylor DJ. Adolescent insomnia as a risk factor for early adult depression and substance abuse. Sleep 2008;31(10):1351–6.

[8] Wong MM, Brower KJ, Fitzgerald HE, Zucker RA. Sleep problems in early childhood and early onset of alcohol and other drug use in adolescence. Alcohol Clin Exp Res 2004;28(4):578–87.

[9] Wong MM, Brower KJ, Nigg JT, Zucker RA. Childhood sleep problems, response inhibition, and alcohol and drug outcomes in adolescence and young adulthood. Alcohol Clin Exp Res 2010;34(6):1033–44. https://doi.org/10.1111/j.1530-0277.2010.01178.x.

[10] Wong MM, Brower KJ, Zucker RA. Childhood sleep problems, early onset of substance use and behavioral problems in adolescence. Sleep Med 2009;10(7):787–96. https://doi.org/10.1016/j.sleep.2008.06.015.

[11] Van Reen E, Roane BM, Barker DH, McGeary JE, Borsari B, Carskadon MA. Current alcohol use is associated with sleep patterns in first-year college students. Sleep 2016;39(6):1321–6. https://doi.org/10.5665/sleep.5862.

[12] Tachibana H, Izumi T, Honda S, Horiguchi I, Manabe E, Takemoto T. A study of the impact of occupational and domestic factors on insomnia among industrial workers of a manufacturing company in Japan. Occup Med (Lond) 1996;46(3):221–7.

[13] Ford DE, Kamerow DB. Epidemiologic study of sleep disturbances and psychiatric disorders. an opportunity for prevention? JAMA 1989;262(11):1479–84.

[14] Weissman MM, Greenwald S, Nino-Murcia G, Dement WC. The morbidity of insomnia uncomplicated by psychiatric disorders. Gen Hosp Psychiatry 1997;19(4):245–50. doi: S016383439700056X [pii].

[15] Breslau N, Roth T, Rosenthal L, Andreski P. Sleep disturbance and psychiatric disorders: a longitudinal epidemiological study of young adults. Biol Psychiatry 1996;39(6):411–8. doi: 0006322395001883 [pii].

[16] Currie SR, Clark S, Rimac S, Malhotra S. Comprehensive assessment of insomnia in recovering alcoholics using daily sleep diaries and ambulatory monitoring. Alcohol Clin Exp Res 2003;27(8):1262–9. https://doi.org/10.1097/01.alc.0000081622.03973.57.

[17] Roehrs T, Papineau K, Rosenthal L, Roth T. Ethanol as a hypnotic in insomniacs: self administration and effects on sleep and mood. Neuropsychopharmacology 1999;20(3):279–86. https://doi.org/10.1016/S0893-133X(98)00068-2.

[18] Kaneita Y, Uchiyama M, Takemura S, Yokoyama E, Miyake T, Harano S, … Ohida T. Use of alcohol and hypnotic medication as aids to sleep among the Japanese general population. Sleep Med 2007;8(7–8):723–32. https://doi.org/10.1016/j.sleep.2006.10.009.

[19] Ancoli-Israel S, Roth T. Characteristics of insomnia in the United States: results of the 1991 National Sleep Foundation Survey. I. Sleep 1999;22(Suppl 2):S347–53.

[20] Johnson EO, Roehrs T, Roth T, Breslau N. Epidemiology of alcohol and medication as aids to sleep in early adulthood. Sleep 1998;21(2):178–86.

[21] Vinson DC, Manning BK, Galliher JM, Dickinson LM, Pace WD, Turner BJ. Alcohol and sleep problems in primary care patients: a report from the AAFP National Research Network. Ann Fam Med 2010;8(6):484–92. https://doi.org/10.1370/afm.1175.

[22] Perlis ML, Kehr EL, Smith MT, Andrews PJ, Orff H, Giles DE. Temporal and stagewise distribution of high frequency EEG activity in patients with primary and secondary insomnia and in good sleeper controls. J Sleep Res 2001;10(2):93–104.

[23] Perlis ML, Smith MT, Andrews PJ, Orff H, Giles DE. Beta/Gamma EEG activity in patients with primary and secondary insomnia and good sleeper controls. Sleep 2001;24(1):110–7.

[24] Chakravorty S, Chaudhary NS, Brower KJ. Alcohol dependence and its relationship with insomnia and other sleep disorders. Alcohol Clin Exp Res 2016;https://doi.org/10.1111/acer.13217.

[25] Thiele TE, Marsh DJ, Ste Marie L, Bernstein IL, Palmiter RD. Ethanol consumption and resistance are inversely related to neuropeptide Y levels. Nature 1998;396(6709):366–9. https://doi.org/10.1038/24614.

[26] Comasco E, Nordquist N, Gokturk C, Aslund C, Hallman J, Oreland L, Nilsson KW. The clock gene PER2 and sleep problems: association with alcohol consumption among Swedish adolescents. Ups J Med Sci 2010;115(1):41–8. https://doi.org/10.3109/03009731003597127.

[27] Kovanen L, Saarikoski ST, Haukka J, Pirkola S, Aromaa A, Lonnqvist J, Partonen T. Circadian clock gene polymorphisms in alcohol use disorders and alcohol consumption. Alcohol Alcohol 2010;45(4):303–11. https://doi.org/10.1093/alcalc/agq035.

[28] Ando H, Ushijima K, Kumazaki M, Eto T, Takamura T, Irie S, … Fujimura A. Associations of metabolic parameters and ethanol consumption with messenger RNA expression of clock genes in healthy men. Chronobiol Int 2010;27(1):194–203. https://doi.org/10.3109/07420520903398617.

[29] Skarupke C, Schlack R, Lange K, Goerke M, Dueck A, Thome J, … Cohrs S. Insomnia complaints and substance use in German adolescents: did we underestimate the role of coffee consumption? Results of the KiGGS study. J Neural Transm 2017;124:69–78. https://doi.org/10.1007/s00702-015-1448-7.

[30] Zhabenko O, Austic E, Conroy DA, Ehrlich P, Singh V, Epstein-Ngo Q, … Walton MA. Substance use as a risk factor for sleep problems among adolescents presenting to the emergency department. J Addict Med 2016;10(5):331–8. https://doi.org/10.1097/ADM.0000000000000243.

[31] Hasler BP, Casement MD, Sitnick SL, Shaw DS, Forbes EE. Eveningness among late adolescent males predicts neural reactivity to reward and alcohol dependence two years later. Behav Brain Res 2017;327:112–20. https://doi.org/10.1016/j.bbr.2017.02.024.

[32] Fucito LM, DeMartini KS, Hanrahan TH, Yaggi HK, Heffern C, Redeker NS. Using sleep interventions to engage and treat heavy-drinking college students: a randomized pilot study. Alcohol Clin Exp Res 2017;41(4):798–809. https://doi.org/10.1111/acer.13342.

[33] Popovici I, French MT. Does unemployment lead to greater alcohol consumption? Ind Relat 2013;52(2):444–66. https://doi.org/10.1111/irel.12019.

[34] Bruck D, Astbury J. Population study on the predictors of sleeping difficulties in young Australian women. Behav Sleep Med 2012;10(2):84–95. https://doi.org/10.1080/15402002.2011.592888.

[35] Lind MJ, Hawn SE, Sheerin CM, Aggen SH, Kirkpatrick RM, Kendler KS, Amstadter AB. An examination of the etiologic overlap between the genetic and environmental influences on insomnia and common psychopathology. Depress Anxiety 2017;34(5):453–62. https://doi.org/10.1002/da.22587.

[36] Haario P, Rahkonen O, Laaksonen M, Lahelma E, Lallukka T. Bidirectional associations between insomnia symptoms and unhealthy behaviours. J Sleep Res 2013;22(1):89–95. https://doi.org/10.1111/j.1365-2869.2012.01043.x.

[37] Canham SL, Kaufmann CN, Mauro PM, Mojtabai R, Spira AP. Binge drinking and insomnia in middle-aged and older adults: the Health and Retirement Study. Int J Geriatr Psychiatry 2015;30(3):284–91. https://doi.org/10.1002/gps.4139.

[38] Wang YM, Chen HG, Song M, Xu SJ, Yu LL, Wang L, … Lu L. Prevalence of insomnia and its risk factors in older individuals: a community-based study in four cities of Hebei Province, China. Sleep Med 2016;19:116–22. https://doi.org/10.1016/j.sleep.2015.10.018.

[39] Taylor DJ, Pruiksma KE, Hale WJ, Kelly K, Maurer D, Peterson AL, … Williamson DE. Prevalence, correlates, and predictors of insomnia in the US army prior to deployment. Sleep 2016;39(10):1795–806. https://doi.org/10.5665/sleep.6156.

[40] Cucciare MA, Darrow M, Weingardt KR. Characterizing binge drinking among U.S. military Veterans receiving a brief alcohol intervention. Addict Behav 2011;36(4):362–7. https://doi.org/10.1016/j.addbeh.2010.12.014.

[41] Casey DA, Northcott C, Stowell K, Shihabuddin L, Rodriguez-Suarez M. Dementia and palliative care. Clin Geriatr 2012;20(1):36–41.

[42] Kaneita Y, Ohida T, Takemura S, Sone T, Suzuki K, Miyake T, … Umeda T. Relation of smoking and drinking to sleep disturbance among Japanese pregnant women. Prev Med 2005;41(5–6):877–82. https://doi.org/10.1016/j.ypmed.2005.08.009.

[43] Blumel JE, Cano A, Mezones-Holguin E, Baron G, Bencosme A, Benitez Z, … Chedraui P. A multinational study of sleep disorders during female mid-life. Maturitas 2012;72(4):359–66. https://doi.org/10.1016/j.maturitas.2012.05.011.

[44] Chakravorty S, Perlis ML, Arnedt JT, Kranzler HR, Sturgis EB, Findley JC, Oslin DW. The efficacy of individual 8-week CBT-I for insomnia in recovering alcohol dependent Veterans. Sleep 2016;39(Abstract Supplement):1.

[45] Brower KJ, Aldrich MS, Robinson EA, Zucker RA, Greden JF. Insomnia, self-medication, and relapse to alcoholism. Am J Psychiatry 2001;158(3):399–404.

[46] Berman AH, Wennberg P, Sinadinovic K. Changes in mental and physical well-being among problematic alcohol and drug users in 12-month internet-based intervention trials. Psychol Addict Behav 2015;29(1):97–105. https://doi.org/10.1037/a0038420.

[47] Pandi-Perumal SR, Smits M, Spence W, Srinivasan V, Cardinali DP, Lowe AD, Kayumov L. Dim light melatonin onset (DLMO): a tool for the analysis of circadian phase in human sleep and chronobiological disorders. Prog Neuropsychopharmacol Biol Psychiatry 2007;31(1):1–11. https://doi.org/10.1016/j.pnpbp.2006.06.020.

[48] Zhou JN, Liu RY, van Heerikhuize J, Hofman MA, Swaab DF. Alterations in the circadian rhythm of salivary melatonin begin during middle-age. J Pineal Res 2003;34(1):11–6.

[49] Paine SJ, Gander PH, Travier N. The epidemiology of morningness/eveningness: influence of age, gender, ethnicity, and socioeconomic factors in adults (30–49 years). J Biol Rhythms 2006;21(1):68–76. https://doi.org/10.1177/0748730405283154.

[50] Taillard J, Philip P, Chastang JF, Bioulac B. Validation of Horne and Ostberg morningness-eveningness questionnaire in a middle-aged population of French workers. J Biol Rhythms 2004;19(1):76–86. https://doi.org/10.1177/0748730403259849.

[51] Pieters S, Burk WJ, Van der Vorst H, Dahl RE, Wiers RW, Engels RC. Prospective relationships between sleep problems and substance use, internalizing and externalizing problems. J Youth Adolesc 2015;44(2):379–88. https://doi.org/10.1007/s10964-014-0213-9.

[52] Saxvig IW, Pallesen S, Wilhelmsen-Langeland A, Molde H, Bjorvatn B. Prevalence and correlates of delayed sleep phase in high school students. Sleep Med 2012;13(2):193–9. https://doi.org/10.1016/j.sleep.2011.10.024.

[53] Urban R, Magyarodi T, Rigo A. Morningness-eveningness, chronotypes and health-impairing behaviors in adolescents. Chronobiol Int 2011;28(3):238–47. https://doi.org/10.3109/07420528.2010.549599.

[54] Adan A. Chronotype and personality factors in the daily consumption of alcohol and psychostimulants. Addiction 1994;89(4):455–62.

[55] Onyper SV, Thacher PV, Gilbert JW, Gradess SG. Class start times, sleep, and academic performance in college: a path analysis. Chronobiol Int 2012;29(3):318–35. https://doi.org/10.3109/07420528.2012.655868.

[56] Wittmann M, Paulus M, Roenneberg T. Decreased psychological well-being in late 'chronotypes' is mediated by smoking and alcohol consumption. Subst Use Misuse 2010;45(1–2):15–30. https://doi.org/10.3109/10826080903498952.

[57] Watson NF, Buchwald D, Harden KP. A twin study of genetic influences on diurnal preference and risk for alcohol use outcomes. J Clin Sleep Med 2013;9(12):1333–9. https://doi.org/10.5664/jcsm.3282.

[58] Martin JS, Gaudreault MM, Perron M, Laberge L. Chronotype, light exposure, sleep, and daytime functioning in high school students attending morning or afternoon school shifts: an actigraphic study. J Biol Rhythms 2016;31(2):205–17. https://doi.org/10.1177/0748730415625510.

[59] Hasler BP, Bootzin RR, Cousins JC, Fridel K, Wenk GL. Circadian phase in sleep-disturbed adolescents with a history of substance abuse: a pilot study. Behav Sleep Med 2008;6(1):55–73. https://doi.org/10.1080/15402000701796049.

[60] Hasler BP, Franzen PL, de Zambotti M, Prouty D, Brown SA, Tapert SF, … Clark DB. Eveningness and later sleep timing are associated with greater risk for alcohol and marijuana use in adolescence: initial findings from the National Consortium on Alcohol and Neurodevelopment in Adolescence Study. Alcohol Clin Exp Res 2017;41(6):1154–65. https://doi.org/10.1111/acer.13401.

[61] Burgess HJ, Rizvydeen M, Fogg LF, Keshavarzian A. A single dose of alcohol does not meaningfully alter circadian phase advances and phase delays to light in humans. Am J Physiol Regul Integr Comp Physiol 2016;310(8):R759–65. https://doi.org/10.1152/ajpregu.00001.2016.

[62] Hasler BP, Smith LJ, Cousins JC, Bootzin RR. Circadian rhythms, sleep, and substance abuse. Sleep Med Rev 2012;16(1):67–81. https://doi.org/10.1016/j.smrv.2011.03.004.

[63] Swanson GR, Gorenz A, Shaikh M, Desai V, Kaminsky T, van Den Berg J, ... Keshavarzian A. Night workers with circadian misalignment are susceptible to alcohol-induced intestinal hyperpermeability with social drinking. Am J Physiol Gastrointest Liver Physiol 2016;311(1):G192–201. https://doi.org/10.1152/ajpgi.00087.2016.

[64] Dorrian J, Heath G, Sargent C, Banks S, Coates A. Alcohol use in shiftworkers. Accid Anal Prev 2017;99(Pt B):395–400. https://doi.org/10.1016/j.aap.2015.11.011.

[65] Danel T, Touitou Y. Chronobiology of alcohol: from chronokinetics to alcohol-related alterations of the circadian system. Chronobiol Int 2004;21(6):923–35.

[66] Danel T, Libersa C, Touitou Y. The effect of alcohol consumption on the circadian control of human core body temperature is time dependent. Am J Physiol Regul Integr Comp Physiol 2001;281(1):R52–5.

[67] Rupp TL, Acebo C, Carskadon MA. Evening alcohol suppresses salivary melatonin in young adults. Chronobiol Int 2007;24(3):463–70. https://doi.org/10.1080/07420520701420675.

[68] Kuhlwein E, Hauger RL, Irwin MR. Abnormal nocturnal melatonin secretion and disordered sleep in abstinent alcoholics. Biol Psychiatry 2003;54(12):1437–43.

[69] Conroy DA, Hairston IS, Arnedt JT, Hoffmann RF, Armitage R, Brower KJ. Dim light melatonin onset in alcohol-dependent men and women compared with healthy controls. Chronobiol Int 2012;29(1):35–42. https://doi.org/10.3109/07420528.2011.636852.

[70] Kline C. Sleep duration. In: Gellman MD, Turner JR, editors. Encyclopedia of behavioral medicine. New York, NY: Springer; 2013. p. 1808–10.

[71] Consensus Conference Panel, Watson NF, Badr MS, Belenky G, Bliwise DL, Buxton OM, Buysse D, Dinges DF, Gangwisch J, Grandner MA, Kushida C, Malhotra RK, Martin JL, Patel SR, Quan SF, Tasali E, Non-Participating O, Twery M, Croft JB, Maher E, American Academy of Sleep Medicine Staff, Barrett JA, Thomas SM, Heald JL. Recommended amount of sleep for a healthy adult: a joint consensus statement of the american academy of sleep medicine and sleep research society. J Clin Sleep Med 2015;11:591–2.

[72] Grandner MA, Chakravorty S, Perlis ML, Oliver L, Gurubhagavatula I. Habitual sleep duration associated with self-reported and objectively determined cardiometabolic risk factors. Sleep Med 2014;15(1):42–50. https://doi.org/10.1016/j.sleep.2013.09.012.

[73] Knutson KL, Van Cauter E, Rathouz PJ, DeLeire T, Lauderdale DS. Trends in the prevalence of short sleepers in the USA: 1975–2006. Sleep 2010;33(1):37–45.

[74] Consensus Conference Panel, Watson NF, Badr MS, Belenky G, Bliwise DL, Buxton OM, ... Tasali E. Joint consensus Statement of the American Academy of Sleep Medicine and Sleep Research Society on the recommended amount of sleep for a healthy adult: methodology and discussion. J Clin Sleep Med 2015;11(8):931–52. https://doi.org/10.5664/jcsm.4950.

[75] Cappuccio FP, Cooper D, D'Elia L, Strazzullo P, Miller MA. Sleep duration predicts cardiovascular outcomes: a systematic review and meta-analysis of prospective studies. Eur Heart J 2011;32(12):1484–92. https://doi.org/10.1093/eurheartj/ehr007.

[76] Kaneita Y, Ohida T, Uchiyama M, Takemura S, Kawahara K, Yokoyama E, ... Fujita T. The relationship between depression and sleep disturbances: a Japanese nationwide general population survey. J Clin Psychiatry 2006;67(2):196–203.

[77] Ryu SY, Kim KS, Han MA. Factors associated with sleep duration in Korean adults: results of a 2008 community health survey in Gwangju metropolitan city, Korea. J Korean Med Sci 2011;26(9):1124–31. https://doi.org/10.3346/jkms.2011.26.9.1124.

[78] Palmer CD, Harrison GA, Hiorns RW. Association between smoking and drinking and sleep duration. Ann Hum Biol 1980;7(2):103–7.

[79] Schuckit MA, Bernstein LI. Sleep time and drinking history: a hypothesis. Am J Psychiatry 1981;138(4):528–30.

[80] Chaput JP, McNeil J, Despres JP, Bouchard C, Tremblay A. Short sleep duration is associated with greater alcohol consumption in adults. Appetite 2012;59(3):650–5. https://doi.org/10.1016/j.appet.2012.07.012.

[81] Krueger PM, Friedman EM. Sleep duration in the United States: a cross-sectional population-based study. Am J Epidemiol 2009;169(9):1052–63. https://doi.org/10.1093/aje/kwp023.

[81a] Miller MB, Janssen T, Jackson KM. The prospective association between sleep and initiation of substance use in young adolescents. J Adolesc Health 2017;60(2):154–60.

[82] Miller MB, DiBello AM, Lust SA, Carey MP, Carey KB. Adequate sleep moderates the prospective association between alcohol use and consequences. Addict Behav 2016;63:23–8. https://doi.org/10.1016/j.addbeh.2016.05.005.

[83] Iber C, American Academy of Sleep Medicine. The AASM manual for the scoring of sleep and associated events: rules, terminology and technical specifications. Westchester, IL: American Academy of Sleep Medicine; 2007.

[84] Berry RB, Bonnet MH, Light RW. Effect of ethanol on the arousal response to airway occlusion during sleep in normal subjects. Am Rev Respir Dis 1992;145(2 Pt 1):445–52. https://doi.org/10.1164/ajrccm/145.2_Pt_1.445.

[85] Krol RC, Knuth SL, Bartlett Jr. D. Selective reduction of genioglossal muscle activity by alcohol in normal human subjects. Am Rev Respir Dis 1984;129(2):247–50.

[86] Issa FG, Sullivan CE. Alcohol, snoring and sleep apnea. J Neurol Neurosurg Psychiatry 1982;45(4):353–9.

[87] Mitler MM, Dawson A, Henriksen SJ, Sobers M, Bloom FE. Bedtime ethanol increases resistance of upper airways and produces sleep apneas in asymptomatic snorers. Alcohol Clin Exp Res 1988;12(6):801–5.

[88] Riemann R, Volk R, Muller A, Herzog M. The influence of nocturnal alcohol ingestion on snoring. Eur Arch Otorhinolaryngol 2010;267(7):1147–56. https://doi.org/10.1007/s00405-009-1163-9.

[89] Sakurai S, Cui R, Tanigawa T, Yamagishi K, Iso H. Alcohol consumption before sleep is associated with severity of sleep-disordered breathing among professional Japanese truck drivers. Alcohol Clin Exp Res 2007;31(12):2053–8. https://doi.org/10.1111/j.1530-0277.2007.00538.x.

[90] Scanlan MF, Roebuck T, Little PJ, Redman JR, Naughton MT. Effect of moderate alcohol upon obstructive sleep apnoea. Eur Respir J 2000;16(5):909–13.

[91] Teschler H, Berthon-Jones M, Wessendorf T, Meyer HJ, Konietzko N. Influence of moderate alcohol consumption on obstructive sleep apnoea with and without AutoSet nasal CPAP therapy. Eur Respir J 1996;9:2371–7.

[92] Kolla BP, Foroughi M, Saeidifard F, Chakravorty S, Wang Z, Mansukhani MP. The impact of alcohol on breathing parameters during sleep: a systematic review and meta-analysis. Sleep Med Rev 2018;42:59–67. 1532–2955 (Electronic).

[93] Le Bon O, Verbanck P, Hoffmann G, Murphy JR, Staner L, De Groote D, ... Pelc I. Sleep in detoxified alcoholics: impairment of most standard sleep parameters and increased risk for sleep apnea, but not for myoclonias—a controlled study. J Stud Alcohol 1997;58(1):30–6.

[94] Aldrich MS, Shipley JE, Tandon R, Kroll PD, Brower KJ. Sleep-disordered breathing in alcoholics: association with age. Alcohol Clin Exp Res 1993;17(6):1179–83.

[95] Phillips B, Young T, Finn L, Asher K, Hening WA, Purvis C. Epidemiology of restless legs symptoms in adults. Arch Intern Med 2000;160(14):2137–41.

[96] Batool-Anwar S, Li Y, De Vito K, Malhotra A, Winkelman J, Gao X. Lifestyle factors and risk of restless legs syndrome: prospective cohort study. J Clin Sleep Med 2016;12(2):187–94. https://doi.org/10.5664/jcsm.5482.

[97] Winter AC, Berger K, Glynn RJ, Buring JE, Gaziano JM, Schurks M, Kurth T. Vascular risk factors, cardiovascular disease, and restless legs syndrome in men. Am J Med 2013;126(3):228–35. 235 e221–222 https://doi.org/10.1016/j.amjmed.2012.06.039.

[98] Li LH, Chen HB, Zhang LP, Wang ZW, Wang CP. A community-based investigation on restless legs syndrome in a town in China. Sleep Med 2012;13(4):342–5. https://doi.org/10.1016/j.sleep.2011.09.008.

[99] Zhang J, Lam SP, Li SX, Li AM, Kong AP, Wing YK. Restless legs symptoms in adolescents: epidemiology, heritability, and pubertal effects. J Psychosom Res 2014;76(2):158–64. https://doi.org/10.1016/j.jpsychores.2013.11.017.

[100] Aldrich MS, Shipley JE. Alcohol use and periodic limb movements of sleep. Alcohol Clin Exp Res 1993;17(1):192–6.

[101] Brower KJ, Hall JM. Effects of age and alcoholism on sleep: a controlled study. J Stud Alcohol 2001;62(3):335–43.

[102] Gann H, Feige B, Fasihi S, van Calker D, Voderholzer U, Riemann D. Periodic limb movements during sleep in alcohol dependent patients. Eur Arch Psychiatry Clin Neurosci 2002;252(3):124–9. https://doi.org/10.1007/s00406-002-0371-8.

[103] Aurora RN, Kristo DA, Bista SR, Rowley JA, Zak RS, Casey KR, Lamm CI, Tracy SL, Rosenberg RS, American Academy of Sleep Medicine. The treatment of restless legs syndrome and periodic limb movement disorder in adults—an update for 2012: practice parameters with an evidence-based systematic review and meta-analyses: an American Academy of Sleep Medicine Clinical Practice Guideline. Sleep 2012;35:1039–62.

[104] Hornyak M, Haas P, Veit J, Gann H, Riemann D. Magnesium treatment of primary alcohol-dependent patients during subacute withdrawal: an open pilot study with polysomnography. Alcohol Clin Exp Res 2004;28(11):1702–9.

[105] Ohayon MM, Li KK, Guilleminault C. Risk factors for sleep bruxism in the general population. Chest 2001;119(1):53–61.

[106] Rintakoski K, Kaprio J. Legal psychoactive substances as risk factors for sleep-related bruxism: a nationwide Finnish Twin cohort study. Alcohol Alcohol 2013;48(4):487–94. https://doi.org/10.1093/alcalc/agt016.

[107] Hojo A, Haketa T, Baba K, Igarashi Y. Association between the amount of alcohol intake and masseter muscle activity levels recorded during sleep in healthy young women. Int J Prosthodont 2007;20(3):251–5.

[108] Ebrahim IO, Shapiro CM, Williams AJ, Fenwick PB. Alcohol and sleep I: effects on normal sleep. Alcohol Clin Exp Res 2013;37:539–49.

[109] Pressman MR, Mahowald MW, Schenck CH, Bornemann MC. Alcohol-induced sleepwalking or confusional arousal as a defense to criminal behavior: a review of scientific evidence, methods and forensic considerations. J Sleep Res 2007;16:198–212.

[110] Cartwright R, Guilleminault C. Slow wave activity is reliably low in sleepwalkers: response to Pressman et al. letter to the editor. J Clin Sleep Med 2014;10(1):113–5. (1550–9397 (Electronic)).

[111] Ioffe S, Chernick V. Development of the EEG between 30 and 40 weeks gestation in normal and alcohol-exposed infants. Dev Med Child Neurol 1988;30(6):797–807.

[112] Alvik A, Torgersen AM, Aalen OO, Lindemann R. Binge alcohol exposure once a week in early pregnancy predicts temperament and sleeping problems in the infant. Early Hum Dev 2011;87:827–33.

[113] Dhand R, Sohal H. Good sleep, bad sleep! The role of daytime naps in healthy adults. Curr Opin Pulm Med 2006;12:379–82.

[114] Furihata R, Kaneita Y, Jike M, Ohida T, Uchiyama M. Napping and associated factors: a Japanese nationwide general population survey. Sleep Med 2016;20:72–9.

[115] Barateau L, Jaussent I, Lopez R, Boutrel B, Leu-Semenescu S, Arnulf I, Dauvilliers Y. Smoking, alcohol, drug use, abuse and dependence in narcolepsy and idiopathic hypersomnia: a case-control study. Sleep 2016;39:573–80.

FURTHER READING

Conroy DA, Hairston IS, Zucker RA, Heitzig MM. Sleep patterns in children of alcoholics and the relationship with parental reports. Austin J Sleep Disord 2015;2(1):01–9.

Crum RM, Storr CL, Chan YF, Ford DE. Sleep disturbance and risk for alcohol-related problems. Am J Psychiatry 2004;161(7):1197–203.

Dahl RE, Williamson DE, Bertocci MA, Stolz MV, Ryan ND, Ehlers CL. Spectral analyses of sleep EEG in depressed offspring of fathers with or without a positive history of alcohol abuse or dependence: a pilot study. Alcohol 2003;30(3):193–200.

Roehrs T, Roth T. Insomnia as a path to alcoholism: tolerance development and dose escalation. Sleep 2018;https://doi.org/10.1093/sleep/zsy091.

Tarokh L, Carskadon MA. Sleep electroencephalogram in children with a parental history of alcohol abuse/dependence. J Sleep Res 2010;19(1 Pt 2):165–74. https://doi.org/10.1111/j.1365-2869.2009.00763.x.

Wong MM, Robertson GC, Dyson RB. Prospective relationship between poor sleep and substance-related problems in a national sample of adolescents. Alcohol Clin Exp Res 2015;39(2):355–62. https://doi.org/10.1111/acer.12618.

Chapter 22

Improved sleep as an adjunctive treatment for smoking cessation☆

Freda Patterson[a], Rebecca Ashare[b]
[a]Department of Behavioral Health and Nutrition, College of Health Sciences, University of Delaware, Newark, DE, United States, [b]Perelman School of Medicine at the University of Pennsylvania, Philadelphia, PA, United States

INTRODUCTION

Despite declines in adult cigarette smoking prevalence during the past 50 years, cigarette smoking remains the leading cause of preventable death and disability in the United States. Data show that cigarette smoking and secondhand smoke exposure are accountable for at least 443,000 premature deaths and up to $289 billion in direct health care expenditures and productivity losses each year [1]. Mortality associated with continued tobacco use is well-documented: 33% of cardiovascular and metabolic diseases, 32% of all cancers (including 87% of lung cancer), and 62% of pulmonary and respiratory diseases are attributable to cigarette smoking [2]. In spite of these adverse health effects, 15.1% of adults in the United States (~36.5 million people) are current smokers, with rates of 33%–48% reported among demographic subgroups including those who are uninsured, low-income, and low-education [3]. Data also show that adults with a mental health disorder (e.g., anxiety disorder) are twice as likely to smoke than those in the general population [4]. Thus, smoking cessation remains a public health priority.

Current FDA-approved treatments for nicotine dependence, including nicotine replacement therapies (e.g., nicotine patch, spray, gum, lozenge) and non-nicotinic treatments (e.g., bupropion, varenicline) are sub-optimally effective. Whereas these treatments do double the odds of 6-month abstinence compared to placebo, less than one quarter remain abstinent [5]. *Healthy People* 2020 has set the national goal of a 12% smoking prevalence rate for all demographic groups; achieving this goal will require the development of more effective treatments for smoking cessation, as well as strategies to optimize current treatments [6].

As a common biologic function that plays a central role in metabolic regulation, emotion regulation, performance, memory consolidation, brain recuperation processes, and learning, sleep may be such an intervention target that could optimize nicotine dependence treatment response. For example, insomnia (difficulty falling and/or staying asleep) is a clinically-recognized nicotine withdrawal symptom [7] that is not addressed in the clinical guidelines for nicotine dependence treatment [5]. On this basis, this chapter will first provide an overview of the epidemiology of cigarette smoking, followed by a review of the differences in sleep quality metrics in smokers versus non-smokers. Next, a review of the effects of smoking abstinence on sleep quality and a brief overview of the possible mechanisms that may link sleep with smoking cessation outcomes will be provided. Following this, a review of evidence-based treatments for sleep disturbances will be considered with the goal of identifying sleep therapies that could be used in the context of smoking cessation. Last, future research directions needed to validate the extent to which poor sleep quality may be a viable target with which to optimize response to standard nicotine dependence treatment, will be considered.

EPIDEMIOLOGY OF CIGARETTE SMOKING

Between 1965 and 2014, the United States adult smoking rate dropped from >42% to about 17% [8]. This monumental public health achievement was driven by several initiatives including enhanced public education about the adverse health effects of smoking, the development of efficacious behavioral treatments and medications, and enhanced public health policies (e.g., clean indoor air laws, cigarette taxes). However, this success appears to have plateaued—approximately one in six adults (15.1%) are current smokers, and these rates climb to about one in two (48%) in high-risk groups (i.e., low-income, low-education, non-Caucasian) [3,9]. Because of these socioeconomic factors, culture,

☆ Sections of Patterson F, Grandner MA, Malone SK, Rizzo A, Davey A, Edwards, DG. Sleep as a target for optimized response to smoking cessation treatment. *Nicotine and Tobacco Research*, 2017, reprinted by permission of Oxford University Press.

policies, and lack of proper healthcare, there are growing health disparities with respect to the impact of tobacco use among those living in rural areas compared to those living in urban and metropolitan areas [10]. Indeed, smoking prevalence varies widely depending on geographic region within the United States with prevalence rates of 25.4%, 24.2%, 21.3%, and 18.0% in the Midwest, South, Northeast, and West regions, respectively [11].

Despite increased awareness of the adverse health consequences, cigarette smoking remains the leading cause of preventable disease and death in the United States and accounts for 1 out of every 5 deaths [12, 13]. Cigarette smoking causes 9 out of 10 lung cancers, and increases the risk of other cancers, cardiovascular disease, lung disease, and infectious diseases [14]. In the context of cardiovascular diseases (CVDs), smokers are twice as likely to have a sudden cardiac death [15], seven times more likely to develop peripheral arterial disease [16], and more than twice as likely to have a stroke [17, 18] than non-smokers. Moreover, tobacco use costs $170 billion in direct medical costs each year [1] representing a significant public health burden.

The increasing availability of effective treatments for nicotine dependence has contributed to the substantial decline in smoking rates. Currently, there are three FDA-approved medications for nicotine dependence: nicotine replacement therapy (NRT), which includes transdermal nicotine (TN), nasal spray, gum, and lozenges; bupropion; and varenicline. Use of these treatments significantly increase the likelihood that a quit attempt will be successful, versus no medication [19]. These medications are safe, with little evidence that serious adverse events are associated with their use [19, 20], even among smokers with psychiatric [21] or medical [22] comorbidities. With respect to behavioral interventions, quit rates are generally low, ranging from 7% to 13% [23]. While recent studies suggest that novel behavioral interventions such as mindfulness treatments [24] or acceptance and commitment therapy [25] have received initial support, the majority of interventions are based on standard cognitive-behavioral and social support models [26, 27]. Despite the availability of these treatments and the fact that most smokers want to quit [28], 75%–90% of smokers are unable to sustain long-term abstinence [29–31]. In order to achieve further reductions in population smoking rates, new strategies or behavioral targets are necessary to optimize current treatments. Sleep health may be such a behavioral target [32].

SLEEP CONTINUITY AND ARCHITECTURE IN SMOKERS VERSUS NON-SMOKERS

Overview of sleep continuity and architecture

Sleep is quantified by metrics of sleep continuity and sleep architecture. Sleep continuity refers to the timeline of when an individual is asleep, compared to the time when they are intending to sleep. For example, key metrics within sleep continuity include the timing of sleep, the total amount of time spent in bed (time in bed, or TIB), sleep latency (time to fall asleep, or SL), number of awakenings, total time awake after sleep onset (also referred to as "wake after sleep onset" or WASO), time of final awakening, total sleep time (computed as TST=TIB—SL—WASO), and sleep efficiency (the proportion of time spent in bed actually asleep, computed as [TST/TIB]*100) [33].

Sleep architecture represents the cyclical pattern of sleep as it shifts among the various sleep stages, including non-rapid eye movement (NREM) and rapid eye movement (REM) sleep. Polysomnography (PSG) provides objective assessment of different sleep stages; the temporal and percentage of time in each of these stages are key markers of individual sleep quality. Briefly, the three NREM stages (N1, N2, N3) roughly parallel a depth-of-sleep continuum, with arousal thresholds generally lowest in N1 and highest in N3 sleep. N1 and N2 sleep stages are associated with minimal or fragmentary neuronal activity. REM sleep is characterized by heart rate, breathing rate and brain wave activity that is similar to waking levels, compared to other stages of sleep [34]. REM sleep (as with N3) is important for cognitive tasks such as memory consolidation and information processing; dreaming predominantly occurs during REM sleep [35]. Throughout the sleep period, adults will cycle between stages of NREM and REM, spending 75%–80% of sleep time in NREM and the remainder in REM sleep [34].

Self-reported perceptions regarding sleep are also valuable metrics. Sleep disruptive events (i.e., sleep walking, night terrors) and daytime sleepiness or dysfunction (i.e., sleepiness, lack of energy, drowsiness that may prevent the completion of daytime tasks) are commonly measured characteristics of sleep [36]. A growing body of literature has compared these and other sleep variables in smokers and non-smokers; a review of this work is provided below and summarized in Table 22.1.

Sleep architecture in smokers versus non-smokers

Five studies were found that used polysomnography to examine sleep architecture in smokers and non-smokers (Table 22.1) [37–39, 46, 47]. Three of the five studies found that, compared to non-smokers, smokers had a significantly higher percentage of time in N1. For example, Zang and colleagues found that among 779 smokers and 2916 never-smokers, current smokers accrued 24% more N1 sleep [46]; this would indicate shallower, more disturbed sleep. In another study of women ($N=63$ smokers and $N=323$ non-smokers) the mean time in minutes in N1 was 31 for smokers and 21 for non-smokers [47]. Similarly, smokers

TABLE 22.1 Objective and subjective sleep metrics in smokers vs non-smokers.

Study	Study design	N, smokers/non-smokers	Sleep metrics	Findings	Comments
Zang et al. [46]	Multicenter, longitudinal study; baseline PSG data	N=779/2916	Sleep architecture (% time spent in each stage)	N1: smokers > non-smokers N2: smokers > non-smokers N3: smokers < non-smokers	
			Sleep onset latency	Smokers > non-smokers	
			Total sleep time	Smokers < non-smokers	
			Sleep efficiency	Smokers < non-smokers	
Sahlin et al. [47]	Population-based study; PSG	N=63/323	Sleep onset latency	Smokers > non-smokers	
			Sleep architecture	N1: smokers > non-smokers	
			Wake after sleep onset	Smokers < non-smokers	
Jaehne et al. [37]	Observational PSG study	N=44/44	REM density	Smokers > non-smokers	
			Sleep period time	Smokers < non-smokers	Time between sleep onset and final awakening
			Sleep onset latency	Smokers > non-smokers	
			Subjective sleep rating	Smokers > non-smokers	Measured via PSQI; higher scores indicate more sleep problems
Soldatos et al. [38]	PSG laboratory study	N=50/50	Sleep onset latency	Smokers > non-smokers	
			Total time awake	Smokers > non-smokers	
Redline et al. [39]	Prospective cohort study; PSG	N=259/1256	Sleep architecture (% time spent in each stage)	N1: smokers > non-smokers N2: smokers > non-smokers N3: smokers < non-smokers	
Cohrs et al. [40]	Population-based, case-control	N=1243/1071	Subjective sleep rating	Smokers > non-smokers	Measured via PSQI; higher scores indicate more sleep problems
Branstetter et al. [41]	NHANES Population-based survey	N=2015/5752	Sleep duration	Smokers < non-smokers	
			Sleep onset latency	Smokers > non-smokers	
			Early awakening	Smokers > non-smokers	
			Nighttime awakening	Smokers > non-smokers	
Phillips et al. [48]	Self-report survey	N=77/308	Daytime sleepiness	Smokers > non-smokers	
			Difficulty falling asleep	Smokers > non-smokers	
			Difficulty staying asleep	Smokers > non-smokers	
McNamara et al. [42]	NHANES population-based survey	N=1023/2294	Difficulty falling asleep	Smokers > non-smokers	
			Difficulty staying asleep	Smokers > non-smokers	
			Early awakening	Smokers > non-smokers	
			Total sleep time	Smokers < non-smokers	
			Sleep onset latency	Smokers > non-smokers	

Continued

TABLE 22.1 Objective and subjective sleep metrics in smokers vs non-smokers.—cont'd

Study	Study design	N, smokers/non-smokers	Sleep metrics	Findings	Comments
Patterson et al. [43]	UK Biobank prospective cohort study; self-report	N=34,401/405,212	Sleep Duration	Smokers < non-smokers	Smokers were also more likely to be long sleepers (≥9h)
			Late chronotype	Smokers > non-smokers	
Riedel et al. [49]	Epidemiological survey; 2 weeks sleep diaries	N=62/606	Self-reported Insomnia	Smokers > non-smokers	Findings for light smokers (<15 cigarettes per day); no significant findings between heavier smokers and non-smokers were found
			Time in bed	Smokers < non-smokers	
			Total sleep time	Smokers < non-smokers	
Grandner et al. [44]	2009 behavioral risk factor surveillance system (BRFSS)	N=57,631/184,234	Perceived insufficient sleep	Smokers > non-smokers	Sample includes daily and occasional smokers
Hayley et al. [45]	2012–2013 National epidemiologic survey on alcohol and related conditions (NESARC-III)	N=7265/28,912	Subjective sleep disturbance (difficulty falling/staying asleep)	smokers > non-smokers	DSM-5 diagnosis of tobacco use disorder in the past year
			Sleep duration	Smokers < non-smokers	

Note: For the purpose of this table, "non-smoker" refers to "never smokers" (i.e., smoked fewer than 100 cigarettes lifetime). For the purpose of this table, we focused on sleep metrics that differed between smokers and non-smokers. Readers are referred to the original articles for additional measures. PSQI=Pittsburgh Sleep Quality Index.

in these studies were reported to have a significantly higher percentage of N2 sleep, but significantly lower percentage of N3 sleep [39, 46]. Jaehne and colleagues reported that in a laboratory-conducted PSG assessment of 44 smokers and 44 matched non-smokers, smokers reported a higher REM density than their counterparts [37]. Collectively, this small body of work suggests that smokers may spend less time in deeper, more restful sleep-states than non-smokers.

Sleep continuity in smokers versus non-smokers

In terms of sleep onset latency, there is consensus across PSG verified studies that smokers (vs non-smokers) have a longer sleep onset latency [37–39, 46, 47], shorter sleep duration [37, 46], and later sleep timing [43]. PSG verified sleep onset latency has been reported to range from 5.4 to 24.9 min [37, 38, 46, 47] minutes longer in current versus non-smokers. Mean total sleep time/duration has also been found to differ between smokers and non-smokers, with smokers having shorter sleep. In one study, smokers reported 13.3 fewer minutes of total sleep time [37] and 14.0 min in another study [46]. Overall, smokers recorded significantly more time awake after sleep onset [38].

Findings from these PSG studies showing longer sleep onset latency and shorter duration in smokers versus non-smokers are consistent with the self-report literature. Using data from the National Health and Nutrition Examination study, Branstetter and colleagues found that current smokers took almost 25.9 (SD=21.3) minutes to fall asleep compared to 21.5 (SD=19.5) minutes in former smokers, and 22.1 (SD=19.3) minutes in never smokers [41]. Other studies have found self-reported sleep latency to be significantly longer in smokers than non-smokers [40, 42, 48].

In terms of differences in sleep duration and sleep timing, smokers report shorter sleep duration, and later sleep timing than non-smokers. For example, population level data from the National Health and Nutrition Examination Survey showed that mean sleep duration in smokers is 6.6 versus 6.9h in non/never smokers [41]. While data from the United Kingdom Biobank showed that in a sample of 34,401 smokers, 30.8% reported short sleep (≤6h), and 9.3% reported long sleep (≥9h) duration [43]. Several other studies found self-reported sleep duration to be significantly shorter in adult smokers than non-smokers [42, 45], with one study showing significance for light smokers (<15 cigarettes per day) versus non-smokers, only [49]. Using data from $N=323,047$ adult respondents of the 2009 Behavioral Risk Factor Surveillance System, Grandner and colleagues found that self-reported insufficient sleep was highest among daily current smokers and lowest among those who never smoked [44]. In terms of sleep timing, data from a national sample of adults showed that current smokers had a more than twofold greater odds of having an evening versus intermediate timing preference [43], as well as a 40% greater odds of waking up too early [42].

When the relationship between smoking status and sleep efficiency is considered, two of the five studies that used PSG assessment reported differences. Jaehne and colleagues found that smokers had poorer sleep efficiency that was not significantly different from non-smoker levels (87.08% vs 89.84%, respectively) [37], whereas Redline and colleagues report that sleep efficiency was significantly lower in smokers than non-smokers [39].

Together, these objective assessments of sleep continuity markers indicate that smokers have poorer sleep continuity than non-smokers as suggested by longer sleep latency and shorter sleep duration. Lower sleep efficiency was indicated by some, but not all studies reviewed.

Sleep fragmentation in smokers versus non-smokers

One PSG study of smokers and non-smokers observed that smokers had significantly more disruptive events such as general leg movements and a higher leg movement index as compared to non-smokers [37]. In one of the more comprehensive studies from the self-report literature examining the relationship between sleep and smoking status, a global disturbed sleep quality index was found to be significantly more prevalent in smokers versus non-smokers (28.1% vs 19.1%) [40]. Other data show only male smokers to have significantly greater prevalence of nightmares and disturbing dreams as compared to non-smokers [50].

Among smokers, nocturnal awakenings to smoke are common, reported in 19%–51% of smokers [51–53]. One study showed that among night smokers, night smoking occurred on one-in-four nights (26%) and averaged two episodes per night [52]. Epidemiological evidence also suggests that current nicotine dependence is also associated with greater subjective sleep disturbance [45]. Night-time smokers are more nicotine dependent [51, 52] and, following a cessation attempt, are more likely to relapse [52]. These studies indicate that smokers may be vulnerable to sleep fragmentation and disruptive events.

Daytime sleepiness in smokers versus non-smokers

Across several longitudinal and cross-sectional studies, smokers are more likely to report daytime sleepiness than non-smokers. In one longitudinal, observational study of 3516 adults, excessive daytime sleepiness was related to current smoking in females and not males [50]. In a study that used self-report NHANES data to examine sleep characteristics of current ($N=2015$), former ($N=2741$), and never smokers ($N=5752$), results showed that current smokers reported significantly more occurrences of feeling unrested and overly sleepy during the day as compared to the comparison groups [41]. Cross-sectional data from the Behavioral Risk Factor Surveillance System also showed that smokers reported significantly more daytime sleepiness [48].

SUMMARY

Together, these data suggest that smokers are vulnerable to deficits in sleep continuity and architecture. From a sleep continuity perspective, smokers are more vulnerable to longer sleep latency, more awakenings, poorer sleep quality, and shorter sleep time. From a sleep architecture perspective, shorter percentage of time in slow wave sleep is more common in smokers than non-smokers while subjective reports indicate that smokers have more restless sleep and greater daytime drowsiness and sleepiness.

SMOKING ABSTINENCE AND SLEEP

As a clinically verified symptom of nicotine withdrawal, insomnia is reported by up to 42% of abstinent smokers [54–56], while up to 80% of smokers habitually experience sleep disturbances [57], that then become exacerbated following cessation [58]. Nicotine withdrawal is a robust predictor of relapse to former smoking practices [59] and as such withdrawal symptoms are primary intervention targets. Elucidating the extent to which insomnia and other sleep deficits change following abstinence, and relate to smoking status and cessation outcome, is critical to quantifying the extent to which sleep may be a valid intervention target to promote cessation (see Fig. 22.1). Please also see Jaehne et al. (2009) and Hayley and Downey (2015) for recent reviews [60, 61].

Changes in sleep following abstinence

Three studies have objectively assessed sleep patterns (using polysomnography [PSG]) following cessation in treatment-seeking smokers. In the larger of the two studies, 33 smokers completed a PSG assessment at baseline, 24–36h, and 3-months following cessation [58]. Results showed a significantly increased percentage of wake time

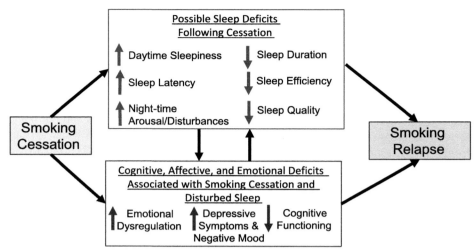

FIG. 22.1 Overview of sleep, cognitive, affective, and emotional deficits associated with smoking cessation and relapse.

after sleep onset and night-time arousal in the first 24–36 h of quitting; no significant differences were seen at the 3-month follow-up [58]. In another study that included an analytic sample of seven treatment-seeking smokers, data showed that sleep duration and efficiency declined significantly in the first month of abstinence, however, by 1 year after cessation, sleep metrics had improved with reductions in latency to REM sleep and stage 1 (light) sleep and increases in REM (deep) sleep [62]. Wetter and colleagues reported on a double-blind randomized trial that compared sleep architecture in 34 treatment-seeking smokers who received either active or placebo nicotine patches [63]. Sleep was PSG monitored for two nights before smoking cessation and three nights afterwards. The results showed that while sleep fragmentation significantly increased among placebo patch users, the active patch users did not demonstrate significant increases in sleep fragmentation following cessation [63]. Converging with these data from treatment-seeking smokers are data from a within-subject laboratory study that objectively compared the effects of smoking abstinence versus smoking as usual on sleep quality, daytime sleepiness and mood in a sample of 18 non-treatment seeking smokers. Results showed that as compared to smoking-as-usual, nicotine abstinence significantly increased relative arousals, sleep stage changes, and awakenings in the first week of abstinence [64]. Collectively, these objective assessments of sleep metrics across the quitting period suggest that sleep deficits in the form of longer sleep latency, decreased sleep duration and efficiency are likely in the first weeks of quitting, but that these deficits are ameliorated 3–12 months after quitting.

Several studies have also examined self-reports of the natural history of withdrawal in abstinent smokers. Cummings and colleagues reported on a sample of 33 smokers who completed withdrawal diaries daily for a 21-day period following cessation [54]. Difficulty sleeping and daytime sleepiness in this sample did not show significant declines across the 21-day observation period as compared to the other withdrawal symptoms measured (i.e., craving, irritability). Meanwhile, heavier smokers reported significantly higher mean scores of difficulty sleeping and daytime sleepiness than light smokers [54]. By contrast, electronic diary assessment of nicotine withdrawal duration and symptom severity showed that in 214 treatment-seeking smokers, sleep disturbances did dissipate in a 21-day monitoring period after abstinence [65]. Data from these self-report studies converge with findings from studies using objective measures of sleep by showing that following nicotine abstinence, smokers experience an exacerbation of insomnia-type symptoms (i.e., longer sleep onset latency, more frequent awakenings) and shorter sleep duration and that cross time, these symptoms may dissipate.

Relationship between sleep and cessation outcome

Ten studies that explicitly examined one or more sleep metrics in relation to smoking cessation outcomes were reviewed [53, 56, 66–73] (Table 22.2). While the range of sleep metrics measured, the use of different tools to measure the same sleep metrics, the variability in smoking cessation treatments used, and time-period of assessment pre- and post-cessation across these studies makes direct comparison challenging, some points of commentary can be raised.

First, eight studies showed that sleep metrics measured immediately before cessation and/or during cessation predicted relapse. For example, Peltier and colleagues reported that in a sample of 139 treatment seeking smokers, increased sleep latency, reduced subjective sleep quality and increased daytime dysfunction in the first week of quitting were predictive of relapse 4-weeks after treatment while increased sleep disturbances were predictive of relapse

TABLE 22.2 Review of literature examining relationship between sleep metrics and smoking cessation outcomes.

Study	Study design and sample	Assessment time points and sleep metrics measured	Treatment (pharmacotherapy, # sessions)	Comments
Foulds et al. [53]	- Cohort study analysis - 1021 smokers or recently quit smokers (59% female)	- Baseline, 4-week, and 6 month follow up - "Sometimes awaken at night to have a cigarette" (yes/no)	- Treatment that was recommended; - Six weekly sessions run by a clinical social worker, clinical psychologist, and intern - Individual counseling - FDA approved smoking cessation drugs: nicotine patch, gum, and lozenges	- At the 6 month follow up, 31.3% reported tobacco abstinence - Participants who reported waking at night to have a cigarette had a 40% increased odds of relapsing by 6-months following treatment even after adjustment for pre-treatment nicotine dependence
Rapp et al. [68]	- Secondary analysis of a cluster randomized trial on smoking cessation - 500 student nurses (82% female)	- Baseline and 13-month follow-up - Average sleep duration (single item)	- Three teaching units delivered to nursing students - No pharmacotherapy	- At 13-month follow-up, 10.6% had quit - Sleep duration positively associated with cessation: every hour of additional sleep increased the relative probability of cessation by 48% (aRR=1.48; CI=1.14–1.93)
Riemerth et al. [71]	- Secondary Analysis of a cohort study - 2884 participants (50.3% female)	- Baseline, week 1, week 2, week 3, week 4, and week 5 were measured - Sleep-disturbing nicotine craving (NSDNC)	- Smoking cessation program with individual counseling - Nicotine replacement therapy	- While looking at NSDNC, 22.4% of patients suffer from symptoms, 77.1% awoke rarely, 9.4% awoke several times per week, 6.8% awoke most days, 6.6% awoke daily - Those with higher rate of NSDNC can be considered high dependent smokers
Okun et al. [56]	- Secondary analysis of a randomized clinical trial - 322 women	- Baseline, 1 month post-quit, and 3 month follow up - Sleep disturbances, insomnia, drowsiness, and sleep quality	- Smoking cessation counseling; concerns or standard - Bupropion hydrochloride or placebo pharmacotherapy	- >25% of women reported sleep disturbances - Smoking cessation outcomes were not related to sleep disturbance ($P=.54$), symptoms of insomnia ($P=.52$), sleep quality scores ($P=.42$), and drowsiness ($P=.14$)
Peters et al. [67]	- Double blind randomized controlled trial - 385 smokers (48% female)	- Baseline before 6-week study duration and smoking prevalence 1, 6, 24, and 48 weeks after quitting - Sleep quality and disturbances (PSQI)	- Nicotine Patches (21 mg) - Naltrexone and placebo pharmacotherapy - 6-weekly counseling	- Participants that were both poor sleepers and night smokers were significantly more likely to be smoking at 6, 24, and 48 weeks - Poor sleepers, only, compared to both poor sleep and night smoking were significantly less likely to be smoking at week 6 (OR=0.44, CI=0.022–0.91)

Continued

TABLE 22.2 Review of literature examining relationship between sleep metrics and smoking cessation outcomes.—cont'd

Study	Study design and sample	Assessment time points and sleep metrics measured	Treatment (pharmacotherapy, # sessions)	Comments
Doner et al. [73]	- Cohort study - 2471 participants (447-two sessions, 421-three sessions, 527-four sessions, 1076-five sessions)	- Baseline then 2, 3 4, or 5 smoking cessation (depending on the amount the participant attended) - CO concentration and withdrawal symptoms measured with DSM-IV	- Four groups: two sessions, three sessions, four sessions, and five sessions - Individual or group based sessions offered - Pharmaceutical therapy	- Participants that attended more sessions has a higher chance of smoking cessation from 12.1% to 61.2% ($P<0.001$) - Baseline nocturnal wakening predicted lower odds of quitting ($P=0.0226$)
Fucito et al. [72]	- Randomized trial - 19 participants (9 with CBT-I + SC and 10 with SC)	- Two weeks prior to treatment (baseline), treatment times, and follow up - Sleep apnea (Berlin Questionnaire), daily sleep (Pittsburgh sleep diaries), Insomnia (Insomnia severity index)	- Two groups: (1) cognitive behavioral therapy for insomnia with smoking cessation counseling or (2) smoking cessation counseling alone - 8 sessions over 10 weeks	- Smoking abstinence at the end of treatment was low (CBT-I+SC:1/7, 14%; SC: 2/10, 20%) and follow-up (CBT-I+SC: 1/7, 14%; SC: 0/10, 0%) - Behavior intervention such as CBT-I might improve sleep for smokers that have insomnia
Ashare et al. [70]	- Secondary analysis of placebo controlled clinical trial - 1136 smokers (46% female)	- Baseline, 1, 4, 8 weeks after target quit date, and 12 months after target quit date - Sleep disturbances calculated from sleep problems, insomnia, and abnormal dreams	- Behavioral counseling through the telephone - Placebo, transdermal nicotine, or varenicline pharmacotherapy	- Treatments do not lessen withdrawal related sleep disturbances. But treatments that focus on sleep disturbances could improve smoking cessation rates - Participants that reported a higher amount of sleep disturbances at baseline testing were less likely to be abstinent (OR=0.79, CI=0.67–0.93, $P=0.004$)
Short et al. [69]	- Randomized control trial - 250 participants (52.8% female)	- Baseline, start of quitting process, and 3 month follow up - Insomnia (single item)	- Active or control group with smoking cessation program led by study staff - No pharmacotherapy	- Pre-quite insomnia measures were indicators and predictors of smoking cessation at month 3 from the quit date - Post-quit insomnia symptoms among patients did not show statistical significance with smoking status at month 3
Peltier et al. [66]	- Randomized control trial - 139 participants (57.6% female)	- Baseline and weeks 1, 4, 12 post-quit date - Sleep quality measured with WSWS and PSQI	- Two groups; usual care (weekly counseling, physician visits, pharmacotherapy) and usual care plus small financial incentives - 4 weeks of weekly sessions	- Participants that reported poor sleep quality the week prior to quitting and the week following the quit date had a reduced smoking cessation at weeks 4 and 12 among those of lower SES - Poor PSQI score was significantly correlated with WSWS measures assessed at quit date ($r=.58$, $P<0.001$) and at 1 week post quit ($r=.48$, $P=0.001$)

12-weeks after treatment [66]. Sleep disturbances alone did not predict relapse in a different sample of 385 treatment-seeking smokers. Instead, pre-cessation sleep disturbances interacted with waking at night to smoke (pre-cessation) to predict relapse 6, 24, and 48 weeks post-quitting [67].

Second, pre-treatment sleep habits are relevant to smoking outcomes. Four of the studies found that pre-cessation (vs abstinence-induced) sleep deficits were predictive of relapse [67, 69, 70, 73]. In a sample of 579 smokers who received a 12-week anxiety-related smoking cessation program versus a control condition, the results showed that smokers who self-reported pre-cessation insomnia symptoms had a greater odds (aOR = 1.11; 95% CI = 1.01–1.22) of relapsing 3-months following cessation than those who did not have pre-cessation insomnia symptoms. Post-quit insomnia was not related to cessation outcome [69]. Likewise, in another study of 1136 smokers who received pharmacotherapy and counseling, data showed that smokers reporting more sleep disturbance pre-treatment were less likely to be quit at the end of treatment (OR = 0.79; 95% CI = 0.67–0.93) [70]. Dorner and colleagues reported that greater nocturnal awakenings at baseline was an independent predictor of relapse 5-weeks following cessation in a sample of 2471 treatment-seeking smokers [73]. The remaining studies reviewed either did not have an assessment of sleep in the first week(s) of cessation [68], found that sleep patterns both before and after cessation predicted cessation outcome [66], did not report results in sufficient detail to ascertain whether sleep quality before or after cessation was related most to cessation [71], or did not find that sleep was related to cessation outcome [56].

Third, only one of the studies reviewed was designed specifically to test the efficacy of a behavioral sleep intervention on cessation outcome in a small sample of 19 smokers with a clinical diagnosis of insomnia [72]. Fucito and colleagues compared quit rates in nine participants who received a cognitive-behavioral treatment for insomnia + smoking cessation counseling + transdermal nicotine versus smoking cessation counseling + transdermal nicotine. The results of this small study showed that participants receiving the experimental insomnia treatment reported better sleep quality and efficiency; they also had more days to relapse [72].

Some of the take-home points from this literature are that sleep deficits (i.e., insomnia type symptoms of longer sleep latency, night-time awakenings, difficulty staying asleep) both before and after a quit attempt may predict relapse in treatment seeking smokers. Importantly, not all studies found these associations, suggesting that there may be subgroups of smokers (i.e., those with higher levels of pretreatment insomnia symptoms) who may be more vulnerable to the exacerbated sleep deficits following cessation. Cognitive behavioral treatment for insomnia as an adjunctive treatment for smoking cessation may be a plausible approach to delaying relapse. The characterization or phenotype of treatment seeking smokers most vulnerable to relapse because of sleep deficits, and the extent to which cognitive behavioral therapy for insomnia increased days of abstinence in this population warrants consideration.

Effects of pharmacotherapy on sleep

Sleep disturbances are a recognized side-effect of the FDA-approved treatments for nicotine dependence including nicotine replacement therapies (patch, spray, gum, lozenge), bupropion and varenicline. One placebo-controlled trial that utilized nicotine patch and varenicline treatment arms showed that these active treatments did not ameliorate withdrawal-related sleep disturbance, thus strategies to address sleep disturbances induced by smoking cessation pharmacologic treatments are needed to promote cessation [70]. Characterizing the sleep disturbances presented by each of the pharmacologic treatments is therefore necessary to informing the design of adjunctive nicotine dependence treatments.

Nicotine replacement therapy

Nicotine replacement therapies (NRTs; transdermal patch, gum, spray, lozenge) provide partial nicotine replacement upon cessation of smoking and in doing so, ameliorate nicotine craving and pharmacologic withdrawal symptoms [74]. Up to 50% of treatment seeking smokers using nicotine replacement therapies report sleep disturbances that start on the day of use [75]. Disturbed sleep, vivid dreams and daytime drowsiness are some of the more commonly reported side effects from using nicotine replacement therapies. In one study, 6.4% of participants reported disturbed sleep, 4.4% reported vivid dreams, and 1.5% reported daytime drowsiness while using NRT [76]. Meta-analytic data of 120 studies involving 177,390 individuals, showed that the prevalence of insomnia among individuals using nicotine replacement therapy for smoking cessation was 11.4% [77]. High levels of pre-treatment nicotine dependence, continued cessation, and female gender were found to significantly predict sleep disturbances 4-weeks after quitting in a sample of 1392 treatment-seeking smokers [75]. Importantly, wearing the patch for 16h (vs 24h) does not reduce its efficacy [78]. Therefore, smokers who experience sleep disruption while using NRT may remove the patch before going to sleep.

Studies examining the trajectory of NRT sleep-related side effects suggest that sleep disturbances among NRT users may take some time to subside. In one cohort study, instances of sleep disturbance (vivid dreams, other sleep disturbances) were still being reported by up to 50% of abstinent smokers after 12-weeks of treatment [75]. This is consistent with another study that showed no change in reports of sleep disturbance in the 21-days following cessation [65], but inconsistent with data showing that use of transdermal nicotine actually ameliorates sleep disturbances following cessation compared to placebo [63]. Collectively, these studies reporting on NRT use and sleep in smokers suggest that up to one-in-ten treatment seeking smokers can experience NRT-induced sleep disturbance following cessation that may last well into the quitting period (i.e., up to 12 weeks).

Bupropion

Sustained release bupropion (bupropion SR) is an aminoketone anti-depressant that is hypothesized to promote smoking cessation and delay relapse [79] to smoking by inhibiting dopamine reuptake in the reward center of the brain. Compared to placebo, bupropion increases the relative risk of cessation by 1.62 (95% CI=1.49–1.76) [80].

Between 4% and 21% of treatment seeking smokers using bupropion SR report disturbed sleep including insomnia, abnormal dreams and daytime fatigue [81]. Some studies show that sleep disturbances associated with bupropion are significantly higher than those found in placebo, and varenicline [82]. Conversely, other studies show no significant increases in sleep disturbances associated with bupropion treatment [83]. Although this evidence reporting on the increases of sleep disturbances following cessation using bupropion is mixed, that up to one-in-five bupropion users report an increase in sleep disturbances is clinically meaningful.

Varenicline

Varenicline is an $\alpha 4 \beta 2$ partial agonist medication indicated for the treatment of nicotine dependence. As a $\alpha 4 \beta 2$ partial agonist, varenicline stimulates sufficient dopamine to reduce craving while simultaneously acting as a partial antagonist by blocking reinforcement from smoked nicotine [84]. Double-blind, randomized trials show varenicline to outperform bupropion, nicotine replacement therapy, and placebo in producing higher quit rates. For example, Gonzales and colleagues report that following a 12-week treatment period, varenicline quit rates were 50.3% as compared to 33.5% in the bupropion arm and 14.5% in the placebo arm [85]. Compared to placebo, meta-analytic data show bupropion to increase the odds of cessation by 1.84 and varenicline by 2.88 [86], thus, varenicline is considered the most effective FDA-approved treatment for nicotine dependence.

Listed side-effects of varenicline include insomnia, vivid or lucid dreams and other sleep disturbances such as difficulty staying asleep. McClure and colleagues reported that 39%–46% of treatment seeking smokers using varenicline reported difficulty sleeping, while 56%–68% reported a change in dreaming, and that these sleep disturbances were retained 21-days after cessation [87]. Meta-analysis of clinical trials that compared the efficacy of varenicline to placebo, show that disturbed sleep, specifically insomnia symptoms of difficulty falling and staying asleep, as well as the incidence of abnormal dreams were between 50% and 70% higher in varenicline recipients [88, 89]. One study that prospectively evaluated changes in sleep insomnia and dreams among treatment seeking smokers using varenicline ($N=38$), showed that, based on daily sleep diaries over a 7-day period, participants retained excellent sleep efficiency (>90%) and that while overall sleep measures did not change significantly, an increased number of awakenings and reports of dreams was observed [90]. Prospective studies suggest that insomnia-related symptoms peak in the first week of quitting and then progressively decline until pre-treatment levels are achieved at 2–12 weeks [91]. Together, these studies reporting on the relationship between varenicline use and sleep disturbances show that while as many as seven-in-ten treatment seeking smokers using varenicline report sleep symptoms, the symptoms dissipate across time.

Take home points: Relationship between sleep and cessation outcome

Poor sleep health as characterized by shorter sleep duration, difficulty falling asleep, difficulty staying asleep, early awakenings and night-time awakenings are more common in smokers than non-smokers. Of particular relevance to smoking cessation efforts, sleep health deteriorates following cessation in many smokers, and this in turn is implicated in relapse. Importantly, FDA-approved treatments for nicotine dependence may also impede healthy sleep. Varenicline, the most effective smoking cessation treatment, in particular produces insomnia symptoms and abnormal dreams as a notable side effect. These different lines of evidence converge to underscore sleep as an intervention target for treatment-seeking smokers, particularly for those using pharmacotherapy. Another question raised by this body of work is whether there are sub-groups of smokers (i.e., those with higher nicotine dependence; those with poorer pre-cessation sleep health; those with conditions associated with smoking and poor sleep health, such as depression) who are particularly vulnerable to sleep deficits and poorer sleep health following cessation, and therefore might be a higher-priority for a sleep health intervention.

POSSIBLE MECHANISMS LINKING POOR SLEEP TO SMOKING CESSATION OUTCOMES

To further understand the possible relationship between sleep and smoking cessation, it is important to consider the different mechanisms through which sleep may impact smoking behavior and vice versa. Plausible mechanisms through which tobacco use and sleep interact include cognitive, affective (i.e., mood, depressive symptoms) and emotional (i.e., emotional dysregulation) states, as well as neurobiological mechanisms (see Fig. 22.1). A better understanding of these mechanisms may also shed light on sub-groups of smokers who may be most likely to experience poor sleep during a quit attempt.

Unhealthy sleep has been associated with cognitive deficits [92–94], and cognitive impairment following smoking cessation predicts relapse [95]. Adverse changes in sleep (either substantial increases or decreases in sleep duration) have been associated with compromised cognitive function [96]. For example, short (≤ 6h) and long (≥ 9h) sleep has predicted poorer cognitive function [92]. Even an extra 6h of wakefulness can produce deficits in alertness and working memory [97]. In studies of experimentally-induced sleep restriction, sleep loss leads to impairments in vigilance and sustained attention [98], as well as executive function and decision-making [99], which could plausibly lead to difficulty making healthy choices. For example, Greer and colleagues [100] showed that sleep loss led to worse food-related decision making. However, studies specifically linking sleep loss due to smoking and decision making around smoking have not yet been conducted.

Disruption in cognitive processing is a common nicotine withdrawal symptom, [55] with up to one-half of abstinent smokers reporting difficulty concentrating [101]. During abstinence, smokers experience specific deficits in sustained attention [102], working memory [103, 104], and executive function [95] which are mitigated upon resumption of nicotine use [105]. Importantly, attention and concentration deficits following a quit attempt increase the risk of smoking relapse in clinical studies [106–108]. Thus, cognitive-deficits and disturbed sleep are both abstinence symptoms in habitual smokers that may interact to increase the likelihood of relapse.

Moreover, comorbid conditions associated with cognitive impairment and high smoking rates also exhibit higher prevalence of poor sleep. For instance, smoking rates among people living with HIV (PLWH) are 50%–74%—about three times higher than in the general population [109–113]. The widespread use of anti-retroviral therapy (ART) has improved survival rates among PLWH [114–116], making addressing modifiable health-risk behaviors, such as tobacco use has become a critical priority. PLWH are also increasingly vulnerable to non-AIDS-related diseases including cardiovascular disease, bone disease, frailty, and HIV-associated neurocognitive disorder (HAND) [117–122]. Moreover, there is evidence that poor sleep is common among PLWH, including increased sleep onset latency and reduced N1 sleep, relative to controls [123]. PLWH who experience poor sleep, either subjectively or objectively measured, report lower quality of life and greater daytime dysfunction [124]. While there is some evidence that certain ART regimens may contribute to poor sleep [125], high rates of tobacco use among PLWH may also exacerbate poor sleep, which may in turn, increase the severity of HAND. These relationships are clearly complex and more research is necessary to evaluate the unique and combined effects of tobacco use and HIV on sleep metrics and cognitive function.

Similar to cognition, there are data to suggest that depressive symptoms and emotional dysregulation are associated with smoking [126, 127] and habitually poor sleep [128–130]. Poor sleep health is considered a central component of mood disorders [131] and there is growing consensus that disruption of the circadian system (sleep-wake cycle) contributes to the pathophysiology of mood disorders [131–133]. Up to 90% of depressed patients self-report difficulty falling or staying asleep [134, 135]. PSG measures of sleep health including decreased REM latency (i.e., interval between sleep onset and the first REM sleep period), increased total REM sleep time and REM density (i.e., the frequency of rapid eye movements per REM period), and diminished slow wave sleep (SWS) production [136–139] predict response to depression treatment and recurrence of depression symptoms [140–142]. Moreover, sleep disturbance often precedes the onset of depression [143–145]. Even among non-depressed adults, poor sleep quality precedes a subsequent increase in depressive symptoms and negative mood [128].

Importantly, tobacco use and depression are highly comorbid. While smoking prevalence continues to decline in the general population, those with psychiatric disorders, including depression, are increasingly overrepresented among smokers [4, 146–148]. Smokers have a higher prevalence of depression than non-smokers [149, 150] and upwards of 43% of individuals with depression are smokers [4, 147]. Smoking increases risk of first incidence, severity and recurrence of depression [151–154], and heavier smokers are at higher risk of depression [147, 155]. Neuroimaging studies have revealed that abstinent smokers [156–158], depressed individuals [159, 160], and individuals with sleep disorders [161] exhibit similar patterns of brain activity during difficult cognitive tasks. This complex interplay between sleep, smoking, and depressive symptomology is likely exacerbated upon smoking cessation when abstinence from nicotine leads to increases in negative mood and insomnia symptoms [128], both of which have been shown to relate to relapse among treatment-seeking smokers [66, 162]. The temporal sequence of changes in depressive symptoms and sleep heath following cessation has yet to be fully understood, but such information would inform upstream intervention targets for smoking behavior.

Likewise, emotion dysregulation, or the ability to regulate emotions and control behavioral responses, has been implicated as a mechanism for how sleep may relate to smoking cognitions and quitting outcomes. From the outset, poor sleep quality has been highly correlated with emotion dysregulation in smokers. Recent evidence suggests that emotion dysregulation mediates the relationship between insomnia symptoms and smoking variables including, negative reinforcement smoking outcome expectancies, negative reinforcement smoking motives and negative reinforcement expectancies from smoking abstinence. Importantly these associations were adjusted for other demographic and smoking behavior variables [129]. Similarly, Filio and colleagues, showing that in a sample of 128 treatment-seeking smokers, greater emotion dysregulation was associated with lower self-efficacy for remaining abstinent, and a lower likelihood of having had a quit attempt of 24 h or greater [127]. This small body of work converges to suggest that emotion regulation may be an important mechanism linking sleep with cigarette smoking behaviors and quitting.

Sleep and smoking behavior also share several neurobiological mechanisms, which may partially explain these associations. For instance, the naturally occurring hormone, melatonin, which plays an essential role in sleep-wake function, has been shown to be lower among smokers compared to non-smokers [163, 164]. In preclinical models, melatonin receptor knockout mice exhibited greater sensitivity to nicotine [165]; enhancing melatonin function may attenuate nicotine withdrawal symptoms [166] and reduce nicotine administration in mice [167]. More recently, the peptide, hypocretin which plays an important role in the sleep/wake cycle through its wake-promoting effects [168] has been shown to be associated with nicotine self-administration in rodents [169]. While there are certainly other neurotransmitters involved, melatonin and hypocretin are reviewed here to emphasize the neurobiological links between smoking and sleep that may shed light on this complex relationship.

Cognitive, affective, and emotional states present plausible pathways through which sleep and tobacco use may interact (see Fig. 22.1). This area of work is severely underdeveloped, and longitudinal studies are needed to quantify the association, and the temporal relationships, between these variables across time. Moreover, mechanistic studies are necessary to better understand the neurobiological pathways that are common and unique to sleep health and tobacco use that may lead to novel targets for interventions. Testing the extent to which improving sleep ameliorates deficits in cognitive, affective, and emotional states in smokers across the smoking cessation process will help determine if sleep improvement is a viable adjunctive therapy for smoking cessation.

PLAUSIBLE ADJUNCTIVE SLEEP THERAPIES TO PROMOTE SMOKING CESSATION

Overview

Smokers typically exhibit sleep patterns consistent with insomnia-type symptoms including difficulty falling asleep (long sleep latency) and difficulty staying asleep (short sleep duration, frequent awakenings and arousal during the night), that are amplified following cessation [58]. Some studies suggest that increases in disturbed sleep following cessation is attributed to the use of pharmacotherapy, whereas others suggest that disturbed sleep following cessation is attributable to nicotine-withdrawal [70]. In both scenarios, disturbed sleep before [67, 69, 70], and after cessation predicts relapse, and as such, warrants treatment as part of the cessation process. There are a range of behavioral and pharmacological treatments for insomnia-type symptoms that may be suitable for use in conjunction with standard nicotine dependence treatment (counseling + pharmacotherapy); an overview is provided here.

Behavioral treatments

Cognitive-behavioral therapy for insomnia (CBT-I) is a first-line treatment for chronic insomnia [170] that improves sleep outcomes for up to 2 years after treatment [171] and is preferred by patients with a clinical diagnosis of insomnia to drug therapy [172]. CBT-I is comprised of two core components (stimulus control and sleep restriction therapy), as well as several optional components including cognitive therapy, sleep hygiene, and relaxation [173]. Stimulus control techniques work to strengthen the association between the bed and bedroom with sleep, and to establish a consistent sleep schedule. Sleep restriction therapy is a specific approach that addresses the mismatch between sleep ability and sleep opportunity by reducing sleep opportunity to match ability and then slowly upwardly titrating sleep opportunity as long as the individual is able to maintain high sleep efficiency. Cognitive therapy seeks to identify and replace dysfunctional beliefs and attitudes about sleep and insomnia. Sleep hygiene works to address environmental factors, physiologic factors, and behavioral components (i.e., regular sleep scheduling, limiting alcohol intake). Relaxation training seeks to address the high levels of physiologic, cognitive, and/or emotional arousal, both at night and during the daytime, which is exhibited by individuals who have difficulty falling and/or staying asleep [170, 174]. Deep breathing, progressive relaxation, and meditation are relaxation techniques that haven been shown to lower pre-sleep arousal (e.g., racing thoughts) and improve sleep metrics [175]. In a recent meta-analysis of 20 studies that examined the efficacy of CBT-I among patients with chronic insomnia, sleep onset latency, wake after sleep onset, total sleep time, and sleep efficiency, were all significantly improved by multimodal CBT-I [175]. This is in the context of several other meta-analyses and systematic reviews of CBT-I showing that not only is it superior to placebo [174] and equivalent or superior to pharmacotherapy for insomnia [176], but it is effective even in the presence of comorbid conditions such as depression and chronic pain [177].

To date, only one study has examined the effects of a CBT-I intervention on smoking cessation outcomes [178]. Nineteen treatment-seeking smokers were randomized to receive eight sessions of CBT-I, transdermal nicotine patch and smoking cessation counseling ($N=9$) versus transdermal nicotine patch and smoking cessation counseling ($N=10$) alone. While the results showed no difference in smoking cessation rates between the groups, participants receiving the CBT-I had a longer time to relapse [178]. A fully-powered examination of the effects of CBT-I on smoking cessation outcomes is warranted.

Pharmacological treatments

Benzodiazepines are a pharmacologic first-line treatment for insomnia. Currently there are five FDA-approved benzodiazepines for this indication: estazolam, flurazepam, quazepam, temazepam, and triazolam [179]. These medications act by increasing the activity of the inhibitory neurotransmitter GABA to inspire drowsiness or sedation. Consistent with this mechanism, sleep latency (time to sleep) and wake after sleep onset, are both significantly reduced while sleep duration and sleep quality are significantly increased using these therapeutics in the short term. However, with increased tolerance of these pharmaceutics, sleep improvements may be curtailed [180]. Of particular relevance to smokers, cigarette smoke contains beta-carbolines that block the actions of benzodiazepines at the GABA-A receptors [181], thus higher doses may be needed in smokers versus non-smokers to observe comparable effects. In addition, benzodiazepines and other hypnotics (e.g., eszopiclone, zolpidem) often produce rebound insomnia and next-day residual effects, such as memory impairment, difficulty concentrating, or mood symptoms [182–184]—all of which may promote smoking relapse [185, 186]. Nevertheless, benzodiazepines are not contraindicated with any of the FDA-approved treatments for nicotine dependence and their role in promoting smoking cessation through improved sleep has yet to be evaluated.

Melatonin is a hormone normally secreted from the pineal gland at night that serves as the signal of darkness in the organism and as such plays a pivotal role in the physiological regulation of circadian rhythms, including sleep. Several melatonin receptor agonists have recently become available for treatment of sleep disorders: ramelteon for the treatment of insomnia characterized by difficulty with sleep onset, prolonged-release melatonin for treatment of primary insomnia characterized by poor quality of sleep in patients who are aged 55 or over, agomelatine for the treatment of depression and associated sleep disorder, and tasimelteon for the treatment of non-24 h sleep-wake disorder in the blind [187]. Given that longer sleep latency (difficulty falling asleep) is a characteristic of smokers (vs non-smokers) that is exacerbated following cessation, the reported reductions in sleep latency in ramelteon users [188] may be particularly beneficial to curbing sleep deficits following smoking cessation. Moreover, these medications have not been shown to produce the adverse next-day effects or rebound insomnia associated with hypnotics [189]. These melatonin receptor agonists are not contraindicated with the FDA-approved smoking cessation medications and their efficacy as adjunctive smoking cessation treatments warrants investigation.

The discovery of the role of the neuropeptide, orexin, in the sleep-wake cycle is thought to be one of the major advances in sleep research in the last two decades [190]. This finding led to the development of several orexin receptor antagonists as potential treatments for insomnia. Suvorexant, a dual orexin 1 and orexin 2 receptor antagonist, was approved by the FDA in 2014 for insomnia and other medications that act exclusively on the orexin 2 receptor are currently being tested. These medications promote sleep by increasing REM sleep with few, if any, effects on slow wave sleep [190]. Given that smokers may spend less time in REM sleep [37], it is plausible that these medications may address this sleep deficit during a smoking cessation attempt. Indeed, the orexin system has been shown to play a role in preclinical studies of nicotine self-administration [191]. However, these medications negative side effects similar to other hypnotics and whether they can be used in conjunction with FDA-approved smoking cessation treatments is unknown.

DIRECTIONS FOR FUTURE RESEARCH

On the basis of the research reviewed in this chapter, we suggest that sleep is an understudied and underutilized intervention target for promoting smoking cessation and preventing relapse in treatment-seeking smokers. As demonstrated, sleep deficits in terms of shorter sleep duration and insomnia symptoms (difficulty getting to sleep and staying asleep) is a sleep phenotype of smokers that may become exacerbated following cessation, both as an abstinence symptom, and, as a side-effect of quit-smoking medications. As such, it could be argued that smoking cessation practitioners have a basis from which to advise treatment-seeking smokers to strive to develop and maintain a healthy sleep schedule. A healthy sleep schedule could be defined as maximizing sleep efficiency through restricting time in bed for sleep or sex, achieving adequate sleep duration of 7–8 h, and achieving an earlier time to bed. Maintaining healthy sleep may facilitate the quitting process and increase abstinence. In addition, because standard pharmacotherapies for smoking cessation may exacerbate sleep disturbance, practitioners might discuss these potential side effects and strategies to mitigate them (e.g., for patients reporting sleep difficulty prior to a quit attempt, advise them to remove the nicotine patches before going to sleep).

To build the empirical basis from which to support these (and potentially other) sleep health recommendations to promote smoking cessation treatment response, there are several directions for future work that are needed. First, the temporal relationship between smoking and sleep needs further consideration. As discussed in this review, sleep may be disrupted because of the physiological effects of nicotine and nicotine withdrawal upon abstinence. Conversely, smokers may use their smoking habit to counter the effects of daytime sleepiness because of poor sleep. Prospective, observational studies examining the temporal relationship underpinning this complex interplay between sleep and smoking are needed. Meta-analytic studies to quantify the relationship between sleep deficits with smoking behaviors and cessation outcomes would also be valuable.

Second, laboratory and clinical studies to examine the effects of pharmacological treatments for insomnia on tobacco consumption in a natural setting and as an adjunctive treatment for smoking cessation are needed. Behavioral pharmacology and laboratory studies also provide a unique opportunity to investigate mechanisms underlying this association including neurobiological mechanisms as well as the mediating role of cognition and affect on the relationship between tobacco use and sleep.

Third, behavioral therapies for insomnia targeting treatment-seeking smokers need to be developed and evaluated in the context of smoking cessation interventions. For example, physical activity and mindfulness based approaches have independently been shown to reduce insomnia symptoms [192] and promote smoking abstinence [193, 194], thus, these approaches may have promise in treating smokers with higher levels of sleep deficits.

Another goal of this line of research is to elucidate the relationship between sleep, tobacco use, and cessation outcomes that will ultimately inform a sleep phenotype of risk for continued smoking. Characterizing smokers according to this phenotype will inform targeted intervention approaches to promote cessation outcomes in smokers most vulnerable to sleep deficits.

ACKNOWLEDGMENTS

Research reported in this publication was supported by an Institutional Development Award (IDeA) Center of Biomedical Research Excellence from the National Institute of General Medical Sciences of the National Institutes of Health under grant number P20GM113125, by grant support from the National Institute on Drug Abuse (R21 DA040902), by the National Institute On Minority Health And Health Disparities (R01MD012734) and by the University of Delaware Research Foundation grant number 16A01366.

CONFLICTS OF INTEREST

Dr. Patterson receives medication free of charge from Pfizer.

REFERENCES

[1] Xu X, Bishop EE, Kennedy SM, Simpson SA, Pechacek TF. Annual healthcare spending attributable to cigarette smoking: an update. Am J Prev Med 2015;48(3):326–33.

[2] USDHHS. The health consequences of smoking—50 years of progress: a report of the surgeon general. Atlanta, GA: Centers for Disease Control; 2014.

[3] Jamal A, King BA, Neff LJ, Whitmill J, Babb SD, Graffunder CM. Current cigarette smoking among adults—United States, 2005-2015. MMWR Morb Mortal Wkly Rep 2016;65(44):1205–11.

[4] Lasser K, Boyd JW, Woolhandler S, Himmelstein DU, McCormick D, Bor DH. Smoking and mental illness: a population-based prevalence study. JAMA 2000;284(20):2606–10.

[5] Fiore MC. Treating tobacco use and dependence: 2008 update: clinical practice guideline. U.S. Department of Health and Human Services; Public Health Service; 2008.

[6] Carpenter MJ, Jardin BF, Burris JL, Mathew AR, Schnoll RA, Rigotti NA, et al. Clinical strategies to enhance the efficacy of nicotine replacement therapy for smoking cessation: a review of the literature. Drugs 2013;73(5):407–26.

[7] Hughes JR, Hatsukami D. Signs and symptoms of tobacco withdrawal. Arch Gen Psychiatry 1986;43(3):289–94.

[8] Centers of Disease Control. Trends in current cigarette smoking among high school students and adults, United States, 1965–2014. Available from: https://www.cdc.gov/tobacco/data_statistics/tables/trends/cig_smoking/index.htm; 2015.

[9] Perkett M, Robson SM, Kripalu V, Wysota C, McGarry C, Weddle D, et al. Characterizing cardiovascular health and evaluating a low-intensity intervention to promote smoking cessation in a food-assistance population. J Community Health 2017;42(3):605–11.

[10] Association AL. Cutting tobacco's rural roots: tobacco use in rural communities Chicago. American Lung Association; 2015.

[11] Substance Abuse and Mental Health Services Administration. Results from the 2014 National survey on drug use and health: detailed tables [PDF–9.48 MB]. Tables 2.56B, 2.61B, 2.66B. Rockville, MD: Substance Abuse and Mental Health Services Administration, Center for Behavioral Health Statistics and Quality; 2015.

[12] Jamal A, Agaku IT, O'Connor E, King BA, Kenemer JB, Neff L. Current cigarette smoking among adults—United States, 2005-2013. MMWR Morb Mortal Wkly Rep 2014;63(47):1108–12.

[13] Mokdad AH, Marks JS, Stroup DF, Gerberding JL. Actual causes of death in the United States, 2000. JAMA 2004;291(10):1238–45.

[14] Services. USDoHaH. Let's make the next generation Tobacco-Free: your guide to the 50th anniversary surgeon General's report on smoking and health. Atlanta: U.S.: Department of Health and Human Services, Centers for Disease Control and Prevention, National Center for Chronic Disease Prevention and Health Promotion, Office on Smoking and Health; 2014.

[15] Wannamethee G, Shaper AG, Macfarlane PW, Walker M. Risk factors for sudden cardiac death in middle-aged British men. Circulation 1995;91(6):1749–56.

[16] Price JF, Mowbray PI, Lee AJ, Rumley A, Lowe GD, Fowkes FG. Relationship between smoking and cardiovascular risk factors in the development of peripheral arterial disease and coronary artery disease: Edinburgh artery study. Eur Heart J 1999;20(5):344–53.

[17] Colditz GA, Bonita R, Stampfer MJ, Willett WC, Rosner B, Speizer FE, et al. Cigarette smoking and risk of stroke in middle-aged women. N Engl J Med 1988;318(15):937–41.

[18] Kawachi I, Colditz GA, Stampfer MJ, Willett WC, Manson JE, Rosner B, et al. Smoking cessation and decreased risk of stroke in women. JAMA 1993;269(2):232–6.

[19] Cahill K, Stevens S, Lancaster T. Pharmacological treatments for smoking cessation. JAMA 2014;311(2):193–4.

[20] Hartmann-Boyce J, Lancaster T, Stead LF. Print-based self-help interventions for smoking cessation. Cochrane Database Syst Rev 2014;6:CD001118.

[21] Anthenelli RM, Benowitz NL, West R, St Aubin L, McRae T, Lawrence D, et al. Neuropsychiatric safety and efficacy of varenicline, bupropion, and nicotine patch in smokers with and without psychiatric disorders (EAGLES): a double-blind, randomised, placebo-controlled clinical trial. Lancet 2016;387(10037):2507–20.

[22] Rigotti NA, Bitton A, Kelley JK, Hoeppner BB, Levy DE, Mort E. Offering population-based tobacco treatment in a healthcare setting: a randomized controlled trial. Am J Prev Med 2011;41(5):498–503.

[23] Patnode CD, Henderson JT, Thompson JH, Senger CA, Fortmann SP, Whitlock EP. Behavioral counseling and pharmacotherapy interventions for tobacco cessation in adults, including pregnant women: a review of reviews for the U.S. preventive services task force. Ann Intern Med 2015;163(8):608–21.

[24] de Souza IC, de Barros VV, Gomide HP, Miranda TC, Menezes Vde P, Kozasa EH, et al. Mindfulness-based interventions for the treatment of smoking: a systematic literature review. J Altern Complement Med 2015;21(3):129–40.

[25] Lee EB, An W, Levin ME, Twohig MP. An initial meta-analysis of acceptance and commitment therapy for treating substance use disorders. Drug Alcohol Depend 2015;155:1–7.

[26] Stead LF, Koilpillai P, Lancaster T. Additional behavioural support as an adjunct to pharmacotherapy for smoking cessation. Cochrane Database Syst Rev 2015;10:CD009670.

[27] Stead LF, Lancaster T. Behavioural interventions as adjuncts to pharmacotherapy for smoking cessation. Cochrane Database Syst Rev 2012;12:CD009670.

[28] U.S. Department of Health and Human Services. The health consequences of smoking—50 years of progress: a report of the surgeon general. Atlanta, U.S.: Department of Health and Human Services, Centers for Disease Control and Prevention, National Center for Chronic Disease Prevention and Health Promotion, Office on Smoking and Health; 2014.

[29] Schnoll RA, Lerman C. Current and emerging pharmacotherapies for treating tobacco dependence. Expert Opin Emerg Drugs 2006;11(3):429–44.

[30] Kotz D, West R. Explaining the social gradient in smoking cessation: it's not in the trying, but in the succeeding. Tob Control 2009;18(1):43–6.

[31] Gilman SE, Martin LT, Abrams DB, Kawachi I, Kubzansky L, Loucks EB, et al. Educational attainment and cigarette smoking: a causal association? Int J Epidemiol 2008;37(3):615–24.

[32] Patterson F, Grandner MA, Malone SK, Rizzo A, Davey A, Edwards DG. Sleep as a target for optimized response to smoking cessation treatment. Nicotine Tob Res 2019;21(2):139–48.

[33] Smith LJ, Nowakowski S, Soeffing JP, Orff HJ, P M. The measurement of sleep. In: Perlis ML, KL L, editors. Treating sleep disorders: principles and practice of behavioral sleep medicine. Hoboken, NJ: Wiley; 2003.

[34] Carskadon MA, Dement WC. Monitoring and staging human sleep. In: Kryger MH, Roth T, Dement WC, editors. Principles and practice of sleep medicine. 5th ed. St. Louis: Elsevier Saunders; 2011. p. 16–26.

[35] Chambers AM. The role of sleep in cognitive processing: focusing on memory consolidation. Wiley Interdiscip Rev Cogn Sci 2017;8(3): https://doi.org/10.1002/wcs.1433.

[36] Krystal AD, Edinger JD. Measuring sleep quality. Sleep Med 2008;9(Suppl 1):S10–7.

[37] Jaehne A, Unbehaun T, Feige B, Lutz UC, Batra A, Riemann D. How smoking affects sleep: a polysomnographical analysis. Sleep Med 2012;13(10):1286–92.

[38] Soldatos CR, Kales JD, Scharf MB, Bixler EO, Kales A. Cigarette smoking associated with sleep difficulty. Science 1980;207(4430):551–3.

[39] Redline S, Kirchner HL, Quan SF, Gottlieb DJ, Kapur V, Newman A. The effects of age, sex, ethnicity, and sleep-disordered breathing on sleep architecture. Arch Intern Med 2004;164(4):406–18.

[40] Cohrs S, Rodenbeck A, Riemann D, Szagun B, Jaehne A, Brinkmeyer J, et al. Impaired sleep quality and sleep duration in smokers-results from the German multicenter study on nicotine dependence. Addict Biol 2014;19(3):486–96.

[41] Branstetter SA, Horton WJ, Mercincavage M, Buxton OM. Severity of nicotine addiction and disruptions in sleep mediated by early awakenings. Nicotine Tob Res 2016;18(12):2252–9.

[42] McNamara JP, Wang J, Holiday DB, Warren JY, Paradoa M, Balkhi AM, et al. Sleep disturbances associated with cigarette smoking. Psychol Health Med 2014;19(4):410–9.

[43] Patterson F, Malone SK, Lozano A, Grandner MA, Hanlon AL. Smoking, screen-based sedentary behavior, and diet associated with habitual sleep duration and chronotype: data from the UK biobank. Ann Behav Med 2016;50(5):715–26.

[44] Grandner MA, Jackson NJ, Izci-Balserak B, Gallagher RA, Murray-Bachmann R, Williams NJ, et al. Social and behavioral determinants of perceived insufficient sleep. Front Neurol 2015;6:112.

[45] Hayley AC, Stough C, Downey LA. DSM-5 tobacco use disorder and sleep disturbance: findings from the National Epidemiologic survey on alcohol and related conditions-III (NESARC-III). Subst Use Misuse 2017;52(14):1859–70.

[46] Zhang L, Samet J, Caffo B, Punjabi NM. Cigarette smoking and nocturnal sleep architecture. Am J Epidemiol 2006;164(6):529–37.

[47] Sahlin C, Franklin KA, Stenlund H, Lindberg E. Sleep in women: normal values for sleep stages and position and the effect of age, obesity, sleep apnea, smoking, alcohol and hypertension. Sleep Med 2009;10(9):1025–30.

[48] Phillips BA, Danner FJ. Cigarette smoking and sleep disturbance. Arch Intern Med 1995;155(7):734–7.

[49] Riedel BW, Durrence HH, Lichstein KL, Taylor DJ, Bush AJ. The relation between smoking and sleep: the influence of smoking level, health, and psychological variables. Behav Sleep Med 2004;2(1):63–78.

[50] Wetter DW, Young TB. The relation between cigarette smoking and sleep disturbance. Prev Med 1994;23(3):328–34.

[51] Bover MT, Foulds J, Steinberg MB, Richardson D, Marcella SW. Waking at night to smoke as a marker for tobacco dependence: patient characteristics and relationship to treatment outcome. Int J Clin Pract 2008;62(2):182–90.

[52] Scharf DM, Dunbar MS, Shiffman S. Smoking during the night: prevalence and smoker characteristics. Nicotine Tob Res 2008;10(1):167–78.

[53] Foulds J, Gandhi KK, Steinberg MB, Richardson DL, Williams JM, Burke MV, et al. Factors associated with quitting smoking at a tobacco dependence treatment clinic. Am J Health Behav 2006;30(4):400–12.

[54] Cummings KM, Giovino G, Jaen CR, Emrich LJ. Reports of smoking withdrawal symptoms over a 21 day period of abstinence. Addict Behav 1985;10(4):373–81.
[55] Hughes JR. Effects of abstinence from tobacco: valid symptoms and time course. Nicotine Tob Res 2007;9(3):315–27.
[56] Okun ML, Levine MD, Houck P, Perkins KA, Marcus MD. Subjective sleep disturbance during a smoking cessation program: associations with relapse. Addict Behav 2011;36(8):861–4.
[57] Zhang L, Samet J, Caffo B, Bankman I, Punjabi NM. Power spectral analysis of EEG activity during sleep in cigarette smokers. Chest 2008;133(2):427–32.
[58] Jaehne A, Unbehaun T, Feige B, Cohrs S, Rodenbeck A, Riemann D. Sleep changes in smokers before, during and 3 months after nicotine withdrawal. Addict Biol 2015;20(4):747–55.
[59] al'Absi M, Hatsukami D, Davis GL, Wittmers LE. Prospective examination of effects of smoking abstinence on cortisol and withdrawal symptoms as predictors of early smoking relapse. Drug Alcohol Depend 2004;73(3):267–78.
[60] Jaehne A, Loessl B, Barkai Z, Riemann D, Hornyak M. Effects of nicotine on sleep during consumption, withdrawal and replacement therapy. Sleep Med Rev 2009;13(5):363–77.
[61] Hayley AC, Downey LA. Quitters never sleep: the effect of nicotine withdrawal upon sleep. Curr Drug Abuse Rev 2015;8(2):73–4.
[62] Moreno-Coutino A, Calderon-Ezquerro C, Drucker-Colin R. Long-term changes in sleep and depressive symptoms of smokers in abstinence. Nicotine Tob Res 2007;9(3):389–96.
[63] Wetter DW, Fiore MC, Baker TB, Young TB. Tobacco withdrawal and nicotine replacement influence objective measures of sleep. J Consult Clin Psychol 1995;63(4):658–67.
[64] Prosise GL, Bonnet MH, Berry RB, Dickel MJ. Effects of abstinence from smoking on sleep and daytime sleepiness. Chest 1994;105(4):1136–41.
[65] Shiffman S, Patten C, Gwaltney C, Paty J, Gnys M, Kassel J, et al. Natural history of nicotine withdrawal. Addiction 2006;101(12):1822–32.
[66] Peltier MR, Lee J, Ma P, Businelle MS, Kendzor DE. The influence of sleep quality on smoking cessation in socioeconomically disadvantaged adults. Addict Behav 2017;66:7–12.
[67] Peters EN, Fucito LM, Novosad C, Toll BA, O'Malley SS. Effect of night smoking, sleep disturbance, and their co-occurrence on smoking outcomes. Psychol Addict Behav 2011;25(2):312–9.
[68] Rapp K, Buechele G, Weiland SK. Sleep duration and smoking cessation in student nurses. Addict Behav 2007;32(7):1505–10.
[69] Short NA, Mathes BA, Gibby B, Oglesby ME, Zvolensky MJ, Schmidt NB. Insomnia symptoms as a risk factor for cessation failure following smoking cessation treatment. Addict Res Theory 2016;25:17–23.
[70] Ashare RL, Lerman C, Tyndale RF, Hawk LW, George TP, Cinciripini PM, et al. Sleep disturbance during smoking cessation: withdrawal or side effect of treatment. J Smok Cessat 2017;12:63–70.
[71] Riemerth A, Kunze U, Groman E. Nocturnal sleep-disturbing nicotine craving and accomplishment with a smoking cessation program. Wien Med Wochenschr 2009;159(1–2):47–52.
[72] Fucito LM, Redeker NS, Ball SA, Toll BA, Ikomi JT, Carroll KM. Integrating a behavioral sleep intervention into smoking cessaiton treatment for smokers with insomnia: a randomised pilot study. J Smok Cessat 2013;9(1):31–8.
[73] Dorner TE, Trostl A, Womastek I, Groman E. Predictors of short-term success in smoking cessation in relation to attendance at a smoking cessation program. Nicotine Tob Res 2011;13(11):1068–75.
[74] Stead LF, Perera R, Bullen C, Mant D, Lancaster T. Nicotine replacement therapy for smoking cessation. Cochrane Database Syst Rev 2008;1:CD000146.
[75] Gourlay SG, Forbes A, Marriner T, McNeil JJ. Predictors and timing of adverse experiences during trandsdermal nicotine therapy. Drug Saf 1999;20(6):545–55.
[76] Stapleton JA, Watson L, Spirling LI, Smith R, Milbrandt A, Ratcliffe M, et al. Varenicline in the routine treatment of tobacco dependence: a pre-post comparison with nicotine replacement therapy and an evaluation in those with mental illness. Addiction 2008;103(1):146–54.
[77] Mills EJ, Wu P, Lockhart I, Wilson K, Ebbert JO. Adverse events associated with nicotine replacement therapy (NRT) for smoking cessation. A systematic review and meta-analysis of one hundred and twenty studies involving 177,390 individuals. Tob Induc Dis 2010;8:8.
[78] Stead LF, Perera R, Bullen C, Mant D, Hartmann-Boyce J, Cahill K, et al. Nicotine replacement therapy for smoking cessation. Cochrane Database Syst Rev 2012;11:CD000146.
[79] Hays JT, Hurt RD, Rigotti NA, Niaura R, Gonzales D, Durcan MJ, et al. Sustained-release bupropion for pharmacologic relapse prevention after smoking cessation. A randomized, controlled trial. Ann Intern Med 2001;135(6):423–33.
[80] Hughes JR, Stead LF, Hartmann-Boyce J, Cahill K, Lancaster T. Antidepressants for smoking cessation. Cochrane Database Syst Rev 2014;(1)CD000031.
[81] Jorenby DE, Leischow SJ, Nides MA, Rennard SI, Johnston JA, Hughes AR, et al. A controlled trial of sustained-release bupropion, a nicotine patch, or both for smoking cessation. N Engl J Med 1999;340(9):685–91.
[82] West R, Baker CL, Cappelleri JC, Bushmakin AG. Effect of varenicline and bupropion SR on craving, nicotine withdrawal symptoms, and rewarding effects of smoking during a quit attempt. Psychopharmacology (Berl) 2008;197(3):371–7.
[83] Shiffman S, Johnston JA, Khayrallah M, Elash CA, Gwaltney CJ, Paty JA, et al. The effect of bupropion on nicotine craving and withdrawal. Psychopharmacology (Berl) 2000;148(1):33–40.
[84] Papke RL, Heinemann SF. Partial agonist properties of cytisine on neuronal nicotinic receptors containing the beta 2 subunit. Mol Pharmacol 1994;45(1):142–9.
[85] Gonzales D, Rennard SI, Nides M, Oncken C, Azoulay S, Billing CB, et al. Varenicline, an alpha4beta2 nicotinic acetylcholine receptor partial agonist, vs sustained-release bupropion and placebo for smoking cessation: a randomized controlled trial. JAMA 2006;296(1):47–55.
[86] Cahill K, Stevens S, Perera R, Lancaster T. Pharmacological interventions for smoking cessation: an overview and network meta-analysis. Cochrane Database Syst Rev 2013;31(5):CD009329. https://doi.org/10.1002/14651858.CD009329.pub2.
[87] McClure JB, Swan GE, Jack L, Catz SL, Zbikowski SM, McAfee TA, et al. Mood, side-effects and smoking outcomes among persons with and without probable lifetime depression taking varenicline. J Gen Intern Med 2009;24(5):563–9.
[88] Thomas KH, Martin RM, Knipe DW, Higgins JP, Gunnell D. Risk of neuropsychiatric adverse events associated with varenicline: systematic review and meta-analysis. BMJ 2015;350:h1109.

[89] Drovandi AD, Chen CC, Glass BD. Adverse effects cause varenicline discontinuation: a meta-analysis. Curr Drug Saf 2016;11(1):78–85.

[90] Polini F, Principe R, Scarpelli S, Clementi F, DeGennaro L. Use of varenicline in smokeless tobacco cessation influences sleep quality and dream recall frequency but not dream affect. Sleep Med 2017;30:1–6.

[91] Foulds J, Russ C, Yu CR, Zou KH, Galaznik A, Franzon M, et al. Effect of varenicline on individual nicotine withdrawal symptoms: a combined analysis of eight randomized, placebo-controlled trials. Nicotine Tob Res 2013;15(11):1849–57.

[92] Xu L, Jiang CQ, Lam TH, Liu B, Jin YL, Zhu T, et al. Short or long sleep duration is associated with memory impairment in older Chinese: the Guangzhou biobank cohort study. Sleep 2011;34(5):575–80.

[93] Shekleton JA, Flynn-Evans EE, Miller B, Epstein LJ, Kirsch D, Brogna LA, et al. Neurobehavioral performance impairment in insomnia: relationships with self-reported sleep and daytime functioning. Sleep 2014;37(1):107–16.

[94] Goel N, Basner M, Rao H, Dinges DF. Circadian rhythms, sleep deprivation, and human performance. Prog Mol Biol Transl Sci 2013;119:155–90.

[95] Patterson F, Jepson C, Strasser AA, Loughead J, Perkins KA, Gur RC, et al. Varenicline improves mood and cognition during smoking abstinence. Biol Psychiatry 2009;65(2):144–9.

[96] Faubel R, Lopez-Garcia E, Guallar-Castillon P, Graciani A, Banegas JR, Rodriguez-Artalejo F. Usual sleep duration and cognitive function in older adults in Spain. J Sleep Res 2009;18(4):427–35.

[97] Smith ME, McEvoy LK, Gevins A. The impact of moderate sleep loss on neurophysiologic signals during working-memory task performance. Sleep 2002;25(7):784–94.

[98] Van Dongen HP, Bender AM, Dinges DF. Systematic individual differences in sleep homeostatic and circadian rhythm contributions to neurobehavioral impairment during sleep deprivation. Accid Anal Prev 2012;45(Suppl):11–6.

[99] Killgore WD, Balkin TJ, Wesensten NJ. Impaired decision making following 49 h of sleep deprivation. J Sleep Res 2006;15(1):7–13.

[100] Greer SM, Goldstein AN, Walker MP. The impact of sleep deprivation on food desire in the human brain. Nat Commun 2013;4:2259.

[101] Ward MM, Swan GE, Jack LM. Self-reported abstinence effects in the first month after smoking cessation. Addict Behav 2001;26(3):311–27.

[102] Myers CS. Nicotine nasal spray dose-dependently enhanced sustained attention as assessed by the continous performance task. Society for Research on Nicotine and Tobacco. 2005. Prague2005.

[103] Jacobsen LK, Krystal JH, Mencl WE, Westerveld M, Frost SJ, Pugh KR. Effects of smoking and smoking abstinence on cognition in adolescent tobacco smokers. Biol Psychiatry 2005;57(1):56–66.

[104] Mendrek A, Monterosso J, Simon SL, Jarvik M, Brody A, Olmstead R, et al. Working memory in cigarette smokers: comparison to non-smokers and effects of abstinence. Addict Behav 2006;31(5):833–44.

[105] Myers C, Taylor RC, Moolchan ET, Heishman SJ. Dose-related enhancement of mood and cognition in smokers administered nicotine nasal spray. Neuropsychopharmacology 2008;33(3):588–98.

[106] Dolan SL, Sacco KA, Termine A, Seyal AA, Dudas MM, Vessicchio JC, et al. Neuropsychological deficits are associated with smoking cessation treatment failure in patients with schizophrenia. Schizophr Res 2004;70(2–3):263–75.

[107] Krishnan-Sarin S, Reynolds B, Duhig AM, Smith A, Liss T, McFetridge A, et al. Behavioral impulsivity predicts treatment outcome in a smoking cessation program for adolescent smokers. Drug Alcohol Depend 2007;88(1):79–82.

[108] Rukstalis M, Jepson C, Patterson F, Lerman C. Increases in hyperactive-impulsive symptoms predict relapse among smokers in nicotine replacement therapy. J Subst Abuse Treat 2005;28(4):297–304.

[109] Crothers K, Griffith TA, McGinnis KA, Rodriguez-Barradas MC, Leaf DA, Weissman S, et al. The impact of cigarette smoking on mortality, quality of life, and comorbid illness among HIV-positive veterans. J Gen Intern Med 2005;20(12):1142–5.

[110] Feldman JG, Minkoff H, Schneider MF, Gange SJ, Cohen M, Watts DH, et al. Association of cigarette smoking with HIV prognosis among women in the HAART era: a report from the women's interagency HIV study. Am J Public Health 2006;96(6):1060–5.

[111] Webb MS, Vanable PA, Carey MP, Blair DC. Cigarette smoking among HIV+ men and women: examining health, substance use, and psychosocial correlates across the smoking spectrum. J Behav Med 2007;30(5):371–83.

[112] Collins RL, Kanouse DE, Gifford AL, Senterfitt JW, Schuster MA, McCaffrey DF, et al. Changes in health-promoting behavior following diagnosis with HIV: prevalence and correlates in a national probability sample. Health Psychol 2001;20(5):351–60.

[113] Burkhalter JE, Springer CM, Chhabra R, Ostroff JS, Rapkin BD. Tobacco use and readiness to quit smoking in low-income HIV-infected persons. Nicotine Tob Res 2005;7(4):511–22.

[114] Deeken JF, Tjen ALA, Rudek MA, Okuliar C, Young M, Little RF, et al. The rising challenge of non-AIDS-defining cancers in HIV-infected patients. Clin Infect Dis 2012;55(9):1228–35.

[115] Brugnaro P, Morelli E, Cattelan F, Petrucci A, Panese S, Eseme F, et al. Non-AIDS definings malignancies among human immunodeficiency virus-positive subjects: epidemiology and outcome after two decades of HAART era. World J Virol 2015;4(3):209–18.

[116] Palella Jr. FJ, Delaney KM, Moorman AC, Loveless MO, Fuhrer J, Satten GA, et al. Declining morbidity and mortality among patients with advanced human immunodeficiency virus infection. HIV outpatient study investigators. N Engl J Med 1998;338(13):853–60.

[117] Rubinstein PG, Aboulafia DM, Zloza A. Malignancies in HIV/AIDS: from epidemiology to therapeutic challenges. AIDS 2014;28(4):453–65.

[118] Vaccher E, Serraino D, Carbone A, De Paoli P. The evolving scenario of non-AIDS-defining cancers: challenges and opportunities of care. Oncologist 2014;19(8):860–7.

[119] Palella Jr. FJ, Baker RK, Moorman AC, Chmiel JS, Wood KC, Brooks JT, et al. Mortality in the highly active antiretroviral therapy era: changing causes of death and disease in the HIV outpatient study. J Acquir Immune Defic Syndr 2006;43(1):27–34.

[120] D'Abramo A, Zingaropoli MA, Oliva A, D'Agostino C, Al Moghazi S, De Luca G, et al. Higher levels of osteoprotegerin and immune activation/immunosenescence markers are correlated with concomitant bone and endovascular damage in HIV-suppressed patients. PLoS One 2016;11(2):e0149601.

[121] Nasi M, De Biasi S, Gibellini L, Bianchini E, Pecorini S, Bacca V, et al. Ageing and inflammation in patients with HIV infection. Clin Exp Immunol 2017;187(1):44–52.

[122] Leng SX, Margolick JB. Understanding frailty, aging, and inflammation in HIV infection. Curr HIV/AIDS Rep 2015;12(1):25–32.

[123] Gamaldo CE, Spira AP, Hock RS, Salas RE, McArthur JC, David PM, et al. Sleep, function and HIV: a multi-method assessment. AIDS Behav 2013;17(8):2808–15.

[124] Low Y, Goforth H, Preud'homme X, Edinger J, Krystal A. Insomnia in HIV-infected patients: pathophysiologic implications. AIDS Rev 2014;16(1):3–13.

[125] Allavena C, Guimard T, Billaud E, De la Tullaye S, Reliquet V, Pineau S, et al. Prevalence and risk factors of sleep disturbance in a large HIV-infected adult population. AIDS Behav 2016;20(2):339–44.

[126] Cooper J, Borland R, Yong HH, Fotuhi O. The impact of quitting smoking on depressive symptoms: findings from the international tobacco control four-country survey. Addiction 2016;111(8):1448–56.

[127] Fillo J, Alfano CA, Paulus DJ, Smits JA, Davis ML, Rosenfield D, et al. Emotion dysregulation explains relations between sleep disturbance and smoking quit-related cognition and behavior. Addict Behav 2016;57:6–12.

[128] Vanderlind WM, Beevers CG, Sherman SM, Trujillo LT, McGeary JE, Matthews MD, et al. Sleep and sadness: exploring the relation among sleep, cognitive control, and depressive symptoms in young adults. Sleep Med 2014;15(1):144–9.

[129] Kauffman BY, Farris SG, Alfano CA, Zvolensky MJ. Emotion dysregulation explains the relation between insomnia symptoms and negative reinforcement smoking cognitions among daily smokers. Addict Behav 2017;72:33–40.

[130] Hall FS, Der-Avakian A, Gould TJ, Markou A, Shoaib M, Young JW. Negative affective states and cognitive impairments in nicotine dependence. Neurosci Biobehav Rev 2015;58:168–85.

[131] Germain A, Kupfer DJ. Circadian rhythm disturbances in depression. Hum Psychopharmacol 2008;23(7):571–85.

[132] McClung CA. Circadian genes, rhythms and the biology of mood disorders. Pharmacol Ther 2007;114(2):222–32.

[133] McClung CA. How might circadian rhythms control mood? Let me count the ways. Biol Psychiatry 2013;74(4):242–9.

[134] Tsuno N, Besset A, Ritchie K. Sleep and depression. J Clin Psychiatry 2005;66(10):1254–69.

[135] Staner L. Comorbidity of insomnia and depression. Sleep Med Rev 2010;14(1):35–46.

[136] Pillai V, Kalmbach DA, Ciesla JA. A meta-analysis of electroencephalographic sleep in depression: evidence for genetic biomarkers. Biol Psychiatry 2011;70(10):912–9.

[137] Winkler D, Pjrek E, Praschak-Rieder N, Willeit M, Pezawas L, Konstantinidis A, et al. Actigraphy in patients with seasonal affective disorder and healthy control subjects treated with light therapy. Biol Psychiatry 2005;58(4):331–6.

[138] Nutt D, Wilson S, Paterson L. Sleep disorders as core symptoms of depression. Dialogues Clin Neurosci 2008;10(3):329–36.

[139] Armitage R, Hoffmann R, Trivedi M, Rush AJ. Slow-wave activity in NREM sleep: sex and age effects in depressed outpatients and healthy controls. Psychiatry Res 2000;95(3):201–13.

[140] Clark C, Dupont R, Golshan S, Gillin JC, Rapaport MH, Kelsoe JR. Preliminary evidence of an association between increased REM density and poor antidepressant response to partial sleep deprivation. J Affect Disord 2000;59(1):77–83.

[141] Mendlewicz J. Sleep disturbances: core symptoms of major depressive disorder rather than associated or comorbid disorders. World J Biol Psychiatry 2009;10(4):269–75.

[142] Troxel WM, Kupfer DJ, Reynolds 3rd CF, Frank E, Thase ME, Miewald JM, et al. Insomnia and objectively measured sleep disturbances predict treatment outcome in depressed patients treated with psychotherapy or psychotherapy-pharmacotherapy combinations. J Clin Psychiatry 2012;73(4):478–85.

[143] Breslau N, Roth T, Rosenthal L, Andreski P. Sleep disturbance and psychiatric disorders: a longitudinal epidemiological study of young adults. Biol Psychiatry 1996;39(6):411–8.

[144] Ford DE, Kamerow DB. Epidemiologic study of sleep disturbances and psychiatric disorders. An opportunity for prevention? JAMA 1989;262(11):1479–84.

[145] Buysse DJ, Angst J, Gamma A, Ajdacic V, Eich D, Rossler W. Prevalence, course, and comorbidity of insomnia and depression in young adults. Sleep 2008;31(4):473–80.

[146] Lawrence D, Mitrou F, Zubrick SR. Smoking and mental illness: results from population surveys in Australia and the United States. BMC Public Health 2009;9:285.

[147] Pratt LA, Brody DJ. Depression and smoking in the U.S. household population aged 20 and over, 2005-2008. NCHS Data Brief 2010;(34)1–8.

[148] Grant BF, Hasin DS, Chou S, Stinson FS, Dawson DA. Nicotine dependence and psychiatric disorders in the United States: results from the National Epidemiologic survey on alcohol and related conditions. Arch Gen Psychiatry 2004;61(11):1107–15.

[149] Glassman AH, Helzer JE, Covey LS, Cottler LB, Stetner F, Tipp JE, et al. Smoking, smoking cessation, and major depression. JAMA 1990;264(12):1546–9.

[150] Breslau N, Kilbey M, Andreski P. Nicotine dependence, major depression, and anxiety in young adults. Arch Gen Psychiatry 1991;48(12):1069–74.

[151] Leventhal AM, Kahler CW, Ray LA, Zimmerman M. Refining the depression-nicotine dependence link: patterns of depressive symptoms in psychiatric outpatients with current, past, and no history of nicotine dependence. Addict Behav 2009;34(3):297–303.

[152] Breslau N, Kilbey MM, Andreski P. Nicotine dependence and major depression. New evidence from a prospective investigation. Arch Gen Psychiatry 1993;50(1):31–5.

[153] Breslau N, Peterson EL, Schultz LR, Chilcoat HD, Andreski P. Major depression and stages of smoking. A longitudinal investigation. Arch Gen Psychiatry 1998;55(2):161–6.

[154] Breslau N, Novak SP, Kessler RC. Daily smoking and the subsequent onset of psychiatric disorders. Psychol Med 2004;34(2):323–33.

[155] Colman I, Naicker K, Zeng Y, Ataullahjan A, Senthilselvan A, Patten SB. Predictors of long-term prognosis of depression. CMAJ 2011;183(17):1969–76.

[156] Ashare RL, Valdez JN, Ruparel K, Albelda B, Hopson RD, Keefe JR, et al. Association of abstinence-induced alterations in working memory function and COMT genotype in smokers. Psychopharmacology (Berl) 2013;230(4):653–62.

[157] Falcone M, Wileyto EP, Ruparel K, Gerraty RT, Laprate L, Detre JA, et al. Age-related differences in working memory deficits during nicotine withdrawal. Addict Biol 2014;19(5):907–17.

[158] Loughead J, Ray R, Wileyto EP, Ruparel K, Sanborn P, Siegel S, et al. Effects of the alpha4beta2 partial agonist varenicline on brain activity and working memory in abstinent smokers. Biol Psychiatry 2010;67(8):715–21.

[159] Harvey PO, Fossati P, Pochon JB, Levy R, Lebastard G, Lehericy S, et al. Cognitive control and brain resources in major depression: an fMRI study using the n-back task. Neuroimage 2005;26(3):860–9.

[160] Norbury R, Godlewska B, Cowen PJ. When less is more: a functional magnetic resonance imaging study of verbal working memory in remitted depressed patients. Psychol Med 2013;1–7.

[161] Drummond SP, Walker M, Almklov E, Campos M, Anderson DE, Straus LD. Neural correlates of working memory performance in primary insomnia. Sleep 2013;36(9):1307–16.

[162] Cooper J, Borland R, McKee SA, Yong HH, Dugue PA. Depression motivates quit attempts but predicts relapse: differential findings for gender from the international tobacco control study. Addiction 2016;111(8):1438–47.

[163] Nogueira LM, Sampson JN, Chu LW, Yu K, Andriole G, Church T, et al. Individual variations in serum melatonin levels through time: implications for epidemiologic studies. PLoS One 2013;8(12):e83208.

[164] Ozguner F, Koyu A, Cesur G. Active smoking causes oxidative stress and decreases blood melatonin levels. Toxicol Ind Health 2005;21(1–2):21–6.

[165] Mexal S, Horton WJ, Crouch EL, Maier SI, Wilkinson AL, Marsolek M, et al. Diurnal variation in nicotine sensitivity in mice: role of genetic background and melatonin. Neuropharmacology 2012;63(6):966–73.

[166] Zhdanova IV, Piotrovskaya VR. Melatonin treatment attenuates symptoms of acute nicotine withdrawal in humans. Pharmacol Biochem Behav 2000;67(1):131–5.

[167] Horton W, Gissel H, Saboy J, Wright Jr. K, Stitzel J. Melatonin administration alters nicotine preference consumption via signaling through high-affinity melatonin receptors. Psychopharmacology (Berl) 2015;232(14):2519–30.

[168] Sutcliffe JG, de Lecea L. The hypocretins: setting the arousal threshold. Nat Rev Neurosci 2002;3(5):339–49.

[169] Plaza-Zabala A, Maldonado R, Berrendero F. The hypocretin/orexin system: implications for drug reward and relapse. Mol Neurobiol 2012;45(3):424–39.

[170] Qaseem A, Kansagara D, Forciea MA, Cooke M, Denberg TD, Clinical Guidelines Committee of the American College of Physicians. Management of chronic insomnia disorder in adults: a clinical practice guideline from the American College of Physicians. Ann Intern Med 2016;165(2):125–33.

[171] National Institutes of Health. National Institutes of Health State of the Science Conference Statement on manifestations and management of chronic insomnia in adults. June 13-15, 2005; Sleep 2005;28(9):1049–57.

[172] Vincent N, Lionberg C. Treatment preference and patient satisfaction in chronic insomnia. Sleep 2001;24(4):411–7.

[173] Perlis M, Aloia M, Kuhn B. Behavioral treatments for sleep disorders: a comprehensive primer of behavioral sleep medicine interventions. San Diego, CA: Elsiver Academic Press; 2011.

[174] Morin CM, Bootzin RR, Buysse DJ, Edinger JD, Espie CA, Lichstein KL. Psychological and behavioral treatment of insomnia: update of the recent evidence (1998-2004). Sleep 2006;29(11):1398–414.

[175] Trauer JM, Qian MY, Doyle JS, Rajaratnam SM, Cunnington D. Cognitive behavioral therapy for chronic insomnia: a systematic review and meta-analysis. Ann Intern Med 2015;163(3):191–204.

[176] Smith MT, Perlis ML, Park A, Smith MS, Pennington J, Giles DE, et al. Comparative meta-analysis of pharmacotherapy and behavior therapy for persistent insomnia. Am J Psychiatry 2002;159(1):5–11.

[177] Xu H, Guan J, Yi H, Yin S. A systematic review and meta-analysis of the association between serotonergic gene polymorphisms and obstructive sleep apnea syndrome. PLoS One 2014;9(1):e86460.

[178] Fucito LM, Redeker NS, Ball SA, Toll BA, Ikomi JT, Carroll KM. Integrating a Behavioural sleep intervention into smoking cessation treatment for smokers with insomnia: a randomised pilot study. J Smok Cessat 2014;9(1):31–8.

[179] Physician Desk Reference Network. 2015 Physician's desk reference. 69th ed. Montvale, NJ: PDR Network; 2014.

[180] Nowell PD, Mazumdar S, Buysse DJ, Dew MA, Reynolds 3rd CF, Kupfer DJ. Benzodiazepines and zolpidem for chronic insomnia: a meta-analysis of treatment efficacy. JAMA 1997;278(24):2170–7.

[181] Cosgrove KP, McKay R, Esterlis I, Kloczynski T, Perkins E, Bois F, et al. Tobacco smoking interferes with GABAA receptor neuroadaptations during prolonged alcohol withdrawal. Proc Natl Acad Sci U S A 2014;111(50):18031–6.

[182] MacFarlane J, Morin CM, Montplaisir J. Hypnotics in insomnia: the experience of zolpidem. Clin Ther 2014;36(11):1676–701.

[183] Michelson D, Snyder E, Paradis E, Chengan-Liu M, Snavely DB, Hutzelmann J, et al. Safety and efficacy of suvorexant during 1-year treatment of insomnia with subsequent abrupt treatment discontinuation: a phase 3 randomised, double-blind, placebo-controlled trial. Lancet Neurol 2014;13(5):461–71.

[184] Leufkens TR, Ramaekers JG, de Weerd AW, Riedel WJ, Vermeeren A. Residual effects of zopiclone 7.5 mg on highway driving performance in insomnia patients and healthy controls: a placebo controlled crossover study. Psychopharmacology (Berl) 2014;231(14):2785–98.

[185] Loughead J, Wileyto EP, Ruparel K, Falcone M, Hopson R, Gur R, et al. Working memory-related neural activity predicts future smoking relapse. Neuropsychopharmacology 2015;40(6):1311–20.

[186] Patterson F, Jepson C, Loughead J, Perkins K, Strasser A, Siegel S, et al. Working memory deficits predict short-term smoking resumption following brief abstinence. Drug Alcohol Depend 2010;106(1):61–4.

[187] Laudon M, Frydman-Marom A. Therapeutic effects of melatonin receptor agonists on sleep and comorbid disorders. Int J Mol Sci 2014;15(9):15924–50.

[188] Kohsaka M, Kanemura T, Taniguchi M, Kuwahara H, Mikami A, Kamikawa K, et al. Efficacy and tolerability of ramelteon in a double-blind, placebo-controlled, crossover study in Japanese patients with chronic primary insomnia. Expert Rev Neurother 2011;11(10):1389–97.

[189] Mayer G, Wang-Weigand S, Roth-Schechter B, Lehmann R, Staner C, Partinen M. Efficacy and safety of 6-month nightly ramelteon administration in adults with chronic primary insomnia. Sleep 2009;32(3):351–60.

[190] Jacobson LH, Chen S, Mir S, Hoyer D. Orexin OX2 receptor antagonists as sleep Aids. Curr Top Behav Neurosci 2017;33:105–36.

[191] Kenny PJ. Tobacco dependence, the insular cortex and the hypocretin connection. Pharmacol Biochem Behav 2011;97(4):700–7.

[192] Wong SY, Zhang DX, Li CC, Yip BH, Chan DC, Ling YM, et al. Comparing the effects of mindfulness-based cognitive therapy and sleep psycho-education with exercise on chronic insomnia: a randomised controlled trial. Psychother Psychosom 2017;86(4):241–53.

[193] Nair US, Patterson F, Rodriguez D, Collins BN. A telephone-based intervention to promote physical activity during smoking cessation: a randomized controlled proof-of-concept study. Transl Behav Med 2017;7(2):138–47.

[194] Maglione MA, Maher AR, Ewing B, Colaiaco B, Newberry S, Kandrack R, et al. Efficacy of mindfulness meditation for smoking cessation: a systematic review and meta-analysis. Addict Behav 2017;69:27–34.

Chapter 23

Sleep and the impact of caffeine, supplements, and other stimulants

Ninad S. Chaudhary[a,b], Priyamvada M. Pitale[c], Favel L. Mondesir[d]

[a]Department of Epidemiology, School of Public Health, University of Alabama at Birmingham, Birmingham, AL, United States, [b]Department of Neurology, University of Alabama School of Medicine, Birmingham, AL, United States, [c]Department of Optometry and Vision Science, School of Optometry, University of Alabama at Birmingham, Birmingham, AL, United States, [d]Division of Cardiovascular Medicine, School of Medicine, University of Utah, Salt Lake City, UT, United States

ABBREVIATIONS

ADA	adenosine deaminase
ADORA2A	adenosine A2A receptor
CYP1A2	cytochrome P450 family 1 subfamily A member 2
DSM	diagnostic statistical manual
ED	energy drinks
EEG	electroencephalogram
FDA	Food and Drug Administration
GBP4	guanylate binding protein 4
GWAS	genome-wide association studies
ICD-9	International Classification of Diseases
IL-6	interleukin 6
KiGGS	German Health Interview and Examination Survey for Children and Adolescents
NHANES	National Health and Nutritional Examination Survey
NHIS	National Health Interview Survey
NICHD	National Institute of Child Health and Human Development
N-REM	non-rapid eye movement
PRIMA1	proline-rich membrane anchor 1
PTSD	post-traumatic stress disorder
REM	rapid eye movement
SNP	single nucleotide polymorphisms
WASO	waking after sleep onset

INTRODUCTION

Caffeine is the most frequently used psychostimulant drugs worldwide [1]. Recently, new categories of caffeine-containing products such as energy drinks supplements are popular and have been increasing rapidly in the past decade [2,3]. The most frequent reason people use caffeine or energy drinks is to counteract the effects of insufficient sleep or sleepiness [4,5]. The use of caffeine as a countermeasure is of grave concern considering the prevalence of sleep problems has been on the rise in the global population [6]. These products also have adverse effects on the cardiovascular, metabolic, and neurological systems [3]. These effects may potentially add to existing comorbid burden due to underlying sleep disturbances. The global impact of the combined health outcomes of caffeine use and sleep problem is overarching and requires close evaluation.

The average intake of caffeine in general population is 165 mg, amount sufficient enough to interfere with sleep [1]. However, the average intake may be underestimated due to varying sources of caffeine consumed throughout the day. Caffeine is also the essential ingredient of energy drinks. In addition to caffeine, energy drinks also contain taurine, L-carnitine, vitamins, carbohydrates, herbal supplements and other additives such as cocoa, guarana, ginseng [2]. The current regulations are not required to include the caffeine content of these additional ingredients which limits the information on the exact amount [7,8]. Most of the research studies examined the physiological effects of these products that improve performance; however, the associated adverse effects are still understudied [8].

There is extensive literature on the association between caffeine and sleep over last few decades [9]. The research is mostly divided between understanding the effect of caffeine after sleep deprivation, and on post-caffeine recovery sleep. The earliest reference to caffeine-sleep mechanisms is observed in a research study by Brezinova et al. [10] who reported that caffeine use is associated with stable periods of wakefulness between episodes of drowsiness which have a similar duration of any of the sleep stages. They coined this phenomenon "Caffeine-insomnia." Subsequent research has made landslide improvement in measuring the associations between caffeine and sleep. The research is still limited in understanding the individual variation of caffeine on sleep effects, time and dose-response relationship of these associations, establishing the causality, and interaction of caffeine, circadian rhythm and sleep regulatory processes. The advent of energy drinks in the last decade places the importance of these associations in a new light.

This chapter focuses on the effect of caffeine, energy drinks and other psychostimulants on the sleep of adolescents and adults. We outline the epidemiology of sleep disorders in light of caffeine and energy drinks consumption, the possible underlying mechanisms and related factors, the health implications of the cumulative effect of sleep and these products and probable recommendations.

EPIDEMIOLOGY

Epidemiology of sleep in caffeine

Caffeine is a by-product obtained from cocoa grown indigenously in South America and is now frequently consumed as a beverage form as a coffee, soft drinks, and energy drinks, gum or concentrated form [2]. Approximately 85% of adult Americans and 30% of American adolescents regularly consume at least one drink of caffeine each day, while only 50% consumes energy drink supplements [1,11]. The use of psychostimulants is prescription-based and is not known for the civilian population. The mean caffeine intake for all adults varies from 164 to 228 mg/day primarily due to the difference in the source of consumption that predominantly vary by age, in addition to gender [1]. The mean daily caffeine intake was 165 ± 1 mg combined for all ages, and highest for the age group of 50–64 years (226 ± 2 mg/day) [1]. The older population is more likely to consume coffee while younger population relies on sodas and energy drink supplements for their caffeine intake [11]. While the mean intake in men is slightly higher than women, the source of caffeine is mostly age dependent for both [9].

Caffeine use on a routine basis or sleep-deprived state improves neurobehavioral functioning in healthy individuals at a dose as low as 32 mg [12]. However, with increasing consumption or dosages, caffeine may have adverse sleep-related consequences on subsequent nights or for a longer duration [4,9]. The contents of caffeine vary in these products which determine the possible sleep disturbances based on the amount consumed [2,13] (Table 23.1). These dose-dependent adverse effects range from minor abnormalities in sleep patterns such as poor sleep quality, or reduced sleep duration to those meeting the diagnosis of insomnia [14].

TABLE 23.1 Caffeine beverages and contents.

	Caffeine content[a]	Sleep effects[c]
Coffees		
Brewed coffee	133 mg/8 oz	SSD, DFA
Instant coffee	93 mg/8 oz	SSD, frequent awakening
Espresso	320 mg/8 oz	Poor SQ, DS, changes in EEG
Soft drinks/energy drinks		
Regular or diet	35–71 mg/12 oz[b]	None to SSD
5 h Energy	215 mg/12 oz	Decreased time in bed, poor SQ, changes in EEG
Monster energy	120 mg/12 oz	DFA, DSA
Red bull	116 mg/12 oz	DFA, increased WASO, poor SS
Others	50–300 mg/8.3 oz	None to poor SQ, DS, changes in EEG
Tea and other	15–40 mg/16 oz	None

[a]FDA limit.
[b]FDA limit for soft drinks is 71 mg/12 oz.
[c]Effects are based on quantity of caffeine obtained from laboratory studies.
SSD, short sleep duration; DFA, difficulty falling asleep; DS, daytime sleepiness; SQ, sleep quality; DFA, difficulty falling asleep; DSA, difficulty staying asleep; SS, sleep satisfaction; EEG, electroencephalograph.
Somogyi, LP. Caffeine intake by the US population. Prepared for The Food and Drug Administration and Oakridge National Laboratory; 2010; Clark I, Landolt HP. Coffee, caffeine, and sleep: a systematic review of epidemiological studies and randomized controlled trials. Sleep Med Rev 2017;31:70–8; Juliano LM GR. Caffeine. In: Lowinson JH RP, Millman RB, Langrod JG, editors. Substance abuse: a comprehensive textbook. 4th ed. Baltimore: Lippincott Williams & Wilkins; 2005.

Caffeine use has primarily been researched to overcome sleep deprivation. While few studies focus on caffeine use for sleep deprived in the general population, most of them were concentrated in particular community, specifically military, and shift workers. Roehrs and Roth [4] reviewed anecdotal studies providing practice guidelines and recommendations for caffeine and sleep-related research. The guidelines emphasized the importance of the comprehensive assessment of dietary sources of caffeine to identify caffeine-related sleep disturbances, to integrate its use in clinical evaluations of insomnia complaints, and determine caffeine dependence in the light of insomnia for appropriate interventions. The exhaustive research in caffeine and sleep-related investigations over past few years led to identifying caffeine use disorder in DSM. The DMS-IV recognizes four caffeine-related diagnoses: Caffeine

Intoxication, Caffeine-Induced Anxiety Disorder, Caffeine Related Disorder not otherwise specified and most important from the perspective of this chapter, Caffeine-Induced Sleep Disorder (DMS IV) [15,16]. This classification was however updated in DSM-V to include Caffeine-induced Sleep Disorder under the subheading of "Other Caffeine-Induced Disorders," and limiting the utility of caffeine use disorder as a condition for further study [17,18] (DSM V). The decision reflects the consensus of the experts that the caffeine use disorder can likely be over-diagnosed since the use of caffeine is extensive worldwide [19]. These recommendations emphasize the dire need for understanding caffeine-related sleep disturbances which are likely etiological or consequential factor of caffeine withdrawal syndrome as well as dependence.

Despite the beneficial effects of caffeine on sleep deprivation, numerous studies have observed adverse sleep-related consequences on subsequent nights [4,9]. However, the epidemiological studies quantifying these associations are limited [20–25]. The complexity of this relationship can be attributed to the non-standardized measurements of caffeine intake or sleep disturbances, age-related differences, and individual variation in dose-dependent effects of caffeine on sleep [9,26–28]. Here we attempt to delineate the studies based on sleep disturbances.

One of the oldest cohort of community-dwelling adults of US Iowa state aged 65 years and more reported that caffeine intake poses trouble falling asleep which was driven by over-the-counter analgesics that contain caffeine [29]. The trends in caffeine consumption have changed drastically over last decade, and additional sources of caffeine now influences sleep. A systematic review on the effects of caffeinated foods and beverages on cognitive functioning in healthy adult population observed that high dose caffeine intake of 200–400 mg more than once a week was independently associated with short sleep duration [30,31]. The interplay of caffeine consumption with sleep duration and insomnia symptoms was recently evaluated in American adults using data from the 2007–08 National Health and Nutrition Examination Survey (NHANES) [22]. The study provided a unique perspective that the caffeine was not only associated with insomnia symptoms of trouble falling asleep, waking after sleep onset (WASO), and daytime sleepiness but the interaction of caffeine intake and habitual sleep duration predicted non-restorative sleep. The relationship between higher caffeine consumption with sleep duration or sleep quality was consistent in studies across the globe despite the heterogeneity in the methods of assessment of sleep duration or sleep quality [24,32–34].

There are more small-scale epidemiological studies performed in adolescents than in adult population owing to research recommendations in past literature and unique consumption pattern in adolescents [21,23–25]. The shortened sleep duration, excessive daytime sleepiness and dose-dependent initial insomnia and poor sleep quality is a consistent pattern observed in adolescents across these epidemiological studies. A large nationally representative school-based NICHD (the National Institute of Child Health and Human Development) study of 15,686 adolescents from grades 6 to 10 found that 75% of adolescents drank at least one soda/day, but coffee use was less prevalent with 75% having coffee less than once a week [24]. The study reported that adolescents with high caffeine intake, defined as >1 drink/day, had almost twice the odds of reporting difficulty in sleeping, or daytime sleepiness. A similar association in a large German-based epidemiological study of 7698 adolescents (11–17 years) recognized that the coffee use is an essential risk factor for insomnia at this age [35]. The dose-dependent effects of caffeine on sleep are more prominent in early age group. Caffeine intake of as high as 77.1 mg in 12–15 years old results in shorter sleep duration, longer WASO, and longer daytime sleep on a two-week sleep diary [25]. The sleep was more interrupted on the night after consumption in those who had 100–150 mg/dl compared to those who had 0–50 mg/day. In another study of 1522 adolescents of 13 years of age, median caffeine intake of 22–27 mg/day via soft drinks increases the odds of shortened sleep duration such that for each 10 mg/day increase in intake there were higher odds of sleep duration 8.5 h or less [23].

The factors that dictate the relationship between caffeine and sleep are different in adolescents compared to other age groups. The short sleep duration reported in adolescents is an interactive effect of caffeine, technology or use of other substances especially alcohol [21,36]. Studies observed that coffee use modifies the relationship between sleep and other substance use (alcohol, nicotine) which in turn disrupt the sleep cycle and perpetuate the cycle of sleep fragmentation and substance use [35,37–39]. These findings however are not always consistent across studies [40,41]. The discrepancies can be due to the different sources of caffeine, time of intake, or duration of intake. Further the current research has not delineated the effect of soda beverages from energy drinks especially in young age groups who consume an excess of energy drinks. The results should not differ much considering the main ingredient in both the products is still caffeine.

The lack of objective measures of sleep disturbances or insomnia complaints has been the challenge in better understanding and generalization of these relationships from the epidemiological view. There is an overwhelming number of studies that attempted to account the limitation of non-standardized measures by studying insomnia symptoms, more closely related to clinical diagnosis, in a controlled environment of randomized clinical trials, laboratory-based intervention studies, or polysomnographic studies [4,9]. The exciting aspect of these studies is that they consistently found that caffeine results in reduced sleep efficiency, increased sleep latency, more episodes of WASO, and reduced sleep duration. However, these studies are limited by small sample size, restriction to younger participants, and use of non-pragmatic concentrated forms of caffeine as a part of the intervention. The randomized polysomnographic study trials, being the gold standard, will be discussed here. A double-blind crossover study on 22–25 years old and 40–60 years old observed that higher dose of caffeine (200–400 mg) was associated with dose-dependent increase in relative stage 1 sleep and reduction in absolute and relative slow wave sleep and absolute REM movements in both age groups [42]. The younger individuals had increased absolute stage 1 sleep while older adults had decreased absolute stage 2 sleep. Overall, the middle-aged adults are more sensitive to the high dose of caffeine [43]. A randomized controlled trial focused on understanding the sleep hygiene patterns compared the effects of 400 mg caffeine and placebo at 0, 3, and 6 h consumed before self-reported habitual bedtime. The polysomnographic findings confirmed that caffeine compared to placebo is associated with reduced total sleep time, delayed onset of sleep, disturbance in the sleep stages, and reduced sleep quality. These changes were prominent for a minimum of 6 h before bedtime supporting the role of metabolic products of caffeine on sleep [44]. From the perspective of dose-dependent effects, many studies observed reduced total sleep time, duration, and sleep stages at a dose of approximately 200 mg. However, the smallest dose at which these changes are observable is still not established. One of the archival studies by Rosenthal et al. observed increased sleep latency and reduced sleep duration at a dose as low as 75 mg/day with no effects on recovery sleep [45]. Ho et al. [46] in a double-blind control group design among 20–22 years old observed that there are no changes in sleep measures at a dose of 60 mg/day x week. While the pragmatic studies will be more convincing, detailed information of these studies provided by Clark et al. [9], Poole et al. [47], and Roehrs et al. [4] can be conclusive for some researchers.

The recovery sleep or physiological sleep debt secondary to prolonged sleep deprivation is more common than acute sleep deprivation [48]. However, the relationship of caffeine with recovery sleep is studied less in the general population and more in those who are at work in high-risk occupation setting such as military, medical professionals, truck drivers, pilots and shift duty workers. Few studies conducted within the general population have shown that caffeine reduces sleep efficiency, and is associated with reduced N-REM sleep EEG synchronization in daytime recovery sleep [42,49]. In addition to occupation, age is another critical demographic as the detrimental effects of caffeine on recovery sleep are more prominent in middle-aged individuals than younger subjects [42].

Epidemiology of sleep in energy drink supplements

Even though the advent of energy drink supplements in the last decade is frequently reported to be alarming, the percent consumers continue to be <10% across all age groups [1,50]. The reasons for concern about these supplements are: (1) Though the consumption is <10%, caffeine intake via these supplements has increased from 0.1–0.3 mg/day in 2001 to 1.9–2.1 mg/day in 2010 in the US population. The data collected by Kantar World Panel provided age-specific rates which were higher in the age groups 13–17 years (5.3–6.9 mg/day) and 18–24 years (6.2–0.8 mg/day), and lower in elderly population (0.7–1.1 mg/day); (2) There is no specific nutritional information available on the contents of all these supplements which limits our knowledge on the consumption and effects of other less-known substances; (3) There is limited data on health effects of these supplements in the general population due to lack of evidence-based research. The studies related to supplements have been small and focused on young adults, college students, athletes or military personnel. The findings from these studies are not generalizable owing to high consumption in these segmented populations, and differential comorbidity pattern compared to other age groups. The above-mentioned points are also valid for the role of these supplements in sleep disturbances.

The causal relationship between energy drinks and sleep disturbances is still debatable despite strong contemporary research. Two recent large epidemiological studies reported that the short sleep duration was associated with energy drinks use. A nationally representative survey based on NHANES data (2005–14) published that short sleep duration was associated with higher intake of beverages in adults >18 years of age and it varies with an hour of sleep lost compared to regular 7–8 h of sleep [32]. A study based on Ontario Student Drug Use and Health Survey observed that 13% of high school students reported energy drinks use, and were more likely to report shortened sleep duration compared to non-users even after accounting for tobacco or alcohol use [51]. The study findings lacked cause-effect directionality, a pattern consistent in most of other related investigations. The research on sleep and energy drinks supplements primarily focus on changes in work performance, cognition or health behaviors following their consumption. In adolescents, the association focuses on academic achievement, and risk-taking behavior, while in adults it is based on a change in cognition or performance. In a small university sample, high-end energy drinks result in sleep disturbances such as later bedtimes, harder time falling asleep, and more all-nighter episodes. The concerning factor was that the majority of students were unaware about the presence of caffeine in these supplements [52]. In another cross-sectional study of 498 random groups of college students, the researchers observed that the students consumed more than one energy drink each month on average, and 67% of them consumed for insufficient sleep or energy to improve overall performance. However, the consumption, in general, was associated with binging episodes [53]. More than 58% of emergency department adolescent visitors reported ED use within last 30 days and had trouble sleeping and work problems [54].

The military members report around three times higher daily energy drinks use (27%) compared to the general population [30]. Studies on the energy drinks use in the military population have demonstrated the association of energy drinks use on sleep continuity disturbances and next day functioning. In a 2008 Air Force Research Laboratory report, 30.8% of active duty Air Force Personnel reported trouble falling asleep, and 10.8% reported trouble staying asleep as common adverse effects related to the consumption of energy drinks [55]. In a similar study among military volunteers of 2007–08 Millennium Cohort Study, energy drinks use was higher in those who reported less than recommended 7–8 h of sleep or had trouble sleeping [56]. Energy drinks use has similar reports of abnormalities of sleep duration and sleepiness among combat-deployed personnel in more contemporary studies. The personnel who had at least 3 or more energy drinks/day were more likely to report sleep <4 h per day, or disturbed sleep for at least half the days in a month. These personnel also reported insomnia or episodes of falling asleep during regular guard duties [57–59]. The findings require special attention considering military personnel are at constant stress from high tempo combats, nighttime duties, personal life, and illness and frequently use these supplements for bodybuilding, energy, improving alertness and counteracting sleep-deprived state [60,61].

The most popular cause of sleep disturbances due to ED use is their caffeine content. However, the role of taurine and other carbohydrates has been recently explored in animal studies [62]. Tyrosine, an ingredient of some caffeinated energy drinks, was investigated in sleep-deprived healthy young male and found to have no independent association on sleep [62]. The additional nutritional information on the other ingredients is warranted for future research.

Epidemiology of sleep in other psychostimulants

Research on other psychostimulants is limited to military personnel, air force and other service members owing to their extended period of sleep deprivation. The substances most frequently researched as a countermeasure to sleep deprivation are modafinil and dexamphetamine. Modafinil and dextroamphetamine improved executive function, objectively measured alertness, and psychomotor vigilance speed [63]. However, modafinil has been reported to be a more promising stimulant [64]. In comparative efficacy of caffeine, modafinil and dextroamphetamine, modafinil dose of 400 mg improve performance and mood symptoms without any other adverse effects compared to other stimulants after sleep deprivation of 40 h or more. Dextroamphetamine is associated with dose-dependent impairment of sleep maintenance and decreased executive functioning, while caffeine-related adverse sleep disturbances have been discussed earlier [65]. Modafinil is also well-tolerated compared to other drugs [66,67]. In a comparison between energy drink ingredients (tyrosine, phentermine), caffeine and dextroamphetamine after 36 h of sleep deprivation, dextroamphetamine was associated with marked decrease in sleep drive, increased alertness on the first-day recovery. Tyrosine did not affect while caffeine and phentermine had similar effects on sleep [68].

Interestingly the most common illicit stimulants used by military personnel were cocaine and amphetamines as reported by latest 2005 report of Army Forensic Laboratories [69,70]. With the ongoing efforts of legislation of marijuana, the association between sleep and these stimulants may play an essential role in the general population in future.

Relationships in specific populations

The findings from general population may not be applicable for individuals with certain occupations such as Military Personnel, Air Force Personnel, Shift Workers, Emergency Duty workers, and others. The daily caffeine consumption in the form of sodas and energy drinks in these high-risk occupational settings is considerably higher than the general population owing to their work expectations even in the sleep-deprived state [30]. Secondly, the effect of caffeine is studied in this population from the perspective of improvement of work performance, recovery sleep, and consequences of physiological sleep debt. The risk and benefits of caffeine use in these populations can be summarized as follows:

(i) *Sleep deprivation and sleepiness*: Caffeine has shown to be effective in maintaining the cognition and performance measures even after 3–4 nights of sleep deprivation [71,72]. Compared to placebo, caffeine intake of 100–400 mg improved subjective exertion and task completion time, maintained vigilance and maintained indices for performance and marksmanship in military personnel who are sleep deprived for 22–27 h [72,73]. These caffeine dosages are high, but caffeine can improve the performance at lower doses [12]. The unique phenomenon observed in these populations is abrupt awakening from naps which results in momentary degradation of performance known as sleep inertia. A double-blind crossover design in non-smoking adults observed that the immediate use of caffeine gum of 100 mg post-inertia improved vigilance at 18 min post-awakening [74]. Shift workers who have fixed shifts at evening or night are more likely to report sleepiness, insomnia symptoms, and excess caffeine consumption compared to those who have rotating shifts [75,76]. The rotating shift workers are more likely to report delayed sleep onset and improved overnight performance after caffeine use [76,77]. Caffeine also improves mood disturbances associated with sleepiness after 48 h of sleep deprivation [78,79].

However, these findings may vary depending on the pattern of consumption and activity involved. A study of pilots who were sleep-deprived for 37 h did not find any significant changes in subjective sleepiness in spite of improvement in performance [80]. Similarly, the truck drivers who were habitual coffee drinkers had no effects on sleep time after prolonged sleep deprivation [81].

(ii) *Recovery sleep:* Chronic sleep restriction is more common and concerning than acute, total sleep deprivation [48]. Individuals in these settings undergo a prolonged period of sleep deprivation, have circadian misalignment, and have a higher impact on recovery sleep. While caffeine shows sustained performance in studies measuring short-term sleep-deprived states, its effects on performance after prolonged sleep deprivation and subsequent recovery sleep are not well-characterized. Habitual caffeine use may interfere with recovery sleep following acute sleep-deprived state. A study of healthy civilian and active-duty military personnel observed that the effect of daily intake of caffeine (400 mg) on subjective alertness were attenuated compared to placebo after the third night of total sleep deprivation. On the maintenance of wakefulness test, caffeine users were no longer able to maintain wakefulness after the second night of sleep deprivation and were also slower to return to baseline in the recovery period [82]. Caffeine use after 23 h of wakefulness in night shift workers was associated with increased WASO, increased stage 1 sleep, decreased slow wave sleep, increased core body temperature and broader distal to proximal skin temperature gradient during recovery sleep [83]. High core body temperature indicates increased alertness which is congruent to other literature that reported no influence on performance in the period after recovery sleep [84].

(iii) *Insomnia*: The overlap of insomnia symptoms with other sleep measures limits the research of caffeine use among those with a clinical diagnosis of insomnia. Further, the treatment strategies of insomnia including abstinence from caffeine will influence the use of caffeine among those with insomnia. In a treatment-seeking active duty military personnel reporting PTSD and insomnia symptoms, those who had elevated insomnia symptom severity avoided coffee use [85]. In a study of patients with primary insomnia, the use of caffeine as a treatment for sleep deprivation was associated with decreased total sleep time and increased subsequent sleep onset latency [86]. There is also a distinct overlap of insomnia symptoms with psychiatric symptoms in these populations. The contribution of psychiatric symptoms in the causation of insomnia could be more significant than the overall effect of caffeine use alone which could potentially explain the negative association between insomnia severity and caffeine consumption across different studies [85,87]. The self-management and educational efforts to reduce caffeine in these work settings once diagnosed with insomnia can be postulated but require further supporting evidence.

In conclusion, caffeine negatively affects sleep quality and sleep duration in these populations. However, it is of paramount importance to find the risk/benefit ratio of use of caffeine or energy drinks for the community members who are expected to maintain high performance in sleep-deprived states which is possibly achieved by use of these products.

PHYSIOLOGY OF CAFFEINE IN SLEEP-WAKE HOMEOSTASIS

The biological effects of caffeine and other psychostimulants on sleep are the acute or chronic reactions at a psychological or physiological level such as anxiety response, rewarding, or reinforcing effects. Though one of these responses is used to improve cognition, and attention span in sleep-deprived individuals, the chronic reactions or tolerance to the doses over time contribute to recovery sleep or physiological sleep debt. To understand this delayed response, we will briefly talk about these underlying biological processes. As caffeine is the essential component of energy drinks supplements compared to taurine, or sugar, we will limit our discussion on physiology effects of caffeine.

Role of adenosine and caffeine in sleep-wake cycle

The biological mechanisms of caffeine involve interactions at multiple sites. Caffeine is a non-selective competitive adenosine receptor antagonist and produces its psychostimulant effects by counteracting the pulsating effects of endogenous adenosine in the central adenosine receptor [4,88]. Adenosine principally modulates the function of neurotransmitter pathways that are involved in motor activation and reward. The dopaminergic systems primarily control these pathways, as well as arousal processes involved in cholinergic, noradrenergic, histaminergic or orexinergic systems [89]. Caffeine is structurally similar to adenosine and binds to adenosine receptors, specifically to adenosine receptors (A1 and A2A) expressed extensively in the brain.

Adenosine is a prime promoter of neural activity of sleep-wake homeostasis. Adenosine is produced from the breakdown of ATP in normal or pathological conditions due to hypoxia or ischemia [90]. The release of adenosine and subsequent mechanisms that balance the intracellular and extracellular levels of adenosine modulates the function of sleep-active neurons. Adenosine primarily causes sleep enhancement via its mechanism on the A1 receptor to inhibit the cholinergic and anti-cholinergic wake responsive neurons resulting in GABAergic neuron inhibition [91]. The second most studied A2a adenosine receptor inhibits the histaminergic neurons and also directly stimulate the sleep-related active neurons in the ventrolateral preoptic nucleus promoting sleep activity. Caffeine, an adenosine antagonist, competes with extracellular adenosine and attenuates the effect of adenosine stimulation by inhibiting the A1 and A2a receptors [89] (see Fig. 23.1). Studies in the rodent models have shown that caffeine also increases the sensitivity of D2/D3 dopamine receptors promoting the wakefulness [92]. The high affinity of caffeine to these adenosine brain receptors is intertwined with physiological extracellular levels of adenosine and is mainly responsible for motor-activating, and reinforcing and arousing properties of caffeine. Fig. 23.1 summarizes the adenosine regulation of sleep-active neurons.

Genetic factors and response to caffeine

There are pronounced individual differences in response to caffeine. Not all develop caffeine-induced sleep disturbances and insomnia [93]. These differences are secondary to pharmacokinetic or pharmacodynamics mechanism of caffeine at the genetic level which is further influenced by the amount of caffeine consumption or other environmental factors such as demographics, health behaviors, medical comorbidities, or comorbid substance use. The cumulative effect of these factors plays a role in determining the magnitude and duration of tolerance development or caffeine sensitivity to sleep. In this section, we will talk about genetic factors responsible for individual variation, while environmental factors are addressed in subsequent section.

One prominent gene polymorphism involved at the pharmacokinetic level is ADORA2A gene. The gene variants modulate the subjective responses of caffeine through its effects on A2A receptors. A genome-wide association study in a community-based sample of Australian twins from the Australian Twin Registry found that the individual sensitivity of caffeine-induced insomnia depends on polymorphism in the ADORA2A gene (rs5751876) [94,95]. Individuals who had C/C genotype at this SNP reported high sensitivity on the questionnaire and increased EEG beta activity on polysomnography. Individuals with low caffeine sensitivity and no activity had T/T genotype. The GWAS study additionally identified two specific SNPs, rs6575353 coding for PRIMA1 and rs521704 coding for GBP4, which were associated with caffeine-induced insomnia. However, the effect of ADORA2A polymorphism may not be similar for each insomnia symptom. A small sample size study ($n=50$) in college students observed that ADORA2A CC genotype did not moderate the effects of caffeine on WASO even though CC genotype was associated with WASO compared to TT phenotype [96]. In another GWAS twin study of 3808 participants, a region on chromosome 2q and chromosome 17q was associated with caffeine attributed sleep disturbances. The chromosome 17q is of particular interest as the gene regulating dopamine, and cAMP-regulated protein pathways reside in this area which can potentially link to caffeine-induced insomnia. However, more research is needed to validate these findings.

Another polymorphism of importance is a gene for adenosine deaminase (ADA), an enzyme regulating intracellular and extracellular adenosine levels. A heterozygous genotype at the gene (rs73598374) results in reduced activity of ADA which increases EEG delta activity in NREM sleep. Based on this pathway, a polysomnography study of 1000 participants assessed the association between caffeine and sleep measures within both heterozygous and homozygous state [97]. Individuals with a heterozygous form of ADA had better sleep quality, a higher proportion of REM sleep and lesser WASO episodes [98]. A CYP1A2 gene polymorphism that interferes with the hepatic metabolism of caffeine can account for variability in caffeine effect on sleep [99]. However, this mechanism affecting the bioavailability of caffeine relies first on the amount of caffeine intake followed by CYP1A2 gene polymorphism.

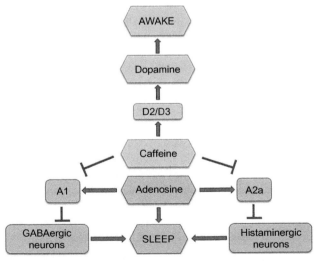

FIG. 23.1 Schematic representation of the role of caffeine in the neurochemistry of sleep-wake homeostasis. *A1/A2a*, adenosine receptors; *D2/D3*, dopamine receptors; *Red arrows* indicate inhibition. Caffeine inhibits adenosine receptors and blocks the action of extracellular levels adenosine on these receptors. Wake pathway is activated, and sleep pathway via adenosine is inhibited in the presence of caffeine.

Environmental factors and response to caffeine

Currently, there are no restrictions on the amount of caffeine intake even though US Food and Drug Administration (FDA) lists 400 mg per day as the safe amount to ingest [1,2]. There is no standard limit for energy drinks yet. The major implication is that the individual threshold at which caffeine or energy drinks will influence sleep is not determined. It inadvertently forces us to consider individual-level variation where this relationship will be vital. We can postulate that individuals regulate their intake to achieve their primary objective of reducing sleepiness or increasing alertness, and secondly to maintain within the safe dose range. While the literature has strongly focused on genetics for individual variation, following other aspects should be noted.

1. *Amount of intake and adverse effects*: A study of US military personnel found that sleep problems are higher in those who drink 3 or more energy drinks per day [57]. A similar pattern is observed in children and adolescents where moderate-level caffeine consumers have more sleep problems [100]. These associations led to hypothesis whether caffeine/energy drinks are used as a countermeasure of sleep deprivation or perpetuate sleep problems. Irrespective of the directionality of the association, such pattern will inadvertently lead to upward trajectory of ingestion. High doses of consumption subsequently cause adverse effects, most notably seen are restlessness, nervousness, cardiovascular instability, and insomnia [100]. Caffeine results in a dose-dependent increase in anxiety, worries, and tremors compared to placebo which was partly shown to vary by individual personality [101]. In another study of healthy non-smoking individuals, a sequential increase of caffeine dosage from none to 600 mg was associated with irritability, headache, anxiety, talkativeness, and sleepiness but the specific threshold was not determined [79]. Caffeine withdrawal among moderate-high caffeine consumers also has similar symptoms. In these individuals, caffeine withdrawal was associated with a headache, impaired mood, impaired cognition, and increased blood pressure [102]. These effects are subjective, and the extent to which they are linked to the caffeine-sleep paradigm is hard to determine.

2. *Chronotype/time of intake*: One of the frequently reported reasons for individual variability is the time of the day when these stimulants are consumed. A Brighton Sleep Study in 20 to 45 years old observed that 1–4 cups per day within 6 h of bedtime was associated with shorter sleep latency, fewer awakenings, and more sleep satisfaction than heavy consumers or those who abstained from caffeine [103]. Early morning caffeine decreases the propensity of wakefulness on a subsequent night [104]. Further, evidence suggests that evening chronotype of individuals has a significant role in the caffeine-sleep relationship. In a study of college students, those who had a circadian preference for evening time were three times more likely to consume high dose caffeinated energy drinks and report daytime sleepiness compared to those with morning chronotype [105–107]. Adolescents are showing a slow transition to evening chronotype with increased use of electronic devices at nighttime. The use of technology at night has been associated with increased consumption of energy drinks and other caffeinated products which cumulatively results in short sleep duration [21,36]. Few studies found no role of chronotype at the caffeine-sleep interface or observed increased WASO episodes and sleep efficiency among morning-type individuals [96,108]. Overall the behavioral aspect of circadian preference has some role to play.

3. *Comorbid burden of disease*: Little research has examined the effect of the comorbid burden of an individual on the sleep and use of caffeine, energy drinks, and other stimulants. In a small case-control study (134 cases, 230 controls) of diabetic participants, diabetic individuals were habitual caffeine users and reported greater daytime sleepiness after an extra cup of coffee [109]. This relationship was attributed to the presence of a highly inducible CYP1A2 genotype in the diabetic population [110].

4. *Sociodemographic pattern*: Another individual aspect that potentially affects both caffeine and sleep is their sociodemographic characteristics. The role of these characteristics as a risk factor for a stimulant-sleep interface is still understudied. Both energy drinks and sleep are associated with low socio-economic status [33,111]. Energy drinks are part of government assistance programs and account for 50% of their budget indicating high use in the low-income population [112]. Sleep also varies by socio-economic status, and individuals with low socioeconomic status are more likely to report insomnia complaints and shorter sleep duration [111]. The consistent pattern of sleep and caffeine usage indicates a potentially stronger role of low income in addressing the outcomes of caffeine-induced sleep disturbances. Based on gender, males are three times more likely to report energy drink use than females [111]. However, in Brazilian sample, women were reported to have a higher proportion of caffeine-induced insomnia than men [113]. There is a distinct age-related pattern in the relationship between sleep and energy drink use. A nationally representative study using NHANES data observed that caffeine intake from energy drinks and dietary supplements is higher in the age group of

19–30 years [114]. In another study using community survey, younger respondents (< 30 years) were eight times more likely to use energy drink compared to older individuals [115]. On the other hand, sleep disturbances are aging phenomenon yet we are observing sleep disturbances in younger population owing to energy drinks. The age-related changes in sleep patterns highlight the importance of monitoring caffeine dependent sleep problems in the younger population. The NHANES study also reported that the use of energy drinks is highest among non-Hispanic White individuals followed by Hispanic and Black respondents [114]. In contrast, the data using NHIS study found that the consumption is highest in Black respondents [116]. The NHIS finding aligns with racial pattern of sleep disturbances where Blacks are more likely to have poor sleep continuity, shorter sleep duration, and less slow-wave sleep [117].

5. *Alcohol use*: The use of alcohol mixed with caffeinated energy drinks is frequently reported, and is associated with vicious cycle of substance use and sleep disturbances. This pattern of alcohol mixed energy drinks is noticed commonly in adolescent's age group. A community based study found that alcohol-mixed energy drink consumptions compared to alcohol only drinks is associated with increased chances of difficulty in falling asleep [118]. Energy drinks consumption is associated with increased alcohol use independently and also in presence of sleep disturbances perpetuating overall sleep problems [33,39,119,120]. Another study in college students observed that caffeinated beer improves sleep quality with no effect on sleep latency or next day hangover symptoms which may promote substance use [41]. In general, the combined use of alcohol and energy drinks distorts the perceptions of adaptive response to excess caffeine intake or alcohol intake which heightens the consequences. Excess alcohol use secondary to energy drinks use increases the probability of daytime sleepiness with binge drinking [121]. The role of other illegal substance use can potentially be similar to alcohol but will not be discussed in this chapter.

Above knowledge will help understand how caffeine interferes with the interplay between circadian and homeostatic regulatory systems to affect the periodic cycle of wakefulness and sleep.

HEALTH IMPLICATIONS OF CAFFEINE(STIMULANT) USE—SLEEP DISTURBANCES MODEL

Although the caffeine use and sleep disturbances have independent associations with health outcomes, the major impact of stimulant-induced insomnia on health status is sleep disturbances. These sleep disturbances are secondary to circadian misalignment where the endogenous circadian pattern could not match with altered sleep-wake cycle owing to caffeine/energy drink consumption resulting in deleterious effects [122]. Though not established we will briefly discuss some of these health implications from the perspective of stimulant use resulting in, sleep disturbances, short sleep duration, and circadian misalignment. The discussion is also primarily focused on caffeine use.

(1) *Short sleep duration*: The associations of short sleep duration with chronic diseases and associated risk factors are extensively researched and are beyond the scope of this chapter. Most notable about the use of stimulant to counteract sleepiness is weight gain or obesity, glucose dysregulation, hypertension, dyslipidemia, inflammation and hormonal dysregulation [117]. Obesity is the most immediate health outcome of concern considering the use of energy drinks as a part of energy supplements. Adolescents with inadequate sleep have poor eating behaviors including excess consumption of caffeinated energy drink which are associated with higher body weight and blood pressure [123]. Community-dwelling adults and military personnel reporting caffeine or energy drink use had increased concentrations of IL-6 which is indicative of low-grade chronic inflammation in the presence of inadequate sleep leading to morbidities [50,124]. A randomized controlled trial of sleep-deprived individuals reported that caffeinated coffee adversely modulated the glucose homeostasis compared to decaffeinated ones [125]. Continuous caffeine intake over a prolonged period after an extended period of sleep deprivation was associated with mental exhaustion, irritability and mood disturbances but not with glucose levels [126]. However, there was increased glucose response to subsequent food intake which could be the potential mechanism to glucose dysregulation. Despite overlap between sleep duration and sleep deprivation, short sleep duration has shown to play a prominent role in obesity.

(2) *Non-specific sleep-disturbances*: The sleep symptoms that form a part of the ICD-9 diagnosis of insomnia but do not meet the criteria will be considered as non-specific sleep-disturbances here. These symptoms include difficulty falling asleep, difficulty staying asleep, waking after sleep onset and daytime sleepiness. There is a strong correlation between these symptoms or insomnia diagnosis and psychiatric comorbidities. Caffeine increases hyperactivity of dopaminergic system through its inhibitory action on adenosine which is further pressurized by sleep disturbances affecting mood, executive functioning, cognition and behavioral disorders [127,128]. Caffeine-induced sleep disturbances increase nighttime restlessness [101]. A KiGGS German study observed that caffeine use influences the

association between insomnia complaints and use of alcohol, marijuana, and tobacco emphasizing the role of common stimulants [35]. A case study of psychiatric patients showed those with bipolar disorder, of which sleep disturbances are component, had some temporal relationship between use of energy drinks and frequent hospitalizations [129]. This finding aligns with the results of a randomized controlled trial where individuals with identified vulnerability to stress-induced sleep disturbances exhibited higher sleep-reactivity to a caffeine challenge [130]. High caffeine consumption is also associated with multiple psychiatric disorders including depression, anxiety and substance dependence [131]. Excess consumption of energy drinks or stimulants by military population is of special concern considering pre-deployment sleep disturbances, or insomnia is associated with subsequent development of anxiety disorder including PTSD [132]. Despite these findings, we preclude any causal interpretation due to cross-sectional nature of these studies as well as lack of data on caffeine use among those who show symptoms of insomnia.

(3) *Circadian misalignment:* The sleep disturbances due to circadian rhythm misalignment occur commonly in the special populations described above. There is strong empirical evidence that excess stimulant use improves their vigilance, cognitive function, alertness, endurance performance, a memory task, and reaction time. However, the benefits associated with the use of these products interfere with underlying biological sleep rhythms. These disturbed physiological processes tend to result in adverse outcomes in the physical or psychological domain. The most pertinent concerning behavior observed in shift workers, healthcare, law enforcement, and drivers are impulsive risk-taking behaviors. The caffeine response to impulsivity is secondary to dose-dependent increase in adrenaline during wakefulness period [133]. Accordingly, the truck drivers or shift workers are likely to report impaired driving performance, commit avoidable errors and increased number of accidents [134–136]. These episodes are more likely to occur during daytime work when the underlying physiological response is to stay awake [137]. Studies researching likelihood of accidents in sleep-deprived truck drivers recommend caffeine use with short naps emphasizing the vital role of sleep in preventing accidents [138,139].

The predominant health consequences of circadian misalignment are related to cardiovascular disease and metabolic disorders. Physiological hyperarousal in the presence of chronic sleep restriction and caffeine use is associated with significant odds of hypertension [140,141]. Shift workers report higher mortality due to diabetes, cardiovascular and stroke. Rotating shift workers have increased the risk of obesity, hypertriglyceridemia, poor dietary patterns, and metabolic syndrome [122]. Bonnet and Arand in their archival study of caffeine as a model for insomnia observed that caffeine use causes significant metabolic changes which are associated with decrease in sleep efficiency [14]. Chen and colleagues found that men who work long hours are twice at risk for coronary heart disease [142]. In adolescents with different chronotype and caffeine sensitivity, poor sleep and situational stress is associated with cardiac changes secondary to sympathetic system activation, a phenomenon consistent with circadian disturbances. These individuals are at subsequent risk of developing insomnia and related comorbidities [143]. In summary, the burden of caffeine use as a countermeasure to sleepiness is associated with an enormous burden on health and functioning.

RECOMMENDATIONS

We recommend the following suggestions to improve the health and functioning of individuals:

1. *Sleep Hygiene practices*: Insufficient sleep is a significant problem, and strict implementation of a sleep management plan to maintain sleep duration should be prioritized in all educational and workplace settings such as schools, colleges, medical facilities, military centers and other training facilities. Stimulants intake before bedtime should be discouraged, and daily variation should be monitored. Physical activity has been shown to alleviate the effects of caffeine use on sleep and should be recommended. A strict sleep schedule should be followed to maintain the circadian rhythm, and the daytime nap should be avoided. In contrast, strategically timed short naps in conjunction with small doses of caffeine may improve mood and performance after sleep deprivation without affecting subsequent sleep patterns in specific populations and should be endorsed. However, the clinical implications of frequent napping in general population is still preliminary and should be further investigated [144].

2. *Better food labeling practices*: Despite the statutory FDA limit on caffeine amount, there are no strict restrictions on the contents of energy drinks and amount of intake. The judicious use of caffeine from different sources may lead to tolerance and create a vicious cycle of insufficient sleep and habitual increasing intake of caffeine over time. This cycle will secondarily amplify health-related outcomes. The content or safety level of other ingredients in energy drinks is not even known to assess whether they show caffeine like tolerance. The issue is further complicated for dietary supplements which may contain adulterants. The FDA should stress upon the labeling options on these ingredients to inform con-

sumers. The labels should standardize the content level of each ingredient, warnings about caffeine toxicity, and age-specific symptom-free doses [8]. Alternatively, the marketing practice for beverages should highlight the caffeine content. The marketing practices directed toward youth should be supervised. The ultimate goal is to develop a culture where caffeine intake is monitored to detect early signs of chronic use and withdrawal symptoms.

3. *Concurrent substance use*: The concurrent use of caffeine and other substance use is discouraged. Adolescents have frequently been involved in the use of caffeine with alcohol. Their simultaneous use facilitates extended drinking sessions, interferes with executive function and masquerades the alcohol intoxication. Such behavior is linked to risk-taking patterns such as intoxicated driving. It also has deleterious effects on sleep maintenance and aggravates insomnia [145]. The sleep education should actively focus on this issue.

4. *Rethinking research methodologies*: The current research on the stimulant use and sleep is based on nonstandardized measures. Though we have sufficient evidence on the qualitative aspect of these relationships, the quantitative parameters are still unclear. The lack of quantitative threshold for caffeine and other ingredients have limited authorities from imposing stronger regulations. There is a distinct gap between cross-sectional population level studies and experimental or laboratory-based studies in this area of research. The next steps in research should focus on bridging that gap. There is a need of a socio-ecological model, as proposed by Grandner and colleagues, that will define the temporal pathway upstream from sociodemographic and behavioral determinants of caffeine-sleep paradigm to downstream adverse consequences of this paradigm [117]. Such structured framework will help identify high-risk populations and plan interventions tailored to these groups. Numerous interventions that have been formulated over time have not shown improvement in the change in caffeine consumption [146]. Recently, a publicly accessible Web tool (2 B Web Alert) was used to determine the impact of sleep-wake schedule, time of day of caffeine use, and performance. Evidence-based tools like 2 B Web Alert will aid in designing effective work schedules, proposing better sleep restriction and caffeine studies, and increasing public awareness of the pattern of caffeine consumption on alertness [147].

CONCLUSION

In conclusion, the research examining connections between sleep and caffeine, energy drinks and other stimulants is extensive but still preliminary. Given the positive and negative effects of caffeine and energy drinks, these beverages may provide temporary relief from sleep disturbances, but their prolonged use may affect the quality of life in general. Future research should investigate the cumulative effect of sleep and stimulants on health outcomes rather than independent associations. Such investigations will provide insights to identify candidates with differential risk of sleep effects secondary to stimulant use. It will also provide guidelines to assess appropriate candidates for caffeine supplementation during sleep deprivation, and derive the optimal range of caffeine dosages that may alleviate sleep deprivation without affecting subsequent recovery sleep or inducting additive behavior. The overarching purpose of this chapter is to contribute toward any policy implications in the field of sleep management of the general population.

REFERENCES

[1] Mitchell DC, Knight CA, Hockenberry J, Teplansky R, Hartman TJ. Beverage caffeine intakes in the U.S. Food Chem Toxicol 2014;63:136–42.

[2] Somogyi LP. Caffeine intake by the US population. Prepared for The Food and Drug Administration and Oakridge National Laboratory; 2010.

[3] Seifert SM, Schaechter JL, Hershorin ER, Lipshultz SE. Health effects of energy drinks on children, adolescents, and young adults. Pediatrics 2011;2009–3592.

[4] Roehrs T, Roth T. Caffeine: sleep and daytime sleepiness. Sleep Med Rev 2008;12(2):153–62.

[5] De Valck E, Cluydts R. Slow-release caffeine as a countermeasure to driver sleepiness induced by partial sleep deprivation. J Sleep Res 2001;10(3):203–9.

[6] Ohayon M, Wickwire EM, Hirshkowitz M, Albert SM, Avidan A, Daly FJ, et al. National Sleep Foundation's sleep quality recommendations: first report. Sleep Health 2017;3(1):6–19.

[7] Martyn D, Lau A, Richardson P, Roberts A. Temporal patterns of caffeine intake in the United States. Food Chem Toxicol 2018;111:71–83.

[8] Pomeranz JL, Munsell CR, Harris JL. Energy drinks: an emerging public health hazard for youth. J Public Health Policy 2013;34(2):254–71.

[9] Clark I, Landolt HP. Coffee, caffeine, and sleep: a systematic review of epidemiological studies and randomized controlled trials. Sleep Med Rev 2017;31:70–8.

[10] Brezinova V, Oswald I, Loudon J. Two types of insomnia: too much waking or not enough sleep. Br J Psychiatry 1975;126:439–45.

[11] Fulgoni 3rd VL, Keast DR, Lieberman HR. Trends in intake and sources of caffeine in the diets of US adults: 2001–2010. Am J Clin Nutr 2015;101(5):1081–7.

[12] Lieberman HR, Wurtman RJ, Emde GG, Roberts C, Coviella IL. The effects of low doses of caffeine on human performance and mood. Psychopharmacology 1987;92(3):308–12.

[13] Juliano LMGR. Caffeine. In: Lowinson JHRP, Millman RB, Langrod JG, editors. Substance abuse: a comprehensive textbook. 4th ed. Baltimore: Lippincott Williams & Wilkins; 2005.

[14] Bonnet MH, Arand DL. Caffeine use as a model of acute and chronic insomnia. Sleep 1992;15(6):526–36.

[15] Reid WH, Wise MG. DSM-IV training guide. New York: Routledge; 2014.

[16] Addicott MA. Caffeine use disorder: a review of the evidence and future implications. Curr Addict Rep 2014;1(3):186–92.

[17] Ágoston C, Urbán R, Richman MJ, Demetrovics Z. Caffeine use disorder: an item-response theory analysis of proposed DSM-5 criteria. Addict Behav 2018;81:109–16.

[18] Meredith SE, Juliano LM, Hughes JR, Griffiths RR. Caffeine use disorder: a comprehensive review and research agenda. J Caffeine Res 2013;3(3):114–30.

[19] Striley MCW, Hughes PJR, Griffiths R, Juliano L, Budney AJ. A critical examination of the caffeine provisions in the diagnostic and statistical manual (DSM-5). J Caffeine Res 2013;3(3):101–7.

[20] Bryant Ludden A, Wolfson AR. Understanding adolescent caffeine use: connecting use patterns with expectancies, reasons, and sleep. Health Educ Behav 2010;37(3):330–42.

[21] Calamaro CJ, Yang K, Ratcliffe S, Chasens ER. Wired at a young age: the effect of caffeine and technology on sleep duration and body mass index in school-aged children. J Pediatr Health Care 2012;26(4):276–82.

[22] Chaudhary NS, Grandner MA, Jackson NJ, Chakravorty S. Caffeine consumption, insomnia, and sleep duration: results from a nationally representative sample. Nutrition 2016;32(11–12):1193–9.

[23] Lodato F, Araujo J, Barros H, Lopes C, Agodi A, Barchitta M, et al. Caffeine intake reduces sleep duration in adolescents. Nutr Res 2013;33(9):726–32.

[24] Orbeta RL, Overpeck MD, Ramcharran D, Kogan MD, Ledsky R. High caffeine intake in adolescents: associations with difficulty sleeping and feeling tired in the morning. J Adolesc Health 2006;38(4):451–3.

[25] Pollak CP, Bright D. Caffeine consumption and weekly sleep patterns in US seventh-, eighth-, and ninth-graders. Pediatrics 2003;111(1):42–6.

[26] Spaeth AM, Goel N, Dinges DF. Cumulative neurobehavioral and physiological effects of chronic caffeine intake: individual differences and implications for the use of caffeinated energy products. Nutr Rev 2014;72(Suppl. 1):34–47.

[27] Loftfield E, Freedman ND, Dodd KW, Vogtmann E, Xiao Q, Sinha R, et al. Coffee drinking is widespread in the United States, but usual intake varies by key demographic and lifestyle factors-3. J Nutr 2016;146(9):1762–8.

[28] Nehlig A. Interindividual differences in caffeine metabolism and factors driving caffeine consumption. Pharmacol Rev 2018;70(2):384–411.

[29] Brown SL, Salive ME, Pahor M, Foley DJ, Corti MC, Langlois JA, et al. Occult caffeine as a source of sleep problems in an older population. J Am Geriatr Soc 1995;43(8):860–4.

[30] Knapik JJ, Austin KG, McGraw SM, Leahy GD, Lieberman HR. Caffeine consumption among active duty United States Air Force personnel. Food Chem Toxicol 2017;105:377–86.

[31] Ding M, Bhupathiraju SN, Chen M, van Dam RM, Hu FB. Caffeinated and decaffeinated coffee consumption and risk of type 2 diabetes: a systematic review and a dose-response meta-analysis. Diabetes Care 2014;37(2):569–86.

[32] Prather AA, Leung CW, Adler NE, Ritchie L, Laraia B, Epel ES. Short and sweet: associations between self-reported sleep duration and sugar-sweetened beverage consumption among adults in the United States. Sleep Health 2016;2(4):272–6.

[33] Grandner MA, Knutson KL, Troxel W, Hale L, Jean-Louis G, Miller KE. Implications of sleep and energy drink use for health disparities. Nutr Rev 2014;72(Suppl. 1):14–22.

[34] Kant AK, Graubard BI. Association of self-reported sleep duration with eating behaviors of American adults: NHANES 2005-2010. Am J Clin Nutr 2014;100(3):938–47.

[35] Skarupke C, Schlack R, Lange K, Goerke M, Dueck A, Thome J, et al. Insomnia complaints and substance use in German adolescents: did we underestimate the role of coffee consumption? Results of the KiGGS study. J Neural Transm (Vienna) 2017;124(Suppl. 1):69–78.

[36] Calamaro CJ, Mason TB, Ratcliffe SJ. Adolescents living the 24/7 lifestyle: effects of caffeine and technology on sleep duration and daytime functioning. Pediatrics 2009;123(6):e1005–10.

[37] Jaehne A, Loessl B, Barkai Z, Riemann D, Hornyak M. Effects of nicotine on sleep during consumption, withdrawal and replacement therapy. Sleep Med Rev 2009;13(5):363–77.

[38] Kozak P, Paer A, Jackson N, Chakravorty S, Grandner M, editors. Alcohol, smoking, caffeine and drug use associated with sleep duration and sleep quality. Sleep; Amer Acad Sleep Medicine One Westbrook Corporate CTR, Westchester, IL; 2011.

[39] Marmorstein NR. Interactions between energy drink consumption and sleep problems: associations with alcohol use among young adolescents. J Caffeine Res 2017;7(3):111–6.

[40] Lund HG, Reider BD, Whiting AB, Prichard JR. Sleep patterns and predictors of disturbed sleep in a large population of college students. J Adolesc Health 2010;46(2):124–32.

[41] Rohsenow DJ, Howland J, Alvarez L, Nelson K, Langlois B, Verster JC, et al. Effects of caffeinated vs. non-caffeinated alcoholic beverage on next-day hangover incidence and severity, perceived sleep quality, and alertness. Addict Behav 2014;39(1):329–32.

[42] Carrier J, Paquet J, Fernandez-Bolanos M, Girouard L, Roy J, Selmaoui B, et al. Effects of caffeine on daytime recovery sleep: a double challenge to the sleep-wake cycle in aging. Sleep Med 2009;10(9):1016–24.

[43] Robillard R, Bouchard M, Cartier A, Nicolau L, Carrier J. Sleep is more sensitive to high doses of caffeine in the middle years of life. J Psychopharmacol 2015;29(6):688–97.

[44] Drake C, Roehrs T, Shambroom J, Roth T. Caffeine effects on sleep taken 0, 3, or 6 hours before going to bed. J Clin Sleep Med 2013;9(11):1195–200.

[45] Rosenthal L, Roehrs T, Zwyghuizen-Doorenbos A, Plath D, Roth T. Alerting effects of caffeine after normal and restricted sleep. Neuropsychopharmacology 1991;4(2):103–8.

[46] Ho SC, Chung JW. The effects of caffeine abstinence on sleep: a pilot study. Appl Nurs Res 2013;26(2):80–4.

[47] Poole R, Kennedy OJ, Roderick P, Fallowfield JA, Hayes PC, Parkes J. Coffee consumption and health: umbrella review of meta-analyses of multiple health outcomes. BMJ 2017;359:j5024.

[48] Basner M, Fomberstein KM, Razavi FM, Banks S, William JH, Rosa RR, et al. American time use survey: sleep time and its relationship to waking activities. Sleep 2007;30(9):1085–95.

[49] Carrier J, Fernandez-Bolanos M, Robillard R, Dumont M, Paquet J, Selmaoui B, et al. Effects of caffeine are more marked on daytime recovery sleep than on nocturnal sleep. Neuropsychopharmacology 2007;32(4):964–72.

[50] Manchester J, Eshel I, Marion DW. The benefits and risks of energy drinks in young adults and military service members. Mil Med 2017;182(7):e1726–33.

[51] Sampasa-Kanyinga H, Hamilton HA, Chaput JP. Sleep duration and consumption of sugar-sweetened beverages and energy drinks among adolescents. Nutrition 2018;48:77–81.

[52] Kelly CK, Prichard JR. Demographics, health, and risk behaviors of young adults who drink energy drinks and coffee beverages. J Caffeine Res 2016;6(2):73–81.

[53] Malinauskas BM, Aeby VG, Overton RF, Carpenter-Aeby T, Barber-Heidal K. A survey of energy drink consumption patterns among college students. Nutr J 2007;6:35.

[54] Jackson DA, Cotter BV, Merchant RC, Babu KM, Baird JR, Nirenberg T, et al. Behavioral and physiologic adverse effects in adolescent and young adult emergency department patients reporting use of energy drinks and caffeine. Clin Toxicol (Phila) 2013;51(7):557–65.

[55] Schmidt RM, McIntire LK, Caldwell JA, Hallman C. Prevalence of energy-drink and supplement usage in a sample of air force personnel. Air Force Research Lab Wright-Patterson AFB OH Human Effectiveness Directorate; 2008.

[56] Jacobson IG, Ryan MA, Hooper TI, Smith TC, Amoroso PJ, Boyko EJ, et al. Alcohol use and alcohol-related problems before and after military combat deployment. JAMA 2008;300(6):663–75.

[57] Centers for Disease C, Prevention. Energy drink consumption and its association with sleep problems among U.S. service members on a combat deployment—Afghanistan, 2010. MMWR Morb Mortal Wkly Rep 2012;61(44):895–8.

[58] McLellan TM, Riviere LA, Williams KW, McGurk D, Lieberman HR. Caffeine and energy drink use by combat arms soldiers in Afghanistan as a countermeasure for sleep loss and high operational demands. Nutr Neurosci 2018;1–10.

[59] Waits WM, Ganz MB, Schillreff T, Dell PJ. Sleep and the use of energy products in a combat environment. US Army Med Dep J 2014;22–8.

[60] Jacobson IG, Horton JL, Smith B, Wells TS, Boyko EJ, Lieberman HR, et al. Bodybuilding, energy, and weight-loss supplements are associated with deployment and physical activity in U.S. military personnel. Ann Epidemiol 2012;22(5):318–30.

[61] Wesensten NJ. Legitimacy of concerns about caffeine and energy drink consumption. Nutr Rev 2014;72Suppl. 1:78–86.

[62] Lin FJ, Pierce MM, Sehgal A, Wu T, Skipper DC, Chabba R. Effect of taurine and caffeine on sleep-wake activity in Drosophila melanogaster. Nat Sci Sleep 2010;2:221–31.

[63] Eliyahu U, Berlin S, Hadad E, Heled Y, Moran DS. Psychostimulants and military operations. Mil Med 2007;172(4):383–7.

[64] Buguet A, Moroz DE, Radomski MW. Modafinil—medical considerations for use in sustained operations. Aviat Space Environ Med 2003;74(6 Pt 1):659–63.

[65] Killgore WD, Rupp TL, Grugle NL, Reichardt RM, Lipizzi EL, Balkin TJ. Effects of dextroamphetamine, caffeine and modafinil on psychomotor vigilance test performance after 44 h of continuous wakefulness. J Sleep Res 2008;17(3):309–21.

[66] Erman MK, Rosenberg R, The USMSWSDSG. Modafinil for excessive sleepiness associated with chronic shift work sleep disorder: effects on patient functioning and health-related quality of life. Prim Care Companion J Clin Psychiatry 2007;9(3):188–94.

[67] Wesensten NJ, Belenky G, Kautz MA, Thorne DR, Reichardt RM, Balkin TJ. Maintaining alertness and performance during sleep deprivation: modafinil versus caffeine. Psychopharmacology 2002;159(3):238–47.

[68] Waters WF, Magill RA, Bray GA, Volaufova J, Smith SR, Lieberman HR, et al. A comparison of tyrosine against placebo, phentermine, caffeine, and D-amphetamine during sleep deprivation. Nutr Neurosci 2003;6(4):221–35.

[69] Grayson JK, Gibson RL, Shanklin SL, Neuhauser KM, McGhee C. Trends in positive drug tests, United States Air Force, fiscal years 1997-1999. Mil Med 2004;169(7):499–504.

[70] Lacy BW, Ditzler TF, Wilson RS, Martin TM, Ochikubo JT, Roussel RR, et al. Regional methamphetamine use among U.S. Army personnel stationed in the continental United States and Hawaii: a six-year retrospective study (2000-2005). Mil Med 2008;173(4):353–8.

[71] McLellan TM, Kamimori GH, Voss DM, Tate C, Smith SJ. Caffeine effects on physical and cognitive performance during sustained operations. Aviat Space Environ Med 2007;78(9):871–7.

[72] McLellan TM, Kamimori GH, Bell DG, Smith IF, Johnson D, Belenky G. Caffeine maintains vigilance and marksmanship in simulated urban operations with sleep deprivation. Aviat Space Environ Med 2005;76(1):39–45.

[73] McLellan TM, Bell DG, Kamimori GH. Caffeine improves physical performance during 24 h of active wakefulness. Aviat Space Environ Med 2004;75(8):666–72.

[74] Newman RA, Kamimori GH, Wesensten NJ, Picchioni D, Balkin TJ. Caffeine gum minimizes sleep inertia. Percept Mot Skills 2013;116(1):280–93.

[75] Geiger-Brown J, Rogers VE, Trinkoff AM, Kane RL, Bausell RB, Scharf SM. Sleep, sleepiness, fatigue, and performance of 12-hour-shift nurses. Chronobiol Int 2012;29(2):211–9.

[76] Walia HK, Hayes AL, Przepyszny KA, Karumanchi P, Patel SR. Clinical presentation of shift workers to a sleep clinic. Sleep Breath 2012;16(2):543–7.

[77] Borland RG, Rogers AS, Nicholson AN, Pascoe PA, Spencer MB. Performance overnight in shiftworkers operating a day-night schedule. Aviat Space Environ Med 1986;57(3):241–9.

[78] Kelly TL, Mitler MM, Bonnet MH. Sleep latency measures of caffeine effects during sleep deprivation. Electroencephalogr Clin Neurophysiol 1997;102(5):397–400.

[79] Penetar D, McCann U, Thorne D, Kamimori G, Galinski C, Sing H, et al. Caffeine reversal of sleep deprivation effects on alertness and mood. Psychopharmacology 1993;112(2–3):359–65.

[80] Lohi JJ, Huttunen KH, Lahtinen TM, Kilpelainen AA, Muhli AA, Leino TK. Effect of caffeine on simulator flight performance in sleep-deprived military pilot students. Mil Med 2007;172(9):982–7.

[81] Heaton K, Griffin R. The effects of caffeine use on driving safety among truck drivers who are habitual caffeine users. Workplace Health Saf 2015;63(8):333–41.

[82] Doty TJ, So CJ, Bergman EM, Trach SK, Ratcliffe RH, Yarnell AM, et al. Limited efficacy of caffeine and recovery costs during and following 5 days of chronic sleep restriction. Sleep 2017;40(12).

[83] McHill AW, Smith BJ, Wright Jr KP. Effects of caffeine on skin and core temperatures, alertness, and recovery sleep during circadian misalignment. J Biol Rhythm 2014;29(2):131–43.

[84] LaJambe CM, Kamimori GH, Belenky G, Balkin TJ. Caffeine effects on recovery sleep following 27 h total sleep deprivation. Aviat Space Environ Med 2005;76(2):108–13.

[85] McLean CP, Zandberg L, Roache JD, Fitzgerald H, Pruiksma KE, Taylor DJ, et al. Caffeine use in military personnel with PTSD: prevalence and impact on sleep. Behav Sleep Med 2017;1–11.

[86] Salin-Pascual RJ, Valencia-Flores M, Campos RM, Castano A, Shiromani PJ. Caffeine challenge in insomniac patients after total sleep deprivation. Sleep Med 2006;7(2):141–5.

[87] Knapik J, Trone D, McGraw S, Steelman R, Austin K, Lieberman H. Caffeine use among active duty Navy and Marine Corps personnel. Nutrients 2016;8(10):620.

[88] Landolt HP. Sleep homeostasis: a role for adenosine in humans? Biochem Pharmacol 2008;75(11):2070–9.

[89] Bjorness TE, Greene RW. Adenosine and sleep. Curr Neuropharmacol 2009;7(3):238–45.

[90] Sperlagh B, Vizi ES. The role of extracellular adenosine in chemical neurotransmission in the hippocampus and Basal Ganglia: pharmacological and clinical aspects. Curr Top Med Chem 2011;11(8):1034–46.

[91] Halassa MM, Florian C, Fellin T, Munoz JR, Lee SY, Abel T, et al. Astrocytic modulation of sleep homeostasis and cognitive consequences of sleep loss. Neuron 2009;61(2):213–9.

[92] Volkow ND, Wang GJ, Logan J, Alexoff D, Fowler JS, Thanos PK, et al. Caffeine increases striatal dopamine D2/D3 receptor availability in the human brain. Transl Psychiatry 2015;5:e549.

[93] Bchir F, Dogui M, Ben Fradj R, Arnaud MJ, Saguem S. Differences in pharmacokinetic and electroencephalographic responses to caffeine in sleep-sensitive and non-sensitive subjects. C R Bioll 2006;329(7):512–9.

[94] Byrne EM, Johnson J, McRae AF, Nyholt DR, Medland SE, Gehrman PR, et al. A genome-wide association study of caffeine-related sleep disturbance: confirmation of a role for a common variant in the adenosine receptor. Sleep 2012;35(7):967–75.

[95] Retey JV, Adam M, Khatami R, Luhmann UF, Jung HH, Berger W, et al. A genetic variation in the adenosine A2A receptor gene (ADORA2A) contributes to individual sensitivity to caffeine effects on sleep. Clin Pharmacol Ther 2007;81(5):692–8.

[96] Nova P, Hernandez B, Ptolemy AS, Zeitzer JM. Modeling caffeine concentrations with the Stanford Caffeine Questionnaire: preliminary evidence for an interaction of chronotype with the effects of caffeine on sleep. Sleep Med 2012;13(4):362–7.

[97] Mazzotti DR, Guindalini C, Pellegrino R, Barrueco KF, Santos-Silva R, Bittencourt LR, et al. Effects of the adenosine deaminase polymorphism and caffeine intake on sleep parameters in a large population sample. Sleep 2011;34(3):399–402.

[98] Mazzotti DR, Guindalini C, de Souza AA, Sato JR, Santos-Silva R, Bittencourt LR, et al. Adenosine deaminase polymorphism affects sleep EEG spectral power in a large epidemiological sample. PLoS One 2012;7(8):e44154.

[99] Sachse C, Brockmoller J, Bauer S, Roots I. Functional significance of a C-->A polymorphism in intron 1 of the cytochrome P450 CYP1A2 gene tested with caffeine. Br J Clin Pharmacol 1999;47(4):445–9.

[100] Temple JL, Bernard C, Lipshultz SE, Czachor JD, Westphal JA, Mestre MA. The safety of ingested caffeine: a comprehensive review. Front Psychiatry 2017;8:80.

[101] Omvik S, Pallesen S, Bjorvatn B, Thayer J, Nordhus IH. Nighttime thoughts in high and low worriers: reaction to caffeine-induced sleeplessness. Behav Res Ther 2007;45(4):715–27.

[102] Rogers PJ, Heatherley SV, Hayward RC, Seers HE, Hill J, Kane M. Effects of caffeine and caffeine withdrawal on mood and cognitive performance degraded by sleep restriction. Psychopharmacology 2005;179(4):742–52.

[103] Mniszek DH. Brighton sleep survey: a study of sleep in 20–45-year olds. J Int Med Res 1988;16(1):61–5.

[104] Landolt HP, Werth E, Borbely AA, Dijk DJ. Caffeine intake (200 mg) in the morning affects human sleep and EEG power spectra at night. Brain Res 1995;675(1–2):67–74.

[105] Suh S, Yang HC, Kim N, Yu JH, Choi S, Yun CH, et al. Chronotype differences in health behaviors and health-related quality of life: a population-based study among aged and older adults. Behav Sleep Med 2017;15(5):361–76.

[106] Tran J, Lertmaharit S, Lohsoonthorn V, Pensuksan WC, Rattananupong T, Tadesse MG, et al. Daytime sleepiness, circadian preference, caffeine consumption and use of other stimulants among thai college students. J Public Health Epidemiol 2014;8(6):202–10.

[107] Whittier A, Sanchez S, Castaneda B, Sanchez E, Gelaye B, Yanez D, et al. Eveningness chronotype, daytime sleepiness, caffeine consumption, and use of other stimulants among peruvian university students. J Caffeine Res 2014;4(1):21–7.

[108] Kerpershoek ML, Antypa N, Van den Berg JF. Evening use of caffeine moderates the relationship between caffeine consumption and subjective sleep quality in students. J Sleep Res 2018;27(5):e12670.

[109] Urry E, Jetter A, Holst SC, Berger W, Spinas GA, Langhans W, et al. A case-control field study on the relationships among type 2 diabetes, sleepiness and habitual caffeine intake. J Psychopharmacol 2017;31(2):233–42.

[110] Urry E, Jetter A, Landolt HP. Assessment of CYP1A2 enzyme activity in relation to type-2 diabetes and habitual caffeine intake. Nutr Metab (Lond) 2016;13:66.

[111] Grandner MA, Martin JL, Patel NP, Jackson NJ, Gehrman PR, Pien G, et al. Age and sleep disturbances among American men and women: data from the U.S. Behavioral Risk Factor Surveillance System. Sleep 2012;35(3):395–406.

[112] Andreyeva T, Luedicke J, Henderson KE, Tripp AS. Grocery store beverage choices by participants in federal food assistance and nutrition programs. Am J Prev Med 2012;43(4):411–8.

[113] Frozi J, de Carvalho HW, Ottoni GL, Cunha RA, Lara DR. Distinct sensitivity to caffeine-induced insomnia related to age. J Psychopharmacol 2018;32(1):89–95.

[114] Bailey RL, Saldanha LG, Dwyer JT. Estimating caffeine intake from energy drinks and dietary supplements in the United States. Nutr Rev 2014;72(Suppl. 1):9–13.

[115] Berger LK, Fendrich M, Chen HY, Arria AM, Cisler RA. Sociodemographic correlates of energy drink consumption with and without alcohol: results of a community survey. Addict Behav 2011;36(5):516–9.

[116] Park S, Onufrak S, Blanck HM, Sherry B. Characteristics associated with consumption of sports and energy drinks among US adults: National Health Interview Survey, 2010. J Acad Nutr Diet 2013;113(1):112–9.

[117] Grandner MA. Addressing sleep disturbances: an opportunity to prevent cardiometabolic disease? Int Rev Psychiatry 2014;26(2):155–76.

[118] Peacock A, Bruno R, Martin FH. The subjective physiological, psychological, and behavioral risk-taking consequences of alcohol and energy drink co-ingestion. Alcohol Clin Exp Res 2012;36(11):2008–15.

[119] Brache K, Stockwell T. Drinking patterns and risk behaviors associated with combined alcohol and energy drink consumption in college drinkers. Addict Behav 2011;36(12):1133–40.

[120] Patrick ME, Griffin J, Huntley ED, Maggs JL. Energy drinks and binge drinking predict college students' sleep quantity, quality, and tiredness. Behav Sleep Med 2018;16(1):92–105.

[121] Chakravorty S, Chaudhary NS, Brower KJ. Alcohol dependence and its relationship with insomnia and other sleep disorders. Alcohol Clin Exp Res 2016;40(11):2271–82.

[122] Rajaratnam SM, Howard ME, Grunstein RR. Sleep loss and circadian disruption in shift work: health burden and management. Med J Aust 2013;199(8):S11–5.

[123] Quan SF, Combs D, Parthasarathy S. Impact of sleep duration and weekend oversleep on body weight and blood pressure in adolescents. Southwest J Pulm Crit Care 2018;16(1):31–41.

[124] Okun ML, Reynolds 3rd CF, Buysse DJ, Monk TH, Mazumdar S, Begley A, et al. Sleep variability, health-related practices, and inflammatory markers in a community dwelling sample of older adults. Psychosom Med 2011;73(2):142–50.

[125] Rasaei B, Talib RA, Noor MI, Karandish M, Karim NA. Simultaneous coffee caffeine intake and sleep deprivation alter glucose homeostasis in Iranian men: a randomized crossover trial. Asia Pac J Clin Nutr 2016;25(4):729–39.

[126] Grant CL, Coates AM, Dorrian J, Paech GM, Pajcin M, Della Vedova C, et al. The impact of caffeine consumption during 50 hr of extended wakefulness on glucose metabolism, self-reported hunger and mood state. J Sleep Res 2018;e12681.

[127] Cauli O, Morelli M. Caffeine and the dopaminergic system. Behav Pharmacol 2005;16(2):63–77.

[128] Lara DR. Caffeine, mental health, and psychiatric disorders. J Alzheimers Dis 2010;20(Suppl. 1):S239–48.

[129] Chelben J, Piccone-Sapir A, Ianco I, Shoenfeld N, Kotler M, Strous RD. Effects of amino acid energy drinks leading to hospitalization in individuals with mental illness. Gen Hosp Psychiatry 2008;30(2):187–9.

[130] Drake CL, Jefferson C, Roehrs T, Roth T. Stress-related sleep disturbance and polysomnographic response to caffeine. Sleep Med 2006;7(7):567–72.

[131] Kendler KS, Myers J, O Gardner C. Caffeine intake, toxicity and dependence and lifetime risk for psychiatric and substance use disorders: an epidemiologic and co-twin control analysis. Psychol Med 2006;36(12):1717–25.

[132] Gehrman P, Seelig AD, Jacobson IG, Boyko EJ, Hooper TI, Gackstetter GD, et al. Predeployment sleep duration and insomnia symptoms as risk factors for new-onset mental health disorders following military deployment. Sleep 2013;36(7):1009–18.

[133] Kamimori GH, Penetar DM, Headley DB, Thorne DR, Otterstetter R, Belenky G. Effect of three caffeine doses on plasma catecholamines and alertness during prolonged wakefulness. Eur J Clin Pharmacol 2000;56(8):537–44.

[134] Barger LK, Lockley SW, Rajaratnam SM, Landrigan CP. Neurobehavioral, health, and safety consequences associated with shift work in safety-sensitive professions. Curr Neurol Neurosci Rep 2009;9(2):155–64.

[135] de Mello MT, Narciso FV, Tufik S, Paiva T, Spence DW, Bahammam AS, et al. Sleep disorders as a cause of motor vehicle collisions. Int J Prev Med 2013;4(3):246–57.

[136] Souza JC, Paiva T, Reimao R. Sleep habits, sleepiness and accidents among truck drivers. Arq Neuropsiquiatr 2005;63(4):925–30.

[137] Akerstedt T, Fredlund P, Gillberg M, Jansson B. A prospective study of fatal occupational accidents—relationship to sleeping difficulties and occupational factors. J Sleep Res 2002;11(1):69–71.

[138] Horne JA, Reyner LA. Counteracting driver sleepiness: effects of napping, caffeine, and placebo. Psychophysiology 1996;33(3):306–9.

[139] Sharwood LN, Elkington J, Meuleners L, Ivers R, Boufous S, Stevenson M. Use of caffeinated substances and risk of crashes in long distance drivers of commercial vehicles: case-control study. BMJ 2013;346:f1140.

[140] Li Y, Vgontzas AN, Fernandez-Mendoza J, Bixler EO, Sun Y, Zhou J, et al. Insomnia with physiological hyperarousal is associated with hypertension. Hypertension 2015;65(3):644–50.

[141] Savoca MR, MacKey ML, Evans CD, Wilson M, Ludwig DA, Harshfield GA. Association of ambulatory blood pressure and dietary caffeine in adolescents. Am J Hypertens 2005;18(1):116–20.

[142] Cheng Y, Du CL, Hwang JJ, Chen IS, Chen MF, Su TC. Working hours, sleep duration and the risk of acute coronary heart disease: a case-control study of middle-aged men in Taiwan. Int J Cardiol 2014;171(3):419–22.

[143] Bonnet MH, Arand DL. Situational insomnia: consistency, predictors, and outcomes. Sleep 2003;26(8):1029–36.

[144] McCrae CS, Rowe MA, Dautovich ND, Lichstein KL, Durrence HH, Riedel BW, et al. Sleep hygiene practices in two community dwelling samples of older adults. Sleep 2006;29(12):1551–60.

[145] Bonnet MH, Balkin TJ, Dinges DF, Roehrs T, Rogers NL, Wesensten NJ, et al. The use of stimulants to modify performance during sleep loss: a review by the sleep deprivation and Stimulant Task Force of the American Academy of Sleep Medicine. Sleep 2005;28(9):1163–87.

[146] Mindell JA, Sedmak R, Boyle JT, Butler R, Williamson AA. Sleep well!: a pilot study of an education campaign to improve sleep of socioeconomically disadvantaged children. J Clin Sleep Med 2016;12(12):1593–9.

[147] Reifman J, Kumar K, Wesensten NJ, Tountas NA, Balkin TJ, Ramakrishnan S. 2B-alert web: an open-access tool for predicting the effects of sleep/wake schedules and caffeine consumption on neurobehavioral performance. Sleep 2016;39(12):2157–9.

Chapter 24

Sleep, stress, and immunity

Aric A. Prather

Department of Psychiatry, University of California, San Francisco, CA, United States

INTRODUCTION

Sleep is a biological imperative, conserved across species, that plays a fundamental role in promoting health and well-being [1, 2]. Epidemiologic evidence consistently links poor sleep, characterized by short sleep duration, poor sleep continuity, and poor subjective sleep quality, with increased rates of a number of age-related conditions, including cardiovascular disease and metabolic conditions, as well as, premature mortality [3–6]; however, the biological mechanisms that underlie these associations are not well understood. The immune system has emerged as one promising pathway [7]. Over the past several decades, researchers have been investigating the sleep-immune connection in humans. As will be summarized below, there is now a compelling literature supporting the role of sleep and sleep loss on immune functioning and risk for immune-related conditions. Moreover, there is growing appreciation that sleep may serve as a predictor of how quickly the immune system ages.

The influence of sleep on physical health cannot be overstated, but it would be a mistake to focus on sleep in isolation. Sleep is malleable, always susceptible to the perturbations of the prior day. Energy expenditure, substances consumed (e.g., caffeine, nutrition), and one's psychological stress acutely affect the duration, continuity, and quality of sleep. Conversely, sleep the night prior can have dramatic effects on how someone thinks and feels the following day. Indeed, data suggest that sleep and stress are reciprocally connected. This is particularly relevant when considering sleep's influence on the immune system because there is a complementary literature demonstrating that psychological stress modulates many of the same immunologic pathways observed in sleep research [8]. Unfortunately, few studies have examined the independent and synergistic roles of sleep and psychological stress on immunity.

The goal of this chapter is to provide a review of the scientific human literature linking sleep and the immune system, including the acquired and innate immune system as well as markers of immune cell aging (i.e., immunosenescence). Next, parallels will be drawn using the psychological stress and immune system literature, which will be followed by a discussion of the reciprocal links between sleep and psychological stress and the pathways through which they can influence immune function. Finally, a review of the limited research examining the synergistic influences of sleep and psychological stress will be presented. First, however, a brief overview of the immune system is provided to help orient the reader.

OVERVIEW OF THE IMMUNE SYSTEM

The immune system comprises cells and soluble molecules that work together to protect the body (i.e., self) from the foreign antigens such as viruses and bacteria (i.e., non-self). Though exquisitely dependent on one another, the immune system is typically separated into two distinct arms: the acquired and innate immune system.

Acquired immune system

The acquired immune system, as the name suggests, develops over time in response to antigen exposure. This arm is slow acting, often requiring days to weeks to produce the intended immune response, and is comprised of various lymphocytes (e.g., helper (CD4+) and cytotoxic (CD8+) T cells and B (CD19+) cells) that have receptors on their cell surfaces that respond to one and only one antigen. In response to an infectious challenge, the antigen is taking up by antigen presenting cells (APCs), such as dendritic cells or macrophages, that then migrate to lymphoid organs (e.g., lymph nodes). The APCs present the antigen to helper T cells. Once activated these begin to divide and proliferate to mount an immune cell army whose role is to clear the body of the invader (i.e., antigen). The primary role of helper T cells is to produce and release cytokines that modulate the rest of the immune system. The role of cytotoxic T cells is to seek out and lyse infected cells (e.g., virally infected cells), while B cells produce antibodies, which are soluble proteins critical in neutralizing bacterial toxins and flagging free-floating viruses and infected cells so as to communicate the need for destruction to the innate arm of the immune system. Antibody levels are clinically meaningful as they are the end product of vaccinations. Activated T and B cells maintain immunological memory and can circulate in the blood for years to provide a rapid response if challenged by the same antigen once again.

Innate immune system

The innate immune system is functional at birth and is activated quickly (e.g., minutes to hours). It is the body's first line of immune defense and includes physical and anatomical barriers such as the skin and, unlike the acquired immune system, is made up of specialized cells that do not require specific recognition of an antigen to become activated [9]. Examples of these cells include natural killer (NK) cells, granulocytes (e.g., neutrophils), and macrophages. NK cells play an important role in halting the early phases of viral infections and attacking cells that are malignant. NK cells release a toxic substance to lyse unwanted cells. Macrophages, in contrast, are phagocytotic, meaning that they eat their targets (i.e., unwanted invaders). They also release cytokines, which are proteins that facilitate inflammation. During an inflammatory response, immune cells congregate at the site of injury, such as a wound, releasing toxic molecules and signaling proteins to neutralize the threat and call other surrounding immune cells to their aid. The inflammatory response is critical to survival; however, prolonged or unregulated inflammation can contribute to inflammatory related diseases, including autoimmune conditions and cardiovascular disease, among others. Key proinflammatory cytokines include interleukin (IL)-6, tumor necrosis factor (TNF)-α, and IL-1β. Additionally, C-reactive protein, which is an acute phase protein produced by the liver in response to increasing levels of IL-6, has emerged as a measure of chronic systemic inflammation and is a clinical risk factor for cardiovascular disease [10].

The aging immune system

There are a several well-recognized changes that occur as the immune system ages. These changes include involution of the thymus, which causes diminution of T cell diversity, and increases in low grade systemic inflammation, known as inflammaging [11]. There is also an accumulation of aged immune cells that are characterized by cellular senescence. Cellular senescence is a state of cell cycle arrest that is marked by an inability to effectively replicate. One pathway to cell cycle arrest that has been studied by a number of sleep researchers is telomere attrition. Telomeres are DNA protein complexes at the ends of chromosomes of eukaryotic cells that protects the DNA that encodes genetic information from genomic instability and damage. Telomeres shorten with each cellular division and if not replaced by the enzyme telomerase, critically short telomeres will send the cell into either apoptosis or cell cycle arrest. With advancing age, the shortening of telomeres primarily occurs in cytotoxic T cells (CD8+), particularly those who have lost CD28 expression. CD28 is a co-stimulatory molecule on the cell surface important for facilitating proliferative capacity.

Beyond an inability to replicate, senescent cells are also characterized by a unique secretory pattern known as the senescence associated secretory phenotype (SASP), which is marked by increased proinflammatory and chemokine activity (e.g., IL-6, IL-8, monocyte chemotactic protein (MCP)-2 and MCP-4, as well as intracellular adhesion molecule (ICAM)-1, among other molecules). Prior work in animals demonstrate the removal of senescent cells reduces age-related pathology [12], potentially due to the reduction in SASP.

Variability is inherent in all of the immune parameters described above. As such, there has been growing interest among sleep researchers to determine whether different aspects of sleep may account for significant levels of variability in these markers, which may illuminate why poor sleep is associated with susceptibility to infectious illness and age-related medical conditions.

SLEEP, ACQUIRED IMMUNITY, AND INFECTIOUS DISEASE RISK

Animal and human studies demonstrate that poor sleep is associated with alterations in aspects of acquired immunity with implications for infectious disease risk [13]. This is certainly consistent with anecdotal experiences where prolonged periods of insufficient sleep tracks with one's increased risk of "catching a cold". However, there is now strong empirical evidence to support this common belief. For example, in a sample taken from the National Health and Nutritional Examinations Surveys (NHANES) spanning form 2005 to 2012, self-reported short sleep duration ($\leq$5h per night), endorsing a physician diagnosis of a sleep disorder, and having told a physician about having a sleep disturbance were associated with increased rates of head and chest colds as well as infection compared to better sleepers [14].

The cross-sectional nature of these data limit inferences regarding the directionality of these associations. However, prospective evidence showed that self-reported short sleep duration ($\leq$5h per night) and long sleep duration ($\geq$9h per night) predicted increased incidence of physician diagnosed pneumonia compared to normal sleepers (8h sleepers) [15]. This study, though intriguing, was limited by a number of factors that typically plague the sleep and health literature, including use of single retrospective self-report item to assess typical sleep duration. In addition, exposure to the pneumococcal virus was not controlled, raising concerns about possible unmeasured, confounding variables that predict likelihood of exposure rather than response to the virus.

The strongest human evidence demonstrating that poor sleep is associated with increased susceptibility to infectious illness comes from a series of experimental studies in which healthy participants are exposed to a known quantity of rhinovirus—the virus responsible for producing an upper respiratory infection (URI) [16–18]. The general design of these studies is as follows: participants are quarantined in a hotel throughout the course of the study, which begins 1 to 2 days prior to being inoculated with the live virus and spans an additional 5 to 7 days of monitoring. On the days prior to inoculation, participants undergo a nasal wash to ensure that they have no baseline infection and have their blood drawn to assess any pre-existing antibodies to the virus. On the third day of quarantine, the participants receive known quantities of the virus, administered intranasally. Subsequent to this, participants are then monitored for subjective and objective signs of illness. In order to verify the presence of infection, daily nasal washings are conducted and assayed for evidence of viral shedding (i.e., replication of the virus). In addition, viral-specific antibody titers are obtained 21 to 28 days after inoculation. Individuals are deemed infected if (1) they show evidence of viral shedding or (2) demonstrate a twofold increase over their baseline levels in the virus-specific antibody titers. Importantly, not all participants infected go on to show signs of objective illness. Physical examinations are carried out each day to monitor signs of objective illness. Specifically, nasal congestion is quantified by measuring the time required for a dye administered in the nose to reach the nasopharynx [19]. Daily mucus secretion is quantified by weighing tissues used throughout the day, subtracting the weight of the tissue. Typically, a baseline adjusted nasal clearance time of >7 min and/or a total adjusted mucus weight of at least 10 g is used as the threshold for the presence of clinical illness in participants who show evidence of infection [20].

This paradigm provides the unique opportunity to prospectively test whether sleep prior to viral exposure predicts who is susceptible to becoming infected *and* developing a biologically-verified cold. In this regard, Cohen and colleagues found that shorter sleep duration and poorer sleep efficiency assessed by 14-night sleep diary significantly predicted increased likelihood for developing a biologically-verified cold [16]. Similarly, in a separate sample, Prather and colleagues demonstrated that shorter sleep duration, this time measured by 7 nights of wrist actigraphy, predicted increased likelihood of developing a cold [18]. As displayed in Fig. 24.1, they found that participants who obtained 6 or fewer hours of sleep on average were four times more likely to develop a cold compared to those who obtained >7 h per night. Importantly, these findings were independent of a bevy of potential confounders, including baseline antibody titers, sociodemographic factors, health behaviors, and psychological processes, such as levels of psychological stress. Furthermore, neither study found that sleep predicted susceptibility to infection.

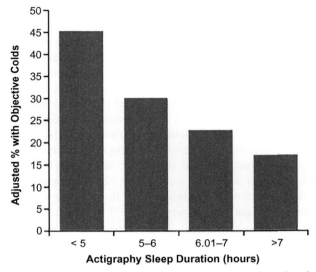

FIG. 24.1 Sleep duration, averaged over 7 nights of wrist actigraphy and measured prior to virus exposure, is associated with percentage of participants who subsequently developed a biologically-verified cold. The percentage of colds is based on predicted values (adjusted for age and pre-challenge viral specific antibody levels). *(From Prather AA, Janicki-Deverts D, Hall MH, Cohen S. Behaviorally assessed sleep and susceptibility to the common cold. Sleep 2015;38(9):1353–9. https://doi.org/10.5665/sleep.4968, used with permission.)*

This experimental paradigm provides important objective, empirical evidence that insufficient sleep confers risk for infectious illness. Another clinically-relevant model used to assess the role of sleep on infectious disease risk is through the use of vaccinations. Prophylactic vaccination is used to simulate infection and induce the formation of memory T and B cells with antibodies to the specific targeted pathogen. Several experimental studies employing total or partial sleep restriction following vaccination suggest that acute sleep loss can impair, albeit transiently, antibody responses compared to undisturbed sleep [21–24]. For example, Spiegel and colleagues examined the effects of partial sleep restriction (i.e., reducing sleep opportunity in the lab from 8 to 4 h per night for six consecutive nights) compared to undisturbed sleepers on response to the influenza vaccination. Antibody titers to the influenza vaccine were measured at baseline, 10-days post-vaccination, and between 21 and 30 days post-vaccination. Analyses revealed that 10-days after the vaccination, those who were randomized to the sleep restriction condition mounted a response half that of the undisturbed sleepers; however, by the later measurement time point there were no group differences.

It is possible that the modest and transient effects of experimental sleep loss may be more substantial among those who experience chronically short sleep. In this regard, Prather and colleagues examined the prospective associations between sleep measures, obtained via wrist actigraphy and averaged over time, and vaccination response to the Hepatitis B vaccination series in a sample of healthy midlife adults [25]. Analyses revealed that shorter average sleep duration was associated with fewer viral specific antibodies to the vaccination, such that for every additional hour of sleep, participants experienced a 56% increase in antibody production. This study also attempted to examine whether the reduction in antibody responses carried any clinical relevance. Using CDC guidelines for determining the threshold for "protection" conferred by the hepatitis B vaccination series, analyses revealed that short sleepers (i.e., those sleeping <6h per night) were nearly 12 times more likely to be left unprotected 6-months after the vaccination series compared to participants who slept >7h per night.

To date, only one study has examined whether insomnia serves as a risk factor for impaired vaccination response. Taylor and colleagues examined whether participants with Insomnia Disorder, assessed by structured clinical interview, showed impaired response to the influenza vaccine as compared to non-insomniac participants [26]. Overall, participants with insomnia showed fewer antibodies to the influenza vaccine compared to controls both before and after the vaccination, suggesting that while the magnitude of the response to the vaccination was similar between those with and without insomnia, antibody titers produced by insomnia participants were lower overall.

What are the underlying immune mechanisms through which poor sleep is linked to infectious disease risk? Studies employing experimental sleep loss support reliable alterations in immune parameters that are thought to underlie host resistance. For example, acute sleep loss is associated with redistribution of T and B cells in peripheral circulation [27–30]. Circulating T and B cells peak early in the evening and then migrate to the lymphoid organs where they may come in contact with antigens, such as viruses. As such, sleep loss may impact the ability of these immune cells to be in the "right place at the right time" [31]. In addition, sleep loss is associated with impaired T cell functioning, including diminished antigen-specific response by helper T cells [22], as well as a decline in the production of IL-12 [32], a cytokine central to T-cell maturation. Finally, sleep loss is associated with impaired proliferative capacity (i.e., cellular replication) of T cells when stimulated in vitro and modulation of the function of antigen presenting cells (APCs), which may affect how well these cells present viruses to the rest of the immune system [27, 32].

Summary. Experimental and observational data support the notion that poor sleep can impair acquired immunity, with clinical implications related to susceptibility to the common cold and vaccination efficacy. While laboratory studies have identified which aspects of the acquired immune response are altered during disturbed sleep, there is still a need for researchers to test sleep interventions to determine whether improvements in sleep can enhance vaccination responses and otherwise protect populations whose acquired immune system is suboptimal, such as the elderly or individuals who are HIV+.

SLEEP, INNATE IMMUNITY, AND INFLAMMATORY DISEASE RISK

The innate immune system is regulated by both sleep and circadian processes. For example, there are nocturnal increases in NK cell number and function during sleep that are markedly impaired under periods of sleep loss [33, 34]. In a small study of healthy volunteers, NK cell cytotoxicity was significantly lower in response to a night of partial sleep restriction (i.e., deprived of sleep from 3AM to 7AM) compared to an undisturbed night of sleep [33]. Inflammatory cytokines increase across the night, which is partially due to circadian rhythmicity, however, there is a substantial literature that shows that sleep also modulate inflammatory activity. This is not surprising given that inflammation appears to be a central pathway in the pathogenesis of several chronic diseases where sleep plays a role, such as cardiovascular disease [35]. Irwin and colleagues recently published a comprehensive meta-analytic review on the associations between measures of sleep and inflammation [36]. In an analysis of 72 studies, greater sleep disturbances, assessed by questionnaires, were associated with higher levels of circulating IL-6 and CRP but not TNF-α. Shorter sleep duration, when measured subjectively by self-report, was unrelated to IL-6 or CRP levels, though when measured objectively, shorter sleep duration was significantly related to higher IL-6. In addition, longer sleep duration was associated with higher CRP and IL-6 but not TNF-α, highlighting the curvilinear risk conferred by both long and short sleep.

There is a large experimental literature examining the effects of partial and total sleep restriction on markers of inflammation. For example, several studies have found that multiple nights of partial sleep restriction (e.g., reducing sleep opportunity from 8 to 4h per night) is associated with elevated levels of CRP and IL-6; however, other studies have failed to find such effects. In fact, in the most recent meta-analysis, Irwin and colleagues failed to observe an aggregated effect of sleep restriction on next day levels of CRP, IL-6, or TNF-α (non-significant effect sizes ranging from -0.43 to 0.61) [36].

The effects of sleep restriction on inflammation have been more consistent when researchers have focused on genomic and cellular measures of the inflammatory response as opposed to protein levels in systemic circulation. The inflammatory response is initiated when an antigen [such as a lipopolysaccharide (LPS), which is an endotoxin that is a major component of the outer membrane of Gram-negative bacteria] binds to a toll-like receptor found on the cell surfaces of macrophages. This initiates a signaling cascade characterized by the activation of intracellular transcriptional factors nuclear factor kB (NF-kB) and activator protein 1 (AP-1). Activation of these transcriptional factors lead to the transcription of inflammatory response genes within the nucleus of the cell, including *TNF* and *IL1*. Studies examining the influence of sleep loss on inflammatory gene expression, transcriptional pathways, and the intracellular production of inflammatory proteins tend to demonstrate that acute sleep loss results in an upregulation in inflammatory activity [37–40]. For example, in a sample of 30 healthy adults, a night of partial sleep loss (i.e., sleep opportunity from 3 AM to 7 AM) was associated with a threefold increase in transcription of IL-6 messenger RNA and a twofold increase in transcription TNF-αRNA the following morning compared to a baseline period [37].

Like was the case for outcomes within the acquired immune system, few studies have examined the associations between inflammation and insomnia. One exception is a study of 22 participants, half of whom were diagnosed with insomnia. Nocturnal levels of IL-6 were sampled across a single night in the sleep laboratory and it was observed that patients with insomnia had significantly higher levels of circulating IL-6 [41]. Three other studies have examined the association between insomnia diagnosis and systemic levels of IL-6 with mixed results [42–44]. As such, when examined in aggregate, insomnia diagnosis does not appear to be associated with higher levels of inflammation relative to patients without insomnia [36]. One explanation for these mixed findings may be that some participants with insomnia also experience short sleep duration while others do not. Insomnia with short sleep duration has emerged as a more severe biological phenotype [45]. In this regard, in an adolescent sample, shorter sleepers ($\leq$7 h per night) with symptoms of insomnia showed higher levels of circulating CRP compared to those with insomnia but sleep >7 h per night [46].

While current data suggests that overall clinical insomnia does not appear to be strongly associated with inflammatory activity, there is intriguing evidence that treating insomnia may result in related regulation of inflammation. In this regard, a recent randomized controlled study examined the effects of cognitive behavioral therapy for insomnia (CBTI), Tai Chi Chih (TCC), and a sleep seminar control condition on cellular and genomic markers of inflammation in an older adult sample with insomnia. Over the course of the 4-month study and 16-month follow-up, CBTI produced a significant decrease in levels of CRP compared to the control condition. The TCC condition also produced an initial reduction in CRP, though this was lost by the 16-month follow up period. CBTI also produced a reduction in inflammatory gene expression over the course of the study, as did the TCC condition [47]. Notably, TCC has also been shown to reduce inflammation in breast cancer patients with insomnia [48]. More intervention research is needed to assess whether behavioral treatments for individuals with sleep disturbances can produce robust, clinically meaningful improvements in aspects of innate immune function.

Summary. Poor sleep is associated with impairment in some aspects of innate immunity (e.g., NK cell cytotoxicity) and enhancement in inflammatory activity. Overall, observational studies demonstrate that systemic inflammation is elevated in individuals reporting sleep disturbances, and those who report (or demonstrate via more objective methods) short and long sleep duration. In contrast, experimental studies of sleep restriction do not support a consistent increase in protein levels of proinflammatory cytokines; however, there are differences in study design that may contribute to heterogeneity in these effects. Moreover, sleep loss is associated with alterations in transcriptional pathways responsible for inflammatory activity, suggesting that acute sleep loss affect cellular processes. Finally, recent intervention data show that in some instances (e.g., in patients with insomnia in late life) behavioral strategies aimed at improving sleep can produce changes in markers of inflammation; however, more work in this area is needed.

SLEEP AND IMMUNOLOGICAL AGING

Research demonstrating that poor sleep is predictive of disease and premature mortality has raised the possibility that sleep may play a role in the rate at which the immune system ages. As noted above, poor sleep, in some but not all studies, is associated with elevated levels of systemic inflammation, which is one aspect of immune system aging. Similarly, short sleep duration is associated with impaired vaccination efficacy—yet another aspect of the immune system that degrades from mid to late life. In addition, a growing body of research has focused on associations between sleep and telomere length, a recognized marker of immunosenescence.

The first evidence supporting an association between sleep and telomere length came from a study of 245 women aged 49 to 66 years [49]. Here, researchers found poorer subjective sleep quality was significantly associated with shorter immune cell telomere length, independent of chronological age, race, body mass index, and income. This association was strongest among participants who endorsed that their reports of poor sleep reflected a more chronic problem, which is consistent with the notion that prolonged poor sleep may promote a "wear and tear" on the immune system. Since this initial finding, several other studies have found that poor subjective quality and short sleep duration, measured both subjectively and by actigraphy, were associated with shorter immune cell telomere length [50–53].

Sleep disorders, such as obstructive sleep apnea (OSA) and insomnia, have also been examined in the context of telomere length. Indeed, a recent meta-analytic review supported a significant association between diagnosis of obstructive sleep apnea and shorter telomeres [54]. Unfortunately, only eight studies were available for review, which indicates a need for more rigorous research in this area. Relatedly, it remains unclear if effective treatment of OSA can slow cellular aging. Only two studies have examined the association between insomnia and telomere length. In one study of women with a prior diagnosis of breast cancer, those who were experiencing insomnia had shorter telomeres than women without insomnia, though this difference was not statistically significant [55]. In a second sample, insomnia status interacted with chronological age to predict immune cell telomere length such that insomnia was associated with shorter telomeres in older participants (aged 70–88 years) but not younger participants (aged 60–69 years) [56].

The majority of studies that have investigated links between sleep and immune cell telomere length relied on telomere samples from either whole blood or peripheral blood mononuclear cells (PBMCs). In either case, the blood sample is comprised of several different immune cell subsets (e.g., B and T cells, monocytes, granulocytes), all of which may have telomeres of differing lengths [57, 58]. This poses a challenge for understanding the extent to which poor sleep promotes accelerated immunosenescence in particular cell types, given that the accrual of senescent immune cells (often driven by short telomere length) occurs disproportionately in CD8+ T cells. To date, only one study has investigated associations between sleep and telomere length in sorted immune cells [59]. In this regard, in a study of 87 obese women, poorer overall sleep quality, as measured using the Pittsburgh Sleep Quality Index (PSQI), was significantly associated with shorter immune cell telomere length in CD8+ and CD4+ T cells, but not in B cells or granulocytes. Self-reported sleep duration, obtained by sleep diary, was not associated with telomere length in any immune cell subset.

Finally, one study has examined the influence of sleep loss on signaling pathways active within senescent immune cells, including the transcription of the SASP pathway and the expression of p16^{INK4a}, a marker of cellular senescence. Here, Carroll and colleagues found that a night of partial sleep deprivation (i.e., sleep opportunity from 3 AM to 7 AM) was associated with an upregulation in gene expression of SASP genes and senescent marker p16^{INK4a} in PBMCs compared to an undisturbed night of sleep [60]. These data raise the possibility that acute sleep loss may drive immune cells toward senescence, though this needs to be examined more comprehensively.

Summary. Accumulating evidence points to the role of sleep in the development and progression of age-related diseases, many of which include alterations in immune functioning. As such, it should not be surprising that poor sleep is associated with markers of immunosenescence. Short sleep duration and poor sleep quality predict shorter immune cell telomere length in several studies. The same can be said for sleep disorders like OSA and insomnia. To date, however, studies of sleep and telomere length have been cross-sectional and prospective designs are needed to determine whether poor sleep promotes accelerated telomere attrition. It should be noted that senescent cells, by their very nature, are inflammatory and also implicated in compromised acquired immunity (e.g., vaccination efficacy), which raises the possibility that accelerated immunosenescence may serve as a pathway through which poor sleep may contribute to inflammatory and infectious immune-related outcomes.

BEYOND SLEEP: DOES STRESS INFLUENCE IMMUNITY?

As will be reviewed below, sleep and stress are bidirectionally linked. Moreover, stress is associated with alterations in immunity. While a comprehensive review of the stress-immunity literature is beyond the scope of this chapter, what is striking is that many of the findings replete in the psychological stress-immunity literature parallel those observed studies of habitual short sleep duration, sleep disturbance, and laboratory studies employing acute sleep loss [8]. With regards to acquired immunity, higher perceptions of psychological stress as well as a greater number of stressful life events (i.e., exposures to stress) have been associated with increased susceptibility to infectious illness using the same experimental viral challenge paradigm described above [61, 62]. Moreover, higher levels of stress, particularly when it is chronic, are associated with impaired vaccination response [63, 64]. Higher stress is also associated with increased likelihood of reactivation of latent viruses (e.g., herpes), which is typically attributed to an inability of the acquired immune system to keep the virus dormant, as well as impaired wound healing [65]. Laboratory paradigms that employ acute stress exposure, such as having a participant give a speech in front of harsh evaluators, provides an ideal context to investigate how acute bouts of stress modulate immunity. Again, similar to what is seen under periods of sleep loss, acute laboratory stress results in redistribution of immune cells in peripheral circulation as well as an impairment in the ability of T cells to proliferate when challenged [66, 67].

The innate immune system is also affected by stress. For instance, chronic stress has been associated with enhanced levels of inflammation and impairments in NK cell activity [68, 69]. There is also consistent evidence that chronic stress, such as serving as a caregiver, is associated with an upregulation in inflammatory genes as well as a downregulation in neuroendocrine pathways that regulate inflammatory activity [70]. Moreover, a recent meta-analysis demonstrated that acute laboratory stress produces consistent stress-related increases in cellular production and circulating levels of several markers of systemic inflammation, including IL-6, TNF-α, and IL-1β [71].

Finally, in contrast to the limited data on sleep and markers of immunological aging, much more work has been done on the influence of psychological stress. For example, Epel and colleagues provided the first evidence that greater perceptions of stress were associated with shorter immune cell telomere length [72]. Since this seminal study, several meta-analytic reviews have supported that higher levels of psychological stress are associated with shorter telomere length [73]. In addition, there is some compelling evidence that stress exposures, in both early life and adulthood, can predict accelerated telomere attrition over time [74, 75].

Summary. The influence of psychological stress on the immune system is well documented, with many of the findings mirroring those observed in the sleep literature, including impairments in acquired (e.g., vaccination efficacy) and innate (e.g., NK cell cytotoxicity) immunity as well as an upregulation in inflammation.

SLEEP AND PSYCHOLOGICAL STRESS: RECIPROCAL PROCESSES

Sleep and psychological stress are bidirectionally linked, where daytime stressors affect one's ability to sleep soundly and poor sleep affects how one responds to the hassles of the day. With respect to the former (i.e., stress affecting sleep), this is no better example than when one considers insomnia. Indeed, stressful life events are routinely identified as precipitating factors in 3P model of insomnia [76], and there is strong evidence that stress exposures often precede insomnia onset [77]. For example, Drake and colleagues prospectively examined whether presence of stressful life events over the past year predicted incidence of insomnia 1-year later in a sample of >2000 participants free of insomnia and other comorbid psychiatric illnesses at baseline. Analyses revealed that more stressful life events predicted greater likelihood of insomnia, with a 13% increase risk of insomnia for every additional stressor reported [78]. Notably, stressors do not always lead to insomnia but may be more likely among those who tend to experience hyperarousal or vigilance in response to stress. In this regard, in the same study described above, Drake and colleagues found that participants who had a propensity for sleep disturbance in response to stress (known as sleep reactivity) were more likely to experience insomnia in the future and this was particularly true among participants who also reported more stress exposures.

In contrast to research examining the role of stress in promoting poor sleep, literature on how poor sleep influences an individual's perception and response to stress is less well developed [79]. One of the challenges inherent in incorporating stress into sleep research is the variability in which the term "stress" is used across disciplines [80]. In an effort to provide a model for testing how sleep influences the experience of stress (Table 24.1), it is important to first separate the stress process from the stress exposure (i.e., a stressor). The stress process can be further partitioned into (1) psychological appraisal of the stressor (i.e., evaluation of whether the stressor is demanding and beyond one's ability to effectively cope [81]), (2) response, including one's affective (i.e., emotional) and physiological reactivity to the stressor, and (3) recovery, including how long one's affective or physiological response lasts following the cessation of the stressor. Though not part of the stress process per se, it is likely that sleep can influences one's situation selection (i.e., the tendency to select into situations where a stressor is more likely to occur).

What is the evidence that sleep can influence the stress process? First starting with situation selection, Gordon and Chen provide some intriguing evidence using romantic couples. In this regard, they found that a poor night of sleep increased the likelihood of an interpersonal conflict (e.g., a stress exposure) within the couple the following day [82]. With respect to the stress process, much of the evidence comes from experimental studies of sleep loss. For example, Dinges and colleagues demonstrated that a night of sleep loss led participants to subjectively report more perceived stress in response to a task than under non-sleep deprived conditions [83]. This raises the possibility that sleep loss may lower one's threshold for what is perceived as stressful (i.e., appraisal). There is also compelling evidence when

TABLE 24.1 Proposed ways in which poor sleep can affect experiences of daily stress.

	Stress process		
Situation selection	Appraisal	Reactivity	Recovery
• Increased likelihood of stress exposure (e.g., interpersonal stress, accident)	• Increase allocation of attention to negative stimuli/threat • Tendency to appraise exposure as more threatening/stressful	• Greater affective and physiologic reactivity to stress exposure	• Prolonged physiologic arousal • Tendency toward maladaptive cognitive strategies (e.g., rumination)

it comes to physiological responses to stress. In a seminal study, Walker and colleagues reported that a night of sleep loss produced greater amygdala activity in response to threatening stimuli compared to an undisturbed sleep control condition [84]. Similarly, Franzen and colleagues found that compared to undisturbed sleep, a night of sleep loss produced greater blood pressure reactivity to a standardized acute laboratory stressor [85].

HOW DOES POOR SLEEP AND PSYCHOLOGICAL STRESS AFFECT IMMUNITY?

As displayed in Fig. 24.2, there are a number of pathways through which poor sleep and psychological stress can affect immunity. The immune system is strongly regulated by outputs from the autonomic nervous system (ANS) and the hypothalamic pituitary adrenal (HPA) axis. The autonomic nervous system is composed of the sympathetic (SNS), parasympathetic (PNS), and enteric nervous system, though much of the research on ANS and immunity in humans has focused exclusively on the SNS and PNS. Catecholamines, norepinephrine and epinephrine, bind to adrenergic receptors on immune cells to influence function. In addition, sympathetic nerve fibers directly innervate lymphoid organs, modulating immunity. Sympathetic activation is routinely observed with acute bouts of stress while decreases in SNS and s related increase in PNS is observed during sleep. In disrupted sleep, there is evidence of increased SNS activity. For instance, patients with insomnia are often characterized by hypervigilance, including enhanced SNS activation. Similarly, even short bouts of sleep loss are associated with increased SNS in some but not all studies [86]. As such, immune changes observed during stress and sleep may be due in part to alterations in SNS activity. There is growing evidence that the PNS, and its substrate acetylcholine, plays an important role in regulating inflammatory activity [87, 88].

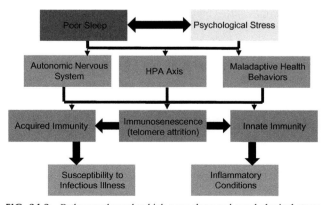

FIG. 24.2 Pathways through which poor sleep and psychological stress may increase one's susceptibility to infectious illness and inflammatory conditions.

PNS activity is reduced both under stress and poor sleep, suggesting that inflammatory activity may be dysregulated in part due to alterations in both SNS and PNS.

The HPA axis is another canonical biological pathway engaged by acute psychological stress and often disturbed by poor sleep. The primary output of the HPA axis is the hormone cortisol, which is a glucocorticoid, that plays a central role in modulating acquired and innate immunity. In particular, cortisol regulates inflammation by down regulating inflammatory gene expression within the cell. However, under chronic stress, immune cells become resistant to the anti-inflammatory effects of cortisol, rendering inflammatory functioning unchecked [89]. It is unclear whether prolonged sleep disturbance promotes a similar phenomenon but is certainly an area of worthy investigation. Insomnia has been shown to be related to elevated levels of cortisol [86, 90], which may accumulate over the 24-h period [44]. In addition, patients with insomnia show alterations in the diurnal rhythm of cortisol compared to participants without insomnia [91]. Cortisol is also implicated in regulating nocturnal immunity, including the movement of immune cells from the bone morrow into circulation and on into lymphoid organs [31]. Thus, nocturnal immunity, and possibly host resistance to infection, could be altered to the extent that sleep disruption alters cortisol rhythms. Finally, there is some basic science data to suggest that cortisol also plays a role in accelerating telomere attrition. Indeed, T cells treated in vitro with cortisol show a downregulation in telomerase activity, which is the enzyme charged with maintaining telomere integrity [92].

In addition to biological mechanisms, maladaptive health behaviors, such as tobacco smoke exposure, insufficient physical activity, excess alcohol consumption, illicit drug use, and poor nutrition can be pathways toward impaired immune functioning. The engagement in such behaviors may serve as coping mechanisms for individuals under prolonged stress or sleep loss. Furthermore, many of these behaviors (e.g., excess alcohol consumption [93]) can directly impair sleep quality, thus perpetuating the cycle between sleep and stress.

STRESS-SLEEP CONNECTION AND IMMUNITY

The parallels between poor sleep and stress on immunity begs the question of whether there may be synergistic effects worth investigation. To date, research in this area has been limited which is unfortunate because the sleep-stress connection presents important intervention opportunities. For example, sleep interventions such as sleep extension or behavioral treatments for insomnia (e.g., CBTI) may not only lead to better sleep but could have "spillover" effects on one's capacity to better regulate negative emotions in response to stress exposure. Similarly, stress reduction programs [e.g., mindfulness-based stress reduction (MBSR)] may not only improve one's experiences of stress but also lead to more restful, consolidated sleep. Moreover, either target (stress or sleep) may confer salubrious effects on immunity.

Though interventions have not been tested to date, the few studies that have examined the synergistic effects of sleep and stress on immunity are promising. In this regard, two studies have examined whether global sleep quality modulated increases in circulating levels of IL-6 in response in healthy adults exposed to an acute laboratory stressor [94, 95]. In both cases, poorer overall sleep quality was associated with stronger inflammatory responses to the acute stressors. Synergistic effects have also been observed in the context of infectious disease risk. Prather and colleagues, using the cold study paradigm described above, found in a sample of over 700 participants that shorter self-reported sleep duration predicted who went on to develop a biologically-verified cold following inoculation with the cold virus. However, this association was only true among individuals who reported average or below average subjective socioeconomic status (SES) [17]. While not a commonly used measure of psychological stress, subjective SES may reflect the chronic burden of feeling "lesser" compared to others. This is speculative but is consistent with other literature that finds that lower subjective SES predicts increased rates of diseases routinely observed at higher rates among those experiencing elevated levels of stress [96, 97]. Finally, one study reported an interaction between global sleep quality and perceived stress in predicting immune cell telomere length, such that the relationship between poorer overall sleep and shorter telomere length was significantly stronger participants who also reported higher levels psychological stress. Moreover, sleep quality was unrelated to telomere length in participants reporting lower levels of psychological stress [59].

CONCLUSION

Poor sleep and psychological stress are associated with alterations in immune system functioning that can have important implications for the development and progression of many age-related conditions. While efforts within in the areas of basic and translational science have focused on the independent effects of sleep and stress on immunity, there is a significant gap in our understanding of the interactions between the two. As was highlighted above, there are parallels regarding the influence of sleep and stress on immunity as well as obvious shared biological and behavioral pathways. Furthermore, there are exciting opportunities for employing interventions that may take advantage of the reciprocal links between sleep and stress and produce important benefits for immune health. From a population health perspective, it is clear that stress exposures and sleep disturbances are not evenly distributed across the population. As such, there is a pressing need to better understand these bidirectional processes so that more appropriate interventions can help those at greatest risk for infectious and inflammatory conditions as well as diseases associated with immunosenescence.

REFERENCES

[1] Buysse DJ. Sleep health: can we define it? Does it matter? Sleep 2014;37(1):9–17. https://doi.org/10.5665/sleep.3298.

[2] Luyster FS, Strollo Jr PJ, Zee PC, Walsh JK, Boards of Directors of the American Academy of Sleep Medicine and the Sleep Research Society. Sleep: a health imperative. Sleep 2012;35(6):727–34. https://doi.org/10.5665/sleep.1846.

[3] Cappuccio FP, Cooper D, D'Elia L, Strazzullo P, Miller MA. Sleep duration predicts cardiovascular outcomes: a systematic review and meta-analysis of prospective studies. Eur Heart J 2011;32(12):1484–92. https://doi.org/10.1093/eurheartj/ehr007.

[4] Cappuccio FP, D'Elia L, Strazzullo P, Miller MA. Sleep duration and all-cause mortality: a systematic review and meta-analysis of prospective studies. Sleep 2010;33(5):585–92.

[5] Cappuccio FP, D'Elia L, Strazzullo P, Miller MA. Quantity and quality of sleep and incidence of type 2 diabetes: a systematic review and meta-analysis. Diabetes Care 2010;33(2):414–20. https://doi.org/10.2337/dc09-1124.

[6] Watson NF, Badr MS, Belenky G, Bliwise DL, Buxton OM, Buysse D, et al. Recommended amount of sleep for a healthy adult: a joint consensus statement of the American Academy of Sleep Medicine and Sleep Research Society. Sleep 2015;38(6):843–4. https://doi.org/10.5665/sleep.4716.

[7] Irwin MR. Why sleep is important for health: a psychoneuroimmunology perspective. Annu Rev Psychol 2015;66:143–72. https://doi.org/10.1146/annurev-psych-010213-115205.

[8] Segerstrom SC, Miller GE. Psychological stress and the human immune system: a meta-analytic study of 30 years of inquiry. Psychol Bull 2004;130(4):601–30. https://doi.org/10.1037/0033-2909.130.4.601. pii:2004-15935-004.

[9] Medzhitov R. Toll-like receptors and innate immunity. Nat Rev Immunol 2001;1(2):135–45. https://doi.org/10.1038/35100529.

[10] Pearson TA, Mensah GA, Alexander RW, Anderson JL, Cannon 3rd RO, Criqui M, et al. Markers of inflammation and cardiovascular disease: application to clinical and public health practice: a statement for healthcare professionals from the Centers for Disease Control and Prevention and the American Heart Association. Circulation 2003;107(3):499–511.

[11] Franceschi C, Campisi J. Chronic inflammation (inflammaging) and its potential contribution to age-associated diseases. J Gerontol A Biol Sci Med Sci 2014;69(Suppl. 1):S4–9. https://doi.org/10.1093/gerona/glu057.

[12] Baker DJ, Wijshake T, Tchkonia T, LeBrasseur NK, Childs BG, van de Sluis B, et al. Clearance of p16Ink4a-positive senescent cells delays ageing-associated disorders. Nature 2011;479(7372):232–6. https://doi.org/10.1038/nature10600.

[13] Opp MR, Born J, Irwin MR. Sleep and the immune system. In: Ader R, editor. Psychoneuroimmunology. 4th ed. New York, NY: Academic Press; 2007. p. 570–618.

[14] Prather AA, Leung CW. Association of insufficient sleep with respiratory infection among adults in the United States. JAMA Intern Med 2016;176(6):850–2. https://doi.org/10.1001/jamainternmed.2016.0787.

[15] Patel SR, Malhotra A, Gao X, Hu FB, Neuman MI, Fawzi WW. A prospective study of sleep duration and pneumonia risk in women. Sleep 2012;35(1):97–101. https://doi.org/10.5665/sleep.1594.

[16] Cohen S, Doyle WJ, Alper CM, Janicki-Deverts D, Turner RB. Sleep habits and susceptibility to the common cold. Arch Intern Med 2009;169(1):62–7. pii:169/1/62, https://doi.org/10.1001/archinternmed.2008.505.

[17] Prather AA, Janicki-Deverts D, Adler NE, Hall M, Cohen S. Sleep habits and susceptibility to upper respiratory illness: the moderating role of subjective socioeconomic status. Ann Behav Med 2017;51:137–46. https://doi.org/10.1007/s12160-016-9835-3.

[18] Prather AA, Janicki-Deverts D, Hall MH, Cohen S. Behaviorally assessed sleep and susceptibility to the common cold. Sleep 2015;38(9):1353–9. https://doi.org/10.5665/sleep.4968.

[19] Doyle WJ, McBride TP, Skoner DP, Maddern BR, Gwaltney Jr JM, Uhrin M. A double-blind, placebo-controlled clinical trial of the effect of chlorpheniramine on the response of the nasal airway, middle ear and eustachian tube to provocative rhinovirus challenge. Pediatr Infect Dis J 1988;7(3):229–38.

[20] Cohen S, Doyle WJ, Skoner DP, Rabin BS, Gwaltney Jr. JM. Social ties and susceptibility to the common cold. JAMA 1997;277(24):1940–4.

[21] Benedict C, Brytting M, Markstrom A, Broman JE, Schioth HB. Acute sleep deprivation has no lasting effects on the human antibody titer response following a novel influenza A H1N1 virus vaccination. BMC Immunol 2012;13(1):https://doi.org/10.1186/1471-2172-13-1.

[22] Lange T, Dimitrov S, Bollinger T, Diekelmann S, Born J. Sleep after vaccination boosts immunological memory. J Immunol 2011;187(1):283–90. https://doi.org/10.4049/jimmunol.1100015.

[23] Lange T, Perras B, Fehm HL, Born J. Sleep enhances the human antibody response to hepatitis A vaccination. Psychosom Med 2003;65(5):831–5.

[24] Spiegel K, Sheridan JF, Van Cauter E. Effect of sleep deprivation on response to immunization. JAMA 2002;288(12):1471–2. pii:jlt0925-3.

[25] Prather AA, Hall M, Fury JM, Ross DC, Muldoon MF, Cohen S, et al. Sleep and antibody response to hepatitis B vaccination. Sleep 2012;35(8):1063–9. https://doi.org/10.5665/sleep.1990.

[26] Taylor DJ, Kelly K, Kohut ML, Song KS. Is insomnia a risk factor for decreased influenza vaccine response? Behav Sleep Med 2017;15(4):270–87. https://doi.org/10.1080/15402002.2015.1126596.

[27] Bollinger T, Bollinger A, Skrum L, Dimitrov S, Lange T, Solbach W. Sleep-dependent activity of T cells and regulatory T cells. Clin Exp Immunol 2009;155(2):231–8. pii:CEI3822, https://doi.org/10.1111/j.1365-2249.2008.03822.x.

[28] Besedovsky L, Lange T, Born J. Sleep and immune function. Arch Eur J Physiol 2012;463(1):121–37. https://doi.org/10.1007/s00424-011-1044-0.

[29] Born J, Lange T, Hansen K, Molle M, Fehm HL. Effects of sleep and circadian rhythm on human circulating immune cells. J Immunol 1997;158(9):4454–64.

[30] Born J, Uthgenannt D, Dodt C, Nunninghoff D, Ringvolt E, Wagner T, et al. Cytokine production and lymphocyte subpopulations in aged humans. An assessment during nocturnal sleep. Mech Ageing Dev 1995;84(2):113–26. https://doi.org/10.1016/0047-6374(95)01638-4.

[31] Lange T, Dimitrov S, Born J. Effects of sleep and circadian rhythm on the human immune system. Ann NY Acad Sci 2010;1193:48–59. pii:NYAS5300, https://doi.org/10.1111/j.1749-6632.2009.05300.x.

[32] Dimitrov S, Lange T, Nohroudi K, Born J. Number and function of circulating human antigen presenting cells regulated by sleep. Sleep 2007;30(4):401–11.

[33] Irwin M, Mascovich A, Gillin JC, Willoughby R, Pike J, Smith TL. Partial sleep deprivation reduces natural killer cell activity in humans. Psychosom Med 1994;56(6):493–8.

[34] Irwin M, McClintick J, Costlow C, Fortner M, White J, Gillin JC. Partial night sleep deprivation reduces natural killer and cellular immune responses in humans. FASEB J 1996;10(5):643–53.

[35] Medzhitov R. Origin and physiological roles of inflammation. Nature 2008;454(7203):428–35. https://doi.org/10.1038/nature07201.

[36] Irwin MR, Olmstead R, Carroll JE. Sleep disturbance, sleep duration, and inflammation: a systematic review and meta-analysis of cohort studies and experimental sleep deprivation. Biol Psychiatry 2016;80:40–52. https://doi.org/10.1016/j.biopsych.2015.05.014.

[37] Irwin M, Wang M, Campomayor CO, Collado-Hidalgo A, Cole S. Sleep deprivation and activation of morning levels of cellular and genomic markers of inflammation. Arch Intern Med 2006;166(16):1756–62. pii:166/16/1756, https://doi.org/10.1001/archinte.166.16.1756.

[38] Irwin MR, Carrillo C, Olmstead R. Sleep loss activates cellular markers of inflammation: sex differences. Brain Behav Immun 2010;24(1):54–7. https://doi.org/10.1016/j.bbi.2009.06.001.

[39] Irwin MR, Wang M, Ribeiro D, Cho HJ, Olmstead R, Breen EC, et al. Sleep loss activates cellular inflammatory signaling. Biol Psychiatry 2008;64(6):538–40. https://doi.org/10.1016/j.biopsych.2008.05.004.

[40] Irwin MR, Witarama T, Caudill M, Olmstead R, Breen EC. Sleep loss activates cellular inflammation and signal transducer and activator of transcription (STAT) family proteins in humans. Brain Behav Immun 2015;47:86–92. https://doi.org/10.1016/j.bbi.2014.09.017.

[41] Burgos I, Richter L, Klein T, Fiebich B, Feige B, Lieb K, et al. Increased nocturnal interleukin-6 excretion in patients with primary insomnia: a pilot study. Brain Behav Immun 2006;20(3):246–53. https://doi.org/10.1016/j.bbi.2005.06.007.

[42] Song C, Lin A, Bonaccorso S, Heide C, Verkerk R, Kenis G, et al. The inflammatory response system and the availability of plasma tryptophan in patients with primary sleep disorders and major depression. J Affect Disord 1998;49(3):211–9.

[43] Okun ML, Reynolds 3rd CF, Buysse DJ, Monk TH, Mazumdar S, Begley A, et al. Sleep variability, health-related practices, and inflammatory markers in a community dwelling sample of older adults. Psychosom Med 2011;73(2):142–50. https://doi.org/10.1097/PSY.0b013e3182020d08.

[44] Vgontzas AN, Tsigos C, Bixler EO, Stratakis CA, Zachman K, Kales A, et al. Chronic insomnia and activity of the stress system: a preliminary study. J Psychosom Res 1998;45:21–31. 1 Spec No.

[45] Vgontzas AN, Fernandez-Mendoza J, Liao D, Bixler EO. Insomnia with objective short sleep duration: the most biologically severe phenotype of the disorder. Sleep Med Rev 2013;17(4):241–54. https://doi.org/10.1016/j.smrv.2012.09.005.

[46] Fernandez-Mendoza J, Baker JH, Vgontzas AN, Gaines J, Liao D, Bixler EO. Insomnia symptoms with objective short sleep duration are associated with systemic inflammation in adolescents. Brain Behav Immun 2017;61:110–6. https://doi.org/10.1016/j.bbi.2016.12.026.

[47] Irwin MR, Olmstead R, Carrillo C, Sadeghi N, Breen EC, Witarama T, et al. Cognitive behavioral therapy vs. Tai Chi for late life insomnia and inflammatory risk: a randomized controlled comparative efficacy trial. Sleep 2014;37(9):1543–52.

[48] Irwin MR, Olmstead R, Breen EC, Witarama T, Carrillo C, Sadeghi N, et al. Tai chi, cellular inflammation, and transcriptome dynamics in breast cancer survivors with insomnia: a randomized controlled trial. J Natl Cancer Inst Monogr 2014;2014(50):295–301. https://doi.org/10.1093/jncimonographs/lgu028.

[49] Prather AA, Puterman E, Lin J, O'Donovan A, Krauss J, Tomiyama AJ, et al. Shorter leukocyte telomere length in midlife women with poor sleep quality. J Aging Res 2011;2011:721390. https://doi.org/10.4061/2011/721390.

[50] Lee KA, Gay C, Humphreys J, Portillo CJ, Pullinger CR, Aouizerat BE. Telomere length is associated with sleep duration but not sleep quality in adults with human immunodeficiency virus. Sleep 2014;37(1):157–66. https://doi.org/10.5665/sleep.3328.

[51] Liang G, Schernhammer E, Qi L, Gao X, De Vivo I, Han J. Associations between rotating night shifts, sleep duration, and telomere length in women. PLoS One 2011;6(8):e23462. https://doi.org/10.1371/journal.pone.0023462.

[52] Jackowska M, Hamer M, Carvalho LA, Erusalimsky JD, Butcher L, Steptoe A. Short sleep duration is associated with shorter telomere length in healthy men: findings from the Whitehall II cohort study. PLoS One 2012;7(10):e47292. https://doi.org/10.1371/journal.pone.0047292.

[53] Tempaku PF, Mazzotti DR, Tufik S. Telomere length as a marker of sleep loss and sleep disturbances: a potential link between sleep and cellular senescence. Sleep Med 2015;16(5):559–63. https://doi.org/10.1016/j.sleep.2015.02.519.

[54] Huang P, Zhou J, Chen S, Zou C, Zhao X, Li J. The association between obstructive sleep apnea and shortened telomere length: a systematic review and meta-analysis. Sleep Med 2018;48:107–12. https://doi.org/10.1016/j.sleep.2017.09.034.

[55] Garland SN, Palmer C, Donelson M, Gehrman P, Johnson FB, Mao JJ. A nested case-controlled comparison of telomere length and psychological functioning in breast cancer survivors with and without insomnia symptoms. Rejuvenation Res 2014;17(5):453–7. https://doi.org/10.1089/rej.2014.1586.

[56] Carroll JE, Esquivel S, Goldberg A, Seeman TE, Effros RB, Dock J, et al. Insomnia and telomere length in older adults. Sleep 2016;39:559–64.

[57] Lin J, Epel E, Cheon J, Kroenke C, Sinclair E, Bigos M, et al. Analyses and comparisons of telomerase activity and telomere length in human T and B cells: insights for epidemiology of telomere maintenance. J Immunol Methods 2010;352(1–2):71–80. https://doi.org/10.1016/j.jim.2009.09.012.

[58] Lin J, Cheon J, Brown R, Coccia M, Puterman E, Aschbacher K, et al. Systematic and cell type-specific telomere length changes in subsets of lymphocytes. J Immunol Res 2016;2016:5371050. https://doi.org/10.1155/2016/5371050.

[59] Prather AA, Gurfein B, Moran P, Daubenmier J, Acree M, Bacchetti P, et al. Tired telomeres: poor global sleep quality, perceived stress, and telomere length in immune cell subsets in obese men and women. Brain Behav Immun 2015;47:155–62. https://doi.org/10.1016/j.bbi.2014.12.011.

[60] Carroll JE, Cole SW, Seeman TE, Breen EC, Witarama T, Arevalo JM, et al. Partial sleep deprivation activates the DNA damage response (DDR) and the senescence-associated secretory phenotype (SASP) in aged adult humans. Brain Behav Immun 2016;51:223–9. https://doi.org/10.1016/j.bbi.2015.08.024.

[61] Cohen S, Frank E, Doyle WJ, Skoner DP, Rabin BS, Gwaltney Jr JM. Types of stressors that increase susceptibility to the common cold in healthy adults. Health Psychol 1998;17(3):214–23.

[62] Cohen S, Williamson GM. Stress and infectious disease in humans. Psychol Bull 1991;109(1):5–24.

[63] Burns VE, Carroll D, Ring C, Drayson M. Antibody response to vaccination and psychosocial stress in humans: relationships and mechanisms. Vaccine 2003;21(19–20):2523–34. pii:S0264410X03000410.

[64] Cohen S, Miller GE, Rabin BS. Psychological stress and antibody response to immunization: a critical review of the human literature. Psychosom Med 2001;63(1):7–18.

[65] Glaser R, Kiecolt-Glaser JK. Stress-induced immune dysfunction: implications for health. Nat Rev Immunol 2005;5(3):243–51. https://doi.org/10.1038/nri1571.

[66] Marsland AL, Herbert TB, Muldoon MF, Bachen EA, Patterson S, Cohen S, et al. Lymphocyte subset redistribution during acute laboratory stress in young adults: mediating effects of hemoconcentration. Health Psychol 1997;vol. 16(4):341–8.

[67] Marsland AL, Kuan DC, Sheu LK, Krajina K, Kraynak TE, Manuck SB, et al. Systemic inflammation and resting state connectivity of the default mode network. Brain Behav Immun 2017;62:162–70. https://doi.org/10.1016/j.bbi.2017.01.013.

[68] Wyman PA, Moynihan J, Eberly S, Cox C, Cross W, Jin X, et al. Association of family stress with natural killer cell activity and the frequency of illnesses in children. Arch Pediatr Adolesc Med 2007;161(3):228–34. https://doi.org/10.1001/archpedi.161.3.228.

[69] Kiecolt-Glaser JK, Preacher KJ, MacCallum RC, Atkinson C, Malarkey WB, Glaser R. Chronic stress and age-related increases in the proinflammatory cytokine IL-6. Proc Natl Acad Sci USA 2003;100(15):9090–5. https://doi.org/10.1073/pnas.1531903100.

[70] Miller GE, Chen E, Sze J, Marin T, Arevalo JM, Doll R, et al. A functional genomic fingerprint of chronic stress in humans: blunted glucocorticoid and increased NF-kappaB signaling. Biol Psychiatry 2008;64(4):266–72. https://doi.org/10.1016/j.biopsych.2008.03.017.

[71] Marsland AL, Walsh C, Lockwood K, John-Henderson NA. The effects of acute psychological stress on circulating and stimulated inflammatory markers: a systematic review and meta-analysis. Brain Behav Immun 2017;64:208–19. https://doi.org/10.1016/j.bbi.2017.01.011.

[72] Epel ES, Blackburn EH, Lin J, Dhabhar FS, Adler NE, Morrow JD, et al. Accelerated telomere shortening in response to life stress. Proc Natl Acad Sci USA 2004;101(49):17312–5. pii:0407162101, https://doi.org/10.1073/PNAS0407162101.

[73] Mathur MB, Epel E, Kind S, Desai M, Parks CG, Sandler DP, et al. Perceived stress and telomere length: a systematic review, meta-analysis, and methodologic considerations for advancing the field. Brain Behav Immun 2016;54:158–69. https://doi.org/10.1016/j.bbi.2016.02.002.

[74] Shalev I, Entringer S, Wadhwa PD, Wolkowitz OM, Puterman E, Lin J, et al. Stress and telomere biology: a lifespan perspective. Psychoneuroendocrinology 2013;38(9):1835–42. https://doi.org/10.1016/j.psyneuen.2013.03.010.

[75] Puterman E, Gemmill A, Karasek D, Weir D, Adler NE, Prather AA, et al. Lifespan adversity and later adulthood telomere length in the nationally representative US health and retirement study. Proc Natl Acad Sci USA 2016;113(42):E6335–42. https://doi.org/10.1073/pnas.1525602113.

[76] Spielman AJ, Caruso LS, Glovinsky PB. A behavioral perspective on insomnia treatment. Psychiatr Clin North Am 1987;10(4):541–53.

[77] Drake CL, Roehrs T, Roth T. Insomnia causes, consequences, and therapeutics: an overview. Depress Anxiety 2003;18(4):163–76. https://doi.org/10.1002/da.10151.

[78] Drake CL, Pillai V, Roth T. Stress and sleep reactivity: a prospective investigation of the stress-diathesis model of insomnia. Sleep 2014;37(8):1295–304. https://doi.org/10.5665/sleep.3916.

[79] Gordon AM, Mendes WB, Prather AA. The social side of sleep: elucidating the links between sleep and social processes. Curr Dir Psychol Sci 2017;26(5):470–5. https://doi.org/10.1177/0963721417712269.

[80] Epel ES, Crosswell AD, Mayer SE, Prather AA, Slavich GM, Puterman E, et al. More than a feeling: a unified view of stress measurement for population science. Front Neuroendocrinol 2018;49:146–69. https://doi.org/10.1016/j.yfrne.2018.03.001.

[81] Lazarus R, Stress FS. Appraisal and coping. New York: Springer; 1984.

[82] Gordon AM, Chen S. The role of sleep in interpersonal conflict: do sleepless nights mean worse fights? Soc Psychol Personal Sci 2014;5:168–75.

[83] Minkel JD, Banks S, Htaik O, Moreta MC, Jones CW, McGlinchey EL, et al. Sleep deprivation and stressors: evidence for elevated negative affect in response to mild stressors when sleep deprived. Emotion 2012;12:1015–20. https://doi.org/10.1037/a0026871.

[84] Yoo SS, Gujar N, Hu P, Jolesz FA, Walker MP. The human emotional brain without sleep—a prefrontal amygdala disconnect. Curr Biol 2007;17(20):R877-8. pii:S0960-9822(07)01783-6. https://doi.org/10.1016/j.cub.2007.08.007.

[85] Franzen PL, Gianaros PJ, Marsland AL, Hall MH, Siegle GJ, Dahl RE, et al. Cardiovascular reactivity to acute psychological stress following sleep deprivation. Psychosom Med 2011;73(8):679–82. https://doi.org/10.1097/PSY.0b013e31822ff440.

[86] Meerlo P, Sgoifo A, Suchecki D. Restricted and disrupted sleep: effects on autonomic function, neuroendocrine stress systems and stress responsivity. Sleep Med Rev 2008;12(3):197–210. https://doi.org/10.1016/j.smrv.2007.07.007.

[87] Pavlov VA, Tracey KJ. Neural regulators of innate immune responses and inflammation. Cell Mol Life Sci 2004;61(18):2322–31. https://doi.org/10.1007/s00018-004-4102-3.

[88] Tracey KJ. The inflammatory reflex. Nature 2002;420(6917):853–9. https://doi.org/10.1038/nature01321.

[89] Miller GE, Cohen S, Ritchey AK. Chronic psychological stress and the regulation of pro-inflammatory cytokines: a glucocorticoid-resistance model. Health Psychol 2002;21(6):531–41.

[90] Meerlo P, Koehl M, van der Borght K, Turek FW. Sleep restriction alters the hypothalamic-pituitary-adrenal response to stress. J Neuroendocrinol 2002;14(5):397–402.

[91] Backhaus J, Junghanns K, Hohagen F. Sleep disturbances are correlated with decreased morning awakening salivary cortisol. Psychoneuroendocrinology 2004;29(9):1184–91. https://doi.org/10.1016/j.psyneuen.2004.01.010. pii:S0306453004000125.

[92] Choi J, Fauce SR, Effros RB. Reduced telomerase activity in human T lymphocytes exposed to cortisol. Brain Behav Immun 2008;22(4):600–5. https://doi.org/10.1016/j.bbi.2007.12.004.

[93] Redwine L, Dang J, Hall M, Irwin M. Disordered sleep, nocturnal cytokines, and immunity in alcoholics. Psychosom Med 2003;65(1):75–85.

[94] Heffner KL, Ng HM, Suhr JA, France CR, Marshall GD, Pigeon WR, et al. Sleep disturbance and older adults' inflammatory responses to acute stress. Am J Geriatr Psychiatry 2012;20(9):744–52. https://doi.org/10.1097/JGP.0b013e31824361de.

[95] Prather AA, Puterman E, Epel ES, Dhabhar FS. Poor sleep quality potentiates stress-induced cytokine reactivity in postmenopausal women with high visceral abdominal adiposity. Brain Behav Immun 2014;35:155–62. https://doi.org/10.1016/j.bbi.2013.09.010.

[96] Adler NE, Boyce T, Chesney MA, Cohen S, Folkman S, Kahn RL, et al. Socioeconomic status and health. The challenge of the gradient. Am Psychol 1994;49(1):15–24.

[97] Cohen S, Janicki-Deverts D, Miller GE. Psychological stress and disease. JAMA 2007;298(14):1685–7. pii:298/14/1685. https://doi.org/10.1001/jama.298.14.1685.

Part VI

Sleep loss and neurocognitive function

Chapter 25

Sleep loss and impaired vigilant attention

Mathias Basner
Unit for Experimental Psychiatry, Division of Sleep and Chronobiology, Department of Psychiatry, University of Pennsylvania Perelman School of Medicine, Philadelphia, PA, United States

Sleep is found widely throughout the animal kingdom [1]. Although its functions are not fully understood, there is substantial evidence that sleep of sufficient duration and quality is necessary to ensure high levels of waking alertness, attention, and cognitive performance [2–4], and to avoid predisposing humans to adverse health outcomes [5, 6]. Despite growing awareness of these negative outcomes [7], recent surveys indicate that 35%–40% of the adult US population chronically curtail their sleep to <7h on weekday nights [8], primarily for lifestyle reasons [9, 10].

NEUROBEHAVIORAL CONSEQUENCES OF ACUTE AND CHRONIC SLEEP LOSS

Both acute total deprivation (i.e., a period without any sleep beyond the typical 16h awake) and chronic partial sleep restriction (i.e., multiple days with reduced sleep duration per 24h) induce neurobehavioral changes in humans beyond subjective sleepiness, despite motivation to prevent these effects. The most reliable changes include increased lapses of sustained attention (i.e., errors of omission) and compensatory response disinhibition (i.e., errors of commission); psychomotor and cognitive slowing; working memory deficits; slow eyelid closures; and reduced sleep latency [3, 4]. Recent publications [11, 12] have challenged the claim that sleep loss primarily impairs executive functions and reasoning. High-order cognitive functions can be affected by sleep loss, but these effects are likely mediated by deficits in the ability to sustain wakefulness and attention, and to accurately respond in a timely manner. The most sensitive indicators of sleep loss seem to be those that precisely track moment-to-moment changes in neural indicators of state (especially EEG and fMRI), or behavioral markers of the stability of sustained attention, such as the psychomotor vigilance test (PVT).

Studies on the effects of chronic partial sleep restriction and recovery indicate that the mechanisms underlying the dynamic neurobehavioral changes induced by chronic sleep restriction may be fundamentally different from those associated with acute total sleep deprivation [13]. Sleep-dose-response experiments found that chronic restriction of sleep to between 3 and 7h time in bed per 24h, resulted in sleep dose-dependent, near-linear declines in vigilant attention across 7–14 days of sleep restriction, reaching levels that were comparable to 2–3 nights without any sleep [14, 15]. The neurobehavioral effects of chronic sleep restriction are modulated by endogenous circadian phase—with the greatest deficits during the circadian night [16–18]. Also, several experiments demonstrate that subjects frequently underestimate sleep loss related decrements in neurobehavioral performance, especially during the biological night [15, 19].

DIFFERENTIAL VULNERABILITY TO SLEEP LOSS

Several studies found trait-like individual differences in the magnitude of neurobehavioral consequences to both acute total [20, 21] and to chronic partial sleep deprivation [15, 22–24] that were highly replicable, suggesting a polygenetic trait. Numerous factors have been investigated as potential predictors of phenotypic vulnerability to sleep loss but none fully account for this phenomenon [25]. A few small studies do suggest that those vulnerable to acute total sleep deprivation may also be vulnerable to chronic partial sleep loss [11, 26–29].

EFFECTS OF SLEEP LOSS ON VIGILANT ATTENTION

As mentioned above, there is extensive evidence that the neurobehavioral consequences of sleep loss can be measured in certain aspects of cognitive functioning [3, 4, 30]. Among the most reliable effects of sleep deprivation is degradation of attention [3, 12], especially vigilant attention as measured by the PVT [2, 31]. The effects of sleep loss on PVT performance appear to be due to variability in maintenance of the alert state (i.e., alerting network) [2], and can include deficits in endogenous selective attention [32, 33], but they may also occur in attention involved in orienting to sensory events (i.e., orienting network), and attention central to regulating thoughts and behaviors (i.e., executive network) [34–36]. These multidimensional features of attention suggest it has a fundamental role in a wide range of cognitive functions, which may be the mechanisms by which sleep loss affects a range of performances, although it remains controversial whether impairment due to sleep deprivation is generic to all cognitive processes subserved by attentional processes [37].

THE PSYCHOMOTOR VIGILANCE TEST (PVT)

The PVT [2, 38–40] has become arguably the most widely used measure of behavioral alertness owing in large part to the combination of its high sensitivity to sleep deprivation [2, 31] and its psychometric advantages over other cognitive tests. The standard 10 min PVT measures sustained or vigilant attention by recording response times (RT) to visual (or auditory) stimuli that occur at random inter-stimulus intervals (ISI, 2–10 s in the standard 10 min PVT, including a 1 s feedback period during which the RT to the last stimulus is displayed) [31, 38, 41, 42]. It is not entirely accurate to describe the PVT as merely simple RT. The latter is a generic phrase historically used to refer to the measurement of the time it takes to respond to a stimulus with one type of response (in contrast a complex RT task can require different responses to different stimuli). A simple RT test assumes no specific number of RTs—in fact, it can be based on a single RT. Similar to simple RT, the PVT relies on a stimulus (typically visual) and an RT (typically a button press), but it also relies on sampling many responses to stimuli that appear at a random ISI within a pre-specified ISI range, and that therefore occur over a period of time (i.e., 10 min in terms of the most commonly used PVT). Therefore, time on task and ISI parameterization instantiate the "vigilance" aspect of the PVT. Response time to stimuli attended to has been used since the late 19th century in sleep deprivation research [41, 43, 44] because it offers a simple way to track changes in behavioral alertness caused by inadequate sleep, without the confounding effects of aptitude and learning [2, 31, 40, 45]. Moreover, the 10 min PVT [38] has been shown to be highly reliable, with intra-class correlations for key metrics such as lapses measuring test-retest reliability above 0.8 [31].

PVT performance also has ecological validity in that it can reflect real-world risks, because deficits in sustained attention and timely reactions adversely affect many applied tasks, especially those in which work-paced or timely responses are essential (e.g., stable vigilant attention is critical for safe performance in all transportation modes, many security-related tasks, and a wide range of industrial tasks). Lapses in attention as measured by the PVT can occur when fatigue is caused by either sleep loss or time on task [41, 46, 47], which are the two factors that make up virtually all theoretical models of fatigue in real-world performance. There is a large body of literature on attentional deficits having serious consequences in applied settings [48–51].

Sleep deprivation (SD) induces reliable changes in PVT performance, causing an overall slowing of response times, a steady increase in the number of errors of omission (i.e., lapses of attention, historically defined as RTs ≥ twice the mean RT or 500 ms), and a more modest increase in errors of commission (i.e., responses without a stimulus, or false starts) [15, 52]. These effects can increase as task duration increases (so-called time-on-task effect or vigilance decrement) [53], and they form the basis of the state instability theory (Fig. 25.1) [2, 31, 38–40, 43]. According to this the-

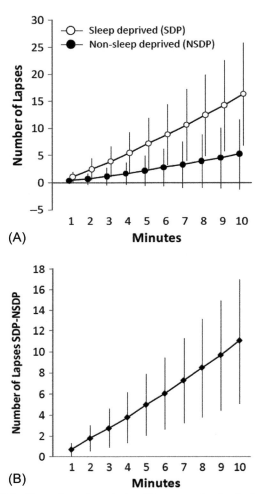

FIG. 25.1 The analyses shown in A and B were restricted to the first 1, 2, 3, 4, 5, 6, 7, 8, 9, and 10 min of the 10 min PVT (abscissa). (A) The number of lapses and their standard deviation are shown for the sleep deprived and the non-sleep deprived state. (B) The within-subject differences between sleep deprived and non-sleep deprived states of the number of lapses and their standard deviations are shown. *(Adapted from Basner M, Dinges DF. Maximizing sensitivity of the psychomotor vigilance test (PVT) to sleep loss. Sleep 2011;34 (5):581–91.)*

ory, several competing systems influence behavior during periods of sleep loss, two of the most important being the involuntary drive to fall asleep and a counteracting top-down drive to sustain alertness [2]. The interaction of these sleep-initiating and wake-maintaining systems leads to unstable sustained attention as manifested in longer RTs occurring stochastically throughout each PVT performance bout [2, 40]. Neuroimaging studies reveal that slowed responses on visual attention tasks—including the PVT—during sleep deprivation are associated with changes in neural activity in distributed brain regions that can include frontal and parietal control regions, visual and insular cortices, cingulate gyrus, and the thalamus [33, 54–57].

The 10 min PVT [2, 31, 38] has been shown to be sensitive to both acute total sleep deprivation (TSD) [15, 40, 58] and chronic partial sleep deprivation (PSD) [14, 15, 39, 58–60]; to be affected both by sleep homeostatic and circadian drives [61, 62]; to reveal large inter-subject variability in the response to sleep loss [20, 22, 63]; to demonstrate the effects of jet lag and shift work [64]; and to reveal improvements in alertness after wake-promoting interventions [65–67] and recovery from sleep loss [68, 69], and after initiation of CPAP treatment in patients with obstructive sleep apnea (OSA) [70]. The PVT is often used as a "gold standard" measure for the neurobehavioral effects of sleep loss, against which other biomarkers or fatigue detection technologies are compared [71, 72].

However, despite its simplicity, the complexity of developing a valid and reliable PVT is often underestimated. In contrast to most other cognitive tests, RT shifts of a few milliseconds can be meaningful on the PVT, and it is therefore important to use calibrated software and hardware. This and other caveats are discussed in greater detail below.

PVT SOFTWARE AND HARDWARE

Numerous versions of the PVT are available, either commercially or for free, across several hardware platforms. Researchers also frequently program their own version of the PVT. It is therefore unclear how data produced on different versions of the PVT compare across studies. This is especially true for PC-based PVTs [73, 74] (that often use a mouse or a keyboard for response input) relative to platforms that use a touchscreen for input (e.g., tablets and smartphones). For example, both the orientation of a smartphone and the input method (e.g., tapping the screen versus swiping) have been shown to influence response times [75, 76]. Also, each system has a certain response latency, i.e., the time it takes for a response (e.g., pushing down the spacebar) to be registered by the system. During calibration, the average response latency is typically determined and then subtracted from response times registered by the system. However, the system latency is typically the result of several serial processes that may sample at different frequencies. Therefore, the variability of response latencies may also be critical for the validity of the PVT and should be small. Our research group recently ran into a problem when switching from a 4th generation iPad to the iPad Air. With the new hardware iteration, Apple had introduced a new power saving feature. The touchscreen polling rate decreased to 60 Hz <2 s after the last touch. This led to a response time binning on the PVT with a bin size of ~17 ms. Although dedicated hardware versions like the PVT-192 exist, both bulkiness and the high price are likely prohibitive for their use in large-scale studies. Valid and reliable smartphone versions of the PVT are urgently needed to facilitate these studies [76].

Another factor affecting cross-study comparability of the PVT is poor standardization of both test parameters (e.g., inclusion of the 1 s feedback interval in the ISI or not?) and outcome variables (e.g., inclusion of timeouts in calculation of average response times?). These parameters can greatly affect PVT performance, and it is important that the field uses the same instrument to assess vigilant attention. In an effort to increase standardization across studies, Basner and Dinges published and encouraged to adopt definitions of test parameters and outcomes variables for the PVT [77].

PVT DURATION

For many applied settings, the standard 10 min version of the PVT is too long. Therefore, several groups have developed 5 min [75, 78–82] and 3 min [83] versions of the PVT. As mentioned above, PVT performance deteriorates with time-on-task, and sleep deprived subjects performing the PVT can likely compensate for brief periods of time. Thus, there is a trade-off between test duration and sensitivity (as shown by Basner et al. [83]), and it is likely not feasible to develop very short versions of the PVT that remain sensitive. Accordingly, both 2-min [78] and 90 s [80] versions of the PVT were deemed to be too insensitive to be used as valid tools for the detection of neurobehavioral effects of fatigue. Obviously, data generated with different duration versions of the PVT are not directly comparable, especially if parameters affecting data sampling and analysis are not changed. In an effort to increase comparability between the 10 min and a 3 min version of the PVT, ISIs were shortened from 2–10 s to 1–4 s (a newer iteration [45] uses 2–5 s) and the lapse threshold was decreased from 500 to 355 ms. Shorter ISIs have been shown to be associated with longer RTs [84]. This 3 min version has been successfully implemented in large-scale field studies and was shown to be sensitive to, e.g., the effects of sleep loss [85] and sleep medication use [86].

Yet another approach to decrease PVT duration was introduced by Basner and Dinges in 2012. They proposed an adaptive duration PVT (PVT-A). The PVT-A algorithm samples data until a certain decision threshold is exceeded, at which point the test determines to have gathered enough information and stop administration. With this approach, it was possible to decrease average test duration from 10 min to <6.5 min, with a minimal test duration of <30 s. The adaptive duration strategy may be superior to a simple reduction of PVT duration where the fixed test duration may be too short to identify subjects with moderate impairment (showing deficits only later during the test), but unnecessary long for those who are either fully alert or severely impaired.

PVT OUTCOME METRIC

Based on the time series of RTs, a number of outcome metrics can be produced for the PVT, including average and median RT, reciprocal transforms of RT (i.e., 1/RT or response speed), the number of lapses (i.e., errors of omission; typically RTs ≥ 500 ms), false starts (i.e., errors of commission; typically RTs <100 ms), the fastest or slowest 10% of RT or 1/RT, the standard deviation of RT, to only name a few. All of these outcome variables are in use but inconsistently reported across studies. Also, it is sometimes unclear whether a primary PVT outcome metric was defined a priori, or whether the researcher engaged in a fishing expedition among available PVT outcomes (often without proper correction for multiple testing). In an effort to determine the most sensitive PVT outcomes, Basner and Dinges compared effect sizes of 10 frequently used PVT outcomes, and suggested that response speed and lapses should be considered as primary PVT outcomes [77]. Several investigators have suggested new and more sensitive PVT outcomes (e.g., Rajaraman et al. [87], Basner et al. [88], Chavali et al. [89]), but it remains to be seen to what degree they will be accepted and used by the research community. Importantly, Basner et al. [88] found that the sensitivity of response speed was comparable to their new PVT metric, and thus corroborates the superiority of response speed as a primary outcome for the PVT.

RESEARCH AGENDA

Further studies are needed to shed light on the biological mechanisms underlying the changes in vigilant attention induced by acute total and chronic partial sleep loss, and the changes observed during recovery from sleep loss. Also, we know very little about the effects of chronic sleep restriction on vigilant attention beyond 3 consecutive weeks, stressing the need for a low-cost but yet valid and reliable version of the PVT that can easily be deployed in large-scale field studies.

REFERENCES

[1] Siegel JM. Sleep viewed as a state of adaptive inactivity. Nat Rev Neurosci 2009;10(10):747–53.

[2] Lim J, Dinges DF. Sleep deprivation and vigilant attention. Ann NY Acad Sci 2008;1129:305–22.

[3] Goel N, Rao H, Durmer JS, Dinges DF. Neurocognitive consequences of sleep deprivation. Semin Neurol 2009;29(4):320–39.

[4] Banks S, Dinges DF. Behavioral and physiological consequences of sleep restriction. J Clin Sleep Med 2007;3(5):519–28.

[5] Buxton OM, Cain SW, O'Connor SP, et al. Adverse metabolic consequences in humans of prolonged sleep restriction combined with circadian disruption. Sci Transl Med 2012;4(129):129ra143.

[6] Watson NF, Badr MS, Belenky G, et al. Joint consensus statement of the American Academy of Sleep Medicine and Sleep Research Society on the recommended amount of sleep for a healthy adult: methodology and discussion. Sleep 2015;38(8):1161–83.

[7] Basner M, Dinges DF. Sleep duration in the United States 2003-2016: first signs of success in the fight against sleep deficiency? Sleep 2018;41(4):1–16.

[8] Centers for Disease C, Prevention. Effect of short sleep duration on daily activities—United States, 2005–2008. MMWR Morb Mortal Wkly Rep 2011;60(8):239–42.

[9] Basner M, Fomberstein KM, Razavi FM, et al. American time use survey: sleep time and its relationship to waking activities. Sleep 2007;30(9):1085–95.

[10] Basner M, Spaeth AM, Dinges DF. Sociodemographic characteristics and waking activities and their role in the timing and duration of sleep. Sleep 2014;37(12):1889–906.

[11] Lo JC, Groeger JA, Santhi N, et al. Effects of partial and acute total sleep deprivation on performance across cognitive domains, individuals and circadian phase. PLoS One 2012;7(9):e45987.

[12] Lim J, Dinges DF. A meta-analysis of the impact of short-term sleep deprivation on cognitive variables. Psychol Bull 2010;136(3):375–89.

[13] Basner M, Rao H, Goel N, Dinges DF. Sleep deprivation and neurobehavioral dynamics. Curr Opin Neurobiol 2013;23(5):854–63.

[14] Belenky G, Wesensten NJ, Thorne DR, et al. Patterns of performance degradation and restoration during sleep restriction and subsequent recovery: a sleep dose-response study. J Sleep Res 2003;12(1):1–12.

[15] Van Dongen HP, Maislin G, Mullington JM, Dinges DF. The cumulative cost of additional wakefulness: dose-response effects on neurobehavioral functions and sleep physiology from chronic sleep restriction and total sleep deprivation. Sleep 2003;26(2):117–26.

[16] Cohen DA, Wang W, Wyatt JK, et al. Uncovering residual effects of chronic sleep loss on human performance. Sci Transl Med 2010;2(14):14ra13.

[17] Zhou X, Ferguson SA, Matthews RW, et al. Sleep, wake and phase dependent changes in neurobehavioral function under forced desynchrony. Sleep 2011;34(7):931–41.

[18] Mollicone DJ, Van Dongen HP, Rogers NL, Banks S, Dinges DF. Time of day effects on neurobehavioral performance during chronic sleep restriction. Aviat Space Environ Med 2010;81(8):735–44.

[19] Zhou X, Ferguson SA, Matthews RW, et al. Mismatch between subjective alertness and objective performance under sleep restriction is greatest during the biological night. J Sleep Res 2012;21(1):40–9.

[20] Van Dongen HP, Baynard MD, Maislin G, Dinges DF. Systematic interindividual differences in neurobehavioral impairment from sleep loss: evidence of trait-like differential vulnerability. Sleep 2004;27(3):423–33.

[21] Van Dongen HP, Maislin G, Dinges DF. Dealing with inter-individual differences in the temporal dynamics of fatigue and performance: importance and techniques. Aviat Space Environ Med 2004;75(3 Suppl):A147–54.

[22] Goel N, Banks S, Mignot E, Dinges DF. PER3 polymorphism predicts cumulative sleep homeostatic but not neurobehavioral changes to chronic partial sleep deprivation. PLoS One 2009;4(6):e5874.

[23] Goel N, Banks S, Mignot E, Dinges DF. DQB1*0602 predicts interindividual differences in physiologic sleep, sleepiness, and fatigue. Neurology 2010;75(17):1509–19.

[24] Goel N, Banks S, Lin L, Mignot E, Dinges DF. Catechol-O-methyltransferase Val158Met polymorphism associates with individual differences in sleep physiologic responses to chronic sleep loss. PLoS One 2011;6(12):e29283.

[25] Tkachenko O, Dinges DF. Interindividual variability in neurobehavioral response to sleep loss: a comprehensive review. Neurosci Biobehav Rev 2018;89:29–48.

[26] Tassi P, Schimchowitsch S, Rohmer O, et al. Effects of acute and chronic sleep deprivation on daytime alertness and cognitive performance of healthy snorers and non-snorers. Sleep Med 2012;13(1):29–35.

[27] Philip P, Sagaspe P, Prague M, et al. Acute versus chronic partial sleep deprivation in middle-aged people: differential effect on performance and sleepiness. Sleep 2012;35(7):997–1002.

[28] Drake CL, Roehrs TA, Burduvali E, Bonahoom A, Rosekind M, Roth T. Effects of rapid versus slow accumulation of eight hours of sleep loss. Psychophysiology 2001;38(6):979–87.

[29] Rupp TL, Wesensten NJ, Balkin TJ. Trait-like vulnerability to total and partial sleep loss. Sleep 2012;35(8):1163–72.

[30] Van Dongen HP, Vitellaro KM, Dinges DF. Individual differences in adult human sleep and wakefulness: Leitmotif for a research agenda. Sleep 2005;28(4):479–96.

[31] Dorrian J, Rogers NL, Dinges DF, Kushida CA. Psychomotor vigilance performance: neurocognitive assay sensitive to sleep loss. In: Sleep deprivation: clinical issues, pharmacology and sleep loss effects. New York, NY: Marcel Dekker; 2005. p. 39–70.

[32] Cordova CA, Said BO, McCarley RW, Baxter MG, Chiba AA, Strecker RE. Sleep deprivation in rats produces attentional impairments on a 5-choice serial reaction time task. Sleep 2006;29(1):69–76.

[33] Lim J, Tan JC, Parimal S, Dinges DF, Chee MW. Sleep deprivation impairs object-selective attention: a view from the ventral visual cortex. PLoS One 2010;5(2):e9087.

[34] Posner MI. Measuring alertness. Ann NY Acad Sci 2008;1129:193–9.

[35] Trujillo LT, Kornguth S, Schnyer DM. An ERP examination of the different effects of sleep deprivation on exogenously cued and endogenously cued attention. Sleep 2009;32(10):1285–97.

[36] Anderson C, Horne JA. Sleepiness enhances distraction during a monotonous task. Sleep 2006;29(4):573–6.

[37] Tucker AM, Whitney P, Belenky G, Hinson JM, Van Dongen HPA. Effects of sleep deprivation on dissociated components of executive functioning. Sleep 2010;33(1):47–57.

[38] Dinges DF, Powell JW. Microcomputer analysis of performance on a portable, simple visual RT task during sustained operations. Behav Res Methods Instrum Comput 1985;17(6):652–5.

[39] Dinges DF, Pack F, Williams K, et al. Cumulative sleepiness, mood disturbance, and psychomotor vigilance performance decrements during a week of sleep restricted to 4–5 hours per night. Sleep 1997;20(4):267–77.

[40] Doran SM, Van Dongen HP, Dinges DF. Sustained attention performance during sleep deprivation: evidence of state instability. Arch Ital Biol 2001;139(3):1–15.

[41] Dinges DF, Kribbs NB. Performing while sleepy: effects of experimentally-induced sleepiness. In: Monk TH, editor. Sleep, sleepiness and performance. Chichester: John Wiley and Sons; 1991. p. 97–128.

[42] Warm JS, Parasuraman R, Matthews G. Vigilance requires hard mental work and is stressful. Hum Factors 2008;50(3):433–41.

[43] Patrick GTW, Gilbert JA. On the effects of sleep loss. Psychol Rev 1896;3(5):469–83.

[44] Dinges DF. Probing the limits of funcional capability: the effects of sleep loss on short-duration task. In: Broughton RJ, Ogilvie RD, editors. Sleep, arousal, and performance. Boston, MA: Birkhäuser; 1992. p. 177–88.

[45] Basner M, Hermosillo E, Nasrini J, et al. Repeated administration effects on psychomotor vigilance test performance. Sleep 2018;41(1):181–6. zsx187.

[46] Lim J, Wu WC, Wang J, Detre JA, Dinges DF, Rao H. Imaging brain fatigue from sustained mental workload: an ASL perfusion study of the time-on-task effect. NeuroImage 2010;49(4):3426–35.

[47] Davies DR, Parasuraman R. The psychology of vigilance. New York, NY: Academic Press; 1982.

[48] Philip P, Akerstedt T. Transport and industrial safety, how are they affected by sleepiness and sleep restriction? Sleep Med Rev 2006;10(5):347–56.

[49] Dinges DF. An overview of sleepiness and accidents. J Sleep Res 1995;4(S2):4–14.

[50] Van Dongen HP, Dinges DF. Sleep, circadian rhythms, and psychomotor vigilance. Clin Sports Med 2005;24(2). 237–249, vii-viii.

[51] Gunzelmann G, Moore LR, Gluck KA, Van Dongen HP, Dinges DF. Individual differences in sustained vigilant attention: insights from computational cognitive modeling. In: Love BC, McRae K, Sloutsky VM, editors. 30th Annual Meeting of the Cognitive Science Society, 2008. Austin, TX: Cognitive Science Society; 2008. p. 2017–22.

[52] Dinges DF, Mallis M. Managing fatigue by drowsiness detection: can technological promises be realized? In: Hartley L, editor. Managing Fatigue in Transportation—Proceedings of the 3rd Fatigue in Transportation Conference, Fremantle, Western Australia, Pergamon; 1998. p. 209–29.

[53] Gunzelmann G, Moore LR, Gluck KA, Van Dongen HP, Dinges DF. Fatigue in sustained attention: generalizing mechanisms for time awake to time on task. In: Ackerman PL, editor. Cognitive fatigue: multidisciplinary perspectives on current research and future applications. Washington, DC: American Psychological Association; 2010. p. 83–101.

[54] Chee MW, Tan JC, Zheng H, et al. Lapsing during sleep deprivation is associated with distributed changes in brain activation. J Neurosci 2008;28(21):5519–28.

[55] Drummond SP, Bischoff-Grethe A, Dinges DF, Ayalon L, Mednick SC, Meloy MJ. The neural basis of the psychomotor vigilance task. Sleep 2005;28(9):1059–68.

[56] Tomasi D, Wang RL, Telang F, et al. Impairment of attentional networks after 1 night of sleep deprivation. Cereb Cortex 2009;19(1):233–40.

[57] Ma N, Dinges DF, Basner M, Rao H. How acute total sleep loss affects the attending brain: a meta-analysis of neuroimaging studies. Sleep 2015;38(2):233–40.

[58] Jewett ME, Dijk DJ, Kronauer RE, Dinges DF. Dose-response relationship between sleep duration and human psychomotor vigilance and subjective alertness. Sleep 1999;22(2):171–9.

[59] Balkin TJ, Bliese PD, Belenky G, et al. Comparative utility of instruments for monitoring sleepiness-related performance decrements in the operational environment. J Sleep Res 2004;13(3):219–27.

[60] Mollicone DJ, van Dongen HPA, Rogers NL, Dinges DF. Response surface mapping of neurobehavioral performance: testing the feasibility of split sleep schedules for space operations. Acta Astronaut 2008;63(7):833–40.

[61] Wyatt JK, Ritz-De Cecco A, Czeisler CA, Dijk DJ. Circadian temperature and melatonin rhythms, sleep, and neurobehavioral function in humans living on a 20-h day. Am J Phys 1999;277(4 Pt 2):R1152–63.

[62] Graw P, Krauchi K, Knoblauch V, Wirz-Justice A, Cajochen C. Circadian and wake-dependent modulation of fastest and slowest reaction times during the psychomotor vigilance task. Physiol Behav 2004;80(5):695–701.

[63] Killgore WD, Grugle NL, Reichardt RM, Killgore DB, Balkin TJ. Executive functions and the ability to sustain vigilance during sleep loss. Aviat Space Environ Med 2009;80(2):81–7.

[64] Neri DF, Oyung RL, Colletti LM, Mallis MM, Tam PY, Dinges DF. Controlled breaks as a fatigue countermeasure on the flight deck. Aviat Space Environ Med 2002;73(7):654–64.

[65] Dinges DF, Orne MT, Whitehouse WG, Orne EC. Temporal placement of a nap for alertness: contributions of circadian phase and prior wakefulness. Sleep 1987;10(4):313–29.

[66] Van Dongen HP, Price NJ, Mullington JM, Szuba MP, Kapoor SC, Dinges DF. Caffeine eliminates psychomotor vigilance deficits from sleep inertia. Sleep 2001;24(7):813–9.

[67] Czeisler CA, Walsh JK, Roth T, et al. Modafinil for excessive sleepiness associated with shift-work sleep disorder. N Engl J Med 2005;353(5):476–86.

[68] Rupp TL, Wesensten NJ, Bliese PD, Balkin TJ. Banking sleep: realization of benefits during subsequent sleep restriction and recovery. Sleep 2009;32(3):311–21.

[69] Banks S, Van Dongen HP, Maislin G, Dinges DF. Neurobehavioral dynamics following chronic sleep restriction: dose-response effects of one night of recovery. Sleep 2010;33(8):1013–26.

[70] Kribbs NB, Pack AI, Kline LR, et al. Effects of one night without nasal CPAP treatment on sleep and sleepiness in patients with obstructive sleep apnea. Am Rev Respir Dis 1993;147(5):1162–8.

[71] Dawson D, Searle AK, Paterson JL. Look before you (s)leep: evaluating the use of fatigue detection technologies within a fatigue risk management system for the road transport industry. Sleep Med Rev 2014;18(2):141–52.

[72] Chua EC, Tan WQ, Yeo SC, et al. Heart rate variability can be used to estimate sleepiness-related decrements in psychomotor vigilance during total sleep deprivation. Sleep 2012;35(3):325–34.

[73] Reifman J, Kumar K, Khitrov MY, Liu J, Ramakrishnan S. PC-PVT 2.0: an updated platform for psychomotor vigilance task testing, analysis, prediction, and visualization. J Neurosci Methods 2018;304:39–45.

[74] Basner M, Savitt A, Moore TM, et al. Development and validation of the cognition test battery. Aerosp Med Hum Perform 2015;86(11):942–52.

[75] Arsintescu L, Mulligan JB, Flynn-Evans EE. Evaluation of a psychomotor vigilance task for touch screen devices. Hum Factors 2017;59(4):661–70.

[76] Grandner MA, Watson NF, Kay M, Ocano D, Kientz JA. Addressing the need for validation of a touchscreen psychomotor vigilance task: important considerations for sleep health research. Sleep Health 2018;4(5):387–9.

[77] Basner M, Dinges DF. Maximizing sensitivity of the psychomotor vigilance test (PVT) to sleep loss. Sleep 2011;34(5):581–91.

[78] Loh S, Lamond N, Dorrian J, Roach G, Dawson D. The validity of psychomotor vigilance tasks of less than 10-minute duration. Behav Res Methods Instrum Comput 2004;36(2):339–46.

[79] Lamond N, Jay SM, Dorrian J, Ferguson SA, Roach GD, Dawson D. The sensitivity of a palm-based psychomotor vigilance task to severe sleep loss. Behav Res Methods 2008;40(1):347–52.

[80] Roach GD, Dawson D, Lamond N. Can a shorter psychomotor vigilance task be used as a reasonable substitute for the ten-minute psychomotor vigilance task? Chronobiol Int 2006;23(6):1379–87.

[81] Lamond N, Dawson D, Roach GD. Fatigue assessment in the field: validation of a hand-held electronic psychomotor vigilance task. Aviat Space Environ Med 2005;76(5):486–9.

[82] Thorne DR, Johnson DE, Redmond DP, Sing HC, Belenky G, Shapiro JM. The Walter Reed palm-held psychomotor vigilance test. Behav Res Methods 2005;37(1):111–8.

[83] Basner M, Mollicone DJ, Dinges DF. Validity and sensitivity of a brief psychomotor vigilance test (PVT-B) to total and partial sleep deprivation. Acta Astronaut 2011;69:949–59.

[84] Yang FN, Xu S, Chai Y, Basner M, Dinges DF, Rao H. Sleep deprivation enhances inter-stimulus interval effect on vigilant attention performance. Sleep 2018;1–12.

[85] Basner M, Dinges DF, Shea JA, et al. Sleep and alertness in medical interns and residents: an observational study on the role of extended shifts. Sleep 2017;40(4):1–8.

[86] Dinges DF, Basner M, Ecker AJ, Baskin P, Johnston S. Effects of zolpidem and zaleplon on cognitive performance after emergent tmax and morning awakenings: a randomized placebo-controlled trial. Sleep 2019;zsy258:1–9.

[87] Rajaraman S, Ramakrishnan S, Thorsley D, Wesensten NJ, Balkin TJ, Reifman J. A new metric for quantifying performance impairment on the psychomotor vigilance test. J Sleep Res 2012;21(6):659–74.

[88] Basner M, McGuire S, Goel N, Rao H, Dinges DF. A new likelihood ratio metric for the psychomotor vigilance test and its sensitivity to sleep loss. J Sleep Res 2015;24(6):702–13.

[89] Chavali VP, Riedy SM, Van Dongen HPA. Signal-to-noise ratio in PVT performance as a cognitive measure of the effect of sleep deprivation on the fidelity of information processing. Sleep 2017;40(3):zsx016.

Chapter 26

Sleep loss, executive function, and decision-making

Brieann C. Satterfield, William D.S. Killgore
Department of Psychiatry, College of Medicine, University of Arizona, Tucson, AZ, United States

ABBREVIATIONS

ACC	Anterior cingulate cortex
ADA	Adenosine deaminase
ADORA2A	Adenosine A2A receptor
BART	Balloon risk analog test
BDNF	Brain derived neurotrophic factor
DAT1	Dopamine transporter
DRD2	Dopamine D2 receptor
DTI	Diffusion tensor imaging
h	Hour
IGT	Iowa gambling task
mg	Milligram
min	Minute
OFC	Orbitofrontal cortex
PER3	PERIOD3
PFC	Prefrontal cortex
PVT	Psychomotor vigilance test
RT	Response time
SD	Standard deviation
SE	Standard error
sec	Second
TNFα	Tumor necrosis factor alpha
VWM	Visual working memory
WCST	Wisconsin card sorting task

INTRODUCTION

We live in a society that operates around the clock, often forgetting that sleep is important. In fact, insufficient sleep has become a public health epidemic and is often overlooked as a serious problem [1]. While the National Sleep Foundation recommends adults sleep >7h per night [2], 35% of adults in the United States sleep less [3]. Most people have suffered from sleep loss, either chronic or acute, at some point in their lives whether it be due to a new baby, stress, studying for an exam, or other circumstances. However, many people fail to realize the negative impact sleep loss has on cognitive functioning and how this has far-reaching real-world implications.

In fact, insufficient sleep is common in several safety-critical occupations, including medical professionals, military personnel, airline pilots, and truck drivers, just to name a few. Thus, it is important to understand how sleep loss impacts various aspects of cognition.

The present chapter provides an overview of the effects of sleep loss on several major cognitive domains. First, it is important to discuss the underlying neurobiological mechanisms that regulate sleep and wake, and thus modulate cognitive performance. We must also appreciate that human cognitive capacities are complex, with higher-order processes (e.g., executive functions, decision-making) building upon a foundation of elementary processes (e.g., attention). Therefore, this chapter will offer a discussion of how sleep loss impairs alertness, sustained attention, and vigilance. Additionally, we will discuss the importance of considering how inter-individual differences are related to relative resistance or vulnerability to cognitive impairment. We will then build upon these elementary capacities and focus on the consequences that sleep loss has on several complex executive function domains including working memory, inhibitory control, cognitive control, problem solving, risk-taking, and decision-making.

NEUROBIOLOGY OF SLEEP AND FATIGUE

There are two fundamental neurobiological processes that drive fatigue and alertness: the *homeostatic process* (Process S) and the *circadian process* (Process C) [4, 5]. The homeostatic process keeps track of prior amounts of sleep and wakefulness, and is conceptualized as an accumulating pressure for sleep with increasing time spent awake. This pressure is then dissipated over the course of a sleep period. The circadian process is the body's natural 24-h rhythm that keeps track of time of day. This process, modulated by the suprachiasmatic nucleus (SCN) of the hypothalamus, oscillates throughout a 24-h period to drive daytime alertness and nighttime sleepiness. During daytime hours, homeostatic sleep pressure accumulates with each hour awake, but is counteracted by the circadian drive for alertness. This interaction between the homeostatic and circadian pressures allows us to maintain normal daytime functioning at a fairly constant level. During nighttime hours, the homeostatic pressure for sleep is high and the circadian drive for alertness is low, promoting the onset and maintenance of sleep. Thus, waking performance is optimal during daytime hours and worst during nighttime hours (Fig. 26.1) [5, 6]. However, perturbations to this system (i.e., shift work, mistimed sleep, travel across time zones) can result in impaired neurobehavioral functioning. For example, the homeostatic and circadian processes become misaligned when an individual works during the night and sleeps during the day. In such a case, the homeostatic pressure for sleep mounts over the course of nighttime waking hours, but the circadian drive for alertness decreases, and hits the nadir during the early morning hours (Fig. 26.1). The net effect of this misalignment is increased fatigue, which can lead to cognitive performance impairment. Further, sleeping during the day is often difficult for a nightshift worker. This is because the circadian process increases the pressure for wakefulness throughout the day, forcing an individual to awaken before the homeostatic pressure is fully dissipated. This type of sleep curtailment can lead to a net accumulation of sleep debt [6].

ALERTNESS, SUSTAINED ATTENTION, AND VIGILANCE

In our day-to-day lives, the ability to maintain focus and attention is essential for effectively completing the task at hand and solving problems that require complex cognitive processing. Our ability to maintain attention and alertness fluctuates throughout the day as a function of the circadian and homeostatic processes, often without notice. However, when these two processes become misaligned due to extended wakefulness or mistimed sleep, attention begins to degrade. When wake is extended beyond 16 h or restricted to <6 h per night, individuals tend to show consistent and profound impairment in sustained attention.

Psychomotor vigilance

Sustained attention is typically measured using the psychomotor vigilance test (PVT) [7, 8]. The PVT is a simple computerized reaction time task that is considered to be the gold standard measure of behavioral alertness. It is sensitive to sleep loss and does not show an appreciable learning effect [9, 10]. In the standard version of the task, a visual stimulus is presented on the screen at random intervals between 2 and 10 s for a total of 10 min in duration. When the stimulus appears, the examinee presses a response button as quickly as possible, while avoiding false starts. Lim and Dinges [11] identified several distinct impacts that sleep deprivation has on PVT performance: (1) slowing of response times (RTs), (2) increases in attentional lapses, (3) exaggerated time-on-task effects, and (4) sensitivity to homeostatic and circadian influences.

In a typical sleep deprivation study, PVT RTs begin to slow around 16 h of wakefulness and degrade further across the night, with impairment being the most prominent during the early morning hours (i.e., the circadian nadir). *Slower responses* on

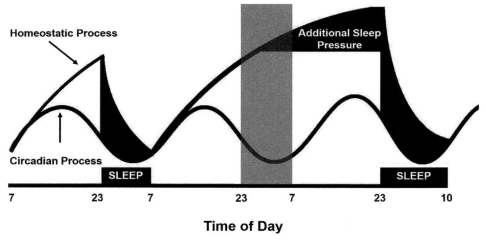

FIG. 26.1 The two process model of sleep regulation. The homeostatic process (S) and circadian process (C) interact to drive daytime alertness and nighttime sleepiness. Homeostatic pressure for sleep increases as a function of time spent awake and dissipates with time spent asleep. At the same time, circadian pressure oscillates across a 24-h period, with pressure for alertness highest in the early evening and lowest during the early morning hours. However, when an individual skips a night of sleep (shaded area) homeostatic pressure continues to build, while the circadian process continues to drive sleepiness during the earlier morning hours. At this point, the net effect of high homeostatic pressure and low circadian pressure is reduced alertness and increased fatigue, resulting in impaired cognitive functioning. Once an individual goes to sleep, homeostatic pressure decreases, and often results in increased sleep duration that is required to fully dissipate the homeostatic buildup. *Modified from File:Two-process model of sleep regulation.jpg [Internet]. WikiMedia. (2007). Available from: https://en.wikipedia.org/wiki/File:Two-process_model_of_sleep_regulation.jpg.*

the PVT are associated with reduced activation in the default mode network, a cortical system that includes the medial frontal and posterior cingulate cortex, regions that are most active when the brain is idle and not involved in complex cognitive processing [12]. While the average RT across trials increases during periods of sleep deprivation, there is also a significant slowing of both the fastest 10% and slowest 10% of RTs on the PVT. Albeit, the slowest RTs are disproportionately affected compared to the fastest 10% RTs (Fig. 26.2) [8, 10, 13, 14]. This indicates that sleep loss impacts not only the typical response, but also the best and worst performance. Decrements in PVT performance are not limited to conditions of total sleep deprivation. PVT performance is also substantially degraded when sleep is restricted by only a few hours each night. Belenky et al. demonstrated that when sleep is restricted to either 7, 5, or 3 h per night over the course of 1 week, response speeds (1/RT*1000) slowed in a cumulative manner across days [15]. Even when participants were allowed three 8 h nighttime recovery sleep periods, performance did not return to baseline levels (Fig. 26.3) [15].

Another characteristic of sleep loss is the increased frequency and duration of *attentional lapses* (RTs ≥ 500 ms) that occur within a single PVT bout, which are also accompanied by increases in errors of commission or false alarms (i.e., responding when no stimulus is present). Van Dongen and colleagues demonstrated that when sleep is restricted to either 6, 4, or 8 h over the course of 2 weeks the number of attentional lapses increases in a cumulative and

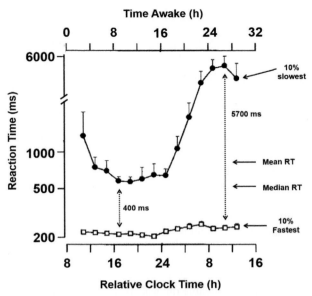

FIG. 26.2 Time course of PVT mean RTs across 32 h of sleep deprivation. PVT performance remained relatively stable until 16 h wake. Up until this point, only 400 ms separated the average 10% slowest (black circles) and 10% fastest (white squares) RTs. However, with increased time awake, RTs slowed dramatically. Just after 24 h wake, approximately 5700 ms separated the the fastest and slowest 10% of RTs. While not displayed in this figure, mean and median RTs are also significantly impacted by sleep loss and fall between the fastest and slowest curves shown here. *Modified from Cajochen C, Khalsa SB, Wyatt JK, Czeisler CA, Dijk DJ. EEG and ocular correlates of circadian melatonin phase and human performance decrements during sleep loss. Am J Physiol 1999;277:R640–9.*

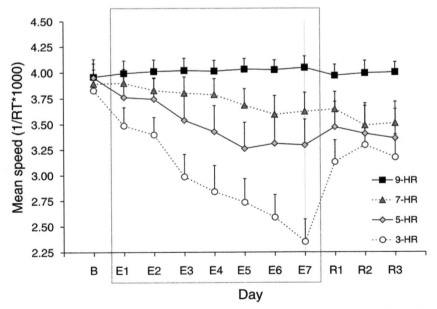

FIG. 26.3 Mean response speed (1/RT*1000) on the PVT over the course of a 7 day sleep restriction protocol as a function of sleep condition group. All groups had similar PVT performance at baseline (B). In the 9 h sleep group, PVT performance remained stable across sleep restriction days (E1–E7) and into the recovery days (R1–R3). When sleep was restricted to either 7, 5, or 3 h, there was a steady decline in PVT mean speed as days progressed. This decline was more pronounced in the 3 h sleep condition compared to the 7 h sleep condition. Additionally, 8 h sleep for three nights (recovery) was not sufficient to return PVT performance back to baseline levels. *Reproduced from Belenky G, Wesensten NJ, Thorne DR, Thomas ML, Sing HC, Redmond DP, Russo MB, Balkin TJ. Patterns of performance degradation and restoration during sleep restriction and subsequent recovery: a sleep dose-response study. J Sleep Res 2003;12(1):1–12, with permission from John Wiley and Sons.*

dose-dependent manner [16]. In fact, when sleep restriction was most severe (i.e., 4 h per night), the average number of lapses at the end of the 2 weeks was similar to the average number of lapses seen at the end of an 88 h sleep deprivation period [16]. Neuroimaging findings suggest that attentional lapses during sleep deprivation are related to reduced neural activation within the frontal, parietal and occipital regions, as well as the thalamus [17]. Together, imaging and behavioral studies have demonstrated how sleep loss disrupts normal functioning within the vigilant attention network, in turn hindering the ability to sustain attention.

The *time-on-task effect* is a phenomenon in which performance degrades as a function of time spent performing a cognitive task. That is, performance progressively declines the longer an individual is required to sustain attention necessary to perform the task [18]. This results in increased performance variability [19]. The time-on-task effect is apparent on several different types of cognitive tasks, however it is especially noticeable on tasks of vigilant attention, like the PVT. [20] On the PVT, the time-on-task effect manifests as a steady increase in the standard deviation of RTs across the task duration [10]. This phenomenon is present even under well-rested baseline conditions. Variability in PVT RTs is also a distinct characteristic of how vigilant attention is affected by sleep loss. Interestingly, the time-on-task effect interacts with sleep loss to amplify performance impairments when homeostatic pressure is high [8, 10, 21]. When faced with sleep loss, the time-on-task effect can be mediated by taking short breaks or switching tasks [6, 22, 23].

Last, PVT performance is sensitive to *homeostatic and circadian influences* [8, 24]. Fig. 26.4 shows the dynamic influence that the two neurobiological processes exert on performance. As described above, homeostatic pressure increases across hours awake, while the circadian process waxes and wanes across a 24-h period (Fig. 26.4, left). When these processes are considered in interaction, the sum of the two processes modulates PVT performance in a distinct manner. Fig. 26.4 (right) shows how the net effect of the two neurobiological processes impact PVT performance during 62 h of total sleep deprivation. Not only does PVT impairment increase with time spent awake, it also oscillates with the circadian process. Performance slightly improves during the early evening hours when the circadian pressure for wake is high, but further deteriorates after the circadian nadir and with mounting homeostatic pressure [25].

Wake state instability

Several aspects of PVT performance impairment, as described above, have been summarized into a single theory: the wake state instability hypothesis [10]. Sleep loss leads to a decrease in RTs, increase in attentional lapses and errors, and an increase in the time-on-task effect, all of which are influenced by mounting homeostatic pressure and manifesting as performance instability [8]. These moment-to-moment variations in performance are not gradual, linear, or predictable, but rather stochastic in nature. For example, Fig. 26.5 shows PVT responses from a single subject

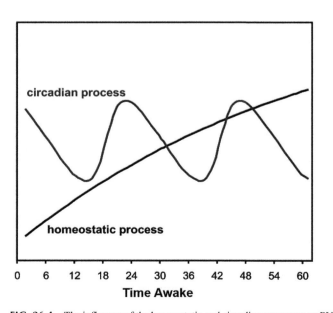
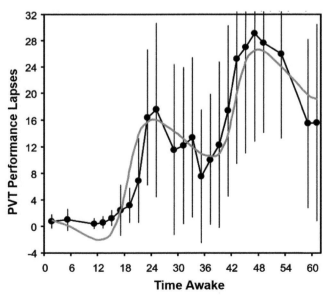

FIG. 26.4 The influence of the homeostatic and circadian processes on PVT performance during 62 h of extended wakefulness. The left panel shows the steady increase in homeostatic pressure across the sleep deprivation period in interaction with the waxing and waning of the circadian process. The right panel shows a mathematical derivation of the sum of the homeostatic and circadian processes (gray curve) overlaid on mean PVT lapses (± SD; *black curve*) collected from 12 healthy adults. *Reproduced from Van Dongen HPA, Belenky G. Individual differences in vulnerability to sleep loss in the work environment. Ind Health 2009;47(5):518–26, with permission.*

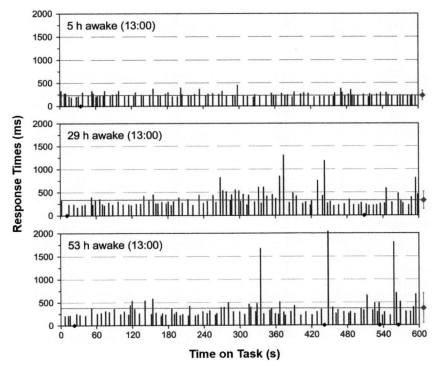

FIG. 26.5 Raw PVT RTs from a single subject collected over the course of a 62 h sleep deprivation period. RTs are plotted against time-on-task. PVT performance is shown at 5 h wakefulness (top panel), again 24 h later (middle panel), and another 24 h later (bottom panel). RTs become longer and more variable as a function of both time-on-task and time awake. Additionally, false starts (*black diamonds*) also increase. *Gray diamonds*: mean RT ± SD. *Reproduced from Satterfield BC, Van Dongen HPA. Occupational fatigue, underlying sleep and circadian mechanisms, and approaches to fatigue risk management. Fatigue Biomed Heal Behav 2013;246(3):118–36, with permission of Taylor & Francis Ltd. (http://www.informaworld.com).*

throughout the course of a 62 h sleep deprivation period. In the early afternoon when the subject has only been awake for 5 h, PVT performance is stable, with minimal variability in RTs and no attentional lapses. However, a different picture emerges 24 h later when the subject has been awake for 29 consecutive hours (Fig. 26.5, middle). At this point, there is moderate variability in RTs as time-on-task increases, and the occasional response exceeds the 500 ms attentional lapse threshold. Another 24 h later (Fig. 26.5, bottom), performance variability is further increased. At 53 h awake, attentional lapses become more frequent, RTs become longer, more errors are made, and the time-on-task effect is amplified. Together, these data illustrate that performance instability is a hallmark of sleep loss [6, 18, 23]. It is this unstable and unpredictable nature that makes fatigue so dangerous, especially in safety-critical operations. It has been posited that the stochastic nature of performance instability is the result of neuronal groups involved in the task expressing a local, use-dependent sleep like state. The local sleep theory suggests that activity from sustained use during a performance task and extended wakefulness pushes local neuronal groups to fall asleep. In turn, information processing in the task-specific pathway is interrupted, causing performance instability and increased attentional lapses [18, 26].

INDIVIDUAL DIFFERENCES

Research has shown that there are varying degrees of cognitive impairment during sleep loss across individuals. That is, not all individuals respond to sleep loss in the same manner [27–30]. These inter-individual differences are substantial and robust across a variety of manipulations, and constitute a trait [21, 31, 32]. As demonstrated by Van Dongen and colleagues [32], there are individuals who are resilient to the effects that sleep deprivation exerts on cognitive performance and individuals who are incredibly vulnerable. Fig. 26.6 shows that resilient individuals (triangles) are able to maintain stable performance across a 40 h sleep deprivation period, while vulnerable individuals (circles) show substantial impairment as wake extends past 16 h [33]. Due to the stable, trait nature of inter-individual differences, a number of biomarkers have been assessed to predict which individuals may be more or less susceptible to cognitive impairment due to sleep loss. These include personality and sensory markers [34, 35], neural markers [36], and genetic markers [37].

Several neuroimaging studies have sought to identify neural predictors of inter-individual differences in cognitive performance by assessing functional activation while performing a cognitive task [17, 38–40], functional

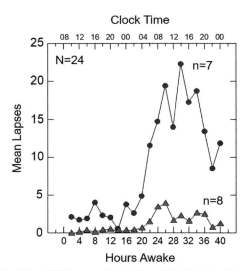

FIG. 26.6 Mean PVT lapses collected from 24 individuals during 40 h of total sleep deprivation. The number of PVT lapses remains low until about 16 h awake. At this point, the number of PVT lapses increases significantly for those most vulnerable ($n=7$) to impairment (*circles*). In contrast, the number of PVT lapses remains relatively stable for those most resilient ($n=8$) to impairment (*triangles*). Performance for the remaining nine individuals falls between these curves. *Modified from Van Dongen HPA, Maislin G, Dinges DF. Dealing with inter-individual differences in the temporal dynamics of fatigue and performance: Importance and techniques. Aviat Space Environ Med 2004;75(Suppl 3):A147–54, with permission.*

connectivity between brain regions [41], and white matter microstructure [42]. Chee and Tan found that sleep loss was associated with lower fronto-parietal activation compared to the rested state, and individuals most vulnerable to impaired selective attention had reduced activation in top-down cognitive bias regions (i.e., frontal and parietal cortices) [17]. In addition to measuring changes in neural activation, individual differences in neuroanatomical connectivity and structure have been identified. For example, our lab used diffusion tensor imaging (DTI) to assess the association between microstructure of the fronto-parietal attention system and PVT performance during a single night of sleep deprivation. We found that indirect measures of higher white matter integrity and higher myelination in the fiber pathways connecting the left frontal and left parietal regions were significantly correlated with resistance to PVT impairment [42]. Neural markers have the potential to help identify those individuals most vulnerable and those most resistant to impaired cognitive performance without having to expose them to any sleep loss paradigm. This affords us with a better understanding of the neural mechanisms underlying sleep loss related cognitive impairment.

Identifying genetic markers of inter-individual differences to performance impairments has become a large area of research in the last several years. Often, genetic polymorphisms are used as a tool to investigate how the functional differences brought about by the polymorphisms influence inter-individual differences in cognitive performance [43]. These studies have focused on polymorphisms associated with circadian pathways, adenosine (a marker of homeostatic pressure) pathways, neurotransmitters, neural signaling pathways, and immune responses [43].

PVT performance during sleep deprivation is mediated by several genetic variants, including those of the adenosine A_{2A} receptor (ADORA2A) gene [44], adenosine deaminase (ADA) gene, dopamine transporter (DAT1) gene [45, 46], and the tumor necrosis factor alpha (TNFα) gene [47]. For example, it was recently found that a variant of DAT1 mediates the time-on-task effect during sleep deprivation [45]. Study participants performed the PVT every 2 h over the course of a 38 h sleep deprivation period. Subjects homozygous for the 10-repeat variant of DAT1 were resilient to the time-on-task effect compared to subjects with the 9-repeat variant of the same gene. As Fig. 26.7 shows, performance between the two DAT1 genotype groups diverged as sleep deprivation progressed, with the most resilient individuals (i.e., the 10/10 group) maintaining stable performance with very little time-on-task effect [45]. Holst et al. also found that DAT1 genotype modulates PVT performance, specifically PVT lapses [46]. Genetic markers have also been found to influence performance on a variety of other cognitive tasks and will be discussed throughout the remainder of the chapter.

EXECUTIVE FUNCTIONS

The term "executive function" is used to describe a group of higher order cognitive processes that are necessary to coordinate and control deliberate actions toward future goals [48]. The term encompasses several cognitive processes including the ability to sustain attention while suppressing distractors, inhibit inappropriate actions, switch tasks, shift mental sets, think flexibly, plan and sequence events, and make appropriate and low-risk decisions, to name a few (Fig. 26.8). While these complex cognitive processes are mediated by several interacting cortical and subcortical regions, they rely heavily on the prefrontal cortex (PFC) which is sensitive to the effects of sleep loss [49]. Notably, the PFC shows reduced glucose metabolism following sleep deprivation (Fig. 26.9) [50], which is not fully reversed following a single night of recovery sleep [51]. This decline in prefrontal metabolic activity is thought to underlie some of the cognitive impairments seen during sleep loss.

To further complicate matters, not only are there inter-individual differences in cognitive performance as discussed in the previous section, these differences are also task-dependent [52], meaning that those most vulnerable to impairment on one task are not necessarily vulnerable to performance impairment on a different task. This is because cognitive performance, including executive functioning, is not a unitary concept. Most tasks designed to measure

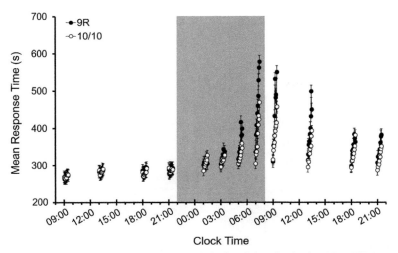

FIG. 26.7 Time-on-task performance for the DAT1 genotype groups across 38 h of total sleep deprivation. Mean RTs (±standard error) from 12 individual test bouts are plotted in 1-min bins for the 10-min PVT. Individuals carrying the 9-repeat allele, as either heterozygous or homozygous, were grouped together (9R). Data are plotted against the start time of the PVT test bout. As sleep deprivation progressed, time-on-task performance diverged between the 9R and 10/10 DAT1 genotype groups, such that those homozygous (10/10) for the DAT1 10-repeat allele were protected against severe time-on-task impairment. *Shaded area*: nighttime test bouts. *Modified from Satterfield BC, Wisor JP, Schmidt MA, Van Dongen HPA. Time-on-task effect during sleep deprivation in healthy young adults is modulated by dopamine transporter genotype. Sleep 2017;40(12):zsx167, with permission from Oxford University Press.*

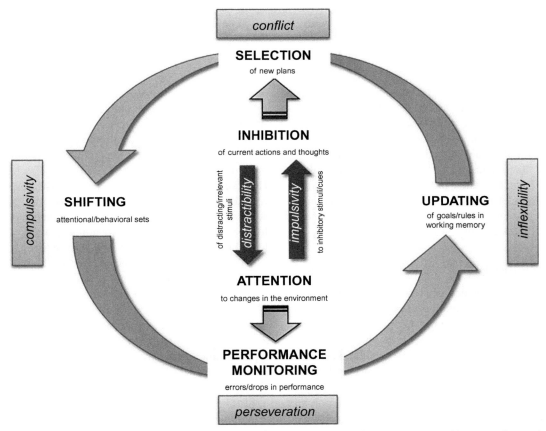

FIG. 26.8 A simplified schematic of the hypothesized relationship between the different executive functions. The ability to sustain attention and inhibit inappropriate responses are thought to be central components of executive function. A well-rested individual will pay *attention* to incoming information in order to respond appropriately to incoming stimuli, such as *inhibiting* the current course of action if needed. At the same time, the brain *monitors performance* based on internal and external feedback and triggers a signal that a new plan of action is required if performance levels drop and the number of errors increase. Then, behaviors are updated to reflect a change in goals and a new plan is *selected*. With the implementation of a new course of action, an individual must then *shift* both behavioral and attentional resources to continue with the new plan. However, sleep loss disrupts several points in this cycle leading to unfavorable actions and outcomes. For example, impairments in attention and inhibition may lead to distractibility and impulsivity, respectively. Further, impaired attention can lead to perseveration, or over focused behavior. In turn, relevant goals and rules are unable to be updated, resulting in inflexibility. Inflexible behavior does not allow for an individual to select a new course of action and could lead to compulsive behavior. Thus, one must be able to effectively integrate attention, inhibition, and flexibility in order to monitor performance and accurately update goals in response to environmental changes. *Reproduced from Bari A, Robbins TW. Inhibition and impulsivity: Behavioral and neural basis of response control. Prog Neurobiol 2013;108:44–79, with permission from Elsevier.*

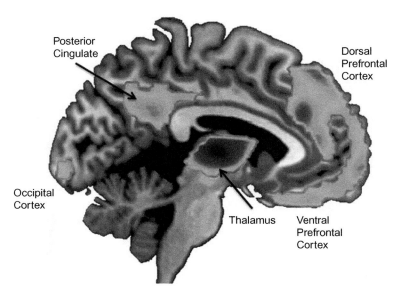

FIG. 26.9 A positron emission tomography (PET) image of regional cerebral glucose metabolism following 24 h of sleep deprivation. Sleep deprivation results in decreased glucose metabolism in areas of the prefrontal cortex, thalamus, and posterior cingulate. Reduced metabolism in these areas is thought to subserve some of the sleep loss induced impairments in cognitive performance we often see. *Reproduced from Killgore WDS, Weber M. Sleep deprivation and cognitive performance. In: Bianchi MT, editor, Sleep deprivation and disease: effects on the body, brain, and behavior. New York: Springer; 2014. p. 209–29.*

executive functions involve several integrated processes that are differentially impacted by sleep loss [53]. The complexity and multiple cognitive processes involved in many executive function tasks introduces a "task impurity problem." As Tucker and colleagues have demonstrated, performance on executive function tasks may not be attributable to global task impairments, but rather impairments in specific cognitive components of the task [54]. Thus, caution should be taken when administering and interpreting performance data from complex executive function tasks.

Working memory

Working memory can be described as the capacity to maintain and manipulate information in immediate memory, and underlies most executive functions. Working memory is conceptualized as having four components that include the storage of information, integration of information, regulation of information, and manipulation of information [55–57]. Working memory is distinct from short-term memory in that it requires both short-term storage of information *and* manipulation of that information [55]. There are several cognitive tasks that are used to measure various aspects of working memory, including digit span, word recall, number generation, serial addition, Sternberg, and N-back tasks [56].

In a meta-analysis, Lim and Dinges found that sleep loss impacts working memory performance with moderate effects sizes. Specifically, they found that both accuracy and RTs are impaired on these tasks [11]. Chee and colleagues conducted a series of studies using two different working memory tasks to investigate how sleep loss disrupts neural signaling specific to maintenance and manipulation of information. Following either 24 or 35 h of total sleep deprivation, both tasks showed reduced functional activation within bilateral parietal regions [58, 59], a common finding in studies of working memory and sleep loss [39, 60, 61]. However, Chee et al. [58, 59] found conflicting results in regard to activity within the PFC. In the latter study of the series [59], activity in the left PFC was *reduced* after sleep loss, while the first study [58] found that activation in the left PFC actually *increased* after 24 h of sleep deprivation. The increase in neural responsiveness of the PFC following sleep loss may reflect the initiation of compensatory mechanisms that are required to maintain stable performance. The compensatory recruitment hypothesis suggests that some individuals are able to sustain cognitive performance during sleep loss by recruiting areas of the cortex that are typically not engaged by the same task during rested wakefulness [62].

Working memory performance during sleep loss appears to also be mediated by a genetic polymorphism of the circadian clock gene PERIOD3 (PER3). Individuals with the 5-repeat allele for PER3 had better working memory performance on an N-back task than those with the 4-repeat allele. The difference in performance was significant only at the circadian nadir in the early morning hours [63]. Taken together, findings from neuroimaging studies show that sleep deprivation influences working memory through disruption to fronto-parietal networks, and performance is mediated by genetic polymorphisms of the circadian system.

While imaging studies have been able to identify brain regions involved in working memory performance during sleep loss, behavioral studies have found that sleep loss differentially impacts specific aspects of working memory performance [54, 59, 64], and in a sex-dependent manner [65, 66]. For example, Chee and Chuah found that sleep deprivation impairs general visual working memory (VWM) capacity, possibly due to degraded perceptual processing [67]. However, others have found that sleep loss does not impair VWM capacity, but rather impairs the ability to filter out VWM distractors [64]. Tucker et al. [54] also demonstrated that while sleep loss may show global decrements in working memory, the impairment is driven by specific working memory components. When dissociating a working memory task into "executive" and "nonexecutive" components, the non-executive working memory components (i.e., RTs) were the only elements impacted. "Executive" working memory scanning efficiency and resistance to proactive interference remained intact [54].

In a recent study, Rångtell and colleagues [65] administered a sequence-type working memory task following a single night of sleep loss, or 8 h sleep, which study participants performed in silence or with an auditory distraction. They were then asked to rate how confident they were in their performance. Overall, sleep loss impaired working memory performance in women, but not in men. Neither sex reported differences in subjective working memory performance. The auditory distraction impaired performance in both conditions and was not impacted by sex [65]. Overall, the accumulating data suggests that working memory impairments may actually be driven by degradation in alertness and vigilance, rather than the specific executive functions such as the ability to maintain and manipulate information, and these impairments are sex-specific.

Inhibitory control

Some actions may be adaptive under one set of circumstances, yet maladaptive in other circumstances. A key aspect of executive functioning is the ability to inhibit inappropriate responses or behaviors in a particular context. For instance, lack of inhibitory control can lead to impulsive decisions that may have negative consequences. Inhibitory control is typically assessed using response inhibition tasks, including the stop signal task or go/no-go paradigms. These tasks are designed to measure the ability to withhold a prepotent (i.e., automatic) response [68]. In a typical go/no-go task, individuals learn to respond to a specific set of stimuli (*go* stimuli) and learn to withhold a response for a different set of stimuli (*no-go* stimuli). Performance is assessed based on correctly responding to *go* stimuli (simple attention and response time) and correctly withholding a response to *no-go* stimuli (inhibitory control).

Neuroimaging studies using the go/no-go paradigm suggest that the task recruits several PFC regions. Specifically, the ability to correctly withhold a response most consistently activates the right lateral PFC and bilateral insula. In contrast, failure to withhold a response engages the right anterior cingulate cortex (ACC), medial frontal gyrus and portions of the parietal lobe. All of which are regions often associated with error detection and behavioral monitoring [69]. These are some of the same regions that show reduced metabolic activity following sleep loss [50]. Thus, it would be expected that sleep loss impairs the ability to withhold inappropriate responses, which has been observed. In fact, sleep deprived individuals who are unable to efficiently inhibit responses on the go/no-go have difficulty recruiting the ventrolateral PFC. Conversely, resilient individuals show increased activation within this region [70].

Drummond and colleagues [71] used the go/no-go to assess the effects of 64 h of sleep deprivation on inhibitory control. As expected, the ability to inhibit inappropriate responses decreased as a function of time awake. Interestingly, hit rates (correct *go* responses) remained unaffected for most of the sleep deprivation period, but rapidly declined at 55 h of wakefulness [71]. Another sleep deprivation study found similar results [72]. Additionally, these findings have been replicated under conditions of partial sleep restriction, where sleep was limited to 6 h per night for four nights. Study participants showed impaired inhibitory actions while maintaining correct responses [73]. Sleep loss causes a steady decline in response withholding with increasing time spent awake, while maintaining the ability to attend to incoming stimuli. These findings emphasize the fact that sleep deprivation does not result in a global degradation of cognitive performance due to impaired basic attention, thus cognitive impairment is task and domain-specific [32, 53, 74].

Cognitive control

A hallmark characteristic of executive function is the ability to modulate cognitive processes. In a broad sense, cognitive control is the ability to regulate and coordinate thoughts and actions in-line with behavioral goals or changes in situational demands [75]. This allows us to balance cognitive *stability*—the ability to actively focus on and maintain task-relevant information—with cognitive *flexibility*—the ability to update information according to changes in situational demands, while also suppressing irrelevant information in order to appropriately adapt behavioral actions to meet new goals [75–77]. For example, you may be driving down a long, straight highway when a large deer jumps out in front of your vehicle. Your current goal of driving down a straight highway is disrupted by the unexpected object in the road. You must update your goal in order to appropriately adapt your response to the situation (i.e., avoid hitting the deer).

Impairments in cognitive control can lead to perseverative, or over-focused, behaviors that can have serious consequences. These types of behaviors are often seen in psychiatric conditions such as obsessive compulsive disorder and schizophrenia [78].

Cognitive control encompasses the interaction of multiple cognitive processes, including working memory, attention, decision-making, response selection, response inhibition, and associated learning [78]. These processes underlie several behaviors such as multi-tasking/task-switching, changing behavior to fit a new rule, or suppressing distractions. Multi-tasking and task-switching are typically assessed using paradigms that require an individual to rapidly switch between response sets. The effect of *interference* (i.e., failure to suppress distractions) is often assessed using task paradigms that involve ignoring irrelevant information presented in order to stay focused on the task goal. Whereas *flexibility* (changing behavior to fit a new rule) is often measured using reversal learning tasks that require an individual to recognize changes in contingencies (changes in stimulus-response patterns) and update behavior accordingly.

In well-rested individuals, these task paradigms have been shown to reliably recruit areas of the PFC, specifically the orbitofrontal cortex (OFC) and dorsolateral PFC. There are also several reciprocal projections between the PFC and subcortical structures such as the ventral striatum, amygdala, and thalamus that are involved in maintaining cognitive control [78, 79]. Additionally, both the cortical and subcortical regions are highly sensitive to disruptions in the neurochemical environment. Dopamine is a primary neuromodulator in the fronto-striatal pathway that is quite sensitive to perturbations such as sleep loss. Even small variations in dopamine levels can result in cognitive impairment [78, 80]. Thus, alteration of functioning within the dopamine system may be one of the primary ways that sleep deprivation can affect cognition.

Multi-tasking and task-switching

In safety-critical operations, the ability to rapidly and efficiently switch between multiple tasks is paramount. Unfortunately, many of the occupations (e.g., airline pilots, truck drivers, medical personnel, military personnel, etc.) that require multi-tasking are also often subjected to chronic sleep loss. During a typical task-switching paradigm, individuals perform two types of tasks in succession in which numerical stimuli are presented. On one task type study participants are asked to identify which of the numbers is even or odd. On another task type study participants are asked to identify if the number presented is smaller or larger than a predetermined value. When two of the same task types are presented one after another, this is considered a repetition trial. When the trial switches from one task to the other this is considered a switch trial. Switch trials are used to calculate *switch cost*, or the change in reaction time and accuracy between the switch and repetition trials. Essentially, switch cost is a measure of the amount of time that is required to reconfigure the cognitive processes needed to perform the new task—a basic executive function.

A recent neuroimaging study found that while performing a task-switching paradigm following sleep deprivation, neural activation increased in the fronto-parietal network and cingulate gyrus as compared to the well-rested state. However, different brain regions were involved in the switch trials. Task-switching was associated specifically with increased activation in the superior temporal gyrus and thalamus. Based on the cerebral metabolic data described earlier, it would seem sensible to expect reduced activation in these key brain regions. However, the fact that sleep deprivation was associated with increased neural activation suggests that compensatory mechanisms may be initiated to maintain some level of information retrieval necessary for the task [81]. Nonetheless, from a behavioral perspective, sleep loss results in slowed RTs, especially during switch trials [81, 82]. A single night of sleep loss also reduces performance accuracy and increases switch costs [82].

Total sleep deprivation studies are extreme cases of sleep loss and often do not translate to real-world scenarios. Haavisto and colleagues [83] investigated how multi-tasking performance is affected by sleep restriction over the course of what some would consider a typical workweek. Individuals in the restricted condition were only allowed to sleep for 4 h per night for five consecutive nights, compared to those in the well-rested condition who were allowed to sleep for 8 h per night. A multi-tasking paradigm was used in which study participants performed a series of subtasks to assess short-term memory, arithmetic skills, and visual and auditory monitoring. Sleep restriction impaired the ability to multi-task as a function of the number of days of sleep restriction, with performance also degrading further as time-on-task increased. Additionally, it took two nights of recovery sleep (8 h) to return to baseline performance levels [83]. Because sleep loss degrades the ability to multi-task or rapidly switch between activities, the potential for errors and accidents significantly increases.

Cognitive interference

Another aspect of cognitive control is being able to suppress irrelevant or distracting information while maintaining focus on relevant task information. When the irrelevant aspects of the task cannot be ignored, this is known as *cognitive interference*. Typically, cognitive interference is measured using various forms of the Stroop paradigm. The goal of a Stroop task is to inhibit a common or "prepotent" response in favor of a less common response. For example, the brain naturally reads printed words it sees without any effort. This automatic tendency to read is known as a prepotent response. During a typical Stroop task, the participant is presented with a series of words depicting color names (e.g., "RED", "GREEN", "BLUE"). In some conditions the words are printed in either congruent (e.g., "RED" in red letters) or incongruent (e.g., "RED" in blue letters) ink colors, while in the neutral condition the words are printed in black ink. Individuals are to state the color of the ink in which the word appears, but not the word itself. The goal is to suppress the prepotent response (i.e., reading the word) in favor of the less common response (i.e., saying the color of the ink in which the word appears). This induces cognitive interference. Interestingly, several studies using the Stroop task have found that sleep deprivation does not affect cognitive interference, but rather only causes a general slowing of RTs [84–86]. A recent study found that resilience to slowed RTs and increased errors on the Stroop during a 30h sleep deprivation period was related to a genetic polymorphism of the brain derived neurotrophic factor (BDNF) gene. Those individuals with the common Val allele had fewer errors compared to those with the Met allele [87], suggesting that some aspects of cognitive interference are predictable by genetic markers.

However, Gevers and colleagues [88] recognized the importance of assessing task performance in relevant components rather than assessing performance across the task as a whole. The Stroop task was administered once after a full night of sleep and again following a night of sleep deprivation. The task was decomposed into three components for analysis: size of the interference effect, bottom-up modulation (facilitated processing after repetitions), and top-down modulation (cognitive control adjustments for incongruent trials). They found that sleep deprivation impaired top-down control such that there was a reduced ability to efficiently recognize and adapt to conflicts (i.e., incongruent trials) [88]. It is possible that earlier studies did not find an effect of cognitive interference because different aspects of the task, as demonstrated by Gevers et al. [88], are differentially impacted by sleep loss, further highlighting the importance of deconstructing a task into specific, well-defined cognitive components.

Flexible attentional control

As with multi-tasking, the ability to think flexibly and quickly adapt to changing environmental circumstances is important in safety-critical operations. Effective attentional control requires an individual to anticipate responses and outcomes based on a predetermined set of expectations. However, real-world situations place individuals in dynamic environments that often challenge set expectations. Thus, individuals must be able to effectively recognize a change in circumstances and appropriately update the behavioral response. Reversal learning paradigms are typically used to assess flexible attentional control. While there are a wide variety of reversal learning paradigms, in the simplest form these tasks involve learning specific stimulus-response mappings that are tied to a reward and those that are tied to an unfavorable outcome. At a point in the middle of the task, the stimulus-response contingencies are reversed, such that previously rewarding stimuli become unfavorable and vice versa. Those that are able to maintain flexible attentional control will quickly recognize a change has occurred and adapt their responses. In contrast, those that do not think flexibly tend to show perseverative behavior by responding to the old stimulus-response mappings.

Interestingly, impairments in cognitive flexibility mimic impairments seen in individuals suffering from damage to the OFC and the basal ganglia [78, 89–91]. Neuroimaging studies in well-rested adults show that reversal learning recruits the ventrolateral PFC when a subject stops responding to the previously correct stimuli and starts responding to the new, relevant stimuli. When a subject makes a reversal error (i.e., responding to the incorrect stimulus following the stimulus-response reversal), there is neural activation within the ventral striatum [92]. Until recently, flexible attentional control had not been thoroughly explored under conditions of sleep deprivation. Whitney and colleagues [93, 94] conducted a series of sleep deprivation studies to assess how sleep loss impacts the ability to maintain flexible attentional control. In the first study, research participants were exposed to a 62h sleep deprivation period. These individuals performed a modified version of the basic go/no-go paradigm during well-rested baseline, after 55h of extended wakefulness, and following recovery sleep. During this novel task, participants were required to respond to a specific set of numeric stimuli (*go* stimuli) and withhold their response from a different set of numeric stimuli (*no-go* stimuli). However, they were required to learn which stimuli were *go* and which were *no-go* based on monetary reward feedback. Halfway throughout the task, the stimulus-response contingencies were reversed. Stimuli that were previously *go* stimuli became *no-go* stimuli and stimuli that were previously *no-go* stimuli became *go* stimuli. Participants were unaware of the reversal, and were again required to use monetary reward feedback to determine the correct stimulus-response mappings [93]. Sleep deprivation degraded pre-reversal performance, and even further degraded post-reversal performance (Fig. 26.10). Importantly, the profound impairment seen on the reversal learning task was distinct from vigilant attention impairment [93]. Further, it was found that the Val165Met genetic polymorphism of catechol-*O*-methyltransferase (i.e., the enzyme that degrades dopamine in the PFC) was associated with resilience to impairment on the go/no-go reversal learning task. Specifically, individuals carrying the Met allele were protected from the post-reversal performance impairment described above [76].

In a follow-up study, Whitney and colleagues [94] used a novel adaptation of the continuous performance task (AX-CPT) in which previous cue-probe contingencies were switched halfway throughout the task. They again

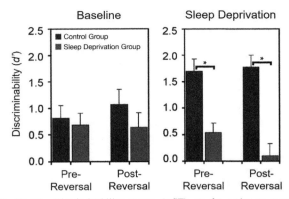

FIG. 26.10 Discriminability scores (±SE) on the go/no-go reversal learning paradigm. At baseline (left), there were no significant differences in pre- or post-reversal performance between the well-rested control and the sleep deprived groups. However, performance was profoundly degraded in the sleep deprived group (right). Impairment was further impaired following the stimulus-response reversal. Asterisks (*) indicate statistically significant pairwise differences. *Modified from Whitney P, Hinson JM, Jackson ML, Van Dongen HPA. Feedback blunting: Total sleep deprivation impairs decision making that requires updating based on feedback. Sleep 2015;38(5):745–54, with permission from Oxford University Press.*

demonstrated that sleep deprivation diminishes flexible attentional control, and also found that top-down control does not efficiently prevent errors [94]. They also found that a polymorphism that affects the binding potential of the dopamine D_2 receptor (DRD2) is associated with protection from impairment. Specifically, individuals homozygous for the C allele were resilient to impaired flexible attentional control. Together, these findings demonstrate that sleep deprivation profoundly degrades the ability to maintain flexible attention control, which can lead to maladaptive behaviors, including perseveration. Further, resilience to cognitive control impairment seems to be mediated by functional dopaminergic polymorphisms involved in the frontostriatal pathways.

Problem solving

The ability to solve problems is a core aspect of nearly any job. Depending on the operational environment, the kinds of problems that workers may encounter can range from simple mundane challenges to those that can be mission critical or even life threatening. Because many occupations require individuals to remain awake for extended periods or during times that are out of phase with their circadian rhythm, it is important to understand how various aspects of problem-solving ability can be impacted by lack of sleep.

Convergent thinking and logical deduction

Different kinds of problems require different kinds of solutions. One type of problem can be solved through the process of *convergent thinking*. This type of problem solving involves the step-by-step application of logical deductive reasoning and the use of established rules to reach a solution. These kinds of problems can be solved by beginning with an established set of information or major premise, adding a second minor premise, and finally arriving at a logical conclusion. For example, given the major premise that "every A is B" and the minor premise that "this C is A," then it is logical to conclude that "therefore this C is B." Concretely, we could apply this to a real-life example such as "all bulldogs are animals. Baxter is a bulldog. Therefore, Baxter is an animal." While this process sounds complex, evidence suggests that this form of convergent thinking is not significantly degraded by sleep deprivation [49]. Most studies that have specifically tested outcome measures such as logical deduction, intellectual functioning, grammatical reasoning, reading comprehension, and nonverbal problem solving have found negligible effects of sleep deprivation on these capacities [11, 49].

Divergent and innovative thinking

In contrast to the convergent thought processes discussed above, the ability to think laterally, innovatively, and flexibly *does* appear to be particularly susceptible to sleep deprivation [49]. In one study, a single night without sleep was associated with fewer creative responses and greater difficulty letting go of unsuccessful strategies [95]. Similarly, sleep deprivation has also been shown to adversely affect the ability to generate lists of novel words and produces slower and less efficient performance on the Tower of London, a task that requires planning, forethought, and cognitive flexibility [49]. The ability to generate and vocalize a series of random numbers is also degraded by sleep deprivation, leading to increased redundancy and stereotypy of responses and frequent rule violations [96]. On the other hand, inconsistent effects of sleep deprivation have been reported for the Wisconsin Card Sorting Test (WCST), a clinically based test of concept formation, set shifting, and mental flexibility [97]. It is important to keep in mind, however, that the WCST is a clinical task that was designed to detect relatively severe brain injuries and may not be sensitive enough to detect the subtle effects produced by sleep loss.

One particularly interesting study attempted to mimic real life decision-making during sleep deprivation by using a complex marketing strategy game. The task required participants to engage in several high level executive function tasks during a prolonged period of sleep loss. In particular, participants had to continuously monitor ongoing activities, revise their marketing strategies in light of periodically appearing new information, and apply available information to develop creative and innovative solutions under severe time constraints [98]. When normally rested, participants were able to think flexibly and innovatively, but once sleep deprived, they showed rigid thinking and perseverated on poor and ineffective strategies. As they reached the end of the game, these sleep deprived individuals had exhausted their financial resources and were in a significantly worse financial position than compared to playing the same game in a rested state [98]. While such tasks can be incredibly ecologically valid and applicable to real-world situations, these types of complex tasks also suffer from the previously described task impurity problem [53, 54]. These types of tasks are not designed to deconstruct the component processes most affected by sleep loss, but do provide important understanding of how lack of sleep may actually be manifested in real-world situations.

RISK-TAKING, JUDGMENT, AND DECISION-MAKING

Can sleep deprivation increase the tendency to engage in high-risk activities or does it affect decisions that involve risk? While these questions seem simple, the answers appear to be complex and depend on a number of factors. We will address several issues, including how sleep loss affects self-reported risk-taking propensity, decision-making under conditions of uncertainty, the role of effort on risky behavior, implicit cognitive biases, aggressive behaviors, and moral decision-making.

Self-rated risk propensity

People can engage in high risk activities for a number of reasons. The construct of *risk-taking* is often confused with the closely related construct of *sensation seeking*, a preference for seeking out novel experiences and other thrilling activities that produce high levels of physiological arousal [99]. In contrast to sensation seeking, *Risk-Taking Propensity* represents the tendency to engage in activities that include a high level of risk, danger, or uncertainty of outcome [100, 101]. Although risk taking can occur because an individual is sensation seeking, risky behavior can also occur for reasons other than the pursuit of thrills or excitement. Accordingly, these two constructs are only modestly correlated with one another [102].

Interestingly, sleep deprivation has been shown to affect scores on measures of both sensation seeking and risk-taking, but the associations are typically inverse. For instance, one night of total sleep deprivation has been shown to significantly reduce scores on measures of self-reported sensation-seeking and self-reported risk-taking propensity [103, 104]. Similar findings have also been reported following two nights without sleep [35, 105]. Such findings are not surprising when considered in light of the fact that one of the most common symptoms of sleep loss is increased fatigue and reduced physical and mental energy [106]. It seems sensible that increased fatigue would lead to a reduction in activities requiring energy expenditure or exertion of mental or physical effort. Interestingly, longer periods of total sleep deprivation (i.e., 75h awake) have been associated with a reversal of this trend, with participants showing greater interest in risky activities by the third night without sleep [35]. It is not entirely clear why this upsurge in risky preferences may occur, but it is conceivable that severe extended sleep deprivation may (1) substantially alter judgment, (2) lead to a burst of hypomanic disinhibition due to altered prefrontal functioning, or (3) be an attempt of participants to seek out stimulation as a means to behaviorally induce arousal. Notably, repeated doses of caffeine (200mg every 2h) appeared to be protective against this sudden surge in self-reported risk-seeking. Overall, these findings suggest that short term sleep deprivation (one or two nights) reduces interest in high-risk sensational activities, whereas that interest may show a rebound when sleep deprivation becomes extreme (≥3 nights).

Risky decision-making

While it is clear that sleep loss can lead to altered risk-related perceptions, it is also important to understand how sleep deprivation can affect actual risk-taking behavior. These types of effects are often revealed by gambling or other similar game-like risk tasks. As discussed below, sleep deprivation can lead to altered perception of risk, which will alter behavioral outcomes.

Cognitive framing

Sleep deprivation can affect how a person responds to the way in which a risk is presented to them, a phenomenon known as "framing." In most circumstances, risks can be framed as a potential gain (e.g., would you rather have an 80% chance of winning $4000, or a 100% chance of winning $3000) or as a potential loss (e.g., would you prefer an 80% chance of losing $4000 or a 100% chance of losing $3000). In such cases, it is well established that most people are risk avoiding when considering possible gains (i.e., they would prefer the "sure thing") and risk seeking when considering possible losses (i.e., they would prefer the "long shot") [107]. Interestingly, sleep deprivation appears to shift this basic cognitive bias. For example, in one study using a gambling game, when possible outcomes were described in terms of potential gains, sleep deprivation produced an increase in risk-taking above baseline. However, when possible outcomes were framed as potential losses, sleep deprivation caused participants to become more risk averse than when normally rested [108]. These findings suggest that sleep deprivation modifies the typical framing effect, increasing risk-taking when gains are emphasized and increasing risk-aversion when losses are emphasized, thus magnifying our typical tendencies.

Altered expectations of reward

Sleep deprivation appears to alter the cognitive assessment of risk by changing functioning within brain regions that assign value to objects or situations. For example, one study examined the effects of sleep deprivation on brain activation while participants completed a roulette-style gambling task [109]. During a neuroimaging session, participants completed a series of roulette gambles that ranged from certain wins to highly risky bets. One night of sleep deprivation led to increased activation within the nucleus accumbens during high-risk decisions. This brain structure is involved in the anticipation of rewards and the increased responsiveness of this area following sleep loss suggests that it may be increasing the expected value of the risky bets. Simultaneously, sleep deprivation also blunted activation within the insular cortex during losses. Together, these findings suggest that sleep deprivation alters brain activation in a way that could bias an individual toward risky-behavior (i.e., increasing expectation of rewards and minimizing responses to losses).

The same research team conducted a follow-up study to examine the effects of sleep deprivation on complex reward-based decision-making [110]. Research participants completed a series of trials, some of which focused on gains and others that focused on losses. For instance, during the gain-focused trials, participants could choose to increase the potential amount of money that could be won or increase the probability of winning a particular amount. On the other hand, loss-focused trials permitted the participant to either reduce the amount of money that could be lost or lower the probability of losing a specified amount. Rested individuals showed a bias toward minimizing losses, but this pattern shifted toward maximizing gain after sleep deprivation. These changes were associated with increased activation of reward processing regions, including the ventromedial PFC, and a decline in activation of the insular cortex, which is generally associated with aversion and negative affective experiences [110]. Together, these findings suggest that sleep deprivation alters functional activation in brain regions associated with reward and punishment, which may increase the expectation that risky decisions will lead to reward.

Reward-based learning

Poor decision-making is often characterized by a preference for short-term gains that ultimately lead to long-term losses. Everyday life is full of choices that involve deciding whether to forgo immediate satisfaction in service of longer lasting benefits. One experimental paradigm that seems to get to the heart of these kinds of decisions is the Iowa Gambling Task (IGT), a computerized gambling game-like task that involves selecting cards from four decks with varied, but unstipulated, payout schedules. Two of the decks are high risk because of their widely variable payouts that lead to a net loss, and two of the decks are low risk because they have very consistent but small payouts that reliably lead to a net gain. When healthy normal individuals play this game, they rapidly learn to maximize long-term profits over short-term gains by sticking with the low risk decks. In contrast, patients with focal lesions to the ventromedial PFC, a region critical to learning from rewards and punishments, tend to become selectively attracted to the short-term gains associated with the high-risk decks, which eventually leads them to progressively lose money throughout the course of the game [111, 112].

The IGT has been studied in several studies of sleep deprivation, which have consistently demonstrated that lack of sleep is associated with a pattern of performance that is qualitatively similar to that of patients with lesions in the ventromedial region of the PFC [113–115]. In short, sleep deprivation leads to a short-term focus on immediate gains to the detriment of longer-term outcomes, a pattern that appears to be more severe with greater durations of sleep deprivation. This effect is mediated, in part, by the DAT1 polymorphism, such that individuals with the 9-repeat allele have elevated responsivity to gain anticipations [116]. Interestingly, stimulant countermeasures such as caffeine, modafinil, and dextroamphetamine have not been effective at restoring performance on the IGT (Fig. 26.11), despite normalizing performance on psychomotor vigilance [113, 114]. This lack of effect of stimulants suggests that the deficits on the IGT are probably not due to problems with attention and vigilance and are brought about by alteration

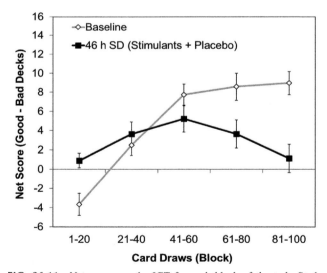

FIG. 26.11 Net scores on the IGT for each block of the task. Study participants performed the task following 46 h of extended wakefulness. Stimulants (600 mg caffeine, 20 mg dextroamphetamine, 400 mg modafinil, or placebo) were administered at 44 h wakefulness. Stimulants did not affect IGT performance, and are thus grouped with the placebo group here. Stimulants (*black squares*) were not effective at restoring IGT performance back to baseline levels (*white diamonds*). *Modified from Killgore WDS, Grugle NL, Balkin TJ. Gambling when sleep deprived: don't bet on stimulants. Chronobiol Int 2012;29(1):43–54, with permission from Taylor & Francis Ltd. (http://www.informaworld.com).*

in the process of integrating information about rewards and punishments with ongoing decision-making processes. Furthermore, we also recently showed that daytime sleepiness reduces the psychological weight that individuals give to more temporally distant versus more recent trials on the IGT in their decision-making strategy [117]. These data suggest that sleepiness may shorten the "time horizon" over which decision information is integrated into the decision-making process. Thus, a sleep deprived individual could also make risky choices because they base decisions on a limited amount of information.

Impulsive behavior

While risk taking often involves deciding between high and low risk options, another form of risk-taking involves "pressing one's luck" beyond the point where the benefits of success are outweighed by the costs of failure. A task that assesses the tendency to push the limits and behave impulsively is known as the Balloon Analog Risk Task (BART). The BART is a computerized task that presents a series of 30 virtual balloons that must be inflated on the screen to win money. To inflate each balloon, the participant presses the spacebar on the keyboard. With each key press, or "pump," the balloon increases in size slightly and gains an additional five cents in value. The larger the balloon becomes, the greater its potential monetary value. The participant can "bank" the accumulated value of a balloon at any time, as long as it has not exploded. If the balloon is inflated too much, the balloon will explode and all accumulated value for that balloon will be lost. Each balloon has a different breaking point that is not known to the participant. In order to win the most money possible, the participant must make a judgment about how much to inflate each balloon and then attempt to cash in before reaching the unknown explosion point. A commonly used output variable from this task is mean number of key presses for the unexploded balloon trials (i.e., those trials that were "banked" without popping the balloon), which is commonly known as the Adjusted Average Number of Pumps. Some studies have also calculated a "Cost/Benefit Ratio," which considers both the Cost (i.e., proportion of exploded balloons) versus the Benefit (i.e., the proportion of all potential money that was actually banked) [35, 104, 105]. Higher Cost/Benefit Ratio scores suggest greater risk-taking.

Several sleep deprivation studies have used the BART to examine the effects of sleep loss on risky behavior. The first published study to examine the BART during sleep deprivation showed that one night without sleep led to a decline in the Cost/Benefit Ratio, suggesting a tendency toward less behavioral risk-taking [104]. A second study published around the same time also found that a single night of sleep deprivation was associated with reduced risk-taking (i.e., lower Adjusted Average Number of Pumps) among women but not men [118]. Later work further confirmed that risk taking on the BART was reduced with two nights of sleep deprivation, but was returned to baseline levels with a 20 mg dose of dextroamphetamine, but not by similarly alerting doses of 400 mg modafinil, or 600 mg of caffeine [105]. In contrast, Killgore and colleagues found that BART Cost/Benefit scores were generally unaffected by two nights of sleep deprivation, but this was followed by a surge in behavioral risk-taking after three nights without sleep [35]. The cause of this surge in risk-taking after extreme sleep deprivation is not clear, but it is possible that inhibitory capacity eventually fails after several nights awake, or that the increased risk taking is simply a way for participants to stimulate arousal [35].

The fact that the BART typically shows reduced risk-taking during sleep loss seems to stand in contradiction to the increased risk-taking that is consistently found on the IGT. One explanation for this discrepancy may involve the difference in effort required by these two tasks [104]. While both the IGT and BART involve risky decision-making, risky choices on the IGT require no more effort than the safe options (i.e., a single button press is required regardless of which option is selected), while greater risk taking on the BART requires the expenditure of additional physical and cognitive effort (i.e., more button presses are required to be risky, while fewer button presses are safer). This explanation was given further support by a study that showed that sleep deprivation leads to "effort discounting," a willingness to accept less reward if it requires only minimal effort rather than expend greater effort to obtain higher value rewards [119]. Sleep deprived individuals appear to be less willing to expend effort to engage in risky activities.

Aggressive/punitive responses

Negative mood states are common during sleep deprivation and evidence suggests that individuals may become more easily frustrated by even minor hassles or interpersonal slights. For instance, sleep deprivation appears to increase the willingness to blame others for frustrating problems and makes people less willing to work with others to achieve mutually satisfying outcomes [120]. Without sleep, people often feel picked on or targeted for persecution [121]. Sometimes, this can even lead to aggressive behaviors [122]. In one study, participants played a series of "bargaining" and "trust" games that required them to interact with other players to earn various levels of money [123]. On these games, sleep deprivation increased the tendency to engage in aggressive exchanges with the other players. Moreover, sleep deprived individuals were less trusting of their partners and more often rejected monetary offers that were perceived as unfair, even when rejecting the offer would come at a financial cost to themselves. Sleep deprivation appears to have an adverse effect on trust and normal social discourse.

Moral judgment

Our stable moral precepts and beliefs dictate our responses to difficult situations where the appropriate decision is not obvious. A few studies have demonstrated moral judgment and moral decision-making can be affected by sleep deprivation. In the earliest published study to examine moral judgment following sleep loss, participants completed a series of moral and non-moral dilemmas when fully rested and again following 53 h of sleep deprivation [124]. The findings showed that most decision-making processes were relatively unaffected by sleep deprivation, including non-moral decisions and moral decisions that were generally low in emotional intensity. However, sleep deprivation appeared to significantly slow responses to difficult moral decisions that involved high levels of emotional conflict. Compared to the speed of decisions at baseline, the time to respond to emotionally challenging situations was much slower, suggesting that sleep deprivation does not affect all decisions equally—sleep deprivation specifically impairs the ability to make emotionally based decisions. Moreover, sleep deprivation also altered the qualitative direction of the judgments. Specifically, sleep deprived individuals were more likely to make utilitarian type judgments that violated their own moral beliefs compared to when they were well rested [124]. However, this effect was not significant in a second study of only a single night of sleep loss [125], suggesting that deficits in moral judgments may only emerge with prolonged periods without sleep. Other evidence also suggests that moral reasoning may be affected by partial sleep restriction as well. For instance, when sleep was restricted to approximately 2.5 h per night over 5 days, military personnel showed significant reductions in principle-oriented moral reasoning [126]. Among this sample of military cadets, their moral decisions became more rules-focused and self-oriented over the course of sleep restriction, and they showed progressively greater difficulty with higher-level principle-oriented reasoning. Overall, it appears clear that sleep deprivation affects the speed and quality of moral decisions and judgments.

PRACTICAL IMPLICATIONS

Extreme cases of acute sleep deprivation often fail to be generalizable to real-word situations. However, a common occurrence in everyday life is that of chronic sleep restriction. It is common for individuals to repeatedly, and regularly, obtain insufficient amounts of sleep. Individuals that are chronically sleep-restricted often have difficulty with day-to-day tasks, and they may not even realize it until it is too late. For example, individuals that are leaving a night shift often drive drowsy. While drowsy driving may not always lead to a direct negative consequence, under the right circumstances the results can be catastrophic due to the unpredictable nature of sleep loss induced impairments. For instance, if an individual experiences even a single lapse in attention at the same moment a stop light turns red, the result could be disastrous. Further, the inability to effectively inhibit responses and make rational decisions can significantly impact work performance and interpersonal relationships. As sleep loss impairs inhibitory control, individuals that are sleep-restricted may make decisions or act in ways that are out of character or inappropriate to the context due to a diminished capacity to inhibit responses. This could prove harmful to an individual's social or professional reputation or could damage close interpersonal relationships. Additionally, sleep loss alters reward expectation in such a way that individuals do not realize the consequences of their actions, as they expect to be rewarded by their choices, regardless of the quality of the decisions. This exaggerated expectation of reward may lead individuals to make risky decisions, which could affect economic choices such as gambling or selecting risky investments. Of course, these same unrealistic expectations of reward could potentially affect other behaviors as well, including social, interpersonal, and professional decisions. Overall, it is important for individuals to recognize the range of cognitive consequences of sleep loss and the downstream effects that insufficient sleep can have on basic day-to-day activities and interpersonal relationships.

Often times, the effects of sleep loss can be mitigated with effective countermeasure strategies, including caffeine and strategic napping [6], although there is some evidence that widely used stimulants, such as caffeine, may improve some executive functions [35, 127, 128] while having no discernable effects on others [113, 114]. However, many of the negative consequences associated with sleep loss can be avoided all together with a proactive approach. Those who are often faced with chronic sleep loss should take the steps necessary to educate themselves on causes and consequences of fatigue. It is also important to become educated about and implement proper sleep hygiene techniques, including making sleep a priority, standardizing sleep schedules, creating a good sleep environment, and "unplugging" from technology and other forms of stimulation at least 30 min before bed [129]. Additionally, individuals should attempt to align lifestyle choices with their work and social schedules to maximize sleep opportunities [6].

CONCLUSIONS

Sleep loss appears to have a multifaceted impact on both neural and behavioral measures. These impacts affect both global and domain-specific aspects of cognition. This, in part, is due to the differential responsiveness of the interconnected brain regions underlying each specific cognitive task. Sleep loss consistently impairs vigilant attention performance, resulting in increased lapses of attention and slowed response times. However, the literature is mixed as to if, and how, insufficient sleep influences performance on higher-order executive function and decision-making tasks. For example, the executive and non-executive components of working memory are differentially impacted by sleep loss. Further, other complex executive functions, such as cognitive control, are negatively impacted by sleep loss, yet impaired vigilant attention does not seem to be the underlying cause. For some higher-level tasks that involve judgment and decision-making, the effects of sleep loss on emotional systems may be particularly important. While research into the underlying mechanisms of cognitive impairment due to sleep loss has increased in recent years, more work is needed in order to fully elucidate how sleep loss specifically impacts various aspects of cognition. Identification of task-specific impairments, and the mechanisms subserving these impairments, can aid in the development of appropriate countermeasures and fatigue risk management strategies for those most at risk for experiencing chronic and acute sleep loss.

REFERENCES

[1] Center for Disease Control and Prevention. Insufficient sleep is a public health epidemic [Internet]. Insufficient sleep is a public health problem. Available from: http://www.cdc.gov/features/dssleep/; 2015.

[2] Bayon V, Leger D, Gomez-Merino D, Vecchierini MF, Chennaoui M. Sleep debt and obesity. Ann Med 2014;46(5):264–72.

[3] Hirshkowitz M, Whiton K, Albert SM, Alessi C, Bruni O, DonCarlos L, Hazen N, Herman H, Katz ES, Kheirandish-Gozal L, Neubauer DN, O'Donnel AE, Ohayon M, Peever J, Rawding R, Sachdeva RC, Setters B, Vitiello MV, Ware JC, Adams Hillard PJ. National sleep foundation's sleep time duration recommendations: methodology and results summary. Sleep Heal 2015;1(1):40–3.

[4] Borbély AA. A two process model of sleep regulation. Hum Neurobiol 1982;1(3):195–204.

[5] Daan S, Beersma DG, Borbély AA. Timing of human sleep: recovery process gated by a circadian pacemaker. Am J Physiol 1984;246:R161–83.

[6] Satterfield BC, Van Dongen HPA. Occupational fatigue, underlying sleep and circadian mechanisms, and approaches to fatigue risk management. Fatigue Biomed Heal Behav 2013;246(3):118–36.

[7] Dinges DF, Powell JW. Microcomputer analyses of performance on a portable, simple visual RT task during sustained operations. Behav Res Methods Instrum Comput 1985;17(6):652–5.

[8] Lim J, Dinges DF. Sleep deprivation and vigilant attention. Ann N Y Acad Sci 2008;1129:305–22.

[9] Dorrian J, Rogers NL, Dinges DF. Psychomotor vigilance performance: neurocognitive assay sensitive to sleep loss. In: Kushida CA, editor. Sleep deprivation: clinical issues, pharmacology and sleep loss effects. New York: Marcell Dekker; 2005. p. 39–70.

[10] Doran SM, Van Dongen HPA, Dinges DF. Sustained attention performance during sleep deprivation: evidence of state instability. Arch Ital Biol 2001;139(3):253–67.

[11] Lim J, Dinges DF. A meta-analysis of the impact of short-term sleep deprivation on cognitive variables. Psychol Bull 2010;136(3):375–89.

[12] Drummond SPA, Bischoff-Grethe A, Dinges DF, Ayalon L, Mednick SC, Meloy MJ. The neural basis of the psychomotor vigilance task. Sleep 2005;28(9):1059–68.

[13] Wesensten NJ, Killgore WDS, Balkin TJ. Performance and alertness effects of caffeine, dextroamphetamine, and modafinil during sleep deprivation. J Sleep Res 2005;14(3):255–66.

[14] Cajochen C, Khalsa SB, Wyatt JK, Czeisler CA, Dijk DJ. EEG and ocular correlates of circadian melatonin phase and human performance decrements during sleep loss. Am J Physiol 1999;277:R640–9.

[15] Belenky G, Wesensten NJ, Thorne DR, Thomas ML, Sing HC, Redmond DP, Russo MB, Balkin TJ. Patterns of performance degradation and restoration during sleep restriction and subsequent recovery: a sleep dose-response study. J Sleep Res 2003;12(1):1–12.

[16] Van Dongen HPA, Maislin G, Mullington JM, Dinges DF. The cumulative cost of additional wakefulness: dose-response effects on neurobehavioral functions and sleep physiology from chronic sleep restriction and total sleep deprivation. Sleep 2003;26(2):117–26.

[17] Chee MWL, Tan JC. Lapsing when sleep deprived: neural activation characteristics of resistant and vulnerable individuals. Neuroimage 2010;51(2):835–43.

[18] Van Dongen HPA, Belenky G, Krueger JM. A local, bottom-up perspective on sleep deprivation and neurobehavioral performance. Curr Top Med Chem 2011;11(19):2414–22.

[19] Bills AG. Blocking: a new principle of mental fatigue. Am J Psychol 1931;43(2):230–45.

[20] Kribbs NB, Dinges DF. Vigilance decrement and sleepiness. In: Ogilvie RD, Harsh JR, editors. Sleep onset: normal and abnormal processes. Washington, DC: American Psychological Association; 1994. p. 113–25.

[21] Van Dongen HPA, Dinges DF. Investigating the interaction between the homeostatic and circadian processes of sleep-wake regulation for the prediction of waking neurobehavioural performance. J Sleep Res 2003;12(3):181–7.

[22] Bills AG. Fatigue in mental work. Physiol Rev 1937;17:436–53.

[23] Van Dongen HPA, Belenky G, Krueger JM. Investigating the temporal dynamics and underlying mechanisms of cognitive fatigue. In: Ackerman PL, editor. Cognitive fatigue: mulitidisciplinary perspectives on current research and future applications. Washington, DC: American Psychological Association; 2011. p. 127–47.

[24] Van Dongen HPA, Dinges DF. Sleep, circadian rhythms, and psychomotor vigilance. Clin Sports Med 2005;24:237–49.

[25] Van Dongen HPA, Belenky G. Individual differences in vulnerability to sleep loss in the work environment. Ind Health 2009;47(5):518–26.

[26] Krueger JM, Rector DM, Roy S, Van Dongen HPA, Belenky G, Panksepp J. Sleep as a fundamental property of neuronal assemblies. Nat Rev Neurosci 2008;9(12):910–9.

[27] Wilkinson RT. Interaction of lack of sleep with knowledge of results, repeated testing, and individual differences. J Exp Psychol 1961;62(3):263–71.

[28] Leproult R, Colecchia EF, Berardi AM, Stickgold R, Kosslyn SM, Van Cauter E. Individual differences in subjective and objective alertness during sleep deprivation are stable and unrelated. Am J Physiol Regul Integr Comp Physiol 2003;284:R280–90.

[29] Tucker AM, Dinges DF, Van Dongen HPA. Trait interindividual differences in the sleep physiology of healthy young adults. J Sleep Res 2007;16(2):170–80.

[30] Grant DA, HPA VD. Individual differences in sleep duration and responses to sleep loss. In: Shaw P, Tafti M, Thorpy M, editors. The genetic basis of sleep and sleep disorders. Cambridge: Cambridge University Press; 2013. p. 189–96.

[31] Rupp TL, Wesensten NJ, Balkin TJ. Trait-like vulnerability to total and partial sleep loss. Sleep 2012;35(8):1163–72.

[32] Van Dongen HPA, Baynard MD, Maislin G, Dinges DF. Systematic interindividual differences in neurobehavioral impairment from sleep loss: evidence of trait-like differential vulnerability. Sleep 2004;27(3):423–33.

[33] Van Dongen HPA, Maislin G, Dinges DF. Dealing with interindividual differences in the temporal dynamics of fatigue and performance: importance and techniques. Aviat Space Environ Med 2004;75(Suppl 3):A147–54.

[34] Killgore WDS, Richards JM, Killgore DB, Kamimori GH, Balkin TJ. The trait of introversion-extraversion predicts vulnerability to sleep deprivation. J Sleep Res 2007;16(4):354–63.

[35] Killgore WDS, Kamimori GH, Balkin TJ. Caffeine protects against increased risk-taking propensity during severe sleep deprivation. J Sleep Res 2011;20(3):395–403.

[36] Chee MWL, Van Dongen HPA. Functional imaging of inter-individual differences in response to sleep deprivation. In: Nofzinger E, Maquet P, Thorpy M, editors. Neuroimaging of sleep and sleep disorders. Cambridge: Cambridge University Press; 2013. p. 154–62.

[37] Goel N. Genetic markers of sleep and sleepiness. Sleep Med Clin 2017;12(3):289–99.

[38] Mu Q, Mishory A, Johnson KA, Nahas Z, Kozel FA, Yamanaka K, Bohning DE, George MS. Decreased brain activation during a working memory task at rested baseline is associated with vulnerability to sleep deprivation. Sleep 2005;28(4):433–46.

[39] Mu Q, Nahas Z, Johnson KA, Yamanaka K, Mishory A, Koola J, Hill S, Horner MD, Bohning DE, George MS. Decreased cortical response to verbal working memory following sleep deprivation. Sleep 2005;28(1):55–67.

[40] Lythe KE, Williams SCR, Anderson C, Libri V, Mehta MA. Frontal and parietal activity after sleep deprivation is dependent on task difficulty and can be predicted by the fMRI response after normal sleep. Behav Brain Res 2012;233(1):62–70.

[41] Yeo BTT, Tandi J, Chee MWL. NeuroImage functional connectivity during rested wakefulness predicts vulnerability to sleep deprivation. Neuroimage 2015;111:147–58.

[42] Cui J, Tkachenko O, Gogel H, Kipman M, Preer LA, Weber M, Divatia SC, Demers LA, Olson EA, Buchholz JL, Bark JK, Rosso IM, Rauch SL, WDS K. Microstructure of frontoparietal connections predicts individual resistance to sleep deprivation. Neuroimage 2015;106:123–33.

[43] Goel N. Neurobehavioral effects and biomarkers of sleep loss in healthy adults. Curr Neurol Neurosci Rep 2017;17(11):89.

[44] Bodenmann S, Hohoff C, Freitag C, Deckert J, Rétey JV, Bachmann V, Landolt H-P. Polymorphisms of ADORA2A modulate psychomotor vigilance and the effects of caffeine on neurobehavioural performance and sleep EEG after sleep deprivation. Br J Pharmacol 2012;165(6):1904–13.

[45] Satterfield BC, Wisor JP, Schmidt MA, Van Dongen HPA. Time-on-task effect during sleep deprivation in healthy young adults is modulated by dopamine transporter genotype. Sleep 2017;40(12):zsx167.

[46] Holst SC, Müller T, Valomon A, Seebauer B, Berger W, Landolt H-P. Functional polymorphisms in dopaminergic genes modulate neurobehavioral and neurophysiological consequences of sleep deprivation. Sci Rep 2017;7:45982.

[47] Satterfield BC, Wisor JP, Field SA, Schmidt MA, Van Dongen HPA. TNFα G308A polymorphism is associated with resilience to sleep deprivation-induced psychomotor vigilance performance impairment in healthy young adults. Brain Behav Immun 2015;47:66–74.

[48] Miller EK, Wallis JD. Executive function and higher-order cognition: definition and neural substrates. Encycl Neurosci 2010;4:99–104.

[49] Harrison Y, Horne JA. The impact of sleep deprivation on decision making: a review. J Exp Psychol Appl 2000;6(3):236–49.

[50] Thomas ML, Sing HC, Belenky G, Holcomb HH, Mayberg HS, Dannals RF, Wagner Jr. J, Thorne D, Popp K, Rowland L, Welsh A, Balwinski S, Redmond D. Neural basis of alertness and cognitive performance impairments during sleepiness I. Effects of 48 and 72 h of sleep deprivation on waking human regional brain activity. J Sleep Res 2000;9:335–52.

[51] Wu JC, Gillin JC, Buchsbaum MS, Chen P, Keator DB, Wu NK, Darnall LA, Fallon JH, Bunney WE. Frontal lobe metabolic decreases with sleep deprivation not totally reversed by recovery sleep. Neuropsychopharmacology 2006;31(12):2783–92.

[52] Van Dongen HPA, Bender AM, Dinges DF. Systematic individual differences in sleep homeostatic and circadian rhythm contributions to neurobehavioral impairment during sleep deprivation. Accid Anal Prev 2012;45(Suppl):11–6.

[53] Jackson ML, Gunzelmann G, Whitney P, Hinson JM, Belenky G, Rabat A, Van Dongen HPA. Deconstructing and reconstructing cognitive performance in sleep deprivation. Sleep Med Rev 2013;17(3):215–25.

[54] Tucker AM, Whitney P, Belenky G, Hinson JM, Van Dongen HPA. Effects of sleep deprivation on dissociated components of executive functioning. Sleep 2010;33(1):47–57.

[55] Reichert C, Maire M, Schmidt C, Cajochen C. Sleep-wake regulation and its impact on working memory performance: the role of adenosine. Biology 2016;5(1):11.

[56] Frenda SJ, Fenn KM. Sleep less, think worse: the effect of sleep deprivation on working memory. J Appl Res Mem Cogn 2016;5(4):463–9.

[57] Baddeley AD. Working memory: theories, models, and controversies. Annu Rev Psychol 2012;63:1–29.

[58] Chee MWL, Choo WC. Functional imaging of working memory after 24 hr of total sleep deprivation. J Neurosci 2004;24(19):4560–7.

[59] Chee MWL, Chuah LYM, Venkatraman V, Chan WY, Philip P, Dinges DF. Functional imaging of working memory following normal sleep and after 24 and 35 h of sleep deprivation: correlations of fronto-parietal activation with performance. Neuroimage 2006;31(1):419–28.

[60] Choo WC, Lee WW, Venkatraman V, Sheu FS, Chee MWL. Dissociation of cortical regions modulated by both working memory load and sleep deprivation and by sleep deprivation alone. Neuroimage 2005;25(2):579–87.

[61] Bell-McGinty S, Habeck C, Hilton HJ, Rakitin B, Scarmeas N, Zarahn E, Flynn J, DeLaPaz R, Basner R, Stern Y. Identification and differential vulnerability of a neural network in sleep deprivation. Cereb Cortex 2004;14(5):496–502.

[62] Drummond SPA, Meloy MJ, Yanagi MA, Orff HJ, Brown GG. Compensatory recruitment after sleep deprivation and the relationship with performance. Psychiatry Res Neuroimaging 2005;140(3):211–23.

[63] Groeger J, Viola A, Lo J, von Schantz M, Archer SN, Dijk D-J. Early morning executive functioning during sleep deprivation is compromised by a PERIOD3 polymorphism. Sleep 2008;31(8):1159–67.

[64] Drummond SPA, Anderson DE, Straus LD, Vogel EK, Perez VB. The effects of two types of sleep deprivation on visual working memory capacity and filtering efficiency. PLoS One 2012;7(4):1–8.

[65] Rångtell FH, Karamchedu S, Andersson P, Liethof L, Olaya Búcaro M, Lampola L, Schiöth HB, Cedernaes H, Benedict C. A single night of sleep loss impairs objective but not subjective working memory performance in a sex-dependent manner. J Sleep Res 2019;28:1–8.

[66] Santhi N, Lazar AS, McCabe PJ, Lo JC, Groeger JA, Dijk D-J. Sex differences in the circadian regulation of sleep and waking cognition in humans. Proc Natl Acad Sci 2016;113(19):E2730–9.

[67] Chee MWL, Chuah YML. Functional neuroimaging and behavioral correlates of capacity decline in visual short-term memory after sleep deprivation. Proc Natl Acad Sci 2007;104(22):9487–92.

[68] Bari A, Robbins TW. Inhibition and impulsivity: behavioral and neural basis of response control. Prog Neurobiol 2013;108:44–79.

[69] Taylor SF, Stern ER, Gehring WJ. Neural systems for error monitoring: recent findings and theoretical perspectives. Neuroscientist 2007;13(2):160–72.

[70] Chuah YML, Venkatraman V, Dinges DF, Chee MWL. The neural basis of interindividual variability in inhibitory efficiency after sleep deprivation. J Neurosci 2006;26(27):7156–62.

[71] Drummond SPA, Paulus MP, Tapert SF. Effects of two nights sleep deprivation and two nights recovery sleep on response inhibition. J Sleep Res 2006;15(3):261–5.

[72] Sagaspe P, Taillard J, Amieva H, Beck A, Rascol O, Dartigues J-F, Capelli A, Philip P. Influence of age, circadian and homeostatic processes on inhibitory motor control: a Go/Nogo task study. PLoS One 2012;7(6):e39410.

[73] Demos KE, Hart CN, Sweet LH, Mailloux KA, Trautvetter J, Williams SE, Wing RR, McCaffery JM. Partial sleep deprivation impacts impulsive action but not impulsive decision-making. Physiol Behav 2016;164:214–9.

[74] Killgore WDS. Effects of sleep deprivation on cognition. In: Kerkhof G, HPA VD, editors. Progress in brain research. Elsevier B.V.; 2010. p. 105–29.

[75] Braver TS. The variable nature of cognitive control: a dual mechanisms framework. Trends Cogn Sci 2012;16(2):106–13.

[76] Satterfield BC, Hinson JM, Whitney P, Schmidt MA, Wisor JP, Van Dongen HPA. Catechol-O-methyltransferase (COMT) genotype affects cognitive control during total sleep deprivation. Cortex 2017;99:179–86.

[77] Cools R. The costs and benefits of brain dopamine for cognitive control. WIREs Cogn Sci 2016;7:317–29.

[78] Klanker M, Feenstra M, Denys D, Lo C, Tsing N. Dopaminergic control of cognitive flexibility in humans and animals. Front Neurosci 2013;7:201.

[79] Botvinick M, Braver TS. Motivation and cognitive control: from behavior to neural mechanism. Annu Rev Psychol 2015;66:83–113.

[80] Cools R, Robbins TW. Chemistry of the adaptive mind. Philos Trans R Soc A Math Phys Eng Sci 2004;362(1825):2871–88.

[81] Nakashima A, Bouak F, Lam Q, Smith I, Vartanian O. Task switching following 24 h of total sleep deprivation. Neuroreport 2018;29(2):123–7.

[82] Couyoumdjian A, Sdoia S, Tempesta D, Curcio G, Rastellini E, De Gennaro L, Ferrara M. The effects of sleep and sleep deprivation on task-switching performance. J Sleep Res 2010;19:64–70.

[83] Haavisto ML, Porkka-Heiskanen T, Hublin C, Härmä M, Mutanen P, Müller K, Virkkala J, Sallinen M. Sleep restriction for the duration of a work week impairs multitasking performance: sleep restriction and multitasking. J Sleep Res 2010;19(3):444–54.

[84] Sagaspe P, Sanchez-Ortuno M, Charles A, Taillard J, Valtat C, Bioulac B, Philip P. Effects of sleep deprivation on color-word, emotional, and specific stroop interference and on self-reported anxiety. Brain Cogn 2006;60(1):76–87.

[85] Cain SW, Silva EJ, Chang AM, Ronda JM, Duffy JF. One night of sleep deprivation affects reaction time, but not interference or facilitation in a stroop task. Brain Cogn 2011;76(1):37–42.

[86] Bratzke D, Steinborn MB, Rolke B, Ulrich R. Effects of sleep loss and circadian rhythm on executive inhibitory control in the stroop and simon tasks. Chronobiol Int 2012;29(1):55–61.

[87] Grant LKL, Cain SW, Chang A-M, Saxena R, Czeisler CA, Andersaon C. Impaired cognitive flexibility during sleep deprivation among carriers of the brain derived neurotrophic factor (BDNF) Val66Met allele. Behav Brain Res 2018;338:51–5.

[88] Gevers W, Deliens G, Hoffmann S, Notebaert W, Peigneux P. Sleep deprivation selectively disrupts top-down adaptation to cognitive conflict in the stroop test. J Sleep Res 2015;24(6):666–72.

[89] Frank MJ, Claus ED. Anatomy of a decision: Striato-orbitofrontal interactions in reinforcement learning, decision making, and reversal. Psychol Rev 2006;113(2):300–26.

[90] Fellows LK. The role of orbitofrontal cortex in decision making: a component process account. Ann N Y Acad Sci 2007;1121(1):421–30.

[91] Pauli WM, Hazy TE, O'Reilly RC.. Expectancy, ambiguity, and behavioral flexibility: separable and complementary roles of the orbital frontal cortex and amygdala in processing reward expectancies. J Cogn Neurosci 2011;24(2):351–66.

[92] Cools R, Clark L, Owen AM, Robbins TW. Defining the neural mechanisms of probabilistic reversal learning using event-related functional magnetic resonance imaging. J Neurosci 2002;22(11):4563–7.

[93] Whitney P, Hinson JM, Jackson ML, Van Dongen HPA. Feedback blunting: total sleep deprivation impairs decision making that requires updating based on feedback. Sleep 2015;38(5):745–54.

[94] Whitney P, Hinson JM, Satterfield BC, Grant DA, Honn KA, Van Dongen HPA. Sleep deprivation diminishes attentional control effectiveness and impairs flexible adaptation to changing conditions. Sci Rep 2017;7(1):16020.

[95] Horne JA. Sleep loss and "divergent" thinking ability. Sleep 1988;11(6):528–36.

[96] Cade BE, Gottlieb DJ, Lauderdale DS, Bennett DA, Buchman AS, Buxbaum SG, De Jager PL, Evans DS, Fülöp T, Gharib SA, Johnson WC, Kim H, Larkin EK, Lee SK, Lim AS, Punjabi NM, Shin C, Stone KL, Tranah GJ, Weng J, Yaffe K, Zee PC, Patel SR, Zhu X, Redline S, Saxena R. Common variants in DRD2 are associated with sleep duration: the CARe consortium. Hum Mol Genet 2016;25(1):167–79.

[97] Jones K, Harrison Y. Frontal lobe function, sleep loss and fragmented sleep. Sleep Med Rev 2001;5(6):463–75.

[98] Harrison Y, Horne JA. One night of sleep loss impairs innovative thinking and flexible decision making. Organ Behav Hum Decis Process 1999;78(2):128–45.

[99] Hoyle RH, Stephenson MT, Palmgreen P, Lorch EP, Donohew RL. Reliability and validity of a brief measure of sensation seeking. Pers Individ Dif 2002;32(3):401–14.

[100] Killgore WDS, Vo A, Castro C, Hoge C. Assessing risk propensity in American soldiers: preliminary reliability and validity of the evaluation of risks (EVAR) scale—english version. Mil Med 2006;171(3):233–9.

[101] Sicard B, Jouve E, Blin O. Risk propensity assessment in military special operations. Mil Med 2001;166(10):871–4.
[102] Killgore WDS, Grugle NL, Killgore DB, Balkin TJ. Sex differences in self-reported risk-taking propensity on the evaluation of risks scale. Psychol Rep 2010;106(3):693–700.
[103] Chaumet G, Taillard J, Sagaspe P, Pagani M, Dinges DF, Pavy-Le-Traon A, Bareille MP, Rascol O, Philip P. Confinement and sleep deprivation effects on propensity to take risks. Aviat Space Environ Med 2009;80(2):73–80.
[104] Killgore WDS. Effects of sleep deprivation and morningness-eveningness traits on risk-taking. Psychol Rep 2007;100:613–26.
[105] Killgore WDS, Grugle NL, Killgore DB, Leavitt BP, Watlington GI, McNair S, Balkin TJ. Restoration of risk-propensity during sleep deprivation: caffeine, dextroamphetamine, and modafinil. Aviat Space Environ Med 2008;79(9):867–74.
[106] Dinges DF, Pack F, Williams K, Gillen K, Powell J, Ott G, Aptowicz C, Pack AI. Cumulative sleepiness, mood disturbance, and psychomotor vigilance performance decrements during a week of sleep restricted to 4–5 hours per night. Sleep 1997;20(4):267–77.
[107] Kahneman D. A perspective on judgment and choice: mapping bounded rationality. Am Psychol 2003;58(9):697–720.
[108] Mckenna BS, Dickinson DL, Orff HJ, Drummond SPA. The effects of one night of sleep deprivation on known-risk and ambiguous-risk decisions. J Sleep Res 2007;16(3):245–52.
[109] Venkatraman V, Chuah LY, Huettel SA, Chee MW. Sleep deprivation elevates expectation of gains and attenuates response to losses following risky decisions. Sleep 2007;30(5):603–9.
[110] Venkatraman V, Huettel SA, Chuah LYM, Payne JW, Chee MWL. Sleep deprivation biases the neural mechanisms underlying economic preferences. J Neurosci 2011;31(10):3712–8.
[111] Bechara A, Damasio H, Tranel D, Damasio AR. Deciding advantageously before knowing the advantageous strategy. Science 1997;275(5304):1293–5.
[112] Bechara A, Damasio AR, Damasio H, Anderson SW. Insensitivity to future consequences following damage to human prefrontal cortex. Cognition 1994;50(1–3):7–15.
[113] Killgore WDS, Grugle NL, Balkin TJ. Gambling when sleep deprived: don't bet on stimulants. Chronobiol Int 2012;29(1):43–54.
[114] Killgore WDS, Lipizzi EL, Kamimori GH, Balkin TJ. Caffeine effects on risky decision making after 75 hours of sleep deprivation. Aviat Sp Environ Med 2007;78(10):957–62.
[115] Killgore WDS, Balkin TJ, Wesensten NJ. Impaired decision making following 49 hours of sleep deprivation. J Sleep Res 2006;15(1):7–13.
[116] Greer SM, Goldstein AN, Knutson B, Walker MP. A genetic polymorphism of the human dopamine transporter determines the impact of sleep deprivation on brain responses to rewards and punishments. J Cogn Neurosci 2016;28(6):803–10.
[117] Olson EA, Weber M, Rauch SL, Killgore WDS. Daytime sleepiness is associated with reduced integration of temporally distant outcomes on the Iowa gambling task. Behav Sleep Med 2016;14(2):200–11.
[118] Acheson A, Richards JB, de Wit H. Effects of sleep deprivation on impulsive behaviors in men and women. Physiol Behav 2007;91(5):579–87.
[119] Libedinsky C, Massar SAA, Ling A, Chee W, Huettel SA, Chee MWL. Sleep deprivation alters effort discounting but not delay discounting of monetary rewards. Sleep 2013;36(6):899–904.
[120] Kahn-Greene ET, Lipizzi EL, Conrad AK, Kamimori GH, Killgore WDS. Sleep deprivation adversely affects interpersonal responses to frustration. Pers Individ Dif 2006;41(8):1433–43.
[121] Kahn-Greene ET, Killgore DB, Kamimori GH, Balkin TJ, Killgore WDS. The effects of sleep deprivation on symptoms of psychopathology in healthy adults. Sleep Med 2007;8(3):215–21.
[122] Kamphuis J, Meerlo P, Koolhaas JM, Lancel M. Poor sleep as a potential causal factor in aggression and violence. Sleep Med 2012;13(4):327–34.
[123] Anderson C, Dickinson DL. Bargaining and trust: the effects of 36-h total sleep deprivation on socially interactive decisions. J Sleep Res 2010;19:54–63.
[124] Killgore WDS, Killgore DB, Day LM, Li C, Kamimori GH, Balkin TJ. The effects of 53 hours of sleep deprivation on moral judgement. Sleep 2007;30(3):345–52.
[125] Tempesta D, Couyoumdjian A, Moroni F, Marzano C, De Gennaro L, Ferrara M. The impact of one night of sleep deprivation on moral judgments. Soc Neurosci 2012;7(3):292–300.
[126] Olsen OK, Pallesen S, Eid J. The impact of partial sleep deprivation on moral reasoning in military officers. Sleep 2010;33(8):1086–90.
[127] Killgore WDS, Kamimori GH, Balkin TJ. Caffeine improves the efficiency of planning and sequencing abilities during sleep deprivation. J Clin Psychopharmacol 2014;34(5):660–2.
[128] Killgore WDS, Kahn-Greene ET, Grugle NL, Killgore DB, Balkin TJ. Sustaining executive functions during sleep deprivation: a comparison of caffeine, dextroamphetamine, and modafinil. Sleep 2009;32(2):205–16.
[129] National Sleep Foundation. Healthy sleep tips. [Internet]. Available from: https://sleepfoundation.org/sleep-tools-tips/healthy-sleep-tips; 2018.

Chapter 27

Sleep and healthy decision making

Kelly Glazer Baron[a,b], Elizabeth Culnan[b]
[a]Division of Public Health, Department of Family and Preventive Medicine, University of Utah, Salt Lake City, UT, United States, [b]Department of Behavioral Sciences, Rush University Medical Center, Chicago, IL, United States

ABBREVIATIONS

CPAP	continuous positive airway pressure
ICSD3	International Classification of Sleep Disorders, 3rd edition
OSA	obstructive sleep apnea

INTRODUCTION

Engaging in healthy behaviors is critical to the prevention of the most prevalent chronic illnesses including cardiovascular disease, diabetes and cancer. For example, a study of adults in the United Kingdom demonstrated adults with poor health behaviors in all four categories studied (smoking, low physical activity, low fruit and vegetable intake and high alcohol intake) had a 3.5-fold mortality rate over a 20 year follow-up compared with those who did not engage in these unhealthy behaviors [1]. The authors suggested that having poor health habits in all four behaviors was equivalent to increasing participants' chronological age by 12 years. Sleep loss and circadian disruption are linked to both poorer health behaviors and the development of chronic illnesses. Therefore, health behaviors are thought to be one of the pathways by which sleep and circadian rhythms influence chronic disease risk. The focus of this chapter is to defining how changes in sleep and circadian rhythms influence the decisions to engage in healthy behaviors (e.g., physical activity) or avoid unhealthy behaviors (e.g., smoking). We will first start with a discussion of how sleep behavior itself is a daily health decision with some controllable components (leisure and social time) and some uncontrollable (sleep disorders, commute times, work hours). Next, we will examine several key behavioral pathways linking sleep loss and circadian disruption to health behaviors. The chapter will focus on examples from four health behaviors: physical activity, diet, smoking and alcohol use. In this chapter we will present what is known about how and why sleep loss and circadian disruption influences these important health behaviors including (1) environmental exposures at night, (2) neurocognitive changes experienced after sleep loss and circadian disruption, (3) affective response to sleep loss and circadian disruption and (4) how sleep loss and circadian disruption affect effort and motivation for health behaviors. Last, we will discuss what is known about whether improving sleep and circadian rhythms may influence the ability to make changes to other health behaviors.

SLEEP AS A HEALTH BEHAVIOR

Influences on sleep and health behaviors

Health behaviors are defined as "overt behavioral patterns, actions and habits that relate to health maintenance, health restoration or health improving" [2]. Sleep is considered a "pillar of health" along with diet and exercise. Sleep is also a unique health behavior in that it involves the physiological ability to sleep as well as volitional processes (opportunity to sleep). According to the 2-process model of sleep regulation [3], sleep propensity, or the physiologic ability to sleep, is determined by the homeostatic pressure to sleep (process S) as well as circadian timing (process C). In order to sleep well, an individual must have a high drive to sleep and sleep is better quality if it occurs in alignment with the internal circadian rhythm. If either of these processes is impaired, sleep is more difficult to achieve (e.g., low homeostatic pressure due to napping or mistimed sleep opportunity, such as day sleep due to shift work).

There are some similarities between sleep and eating behavior, which is driven by both hunger (desire to eat) and appetite (desire for particular food categories). More specifically, eating behavior is controlled by an interaction of homeostatic and hedonic influences [4]. The homeostatic drive refers to an increased drive to eat as a response to depleted energy stores. However, individuals can override these messages regarding energy stores to consume palatable foods when there is no caloric need, which is called hedonic eating (eating for pleasure) [5]. It is thought that hedonic eating is one of the main drivers of the obesity epidemic, due to the abundance of readily available palatable and high caloric food. Like sleep, there are also circadian influences on eating. Studies have been conducted using a forced desynchrony protocol, a laboratory sleep schedule in which individuals are on a 22 or 28 h "day," which progressively moves sleep around the clock and allows researchers to conduct assessments at multiple circadian phases while controlling for sleep duration. Research conducted by Scheer et al. [6] using this protocol demonstrated that hunger as well as appetite for sweet, salty, starchy foods, fruits, meat/poultry and food overall and ability to eat demonstrated a robust circadian rhythm, with the lowest values around the biological morning (8 am) and peak values around 8 pm.

Short sleep duration is highly prevalent in the population

According to data from the National Health Interview Survey, 70.1 million US adults (29.2%) sleep <6 h per 24 h period [7]. Short sleep duration is highly prevalent and linked to negative mental and physical health consequences, including increased cardiovascular disease risk [8]. These statistics are a stark contrast to recommendations made by a recent consensus panel of sleep experts that concluded "at least 7 hours" as the amount of sleep needed for health and performance among adults [9]. Therefore, a high number of US adults could benefit from extending sleep duration. The high prevalence of short sleep duration is leading researchers to consider how sleep behaviors can be influenced, through intervention, public health education and occupational policies.

What predicts the decision to sleep or not to sleep?

In the epidemiologic literature, short sleep duration is more prevalent in males than females and more prevalent among blacks compared to whites [10–13]. Short sleep duration is also more common in urban areas compared with rural areas and in areas with lower safety and social cohesion [10, 14]. However, short sleep duration is present across the SES spectrum. Basner et al. [15] conducted an analysis of the American Time Use Survey and pooled data from 3 years of telephone surveys that asked participants (all age > 14) to report on their activities from the past 24 h. The data demonstrated the most common trade offs for sleep were work and commute time. Shorter sleepers engaged in more social and leisure activities, whereas both shorter and longer sleepers watched more TV than the average sleeper.

A later study found similar results and demonstrated for every hour that work or educational activities started earlier, sleep duration was 20 min shorter [16]. Furthermore, individuals working more than one job had the highest risk for short sleep duration.

Several studies have examined the role of social cognitive factors (knowledge, beliefs, social norms, etc.) in sleep duration. These studies have demonstrated that attitudes toward the importance of sleep and social norms are associated with sleep duration. This suggests that individuals' beliefs about sleep influence sleep behaviors, which is similar to what has been documented with other health behaviors [17]. In addition, perceived control over sleep, as well as self-efficacy (the belief that one can control and overcome barriers to sleep), are strongly associated with sleep duration. Another potentially modifiable correlate of short sleep duration is "bedtime procrastination," which is based in self-control theory, and refers to delaying bedtime due to not wanting to stop other activities (e.g., working, watching TV) [18]. Bedtime procrastination has been associated with lower overall self-control as well as poorer sleep habits and lower self reported sleep duration [18]. These studies suggest that cognitive and behavioral techniques such as education, goal setting and self-monitoring may be successful for extending sleep duration, as has been demonstrated in improving other domains of health behaviors, such as diet and physical activity.

Some individuals make time to sleep but cannot sleep

For individuals with sleep disorders, desire to sleep may be present but they are unable to achieve adequate sleep quality or quantity. Multiple studies demonstrating that psychosocial factors such as stress and perceived discrimination are associated with short sleep duration [19–21]. Insomnia is diagnosed if an individual has difficulty falling asleep, staying asleep or waking up too early despite adequate opportunity for sleep (ICSD3). It has been estimated that 9%–12% of the population has chronic insomnia. Often is the case in insomnia, the individual will engage in what would typically be a good health behavior (spending time in bed trying to sleep) but this can serve to exacerbate their insomnia because it provides greater opportunity for worry, rumination and frustration. Data suggest that having both short sleep time and an insomnia diagnosis has the greatest impact on health [22]. This combination has been considered "most severe phenotype" of insomnia. It is unclear whether this subtype of sleeper (insomnia with short sleep duration) would respond to the same interventions as individuals with short sleep time but not insomnia. For example, it has been demonstrated that individuals with short sleep time and insomnia respond poorly to cognitive behavioral therapy for insomnia, the recommended behavioral insomnia treatment [23].

PROPOSED PATHWAYS LINKING SLEEP TO OTHER HEALTH BEHAVIORS

Data from many studies have demonstrated that sleep loss and/or circadian disruption are associated with poorer health behaviors, including poorer dietary patterns, lower physical activity, higher alcohol intake and more smoking. However, only a few studies have examined the mechanisms driving these effects. We are going to discuss these proposed pathways in the next section including environmental exposure, neuropsychological, affective responses to sleep loss and circadian disruption, and the impact of sleep loss and circadian disruption to effort and motivational processes.

Exposure

The physical and social environment contributes to health behaviors. Additional exposure to certain social scenarios and environments late at night also likely interacts with the difficulty of making decisions later in the evening to ultimately impact engagement in healthy or unhealthy behaviors. For instance, Campbell et al. [24] found availability of junk food in the home environment to be associated with consumption of unhealthy foods and beverages among adolescents. Similarly, exposure to cues for smoking (e.g., seeing a picture of a cigarette; seeing a picture of a bar) also leads to increased cravings to smoke a cigarette when compared to scenarios not associated with smoking (e.g., seeing a picture of a gym) [25]. Exposure to environments that offer unhealthy choices may be higher in the evening. For instance, individuals may be more likely to go to bars or clubs and be exposed to alcohol and tobacco in the late evening hours.

One hypothesis is that the ability to self-regulate is a finite resource, and that repeated decision-making depletes the ability to self-regulate [26]. Thus, it may be that the longer one is awake, the more opportunities one has to make decisions regarding health behaviors and the more difficult it becomes to self-regulate. This may make it more likely that an individual engages in an unhealthy behavior (e.g., smoking a cigarette, eating more calories than planned, consuming junk food). Research suggests that individuals who remain awake engage in more unhealthy behaviors when compared to those who do not remain awake late into the night. For instance, Onyper et al. [27] found that college students who stayed awake all night at least once during a 2 week time period were more likely to report engaging in binge drinking and using stimulants. Similarly, participants with experimentally restricted sleep demonstrated increased snacking after dinner [28].

Evening chronotypes, who likely delay bedtime due to their preference for engaging in evening activity, have also been noted to have worse health behaviors. This may be due to a variety of factors (e.g., impulsivity, emotional dysregulation) including having additional time awake in the evening [29, 30]. For example, evening types have been found to use more nicotine and alcohol than their morning type counterparts [31]. Further, evening types have been noted to have higher scores on a measure of hazardous alcohol use [31], indicating that they are not only consuming more alcohol, but that the manner in which they consume alcohol may be more problematic than morning and neither types. Evening types have also been shown to consume fewer servings of fruits and vegetables [32] and more fast food [33].

In sum, environmental exposure likely influences decision-making processes. Staying up later in the evening likely influences not only the environments an individual is exposed to (e.g., a bar), but also the ability to self-regulate and make healthy choices. Further research is needed to fully understand the associations between decision-making processes, sleep and circadian factors, and environmental exposure.

Neurocognitive factors

Sleep loss and/or circadian disruption has an impact on health behaviors, such as increases in dietary intake [34] and decreases in physical activity [35]. One pathway that may explain these behavior changes is the effects of sleep on neurocognitive functioning in that the way individuals process information in their environment, such as salience of health-related cues or their ability to think about short and long-term consequences before acting, is affected by sleep and circadian factors. In fact, sleep and circadian rhythms are critical to cognitive processes including attention, memory and executive function, domains of cognition that support decision-making processes as they allow individuals to attend to information, remember that information, and subsequently plan or inhibit behaviors. There is a large literature examining the neurocognitive changes with sleep loss (see Refs. 36, 37 for review). Short term experiments have demonstrated that both partial (decreased sleep duration over days or weeks) and total (e.g., 24 h or more for a few days) sleep loss are associated with changes in vigilance, processing speed, executive function, decision making and impulsivity/risk taking (see Ref. 38, for review). These cognitive changes have clear indications for health behaviors but only a few studies have examined these changes in applied settings.

Fig. 27.1 demonstrates some of the possible neurocognitive processes that may link sleep loss and circadian disruption to health behaviors.

Attention: The ability to attend to information is crucial for cognitive functioning and subsequent decision-making. Sustained attention decreases with chronic sleep restriction, and is further worsened during the circadian night [39]. Given this, sleep loss may or alter attention or attentional

FIG. 27.1 Neurocognitive processes affected by sleep loss and circadian disruption.

bias for health-related cues. Similarly, attempting to make decisions during the circadian night would be impaired due to the difficulty with attending at this time. There is much research on the attentional biases associated with food cues and it has been demonstrated that individuals are faster to detect food cues and have greater difficulty disengaging on food cues [40]. It is currently unknown how sleep loss affects food related or other heath related attentional biases.

Executive function: Executive function comprises a variety of cognitive skills including working memory, inhibition and cognitive flexibility [41]. These cognitive skills have been linked to eating behaviors, exercise and weight gain. It is thought that these cognitive processes are important to maintaining healthy behaviors, such as navigating the obesogenic food environment and planning to fit in a workout. Binks et al. [42] reported no difference in higher order cortical functions such as sustained attention or cognitive flexibility after 34–36h of wakefulness compared to non-sleep deprived controls. However, this study was conducted under controlled conditions. Killgore [43] suggests that the interaction of declines in attention and emotional control may have an affect on these higher order processes as well.

Delay (temporal) discounting: Delay discounting refers to the relative preference for immediate versus delayed rewards. This measure has been associated with behaviors such as propensity for drug abuse [44–47]; as well as eating behaviors and weight gain over time [48]. Current data suggest that sleep loss itself is not associated with delay discounting [49]. Other studies have suggested that delay discounting may be trait-like and related to chronotype rather than sleep duration. One study demonstrated that evening types have a steeper discounting curve, which indicates a greater preference for more immediate rewards and are thought to be more "present oriented" than "future oriented" [50]. It is unclear whether circadian rhythm disruption affects delay discounting.

Effort discounting: Effort discounting refers to the effort an individual will put forth to achieve a reward. Data has demonstrated that sleep loss is associated with decreased effort discounting (or less effort to earn a reward) [49]. This may suggest that sleep loss could influence health behaviors because individuals put less effort toward them (e.g., effort to prepare healthy meals).

Impulsivity: Can be conceptualized as impulsive action (responding quickly/carelessly) or impulsive decision making, which involves weighing risks vs rewards [51]. Studies have demonstrated sleep loss affects impulsive behavior and possibly decision making in cognitive tasks [52–54]. Given the number of health behavior decisions made each and every day, increased impulsivity could affect a multitude of choices in daily life. There are several studies demonstrating individuals with evening chronotype have greater impulsivity.

Linking sleep related changes in neurocognitive function to health behaviors

Despite the many studies linking these neurocognitive processes to sleep and circadian rhythms, few studies have applied these concepts to healthy behaviors. Cedernaes et al. [55] demonstrated greater impulsivity in a go/no go task that asked participants to respond to food related cues after total sleep deprivation. Participants committed more errors of commission (i.e. responded to non-food related cues) after total sleep deprivation when compared to a control condition. Additionally, participants reported greater hunger after the sleep deprivation condition. This study did not have a condition in which individuals had to inhibit responses to food related cues, and therefore it is not clear whether these results were due to general impulsivity or specific to food related impulsivity. In another study, Chan [56] found that circadian disruption, as measured by inconsistent bedtimes, rather than sleep duration was associated with high BMI only in individuals with low delay discounting. This study demonstrates the potential that individual differences in trait-like cognitive factors rather than a simple cause-effect relationship.

Neuroimaging data

Several studies using neuroimaging have found that sleep loss is correlated with changes in brain regions implicated in healthy decision making. Greer et al. [57] found that after one night of total sleep deprivation, individuals had increases in appetitive evaluation regions within the prefrontal cortex and insular cortex during food desirability choices, combined with increased activity in the amygdala.

Similarly, St-Onge et al. [58] found that after receiving just 4 h per night in bed, participants displayed greater activation in the thalamus and areas of the orbitofrontal cortex (part of the prefrontal cortex) in response to food-related images when compared to non-food images. These relationships were attenuated in the group receiving a 9 h sleep opportunity. In addition, St-Onge et al. [59] conducted a partial sleep deprivation study and exposed individuals to 5 vs 9 h time in bed for five nights. After a period of restricted sleep, viewing unhealthy foods led to greater activation in the superior and middle temporal gyri, middle and superior frontal gyri, left inferior parietal lobule, orbitofrontal cortex, and right insula compared with healthy foods. These same stimuli presented after a period of habitual sleep did not produce marked activity patterns specific to unhealthy foods. Further, food intake during restricted sleep increased in association with a relative decrease in brain oxygenation level-dependent (BOLD) activity observed in the right insula. In sum, the above studies demonstrate that sleep loss may impact cognitive processes in areas of the brain such as the prefrontal cortex, which assists in planning, complex decision-making, and moderating behavior. These studies also demonstrate that longer sleep opportunities may protect an individual from experiencing these effects.

Most of the above studies were conducted in a highly controlled environment with scheduled meals. In contrast, Fang et al. [60] conducted a total sleep deprivation study to examine the impact that sleep deprivation may have on the salience network, which is thought play a role in reward-processing and homeostatic regulation [61]. Throughout the course of the study, participants were able to consume food ad libitum, with the exception of when receiving fMRI scans. Results indicated that after a night of total sleep deprivation, participants consumed a greater percentage of calories from fat and lower percentage of calories from carbohydrates than they had after a baseline night of sleep (9 h of time in bed). Further, after total sleep deprivation, enhanced functional connectivity within regions of the salience network was noted, and these changes appeared to predict the increased fat and decreased carbohydrate intake. This study indicates that changes in neuronal activity and functional connectivity that affect reward processing associated with sleep loss may lead to changes in behavior. More specifically, the neurocognitive changes may have directly impacted decision-making surrounding food choices and may have lead to the increased consumption of fat and carbohydrates.

There is less research on the changes in neuronal activity associated with circadian rhythm disruption. Much of this research has focused on chronotype, the self-reported preference for timing of sleep and activity. Hasler has explored the associations between eveningness and alcohol among late adolescents and young adults. Hasler and Clark [62] explored brain related pathways (medial prefrontal cortex and ventral striatum) and alcohol intake in late adolescents (age 20). He reported evening types had an altered response to reward stimuli compared to morning types. Furthermore this decreased activation in response to rewards was associated with more symptoms of alcohol dependence. He later reported in a longitudinal study that evening chronotype at age 20 predicted alcohol dependence at age 22 via these brain reward pathways [63].

In summary, there are many studies of the influence of sleep and circadian factors on neurocognitive function but only a handful of studies have examined these in reference to health behaviors. Data suggests that sleep loss affects many of the cognitive processes needed to navigate the multitude of healthy decisions needed in daily life.

AFFECTIVE RESPONSE TO SLEEP LOSS

Sleep loss and circadian disruption may also affect health decision making through their impact on emotional functioning. There are well known effects of sleep loss on emotional functioning. In the short term, there is an antidepressant effect of sleep deprivation but the antidepressant effects are reversed after sleep duration is restored [64]. In contrast, chronic sleep restriction can lead to depression and irritability [65]. The impact of sleep loss on psychopathology is not just limited to depression. Kahn-Greene et al. [66] reported that after 56 h of wakefulness, Personality Assessment Inventory clinical scales of somatic complaints, anxiety, depression, and paranoia were higher when compared to assessments completed prior to sleep deprivation. Post-hoc analyses revealed that the increase in somatic complaints was associated with an increase in the score on the health concerns subscale, the increase in anxiety was associated with an increase in the score on the physiological subscale, the increase in depression was associated with an increase in both the cognitive and affective subscales, and the increase in paranoia was associated with an increase in both the persecution and the resentment subscales [66]. Negative emotions may impact decision-making processes [67], thus, these mood changes may have an impact on health behaviors. There is evidence that the mood effects of sleep loss may interfere with healthy behaviors. One study demonstrated that when individuals were required to exercise during 30 h of sleep loss had greater negative mood than those who did not exercise [68].

The role of circadian factors in mood and mood disorders has been widely studied with far reaching implications for mood disorders and health behaviors [69]. There is a well known circadian rhythm of mood, with lowest mood in the morning hours and higher mood around 8 pm [70]. Disruption of the circadian rhythm via travel or shift work may affect health behaviors via mood. Silva et al. [71] demonstrated that among shift workers, anxiety ratings were increased the morning after night shift work when compared to the morning after a night of sleep. Further, anxiety scores were negatively associated with hunger ratings, indicating that those with greater anxiety reported less hunger.

EFFORT AND MOTIVATION

"I want to exercise but I am just too tired." This is a common statement that health care providers hear from their patients, who report poor sleep, fatigue, long work hours and exhaustion interfere with the ability to participate in regular exercise. Yet, on the other hand, data suggest that sleep loss itself does not affect aerobic capacity. It does impact their perception and emotional responses to physical activity, however. Even 24-h sleep loss has a relatively small effect on aerobic capacity, although this study did report there was significant variability in response to sleep loss [72]. For example, half of the participants demonstrated a very small change in aerobic capacity after sleep loss (5%), while the other half showed larger decrements in exercise tolerance (15%–40%). Notably, participants reported perceptions of greater effort following sleep loss, regardless of change in exercise tolerance, suggesting that exercise felt more difficult for everyone. Baron and colleagues also reported in analyses of an exercise intervention for older adults with insomnia, although the intervention overall improved aerobic capacity, sleep quality and daytime sleepiness over the 16 week intervention period, individuals who had higher self-reported sleepiness had lower exercise participation in the trial [73]. Therefore, an important message to patients is that losing sleep does make exercise feel harder but their physical performance is still relatively preserved in most cases.

In addition to sleep loss, circadian factors play a role in physical activity [74]. Studies have reported that evening chronotypes have lower physical activity, particularly in the morning [75, 76]. Individuals with evening chronotype also report higher perceived effort and demonstrate a higher heart rate to exercise in the morning [77, 78]. Therefore, time of day and chronotype both play a role in the perception of physical activity. There is little research about whether circadian rhythm disruption affects physical activity.

There is a dearth of research examining the relationships between sleep, circadian factors, and motivation, which can be defined as "Wanting. A condition of an organism that includes a subjective sense (not necessarily conscious) of desiring some change in self and/or environment" [79, p. 1]. Motivation is influenced by cognitive processes and emotion [79], which as reviewed earlier, are factors that can be impacted by sleep loss and circadian dysfunction. Thus, it might be postulated that disruptions to cognitive processes and to emotional functioning may in turn lead to decreased motivation to engage in health behaviors, and may ultimately result in less healthy decision-making.

To our knowledge, there have been no studies directly examining the links between sleep loss, subjectively reported motivation, and health behaviors such as physical activity and diet. However, there have been studies examining the impact of sleep loss on motivation and studies that indirectly measure motivation (e.g., wanting), sleep loss, and health behaviors. Motivation to engage in cognitive tasks has been found to decrease across a night of total sleep deprivation [80]. It might be hypothesized that a similar relationship may exist between sleep deprivation and motivation to engage in healthy decision-making, where the more sleep deprived an individual becomes, the less motivated they feel to engage in health behaviors such as physical activity. Of note, evidence has shown that rewards can influence the ability to attend after sleep deprivation [81]. More specifically, when asked to complete a task of sustained attention after sleep deprivation, those who are offered low rewards for doing so have more attentional lapses and slowed reaction time when compared with those who are offered high rewards for doing so. These findings demonstrate that a reward may increase motivation to perform. Thus, rewards may help to protect against some of the effects of sleep loss on motivational processes.

Although not measured directly, sleep loss may result in increased motivation to obtain unhealthy foods. Participants have indicated wanting more high-fat than low-fat foods after partial sleep restriction [82], which may translate into increased motivation to obtain these foods under conditions of sleep loss. Participants have also rated their desire to eat as higher after partial sleep restriction when compared to a control condition [82]. However, the literature has been mixed regarding the degree to which individuals may increase intake in association with increased hunger ratings [82–84]. Thus, it is unclear to what extent increased motivation to obtain highly palatable food results in a clinically significant change in behavior.

There is also some limited evidence that circadian disruption influences motivation. For instance, circadian disruption in the form of jetlag has been associated with temporary decreases in ratings of motivation to engage in physical activity [85]. Further, motivation has been shown to be lower after eastward travel when compared to westward travel [85].

Given that an individual's level of motivation may ultimately impact engagement in health behaviors, it is crucial to consider these factors when examining the pathways through which sleep and circadian functioning may impact health. Further research directly examining the relationships between sleep, circadian rhythms and motivation for different health behaviors is needed. Additionally, research is needed to assess how these relationships ultimately impact decision-making regarding health behaviors. For instance, if sleep loss leads to reduced motivation, and the reduced motivation leads to difficulty selecting healthy foods, interventions to increase motivation among those not receiving enough sleep may help to improve cardiometabolic health.

DOES CHANGING SLEEP MAKE IT EASIER TO MAKE HEALTHY DECISIONS?

Despite the data linking sleep loss and circadian disruption to poorer health behaviors, there are few studies that evaluate whether extending sleep or aligning circadian rhythms leads to improved health behaviors. There have been several studies that have looked at whether treating obstructive sleep apnea (OSA) with continuous positive airway pressure (CPAP) improves other health behaviors. OSA is a disorder characterized by repeated airway collapse that leads to intermittent hypoxia and sleep fragmentation. It was hypothesized that individuals with OSA, once treated, would have more energy to engage in better health behaviors. However, this has not been supported by the literature. In general, individuals with OSA who are treated with CPAP actually have an increase in BMI [86] rather than a decrease. It has also been shown that individuals in general, do not change their diet or become more physically active after treatment [87]. On the other hand, multiple studies have evaluated intensive lifestyle interventions in patients with OSA and have demonstrated that these interventions can achieve weight loss in this population [88].

There has also been research conducted to examine the effects of insomnia treatment (cognitive behavioral therapy for insomnia or CBT-I) among individuals with alcohol dependence. It is well known that alcohol use and abuse is implicated in the development and maintenance of insomnia [89]. For example, individuals may consume alcohol to cope with sleep problems. Furthermore, if alcohol dependent, they also experience sleep problems in response to abstaining from alcohol. There are multiples studies that demonstrate that CBT-I is effective at improving sleep among individuals with alcohol dependence [90]. However, a recent clinical trial did not demonstrate an improvement in relapse.

A handful of studies have evaluated effects of extending sleep duration on eating and weight related outcomes. One study by Cizza et al. [91] conducted sleep extension intervention however they found improvements in the intervention and control group in metabolic factors. It is thought that a Hawthorne effect occurred in the lengthy period between the screening and intervention period, in that the act of completing sleep logs and enrolling in a study lead to improvements in both groups. Two more recent studies have demonstrated changes in eating behaviors with sleep extension. A study of sleep extension in adolescents (bedtime advancement) demonstrated adolescents with earlier bedtimes increased low GI fruit and dairy at post treatment [92]. Another study demonstrated 4 weeks of sleep extension in adults reduced appetite, desire for sweet and salty foods [93]. In this study, individuals extended their sleep by an enormous 1.6 h. It is not clear whether this amount of sleep extension is possible in most participants. Two other sleep extension studies were published that demonstrated more modest improvements in sleep duration (about 40 min) [94, 95]. More studies are needed to evaluate whether changing sleep will improve other health behaviors. Therefore, these studies suggest that it is possible that extending sleep can have a broader reaching effect on other health behaviors. The limitation is that existing studies are small, short term and intensive interventions. More research is needed to test whether these interventions can lead to sustained sleep behavior change as well as sustained changes to health behaviors.

SUMMARY

In summary, we have presented multiple environmental, social, neurocognitive and behavioral pathways by which sleep and circadian rhythms can influence healthy decision making. We have depicted these pathways in a conceptual diagram (Fig. 27.2). Environmental factors, changes to

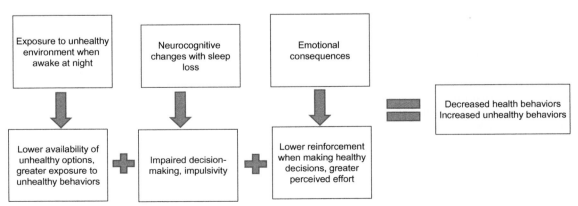

FIG. 27.2 Conceptual diagram demonstrating the complex interaction of environment, neurocognitive function, and emotional changes associated with sleep loss and circadian disruption.

cognitive function and emotional changes are all altered by sleep loss and can influence desire and self-control, motivation and energy to engage in health behaviors. Although a great deal of research has focused on how sleep loss affects these basic cognitive processes, only a few studies have evaluated how these changes affect health behavior and decision making. In addition, translational research is needed to understand how sleep effects behavioral decision making in naturalistic settings. Finally, given that vulnerabilities exist, interventions are needed to (1) determine if improving sleep can help other health behaviors, and (2) when changing sleep is not possible, if addressing the underlying vulnerability in those at risk for sleep loss (e.g., cognitive interventions).

REFERENCES

[1] Kvaavik E, Batty GD, Ursin G, Huxley R, Gale CR. Influence of individual and combined health behaviors on total and cause-specific mortality in men and women: the United Kingdom health and lifestyle survey. Arch Intern Med 2010;170(8):711–8.

[2] Gochman DS. Handbook of health behavior research. vol. 1–4. New York, NY: Plenum; 19973.

[3] Borbély AA. A two process model of sleep regulation. Hum Neurobiol 1982;1(3):195–204.

[4] Lutter M, Nestler EJ. Homeostatic and hedonic signals interact in the regulation of food intake. J Nutr 2009;139(3):629–32.

[5] Lowe MR, Butryn ML. Hedonic hunger: a new dimension of appetite? Physiol Behav 2007;91(4):432–9.

[6] Scheer FA, Morris CJ, Shea SA. The internal circadian clock increases hunger and appetite in the evening independent of food intake and other behaviors. Obesity 2013;21(3):421–3.

[7] Ford ES, Cunningham TJ, Croft JB. Trends in self-reported sleep duration among US adults from 1985 to 2012. Sleep 2015;38(5):829–32.

[8] Grandner MA, Chakravorty S, Perlis ML, Oliver L, Gurubhagavatula I. Habitual sleep duration associated with self-reported and objectively determined cardiometabolic risk factors. Sleep Med 2014;15(1):42–50.

[9] Watson NF, Badr MS, Belenky G, Bliwise DL, Buxton OM, Buysse D, et al. Joint consensus statement of the American Academy of Sleep Medicine and Sleep Research Society on the recommended amount of sleep for a healthy adult: methodology and discussion. Sleep 2015;38(8):1161–83.

[10] Grandner MA, Jackson NJ, Izci-Balserak B, Gallagher RA, Murray-Bachmann R, Williams NJ, et al. Social and behavioral determinants of perceived insufficient sleep. Front Neurol 2015;6:112.

[11] Hale L, Do DP. Racial differences in self-reports of sleep duration in a population-based study. Sleep 2007;30(9):1096–103.

[12] Whinnery J, Jackson N, Rattanaumpawan P, Grandner MA. Short and long sleep duration associated with race/ethnicity, sociodemographics, and socioeconomic position. Sleep 2014;37(3):601–11.

[13] Williams NJ, Grandner MA, Wallace DM, Cuffee Y, Airhihenbuwa C, Okuyemi K, et al. Social and behavioral predictors of insufficient sleep among African Americans and Caucasians. Sleep Med 2016;18:103–7.

[14] DeSantis AS, Diez Roux AV, Moore K, Baron KG, Mujahid MS, Nieto FJ. Associations of neighborhood characteristics with sleep timing and quality: the multi-ethnic study of atherosclerosis. Sleep 2013;36(10):1543–51.

[15] Basner M, Fomberstein KM, Razavi FM, Banks S, William JH, Rosa RR, et al. American time use survey: sleep time and its relationship to waking activities. Sleep 2007;30(9):1085–95.

[16] Basner M, Spaeth AM, Dinges DF. Sociodemographic characteristics and waking activities and their role in the timing and duration of sleep. Sleep 2014;37(12):1889–906.

[17] Knowlden AP, Sharma M. Health belief structural equation model predicting sleep behavior of employed college students. Fam Community Health 2014;37(4):271–8.

[18] Kroese FM, De Ridder DT, Evers C, Adriaanse MA. Bedtime procrastination: introducing a new area of procrastination. Front Psychol 2014;5:611.

[19] Hoggard LS, Hill LK. Examining how racial discrimination impacts sleep quality in African Americans: is perseveration the answer? Behav Sleep Med 2018;16(5):471–81.

[20] Johnson DA, Lisabeth L, Lewis TT, Sims M, Hickson DA, Samdarshi T, et al. The contribution of psychosocial stressors to sleep among African Americans in the Jackson Heart Study. Sleep 2016;39(7):1411–9.

[21] Sims M, Lipford KJ, Patel N, Ford CD, Min Y-I, Wyatt SB. Psychosocial factors and behaviors in African Americans: the jackson heart study. Am J Prev Med 2017;52(1):S48–55.

[22] Vgontzas AN, Fernandez-Mendoza J, Liao D, Bixler EO. Insomnia with objective short sleep duration: the most biologically severe phenotype of the disorder. Sleep Med Rev 2013;17(4):241–54.

[23] Bathgate CJ, Edinger JD, Krystal AD. Insomnia patients with objective short sleep duration have a blunted response to cognitive behavioral therapy for insomnia. Sleep 2017;40(1):zsw012.

[24] Campbell KJ, Crawford DA, Salmon J, Carver A, Garnett SP, Baur LA. Associations between the home food environment and obesity-promoting eating behaviors in adolescence. Obesity 2007;15(3):719–30.

[25] Conklin CA, Robin N, Perkins KA, Salkeld RP, McClernon FJ. Proximal versus distal cues to smoke: the effects of environments on smokers' cue-reactivity. Exp Clin Psychopharmacol 2008;16(3):207.

[26] Vohs KD, Baumeister RF, Schmeichel BJ, Twenge JM, Nelson NM, Tice DM. Making choices impairs subsequent self-control: a limited-resource account of decision making, self-regulation, and active initiative. J Pers SocPsychol 2008;94(5):883–98.

[27] Onyper SV, Thacher PV, Gilbert JW, Gradess SG. Class start times, sleep, and academic performance in college: a path analysis. Chronobiol Int 2012;29(3):318–35.

[28] Markwald RR, Melanson EL, Smith MR, Higgins J, Perreault L, Eckel RH, et al. Impact of insufficient sleep on total daily energy expenditure, food intake, and weight gain. Proc NatlAcad Sci USA 2013;110(14):5695–700.

[29] Caci H, Robert P, Boyer P. Novelty seekers and impulsive subjects are low in morningness. Eur Psychiatry 2004;19(2):79–84.

[30] Chelminski I, Ferraro FR, Petros TV, Plaud JJ. An analysis of the "eveningness–morningness" dimension in "depressive" college students. J Affect Disord 1999;52(1–3):19–29.

[31] Prat G, Adan A. Influence of circadian typology on drug consumption, hazardous alcohol use, and hangover symptoms. Chronobiol Int 2011;28(3):248–57.

[32] Patterson F, Malone SK, Lozano A, Grandner MA, Hanlon AL. Smoking, screen-based sedentary behavior, and diet associated with habitual sleep duration and chronotype: data from the UK Biobank. Ann Behav Med 2016;50(5):715–26.

[33] Fleig D, Randler C. Association between chronotype and diet in adolescents based on food logs. Eat Behav 2009;10(2):115–8.

[34] Brondel L, Romer MA, Nougues PM, Touyarou P, Davenne D. Acute partial sleep deprivation increases food intake in healthy men. Am J Clin Nutr 2010;91(6):1550–9.

[35] Schmid SM, Hallschmid M, Jauch-Chara K, Wilms B, Benedict C, Lehnert H, et al. Short-term sleep loss decreases physical activity under free-living conditions but does not increase food intake under time-deprived laboratory conditions in healthy men. Am J Clin Nutr 2009;90(6):1476–82.

[36] Durmer JS, Dinges DF, editors. Neurocognitive consequences of sleep deprivation. Seminars in neurology, New York, NY; 2005.

[37] Kilgore W. Socio-emotional and neurocognitive effects of sleep loss. In: The handbook of operator fatigue. Boca Raton, FL: Taylor & Francis Group; 2012. p. 173–85.

[38] Goel N, Rao H, Durmer JS, Dinges DF, editors. Neurocognitive consequences of sleep deprivation. Seminars in neurology; NIH Public Access; 2009.

[39] McHill AW, Hull JT, Wang W, Czeisler CA, Klerman EB. Chronic sleep curtailment, even without extended (> 16-h) wakefulness, degrades human vigilance performance. Proc Natl Acad Sci 2018;115(23):6070–5.

[40] Pool E, Brosch T, Delplanque S, Sander D. Where is the chocolate? Rapid spatial orienting toward stimuli associated with primary rewards. Cognition 2014;130(3):348–59.

[41] Diamond A. Executive functions. Annu Rev Psychol 2013;64:135–68.

[42] Binks PG, Waters WF, Hurry M. Short-term total sleep deprivations does not selectively impair higher cortical functioning. Sleep 1999;22(3):328–34.

[43] Killgore WD. Effects of sleep deprivation on cognition. Prog Brain Res 2010;185:105–29.

[44] Amlung M, MacKillop J. Delayed reward discounting and alcohol misuse: the roles of response consistency and reward magnitude. J Exp Psychol 2011;2(3). jep. 017311.

[45] Audrain-McGovern J, Rodriguez D, Epstein LH, Cuevas J, Rodgers K, Wileyto EP. Does delay discounting play an etiological role in smoking or is it a consequence of smoking? Drug Alcohol Depend 2009;103(3):99–106.

[46] Bickel WK, Odum AL, Madden GJ. Impulsivity and cigarette smoking: delay discounting in current, never, and ex-smokers. Psychopharmacology 1999;146(4):447–54.

[47] Fernie G, Peeters M, Gullo MJ, Christiansen P, Cole JC, Sumnall H, et al. Multiple behavioural impulsivity tasks predict prospective alcohol involvement in adolescents. Addiction 2013;108(11):1916–23.

[48] Appelhans BM, Waring ME, Schneider KL, Pagoto SL, DeBiasse MA, Whited MC, et al. Delay discounting and intake of ready-to-eat and away-from-home foods in overweight and obese women. Appetite 2012;59(2):576–84.

[49] Libedinsky C, Massar SA, Ling A, Chee W, Huettel SA, Chee MW. Sleep deprivation alters effort discounting but not delay discounting of monetary rewards. Sleep 2013;36(6):899–904.

[50] Milfont TL, Schwarzenthal M. Explaining why larks are future-oriented and owls are present-oriented: self-control mediates the chronotype–time perspective relationships. Chronobiol Int 2014;31(4):581–8.

[51] Krishnan-Sarin S, Reynolds B, Duhig AM, Smith A, Liss T, McFetridge A, et al. Behavioral impulsivity predicts treatment outcome in a smoking cessation program for adolescent smokers. Drug Alcohol Depend 2007;88(1):79–82.

[52] Acheson A, Richards JB, de Wit H. Effects of sleep deprivation on impulsive behaviors in men and women. Physiol Behav 2007;91(5):579–87.

[53] Demos K, Hart C, Sweet L, Mailloux K, Trautvetter J, Williams S, et al. Partial sleep deprivation impacts impulsive action but not impulsive decision-making. Physiol Behav 2016;164:214–9.

[54] Mckenna BS, Dickinson DL, Orff HJ, Drummond SP. The effects of one night of sleep deprivation on known-risk and ambiguous-risk decisions. J Sleep Res 2007;16(3):245–52.

[55] Cedernaes J, Brandell J, Ros O, Broman JE, Hogenkamp PS, Schiöth HB, et al. Increased impulsivity in response to food cues after sleep loss in healthy young men. Obesity 2014;22(8):1786–91.

[56] Chan WS. Delay discounting and response disinhibition moderate associations between actigraphically measured sleep parameters and body mass index. J Sleep Res 2017;26(1):21–9.

[57] Greer SM, Goldstein AN, Walker MP. The impact of sleep deprivation on food desire in the human brain. Nat Commun 2013;4:ncomms3259.

[58] St-Onge M-P, McReynolds A, Trivedi ZB, Roberts AL, Sy M, Hirsch J. Sleep restriction leads to increased activation of brain regions sensitive to food stimuli. Am J Clin Nutr 2012;95(4):818–24.

[59] St-Onge M, Wolfe S, Sy M, Shechter A, Hirsch J. Sleep restriction increases the neuronal response to unhealthy food in normal-weight individuals. Int J Obes 2014;38(3):411.

[60] Fang Z, Spaeth AM, Ma N, Zhu S, Hu S, Goel N, et al. Altered salience network connectivity predicts macronutrient intake after sleep deprivation. Sci Rep 2015;5:8215.

[61] Seeley WW, Menon V, Schatzberg AF, Keller J, Glover GH, Kenna H, et al. Dissociable intrinsic connectivity networks for salience processing and executive control. J Neurosci 2007;27(9):2349–56.

[62] Hasler BP, Clark DB. Circadian misalignment, reward-related brain function, and adolescent alcohol involvement. Alcohol Clin Exp Res 2013;37(4):558–65.

[63] Hasler BP, Casement MD, Sitnick SL, Shaw DS, Forbes EE. Eveningness among late adolescent males predicts neural reactivity to reward and alcohol dependence 2 years later. Behav Brain Res 2017;327:112–20.

[64] Giedke H, Schwärzler F. Therapeutic use of sleep deprivation in depression. Sleep Med Rev 2002;6(5):361–77.

[65] Zhai L, Zhang H, Zhang D. Sleep duration and depression among adults: a meta-analysis of prospective studies. Depress Anxiety 2015;32(9):664–70.

[66] Kahn-Greene ET, Killgore DB, Kamimori GH, Balkin TJ, Killgore WD. The effects of sleep deprivation on symptoms of psychopathology in healthy adults. Sleep Med 2007;8(3):215–21.

[67] Raghunathan R, Pham MT. All negative moods are not equal: Motivational influences of anxiety and sadness on decision making. Organ Behav Hum Decis Process 1999;79(1):56–77.

[68] Scott JP, McNaughton LR, Polman RC. Effects of sleep deprivation and exercise on cognitive, motor performance and mood. Physiol Behav 2006;87(2):396–408.

[69] Li JZ. Circadian rhythms and mood: Opportunities for multi-level analyses in genomics and neuroscience: circadian rhythm

[69] ... dysregulation in mood disorders provides clues to the brain's organizing principles, and a touchstone for genomics and neuroscience. Bioessays 2014;36(3):305–15.
[70] Monk T, Buysse D, Reynolds III C, Berga S, Jarrett D, Begley A, et al. Circadian rhythms in human performance and mood under constant conditions. J Sleep Res 1997;6(1):9–18.
[71] Silva AASC, TdVC L, Teixeira KR, Mendes JA, de Souza Borba ME, Mota MC, et al. The association between anxiety, hunger, the enjoyment of eating foods and the satiety after food intake in individuals working a night shift compared with after taking a nocturnal sleep: a prospective and observational study. Appetite 2017;108:255–62.
[72] Martin BJ. Effect of sleep deprivation on tolerance of prolonged exercise. Eur J Appl Physiol Occup Physiol 1981;47(4):345–54.
[73] Baron KG, Reid KJ, Zee PC. Exercise to improve sleep in insomnia: exploration of the bidirectional effects. J Clin Sleep Med 2013;9(08):819–24.
[74] Vitale JA, Weydahl A. Chronotype, physical activity, and sport performance: a systematic review. Sports Med 2017;47(9):1859–68.
[75] McNeil J, Doucet É, Brunet J-F, Hintze LJ, Chaumont I, Langlois É, et al. The effects of sleep restriction and altered sleep timing on energy intake and energy expenditure. Physiol Behav 2016;164:157–63.
[76] Shechter A, St-Onge M-P. Delayed sleep timing is associated with low levels of free-living physical activity in normal sleeping adults. Sleep Med 2014;15(12):1586–9.
[77] Rae DE, Stephenson KJ, Roden LC. Factors to consider when assessing diurnal variation in sports performance: the influence of chronotype and habitual training time-of-day. Eur J Appl Physiol 2015;115(6):1339–49.
[78] Rossi A, Formenti D, Vitale JA, Calogiuri G, Weydahl A. The effect of chronotype on psychophysiological responses during aerobic self-paced exercises. Percept Mot Skills 2015;121(3):840–55.
[79] Baumeister RF. Toward a general theory of motivation: problems, challenges, opportunities, and the big picture. Motiv Emot 2016;40(1):1–10.
[80] Odle-Dusseau HN, Bradley JL, Pilcher JJ. Subjective perceptions of the effects of sustained performance under sleep-deprivation conditions. Chronobiol Int 2010;27(2):318–33.
[81] Massar SA, Lim J, Sasmita K, Chee MW. Sleep deprivation increases the costs of attentional effort: performance, preference and pupil size. Neuropsychologia 2019;123:169–77.
[82] McNeil J, Forest G, Hintze LJ, Brunet J-F, Finlayson G, Blundell JE, et al. The effects of partial sleep restriction and altered sleep timing on appetite and food reward. Appetite 2017;109:48–56.
[83] Nedeltcheva AV, Kilkus JM, Imperial J, Kasza K, Schoeller DA, Penev PD. Sleep curtailment is accompanied by increased intake of calories from snacks. Am J Clin Nutr 2008;89(1):126–33.
[84] St-Onge M-P, Roberts AL, Chen J, Kelleman M, O'Keeffe M, RoyChoudhury A, et al. Short sleep duration increases energy intakes but does not change energy expenditure in normal-weight individuals. Am J Clin Nutr 2011;94(2):410–6.
[85] Fowler PM, Knez W, Crowcroft S, Mendham AE, Miller J, Sargent C, et al. Greater effect of east vs. west travel on jet lag, sleep, and team-sport performance. Med Sci Sports Exerc 2017;49(12):2548–61.
[86] Drager LF, Brunoni AR, Jenner R, Lorenzi-Filho G, Bensenor IM, Lotufo PA. Effects of CPAP on body weight in patients with obstructive sleep apnoea: a meta-analysis of randomised trials. Thorax 2015;70(3):258–64.
[87] Batool-Anwar S, Goodwin JL, Drescher AA, Baldwin CM, Simon RD, Smith TW, et al. Impact of CPAP on activity patterns and diet in patients with obstructive sleep apnea (OSA). J Clin Sleep Med 2014;10(05):465–72.
[88] Thomasouli M-A, Brady EM, Davies MJ, Hall AP, Khunti K, Morris DH, et al. The impact of diet and lifestyle management strategies for obstructive sleep apnoea in adults: a systematic review and meta-analysis of randomised controlled trials. Sleep Breath 2013;17(3):925–35.
[89] Brower KJ. Insomnia, alcoholism and relapse. Sleep Med Rev 2003;7(6):523–39.
[90] Brooks AT, Wallen GR. Sleep disturbances in individuals with alcohol-related disorders: a review of cognitive-behavioral therapy for insomnia (CBT-I) and associated non-pharmacological therapies. Subst Abuse 2014;8:55–62.
[91] Cizza G, Piaggi P, Rother KI, Csako G. Hawthorne effect with transient behavioral and biochemical changes in a randomized controlled sleep extension trial of chronically short-sleeping obese adults: implications for the design and interpretation of clinical studies. PLoS One 2014;9(8):e104176.
[92] Asarnow LD, Greer SM, Walker MP, Harvey AG. The impact of sleep improvement on food choices in adolescents with late bedtimes. J Adolesc Health 2017;60(5):570–6.
[93] Tasali E, Chapotot F, Wroblewski K, Schoeller D. The effects of extended bedtimes on sleep duration and food desire in overweight young adults: a home-based intervention. Appetite 2014;80:220–4.
[94] Haack M, Serrador J, Cohen D, Simpson N, Meier-Ewert H, Mullington JM. Increasing sleep duration to lower beat-to-beat blood pressure: a pilot study. J Sleep Res 2013;22(3):295–304.
[95] Leproult R, Deliens G, Gilson M, Peigneux P. Beneficial impact of sleep extension on fasting insulin sensitivity in adults with habitual sleep restriction. Sleep 2015;38(5):707–15.

GLOSSARY

Bedtime procrastination A lack of self-regulation at bedtime that involves avoiding interrupting a more pleasurable activity to go to bed.

Circadian rhythm The approximately 24 h rhythm generated by the suprachiasmatic nucleus, in humans. Circadian rhythms are observed in sleep wake patterns, mood, cognitive performance and many hormones (cortisol) and physiological processes (heart rate, blood pressure).

Chronotype Self-reported preference for timing of sleep/wake schedule and activities.

Delay discounting The ability to delay immediate rewards for later rewards.

Forced desynchrony A laboratory protocol that examines the circadian rhythm while controlling for sleep duration. Individuals are put on a shorter (22 h) or longer (28 h) "day" and sleep is moved progressively around the clock in order to examine performance and physiology at different phases of the circadian rhythm.

Eveningness Self-reported preference for delayed timing of activity and sleep wake schedule.

Executive function A collection of top down cognitive processes including inhibitory control, multi-tasking, planning, switching tasks.

Effort discounting The willingness to exert effort to obtain a reward.

Hawthorne effect When participants change based on being observed, rather than the intervention.

Hedonic eating Eating for pleasure.

Homeostatic Refers to the system of balance, discussed in terms of both sleep propensity and hunger. The build-up of drive to eat or sleep is in part driven by the time since the last sleep or meal.

Impulsivity Refers to impulsive action or decision making made without weighing the pros and cons.

Part VII

Public health implications of sleep disorders

Chapter 28

Insomnia and psychiatric disorders

Ivan Vargas[a,b], Sheila N. Garland[c,d], Jacqueline D. Kloss[a], Michael L. Perlis[a,b]

[a]Behavioral Sleep Medicine Program, University of Pennsylvania, Philadelphia, PA, United States, [b]Center for Sleep and Circadian Neurobiology, University of Pennsylvania, Philadelphia, PA, United States, [c]Department of Psychology, Faculty of Science, Memorial University, St. John's, NL, Canada, [d]Division of Oncology, Faculty of Medicine, Memorial University, St. John's, NL, Canada

ABBREVIATIONS

ACP	American College of Physicians
ADHD	Attention-Deficit/Hyperactivity Disorder
APA	American Psychiatric Association
ASD	Autism Spectrum Disorder
AUD	Alcohol Use Disorder
CBT-I	Cognitive Behavioral Therapy for Insomnia
DLMO	Dim Light Melatonin Onset
DSM	Diagnostic and Statistical Manual of Mental Disorders
EEG	Electroencephalography
GABA	Gamma-Aminobutyric acid
GERD	Gastroesophageal reflux disease
MDD	Major Depressive Disorder
MDE	Major Depressive Episode
NHANES	National Health and Nutrition Examination Survey
PSG	Polysomnography
PTSD	Post-traumatic Stress Disorder
REM	Rapid Eye Movement

INTRODUCTION

Sleep continuity disturbance (i.e., insomnia) is ubiquitous among psychiatric conditions, and is a diagnostic feature and/or correlate of most, if not all, "Axis 1" disorders [1]. Approximately 40% of patients with insomnia report at least one other comorbid psychiatric disorder [2]. Of those reporting insomnia complaints, only 16% met criteria for primary insomnia (i.e., no other comorbidities), whereas, 36% met criteria for another Diagnostic and Statistical Manual of Mental Disorders (DSM) diagnoses (e.g., major depression, bipolar, etc.) [3]. Therefore, the study and treatment of insomnia must take into consideration the relative impact of psychiatric comorbidities, and vice versa. In the present chapter we aim to: (1) review the definition, incidence, prevalence and theoretical perspectives on the etiology of insomnia; (2) summarize the literature on the relationship between insomnia and various psychiatric conditions, particularly with respect to comorbidity rates, insomnia as a risk factor, and the potential mechanisms that may explain the association; (3) briefly introduce the theoretical components of Cognitive Behavioral Therapy for Insomnia (CBT-I); and (4) examine the evidence for the efficacy of CBT-I among patients with comorbid psychiatric conditions.

DEFINITION, INCIDENCE, AND PREVALENCE

Definition

Insomnia, broadly defined, refers to difficulty initiating and maintaining sleep. Insomnia is often conceptualized in terms of frequency, chronicity, type, and subtype. Frequency refers to how often an individual experiences insomnia symptoms (typically days per week). Chronicity refers to whether the insomnia is acute or chronic. Type refers to the forms of insomnia that have been historically identified as distinct entities previously included in the International Classification of Sleep Disorders [4] nosology including idiopathic insomnia, psychophysiologic insomnia, paradoxical insomnia, insomnia due to inadequate sleep hygiene, and insomnia comorbid with medical or psychiatric illness; however, these types are no longer included in the most current version [5]. Subtype refers to the insomnia phenotype (initial, middle, late, or mixed insomnia). The formal definition or diagnostic criteria for insomnia disorder

(DSM-5, 6) is outlined below, but what is relevant for this chapter is that these distinctions with regard to chronicity, type, and subtype exist and should be taken into account. Notably, while frequency and chronicity are defined in the diagnostic criteria, severity is not. It is, however, common in research criteria to use 30 min as a severity threshold (e.g., a sleep latency greater than or equal to 30 min is considered clinically significant). Why it is not adopted into the clinical nosology is still unclear. This may be related to the relative difference in how one interprets the severity criteria (i.e., for one individual, 30 min may be functionally impairing or distressing, but for someone else it may not be). Taylor and colleagues reported, however, that greater than 30 min was reliably endorsed as a "problem" [6].

Historically, insomnia was considered "just a symptom" of a medical or psychiatric disease and it was believed that the treatment of the underlying disorder was sufficient and would consequently ameliorate the insomnia as well. Long-term management of insomnia, therefore, was thought to be unnecessary. This perspective has since changed, to where chronic insomnia is now conceptualized as an independent disorder [7]. Adopted by the American Psychiatric Association's (APA) diagnostic nomenclature (i.e., DSM-5), "insomnia disorder" is used to distinguish insomnia, what is considered to be a distinct diagnostic entity (see diagnostic criteria below), from the sleep continuity disturbance that is a symptom of an underlying medical and/or psychiatric condition, eliminating the need for the "primary" or "secondary" distinctions. Of note, sleep continuity refers to a class, or set, of variables, that we use to talk about "sleep performance" in the context of insomnia. Polysomnography (PSG) recorded sleep has the class term "sleep architecture" which refers to a group of variables that we use to discuss differences in PSG-recorded sleep (e.g., sleep stages, REM onset latency, REM density, K-complexes). Our field, however, has yet to "doctrinize" a class term that refers to all the sleep variables relevant to assessing insomnia (e.g., sleep latency, wake after sleep onset, total sleep time, sleep efficiency). When one or more of these variables are pathological, we refer to this as *sleep continuity disturbance*.

The specific DSM-5 criteria for insomnia disorder are [8]:

- A predominant complaint of dissatisfaction with sleep quantity or quality, associated with one or more of the following symptoms:
  - Difficulty initiating sleep;
  - Difficulty maintaining sleep, characterized by frequent awakenings or problems returning to sleep after awakenings; or
  - Early-morning awakening with inability to return to sleep.
- The sleep disturbance causes clinically significant distress or impairment in social, occupational, educational, academic, behavioral, or other important areas of functioning.
- The sleep difficulty occurs at least three nights per week.
- The sleep difficulty is present for at least 3 months.
- The sleep difficulty occurs despite adequate opportunity for sleep.
- The insomnia is not better explained by and does not occur exclusively during the course of another sleep-wake disorder (e.g., narcolepsy, a breathing-related sleep disorder, a circadian rhythm sleep-wake disorder), the physiological effects of a substance (e.g., a drug of abuse, a medication) or coexisting mental disorders and medical conditions.

Incidence and prevalence

Approximately 30%–50% of the U.S. population experience acute sleep continuity disturbance per annum [9], and approximately 6%–10% of the population report chronic levels of insomnia [10]. Recent data from our group showed that the incident rate of acute insomnia was approximately 25%. Of those 357 individuals who developed new onset acute insomnia, 72% recovered (i.e., resumed good sleep) and 7% developed insomnia disorder. Interestingly, the other 21% of subjects experienced a form of persistent poor sleep that was not consistent with good sleep but also did not meet diagnostic criteria for insomnia disorder [11]. With respect to insomnia subtypes (i.e., initial, middle, and late insomnia), data from the 2007–08 National Health and Nutrition Examination Survey (NHANES), which included a nationally-representative sample of US adults, found that self-reported difficulty falling asleep (initial insomnia) was reported by about 19% of the US population [12]. Other insomnia subtypes, such as difficulty resuming sleep during the night (middle insomnia) and early morning awakenings (late insomnia) were endorsed at similar rates. These high prevalence and incidence estimates have important public health implications because insomnia is associated with significant daytime impairment, including mood dysregulation, cognitive deficits, and fatigue [13]. Just as or more importantly, insomnia is a risk factor for multiple psychiatric and medical disorders, including depression [14–18], hypertension [19–21], diabetes [22, 23], and cardiovascular disease [24]. Taken together, efforts to identify the factors and/or mechanisms that explain the transition from acute to chronic insomnia and/or increase risk for chronic insomnia (in general) are an important research and public health priority.

THEORETICAL PERSPECTIVES ON THE ETIOLOGY OF INSOMNIA

No matter how important sleep may be, it was adaptively deferred when the mountain lion entered the cave.

-Spielman and Glovinsky [24a].

In keeping with this statement is the expression: "we live with insomnia today because at some point in our evolutionary history, insomnia allowed us to live" (Dean Handley, Sepracor, c.2005). Both of these quotes suggest that acute insomnia is adaptive. Most would argue that stress reactivity is adaptive. Physiological, cognitive, and behavioral responses to environmental challenges are not only necessary for survival, they directly bear on the individual's health and wellbeing [25, 26]. This said, altered stress responses are also risk factors for chronic disease, such as insomnia and depression [27–29]. Given this point of view, an essential question is "how does something that is inherently adaptive become maladaptive?" That is, how does acute sleeplessness in the face of a threat become pathological over time? While many things may precipitate acute sleeplessness, chronic insomnia is thought to be an independent disorder with a unique etiology (i.e., insomnia disorder). While the exact pathogenesis of insomnia disorder is likely to be multifactorial, five models are presented below. In addition to putting forward a transdiagnostic model, there is also one trans-theoretical model summarized as well (see also Fig. 28.1). It is important to review these models and provide some context, as it may help explain the high comorbidity and shared pathophysiology between insomnia and psychiatric conditions. That is, understanding the factors that are related to the transition from sleep continuity disturbance that is acute (and likely adaptive) to chronic and dysfunctional may shed some light on how insomnia subsequently increases the risk for psychiatric problems.

Stimulus control model

As originally described by Bootzin in 1972 [30], stimulus control is based on the behavioral principle that one stimulus may elicit a variety of responses, depending on the conditioning history. In the instance where one stimulus is always paired with a single behavior, there is a high probability that the stimulus will yield only one response. In good sleepers, the stimuli typically associated with sleep (e.g., bed, bedroom, etc.) are paired with (and subsequently elicit the response of) sleep. In the instance where there is a complex conditioning history, as typically occurs in patients with insomnia, this is often not the case. When a stimulus is paired with a variety of behaviors, there is a low probability that the stimulus will elicit only one response. In patients with insomnia, stimuli typically associated with sleep are often paired with activities other than sleep, such as reading, and watching television in bed. The Stimulus Control Model of Insomnia suggests that engaging in these other behaviors sets the stage for a complex conditioning relationship, or stimulus dyscontrol, that is, reduced probability that sleep-related stimuli will elicit the desired response of sleepiness and sleep. Put differently, lying awake in bed (whether one is engaging in sleep effort or not) only further reduces the probability that lying in bed (stimulus) will result in sleep (response). More recently, this model has been expanded to suggest that part of stimulus dyscontrol is that while individuals are lying in bed, they are *microsleeping*. During these microsleeps, the sleep may not be perceptible to the individual but also the sleep is not of sufficient quality to be sating. Microsleeps are, however, enough sleep to make it difficult to fall asleep later on (i.e., they reduce the homeostatic sleep drive).

Behavioral model (Spielman's 3P model)

In contrast to the stimulus control model, Spielman's 3P model proposes there are certain factors associated with both the onset of acute insomnia and the transition from acute to chronic insomnia. The first two sets of factors (predisposing and precipitating factors) represent a stress-diathesis conceptualization of how insomnia comes to be expressed, whereas the third set of factors (perpetuating factors) explain the mechanisms by which insomnia can become chronic. Predisposing factors increase the underlying vulnerability to develop insomnia and comprise biological features such sex or a genetic predisposition to develop insomnia and psychological traits such as the tendency to worry and/or ruminate [31, 32]. A predisposition to sleep disturbance, however, requires a sufficiently stressful *precipitant*, or combination of precipitants, before it may be expressed [4]. Perpetuating factors refer to the behaviors that an individual engages in while attempting to manage sleep continuity disturbance, which in turn actually contribute to the persistence of insomnia. Examples include: going to bed earlier, napping during the day, and delaying time out of bed. Collectively, these behaviors are referred to as "sleep extension." While typically used as an attempt to recover lost sleep, sleep extension instead results in a mismatch between sleep opportunity (i.e., how much time the person spends in bed) and sleep ability (i.e., how much time the person actually sleeps), thereby, increasing the likelihood of stimulus dyscontrol. Sleep extension also perpetuates insomnia by attenuating the homeostatic drive (i.e., pressure) to sleep, thus making it harder to fall and/or stay asleep. Taken together, addressing these two problems (i.e., stimulus dyscontrol and a reduced homeostatic sleep drive) are the primary therapeutic targets of cognitive behavioral therapy for insomnia (CBT-I; see "What is CBT-I" for more information).

Neurocognitive model

The neurocognitive model extends the behavioral models of insomnia by suggesting that the repeated pairing of sleep-related stimuli with insomnia-related wakefulness leads to conditioned cortical hyperarousal. While hyperarousal is widely considered the underlying factor that gives rise to

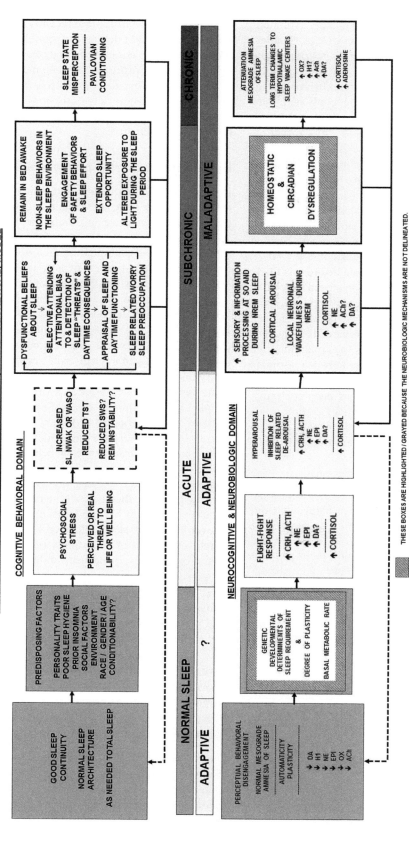

FIG. 28.1 Parallel process (Trans-theoretical) model—The parallel process model is provided to illustrate how the cognitive and behavioral domains may be viewed as parallel processes to the neurocognitive and neurobiologic domains. *ACH*, Acetylcholine; *ACTH*, adrenocorticotropic hormone; *CRH*, corticotrophin-releasing hormone; *DA*, dopamine; *EPI*, epinephrine; *H1*, histamine-1 receptor antagonist; *NE*, norepinephrine; *NWAK*, number of awakenings; *OX*, orexin; *SL*, sleep latency; *SO*, sleep onset; *TST*, total sleep time; *WASO*, wake after sleep onset. Adapted with permission from Perlis ML, Ellis JG, Kloss JD, Riemann D. Etiology and pathophysiology of insomnia. In: *Principles and practice of sleep medicine*. (2016) [Cited 24 July 2018]. p. 769–84.

insomnia [33], a strength of the Neurocognitive Model is that it proposes a pluralistic perspective of hyperarousal, such that there are several forms of hyperarousal (e.g., cortical, cognitive, and somatic arousal). While the model suggests that hyperarousal may be construed in at least three dimensions, it is cortical hyperarousal (which may be indexed by increases in high frequency EEG activity [beta/Gamma activity between 16 and 45 Hz]), that is central to the etiology and pathophysiology of insomnia [8]. Heightened cortical arousal is hypothesized to (1) allow for increased levels of sensory and information processing at and around sleep onset and during NREM sleep or (2) for the attenuation of the normal mesograde amnesia (middle of the night) that occurs in association with sleep. These two phenomena are hypothesized to increase the probability of difficulties falling and staying asleep and to contribute to sleep state misperception [34].

Cognitive model

Harvey's cognitive model of insomnia posits that, in chronic insomnia, increased negative cognitive biases (e.g., sleep-related worry, selective attention and monitoring, and the detection of sleep-related threats) perpetuate a level of physiologic arousal that interferes with sleep initiation or sleep maintenance [35]. In turn, this increased physiological arousal both during the day and at night initiates an attentional bias and a monitoring of perceived internal (e.g., body sensations for signs of fatigue) and external (e.g., the alarm clock) sleep-related threats that might indicate to a person that they did (or will) not receive enough sleep. Taken together, these processes lead to an exaggerated perception of sleep continuity disturbance and its potentially negative impact on daytime performance. Adding to the daytime dysfunction is the tendency to hold incorrect beliefs about the impact of sleep disruption and the utility of worrying and/or engaging in safety behaviors (e.g., cancelling appointments or taking a nap during the day). The cognitive model highlights the importance of targeting specific factors that maintain the disorder (i.e., attentional bias) and eliminating the use of safety behaviors in the successful treatment of insomnia.

Psychobiological inhibition model

The psychobiological inhibition model suggests that difficulty with sleep initiation and maintenance is caused by the failure to inhibit wakefulness [36], as opposed to the conditioned hyperarousal [37]. Under normal circumstances, sleep occurs passively (without attention, intention, or effort). In acute insomnia, acute stress precipitates both physiologic and psychological arousal, which can result in the inhibition of sleep-related dearousal and the occurrence of selective attending to the life stressors, and ultimately, interfere with the normal homeostatic and circadian regulation of sleep. Acute insomnia may, in turn, resolve or be perpetuated based on whether the stressor resolves or if the individual instead attends to the insomnia symptoms that occur with the acute insomnia. In chronic insomnia, failure to inhibit wakefulness is thought to occur from an activation of the cognitive attention-intention-effort (A-I-E) pathway. When an individual experiences sleeplessness, their attention shifts towards the process of sleep, something that is typically an automatic and passive event. This shift in attention prevents the normal disengagement from wakefulness and makes the acquisition of sleep an intentional activity, where the person begins to demonstrate active effort to sleep and further weakens the processes related to the inhibition of wakefulness.

Parallel process (trans-theoretical) model

Each of the models presented above provides a unique perspective, and for the most part, none are mutually exclusive. In recognition of this, we provide in Fig. 28.1 an integrative perspective, parallel process model [38]. This model is intended to illustrate how [1] all of the identified factors may be contributory and [2] the cognitive and behavioral domains may be viewed as parallel processes to the neurocognitive and neurobiologic domains. That is, the perspective that the cognitive-behavioral and the neurocognitive-neurobiologic domains represent two sides of the same phenomena.

INSOMNIA AND PSYCHIATRIC MORBIDITY

Insomnia (both the symptom and the disorder) is a substantial risk factor for psychiatric morbidity [39] (Fig. 28.2). While sleep continuity disturbance is associated with a number of psychiatric conditions, it is a diagnostic feature for multiple disorders, such as depression, generalized anxiety disorder, and post-traumatic stress disorder (PTSD) [8]. Insomnia may be common in multiple psychiatric disorders because they share: (1) a similar trigger (e.g., stress) and/or underlying pathophysiology (e.g., serotonin deficiency) [40–42] and (2) similar functional consequences (e.g., dysfunctional beliefs) [43]. With respect to similar trigger, most research suggests that an increase in stress is the most common precipitating event observed in insomnia, but also in depression, anxiety, and by nature of the disorder, PTSD. More recent literature has also suggested that a potential underlying pathophysiology common among these disorders may be a decrease in serotoninergic neurotransmission [42]. With respect to similar functional consequences, insomnia and some of these other psychiatric conditions share many of the same symptoms or clinical presentation, such as depressed

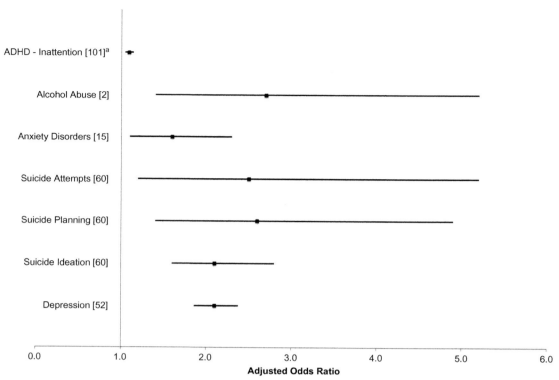

FIG. 28.2 Insomnia as a risk factor for multiple psychiatric conditions. The forest plot below provides the adjusted odds ratio estimates for insomnia by psychiatric condition or phenomena. [a]Represents adjusted odds ratio for ADHD-Inattention with Insomnia as predictor.

mood, worry, fatigue, and difficulty concentrating. For example, a core feature of insomnia is somatic and cognitive hyperarousal [44], which may increase severity of symptoms for both mood and anxiety disorders. Hyperarousal is also a core feature of PTSD, and is known to interact with the hyperarousal present in insomnia [45]. Another consequence of insomnia is emotional dysregulation. Studies show that negative emotional experiences (i.e., mood and affect) are more common in insomnia, especially at night [46, 47]. Insomnia has also been implicated in functional deficits across a wide range of domains [48] as well as reduced quality of life [49]. Taken together, an important question for the field is whether insomnia and these other disorders represent distinct clinical phenomena that are inter-related, such that they increase risk for one another or whether they represent a common disorder/disease with an array of possible clinical sequelae. To this end, the sections below review the relationships between insomnia and a number of psychiatric disorders, particularly with respect to prevalence, directionality of the association, and shared risk factors. Please note that while attention-deficit/hyperactivity disorder (ADHD) and autism spectrum disorder (ASD) may be better considered neurodevelopmental disorders, they were included given that they are a part of the psychiatric nosology (i.e., DSM-5) and are highly comorbid with insomnia.

Depressive disorders

Major depressive disorder (MDD) is a common and heterogeneous disorder that is primarily characterized by episodes (of at least 2 weeks) of depressed mood and/or anhedonia (a loss of interest or pleasure in daily activities) [8]. A major depressive episode (MDE) may also consist of significant changes in weight, sleep, psychomotor activity, fatigue, feelings of worthlessness/guilt, concentration, and/or suicidality. Unlike normal fluctuations in mood, MDD causes significant distress and/or functional impairment. Among all psychiatric disorders, the comorbidity between insomnia and depression is highest. Up to 90% of individuals with MDD experience insomnia [17]. The challenge in understanding the association between insomnia and depression is that insomnia can represent both a risk factor (or prodrome) and a consequence of depression (i.e., the relationship is bidirectional). For example, residual insomnia after treatment for depression [50], is the largest predictor of a subsequent MDE [51]. A meta-analysis of 21 longitudinal studies also identified insomnia as a significant predictor of the onset of an MDE, such that those with insomnia, compared to those without, were twice as likely to develop depression [52]. A more recent meta-analysis of 34 cohort studies showed similar results (relative risk of developing depression was 2.3 among those with insomnia; [53]). These rates have been shown to be four times greater in adolescent samples [54]. Similarly, persistent poor sleep is a known risk factor for relapse after pharmacological [55] and nonpharmacological [56] treatment for MDD.

Despite research suggesting that insomnia is more than just a symptom of depression, and indeed, a risk factor, the mechanisms that explain the association between insomnia and depression are still unclear [17]. Staner [43] proposed that insomnia may directly and/or indirectly cause a depressive episode. For example, insomnia and/or insomnia-related daytime consequences may directly increase depressed mood and other depressive symptoms (e.g., increases in anhedonia, fatigue, and concentration problems). The severity and persistence of these symptoms may ultimately reach a point that is consistent with diagnostic criteria for an MDE. Alternatively, insomnia may cause and/or be caused by a common factor that also contributes to the development of MDD (i.e., indirect causality). The most obvious example of this is stress. Acute stressful life events can simultaneously lead to the development of both insomnia and depressive symptoms (albeit the insomnia symptoms may manifest first), but also, insomnia or insomnia-related daytime consequences may produce additional stress, which may culminate in the development of depression. A number of biological factors have also been identified as potential mechanisms that explain how insomnia may be a risk factor for MDD or at the very least explain why these two disorders are highly comorbid. These biological mechanisms include: monoaminergic neurotransmission, abnormalities in circadian genes, overactivity of the hypothalamic-pituitary-adrenal (HPA) axis, and impaired functioning of plasticity-related gene cascades [42, 57]. The nature and/or role that theses biological mechanisms have on the association between insomnia and depression are unknown, but what is clear is that these are potential targets for future research.

Suicide

Insomnia has also been implicated in suicide. A meta-analysis indicated that the presence of insomnia is associated with an approximately threefold likelihood of suicide ideation, attempts, and death by suicide [58]. This research has been supported by population-level studies that show that suicidal ideation, even in the general population is predicted by insomnia symptoms [59, 60]. While research consistently supports the association between insomnia and suicidal ideation, the specific mechanisms by which insomnia confers risk for suicidal ideation are unknown. A number of psychological and physiological mechanisms have been proposed [61, 62]. Potential psychological mechanisms include: insomnia-related psychosocial impairments (that may or may not be related to loneliness and lack of belonging); activation of hopelessness and/or helplessness schema; and/or diminished executive function. Potential physiological mechanisms, which may occur as a result of insomnia or a common neurobiologic substrate, include: serotonin deficiency; hypercortisolemia; and/or elevated basal metabolic rate. There has also been additional evidence that suicides are disproportionately likely to occur at night [63] suggesting that nocturnal wakefulness itself may represent a risk factor for suicide [64]. That is, being awake at night, and the associated hypofrontality that occurs during the night and/or with sleep loss (i.e., decreased frontal lobe function), may be another mechanism by which insomnia increases risk for suicidal ideation [64]. Insomnia increases the likelihood of being awake at night, the time of day in which one's ability to reason, think rationally, and to engage in impulse control may be at its lowest [65, 66]. Being awake at night, especially during times of increased stress or mood disturbance, may therefore increase risk for suicidal ideation.

Bipolar disorder (BPD)

In contrast to depression, bipolar disorder is characterized by episodes of both mania (and/or hypomania) and depression. With respect to the depressive episodes in bipolar disorder, it is not surprising that similar challenges arise when teasing apart whether insomnia is a risk factor or a consequence (or symptom) of the depression. With respect to the manic episodes, sleeplessness is also a symptom of mania, but qualitatively differs from insomnia in that patients experiencing manic episodes think and feel as though they do not *need* sleep [67]. Of note, the diagnostic criteria for insomnia states that the patient must experience difficulty initiating and/or maintaining sleep despite "adequate opportunity for sleep" [8], a condition that is often not met in patients experiencing manic episodes. Despite these nuances in diagnostic criteria, sleep continuity disturbance is highly prevalent among patients with bipolar disorder [68, 69], and the most common prodromal symptom of a manic episode [70]. Sleep continuity disturbance is greater among patients with bipolar disorder, relative to controls, and nearly as severe as in patients with insomnia (without bipolar disorder). For example, lower sleep efficiency, night-to-night variability, and longer sleep latency have also been shown to be significantly associated with a history of depression in patients with bipolar disorder [71]. Talbot et al. [72] also reported that, in subjects with bipolar disorder, morning negative mood varied as a function of total sleep time during the preceding night [72]. Similarly, sleep loss is highly correlated with daily manic symptoms among patients with bipolar [70]. These findings support the notion that there is a reciprocal relationship between sleep continuity disturbance and daily changes in mood [72]. As with all of the other psychiatric disorders discussed in this chapter, the nature of the association between insomnia and bipolar disorder is elusive. There has been some evidence for a possible genetic linkage between sleep/circadian dysregulation and bipolar disorder. A number of "clock" genes (e.g., per3 and gsk3) are associated with sleep and mood regulation, and may have implications for the development and/or treatment of both insomnia and bipolar [69, 73, 74].

Anxiety disorders

Like depressive disorders, anxiety disorders are heterogeneous in their phenotypic presentation. So much so, that the literature on the association between insomnia and each individual type of anxiety disorder is limited and therefore beyond the scope of the present chapter. Here, we discuss the relationship between insomnia and anxiety disorders, broadly defined, and go into more detail regarding the link between insomnia and PTSD in the next section. Common among anxiety disorders are feelings of intense fear or worry, avoidance-related behaviors, and that these feelings and behaviors are triggered by stress [75]. Not surprisingly, these are also core features of insomnia, and may possibly explain the high comorbidity between the two phenomena [54, 76]. According to one study, the prevalence rate of clinically significant anxiety among subjects with insomnia is nearly 20%, as compared to 3% in subjects without comorbid insomnia [77]. A large prospective study (approximately 25,000 Norwegian adults) also showed that subjects with insomnia were significantly more likely to endorse an anxiety disorder at follow-up. Similarly, those that endorsed insomnia were more than three times as likely to also report a concurrent anxiety disorder [15]. The evidence for anxiety as a potential risk factor for insomnia is even more compelling. An epidemiological study among Swedish adults supported that clinically significant anxiety increased the risk for developing clinically significant insomnia more than fourfold [78]. Another study that surveyed nearly 15,000 Europeans found that in new onset cases of anxiety disorders (that were also comorbid with insomnia), insomnia preceded the anxiety disorder 18% of the time, the two disorders appeared simultaneously approximately 39% of the time, and the anxiety disorder came first approximately 44% of the time [50]. A separate study found that, in contrast to depression where insomnia occurred first in 69% of comorbid cases, anxiety disorders preceded insomnia 73% of the time [54]. Anxiety disorders and insomnia have often been thought of as disorders sharing certain vulnerabilities and characteristics and are often treated with similar pharmacological and behavioral interventions [79]. Because of the high comorbidity and the bidirectional relationship between insomnia and anxiety, Uhde et al. [80] have proposed two potential explanatory models [80]. In the first model, anxiety and insomnia represent different dimensions of a common underlying disorder, whereby different clinical symptoms may emerge as a result of repeated stress. In the second model, anxiety and insomnia represent different neurobiological disorders whereby each separately causes remarkably similar symptoms or that they are both produced by another widely prevalent third factor. How these disorders are conceptualized has important treatment implications; namely, will the treatment of one diminish symptoms of the other or should they be treated simultaneously?

Post-traumatic stress disorder (PTSD)

PTSD refers to a disorder characterized by a complex set of symptoms that arise following a life-threatening or traumatic event and typically include: [1] re-experiencing the traumatic event; [2] avoidance of stimuli that resemble or remind one of the event; [3] negative thoughts or feelings; and/or [4] increased arousal and/or reactivity (including difficulty sleeping). Studies suggest that there is a 60%–90% chance of experiencing insomnia following a traumatic event [81, 82]. Sleep disturbance, in general, and insomnia and nightmares, in specific, has been referred to as the cardinal symptom of post-traumatic stress disorder [83, 84]. Moreover, insomnia symptoms following a traumatic event may increase the likelihood of developing PTSD and/or PTSD severity [85–87]. Other studies have also shown that treating insomnia among patients with PTSD may indirectly reduce the severity of PTSD symptoms [88–90]. Insomnia may therefore represent a risk factor and/or prodromal symptom of PTSD. In fact, insomnia is the most frequently reported symptom among individuals with PTSD and does not remit with otherwise successful first line interventions [91, 92]. With respect to sleep continuity disturbance, difficulty initiating and maintaining sleep at least "sometimes" was reported in 44% and 91% (respectively) of a sample comprised of Vietnam veterans with PTSD [81]. Similarly, Pigeon et al. [93] found that clinical levels of insomnia were significantly associated with greater baseline PTSD severity, but also predicted increases in PTSD 6-months later, such that approximately 38% of subjects with insomnia endorsed PTSD at follow-up (as compared to 5% of subjects without insomnia) [93]. Consistent with other psychiatric disorders, the association between insomnia and PTSD also appears to be bidirectional, in that insomnia is a probable consequence of PTSD, yet insomnia may also further perpetuate the PTSD symptoms. The mechanisms that explain this bidirectional relationship are unknown, but relevant targets have been identified, such as, hyperarousal related to increased noradrenergic activity [94, 95], sleep-related anxiety [96], comorbid depression [97, 98].

Attention-deficit/hyperactivity disorder (ADHD)

Attention-deficit/hyperactivity disorder (ADHD) is a neurodevelopmental disorder that emerges in youth (before age 12) and typically runs a lifelong course [99]. ADHD is characterized by a "persistent pattern of inattention and/

or hyperactivity that interferes with functioning and/or development" (DSM-5). While ADHD is more commonly studied in youth, ADHD in adulthood is also common (an estimated prevalence of 14 million adults in the United States; [100]). Common among patients with ADHD is comorbid insomnia. According to results from a recent study, 67% of adult patients with ADHD also met DSM-based criteria for insomnia. This is compared to 28% of adults in the sample who did not have a diagnosis of ADHD [101]. The most common type of sleep continuity disturbance reported in ADHD patients is difficulty initiating sleep (also referred to as chronic sleep-onset insomnia or initial insomnia) [102–104]. While a prolonged sleep latency has often been thought to be a side effect of stimulant medication [105–107], other studies have hypothesized that initial insomnia may be instead related to abnormalities in circadian functioning among patients with ADHD. Specifically, Van Veen et al., (2010) assessed dim-light melatonin onset (DLMO) and rest-activity patterns in ADHD patients with initial insomnia, as compared to ADHD patients without initial insomnia and healthy controls [108]. Their data supported that ADHD patients with initial insomnia had a delayed DLMO and a reduced 24-h amplitude in their rest-activity cycle. Insomnia, in patients with ADHD, may therefore instead be a natural consequence of the mismatch between a patient's [delayed] circadian rhythm and their attempt to adhere to a 'normal' sleep schedule (due to obligations at home, school, or work). For example, as the research noted above suggests, patients with ADHD may be more likely to have a delayed circadian rhythm (i.e., natural tendency to want to go to sleep later), and consequently, experience insomnia when they attempt to go to sleep early. A delayed circadian rhythm in ADHD may be related to age (patients with ADHD are typically younger) or may be a byproduct of the disorder. This said, future work evaluating the association between insomnia and ADHD, should consider the effect of age and how age-related differences in chronobiology may account for the presence and/or absence of insomnia. With regard to treatment, subjects with ADHD who are currently being treated with stimulant medication report less severe insomnia [101]. This suggests that the pharmacological treatment of ADHD does not exacerbate sleep continuity disturbance, and that sleep continuity disturbance may be a consequence of ADHD given that it subsides with the effective treatment of ADHD symptoms.

Alcohol use disorder (AUD)

In contrast to prior versions of DSM criteria that divided alcohol use problems into abuse and dependence, DSM-5's Alcohol Use Disorder (AUD) makes no such distinction. AUD refers to, among others, problems related to how or how long one drinks, difficulty stopping or cutting back one's drinking, and functional/emotional impairment as a result of drinking [8]. Alcohol use problems and insomnia, not surprisingly, are highly comorbid [109, 110]. While specific prevalence rates vary by how insomnia and/or alcohol use disorders are defined, Brower et al. [111] showed that 18% of subjects who met criteria for an alcohol use disorder also endorsed significant levels of insomnia (this is compared to 10% of subjects without alcohol problems). In a study of subjects being treated for alcohol dependence, 61% of subjects endorsed insomnia symptoms during the pre-treatment phase. Sleep continuity disturbance has also been shown to persist through the early stages of alcohol recovery (up to 5 weeks following the abstinence of alcohol) [111]. In addition, subjects who endorsed insomnia symptoms were more likely to report using alcohol to self-medicate for their sleep problems, had more severe alcohol dependence, and were more likely to relapse to alcohol use (60% of subjects), as compared to subjects without insomnia (30% of subjects) [111]. Alternatively, in subjects with insomnia, 7%–19% of subjects also reported significant alcohol use problems (this is compared to 4%–9% of subjects without insomnia) [2, 112]. The economic burden of alcohol-related insomnia is also considerable given that insomnia accounts for approximately 10% of all alcohol-related costs. That is the equivalent of about 28 billion dollars each year in the United States alone (please note the cost was converted to account for inflation) [113, 114]. It is no surprise that alcohol is one of the most common self-medicating substances used among patients with insomnia [115, 116], given its sedating effects [109]. This said, alcoholism is a significant predictor (i.e., risk factor) of insomnia [111, 112]. The prevalence and economic/societal burden of comorbid insomnia and alcohol use disorders is relatively well defined, however, the neurobiological and psychosocial factors that explain this association are less clear. As indicated above, alcohol is considered a sedative, and therefore has the potential to promote the initiation of sleep. The sleep-promoting effects of alcohol, however, are dose-dependent and do not persist with continued alcohol use (i.e., after 3 days of continued use), and more importantly, alcohol can significantly disrupt both sleep continuity and sleep architecture (i.e., increased likelihood of REM sleep inhibition, nocturnal arousals, and rebound insomnia during the second part of the night). Taken together, for individuals with insomnia, especially initial insomnia, alcohol might be appealing as a convenient, low-cost hypnotic, yet the overall effects are detrimental and can ultimately lead to greater sleep continuity disturbance (particularly the perpetuation of middle and late insomnia) and more severe alcohol dependence (increased amounts of alcohol are required to achieve the same sedating effect). In conclusion, comorbid insomnia and alcohol use disorders are prevalent, costly to society, and associated with worse overall outcomes.

Autism spectrum disorder (ASD)

The association between insomnia and autism spectrum disorder (ASD) has recently become a topic of interest given the increasing literature on sleep problems in children with ASD and the subsequent negative effects those sleep problems have on their overall functioning and quality of life [117]. Prevalence rates for sleep problems range from 50 to 80% of children with ASD [118, 119]. While children with ASD experience a multitude of sleep problems, the most commonly reported sleep problem among parents of children with ASD is insomnia, and more specifically, an extended sleep latency (i.e., initial insomnia) [120–122]. The link between ASD and insomnia is multifactorial. ASD is associated with abnormalities in several neurotransmitters that are also implicated in insomnia (e.g., GABA, serotonin, and melatonin) [123–126], and therefore, the two disorders may share a common neurobiological core. Alternatively, insomnia in individuals with ASD may be related to medical (e.g., epilepsy and GERD) and psychiatric comorbidities (depression and ADHD), the medications used to treat these comorbidities or the behavioral/emotional consequence of ASD (e.g., difficulties with transitions and emotion regulation) [117, 127]. While it is unknown whether insomnia is a risk factor for a more severe course of ASD, there is some evidence to support that sleep continuity disturbance may exacerbate ASD symptoms and/or the functional consequences of ASD (e.g., increases in emotion dysregulation, inattention, and family stress) [117, 128, 129].

Schizophrenia

The association between schizophrenia and insomnia has also been documented [130]. Similar to most other psychiatric disorders, their co-occurrence is prevalent and their relationship bidirectional. Significant sleep disturbances are present in roughly 35%–50% of individuals with schizophrenia and other psychotic disorders [131–133], and are associated with exacerbated positive symptoms [134] and a reduced quality of life [131, 135–138]. Past research supports that sleep continuity disturbance is a risk factor for relapse and may even be a prodromal sign of a psychotic relapse [139]. Moreover, insomnia symptoms increase over time following anti-psychotic medication withdrawal, which supports the notion that insomnia is also a consequence of schizophrenia that can be resolved with the successful treatment of the positive symptoms of schizophrenia [140]. While insomnia as a natural consequence of schizophrenia makes sense (i.e., patients with schizophrenia often have irregular schedules, increased depression and anxiety, extensive medication regimens), how insomnia increases risk for psychotic symptoms is much less clear. There are, however, some studies indicating that low melatonin levels at night may be a common neurobiological feature of both insomnia and schizophrenia, and exogenous melatonin administration may improve the overall quality and quantity of sleep in patients with schizophrenia [141]. Like ADHD, age and age-related differences in chronobiology may also explain part of the association between insomnia and schizophrenia, and thus, should be the focus of future work.

BEHAVIORAL TREATMENT OF INSOMNIA

What is CBT-I?

Cognitive behavioral therapy for insomnia (CBT-I) combines principles from stimulus control and sleep restriction therapy with formal cognitive restructuring in order to target hyperarousal, dysfunctional behaviors and maladaptive thoughts, beliefs, and attitudes about sleep [142]. Stimulus control targets a person's tendency to engage in behaviors other than sleep in the bedroom (e.g., reading or watching television in bed) thereby weakening the association between the sleep environment and the physiologic state of sleep. Stimulus control instructions, in general, are simple. Patients are to avoid using the bed for activities other than sleep or intimacy and get out of bed if unable to sleep within 15–20 min and return to bed only when sleepy [143]. The primary goal of sleep restriction is to address the mismatch between sleep opportunity (time in bed) and sleep ability (time asleep) by limiting the amount of time a patient spends in bed to the amount of time that they are actually sleeping. Sleep restriction has the following important objectives: (1) it increases the homeostatic sleep drive (i.e., sleep "pressure") and reduces the heightened arousal caused by an individuals' effort to force sleep, and (2) it reduces time spent awake during the night by consolidating sleep into longer, more restorative periods. As mentioned above, cognitive restructuring is also commonly used in conjunction with the more behavioral interventions. The primary goal of cognitive restructuring is to identify problematic thoughts that may contribute to the development of, or reinforce, behaviors that produce pre-sleep worry, examine these thoughts for accuracy and if necessary, modify them to be more rational and/or realistic.

CBT-I was recently endorsed by the American College of Physicians [144]. Specifically, the official position of the ACP is that not only is CBT-I recommended as the first-line therapy of choice, but that pharmacotherapy is only recommended in cases where its use is short-term, and/or in combination with behavioral treatment, and/or after discussion with patients regarding the limitations of this approach [144]. More specific information regarding the delivery of CBT-I can be found elsewhere [143].

CBT-I in the context of psychiatric disorders

A recent meta-analysis of 37 randomized controlled trials evaluated the impact of CBT-I not only on insomnia severity and sleep continuity disturbance but also on the symptoms of the comorbid disorder in a sample of 2189 participants [145]. Ten of the studies included in this meta-analysis were conducted in patients with comorbid psychiatric conditions (i.e., substance use disorders, depressive disorders, and PTSD), 26 with medical comorbidities, and 1 with a mixed sample. While patients with comorbid medical conditions, as compared to those with comorbid psychiatric conditions, improved equally on symptoms of insomnia, a larger effect of CBT-I was found for reducing symptoms of the comorbid psychiatric conditions ($g=0.76$) than symptoms of medical comorbidities ($g=0.20$). These data suggest that psychiatric symptoms may be more responsive to CBT-I than those associated with a medical condition. Consistent with other studies, the improvement in sleep continuity was maintained for 3–12 months after completing CBT-I. A number of recent trials of CBT-I have been conducted in patients with comorbid insomnia and depression [146, 147]. These RCTs provide strong evidence that treatment of insomnia (alone or in combination with pharmacotherapy) in patients with depression can produce comparable effects for both depression and insomnia symptoms. This opens treatment options for patients who have not adequately responded to antidepressant medication or who would prefer a non-pharmacological option. Similarly, A meta-analytic review of the effect of CBT for anxiety on comorbid sleep continuity disturbance concluded that CBT for anxiety could be expected to have a moderate effect on sleep outcomes (effect size = 0.52) but that residual sleep problems should be expected [148].

There is a growing interest in the application of CBT-I with patients who have other potentially more serious psychiatric conditions such as PTSD, bipolar mood disorders, and psychotic disorders, but these areas are relatively underdeveloped compared to depressive and anxiety disorders. For example, there is growing evidence that CBT-I, alone or in combination with PTSD-specific treatment components, can significantly improve both subjective and objective sleep outcomes in patients with PTSD [89, 149]. While treatment for insomnia comorbid with psychotic disorders is still largely pharmacological [150], some early work has tested the ability of CBT-I to improve sleep and reduce delusions and hallucinations [151]. Similarly, a modified version of cognitive behavioral therapy for insomnia for patients with bipolar disorder (CBTI-BP) has demonstrated efficacy for both reducing sleep continuity problems but also reducing mood symptoms. Importantly, CBTI-BP takes a more conservative approach to sleep restriction (time in bed is restricted to no less than 6.5 h), given that it has the potential to increase the risk for a hypomanic/manic episode [152]. This said, subjects that underwent CBTI-BP were at a reduced risk for a mood episode relapse as compared to a psychoeducation control group (27% reduction in probability for manic/hypomanic episodes and 28.5% reduction for depressive episodes) [153].

CONCLUSION

Sleep has clear importance for the maintenance of physical and psychological health, making insomnia a serious public health concern. Insomnia increases the risk for, and severity of, a number of psychiatric conditions. When left unaddressed, insomnia negatively impacts the ability of the individual to completely recover from their disorder. Evidence based treatment for insomnia, CBT-I, exists and is effective when delivered in individuals with comorbid psychiatric conditions, but it is currently underutilized. Increased attention and awareness of the importance of treating insomnia is needed. There is a clear and profound association between poor sleep and poor mental health. Apart from this association, there is the possibility that good sleep continuity may not only ward off new onset disease but that it may also promote good mental health.

REFERENCES

[1] Harvey G. Insomnia, psychiatric disorders and the transdiagnostic disorders perspective. Assoc Psychol Sci 2008;17(5):299–303.

[2] Ford DE, Kamerow DB. Epidemiologic study of sleep disturbances and psychiatric disorders an opportunity for prevention? JAMA 1989;262(11):1479–84.

[3] American Psychiatric Association. Diagnostic and statistical manual of mental disorders (DSM). Washington, DC: APA; 1994. 866. Available from: https://books.google.com/books?hl=en&lr=&id=-JivBAAAQBAJ&oi=fnd&pg=PT18&dq=diagnostic+and+statistical+manual+&ots=cdWQ04LKx7&sig=wBBV2FDDB-n4dI3p-wQFUyYTyW2A. [Cited 3 April 2017].

[4] The International classification of sleep disorders. In: American Academy of Sleep Medicine, editors. American Academy of Sleep Medicine. 2nd ed. Darien, IL: American Academy of Sleep Medicine; 2005. Available from: http://scholar.google.com/scholar?hl=en&btnG=Search&q=intitle:THE+INTERNATIONAL+CLASSIFICATION+OF+SLEEP+DISORDERS,+REVISED#0. [Cited 25 June 2018].

[5] American Academy of Sleep Medicine. International classification of sleep disorders. 3rd ed. Darien, IL: American Academy of Sleep Medicine; 2014.

[6] Taylor DJ, Lichstein KL, Durrence HH, Reidel BW, Bush AJ. Epidemiology of insomnia, depression, and anxiety | sleep | Oxford Academic. Sleep 2005;28(11):1457–64.

[7] Lichstein KL. Secondary insomnia: a myth dismissed. Sleep Med Rev 2006;10(1):3–5. Available from: http://www.ncbi.nlm.nih.gov/pubmed/16380276. [Cited 12 June 2018].

[8] Association AP. Diagnostic and Statistical Manual of Mental Disorders (DSM-5®)—American Psychiatric Association—Google Books. 5th ed. 2013. Available from: https://books.google.com/books?hl=en&lr=&id=-JivBAAAQBAJ&oi=fnd&pg=PT22&dq=dsm-5&ots=cePS70QFAc&sig=5N4CULNv6AcocfgcLDTqOq0uH5Y#v=onepage&q=dsm-5&f=false. [Cited 12 June 2018].

[9] Ellis JG, Perlis ML, Neale LF, Espie CA, Bastien CH. The natural history of insomnia: focus on prevalence and incidence of acute insomnia. J Psychiatr Res 2012;46(10):1278–85. Available from: https://doi.org/10.1016/j.jpsychires.2012.07.001.

[10] NIH. NIH State-of-the- Science Conference Statement on Manifestations and Management of Chronic Insomnia in Adults. Bethesda (MD): NIH; 2005.

[11] Gencarelli A, Khader W, Morales K, et al. A one year study of 1,069 good sleepers: the incidence of acute and chronic insomnia. Sleep 2018;41:A137. Available from: https://scholar.google.com/scholar?hl=en&as_sdt=0,39&cluster=9092140357354863201. [Cited 25 June 2018].

[12] Grandner M, Petrov ME, Rattanaumpawan P, Jackson N, Platt A, Patel NP. 2013 Sleep symptoms, race/ethnicity, and socioeconomic position, 9(9):897-905 ncbi.nlm.nih.gov. Available from: https://www.ncbi.nlm.nih.gov/pmc/articles/PMC3746717/ [Cited 12 June 2018].

[13] Riedel BW, Lichstein KL. Insomnia and daytime functioning. Sleep Med Rev 2000;4(3):277–98. Available from: http://www.ncbi.nlm.nih.gov/pubmed/12531170. [Cited 6 May 2014].

[14] Benca RM, Peterson MJ. Insomnia and depression. Sleep Med 2008;9(Suppl. 1):3–9.

[15] Neckelmann D, Mykletun A, Dahl A. Chronic insomnia as a risk factor for developing anxiety and depression. Sleep 2007;30:873–80. Available from: https://www.researchgate.net/profile/Alv_Dahl/publication/6156199_Neckelmann_D_Mykletun_A_Dahl_AA_Chronic_insomnia_as_a_risk_factor_for_developing_anxiety_and_depression_Sleep_30_873-80/links/00b4953a12bf18dafd000000.pdf. [Cited 29 May 2017].

[16] Perlis ML, Smith LJ, Lyness JM, Matteson SR, Pigeon WR, Jungquist CR, et al. Insomnia as a risk factor for onset of depression in the elderly. Behav Sleep Med 2010;4(2):85–103. Available from: http://www.tandfonline.com/doi/abs/10.1207/s15402010bsm0402_3. Cited 29 May 2017.

[17] Riemann D, Voderholzer U. Primary insomnia: a risk factor to develop depression? J Affect Disord 2003;76(1–3):255–9. Available from: http://linkinghub.elsevier.com/retrieve/pii/S0165032702000721. [Cited 18 May 2014].

[18] Roane BM, Taylor DJ. Adolescent insomnia as a risk factor for early adult depression and substance abuse. Sleep 2008;31(10):1351–6. Available from: http://www.ncbi.nlm.nih.gov/pubmed/18853932%5Cnhttp://www.pubmedcentral.nih.gov/articlerender.fcgi?artid=PMC2572740. [Cited 29 May 2017].

[19] Bathgate CJ, Edinger JD, Wyatt JK, Krystal AD. Objective but not subjective short sleep duration associated with increased risk for hypertension in individuals with insomnia. Sleep 2016;39(5):1037–45. Available from: http://www.ncbi.nlm.nih.gov/pubmed/26951399%5Cnhttp://www.pubmedcentral.nih.gov/articlerender.fcgi?artid=PMC4835301.

[20] Suka M, Yoshida K, Sugimori H. Persistent insomnia is a predictor of hypertension in Japanese male workers. J Occup Health 2003;45(6):344–50. Available from: https://www.jstage.jst.go.jp/article/joh/45/6/45_6_344/_article/-char/ja/. [Cited 29 May 2017].

[21] Vgontzas A, Liao D, Bixler E, Chrousos G. Insomnia with objective short sleep duration is associated with a high risk for hypertension. Sleep 2009;32(4):491–7. Available from: http://ctsi.psu.edu/wp-content/uploads/2014/09/Handout-for-Sept-8.pdf. [Cited 29 May 2017].

[22] Mallon L, Broman JE, Hetta J. High incidence of diabetes in men with sleep complaints or short sleep duration: a 12-year follow-up study of a middle-aged population 43. Diabetes Care 2005;28:2762–7. (0149–5992 (Print)). Available from: http://care.diabetesjournals.org/content/28/11/2762.short. [Cited 29 May 2017].

[23] Vgontzas AN, Liao D, Pejovic S, Calhoun S, Karataraki M, Bixler EO. Insomnia with objective short sleep duration is associated with type 2 diabetes: a population-based study. Diabetes Care 2009;32(11):1980–5. Available from: http://care.diabetesjournals.org/content/32/11/1980.short. [Cited 29 May 2017].

[24] Sofi F, Cesari F, Casini A, Macchi C, Abbate R, Gensini GF. Insomnia and risk of cardiovascular disease: a meta-analysis. Eur J Prev Cardiol 2014;21(1):57–64. Available from: http://journals.sagepub.com/doi/10.1177/2047487312460020. [Cited 29 May 2017]..

[24a] Spielman AJ, Glovinsky PB. The varied nature of insomnia. Case Stud Insomnia 1991;1–15.

[25] Cannon W. Bodily changes in pain, hunger, fear and rage. New York: Appleton; 1929. 404. Available from: http://psycnet.apa.org/psycinfo/1929-04389-000. [Cited 29 May 2017].

[26] Jansen ASP, Van NX, Karpitskiy V, Mettenleiter TC, Loewy AD. Central command neurons of the sympathetic nervous system: basis of the fight-or-flight response. Science 1995;270(5236):644–6. Available from: http://www.jstor.org/stable/2888338. [Cited 29 May 2017].

[27] Harvey C-J, Gehrman P, Espie CA. Who is predisposed to insomnia: a review of familial aggregation, stress-reactivity, personality and coping style. Sleep Med Rev 2014;18(3):237–47. Available from: https://doi.org/10.1016/j.smrv.2013.11.004.

[28] Drake CL, Pillai V, Roth T. Stress and sleep reactivity: a prospective investigation of the stress-diathesis model of insomnia. Sleep 2014;37:1295–304. Available from: http://www.ncbi.nlm.nih.gov/pubmed/25083009.

[29] Burke H, Davis M, Otte C, Mohr D. Depression and cortisol responses to psychological stress: a meta-analysis. Psychoneuroendocrinology 2005;30:846–56. Available from: http://www.sciencedirect.com/science/article/pii/S0306453005000831. [Cited 29 April 2014].

[30] Bootzin RR. Stimulus control treatment for insomnia. Proc Am Psychol Assoc 1972;7:395.

[31] Singareddy R, Vgontzas AN, Fernandez-Mendoza J, Liao D, Calhoun S, Shaffer ML, et al. Risk factors for incident chronic insomnia: a general population prospective study. Sleep Med 2012;13(4):346–53. Available from: https://www.sciencedirect.com/science/article/pii/S138994571200038X. [Cited 24 July 2018].

[32] van de Laar M, Verbeek I, Pevernagie D, Aldenkamp A, Overeem S. The role of personality traits in insomnia. Sleep Med Rev 2010;14(1):61–8. Available from: https://www.sciencedirect.com/science/article/pii/S1087079209000732. [Cited 24 July 2018].

[33] Riemann D, Spiegelhalder K, Feige B, Voderholzer U, Berger M, Perlis M, et al. The hyperarousal model of insomnia: a review of the concept and its evidence. Sleep Med Rev 2010;14:19–31. Available from: https://www.sciencedirect.com/science/article/pii/S1087079209000410. [Cited 24 July 2018].

[34] Bastien CH, Ceklic T, St-Hilaire P, Desmarais F, Pérusse AD, Lefrançois J, et al. Insomnia and sleep misperception. Pathol Biol 2014;62:241–51. Available from: http://www.ncbi.nlm.nih.gov/pubmed/25179115. [Cited 24 July 2018].

[35] Harvey A. A cognitive model of insomnia. Behav Res Ther 2002;40(8):869–93. Available from: http://www.sciencedirect.com/science/article/pii/S0005796701000614. [Cited 25 July 2015].

[36] Espie CA, Broomfield NM, MacMahon KMA, Macphee LM, Taylor LM. The attention-intention-effort pathway in the development of psychophysiologic insomnia: a theoretical review. Sleep Med Rev 2006;10:215–45. Available from: https://www.sciencedirect.com/science/article/pii/S1087079206000219. [Cited 24 July 2018].

[37] Perlis ML, Giles DE, Mendelson WB, Bootzin RR, Wyatt JK. Psychophysiological insomnia: the behavioural model and a neurocognitive perspective. J Sleep Res 1997;6:179–88. Available from: https://onlinelibrary.wiley.com/doi/abs/10.1046/j.1365-2869.1997.00045.x. [Cited 24 July 2018].

[38] Perlis ML, Ellis JG, Kloss JD, Riemann D. Etiology and pathophysiology of insomnia. Available from: In: Principles and practice of sleep medicine. Philadelphia, PA: Elsevier Health Sciences; 2016. p. 769–84. [Cited 24 July 2018] http://www.med.upenn.edu/cbti/assets/user-content/documents/PPSM 2016 Insomnia Models Chapter.pdf.

[39] Spiegelhalder K, Regen W, Nanovska S, Baglioni C, Riemann D. Comorbid sleep disorders in neuropsychiatric disorders across the life cycle. Curr Psychiatry Rep 2013;15(6):364. Available from: http://link.springer.com/10.1007/s11920-013-0364-5. [Cited 24 July 2018].

[40] Morin CM, Rodrigue S, Ivers H. Role of stress, arousal, and coping skills in primary insomnia. Psychosom Med 2003;65(2):259–67. Available from: https://insights.ovid.com/crossref?an=00006842-200303000-00012. [Cited 24 July 2018].

[41] Monroe SM, Harkness KL. Life stress, the "kindling" hypothesis, and the recurrence of depression: considerations from a life stress perspective. Psychol Rev 2005;112(2):417–45. Available from: http://www.ncbi.nlm.nih.gov/pubmed/15783292. [Cited 29 April 2014].

[42] Adrien J. Neurobiological bases for the relation between sleep and depression. Sleep Med Rev 2002;6:341–51. Available from: http://www.ncbi.nlm.nih.gov/pubmed/12531125. [Cited 12 June 2018].

[43] Staner L. Comorbidity of insomnia and depression. Sleep Med Rev 2010;14:35–46. W.B. Saunders. Available from: https://www.sciencedirect.com/science/article/pii/S108707920900094X. [Cited 24 July 2018].

[44] Bonnet MH, Arand DL. Hyperarousal and insomnia: state of the science. Sleep Med Rev 1997;1(2):97–108. Available from: https://doi.org/10.1016/j.smrv.2009.05.002.

[45] Spoormaker VI, Montgomery P. Disturbed sleep in post-traumatic stress disorder: secondary symptom or core feature? Sleep Med Rev 2008;12:169–84. Available from: https://www.sciencedirect.com/science/article/pii/S1087079207001219. [Cited 24 July 2018].

[46] Buysse DJ, Thompson W, Scott J, Franzen PL, Germain A, Hall M, et al. Daytime symptoms in primary insomnia: a prospective analysis using ecological momentary assessment. Sleep Med 2007;8(3):198–208. Available from: https://www.sciencedirect.com/science/article/pii/S1389945706006290. [Cited 24 July 2018].

[47] Mccrae CS, Mcnamara JPH, Rowe MA, Dzierzewski JM, Dirk J, Marsiske M, et al. Sleep and affect in older adults: using multilevel modeling to examine daily associations. J Sleep Res 2008;17(1):42–53. Available from: http://doi.wiley.com/10.1111/j.1365-2869.2008.00621.x. [Cited 24 July 2018].

[48] Fortier-Brochu E, Beaulieu-Bonneau S, Ivers H, Morin CM. Insomnia and daytime cognitive performance: a meta-analysis. Sleep Med Rev 2012;16(1):83–94. Available from: http://www.smrv-journal.com/article/S1087079211000372/fulltext. [Cited 8 February 2016].

[49] Matteson-Rusby SE, Pigeon WR, Gehrman P, Perlis ML. Why treat insomnia? Prim Care Companion J Clin Psychiatry 2010;12(1). PCC.08r00743. Available from: http://mbldownloads.com/0806PP_Neubauer-Smith_CME.pdf%5Cnhttp://www.pubmedcentral.nih.gov/articlerender.fcgi?artid=2882812&tool=pmcentrez&rendertype=abstract.

[50] Ohayon MM, Roth T. Place of chronic insomnia in the course of depressive and anxiety disorders. J Psychiatr Res 2003;37(1):9–15.

[51] Perlis ML, Giles DE, Buysse DJ, Tu X, Kupfer DJ. Self-reported sleep disturbance as a prodromal symptom in recurrent depression—Google Scholar. J Affect Disord 1997;42(2–3):209–12. Available from: http://psycnet.apa.org/record/1997-06526-013. [Cited 24 July 2018].

[52] Baglioni C, Battagliese G, Feige B, Spiegelhalder K, Nissen C, Voderholzer U, et al. Insomnia as a predictor of depression: a meta-analytic evaluation of longitudinal epidemiological studies. J Affect Disord 2011;135(1–3):10–9. Available from: http://www.sciencedirect.com/science/article/pii/S0165032711000292. [Cited 28 April 2014].

[53] Li L, Wu C, Gan Y, Qu X, Lu Z. Insomnia and the risk of depression: a meta-analysis of prospective cohort studies. BMC Psychiatry 2016;16(1):375. Available from: https://doi.org/10.1186/s12888-016-1075-3.

[54] Johnson EO, Roth T, Breslau N. The association of insomnia with anxiety disorders and depression: exploration of the direction of risk. J Psychiatr Res 2006;40(8):700–8. Available from: http://www.sciencedirect.com/science/article/pii/S0022395606001440. [Cited 16 February 2016].

[55] Gulec M, Selvi Y, Boysan M, Aydin A, Besiroglu L, Agargun MY. Ongoing or re-emerging subjective insomnia symptoms after full/partial remission or recovery of major depressive disorder mainly with the selective serotonin reuptake inhibitors and risk of relapse or recurrence: a 52-week follow-up study. J Affect Disord 2011;134(1–3):257–65. Available from: https://www.sciencedirect.com/science/article/pii/S0165032711003181. [Cited 24 July 2018].

[56] Dombrovski AY, Cyranowski JM, Mulsant BH, Houck PR, Buysse DJ, Andreescu C, et al. Which symptoms predict recurrence of depression in women treated with maintenance interpersonal psychotherapy? Depress Anxiety 2008;25(12):1060–6. Available from: http://doi.wiley.com/10.1002/da.20467. [Cited 24 July 2018].

[57] Pigeon WR, Perlis ML. Insomnia and depression: birds of a feather. Int J Sleep Disord 2007;1(3):82–91.

[58] Pigeon WR, Pinquart M, Conner K. Meta-analysis of sleep disturbance and suicidal thoughts and behaviors. J Clin Psychiatry 2012;73(9):1160–7.

[59] Chakravorty S, Siu HYK, Lalley-Chareczko L, Brown GK, Findley JC, Perlis ML, et al. Sleep duration and insomnia symptoms as risk factors for suicidal ideation in a nationally representative sample. Prim Care Companion CNS Disord 2015;17(6). Available from: http://www.ncbi.nlm.nih.gov/pubmed/27057399. e1–9, [Cited 24 July 2018].

[60] Wojnar M, Ilgen MA, Wojnar J, McCammon RJ, Valenstein M, Brower KJ. Sleep problems and suicidality in the national comorbidity survey replication. J Psychiatr Res 2009;43(5):526–31. Available from: https://doi.org/10.1016/j.jpsychires.2008.07.006.

[61] McCall WV, Black CG. The link between suicide and insomnia: theoretical mechanisms. Curr Psychiatry Rep 2013;15(9):389.

[62] Woosley JA, Lichstein KL, Taylor DJ, Riedel BW, Bush AJ. Hopelessness mediates the relation between insomnia and suicidal ideation. J Clin Sleep Med 2014;10(11):1223–30.

[63] Perlis ML, Grandner MA, Brown GK, Basner M, Chakravorty S, Morales K, et al. Nocturnal wakefulness as a previously unrecognized risk factor for suicide. J Clin Psychiatry 2016;77(6):726–33.

[64] Perlis ML, Grandner MA, Chakravorty S, Bernert RA, Brown GK, Thase ME. Suicide and sleep: is it a bad thing to be awake when reason sleeps? Sleep Med Rev 2016;29:101–7. Available from: https://doi.org/10.1016/j.smrv.2015.10.003.

[65] Blatter K, Cajochen C. Circadian rhythms in cognitive performance: methodological constraints, protocols, theoretical underpinnings. Physiol Behav 2007;90(2–3):196–208.

[66] Schmidt C, Collette F, Cajochen C, Peigneux P. A time to think: circadian rhythms in human cognition. Cogn Neuropsychiatry 2007;24(7):755–89. Available from: http://www.ncbi.nlm.nih.gov/entrez/query.fcgi?db=pubmed&cmd=Retrieve&dopt=AbstractPlus&list_uids=18066734.

[67] Lewinsohn P, Klein D, Seeley J. Bipolar disorders in a community sample of older adolescents: prevalence, phenomenology, comorbidity, and course. J Am Acad Child Adolesc Psychiatry 1995;34(4):454–63. Available from: https://www.sciencedirect.com/science/article/pii/S089085670963731X. [Cited 24 July 2018].

[68] Hudson JI, Lipinski JF, Keck Jr. PE, et al. Polysomnographic characteristics of young manic patients: comparison with unipolar depressed patients and normal control subjects. Arch Gen Psychiatry 1992;49(5):378–83. Available from: https://doi.org/10.1001/archpsyc.1992.01820050042006%5Cnhttp://archpsyc.jamanetwork.com/data/Journals/PSYCH/12524/archpsyc_49_5_006.pdf.

[69] Peterson MJ, Rumble ME, Benca RM. Insomnia and psychiatric disorders. Psychiatr Ann 2008;38(8):597–605.

[70] Harvey AG, Mullin BC, Hinshaw SP. Sleep and circadian rhythms in children and adolescents with bipolar disorder. Dev Psychopathol 2006;18(4):1147–68. Available from: http://www.journals.cambridge.org/abstract_S095457940606055X. [Cited 24 July 2018].

[71] Eidelman P, Talbot LS, Gruber J, Harvey AG. Sleep, illness course, and concurrent symptoms in inter-episode bipolar disorder. J Behav Ther Exp Psychiatry 2010;41(2):145–9. Available from: https://doi.org/10.1016/j.jbtep.2009.11.007.

[72] Talbot LS, Stone S, Gruber J, Hairston IS, Eidelman P, Harvey AG. A test of the bidirectional association between sleep and mood in bipolar disorder and insomnia. J Abnorm Psychol 2012;121(1):39–50.

[73] Artioli P, Lorenzi C, Pirovano A, Serretti A, Benedetti F, Catalano M, et al. How do genes exert their role? Period 3 gene variants and possible influences on mood disorder phenotypes. Eur Neuropsychopharmacol 2007;17(9):587–94. Available from: https://www.sciencedirect.com/science/article/pii/S0924977X07000818. [Cited 24 July 2018].

[74] Benedetti F, Dallaspezia S, Fulgosi MC, Lorenzi C, Serretti A, Barbini B, et al. Actimetric evidence that CLOCK 3111 T/C SNP influences sleep and activity patterns in patients affected by bipolar depression. Am J Med Genet B Neuropsychiatr Genet 2007;144(5):631–5. Available from: http://doi.wiley.com/10.1002/ajmg.b.30475. [Cited 24 July 2018].

[75] Barlow DH. Unraveling the mysteries of anxiety and its disorders from the perspective of emotion theory. Am Psychol 2000;55(11):1247–63. Available from: http://doi.apa.org/getdoi.cfm?doi=10.1037/0003-066X.55.11.1247. [Cited 24 July 2018].

[76] Mellman TA. Sleep and anxiety disorders. Sleep Med Clin 2008;3(2):261–8.

[77] Taylor DJ, Lichstein KL, Durrence HH, Reidel BW, Bush AJ. Epidemiology of insomnia, depression, and anxiety. Sleep 2005;28(11):1457–64. Available from: https://academic.oup.com/sleep/article-lookup/doi/10.1093/sleep/28.11.1457.

[78] Jansson-Fröjmark M, Lindblom K. A bidirectional relationship between anxiety and depression, and insomnia? A prospective study in the general population. J Psychosom Res 2008;64(4):443–9.

[79] Glidewell RN, McPherson Botts E, Orr WC. Insomnia and anxiety: diagnostic and management implications of complex interactions. Sleep Med Clin 2015;10:93–9. Elsevier. Available from: http://linkinghub.elsevier.com/retrieve/pii/S1556407X14001258. [Cited 24 July 2018].

[80] Uhde TW, Cortese BM, Vedeniapin A. Anxiety and sleep problems: emerging concepts and theoretical treatment implications. Curr Psychiatry Rep 2009;11:269–76. Current Science Inc. Available from: http://link.springer.com/10.1007/s11920-009-0039-4. [Cited 24 July 2018].

[81] Neylan TC, Marmar CR, Metzler TJ, Weiss DS, Zatzick DF, Delucchi KL, et al. Sleep disturbances in the Vietnam generation: findings from a nationally representative sample of male Vietnam Veterans. Am J Psychiatry 1998;155(7):929–33. Available from: http://psychiatryonline.org/doi/abs/10.1176/ajp.155.7.929. [Cited 24 July 2018].

[82] Ohayon MM, Shapiro CM, Ohayon MM, Shapiro CM, et al. Compr Psychiatry 2000;41(6):469–78. Available from: http://linkinghub.elsevier.com/retrieve/pii/S0010440X00330498. [Cited 24 July 2018].

[83] Ross RJ, Ball WA, Sullivan KA, Caroff SN. Sleep disturbance as the hallmark of posttraumatic stress disorder. Am J Psychiatry 1989;146(6):697–707. Available from: http://search.proquest.com/openview/27835554b9cff8fdbbee618d96397f57/1?pq-origsite=gscholar&cbl=40661. [Cited 24 July 2018].

[84] Lamarche LJ, De Koninck J. Sleep disturbance in adults with posttraumatic stress disorder: a review. J Clin Psychiatry 2007;68(August):1257–70.

[85] McLay RN, Klam WP, Volkert SL. Insomnia is the most commonly reported symptom and predicts other symptoms of posttraumatic stress disorder in U.S. service members returning from military deployments. Mil Med 2010;175(10):759–62. Available from: http://publications.amsus.org/doi/abs/10.7205/MILMED-D-10-00193.

[86] Lavie P. Sleep disturbances in the wake of traumatic events. N Engl J Med 2001;345(25):1825–32. Available from: http://www.nejm.org/doi/abs/10.1056/NEJMra012893. [Cited 24 July 2018].

[87] Koren D, Arnon I, Lavie P, Klein E. Sleep complaints as early predictors of posttraumatic stress disorder: a 1-year prospective study of injured survivors of motor vehicle accidents. Am J Psychiatry 2002;159(5):855–7. Available from: http://psychiatryonline.org/doi/abs/10.1176/appi.ajp.159.5.855. [Cited 24 July 2018].

[88] Krakow B, Melendrez D, Pedersen B, Johnston L, Hollifield M, Germain A, et al. Complex insomnia: insomnia and sleep-disordered breathing in a consecutive series of crime victims with nightmares and PTSD. Biol Psychiatry 2001;49(11):948–53.

[89] Margolies SO, Rybarczyk B, Vrana SR, Leszczyszyn DJ, Lynch J. Efficacy of a cognitive-behavioral treatment for insomnia and nightmares in afghanistan and iraq veterans with PTSD. J Clin Psychol 2013;69(10):1026–42.

[90] Ulmer CS, Edinger JD, Calhoun PS. A multi-component cognitive-behavioral intervention for sleep disturbance in Veterans with PTSD: a pilot study. J Clin Sleep Med 2011;7(1):57–68. Available from: http://jcsm.aasm.org/Articles/07_01_57.pdf. [Cited 24 July 2018].

[91] Green MM, McFarlane AC, Hunter CE, Griggs WM. Undiagnosed post-traumatic stress disorder following motor vehicle accidents. Med J Aust 1993;159(8):529–34. Available from: http://www.ncbi.nlm.nih.gov/pubmed/8412952. [Cited 24 July 2018].

[92] Cox RC, McIntyre WA, Olatunji BO. Interactive effects of insomnia symptoms and trauma exposure on PTSD: examination of symptom specificity. Psychol Trauma Theory Res Pract Policy 2017;. Available from: http://doi.apa.org/getdoi.cfm?doi=10.1037/tra0000336. [Cited 24 July 2018].

[93] Pigeon WR, Campbell CE, Possemato K, Ouimette P. Longitudinal relationships of insomnia, nightmares, and PTSD severity in recent combat veterans. J Psychosom Res 2013;75(6):546–50. Available from: https://doi.org/10.1016/j.jpsychores.2013.09.004.

[94] Mellman TA, Kumar A, Kulick-Bell R, Kumar M, Nolan B. Nocturnal/daytime urine noradrenergic measures and sleep in combat-related PTSD. Biol Psychiatry 1995;38(3):174–9. Available from: https://www.sciencedirect.com/science/article/pii/000632239400238X. [Cited 12 June 2018].

[95] Hendrickson RC, Raskind MA. Noradrenergic dysregulation in the pathophysiology of PTSD. Exp Neurol 2016;284:181–95. Available from: https://www.sciencedirect.com/science/article/pii/S001448861630125X. [Cited 12 June 2018].

[96] Inman DJ, Silver SM, Doghrarnji K. Sleep disturbance in post-traumatic stress disorder: a comparison with non-PTSD insomnia. J Trauma Stress 1990;3(3):429–37. [Cited 12 June 2018]. Available from: http://doi.wiley.com/10.1002/jts.2490030311.

[97] Dow BM, Kelsoe JR, Gillin JC. Sleep and dreams in Vietnam PTSD and depression. Biol Psychiatry 1996;39(1):42–50. Available from: https://www.sciencedirect.com/science/article/pii/0006322395001034. [Cited 12 June 2018].

[98] Wright KM, Britt TW, Bliese PD, Adler AB, Picchioni D, Moore D. Insomnia as predictor versus outcome of PTSD and depression among Iraq combat veterans. J Clin Psychol 2011;67(12):1240–58.

[99] Brod M, Schmitt E, Goodwin M, Hodgkins P, Niebler G. ADHD burden of illness in older adults: a life course perspective. Qual Life Res 2012;21:795–9. Springer Netherlands. Available from: http://link.springer.com/10.1007/s11136-011-9981-9. [Cited 24 July 2018].

[100] Kessler RC, Adler L, Barkley R, Biederman J, Conners CK, Demler O, et al. The prevalence and correlates of adult ADHD in the United States: results from the National Comorbidity survey replication. Am J Psychiatry 2006;163(4):716–23. Available from: http://psychiatryonline.org/doi/abs/10.1176/ajp.2006.163.4.716. [Cited 24 July 2018].

[101] Brevik EJ, Lundervold AJ, Halmøy A, Posserud M-B, Instanes JT, Bjorvatn B, et al. Prevalence and clinical correlates of insomnia in adults with attention-deficit hyperactivity disorder. Acta Psychiatr Scand 2017;136(2):220–7. Available from: http://doi.wiley.com/10.1111/acps.12756.

[102] Corkum P, Moldofsky H, Hogg-Johnson S, Humphries T, Tannock R. Sleep problems in children with attention-deficit/hyperactivity disorder: impact of subtype, comorbidity, and stimulant medication. J Am Acad Child Adolesc Psychiatry 1999;38(10):1285–93. Available from: http://www.ncbi.nlm.nih.gov/pubmed/10517062. [Cited 24 July 2018].

[103] Kooij JJS, Aeckerlin LP, Buitelaar JK. Functioning, comorbidity and treatment of 141 adults with attention deficit hyperactivity disorder (ADHD) at a Psychiatric Outpatients' Department. Ned Tijdschr Geneeskd 2001;145(31):1498–501. Available from: http://www.ncbi.nlm.nih.gov/pubmed/11512422. [Cited 24 July 2018].

[104] Boonstra AM, Kooij JJS, Oosterlaan J, Sergeant JA, Buitelaar JK, Van Someren EJW. Hyperactive night and day? Actigraphy studies in adult ADHD: a baseline comparison and the effect of methylphenidate. Sleep 2007;30(4):433–42. Available from: https://academic.oup.com/sleep/article-lookup/doi/10.1093/sleep/30.4.433. [Cited 24 July 2018].

[105] Barrett JR, Tracy DK, Giaroli G. To sleep or not to sleep: a systematic review of the literature of pharmacological treatments of insomnia in children and adolescents with attention-deficit/hyperactivity disorder. J Child Adolesc Psychopharmacol 2013;23(10):640–7. Available from: http://online.liebertpub.com/doi/abs/10.1089/cap.2013.0059.

[106] Giblin JM, Strobel AL. Effect of lisdexamfetamine dimesylate on sleep in children with ADHD. J Atten Disord 2011;15(6):491–8. Available from: http://journals.sagepub.com/doi/10.1177/1087054710371195. [Cited 24 July 2018].

[107] Huang Y-S, Tsai M-H. Long-term outcomes with medications for attention-deficit hyperactivity disorder. CNS Drugs 2011;25(7):539–54. Available from: http://link.springer.com/10.2165/11589380-000000000-00000. [Cited 24 July 2018].

[108] Van Veen MM, Kooij JJS, Boonstra AM, Gordijn MCM, Van Someren EJW. Delayed circadian rhythm in adults with attention-deficit/hyperactivity disorder and chronic sleep-onset insomnia. Biol Psychiatry 2010;67(11):1091–6. Available from: https://doi.org/10.1016/j.biopsych.2009.12.032.

[109] Stein M, Friedmann P. Disturbed sleep and its relationship to alcohol use. Subst Abus. 2005;26(1): 1-13. Available from: http://www.ncbi.nlm.nih.gov/pubmed/16492658

[110] Brower KJ. Insomnia, alcoholism and relapse. Sleep Med Rev 2003;7(6):523–39.

[111] Brower KJ, Aldrich MS, Robinson EAR, Zucker RA, Greden JF. Insomnia, self-medication, and relapse to alcoholism. Am J Psychiatry 2001;158(3):399–404.

[112] Janson C, Lindberg E, Gislason T, Elmasry A, Boman G. Insomnia in men—A 10-year prospective population based study. Sleep 2001;24(4):425–30. Available from: https://academic.oup.com/sleep/article-lookup/doi/10.1093/sleep/24.4.425. [Cited 24 July 2018].

[113] Brower KJ. Alcohol's effects on sleep in alcoholics. Alcohol Res Health 2001;25(2):110–25. Available from. http://www.ncbi.nlm.nih.gov/pubmed/11584550; . [Cited 24 July 2018].

[114] Stoller MK. Economic effects of insomnia. Clin Ther 1994;16(5):873–97. discussion 854. Available from: http://www.ncbi.nlm.nih.gov/pubmed/7859246. [Cited 24 July 2018].

[115] Ancoli-Israel S, Roth T. Characteristics of insomnia in the United States: results of the 1991 National Sleep Foundation Survey. I. Sleep 1999;22(Suppl 2):S347–53. Available from: http://www.ncbi.nlm.nih.gov/pubmed/10394606. [Cited 24 July 2018].

[116] Johnson EO, Roehrs T, Roth T, Breslau N. Epidemiology of medication as aids to alertness in early adulthood. Sleep 1999;22(4):485–8. Available from: https://academic.oup.com/sleep/article-abstract/21/2/178/2731635. [Cited 24 July 2018].

[117] Reynolds AM, Malow BA. Sleep and autism spectrum disorders. Pediatr Clin N Am 2011;58(3):685–98. Available from: https://doi.org/10.1016/j.pcl.2011.03.009.

[118] Richdale AL, Baker E, Short M, Gradisar M. The role of insomnia, pre-sleep arousal and psychopathology symptoms in daytime impairment in adolescents with high-functioning autism spectrum disorder. Sleep Med 2014;15(9):1082–8. Available from: https://doi.org/10.1016/j.sleep.2014.05.005.

[119] Souders MC, Mason TBA, Valladares O, Bucan M, Levy SE, Mandell DS, et al. Sleep behaviors and sleep quality in children

with autism spectrum disorders. Sleep 2009;32(12):1566–78. Available from: https://academic.oup.com/sleep/article-lookup/doi/10.1093/sleep/32.12.1566. [Cited 24 July 2018].

[120] Krakowiak P, Goodlin-Jones B, Hertz-Picciotto I, Croen LA, Hansen RL. Sleep problems in children with autism spectrum disorders, developmental delays, and typical development: a population-based study. J Sleep Res 2008;17(2):197–206. Available from: http://doi.wiley.com/10.1111/j.1365-2869.2008.00650.x. [Cited 24 July 2018].

[121] Richdale AL. Sleep problems in autism:Prevalence, cause, and intervention. Dev Med Child Neurol 1999;41(1):60–6. Available from: https://www.cambridge.org/core/journals/developmental-medicine-and-child-neurology/article/sleep-problems-in-autism-prevalence-cause-and-intervention/15D37F5D3F536E2E271F6C32566AAECC. [Cited 24 July 2018].

[122] Patzold LM, Richdale AL, Tonge BJ. An investigation into sleep characteristics of children with autism and Asperger's disorder. J Paediatr Child Health 1998;34(6):528–33. Available from: http://doi.wiley.com/10.1046/j.1440-1754.1998.00291.x. [Cited 24 July 2018].

[123] Levitt P, Eagleson KL, Powell EM. Regulation of neocortical interneuron development and the implications for neurodevelopmental disorders. Trends Neurosci 2004;27:400–6. Elsevier Current Trends. Available from: https://www.sciencedirect.com/science/article/pii/S0166223604001626. [Cited 24 July 2018].

[124] McCauley JL, Olson LM, Delahanty R, Amin T, Nurmi EL, Organ EL, et al. A linkage disequilibrium map of the 1-Mb 15q12 GABAA receptor subunit cluster and association to autism. Am J Med Genet B Neuropsychiatr Genet 2004;131 B(1):51–9. Available from: http://doi.wiley.com/10.1002/ajmg.b.30038. [Cited 24 July 2018].

[125] Lin-Dyken D, Dyken E. Use of melatonin in young children for sleep disorders. Infants Young Child 2002;15(2):20–37. Available from: https://journals.lww.com/iycjournal/Abstract/2002/10000/Use_of_Melatonin_in_Young_Children_for_Sleep.5.aspx. [Cited 24 July 2018].

[126] Rapin I, Katzman R. Neurobiology of autism. Ann Neurol 1998;43:7–14. Available from: http://doi.wiley.com/10.1002/ana.410430106. [Cited 24 July 2018].

[127] Kotagal S, Broomall E. Sleep in children with autism spectrum disorder. Pediatr Neurol 2012;47(4):242–51. Available from: https://doi.org/10.1016/j.pediatrneurol.2012.05.007.

[128] Honomichl RD, Goodlin-Jones BL, Burnham M, Gaylor E, Anders TF. Sleep patterns of children with pervasive developmental disorders. J Autism Dev Disord 2002;32(6):553–61. Available from: http://link.springer.com/10.1023/A:1021254914276. [Cited 24 July 2018].

[129] Schreck KA, Mulick JA, Smith AF. Sleep problems as possible predictors of intensified symptoms of autism. Res Dev Disabil 2004;25(1):57–66. Available from: https://www.sciencedirect.com/science/article/pii/S0891422203000933. [Cited 24 July 2018].

[130] Monti JM, Monti D. Sleep disturbance in schizophrenia. Int Rev Psychiatry 2005;17(4):247–53. Available from: http://www.tandfonline.com/doi/full/10.1080/09540260500104516. [Cited 24 July 2018].

[131] Palmese LB, DeGeorge PC, Ratliff JC, Srihari VH, Wexler BE, Krystal AD, et al. Insomnia is frequent in schizophrenia and associated with night eating and obesity. Schizophr Res 2011;133(1-3):238–43. Available from: https://doi.org/10.1016/j.schres.2011.07.030.

[132] Xiang YT, Weng Y-Z, Leung CM, Tang WK, Lai KYC, Ungvari GS. Prevalence and correlates of insomnia and its impact on quality of life in Chinese schizophrenia patients. [References] Sleep 2009;32(1):105–9.

[133] Freeman D, Pugh K, Vorontsova N, Southgate L. Insomnia and paranoia. Schizophr Res 2009;108(1–3):280–4. Available from: https://www.sciencedirect.com/science/article/pii/S092099640800532X. [Cited 24 July 2018].

[134] Afonso P, Brissos S, Figueira ML, Paiva T. Schizophrenia patients with predominantly positive symptoms have more disturbed sleep-wake cycles measured by actigraphy. Psychiatry Res 2011;189(1):62–6. Available from: https://www.sciencedirect.com/science/article/pii/S0165178111000023. [Cited 24 July 2018].

[135] Benca RM, Obermeyer WH, Thisted RA, Gillin JC. Sleep and psychiatric disorders: a meta-analysis. Arch Gen Psychiatry 1992;49(8):651–68. Available from: http://archpsyc.jamanetwork.com/article.aspx?doi=10.1001/archpsyc.1992.01820080059010. [Cited 24 July 2018].

[136] Haffmans P, Hoencamp E, Knegtering HJ, van Heycop ten Ham BF. Sleep disturbance in schizophrenia. Br J Psychiatry 1994;165(5):697–8. Available from: https://www.cambridge.org/core/product/identifier/S0007125000073414/type/journal_article. [Cited 24 July 2018].

[137] Chouinard S, Poulin J, Stip E, Godbout R. Sleep in untreated patients with schizophrenia: a meta-analysis. Schizophr Bull 2004;30(4):957–67. Available from: https://academic.oup.com/schizophreniabulletin/article-lookup/doi/10.1093/oxfordjournals.schbul.a007145. [Cited 24 July 2018].

[138] Ritsner M, Kurs R, Ponizovsky A, Hadjez J. Perceived quality of life in schizophrenia: relationships to sleep quality. Qual Life Res 2004;13(4):783–91. Available from: http://link.springer.com/10.1023/B:QURE.0000021687.18783.d6. [Cited 24 July 2018].

[139] Herz MI, Melville C. Relapse in schizophrenia. Am J Psychiatry 1980;137(7):801–5. Available from: http://psychiatryonline.org/doi/abs/10.1176/ajp.137.7.801. [Cited 24 July 2018].

[140] Chemerinski E, Ho BC, Flaum M, Arndt S, Fleming F, Andreasen NC. Insomnia as a predictor for symptom worsening following antipsychotic withdrawal in schizophrenia. Compr Psychiatry 2002;43(5):393–6.

[141] Kumar PNS, Andrade C, Bhakta SG, Singh NM. Melatonin in schizophrenic outpatients with insomnia: a double-blind, placebo-controlled study. J Clin Psychiatry 2007;68(2):237–41. Available from: http://www.ncbi.nlm.nih.gov/pubmed/17335321. [Cited 24 July 2018].

[142] Morin CM. Cognitive-behavioral approaches to the treatment of insomnia. J Clin Psychiatry 2004;6565(16):33–40. Available from: https://www.betrisvefn.is/wp-content/uploads/2017/10/cbt_for_insomnia_morin_2004.pdf. [Cited 24 July 2018].

[143] Perlis ML, Jungquist CR, Smith MT, Posner DA. Cognitive behavioral treatment of insomnia. New York: Springer Science+Business Media, Inc.; 2005. 1–182. Available from: https://books.google.com/books?hl=en&lr=&id=xEeRnTkyPKMC&oi=fnd&pg=PR7&dq=perlis+cbti+manual&ots=eCkx87CJN1&sig=go2SEjuU0w_FfCgCgxxf_o0gd1o. [Cited 24 July 2018].

[144] Qaseem A, Kansagara D, Forciea MA, Cooke M, Denberg TD, Barry MJ, et al. Management of chronic insomnia disorder in adults: a clinical practice guideline from the American college of physicians. Ann Intern Med 2016;165(2):125–33.

[145] Wu JQ, Appleman ER, Salazar RD, Ong JC. Cognitive behavioral therapy for insomnia comorbid with psychiatric and medical conditions. JAMA Intern Med 2015;175(9):1461. Available from: http://archinte.jamanetwork.com/article.aspx?doi=10.1001/jamainternmed.2015.3006.

[146] Manber R, Buysse DJ, Edinger J, Krystal A, Luther JF, Wisniewski SR, et al. Efficacy of cognitive-behavioral therapy for insomnia combined with antidepressant pharmacotherapy in patients with comorbid depression and insomnia: a randomized controlled trial. J Clin Psychiatry 2016;77:e1316–23. Available from: http://www.ncbi.nlm.nih.gov/pubmed/27788313. [Cited 12 June 2018].

[147] Carney CE, Edinger JD, Kuchibhatla M, Lachowski AM, Bogouslavsky O, Krystal AD, et al. Cognitive behavioral insomnia therapy for those with insomnia and depression: a randomized controlled clinical trial. Sleep 2017;40(4):1–13.

[148] Belleville G, Cousineau H, Levrier K, St-Pierre-Delorme ME. Meta-analytic review of the impact of cognitive-behavior therapy for insomnia on concomitant anxiety. Clin Psychol Rev 2011;31:638–52. Available from: https://www.sciencedirect.com/science/article/pii/S0272735811000298. [Cited 24 July 2018].

[149] Talbot LS, Maguen S, Metzler TJ, Schmitz M, McCaslin SE, Richards A, et al. Cognitive behavioral therapy for insomnia in posttraumatic stress disorder: a randomized controlled trial. Sleep. 2014;37(2): 327–41 Available from: https://www.ncbi.nlm.nih.gov/pubmed/24497661

[150] Cole K, Tabbane K, Boivin DB, Joober R. An algorithmic approach to the management of insomnia in patients with schizophrenia. Ann Clin Psychiatry 2017;29(2):133–44. Available from: https://www.ingentaconnect.com/contentone/fmc/acp/2017/00000029/00000002/art00007. [Cited 24 July 2018].

[151] Freeman D, Dunn G, Startup H, Pugh K, Cordwell J, Mander H, et al. Effects of cognitive behaviour therapy for worry on persecutory delusions in patients with psychosis (WIT): a parallel, single-blind, randomised controlled trial with a mediation analysis. Lancet Psychiatry 2015;2(4):305–13. Available from: https://www.sciencedirect.com/science/article/pii/S2215036615000395. [Cited 24 July 2018].

[152] Colombo C, Benedetti F, Barbini B, Campori E, Smeraldi E. Rate of switch from depression into mania after therapeutic sleep deprivation in bipolar depression. Psychiatry Res 1999;86(3):267–70. Available from: https://www.sciencedirect.com/science/article/pii/S0165178199000360. [Cited 24 July 2018].

[153] Harvey AG, Soehner AM, Kaplan KA, Hein K, Lee J, Kanady J, et al. Treating insomnia improves mood state, sleep, and functioning in bipolar disorder: a pilot randomized controlled trial. J Consult Clin Psychol 2015;83(3):564–77.

Chapter 29

Insomnia and cardiometabolic disease risk

Julio Fernandez-Mendoza
Sleep Research & Treatment Center, Department of Psychiatry, Penn State Health Milton S. Hershey Medical Center, Pennsylvania State University College of Medicine, Hershey, PA, United States

ABBREVIATIONS

BMI	body mass index
BP	blood pressure
CBVD	cerebrovascular disease
CHD	coronary heart disease
CMR	cardiometabolic risk factors
CRP	C-reactive protein
CVD	cardiovascular disease
DBP	diastolic blood pressure
HPA	hypothalamic-pituitary-adrenal
HR	heart rate
HRV	heart rate variability
HTN	hypertension
IL-6	interleukin 6
MetS	metabolic syndrome
MI	myocardial infarction
MSLT	multiple sleep latency test
PSG	polysomnography
SAM	sympatho-adrenal-medullary
SBP	systolic blood pressure
SDB	sleep disordered breathing
T2D	type 2 diabetes
TNF-α	tumor necrosis factor alpha

INTRODUCTION

Insomnia has been traditionally viewed either as a symptom of underlying medical or psychiatric disorders or as a symptom of "otherwise healthy but worried" individuals. However, accumulating evidence indicates that insomnia, particularly when chronic and coupled with objective short sleep duration, is associated with cardiometabolic risk factors (CMR) such as hypertension (HTN) and type 2 diabetes (T2D) and increased risk of cardiovascular disease (CVD) and cerebrovascular disease (CBVD) morbidity and mortality. We will review herein the evidence linking insomnia with these cardiometabolic disease outcomes as well as with the pathophysiologic mechanisms (e.g., stress system activation, cardiac autonomic dysregulation, chronic low-grade inflammation) and poor health behaviors (e.g., smoking, physical inactivity) that potentially mediate such increased risk. Finally, we will discuss the public health and clinical implications of these associations, including the need for efficacy and population-level interventions testing whether insomnia therapies improve clinical and subclinical biomarkers of cardiometabolic disease risk.

INSOMNIA: A SYMPTOM AND A CHRONIC DISORDER

Insomnia is the most prevalent sleep disorder, much more than sleep disordered breathing (SDB), and is a major public health problem. About 20%–30% of people from the general population report insomnia symptoms and another 8%–10% fulfill criteria for a chronic insomnia disorder. Insomnia symptoms consist of self-reported difficulties initiating sleep, difficulties maintaining sleep or difficulties waking up too early and being unable to resume sleep (i.e., early morning awakening) without any chronicity or daytime impairment criteria [1, 2]. A self-report of nonrestorative sleep has traditionally been included in experimental and epidemiological studies as a core insomnia symptom [1, 2], however, it has recently been dropped in diagnostic nosologies [3]. Diagnostic criteria for chronic insomnia disorder include the self-report of at least one insomnia symptom, occurring at least 3 nights per week, for at least 3 months that is associated with significant daytime functioning impairment [3]. Both insomnia symptoms and chronic insomnia disorder are associated with impaired quality of life and are well-established correlates and risk factors of mental health problems, such as depression, anxiety and other psychiatric disorders (see Chapter 30). Natural history studies have shown that insomnia disorder is indeed a highly chronic condition with a remission rate as low as 25%, whereas the course of insomnia symptoms is characterized by a waxing-and-waning pattern with high remission rates (about 50%) [4–8]. These longitudinal, epidemiological data have supported that chronic insomnia is a disorder in its own right, whereas insomnia symptoms may occur in relation to the course of an underlying physical or mental health disorder, including depression, pain or SDB [4–6, 8]. As reviewed below, not many of the existing large, population-based studies have been able to examine these two mutually-exclusive categories of insomnia symptoms and chronic insomnia disorder separately. This distinction, however, is critical in population science and clinical practice (Table 29.1).

Another important issue unique to insomnia is the absence of objective and/or quantitative criteria. Insomnia symptoms are, by definition, subjective complaints and, thus, chronic insomnia disorder is a diagnosis reached solely based on self-reports. While there are guiding thresholds to help define what is a clinically significant difficulty initiating or resuming sleep (> 30 min), these quantitative criteria (Table 29.1) are not used alone to identify chronic insomnia disorder if there are no subjective complaints [3]. Furthermore, although the polysomnography (PSG) or actigraphy (ACT) measured objective sleep of people with chronic insomnia is significantly impaired compared to good sleepers [3], and that measurements of objective short sleep duration are a good predictor of the persistent course of insomnia disorder [8] and of who among those with insomnia symptoms are at risk of developing a chronic insomnia disorder [6], PSG or ACT are not required for the diagnosis of chronic insomnia [3]. This is in sharp contrast with the current use of PSG data for the evaluation of patients at risk of SDB (e.g., obese snorers), to establish the severity of SDB [e.g., apnea/hypopnea index (AHI)], and

TABLE 29.1 Most frequent definitions used in insomnia studies.

Domain	Criteria	Method of measurement
Self-reported		
Sleep difficulties	Difficulty initiating sleep Difficulty maintaining sleep Early morning awakening Nonrestorative sleep	Retrospective questionnaire
Insomnia symptoms	At least one sleep difficulty Usually based on severity (moderate-to-severe) and/or frequency (≥ 3 nights per week)	Retrospective questionnaire
Chronic insomnia	Duration (≥ 3 months)	Retrospective questionnaire
Chronic insomnia disorder	ICD/DSM/ICSD (frequency, duration, daytime impairment and adequate sleep opportunities)	Retrospective questionnaire, clinical interview
Nighttime sleep continuity	Sleep onset latency (≥30 min) Wake after sleep onset (≥30 min) Total sleep time (<7 or 6 h) Sleep efficiency (<85%)	Retrospective questionnaire, clinical interview or prospective sleep diary
Objective		
Nighttime sleep continuity	Sleep onset latency (≥30 min) Wake after sleep onset (≥30 min) Total sleep time (<7 or 6 h) Sleep efficiency (<85%)	In-lab PSG, at-home PSG or at-home ACT
Daytime sleep propensity	Daytime sleep latency (>14 min)	In-lab MSLT
Combined/phenotypes		
Self-reported + self-reported	Insomnia symptoms + total sleep time Insomnia disorder + total sleep time	Retrospective questionnaires Retrospective questionnaire + sleep diary
Self-reported + objective	Insomnia symptoms + total sleep time Chronic insomnia + total sleep time Insomnia disorder + total sleep time Insomnia disorder + sleep efficiency Insomnia disorder + total sleep time Insomnia disorder + daytime sleep latency	Retrospective questionnaire + PSG Retrospective questionnaire + PSG Clinical interview + PSG Clinical interview + PSG Clinical interview + ACT Clinical interview + MSLT

ACT, actigraphy; *DSM*, diagnostic and statistical manual of mental disorders; *ICD*, international classification of disease; *ICSD*, international classification of sleep disorders; *MSLT*, multiple sleep latency test; *PSG*, polysomnography.

to predict its associated risk of CVD and CBVD [3]. More importantly, the association of insomnia, as measured by self-reports or objectively, with specific CMR, CVD or CBVD (Table 29.2) has been largely ignored until the past decade [9].

HYPERTENSION AND BLOOD PRESSURE

Since the 1970s, clinical studies had observed a high co-morbidity of insomnia with clinical hypertension, as per patient's clinical history [10]. However, the association of

TABLE 29.2 Most frequent cardiometabolic disease risk outcomes used in insomnia studies.

Domain	Criteria	Method of measurement
Self-reported		
Hypertension	Use of antihypertensive medication Treatment for high blood pressure Medical history of hypertension	Retrospective questionnaire or clinical interview
Type 2 diabetes	Use of insulin medication Treatment for type 2 diabetes Medical history of type 2 diabetes	Retrospective questionnaire or clinical interview
Cardiovascular disease	Treatment for heart disease Medical history of heart disease (myocardial infarction, coronary heart disease, heart failure)	Retrospective questionnaire or clinical interview
Cerebrovascular disease	Treatment for stroke Medical history of stroke	Retrospective questionnaire or clinical interview
Objective		
High blood pressure	Systolic ($\geq$ 130 or 140 mmHg) Diastolic ($\geq$ 80, 85 or 90 mmHg) Mean arterial pressure	In-lab or at-home automatic blood pressure monitoring
Blood pressure regulation	Blood pressure dipping (day-to-night) Nighttime blood pressure (sleep stage-related) Blood pressure reactivity (stress test)	In-lab automatic blood pressure monitoring
Metabolic regulation	Fasting glucose levels ($\geq$ 100 or 110 mg/dL) Fasting insulin levels Homeostatic model assessment Oral glucose tolerance test Hyperinsulinemic-euglycemic clamp	In-lab blood draw In-lab tests
Dyslipidemia	Fasting total cholesterol levels Fasting LDL cholesterol levels Fasting HDL cholesterol levels Fasting total triglycerides levels	In-lab blood draw
Metabolic syndrome	High blood pressure, insulin resistance, dyslipidemia and/or central obesity	In-lab blood draw and physical examination (waist circumference)
Type 2 diabetes	Fasting glucose levels ($\geq$ 126 mg/dL)	In-lab blood draw
Cardiac autonomic modulation	Heart rate Heart rate variability (frequency, time) Pre-ejection period Rate pressure product	In-lab nighttime EKG (polysomnography) In-lab or at-home 24-h EKG
Neuroendocrine regulation	Adrenocorticotropic hormone levels Cortisol levels Overnight catecholamine secretion	In-lab blood draw In-lab or at-home salivary sampling
Inflammation	C-reactive protein levels Interleukin-6 levels Tumor necrosis factor alpha levels	In-lab blood draw
Mortality	Physician-confirmed death Physician-diagnosed cause of death	Death records (social security, national death index in the United States)

insomnia with clinical and subclinical measures of elevated blood pressure (BP) or BP dysregulation remained largely unexplored in favor of mental health comorbidities (see Chapter 30). In the past decade, there has been a renewed interest and multiple studies, including several reviews [9, 11–14] and a meta-analysis [15], have been published on the association of insomnia with HTN. Population-based studies using self-reported data have shown a significant relationship between insomnia, either defined as a symptom or as a disorder, and HTN [15–31]. As shown in Table 29.3, the only available meta-analysis performed by Meng and colleagues [15] on findings from 11 large longitudinal studies comprising a total of 42,636 subjects, estimated that the risk of incident HTN associated with insomnia ranges between 5% and 20%. Most recent studies that have examined the association of insomnia with HTN since the publication of this meta-analysis, have primarily relied on self-reported insomnia symptoms [20–24, 28], a couple on insomnia symptoms associated with impaired daytime functioning [25, 26] and only three using insomnia disorder criteria [27, 29, 30]. These more recent studies used either self-reported data on HTN as a current medical problem or antihypertensive medication use as well as pre-hypertensive (SBP $\geq$ 130 mmHg or DDP $\geq$ 85 mmHg) or hypertensive BP levels (SBP $\geq$ 140 mmHg or DBP $\geq$ 90 mmHg). Most of these studies did not include a PSG study and were, thus, unable to control for the presence of SDB. This is a significant caveat in these studies as SDB has long been associated with HTN (see Chapter 32). To add to this, one cross-sectional study from Norway (HUNT-3; $N=50,806$) has shown that insomnia symptoms were associated with lower SBP or DBP levels [24]. Taken together, the degree of association found in previous reviews [38] and meta-analysis [15] have been regarded as preliminary.

Many of the caveats mentioned above were addressed by the work with the Penn State Adult Cohort (PSAC) conducted almost 10 years ago. These studies showed a synergistic effect between insomnia and PSG-measured short sleep duration (i.e., < 6 h of sleep) on the risk of HTN, while adjusting for the potential effect of multiple demographic, lifestyle and clinical factors, including SDB, T2D, smoking, alcohol or depression [16, 17]. In the first study, Vgontzas and colleagues [16] showed that individuals with chronic insomnia who slept objectively between 5 and 6 h and those with chronic insomnia who slept <5 h were 3.5-fold and 5.1-fold times more likely to have HTN

TABLE 29.3 Findings of available meta-analyses on the association between insomnia and cardiometabolic disease risk.

First author, year (design, vital status)	N (# of studies)	Insomnia definitions	Outcome	Findings
Cappuccio, 2010 [32] (longitudinal)	24,812 (6)	Difficulty initiating sleep Difficulty maintaining sleep	T2D	RR = 1.57* RR = 1.84*
Meng, 2013 [15] (longitudinal)	42,636 (7)	Difficulty initiating sleep Difficulty maintaining sleep Early morning awakening Insomnia symptoms	HTN	RR = 1.17 RR = 1.20* RR = 1.14* RR = 1.05*
Li, 2014 [33] (longitudinal, mortality)	311,260 (17)	Insomnia symptoms	MI CHD CBVD Mortality	RR = 1.41* RR = 1.28* RR = 1.55* RR = 1.33*
Sofi, 2014 [34] (longitudinal, mortality)	122,501 (10)	Insomnia symptoms	CVD/CBVD	RR = 1.45*
Li, 2014 [35] (longitudinal, mortality)	110,530 (10)	Difficulty initiating sleep Difficulty maintaining sleep Early morning awakening Nonrestorative sleep	CVD/CBVD Mortality	RR = 1.45* RR = 1.02 RR = 1.00 RR = 1.30*
Anothaisintawee, 2016 [36] (longitudinal)	289,588 (11)	Difficulty initiating sleep Difficulty maintaining sleep Insomnia symptoms	T2D	RR = 1.55* RR = 1.74* RR = 1.40*
Irwin, 2016 [37] (cross-sectional and longitudinal)	34,943 (31)	Sleep disturbance	CRP IL-6 TNF-α	ES = 0.12* ES = 0.20* ES = 0.07

Note that some studies included death from CVD or CBVD as part of the outcome definition and this is noted as vital status. *CVD*, incident or death from cardiovascular disease, including myocardial infarction (MI), coronary heart disease (CHD) and/or heart failure (HF); *CBVD*, incident or death from cerebrovascular disease, including ischemic stroke; *ES*, effect size; *HTN*, incident hypertension; *NS*, not statistically significant; *RR*, relative risk; *T2D*, incident type 2 diabetes.
*Statistically significant.

(defined as SBP ≥ 140 mmHg or DBP ≥ 90 mmHg and/or antihypertensive medication use), respectively, as compared to good sleepers. A similar but smaller association with prevalent HTN was found in individuals with insomnia symptoms who also slept objectively <6h (OR = 1.5 if sleeping 5–6h and OR = 2.4 if sleeping <5h) [16]. In contrast, individuals with insomnia symptoms or chronic insomnia who slept objectively >6h were not significantly associated with increased odds of prevalent HTN (OR = 0.8 and OR = 1.3, respectively) [16]. In a follow-up study of the PSAC, Fernandez-Mendoza and colleagues showed that, compared to good sleepers, individuals with chronic insomnia who slept <6h were 3.8 times more likely to develop HTN after 7.5 years of follow-up (defined as a report of being treated for HTN adjusted for pre-hypertensive and hypertensive BP levels at baseline) [17]. Similarly to the cross-sectional study above [16], individuals with insomnia symptoms or chronic insomnia who slept objectively >6h were not significantly associated with increased odds of incident HTN (OR = 0.50 and OR = 0.85, respectively) [17]. The findings of these seminal studies are included in Table 29.4.

Following these studies, several other studies have cemented the significance of short sleep duration in the association of insomnia with HTN, while the smaller effect sizes in some of these more recent studies can be explained by the use of subjective instead of objective sleep measures [18, 28–30, 58–61]. The findings of other studies using objective sleep measures are summarized in Table 29.4. Specifically, in a well-characterized sample of adults with chronic insomnia disorder ($N=255$), Bathgate and colleagues [30] used average total sleep time across two consecutive nights of PSG to classify individuals with chronic insomnia disorder into those who slept objectively <6h and >6h. After controlling for numerous potential confounders, including SDB, T2D, hypercholesterolemia or depression, adults with chronic insomnia disorder who slept objectively <6h were 3.6 times more likely to have HTN (defined as a report of HTN as a current problem) than those with chronic insomnia disorder who slept >6h. Interestingly, Bathgate and colleagues [30] did not observe a significant association with prevalent HTN using short sleep duration (<6h) derived from self-reported 2-week sleep diaries (OR = 1.13). In a large, community study ($N=3911$), Kalmbach and colleagues found that individuals with insomnia disorder who reported sleeping <6h were 2.13 times more likely to have HTN (defined as a report of being treated with antihypertensive medication) than good sleepers, even after controlling for sex, age and obesity, after excluding individuals at high risk for SDB or defining insomnia disorder as current or remitted [29]. Furthermore, African Americans were more likely to report insomnia disorder and sleeping <6h compared to their non-Hispanic White counterparts [29]. In a large prospective study from Taiwan ($N=162{,}121$), Deng and colleagues stratified their analyses by the presence of self-reported insomnia symptoms and found that the risk of HTN was significantly elevated for individuals reporting insomnia symptoms and sleeping <6h (HR = 1.06), an effect size smaller than studies relying on objective sleep measures or on a chronic insomnia definition [28]. However, this study adds to the literature in other geographically, racially and ethnically diverse populations [28, 29].

In contrast to these recent studies, a study conducted in a sample of clinically-referred patients with chronic insomnia disorder in Germany ($N=328$) did not find a significant association between patients with chronic insomnia disorder who slept objectively <6h with prevalent HTN (defined as SBP ≥ 140 mmHg or DBP ≥ 90 mmHg and/or antihypertensive medication use) [31]. There are methodological differences with previous studies that need to be considered when interpreting these results. Although participants underwent two consecutive nights of PSG, each night was treated as an independent phenomenon and the two nights were not averaged to account for inter-individual and intra-individual variability across nights [30]. Additionally, the method by which BP was ascertained changed across time, which may have influenced the precision of BP measurement and introduced variability across subjects. Importantly, this study relied on an opportunistic clinical sample of patients referred to a specialty sleep clinic housed within psychiatry, rather than randomly-selected from the general population or from more diverse outpatient clinics. In fact, population-based studies in Germany have suggested a downward trend in BP over the past decade among middle age and older adults [62], which is at odds with the increasing trajectory in the US population [63]. The authors were able to exclude, based on clinical interview and PSG, individuals with occult sleep disorders such as SDB, however, they also excluded those with psychiatric disorders or "serious medical conditions." Thus, the sample included otherwise physically healthy individuals with "primary insomnia," a diagnostic category that has been abandoned in current nosologies given the high comorbidity of insomnia with other medical and psychiatric disorders at any time point during its natural course. Nevertheless, the authors did replicate the finding that individuals with chronic insomnia disorder who slept objectively <6h have a longer chronic course (specifically 3.7 years longer in this study) than those with chronic insomnia and normal sleep duration [6, 8].

Other studies have also examined whether the association of insomnia with HTN is stronger on specific insomnia phenotypes using other sleep/wake-related measures. The multiple sleep latency (MSLT) is an in-lab measure of a person's sleep propensity and, thus, daytime alertness. Interestingly, individuals with chronic insomnia show longer sleep latencies on the MSLT as compared to good sleepers. In fact, this increased alertness (MSLT >14 min) is primarily found in individuals with chronic insomnia and objective short sleep

TABLE 29.4 Cardiometabolic disease risk associated with insomnia based on objective sleep measures.

First author, year (design, sample)	N (men, age)	Insomnia phenotype	Outcome	Findings
Stepanski, 1994 [39] (cross-sectional, research)	49 (100%, 21–50)	Chronic insomnia +1-night PSG SE<85%	Nighttime HR Stress-task HR	↑HR ↑HR reactivity
Bonnet, 1995 [40] (cross-sectional, research)	20 (NA, 18–50)	Chronic insomnia +2-nights PSG SOL >30 min or SE<85%	24-h VO$_2$	↑VO$_2$
Bonnet, 1998 [41] (cross-sectional, research)	37 (NA, 18–50)	Chronic insomnia +2-nights PSG SOL >30 min or SE<85%	Nighttime HR Nighttime HRV	↑HR ↓SDNN ↑LF ↓HF
Vgontzas, 2001 [42] (cross-sectional, research)	24 (63%, 31.4±6.7)	Chronic insomnia +1-night PSG<80%	24-h HPA axis	↑Cortisol ↑ACTH 21:30–0:30
Shaver, 2002 [43] (cross-sectional, research)	53 (0%, 46.2±3.3)	Chronic insomnia +5-nights PSG<85%	HPA axis HR MAP NE	↑Cortisol NS NS NS
Vgontzas, 2002 [44] (cross-sectional, research)	22 (64%, 31.6±6.7)	Chronic insomnia +1-night PSG<80%	IL-6 TNF-α	↑IL-6 14:00–21:00
Vgontzas, 2009 [16] (cross-sectional, population-based)	1741 (48%, 20–88)	Insomnia symptoms +1-night PSG 5–6 h Insomnia symptoms +1-night PSG ≤5 h Chronic insomnia +1-night PSG 5–6 h Chronic insomnia +1-night PSG ≤5 h	HTN	OR=1.48 OR=2.43* OR=3.53* OR=5.12*
Vgontzas, 2009 [45] (cross-sectional, population-based)	1741 (48%, 20–88)	Insomnia symptoms +1-night PSG 5–6 h Insomnia symptoms +1-night PSG ≤5 h Chronic insomnia +1-night PSG 5–6 h Chronic insomnia +1-night PSG ≤5 h	T2D	OR=1.55 OR=1.06 OR=2.07 OR=2.95*
Vgontzas, 2010 [46] (longitudinal, population-based)	1741 (48%, 20–88)	Chronic insomnia +1-night PSG<6 h Chronic insomnia +1-night PSG<6 h	Mortality Men Women	OR=4.00* OR=0.36
Knutson, 2011 [47] (cross-sectional, population-based)	571 (37%, 37–52)	Insomnia symptoms +6-days ACT <80%	Glucose Insulin HOMA	↑Glucose ↑Insulin ↑HOMA if T2D
Fernandez-Mendoza, 2012 [17] (longitudinal, population-based)	786 (49%, 20–84)	Insomnia symptoms +1-night PSG<6 h Chronic insomnia +1-night PSG<6 h	HTN	OR=1.34 OR=3.75*
Spiegelhalder, 2011 [48] (cross-sectional, research)	104 (39%, 39.5±11.8)	Insomnia disorder +2nd-night PSG<85%	Nighttime HR Nighttime HRV	↓SDNN ↓RMSSD ↓pNN50 ↓HF

TABLE 29.4 Cardiometabolic disease risk associated with insomnia based on objective sleep measures.—cont'd

First author, year (design, sample)	N (men, age)	Insomnia phenotype	Outcome	Findings
Vasisht, 2013 [49] (cross-sectional, research)	28 (39%, 30–64)	Insomnia disorder +1-night PSG ≤ 6 h	Glucose 2-h glucose HA1C Insulin HOMA-B HOMA-IR 2-h insulin 2nd-phase ins. resp. IS	NS NS NS ↓Insulin ↓HOMA-B ↓HOMA-IR ↓2-h insulin ↓2nd-phase ins. resp. ↓IS
Fernandez-Mendoza, 2014 [50] (cross-sectional, population-based)	327 (46%, 5–12)	Insomnia symptoms +1-night PSG < 7.7 h	HPA axis	↑Cortisol 19:00 & 7:00
Li, 2015 [51] (cross-sectional, clinical)	315 (33%, 40.0 ± 10.2)	Insomnia disorder + MSLT > 14 min Insomnia disorder + MSLT > 17 min	HTN	OR = 3.27* OR = 4.33*
D'Aurea, 2015 [52] (cross-sectional, research)	30 (17%, 30–55)	Insomnia disorder +2-nights PSG ≤ 5 h	Glucose Cortisol Insulin HOMA ACTH GH	↑Glucose ↑Cortisol NS NS NS NS
Bathgate, 2016 [30] (cross-sectional, clinical)	255 (35%, 46.2 ± 13.7)	Insomnia disorder +2-nights PSG < 6 h	HTN	OR = 3.59*
Castro-Diehl, 2016 [53] (cross-sectional, population-based)	527 (46%, 45–84)	Insomnia symptoms +7-days ACT < 7 h	HR HF-HRV reactivity	↑HR ↑HF-HRV reactivity
Johann, 2017 [31] (cross-sectional, clinical)	328 (38%, 44.3 ± 12.2)	Insomnia disorder +1st-night PSG < 6 h Insomnia disorder +2nd-night PSG < 6 h	HTN	OR = 0.79 OR = 1.21
Fernandez-Mendoza, 2017 [54] (cross-sectional, population-based)	378 (54%, 12–23)	Insomnia symptoms +1-night PSG ≤ 7 h	CRP IL-6 TNF-α	↑CRP NS NS
Jarrin, 2018 [55] (cross-sectional, research)	180 (37%, 49.9 ± 11.3)	Insomnia disorder +2-nights PSG < 6 h	Nighttime HR & HRV	↑HR ↓HF ↑LF/HF
Fernandez-Mendoza, 2018 [56] (cross-sectional, population-based)	1741 (48%, 20–88)	Chronic insomnia or symptoms +1-night PSG < 6 h	CVD/CBVD	OR = 2.00*
Bertisch, 2018 [57] (longitudinal, population-based)	4994 (47%, 64.0 ± 11.1)	Insomnia symptoms +1-night PSG < 6 h	CVD/CBVD Mortality	HR = 1.29* HR = 1.07

Note that when the age range was not available, the mean and standard deviation for the insomnia group is reported. *ACT*, actigraphy; *CRP*, C-reactive protein; *CVD*, prevalent or incident cardiovascular diseases in cross-sectional and longitudinal studies, respectively, including myocardial infarction (MI), coronary heart disease (CHD) and/or heart failure (HF); *CBVD*, prevalent or incident cerebrovascular diseases in cross-sectional and longitudinal studies, respectively, including ischemic stroke; *HTN*, prevalent or incident hypertension in cross-sectional and longitudinal studies, respectively; *HR*, hazard ratio or heart rate, depending on the study; *HRV*, heart rate variability; *IL-6*, interleukin 6; *IS*, insulin sensitivity; *MSLT*, multiple sleep latency test; *NA*, not available; *NE*, norepinephrine; *NS*, not statistically significant; *OR*, odds ratio; *PSG*, polysomnography; *T2D*, prevalent or incident type 2 diabetes in cross-sectional and longitudinal studies, respectively; *TNF-α*, tumor necrosis factor alpha.

*Statistically significant.

duration [12, 64–67]. Based on this observation, Li and colleagues [51] showed in an in-lab study of 315 patients that chronic insomnia combined with an MSLT >14 min, was associated with 3.3-fold increased prevalence of HTN (defined as SBP≥140 mmHg or DBP≥90 mmHg and/or antihypertensive medication use), whereas chronic insomnia with MSLT <14 min was not (OR=1.17). This association with HTN was 4.3-fold for chronic insomnia combined with an MSLT >17 min [51]. This study also showed that chronic insomnia was associated with increasing SBP and DBP levels in a dose-response manner as a function of increasing MSLT levels [51]. Taken together, there appears to be methodologically stronger evidence that supports the hypothesis that the association between chronic insomnia and HTN is primarily found in those with short sleep duration (< 6 h) or other markers of physiologic hyperarousal (MSLT >14 min). Nevertheless, more studies are needed to examine the synergistic effect of having both insomnia disorder and objective short sleep duration on BP regulation and different levels of HTN using longitudinal design. Furthermore, previous studies using objective sleep or alertness measures could not examine other subclinical markers of BP regulation such as nighttime BP or non-dipping BP levels.

The association of insomnia with subclinical markers of BP regulation has been examined primarily in laboratory-based studies. An elevated nighttime SBP and a lower day-to-night SBP dipping have been found in a study of otherwise normotensive patients with chronic insomnia disorder [68]. Interestingly, beta (15–35 Hz) activity in the electroencephalogram (EEG) during the PSG study was positively correlated with concomitant nighttime SBP [68]. Also, BP dysregulation has been observed in individuals with chronic insomnia disorder in response to a psychosocial stress challenge test [69]. Another study has shown that morning-to-evening and day-to-day ambulatory BP variability was significantly higher in patients with chronic insomnia compared to good sleeping control, an association that was greatest in chronic insomnia combined with short sleep duration [70]. In a recent systematic review of 26 observational studies comprising 1484 subjects, Nano and colleagues [71] have reported that 80% of these studies found significant differences in cardiovascular activity, including impaired heart rate variability (HRV) or BP dysregulation, and that the insomnia with objective short sleep duration phenotype [12] presented most consistent findings across markers of impaired cardiovascular regulation [71]. The findings of several of these studies using objective sleep measures are summarized in Table 29.4.

In summary, epidemiologic and experimental studies support an association of insomnia with clinical HTN and subclinical BP dysregulation. However, most epidemiologic evidence pertains to insomnia symptoms, rather than chronic insomnia disorder, was not able to control for SDB, and reported modest effect sizes. The interest in understanding the dynamics of the association of chronic insomnia with HTN in those with objective short sleep duration is growing, as seen by the spurt of literature examining these associations in recent years.

TYPE 2 DIABETES AND INSULIN RESISTANCE

The relationship between insomnia and metabolic dysfunction has been examined using either subclinical markers such as insulin resistance, increased fasting glucose, dyslipidemia or the clustering of these markers into the metabolic syndrome (MetS) as well as using clinically-identified T2D. Multiple cross-sectional and longitudinal studies have reported significant associations between the presence of insomnia symptoms and insulin resistance, increased fasting glucose levels and prevalent or incident T2D [29, 32, 36, 45, 52, 72]. As shown in Table 29.3, a meta-analysis performed by Cappuccio and colleagues [32] on findings from 10 large longitudinal studies comprising a total of 107,756 subjects, estimated that the risk of incident T2D associated with the insomnia symptoms of difficulty initiating sleep and difficulty maintaining sleep was about 57% and 84%, respectively. The vast majority of the epidemiologic studies relied on broad self-reported definitions of sleep disturbance/sleep quality and, in some studies, of insomnia symptoms. However, two experimental studies did not find an association between chronic insomnia disorder and impaired glucose metabolism in the oral glucose tolerance test [73] or hyperinsulinemic-euglycemic clamp [74] under laboratory conditions. The relationship between insomnia and MetS—the clustering of central obesity, dyslipidemia, insulin resistance or glucose dysregulation, and/or elevated BP—has revealed to be much more complex, if not limited and inconsistent [75–82]. The only longitudinal study to date found that while insomnia symptoms were associated with incident MetS over a 3-year follow-up, chronic insomnia disorder was not [80]. In contrast to these inconsistent findings, it is well-established that SDB is associated with the MetS (see Chapter 32). A potential explanation for the stronger association of SDB with MetS as compared to chronic insomnia disorder, may be related to differing underlying mechanisms. A core feature of MetS is central obesity, which is a strong risk factor for the development of SDB (see Chapter 32) but not chronic insomnia [5, 6, 8]. Individuals with chronic insomnia disorder are typically non-obese, not significantly heavier than healthy controls, and not likely to develop obesity despite sleeping objectively shorter than controls [83–85], findings consistent with the presence of increased whole-body metabolic rate [40]. Importantly, PSG sleep duration does not significantly correlate with body mass index in individuals with chronic insomnia disorder [84]. These data indicate that chronic insomnia may be linked to impaired glucose levels and T2D through underlying mechanisms other than central obesity.

As with HTN above, the risk of T2D has also been examined from the perspective of an interplay between insomnia and short sleep duration that can explain the heterogeneity and small effect sizes of previous findings (see Table 29.4). In another study of the PSAC, Vgontzas and colleagues showed that individuals with chronic insomnia who slept objectively between 5 and 6h and those with chronic insomnia who slept <5h were 2.1 and 3.0 times more likely to have T2D (defined as fasting plasma glucose ≥126 and/or use of T2D medication), respectively, as compared to good sleepers [45]. In contrast, individuals with chronic insomnia who slept objectively >6h were not significantly associated with increased odds of prevalent T2D (OR=1.10) [45]. The few other epidemiologic studies that have examined the combined effect of insomnia symptoms and short sleep duration on T2D risk have reported similar findings. For example, a large study showed that individuals who reported poor sleep quality and sleeping <6h had 6.4-fold odds of impaired glucose tolerance [86]. In another small study, individuals with chronic insomnia disorder who slept objectively <6h (as measured by PSG) responded to an oral glucose tolerance test with lower insulin secretion in conjunction with greater insulin sensitivity, and no difference in glycemic control, compared to those with chronic insomnia disorder who slept objectively >6h [49]. These findings were consistent with an epidemiologic study that found that individuals with insomnia symptoms who slept objectively short (ACT sleep efficiency <80%) were associated with lower insulin levels and greater insulin sensitivity if they did not have T2D, while individuals with insomnia symptoms who slept objectively short were associated with worse insulin sensitivity if they also had T2D [47]. In a large, community study ($N=3911$), Kalmbach and colleagues found that individuals with insomnia disorder who reported sleeping <6h were 1.83 times more likely to report a history of T2D than good sleepers, even after adjusting for sex, age and obesity [29]. Finally, in a recent study of 30 adults, D'Aurea and colleagues showed that individuals with chronic insomnia who slept objectively <5h (as measured by PSG) had significantly higher fasting glucose levels than individuals with insomnia who slept objectively longer than 5h [52]. The findings of these and other studies using objective sleep measures and metabolic function indices are summarized in Table 29.4. There are no studies, however, that have examined the synergistic effect of insomnia and objective short sleep duration on prevalent or incident MetS.

In summary, most research suggests a significant association between insomnia symptoms and metabolic dysfunction, including increased risk of T2D. The available evidence suggests that this risk is stronger in chronic insomnia disorder when coupled with objective short sleep duration. The association of chronic insomnia with MetS is not clear and deserves further investigation in well-defined samples by assessing the severity and chronicity of insomnia, including objective measures of relevant confounders such as SDB as well as of nighttime sleep (e.g., PSG or ACT) or daytime alertness (e.g., MSLT).

HEART DISEASE AND STROKE

As mentioned above, clinical observation indicated early on a high comorbidity between insomnia and CVD and CBVD. Clinical patients with chronic insomnia were observed to have a history of heart disease or stroke to a greater rate than expected for the general population or similar outpatients [10, 11, 38]. However, whether insomnia itself conferred a significant increased risk of developing CVD or CBVD remained largely unstudied until the past decade, in which an abundance of data has flourished. Epidemiologic studies have examined whether insomnia is a risk factor for coronary heart disease (CHD), myocardial infarction (MI), heart failure, or, to a lesser extent, ischemic stroke [33–35]. As shown in Table 29.3, several systematic reviews and meta-analyses have been published in the past few years on the association of insomnia with these CVD and CBVD [33, 34]. The available longitudinal studies have estimated that individuals with insomnia symptoms have a 45% higher risk of incident CVD [34], a 41% risk of incident MI [33], and 55% of incident CHD or CBVD [33]. Once again, the available evidence on increased CVD/CBVD risk pertains primarily to insomnia symptoms, rather than chronic insomnia disorder, and most studies did not include PSG and could not control for SDB (see Chapter 32).

Following the seminal findings in the PSAC of a synergistic effect between insomnia and objective short sleep duration on the risk for HTN and T2D [16, 17, 45], other independent investigators have examined whether this insomnia phenotype is also more strongly associated with CVD/CBVD than insomnia with normal sleep duration. This hypothesis has been examined in several epidemiologic studies with the limitation that most of them relied on self-reported sleep duration, which may account for some of the inconsistent findings [29, 56–60, 87]. Specifically, Kalmbach and colleagues found in a large community study ($N=3911$) that individuals with insomnia disorder who reported sleeping <6h were 3.2 and 3.8 times more likely than good sleepers to report a history of myocardial infarction and stroke, respectively, even after adjusting for sex, age and obesity [29]. Recent work from the PSAC has also shown a cross-sectional association between insomnia with objective short sleep duration and prevalent CVD/CBVD [56]. This study showed that compared to normal sleepers who slept objectively >6h (as measured by PSG), individuals with insomnia who slept objectively <6h were 2.0 times more likely to have a history of CVD/CBVD, while individuals with insomnia who slept >6h of sleep were not significantly associated with prevalent CVD/CBVD

(OR=1.3) [56]. Importantly, Bertisch and colleagues [57] examined whether insomnia symptoms with PSG-measured short sleep duration were associated with increased risk of incident CVD/CBVD in 4437 subjects from the Sleep Heart Health Study (SHHS). Individuals with insomnia symptoms who slept objectively <6h showed a 29% higher risk of developing CVD/CBVD after a median of 11.4 years of follow-up, while individuals with insomnia symptoms who slept objectively >6h were not at significantly increased risk of incident CVD/CBVD (HR=0.99) [57]. Thus, the available epidemiologic studies support an association of insomnia with CVD/CBVD and that this association is found when insomnia is defined based on objectively-measured short sleep duration. The findings of studies using objective sleep measures are summarized in Table 29.4.

Finally, several large epidemiologic studies have examined the association of insomnia with all-cause and CVD/CBVD mortality. Early studies reported either a lack of mortality risk associated with insomnia or even a protective effect [9], however, two recent meta-analyses including 13 and 17 studies and 122,501 and 311,260 subjects, respectively, estimated that individuals with insomnia symptoms had a 33%–45% increased risk of CVD/CBVD mortality [33, 34] as compared to good sleepers (see Table 29.3). Following the approach of a synergistic effect between insomnia and objective short sleep duration on the risk of HTN and T2D [16, 17, 45], Vgontzas and colleagues [46] found that men from the PSAC with chronic insomnia who slept objectively <6h (as measured by PSG) were associated with a fourfold increased odds of all-cause mortality after 10 years of follow-up, a risk that was elevated to sevenfold among men with this insomnia phenotype and comorbid HTN or T2D at baseline. In contrast, no significant association was found in men who slept objectively >6h or in women [46]. However, this study was limited by the small number of deceased men (21%) and women (5%), which precluded estimating the risk of CVD/CBVD mortality in either gender. Only three studies have further examined whether insomnia is associated with increased mortality in men, but not women, or when combined with short sleep duration. In a cohort study of 6236 adults (40–45 years old) followed-up after 13–15 years, Sivertsen and colleagues [88] found that the risk of all-cause mortality associated with insomnia symptoms was 4.7-fold in men, a gender difference that has been replicated in another large study [89]. Furthermore, the authors found that individuals with insomnia symptoms who reported sleeping <6.5h had a 2.8-fold risk of all-cause mortality, whereas individuals with insomnia symptoms who reported sleeping >6.5h were not associated with a significant mortality risk (HR=1.8) [88]. In contrast, the recent study by Bertisch and colleagues [57] could not replicate these findings and found neither gender differences in mortality risk nor increased all-cause or CVD/CBVD mortality in individuals with insomnia symptoms who slept objectively <6h (HR=1.07) in the SHHS. Thus, the association between insomnia and mortality is rather modest and inconsistent, particularly when chronicity criteria is not used. Recent studies that have focused on chronic insomnia or used PSG measures have suggested an increased mortality risk, particularly in men and in those with short sleep duration. However, these findings have not been consistently replicated (see Table 29.4). It is likely that the association of insomnia with mortality is rather complex and multiple methodological (i.e., definitions used, cohort effects, critical length of survival time) and developmental factors (i.e., aging in normal sleepers) are at play. Future studies are needed with larger cohorts, uniform criteria for insomnia disorder, and objective sleep data to identify the population with the greatest mortality risk.

STRESS, IMMUNITY AND HEALTH BEHAVIORS

Regardless of the theoretical model adopted to explain the pathophysiology of insomnia (e.g., neurobiological, behavioral), the etiology of insomnia has been conceptualized from a diathesis-stress perspective since the 1980s. It is posited that the joint effects of stressful life events [90] and cognitive-emotional factors are central to the etiopathogenesis of insomnia [91]. In other words, certain individuals with predisposing traits, when faced with common, unexpected or traumatic precipitating events, experience stress-related insomnia symptoms. To cope with this transient insomnia, individuals use maladaptive cognitive and behavioral resources that ultimately lead to developing a chronic insomnia disorder [91]. From a neurobiological perspective, however, it is hypothesized that physiologic changes in the stress and immune system are also responsible for the perpetuation of chronic insomnia [12, 92, 93].

Early studies focused on the neuroendocrine stress response and were driven by the seminal study on psychophysiologic hyperarousal in insomnia by Monroe [94]. The majority of in-lab studies, with three exceptions, have reported increased hypothalamic-pituitary-adrenal (HPA) axis activation in individuals with chronic insomnia, including increased cortisol secretion and alterations in the diurnal cortisol profile [12, 92, 93]. Other neuroendocrine studies found that overnight norepinephrine and catecholamine metabolite levels were increased in individuals with chronic insomnia or were correlated with PSG indices of sleep disturbance in insomnia patients [12, 93]. Also, other studies, with a few exceptions, have found that insomnia is associated with increased nocturnal HR, impaired HRV, altered sympathovagal balance, as measured by impedance cardiography, increased whole-body metabolic rate, increased pupil size (indicative of sympathetic activation), increased or altered systemic inflammation, and increased central nervous system activation during wake and sleep [12, 37, 71, 92, 93, 95].

Physiologic changes in cardiac autonomic modulation and inflammation are of particular relevance to the increased cardiometabolic disease risk observed in individuals with insomnia (see Chapter 26). Only in the past few years two systematic reviews and meta-analyses have been published on the association of insomnia with HR, HRV and other indices of cardiac autonomic modulation [71, 95]. The multiple studies to date suggest a shift toward a predominance of sympathetic modulation during both wake and nighttime periods in individuals with chronic insomnia disorder, given the observed decreased parasympathetic activity (as measured by high-frequency HRV) during NREM sleep and increased sympathetic nervous system activity, as measured by impedance cardiography and the low-to high-frequency (LF/HF) ratio in HRV [71]. Importantly, studies in which a positive association was reported, individuals with chronic insomnia were carefully screened and showed objective sleep disturbances (as measured by PSG or ACT), while studies that did not find an association between chronic insomnia and HRV parameters defined the disorder solely based on subjective reports [12, 71]. For example, Spiegelhalder and colleagues [48] found that when objective measures were not used, chronic insomnia patients did not differ significantly from their good sleeper counterparts in either resting HR or nighttime HRV [48]. On the other hand, chronic insomnia subjects with objective short sleep duration (PSG sleep efficiency <85%) had reduced parasympathetic activity compared to good sleepers, while chronic insomnia subjects with normal sleep duration (PSG sleep efficiency >85%) had similar measures of HR and HRV as good sleeping controls [48]. The findings of the studies that used objective sleep measures in insomnia subjects are presented in Table 29.4.

Studies focusing on immune system activity in individuals with insomnia have been more limited and have reported modest associations [37]. Epidemiologic studies in adults reported no significant association between insomnia symptoms and C-reactive protein (CRP) levels, an acute-phase inflammatory protein of hepatic origin that increases following interleukin-6 (IL-6) secretion [96]. However, in-lab controlled studies found increased inflammation in individuals with chronic insomnia compared to good sleepers, as measured by the secretion [97] and diurnal profiles of IL-6 [44]. In these latter studies, individuals with chronic insomnia were carefully-screened and showed objective sleep disturbances (as measured by PSG). Indeed, evidence accumulates that biomarkers of stress and immune system hyperarousal are primarily present in individuals with insomnia and short sleep duration [12, 52, 53, 98–101] (see Table 29.4). For example, in a controlled, in-lab study Floam and colleagues [98] showed that young adults with insomnia disorder slept objectively shorter (as measured by ACT) and had elevated HPA-axis and pro-inflammatory activity than good sleeping controls and that ACT-measured wake after sleep onset (WASO) was positively associated with HPA-axis activity among insomnia participants. In another in-lab study, D'Aurea and colleagues [52] showed that middle-aged adults with chronic insomnia who slept objectively <5h (as measured by PSG) had significantly higher cortisol levels than individuals with chronic insomnia who slept objectively >5h and that cortisol and adrenocorticotropic hormone (ACTH) levels were inversely correlated with PSG-measured sleep duration among individuals with chronic insomnia. Castro-Diehl and colleagues [53] found in a population-based study of 527 adults from the MultiEthnic Study of Atherosclerosis (MESA), that individuals with insomnia symptoms who slept objectively <7h (as measured by ACT) had greater HR orthostatic reactivity and high-frequency HRV mental reactivity compared to good sleepers. The authors concluded that insomnia with objective short sleep duration was associated with lower levels of cardiac parasympathetic tone and/or higher levels of sympathetic activity [53], a finding consistent with de Zambotti and colleagues [100] data supporting dysfunctional sympathetic activity but normal parasympathetic modulation before and during sleep in young adults with chronic insomnia who sleep objectively shorter compared to controls. Finally, in a recent study of 378 adolescents from the Penn State Child Cohort (PSCC), Fernandez-Mendoza and colleagues [54] found that adolescents with insomnia symptoms who slept objectively <7h (as measured by PSG) had significantly higher CRP levels as compared to good sleepers. In contrast, adolescents with insomnia symptoms who slept objectively >7h were not associated with significantly increased CRP levels [54].

In summary, peripheral and central markers of increased HPA axis activity and impaired cardiac autonomic modulation are primarily found in individuals with chronic insomnia and objective short sleep duration, a phenotype in which 24-h hyperarousal is the main pathophysiologic mechanism [12, 102]. Thus, neuroendocrine dysregulation, cardiac autonomic imbalance, and chronic low-grade inflammation are believed to underlie the increased cardiometabolic disease risk associated with chronic insomnia with objective short sleep duration. A diagrammatic representation of this conceptual model is presented in Fig. 29.1.

Another potential pathway by which insomnia is linked to greater cardiometabolic disease risk that deserves separate consideration is the presence of inadequate health-related behaviors (see Chapters 21–25). Individuals with insomnia are indeed more likely to report smoking, excessive alcohol or caffeine use and lack of physical activity [103], which are all well-established lifestyle CMR. Individuals with insomnia have also been found to have lower cardiorespiratory fitness, which is an independent risk factor for CVD/CBVD [104]. Another health-related behavior recently examined as a potential link between insomnia and cardiometabolic disease risk has been poor diet [105, 106]. A recent large, general population study, for example, found that individuals

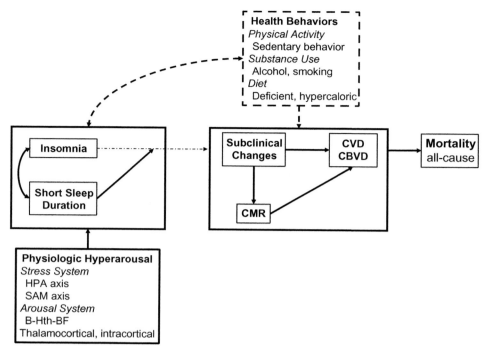

FIG. 29.1 Insomnia with short sleep duration and cardiometabolic disease risk. The *dotted arrow* represents the synergistic effect between insomnia and objective short sleep duration on cardiometabolic disease risk. This diagram depicts how objective short sleep duration in individuals with insomnia is the result of physiologic hyperarousal and identifies those with increased risk of cardiometabolic disease morbidity and mortality. The arousal system includes ascending projections from the brainstem and hypothalamus to the diencephalon, limbic system, basal forebrain, and neocortex (B-Hth-BF) as well as descending projections regulating the autonomic nervous system. These ascending and descending projections interact with the acute or chronic activation of the stress system. This diagram includes potential subclinical changes such as chronic low-grade inflammation, impaired cardiac autonomic modulation, endothelial dysfunction, carotid intima media thickness or coronary artery calcification. It also depicts the potential relative role (discontinuous boxes) of inadequate health behaviors in increasing cardiometabolic disease risk. *CBVD*, cerebrovascular disease; *CMR*, cardiometabolic risk factors; *CVD*, cardiovascular disease; *HPA*, hypothalamic-pituitary-adrenal; *SAM*, sympatho-adrenal-medullary.

who reported insomnia symptoms were associated with an inadequate intake of alpha-carotene, calcium, selenium, salt, carbohydrates, vitamin D, lycopene, or dodecanoic, hexadecanoic, butanoic, or hexanoic acids [106]. It is likely that the relationship between insomnia and many of these health-related behaviors is bidirectional [5, 6, 8, 103]. Interestingly, however, sleep hygiene therapy alone, which targets most of these inadequate health behaviors, is not an effective treatment for chronic insomnia disorder [107]. Importantly, most studies reviewed above on the association of insomnia with CMR or CVD/CBVD adjusted for the potential confounding effect of alcohol and smoking but not diet or physical activity. More work is needed to establish the relative contribution and potential causal role of inadequate health behaviors to the increased cardiometabolic disease risk associated with insomnia, above and beyond the other putative stress-, autonomic-, and immune-related mechanisms (see Fig. 29.1).

PUBLIC HEALTH AND CLINICAL IMPLICATIONS

If insomnia symptoms and chronic insomnia account for about 40% of the population and are significantly associated with increased cardiometabolic disease risk, there is no doubt then that they should become a target of public health policies [108]. Better screening of insomnia in the general population is needed, including in family medicine, occupational and school settings. There are existing brief, reliable, valid and easy to use self-reported tools that can help identify individuals from the general population with subthreshold insomnia symptoms and with clinically significant insomnia, such as the insomnia severity index (ISI), as well as objective measures of sleep duration such as ACT or home-based PSG. This improved detection of insomnia should lead to early phenotyping of those at greatest risk and early targeted treatment, and may help prevent downstream adverse health outcomes, including cardiometabolic risk [102]. Such preventive interventions should also be conducted outside the clinical office at the community level, mimicking those that are applied for cardiometabolic disease risk reduction.

Chronic insomnia disorder has become the focus of many professional health organizations. The American College of Physicians (ACP) has recently released a clinical guideline recommending that "all adult patients receive cognitive-behavioral therapy for insomnia (CBT-I) as the initial treatment for chronic insomnia disorder" and that "clinicians use a shared decision-making approach, including a discussion of the benefits, harms, and costs of short-term use of [sleep] medications, to decide whether to add pharmacological therapy in adults with chronic insomnia disorder in whom CBT-I alone was unsuccessful" [107]. This clinical guideline represents a shift in the current treatment of insomnia and disseminates CBT-I as a well-established, first-line, effective treatment. However, the introduction of pharmacological therapy, which is the current norm, is left up to a failure to respond to CBT-I alone, which indicates that the field still has difficulties matching insomnia treatments to specific phenotypes. This issue pertains to the relationship between insomnia and cardiometabolic disease risk, as it is unknown whether insomnia therapies are associated with concomitant improvements in physiology (e.g., improved inflammation, cortisol, HRV) or health behaviors (e.g., increased physical activity, improved diet). It is also unclear whether any improvements in physiology and health behaviors would attenuate the long-term cardiometabolic disease risk associated with chronic insomnia disorder.

There are no randomized clinical trials (RCT) that have systematically assessed whether widely-used pharmacological therapies (e.g., zolpidem, trazodone) improve CMR. A recent clinical trial found that the addition of the intermediate-acting benzodiazepine estazolam to usual antihypertensive treatment in individuals with insomnia produced significant mean decreases in BP levels of −8 to −11 mmHg as compared to placebo with antihypertensive treatment as usual, which produced mean decreased in BP levels of about −3 mmHg [109]. In contrast, a recent clinical trial that examined the effect of CBT-I, administered via a web platform (SLEEPIO) for 8 weeks, on ambulatory BP levels showed no statistically significant or clinically meaningful changes as compared to the standard of care of education on vascular risk factors (−0.9 vs 0.8 mmHg, respectively) [110]. Interestingly, Bathgate and colleagues [111] have recently reported that individuals with chronic insomnia who slept objectively <6 h (as measured by ACT) had a blunted response to CBT-I, while individuals with chronic insomnia who slept objectively >6 h showed high response and remission rates after undergoing CBT-I.

Thus, future RCTs should be designed to test the effectiveness of pharmacological and cognitive-behavioral therapies for insomnia in improving subclinical and prognostic markers of cardiometabolic disease risk, particularly in individuals with chronic insomnia and objective short sleep duration. It is likely that combined CBT-I plus pharmacological treatment in the most severe insomnia phenotype would show the most significant and clinically-relevant effects on cardiometabolic outcomes, however, this hypothesis needs to be tested.

CONCLUSION

Insomnia is a premorbid risk factor for CVD/CBVD, including mortality. Accumulating evidence suggests that the association of insomnia with CMR is more pronounced when insomnia is defined as a chronic disorder and is associated with objective short sleep duration (i.e., < 6 h) or other measures of physiologic hyperarousal (e.g., such as longer latencies in the MSLT). Evidence suggests that measures of insomnia, and especially objective sleep measures, should be included in the estimation of cardiometabolic disease risk in the clinical and general population. Future studies should make use of sophisticated in-lab and ambulatory study designs in order to better understand the cardiovascular and metabolic profiles of individuals with chronic insomnia in an ecologically-valid manner. Also, longitudinal studies with a lifespan perspective are needed in order to better understand in which critical developmental stages insomnia starts to have an impact on cardiometabolic disease risk. As mentioned above, well-designed RCTs are needed, as they will not only provide therapeutic evidence but also proof-of-concept that improving insomnia and sleep duration improves cardiometabolic biomarkers. Based on the current evidence, personalized medicine approaches using treatment-matching to specific insomnia phenotypes in large RCTs are essential.

REFERENCES

[1] Ohayon MM. Epidemiology of insomnia: what we know and what we still need to learn. Sleep Med Rev 2002;6(2):97–111.
[2] Bixler EO, Vgontzas AN, Lin HM, Vela-Bueno A, Kales A. Insomnia in central Pennsylvania. J Psychosom Res 2002;53(1):589–92.
[3] American Academy of Sleep Medicine. International classification of sleep disorders. 3rd ed. Darien, IL: American Academy of Sleep Medicine; 2014.
[4] LeBlanc M, Merette C, Savard J, Ivers H, Baillargeon L, Morin CM. Incidence and risk factors of insomnia in a population-based sample. Sleep 2009;32(8):1027–37.
[5] Singareddy R, Vgontzas AN, Fernandez-Mendoza J, Liao D, Calhoun S, Shaffer ML, et al. Risk factors for incident chronic insomnia: a general population prospective study. Sleep Med 2012;13(4):346–53.
[6] Fernandez-Mendoza J, Vgontzas AN, Bixler EO, Singareddy R, Shaffer ML, Calhoun SL, et al. Clinical and polysomnographic predictors of the natural history of poor sleep in the general population. Sleep 2012;35(5):689–97.
[7] Morin CM, Belanger L, LeBlanc M, Ivers H, Savard J, Espie CA, et al. The natural history of insomnia: a population-based 3-year longitudinal study. Arch Intern Med 2009;169(5):447–53.
[8] Vgontzas AN, Fernandez-Mendoza J, Bixler EO, Singareddy R, Shaffer ML, Calhoun SL, et al. Persistent insomnia: the role of objective short sleep duration and mental health. Sleep 2012;35(1):61–8.
[9] Hall M, Fernandez-Mendoza J, Kline C, Vgontzas A. Insomnia and health. In: Kryger M, Roth T, Dement W, editors. Principles and practice of sleep medicine. 6th ed. New York, NY: Elsevier; 2016. p. 794–803.

[10] Kales A, Kales J. Evaluation and treatment of insomnia. New York, NY: Oxford University Press; 1984.

[11] Bonnet M, Arand D. Cardiovascular implications of poor sleep. Sleep Med Clin 2007;2:529–38.

[12] Vgontzas AN, Fernandez-Mendoza J, Liao D, Bixler EO. Insomnia with objective short sleep duration: the most biologically severe phenotype of the disorder. Sleep Med Rev 2013;17(4):241–54.

[13] Spiegelhalder K, Scholtes C, Riemann D. The association between insomnia and cardiovascular diseases. Nat Sci Sleep 2010;2:71–8.

[14] Thomas SJ, Calhoun D. Sleep, insomnia, and hypertension: current findings and future directions. J Am Soc Hypertens 2017;11(2):122–9.

[15] Meng L, Zheng Y, Hui R. The relationship of sleep duration and insomnia to risk of hypertension incidence: a meta-analysis of prospective cohort studies. Hypertens Res 2013;36(11):985–95.

[16] Vgontzas AN, Liao D, Bixler EO, Chrousos GP, Vela-Bueno A. Insomnia with objective short sleep duration is associated with a high risk for hypertension. Sleep 2009;32(4):491–7.

[17] Fernandez-Mendoza J, Vgontzas AN, Liao D, Shaffer ML, Vela-Bueno A, Basta M, et al. Insomnia with objective short sleep duration and incident hypertension: the Penn State Cohort. Hypertension 2012;60(4):929–35.

[18] Vozoris NT. The relationship between insomnia symptoms and hypertension using United States population-level data. J Hypertens 2013;31(4):663–71.

[19] Vozoris NT. Insomnia symptom frequency and hypertension risk: a population-based study. J Clin Psychiatry 2014;75(6):616–23.

[20] Shivashankar R, Kondal D, Ali MK, Gupta R, Pradeepa R, Mohan V, et al. Associations of sleep duration and disturbances with hypertension in metropolitan cities of Delhi, Chennai, and Karachi in South Asia: cross-sectional analysis of the CARRS study. Sleep 2017;40(9).

[21] Wang YM, Song M, Wang R, Shi L, He J, Fan TT, et al. Insomnia and multimorbidity in the community elderly in China. J Clin Sleep Med 2017;13(4):591–7.

[22] Clark AJ, Salo P, Lange T, Jennum P, Virtanen M, Pentti J, et al. Onset of impaired sleep and cardiovascular disease risk factors: a longitudinal study. Sleep 2016;39(9):1709–18.

[23] Leigh L, Hudson IL, Byles JE. Sleep difficulty and disease in a cohort of very old women. J Aging Health 2016;28(6):1090–104.

[24] Hauan M, Strand LB, Laugsand LE. Associations of insomnia symptoms with blood pressure and resting heart rate: the HUNT study in Norway. Behav Sleep Med 2016;1–21.

[25] Wang Y, Jiang T, Wang X, Zhao J, Kang J, Chen M, et al. Association between insomnia and metabolic syndrome in a Chinese Han population: a cross-sectional study. Sci Rep 2017;7(1):10893.

[26] Ramos AR, Weng J, Wallace DM, Petrov MR, Wohlgemuth WK, Sotres-Alvarez D, et al. Sleep patterns and hypertension using actigraphy in the hispanic community health study/study of Latinos. Chest 2018;153(1):87–93.

[27] Lin CL, Liu TC, Lin FH, Chung CH, Chien WC. Association between sleep disorders and hypertension in Taiwan: a nationwide population-based retrospective cohort study. J Hum Hypertens 2017;31(3):220–4.

[28] Deng HB, Tam T, Zee BC, Chung RY, Su X, Jin L, et al. Short sleep duration increases metabolic impact in healthy adults: a population-based cohort study. Sleep 2017;40:1–11.

[29] Kalmbach DA, Pillai V, Arnedt JT, Drake CL. DSM-5 insomnia and short sleep: comorbidity landscape and racial disparities. Sleep 2016;39(12):2101–11.

[30] Bathgate CJ, Edinger JD, Wyatt JK, Krystal AD. Objective but not subjective short sleep duration associated with increased risk for hypertension in individuals with insomnia. Sleep 2016;39(5):1037–45.

[31] Johann AF, Hertenstein E, Kyle SD, Baglioni C, Feige B, Nissen C, et al. Insomnia with objective short sleep duration is associated with longer duration of insomnia in the Freiburg insomnia Cohort compared to insomnia with normal sleep duration, but not with hypertension. PLoS One 2017;12(7):e0180339.

[32] Cappuccio FP, D'Elia L, Strazzullo P, Miller MA. Quantity and quality of sleep and incidence of type 2 diabetes: a systematic review and meta-analysis. Diabetes Care 2010;33(2):414–20.

[33] Li M, Zhang XW, Hou WS, Tang ZY. Insomnia and risk of cardiovascular disease: a meta-analysis of cohort studies. Int J Cardiol 2014;176(3):1044–7.

[34] Sofi F, Cesari F, Casini A, Macchi C, Abbate R, Gensini GF. Insomnia and risk of cardiovascular disease: a meta-analysis. Eur J Prev Cardiol 2014;21(1):57–64.

[35] Li Y, Zhang X, Winkelman JW, Redline S, Hu FB, Stampfer M, et al. Association between insomnia symptoms and mortality: a prospective study of U.S. men. Circulation 2014;129(7):737–46.

[36] Anothaisintawee T, Reutrakul S, Van Cauter E, Thakkinstian A. Sleep disturbances compared to traditional risk factors for diabetes development: systematic review and meta-analysis. Sleep Med Rev 2016;30:11–24.

[37] Irwin MR, Olmstead R, Carroll JE. Sleep disturbance, sleep duration, and inflammation: a systematic review and meta-analysis of cohort studies and experimental sleep deprivation. Biol Psychiatry 2016;80(1):40–52.

[38] Schwartz S, McDowell Anderson W, Cole SR, Cornoni-Huntley J, Hays JC, Blazer D. Insomnia and heart disease: a review of epidemiologic studies. J Psychosom Res 1999;47(4):313–33.

[39] Stepanski E, Glinn M, Zorick F, Roehrs T, Roth T. Heart rate changes in chronic insomnia. Stress Med 1994;10(4):261–6.

[40] Bonnet MH, Arand DL. 24-Hour metabolic rate in insomniacs and matched normal sleepers. Sleep 1995;18(7):581–8.

[41] Bonnet MH, Arand DL. Heart rate variability in insomniacs and matched normal sleepers. Psychosom Med 1998;60(5):610–5.

[42] Vgontzas AN, Bixler EO, Lin HM, Prolo P, Mastorakos G, Vela-Bueno A, et al. Chronic insomnia is associated with nyctohemeral activation of the hypothalamic-pituitary-adrenal axis: clinical implications. J Clin Endocrinol Metab 2001;86(8):3787–94.

[43] Shaver JL, Johnston SK, Lentz MJ, Landis CA. Stress exposure, psychological distress, and physiological stress activation in midlife women with insomnia. Psychosom Med 2002;64(5):793–802.

[44] Vgontzas AN, Zoumakis M, Papanicolaou DA, Bixler EO, Prolo P, Lin HM, et al. Chronic insomnia is associated with a shift of interleukin-6 and tumor necrosis factor secretion from nighttime to daytime. Metabolism 2002;51(7):887–92.

[45] Vgontzas AN, Liao D, Pejovic S, Calhoun S, Karataraki M, Bixler EO. Insomnia with objective short sleep duration is associated with type 2 diabetes: a population-based study. Diabetes Care 2009;32(11):1980–5.

[46] Vgontzas AN, Liao D, Pejovic S, Calhoun S, Karataraki M, Basta M, et al. Insomnia with short sleep duration and mortality: the Penn State cohort. Sleep 2010;33(9):1159–64.

[47] Knutson KL, Van Cauter E, Zee P, Liu K, Lauderdale DS. Cross-sectional associations between measures of sleep and markers of glucose metabolism among subjects with and without diabetes: the Coronary Artery Risk Development in Young Adults (CARDIA) Sleep Study. Diabetes Care 2011;34(5):1171–6.

[48] Spiegelhalder K, Fuchs L, Ladwig J, Kyle SD, Nissen C, Voderholzer U, et al. Heart rate and heart rate variability in subjectively reported insomnia. J Sleep Res 2011;20 (1 Pt 2):137–45.

[49] Vasisht KP, Kessler LE, Booth 3rd JN, Imperial JG, Penev PD. Differences in insulin secretion and sensitivity in short-sleep insomnia. Sleep 2013;36(6):955–7.

[50] Fernandez-Mendoza J, Vgontzas AN, Calhoun SL, Vgontzas A, Tsaoussoglou M, Gaines J, et al. Insomnia symptoms, objective sleep duration and hypothalamic-pituitary-adrenal activity in children. Eur J Clin Investig 2014;44(5):493–500.

[51] Li Y, Vgontzas AN, Fernandez-Mendoza J, Bixler EO, Sun Y, Zhou J, et al. Insomnia with physiological hyperarousal is associated with hypertension. Hypertension 2015;65(3):644–50.

[52] D'Aurea C, Poyares D, Piovezan RD, Passos G, Tufik S, Mello MT. Objective short sleep duration is associated with the activity of the hypothalamic-pituitary-adrenal axis in insomnia. Arq Neuropsiquiatr 2015;73(6):516–9.

[53] Castro-Diehl C, Diez Roux AV, Redline S, Seeman T, McKinley P, Sloan R, et al. Sleep duration and quality in relation to autonomic nervous system measures: the multi-ethnic study of atherosclerosis (MESA). Sleep 2016;39(11):1927–40.

[54] Fernandez-Mendoza J, Baker JH, Vgontzas AN, Gaines J, Liao D, Bixler EO. Insomnia symptoms with objective short sleep duration are associated with systemic inflammation in adolescents. Brain Behav Immun 2017;61:110–6.

[55] Jarrin DC, Ivers H, Lamy M, Chen IY, Harvey AG, Morin CM. Cardiovascular autonomic dysfunction in insomnia patients with objective short sleep duration. J Sleep Res.e12663.

[56] Fernandez-Mendoza J, He F, Liao D, Vgontzas AN, Bixler EO, editors. Insomnia with objective short sleep duration is associated with increased risk of cardiocerebrovascular disease. 32nd Annual Meeting of the Associated Professional Sleep Societies; Baltimore, MD: Associated Professional Sleep Societies; 2018.

[57] Bertisch SM, Pollock BD, Mittleman MA, Buysse DJ, Bazzano LA, Gottlieb DJ, et al. Insomnia with objective short sleep duration and risk of incident cardiovascular disease and all-cause mortality: sleep heart health study. Sleep 2018;. zsy047-zsy.

[58] Chandola T, Ferrie JE, Perski A, Akbaraly T, Marmot MG. The effect of short sleep duration on coronary heart disease risk is greatest among those with sleep disturbance: a prospective study from the Whitehall II cohort. Sleep 2010;33(6):739–44.

[59] Sands-Lincoln M, Loucks EB, Lu B, Carskadon MA, Sharkey K, Stefanick ML, et al. Sleep duration, insomnia, and coronary heart disease among postmenopausal women in the Women's Health Initiative. J Womens Health (Larchmt) 2013;22(6):477–86.

[60] Canivet C, Nilsson PM, Lindeberg SI, Karasek R, Ostergren PO. Insomnia increases risk for cardiovascular events in women and in men with low socioeconomic status: a longitudinal, register-based study. J Psychosom Res 2014;76(4):292–9.

[61] Lu K, Chen J, Wu S, Chen J, Hu D. Interaction of sleep duration and sleep quality on hypertension prevalence in adult Chinese males. J Epidemiol 2015;25(6):415–22.

[62] Neuhauser H, Diederichs C, Boeing H, Felix SB, Junger C, Lorbeer R, et al. Hypertension in Germany. Dtsch Arztebl Int 2016;113(48):809–15.

[63] Benjamin EJ, Virani SS, Callaway CW, Chang AR, Cheng S, Chiuve SE, et al. Heart disease and stroke statistics-2018 update: a report from the American Heart Association. Circulation 2018;137(12):e67-492.

[64] Lichstein KL, Wilson NM, Noe SL, Aguillard RN, Bellur SN. Daytime sleepiness in insomnia: behavioral, biological and subjective indices. Sleep 1994;17(8):693–702.

[65] Seidel WF, Ball S, Cohen S, Patterson N, Yost D, Dement WC. Daytime alertness in relation to mood, performance, and nocturnal sleep in chronic insomniacs and noncomplaining sleepers. Sleep 1984;7(3):230–8.

[66] Stepanski E, Zorick F, Roehrs T, Young D, Roth T. Daytime alertness in patients with chronic insomnia compared with asymptomatic control subjects. Sleep 1988;11(1):54–60.

[67] Sugerman JL, Stern JA, Walsh JK. Daytime alertness in subjective and objective insomnia: some preliminary findings. Biol Psychiatry 1985;20(7):741–50.

[68] Lanfranchi PA, Pennestri MH, Fradette L, Dumont M, Morin CM, Montplaisir J. Nighttime blood pressure in normotensive subjects with chronic insomnia: implications for cardiovascular risk. Sleep 2009;32(6):760–6.

[69] Chen IY, Jarrin DC, Ivers H, Morin CM. Investigating psychological and physiological responses to the Trier Social Stress Test in young adults with insomnia. Sleep Med 2017;40:11–22.

[70] Johansson JK, Kronholm E, Jula AM. Variability in home-measured blood pressure and heart rate: associations with self-reported insomnia and sleep duration. J Hypertens 2011;29(10):1897–905.

[71] Nano MM, Fonseca P, Vullings R, Aarts RM. Measures of cardiovascular autonomic activity in insomnia disorder: a systematic review. PLoS One 2017;12(10):e0186716.

[72] Cespedes EM, Dudley KA, Sotres-Alvarez D, Zee PC, Daviglus ML, Shah NA, et al. Joint associations of insomnia and sleep duration with prevalent diabetes: the Hispanic Community Health Study/Study of Latinos (HCHS/SOL). J Diabetes 2016;8(3):387–97.

[73] Keckeis M, Lattova Z, Maurovich-Horvat E, Beitinger PA, Birkmann S, Lauer CJ, et al. Impaired glucose tolerance in sleep disorders. PLoS One 2010;5(3):e9444.

[74] Seelig E, Keller U, Klarhofer M, Scheffler K, Brand S, Holsboer-Trachsler E, et al. Neuroendocrine regulation and metabolism of glucose and lipids in primary chronic insomnia: a prospective case-control study. PLoS One 2013;8(4):e61780.

[75] Jennings JR, Muldoon MF, Hall M, Buysse DJ, Manuck SB. Self-reported sleep quality is associated with the metabolic syndrome. Sleep 2007;30(2):219–23.

[76] Okubo N, Matsuzaka M, Takahashi I, Sawada K, Sato S, Akimoto N, et al. Relationship between self-reported sleep quality and metabolic syndrome in general population. BMC Public Health 2014;14:562.

[77] Hall MH, Okun ML, Sowers M, Matthews KA, Kravitz HM, Hardin K, et al. Sleep is associated with the metabolic syndrome in a multi-ethnic cohort of midlife women: the SWAN Sleep Study. Sleep 2012;35(6):783–90.

[78] Kazman JB, Abraham PA, Zeno SA, Poth M, Deuster PA. Self-reported sleep impairment and the metabolic syndrome among African Americans. Ethn Dis 2012;22(4):410–5.

[79] Ikeda M, Kaneita Y, Uchiyama M, Mishima K, Uchimura N, Nakaji S, et al. Epidemiological study of the associations between sleep complaints and metabolic syndrome in Japan. Sleep Biol Rhythms 2014;12(4):269–78.

[80] Troxel WM, Buysse DJ, Matthews KA, Kip KE, Strollo PJ, Hall M, et al. Sleep symptoms predict the development of the metabolic syndrome. Sleep 2010;33(12):1633–40.

[81] Lin CL, Tsai YH, Yeh MC. The relationship between insomnia with short sleep duration is associated with hypercholesterolemia: a cross-sectional study. J Adv Nurs 2016;72(2):339–47.

[82] Lin SC, Sun CA, You SL, Hwang LC, Liang CY, Yang T, et al. The link of self-reported insomnia symptoms and sleep duration with metabolic syndrome: a Chinese population-based study. Sleep 2016;39(6):1261–6.

[83] Huang L, Zhou J, Sun Y, Li Z, Lei F, Zhou G, et al. Polysomnographically determined sleep and body mass index in patients with insomnia. Psychiatry Res 2013;209(3):540–4.

[84] Cronlein T, Langguth B, Busch V, Rupprecht R, Wetter TC. Severe chronic insomnia is not associated with higher body mass index. J Sleep Res 2015;24(5):514–7.

[85] Vgontzas AN, Fernandez-Mendoza J, Miksiewicz T, Kritikou I, Shaffer ML, Liao D, et al. Unveiling the longitudinal association between short sleep duration and the incidence of obesity: the Penn State Cohort. Int J Obes 2014;38(6):825–32.

[86] Lou P, Chen P, Zhang L, Zhang P, Chang G, Zhang N, et al. Interaction of sleep quality and sleep duration on impaired fasting glucose: a population-based cross-sectional survey in China. BMJ Open 2014;4(3):e004436.

[87] Westerlund A, Bellocco R, Sundstrom J, Adami HO, Akerstedt T, Trolle LY. Sleep characteristics and cardiovascular events in a large Swedish cohort. Eur J Epidemiol 2013;28(6):463–73.

[88] Sivertsen B, Pallesen S, Glozier N, Bjorvatn B, Salo P, Tell GS, et al. Midlife insomnia and subsequent mortality: the Hordaland health study. BMC Public Health 2014;14:720.

[89] Lallukka T, Podlipskytė A, Sivertsen B, Andruškienė J, Varoneckas G, Lahelma E, et al. Insomnia symptoms and mortality: a register-linked study among women and men from Finland, Norway and Lithuania. J Sleep Res 2016;25(1):96–103.

[90] Healey ES, Kales A, Monroe LJ, Bixler EO, Chamberlin K, Soldatos CR. Onset of insomnia: role of life-stress events. Psychosom Med 1981;43(5):439–51.

[91] Spielman AJ. Assessment of insomnia. Clin Psychol Rev 1986;6(1):11–25.

[92] Riemann D, Spiegelhalder K, Feige B, Voderholzer U, Berger M, Perlis M, et al. The hyperarousal model of insomnia: a review of the concept and its evidence. Sleep Med Rev 2010;14(1):19–31.

[93] Bonnet MH, Arand DL. Hyperarousal and insomnia: state of the science. Sleep Med Rev 2010;14(1):9–15.

[94] Monroe LJ. Psychological and physiological differences between good and poor sleepers. J Abnorm Psychol 1967;72(3):255–64.

[95] Tobaldini E, Nobili L, Strada S, Casali KR, Braghiroli A, Montano N. Heart rate variability in normal and pathological sleep. Front Physiol 2013;4:294.

[96] Laugsand LE, Vatten LJ, Bjorngaard JH, Hveem K, Janszky I. Insomnia and high-sensitivity C-reactive protein: the HUNT study. Norway Psychosom Med 2012;74(5):543–53.

[97] Burgos I, Richter L, Klein T, Fiebich B, Feige B, Lieb K, et al. Increased nocturnal interleukin-6 excretion in patients with primary insomnia: a pilot study. Brain Behav Immun 2006;20(3):246–53.

[98] Floam S, Simpson N, Nemeth E, Scott-Sutherland J, Gautam S, Haack M. Sleep characteristics as predictor variables of stress systems markers in insomnia disorder. J Sleep Res 2015;24(3):296–304.

[99] Clark AJ, Dich N, Lange T, Jennum P, Hansen AM, Lund R, et al. Impaired sleep and allostatic load: cross-sectional results from the Danish Copenhagen Aging and Midlife Biobank. Sleep Med 2014;15(12):1571–8.

[100] de Zambotti M, Cellini N, Baker FC, Colrain IM, Sarlo M, Stegagno L. Nocturnal cardiac autonomic profile in young primary insomniacs and good sleepers. Int J Psychophysiol 2014;93(3):332–9.

[101] Miller CB, Bartlett DJ, Mullins AE, Dodds KL, Gordon CJ, Kyle SD, et al. Clusters of insomnia disorder: an exploratory cluster analysis of objective sleep parameters reveals differences in neurocognitive functioning, quantitative EEG, and heart rate variability. Sleep 2016;39(11):1993–2004.

[102] Fernandez-Mendoza J. The insomnia with short sleep duration phenotype: an update on it's importance for health and prevention. Curr Opin Psychiatry 2017;30(1):56–63.

[103] Buysse DJ. Sleep health: can we define it? Does it matter? Sleep 2014;37(1):9–17.

[104] Strand LB, Laugsand LE, Wisloff U, Nes BM, Vatten L, Janszky I. Insomnia symptoms and cardiorespiratory fitness in healthy individuals: the Nord-Trondelag Health Study (HUNT). Sleep 2013;36(1):99–108.

[105] Grandner MA, Kripke DF, Naidoo N, Langer RD. Relationships among dietary nutrients and subjective sleep, objective sleep, and napping in women. Sleep Med 2010;11(2):180–4.

[106] Grandner MA, Jackson N, Gerstner JR, Knutson KL. Sleep symptoms associated with intake of specific dietary nutrients. J Sleep Res 2014;23(1):22–34.

[107] Qaseem A, Kansagara D, Forciea MA, Cooke M, Denberg TD, Clinical Guidelines Committee of the American College of Physicians. Management of chronic insomnia disorder in adults: a clinical practice guideline from the American College of Physicians. Ann Intern Med 2016;165(2):125–33.

[108] St-Onge MP, Grandner MA, Brown D, Conroy MB, Jean-Louis G, Coons M, et al. Sleep duration and quality: impact on lifestyle behaviors and cardiometabolic health: a scientific statement from the American Heart Association. Circulation 2016;134(18):e367–86.

[109] Li Y, Yang Y, Li Q, Yang X, Wang Y, Ku WL, et al. The impact of the improvement of insomnia on blood pressure in hypertensive patients. J Sleep Res 2017;26(1):105–14.

[110] McGrath ER, Espie CA, Power A, Murphy AW, Newell J, Kelly C, et al. Sleep to lower elevated blood pressure: a randomized controlled trial (SLEPT). Am J Hypertens 2017;30(3):319–27.

[111] Bathgate CJ, Edinger JD, Krystal AD. Insomnia patients with objective short sleep duration have a blunted response to cognitive behavioral therapy for insomnia. Sleep 2017;40(1):1–12.

GLOSSARY

Actigraphy An ambulatory method to estimate an individual's sleep and wake using an accelerometer attached to the non-dominant wrist.

Cardiometabolic risk factors Specific clinical and subclinical factors that put individuals at risk of cardiovascular or cerebrovascular disease. Traditionally, elevated blood pressure, including hypertension, insulin resistance, including type 2 diabetes, dyslipidemia and obesity are included. Lifestyle risk factors such as smoking, alcohol abuse, physical inactivity, and inadequate diet are also included within this category.

Cardiovascular disease A broad category that includes all disorders of the circulatory system, including coronary heart disease such as myocardial infarction, peripheral artery disease, cardiomyopathy, heart failure or cardiac arrhythmias.

Cerebrovascular disease A broad category that includes all disorders of the cerebral circulatory system and blood vessels in the brain, including ischemic stroke, transient ischemic attack, and subarachnoid or intracerebral hemorrhage.

Difficulty initiating sleep The report of inability to fall asleep at the desired bedtime.

Difficulty maintaining sleep The report of inability to resume sleep in the middle of the night or sleep period.

Early morning awakening The report of waking up too early before desired and having difficulty resuming sleep.

Nonrestorative sleep The report of unrefreshing sleep upon awakening regardless of the amount of sleep obtained. An insomnia symptom included in most epidemiological studies and previous diagnostic nosologies.

Insomnia symptoms The report of difficulties initiating sleep, difficulties maintaining sleep, early morning awakening or, in some studies, nonrestorative sleep without any duration (chronicity) or impairment (daytime) criteria. Also referred to as "poor sleep" or "sleep difficulties" in some studies.

Insomnia disorder The report of difficulties initiating sleep, difficulties maintaining sleep or early morning awakening that occur at least 3 nights per week for at least 3 months and are associated with significant daytime functioning impairment such as daytime fatigue, poor concentration, mood problems, or worry about insomnia itself, among others.

Metabolic syndrome The clustering of three or more of the cardiometabolic risk factors central obesity, elevated blood pressure, insulin resistance, hypercholesterolemia and/or hypertriglyceridemia.

Multiple sleep latency test The simultaneous recording of multiple physiologic measures (i.e., electroencephalography, electrooculography, electromyography and electrocardiography, at minimum) during five daytime nap opportunities to ascertain a person's daytime alertness or sleep propensity. It can only be performed in the sleep lab attended by registered polysomnography technicians.

Polysomnography The simultaneous recording of multiple physiologic measures (i.e., electroencephalography, electrooculography, electromyography and electrocardiography, at minimum) during the nighttime period to ascertain whether a person is asleep or awake and/or suffers from a sleep disorder. It can be performed in the sleep lab attended by registered polysomnography technicians or at home unattended using ambulatory monitoring.

Sleep fragmentation The report of multiple, brief (e.g., lasting <30 min) awakenings in the middle of the night or sleep period. A symptom typical of sleep disordered breathing, movement-related disorders and other sleep disorders.

Sleep disordered breathing A cluster of sleep disorders in which respiratory function during sleep is compromised and presence in the form of loud upper-airway related sounds (snoring), breathing pauses and/or hypoventilation. It ranges from simple snoring to obstructive sleep apnea and central sleep apnea.

Chapter 30

Sleep apnea and cardiometabolic disease risk

Andrew Kitcher[a,b,c], Atul Malhotra[a,b,c], Bernie Sunwoo[c]

[a]Chief of Pulmonary and Critical Care Medicine, University of California San Diego, La Jolla, CA, United States, [b]University of California, San Diego School of Medicine, La Jolla, CA, United States, [c]Department of Medicine, Division of Pulmonary, Critical Care and Sleep Medicine, University of California San Diego, San Diego, CA, United States

WHAT IS OSA?

Obstructive sleep apnea (OSA) is the most common form of sleep disordered breathing. OSA is characterized by recurrent collapse of the soft tissues of the upper airway resulting in partial or complete cessation of airflow. Importantly, OSA is distinguished from central sleep apnea in that OSA patients continue to make respiratory efforts throughout the apnea. The mechanisms that cause obstructive sleep apnea have been the subject of intense investigation over the last 20 years, during a surge in sleep apnea diagnoses, evolving treatment strategies, and a better understanding of sleep apnea's contribution to cardiovascular disease.

The upper airways, characterized by the oro- and nasopharyngeal spaces, are flexible multipurpose structures that accommodate mastication, deglutition of solids and liquids, clearance of secretions, phonation and respiration. Instead of the rigid cartilaginous support structure that characterizes the first few generations of lower airways, the upper airway relies on a complex array of muscles for support. Complete and partial airway closure is an essential part of swallowing and phonation. However, this capacity becomes pathological when closure occurs repetitively during sleep. Although the threshold for normal versus abnormal is debated, patients with obstructive sleep apnea have respiratory events in sufficient frequency to cause clinical consequences.

Patients with obstructive sleep apnea have one or more anatomical or physiological predisposition to increase propensity for airway collapse. OSA patients may have crowded anatomy, with narrowing at different points of the upper airway. Obesity, with resultant increase in neck circumference and neck soft tissues, is associated with sleep apnea as well. Upper airway dilator muscle function is also important since OSA patients have increased activity of these muscles during wakefulness ostensibly in compensation for anatomical deficiency [1], but a fall in dilator activity at sleep onset yields collapse of a vulnerable airway. Finally, sedative/hypnotic drugs and alcohol are all positively associated with worsening sleep disordered breathing, likely due to their depressive effect on airway muscle tone (Fig. 30.1).

Apnea itself is characterized by a steady rise in carbon dioxide and fall in oxygen as the patient makes ineffective respiratory efforts that do not result in adequate air flow. Were apnea to continue indefinitely, the patient would expire. However, by themselves, neither hypoxia nor hypercapnia is the primary trigger for the body to intervene during these failed efforts. Rescue from apnea, in the form of cortical arousal and engagement of the dilator muscles of the upper airway, is instead triggered by increasingly negative pleural pressures, generated by the actions of the muscles of inspiration to expand the lung in the absence of airflow [2]. Patients are not typically aware of these arousals, but the requisite muscles are engaged for airway opening and restoring adequate ventilation (Fig. 30.2).

Treatment of obstructive sleep apnea has focused on correction of the anatomical and physiological predispositions listed above. Both dietary weight loss and exercise have been shown to improve and even cure sleep apnea, with weight loss surgery studies often using correction of sleep disordered breathing as an important outcome. Avoidance of sedating medications and abstinence from alcohol are encouraged in all patients with OSA. In positional therapy, a wedge pillow or some other prop is used to maintain side sleep and thus lessen the degree of airway collapse due to gravity. Mandibular advancement devices, dental orthotics that push the mandible forward and thus tether open the airway, are used in a subset of patients. Their use is limited by patient anatomy and comfort during sleep. Finally, surgical

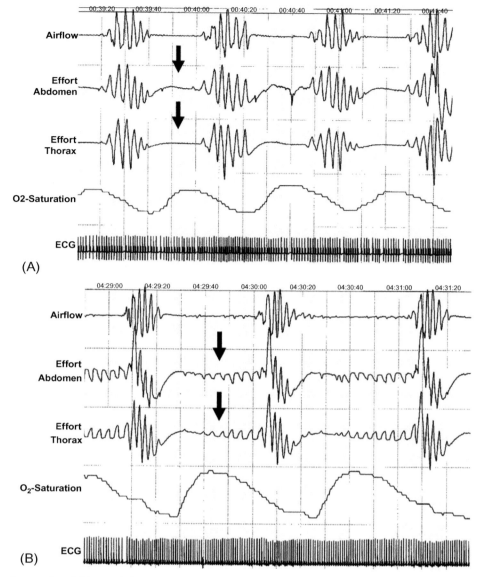

FIG. 30.1 Sleep study examples of (A) central sleep apnea and (B) obstructive sleep apnea, differentiated by respiratory effort signals.

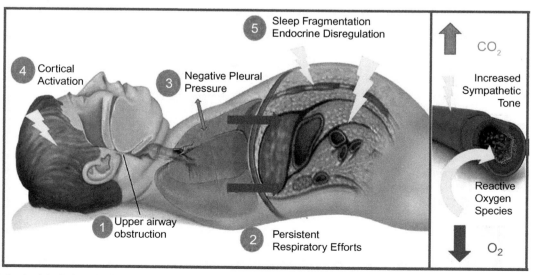

FIG. 30.2 Pathophysiological mechanisms and consequences of obstructive sleep apnea.

treatments, including the uvulopalatopharyngoplasty (or UPPP), have mostly shown variable efficacy [3]. In more recent years, nocturnal hypoglossal nerve stimulation has shown promise as a method to maintain airway patency without triggering cortical arousal.

By far the most common and well-studied OSA treatment is the use of positive airway pressure (PAP) devices to stent open the airway by providing inspiratory and expiratory airway transmural pressure. Continuous positive airway pressure (CPAP) devices have evolved considerably in recent years, and now are capable of collecting data on apneic events, and adjusting applied pressure based on physiological needs. Modern CPAP machines also standardly heat and humidify the air being provided to the patient. Three common masks are used: an oronasal mask sealing over the nose and mouth, a nasal mask that seals around the nose, and nasal "pillows," a small chamber that rests on the upper lip, with fitted prongs that maintain a seal with the nares.

WHO GETS OSA?

Symptomatic obstructive sleep apnea with daytime somnolence affects 3%–7% of adult men and 2%–5% of adult women. The most commonly recognized risk factors are weight, age, and male gender, with a male to female ratio of around 2–3:1 when population surveys are performed [4]. Because men are referred much more often for sleep apnea testing than women, the prevalence of sleep apnea diagnoses has a sex ratio between 5 and 8:1 [5]. It is unclear if this difference in referral patterns is due to differences in clinical suspicion in referring providers, differences in disease presentation, or other factors.

Both overweight and obese status are associated with increased risk of obstructive sleep apnea, as is weight gain regardless of final BMI, with a 32% increase in AHI and sixfold increase in moderate/severe OSA symptoms with a weight gain of 10% [6]. Studies that have attempted to demonstrate an independent association of waist circumference or neck circumference have had mixed results [7, 8]. Sleep apnea increases significantly with age; in men, an AHI >10 events per hour was present in 3.2% of patients aged 20–44 years, 11.3% of those ages 45–64 years, and 18.1% of those >65 years of age [9]. In women, the progression is equally dramatic: 0.6%, 2.0% and 7.0% in the corresponding cohorts [10]. Interestingly, estrogen, progesterone and testosterone may all play roles in the difference in sleep apnea prevalence between the sexes [10]. In general, both men and women have worsening sleep quality with age, with more interruptions and a decrease in the amount of sleep per night. With respect to sleep apnea, it is hypothesized that increased neck fat deposition and tissue laxity may additionally contribute [11, 12].

In addition to the above, several medical conditions make sleep apnea more likely. Both snoring and apneas increase during pregnancy especially in the third trimester [13], despite a decrease in supine sleep. Alcohol use can induce apneas in otherwise healthy sleepers [14], and worsens apneas and hypoxia in those with existing sleep apnea [15]. Current smokers are more likely to snore and have OSA, at least in some studies [16]. Finally, PCOS in women, and Down's Syndrome in both sexes are associated with very high (>60%) prevalence of obstructive sleep apnea [17, 18].

DOES HAVING OSA MAKE YOU MORE LIKELY TO HAVE CARDIOVASCULAR DISEASE?

Obstructive sleep apnea has been associated with hypertension, coronary artery disease, stroke, heart failure and arrhythmias. In men with severe sleep disordered breathing, increased coronary and cerebrovascular morbidity and mortality have been reported [19, 20]. Additionally, there is a dose-response to this association, with worsening apnea-hypopnea indices being associated with greater cardiovascular morbidity, further suggesting a potential causal link. Nevertheless, the precise mechanisms by which obstructive sleep apnea brings about cardiovascular disease remain unclear, and the presence of many common risk factors, such as obesity, age and alcohol/cigarette use, further complicates distinguishing association from causation. Studies have explored the influence of OSA treatment on various cardiovascular end points to delineate better this relationship between OSA and cardiovascular disease.

Hypertension

An increased prevalence of hypertension has been consistently shown in OSA [21–24], and a dose-response effect between severity of OSA and blood pressure has been demonstrated [21]. 71% of patients with resistant hypertension, defined as uncontrolled hypertension requiring at least three antihypertensive agents, have been shown to have OSA compared to 38% of patients with controlled hypertension, and OSA is a common cause of secondary hypertension [25–27].

A causal relationship between OSA and hypertension is supported by multiple studies now showing modest but clinically relevant reductions in blood pressure with OSA treatment. CPAP has been shown to improve nocturnal nondipping and daytime hypertension. A metaanalysis of 4888 patient showed CPAP reduced systolic blood pressure on average by 2.5 mmHg and diastolic blood pressure by 2.0 mmHg. Similar reductions in blood pressure have been shown with oral appliance therapies [28]. Studies have shown greater blood pressure improvement in patients with more severe OSA and in those with resistant hypertension [29, 30]. This reduction in blood pressure has also been associated with increased CPAP adherence further strengthening the relationship between OSA and hypertension.

Coronary artery disease

OSA has been associated with cardiovascular events related to coronary artery disease [20, 31]. Participants in the Wisconsin Sleep Cohort Study with severe sleep disordered breathing were 2.6 times more likely to have an incident coronary heart disease as defined by new reports of myocardial infarction, coronary revascularization procedures, congestive heart failure and cardiovascular death, compared to participants without [31]. Similarly Marin et al. demonstrated a higher incidence of fatal cardiovascular events and nonfatal cardiovascular events defined as nonfatal myocardial infarction, nonfatal stroke, coronary artery bypass surgery and percutaneous transluminal coronary angiography in patients with severe untreated obstructive sleep apnea compared to milder OSA and healthy participants. CPAP treatment of OSA was associated with reduced cardiovascular risk compared to severe untreated OSA [19]. However, these observational studies fall short of proving cardiovascular benefits of CPAP.

Randomized trials exploring the effects of CPAP treatment on cardiovascular events including coronary artery disease are mixed. The SAVE trial was a multicenter randomized trial of 2717 adults with moderate to severe obstructive sleep apnea and coronary or cerebrovascular disease randomized to CPAP therapy plus usual care or usual care alone. After a mean follow up 3.7 years CPAP did not significantly reduce the composite end point from death from cardiovascular causes, myocardial infarction, stroke or hospitalization for unstable angina, heart failure of transient ischemic attack. No significant differences were observed in any of the cause-specific cardiovascular end points including myocardial infarction or hospitalization for unstable angina [32]. The RICCADSA trial randomized 244 nonsleepy patients with newly revascularized coronary artery disease and moderate to severe OSA to CPAP or usual care. Over a median follow-up of 57 months no statistically significant difference was observed in the incidence of the primary end point of repeat vascularization, myocardial infarction, stroke and cardiovascular mortality [33]. This study like the SAVE trial was limited by poor CPAP adherence and when adjusted, on-treatment analysis showed a significant cardiovascular risk reduction in those who used CPAP ≥4 compared to <4h per night.

A reverse causal pathway whereby coronary artery disease and myocardial infarction worsens sleep disordered breathing has also been proposed. In the Sleep Heart Health Study, participants without cardiovascular disease completed two polysomnograms 5 years apart. Participants who developed incident cardiovascular events including myocardial infarction, heart failure and stroke, had greater increases in both mean obstructive and central apnea indices compared with patients without incident cardiovascular disease [34].

Others have examined the association between OSA and identified atherogenic risk factors such as dyslipidemia, endothelial dysfunction and inflammatory markers to try and better elucidate the relationship between OSA and coronary artery disease [35]. Despite experimental studies on mouse models showing increasing levels of triglyceride-rich lipoproteins with intermittent hypoxemia, cross-sectional human studies have not consistently supported a relationship between OSA and dyslipidemia. Studies on dyslipidemia are particularly challenged by the confounding effects of obesity [36, 37]. Similarly studies looking at CPAP treatment on fasting lipid profiles have not shown a consistent improvement in the lipid profile but are methodologically limited. Studies on endothelial dysfunction, another recognized marker of atherosclerosis, largely support an association between OSA and endothelial dysfunction, strengthened by demonstration of improvement in endothelial dysfunction with CPAP therapy [35, 38–42].

Cerebrovascular disease

Given shared atherogenic risk factors in cerebrovascular disease it is not surprising that a high prevalence of sleep disordered breathing has been reported in stroke patients. Sleep disordered breathing has been identified as both an independent risk factor for stroke and a consequence of strokes [43–46]. In one observational cohort study of 2011 consecutive adults who underwent polysomnography, OSA was significantly associated with incident stroke with a hazard ratio of 1.97 after adjusting for body mass index and cardiovascular risk factors [47]. A higher risk of stroke has been shown with higher apnea-hypopnea indices suggesting a dose response relationship [44]. A dose response relationship has been also been demonstrated between severity of sleep disordered breathing and risk of recurrent vascular events and all-cause mortality in stroke and TIA patients [48]. Following strokes, sleep disordered breathing has been associated with worse cognitive and functional outcomes [49]. Consequently a sleep study is recommended in TIA and stroke patients [50]. This is especially relevant given studies showing improvements in neurologic recovery and recurrent vascular events with CPAP treatment post stroke. There remains however conflicting evidence on the effects of CPAP therapy in reducing stroke risk and further studies are needed [45].

Heart failure

Sleep disordered breathing, including obstructive sleep apnea and central sleep apnea, is found in at least 50% of patients with heart failure [51]. A bidirectional relationship between sleep disordered breathing and heart failure exists. Gottlieb et al. followed a total of 1927 men and 2495 women aged ≥40 years free of heart failure over a median of 8.7 years. Men with severe OSA were 58% more likely to develop heart failure compared to those without OSA. This association was not observed in females [52]. In another cohort of community dwelling older men, central sleep apnea was also shown to be associated significantly with development of clinical heart failure and increased risk of decompensated heart failure [53]. Conversely central sleep apnea and Cheyne-Stokes respiration have been shown to be a poor prognostic factor in heart failure, associated with increased postdischarge mortality and hospital readmission in acute heart failure [54–57].

CPAP treatment of OSA has been shown to improve symptoms and cardiac function and is the treatment of choice for OSA in heart failure [58–61]. However, the optimal management of central sleep apnea in heart failure, if any beyond medical optimization of the heart failure, is uncertain. The CANPAP trial randomized 258 heart failure patients (mean ejection fraction 24.5%) with CSA to CPAP or medical therapy alone and was stopped early due to lack of difference in transplant-free survival observed between the two groups [62]. This study was limited by variable CPAP adherence and posthoc subgroup analysis suggested a survival advantage in those effectively treated with CPAP to reduce the apnea-hypopnea index to below 15 events/h [63]. Adaptive servoventilation is a mode of positive airway pressure that adjusts the level of inspiratory support above an expiratory positive airway pressure with the goal of stabilizing ventilator instability. Smaller studies on ASV in heart failure with CSA have shown improvement in the apnea-hypopnea index and cardiac function but a recent randomized controlled trial, the SERVE-HF trial ASV, has questioned the benefits of ASV in systolic heart failure patients with CSA [64, 65].

SERVE-HF randomized 1325 heart failure with reduced ejection fraction ≤45% patients with moderate to severe, predominantly central sleep apnea to ASV or medical management. While ASV improved the apnea-hypopnea index there was no difference in the incidence of the primary endpoint of all-cause mortality, lifesaving cardiovascular intervention or unplanned hospitalization for heart failure with ASV. An unexpected 34% increase in all-cause and cardiovascular mortality was observed in the ASV group [66]. Various explanations have been postulated for this unexpected finding including methodological limitations but the SERVE-HF resulted in an abrupt paradigm shift in the treatment of central sleep apnea in systolic heart failure including guidelines recommending against the use of ASV in systolic heart failure with predominantly central sleep apnea. Due to the results of SERVE-HF a study of 126 hospitalized heart failure patients with moderate to severe SDB randomized to ASV and optimized medical therapy or medical therapy alone was discontinued prematurely. A prespecified subgroup analysis suggested a positive effect of ASV in patients with heart failure with preserved ejection fraction but this study was again limited by variable ASV adherence [67]. Ongoing studies [68] including the ADVENT-HF trial are in progress to try to delineate better the role of positive airway pressure in central sleep apnea.

Arrhythmias

While coronary artery disease and heart failure can result in arrhythmias, OSA itself has been associated with a higher frequency of arrhythmias. In the Sleep Heart Health Study, a multicenter cohort of approximately 6400 patients aged over 40 years with and without severe sleep disordered breathing, individuals with severe sleep disordered breathing had four times the odds of atrial fibrillation, three times the odds of nonsustained ventricular tachycardia and almost twice the odds of complex ventricular ectopy [69]. Additional studies have supported this association between OSA and atrial fibrillation with both an increased prevalence of atrial fibrillation described in patients with OSA and an increased prevalence of OSA described in patients with atrial fibrillation [69–75]. OSA has also been identified as a predictor for recurrence of atrial fibrillation following cardioversion or ablation. In a metaanalysis of 3995 patients who underwent pulmonary vein isolation, patients with OSA had a 25% greater risk of atrial fibrillation recurrence than those without OSA [76]. Treatment of OSA has been shown to reduce this risk of atrial fibrillation recurrence [77–79]. Studies exploring the association between OSA and bradyarrhythmias and sinus pauses are fewer and more mixed. However, improvement in bradycardia was demonstrated in a small subset of patients with moderate or severe OSA implanted with a loop recorder following CPAP treatment [80].

WHY DOES OSA MAKE YOU MORE LIKELY TO HAVE CARDIOVASCULAR DISEASE?

The strong association between the presence of obstructive sleep apnea and cardiovascular comorbidity prompts the question of how. Proposed mechanisms focus on the apneic event as the key stressor, with the hope of distinguishing the unique risks posed by OSA, independent of its common comorbidities. Each component of the apneic event (adrenergic response, hypoxia, and sleep fragmentation) has a body of evidence linking that element with worsening cardiovascular disease.

Apneas trigger a significant adrenergic response that has been correlated with daytime systemic hypertension. Patients with sleep apnea have heightened sympathetic tone throughout the day which only increases with sleep, in contrast to normal controls. Apneic events trigger further increases in sympathetic tone, yielding marked surges in blood pressure [81]. CPAP treatment abates these spikes in sympathetic activity, and decreases the resting daytime and sleeping sympathetic tone [81, 82].

The second deleterious consequence of apnea is pathological hypoxia. Most human beings have a slight decrease in oxygen levels during sleep, which corresponds with a slight drop in body temperature and oxygen consumption and is not deleterious. However, patients with moderate or severe sleep apnea or mild sleep apnea together with underlying pulmonary comorbidities can have profound desaturations during their apneic events. This intermittent hypoxia generates reactive oxygen species (ROS) that cause endothelial dysfunction through a variety of mechanisms. First, reactive oxygen species lead to increased lipid peroxidation [68], and oxidation of the protein side chains of endothelial cells [83]. The resulting endothelial dysfunction makes blood vessels less responsive to usual vasodilatory stimuli, such as nitric oxide [84], a phenomenon that partially reverses with CPAP therapy [85, 86], and prolonged resistance to endothelial-mediated vasodilation leads to atherosclerotic changes [87], ultimately contributing to coronary artery disease and cerebrovascular disease. Reactive oxygen species also trigger endothelin release, which is associated with systemic hypertension and abates with treatment with CPAP [88].

A third consequence of obstructive apnea is sleep fragmentation, which in turn results in daytime symptoms and neurocognitive dysfunction. Arousals combined with hypoxemia and other stimuli are also thought to contribute to raised blood glucose and predisposition to diabetes. In healthy sleep, cortisol levels decline along with blood glucose levels, but when sleep periods are shortened or total sleep curtailed, multiple studies have demonstrated worsening glycemic control. Further studies have extended this observation to demonstrate similar findings in patients with obstructive sleep apnea [89]. Worsening apnea, as measured by hypoxic events and AHI, is correlated with both increased hyperglycemia on average [90] and glycemic variability [91]. These effects may be mediated by a reduction in glucagon-like peptide 1 response to feeding [90] or elevated fasting incretin levels (either glucagon-like peptide 1 or gastric inhibitory polypeptide/glucose-dependent insulinotropic polypeptide) [92]. However, interventional studies using CPAP in OSA have shown somewhat variable effect on glucose control, perhaps related to variable PAP adherence in these studies.

WHAT HAPPENS IF WE REDUCE APNEIC EVENTS?

The mainstay of OSA treatment is the use of continuous positive airway pressure devices to maintain upper airway patency by providing a pneumatic splint. Patients beginning CPAP therapy often report getting their first restful night of sleep in years, and the data tend to back up these anecdotal reports. Sleep studies performed on CPAP demonstrate significant reduction in the apnea-hypopnea index and improvement in oxygenation, across the spectrum of OSA severity, as well as elimination of snoring and improvement in sleep fragmentation. Daytime sleepiness significantly improves, especially in severe OSA, as measured by patient self-assessment and/or by objective testing. Less subjectively, a reduction in road-traffic accidents has also been demonstrated, a major endpoint given the contribution of sleepiness to accidents. Patients also report improvement in their thinking and coordination, but the data here are more mixed, whether assessing via neuropsychological testing or functional neuroimaging. The Apnea Positive Pressure Long-term Efficacy Study (APPLES) was a 6-month randomized, double-blind, sham-controlled study assessing the effects of CPAP on neurocognitive variables in OSA patients and at 2 months significant improvements were seen only in the executive and frontal lobe function variable [93]. However, the study had a number of dropouts which limited power and the endpoints were questioned by some investigators, leading the results to be regarded as not definitive.

As discussed, studies exploring the impact on secondary cardiovascular endpoints, such as lipid metabolism, glycemic control, inflammatory markers or blood pressure control, have all demonstrated significant improvement with CPAP use. The success of CPAP in improving these secondary markers, which are also part of the mechanistic explanation for the OSA/cardiovascular link, has led physicians and researchers to be hopeful for the impact of CPAP on cardiovascular morbidity and mortality. Until recently, this improvement was assumed, in the absence of large prospective studies to confirm it.

Unfortunately, with the reporting of the SAVE and RICCADSA trials, these expectations have been questioned [33, 94, 95]. The SAVE and RICCADSA trials showed no beneficial effect of CPAP on risk of cardiovascular disease in nonsleepy patients with moderate or severe OSA and established cardiovascular disease. This failure prompts the question of why. Was adherence to PAP too low? Was PAP therapy too late in this population with preexisting cardiovascular disease, the endovascular damage having already been done? Is it necessary to eliminate apneas and hypopneas, rather than significantly reduce them? Regardless, as of yet, no randomized trial has demonstrated a mortality benefit with PAP therapy in OSA. Further well-designed studies are required to clarify the relationship between OSA and cardiovascular disease and the impact of CPAP therapy on overall cardiovascular disease. Until then, current clinical guidelines recommend CPAP for moderate to severe OSA with or without symptoms or mild obstructive sleep apnea if accompanied by associated symptoms and/or cardiovascular disorders including hypertension, ischemic heart disease or stroke [96].

CONCLUSION

OSA is an exciting topic with data rapidly evolving regarding its pathogenesis and potential treatment. In the future, randomized trials may need to stratify carefully which patients are likely to benefit from a particular intervention perhaps based on biomarkers. Personalized medicine approaches are also being developed to guide interventions for OSA based on the mechanism underlying apnea in a given individual. Only by further clinical and basic research are new therapies and approaches for OSA likely to emerge.

REFERENCES

[1] Mezzanotte WS, Tangel DJ, White DP. Waking genioglossal electromyogram in sleep apnea patients versus normal controls (a neuromuscular compensatory mechanism). J Clin Invest 1992;89(5):1571–9.

[2] Gleeson K, Zwillich CW, White DP. The influence of increasing ventilatory effort on arousal from sleep. Am Rev Respir Dis 1990;142(2):295–300.

[3] Choi JH, et al. Predicting outcomes after Uvulopalatopharyngoplasty for adult obstructive sleep apnea: a meta-analysis. Otolaryngol Head Neck Surg 2016;155(6):904–13.

[4] Young T, et al. The occurrence of sleep-disordered breathing among middle-aged adults. N Engl J Med 1993;328(17):1230–5.

[5] Strohl KP, Redline S. Recognition of obstructive sleep apnea. Am J Respir Crit Care Med 1996;154(2 Pt 1):279–89.

[6] Peppard PE, et al. Longitudinal study of moderate weight change and sleep-disordered breathing. JAMA 2000;284(23):3015–21.

[7] Young T, Peppard PE, Taheri S. Excess weight and sleep-disordered breathing. J Appl Physiol (1985) 2005;99(4):1592–9.

[8] Young T, et al. Predictors of sleep-disordered breathing in community-dwelling adults: the Sleep Heart Health Study. Arch Intern Med 2002;162(8):893–900.

[9] Bixler EO, et al. Effects of age on sleep apnea in men: I. Prevalence and severity. Am J Respir Crit Care Med 1998;157(1):144–8.

[10] Bixler EO, et al. Prevalence of sleep-disordered breathing in women: effects of gender. Am J Respir Crit Care Med 2001;163(3 Pt 1):608–13.

[11] Eikermann M, et al. The influence of aging on pharyngeal collapsibility during sleep. Chest 2007;131(6):1702–9.

[12] Malhotra A, et al. Aging influences on pharyngeal anatomy and physiology: the predisposition to pharyngeal collapse. Am J Med 2006;119(1):72. e9–14.

[13] Pien GW, et al. Changes in symptoms of sleep-disordered breathing during pregnancy. Sleep 2005;28(10):1299–305.

[14] Taasan VC, et al. Alcohol increases sleep apnea and oxygen desaturation in asymptomatic men. Am J Med 1981;71(2):240–5.

[15] Remmers JE. Obstructive sleep apnea. A common disorder exacerbated by alcohol. Am Rev Respir Dis 1984;130(2):153–5.

[16] Stradling JR, Crosby JH. Predictors and prevalence of obstructive sleep apnoea and snoring in 1001 middle aged men. Thorax 1991;46(2):85–90.

[17] Fogel RB, et al. Increased prevalence of obstructive sleep apnea syndrome in obese women with polycystic ovary syndrome. J Clin Endocrinol Metab 2001;86(3):1175–80.

[18] Trois MS, et al. Obstructive sleep apnea in adults with down syndrome. J Clin Sleep Med 2009;5(4):317–23.

[19] Marin JM, et al. Long-term cardiovascular outcomes in men with obstructive sleep apnoea-hypopnoea with or without treatment with continuous positive airway pressure: an observational study. Lancet 2005;365(9464):1046–53.

[20] Punjabi NM, et al. Sleep-disordered breathing and mortality: a prospective cohort study. PLoS Med 2009;6(8):e1000132.

[21] Hou H, et al. Association of obstructive sleep apnea with hypertension: a systematic review and meta-analysis. J Glob Health 2018;8(1):010405.

[22] Nieto FJ, et al. Association of sleep-disordered breathing, sleep apnea, and hypertension in a large community-based study. Sleep Heart Health Study. JAMA 2000;283(14):1829–36.

[23] O'Connor GT, et al. Prospective study of sleep-disordered breathing and hypertension: the Sleep Heart Health Study. Am J Respir Crit Care Med 2009;179(12):1159–64.

[24] Young T, et al. Population-based study of sleep-disordered breathing as a risk factor for hypertension. Arch Intern Med 1997;157(15):1746–52.

[25] Chobanian AV, et al. The seventh report of the Joint National Committee on prevention, detection, evaluation, and treatment of high blood pressure: the JNC 7 report. JAMA 2003;289(19):2560–72.

[26] Pedrosa RP, et al. Obstructive sleep apnea the Most common secondary cause of hypertension associated with resistant hypertension. Hypertension 2011;58(5):811–7.

[27] Goncalves SC, et al. Obstructive sleep apnea and resistant hypertension: a case-control study. Chest 2007;132(6):1858–62.

[28] Bratton DJ, et al. CPAP vs mandibular advancement devices and blood pressure in patients with obstructive sleep apnea: a systematic review and meta-analysis. JAMA 2015;314(21):2280–93.

[29] Pepperell JC, et al. Ambulatory blood pressure after therapeutic and subtherapeutic nasal continuous positive airway pressure for obstructive sleep apnoea: a randomised parallel trial. Lancet 2002;359(9302):204–10.

[30] Liu L, et al. Continuous positive airway pressure in patients with obstructive sleep apnea and resistant hypertension: a meta-analysis of randomized controlled trials. J Clin Hypertens (Greenwich) 2016;18(2):153–8.

[31] Hla KM, et al. Coronary heart disease incidence in sleep disordered breathing: the Wisconsin Sleep Cohort Study. Sleep 2015;38(5):677–84.

[32] McEvoy RD, et al. CPAP for prevention of cardiovascular events in obstructive sleep apnea. N Engl J Med 2016;375(10):919–31.

[33] Peker Y, et al. Effect of positive airway pressure on cardiovascular outcomes in coronary artery disease patients with nonsleepy obstructive sleep apnea. The RICCADSA randomized controlled trial. Am J Respir Crit Care Med 2016;194(5):613–20.

[34] Chami HA, et al. Association of incident cardiovascular disease with progression of sleep-disordered breathing. Circulation 2011;123(12):1280–6.

[35] Drager LF, Polotsky VY, Lorenzi-Filho G. Obstructive sleep apnea: an emerging risk factor for atherosclerosis. Chest 2011;140(2):534–42.

[36] Newman AB, et al. Relation of sleep-disordered breathing to cardiovascular disease risk factors: the Sleep Heart Health Study. Am J Epidemiol 2001;154(1):50–9.

[37] Trzepizur W, et al. Independent association between nocturnal intermittent hypoxemia and metabolic dyslipidemia. Chest 2013;143(6):1584–9.

[38] Chami HA, et al. Brachial artery diameter, blood flow and flow-mediated dilation in sleep-disordered breathing. Vasc Med 2009;14(4):351–60.

[39] Ip MS, et al. Endothelial function in obstructive sleep apnea and response to treatment. Am J Respir Crit Care Med 2004;169(3):348–53.

[40] Kohler M, et al. CPAP improves endothelial function in patients with minimally symptomatic OSA: results from a subset study of the MOSAIC trial. Chest 2013;144(3):896–902.

[41] Nieto FJ, et al. Sleep apnea and markers of vascular endothelial function in a large community sample of older adults. Am J Respir Crit Care Med 2004;169(3):354–60.

[42] Yeboah J, et al. Brachial flow-mediated dilation predicts incident cardiovascular events in older adults: the Cardiovascular Health Study. Circulation 2007;115(18):2390–7.

[43] Arzt M, et al. Association of sleep-disordered breathing and the occurrence of stroke. Am J Respir Crit Care Med 2005;172(11):1447–51.

[44] Redline S, et al. Obstructive sleep apnea-hypopnea and incident stroke the sleep heart health study. Am J Respir Crit Care Med 2010;182(2):269–77.

[45] Hermann DM, Bassetti CL. Role of sleep-disordered breathing and sleep-wake disturbances for stroke and stroke recovery. Neurology 2016;87(13):1407–16.

[46] Marshall NS, et al. Sleep apnea and 20-year follow-up for all-cause mortality, stroke, and cancer incidence and mortality in the Busselton Health Study cohort. J Clin Sleep Med 2014;10(4):355–62.

[47] Yaggi HK, et al. Obstructive sleep apnea as a risk factor for stroke and death. N Engl J Med 2005;353(19):2034–41.

[48] Birkbak J, Clark AJ, Rod NH. The effect of sleep disordered breathing on the outcome of stroke and transient ischemic attack: a systematic review. J Clin Sleep Med 2014;10(1):103–8.

[49] Aaronson JA, et al. Obstructive sleep apnea is related to impaired cognitive and functional status after stroke. Sleep 2015;38(9):1431–7.

[50] Kernan WN, et al. Guidelines for the prevention of stroke in patients with stroke and transient ischemic attack: a guideline for healthcare professionals from the American Heart Association/American Stroke Association. Stroke 2014;45(7):2160–236.

[51] Arzt M, et al. Prevalence and predictors of sleep-disordered breathing in patients with stable chronic heart failure: the SchlaHF registry. JACC Heart Fail 2016;4(2):116–25.

[52] Gottlieb DJ, et al. Prospective study of obstructive sleep apnea and incident coronary heart disease and heart failure: the sleep heart health study. Circulation 2010;122(4):352–60.

[53] Javaheri S, et al. Sleep-disordered breathing and incident heart failure in older men. Am J Respir Crit Care Med 2016;193(5):561–8.

[54] Javaheri S, et al. Central sleep apnea, right ventricular dysfunction, and low diastolic blood pressure are predictors of mortality in systolic heart failure. J Am Coll Cardiol 2007;49(20):2028–34.

[55] Khayat R, et al. Sleep disordered breathing and post-discharge mortality in patients with acute heart failure. Eur Heart J 2015;36(23):1463–9.

[56] Lanfranchi PA, et al. Prognostic value of nocturnal Cheyne-Stokes respiration in chronic heart failure. Circulation 1999;99(11):1435–40.

[57] Khayat R, et al. Central sleep apnea is a predictor of cardiac readmission in hospitalized patients with systolic heart failure. J Card Fail 2012;18(7):534–40.

[58] Colish J, et al. Obstructive sleep apnea: effects of continuous positive airway pressure on cardiac remodeling as assessed by cardiac biomarkers, echocardiography, and cardiac MRI. Chest 2012;141(3):674–81.

[59] Kaneko Y, et al. Cardiovascular effects of continuous positive airway pressure in patients with heart failure and obstructive sleep apnea. N Engl J Med 2003;348(13):1233–41.

[60] Kasai T, et al. Prognosis of patients with heart failure and obstructive sleep apnea treated with continuous positive airway pressure. Chest 2008;133(3):690–6.

[61] Mansfield DR, et al. Controlled trial of continuous positive airway pressure in obstructive sleep apnea and heart failure. Am J Respir Crit Care Med 2004;169(3):361–6.

[62] Bradley TD, et al. Continuous positive airway pressure for central sleep apnea and heart failure. N Engl J Med 2005;353(19):2025–33.

[63] Arzt M, et al. Suppression of central sleep apnea by continuous positive airway pressure and transplant-free survival in heart failure: a post hoc analysis of the Canadian Continuous Positive Airway Pressure for Patients with Central Sleep Apnea and Heart Failure Trial (CANPAP). Circulation 2007;115(25):3173–80.

[64] Sharma BK, et al. Adaptive servoventilation for treatment of sleep-disordered breathing in heart failure: a systematic review and meta-analysis. Chest 2012;142(5):1211–21.

[65] Teschler H, et al. Adaptive pressure support servo-ventilation: a novel treatment for Cheyne-Stokes respiration in heart failure. Am J Respir Crit Care Med 2001;164(4):614–9.

[66] Cowie MR, et al. Adaptive servo-ventilation for central sleep apnea in systolic heart failure. N Engl J Med 2015;373(12):1095–105.

[67] O'Connor CM, et al. Cardiovascular outcomes with minute ventilation-targeted adaptive servo-ventilation therapy in heart failure: the CAT-HF trial. J Am Coll Cardiol 2017;69(12):1577–87.

[68] Kizawa T, et al. Pathogenic role of angiotensin II and oxidised LDL in obstructive sleep apnoea. Eur Respir J 2009;34(6):1390–8.

[69] Mehra R, et al. Association of nocturnal arrhythmias with sleep-disordered breathing: the Sleep Heart Health Study. Am J Respir Crit Care Med 2006;173(8):910–6.

[70] Braga B, et al. Sleep-disordered breathing and chronic atrial fibrillation. Sleep Med 2009;10(2):212–6.

[71] Caples SM, Somers VK. Sleep-disordered breathing and atrial fibrillation. Prog Cardiovasc Dis 2009;51(5):411–5.

[72] Gami AS, et al. Therapy insight: interactions between atrial fibrillation and obstructive sleep apnea. Nat Clin Pract Cardiovasc Med 2005;2(3):145–9.

[73] Gami AS, et al. Association of atrial fibrillation and obstructive sleep apnea. Circulation 2004;110(4):364–7.

[74] Mehra R, et al. Nocturnal arrhythmias across a spectrum of obstructive and central sleep-disordered breathing in older men: outcomes of sleep disorders in older men (MrOS sleep) study. Arch Intern Med 2009;169(12):1147–55.

[75] Monahan K, et al. Triggering of nocturnal arrhythmias by sleep-disordered breathing events. J Am Coll Cardiol 2009;54(19):1797–804.

[76] Ng CY, et al. Meta-analysis of obstructive sleep apnea as predictor of atrial fibrillation recurrence after catheter ablation. Am J Cardiol 2011;108(1):47–51.

[77] Fein AS, et al. Treatment of obstructive sleep apnea reduces the risk of atrial fibrillation recurrence after catheter ablation. J Am Coll Cardiol 2013;62(4):300–5.

[78] Kanagala R, et al. Obstructive sleep apnea and the recurrence of atrial fibrillation. Circulation 2003;107(20):2589–94.

[79] Naruse Y, et al. Concomitant obstructive sleep apnea increases the recurrence of atrial fibrillation following radiofrequency catheter ablation of atrial fibrillation: clinical impact of continuous positive airway pressure therapy. Heart Rhythm 2013;10(3):331–7.

[80] Simantirakis EN, et al. Severe bradyarrhythmias in patients with sleep apnoea: the effect of continuous positive airway pressure treatment: a long-term evaluation using an insertable loop recorder. Eur Heart J 2004;25(12):1070–6.

[81] Somers VK, et al. Sympathetic neural mechanisms in obstructive sleep apnea. J Clin Invest 1995;96(4):1897–904.

[82] Narkiewicz K, et al. Nocturnal continuous positive airway pressure decreases daytime sympathetic traffic in obstructive sleep apnea. Circulation 1999;100(23):2332–5.

[83] Vatansever E, et al. Obstructive sleep apnea causes oxidative damage to plasma lipids and proteins and decreases adiponectin levels. Sleep Breath 2011;15(3):275–82.

[84] Yamauchi M, et al. Oxidative stress in obstructive sleep apnea. Chest 2005;127(5):1674–9.

[85] Jelic S, Le Jemtel TH. Inflammation, oxidative stress, and the vascular endothelium in obstructive sleep apnea. Trends Cardiovasc Med 2008;18(7):253–60.

[86] Kato M, et al. Impairment of endothelium-dependent vasodilation of resistance vessels in patients with obstructive sleep apnea. Circulation 2000;102(21):2607–10.

[87] Li J, et al. Hyperlipidemia and lipid peroxidation are dependent on the severity of chronic intermittent hypoxia. J Appl Physiol (1985) 2007;102(2):557–63.

[88] Phillips BG, et al. Effects of obstructive sleep apnea on endothelin-1 and blood pressure. J Hypertens 1999;17(1):61–6.

[89] Hui P, et al. Nocturnal hypoxemia causes hyperglycemia in patients with obstructive sleep apnea and type 2 diabetes mellitus. Am J Med Sci 2016;351(2):160–8.

[90] Reutrakul S, et al. The relationship between sleep and glucagon-like peptide 1 in patients with abnormal glucose tolerance. J Sleep Res 2017;26(6):756–63.

[91] Nakata K, et al. Distinct impacts of sleep-disordered breathing on glycemic variability in patients with and without diabetes mellitus. PLoS ONE 2017;12(12):e0188689.

[92] Matsumoto T, et al. Plasma incretin levels and dipeptidyl peptidase-4 activity in patients with obstructive sleep apnea. Ann Am Thorac Soc 2016;13(8):1378–87.

[93] Kushida CA, et al. Effects of continuous positive airway pressure on neurocognitive function in obstructive sleep apnea patients: the Apnea Positive Pressure Long-term Efficacy Study (APPLES). Sleep 2012;35(12):1593–602.

[94] Barbe F, et al. Effect of continuous positive airway pressure on the incidence of hypertension and cardiovascular events in nonsleepy patients with obstructive sleep apnea: a randomized controlled trial. JAMA 2012;307(20):2161–8.

[95] Medeiros AK, et al. Obstructive sleep apnea is independently associated with subclinical coronary atherosclerosis among middle-aged women. Sleep Breath 2017;21(1):77–83.

[96] Epstein LJ, et al. Clinical guideline for the evaluation, management and long-term care of obstructive sleep apnea in adults. J Clin Sleep Med 2009;5(3):263–76.

Part VIII

Sleep health in children and adolescents

Chapter 31

Sleep, obesity and cardiometabolic disease in children and adolescents

Teresa Arora[a], Ian Grey[b]
[a]Zayed University, College of Natural and Health Sciences, Department of Psychology, Abu Dhabi, United Arab Emirates, [b]School of Social Sciences, Lebanese American University, Beirut, Lebanon

INTRODUCTION

The worldwide prevalence of obesity, cardiovascular and metabolic disease has continued to rise in the 21st century. Previously, these diseases were mainly observed in adults but they are now increasingly present in children. The abrupt increase in cardiovascular and metabolic disease as well as obesity has been largely attributed to noxious lifestyle behaviors. These include, but are not limited to, unhealthy diet, inadequate energy expenditure, increased levels of sedentariness and excessive screen-time. Very little was known about sleep behavior as a potential contributor to cardiometabolic health and obesity outcomes until the 1990s. This initial discovery has resulted in increased sleep research attention relating to multiple adverse mental and physical health consequences. There is a growing body of evidence surrounding the link between sleep and obesity, diabetes mellitus and markers of cardiovascular disease across all age groups and in multiple geographic locations. This chapter will focus on highlighting and discussing historical and contemporary evidence surrounding the associations between several important sleep features in relation to obesity and type 2 diabetes mellitus, both of which are strongly correlated with cardiovascular disease, in pediatric populations. The possibility of improving sleep as a novel and contemporary approach to tackle the current epidemic of chronic health issues will also be discussed.

DEFINING OVERWEIGHT AND OBESITY IN CHILDREN

Obesity is characterized as excessive adiposity and is one of the most concerning and serious public health concerns in the 21st century. The condition is complex and multifactorial. Both genetic and environmental factors contribute to the onset and progression of obesity. Statistics from the World Health Organization (WHO) in 2016 estimated that 39% of men and women, aged 18 and over were overweight or obese worldwide. Estimates for children and adolescents (5–19 years) were 18% in 2016. Data collected from 450 nationally representative surveys from 144 countries in the United Nations revealed a stark increase in obesity in preschool children (0–5 years) from 1990 to 2010 [1]. Fig. 31.1 highlights the prevalence of overweight and obesity among children (2–19 years) according to continent over time. In adults, obesity is usually characterized by body mass index (BMI; kg/m^2). This method is used as a diagnostic tool to categorize individuals into underweight, healthy weight, overweight or obese. It is a well utilized method given the ease of assessment for height and weight but it has been criticized due to not directly assessing adiposity which misrepresents the body weight outcome grouping. With children and adolescents, the application of BMI to compartmentalize body weight presents different challenges due to growth spurts and alterations in body fat and muscle. The tool is not usually used as a diagnostic tool in pediatric populations but is instead used to screen for potential weight and health-related conditions because BMI cannot determine if a child has excess adiposity unless assessed by a healthcare professional. The Centers for Disease Control and Prevention (CDC) state that BMI in children and adolescents can be calculated and then expressed as a percentile which is plotted and directly compared, relative to other children in the United States who previously participated in national surveys conducted 1963–65 and 1988–94 [2]. These are based on age and gender-specific percentiles with obesity, overweight, healthy weight and underweight being defined as ≥95th percentile, >85th–94th percentile, 5th–85th percentile, and <5th percentile, respectively. Arguably, these data are now outdated, require revision and are limited to US

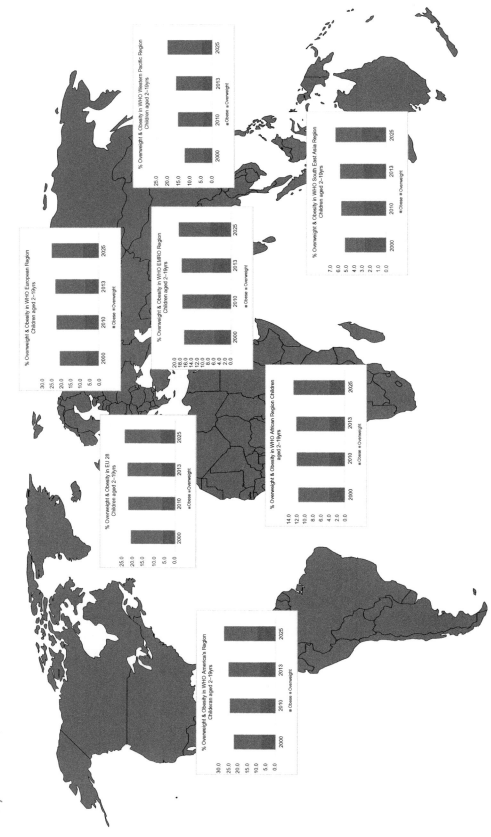

FIG. 31.1 Prevalence of childhood overweight and obesity over time.

children. If the prevalence of childhood obesity continues to rise then the condition could, over time, become the norm. In 2000, Cole and colleagues first proposed an alternative solution to classifying childhood overweight and obesity by developing a globally accepted definition taking into account age and sex-specific cut points [3].

CAUSES AND CONSEQUENCES OF CHILDHOOD OBESITY

Causes

The causes of obesity, both in adults and children, are well established. Weight gain and subsequent obesity onset is due to energy homeostasis imbalance and there are multiple factors which drive this. Excessive energy intake and inadequate energy expenditure are driving factors of obesity but other peripheral variables influence energy balance. These include poverty, socio-economic status, exposure and accessibility to fast food chains, mass media advertising, parental education, behavior and weight status, crime rates, excessive electronic device use, and sleep. While obesity does have a genetic component, it is a disease which is largely driven by a combination of lifestyle behaviors and socio-demographics.

Consequences

Excess adiposity is particularly concerning in pediatric populations given that obesity now presents at a younger age. Childhood obesity has been associated with a wide range of adverse physiological and psychological health comorbidities at the individual level as well as severe social and economic implications. For example, obesity in children effects almost every major organ and obesity at a young age is resulting in earlier onset of type 2 diabetes mellitus and other metabolic dysfunction, cardiovascular disease, some cancer types, respiratory disease, and sleep-disordered breathing, some of which will be discussed in this chapter. The risk for noncommunicable diseases is positively correlated with BMI. Obesity in childhood is also associated with an increased risk of obesity and disability in adulthood as well as premature mortality. The psychological health of those with obesity occurring in childhood is also disquieting and includes depression, anxiety, social isolation, bullying, low self-esteem, confidence and self-image. Suicide ideation is also common in those with obesity, with an increased number of attempted suicides documented among obese teenagers as compared to nonobese [4]. A systematic review which was published in 2009 revealed impaired health-related quality of life (HRQoL) in those with childhood obesity (<21 years old). Interestingly, this review also highlighted that the parent's perception was rated lower for HRQoL than the child's, suggesting that children may be learning to live with the condition from a young age [5]. Moreover, there are a host of neurocognitive impairments evident in obese children including reduced attention span, poorer memory recall, problem-solving abilities and decision-making processes. Perhaps it is not surprising then, that obesity is linked to poorer academic attainment and aspiration [6]. Educational outcomes are among the strongest predictors of life and work satisfaction in adulthood [7]. Obesity also results in hypothalamic inflammation, which interestingly, is the neurological region where sleep-wake behaviors are regulated.

ATTEMPTS TO REDUCE THE OBESITY EPIDEMIC

Obesity is a global concern and childhood obesity is one of the leading causes of premature mortality. With this in mind, and taking into consideration the constellation of comorbidities, raising awareness and educating the public surrounding lifestyle-driven obesity-driven behaviors has been a major global priority. Efforts to reduce the childhood obesity epidemic have largely focused on educating children and parents about the causes and effects of obesity with a strong emphasis on energy homeostasis. The majority of attention has been concentrated on dietary behaviors including controlling portion sizes, consumption of healthy balanced meals, healthy food selection, reduced snacking behavior and more. There have also been multiple interventions designed to promote adherence to healthy lifestyles, including increased physical activity and reduced screen-time with the application of several innovative methods applied in school-based trials [8]. A recent review of the global literature surrounding trials aimed at tackling childhood obesity found that despite interventions being financially demanding, many have an inadequate follow up duration, have not involved parents, have largely focused on elementary school-aged children, have not targeted low-income populations and 50% the studies were conducted in the United States [9]. While some of the US studies generated positive outcomes (reduction in BMI), it was noted that different approaches were used, suggesting that what may work for one individual may not be successful for another. This emphasizes the need for tailored made interventions, based on individual circumstances, which will undoubtedly incur further financial penalties. National campaigns and intervention trials which raise awareness of the causes and consequences of childhood obesity have, in part, been successful in hampering a further rise with a plateau reported in some countries [10, 11] but not in others [12]. The latest statistics gathered in the United States from a nationally representative sample of 2–19 year olds in the United States showed a 4.7% increase in overweight among children in a 2-year period (2014–16) [13].

What then, is not being incorporated into these obesity reduction trials that may be contributing to the ineffectiveness of such costly interventions? Sleep is an overlooked yet key factor and once the extensive evidence surrounding the relationship between sleep and obesity is better understood and integrated into obesity-reduction programs, this will enhance the efficacy of obesity interventions to tackle the current epidemic.

THE IMPORTANCE OF SLEEP IN RELATION TO HEALTH

The importance of sleep in relation to overall health, wellbeing and extended mortality was recognized in 1972 in, what is now known as, the Alameda 7 study [14]. The study showed that acquiring 7–8 h of sleep per night, as well as other lifestyle behaviors, was protective against mortality. This early evidence triggered a number of further studies which produced consistency across findings. For example, one of the largest US population studies reported a U-shaped relationship between number of sleeping hours and mortality as well as body mass index [15]. Mounting evidence has resulted in multiple systematic reviews and metaanalyses that consistently confirm the contribution of sleep duration in relation to mortality [16–20], obesity [20–25], as well as cardiovascular [17, 20, 26, 27] and metabolic disease [20, 28–32] across all age groups. So what then, is the evidence and the link between sleep and obesity? How then does sleep, a behavior associated with reduced consciousness, contribute to weight gain and subsequent obesity?

EVIDENCE FOR A LINK BETWEEN SLEEP DURATION AND OBESITY IN PEDIATRIC POPULATIONS

One of the first studies to report a link between sleep duration and obesity in children was documented in France [33]. A total of 704 controls were recruited and 327 cases of 5-year-old school children with obesity. Anthropometric measurements were obtained within the school setting and parents were interviewed to obtain information on multiple environmental factors, including sleep duration. The study showed that children with short sleep duration, had an almost fivefold estimated relative risk of obesity [33]. Other demographic and lifestyle factors that predicted childhood obesity included the mother's origin, excessive television viewing and snacking but short sleep duration was, by far, the strongest predictor.

There is no shortage of population-based studies that have documented a dose-dependent association between the number of sleep hours and body weight in children. For example, a study of 6862 German children aged 5–6 years old, showed that 10.5–11 h sleep duration per night was protective against overweight and obesity as well as high body fat content with an estimated 23% reduced risk. Furthermore, those sleeping ≥11.5 h per night exhibited a further reduced risk (46%), compared to children who slept for 10 h or less per night [34]. These findings extend well beyond European children with similar observations noted in Japan [35], China [36], Turkey [37], America [38], Australia [39], the United Kingdom [40], and more. The majority of population studies in were initially cross-sectional and relied upon subjective sleep reports. Thus, the arguments were that there were no causal evidence and that sleep data may be inaccurate and subject to various biases. Over time, these limitations have been challenged and overcome. There is now considerable evidence from longitudinal studies which demonstrate that children who have inadequate sleep length, whatever the cause, gain more weight over time, compared to those who obtain sufficient sleep [40, 41]. One of these studies, also obtained objectively estimated sleep measures using a waist-worn accelerometer in free-living environment for 5 days and nights [40]. Moreover, recent systematic reviews and metaanalyses of prospective cohort studies verify that short sleep duration consistently contributes to an increased risk of the onset and progression of obesity among infants, children and adolescents [21, 22, 24, 25].

OTHER SLEEP PARAMETERS AND CHILDHOOD OBESITY

Just as obesity is a complex condition, sleep is also multifaceted and goes beyond duration. Historically, the majority of research has focused on the effects of sleep quantity in relation to health outcomes but sleep has many features including sleep quality, sleep onset latency (time taken to initiate sleep), wake after sleep onset (WASO; a term used to quantify the amount of time spent awake throughout a sleep episode), sleep efficiency (an indicator of sleep quality which is based on the proportion of time spent asleep and awake during a sleep episode), daytime napping, sleep architecture (proportion of specific sleep stages) and sleep-wake timings (onset and offset). Following the plethora of evidence surrounding the sleep-obesity link, the focus of research attention has recently shifted and researchers have begun to investigate other components of sleep in relation to the condition.

A recent systematic review and metaanalysis was conducted to assess the contribution of sleep duration and sleep quality in relation to childhood obesity [21]. The pooled estimates revealed that sleep quality was a stronger predictor of the condition compared to sleep duration. All studies included in the metaanalysis contained 26,533 participants and showed that those with insufficient sleep duration and poor sleep quality combined had a 27% increased risk of overweight/obesity. Poor sleep quality alone was associated with a 46% increased risk of the disease in children, adolescents and young adults. This suggests that other sleep features are important when attempting to understand the sleep-obesity relationship.

Circadian rhythms, also referred to as biological clocks, are controlled by the suprachiasmatic nucleus (SCN) located in the hypothalamus region of the brain. The SCN regulates sleep-wake timings. Interestingly, the timing of sleep, which is delayed in adolescents as a consequence of pubertal transition and exacerbated by lifestyle behaviors, has also been traced to obesity as well as poor dietary habits. A cross-sectional study of 511 young adolescents showed that late circadian preference was associated with higher BMI z-score. Interestingly, evening circadian preference (subjectively estimated and verified by wrist actigraphy) was also associated with a higher frequency of consuming unhealthy snacks and evening caffeine consumption as well as insufficient daily intake of fruit and vegetables [42]. This provides some mechanistic insight into how sleep may be contributing to weight gain and subsequent obesity by influencing energy balance.

As previously mentioned, adolescents have a shifted circadian pattern which is characterized by delayed sleep initiation, and later sleep onset. The recommended sleep duration for adolescents (13–18 years), based on consensus of the American Academy of Sleep Medicine (AASM), is 8–10h per 24-h period [43]. Given that a typical adolescent experiences sleep delays yet has academic attendance commitments on weekdays, this can significantly reduce sleep length in this age group. Many adolescents do not obtain 8–10h of sleep during the week and weekend "catch-up" sleep is common. This pattern of compensatory sleep at weekends to repay "sleep debt" accumulated across the week, however, has been associated with poorer neurocognitive capabilities [44], academic performance and psychological health consequences including depressive symptoms, suicide attempts and self-injury [45]. Thus, sleep consistency seems to be crucial but challenging to achieve, particularly among adolescents.

Two recent studies by the same group have begun to explore the relationship between sleep inconsistency and energy intake [46] as well as abdominal obesity [46, 47], which plays an important role in the development of insulin resistance and type 2 diabetes mellitus. The two studies recruited 305 adolescents and monitored their sleep with wrist actigraphy. They compared the effect of sleep duration versus sleep variability in relation to subjective food and macronutrient intake as well as abdominal obesity, measured using dual-energy-X-ray absorptiometry (DEXA) scan. Those with higher sleep variability had more abdominal obesity which was purported to be the result of increased energy intake, particularly from foods that are high in carbohydrates [47]. These novel approaches provide a better understanding of how different features of sleep are partially contributing to the global epidemic of childhood obesity. Detailed mechanistic studies have also been conducted, which have revealed more clues about the involvement of sleep in obesity.

SLEEP AND ENERGY HOMEOSTASIS

Extensive research has been undertaken to better understand the influence that sleep has upon energy intake and expenditure, given that positive energy balance is known to cause obesity. There have been multiple experimental studies that have been conducted where sleep duration has been manipulated to examine the effect of this behavior upon energy balance. A recent systematic review identified 18 randomized controlled trials (RCTs), 4 of which assessed the effect of sleep on food intake and another 4 which explored total energy expenditure [48]. The authors concluded that increases in energy intake as well as total energy expenditure were observed with sleep restriction. So why, if sleep restriction results in increased energy intake as well as expenditure do people gain weight? The answer is simple. Energy expenditure does not adequately compensate for the increased amount of energy consumption, which results in positive energy and gradual weight gain when persistent.

It should be noted that the majority of experimental sleep studies to investigate alterations in body weight and metabolism have been conducted in adults with limited evidence available in pediatric populations. Sleep restriction studies that explore the mechanisms of obesity and metabolism are limited in children and adolescents. There have been multiple studies conducted in young, healthy adults which have explored the effect of acute partial sleep restriction upon metabolic outcomes as a mechanistic approach to understanding the sleep-obesity link. One of the earliest studies showed that restricting adults sleep to 4h per night for two consecutive nights resulted in alterations to appetite and hunger-regulating hormones, leptin and ghrelin. Specifically, leptin, which is a hormone that indicates satiety, was significantly reduced with sleep restriction, suggesting a delay in the hypothalamic signaling of satiety that can cause overeating. Ghrelin is commonly referred to as a hunger hormone, given that levels rise prior to food consumption then decline afterwards. Levels of ghrelin were increased following sleep restriction in this study suggesting that individuals may be hungrier. The authors obtained subjective reports of hunger in the study which confirmed this notion [49]. Moreover, participants were also asked to rate their appetite for different macronutrients (protein, fats, carbohydrates, sweet/salty foods and more). Interestingly, not only did participants report higher levels of hunger but they also indicated a stronger preference for carbohydrate and calorie-dense foods.

One of the few sleep-metabolism studies to be conducted in children investigated the effects of sleep restriction upon 24-h food recall, levels of leptin and ghrelin as well as body weight [50]. The authors found that sleep restriction (1.5h less than habitual sleep duration) for 1-week was associated with a significant increase in calorie intake per day as well as alterations to leptin. It should be noted that there were only 37 children in the sample and thus generalizability to other populations is problematic. There are many adult studies that have consistently highlighted the relationship between sleep restriction and metabolic disruption. Leptin and ghrelin are neuroendocrine mediators in the sleep-obesity pathway but there are fewer studies in children and the findings are less consistent, as noted in a recent review [51].

A clear picture has emerged surrounding the importance of sleep in relation to obesity and metabolic regulation. The role for sleep curtailment-induced hormonal changes is a plausible mechanistic explanation for obesity in children and adults. Additional work is, however, needed to elucidate the precise role that leptin, ghrelin and other appetite-controlling hormones play in metabolic dysfunction that is associated with sleep alteration/imbalance. Unequivocally, sleep is fundamental to optimal metabolic regulation and provides a robust mechanistic explanation for short, or disrupted, sleep as a driver for obesity onset. Multiple causal pathways between sleep and obesity have been documented which have highlighted a combination of behavioral and physiological drivers.

FUTURE DIRECTIONS

Given that there is now consistent, convincing evidence about the importance of sleep for body weight and hormone regulation, the next logical step seems to apply and incorporate this knowledge into interventions that target obesity reduction and prevention programs. Educating both parents and children about the downstream effects of sleep upon health outcomes is the first step.

There have been multiple school-based interventions which have specifically attempted to optimize sleep in children and adolescents. Most of these trials have shown an increase in sleep knowledge when comparing baseline to follow up. This is good news as raising awareness among children and their parents is the first step, but knowledge alone is not sufficient to elicit behavior change. Unfortunately, improvements to sleep knowledge do not always translate into positive sleep behavior change. Some studies were designed to assess pre and post sleep knowledge only, without exploring possible sleep alteration. Others have shown that improved sleep knowledge does not positively influence subsequent behavior, although the evidence is somewhat heterogeneous. A recent narrative review surrounding sleep improvement as a possible tool for tackling obesity concluded that there are an insufficient number of adolescent studies to support this as a feasible method [52]. Thus, carefully designed studies which include longitudinal and objective assessments of sleep, as well as parental education and involvement, paired with intentional sleep behavior modification is now needed, particularly given the lack of long-term success and/or follow-up that other obesity-reduction trials have shown. When obesity researchers begin to recognize the prominence of sleep behavior, incorporating sleep improvement into future trials may enhance their effectiveness.

METABOLIC DISEASE

The rising incidence of chronic health conditions in recent decades has resulted in a global public health concern. The occurrence of these diseases have been repeatedly and strongly linked to the presence of specific behavioral factors. The Centers for Disease Control and Prevention (CDC) estimated that eliminating three specific behavioral risk factors alone (poor diet, inactivity and smoking) would prevent 80% of heart disease and stroke, 80% of type 2 diabetes and 40% of cancer. In respect of diabetes, the prevalence of the disease has not only been rising steadily among the adult population but also more worryingly among children. The monetary cost alone of diabetes is astounding with estimates upwards of $375 billion per year. Diabetes is not just expensive to treat but it also reduces life expectancy by an average of 10 years. The rising prevalence of diabetes among children has intensified the search for a better understanding of the variables associated with its onset and progression as well as how to best manage the disease. One promising line of research that has begun to gain traction concerns the nature of the relationship between aspects of sleep and heightened risk of diabetes. Research in this domain remains largely in its infancy and the available research base is largely limited to primarily observational studies with limited experimental studies. Though research on the relationship between sleep and diabetes and children has followed on the coat-tails of research with adults, collectively it points to the role of sleep as a modifiable risk factor for the onset of diabetes in children [53]. As the quality and duration of sleep in children and adolescents appears to be decreasing similar to their adult counterparts, the implications for children may be as profound, if not more so, than for adults.

MECHANISMS OF DIABETES

Diabetes Mellitus (DM) is characterized by glucose dysregulation. The most frequently occurring are type 1 and type 2. Type 1 diabetes (T1D) occurs when the pancreas does not produce insulin and type 2 diabetes (T2D) is driven by insulin resistance (IR) or insufficient insulin secretion. The key distinction between these two main types of DM revolves around the degree of insulin produced. In healthy individuals, the hormone insulin has a triggering effect on the cells of the body to absorb elevated levels of glucose in the bloodstream and therefore has a homeostatic type function. These cells of the body open special channels on their surface to absorb an influx of glucose, typically occurring after food intake. When cells do not effectively respond to insulin, or an insufficient amount is secreted, blood glucose levels become elevated. Interestingly, T2D was initially referred to as late onset or adult onset diabetes though this nosology has subsequently been dropped largely in response to the observation that an ever increasing number of children have developed the disease in recent years. Similar to diagnosis of the disease in adults, the criteria for disease in children are polydipsia, polyuria and unexplained weight loss plus casual glucose concentration > equal ≥200 mg/dL (11.1 mmol/L) in venous plasma, fasting glucose ≥126 mg/dL (7.0 mmol/L) in venous or capillary plasma, or 2-h glucose during OGTT ≥200 mg/dL (11.1 mmol/L) in venous plasma or capillary whole blood sample.

The consequences of diabetes can be severe and it remains a major cause of blindness, kidney failure, heart attacks, stroke and lower limb amputation. The associated health, mortality and quality of life costs associated with diabetes have provided the impetus for a, now substantial, research-base. Up until recently, among the most common causes identified have been poor quality diet, excessive body weight and insufficient exercise. As of 2017, the World Health Organization estimated that 422 million people had diabetes worldwide, which is approximately 8.5% of the adult population. According to the Global Burden of Disease (GBD) report for 2015, the prevalence of diabetes increased by approximately 30% between 2005 and 2015 and during the same interval, the annual number of deaths from diabetes rose from 1.2 million to 1.5 million. Up until the 1990s, T2D had been considered a relatively rare occurrence in children and adolescents but since that time a number of industrialized countries such as the United States, Canada, Japan and Germany have witnessed an increasing incidence of the disorder [54]. In the United States alone, overall unadjusted incidence rates of T2D is reported to have increased by 7.1% annually between 2002 and 2012 [55]. The increase in prevalence has unsurprisingly occurred in conjunction with an increase in prevalence and degree of obesity also among children and adolescents. However, some authors have noted a curious anomaly in the case of T2D in children—while incident obesity has stabilized in some countries, the prevalence of diabetes has increased threefold. It also appears that prevalence statistics may be an underestimation of the true rates as several studies have reported that children were asymptomatic at diagnosis. Therefore, it is likely that, as with adults, undiagnosed T2D is a common condition in childhood. However, though the symptoms of the disease may be similar across children and adults, it appears the health consequences are not [56]. For example, one notable outcome of the UK Prospective Diabetes Mellitus Study (UKPDS) was the observation that children and adolescents with T2D have a higher risk for disease-related complications, compared to adults with the same condition. In line with this, developing T2D at a younger age appears to be associated with a much higher risk of long-term cardiovascular disease than those who develop T2D in middle age [56]. Furthermore, young people with T2D appear to be at a much higher risk of developing associated complications than those with T1D. This higher level of risk does not appear to be related to overall levels of glycemic control or disease duration but to the occurrence of hypertension and dyslipidemia [57]. In addition, the negative effects of T2D extend beyond the traditional quantification of medical symptoms to quality of life; adolescents with T2D report poorer health-related quality of life scores compared to T1D counterparts [58]. Furthermore, the burden of psychological disorders in young people with T2D is high, with as many as one in five experiencing either psychological disorders or behavioral problems [59].

It is critical then for researchers to identify, as precisely as possible, all factors implicated in the onset of the disorder. In particular, emphasis should be placed upon the identification of preventative factors and early identification. Principal existing strategies for the prevention of T2D in children revolve around the role of the known factors such as obesity and therefore reflect content surrounding teaching principles of good nutrition and the importance adequate exercise. Family-based treatments for childhood obesity as a specific risk marker have also have been subjected to empirical analysis and support for their effectiveness is evident, particularly in the context of motivated families [54]. However, based on recent sleep research findings which will be discussed, the content of programs aimed at improving the management and reducing the current epidemic of T2D in children is likely to need rethinking with incorporation of effective sleep management strategies.

It is also worth noting that a number of existing reviews and guidelines for the prevention and management of T2D in children fail to mention sleep as a risk factor for the development, or management, of diabetes with the majority focusing on weight control. This is largely attributable to the recent emergence of research on the topic. While sleep may be implicated as a risk factor, it remains currently unclear about what precisely is the role of sleep in children in the development of T2D, and whether optimal sleep functioning relates to improved quality of life and/or better symptom management. Furthermore, self-management programs may require modification in light of emerging research on the role of sleep. The role of sleep may add incremental benefits considering intervention studies have convincingly demonstrated that adoption of a healthy lifestyle characterized by healthy eating, regular physical activity and subsequent modest weight loss can prevent the progression of impaired glucose tolerance to clinical diabetes mellitus [60].

SLEEP AND TYPE 2 DIABETES MELLITUS

Healthy sleep is gaining increasing recognition as an important lifestyle habit in the prevention of chronic diseases, and the evidence is striking. In addition to maintaining normal brain functioning, the importance of the role of sleep in controlling functions of multiple body systems is now well established. It is becoming clear that reducing the total hours of nocturnal sleep can lead to serious negative consequences for almost all bodily organs and systems. For example, immune function is compromised [61], systemic inflammation with increased inflammatory markers becomes apparent [62] and metabolic regulation is significantly disrupted [63]. Sleep deprivation has a profound effect on metabolic health. Specifically, sleep disturbance, insufficient or excessive sleep, and irregular sleep wake patterns have been associated with adverse outcomes such as obesity and impaired glucose metabolism [64] and emerging evidence has assigned an important role to sleep as a modulator of metabolic homeostasis [65]. In a healthy individual, insulin is produced in response to glucose, which trigger the cells of the body to swiftly absorb glucose and regulate its level. Early warning signs of a link between sleep loss and abnormal glucose levels in the blood emerged through several large epidemiological studies. These indicated that sleeping less than 6 h a night was associated with an increased risk of T2D. More importantly, this relationship remained when adjusted for previously identified risk factors such as body weight, alcohol, smoking age and even race. A recent metaanalysis of adult studies concluded that the risk of developing T2D was associated with insufficient sleep and that this was comparable to that of previously identified risk factors including family history of diabetes, excessive weight and reduced levels of physical activity [32]. The authors concluded, on the basis of their findings, that effective sleep should be considered in clinical guidelines for the management of T2D [32].

The relationship between sleep and T2D is more firmly established in adults with a U-shaped relationship being observed, indicating that both short and long sleep duration carry a risk for developing the disease. However, the specific mechanisms underlying this relationship are still not yet fully understood. The issue that arises with the evidence surrounding the link between sleep, T2D and metabolic dysfunction is causality—does diabetes result in sleep reduction or does short sleep interfere with glucose regulation? The emerging consensus is that shortened sleep duration impairs glucose tolerance. For example, Keckeis and colleagues suggest that that some, but not all, sleep disorders considerably compromise glucose metabolism [66]. Specifically, they report that obstructive sleep apnea (OSA) is associated with and increased risk of impaired glucose tolerance and suggests that OSA is a likely risk factor for diabetes, regardless of concomitant obesity [66]. Interestingly, they report that insomnia was not associated with increased risk, although the pathophysiology of insomnia patients is likely to be very different to those who are habitually short sleepers.

Independent of the presence of specific sleep disorders, it also appears that shortened sleep duration is associated with impaired glucose tolerance. Several previous studies have demonstrated that difficulties in maintaining sleep or short sleep duration were associated with an increased incidence of diabetes [67, 68]. A concise narrative review suggested that chronic sleep loss (behavioral or sleep disorder-related) might represent a novel risk factor for weight gain, insulin resistance and T2D [69]. In accounting for this relationship, one mechanism by which sleep deprivation might result in increased risk of insulin resistance and diabetes may be either by directly affecting parameters of glucose tolerance or indirectly through a disturbance in appetite regulation, leading to increased food intake and weight gain. With regard to short sleep duration and metabolic disorders, or weight gain, some have argued that sleep deprivation might simply be the result of increased wakefulness which optimizes the opportunity to eat. Furthermore, tiredness is likely to promote sedentary activities, such as watching television [70].

A number of experimental studies suggest that one pivotal mechanism in the relationship between sleep and T2D in adults is reduced insulin sensitivity. One key experimental study showed that, even after just 4 h of sleep per night over six nights, individuals were 40% less effective at absorbing a standard dose of glucose compared to when they were fully rested. Two possible lines of questions emerge from this finding. First, is it insufficient insulin or suppression of its release that is the culprit? Second, do the cells become unresponsive to an otherwise normal and present message of insulin? The evidence suggests that cells become less responsive to insulin with the subsequent result of impaired glucose tolerance. In addition, sleep restriction studies lend credence to this, beginning with the landmark study published by Spiegel and colleagues in the late 1990s [71]. They found that, in 11 healthy young men, restricting sleep from a baseline of 9 to 4 h for 6 consecutive nights, led to a significant decrease in glucose clearance [71]. A number of subsequent experimental sleep restriction laboratory-based adult studies have found that partial sleep restriction, or changes to sleep-wake timings (circadian desynchronization), leads to impaired glucose sensitivity without compensatory increases in insulin secretion, lower glucose effectiveness and increased glucose levels [72]. While most sleep restriction studies have been laboratory-based, which facilitates the control of relevant variables such as diet, they do not represent a natural environment. However, one recent study found that just 3 weeks of mild sleep restriction (1.5 h less than baseline sleep) in the home setting led to transient impaired insulin sensitivity [73].

The majority of research to date has explored sleep duration rather than sleep architecture, albeit the latter appears to have an impact. Several studies have addressed whether aspects of sleep architecture rather that quantity of sleep affect glucose sensitivity. A small number of studies suggest associations between various sleep stages and insulin sensitivity as well as insulin secretion. These point to positive associations between percentage of sleep time spent in slow wave sleep (SWS; stage 3) and insulin secretory measures as well as insulin sensitivity [74]. Another small study revealed an inverse association between percentage of total sleep time in stage 1 sleep (NREM1) and insulin sensitivity in adolescents, independent of total sleep duration [75]. Another study showed that SWS was positively associated with insulin sensitivity, whereas stage 1 sleep exerted the opposite effect on insulin resistance, after adjustment for age, gender, body mass index z-score, pubertal status, and apnea hypopnea index [76]. Greater sleep efficiency, and longer total sleep, were independently associated with lower glucose levels. This suggests that SWS, sleep efficiency and total sleep duration are protective factors in maintaining glucose and insulin homeostasis.

A growing body of observational evidence points to a relationship between sleep duration and sleep quality and metabolic functioning in adults. These conclusions are supported by several metaanalyses [26, 29, 31, 32] and some of the strongest evidence for sleep and glucose dysregulation come from experimental studies. In summary, the primary mechanisms in the development of impaired glucose metabolism appear to be changes in insulin secretion—the ability of the pancreatic beta-cells to respond to a glucose stimulus, and insulin sensitivity, and the ability of peripheral tissues to respond to an insulin signal. Therefore, it seems that the results from these experimental studies serve to confirm the longitudinal epidemiological and cross-sectional associations observed between chronic sleep restriction and incidence of T2D are causally related—sleep restriction leads to increased insulin resistance, which without increased insulin secretion to compensate can cause hyperglycemia eventually culminating in T2D even when traditional risk factors are accounted for and appear to be independent of age and gender.

SLEEP AND CHILDREN

The consensus of a substantial body of research is that sleep quality and corresponding habits in children and adolescents have undergone a deterioration over recent decades. For example, data generated from the National Survey of Children's Health between 2003 and 2012, indicated that the prevalence of inadequate sleep (defined as 0–6 days of not getting enough sleep), increased across all age groups between 6 and 19 years of age [77]. For instance, inadequate sleep duration increased from 23% to 35% for 6–9 year olds and from 30% to 41% percent for 10–13 year olds [77]. These figures are similar to those reported elsewhere. For instance, it has been reported that sleep deprivation effects 16% of children aged 11 years versus 40.5% of those of 15 year olds [78].

These numbers contrast sharply with the most recent recommendations for sleep duration by the American Academy of Sleep Medicine (AASM) [43]. Based on a review of 864 published articles, they recommended, by

TABLE 31.1 Recommended sleep duration for children and adolescents per 24-h period, as proposed by the American Academy of Sleep Medicine.

Age group	Recommended sleep duration
Infants (4–12 months)	12–16 h
Children (1–2 years)	11–14 h
Children (3–5 years)	10–13 h
Children (6–12 years)	9–12 h
Adolescents (13–18 years)	8–10 h

consensus, sleep duration in infants, children and adolescents, which are depicted in Table 31.1.

The AASM further stated that sleeping the number of recommended hours on a regular basis is associated with better health outcomes including: improved attention, behavior, learning, memory, emotional regulation, quality of life, and mental and physical health. Furthermore, regularly sleeping fewer than the number of recommended hours is associated with attention, behavior, and learning problems. Of particular relevance to this chapter is that insufficient sleep also increases the risk of accidents, injuries, hypertension, obesity, depression and diabetes.

SLEEP, DIABETES AND CHILDREN

Despite the widely held perception regarding an increase of T2D among children and adolescents, difficulties with reliably identifying the prevalence of the disease have been noted. Some have suggested that there are methodological problems surrounding reporting which undermine the validity of current statistics [79]. Nonetheless, others report substantial increases in the prevalence of diabetes among this population. For instance, one study of 10–19 year olds in the United States, highlighted a relative increase of 35% for T2D between 2001 and 2009 [80]. Leaving aside the issue of causality in relation to sleep and T2D, children with diabetes are more likely to experience poorer quality sleep related to glycemic control. For example, one study reported that 67% of children with the T2D experienced poor sleep quality in contrast to only 8% for those with better glycemic control. It is not just older children that may be at potential risk as the relationship between sleep and risk appears to be present for children as young as 4–10 years. For instance, it has been reported that children in this age range routinely sleep <8 h per night and that variations in sleep may result in metabolic dysregulation [81]. In other words, the longer and more stable sleep duration is, the less likely a child is to exhibit metabolic dysfunction.

Although the pathophysiological mechanism of T2D in children is not completely understood, it is clear from adult studies that insulin resistance plays an important role. Evidence of this comes from cross-sectional and longitudinal studies demonstrating that insulin resistance occurs 10–20 years before the onset of the disease in adults, and that it is the best predictor of whether or not an individual will later become diabetic [82]. In contrast, it appears that the disease in children is characterized by more rapid development of glucose homeostasis dysregulation. Therefore, identifying children at risk for T2D is of great importance in order to interrupt its progression and diabetes-related health complications. Another implication of an extended period (though shorter than in adults) of insulin resistance is that children and adolescents with T2D may remain asymptomatic and undiagnosed for a long period of time. The consequences of the disease are as profound as for adults including accelerated development of cardiovascular disease, renal disease, and impaired visual functioning [54].

To date, a small number of large scale cross-sectional studies have been conducted with children. The majority largely mirror results from the few experimental studies that have been performed. In one large, cross-sectional study of 4525 children, aged 9–10 years in the United Kingdom, strong inverse relationships between sleep duration, adiposity and diabetes risk markers were observed. Each additional hour of sleep was associated with less insulin resistance and associations between insulin and glucose remained after an adjustment for adiposity markers [83]. Another very recent large cross-sectional study involving >2700 children conducted in Columbia produced similar results. Boys who met the recommended amount of sleep had a decreased risk of elevated blood glucose levels compared to boys who had a short-sleep duration and the effect remained when potential confounding variables such as adiposity were controlled for [84]. However, the precise mechanisms by which short sleep duration may lead to impaired glucose allostasis in children are still not fully delineated. Furthermore, extrapolating adult findings to a pediatric population is made somewhat complicated by a number of factors, including developmental factors and also key differences in the architecture of adult and child/adolescent sleep. For example, sleep architecture is different between adults and children with slow wave sleep occupying 25%–30% of total sleep for children.

Pediatric studies have also shown associations between sleep duration, sleep architecture and insulin sensitivity and glucose levels [85]. One study showed an association between short sleep duration and insulin resistance [74, 85]. Another found a U-shaped relationship between sleep duration and glycemic measures, with increased glucose levels at both higher and lower sleep durations independent of obesity [74]. One further sleep restriction study, which was performed in lean adolescent males, found that while sleep restriction increased insulin resistance, fasting and postprandial glucose remained unchanged between the two conditions [86].

One proposed pathway from sleep imbalance to T2D relates to the role of endocrine stress, specifically increased cortisol and catecholamine levels. Elevated levels may result in impaired glucose metabolism. Sleep is a refractory period for three stress hormones—cortisol, norepinephrine and epinephrine. These stress hormones are downregulated at night; cortisol in particular, is known to inhibit insulin production and increased levels are related to insulin resistance [87]. Thus, delays to sleep or shortened sleep may cause cortisol levels to rise, which could result in insulin resistance. However, experimental data surrounding the respective roles of these hormones remains unclear in pediatric populations. In one key experimental study, three nights of moderate sleep restriction decreased insulin sensitivity in boys but was unrelated to endocrine stress markers [86].

Another second proposed mechanism concerns the role of disinhibited eating behaviors and risk factors for T2D. For instance, alterations in appetitive hormones are reported to be common following sleep deprivation [49, 88] and may trigger disinhibited eating in adolescents and young adults. Other experimental evidence is, however, somewhat ambiguous. For example, Kelly and colleagues evaluated associations between sleep duration, daytime sleepiness and eating patterns in 119 adolescent girls at risk for T2D [89]. The findings from this study indicated that subjective sleep duration and objectively determined energy intake were positively related, contrary to what was expected. Accounting for age, race, puberty, body composition, depressive symptoms, and perceived stress, sleep duration was positively related to total energy intake [89]. Adjusting for the same covariates, daytime sleepiness was associated with a greater likelihood of binge eating in the previous month. Despite these results, the role of obesity in the relationship between sleep and metabolic disorders, such as T2D, is not completely clear; particularly given the findings from a cross-sectional study, which demonstrated that shorter sleep duration was associated with increased insulin resistance but that the relationship was mediated by abdominal obesity [90].

A third potential mechanism concerns sleep architecture—specifically the role of slow wave sleep (SWS). Existing evidence highlights that the suppression of SWS in adults is related to insulin resistance in adults but the findings are heterogeneous in children. In a landmark study, children's SWS was suppressed, and although the number of participants in the study was limited to 15 children aged 11–14 years, no association was found between SWS suppression and adverse metabolic effects [91]. In contrast, Klingenberg and colleagues showed that SWS suppression was related to insulin resistance [86]. These discrepant findings are mirrored in observational studies suggesting that it is premature to make any definite statement about sleep architecture and metabolic function. It is, however, reasonable to predict that adolescents have the highest risk for insulin resistance and T2D; and that this is probably due to the combination of persistent sleep insufficiency, physiologic declines in SWS, and a peak in insulin resistance during puberty [92].

CONCLUSION

Global statistics indicate an increasing prevalence of T2D among children and adolescents, which are, in part, attributable to traditional risk factors. However, poor sleep habits, whether lifestyle driven or physiological, may also be contributing to the rising levels of childhood diabetes and obesity. Research to date indicates that we are just beginning to better understand the extent to which sleep deficiency impairs glucose metabolism in children. Additional research is, however, needed regarding about the extent, mechanisms, and dynamics of the relationship sleep has upon obesity and T2D, which are comorbid with cardiovascular disease. Overall, the data regarding sleep and chronic health conditions in children are more limited than for adults; short sleep duration may indeed predispose the individual to insulin resistance and hyperglycemia. The magnitude (hours of sleep per night) and duration of sleep restriction (days to weeks) are likely to be important factors in determining the speed and extent of any diabetogenic changes, including elevations of circulating glucose levels caused by reduced insulin sensitivity of peripheral tissues and/or insufficient insulin secretion by the pancreas. So far, the focus has been on typical or average sleep but intra-individual variability may also be an important factor to consider, particularly given the recent emerging evidence surrounding sleep variability. Current experimental findings are consistent with recent epidemiological work which demonstrate that lifestyle factors, including habitually short sleep duration, increase the risk of weight gain over the life course. It is noteworthy that no studies to date have attempted to determine whether improving the quality of sleep of children with T2D brings with it positive changes in glucose homeostasis. In respect of preventative efforts, sleep habits could be integrated into existing management programs that apply an evidence-based approach such as cognitive behavior therapy to provide maximum efforts.

REFERENCES

[1] de Onis M, Blossner M, Borghi E. Global prevalence and trends of overweight and obesity among preschool children. Am J Clin Nutr 2010;92(5):1257–64.

[2] Kuczmarski RJ, Ogden CL, Guo SS, Grummer-Strawn LM, Flegal KM, Mei Z, et al. CDC growth charts for the United States: methods and development. Vital Health Stat 11 2000;2002(246):1–190.

[3] Cole TJ, Bellizzi MC, Flegal KM, Dietz WH. Establishing a standard definition for child overweight and obesity worldwide: international survey. BMJ 2000;320(7244):1240–3.

[4] van Wijnen LG, Boluijt PR, Hoeven-Mulder HB, Bemelmans WJ, Wendel-Vos GC. Weight status, psychological health, suicidal thoughts, and suicide attempts in Dutch adolescents: results from the 2003 E-MOVO project. Obesity (Silver Spring) 2010;18(5):1059–61.

[5] Tsiros MD, Olds T, Buckley JD, Grimshaw P, Brennan L, Walkley J, et al. Health-related quality of life in obese children and adolescents. Int J Obes 2009;33(4):387–400.

[6] Arora T, Hosseini-Araghi M, Bishop J, Yao GL, Thomas GN, Taheri S. The complexity of obesity in U.K. adolescents: relationships with quantity and type of technology, sleep duration and quality, academic performance and aspiration. Pediatr Obes 2013;8(5):358–66.

[7] Stenze T. Intelligence and socioeconomic success: a meta-analytic review of longitudinal research. Dermatol Int 2007;35:401–26.

[8] Verrotti A, Penta L, Zenzeri L, Agostinelli S, De Feo P. Childhood obesity: prevention and strategies of intervention. A systematic review of school-based interventions in primary schools. J Endocrinol Investig 2014;37(12):1155–64.

[9] Ickes MJ, McMullen J, Haider T, Sharma M. Global school-based childhood obesity interventions: a review. Int J Environ Res Public Health 2014;11(9):8940–61.

[10] Schmidt Morgen C, Rokholm B, Sjoberg Brixval C, Schou Andersen C, Geisler Andersen L, Rasmussen M, et al. Trends in prevalence of overweight and obesity in danish infants, children and adolescents—are we still on a plateau? PLoS ONE 2013;8(7):e69860.

[11] Keane E, Kearney PM, Perry IJ, Kelleher CC, Harrington JM. Trends and prevalence of overweight and obesity in primary school aged children in the Republic of Ireland from 2002–2012: a systematic review. BMC Public Health 2014;14:974.

[12] Ranjani H, Mehreen TS, Pradeepa R, Anjana RM, Garg R, Anand K, et al. Epidemiology of childhood overweight & obesity in India: a systematic review. Indian J Med Res 2016;143(2):160–74.

[13] Skinner AC, Ravanbakht SN, Skelton JA, Perrin EM, Armstrong SC. Prevalence of obesity and severe obesity in US children, 1999–2016. Pediatrics 2018;141(3). https://doi.org/10.1542/peds.2017-3459. e20173459.

[14] Belloc NB, Breslow L. Relationship of physical health status and health practices. Prev Med 1972;1(3):409–21.

[15] Kripke DF, Garfinkel L, Wingard DL, Klauber MR, Marler MR. Mortality associated with sleep duration and insomnia. Arch Gen Psychiatry 2002;59(2):131–6.

[16] da Silva AA, de Mello RG, Schaan CW, Fuchs FD, Redline S, Fuchs SC. Sleep duration and mortality in the elderly: a systematic review with meta-analysis. BMJ Open 2016;6(2):e008119.

[17] Yin J, Jin X, Shan Z, Li S, Huang H, Li P, et al. Relationship of sleep duration with all-cause mortality and cardiovascular events: a systematic review and dose-response meta-analysis of prospective cohort studies. J Am Heart Assoc 2017;6(9). https://doi.org/10.1161/JAHA.117.005947. e005947.

[18] Cappuccio FP, D'Elia L, Strazzullo P, Miller MA. Sleep duration and all-cause mortality: a systematic review and meta-analysis of prospective studies. Sleep 2010;33(5):585–92.

[19] Gallicchio L, Kalesan B. Sleep duration and mortality: a systematic review and meta-analysis. J Sleep Res 2009;18(2):148–58.

[20] Itani O, Jike M, Watanabe N, Kaneita Y. Short sleep duration and health outcomes: a systematic review, meta-analysis, and meta-regression. Sleep Med 2017;32:246–56.

[21] Fatima Y, Doi SA, Mamun AA. Longitudinal impact of sleep on overweight and obesity in children and adolescents: a systematic review and bias-adjusted meta-analysis. Obes Rev 2015;16(2):137–49.

[22] Ruan H, Xun P, Cai W, He K, Tang Q. Habitual sleep duration and risk of childhood obesity: systematic review and dose-response meta-analysis of prospective cohort studies. Sci Rep 2015;5:16160.

[23] Chen X, Beydoun MA, Wang Y. Is sleep duration associated with childhood obesity? A systematic review and meta-analysis. Obesity (Silver Spring) 2008;16(2):265–74.

[24] Li L, Zhang S, Huang Y, Chen K. Sleep duration and obesity in children: a systematic review and meta-analysis of prospective cohort studies. J Paediatr Child Health 2017;53(4):378–85.

[25] Miller MA, Kruisbrink M, Wallace J, Ji C, Cappuccio FP. Sleep duration and incidence of obesity in infants, children and adolescents: a systematic review and meta-analysis of prospective studies. Sleep 2018;41(4). https://doi.org/10.1093/sleep/zsy018.

[26] Cappuccio FP, Cooper D, D'Elia L, Strazzullo P, Miller MA. Sleep duration predicts cardiovascular outcomes: a systematic review and meta-analysis of prospective studies. Eur Heart J 2011;32(12):1484–92.

[27] Krittanawong C, Tunhasiriwet A, Wang Z, Zhang H, Farrell AM, Chirapongsathorn S, et al. Association between short and long sleep durations and cardiovascular outcomes: a systematic review and meta-analysis. Eur Heart J Acute Cardiovasc Care 2017. https://doi.org/10.1177/2048872617741733.

[28] Shan Z, Ma H, Xie M, Yan P, Guo Y, Bao W, et al. Sleep duration and risk of type 2 diabetes: a meta-analysis of prospective studies. Diabetes Care 2015;38(3):529–37.

[29] Cappuccio FP, D'Elia L, Strazzullo P, Miller MA. Quantity and quality of sleep and incidence of type 2 diabetes: a systematic review and meta-analysis. Diabetes Care 2010;33(2):414–20.

[30] Upala S, Sanguankeo A, Congrete S, Romphothong K. Sleep duration and insulin resistance in individuals without diabetes mellitus: a systematic review and meta-analysis. Diabetes Res Clin Pract 2015;109(3):e11–2.

[31] Holliday EG, Magee CA, Kritharides L, Banks E, Attia J. Short sleep duration is associated with risk of future diabetes but not cardiovascular disease: a prospective study and meta-analysis. PLoS ONE 2013;8(11):e82305.

[32] Anothaisintawee T, Reutrakul S, Van Cauter E, Thakkinstian A. Sleep disturbances compared to traditional risk factors for diabetes development: systematic review and meta-analysis. Sleep Med Rev 2016;30:11–24.

[33] Locard E, Mamelle N, Billette A, Miginiac M, Munoz F, Rey S. Risk factors of obesity in a five year old population. Parental versus environmental factors. Int J Obes Relat Metab Disord 1992;16(10):721–9.

[34] von Kries R, Toschke AM, Wurmser H, Sauerwald T, Koletzko B. Reduced risk for overweight and obesity in 5- and 6-y-old children by duration of sleep—a cross-sectional study. Int J Obes Relat Metab Disord 2002;26(5):710–6.

[35] Sekine M, Yamagami T, Handa K, Saito T, Nanri S, Kawaminami K, et al. A dose-response relationship between short sleeping hours and childhood obesity: results of the Toyama Birth Cohort Study. Child Care Health Dev 2002;28(2):163–70.

[36] Cao M, Zhu Y, He B, Yang W, Chen Y, Ma J, et al. Association between sleep duration and obesity is age- and gender-dependent in Chinese urban children aged 6–18 years: a cross-sectional study. BMC Public Health 2015;15:1029.

[37] Ozturk A, Mazicioglu M, Poyrazoglu S, Cicek B, Gunay O, Kurtoglu S. The relationship between sleep duration and obesity in Turkish children and adolescents. Acta Paediatr 2009;98(4):699–702.

[38] Bell JF, Zimmerman FJ. Shortened nighttime sleep duration in early life and subsequent childhood obesity. Arch Pediatr Adolesc Med 2010;164(9):840–5.

[39] Shi Z, Taylor AW, Gill TK, Tuckerman J, Adams R, Martin J. Short sleep duration and obesity among Australian children. BMC Public Health 2010;10:609.

[40] Carter PJ, Taylor BJ, Williams SM, Taylor RW. Longitudinal analysis of sleep in relation to BMI and body fat in children: the FLAME study. BMJ 2011;342:d2712.

[41] Reilly JJ, Armstrong J, Dorosty AR, Emmett PM, Ness A, Rogers I, et al. Early life risk factors for obesity in childhood: cohort study. BMJ 2005;330(7504):1357.

[42] Arora T, Taheri S. Associations among late chronotype, body mass index and dietary behaviors in young adolescents. Int J Obes 2015;39(1):39–44.

[43] Paruthi S, Brooks LJ, D'Ambrosio C, Hall WA, Kotagal S, Lloyd RM, et al. Consensus statement of the American Academy of sleep medicine on the recommended amount of sleep for healthy children: methodology and discussion. J Clin Sleep Med 2016;12(11):1549–61.

[44] Kim SJ, Lee YJ, Cho SJ, Cho IH, Lim W, Lim W. Relationship between weekend catch-up sleep and poor performance on attention tasks in Korean adolescents. Arch Pediatr Adolesc Med 2011;165(9):806–12.

[45] Kang SG, Lee YJ, Kim SJ, Lim W, Lee HJ, Park YM, et al. Weekend catch-up sleep is independently associated with suicide attempts and self-injury in Korean adolescents. Compr Psychiatry 2014;55(2):319–25.

[46] He F, Bixler EO, Berg A, Imamura Kawasawa Y, Vgontzas AN, Fernandez-Mendoza J, et al. Habitual sleep variability, not sleep duration, is associated with caloric intake in adolescents. Sleep Med 2015;16(7):856–61.

[47] He F, Bixler EO, Liao J, Berg A, Imamura Kawasawa Y, Fernandez-Mendoza J, et al. Habitual sleep variability, mediated by nutrition intake, is associated with abdominal obesity in adolescents. Sleep Med 2015;16(12):1489–94.

[48] Capers PL, Fobian AD, Kaiser KA, Borah R, Allison DB. A systematic review and meta-analysis of randomized controlled trials of the impact of sleep duration on adiposity and components of energy balance. Obes Rev 2015;16(9):771–82.

[49] Spiegel K, Tasali E, Penev P, Van Cauter E. Brief communication: sleep curtailment in healthy young men is associated with decreased leptin levels, elevated ghrelin levels, and increased hunger and appetite. Ann Intern Med 2004;141(11):846–50.

[50] Hart CN, Carskadon MA, Considine RV, Fava JL, Lawton J, Raynor HA, et al. Changes in children's sleep duration on food intake, weight, and leptin. Pediatrics 2013;132(6):e1473–80.

[51] Hagan EWS, Starke SJ, Peppard PE. The association between sleep duration and leptin, ghrelin, and adiponectin among children and adolescents. Curr Sleep Med Rep 2015;1(4):185–94.

[52] Arora T, Taheri S. Is sleep education an effective tool for sleep improvement and minimizing metabolic disturbance and obesity in adolescents? Sleep Med Rev 2017;36:3–12.

[53] Dutil C, Chaput JP. Inadequate sleep as a contributor to type 2 diabetes in children and adolescents. Nutr Diabetes 2017;7(5):e266.

[54] Reinehr T. Type 2 diabetes mellitus in children and adolescents. World J Diabetes 2013;4(6):270–81.

[55] Mayer-Davis EJ, Lawrence JM, Dabelea D, Divers J, Isom S, Dolan L, et al. Incidence trends of type 1 and type 2 diabetes among youths, 2002–2012. N Engl J Med 2017;376(15):1419–29.

[56] Hillier TA, Pedula KL. Complications in young adults with early-onset type 2 diabetes: losing the relative protection of youth. Diabetes Care 2003;26(11):2999–3005.

[57] Eppens MC, Craig ME, Cusumano J, Hing S, Chan AK, Howard NJ, et al. Prevalence of diabetes complications in adolescents with type 2 compared with type 1 diabetes. Diabetes Care 2006;29(6):1300–6.

[58] Naughton MJ, Ruggiero AM, Lawrence JM, Imperatore G, Klingensmith GJ, Waitzfelder B, et al. Health-related quality of life of children and adolescents with type 1 or type 2 diabetes mellitus: SEARCH for Diabetes in Youth Study. Arch Pediatr Adolesc Med 2008;162(7):649–57.

[59] Levitt Katz LE, Swami S, Abraham M, Murphy KM, Jawad AF, McKnight-Menci H, et al. Neuropsychiatric disorders at the presentation of type 2 diabetes mellitus in children. Pediatr Diabetes 2005;6(2):84–9.

[60] Tuomilehto J, Lindstrom J, Eriksson JG, Valle TT, Hamalainen H, Ilanne-Parikka P, et al. Prevention of type 2 diabetes mellitus by changes in lifestyle among subjects with impaired glucose tolerance. N Engl J Med 2001;344(18):1343–50.

[61] Aldabal L, Bahammam AS. Metabolic, endocrine, and immune consequences of sleep deprivation. Open Respir Med J 2011;5:31–43.

[62] Irwin MR. Why sleep is important for health: a psychoneuroimmunology perspective. Annu Rev Psychol 2015;66:143–72.

[63] Kotronoulas G, Stamatakis A, Stylianopoulou F. Hormones, hormonal agents, and neuropeptides involved in the neuroendocrine regulation of sleep in humans. Hormones (Athens) 2009;8(4):232–48.

[64] Lee SWH, Ng KY, Chin WK. The impact of sleep amount and sleep quality on glycemic control in type 2 diabetes: a systematic review and meta-analysis. Sleep Med Rev 2017;31:91–101.

[65] Koren D, Dumin M, Gozal D. Role of sleep quality in the metabolic syndrome. Diabetes Metab Syndr Obes 2016;9:281–310.

[66] Keckeis M, Lattova Z, Maurovich-Horvat E, Beitinger PA, Birkmann S, Lauer CJ, et al. Impaired glucose tolerance in sleep disorders. PLoS ONE 2010;5(3):e9444.

[67] Gottlieb DJ, Punjabi NM, Newman AB, Resnick HE, Redline S, Baldwin CM, et al. Association of sleep time with diabetes mellitus and impaired glucose tolerance. Arch Intern Med 2005;165(8):863–7.

[68] Katano S, Nakamura Y, Nakamura A, Murakami Y, Tanaka T, Takebayashi T, et al. Relationship between sleep duration and clustering of metabolic syndrome diagnostic components. Diabetes Metab Syndr Obes 2011;4:119–25.

[69] Spiegel K, Knutson K, Leproult R, Tasali E, Van Cauter E. Sleep loss: a novel risk factor for insulin resistance and Type 2 diabetes. J Appl Physiol (1985) 2005;99(5):2008–19.

[70] Sivak M. Sleeping more as a way to lose weight. Obes Rev 2006;7(3):295–6.

[71] Spiegel K, Leproult R, Van Cauter E. Impact of sleep debt on metabolic and endocrine function. Lancet 1999;354(9188):1435–9.

[72] Leproult R, Holmback U, Van Cauter E. Circadian misalignment augments markers of insulin resistance and inflammation, independently of sleep loss. Diabetes 2014;63(6):1860–9.

[73] Robertson MD, Russell-Jones D, Umpleby AM, Dijk DJ. Effects of three weeks of mild sleep restriction implemented in the home environment on multiple metabolic and endocrine markers in healthy young men. Metabolism 2013;62(2):204–11.

[74] Koren D, Levitt Katz LE, Brar PC, Gallagher PR, Berkowitz RI, Brooks LJ. Sleep architecture and glucose and insulin homeostasis in obese adolescents. Diabetes Care 2011;34(11):2442–7.

[75] Armitage R, Lee J, Bertram H, Hoffmann R. A preliminary study of slow-wave EEG activity and insulin sensitivity in adolescents. Sleep Med 2013;14(3):257–60.

[76] Zhu Y, Fenik P, Zhan G, Xin R, Veasey SC. Degeneration in arousal neurons in chronic sleep disruption modeling sleep apnea. Front Neurol 2015;6:109.

[77] Hawkins SS, Takeuchi DT. Social determinants of inadequate sleep in US children and adolescents. Public Health 2016;138:119–26.

[78] Leger D, Beck F, Richard JB, Godeau E. Total sleep time severely drops during adolescence. PLoS ONE 2012;7(10):e45204.

[79] Fazeli Farsani S, van der Aa MP, van der Vorst MM, Knibbe CA, de Boer A. Global trends in the incidence and prevalence of type 2 diabetes in children and adolescents: a systematic review and evaluation of methodological approaches. Diabetologia 2013;56(7):1471–88.

[80] Dabelea D, Mayer-Davis EJ, Saydah S, Imperatore G, Linder B, Divers J, et al. Prevalence of type 1 and type 2 diabetes among children and adolescents from 2001 to 2009. JAMA 2014;311(17):1778–86.

[81] Spruyt K, Molfese DL, Gozal D. Sleep duration, sleep regularity, body weight, and metabolic homeostasis in school-aged children. Pediatrics 2011;127(2):e345–52.

[82] D'Adamo E, Caprio S. Type 2 diabetes in youth: epidemiology and pathophysiology. Diabetes Care 2011;34(Suppl. 2):S161–5.

[83] Rudnicka AR, Nightingale CM, Donin AS, Sattar N, Cook DG, Whincup PH, Owen CG. Sleep duration and risk of type 2 diabetes. Pediatrics 2017;140(3). https://doi.org/10.1542/peds.2017-0338. e20170338.

[84] Pulido-Arjona L, Correa-Bautista JE, Agostinis-Sobrinho C, Mota J, Santos R, Correa-Rodriguez M, et al. Role of sleep duration and sleep-related problems in the metabolic syndrome among children and adolescents. Ital J Pediatr 2018;44(1):9.

[85] Flint J, Kothare SV, Zihlif M, Suarez E, Adams R, Legido A, et al. Association between inadequate sleep and insulin resistance in obese children. J Pediatr 2007;150(4):364–9.

[86] Klingenberg L, Chaput JP, Holmback U, Visby T, Jennum P, Nikolic M, et al. Acute sleep restriction reduces insulin sensitivity in adolescent boys. Sleep 2013;36(7):1085–90.

[87] Plat L, Byrne MM, Sturis J, Polonsky KS, Mockel J, Fery F, et al. Effects of morning cortisol elevation on insulin secretion and glucose regulation in humans. Am J Phys 1996;270(1 Pt 1):E36–42.

[88] Schmid SM, Hallschmid M, Jauch-Chara K, Born J, Schultes B. A single night of sleep deprivation increases ghrelin levels and feelings of hunger in normal-weight healthy men. J Sleep Res 2008;17(3):331–4.

[89] Kelly NR, Shomaker LB, Radin RM, Thompson KA, Cassidy OL, Brady S, et al. Associations of sleep duration and quality with disinhibited eating behaviors in adolescent girls at-risk for type 2 diabetes. Eat Behav 2016;22:149–55.

[90] Javaheri S, Caref EB, Chen E, Tong KB, Abraham WT. Sleep apnea testing and outcomes in a large cohort of Medicare beneficiaries with newly diagnosed heart failure. Am J Respir Crit Care Med 2011;183(4):539–46.

[91] Shaw ND, McHill AW, Schiavon M, Kangarloo T, Mankowski PW, Cobelli C, et al. Effect of slow wave sleep disruption on metabolic parameters in adolescents. Sleep 2016;39(8):1591–9.

[92] Amiel SA, Sherwin RS, Simonson DC, Lauritano AA, Tamborlane WV. Impaired insulin action in puberty. A contributing factor to poor glycemic control in adolescents with diabetes. N Engl J Med 1986;315(4):215–9.

Chapter 32

Sleep and mental health in children and adolescents

Michelle A. Short[a], Kate Bartel[a], Mary A. Carskadon[b]
[a]School of Psychology, Flinders University, Adelaide, SA, Australia, [b]E.P. Bradley Hospital, Brown University, Providence, RI, United States

INTRODUCTION

Mental illness poses one of the largest disease burdens of all conditions [1]. While the burden of diseases is felt most acutely among older adults, such psychiatry illnesses as depression and anxiety are prevalent across much of the human lifespan. Indeed, late childhood and adolescence are important developmental periods in terms of mental health, with a notable acceleration of the incidence of mood disorders, anxiety disorders, eating disorders, and psychosis occurring during this time [2]. Furthermore, these illnesses are frequently chronic and recurrent, and earlier age of onset is associated with a more severe and unremitting course, substantial impairments to educational and social functioning, and reduced quality of life [2]. Anxiety is the most common psychiatric disorder of childhood, affecting between 3.9% and 17.5% of children and adolescents, while depression affects between 2% and 8% [3, 4]. Depression and anxiety disorders are frequently comorbid, with anxiety typically preceding depression [2]. Another affliction that often co-occurs with both conditions is sleep problems, including insufficient sleep, trouble falling asleep, and unrefreshing sleep, among others. As many as 90% of children with anxiety and/or depression report problems with their sleep [5]. As reviewed below, strong evidence indicates that sleep causally impacts a range of factors relating to mental health, including mood, emotion dysregulation, depression, anxiety, and suicide [6–8].

The importance of prevention and early intervention for mental health in youth is indisputable. Identifying sleep disturbances as a factor that contributes to deleterious alteration in mood, emotion regulation, and psychopathology is key because sleep is a target amenable to change. The amount of sleep children and adolescents obtain and the quality of sleep, affect their mood and mental health [7, 9–11]. This chapter focuses on the evidence linking sleep and mental health with the aim to (a) review and summarize the literature regarding the impact of sleep duration and sleep quality in healthy school-age children and adolescents, and those with depression and anxiety and (b) identify approaches for families, school leaders, clinicians, and policy makers to improve child and adolescent sleep for optimal mental health functioning. It is important to note that this chapter focuses on the spectrum of mental health and not solely on mental illness. This is to acknowledge that mental health occurs on a spectrum that is much broader than simply the presence or absence of a diagnosable mental conditions. Thus, the included literature includes both healthy and clinical populations, and mood outcome measures include positive and negative mood states, emotion regulation, symptoms of depression and anxiety, and diagnoses of depression and anxiety.

Another important distinction to make at the outset is regarding the measures used to characterize sleep, which are varied across studies. Most studies use subjective self- or parent-reports of sleep, which are often included as items within a larger survey. These have the advantage of being time- and cost-effective and can be used for larger, epidemiological studies. The limitations of subjective survey measures of sleep include inaccuracy—especially when the reporter is a parent [12]—and reporting biases. Sleep diaries have widespread clinical use and involve recording sleep patterns each day. While this also relies upon self- or parent-report, sleep diaries are not as susceptible to the reporting inaccuracies and biases as survey measures, as the reporting is anchored to the sleep of the previous night. To overcome some of the limitations of subjective reports, objective measures of sleep are also used, often actigraphy (using activity monitors) or polysomnography. Activity monitors are usually worn on the wrist like a wristwatch uses contain an accelerometer to measure movement. Algorithms are then applied to these movement data to estimate sleep and wake. While this is an objective method, limitations include the reliance of concurrent sleep diaries to identify the time in bed period and any times that the device was not worn. In addition, there have been concerns regarding the accuracy of actigraphic algorithms when used with adolescents [13]. The gold standard of sleep measurement is polysomnography. Polysomnography uses electrodes applied to the scalp to directly measure brain activity and thus identify either wake or stage of sleep. While this method is the most accurate, it is time- and cost-intensive and is not often not feasible for many studies. As such, few studies routinely include polysomnography to measure sleep.

SLEEP DURATION AND MENTAL HEALTH

Much of the research on sleep and mental health has focused on how *much* sleep children and adolescents obtain. The recommended sleep duration in children ages 6–13 years is 9–11 h, while adolescents are recommended to sleep 8–10 h per night [14, 15]. Recent studies with mood symptoms as an outcome have estimated that adolescents require between 7.5 and 9.5 h sleep per night for optimal mood [16, 17]; however, many—if not most—teens sleep less [18]. A recent metaanalysis pooled data regarding normative sleep estimates from studies that included actigraphic estimates of sleep on school nights. Pooled results showed that most children and adolescents typically obtain sleep below these recommended amounts [19].

Much of the extant research linking sleep and mental health in children is cross-sectional [20–22], with short sleep linked to increased emotional lability [23]. For example, short sleep duration, objectively measured by actigraphy in a sample of 7-year old children, was associated with heightened emotional reactivity [24]. Similarly, among 8–12-year old children, shorter sleep durations were associated with heightened affective responses in domains including sadness, anger, fear and disgust [25]. Of interest, sleep duration was not correlated with positive affective responses [25].

Sleep duration has also been linked to mental health symptomology and disorders such as anxiety and depression [20]. A telephone survey of parents of children aged 6–17 years revealed that as the number of nights per week of inadequate sleep increased, so did *symptoms* of depression. However, depression and anxiety *diagnoses* were not related to the number of nights of inadequate sleep [26]. Another study reported that, according to sleep diary measures, children with anxiety slept less than those without anxiety [27]. Sleep complaints, including insufficient sleep, also appear to feature highly among children with high levels of depression symptoms or a diagnosis of depression [28]. On the other hand, findings about the association of *objectively* measured sleep to childhood depression are less consistent [20, 28].

Experimental studies have shown a causal relation between sleep duration and a wide range of mood outcomes in children and adolescents [6, 8, 18, 25, 29–33], though experimental studies involving children are less abundant [21, 22]. Nonetheless, these studies provide evidence that children with sufficient sleep experience better mood and are able to regulate their emotions better than those who are sleep restricted [34]. Metaanalytic data from studies of children aged 5 to 12 years indicate that sleep restriction is related to increased internalizing behavior problems, especially when experimentally shortened for 2 or more nights [22]. This association was demonstrated by Gruber and colleagues [32], who after a baseline of 5 days of habitual sleep (as measured via actigraphy), assigned 34 children, aged 7–11 years, to either 1 week of 1 h less time in bed per night or 1 week of 1 h more time in bed per night. When sleep was extended (by an average of 27 min per night), teacher ratings of the children's emotional lability decreased, whereas restriction of sleep (by an average of 54 min per night) led to increased teacher-reported emotional lability [32].

Another experimental study investigated the impact of sleep duration on teacher ratings of internalizing symptoms (e.g., anxious/sad affect and emotional lability) in children aged 6–12 years. Three sleep conditions were compared: 1 week of typical sleep (average 9.5 h' time in bed), 1 week of optimized sleep (minimum 10 h' time in bed), and 1 week of restricted sleep (8 h' time in bed for first and second graders; 6.5 h' time in bed for children in third grade or older) [33]. While attention and academic problems were negatively affected by restricting sleep, internalizing symptoms remained similar across the 3 weeks [33].

Vriend and colleagues used a similar experimental protocol to restrict and extend sleep in 32 children aged 8–12 years [25]. In this study reports of emotion regulation were obtained from parents and children, rather than teachers. All children experienced sleep restriction and sleep extension conditions in a counterbalanced order after 1 week of baseline sleep. Sleep period was extended by going to bed 1 h earlier and restricted by going to bed 1 h later, each condition for 4 consecutive nights; a 3-night "washout" period occurred in between conditions [35]. Children slept an average of 74 min longer during the extended sleep condition compared to the short sleep condition. Positive affective in response to positive emotional images and parent-reported emotion regulation decreased following nights of short sleep. By contrast, negative affective response to negatively valenced images and child-reported emotion regulation did not differ between conditions [35].

While these experimental studies including children were all home-based studies, many studies of adolescents have been laboratory based, thus enabling better adherence to study protocol regarding sleep and avoidance of countermeasures, such as caffeine. Measures of positive and negative affect are often used among adolescent studies.

For example, adolescents consistently report feeling reduced positive affect following sleep restriction [18, 29, 30, 36], suggesting that sleep loss diminishes their ability to feel positive affective states such as enthusiasm and excitement. The effects of restricted sleep on negative affect, however, are less consistent, with only one study showing significantly increased negative affect [36]. When considering discrete mood states, results from adolescent total sleep deprivation and chronic sleep restriction paradigms have shown reports of significantly increased anxiety, anger, confusion, and fatigue following sleep loss [6, 8].

As well as affecting mood states, sleep loss modifies the ability to regulate mood and emotion. Emotion regulation refers to the ability of an individual to monitor, evaluate, and modulate emotional reactions in a way that helps individuals to achieve goals and function effectively across different contexts [37, 38]. One hypothesis as to why emotions are affected by sleep duration proposes that short sleep disrupts the limbic system, which helps maintain emotion regulation [22]. Thus, impaired sleep renders a child vulnerable to emotional instability at a physiological level. This is clinically relevant, as emotion dysregulation is an important transdiagnostic factor that heightens the risk of a wide range of psychopathology outcomes [37]. Sleep loss is also implicated in suicidal ideation and suicide attempts, with one study finding a threefold increased risk of suicide attempt in adolescents who slept less than 8 h per night [22].

Experimental studies have found that adolescents' abilities to regulate their emotions is worsened with sleep loss [8, 35, 39]. For example, one study exposed adolescents aged 10–16 years to two night of sleep restriction (6.5 h on the first night and 2 h on the second night) and two nights with 7–8 h sleep per night. Conditions were separated by 1 week and the order was counterbalanced [39]. Following each sleep condition, participants completed an affective measurement battery. Adolescents reported increased anxiety during a catastrophizing task and rated the likelihood of potential catastrophes as higher following two night of restricted sleep when compared to when they had longer sleep opportunities. Furthermore, the younger adolescents, aged 10–13 years, found their main worry as more threatening when they were sleep deprived.

While experimental studies have been invaluable in demonstrating a causal relationship between sleep loss and many aspects of mood and emotion regulation, such brief sleep manipulations have not consistently shown direct effects of sleep loss on either depressed mood or anxiety symptoms. For example, although self-reported depressed mood symptoms increased with total sleep deprivation, sleep restriction in two studies with adolescent participants did not show the same response [6, 8]. Among intervention studies, adolescents randomly allocated to sleep extension, as well as adolescents who extended their sleep by 45 min following a delay to school start times, reported significantly fewer depressed mood symptoms [31, 40], but not reduced anxiety [41].

While experimental studies have the advantage of experimental control, they often include highly screened, healthy participants without elevated depressed mood or anxiety symptoms, who are exposed to long sleep opportunities prior to sleep restriction, and their sleep is restricted over relatively short periods. Ecologically, children and adolescents typically restrict their sleep over multiple weeks and months during the school term. Thus, short-term in laboratory studies may not be able to capture effects of sleep that may appear only when sleep is chronically restricted over long periods. As a result, longitudinal studies are instrumental to indicating whether chronically restricted sleep is related to subsequent long-term deficits in mood and/or psychopathology and more effective for elucidating relationships between sleep, depression, and anxiety that are difficult to elicit in time-limited experimental studies. This approach often has the advantage of greater ecological validity by including a broader cross-section of participants. Among healthy adolescents, for example, evidence for a longitudinal association between sleep and subsequent depressed mood is reported in most [42, 43], but not all [44], studies. One study of 12- to 15-year old Dutch adolescents found that less time in bed at baseline was associated with greater severity of symptoms of depression/anxiety at follow-up, but not vice-versa [10]. Similarly, a study of 2259 US adolescents aged 12–15 years reported both concurrent and longitudinal associations between short sleep duration and lower self-esteem and higher depressive symptoms [43].

Another approach taken by recent studies is to examine the temporal relationship between sleep duration and next day mood in healthy samples and in adolescents with anxiety and depression disorders [16, 45, 46]. Fuligni and colleagues [16], for example, collected data on nightly sleep and daily mood over a 2-week period from 419 adolescents in grades 9 and 10. The analysis of nightly sleep and next-day depressed mood and anxiety symptoms determined the duration of sleep required for optimal next-day mood. This optimal sleep duration was estimated at 9.03 h (SD=0.86) of sleep per night, similar to estimates of sleep need required for optimal daytime alertness and sustained attention [15, 47]. The association between sleep duration and mood was U-shaped, with both long and short sleep associated with worse anxiety and depressed mood. These findings may explain why some studies do not find significant linear associations between sleep duration and mood, as nonlinear relationships are not always tested. In addition, Fuligni and colleagues found that the relation of sleep duration to subsequent mood was not uniform but varied depending on sex and mental health status. Indeed, this study estimated that girls require more sleep than boys for optimal mood, and adolescents experiencing clinically significant internalizing symptoms require more sleep than adolescents below the clinical range [16]. These findings indicate differential vulnerability to the effect of short sleep in some populations and may explain why such associations are stronger in clinical groups compared to healthy controls [45]. A relationship between sleep duration and anxiety and depression symptoms has similarly been found in children and adolescents, aged 5–18 years (M age = 10.5 years), diagnosed with various psychiatric disorders. Among these clinical groups, parent-reported shorter sleep was associated with increased anxiety and depression symptoms [48].

SLEEP QUALITY AND MENTAL HEALTH

Sleep quality refers to a wide range of factors associated with the ease or difficulties with initiation and maintenance of sleep, as well as how subjectively refreshing or satisfying sleep is to the individual. A number of studies show that sleep quality and mental health are related, with longer sleep onset latencies, more frequent awakenings during the night, longer time spent awake after sleep onset, greater sleep disturbances, and poor subjective sleep quality predicting worse mood, poorer emotion regulation, and increased likelihood of mood disorders [9, 49–51]. Sleep complaints pertaining to the quality of sleep are also common in children with anxiety, including difficulty falling asleep or staying asleep, refusing to go to bed, nightmares and nighttime fears [52]. Similar to studies assessing sleep duration and mental health in children, those examining sleep quality are often cross-sectional [22].

Poor sleep quality has been shown to have a deleterious effect on mood beyond the effect that sleep quality variables may have on sleep duration [53]. Indeed, sleep quality shows unique associations with mental health functioning independent of sleep duration [54, 55]. For example, in a cross-sectional study of nearly 100,000 Japanese high school students, results indicated that difficulty falling asleep, difficulty staying asleep, and subjective sleep quality showed dose-dependent relationships to mental health status, with worse sleep predicting worse mental health [50]. A review of 10–13-year old children regarding sleep and anxiety demonstrated that subjective sleep complaints were common among children with anxiety, especially when parent report was used [27, 56]. Furthermore, based on self and parent report, children who reported sleeping difficulties were more likely to have a diagnosis of anxiety than children who did not report trouble sleeping. Moreover, sleep issues are more likely to persist as the child ages in those who experience anxiety [27]. Conversely, a recent metaanalysis found that decreased sleep efficiency was not associated with internalizing behavior problems [22].

Regarding specific components of sleep quality, children who experience anxiety may exhibit longer sleep latency. However, literature is inconsistent, as when objective measures of sleep are employed (actigraphy or polysomnography), some studies demonstrate this difference, whereas others find no difference between anxious children and controls [27, 56, 57]. Anxious children may have less slow wave sleep and more nightly awakenings than those with depression or no psychiatric diagnoses [28]. The type of anxiety disorder may also relate to the sleep issue, with increased sleep latency, as measured by polysomnography, among children with general anxiety compared to those without any diagnosis. Furthermore, nightmares may be exhibited more frequently among children with separation anxiety [28].

Regarding sleep efficiency, data from a week's actigraphy measurements were not correlated with positive or negative affect [25]. In fact, anxious children may have higher objective sleep efficiency than controls [27]. Of note, when sleep is experimentally restricted, thus decreasing sleep fragmentation (i.e., improving sleep quality), the effects of shorted sleep are still evident—that is, despite improved sleep quality, emotional liability increased during a period of sleep restriction [32].

Overall, it appears that subjective sleep complaints are high among anxious children, yet objective sleep difficulties show less consistent evidence [56, 57]. In part, the study's environment may play a role in different findings. That is, a comfortable home environment may facilitate sleep, compared to a novel laboratory environment exacerbating sleep issues in anxious children [27]. Moreover, some laboratory studies may not capture the home sleep environment, which enables difference in sleep patterns. For example, room sharing and changing beds during the night may occur at home in both anxious and nonanxious children, yet, these behaviors are not practiced in a laboratory [56].

Associations between aspects of sleep quality and mental health are also borne out longitudinally. Longitudinal studies demonstrate that sleep issues in childhood predict later anxiety and depression, in most, but not all studies [20]. In one study of 5-year-old children, sleep was measured at the age of 5 years using polysomnography and internalizing problems were measured 1 year later. Results showed that children who had poor sleep quality (indicated by sleep latency, sleep period time and number of awakenings after sleep onset), reported more internalizing problems in their child 1 year later, when compared to parents of children with good/normal baseline sleep [58].

Longitudinal studies also indicate that sleep disturbances in early childhood increase the likelihood of development of anxiety in adolescence and adulthood [28]. Stronger evidence suggests that sleep issues precede anxiety, however, the inverse relationship may also hold true [28, 59]. That is, sleep issues and emotional disturbances are interrelated, and may predispose a child to future anxiety. It is likely that sleep and emotional functioning hold a bi-directional relationship, thus those with poor sleep are more likely to exhibit symptoms of anxiety and depression, and vice-versa, with each issue exacerbating the other [27, 59, 60]. More research employing objective measures is needed to clarify the strength and direction of associations [59]. When examining bidirectional relationships between sleep quality and anxiety disorders, one review found support for the role of sleep problems predicting anxiety disorders, but limited support for the role of anxiety as a predictor of sleep problems [11]. For example, Gregory and O'Connor assessed sleep problems and behavioral/emotional problems in a sample of 490 young people, assessed at age 4 years and again at mid-adolescence [9]. Sleep problems at age 4 predicted more attention problems, aggression, and depression/anxiety during mid-adolescence. Of note, the reverse relationship was not supported, as behavioral/emotional problems in early childhood did not predict sleep problems during adolescence. The authors also found that the concurrent association between sleep problems and anxiety/depression grew significantly stronger across this developmental period, increasing from $r=0.39$ at 4 years to $r=0.52$ during mid-adolescence [9]. This may be due to the greater prevalence of sleep problems among 4-year-olds, with sleep problems scores decreasing by approximately 50% between early childhood and mid-adolescence. As sleep problems are highly prevalent in young children, the presence of sleep problems may be less sensitive as a predictor of mental health in the very young.

In a study of 516 Japanese adolescents, Kaneita and colleagues assessed sleep and mental health at age 13 years and again after 2 years [51]. Concurrent with the reduction in sleep quality over this time was a reduction in mental health status. A new onset of poor sleep quality and chronically poor sleep quality both significantly predicted the development of poor mental health. Similar findings were reported from a longitudinal study of 3134 US adolescents, aged 11 to 18 years, who were assessed at baseline and approximately 1-year later [61]. Poor quality sleep was highly prevalent, with 60% experiencing nonrestorative sleep, 17% reporting difficulty falling asleep and 12% waking frequently during the night either often or almost every day. After controlling for covariates, there was a dose-response relationship between insomnia symptoms at time 1 and depression at time 2. Specifically, greater insomnia symptom severity at baseline predicted worse mental health 1 year later [61].

One limitation in this literature is the reliance upon subjective self-report measures of sleep and mood. These associations may be inflated due to rater biases, whereby an adolescent who reports poor sleep may be more likely to report poor mood and vice versa. Objective measures of sleep are needed to mitigate against this and to determine whether subjectively short or poor-quality sleep is paralleled by objectively short or poor-quality sleep, or alternatively, whether youth with poorer mental health misperceive their sleep as being worse than it objectively is. Among the limited literature that has examined sleep and mental health using polysomnography, one study assessed objective sleep over two consecutive nights in youth aged 7–17 years with either anxiety disorders ($N=24$), major depressive disorder without comorbid anxiety disorders ($N=128$), or no history of psychiatric disorder ($N=101$) [62]. Youth with anxiety disorders took longer to fall asleep on the second night than controls or youth with depression (longer sleep onset latencies are common on the first night of polysomnography, however, this typically resolves on subsequent nights among most individuals), and they experienced more awakenings than the depressed group [62]. Overall, it appears that sleep disturbances are related to childhood anxiety, possibly in a reciprocal fashion [63]. Similarly, children who experience depression symptoms to a large extent also experience sleep disturbances, such as insomnia [20, 28]. However, this association is likely to be stronger in adolescents and adults [20, 63].

A similar pattern of results was reported in a recent meta-analysis examining bidirectional relationships between sleep and adolescent depression. Specifically, they found that adolescents with depression took longer to fall asleep, had more frequent and longer awakenings during the night, had objectively lighter sleep (more stage 1 sleep) and reported worse sleep quality [7]. When examining prospective relationships over time, poor sleep quality was a predictor of subsequent major depression and suicide attempts, but not vice versa [7]. Among the various subjective and objective sleep predictors of concurrent and future depression, those variables associated with wakefulness in bed were most consistent in predicting depression. The authors propose a model of the relationship between sleep disturbances and depression. They suggest that increased time spent awake in bed due to long sleep onset latencies and wake periods during the night, coupled with poor subjective sleep quality leads to increased night time rumination, where adolescents lie in quiet wakefulness in bed and engage in negative repetitive thoughts focused on the symptoms, causes and consequences of their distress. This focus may be on their distress about their sleep, as is commonly witnessed among individuals with insomnia, or it may be about more general factors, such as relationships with others, issues related to school, mood and tiredness [64]. This model is supported by research showing that such negative thoughts are associated with poor sleep quality in both healthy adolescents, sleep-disordered adolescents, and children and adolescents with mood disorders [64–66]. Taken together, these findings highlight that the presence of sleep problems during adolescence are a "red flag" that teens are at heightened risk of psychopathology [36].

IMPROVING SLEEP AND MENTAL HEALTH IN CHILDREN AND ADOLESCENTS

While insufficient and poor-quality sleep are extremely common among children and adolescents, the opportunities for change are many. Given the contribution of sleep to mental health, simple interventions to target sleep are likely to have broad beneficial impacts on how they experience and regulate mood and emotion, as well as the likelihood of developing mood and/or anxiety disorders. Bronfenbrenner's ecological systems theory [67] posits that children's development occurs in the context of several interacting ecosystems that include the self, the family, peers, school, community and public policy. The sleep of children and adolescents is nested among, and impacted by, these different ecological systems. Thus, suggestions and strategies on how to improve sleep in children and adolescents are provided across four levels: families, schools, clinicians and public policy makers.

Families

There are many ways that families can support better sleep. Limiting technology use, especially in the hour before bed, and removing access to technology overnight, helps to limit exposure to blue light, and allows the opportunity for sleep that is not broken by incoming calls and/or messages [68, 69]. Reducing evening light and limiting or eliminating caffeine can help to ensure that children and adolescents are not being unnecessarily alerted by these exogenous alerting factors [69]. Exercising during the daytime can help children and adolescents to get to sleep faster and have more consolidated and refreshing sleep [70], as can maintaining a comfortable sleeping environment that is dark, cool, and quiet [69].

While adolescents can implement some of these behavioral changes, family involvement to implement, support, and model positive sleep habits to children and adolescents is beneficial [71, 72]. Across the pediatric age range, families have an important role in supporting or harming sleep health. For example, setting limits around bedtime is associated with better sleep, better daytime functioning, and less depression and suicidal ideation [69, 72, 73]. Despite the numerous benefits to regulated bedtimes, research indicates that, even though many parents set limits around the bedtimes of their young children, they relinquish limit-setting at a very early age [72, 74]. A study of North American children and adolescents found that less than 1 in 5 children have a parent-set bed time at age 10 years, while less than 1 in 20 had a parent-set bedtime at age 13 [74]. Of note, however, this developmental shift did not reflect less parental involvement in regulating sleep patterns, overall. Rather, the focus of parental involvement shifted, with the reduction in parent-set bedtimes associated with a concurrent increase in the proportion of parents waking their children up for school in the morning [74].

While a small proportion of children and adolescents have a parent-set bedtime on school nights, this is largely not maintained across weekends. Thus, in older children and adolescents, even less regulation of bedtimes on weekends, coupled with sleep debt accrued across the school week and a delayed body clock, result in a pattern of even later bedtimes on weekends and wake times. Regular bedtimes and waketimes across school nights and weekends are important for good sleep and for entrainment of circadian rhythms, or the body clock [75]. Children and adolescents who obtain sufficient sleep across the school week do not show the same pattern of "sleeping in" on weekends [76]. This catch up sleep extends weekend wake time until later in the morning, or, for some, even into the afternoon. This pattern makes it very difficult for children and adolescents to then fit back into a healthy sleep pattern for school, as their body clocks are shifted later with this weekend catch-up sleep [75]. Maintaining a regular sleep pattern that allows for sufficient sleep across the week avoids these problems.

In addition to sleep-focused behaviors, general environmental factors impact sleep among families. As sleep requires the individual to disengage vigilance to the outside environment, it is important for children and adolescents to feel secure and safe at bedtime. Families can support sleep by maintaining a warm, supportive, and predictable family environment [71, 77]. One study of adolescents found that adolescents who self-reported their families as being more disorganized had worse sleep hygiene, took longer to fall asleep, obtained less sleep and were more sleepy during the day [71]. Conversely, among younger children, increased parental warmth was associated with more sleep [77]. These findings highlight the invaluable role of families in providing a home environment and family culture that supports good sleep.

Schools

Where schools are able to determine their start time, ensuring that the school day does not start before 8:30 a.m. is likely to have wide-ranging benefits to students in terms of enabling them to obtain more sleep, maintain alertness during the day, perform better in the classroom, have fewer motor vehicle accidents, and have improved mood and less psychopathology [40, 77, 78]. A cross-cultural comparison between US adolescents, whose school days began at approximately 7:45 a.m., and Australian adolescents, whose school days started around 8:30 a.m., found that Australian adolescents obtained an average of 47 min more sleep per night sleep [79]. While several factors predicted this cross-cultural difference, the largest predictor was school start time. The American Medical Association and the American Academy of Pediatrics both recommend that schools start no earlier than 8:30 a.m. [80]. The RAND Corporation estimated that if all US schools were to delay school start time until 8:30 a.m., this would add $US83 billion to the economy over the next decade due to higher high school graduation rates and thus better jobs, fewer costs associated with sleep-related car crashes, reduced obesity, and improved mental health [78].

While school start times receive most public attention, any school-related activities that require children and adolescents to wake earlier or stay up later can impinge on their ability to obtain sufficient sleep. Resultingly, school sport and school activities should not be scheduled before 8:30 a.m. Similarly, high academic workloads with large homework volumes and/or attending night school, and high levels of academic pressure can push bedtimes later and negatively impact sleep and mental health [77, 81]. Thus, schools are urged to consider their policies regarding homework, and to question the assumption that homework *quantity* is important for academic achievement. Indeed, recent PISA (Programme for International Student Assessment) results show that, while many of the countries ranked highest in academic performance are countries in which students start school early and have high academic workloads, such as Singapore, Japan, and Taiwan, countries like Finland, who consistently rank among the top academically performing countries, provide evidence that early start times and high volumes of homework are not necessary for academic success. Finnish schools do not start early, their students do not typically attend night school, nor do they have large homework volumes, yet their students rank among the highest in the world when it comes to academic achievement. Most tellingly, students from high achieving countries with early start times and high academic workloads reported among the highest levels of school-work related anxiety, while Finnish students reported very low levels of school-work related anxiety [81]. Thus, high volumes of homework and night school are not necessary for high academic achievement and may come at a high cost to the children in those systems.

Clinicians

Clinicians can work with children and families to improve both sleep and mental health. It is imperative that basic screening for sleep and mental health problems is routinely implemented. If children and adolescents are identified as having a sleep problem, the type of treatment used to improve sleep will depend largely on the sleep issue. Pharmacological interventions are not recommended as a first treatment approach [82], and are thus not reviewed here. Clinically, it is important to first assess the sleep issue. As well as a clinical interview, one means by which to do this is through a sleep diary, kept for at least 1 week [82]. Objective sleep measures, such as actigraphy, are desirable, but not always feasible [82, 83]. Review of both the clinical interview and sleep diary allow definition of the sleep issue (e.g., whether issues falling asleep are due to insomnia or a circadian phase delay) and the goals the client wishes to achieve, both of which inform treatment. Regardless of the specific treatment chosen, psychoeducation for the parent and child regarding sleep is valuable [82, 84]. Psychoeducation typically includes information about developmentally appropriate sleep duration recommendations, sleep hygiene, sleep pressure, sleep architecture, circadian rhythms, and the effects of light and darkness on the circadian rhythm.

Children benefit from sleep hygiene techniques, such as a consistent bedtime routine, and a safe, comfortable sleep environment [83]. Children who have difficulty initiating or maintaining sleep, and negative daytime consequences, in the absence of medical conditions, may have insomnia [82]. As well as good sleep hygiene, insomnia interventions may involve further behavioral treatment, such as cognitive behavior therapy. Both night awakenings and sleep efficiency are improved through behavioral interventions [84]. Adjusting a child's bedtime to a later time, to facilitate faster sleep initiation, may also be used. Once the child has associated bedtime with falling asleep, the bedtime is then moved earlier [82]. A list of resources for clinicians treating childhood insomnia can be found here [82] and here [85].

Concerning adolescents, there are two main psychological therapies which are implemented, again depending on the sleep disorder [82]. The first, bright light therapy, focuses on shifting the body clock of adolescents who have a delayed circadian rhythm. The adolescent is exposed to bright light for 30 min each morning, starting at the adolescents desired waking time, i.e., the time at which they would naturally wake up if they didn't have to go to school. The timing of this light exposure is then shifted earlier by 30 min each day until the adolescent is waking at their desired time. Appropriately timed morning bright light is prescribed in the morning, and dim light is used in the evening, to shift the circadian rhythm. As the light advances each day, so too does the circadian phase [86, 87].

The other treatment widely used among adolescents is cognitive behavior therapy for insomnia, which is used to treat adolescents who experience insomnia. Cognitive behavior therapy for insomnia is a multimodal treatment, which aims to reduce difficulties initialing or maintaining sleep, and the clinical distress associated with such sleep difficulties [86]. It incorporates sleep hygiene, relaxation, stimulus control, sleep restriction and cognitive therapy, which all aim to reduced arousal associated with bedtime and sleep [86]. These treatments have effectively improved sleep latency, sleep duration and awakenings after sleep onset in adolescents. Both depression and anxiety symptoms may also be diminished through this therapy [86].

Many other low-intensity treatment options exist, to improve the sleep of adolescents, especially those who do not experience disordered sleep. These include brief mindfulness and relaxation, and prebed phone restriction [82, 88–90].

Policy makers

While delaying school start times until at least 8:30 a.m. and implementation of lighter extra-curricular homework loads are both discussed in greater detail in relation to how schools can improve sleep in children and adolescents, not all schools can act individually. Thus, public policy regarding these guidelines can have even wider benefit in supporting mental health of children and adolescents.

Health promotion and education regarding child and adolescent sleep may also help to improve sleep and mental health [42]. Health promotion could include information about how much sleep children and adolescents need, indicators of good or problematic sleep, sleep hygiene, and tips on how to improve sleep and where to seek resources. While the present focus has been on how sleep impacts mental health, sleep also affects cognitive performance, risk-taking, drug use, road safety, and delinquency, and so the potential benefits of sleep promotion are widespread.

Just as children sleep better in a safe home environment they also sleep better when they feel safe in their communities [91]. Children and adolescents who are exposed to community violence frequently experience sleep disturbances, nightmares and reduced mental health [92]. Even if children and adolescents are not victims of community violence, the perception of safety in their neighborhoods can impact sleep [91]. One study of 252 adolescents from a wide range of socioeconomic backgrounds in the Southeastern United States asked adolescents about their concern regarding community violence and measured their sleep using actigraphy and self-report. Adolescents who were more concerned about community violence had lower sleep efficiency, woke more during the night, and reported more sleep-wake problems and daytime sleepiness. This effect was stronger among adolescent girls [91]. Thus, policy targeting community safety, especially in areas containing a large proportion of families, is beneficial.

Lastly, paid employment is a factor that can negatively impact the sleep of adolescents [74, 93]. High school students who work more than 20h per week have later bedtimes and less sleep across the week. They also report more daytime sleepiness, are more likely to arrive late for school, have trouble staying awake at school, and are more likely to use caffeine, tobacco and alcohol [94]. These deficits of sleep and daytime functioning were even more pronounced among adolescents who coupled more than 20h of paid work with more than 20h of extra-curricular activity [94]. Australian research indicates that, while most adolescents are aware of their rights at work in terms of declining shifts, they often report feeling pressured to accept shifts, or believe that declining shifts will result in less work being offered to them in the future [95]. Thus, there is scope for policy regarding paid employment for school students. For example, by placing limits around the finishing time of shifts offered to high school workers during school term time. This would mitigate the problem of adolescents finishing shifts so late that they are unable to obtain sufficient sleep before having to rise for school the next day to help support adolescent workloads, sleep and mental health.

CONCLUSION
Summary

Overall, these findings suggest that while estimates of sleep need for optimal mood functioning are between 9 and 11h per night for children and 8–10h per night for adolescents, many young people chronically obtain sleep that is below this amount. Studies examining the relationship between sleep duration and mood suggest that sleep loss produces reductions in positive mood states and affect and increases in negative mood states and negative affect. Longitudinal studies support bidirectional relationships between sleep duration and mood and mood disorders, however, the link from sleep to mood/mental health has been reported more consistently than link from mood/mental health to sleep. Studies focusing on sleep quality have found similar results, with multiple aspects of sleep quality found to impact mental health, including how long it takes to fall asleep, waking during the night, and subjective sleep quality all impacting mental health.

Using Bronfenbrenner's ecological systems theory as a guiding explanatory framework, suggestions and recommendations for improving sleep, and thus mental health, were provided. These recommendations targeted families, schools, clinicians and public policy makers, thus highlighting the importance of a multifaceted approach to support sleep and mental health in young people.

Limitations and future research directions

Overall, the present chapter provides an overview of the relationship between sleep duration and sleep quality and how they either support or diminish mental health in children and adolescents. In addition, we provide suggestions for how sleep (and thus mental health) can be supported across multiple ecological levels. It is important to note, however, that this overview is not exhaustive, and other factors, such as sleep regularity and the timing of circadian rhythms (or the body clock) affect mental health both directly or indirectly, through their influence on sleep duration and sleep quality [54, 96].

While present findings support the importance of sleep for optimal mental health in children and adolescents, there remain gaps and limitations in the literature. First, multimodal assessment of sleep using both subjective and objective measures is needed to determine the degree to which subjectively reported insufficient or poor-quality sleep is also mirrored by objective data. While this is an important point across all developmental stages, it is particularly important among younger children, for whom research typically relies upon parent reports of sleep. Second, while there is a sizeable body of work examining sleep quality and mental health outcomes, sleep quality is not unitary and the conceptual and operational definitions of sleep quality are often ill-defined. Research examining clearly defined and operationalized aspects of sleep quality, such as sleep onset latency, wake after sleep onset, number of night time awakenings, and subjective and objective sleep quality, is needed to determine how different facets of sleep quality affect mental health. Third, the literature consists largely of cross-sectional studies which cannot address the significant issue of causation or causal direction. This is important given the relationship between sleep and mental health is likely bidirectional. Future research would profit from more experimental studies to determine causation and directionality.

Several important areas remain underinvestigated. For example, studies to evaluate the efficacy of health promotion strategies aimed at sleep are needed to determine whether population-level interventions are effective. On a smaller scale, sleep intervention studies are urged to include mood and mental health outcomes to see how improving sleep on an individual level improves mental health. Finally, mood outcome measures need to include positive mood outcomes, such as happiness, as well as negative mood outcomes, such as depressed mood and anxiety. Emerging research suggests that positive emotion may be more sensitive to sleep loss and poor quality sleep than negative emotion [97]. While mood states relevant to mood disorders, such as depression and anxiety are important to measure, reduction in positive mood, or anhedonia, is also a clinically relevant.

Concluding remarks

Sleep plays a crucial role in maintaining optimal mental health across the lifespan. Childhood and adolescence are critical developmental periods when the trajectories of many mental health conditions are begun and thus provides an optimal period for early intervention regarding sleep. Simple interventions to improve and safeguard sleep are thus important to benefit youth mental health and reduce the likelihood or severity of many mental health conditions.

REFERENCES

[1] Vigo D, Thornicroft G, Atun R. Estimating the true global burden of mental illness. Lancet Psychiatry 2016;3(2):171–8.

[2] Giedd JN, Keshavan M, Paus T. Why do many psychiatric disorders emerge during adolescence? Nat Rev Neurosci 2008;9(12):947–57.

[3] Beesdo K, Knappe S, Pine DS. Anxiety and anxiety disorders in children and adolescents: developmental issues and implications for DSM-V. Psychiatr Clin N Am 2009;32(3):483–524.

[4] Son SE, Kirchner JT. Depression in children and adolescents. Am Fam Physician 2000;62(10):2297–308. [311–2].

[5] Alfano CA, Gamble AL. The role of sleep in childhood psychiatric disorders. Child Youth Care Forum 2009;38(6):327–40.

[6] Short MA, Louca M. Sleep deprivation leads to mood deficits in healthy adolescents. Sleep Med 2015;16(8):987–93.

[7] Lovato N, Gradisar M. A meta-analysis and model of the relationship between sleep and depression in adolescents: recommendations for future research and clinical practice. Sleep Med Rev 2014;18(6):521–9.

[8] Baum KT, Desai A, Field J, Miller LE, Rausch J, Beebe DW. Sleep restriction worsens mood and emotion regulation in adolescents. J Child Psychol Psychiatry 2014;55(2):180–90.

[9] Gregory AM, O'Connor TG. Sleep problems in childhood: a longitudinal study of developmental change and association with behavioral problems. J Am Acad Child Adolesc Psychiatry 2002;41(8):964–71.

[10] Meijer AM, Reitz E, Dekovic M, van den Wittenboer GL, Stoel RD. Longitudinal relations between sleep quality, time in bed and adolescent problem behaviour. J Child Psychol Psychiatry 2010;51(11):1278–86.

[11] Leahy E, Gradisar M. Dismantling the bidirectional relationship between paediatric sleep and anxiety. Clin Psychol 2012;16(1):44–56.

[12] Short MA, Gradisar M, Lack LC, Wright HR, Chatburn A. Estimating adolescent sleep patterns: parent reports versus adolescent self-report surveys, sleep diaries, and actigraphy. Nat Sci Sleep 2013;5:23.

[13] Short MA, Gradisar M, Lack LC, Wright H, Carskadon MA. The discrepancy between actigraphic and sleep diary measures of sleep in adolescents. Sleep Med 2012;13(4):378–84.

[14] Hirshkowitz M, Whiton K, Albert S, Alessi C, Bruni O, DonCarlos L, et al. National Sleep Foundation's sleep time duration recommendations: methodology and results summary. Sleep Health 2015;1(1):40–3.

[15] Short MA, Weber N, Reynolds C, Coussens S, Carskadon MA. Estimating adolescent sleep need using dose-response modelling. Sleep 2018; https://doi.org/10.1093/sleep/zsy011.

[16] Fuligni AJ, Bai S, Krull JL, Gonzales NA. Individual differences in optimum sleep for daily mood during adolescence. J Clin Child Adolesc Psychol 2017;53:1–11.

[17] Ojio Y, Nishida A, Shimodera S, Togo F, Sasaki T. Sleep duration associated with the lowest risk of depression/anxiety in adolescents. Sleep 2016;39(8):1555–62.

[18] Lo JC, Ong JL, Leong RL, Gooley JJ, Chee MW. Cognitive performance, sleepiness, and mood in partially sleep deprived adolescents: the Need for Sleep study. Sleep 2016;39(3):687–98.

[19] Galland BC, Short MA, Terrill P, Rigney G, Haszard JJ, Coussens S, et al. Establishing normal values for pediatric nighttime sleep measured by actigraphy: a systematic review and meta-analysis. Sleep 2018; https://doi.org/10.1093/sleep/zsy017.

[20] Gregory AM, Sadeh A. Sleep, emotional and behavioral difficulties in children and adolescents. Sleep Med Rev 2012;16(2):129–36.

[21] Sadeh A. Consequences of sleep loss or sleep disruption in children. Sleep Med Clin 2007;2(3):513–20.

[22] Astill RG, Van der Heijden KB, Van Ijzendoorn MH, Van Someren EJ. Sleep, cognition, and behavioral problems in school-age children: a century of research meta-analyzed. Psychol Bull 2012;138(6):1109–38.

[23] Baglioni C, Spiegelhalder K, Lombardo C, Riemann D. Sleep and emotions: a focus on insomnia. Sleep Med Rev 2010;14(4):227–38.

[24] Nixon GM, Thompson JM, Han DY, Becroft DM, Clark PM, Robinson E, et al. Short sleep duration in middle childhood: risk factors and consequences. Sleep 2008;31(1):71–8.

[25] Vriend JL, Davidson FD, Corkum PV, Rusak B, McLaughlin EN, Chambers CT. Sleep quantity and quality in relation to daytime functioning in children. Child Health Care 2012;41(3):204–22.

[26] Smaldone A, Honig JC, Byrne MW. Sleepless in America: inadequate sleep and relationships to health and well-being of our nation's children. Pediatrics 2007;119(Suppl. 1):S29–37.

[27] Brown WJ, Wilkerson AK, Boyd SJ, Dewey D, Mesa F, Bunnell BE. A review of sleep disturbance in children and adolescents with anxiety. J Sleep Res 2017;27:e12635.

[28] Gregory AM, Sadeh A. Annual research review: sleep problems in childhood psychiatric disorders—a review of the latest science. J Child Psychol Psychiatry 2016;57(3):296–317.

[29] Lo JC, Lee SM, Teo LM, Lim J, Gooley JJ, Chee MW. Neurobehavioral impact of successive cycles of sleep restriction with and without naps in adolescents. Sleep 2016;40(2):1–13.

[30] Reddy R, Palmer CA, Jackson C, Farris SG, Alfano CA. Impact of sleep restriction versus idealized sleep on emotional experience, reactivity and regulation in healthy adolescents. J Sleep Res 2017;26(4):516–25.

[31] Dewald-Kaufmann J, Oort F, Meijer A. The effects of sleep extension and sleep hygiene advice on sleep and depressive symptoms in adolescents: a randomized controlled trial. J Child Psychol Psychiatry 2014;55(3):273–83.

[32] Gruber R, Cassoff J, Frenette S, Wiebe S, Carrier J. Impact of sleep extension and restriction on children's emotional lability and impulsivity. Pediatrics 2012;130(5):e1155–61.

[33] Fallone G, Acebo C, Seifer R, Carskadon MA. Experimental restriction of sleep opportunity in children: effects on teacher ratings. Sleep 2005;28(12):1561–7.

[34] Chaput JP, Gray CE, Poitras VJ, Carson V, Gruber R, Olds T, et al. Systematic review of the relationships between sleep duration and health indicators in school-aged children and youth. Appl Physiol Nutr Metab 2016;41(6 Suppl 3):S266–82.

[35] Vriend J, Davidson F, Corkum P, Rusak B, Chambers C, McLaughlin E. Manipulating sleep duration alters emotional functioning and cognitive performance in children. J Pediatr Psychol 2013; https://doi.org/10.1093/jpepsy/jst033.

[36] McMakin DL, Dahl RE, Buysse DJ, Cousins JC, Forbes EE, Silk JS, et al. The impact of experimental sleep restriction on affective functioning in social and nonsocial contexts among adolescents. J Child Psychol Psychiatry 2016;57(9):1027–37.

[37] McLaughlin KA, Hatzenbuehler ML, Mennin DS, Nolen-Hoeksema S. Emotion dysregulation and adolescent psychopathology: a Prospective Study. Behav Res Ther 2011;49(9):544–54.

[38] Thompson RA. Emotion regulation: a theme in search of definition. Monogr Soc Res Child Dev 1994;59(2–3):25–52.

[39] Talbot LS, McGlinchey EL, Kaplan KA, Dahl RE, Harvey AG. Sleep deprivation in adolescents and adults: changes in affect. Emotion 2010;10(6):831–41.

[40] Owens JA, Belon K, Moss P. Impact of delaying school start time on adolescent sleep, mood, and behavior. Arch Pediatr Adolesc Med 2010;164(7):608–14.

[41] Hasler JC. The effect of sleep extension on academic performance, cognitive functioning and psychological distress in adolescents. The University of Arizona; 2008.

[42] Bonnar D, Gradisar M, Moseley L, Coughlin A-M, Cain N, Short MA. Evaluation of novel school-based interventions for adolescent sleep problems: does parental involvement and bright light improve outcomes? Sleep Health 2015;1(1):66–74.

[43] Fredriksen K, Rhodes J, Reddy R, Way N. Sleepless in Chicago: tracking the effects of adolescent sleep loss during the middle school years. Child Dev 2004;75(1):84–95.

[44] Lovato N, Short MA, Micic G, Hiller RM, Gradisar M. An investigation of the longitudinal relationship between sleep and depressed mood in developing teens. Nat Sci Sleep 2017;9:3–10.

[45] Mullin BC, Pyle L, Haraden D, Riederer J, Brim N, Kaplan D, et al. A preliminary multimethod comparison of sleep among adolescents with and without generalized anxiety disorder. J Clin Child Adolesc Psychol 2017;46(2):198–210.

[46] Cousins JC, Whalen DJ, Dahl RE, Forbes EE, Olino TM, Ryan ND, et al. The bidirectional association between daytime affect and nighttime sleep in youth with anxiety and depression. J Pediatr Psychol 2011;36(9):969–79.

[47] Carskadon MA, Harvey K, Duke P, Anders TF, Litt IF, Dement WC. Pubertal changes in daytime sleepiness. Sleep 1980;2(4):453–60.

[48] Ivanenko A, Crabtree VM, Obrien LM, Gozal D. Sleep complaints and psychiatric symptoms in children evaluated at a pediatric mental health clinic. J Clin Sleep Med 2006;2(1):42–8.

[49] Short MA, Gradisar M, Lack LC, Wright HR, Dohnt H. The sleep patterns and well-being of Australian adolescents. J Adolesc 2013;36(1):103–10.

[50] Kaneita Y, Ohida T, Osaki Y, Tanihata T, Minowa M, Suzuki K, et al. Association between mental health status and sleep status among adolescents in Japan: a nationwide cross-sectional survey. J Clin Psychiatry 2007;68(9):1426–35.

[51] Kaneita Y, Yokoyama E, Harano S, Tamaki T, Suzuki H, Munezawa T, et al. Associations between sleep disturbance and mental health status: a longitudinal study of Japanese junior high school students. Sleep Med 2009;10(7):780–6.

[52] Ivanenko A, Johnson K. Sleep disturbances in children with psychiatric disorders. Semin Pediatr Neurol 2008;15(2):70–8.

[53] Kahn M, Fridenson S, Lerer R, Bar-Haim Y, Sadeh A. Effects of one night of induced night-wakings versus sleep restriction on sustained attention and mood: a pilot study. Sleep Med 2014;15(7):825–32.

[54] Short MA, Gradisar M, Lack LC, Wright HR. The impact of sleep on adolescent depressed mood, alertness and academic performance. J Adolesc 2013;36(6):1025–33.

[55] Warner S, Murray G, Meyer D. Holiday and school-term sleep patterns of Australian adolescents. J Adolesc 2008;31(5):595–608.

[56] McMakin DL, Alfano CA. Sleep and anxiety in late childhood and early adolescence. Curr Opin Psychiatry 2015;28(6):483.

[57] Ramtekkar U, Ivanenko A, editors. Sleep in children with psychiatric disorders. Seminars in pediatric neurology; Elsevier; 2015.

[58] Hatzinger M, Brand S, Perren S, Von Wyl A, Stadelmann S, von Klitzing K, et al. In pre-school children, sleep objectively assessed via sleep-EEGs remains stable over 12 months and is related to psychological functioning, but not to cortisol secretion. J Psychiatr Res 2013;47(11):1809–14.

[59] Sadeh A, Tikotzky L, Kahn M. Sleep in infancy and childhood: implications for emotional and behavioral difficulties in adolescence and beyond. Curr Opin Psychiatry 2014;27(6):453–9.

[60] Meltzer LJ, Mindell JA. Sleep and sleep disorders in children and adolescents. Psychiatr Clin 2006;29(4):1059–76.

[61] Roberts RE, Roberts CR, Chen IG. Impact of insomnia on future functioning of adolescents. J Psychosom Res 2002;53(1):561–9.

[62] Forbes EE, Bertocci MA, Gregory AM, Ryan ND, Axelson DA, Birmaher B, et al. Objective sleep in pediatric anxiety disorders and major depressive disorder. J Am Acad Child Adolesc Psychiatry 2008;47(2):148–55.

[63] Chorney DB, Detweiler MF, Morris TL, Kuhn BR. The interplay of sleep disturbance, anxiety, and depression in children. J Pediatr Psychol 2007;33(4):339–48.

[64] Noone DM, Willis TA, Cox J, Harkness F, Ogilvie J, Forbes E, et al. Catastrophizing and poor sleep quality in early adolescent females. Behav Sleep Med 2014;12(1):41–52.

[65] Hiller RM, Lovato N, Gradisar M, Oliver M, Slater A. Trying to fall asleep while catastrophising: what sleep-disordered adolescents think and feel. Sleep Med 2014;15(1):96–103.

[66] Alfano CA, Pina AA, Zerr AA, Villalta IK. Pre-sleep arousal and sleep problems of anxiety-disordered youth. Child Psychiatry Hum Dev 2010;41(2):156.

[67] Bronfenbrenner U. Ecological systems theory. Six theories of child development: revised formulations and current issues. London: Jessica Kingsley Publishers; 1992:187–249.

[68] Gradisar M, Wolfson AR, Harvey AG, Hale L, Rosenberg R, Czeisler CA. The sleep and technology use of Americans: findings from the National Sleep Foundation's 2011 sleep in America poll. J Clin Sleep Med 2013;9(12):1291–9.

[69] Bartel KA, Gradisar M, Williamson P. Protective and risk factors for adolescent sleep: a meta-analytic review. Sleep Med Rev 2015;21:72–85.

[70] Brand S, Gerber M, Beck J, Hatzinger M, Pühse U, Holsboer-Trachsler E. High exercise levels are related to favorable sleep patterns and psychological functioning in adolescents: a comparison of athletes and controls. J Adolesc Health 2010;46(2):133–41.

[71] Billows M, Gradisar M, Dohnt H, Johnston A, McCappin S, Hudson J. Family disorganization, sleep hygiene, and adolescent sleep disturbance. J Clin Child Adolesc Psychol 2009;38(5):745–52.

[72] Short MA, Gradisar M, Wright H, Lack LC, Dohnt H, Carskadon MA. Time for bed: parent-set bedtimes associated with improved sleep and daytime functioning in adolescents. Sleep 2011;34(6):797–800.

[73] Gangwisch JE, Babiss LA, Malaspina D, Turner JB, Zammit GK, Posner K. Earlier parental set bedtimes as a protective factor against depression and suicidal ideation. Sleep 2010;33(1):97–106.

[74] Carskadon MA. Patterns of sleep and sleepiness in adolescents. Pediatrician 1990;17(1):5–12.

[75] Crowley SJ, Carskadon MA. Modifications to weekend recovery sleep delay circadian phase in older adolescents. Chronobiol Int 2010;27(7):1469–92.

[76] Short MA, Arora T, Gradisar M, Taheri S, Carskadon MA. How many sleep diary entries are needed to reliably estimate adolescent sleep? Sleep 2017;40(3). zsx006.

[77] Adam EK, Snell EK, Pendry P. Sleep timing and quantity in ecological and family context: a nationally representative time-diary study. J Fam Psychol 2007;21(1):4–19.

[78] Hafner M, Stepanek M, Troxel WM. The economic implications of later school start times in the United States. Sleep Health 2017;3(6):451–7.

[79] Short MA, Gradisar M, Lack LC, Wright HR, Dewald JF, Wolfson AR, et al. A cross-cultural comparison of sleep duration between US and Australian adolescents: the effect of school start time, parent-set bedtimes, and extracurricular load. Health Educ Behav 2013;40(3):323–30.

[80] Group ASW. School start times for adolescents. Pediatrics 2014;134(3):642–9.

[81] OECD. PISA 2015 Results (Volume I). OECD Publishing; 2015.

[82] Vriend J, Corkum P. Clinical management of behavioral insomnia of childhood. Psychol Res Behav Manag 2011;4:69.

[83] Hill C. Practitioner review: effective treatment of behavioural insomnia in children. J Child Psychol Psychiatry 2011;52(7):731–40.

[84] Meltzer LJ, Mindell JA. Systematic review and meta-analysis of behavioral interventions for pediatric insomnia. J Pediatr Psychol 2014;39(8):932–48.

[85] Hiller R, Gradisar M. In: Cooper P, Waite P, editors. Helping Your Child with Sleep Problems: a self-help guide for parents. Robinson; 2018.

[86] Blake MJ, Sheeber LB, Youssef GJ, Raniti MB, Allen NB. Systematic review and meta-analysis of adolescent cognitive–behavioral sleep interventions. Clin Child Fam Psychol Rev 2017;20(3):227–49.

[87] Richardson C, Cain N, Bartel K, Micic G, Maddock B, Gradisar M. A randomised controlled trial of bright light therapy and morning activity for adolescents and young adults with delayed sleep-wake phase disorder. Sleep Med 2018;45:114–23.

[88] Bartel K, Richardson C, Gradisar M. Rapid review: sleep and mental wellbeing: exploring the links. Melbourne; 2018. p. 1–51 [in press].

[89] Bartel K, Scheeren R, Gradisar M. Altering adolescents' pre-bedtime phone use to achieve better sleep health. Health Commun 2018;9:1–7.

[90] Bartel K, Huang C, Maddock B, Williamson P, Gradisar M. Brief school-based interventions to assist adolescents' sleep-onset latency: comparing mindfulness and constructive worry versus controls. J Sleep Res 2018;27:e12668.

[91] Bagley EJ, Tu KM, Buckhalt JA, El-Sheikh M. Community violence concerns and adolescent sleep. Sleep Health 2016;2(1):57–62.

[92] Duncan DF. Growing up under the gun: children and adolescents coping with violent neighborhoods. J Prim Prev 1996;16(4):343–56.

[93] Laberge L, Ledoux E, Auclair J, Thuilier C, Gaudreault M, Gaudreault M, et al. Risk factors for work-related fatigue in students with school-year employment. J Adolesc Health 2011;48(3):289–94.

[94] Carskadon M. Adolescent sleepiness: increased risk in a high-risk population. J Saf Res 1990;21(4):169.

[95] House of Representatives Standing Committee on Education and Training. Adolescent overload? Canberra: Commonwealth of Australia; 2009.

[96] Pesonen A-K, Räikkönen K, Paavonen EJ, Heinonen K, Komsi N, Lahti J, et al. Sleep duration and regularity are associated with behavioral problems in 8-year-old children. Int J Behav Med 2010;17(4):298–305.

[97] Watling J, Pawlik B, Scott K, Booth S, Short MA. Sleep loss and affective functioning: more than just mood. Behav Sleep Med 2017;15(5):394–409.

Chapter 33

Delayed school start times and adolescent health

Aaron T. Berger[a], Rachel Widome[a], Wendy M. Troxel[b]

[a]Division of Epidemiology and Community Health, University of Minnesota, Minneapolis, MN, United States, [b]RAND Corporation, Santa Monica, CA, United States

Adolescence is a development period characterized by dramatic changes in neurobiological, physical and socio-emotional development as well as changes in sleep-wake patterns. Most adolescents should optimally sleep a minimum of 8–9.5 h per night [1, 2]. However, the majority fall substantially short of that mark as national surveillance shows that only 7% of US adolescents report getting at least 8.5 h of sleep on school nights [3]. Between ages 12 and 18, the probability of getting at least 7 h of sleep per night drops by about half [4]. The magnitude of the short sleep duration epidemic among adolescents is even larger than what we have observed in adults, with US high school-aged youth being more than twice as likely to report getting insufficient sleep compared to adults [2]. Additionally, there are sleep disparities among adolescents, with girls, non-whites, and those from lower socioeconomic status backgrounds, being more likely to report short sleep duration compared to other groups [4]. A particularly worrisome trend is the secular decline in adolescent sleep over the past 20 years, a trend that has been called a "great sleep recession" [4].

A variety of social and environmental factors pose significant obstacles to sufficient sleep for most US adolescents [5–9], including social pressures to stay up late, bright light from screens at and after bedtime, after school homework and employment, and caffeine use, to name a few. Even in the absence of these social and environmental contexts, there are powerful, hardwired biological factors that push teenagers toward later bedtimes and delayed wake-up times [10, 11].

The circadian biological clock and sleep/wake homeostasis are two body systems that regulate sleep [12]. Early in puberty, in most adolescents, there is a neurobiological change to children's circadian clocks that results in a two- to three-hour delay in the release of the sleep-promoting hormone, melatonin. This circadian timing persists through adolescence [13], and means that the time adolescents naturally fall asleep is deferred to a later hour. After this shift it is challenging, even for an adolescent who tries to get to bed at an early hour, to fall asleep prior to 11 p.m. and it follows that they will struggle to wake before 8 a.m. [11]. Additionally, "sleep drive," which accumulates over the waking hours and diminishes wakefulness as the day progresses, builds slower once children reach their teen years [14]. As adolescents get older, bedtime is delayed on both school and non-school days [5]. This adolescent sleep timing shift has been observed both in the US and around the world, adding confirmatory evidence to this delay of sleep onset being a largely biological, and not purely social, normative or cultural phenomenon [5, 11, 15]. Meanwhile, contrary to popular belief, adolescents require just as much sleep as they did when they were a few years younger [11] with 9.25 h of nightly sleep being considered optimal through the teen years [16].

Early school start times are perhaps the most potent and salient environmental constraint on adolescent sleep [1, 17, 18]. Although delayed sleep onset has many biological and social causes, delayed sleep onset does not necessarily result in truncated sleep duration, unless schedules dictate an untenable wake-up time. Unfortunately, this is nearly always the case in the US, where high schools, almost without exception, tend to start very early [18], leading adolescents to need to wake before they have finished sleeping. In order to meet the biological sleep needs of adolescents, over two dozen medical organizations, including the American Academy of Pediatrics and the Centers for Disease Control and Prevention, have recommended that middle and high schools start at 8:00 a.m., 8:30 a.m., or later [19]. Numerous examples in school districts from across the country have shown that high schools can feasibly shift their start times to 8:30 a.m. or later [20]. Both cross-sectional [21] and longitudinal [22–24] evaluations of start time differences show that later start times allows teens to get more sleep, with each one-hour start time delay associated with 30–90 additional minutes of school night sleep duration. Yet <18% of US middle and high schools start at or after 8:30 a.m. and

42% start at 8 a.m. or earlier [18, 25], start times which appear to most severely curtail sleep [26].

In this chapter we aim to summarize the documented effects of delaying school start time on adolescent health. Given the strong evidence linking short sleep duration with numerous physical and mental health risks (covered in depth in the preceding chapters), delaying start times could be an effective population-level strategy to promote sleep and physical and mental health in adolescents. Our review covers those health outcomes that have been included in peer-reviewed literature, in relation to a K-12 school start time contrast (e.g. studies that have compared two or more schools with different start times, or one school before and after a start time change). We review the effects of school start times on sleep duration, academic outcomes and truancy, mental health and risky behavior, and unintentional injury. We conclude by discussing the obstacles communities face when delaying school start times.

DELAYING HIGH SCHOOL START TIME IMPROVES SLEEP

Delaying school start time is a population-level intervention with the potential to affect sleep duration and sleep-related health during adolescence, a critical developmental period. Studies that have evaluated whether school start times are associated with more or better-quality sleep have overwhelmingly concluded that later school start times are associated with significantly longer school-night sleep duration [27, 28]. Later school start times are associated with longer school-night sleep duration in all eight studies eligible for inclusion in a recent Cochrane review, and in 29 of 31 studies included in another recent systematic review [27, 28]. Fig. 33.1 illustrates the distributions of school-night sleep duration for students at eight US high schools, with start times ranging from 7:35 a.m. to 8:55 a.m. [24]. Compared to students in the earliest-starting school, the entire distribution of sleep duration at later-starting schools shifts progressively toward longer sleep duration. Notably, approximately 10% of US high schools start before the earliest-starting high school in this sample [18].

Some have questioned whether a delayed school start time might be counterproductive, and only serve to delay bedtimes such that sleep duration is not lengthened. Contrary to this belief, most of the available evidence has shown that later school start time leads to longer sleep duration because adolescent bedtimes are stable and likely not influenced much by artificially imposed wake-up times.

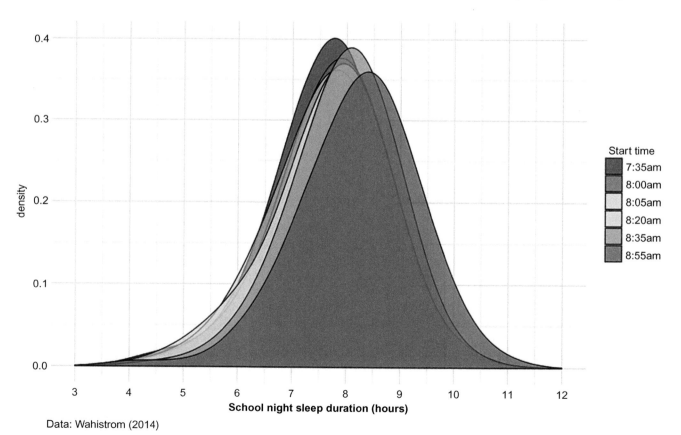

Data: Wahlstrom (2014)

FIG. 33.1 Wahlstrom [24] collected self-reported sleep duration in eight US high schools with a range of start times. At later-starting schools, the entire distribution of school-night sleep duration is progressively shifted to the right, toward greater sleep.

In before-after studies of school start time changes, students typically maintain the same bedtimes after starting later that they had when their school started early [21, 29, 30], although one study showed that bedtimes actually shifted earlier when schools started later [22]. One recent cross-sectional study showed that students had modestly later bedtimes at schools with later start times; however this study did not compare bedtimes before and after start time changes [26]. The stability of adolescent bedtimes across various school start times, early and late, is yet another piece of evidence reinforcing adolescents' circadian clocks are biologically set and rather inflexible.

ACADEMIC ACHIEVEMENT, ATTENTION, AND TRUANCY

Students who do not get enough sleep have more difficulty concentrating in class, and are less likely to graduate high school or college than their better-rested peers [31]. Many sleepy adolescents find themselves unable to stay awake even while they are in class. A 2006 survey by the National Sleep Foundation found that 28% of high school students report falling asleep in school at least once a week [32]. Among high school students in a nationally representative survey, each additional hour of sleep was associated with 16% lower probability of attention problems in school, 15% lower probability of trouble completing homework, and a grade point average (GPA) increase of 0.2 points [31]. Moreover, when those students were followed through early adulthood, each additional hour of sleep was associated with 13% greater probability of graduating high school and a nearly 10% increase in probability of attending college [31].

Because delaying high school start time is recognized as an important strategy for promoting health sleep duration [1, 19], researchers have worked to understand if later start times will improve adolescents' academic achievement. However, there are challenges in interpreting the effect of school start time on academic performance due to the nature of letter grades and standardized tests. Grade point averages may reflect the difficulty of a student's class schedule or the whims of a teacher, while standardized college entrance exams are not taken universally and may reflect only the population who intends to go to college [27]. Additionally, students who struggle in school due to sleepiness might opt for, or be tracked into, less difficult courses.

Studies that have considered the effect of a one-hour school start time delay on standardized test scores have had mixed results [33, 34]. Hinrichs compared ACT test scores of high school students in the Minneapolis and Saint Paul, Minnesota, public school districts before and after a start time delay in Minneapolis schools [33]. Hinrichs found no significant effects of school start time on ACT scores ($\beta=-0.024$, 95% confidence interval [CI] −0.23 to 0.18).

In addition, Hinrichs conducted cross-sectional comparisons of high school student scores on the Kansas Reading Assessment by school start time. As with ACT scores, there was no association between school start time and reading score ($\beta=0.95$, 95% CI −2.67 to 4.58). Meanwhile, Edwards conducted a similar study of the math and reading exam scores of middle school students in neighboring communities in North Carolina, one of which delayed its school start time during the study period [34]. Edwards found that students whose school was delayed by 1h experienced a 1.8% point improvement in math scores (95% CI 1.19–2.37) and 0.98-point improvement in reading scores (95% CI 0.28–1.68) [34].

A few studies have also considered the effect of school start time on students' letter grades. Wahlstrom conducted a pre-post study of student grades before and after start time delays in six school districts [24]. In four districts, the combined pre-post change in all core courses (mathematics, science, social studies, and/or English), and in two districts the core course grades were available for individual analysis. There was significant improvement in grades in three of the four school districts where all courses were analyzed collectively, and no significant difference in one school. In the two districts in which core course grades were analyzed separately, some courses had significant increases and others had significant decreases (although Wahlstrom did not report which specific courses showed letter grade improvements or setbacks) [24].

A considerable problem in causal inference for studies of school start times is that, due to the logistical complexity and multiple competing interests in setting school start times, researchers may never be able to randomize the start times that students receive. One noteworthy study, among first-semester freshmen at the US Air Force Academy (USAFA), was able to overcome that limitation [35]. Because the service academy randomly assigned students' academic schedules, including start time, course, and instructor, this study effectively recreates a randomized trial. The study found that when students are randomly assigned to take a class before 8:00 a.m., their academic performance in their first hour class was significantly worse than those whose first hour class was assigned to 8:00 a.m. or later. In addition, having an early start to the academic day reduced performance throughout the entire day [35]. This study provides the highest-quality evidence that later school start times may play a causal role in improving academic outcomes. While the students at the USAFA are slightly older than high school students, most first-semester freshmen students are still biologically adolescents [35]. The average age of freshman cadets was not reported, but USAFA cadets are required to be between 17 and 23 years old in the year they enter [36], and 33% of all USAFA students are ages 18–19 years old, suggesting that a substantial majority of freshmen enter at age 18 [37].

Later school start times may improve learning and academic achievement by reducing sleepiness during the school day. Indeed, in a recent review of the school start time literature, 10 of the 12 studies that measured sleepiness found significantly less sleepiness in schools with later start times [27]. For example, Danner and Phillips used the Epworth Sleepiness Scale, a validated measure of daytime sleepiness [38], to measure changes in daytime sleepiness before and after a start time change in Kentucky [30]. Mean sleepiness declined from 8.9 to 8.2 ($P < 0.001$). Other longitudinal studies have also found lower levels of sleepiness following delays in start times [22, 23]. Reduced sleepiness may also improve academic achievement by facilitating improved concentration and attention to school and after-school homework. An Israeli study used two different assessments to compare sustained attention in middle school students whose school start time was experimentally delayed by 1 h, compared to a group with consistently early start times [39]. The delayed start time group has a significantly better attention level, and made significantly fewer errors, than the early start time group.

Oversleeping students may be absent for all or part of a school day due to early start times. Edwards estimated that students at schools starting 1 h later average 1.3 fewer absences per year, with a median of five absences per year [34]. Wahlstrom identified the attendance of students who change schools more frequently as experiencing a particularly large benefit from later school start time [21]. While school attendance for continuously enrolled students was consistently high before and after a start time change, attendance for students who had changed schools during high school improved significantly following a start time delay, from 72% to 76% for 9th grade students, and from 73.7% to 77.5% for 10th and 11th grade students [21]. This suggests that the largest gains in school attendance may be observed for the students at highest risk of absenteeism. Combined with reduced daytime sleepiness and better attention to school and homework, improved attendance may account for the increased academic achievement observed in later starting schools.

MENTAL HEALTH AND RISKY BEHAVIOR

Adolescence is known to be a highly vulnerable period for the onset of mental health issues and substance use disorders. For instance, 50% of all lifetime cases of depression begin by age 14 [40]. Furthermore, risky behavior in adolescence, including alcohol, tobacco and other substance use and risky sexual behavior, are a cause of substantial morbidity, mortality, and social problems for youth and can lead to chronic lifelong health issues [41]. Thus, adolescence may be a critical period for preventing the lasting consequences of mental health and substance use-related morbidity and improving sleep health may be a key tactic in the armament of psychological and behavioral health promotion strategies.

Sleep loss and sleep disorders are commonly associated with reduced mental well-being and risky behaviors. Adults with psychiatric mood disorders such as major depression and post-traumatic stress disorder commonly experience sleep disturbances [42], inadequate or poor-quality sleep also co-occur with symptoms of depression [43], hopelessness and suicidal thoughts and attempts [43–45], irritability and impaired emotional regulation [46]. Sleep problems in childhood and adolescence may predict future mental health problems [47, 48] and even suicide [48].

Experimental studies of sleep restriction provide further causal support for a role of sleep disturbance in contributing to mental health problems [46, 49, 50]. Experimentally sleep restricted adolescents exhibit higher levels of anxiety, anger, fatigue and confusion compared to periods of sleep extension [46]. Additionally, experimental sleep deprivation has been associated with reduced positive affect [49, 50], and increased negative affect [49] in adolescents, which may increase adolescent vulnerability to depression. The role played by REM sleep in emotional processing [42] may explain these changes in affective response. When adolescents are woken prematurely they are deprived of rapid eye movement (REM) sleep (the stage of sleep known to be associated with emotional processing). This deficit hampers recovery from emotional conflict, and reduces emotional control by increasing reactivity to negative emotional stimuli [42].

Insufficient sleep may contribute generally to increased adolescent risk behaviors by diminishing teen's executive cognitive functioning and emotional regulation [51]. Early school start times cause many adolescents to live in a state of constant "circadian misalignment" due to the discrepancy between the school-imposed schedule and their own internal clock. Tired from the week, adolescents will sleep dramatically more hours on weekends to make up for school day sleep loss. Differences in weekend and weekday sleep hours have been associated with regulation of reward processing [52, 53] which can manifest as increased sensation-seeking and diminished regulatory control [51, 54]. Sleep problems, including insomnia, short sleep duration, and inconsistencies in weekend versus weekday sleep, are associated cross-sectionally and longitudinally with increased use of alcohol, marijuana, and other drugs in adolescent samples [43, 45].

Although much literature has been published on the associations between sleep and mental health in adolescence, comparatively little work has been done to identify the psychological effects of delayed school start time [55]. We recently conducted a recent systematic review of literature on school start time and psychological health, including substance use. We identified eight eligible studies conducted in the past 20 years. The most commonly studied outcome, symptoms of depression and anxiety, was assessed in four of the studies [21–23, 56]. Positive and negative affect or attitudes were included in two more studies [57, 58], and two other studies included nonspecific measures of mental health [59, 60]. In both cross-sectional [21, 56] and longitudinal studies [22, 23], later school start time is consistently associated with fewer symptoms of depression and anxiety. One additional longitudinal study, with a control group, found that school start time was associated with improved general mental health and reduced psychologically-relevant behavior problems, such as emotional problems and hyperactivity/inattention [60]. Although both studies assessing the effect of school start time on positive and negative affect [57, 58] and one of the two studies of general mental health [59] did not find significant differences between earlier and later-starting schools, these studies were of short, 15-min differences in start times, a school with a late start on only 1 day of the week, and a highly unusual situation where a school was hosted during afternoon hours in another school building due to a fire. The highest quality studies uniformly found that later school start times are associated with better mental health in adolescents [22, 23, 60]. Notably lacking from the literature were studies of the impact of later start times on substance use or other risk-taking behaviors.

UNINTENTIONAL INJURY

Sleep-deprived adolescents are more likely to be involved in car crashes, work-related injuries, and sports injuries [61]. High school students who get 7h of sleep or less each night are more likely to take risks that can lead to serious injury or death, such as failing to wear bike helmets and seat belts, drinking and driving or riding with a drunk driver, and texting while driving, compared to students who get 9 or more hours of sleep [61]. Being in a car crash is the leading cause of death for US adolescents, resulting in nearly 4000 teen deaths in the US every year [62]. Nationally, one-fifth of fatal car crashes between 2009 and 2013 involved a drowsy driver [63]. Half of drowsy driving crashes involve a driver aged 25 or younger [64]. By addressing widespread chronic sleep shortages among adolescents, delaying school start times has a clear link to reducing adolescent injuries and fatalities.

Several studies have analyzed the effect of school start time on car crash rates [24, 30, 65, 66]. One longitudinal study compared car crash rates for 17- and 18-year old drivers in a Kentucky school district to those in the state as a whole over the two years before and two years after a one-hour start time delay. Teens in the county with delayed start times had a 16.5% reduction in car crash rate over the study period, while teen crashes in the rest of the state increased by 7.8% [30]. Two cross-sectional studies compared crash rates for high school-aged drivers in adjacent communities in central [66] and eastern Virginia [65] with 85-min and 75- to 80-min start time differences, respectively. In each comparison, teen drivers in the communities with later start times were significantly less likely to be in crashes. As predicted, the teen crash peaks occurred during high school commute times in both communities, and teen drivers in the central Virginia community with earlier starting times were more likely to be in crashes where the car veered off the road to the right, a commonly sleep-related type of crash [66]. Finally, Wahlstrom compared the number of crashes involving 16- to 18-year old drivers in four communities before and after a start time delay [24]. She found that there were fewer adolescent-involved crashes in three of the four communities following a start time delay, while one community had a slight increase in crashes after a start time delay. However, because the communities were not compared to a reference population, it is not possible to say if the observed trends were unique to communities in which start time was delayed.

Lengthening sleep duration could plausibly reduce these types of injuries by improving concentration, attention, and reaction time, and reducing fatigue and adolescent risk taking. For example, a Norwegian study compared reaction times among students whose start time is delayed one day of the week ($N=33$), to high school students with consistently early start times ($N=45$) [57]. Compared to the consistently early start students, students had significantly fewer reaction time lapses (response delays of over 500 milliseconds) and faster average reaction times, on the late start day. These improvements in reaction time could reduce the risk of multiple types of injury. However, we are aware of no research to date assessing the effect of school start time on other types of accidental injury, including occupational and sports injuries. Another way school start times could be used to reduce the risk of child pedestrian injury is by staggering school start times with periods of high motor vehicle traffic to reduce exposure to traffic [67]. Future research is needed to identify what effect school start time may have on these and other injuries.

CONCLUSIONS

The evidence, which is based on pragmatic observational studies, some of which were natural experiment evaluations, suggests that delaying school start times can promote adolescent sleep and that this can have far reaching effects on healthy youth development. Numerous studies have overwhelmingly demonstrated that adolescents at later-starting schools are more likely get a healthy amount of sleep than their peers at earlier-starting schools. Limited evidence suggests that students at later-starting schools may learn more, and at worst perform no worse than their peers at earlier-starting schools, while an early start may hamper performance throughout the entire school day. There is also limited, but consistent, evidence that later school start times benefit both mental health and physical safety. Students starting school later demonstrate fewer symptoms of anxiety and depression and fewer behavioral problems. The risk of an adolescent being involved in a car crash appears to be reduced in communities with later-starting schools, possibly because or reduced sleepiness and improved reaction time among better-rested teens.

Given the consistency of these findings, with most studies showing clear benefits of later start times for adolescents and the absence of any studies showing harms for adolescents, it is striking that <20% of US middle and high schools start at 8:30 a.m. or later [18]. Although an increasing number of schools have made the change toward later start times or are currently in the process of considering the issue, there are many logistical challenges that deter school districts from making such a change, despite the robust evidence in support. Primary areas of concern include the potential impact on elementary school students (particularly if the schedules are "flipped" with elementary schools starting first and middle/high schools starting later to accommodate bussing), the impact on sports/extracurricular activities, and the impact on before or after-school childcare. Underlying many of these concerns is also a concern about the potential cost implications to school districts—a genuine concern in light of increasingly tightening school budgets. Countering this, however, the RAND Corporation recently published the first comprehensive investigation of the potential economic costs and benefits of a hypothetical state-wide shift in school start times to 8:30 a.m. or later across the US [68]. The study found that, even after just 2 years of such a policy change, the US economy would see an $8.6 billion dollar gain, which would already outweigh the costs per student from delaying school start times to 8.30 a.m. The study projected even larger gains over a more protracted (e.g., 10-year) period of time, with benefits accrued through the improvement in academic outcomes and subsequent lifetime earnings of well-rested students as well as a reduction in adolescent drowsy driving motor vehicle crashes.

There are still important avenues for inquiry and a need to gain additional broad insight on the impacts of the timing of school days, as far as health and social impacts both in adolescence and later in the life course. In particular, there is a critical need for longitudinal studies with longer term follow-up periods, as many potential benefits of later start times are likely to manifest over a longer period of time. Further, there is a need to study the potential impact of start times changes on other members of the community, including parents, teachers, and elementary school students. Additionally, there is a need to increase both the resolution the data on this topic via more objective measures as well as conduct research that would strengthen causal inference in this area. For this, innovative natural experiment evaluations are perhaps the best option given that randomizing schools to start time is highly infeasible. Strengthening the scientific base and disseminating such evidence to school districts and policy-makers is critical because school start times are likely one of the most readily modifiable major determinants of adolescent sleep and this relatively straightforward intervention can reach youth from various backgrounds.

REFERENCES

[1] American Academy of Pediatrics. School start times for adolescents. Pediatrics 2014;134(3):642–9. [Internet]. Available from: http://www.ncbi.nlm.nih.gov/pubmed/25156998%5Cnhttp://pediatrics.aappublications.org/cgi/doi/10.1542/peds.2014-1697.

[2] US Department of Health and Human Services. Healthy people 2020 sleep health objectives [Internet]. [cited 2017 Mar 20]. Available from: https://www.healthypeople.gov/2020/topics-objectives/topic/sleep-health/objectives?topicId=38.

[3] Eaton DK, LR MK-E, Lowry R, Perry GS, Presley-Cantrell L, Croft JB. Prevalence of insufficient, borderline, and optimal hours of sleep among high school students—United States, 2007. J Adolesc Health 2010;46(4):399–401. [Internet]. Elsevier Inc; Available from: https://doi.org/10.1016/j.jadohealth.2009.10.011.

[4] Keyes KM, Maslowsky J, Hamilton A, Schulenberg J. The great sleep recession: changes in sleep duration among US adolescents, 1991–2012. Pediatrics 2015;135(3):460–8. [Internet]. Available from: http://www.ncbi.nlm.nih.gov/pubmed/25687142%5Cnhttp://www.pubmedcentral.nih.gov/articlerender.fcgi?artid=PMC4338325.

[5] Millman RP, Working Group on Sleepiness in Adolescents/Young Adults, AAP Committee on Adolescence. Excessive sleepiness in adolescents and young adults: causes, consequences, and treatment strategies. Pediatrics 2005;115(6):1774–86. [Internet]. United States; Available from: http://ovidsp.ovid.com/ovidweb.cgi?T=JS&PAGE=reference&D=med5&NEWS=N&AN=15930245.

[6] Maume DJ. Social ties and adolescent sleep disruption. J Health Soc Behav 2013;54(4):498–515. [Internet]. Available from: http://www.ncbi.nlm.nih.gov/pubmed/24311758.

[7] Maslowsky J, Ozer EJ. Developmental trends in sleep duration in adolescence and young adulthood: evidence from a national United States sample. J Adolesc Health 2014;54(6):691–7. [Internet]. Elsevier Ltd; Available from: https://doi.org/10.1016/j.jadohealth.2013.10.201.

[8] Owens J, Adolescent Sleep Working Group, Committee on Adolescence. Insufficient sleep in adolescents and young adults: an update on causes and consequences. Pediatrics 2015;134(3):e921–32. Au R, Millman R, Wolfson A, Braverman PK, Adelman WP, Breuner CC, Levine DA, Marcell AV, Murray PJ, O'Brien RF, Carskadon M, editors. [Internet]. United States; Available from: http://ovidsp.ovid.com/ovidweb.cgi?T=JS&PAGE=reference&D=med8&NEWS=N&AN=25157012.

[9] Wing YK, Chan NY, Man Yu MW, Lam SP, Zhang J, Li SX, et al. A school-based sleep education program for adolescents: a cluster randomized trial. Pediatrics 2015;135(3):e635–43. [Internet]. Available from: http://pediatrics.aappublications.org/cgi/doi/10.1542/peds.2014-2419.

[10] Carskadon MA, Vieira C, Acebo C. Association between puberty and delayed phase preference. Sleep 1993;16(3):258–62.

[11] Carskadon MA, Acebo C, Jenni OG. Regulation of adolescent sleep: implications for behavior. Ann N Y Acad Sci 2004;1021(1):276–91. [Internet]. Available from: http://doi.wiley.com/10.1196/annals.1308.032.

[12] O'Malley EB, O'Malley MB. 7 School start time and its impact on learning and behavior. Sleep Psychiatr Disord Child Adolesc 2008;5:79. [Internet]. Available from: http://www.neurofeedback.ch/downloads/Schlaf_und_Schule.pdf.

[13] Frey S, Balu S, Greusing S, Rothen N, Cajochen C. Consequences of the timing of menarche on female adolescent sleep phase preference. PLoS ONE 2009;4(4):e5217.

[14] Taylor DJ, Jenni OG, Acebo C, Carskadon MA. Sleep tendency during extended wakefulness: insights into adolescent sleep regulation and behavior. J Sleep Res 2005;14(3):239–44.

[15] Sanya EO, Kolo PM, Desalu OO, Bolarinwa OA, Ajiboye PO, Tunde-Ayinmode MF. Self-reported sleep parameters among secondary school teenagers in middle-belt Nigeria. Niger J Clin Pract 2015;18(3):337–41.

[16] National Sleep Foundation, editors. Adolescent sleep needs and patterns: research report and resource guide. 2000.

[17] Barnes M, Davis K, Mancini M, Ruffin J, Simpson T, Casazza K. Setting adolescents up for success: promoting a policy to delay high school start times. J Sch Health 2016;86(7):552–7. [Internet]. United States; Available from: http://ovidsp.ovid.com/ovidweb.cgi?T=JS&PAGE=reference&D=prem&NEWS=N&AN=27246680.

[18] Wheaton AG, Ferro GA, Croft JB. School start times for middle school and high school students—United States, 2011–12 school year. MMWR Morb Mortal Wkly Rep 2015;64(30):809–13. [Internet]. United States; Available from: http://ovidsp.ovid.com/ovidweb.cgi?T=JS&PAGE=reference&D=med8&NEWS=N&AN=26247433.

[19] Position statements—Start school later [Internet]. [cited 2018 May 15]. Available from: http://www.startschoollater.net/position-statements.html.

[20] Wahlstrom K. School start time and sleepy teens. Arch Pediatr Adolesc Med 2010;164(7):676–7.

[21] Wahistrom K, Wahlstrom K. Changing times: findings from the first longitudinal study of later high school start limes. NASSP Bull 2002;86(633):3–22. [Internet]. Available from: http://bul.sagepub.com/content/86/633/3%5Cnhttp://bul.sagepub.com/content/86/633/3.full.pdf%5Cnhttp://bul.sagepub.com/content/86/633/3.short.

[22] Owens JA, Belon K, Moss P. Impact of delaying school start time on adolescent sleep, mood, and behavior. Arch Pediatr Adolesc Med 2010;164(7):608–14. [Internet]. United States; Available from: http://ovidsp.ovid.com/ovidweb.cgi?T=JS&PAGE=reference&D=med6&NEWS=N&AN=20603459.

[23] Boergers J, Gable CJ, Owens JA. Later school start time is associated with improved sleep and daytime functioning in adolescents. J Dev Behav Pediatr 2014;35(1):11–7. [Internet]. Available from: http://www.ncbi.nlm.nih.gov/pubmed/24336089.

[24] Wahlstrom KL, Dretzke BJ, Gordon MF, Peterson K, Edwards K, Gdula J. Examining the impact of later high school start times on the health and academic performance of high school students: a multi-site study. 2014. 72. Available from: http://www.ccsdschools.com/Community/documents/ImpactofLaterStartTime.pdf.

[25] National Center for Education Statistics. Schools and Staffing Survey (SASS) [Internet]. Available from: https://nces.ed.gov/surveys/sass/.

[26] Paksarian D, Rudolph KE, He J-P, Merikangas KR. School start time and adolescent sleep patterns: results from the US National Comorbidity Survey-Adolescent Supplement. Am J Public Health 2015;105(7):1351–7.

[27] Wheaton AG, Chapman DP, Croft JB. School start times, sleep, behavioral, health, and academic outcomes: a review of the literature. J Sch Health 2016;86(5):363–81.

[28] Marx R, Tanner-Smith EE, Davison CM, Ufholz L-A, Freeman J, Shankar R, et al. Later school start times for supporting the education, health, and well-being of high school students. Cochrane Database Syst Rev 2017;7:CD009467. [Internet]. England; Available from: http://ovidsp.ovid.com/ovidweb.cgi?T=JS&PAGE=reference&D=mesx&NEWS=N&AN=28670711.

[29] Wahlstrom KL. Accommodating the sleep patterns of adolescents within current educational structures: an uncharted path. In: Adolescent sleep patterns. Cambridge University Press; 2002.

[30] Danner F, Phillips B. Adolescent sleep, school start times, and teen motor vehicle crashes. J Clin Sleep Med 2008;4(6):533–5.

[31] Wang K, Sabia JJ, Cesur R. Sleepwalking through school: new evidence on sleep and academic performance [Internet]. [cited 2018 Apr 12]. (IZA Discussion Paper). Report No.: 9829. Available from: https://poseidon01.ssrn.com/delivery.php?ID=483078123005000094028084119007012126063069030050059041065016107047113058003118121024108087091006073082122082094073070126084026070087092023094121071-02910212300010508811512509711 4&EXT=pdf; 2016.

[32] National Sleep Foundation. 2006 Sleep in America Poll—Teens and Sleep [Internet]. [cited 2018 Apr 12]. Available from: https://sleepfoundation.org/sleep-polls-data/sleep-in-america-poll/2006-teens-and-sleep; 2006.

[33] Hinrichs P. When the bell tolls: the effects of school starting times on academic achievement. Educ Financ Policy 2011;6(4):486–507. [Internet]. [cited 2018 Apr 12]; Available from: https://www.mitpressjournals.org/doi/pdf/10.1162/EDFP_a_00045.

[34] Edwards F. Early to rise? The effect of daily start times on academic performance. Econ Educ Rev 2012;31(6):970–83. [Internet]. [cited 2018 Apr 12]; Available from: https://teensneedsleep.files.wordpress.com/2011/04/edwards-early-to-rise-the-effect-of-daily-start-times-on-academic-performance-published-version.pdf.

[35] Carrell SE, Maghakian T, West JE. A's from Zzzz's? The causal effect of school start time on the academic achievement of adolescents. Am Econ J Econ Pol 2011;3(3):62–81.

[36] Admissions requirements—United States Air Force Academy [Internet]. [cited 2018 May 16]. Available from: https://www.usafa.edu/prep-school/admissions-requirements/.

[37] Undergraduate age diversity at United States Air Force Academy [Internet]. [cited 2018 May 16]. Available from: https://www.collegefactual.com/colleges/united-states-air-force-academy/student-life/diversity/chart-age-diversity.html.

[38] About the ESS—Epworth Sleepiness Scale [Internet]. [cited 2018 Apr 16]. Available from: http://epworthsleepinessscale.com/about-the-ess/.

[39] Lufi D, Tzischinsky O, Hadar S. Delaying school starting time by one hour: some effects on attention levels in adolescents. J Clin Sleep Med 2011;7(2):137–43. [Internet]. Available from: https://www.ncbi.nlm.nih.gov/pmc/articles/PMC3077340/pdf/jcsm.7.2.137.pdf.

[40] Kessler RC, Chiu WT, Demler O, Walters EE. Prevalence, severity, and comorbidity of 12-Month DSM-IV disorders in the national comorbidity survey replication. Arch Gen Psychiatry 2005;62(6):617–27.

[41] Kann L, Kinchen S, Shanklin SL, Flint KH, Kawkins J, Harris WA, et al. Youth risk behavior surveillance—United States, 2013. Morb

[42] Van Der HE, Walker MP. Overnight therapy? The role of sleep in emotional. Psychol Bull 2010;135(5):731–48. [Internet]. Available from: https://www.ncbi.nlm.nih.gov/pmc/articles/PMC2890316/pdf/nihms206917.pdf.

[43] Wahlstrom KL, Berger AT, Widome R. Relationships between school start time, sleep duration, and adolescent behaviors. Sleep Heal J Natl Sleep Found 2017;https://doi.org/10.1016/j.sleh.2017.03.002. (In press). [Internet]. Elsevier Inc.; Available from:.

[44] Daly BP, Jameson JP, Patterson F, McCurdy M, Kirk A, Michael KD. Sleep duration, mental health, and substance use among rural adolescents: developmental correlates. J Rural Ment Health 2015;39(2):108–22. [Internet]. Available from: http://search.ebscohost.com/login.aspx?direct=true&db=pdh&AN=2015-28960-005&site=ehost-live%5Cnbrian.daly@drexel.edu.

[45] Winsler A, Deutsch A, Vorona RD, Payne PA, Szklo-Coxe M. Sleepless in fairfax: the difference one more hour of sleep can make for teen hopelessness, suicidal ideation, and substance use. J Youth Adolesc 2015;44(2):362–78. [Internet]. United States; Available from: http://ovidsp.ovid.com/ovidweb.cgi?T=JS&PAGE=reference&D=med8&NEWS=N&AN=25178930.

[46] Baum KT, Desai A, Field J, Miller LE, Rausch J, Beebe DW. Sleep restriction worsens mood and emotion regulation in adolescents. J Child Psychol Psychiatry Allied Discip 2014;55(2):180–90.

[47] Gregory AM, Caspi A, Eley TC, Moffitt TE, O'Connor TG, Poulton R. Prospective longitudinal associations between persistent sleep problems in childhood and anxiety and depression disorders in adulthood. J Abnorm Child Psychol 2005;33(2):157–63. [Internet]. Available from: https://link.springer.com/content/pdf/10.1007%2Fs10802-005-1824-0.pdf.

[48] Clarke G, Harvey AG. The complex role of sleep in adolescent depression. Child Adolesc Psychiatr Clin N Am 2012;21(2):385–400. [Internet]. Elsevier; Available from: https://doi.org/10.1016/j.chc.2012.01.006.

[49] McMakin DL, Dahl RE, Buysse DJ, Cousins JC, Forbes EE, Silk JS, et al. The impact of experimental sleep restriction on affective functioning in social and nonsocial contexts among adolescents. J Child Psychol Psychiatry Allied Discip 2016;57(9):1027–37.

[50] Lo JC, Ong JL, RLF L, Gooley JJ, MWL C. Cognitive performance, sleepiness, and mood in partially sleep deprived adolescents: the need for sleep study. Sleep 2016;39(3):687–98. [Internet]. Available from: https://academic.oup.com/sleep/article-lookup/doi/10.5665/sleep.5552.

[51] Edwards S, Reeves GM, Fishbein D. Integrative model of the relationship between sleep problems and risk for youth substance use. Curr Addict Rep 2015;2(2):130–40.

[52] Hasler BP, Smith LJ, Cousins JC, Bootzin RR. Circadian rhythms, sleep, and substance abuse. Sleep Med Rev 2012;16(1):67–81.

[53] Hasler BP, Clark DB. Circadian misalignment, reward-related brain function, and adolescent alcohol involvement. Alcohol Clin Exp Res 2013;37(4):558–65.

[54] Hasler BP, Soehner AM, Clark DB. Sleep and circadian contributions to adolescent alcohol use disorder. Alcohol 2015;49(4):377–87.

[55] Berger AT, Widome R, Troxel WM. School start time and psychological health in adolescents. Curr Sleep Med Rep 2018;. Available from: https://link.springer.com/content/pdf/10.1007/s40675-018-0115-6.pdf.

[56] Wahlstrom KL. Technical report, volume II: analysis of student survey data. 1998. II(November).

[57] Vedaa Ø, Saxvig IW, Wilhelmsen-Langeland A, Bjorvatn B, Pallesen S. School start time, sleepiness and functioning in Norwegian adolescents. Scand J Educ Res 2012;56(1):55–67.

[58] Perkinson-Gloor N, Lemola S, Grob A. Sleep duration, positive attitude toward life, and academic achievement: the role of daytime tiredness, behavioral persistence, and school start times. J Adolesc 2013;36(2):311–8. [Internet]. England; Available from: http://ovidsp.ovid.com/ovidweb.cgi?T=JS&PAGE=reference&D=med8&NEWS=N&AN=23317775.

[59] Martin JS, Gaudreault MM, Perron M, Laberge L. Chronotype, light exposure, sleep, and daytime functioning in high school students attending morning or afternoon school shifts: an actigraphic study. J Biol Rhythm 2016;31(2):205–17. [Internet]. United States; Available from: http://ovidsp.ovid.com/ovidweb.cgi?T=JS&PAGE=reference&D=med8&NEWS=N&AN=26825618.

[60] Chan NY, Zhang J, MWM Y, Lam SP, Li SX, APS K, et al. Impact of a modest delay in school start time in Hong Kong school adolescents. Sleep Med 2017;30:164–70. [Internet]. Elsevier Ltd; Available from: https://doi.org/10.1016/j.sleep.2016.09.018.

[61] Wheaton AG, Olsen EO, Miller GF, Croft JB. Sleep duration and injury-related risk behaviors among high school students—United States, 2007–2013 [Internet]. Vol. 65, Morbidity and mortality weekly report. Atlanta, GA; [cited 2018 Apr 9]. Available from: https://www.cdc.gov/mmwr/volumes/65/wr/mm6513a1.htm; 2016.

[62] Governors Highway Safety Association. Mission not accomplished: teen safe driving, the next chapter [Internet]. [cited 2018 Apr 6]. Available from: https://www.ghsa.org/sites/default/files/2016-12/FINAL_TeenReport16.pdf; 2017.

[63] Tefft BC. Prevalence of motor vehicle crashes involving drowsy drivers, United States title prevalence of motor vehicle crashes involving drowsy drivers, United States. [cited 2018 Apr 6]; Available from: www.aaafoundation.org; 2009.

[64] Knipling RR, Wang J-S. Crashes and fatalities related to driver drowsiness/fatigue. Research Note 1996.

[65] Vorona RD, Szklo-Coxe M, Wu A, Dubik M, Zhao Y, Ware JC. Dissimilar teen crash rates in two neighboring southeastern Virginia cities with different high school start times. J Clin Sleep Med 2011;7(2):145–51.

[66] Vorona RD, Szklo-Coxe M, Lamichhane R, Ware JC, McNallen A, Leszczyszyn AD, et al. Adolescent crash rates and school start times in two Central Virginia Counties, 2009–2011: a follow-up study to a Southeastern Virginia study, 2007–2008. J Clin Sleep Med 2014;10(11):1169E–77E.

[67] Yiannakoulias N, Bland W, Scott DM. Altering school attendance times to prevent child pedestrian injuries. Traffic Inj Prev 2013;14(4):405–12. [Internet]. [cited 2018 Apr 11]; Available from: http://www.tandfonline.com/doi/abs/10.1080/15389588.2012.716879

[68] Hafner M, Stepanek M, Troxel WM. The economic implications of later school start times in the United States. Sleep Heal 2017;3(6):451–7. [Internet]. National Sleep Foundation; Available from: https://doi.org/10.1016/j.sleh.2017.08.007.

Part IX

Economic and public policy implications of sleep health

Chapter 34

Sleep health and the workplace

Soomi Lee[a,b], Chandra L. Jackson[c,d], Rebecca Robbins[e], Orfeu M. Buxton[b,f,g,h,i]

[a]School of Aging Studies, University of South Florida, Tampa, FL, United States, [b]Center for Healthy Aging, Pennsylvania State University, State College, PA, United States, [c]Epidemiology Branch, National Institute of Environmental Health Sciences, National Institutes of Health, Department of Health and Human Services, Research Triangle Park, NC, United States, [d]Intramural Program, National Institute on Minority Health and Health Disparities, National Institutes of Health, Department of Health and Human Services, Research Triangle Park, NC, United States, [e]NYU Langone Health, Department of Population Health, New York, NY, United States, [f]Department of Biobehavioral Health, Pennsylvania State University, State College, PA, United States, [g]Division of Sleep Medicine, Harvard Medical School, Boston, MA, United States, [h]Department of Social and Behavioral Sciences, Harvard Chan School of Public Health, Boston, MA, United States, [i]Department of Medicine, Brigham and Women's Hospital, Boston, MA, United States

INTRODUCTION
Work factors impacts nighttime sleep

For most adults, the total time spent asleep or at work together represent the majority of time use in a typical week. Work is necessary to sustain an economy and provide for the well-being of individuals and their families, including essential sustenance and shelter. Ideally, work can also contribute to a sense of purpose, and is an important overall contributor to health and wellbeing. Work can also determine exposure to toxins, pollutants, workplace hazards, psychosocial risks, and other exposures. Work can shape health behaviors, including sleep, which is important to organizations because of its relationship with subsequent employee performance, safety, attitudes, and health.

Most workers need to follow a "cycle of work and rest" in contemporary society, meaning that times for rest (including sleep) are primarily determined by work hours rather than workers' own needs for rest and recovery [1]. Long work hours, shift work, and stress and worry from work are all factors contributing to sleep deficiency among workers. Modifying risk factors for workers' sleep deficiency is a challenge when structural work processes and pressures are difficult to change. This raises a serious concern for workers' health because sleep deficiency over an extended period of time may lead to increased allostatic load [2], the excessive wear and tear of continual adaptation.

Sleep impacts work function and productivity

As noted below and elsewhere in this volume, poor sleep health can directly influence productivity by slowing mental activity, degrading cognition and decision-making, and increasing mistakes on the job. Sleep deficiency may also affect workers' affective and social functioning. For example, an experimental laboratory study has found that sleep-deprived individuals have more difficulty recognizing non-verbal cues (e.g., happy and angry faces) than non-sleep-deprived individuals [3]. Moreover, sleep loss leads to increased mistrust in others [4], which may decrease workplace morale and degrade work team dynamics. Substantial economic costs of untreated insomnia include absenteeism, or missing work, [5, 6] and presenteeism, or presence at work with low productivity [7]. Furthermore, poor sleep health, such as short duration and poor quality sleep, circadian rhythm disruption, and sleep disorders may lead to withdrawal from work (i.e., lateness, absence, and turnover) and withdrawal or disengagement while at work (i.e., cognitive and emotional distraction, and work neglect) through daytime sleepiness [8]. As poor sleep reflects poor recovery [9], poor sleep may lead to compromised work productivity and even unethical behaviors at work [10, 11]. Sleep affects self-regulatory functioning needed for "right and ethical" decision making and judgment [12], and thus poor sleep may result in undesirable work outcomes that can impact many other people, and the workplace.

What theories of work can tell us about modifiable work factors influencing sleep

Despite the recently increasing focus on the work-sleep relationship, a comprehensive understanding of the work-sleep relationship is lacking. Early theories, described in greater detail below, including the Job Demand Control (JD-C) model [13] and the Job Demand-Resources (JD-R) model [14, 15] explain work characteristics influencing workers' stress; yet do not focus on workers' sleep. Several theoretical frameworks, such as the Effort-Recovery Model [16], Cognitive Activation Theory of Stress [17, 18], and Allostatic Load Model [2, 19], have been invoked to explain the link between work experiences and sleep. These theories suggest that time-based (e.g., work hours), thought-based (e.g., rumination), and arousal-based (e.g., anger, stress hormones) processes influence employees' sleep quantity and quality [12]. However, less attention has been paid to the characteristics of work and non-work (family) domains and broader contextual factors (e.g., socioeconomic status) that influence the work-sleep relationship. Combining these relevant theories together, in this chapter, we present a novel, interdisciplinary, and comprehensive theoretical model for the mechanisms of stress and coping that influence workers' sleep health (Fig. 34.2).

This chapter highlights the impact of increasing work demands and decreasing opportunities for good sleep, microlevel effects of work stressors on sleep, workplace intervention effects on sleep, sleep health and workers' future health risks, work characteristics as a potential contributor to socioeconomic and racial/ethnic disparities in sleep health, potential reasons for and consequences of racial/ethnic disparities in work-sleep relationship, sleep-related workplace interventions, and future research directions.

EPIDEMIOLOGY OF SLEEP AND WORK

Epidemiological studies have shown the links between short self-reported sleep duration and negative health outcomes, including obesity and metabolism [20, 21], cardiovascular health [22–24], and mortality risk [25]. These topics are addressed in detail in Chapters 15–18. Most of these studies, however, have examined sleep and health in general adult population, rather than focusing on working population. By doing so, there is lack of knowledge about the prevalence and consequences of sleep deficiency among workers whose sleep may be particularly vulnerable. About 40% of workers report a few unwanted awakenings at night per week [26]. Grandner et al. [27] and Chapter 9 points to directions for future research on sleep to bridge the gap between laboratory and epidemiological studies. While laboratory studies on sleep have tested the effects of sleep deprivation (or sleep restriction) on neurobehavioral performance, metabolism, and psychological health, epidemiological studies have examined the associations of habitual short sleep with mortality risk, obesity and metabolism, cardiovascular disease, and general health and psychosocial stress.

Focusing on sleep duration, there has been a lack of consensus until recently about the criteria for determining "short sleepers" among workers. The American Academy of Sleep Medicine (AASM) and Sleep Research Society (SRS) recommend that "adults obtain 7 or more hours of sleep per night *on a regular basis* to promote optimal health and functioning" [28]. However, according to the National Health Interview Survey in 2004–07, only about 30% of civilian employed workers had 6 h or less sleep per night [29]. Many epidemiological studies have considered short sleep as <6 h per night.

There are hints that workers' sleep varies by industry and occupation. In a rare study on this topic, Luckhaupt et al. [29] found that self-reported short sleep duration (≤6 h per night) among US workers varied by industry and occupation within industry. Specifically, the prevalence of short sleep duration was greatest for management of companies and enterprises (40.5%), followed by transportation/warehousing (37.1%) and manufacturing (34.8%). Occupational categories with the highest prevalence of short sleep duration included production occupations in the transportation/warehousing industry, and installation, maintenance, and repair occupations in both the transportation/warehousing industry and the manufacturing industry. Working in industries and occupations where nonstandard work schedules and long work hours are prevalent may increase the odds of short sleep duration.

Although not based on a national sample, another line of studies focusing on industry-specific implications of work and sleep also shows that the prevalence and nature of poor sleep health differs by industry (Fig. 34.1). In the Work, Family, and Health Study, Olson et al. [30] examined diverse sleep characteristics contributing to poor sleep health among employees in a fortune 500 firm in information technology (IT) division. They defined poor sleep health as having at least one of the following three components: actigraphy total sleep time <6.5 h/day, actigraphy wake-after-sleep-onset (WASO) >45 min/day, and self-reported sleep insufficiency (feeling never or rarely rested upon waking). As shown on the Panel A of Fig. 34.1, 65% of the IT workers exhibited poor sleep health, including 18% with short sleep duration, 41% with long WASO (disturbed sleep), and 22% perceived sleep insufficiency. When examining these sleep characteristics among employees in the extended-care industry who provide direct care to older residents (Panel B, Fig. 34.1), 67% exhibited poor sleep health [31]. The prevalence of perceived sleep insufficiency (31%) among the extended-care workers was higher than among the IT workers (22%), which may suggest the qualitative aspect of their sleep is poor due to nonstandard and varying work shifts, less schedule control, and high work demands.

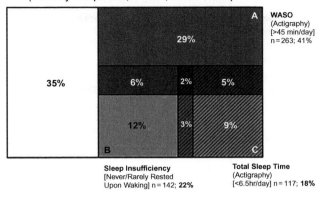
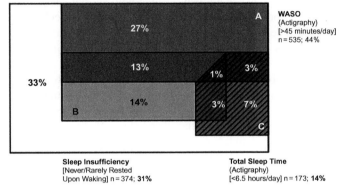

FIG. 34.1 The prevalence and nature of poor sleep health in the IT [30] and extended-care [31] industry samples of employed adults.

The relationship between sleep and work: Results from a meta-analysis (2017)

Well-designed longitudinal studies reveal that work stress predicts changes in sleep over time, rather than vice versa. Van Laethem et al. [32] conducted a meta-analysis on workplace factors and sleep quality based on 16 longitudinal studies. They found strong evidence that high job demands were associated with decreases in sleep quality at follow-up after adjusting for baseline sleep quality. Evidence for high job demands predicting poor sleep (e.g., future sleep disturbances) was also observed in a more recent meta-analysis by Linton et al. [33]. These studies suggest more adverse effects of effort-reward imbalance [34] on employee sleep. For example, employees with higher effort-reward imbalance (i.e., effort invested into work greatly exceeds the rewards received) were more likely to experience sleep disturbances compared to employees with lower effort-reward imbalance [33]. These findings highlight the importance of balance between demands and resources on the job for employees' sleep health, which closely relates to their stress and coping mechanisms.

Results from the American Time Use Study (ATUS) identify work as a main component of US adult time use, and as a factor directly related to shorter sleep duration. "Overwhelmingly, our analyses point to work as the dominant waking activity that is performed instead of sleep in short sleepers (1.55 h more on weekdays and 1.86 h more on weekends/holidays compared to normal sleepers), while long sleepers spent much less time working compared to normal sleepers (2.66 h less on weekdays and 0.90 h less on week- ends/holidays). Working ranked as the primary waking activity that was performed instead of sleep across all sociodemographic strata, with the exception of respondents retired, unemployed, or otherwise not in the labor force." [35]

THEORIES OF WORK AND WORK STRESS THAT INFLUENCE SLEEP

Work stress

Several theories attempt to explain the influence of work stress on workers' health and well-being. The Job Demand Control (JD-C) model [13] suggests that a workers' psychological and physical stress is a function of both job demands and job control (e.g., discretion, decision latitude). Specifically, the quadrant of high job demands and low job control may result in higher strain jobs. The Job Demand-Resources (JD-R) model [14, 15] further includes job resources as motivational process and posits that job resources may buffer the impact of job demands on workers' stress. These perspectives suggest the importance of considering both positive and negative aspects of work that may influence workers' stress and thereby their sleep health.

What is also important, but often forgotten, is the influence of the family home domain. The Work-Home Resources (W-HR) model [36] describes that work and family are inter-related domains and work-family conflict (see the next section for this concept) may occur as a process whereby demands in one domain deplete personal resources (e.g., time for sleep; energy levels) and impede accomplishments in the other domain. Extending upon previous theoretical models on work stress, Fig. 34.2 delineates inclusive mechanisms of stress and coping across work and family domains that influence workers' sleep.

When work demands exceed work resources, it may create stress that may negatively affect sleep health. This process may also be similar when family demands exceed family resources. On the contrary, when work resources exceed work demands, it may create coping mechanisms to stressful work conditions and thus workers' sleep health may be protected. These processes may affect and be affected by daily micro-level contexts, as well as by broader life-course contexts. Environmental, social, individual, and genetic characteristics may also play roles in these mechanisms of stress and coping that influence workers' sleep health (Fig. 34.2).

Work demands and work-family conflict influence sleep

- *Longer work hours and shorter sleep*

As time is a finite resource, it is not surprising to find the negative association between work hours and sleep hours. In the analysis of the American Time Use Survey (ATUS) with a representative cohort of Americans 15+ years between 2003 and 2011, long hours at paid work were associated with shorter sleep hours on weekdays [37]. Further, in a study with medical interns who usually work extended work shifts, an intervention schedule that limited scheduled work hours to 16 or fewer consecutive hours (and encouraged pre-nightshift naps) increased interns' sleep duration on average by 5.8h per week [38]. This intervention also resulted in fewer attentional failures while working during on-call nights. The timing of work also can affect workers' sleep. Shift work and variable work schedules involve disruptions of circadian rhythms and insufficient sleep during the day [39, 40] when it is more difficult to sleep to the wake-promoting signals from the central circadian

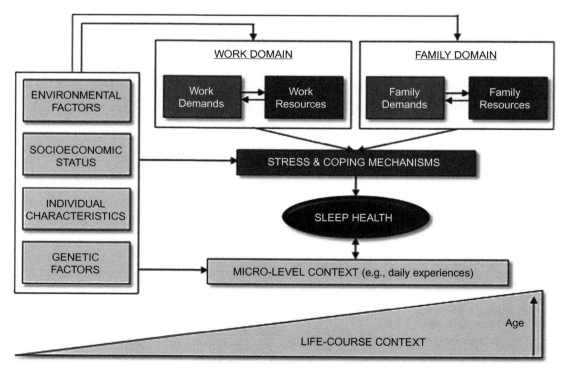

FIG. 34.2 A theoretical model for the mechanisms of stress and coping that influence workers' sleep health. *Note*: Life course stage influences all aspects of the model.

pacemaker [41]. Of note, it is often the case for hourly paid employees on night shifts to be of lower socioeconomic status and minorities, reflecting disparities due to the selection process to these less favorably-timed shifts [42, 43]. Workers with multiple jobs (and multiple shifts) are more likely to be short sleepers (≤6h per night); self-employed workers and those who can start work later in the morning are less likely to be short sleepers [37]. However, there is no linear relationship between socioeconomic status and work hours. Employees with higher education and higher income tend to work longer hours by exchanging their sleep time with economic incentives [37].

- *Higher work-to-family conflict and poorer sleep*

Work-nonwork "balance" has been shown to be cross-sectionally related to both workplace factors and employee health outcomes (for a systematic review, see Nijp et al. [44]). Demands from both the work domain and from the family domain can restrict the time available for sleep. Studies describe sleep as the "victim" in time-based conflict between work and family roles [45]. In particular, the extent to which demands from work permeate into the family domain and interfere with family and personal activities is called work-to-family conflict [46, 47]. Note that the family domain includes any non-work domains, so work-to-family conflict is not just the issue experienced by midlife, married, working parents [48]. Several studies consistently report that workers who experience higher work-to-family conflict have poorer sleep health measured by both self-reports and actigraphy [49–52].

- *Interpersonal stressors at work*

Some work-related stressors may have sustained effects on workers' sleep health when they leave cognitive stress residues, and invoke rumination that may lengthen sleep onset latency by increasing bedtime stress and worries [53] or cognitive hyperarousal [54]. Interpersonal stressors are among the most frequent stressors for an average adult [55–58] and predictive of both affective and physical well-being [59]. Interpersonal stressors may arise in the workplace, for example, in conflicts with colleagues or others, including supervisors [59]. These stressors may linger in workers' mind after work and interfere with their sleep at home. How workers' sleep health is associated with a variety of interpersonal stressors at work (e.g., perceived work inequality, workplace bullying, job discrimination, etc.) is a key avenue for future research, especially with the rapidly changing nature of work as often remote (or not as much at one place as may have been common in the past). Work has also become more "always on," with electronic device-based communications and intrusive social media related to work invading most any hour of the day and night [60]. There may also be racial/ethnic differences in the experience of interpersonal stressors at work and in sleep health.

Later in this chapter, we discuss racial/ethnic disparities in the work-sleep relationship (see the "Racial ethnic disparities in sleep health and sleep disorders" section).

Micro-longitudinal (daily level) effects of work stressors on sleep

Most prior studies examining the associations between work stressors and sleep used cross-sectional data or prospective data, focusing on between-person differences. For example, individuals who experience more work stressors overall may report poorer sleep than others, on average [49, 50, 61]. Examining between-person differences is meaningful, but cannot capture within-person differences, which may provide better insights on modifiable factors appropriate for future intervention strategies. For example, within an individual, some days may be more stressful than other days that may affect nightly sleep behavior. If we know the direction of effect, we may be able to more effectively intervene on the causes and consequences of sleep deficiency in workers' daily context. Crucially, these within-person effects directly convey the range of possible "days" for a given individual, providing both insights and expectations about how and when a workplace intervention or a lifestyle change, for example, might be expected to improve an individual's sleep health.

A daily diary design allows researchers to examine temporal associations, such as day-to-night or night-to-next day associations between work stressors and sleep [62]. Studies using multiple days' data report bidirectional associations between daily psychosocial experiences (positive and negative affect, positive events, and stressors) and nightly sleep [63]. A recent study focusing on work-derived stressors in midlife workers reveal more specific directions of the associations that differ across sleep measures [9]. We conceptualized sleep duration and sleep quality as previous night's sleep recovery, and sleep latency as bedtime rumination and worries that reflect today's stressors and conflicts. We found that shorter sleep duration and lower sleep quality were associated with next-day consequences of more work-to-family conflict and time-based stressors, whereas longer sleep latencies were predicted by more stressors on that day.

An emerging line of research combines daily diary with actigraphy sleep assessment [64, 65] There are still many unanswered questions that can be tested with micro-longitudinal data. More intensive data collection with ecological momentary assessment (EMA) may provide opportunities to test the mechanisms linking work stressors and sleep. Daily physical activity, momentary dietary choices, and emotional residue are potential factors that may mediate the effect of work stressors on nightly sleep health in workers' everyday life.

Sleep health and workers' future health risks

Workers' sleep may influence their future health outcomes. Work environments can impose different sleep burdens on the workers, which may contribute to health disparities in a long–run context. A recent study shows that workers in extended-care settings (i.e., low-wage workers providing direct care in nursing homes) have poorer sleep health than more advantaged workers in the information technology fortune 500 firms [66]. Poorer sleep health is associated with higher cardiovascular disease risk [22, 23, 67], and such associations seem more evident for low-wage workers than for higher-wage workers [66]. This is in line with epidemiological evidence suggesting disparities in sleep and cardiovascular health by socioeconomic status [68–70].

Workers' poor sleep health may also predict their functional limitations in later life. Sleep deficiency is found to be associated with higher rates of pain and functional limitations among health care workers, after controlling for socioeconomic, individual, and workplace characteristics [71]. Note that poor sleep habits persist over time [72], and sleep problems usually increase with age [73]. In other words, disrupted sleep behaviors due to work stress may be sustained after retirement and continue to degrade health and functioning in later life. For example, having more insomnia symptoms and taking sleep medications (even physician-prescribed medications) at baseline predict risk of falling at follow-up, after adjusting for known risk factors of falls [74]. These findings suggest that sleep-related behaviors are critical to understanding age-related changes in health and suggest possible ways to intervene on individuals' sleep and health problems during the second half of life.

WORKPLACE INTERVENTION EFFECTS ON SLEEP

Business case for sleep: Considering the evidence from the employer point of view

Poor sleep, both insufficient sleep duration and sleep disorders, is a pressing public health issue. The majority of adults in the US report sleeping for less than the recommended 7-h duration [28, 75]. Further, research suggests approximately 30% of adults in the US present with insomnia symptoms, and approximately 10% have received a clinical diagnosis for insomnia [76–78]. Estimates suggests total direct and indirect healthcare costs associated with insomnia to be between $30 and $40 billion annually [79, 80]. Research also shows significant healthcare burden of sleep deprivation, estimating total costs associated with insufficient sleep via direct and indirect healthcare cost expenditure to be between 250 and 415 billion US dollars annually [81]. One study found the presenteeism, productivity losses, and safety issues associated with insomnia and insufficient sleep cost on average $1967 per employee [82] while other research found presenteeism and productivity losses per each employees with untreated insomnia to be equivalent to 11.3 days of lost work or $2280 lost each year [7]. Data from a state-wide employee health program shows employee sleep difficulty is linked with absenteeism, lower workplace productivity, and increased healthcare costs (in fully adjusted models, approximately $725 higher healthcare costs with each incremental unit of increased sleep difficulty on a 5-point scale) [83]. Thus, poor sleep health is directly linked to adverse workplace outcomes of consequence for employers, including lower productivity and higher healthcare costs.

In addition to insufficient sleep and insomnia, obstructive sleep apnea (OSA) is an increasingly prevalent and costly disorder. According to the National Sleep Foundation, approximately one third of adults in the US are at high risk for obstructive sleep apnea (OSA) [84], but it is estimated that approximately 80% of individuals with OSA are undiagnosed and untreated [85]. It is estimated individuals with untreated OSA cost approximately 3.4 billion in medical costs each year [86]. In addition to the healthcare costs associated with untreated OSA, workers with untreated OSA are at significantly greater risk of accident or injury (OR 7.2, 95% confidence interval 2.4–21.8) [87]. Thus, it is evident that employee sleep duration, sleep quality, and treatment of existing sleep disorders are topics deserving attention of employers from the standpoint of workplace productivity, presenteeism and healthcare expenditure.

Healthcare spending was reported at 3.3 trillion US dollars and is estimated to continue to grow approximately 5.5% annually to over 5.5 billion US dollars by 2026 [88]. In the US, approximately 60% of the workforce has employer-based health insurance, creating an economic incentive for employers to develop thoughtful, effective workplace-based interventions to address sleep health and sleep disorder screening and care.

Worksite wellness, and the need for more attention to sleep

Most adults spend a large proportion of their waking lives at work, or seeking work. For this reason, worksites represent a ready platform for reaching a potentially wide audience with sleep health promotional activities. Accordingly, the American Heart Association, American Cancer Society, *Healthy People 2020*, National Institute for Occupational Safety and Health, the Centers for Disease Control (CDC) have issued recommendations for comprehensive worksite health promotion efforts [89–93].

Meta-analysis suggests healthcare spending falls by approximately 3 dollars for every 1 dollar spent on wellness programs, and absenteeism and presenteeism costs fall approximately 2 dollars for every 1 dollar spent on wellness programs [94]. With the growing attention to sleep as an important factor in overall health and well-being, and the role sleep plays in daytime productivity and cognition, incorporating sleep into employee health behavior change programs may be beneficial for employers.

In addition to reduced costs, worksite wellness efforts to address sleep are critical from the standpoint of injury and accident prevention. Employees on shift work are at particularly high risk for motor vehicle crashes due to drowsiness and circadian misalignment imposed by shift schedules [95]. After an extended work shift, research with medical residents showed a nearly three times greater risk for a car accident, and almost six times more likely to report a near miss incident on their way home from an extended shift compared to a regular shift [96]. According to meta-analysis, sleepiness behind the wheel among non-shift workers was associated with approximately 2.5 times greater risk of a motor vehicle crash [97]. Accidents, such as motor vehicle crashes, could happen either at the workplace or on the way to or from work, and are a particularly pressing concern for employees in driving-related occupations. Consequently, reducing fatigue among workers, particularly those at greater risk for insufficient sleep and circadian misalignment (such as shift workers) is a critical area of employee health promotion.

Interventions in the workplace to improve employee health outcomes are increasingly common. According to survey data, 90% of worksites in the US featured wellness offerings for employees [98]. According to the meta-analysis conducted by Baicker and colleagues [94], the most common worksite wellness program targets were weight loss and smoking cessation, suggesting attention to sleep in health behavior change efforts at worksites to be an under addressed area in workplace-based behavior change. However, in research assessing readiness to change a variety of health behaviors (e.g., such as quitting smoking, implementing stress management practices), sleep quality was a strong predictor of readiness to change, suggesting sleep may play a role in the initiation of other health practices [83]. Consequently, while underexplored in worksite-based health promotion, employers stand to benefit from incorporating sleep in efforts to improve employee health and well-being. While sleep has many favorable associations with positive health outcomes, sleep is a relatively new component in some workplace-based health programming, especially because of motivating links to weight management, improved physical activity, and other health targets, as well as productivity and injury/accident prevention, along with the subsequent increase in liabilities for the workplace policies, practices and conditions that contribute to fatigue-related errors and mishaps.

To be completely clear, chronic sleep loss directly contributes to sleepiness and "fatigue". Thus, employee and employer motivations can be in alignment with the understanding that adequate restorative sleep is central to workplace readiness. Conversely, failure to enable adequate sleep or to obtain adequate sleep directly contribute to excessive sleepiness and fatigue, and thereby greater risks of accidents, mistakes, and reduced productivity. As one example, "drowsy driving," which sounds potentially innocuous, is becoming more legally similar to drunken driving, and constitutes a source of potential employer liability if fatigue and sleepiness management practices are not in place to counter elevated risks of accidents in sleep-deprived employees.

Worksite programs targeting sleep and sleep related outcomes

According to the National Institutes of Health and Centers for Disease, comprehensive wellness programs play an important role in advancing population health and can take one of several forms, including efforts to address (1) the work environment physical, organizational, or psychosocial components; (2) the work-family-community interface; or (3) individual health-related behaviors [99]. Whereas the health-related behavioral contexts that have been focused on largely have included exercise, smoking, and weight control, there is growing opportunity to integrate sleep and sleep-related interventions.

Sleep-related workplace initiatives are described in a growing body of literature. The taxonomy offered by Sorensen and colleagues [99] falls among the three different types of evidence-based workplace wellness programs of work environment, work-family-community interface, or individual health-related initiatives, listed in examples below, not all of which have shown a favorable or replicated outcome.

Workplace interventions	Intervention duration	Intervention components	Findings
Yoga/mindfulness			
de Bruin et al. [100]	1.5-month intervention	Weekly educational and practice sessions combined physical exercise, restorative yoga, and mindfulness meditation	Results showed exposure to the intervention was associated with positive effects for anxiety, depression, sleep quality
Fang and Li [101]	6-month intervention	Weekly yoga educational intervention	Results from participants exposed to the intervention showed improvements in self-reported sleep quality and work stress

Continued

Workplace interventions	Intervention duration	Intervention components	Findings
Klatt et al. [101a]	8-week intervention	Weekly sessions on mindfulness-based yoga and meditation	Results showed significant reductions in self-reported stress, overall sleep, and improved sleep quality
Blue light			
Jensen et al. [101b]	10-day intervention	Employees were assigned to receive dynamic light exposure compared to ordinary institution light	Results revealed no difference between intervention and control in monitored sleep efficiency and melatonin levels, but intervention nurses subjectively rated their sleep as more effective
Rahman et al. [101c]	2 nights	Employees were assigned to either control lighting or intervention light	Results suggest objective sleep time was improved, sleep efficiency improved and wake after sleep onset decreased
Other Interventions			
Takahashi et al. (Napping) [101d]	1-week intervention	Workers provided several nap sessions	Results show no significant improvement in nocturnal sleep, but did demonstrate improved waking reaction time
Härmä et al. [101e] (shift timing)	10-day intervention	Shift workers were assigned to either backward rotating or forward rotating shift schedule for one shift schedule	Results suggest subjective and objective quality of sleep improved among older adults, and evidence for forward rotating shift was evident
Olson et al. (work/family conflict) [30]	3-month intervention	Focus group discussion, role-playing, and games intended to decrease work-family conflict by improving schedule control of employees and family supportive supervisor behaviors in manager	Results show sleep duration and sleep sufficiency improved in the intervention condition
Järnefelt et al. (CBTI) [101f]	1-month intervention	Educational sessions on CBTI principles	participants reported better subjective sleep quality, although no improvements were identified in sleep quality

RACIAL ETHNIC DISPARITIES IN SLEEP HEALTH AND SLEEP DISORDERS

Health disparities have been defined as differences in health between groups that "are not only unnecessary and avoidable but, in addition, are considered unfair and unjust." [102] Many racial/ethnic minorities as well as immigrant groups to the US are segregated within the labor market into lower-wage and lower-skilled jobs [43, 103–106], and racial/ethnic disparities in the work-sleep relationship have been observed. Understanding the impact of occupational characteristics on sleep as a modifiable potential source of health inequity among ethnically diverse groups can help both identify drivers of poor sleep in the overall population and enable the development of more effective sleep-related interventions that improve population health while addressing health disparities. For instance, under-resourced groups may have greater exposure to both traditional and unique job-related stressors/hazards (whether physical or social) that could help illuminate pathways linking suboptimal work-related factors to poor sleep. However, few studies have investigated the disparate impact of the apparent bi-directional relationship between sleep and work.

In addition to occupational segregation, racial differences in work-related factors that can affect sleep could arise for myriad reasons that may overlap or prove distinct across racial/ethnic minority groups. For instance, blacks – who may be at particularly high risk for insufficient sleep-related morbidity and mortality [107] – are more likely than whites to work non-traditional shifts with non-standard work schedules (especially night shifts) and to have longer work hours, which can negatively affect health through insufficient sleep duration by, for example, disrupting circadian rhythms and increasing one's appetite for sweet and salty foods or causing the brain's effort-reward system to increase pleasure-seeking behaviors that disrupt sleep [108, 109]. Compared with whites, blacks are also more likely to be employed in positions with low control/high demand and that involve low decision-making power. In addition to being more likely to report racial discrimination [105, 110, 111], Blacks are more likely to be among the working poor as defined by individuals with incomes that fell below the official poverty level despite spending at least 27 weeks in the labor force [112]. Among racial/ethnic minorities (especially Hispanics/Latinos), assimilation and acculturation to standard US culture (work-related and beyond) also likely influence health beliefs and behaviors that can affect sleep quantity and quality [113].

Experiences with voluntary or involuntary extended work hours likely vary across all racial/ethnic groups. For instance, a study using 2010 National Health Interview Study data found that whites (20.9% [20.0%–22.0%]) were more likely than Asians (16.6% [13.9%–19.9%]) to formally work at least 48 h per week with a similar percentage (6.2% [5.6%–6.8%] for whites and 6.7% [5.0%–8.8%] for Asians) working in temporary positions, and slightly more likely to work at least 60 h per week as well as engage in alternative shift work although these differences were nonsignificant [114]. With the potential for a differential impact by race/ethnicity, technology (e.g., internet with email capabilities, cellular phones) may have also increased the virtual accessibility of employees in ways that increase job strain as well as disrupt sleep [115, 116]. For cultural reasons, some groups could be more likely than others to feel a particular pressure to be more responsive in order to succeed in workplace settings, which can conceivably increase psychosocial stress and displace sleep.

A prior study investigated whether short sleepers who worked non-standard shifts other than day-shift only were more likely to report hypertension and if the relationship varied by race among blacks and whites where 11.0% reported rotating shift work and 4.0% reported night shift work. Shift work was associated with a 35% increased odds of hypertension among blacks [Odds Ratio = 1.35, Confidence Interval: 1.06–1.72], but not among Whites [OR = 1.01, CI: 0.85–1.20], and black shift workers sleeping <6 h had an 81% increased odds of reporting hypertension [OR = 1.81, CI: 1.29–2.54, $P < 0.01$], while white shift workers did not [OR = 1.17, CI: 0.90–1.52, NS] [117]. Furthermore, Jackson et al. [118] found that *blacks* compared to whites were more likely to be short sleepers across occupational classes within various categories of employment industries such as manufacturing, education, and healthcare [119]. The prevalence of short sleep duration was widest among professionals due to short sleep generally increasing with increasing professional roles based on occupational class (i.e., professional, support services, and laborers) within a given industry among blacks, whereas short sleep prevalence decreased with increasing professional roles for whites. Although short sleep duration among Hispanics/Latinos appeared generally similar to Whites, the occupational pattern observed among blacks was, on average, similar among Hispanics/Latinos [120]. Perhaps, the high prevalence of short sleep duration among professional blacks and Hispanics/Latinos can be attributed, in part, to limited professional/social networks that can provide emotional and financial support, discrimination in the workplace, the perceived high work ethic needed to succeed, and/or greater work-to-family conflict. When effort is not supported by potentially mitigating resources (e.g., financial and emotional support), a strong work ethic among marginalized groups may emerge as a coping strategy that causes strain in response to psychosocial and environmental stressors (e.g., career concerns, racism). This phenomenon, referred to as John Henryism, may be damaging to health through, among other factors, poor sleep that causes physiological dysregulation and adverse health outcomes [121–123].

Furthermore, the 2010 *Sleep in America* poll found that Hispanics/Latinos (38%) were the most likely compared to Blacks (33%), Whites (28%), and Asians (25%) to report being kept awake due to concerns related to employment, finances, personal relationship or health-related concerns [124]. Among 147 Latino farmworkers in North Carolina, most (83%) reported good sleep quality, and the association between working >40 h per week and reporting poor sleep quality approached statistical significance. A previous study has also found that short sleep was similar between Mexican-Americans and Whites but that non-Mexican Latinos were more likely to be habitual short sleepers [125]. Similar to the "healthy worker" effect, Latino laborers could represent a highly select group of particularly healthy and young individuals with minimal sleep disturbances. Acculturation and cultural factors (e.g., religious beliefs and practices, strong work ethic) may influence factors (e.g., stress levels) that have been shown to influence sleep [126]. For some Latino heritages, traditional sleep habits such as "siestas," may still be practiced; thus, increasing the quantity of total sleep [127]. Heterogeneity in sleep patterns of distinct Latino ethnic groups may also confound observed associations [128].

Another study that was nationally representative of the US and conducted among Asians and Whites showed that Asian Americans had an overall age-adjusted prevalence of short sleep duration that was higher than Whites, that varied importantly by both industry and occupation with the largest gap observed in the Finance/Information industry, and that the socioeconomic pattern was similar to whites with generally lower levels of short sleep among professionals [120]. A study among 3510 employees (2371 males and 1139 females) aged 20–65 years working in a local Japanese government evaluated whether work, family, behavioral, and sleep quality characteristics differed with varying time in bed (TIB). They found that high job demands, long work hours, and high work-to-family conflict were more prevalent among those with short TIB while those with long TIB had daily drinking habits. Participants with short TIB had poor sleep largely attributed to poor subjective sleep quality and daytime dysfunction, and those with long TIB had poor sleep largely due to long sleep latency, poor sleep efficiency and sleep disturbances.

Previous studies also suggest that sleep patterns among employed immigrants to the US, regardless of race, may differ importantly from individuals born in the US [43, 104, 120, 129, 130] with immigrant status generally shown to be independently associated with a higher likelihood of short sleep [130]. For instance, a nationally representative sample of the US found that white immigrant workers had longer sleep durations than US-born white workers and black immigrant workers had a higher prevalence of short sleep than US-born Black workers, for whom prevalence of short sleep was higher than other US-born groups. Certain immigrant population could forgo sleep to work longer hours to, for instance, send remittances to their home country.

Overall, occupational factors may contribute to racial/ethnic disparities in sleep health because of differential access to power, prestige, and tangible/intangible resources across racial/ethnic groups. For instance, occupational status is influenced by factors that have been shown to vary substantially by race. For example, differences in educational attainment create differential access to and use of information and knowledge (health related and beyond), and income creates differences through differential access to quality education, as well as to material goods and services. Racial/ethnic disparities in sleep may also be propagated through differential exposure to social hazards in the workplace that produce or further exacerbate stressors that impair sleep on a daily basis. For instance, exposure to everyday and major forms of racial/ethnic discrimination/harassment in the workplace and in society may play an important role in producing psychosocial stress related to job strain or limited control over job demands/prestige in ways that affect sleep [131, 132]. Of note, most prior studies that include racial/ethnic minorities have been cross-sectional and disadvantaged groups may be generally more likely to have comorbid conditions (e.g., obesity, type 2 diabetes, sleep apnea) that could also result in less, poorer quality sleep, and these health conditions have been shown to influence one's working conditions [133].

In conclusion, poor sleep (mainly based on short sleep duration) has been shown to vary importantly by race, socioeconomic status, and immigrant status across various occupations and industries. These complex and preventable differences reflect the need to identify as well as understand structural and sociocultural factors that may influence differences in the work-sleep relationship to effectively address disparities in sleep or optimal health and productivity among workers in the US.

FUTURE RESEARCH TOPICS AND DIRECTIONS

We propose four key directions for future research on Sleep Health and the Workplace: [1] identifying how work characteristics and workers' sleep health are linked over time, [2] applying advanced research designs and methods for examining sleep health and work, [3] developing sleep-focused interventions for workers, and [4] examining sleep health disparities in the working population by industry, occupation, and socioeconomic status. As Litwiller and colleagues noted in a recent meta-analysis linking sleep duration and quality to a wide range of work-related antecedents and outcomes: "the demonstrated importance of sleep indicates that it should be a critical part of theory being developed about the biggest organizational challenges of our time [134]."

First, as many of the authors in this book have suggested, more longitudinal studies are needed to understand how work and sleep health are associated over time. Although non-experimental longitudinal studies cannot determine causality, rigorously designed studies can provide information on the temporal association between a predictor and an outcome and rule out confounding variables. For example, when baseline level of sleep is controlled, we can test whether work demands predict *changes* in worker's sleep at a follow-up [12]. Longitudinal studies that encompass multiple time points over a long period of time can also answer to life-course related questions, whether and how new employment, unemployment, and retirement affect individual sleep patterns. In order to better understand workers' sleep, it is also necessary to consider factors in the family and personal domains. Family researchers have begun to examine the interplay between sleep and family life [135]. Work and family are both the most salient contexts for adult health and well-being [136]. In particular, in couple relationships, a worker's sleep may affect and be affected by his/her partner's sleep [137, 138]. As such, researchers are encouraged to take interdisciplinary perspectives to comprehensively capture work, family, and personal factors contributing to workers' sleep health.

Second, future research on work and sleep health may benefit from applying advanced research designs and methods. Most previous studies have used self-reported sleep measures. Self-reported sleep variables provide information on perceived sleep, however, they may be prone to self-report bias. This raises a concern about common-method bias when examining the relationship between reported work characteristics and reported sleep [139]. Actigraphy is an objective and non-intrusive method of monitoring sleep and activity patterns in population-based studies. Actigraphy has been increasingly used in sleep-related studies since the early 1990s [140]. Actigraphy offers valid and reliable results across studies that are highly correlated with those of polysomnography (PSG) that has been used as a gold standard in the diagnosis of sleep disorders in laboratory settings [141]. Actigraphy is easy to use (e.g., wearing an accelerometer device on the non-dominant wrist), interfering less with usual activities by participants. Thus, this method may be particularly valuable in studying the sleep health of workers who generally have a busy schedule around the clock. Moreover, actigraphy data can provide in-depth information on multidimensional sleep health (e.g., sleep duration, the amount of wake-after-sleep-onset, sleep timing, daytime napping, and variability in sleep) [142]; such information may be more useful when diary data on participants' daily work and nonwork experiences are also available [135]. Taken together, the use of multiple sophisticated methods may advance future research on work and sleep health.

Third, developing sleep-focused interventions for employees that draw on best practices in workplace-based health promotion and behavior change represents a promising future direction for research and practice. Directly considering the role of sleep in employee health and well-being, stress management and personal impact on non-work life is an engaging topic, fosters community and a commitment to the organization and co-workers, and demonstrates employers "get it." For productivity-focused employers (and who isn't and likely to stay in business?), and especially those in safety-critical sectors, the essential role of adequate, restorative sleep on a regular basis can be an important component of workplace programs and policies to promote a safe workplace with fewer accidents and mistakes, and ultimately greater productivity. As noted by several national associations, including the American Heart Association and Centers for Disease Control, the workplace is a critical context for promoting health. Work is a place where many adults spend a large proportion of their waking lives, and employee sleep health is directly related to workplace-related outcomes, including alertness and productivity, making sleep a potentially useful outcome for interventions. Future research drawing upon evidence-based approaches, such as cognitive behavioral therapy for insomnia, could improve employee sleep health and associated workplace productivity. It may be interesting to also consider approaches such as mindfulness-based stress reduction or yoga for improving health while also improving sleep. Targeting interventions toward improved employee sleep is a promising area for future research and practice.

REFERENCES

[1] Zijlstra FRH, Sonnentag S. After work is done: psychological perspectives on recovery from work. Eur J Work Organ Psychol 2006;15(2):129–38.

[2] Ganster DC, Rosen CC. Work stress and employee health: a multidisciplinary review. J Manag 2013;39(5):1085–122. https://doi.org/10.1177/0149206313475815.

[3] van der Helm E, Gujar N, Walker MP. Sleep deprivation impairs the accurate recognition of human emotions. Sleep 2010;33:335–42.

[4] Anderson C, Dickinson DL. Bargaining and trust: the effects of 36-h total sleep deprivation on socially interactive decisions. J Sleep Res 2010;19(1-Part-I):54–63.

[5] Driver HS. Sleep disorders at work. In: Barling J, Barnes CM, Carleton EL, Wagner DT, editors. Work and sleep: research insights for the workplace. New York: Oxford University Press; 2016. p. 31–51.

[6] Ozminkowski RJ, Wang S, Walsh JK. The direct and indirect costs of untreated insomnia in adults in the United States. Sleep 2007;30(3):263–73.

[7] Kessler RC, Berglund PA, Coulouvrat C, Hajak G, Roth T, Shahly V, et al. Insomnia and the performance of US workers: results from the America insomnia survey. Sleep 2011;34(9):1161–71.

[8] Mullins HM, Cortina JM, Drake CL, Dalal RS. Sleepiness at work: a review and framework of how the physiology of sleepiness impacts the workplace. J Appl Psychol 2014;99(6):1096–112.

[9] Lee S, Crain TL, SM MH, Almeida DM, Buxton OM. Daily antecedents and consequences of nightly sleep. J Sleep Res 2017;26(4):498–509. https://doi.org/10.1111/jsr.12488.

[10] Barnes CM, Schaubroeck J, Huth M, Ghumman S. Lack of sleep and unethical conduct. Organ Behav Hum Decis Process 2011;115(2):169–80.

[11] Christian MS, Ellis APJ. Examining the effects of sleep deprivation on workplace deviance: a self-regulatory perspective. Acad Manag J 2011;54(5):913–34.

[12] Barling J, Barnes CM, Carleton EL, Wagner DT, editors. Work and sleep: research insights for the workplace. New York, NY: Oxford University Press; 2016.

[13] Karasek R. Job demands, job decision latitude and mental strain: implications for job redesign. Adm Sci Q 1979;24:285–308.

[14] Bakker AB, Demerouti E. The job demands-resources model: state of the art. J Manag Psychol 2007;22(3):309–28.

[15] Demerouti E, Bakker AB, Nachreiner F, Schaufeli WB. The job demands-resources model of burnout. J Appl Psychol 2001;86(3):499–512.

[16] Meijman TF, Mulder G. Psychological aspects of workload. In: PJD D, Thierry H, de Wolff CJ, editors. Handbook of work and organizational psychology. vol. 2. Hove, East Sussex: Psychology Press Ltd; 1998.

[17] Meurs JA, Perrewé PL. Cognitive activation theory of stress: an integrative theoretical approach to work stress. J Manag 2010;37(4):1043–68.

[18] Ursin H, Eriksen HR. Cognitive activation theory of stress (CATS). Neurosci Biobehav Rev 2010;34(6):877–81.

[19] McEwen BS. Stress, adaptation, and disease. Allostasis and allostatic load. Ann N Y Acad Sci 1998;840:33–44.

[20] Hall MH, Muldoon MF, Jennings JR, Buysse DJ, Flory JD, Manuck SB. Self-reported sleep duration is associated with the metabolic syndrome in midlife adults. Sleep 2008;31(5):635–43.

[21] Knutson KL, Spiegel K, Penev P, Van Cauter E. The metabolic consequences of sleep deprivation. Sleep Med Rev 2007;11(3):163–78.

[22] Gangwisch JE, Heymsfield SB, Boden-Albala B, Buijs RM, Kreier F, Pickering TG, et al. Short sleep duration as a risk factor for hypertension: analyses of the first National Health and Nutrition Examination Survey. Hypertension 2006;47(5):833–9.

[23] Gottlieb DJ, Redline S, Nieto FJ, Baldwin CM, Newman AB, Resnick HE, et al. Association of usual sleep duration with hypertension: the Sleep Heart Health Study. Sleep 2006;29(8):1009–14.

[24] Meisinger C, Heier M, Löwel H, Schneider A, Döring A. Sleep duration and sleep complaints and risk of myocardial infarction in middle-aged men and women from the general population: The MONICA/KORA Augsburg Cohort study. Sleep 2007;30(9):1121–7.

[25] Grandner MA, Drummond SP. Who are the long sleepers? Towards an understanding of the mortality relationship. Sleep Med Rev 2007;11(5):341–60.

[26] Swanson LM, Arnedt JT, Rosekind MR, Belenky G, Balkin TJ, Drake C. Sleep disorders and work performance: findings from the 2008 National Sleep Foundation Sleep in America poll. J Sleep Res 2011;20(3):487–94.

[27] Grandner MA, Patel NP, Gehrman PR, Perlis ML, Pack AI. Problems associated with short sleep: bridging the gap between laboratory and epidemiological studies. Sleep Med Rev 2010;14(4):239–47.

[28] Watson NF, Badr MS, Belenky G, Bliwise DL, Buxton OM, Buysse D, et al. Recommended amount of sleep for a healthy adult: a joint consensus statement of the American Academy of Sleep Medicine and Sleep Research Society. J Clin Sleep Med 2015;11(6):591–2. https://doi.org/10.5664/jcsm.4758.

[29] Luckhaupt SE, Tak S, Calvert GM. The prevalence of short sleep duration by industry and occupation in the National Health Interview Survey. Sleep 2010;33(2):149–59.

[30] Olson R, Crain TL, Bodner TE, King R, Hammer LB, Klein LC, et al. A workplace intervention improves sleep: results from the randomized controlled Work, Family, and Health Study. Sleep Health 2015;1(1):55–65.

[31] Marino M, Killerby M, Lee S, Klein LC, Moen P, Olson R, et al. The effects of a cluster randomized controlled workplace intervention on sleep and work-family conflict outcomes in an extended care setting. Sleep Health 2016;2(4):297–308.

[32] Van Laethem M, Beckers DGJ, Kompier MAJ, Dijksterhuis A, Geurts SAE. Psychosocial work characteristics and sleep quality: a systematic review of longitudinal and intervention research. Scand J Work Environ Health 2013;39(6):535–49.

[33] Linton SJ, Kecklund G, Franklin KA, Leissner LC, Sivertsen B, Lindberg E, et al. The effect of the work environment on future sleep disturbances: a systematic review. Sleep Med Rev 2015;23:10–9.

[34] Siegrist J, Starke D, Chandola T, Godin I, Marmot M, Niedhammer I, et al. The measurement of effort–reward imbalance at work: European comparisons. Soc Sci Med 2004;58(8):1483–99.

[35] Basner M, Fomberstein KM, Razavi FM, Banks S, William JH, Rosa RR, et al. American time use survey: sleep time and its relationship to waking activities. Sleep 2007;30(9):1085–95.

[36] ten Brummelhuis LL, Bakker AB. A resource perspective on the work-home interface: the work-home resources model. Am Psychol 2012;67(7):545–56.

[37] Basner M, Spaeth AM, Dinges DF. Sociodemographic characteristics and waking activities and their role in the timing and duration of sleep. Sleep 2014;37(12):1889–906.

[38] Lockley SW, Cronin JW, Evans EE, Cade BE, Lee CJ, Landrigan CP, et al. Effect of reducing interns' weekly work hours on sleep and attentional failures. N Engl J Med 2004;351(18):1829–37.

[39] Akerstedt T. Shift work and disturbed sleep/wakefulness. Occup Med (Lond) 2003;53(2):89–94.

[40] Sallinen M, Kecklund G. Shift work, sleep, and sleepiness—differences between shift schedules and systems. Scand J Work Environ Health 2010;36(2):121–33.

[41] Czeisler CA, Buxton OM. Human circadian timing system and sleep-wake regulation. In: Kryger MH, Roth T, Dement WC, editors. Principles and practice of sleep medicine. 6th ed. Elsevier; 2017. https://doi.org/10.1016/B978-0-323-24288-2.00035-0. p. 362–76.e5.

[42] Buxton OM, Okechukwu CA. Long working hours can be toxic. Lancet Diabetes Endocrinol 2015;3(1):3–4.

[43] Ertel KA, Berkman LF, Buxton OM. Socioeconomic status, occupational characteristics, and sleep duration in African/Caribbean immigrants and US white health care workers. Sleep 2011;34(4):509–18.

[44] Nijp HH, Beckers DG, Geurts SA, Tucker P, Kompier MA. Systematic review on the association between employee worktime control and work-non-work balance, health and well-being, and job-related outcomes. Scand J Work Environ Health 2012;38(4):299–313. https://doi.org/10.5271/sjweh.3307.

[45] Barnes CM, Wagner DT, Ghumman S. Borrowing from sleep to pay work and family: expanding time-based conflict to the broader non-work domain. Pers Psychol 2012;65(4):789–819.

[46] Netemeyer RG, Boles JS, McMurrian R. Development and validation of work-family conflict and family-work conflict scales. J Appl Psychol 1996;81(4):400–10.

[47] Voydanoff P. Work demands and work-to-family and family-to-work conflict: direct and indirect relationships. J Fam Issues 2005;26(6):707–26.

[48] Barnes CM. Working in our sleep: sleep and self-regulation in organizations. Organ Psychol Rev 2012;2:234–57.

[49] Berkman LF, Liu SY, Hammer L, Moen P, Klein LC, Kelly E, et al. Work-family conflict, cardiometabolic risk, and sleep duration in nursing employees. J Occup Health Psychol 2015;20(4):420–33.

[50] Buxton OM, Lee S, Beverly C, Berkman L, Moen P, Kelly E, et al. Work-family conflict and employee sleep: evidence from IT workers in the work, family and health study. Sleep 2016;39(10):1871–82.

[51] Jacobsen HB, Reme SE, Sembajwe G, Hopcia K, Stoddard AM, Kenwood C, et al. Work-family conflict, psychological distress, and sleep deficiency among patient care workers. Workplace Health Saf 2014;62(7):282–91.

[52] Jacobsen HB, Reme SE, Sembajwe G, Hopcia K, Stiles TC, Sorensen GP, Porter JH, et al. Work stress, sleep deficiency and predicted 10-year cardiometabolic risk in a female patient care worker population. Am J Ind Med 2014;57(8):940–9. https://doi.org/10.1002/ajim.22340. [Epub 8 May 2014]. Erratum in: Am J Ind Med 2015;58(1):112.

[53] Akerstedt T, Kecklund G, Axelsson J. Impaired sleep after bedtime stress and worries. Biol Psychol 2007;76(3):170–3.

[54] Espie CA. Insomnia: conceptual issues in the development, persistence, and treatment of sleep disorder in adults. Annu Rev Psychol 2002;53(1):215–43.

[55] Birditt KS, Fingerman KL, Almeida DM. Age differences in exposure and reactions to interpersonal tensions: a daily diary study. Psychol Aging 2005;20(2):330–40.

[56] Bolger N, DeLongis A, Kessler RC, Schilling EA. Effects of daily stress on negative mood. J Pers Soc Psychol 1989;57(5):808–18.

[57] Clark LA, Watson D. Mood and the mundane: relations between daily life events and self-reported mood. J Pers Soc Psychol 1988;54(2):296–308.

[58] Repetti RL. Short-term effects of occupational stressors on daily mood and health complaints. Health Psychol 1993;12(2):125–31.

[59] Potter PT, Smith BW, Strobel KR, Zautra AJ. Interpersonal workplace stressors and well-being: a multi-wave study of employees with and without arthritis. J Appl Psychol 2002;87(4):789–96.

[60] Pfeffer J. Dying for a paycheck: how modern management harms employee health and company performance—and what we can do about it. 1st ed. HarperBusiness: New York; 2018.

[61] Berkman LF, Buxton OM, Ertel K, Okechukwu C. Manager's practices related to work-family balance predict employee cardiovascular risk and sleep duration in extended care settings. J Occup Health Psychol 2010;115(3):316–29.

[62] Lee S, Almeida DM. Daily diary design. In: Whitbourne SK, editor. Encyclopedia of adulthood and aging. Oxford, UK: Wiley-Blackwell; 2016. p. 297–300.

[63] Sin NL, Almeida DM, Crain TL, Kossek EE, Berkman LF, Buxton OM. Bidirectional, temporal associations of sleep with positive events, affect, and stressors in daily life across a week. Ann Behav Med 2017;51(3):402–15.

[64] Baron KG, Reid KJ, Zee PC. Exercise to improve sleep in insomnia: exploration of the bidirectional effects. J Clin Sleep Med 2013;9(8):819–24.

[65] Russell C, Wearden AJ, Fairclough G, Emsley RA, Kyle SD. Subjective but not actigraphy-defined sleep predicts next-day fatigue in chronic fatigue syndrome: a prospective daily diary study. Sleep 2016;39(4):937–44.

[66] Buxton OM, Lee S, Marino M, Beverly C, Almeida DM, Berkman L. Sleep health and predicted cardiometabolic risk scores in employed adults from two industries. J Clin Sleep Med 2018;14(3):371–83.

[67] Cappuccio FP, Stranges S, Kandala NB, Miller MA, Taggart FM, Kumari M, et al. Gender-specific associations of short sleep duration with prevalent and incident hypertension: the Whitehall II Study. Hypertension 2007;50(4):693–700.

[68] Canivet C, Nilsson PM, Lindeberg SI, Karasek R, Ostergren PO. Insomnia increases risk for cardiovascular events in women and in men with low socioeconomic status: a longitudinal, register-based study. J Psychosom Res 2014;76(4):292–9.

[69] St-Onge MP, Grandner MA, Brown D, Conroy MB, Jean-Louis G, Coons M, et al. Sleep duration and quality: impact on lifestyle behaviors and cardiometabolic health: a scientific statement from the American Heart Association. Circulation 2016;134(18):e367–86.

[70] Whinnery J, Jackson N, Rattanaumpawan P, Grandner MA. Short and long sleep duration associated with race/ethnicity, sociodemographics, and socioeconomic position. Sleep 2014;37(3):601–11.

[71] Buxton OM, Hopcia K, Sembajwe G, Porter JH, Dennerlein JT, Kenwood C, et al. Relationship of sleep deficiency to perceived pain and functional limitations in hospital patient care workers. J Occup Environ Med 2012;54(7):851–8.

[72] Breslow L, Enstrom JE. Persistence of health habits and their relationship to mortality. Prev Med 1980;9(4):469–83.

[73] Ohayon MM, Carskadon MA, Guilleminault C, Vitiello MV. Meta-analysis of quantitative sleep parameters from childhood to old age in healthy individuals: developing normative sleep values across the human lifespan. Sleep 2004;27(7):1255–73.

[74] Chen TY, Lee S, Buxton OM. A greater extent of insomnia symptoms and physician-recommended sleep medication use predict fall risk in community-dwelling older adults. Sleep 2017;40(11).

[75] Liu Y, Wheaton AG, Chapman DP, Cunningham TJ, Lu H, Croft JB. Prevalence of healthy sleep duration among adults—United States, 2014. Morb Mortal Wkly Rep 2016;65(6):137–41.

[76] Morin CM, Bootzin RR, Buysse DJ, Edinger JD, Espie CA, Lichstein KL. Psychological and behavioral treatment of insomnia:update of the recent evidence (1998–2004). Sleep 2006;29(11):1398–414.

[77] Morin CM, Jarrin DC. Insomnia and healthcare-seeking behaviors: impact of case definitions, comorbidity, sociodemographic, and cultural factors. Sleep Med 2013;14(9):808–9.

[78] Ohayon MM. Epidemiology of insomnia: what we know and what we still need to learn. Sleep Med Rev 2002;6(2):97–111.

[79] Chilcott LA, Shapiro CM. The socioeconomic impact of insomnia. An overview. PharmacoEconomics 1996;10(Suppl 1):1–14.

[80] Hafner M, Stepanek M, Taylor J, Troxel WM, van Stolk C. Why sleep matters-the economic costs of insufficient sleep: a cross-country comparative analysis. Rand Health Quart 2017;6(4):11.

[81] Rosekind MR, Gregory KB. Insomnia risks and costs: health, safety, and quality of life. Am J Manag Care 2010;16(8):617–26.

[82] Rosekind MR, Gregory KB, Mallis MM, Brandt SL, Seal B, Lerner D. The cost of poor sleep: workplace productivity loss and associated costs. J Occup Environ Med 2010;52(1):91–8.

[83] Hui SK, Grandner MA. Trouble sleeping associated with lower work performance and greater health care costs: longitudinal data from Kansas State Employee Wellness Program. J Occup Environ Med 2015;57(10):1031–8.

[84] Hiestand DM, Britz P, Goldman M, Phillips B. Prevalence of symptoms and risk of sleep apnea in the US population: results from the national sleep foundation sleep in America 2005 poll. Chest 2006;130(3):780–6.

[85] Young T, Evans L, Finn L, Palta M. Estimation of the clinically diagnosed proportion of sleep apnea syndrome in middle-aged men and women. Sleep 1997;20(9):705–6.

[86] Kapur V, Blough DK, Sandblom RE, Hert R, de Maine JB, Sullivan SD, et al. The medical cost of undiagnosed sleep apnea. Sleep 1999;22(6):749–55.

[87] Connor J, Whitlock G, Norton R, Jackson R. The role of driver sleepiness in car crashes: a systematic review of epidemiological studies. Accid Anal Prev 2001;33(1):31–41.

[88] Cuckler GA, Sisko AM, Poisal JA, Keehan SP, Smith SD, Madison AJ, et al. National health expenditure projections, 2017–26: despite uncertainty, fundamentals primarily drive spending growth. Health Aff (Millwood) 2018;37(3):482–92.

[89] Katz DL, O'Connell M, Yeh MC, Nawaz H, Njike V, Anderson LM, et al. Public health strategies for preventing and controlling overweight and obesity in school and worksite settings: a report on recommendations of the Task Force on Community Preventive Services. Morb Mortal Wkly Rep Recomm Rep 2005;54(RR-10):1–12.

[90] Sorensen G, Barbeau E, editors. Steps to a healthier US workforce: integrating occupational health and safety and worksite health promotion: state of the science. National Institute of Occupational Safety and Health Steps to a Healthier US Workforce Symposium; 2004.

[91] Kumanyika SK, Obarzanek E, Stettler N, Bell R, Field AE, Fortmann SP, et al. Population-based prevention of obesity: the need for comprehensive promotion of healthful eating, physical activity, and energy balance: a scientific statement from American Heart Association Council on Epidemiology and Prevention, Interdisciplinary Committee for Prevention (formerly the expert panel on population and prevention science). Circulation 2008;118(4):428–64.

[92] Koh HK. A 2020 vision for healthy people. N Engl J Med 2010;362(18):1653–6.

[93] Carnethon M, Whitsel LP, Franklin BA, Kris-Etherton P, Milani R, Pratt CA, et al. Worksite wellness programs for cardiovascular disease prevention: a policy statement from the American Heart Association. Circulation 2009;120(17):1725–41.

[94] Baicker K, Cutler D, Song Z. Workplace wellness programs can generate savings. Health Affairs (Millwood) 2010;29(2):304–11.

[95] Garbarino S, De Carli F, Nobili L, Mascialino B, Squarcia S, Penco MA, et al. Sleepiness and sleep disorders in shift workers: a study on a group of italian police officers. Sleep 2002;25(6):648–53.

[96] Barger LK, Cade BE, Ayas NT, Cronin JW, Rosner B, Speizer FE, et al. Extended work shifts and the risk of motor vehicle crashes among interns. N Engl J Med 2005;352:125–34.

[97] Bioulac S, Franchi JM, Arnaud M, Sagaspe P, Moore N, Salvo F, et al. Risk of motor vehicle accidents related to sleepiness at the wheel: a systematic review and meta-analysis. Sleep 2017;40(10).

[98] Linnan L, Bowling M, Childress J, Lindsay G, Blakey C, Pronk S, et al. Results of the 2004 National Worksite Health Promotion Survey. Am J Public Health 2008;98(8):1503–9.

[99] Sorensen G, Stoddard AM, Stoffel S, Buxton OM, Sembajwe G, Hashimoto D, et al. The role of the work context in multiple wellness outcomes for hospital patient care workers. J Occup Environ Med 2011;53(8):899–910.

[100] de Bruin EI, Formsma AR, Frijstein G, Bogels SM. Mindful2Work: effects of combined physical exercise, yoga, and mindfulness meditations for stress relieve in employees. A proof of concept study. Mindfulness 2017;8(1):204–17.

[101] Fang R, Li X. A regular yoga intervention for staff nurse sleep quality and work stress: a randomised controlled trial. J Clin Nurs 2015;24(23–24):3374–9. https://doi.org/10.1111/jocn.12983.

[101a] Klatt M, Steinberg B, Duchemin AM. Mindfulness in motion (MIM): an onsite mindfulness based intervention (MBI) for chronically high stress work environments to increase resiliency and work engagement. J Vis Exp 2015;101:e52359.

[101b] Jensen HI, Markvart J, Holst R, Thomsen TD, Larsen JW, Eg DM, et al. Shift work and quality of sleep: effect of working in designed dynamic light. Int Arch Occup Environ Health 2015;89(1):49–61.

[101c] Rahman SA, Shapiro CM, Wang F, Ainlay H, Kazmi S, Brown TJ, et al. Effects of filtering visual short wavelengths during nocturnal shiftwork on sleep and performance. Chronobiol Int 2013;30(8):951–62.

[101d] Takahashi M, Nakata A, Haratani T, Ogawa Y, Arito H. Post-lunch nap as a worksite intervention to promote alertness on the job. Ergonomics 2004;47(9):1003–13.

[101e] Härmä M, Tarja H, Irja K, et al. A controlled intervention study on the effects of a very rapidly forward rotating shift system on sleep–wakefulness and well-being among young and elderly shift workers. Int J Psychophysiol 2006;59(1):70–9. https://doi.org/10.1016/j.ijpsycho.2005.08.005.

[101f] Järnefelt H, Lagerstedt R, Kajaste S, Sallinen M, Savolainen A, Hublin C. Cognitive behavior therapy for chronic insomnia in occupational health services. J Occup Rehabil 2012;22:511–21.

[102] Whitehead M. The concepts and principles of equity and health. Health Promot Int 1991;6(3):217–28.

[103] Chung-Bridges K, Muntaner C, Fleming LE, Lee DJ, Arheart KL, LeBlanc WG, et al. Occupational segregation as a determinant of US worker health. Am J Ind Med 2008;51(8):555–67.

[104] Hurtado DA, Sabbath EL, Ertel KA, Buxton OM, Berkman LF. Racial disparities in job strain among American and immigrant long-term care workers. Int Nurs Rev 2012;59(2):237–44.

[105] Krieger N, Waterman PD, Hartman C, Bates LM, Stoddard AM, Quinn MM, et al. Social hazards on the job: workplace abuse, sexual harassment, and racial discrimination—a study of Black, Latino, and White low-income women and men workers in the United States. Int J Health Serv Plann Admin Eval 2006;36(1):51–85.

[106] Orrenius PM, Zavodny M. Do immigrants work in riskier jobs? Demography 2009;46(3):535–51.

[107] Nunes J, Jean-Louis G, Zizi F, Casimir GJ, von Gizycki H, Brown CD, et al. Sleep duration among black and white Americans: results of the National Health Interview Survey. J Natl Med Assoc 2008;100(3):317–22.

[108] Sharma S, Kavuru M. Sleep and metabolism: an overview. Int J Endocrinol 2010;2010.

[109] St-Onge MP, McReynolds A, Trivedi ZB, Roberts AL, Sy M, Hirsch J. Sleep restriction leads to increased activation of brain regions sensitive to food stimuli. Am J Clin Nutr 2012;95(4):818–24.

[110] Williams DR, Mohammed SA. Discrimination and racial disparities in health: evidence and needed research. J Behav Med 2009;32(1):20–47.

[111] Okechukwu CA, Souza K, Davis KD, de Castro AB. Discrimination, harassment, abuse, and bullying in the workplace: contribution of workplace injustice to occupational health disparities. Am J Ind Med 2014;57(5):573–86.

[112] Statistics USBoL. A profile of the working poor, 2015. 2017. Contract No.: Report 1068.

[113] Lara M, Gamboa C, Kahramanian MI, Morales LS, Bautista DE. Acculturation and Latino health in the United States: a review of the literature and its sociopolitical context. Annu Rev Public Health 2005;26:367–97.

[114] Alterman T, Luckhaupt SE, Dahlhamer JM, Ward BW, Calvert GM. Prevalence rates of work organization characteristics among workers in the U.S.: data from the 2010 National Health Interview Survey. Am J Ind Med 2013;56(6):647–59.

[115] Costa G. The 24-hour society between myth and reality. J Hum Erol (Tokyo) 2001;30(1–2):15–20.

[116] Presser HB. Toward a 24-hour economy. Science 1999;284(5421):1778–9.

[117] Ceide ME, Pandey A, Ravenell J, Donat M, Ogedegbe G, Jean-Louis G. Associations of short sleep and shift work status with hypertension among black and white Americans. Int J Hypertens 2015;2015:697275.

[118] Jackson CL, Kawachi I, Redline S, Juon HS, Hu FB. Asian-White disparities in short sleep duration by industry of employment and occupation in the US: a cross-sectional study. BMC Public Health 2014;14:552.

[119] Jackson CL, Redline S, Kawachi I, Williams MA, Hu FB. Racial disparities in short sleep duration by occupation and industry. Am J Epidemiol 2013;178(9):1442–51. https://doi.org/10.1093/aje/kwt159.

[120] Jackson CL, Hu FB, Redline S, Williams DR, Mattei J, Kawachi I. Racial/ethnic disparities in short sleep duration by occupation: the contribution of immigrant status. Soc Sci Med 2014;118:71–9.

[121] James SA. John Henryism and the health of African-Americans. Cult Med Psychiatry 1994;18(2):163–82.

[122] Markovic N, Bunker CH, Ukoli FA, Kuller LH. John Henryism and blood pressure among Nigerian civil servants. J Epidemiol Community Health 1998;52(3):186–90.

[123] Flaskerud JH. Coping and health status: John Henryism. Issues Ment Health Nurs 2012;33(10):712–5.

[124] Summary findings of 2010. Sleep in America Poll [press release]. March 8, 2010, 2010.

[125] Hale L, Do DP. Racial differences in self-reports of sleep duration in a population-based study. Sleep 2007;30(9):1096–103.

[126] Hale L, Rivero-Fuentes E. Negative acculturation in sleep duration among Mexican immigrants and Mexican Americans. J Immigr Minor Health 2011;13(2):402–7.

[127] Loredo JS, Soler X, Bardwell W, Ancoli-Israel S, Dimsdale JE, Palinkas LA. Sleep health in U.S. Hispanic population. Sleep 2010;33(7):962–7.

[128] Redline S, Sotres-Alvarez D, Loredo J, Hall M, Patel SR, Ramos A, et al. Sleep-disordered breathing in Hispanic/Latino individuals of diverse backgrounds. The Hispanic Community Health Study/Study of Latinos. Am J Respir Crit Care Med 2014;189(3):335–44.

[129] Seicean S, Neuhauser D, Strohl K, Redline S. An exploration of differences in sleep characteristics between Mexico-born US immigrants and other Americans to address the Hispanic Paradox. Sleep 2011;34(8):1021–31.

[130] Voss U, Tuin I. Integration of immigrants into a new culture is related to poor sleep quality. Health Qual Life Outcomes 2008;6:61.

[131] Thomas KS, Bardwell WA, Ancoli-Israel S, Dimsdale JE. The toll of ethnic discrimination on sleep architecture and fatigue. Health Psychol 2006;25(5):635–42.

[132] Tomfohr L, Pung MA, Edwards KM, Dimsdale JE. Racial differences in sleep architecture: the role of ethnic discrimination. Biol Psychol 2012;89(1):34–8.

[133] Mensah GA, Mokdad AH, Ford ES, Greenlund KJ, Croft JB. State of disparities in cardiovascular health in the United States. Circulation 2005;111(10):1233–41.

[134] Litwiller B, Snyder LA, Taylor WD, Steele LM. The relationship between sleep and work: a meta-analysis. J Appl Psychol 2017;102(4):682–99.

[135] Lee S, Lemmon M. Dynamic interplay between sleep and family life: review and directions for future research. In: SM MH, King V, Buxton OM, editors. Family contexts of sleep and health across the life course. vol. 8. New York: Springer; 2017. p. 201–9.

[136] Moen P, Wethington E. Midlife development in a life course context. In: Willis SL, Reid JD, editors. Life in the middle. San Diego: Academic Press; 1999. p. 3–23.

[137] Gunn HE, Buysse DJ, Hasler BP, Begley A, Troxel WM. Sleep concordance in couples is associated with relationship characteristics. Sleep 2015;38(6):933–9.

[138] Lee S, Martire LM, Damaske SA, Mogle JA, Zhaoyang R, Almeida DM, et al. Covariation in couples' nightly sleep and gender differences. Sleep Health 2018;4(2):201–8.

[139] Podsakoff PM, MacKenzie SB, Lee JY, Podsakoff NP. Common method biases in behavioral research: a critical review of the literature and recommended remedies. J Appl Psychol 2003;88(5):879–903.

[140] Knutson KL, Buxton OM. Actigraphy as a tool for measuring sleep: pros, cons and secrets of the trade. Sleep Res Soc Bull 2011;17(2):14–5.

[141] Jean-Louis G, Kripke DF, Cole RJ, Assmus JD, Langer RD. Sleep detection with an accelerometer actigraph: comparisons with polysomnography. Physiol Behav 2001;72:21–8.

[142] Buysse D. Sleep health: can we define it? Does it matter? Sleep 2014;37(1):9–17.

FURTHER READING

Li I, Mackey MG, Foley B, Pappas E, Edwards K, Chau JY, et al. Reducing office workers' sitting time at work using sit-stand protocols: results from a pilot randomized controlled trial. J Occup Environ Med 2017;59(6):543–9.

Owens J, Adolescent Sleep Working Group, Committee on Adolescence. Insufficient sleep in adolescents and young adults: an update on causes and consequences. Pediatrics 2014;134(3):e921–32.

Chapter 35

Sleep health equity

Judite Blanc[a], Jao Nunes[b], Natasha Williams[c], Rebecca Robbins[d], Azizi A. Seixas[d,e], Girardin Jean-Louis[d,e]

[a]NYU Langone Health, Department of Population Health, Center for Healthful Behavior Change, New York, NY, United States, [b]The City College of New York, New York, NY, United States, [c]NYU Langone Health, Division of Health and Behavior, Department of Population Health, Center for Healthful Behavior Change, New York, NY, United States, [d]NYU Langone Health, Department of Population Health, New York, NY, United States, [e]NYU Langone Health, Department of Psychiatry, New York, NY, United States

INTRODUCTION: SLEEP AND PUBLIC HEALTH

Sleep is the "reversible behavioral state of perceptual disengagement from and unresponsiveness to the environment" [1, p. 15]. Sleep is vital for optimal mental, emotional, and physical well-being, and therefore has emerged as a pillar of health alongside nutrition, exercise, and smoking cessation. Leading scientific, clinical, and governmental organizations in the US and internationally recognize the importance of sleep [2–4].

Despite this wide-scale recognition, schedules in America are inconsistent with healthy sleep habits [5]. For instance, career, social and lifestyle demands represent barriers to adequate sleep that have generated a sleep crisis of significant proportion with impact on individual outcomes (e.g., cognition, mental and physical health) and societal outcomes (e.g., productivity and safety). The Centers for Disease Control and Prevention (CDC) has made specific recommendations to meet daily needs: 9–12h of sleep for school-aged children, 8–10h of sleep for adolescents and a minimum of 7h for adults. However, the 2015 Youth Risk Behavior Surveillance System [6] shows an estimated nationwide prevalence of short sleep duration of 57.8% among middle school students and of 72.7% among high school students. Among adult respondents to the 2009 Behavioral Risk Factor Surveillance System (BRFSS), 41.3% reported 1–13 days of insufficient rest or sleep before the survey [7].

Poor sleep health has significant individual health and societal consequences. Poor sleep is associated with poor mental, emotional (e.g., anxiety, mood disturbance, suicidal ideation) and physical health consequences (e.g., accidents, illness, pain). Additionally, poor sleep health increases the risk for chronic conditions (e.g., high blood pressure, high body mass index (BMI) and obesity) and mortality. Poor sleep health extracts a significant cost in terms of dollars annually in medical expenses such as doctor visits, hospital services, prescriptions, and over-the-counter medications [8].

Conversely, research suggests that adequate sleep health has a protective effect for individual health and societal outcomes. Specifically, individuals who consistently report dimensions of sleep health, including good sleep quality, sufficient sleep duration, and absence of sleep disorder, are more likely to have better measures of mental, emotional, and physical health and longevity compared to those who cut their sleep short. But research also documents differences in sleep health along socio-economic lines, sleep health being out of reach for most individuals living in deprived social economic areas. Specifically, minority groups and individuals from disadvantaged economic backgrounds report lower sleep quality. Sleep, a precious resource in our society, this research suggests, may actually constitute a luxury most commonly practiced by majority race/ethnic groups and socio-economically advantaged individuals.

In order to tackle the sleep health crisis, the past 10 years has seen a dramatic increase in research that articulates these socio-economic limiting factors that affect sleep.

We argue there is a need for a paradigm shift in the way sleep medicine approaches this public health matter. We agree with Buysse [9] that the current focus mainly on sleep disorders should give way to a stronger emphasis on the notion of sleep health. This is crucial to the health of individuals and of the population, and stands to benefit sleep medicine itself.

WHAT IS SLEEP HEALTH?

While the body of knowledge on sleep patterns and associated public health outcomes is growing, there is no precise and specific definition of the concept of sleep health for use by sleep researchers and experts. When *sleep* and *health* are entered together in databases such as *PubMed*, and *Google Scholar*, between 43,714 and 676,000,000 results *are identified (September, the 8th of 2018)*. Authors appear to use "sleep problems" and "sleep health" interchangeably in their titles. Buysse [9] noticed that not even in the 2006 Institute of Medicine Report and in the mission statement for sleep of The Centers for Disease Control and Prevention, an explicit definition of sleep health was included. To fill the gap, in his recently published paper, "Sleep health: can we define it? Does it matter," Buysse [9] articulated this comprehensive definition:

> *Sleep health is a multidimensional pattern of sleep-wakefulness, adapted to individual, social, and environmental demands, that promotes physical and mental well-being. Good sleep health is characterized by subjective satisfaction, appropriate timing, sufficient duration, high efficiency, and sustained alertness during waking hours. [9, p. 12]*

And in contrast to the deficit model, this definition focuses on positive characteristics of sleep health which are physiologically quantifiable. Although adult-centered, it could be easily extrapolated to youth and cannot be conceived outside of its individual, social and environmental components, and offers specific anchors for these five dimensions of sleep health:

- Sleep duration: The number of sleep hours per day
- Sleep continuity or efficiency: The ability of falling asleep and staying asleep
- Timing: The placement of sleep within the 24-h day
- Alertness/sleepiness: The capacity of staying awake
- Satisfaction/Quality: The individual perception of "good" or "poor" sleep.

The following are questions of interest in the context of the present analysis: What is the relationship between the sleep health dimensions and physical and mental well-being? What are the mechanisms that influence sleep dimensions and related health outcomes? Overall, does the *presence of* sleep impairment affect all groups at the same level?

SOCIAL DETERMINANTS OF SLEEP HEALTH DIMENSIONS AND ASSOCIATED HEALTH OUTCOMES

During the last decades, factors affecting the five dimensions of sleep health, including age, sex, body mass index (BMI), race/ethnicity, education, environment, workplace and economic position have been widely documented [7, 10–12]. In Table 35.1, we present a list of problems related to patients, providers and the healthcare system that potentially undermine optimal sleep in disadvantaged communities.

As described in Table 35.1, there is a growing body of studies that show similar findings, not all categories are equally impacted by the sleep health crisis. How does this sleep deficiency vary across racial and ethnic groups? Are other individual, social and environmental components contributing to this variation of sleep patterns among different social groups? *These are valid questions that current research findings cannot yet answer in full*. At this point, we intend to explore another valid question: Can any role be attributed to the historical context, specifically to the race-based slavery system that underpins the birth of the American society?

HEALTH DIFFERENCES AND THE HISTORICAL SLEEP GAP BETWEEN BLACKS AND WHITES

Identifying determinants of health differences

The debate is not a new one about the origin of historically ubiquitous health disparities/inequities and of the elusive health equity, but the tools of the debate remain faulty. Until now, there is little consensus regarding the meaning of the terms "health disparities/inequities" and "health equity." An early articulated and widespread conceptualization of these notions is attributed to Margaret Whitehead in her famous article, *"The concepts and principles of equity and health"* [13]. She stated: "inequality in health is a term commonly used in some countries to indicate systematic, avoidable and important differences." In her discussion, she described seven critical determinants of health differentials: (1) Natural, biological variation; (2) Health-damaging behavior if freely chosen, such as participation in certain sports and pastimes; (3) The transient health advantage of one group over another when that group is first to adopt a health-promoting behavior (as long as other groups have the means to catch up fairly soon); (4) Health-damaging behavior where the degree of choice of lifestyles is severely restricted; (5) Exposure to unhealthy, stressful living and working conditions; (6) Inadequate access to essential health and other public services; and (7) Natural selection or health-related social mobility involving the tendency for sick people to move down the social scale.

According to Whitehead, these seven determinants of health differences are all interacting. Although the impact of biological factors and the effects of sick people moving down the social scale have been demonstrated, she underlined that the major role is to be attributed to socioeconomic and environmental factors, including lifestyles. Consequently, how may this fact hold explanations for the actual sleep health disparities?

History behind the black–white "sleep gap"

Biological imperatives to maintain homeostatic processes and social factors determine sleep in humans. The National Heart Lung and Blood Institute (NHLBI) of the National Institutes of Health (NIH) describes sleep as a fundamental requirement of living. Yet extensive scientific evidence revealed

TABLE 35.1 Summary of social determinants of sleep health and potential health outcomes.

Dimensions of sleep measured	Population studied	Social determinants/ risk factors	Health outcomes	Source
Duration	BRFSS 2014 $N=444,306$ American men and women aged ≤18 from 50 states and the District of Columbia	Location, employment status, education, Black race, Hawaiians or other non-White/ Hispanic ethnicities	N/A	Liu et al. [14]
	$N=474,684$ participants of multiple studies from USA, Japan, UK, Sweden, Germany, Singapore, Israel, and Taiwan	N/A	Mortality of Coronary Heart Disease, Stroke CVD,	Cappuccio et al. [14a]
	$N=29,818$, aged 18–85 Cross-sectional household interview survey 2005 National Health Interview Survey (NHIS)	Black race	Obesity	Donat et al. [14b]
	$N=578$ Chicago residents, aged 33–45 Wrist Actigraphy for three consecutive days from 2003 to 2005	Black race	Incident hypertension difference in diastolic and systolic blood pressure	Knutson et al. [14c]
Sleep continuity/ efficiency	$N=812$ participants (36% African American; 67% female) Longitudinal and cross-sectional	Black race, sedentary life	Metabolic syndrome (increased blood pressure, high blood sugar, excess body fat around the waist, and abnormal cholesterol or triglyceride levels)	Troxel et al. [14d]
Timing (shift work)	37 African American women and 62 women of other races. Day shift ($n=61$), evening shift ($n=11$), and night shift ($n=27$)	Black race, evening and night shift	Nondipping blood pressure	Yamasaki et al. [14e]
Alertness/ sleepiness	$N=84,003$ Multi-ethnic female registered nurses aged 37–54 in 14 US states. Longitudinal analyses of data from the Nurses' Health Study II	Shift work	Hypertension, diabetes, hypercholesterolemia, obesity, and depression	Gangwisch et al. [14f]
Satisfaction/ quality	$N=1139$ including 520 whites, 586 African Americans, and 33 of Asian, Native American, or Hispanic ethnicity	Female gender, age, education, income	Overall self-reported physical health	Moore et al. [14g]

that social categories such as racial/ethnic minorities and *socioeconomically* disadvantaged groups do not attain the adequate and recommended amount, quality and consistency of sleep. The associations between sleep quantity/quality and both demographic and socioeconomic factors have been widely reported in the literature [12, 14, 15]. Also, similar observations have been made for race/ethnicity that emerged as a significant determinant of individual variation in sleep phenotype. For instance, African Americans are more burdened with sleep health disparities [16]. White youth generally have better sleep than minority youth, Hispanics have more than Blacks, and there is inconclusive evidence for Asians and other minorities [17]. Depending on the definition adopted, sleep-disorder-breathing (SDB) was 4–6 times more likely in 8- to 11-year-old black children compared with white children, and almost 3–5 times more likely in those born preterm than term children [18]. Moreover, on average, African–American adults report shorter sleep duration compared to other racial/ethnic groups [14].

This critical prevalence of suboptimal sleep in communities of African descent compared to those of European descent is of interest to this chapter. Traditionally and according to historians and national archives, people of African descent were the most affected compared to native-Americans and Hispanics by the slave trade and chattel enslavement in the US. Sleep health disparities require to be investigated from a historical perspective, considering the birth context of this nation. The roots of the "blacks and whites sleep gap" is suspected by some historians to stretch back to the history of chattel enslavement and colonization in the United States. During this period, after people were kidnapped from the African continent and considered as chattel slaves by white European settlers, resting time control became an efficient weapon in the hands of the European masters to strengthen the race-based slavery system and allowed the maximum exploitation of the enslaved workforce [19, 20]. Report on the sleep condition by the former enslaved abolitionist leader, Frederick Douglas echoes historian Benjamin Reiss's analysis:

> There were no beds given the slaves unless one coarse blanket be considered such, and none but the men and women had these. This, however, is not considered a very great privation. They find less difficulty from the want of beds, than from the want of time to sleep; for when their day's work in the field is done, the most of them having their washing, mending, and cooking to do, and having few or none of the ordinary facilities for doing either of these, very many of their sleeping hours are consumed in preparing for the field the coming day; and when this is done, old and young, male and female, married and single, drop down side by side, on one common bed,—the cold, damp floor,—each covering himself or herself with their miserable blankets; and here they sleep till they are summoned to the field by the driver's horn. At the sound of this, all must rise, and be off to the field. There must be no halting; everyone must be at his or her post, and woe betides them who hear not this morning summons to the field; for if they are not awakened by the sense of hearing, they are by the sense of feeling: no age nor sex finds any favor. [19, pp. 8–9]

Reiss's conception of the structure, ideology, practices, and policies that governed slave plantations and therefore enslaved people and their sleep phenotype might be very relevant for investigating sleep health disparities over a century later. All in all, socio-historical context has always been a significant determinant of who gets the best sleep, for:

> If slaves helped build the modern world, they were never afforded sufficient rest from the toils involved. Nor were they afforded the privacy that, according to the sociologist Norbert Elias, was becoming a hallmark of Western bourgeois sleep. Once they were excluded from normal sleep, they were punished for failures to maintain alertness and productivity and branded as constitutionally lazy for any sign of exhaustion. Their supposedly different sleep patterns- those that marked them as belonging to an inferior race- were actually taken as a justification for race-based slavery by medical authorities and slavery propagandists [20, p. 122].

Sleep health as a contributor to health disparities in modern days

Regarding health differences in our contemporary society, sleep may play an important role among the factors that contribute to health and health care disparities [21], particularly cardiovascular health in the United States [22]. This idea of a mediating effect of sleep on health disparities can be analyzed based on the conceptual framework proposed by Jackson et al. [22], in their review of the multilevel determinants of sleep-cardiovascular health disparities via proximal, intermediate and distal pathways.

(1) *Proximal factors* include individual risks behaviors, biological/genetic pathways, and biological responses, personal demographics such as acculturation, age, and sex that are recognized risks factors for Cardiovascular Diseases and impact sleep quality and quantity.

(2) *Intermediate factors* comprise physical context, built environments, neighborhood and housing disadvantages, social relationships through family influences and social context such as racism. Data from the 2014 Census estimate the portion of Racial/ethnic minorities of US population at 37.8% [22a]. Approximately, ¼ of Blacks (mostly descendants of the former enslaved Africans, Africans, and Afro-Caribbean immigrants) (21.2%) and Hispanic persons (18.3%) lived in Poverty compared with non-Hispanic white (7.8%) (Descendants of the settlers and European immigrants) and Asian (10%) [22b]. Research has demonstrated that people living in disadvantaged neighborhoods have increased exposure to sleep disturbance risk factors such as inappropriate light, noise, allergens, tobacco or air pollution. In addition to the higher rate of poverty, racial/ethnic minorities report more frequently objective and perceived discrimination. Results reported by Jackson et al. [22] have shown that experiences of racial discrimination and internalization of negative racial bias contribute to the acceleration of vascular aging in Black males.

(3) *Distal factors* are occupational patterns, treatment access, and adherence, social conditions, and policies. For example, shift work is more current in African–Americans workers compared to their white counterparts and was reported to play a role in racial differences in sleep quantity. The proportion of job-related stress, low-wage jobs, and discrimination experiences was higher in black workers compared to whites ones.

FROM SLEEP HEALTH DISPARITIES TOWARD SLEEP HEALTH EQUITY

The CDC's Health Report *from 1999 to* 2014 indicates that trends in health were generally progressive for the overall population in the US. Differences in life expectancy, infant mortality, cigarette smoking among women, influenza vaccinations among those aged 65 and over, and health insurance *coverage narrowed among the racial and ethnic groups*. Nonetheless, during 1980–2014, life expectancy at birth for males and females was longest for white persons and shortest for black persons. For both males and females, racial differences in life expectancy at birth lessened, but persisted during 1980–2014. Furthermore, disparities by racial and ethnic group in the rate of high blood pressure and smoking among adult men persisted throughout the study period, with non-Hispanic black adults more likely to have high blood pressure than adults in other racial and ethnic groups throughout the period, and non-Hispanic black and non-Hispanic white males more likely to be current smokers than Hispanic and non-Hispanic Asian men. In summary, the authors of the report concluded that:

> *Despite improvements over time in many of the health measures presented in this Special Feature, disparities by race and ethnicity were found in the most recent year for all 10 measures,[a] indicating that although progress has been made in the 30 years since the Heckler Report, elimination of disparities in health and access to health care has yet to be achieved. [23, p. 21]*

Meanwhile, in the field of sleep medicine, although extensive efforts are being deployed by concerned clinicians and sleep researchers along with recommendations to develop research agenda and implement programs to decrease disparities in sleep health, the data presented previously suggest that there are still miles to go until society de facto attains sleep health equity [24]. Therefore we conclude the chapter proposing a conceptual framework intended as a roadmap toward sleep health which incorporates findings from our group's research initiatives in the field of behavioral sleep-research.

At the beginning of this chapter we adopted Buysse's [9] definition of *sleep health* as a multidimensional pattern of sleep-wakefulness, adapted to the individual, social, and environmental demands, that promotes physical and mental well-being. Moreover, equity is the absence of avoidable, unfair, or remediable differences among groups of people, whether those groups are defined socially, economically, demographically or geographically or by other means of stratification [13]. Pursuing Equity in Health implies pursuing the elimination of health disparities/inequities [25]. To effectively move toward sleep equity, we believe a first step would be to define such a concept. Thus borrowing from the definitions of sleep health and health equity, we define *sleep health equity* as:

---

a. Measures of mortality, natality, health conditions, health behaviors, and health care access and utilization, by race, race and ethnicity, or by detailed Hispanic origin.

Equal opportunities that are given to each individual and/or communities based on their need, no matter their age, sex, race/ethnicity, geographic location, and socio-economic status, to obtain recommended, satisfactory, efficient amount of sleep with appropriate timing that promotes physical and mental well-being.

Similarly to the analogy of sacred circle that has the power to generate unity and to heal, we propose that sleep health equity practice is the constant effort to provide adequate sleep health resources to each group and individual, and to avoid sleep health disparities inherent to socio-economic and environmental factors. Thus, moving toward sleep health equity implies to pay close attention to the importance of contextual factors such as culture, education, policies, funding, governance, institution, historic events, historic collaboration, community capacity and readiness, university capacity and readiness and the dynamic between them. Additionally, the question of mutual respect and trust, cultural relevance and sustained partnerships should receive as much attention in the process (Fig. 35.1).

We identify a list of problems related to patients, providers and the health-care-system that are undermining optimal sleep in disadvantaged communities, and then propose scientifically-informed potential policy examples that may contribute to decrease sleep health disparities [5, 22, 24, 26, 27]. See Table 35.2.

CONCLUSION

Although sleep is fundamental to general health, the American lifestyle and societal schedule are not consistent with healthy sleep patterns. Unfortunately, racial/ethnic minorities are the most affected by the sleep health crisis, which, for Blacks echoes of the historical context of the birth of America. Sleep medicine experts, specialists, practitioners, representatives and policymakers have an ethical responsibility to help to eliminate sleep health inequities. Although recent data demonstrated that in general there was improvement in sleep health parameters in both privileged groups and disadvantaged ones, there is still a lot to be accomplished in order to eliminate health disparities. Youth and adults from disadvantaged communities would benefit if sleep medicine emphasizes the definition of sleep health dimensions and encourages the changes in practice they embody, in addition to identifying and treating sleep disorders. The positive aspect of sleep health as defined previously, and the ideal of equity in health comprise the anchors for our proposed definition of sleep health equity, and its associated conceptual framework toward the elimination of sleep health inequities.

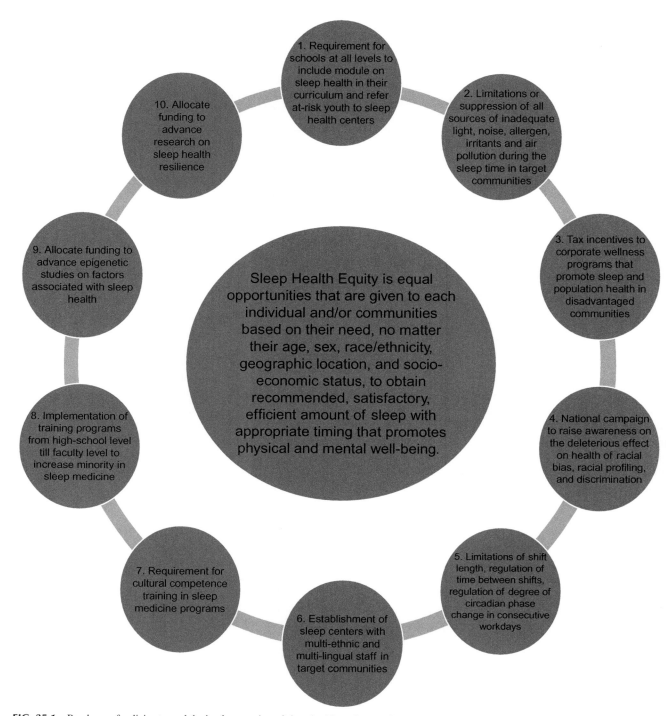

FIG. 35.1 Roadmap of policies toward the implementation of sleep health equity practices.

TABLE 35.2 Barriers to sleep health equity and policy solutions for reducing health disparities and advancing sleep health equity.

Sleep health equity barrier	Solution advancing sleep health equity
1. Higher prevalence of Sleep-Disordered-Breathing (SBD) among African–American children	Implementation of programs for sleep health literacy, early screening and treatment for sleep disorders since elementary schools. Requirement for schools at all levels to include module on sleep health in their curriculum and refer at-risk youth to sleep health centers
2. Greater exposure to environmental risk factors for poor sleep among racial/ethnic minorities living in disadvantaged neighborhoods	Implementation of a multilevel approach to reducing environmental factors that disturb sleep such as inadequate light, noise, allergen and irritants, and air pollution. Limitations or suppression of all sources of inadequate light, noise, allergen, irritants and air pollution during the sleep time in identified communities
3. A higher rate of short sleep duration that increases cardiovascular risk among individuals of African-descents and other minorities	Adopt a multilevel community-oriented sleep health and promotion education campaign (Ex: PEERS-ED, TASHE, and MetSO). Provide incentives to corporate wellness programs that promote sleep and population health among racial/ethnic communities
4. Racial/ethnic minorities, particularly Blacks are exposed to higher racial discrimination, which induces stress that undermines sleep	National campaign to raise awareness on the deleterious effects of racial bias, racial profiling, and discrimination on health. Increase severity of sanctions against racial discrimination nationwide
5. Blacks are disproportionately concerned with effects of shift work. Need stronger work schedule regulations	Limitations of shift length, regulation of time between shifts, regulation of degree of circadian phase changes in consecutive workdays
6. Culture and language barriers limit access to sleep health literacy among racial/ethnic minorities	Establishment of sleep centers with multi-ethnic and multi-lingual staff in vulnerable communities. Requirement for healthcare facilities in vulnerable communities to have a multi-ethnic and multi-lingual staff
7. Poor adherence to treatment of sleep disorders among minorities, particularly Blacks at risk for Obstructive Sleep Apnea (OSA)	A tailored behavioral intervention to increase adherence to physician recommendation (Ex: MetSO and PEERS-ED studies). Requirement for cultural competency training in sleep medicine programs
8. Lack of minority in the field of sleep medicine	Implementation of training programs from the high-school to faculty level to increase minority representation in sleep medicine (Ex: PRIDE and COMRADE programs). Requirement for a specific quota of racial/ethnic minorities in the recruitment of future sleep specialists
9. Lack of research on epigenetic factors associated with sleep problems among children and adults	Implementation of multi-level research that explores links of individual and household/neighborhood factors with poor sleep. Allocate funding to advance epigenetic studies on factors associated with poor sleep health
10. Lack of research on psychological resilience factors that are protective against factors that negatively affect sleep and CVD	Implementation of multi-level research that explores links among, stress exposure, individual, social, cultural and physical factors that affect sleep. Allocate funding to advance research on sleep health resilience

REFERENCES

[1] Carskadon MA, Dement WC. Normal human sleep: an overview. In: Kryger TRM, editor. Principles and practice of sleep medicine. Elsevier; 2017. p. 15–24. e3.

[2] Hirshkowitz M, Whiton K, Albert SM, Alessi C, Bruni O, DonCarlos L, et al. National Sleep Foundation's sleep time duration recommendations: methodology and results summary. Sleep Health 2015;1(1):40–3.

[3] Watson NF, Badr MS, Belenk G, Bliwise DL. Recommended amount of sleep for a healthy adult. Am Acad Sleep Med Sleep Res Soc 2015;38(6):843–4. https://doi.org/10.5665/sleep.4716.

[4] Somers VK, White DP, Amin R, Abraham WT, Costa F, Culebras A, et al. Sleep apnea and cardiovascular disease: an American Heart Association/American College of Cardiology Foundation Scientific Statement from the American Heart Association Council for High Blood Pressure Research Professional Education Committee, Council on Clinical Cardiology, Stroke Council, and Council on Cardiovascular Nursing in collaboration with the National Heart, Lung, and Blood Institute National Center on Sleep Disorders Research (National Institutes of Health). J Am Coll Cardiol 2008;52(8):686–717.

[5] Barnes CM, Drake CL. Prioritizing sleep health: public health policy recommendations. Perspect Psychol Sci 2015;10(6):733–7. https://doi.org/10.1177/1745691615598509.

[6] Wheaton AG, Jones SE, Cooper AC, Croft JB. Short sleep duration among middle school and high school students—United States, 2015. MMWR Morb Mortal Wkly Rep 2018;67:85–90. https://doi.org/10.15585/mmwr.mm6703a1.

[7] Centers for Disease Control and Prevention CDC. Perceived insufficient rest or sleep among adults—United States, 2008. MMWR; 2009. October 30, 2009/58(42);1175–1179 https://www.cdc.gov/mmwr/preview/mmwrhtml/mm5842a2.htm.

[8] Institute of Medicine. Sleep disorders and sleep deprivation: an unmet public health problem. Washington, DC: The National Academies Press; 2006.

[9] Buysse D. Sleep health: can we define it? Does it matter? Sleep 2014;37(1):9–17. https://doi.org/10.5665/sleep.3298.

[10] Johnson DA, Billings ME, Hale L. Curr Epidemiol Rep 2018;5:61. https://doi.org/10.1007/s40471-018-0139-y.

[11] Grandner MA, Williams N, Knutson KL, Roberts D, Jean-Louis G. Sleep disparity, race/ethnicity, and socioeconomic position. Sleep Med 2016;18:7–18.

[12] Grandner MA, Patel NP, Gehrman PR, Xie D, Sha D, Weaver T, Gooneratne N. Who gets the best sleep? ethnic and socioeconomic factors related to sleep complaints. Sleep Med 2010;11(5):470–8. https://doi.org/10.1016/j.sleep.2009.10.006.

[13] Whitehead M. The concepts and principles of equity and health. Health Promot Int 1991;6(3):217–28. https://doi.org/10.1093/heapro/6.3.217.

[14] Liu Y, Wheaton AG, Chapman DP, Cunningham TJ, Lu H, Croft JB. Prevalence of healthy sleep duration among adults—United States, 2014. MMWR Morb Mortal Wkly Rep 2016;65(6):137–41. https://doi.org/10.15585/mmwr.mm6506a1.

[14a] Cappuccio FP, Cooper D, D'Elia L, Strazzullo P, Miller MA. Sleep duration predicts cardiovascular outcomes: a systematic review and meta-analysis of prospective studies. Eur Heart J 2011;32(12):1484–92. https://doi.org/10.1093/eurheartj/ehr007.

[14b] Donat M, Brown C, Williams N, Pandey A, Racine C, McFarlane SI, Jean-Louis G. Linking sleep duration and obesity among black and white US adults. Clin Pract (Lond, Engl) 2013;10(5):https://doi.org/10.2217/cpr.13.47.

[14c] Knutson KL, Van CE, Rathouz PJ, Yan LL, Hulley SB, Liu K, et al. Association between sleep and blood pressure in midlife: the CARDIA sleep study. Arch Intern Med 2009;169:1055–61.

[14d] Troxel WM, Buysse DJ, Matthews KA, Kip KE, Strollo PJ, Hall M, Drumheller O, et al. Sleep symptoms predict the development of the metabolic syndrome. Sleep 2010;33(12):1633–40.

[14e] Yamasaki F, Schwartz JE, Gerber LM, Warren K, Pickering TG. Impact of shift work and race/ethnicity on the diurnal rhythm of blood pressure and catecholamines. Hypertension 1998;32:417–23.

[14f] Gangwisch JE, Rexrode K, Forman JP, Mukamal K, Malaspina D, Feskanich D. Daytime sleepiness and risk of coronary heart disease and stroke: results from the Nurses' Health Study II. Sleep Med 2014;15(7):782–8.

[14g] Moore PJ, Adler NE, Williams DR, Jackson JS. Socioeconomic status and health: the role of sleep. Psychosom Med 2002;64(2):337–44.

[15] Adenekan B, Pandey A, McKenzie S, Zizi F, Casimir GJ, Jean-Louis G. Sleep in America: role of racial/ethnic differences. Sleep Med Rev 2013;17(4):255–62.

[16] Petrov ME, Lichstein KL. Differences in sleep between black and white adults: An update and future directions. Sleep Med 2016; https://doi.org/10.1016/j.sleep.2015.01.011. Elsevier.

[17] Guglielmo D, Gazmararian JA, Chung J, Rogers AE, Hale L. Racial/ethnic sleep disparities in US school-aged children and adolescents: a review of the literature. Sleep Health 2018; https://doi.org/10.1016/j.sleh.2017.09.005. Elsevier Inc.

[18] Rosen GM, Bendel AE, Neglia JP, Moertel CL, Mahowald M. Sleep in children with neoplasms of the central nervous system: case review of 14 children. Pediatrics 2003;112(1 Pt 1):e46–54. https://doi.org/10.1542/peds.112.1.e46.

[19] Douglas F. Narrative of the life of Frederick Douglass, an American Slave. New York: Penguin Classics; 1845.

[20] Reiss B. Wild nights: how taming sleep created our restless world. Basic Books; 2017.

[21] Williams NJ, Grandner MA, Snipes A, Rogers A, Williams O, Airhihenbuwa C, Jean-Louis G. Racial/ethnic disparities in sleep health and health care: Importance of the sociocultural context. Sleep Health 2015;1(1):28–35.

[22] Jackson C, Redline S, Emmons KM. Sleep as a potential fundamental contributor to disparities in cardiovascular health. SSRN; 2015 https://doi.org/10.1146/annurev-publhealth-031914-122838.

[22a] Colby SL, Ortman JM. Projections of the size and composition of the U.S. population: 2014 to 2060. Current Population Reports, P25-1143. Washington, DC: US Census Bureau; 2014. https://doi.org/P25-1143.

[22b] US Census Bureau. Historical poverty tables: people and families—1959 to 2017; 2018. Available at: https://www.census.gov/data/tables/time-series/demo/income-poverty/historical-poverty-people.html.

[23] National Center for Health Statistics. Health, United States, 2015: with special feature on racial and ethnic health disparities. Hyattsville, MD: National Center for Health Statistics; 2016.

[24] Jean-Louis G, Grandner M. Importance of recognizing sleep health disparities and implementing innovative interventions to reduce these disparities. Sleep Med 2016;18:1–2. https://doi.org/10.1016/j.sleep.2015.08.001.

[25] Braveman P. Health disparities and health equity: concepts and measurement. Annu Rev Public Health 2006;27(1):167–94. https://doi.org/10.1146/annurev.publhealth.27.021405.102103.

[26] Jean-Louis G, Newsome V, Williams NJ, Zizi F, Ravenell J, Ogedegbe G. Tailored behavioral intervention among blacks with metabolic syndrome and sleep apnea: results of the MetSO trial. Sleep 2017;40(1):https://doi.org/10.1093/sleep/zsw008.

[27] Williams NJ, Robbins R, Rapoport D, Allegrante JP, Cohall A, Ogedgebe G, Jean-Louis G. Tailored approach to sleep health education (TASHE): study protocol for a web-based randomized controlled trial. Trials 2016;17(1):585. https://doi.org/10.1186/s13063-016-1701-x.

Chapter 36

Obstructive sleep apnea in commercial motor vehicle operators

Indira Gurubhagavatula[a,b], Aesha M. Jobanputra[c], Miranda Tan[d]

[a]Department of Medicine, Division of Sleep Medicine, Perelman School of Medicine at the University Hospital of Pennsylvania Medical Center, Philadelphia, PA, United States, [b]Sleep Disorders Clinic, Philadelphia VA Medical Center, Philadelphia, PA, United States, [c]Department of Medicine, Division of Pulmonary and Critical Care Medicine, Rutgers Robert Wood Johnson Medical School, New Brunswick, NJ, United States, [d]Department of Medicine, Pulmonary Service, Section of Sleep Medicine, Memorial Sloan Kettering Cancer Center, New York, NY, United States

Drowsy driving has been likened to driving while intoxicated, with similar impairment in neurocognitive performance and driving capability [1]. Crash severity related to drowsy driving tends to be severe because the sleep-impaired driver is unable to attenuate impact by braking or steering away, resulting in heavier damages, serious injuries or death [2,3]. The National Highway Traffic Safety Administration (NHTSA) estimates that drowsy driving was responsible for 72,000 crashes, 41,000 injuries, and 800 deaths in 2013 [4]. These statistics, however, may be an underestimate [5,6] and recent data suggest that drowsy driving causes up to 9.5% of all motor-vehicle crashes [7], up to 10.8% of crashes that resulted in significant property damage, airbag deployment, or injury [7], and 16.5% of fatal crashes [6]. The key causes of drowsy driving include insufficient sleep, shift work disorder, medications, and untreated sleep disorders such as obstructive sleep apnea (OSA), chronic insomnia and narcolepsy [8,9].

The association between OSA and daytime sleepiness increases the risk of collision by two- to fourfold [10–12]. OSA is characterized by repetitive collapse of the upper airway, causing breathing to stop or obstruct partially during sleep. As a result, oxyhemoglobin saturation falls briefly, which triggers a surge in sympathetic activity and arousal from sleep. Disruptions during sleep can confer symptoms of daytime sleepiness, fatigue, and inability to sustain attention—these deficits are amplified under mundane conditions or while doing overlearned activities that require sustained attention, such as, for example, driving long distances on a rural highway [13].

Untreated OSA has been associated with an increased risk of multiple health conditions, including cardiovascular diseases, neurocognitive impairment, and metabolic syndrome [14]. OSA tends to be more common in obese individuals, men, post-menopausal women and middle-aged or older individuals [15]. Diagnosis can pose its own challenges, with the first obstacle being recognition of risk for the disease; it is estimated that up to 80% of people remain undiagnosed in the community [16]. Recent work has shown that OSA is highly prevalent in the population of commercial motor vehicle (CMV) operators [8]. This chapter will therefore focus on the prevalence, risks, and effects of OSA in the CMV operator.

PREVALENCE

OSA is much more common in CMV operators than in the general population [8, 11, 17, 18]. A higher proportion of CMV operators carry the common risk factors for the disorder, including middle age, male gender, and central obesity [8,11]. Studies among CMV operators suggest OSA prevalence ranging from 28% [17,19] to 78% [8,11,17,18]; in contrast, the prevalence of OSA in the general population of employed workers is 10–17% in men, and 9% in women aged 30–70 years [20]. Due to the rising prevalence of obesity, the overall prevalence of OSA continues to increase [21].

The FMCSA commissioned a study of 4280 CMV operators in Philadelphia, Pennsylvania; 1392 individuals responded and 407 of the at-risk respondents underwent an in-laboratory polysomnography (PSG) [17]. This study estimated the OSA prevalence among commercial vehicle operators to be approximately 28% [17]. A similar study was conducted in Australia and a higher prevalence of 60% was found, compared to the American study, which was not explained by modestly lower BMIs in the latter group, but may be due to participant bias [8, 15]. In a third study conducted by a large trucking company, 19,371 commercial drivers had employer-mandated screening for OSA with an on-line questionnaire [11]. Screening identified 30% (5908) drivers at high risk; of the drivers tested with polysomnography (PSG) 80% had OSA [11]. These data support the value of systematic screening in identifying latent cases of OSA in this safety-sensitive population [22].

HISTORY OF FEDERALLY-FUNDED RESEARCH AND REGULATORY ACTIVITY

The current regulation regarding OSA in CMV operators is vague and offers medical examiners little detail or specificity in evaluating for OSA. The rule states that the person being evaluated "has no established medical history or clinical diagnosis of [any] condition which is likely to cause loss of consciousness or any loss of ability to control a commercial motor vehicle." [23] This rule has remained unchanged since 1970 [23], despite a series of meetings and publications to address regulation by the administration (Fig. 36.1).

In July 1988, the Department of Transportation (DoT) and Federal Highway Administration (FHA) convened at the Conference on Neurological Disorders and Commercial Drivers in Washington, D.C. and released the recommendation that anyone who would lose consciousness while driving should be excluded from operating a commercial vehicle [24]. Specific conditions that could contribute to such loss of consciousness were cited, including sleep disorders such as sleep apnea syndrome, which could cause excessive daytime sleepiness [24].

The Federal Motor Carrier Safety Administration (FMCSA) was later established, with the mission of reducing injuries and fatalities involving large trucks and buses [22]. The FMCSA authorized research to estimate the prevalence of OSA in CMV operators, which was published in 2002 [17]. In 2001, the FMCSA listed a single question on the Fitness for Duty Evaluation Form for OSA, which combined four items: sleep disorders, pauses in breathing while asleep, daytime sleepiness and loud snoring [25]. Empiric evidence showed that over the course of the first year, responses to this question were rarely affirmative, raising questions about the accuracy of self-reported symptoms to identify risk during fitness-for-duty evaluations. To address this and other questions, the FMCSA convened a Medical Expert Panel to offer guidance on screening and management of OSA in CMV operators in 2008, followed by additional meetings to update this information from its Medical Review Board (FMCSA-MRB) and its Motor Carrier Safety Advisory Committee in 2011 [19,26].

These meetings identified the lack of uniformity in fitness for duty evaluations, and the tendency of CMV operators to "doctor-shop" in gaining medical certification. To address this issue, in 2015, the administration required medical examiners to undergo training and certification to perform screening for OSA during fitness for duty evaluations of CMV operators (before and after hire), and created a National Registry for Certified Medical Examiners [27]. In March 2016, the FMCSA and Federal Road Administration (FRA) issued a Notice of Proposed Rulemaking (NPRM) requesting prevalence data, cost and benefits of OSA evaluation and treatment in CMV operators [28]. The American Academy of Sleep Medicine (AASM) convened a task force in response to address sleep and transportation safety awareness and published these recommendations for access to the larger sleep medicine community [29]. Based on this input and suggestions from other stakeholders, the FMCSA-MRB met in August 2016 to issue its latest recommendations [30]; these recommendations would apply to all truck and rail CMV operators with moderate to severe apnea and were intended to become law [31]. However, the plan to mandate the screening of truck and rail operators for OSA was abandoned in August 2017 when the NPRM was withdrawn by the new administration [32].

SCREENING

Guidance for screening has been offered by several groups [19,29,30]. In general, guidance documents emphasize the use of objective rather than subjective measures when assessing risk during the initial evaluation. Self-reported subjective measures have been shown to be unreliable in several studies, with a preponderance of negative responses [33–35]. Although some operators may admit to subjective symptoms of OSA (e.g., history of sleepiness-related accidents, fatigue, sleepiness during duty hours) and typical symptoms of sleep apnea (e.g., snoring, gasping during sleep), research shows that such reporting is more commonly absent or unreliable [33–35]. In the study conducted by Dagan et al., OSA was detected in 77.7% of the drivers screened by PSG and 47.1% of them were sleepy according to the Multiple Sleep Latency Test (MSLT) [35]. None of the drivers, however, complained about sleep problems including snoring or excessive daytime sleepiness [35]. The absence of subjective reporting in the setting of objective findings consistent with OSA and sleepiness may be due to differential vulnerability to sleepiness based on genetic factors [36], or may be the result of underreporting due to concerns about employment. Therefore, emphasis should be placed on objective criteria.

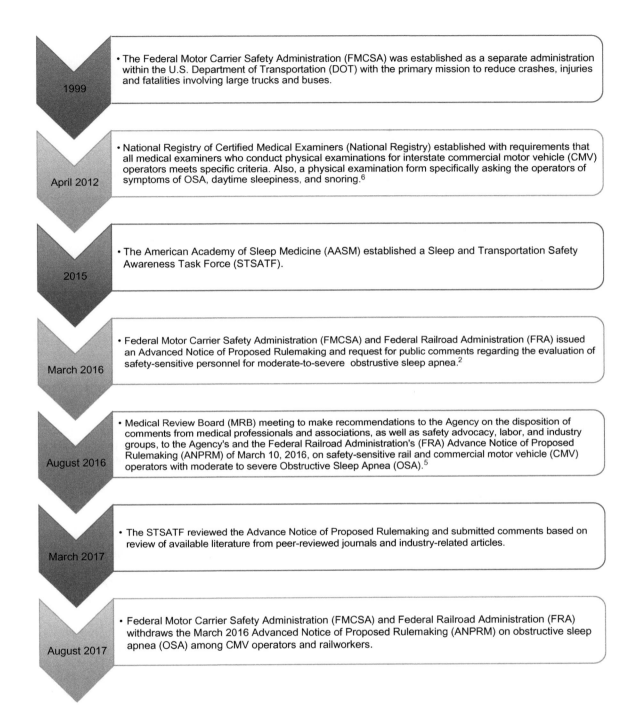

FIG. 36.1 Timeline leading up to current regulations on commercial motor vehicle drives.

Due to the high prevalence of OSA and unreliability of subjective criteria, the AASM and Transportation Safety Awareness Task Force (TSATF) have put forth primary and secondary screening criteria that rely heavily on objective measures (Table 36.1) [29]. Primary criteria refer to high body mass index (BMI), the presence of resistant hypertension or diabetes, or a history of a drowsiness-related crash. If a CMV operator meets primary criteria, they should be referred to a board-certified sleep medicine physician for a thorough evaluation.

The use of BMI as a primary screening tool is well-supported. In a population-based prospective study conducted by Peppard et al., a 10% weight gain was associated with sixfold increase in risk of developing

TABLE 36.1 Screening criteria for OSA recommended by AASM sleep and transportation safety awareness task force [11]

Primary criteria	Secondary criteria
1. BMI ≥ 40 kg/m^2	1. Symptoms of OSA

2. BMI ≥ 33 kg/m^2 and either
   a. Hypertension requiring ≥2 medications for control or
   b. Type 2 diabetes mellitus
	2. BMI 28–33 kg/m^2 with any of the following risk factors
	• Small or recessed jaw
	• Modified Mallampati classification 3 or 4
	• Neck size ≥17 in. (men), ≥15.5 in. (women)
	• Hypertension
	• Type 2 diabetes mellitus (especially if BMI > 30 kg/m^2)
	• Cardiovascular disease
	• Untreated hypothyroidism
	• Age ≥ 42 years
	• Family history of OSA
	• Male or postmenopausal female

3. Sleepiness-related crash or accident, off-road deviation, or rear-ending another vehicle by report or observation	
4. Fatigue or sleepiness during the duty period	

moderate-to-severe OSA [37]. In another prospective study conducted in Brazil, 34.5% of those with BMI of 25 to <30 kg/m^2 and 64.1% of those with BMI ≥ 30 kg/m^2 had OSA [38]. Lower BMI thresholds (such as 30 kg/m^2) [39] for screening would require a larger number of operators to be tested and may result in fewer missed cases, but may result in higher costs. Most organizations recognize BMI ≥ 35 kg/m^2 as increased risk for OSA and advocate for testing using this parameter [19,35,40,41]; the higher BMI threshold limits the number of operators tested, but results in missed cases.

INITIAL EVALUATION

If a CMV operator presents with two or more of the factors listed in Table 36.1, (s)he should be referred for a comprehensive evaluation by a board-certified sleep physician for diagnostic testing. The likelihood of having OSA with these relatively high BMI thresholds may be >80% [28]—if for any reason an initial diagnostic test (such as home-based testing or polysomnography, PSG) is found to be inconclusive or negative, additional evaluation may be warranted.

DIAGNOSIS

In 2008, the Medical Expert Panel advised the use of in-lab polysomnogram (PSG) to confirm the diagnosis of OSA in those who were at high risk for the condition upon screening [19]. PSG is the gold standard test for diagnosis of OSA [42] and can also help identify other sleep disorders, but is expensive and time-consuming. More recently, rapid advancements in diagnostic technologies have allowed home sleep apnea testing (HSAT) to become routine [43]. Typically, HSAT relies on three–four channels, rather than 12–16 used in PSG, to assess respiratory effort, airflow, oxygen saturation, and heart rate [43]. HSAT has many advantages, including portability, lower expense, convenience and accessibility. HSAT can be performed directly in the patient's preferred sleeping environment, rather than in a laboratory.

Despite the allure associated with HSAT, several limitations exist. HSAT is useful for confirming OSA in CMV operators with high preclinical suspicion, but is less useful for ruling out OSA and may be inconclusive [29]. In other words, a negative result is less useful than a positive, confirmatory result because of the high pre-test probability of the chosen sample. In addition to inconclusive results on HSAT, AHI severity is underestimated and thus OSA may also be underestimated or missed [44]. Finally, this type of study is done in a home setting where there is a risk of a person other than the intended worker being evaluated wearing the device. HSAT should be performed in conjunction with a comprehensive sleep evaluation under a board-certified sleep physician [43].

Sleep studies (both PSG and HSAT) are considered positive for sleep apnea if the number of apneas (cessation in airflow) or hypopneas (reduction in airflow) per hour of sleep (if PSG) or test time (if HSAT) exceed a certain threshold. This measure, known as the apnea-hypopnea index (AHI, for PSG) or respiratory event index (REI, for HSAT), is used to grade the severity of illness according to the following thresholds: normal <5 events/h; mild = [5–15) events/h; moderate = [15–30) events/h; and severe AHI ≥30 events/h [42,43].

TREATMENT

CMV operators with AHI ≥20 events/h (i.e., in the moderate to severe range) should receive prompt, definitive treatment with positive airway pressure (PAP) therapy, as this cohort is more likely to experience a sleepiness-related crash [19,41]. Although data linking crashes to those with AHI between 5 and 15 events/h are less reliable, this group may also benefit from treatment with PAP or alternative modalities, such as a mandibular advancement device, weight management, position therapy, or upper airway surgery [10].

MONITORING PAP THERAPY

If PAP is prescribed, treatment should be monitored in an ongoing fashion to assess for efficacy. Efficacy indicates adherence and effectiveness, both of which can be tracked using downloaded data from PAP devices [45–47]. Such data can now be obtained via wireless mechanisms [48], enabling prompt retrieval of information and intervention to improve efficacy, including correction of mask fit and addressing any side effects of PAP therapy (e.g., dry mouth, mask discomfort).

The three main criteria that must be addressed when evaluating the efficacy of treatment for OSA are: [1] resolution of OSA or residual AHI < 5 events/h; [2] adherence to therapy for ≥4 h of PAP use per night for ≥70% of days; and [3] improvement in the symptoms of the intended CMV operator [29]. The assessment of sleepiness in CMV operators post-therapy can be challenging; symptom reporting is unreliable and objective measures of sleepiness (e.g., MSLT and Maintenance of Wakefulness Testing) have not been shown to correlate with on-the-road performance [49]. Therefore, most centers rely heavily on downloaded data from PAP devices to evaluate treatment efficacy of OSA.

BENEFITS OF PAP THERAPY

Treatment of OSA with PAP improves daytime sleepiness within 2–7 days of treatment [50] and, more importantly, decreases the risk of motor vehicle accidents [8, 10, 51–55] based on simulated driving tests. A detailed cost analysis also found PAP therapy to be a cost-effective approach that reduces overall health care costs for individuals suffering from OSA [56]. Health payers and trucking companies using PAP to treat CMV operators with OSA save approximately $2.88 billion and up to $1.3–13.8 million annually, respectively [16,57]. Some studies contend that screening for and treating OSA with PAP results in major savings by reducing absenteeism and comorbidities, while increasing overall productivity [16]. Others have shown that the estimated costs of screening, diagnosis and treatment for OSA are justified by the reduction in costs related to prevented crashes, provided that a high proportion of patients adhere to treatment [56]. A program to screen and diagnose commercial drivers for OSA using PSG was found to cost only half as much as not screening at all, or $358–372 per driver [56]. This favorable cost analysis was attributed to the high cost of crashes, which would have otherwise occurred where screening, diagnosis and treatment were not performed [56].

EDUCATION

CMV operators should be educated regarding symptoms, risk factors and potential consequences of untreated OSA at the time of evaluation. Those who are found to be at low risk for OSA should be advised to seek re-evaluation if they experience a weight gain of at least 10%, as such an increase has been correlated with an increase in AHI by 32% [37] Rescreening is also advised if the CMV operator develops symptoms suggestive of OSA, or downstream consequences, such as hypertension or type 2 diabetes mellitus [29]. They should also be advised to avoid driving while sleepy for any reason, including insufficient sleep, circadian disorder or medications.

CONCLUSION

OSA is more common among CMV operators than in general groups because individuals in this cohort have a higher prevalence of the three common risk factors for OSA: obesity, male gender, and middle age. If left unaddressed, OSA can lead to daytime sleepiness—often a contributing factor in vehicular crashes—as well as other health conditions, ultimately creating tremendous economic costs. Therefore, providing a permissive environment for CMV operators to seek help without fear of employment loss is paramount. Implementation of comprehensive case identification and treatment programs can facilitate management of OSA while maintaining employment. Sleep medicine clinicians should be aware of the high prevalence, potential risks and specific challenges in identifying and treating OSA in this unique population. More specific guidance and a mandate from the federal government to diagnose and treat the condition are long overdue and may help mitigate the loss of life, injuries, property damage and large financial costs that occur annually due to preventable accidents from untreated OSA in the transportation industry.

REFERENCES

[1] Williamson AM, Feyer AM. Moderate sleep deprivation produces impairments in cognitive and motor performance equivalent to legally prescribed levels of alcohol intoxication. Occup Environ Med 2000;57:649.

[2] Strohl KP, Brown DB, Collop N, George C, Grunstein R, Han F, Kline L, Malhotra A, Pack A, Phillips B, Rodenstein D. An official American Thoracic Society clinical practice guideline: sleep apnea, sleepiness, and driving risk in noncommercial drivers. An update of a 1994 statement. Am J Respir Crit Care Med 2013;187(11):1259–66.

[3] Mukherjee S, Patel SR, Kales SN, Ayas NT, Strohl KP, Gozal D, Malhotra A. An official American Thoracic Society statement: the importance of healthy sleep. Recommendations and future priorities. Am J Respir Crit Care Med 2015;191(12):1450–8.

[4] National Highway Traffic Safety Administration. NHTSA Drowsy Driving Research and Program Plan. Available at https://www.nhtsa.gov/sites/nhtsa.dot.gov/files/drowsydriving_strategicplan_030316.pdf; 2016. [Accessed 1 July, 2018].

[5] Masten SV, Stutts JC, Martell CA. Predicting daytime and nighttime drowsy driving crashes based on crash characteristic models. In: 50th Annual Proceedings of the Association for the Advancement of Automotive Medicine; 2006.

[6] Tefft BC. Prevalence of motor vehicle crashes involving drowsy drivers, United States, 1999–2008. Accid Anal Prev 2012;45:180–6.

[7] Owens JM, Dingus TA, Guo F, Fang Y, Perez M, McClafferty J, Tefft BC. Prevalence of drowsy driving crashes: estimates from a large-scale naturalistic driving study. 2018. [AAA Foundation for Traffic Safety].

[8] Howard ME, Desai AV, Grunstein RR, et al. Sleepiness, sleep-disordered breathing, and accident risk factors in commercial vehicle drivers. Am J Respir Crit Care Med 2004;170(9):1014–21.

[9] Pack AI, Pack AM, Rodgman E, Cucchiara A, Dinges DF, Schwab CW. Characteristics of crashes attributed to the driver having fallen asleep. Accid Anal Prev 1995;27(6):769–75.

[10] Tregear S, Reston J, Schoelles K, Phillips B. Obstructive sleep apnea and risk of motor vehicle crash: systematic review and meta-analysis. J Clin Sleep Med 2009;5(6):573–81.

[11] Berger M, Varvarigou V, Rielly A, Czeisler CA, Malhotra A, Kales SN. Employer-mandated sleep apnea screening and diagnosis in commercial drivers. J Occup Environ Med 2012;54(8):1017–25.

[12] Burks SV, Anderson JE, Bombyk M, Haider R, Ganzhorn D, Jiao X, Lewis C, Lexvold A, Liu H, Ning J, Toll A. Nonadherence with employer-mandated sleep apnea treatment and increased risk of serious truck crashes. Sleep 2016;39(5):967–75.

[13] Gurubhagavatula I. Consequences of obstructive sleep apnoea. Indian J Med Res 2010;131(2):188.

[14] Dewan NA, Nieto FJ, Somers VK. Intermittent hypoxemia and OSA: implications for comorbidities. Chest 2015;147(1):266–74.

[15] Senaratna CV, Perret JL, Lodge CJ, Lowe AJ, Campbell BE, Matheson MC, Hamilton GS, Dharmage SC. Prevalence of obstructive sleep apnea in the general population: a systematic review. Sleep Med Rev 2017;34:70–81.

[16] American Academy of Sleep Medicine. "Hidden health crisis costing America billions." Underdiagnosing and undertreating obstructive sleep apnea draining healthcare system. Mountain View, CA: Frost & Sullivan; 2016. Available at: https://aasm.org/resources/pdf/sleep-apnea-economic-crisis.pdf. [Accessed 28 February 2019].

[17] Pack A, Dinges D, Maislin GA. Study of prevalence of sleep apnea among commercial truck drivers. Washington, DC: Federal Motor Carrier Safety Administration; 2002. Publication no. DOT-RT-02-030 https://ntl.bts.gov/lib/55000/55400/55447/DOT-RT-02-030.pdf. Published May 2002.

[18] Stoohs RA, Bingham LA, Itoi A, Guilleminault C, Dement WC. Sleep and sleep-disordered breathing in commercial long-haul truck drivers. Chest 1995;107(5):1275–82.

[19] Ancoli-Israel S. CC, Geroge CFP, Guilleminault C, Pack AI. Expert panel recommendations: obstructive sleep apnea and commercial vehicle driver safety January vol. 14, 2008.

[20] Peppard PE, Young T, Barnet JH, Palta M, Hagen EW, Hla KM. Increased prevalence of sleep-disordered breathing in adults. Am J Epidemiol 2013;177(9):1006–14.

[21] Young T, Palta M, Dempsey J, Skatrud J, Weber S, Badr S. The occurrence of sleep-disordered breathing among middle-aged adults. N Engl J Med 1993;328(17):1230–5.

[22] Federal Motor Carrier Safety Administration History. Available at: https://www.fmcsa.dot.gov/mission/about-us [Accessed 28 February 2019].

[23] Qualification of drivers; exemption applications; epilepsy and seizure disorders. Available at: https://www.fmcsa.dot.gov/regulations/notices/2018-03940 [Accessed 28 February 2019].

[24] Booker HE. Conference on neurological disorders and commercial drivers. Administration UDoTaFH, ed. Washington DC July 1988.

[25] Medical examination report for commercial driver fitness determination. Available at: https://www.fmcsa.dot.gov/sites/fmcsa.dot.gov/files/docs/Medical_Examination_Report_for_Commercial_Driver_Fitness_Determination_649-F%28604529.pdf [Accessed 28 February 2019].

[26] Williams JR AA, Tregear SJ. n.d.Evidence report: obstructive sleep apnea and commercial motor vehicle driver safety, updated review: 11/30/2011. November 30, 2011.

[27] National Registry of Certified Medical Examiners. Available at: https://www.fmcsa.dot.gov/regulations/national-registry/national-registry-certified-medical-examiners [Accessed 28 February 2019].

[28] Evaluation of Safety Sensitive Personnel for Moderate-to-Severe Obstructive Sleep Apnea n.d.. In: Transportation Do, ed. Vol 81 March 10, 2016:6.

[29] Gurubhagavatula I, Sullivan S, Meoli A, et al. Management of Obstructive Sleep Apnea in commercial motor vehicle operators: recommendations of the AASM sleep and transportation safety awareness task force. J Clin Sleep Med 2017;13(5):745–58.

[30] American Academy of Sleep Medicine. American Academy of Sleep Medicine—comments; 2016. Available at: https://www.regulations.gov/document?D=FRA-2015-0111-0065 [Accessed 28 February 2019].

[31] Federal Motor Carrier Safety Administration. Advance Notice of Proposed Rulemaking (ANPRM) on Obstructive Sleep Apnea Medical Review Board (MRB) Meeting, August 22–23, 2016. Available at: https://www.fmcsa.dot.gov/medical-review-board-mrb-meeting-topics [Accessed 28 February 2019].

[32] Federal Motor Carrier Safety Administration, Federal Railroad Administration. Evaluation of safety sensitive personnel for moderate-to-severe obstructive sleep apnea. Federal Register; 2017. Available at: https://www.federalregister.gov/documents/2017/08/08/2017-16451/evaluation-of-safety-sensitive-personnel-for-moderate-to-severe-obstructive-sleep-apnea [Accessed 28 February 2019].

[33] Talmage JB, Hudson TB, Hegmann KT, Thiese MS. Consensus criteria for screening commercial drivers for obstructive sleep apnea: evidence of efficacy. J Occup Environ Med 2008;50(3):324–9.

[34] Parks P, Durand G, Tsismenakis AJ, Vela-Bueno A, Kales S. Screening for obstructive sleep apnea during commercial driver medical examinations. J Occup Environ Med 2009;51(3):275–82.

[35] Dagan Y, Doljansky JT, Green A, Weiner A. Body mass index (BMI) as a first-line screening criterion for detection of excessive daytime sleepiness among professional drivers. Traffic Inj Prev 2006;7(1):44–8.

[36] Van Dongen HP, Bender AM, Dinges DF. Systematic individual differences in sleep homeostatic and circadian rhythm contributions to neurobehavioral impairment during sleep deprivation. Accid Anal Prev 2012;45(Suppl):11–6.

[37] Peppard PE, Young T, Palta M, Dempsey J, Skatrud J. Longitudinal study of moderate weight change and sleep-disordered breathing. JAMA 2000;284(23):3015–21.

[38] Tufik S, Santos-Silva R, Taddei JA, Bittencourt LR. Obstructive sleep apnea syndrome in the Sao Paulo epidemiologic sleep study. Sleep Med 2010;11(5):441–6.

[39] Gurubhagavatula I, Maislin G, Nkwuo JE, Pack AI. Occupational screening for obstructive sleep apnea in commercial drivers. Am J Respir Crit Care Med 2004;170(4):371–6.

[40] Hartenbaum N, Collop N, Rosen IM, et al. Sleep apnea and commercial motor vehicle operators: statement from the joint Task Force of the American College of Chest Physicians, American College of Occupational and Environmental Medicine, and the National Sleep Foundation. J Occup Environ Med 2006;48(9 Suppl):S4–37.

[41] Parker DR HB. n.d. Motor carrier safety advisory committee and medical review board task: final report on obstructive sleep apnea (OSA). In: Transportation Do, ed2012.

[42] Kapur VK, Auckley DH, Chowdhuri S, et al. Clinical practice guideline for diagnostic testing for adult obstructive sleep apnea: an American Academy of sleep medicine clinical practice guideline. J Clin Sleep Med 2017;13(3):479–504.

[43] Collop NA, Anderson WM, Boehlecke B, et al. Clinical guidelines for the use of unattended portable monitors in the diagnosis of obstructive sleep apnea in adult patients. Portable monitoring task force of the American Academy of sleep medicine. J Clin Sleep Med 2007;3(7):737–47.

[44] Bianchi MT, Goparaju B. Potential underestimation of sleep apnea severity by at-home kits: rescoring in-laboratory polysomnography without sleep staging. J Clin Sleep Med 2017;13(4):551–5.

[45] Berry RB, Kushida CA, Kryger MH, Soto-Calderon H, Staley B, Kuna ST. Respiratory event detection by a positive airway pressure device. Sleep 2012;35(3):361–7.

[46] Cilli A, Uzun R, Bilge U. The accuracy of autotitrating CPAP-determined residual apnea-hypopnea index. Sleep Breath 2013;17(1):189–93.

[47] Desai H, Patel A, Patel P, Grant BJ, Mador MJ. Accuracy of autotitrating CPAP to estimate the residual apnea-hypopnea index in patients with obstructive sleep apnea on treatment with autotitrating CPAP. Sleep Breath 2009;13(4):383–90.

[48] Schwab RJ, Badr SM, Epstein LJ, et al. An official American Thoracic Society statement: continuous positive airway pressure adherence tracking systems. The optimal monitoring strategies and outcome measures in adults. Am J Respir Crit Care Med 2013;188(5):613–20.

[49] Antic NA, Catcheside P, Buchan C, et al. The effect of CPAP in normalizing daytime sleepiness, quality of life, and neurocognitive function in patients with moderate to severe OSA. Sleep 2011;34(1):111–9.

[50] Tregear S, Reston J, Schoelles K, Phillips B. Continuous positive airway pressure reduces risk of motor vehicle crash among drivers with obstructive sleep apnea: systematic review and meta-analysis. Sleep 2010;33(10):1373–80.

[51] Shiomi T, Arita AT, Sasanabe R, et al. Falling asleep while driving and automobile accidents among patients with obstructive sleep apnea-hypopnea syndrome. Psychiatry Clin Neurosci 2002;56(3):333–4.

[52] Horstmann S, Hess CW, Bassetti C, Gugger M, Mathis J. Sleepiness-related accidents in sleep apnea patients. Sleep 2000;23(3):383–9.

[53] George CF. Reduction in motor vehicle collisions following treatment of sleep apnoea with nasal CPAP. Thorax 2001;56(7):508–12.

[54] Teran-Santos J, Jimenez-Gomez A, Cordero-Guevara J. The association between sleep apnea and the risk of traffic accidents. Cooperative group Burgos-Santander. N Engl J Med 1999;340(11):847–51.

[55] Young T, Blustein J, Finn L, Palta M. Sleep-disordered breathing and motor vehicle accidents in a population-based sample of employed adults. Sleep 1997;20(8):608–13.

[56] Gurubhagavatula I, Nkwuo JE, Maislin G, Pack AI. Estimated cost of crashes in commercial drivers supports screening and treatment of obstructive sleep apnea. Accid Anal Prev 2008;40(1):104–15.

[57] Watson NF. Health care savings: the economic value of diagnostic and therapeutic care for obstructive sleep apnea. J Clin Sleep Med 2016;12(8):1075–7.

Chapter 37

Sleep health as an issue of public safety

Matthew D. Weaver[a,b], Laura K. Barger[a,b]
[a]Division of Sleep and Circadian Disorders, Brigham and Women's Hospital, Boston, MA, United States,
[b]Division of Sleep Medicine, Harvard Medical School, Boston, MA, United States

INTRODUCTION

The public safety net is comprised of three principal components: Police (law enforcement), Fire (rescue services), and Emergency Medical Services (medical services). Highly reliable 24×7 operations are critical to protect the health and safety of the public and effectively assist them in their time of need. The occupation is uniquely challenging. The work environment is uncontrolled, high-stress, and unpredictable. Rapid, risk-averse decision-making is critical. Tasks are often physically demanding. Public safety professionals also must operate motor vehicles in an unconventional manner, using lights and sirens on crowded roadways.

As is common in many occupations, work schedules are often based on tradition rather than sleep or circadian principles. Rapid backward rotation of shifts persists [1]. Police officers often work extended weekly work hours which are compounded by court appearances and special details outside of their scheduled work shifts. Extended duration shifts (≥24h) are the most commonly scheduled shift among firefighters, and 12 or 24h shifts are most common in the Emergency Medical Services (EMS) setting [2–4]. Seniority drives competition for planned overtime shifts, and unplanned overtime is a regular occurrence. Opportunities for rest are largely unpredictable. Secondary employment is also routine [5]. Furthermore, the prevalence of sleep disorders in this population approaches 40% [6, 7].

One can imagine the myriad of ways that sleep health can impact this occupational group. Sleep disorders drive up the risk of crashes, injuries, and chronic health conditions [6, 7]. Emerging research is beginning to yield even more insights about the pervasive impact of fatigue. Police officers who slept less prior to a test were more likely to exhibit implicit racial bias by associating Black Americans with weapons [8], and were more likely to make errors in shoot/don't shoot situations [9]. Fatigue impairs balance and increases gait variability, which are critical to prevent firefighters from falling in active fire situations [10]. EMS providers who are fatigued are more likely to report occupational injuries, medical errors, and actions which compromise the safety of themselves and their patients [3]. Less than optimal health, safety and performance of police officers, firefighters and EMS providers may negatively impact public safety through crashes, errors, and other mishandling of situations, ultimately compromising the integrity of the public safety net.

The purpose of this chapter is to outline our current understanding of sleep health in the public safety setting. We discuss work hours and scheduling characteristics of public safety personnel, the consequences of work hours on sleep, including the physiological sleep factors that determine alertness and performance, sleep disorders in public safety personnel and finally, the potential of fatigue risk management programs to improve the health and safety of first responders.

DEMOGRAPHICS

Organizational structure

Police are most often employed by the local or city government and their coverage area is arranged by precincts. It is common practice for Police to actively patrol their service area throughout their work shifts. Fire and EMS services are less consistent in their organization and structure. Fire departments service a clearly defined geographic area. They tend to maintain operations at the fire station and respond to calls as they are received. Fire personnel are often volunteers, and as such, would respond to the station from their current location in the event of a call, staff the emergency response vehicle, and then respond to the scene of the emergency. Approximately 40% of fire departments also employ cross-trained EMS personnel. These departments deliver both rescue and medical services. However in many communities, EMS and fire are separate entities. EMS agencies may be government, private, or hospital-based. EMS agencies are increasingly adopting system status management practices, which involves positioning ambulances throughout the service area to minimize emergency response times [11]. These industry practices are relevant when considering the opportunity for and applicability of various fatigue risk management strategies.

Individuals

There are approximately 650,000 police officers, 1.2 million firefighters, and 825,000 EMS providers in the United States (Table 37.1). Approximately 90% of police officers are male, as are 95% of firefighters, while one in four EMS providers are female [12]. Racial and ethnic minorities are often underrepresented in these occupational groups [12]. EMS professionals tend to be younger than police or firefighters. The vast majority of public safety personnel are overweight or obese [13, 14].

Many firefighters serve as volunteers (70%) and scheduling practices often provide 48 or 96 h off between shifts. Likewise, 70% of EMS providers are either volunteers or part-time employees. Consequently, it is common for public safety professionals to work more than one job. Approximately 40% of EMS providers work for more than one EMS agency [15], logging an average of 25 h per month at the second job [16], while one in three firefighters holds multiple jobs [17].

TABLE 37.1 Demographic characteristics of public safety personnel.

	Police	Fire	EMS
Workforce Size[a]	657,690	1,160,450	826,000
Age (mean years)	39	38	35
Gender			
Male	88%	95%	72%
Female	12%	5%	28%
Racial or Ethnic Minority	33%	21%	22%
Overweight or Obese	80%	80%	71%
Most Common Shift	8 h	24 h	12 or 24 h
2016 Nonfatal Occupational Injury Rate (per 100 FTE)[a]	10.2	9.5	7.8

[a]Source: U.S. Bureau of Labor Statistics, U.S. Department of Labor.

WORK HOURS AND SCHEDULING CHARACTERISTICS

Public safety professionals must be available for duty 24×7. This requires employment outside of regular daylight hours to fulfill workforce needs. Although federal regulations strictly limit the number of consecutive hours that truck drivers can drive and that pilots can fly, there are no standardized regulations which limit work hours among public safety personnel.

The optimal timing and duration of work hours is an interesting and multifaceted problem. Shorter shifts may permit employees to maintain high levels of vigilance throughout the duration of the shift and thus may be safer. In industry, short shifts have been associated with greater individual productivity and job satisfaction [18]. However, these schedules require more workers to be hired and trained to fulfill workforce needs. Longer shifts in general allow for a smaller overall workforce, thus lowering overhead benefit costs [19]. However, longer shifts may introduce a greater risk of fatigue-related performance deficiency [20]. Extended shifts likely require more preparation on the part of the worker to arrive capable of working for an extended period. Extended shifts also require a comparatively longer duration of downtime after the shift for recovery purposes, particularly if the shift involves nighttime work [21].

Shift duration

Police tend to work shifts of <12 h duration [22], while extended duration (≥24 h) shifts are often utilized by fire and EMS services. The most commonly scheduled shift duration is 24 or 48 h in U.S. fire departments [2]. Particularly in the western United States, the 48/96 schedule is becoming increasingly popular [23]. Firefighters working these very long shifts often commute extended distances. The most common shift lengths in EMS are 12 or 24 h in duration. In a national study of 511 EMS workers, approximately 50% (48.5%) of respondents reported working 24-h shifts, while 38.4% reported working 12-h shifts [3]. It is commonly believed that rural agencies are more likely to schedule shifts of 24 h or longer duration. An increased prevalence of extended shifts in rural areas may be necessary to provide 24-h coverage with smaller workforces.

Weekly work hours

There are few reliable estimates of work hours among police. There are multiple examples of police officers averaging >80 weekly work hours over the course of year [24]. Many of these hours are accumulated through overtime. Vila found that police in large, urban departments average between 15 and 40 h of overtime per month, though some officers exceeded 80 h of overtime monthly [25]. In our survey of nearly

7000 firefighters employed by 66 departments nationally, respondents reported working an average of 64 h per week after accounting for overtime and secondary employment [6].

The Longitudinal EMT Attribute Demographic Study (LEADS) is a 10-year, longitudinal survey of nationally-registered EMS providers. The LEADS survey found that EMS personnel were available for response approximately 50 h per week [5]. The high prevalence of part-time work may lead to a subset of providers covering the majority of shifts. One national effort found that 1/3 of all shifts are worked by EMS providers who have already worked at least 48 h in the 7-day preceding the shift, and 10% of shifts are staffed by providers who have already exceeded 60 h of work in that week [26]. Recovery between shifts, measured by the Occupational Fatigue Exhaustion Recovery (OFER) scale, is highest for providers who work extended duration shifts, and lowest for EMS providers who work 12 h shifts, likely as a result of schedule compression [27].

The association between work schedules and health and safety outcomes

The bulk of evidence to inform the association between work schedules and health and safety outcomes among police was generated through the Buffalo Cardio-Metabolic Occupational Police Stress (BCOPS) study. BCOPS was a cross-sectional study of Buffalo police which enrolled 65% of the urban police force between 2004 and 2009. BCOPS found that night shift work was associated with poor sleep quality [28], as well as extended absence for sick leave [29]. When combined with short sleep duration or overtime hours, the officers who worked night shifts were also more likely to meet criteria for metabolic syndrome [30]. Furthermore, the night shift is also associated with injury risk, conferring a 72% increased risk of injury [31].

Similarly, temporal patterns of work-related injury among firefighters are closely aligned with the circadian rhythm of alertness, with the highest risk of injury for calls occurring at 0200 h [32]. The work schedules of firefighters have also been associated with health outcomes, though not in a consistent manner. Firefighters who worked 48 h shifts were significantly more likely to have excessive daytime sleepiness relative to firefighters on 24-h or 10/14-h schedules, and those with excessive daytime sleepiness were twice as likely to report depressive symptoms [33]. However, cardiovascular disease is the leading cause of on-duty death among firefighters, and 24-h shifts may increase the risk of elevated diastolic blood pressure in this group [34, 35].

Evaluations of shift schedules in EMS have focused on safety and clinical outcomes. Allen et al. compared the endotracheal intubation success rates of Air Medical providers for 12-h and 24-h shifts after an organization wide change in shift length [36]. They concluded that since success rates were not different before and after the change, the psychomotor agility of providers was not affected by increasing shift length from 12 to 24 h. Similar studies found no difference in cognitive performance for 12- vs. 18-h shifts in a population of 10-flight nurses [37], or 12- vs. 24-h shifts in a population of helicopter EMS providers [38].

LEADS data have also been utilized to evaluate the prevalence of sleep problems in EMS workers nationally [39]. Among respondents to this survey, sleep maintenance disorder was more common in providers working 24-h shifts, those working >40-h in a week, and those working in rural areas. These findings are aligned with those from 30 EMS agencies nationally who administered the Pittsburgh Sleep Quality Index (PSQI) and Chalder Fatigue Questionnaire (CFQ) to determine the prevalence of poor sleep quality and severe fatigue in their workforces [3]. Nearly 60% of respondents reported poor sleep quality (PSQI>5), and 55% were found to have severe mental or physical fatigue. The prevalence of fatigue was highest among those who worked 24-h shifts. Fatigue was associated with nearly twice the odds of injury, 2.2 times the odds of medical error, and >threefold increased odds of safety-compromising behaviors.

The demand for medical emergencies far exceeds that of fire or rescue services [40]. As such, opportunities for sleep on-shift may be less common for EMS providers compared to firefighters. While firefighters have adopted increasingly longer shift durations with little evidence of adverse safety outcomes, these schedules are hazardous in the EMS setting. An observational study of nearly 1 million shifts over a three period determined that the risk of occupational injury or illness was increased for shifts ≥10 h duration [4]. Relative to 8-h shifts, 12-h shifts were associated with a 43% increased risk and 16-h shifts an 82% increased risk, while 24-h shifts more than doubled the risk of an occupational injury or illness. The National Highway Traffic Safety Administration subsequently supported The Fatigue in EMS Research Project—a series of systematic reviews and meta-analyses designed to develop evidence-based guidelines for Fatigue Risk Management in EMS [41]. While the overall evidence base was considered to be low quality, the expert-panel recommended that EMS shift duration should be <24 h [11].

Implementation of schedules based on sleep and circadian principles

There have been few efforts to optimize work schedules based on sleep and circadian principles, though experimental evidence has demonstrated that matching shifts with chronotype (with early chronotypes working morning shifts, and evening chronotypes on evening or night shifts) improves sleep duration, sleep quality, and measures of well-being [42]. In one of the few investigations of chronotype and scheduling in public safety, police officers reporting an evening chronotype actually had lower sleep quality and shorter sleep duration on night shift schedules compared to morning types [1]. However, this design may have been compromised by the rapid, counterclockwise rotation of shifts worked by the police officers under study. There is an opportunity for vast improvements over the current scheduling paradigms. Future efforts should seek to identify and evaluate work schedules that optimize alertness and may be utilized by public safety personnel.

PHYSIOLOGICAL DETERMINANTS OF ALERTNESS

There are four major physiological determinants of fatigue, alertness and performance: (1) circadian phase (biological time of day); (2) number of hours awake; (3) nightly sleep duration; and (4) sleep inertia (impaired performance upon waking). Circadian misalignment [43–46], acute sleep deprivation [47–51], chronic sleep deficiency [52–57], and abrupt awakening [58–61], often inherent to police, firefighter and EMS schedules, have each been independently associated with decrements in performance, and an increased risk of errors and accidents.

Physiological determinants of fatigue in public safety

The detrimental effects of each of these four factors are exacerbated by the long work hours, night and rotating shifts, extended-duration shifts and quick turn-arounds inherent in public safety schedules that are required to cover 24 h per day, 365 days per year (Fig. 37.1).

Alertness and performance vary rhythmically with a period of roughly 24 h [62, 63] driven by an endogenous circadian pacemaker located in the suprachiasmatic nucleus of the hypothalamus [64]. The largest performance decrements are seen when participants are awake during the biological night, with the worst performance several hours before normal wake time (e.g., ~3:00–6:00 am) [43–46]. Further, not only is the ability to stay alert dependent on the time of day, but the quality and quantity of sleep also vary with circadian phase such that sleep during the day is shorter and of poorer quality than sleep during the night [65–67]. Thus, night shift workers are commonly unable to sleep during daytime hours and are fatigued at night [68–70]. Not surprisingly, there is an increased rate of industrial and driving accidents during the night as compared to the day [71]. Public safety personnel regularly work during the biological night when the endogenous drive for alertness is lowest. As mentioned previously, firefighters and police have increased rates of occupational injury on the night shift [2, 31, 32].

Extended-duration shifts are common among public safety personnel, requiring long continuous episodes of wakefulness that induce fatigue. Acute sleep deprivation causes decrements in human alertness and performance, independent of the circadian system [47–51]. Every hour that one is awake, the homeostatic drive to sleep increases resulting in deteriorating performance. This deterioration results in an increase in the risk of fatigue-related fatal truck crashes with increased hours driving and awake [72]. Compared with the first hour, there is more than a 15-fold increase in the risk of a fatigue-related fatal crash after 13 h of driving. Long transports are common for fire and EMS in rural areas or in areas where specialized care

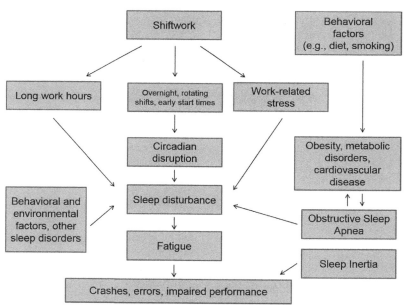

FIG. 37.1 Factors which interact to impact safety and performance in public safety.

is not available locally. In one national study, 4% of EMS providers reported tiredness-related difficulties operating the ambulance for short distances in the past month, while more than twice as many (10%) reported such difficulties operating the ambulance for long distances [39].

Fatigue-related impairments also manifest in increased injury risk. In a review of newly implemented 48-h shifts, it was reported that firefighters had significantly more injuries in the second day as compared to the first day of the shift [2]. In a cross-sectional study of 511 EMS providers nationally, the proportion of providers considered to be fatigued was highest among those working 24h shifts [3].

Public safety personnel are regularly exposed to chronic partial sleep deprivation when they fail to obtain adequate recovery sleep after working long shifts or shifts scheduled too close together. The history of nightly sleep duration has also been demonstrated to affect performance. Sleep loss on a nightly basis, chronic sleep deficiency, results in a sleep "debt." The consequences of the sleep debt are cumulative and affect health and performance [52–56]. Participants restricted to approximately 5h of sleep per night for seven nights exhibit significantly more lapses on a vigilance task [57]. Loss of even 2h of nightly sleep for five to seven consecutive nights causes decrements in performance comparable to those seen after 24h of continuous sleep deprivation. After 12–14 consecutive nights at this level of sleep restriction, lapses of attention on the task were comparable to those observed after 48h of total sleep deprivation [57]. Nearly 60% of EMS providers report poor sleep quality [3], and >1/3 have excessive daytime sleepiness [39]. Nearly 30% of police officers also have excessive daytime sleepiness [7].

Public safety personnel who routinely obtain inadequate nightly sleep and also work extended duration shifts experience even worse decrements in performance as there is synergy between acute and chronic sleep loss. The rate of deterioration in performance during extended (>16h) wakefulness is greatly increased, particularly during the circadian night, when accompanied by the chronic sleep deficiency that often builds up when working 24h shifts [73]. When acute sleep deprivation occurs on a background of chronic sleep deficiency, performance during 28h of wakefulness was 10-fold worse following 3 weeks of chronic sleep restriction, even when participants were tested after 10h of recovery sleep [73].

Public safety personnel who do manage to sleep when on-shift overnight are often asked to perform emergent actions immediately upon awakening (e.g., firefighters driving immediately after being awakened by an alarm). In fact, the time it takes to leave the station following a call for service is a quality measure, with "chute time" expected to be 1 min or less. This can be dangerous as alertness and performance are markedly impaired immediately following awakening. This impairment, known as sleep inertia, is more profound when workers are sleep deprived or have been awakened at an adverse circadian phase (e.g., during the night shift) [61]. Chronic sleep deficiency, which increases the depth of subsequent sleep, also worsens the adverse effects of sleep inertia. The effects of sleep inertia dissipate over time in an asymptotic manner [61]. The consequences of its impact upon awakening are particularly relevant to first responders and present an additional challenge for fatigue risk management programs. Sensory activations which accompany calls for service (such as alarms, lights, and sirens) may promote wakefulness during this period of vulnerability, but their effects remain understudied.

Sleep deficiency and health

Sleep deficiency is an underlying cause of many short- and long-term health problems. Sleep deficiency and working during an adverse circadian phase have been linked with increased risks of weight gain, obesity, cardiovascular disease, stroke, myocardial infarction, and cancer [74]. Workers who routinely work extended hours and night shifts are at particularly high risk of suffering adverse health consequences. Nearly half (45%) of the deaths that occur among US firefighters while at work are attributed to heart disease [2]. Police officers also have an increased risk of cardiovascular disease [75].

In addition to the other common risk factors (e.g., stress, burnout) that first responders face in their jobs, sleep deficiency can exacerbate their risk of poor health. Increased stress experienced by first responders may exacerbate sleep disruption and consequently sleepiness. Sleep deficiency, sleep disorders and shift work interact with the processes controlling appetite and metabolism, increasing the risk of weight gain, which is a risk factor for sleep apnea, and long term increases the risk of cardiovascular disease and diabetes [74]. The vast majority of public safety personnel are overweight or obese (Table 37.1). Public safety personnel in less than optimal health are at risk for adverse events at work. In this setting, adverse events can have far-reaching implications for the safety of the public. Safety-conscious scheduling, along with sleep health interventions, may help to reduce these risks. Importantly, sleep health interventions must address undiagnosed and untreated sleep disorders.

SLEEP DISORDERS

An Institute of Medicine report declared sleep deficiency and untreated sleep disorders an unmet public health problem, estimating 50–70 million people in the United States are living with a sleep disorder [76]. In addition to the degradation in an individual's health, alertness, performance, safety and quality of life, untreated sleep disorders are responsible for substantial costs to employers and society. Sleep deficient individuals and those who have a sleep disorder have up to a 20% increased utilization of the health care system [77]. In the year before diagnosis with obstructive sleep apnea (OSA), a common sleep disorder characterized by repetitive pharyngeal collapse during sleep [78], individual's medical costs were almost twice as much as those without OSA [79]. Costs to employers and society are systemic and are revealed through increased rates of absenteeism, disability day usage, reduced productivity (presenteeism), injuries, accidents and even increased alcohol consumption [80]. These indirect costs have been estimated in the hundreds of billions of dollars [81]. In the case of public safety personnel, untreated sleep disorders are not only a threat to personal health, but may also endanger the public in their role as a public safety workforce. Furthermore, costs associated with sleep disorders are avoidable, but are often borne by taxpayer support.

In a cross-sectional survey of 4957 police officers, 40.4% screened positive for at least one sleep disorder [7] (Fig. 37.2). The most common sleep disorder was obstructive sleep apnea (33.6%), followed by shift work disorder (14.5%), and insomnia (6.5%). Similarly, in a nationwide survey of 6933 firefighters, 37.2% firefighters screened positive for a sleep disorder [6]. Again, the most common sleep disorder was obstructive sleep apnea (28.4%), followed by shift work disorder (9.1%) and insomnia (6.0). In the subset of firefighters who reported their primary responsibility as medical care (fire-based EMS), 45% screened positive for at least common sleep disorder, with 33.9% screening positive for obstructive sleep apnea, 10.1% for shift work disorder, and 7.5% for insomnia [82]. Across the survey studies, >80% of those who screened positive for a sleep disorder were previously undiagnosed and untreated.

In a widespread cross-sectional survey of police officers ($n=4957$) from the United States and Canada, positive sleep disorder screening was associated with adverse health and safety outcomes [7]. Compared with those officers who did not screen positive for a sleep disorder, positive screening was associated with more than twice the prevalence of reported depression and burnout-emotional exhaustion and three times the risk of anxiety. Compared to officers who did not screen positive for OSA, positive screening was also associated with approximately twice the risk of diabetes and cardiovascular disease.

In a 2-year follow-up period, 3545 officers completed 15,735 online monthly surveys (6587 person-months with positive screens and 9148 with negative screens for sleep disorders) to capture performance and safety outcomes. Each officer completed approximately four monthly surveys. Compared to those officers who did not screen positive for a sleep disorder, officers who were prospectively identified as screening positive for a sleep disorder had higher risk of reporting a serious administrative error, falling asleep while driving, an error or safety violation attributed to fatigue, occupational injury and other adverse work-related performance measures, including uncontrolled anger toward a suspect, absenteeism, and falling asleep during meetings (Fig. 37.3).

Similarly, in a nationwide survey of firefighters, firefighters who screened positive for OSA were twice as likely to report an MVC and falling asleep while driving; 85% of MVCs were documented with police reports or detailed descriptions and 48% occurred at work or during commutes. Similarly, near miss crashes and injuries were elevated in those screening positive for OSA (Fig. 37.3).

Positive screening for OSA was also associated with adverse health outcomes. Firefighters who screened positive for a sleep disorder were 106% more likely to report having cardiovascular disease, 84% more likely to report diabetes, 195% more likely to report depression and 163% more likely to report anxiety and to report poorer health status ($P<0.0001$), compared with those who did not screen positive. Safety and health outcomes were similarly significantly increased in those who screened positive for any sleep disorder compared to those who did not screen positive, including among fire-based EMS personnel (Fig. 37.3).

529 firefighters reported a current diagnosis of depression and/or anxiety (9% depression and/or anxiety; 6% depression; 4% anxiety). Although these rates are similar to those in the general population [83], our data revealed an almost threefold increase in the odds of the diagnosis for those who screened positive for OSA. The BCOPS study also reported on the association between sleep (sleep quality in this case) and depression among police officers. The investigators found that depression severity scores increased as sleep quality scores increased [84]. Sleep quality was independently associated with depressive symptoms.

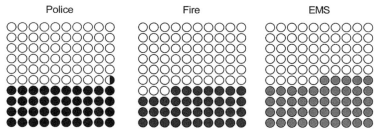

FIG. 37.2 The prevalence of positive sleep disorder screening across branches of public safety.

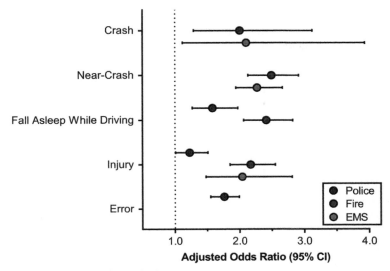

FIG. 37.3 Adjusted associations between sleep disorders and safety outcomes.

Sleep disorders and mood disorders are closely intertwined. Sleep disturbance and fatigue are two of the diagnostic criteria for depression in the Diagnostic and Statistical Manual of Mental Disorders [85]. Treatment for sleep disorders has been shown to reduce symptoms of depression and anxiety [86], and likewise, treatment for depression can reduce symptoms of insomnia [87]. Soldiers with insomnia reported more difficulty with social functioning, lower morale, perceived less support, and reported being less able to cope with stress following deployment [88]. Depression leads to reduced productivity at work and adversely impacts social relationships [81], and may impact the on-demand productivity in the work of public safety personnel and the occupational culture of living as a group in the fire department, EMS base, and police barracks.

Sleep disorders are endemic in police and firefighters. Given the similarities in demographic characteristics and occupational demands, it is likely that there is a similar prevalence of sleep disorders among EMS providers. However, there has been little research in this area. One national survey found that the odds of involvement in an ambulance crash are significantly higher for EMS providers with sleep problems [89]. A separate national survey determined that 70% of actively working EMS providers had a sleep problem [39]. This effort found that 10% of respondents self-reported snoring and pauses in breathing, while 5% self-reported snoring, pauses in breathing, and excessive daytime sleepiness, suggestive of sleep apnea. This approach likely results in under-reporting, as respondents would often be unaware of these events happening during their sleep periods. Among the subset of firefighters in our nationwide survey [6] who were cross-trained as EMS providers and reported their primary responsibility as medical care, the prevalence of positive sleep disorder screening was 45%, higher than what we observed among police or fire-only personnel (Fig. 37.2). Sleep disorders were also associated with adverse outcomes in this group. After controlling for age, gender, body mass index, exercise frequency, years of experience, shift schedule, work at multiple jobs, and call volume, positive sleep disorder screening was independently associated with more than twice the odds of an occupational injury, motor vehicle crash and near-crash [82]. Additional research is needed to evaluate the prevalence of sleep disorders and the impact on health and safety among the third component of the public safety net, EMS clinicians.

Sleep disorders are highly treatable and treatment can reduce associated health and safety risks. For example, in the case of OSA, patients adherent to continuous positive airway pressure (CPAP) therapy have better cardiovascular health outcomes compared to those who are non-adherent [90]. Successful treatment of OSA with CPAP therapy has resulted in a significant decrease in motor vehicle crash (MVC) rates [91–93] and reverses the trend in increased health care costs seen prior to treatment [94, 95]. Treatment for insomnia involves cognitive behavioral therapy (CBT-I), medication, or a combination of the two [96]. CBT-I is effective at improving sleep and reducing fatigue [97]. Recent efforts have even shown that the use of fully automated web-based CBT-I is as effective as in-person therapy sessions [98].

FATIGUE RISK MANAGEMENT

In an effort to improve health and safety, industries such as aviation, railroad and trucking are mandated or encouraged to institute fatigue risk management programs [99, 100]. To address sleep deficiency and sleep disorders, police, fire departments and EMS agencies should consider programs to address sleep health and fatigue risk management. The key components of a comprehensive fatigue risk management program should ideally include: a sleep health education program, recurrent and with certification testing; work schedule policies that are grounded in sleep and circadian science with monitoring of compliance and enforcement; and mandatory screening for sleep disorders with follow-up on effectiveness and compliance with any treatment [101].

The Royal Canadian Mounted Police (RCMP) implemented and evaluated a fatigue management program in a pilot study using a before-after design [102]. Following a train-the-trainer approach, the approximately 4-h program, emphasizing the science of sleep, sleep apnea and other sleep disorders and fatigue countermeasures, was presented to 61 RCMP members. On surveys completed 4 weeks following the training, members reported an increased satisfaction with sleep, reduced symptoms of insomnia and reduced incidence of headaches. The authors stressed that this training program should continue to be tested in larger police organizations to confirm the sleep health benefits for police officers.

A station-randomized trial of a sleep health education and sleep disorders screening program was conducted in a large municipal fire department. Of 1211 active firefighters identified at study onset, 604 were assigned to the intervention group and 607 to the control group. In an intention-to-treat analysis, firefighters assigned to intervention stations which participated in education sessions and had the opportunity to complete sleep disorders screening reported half the number of disability days on average than those assigned to control stations, as recorded by payroll records. In post-hoc analysis accounting for exposure to the intervention, firefighters who attended education sessions were 24% less likely to file at least one official injury report during the study duration than those firefighters who did not attend regardless of randomization [103].

Most recently, the National Highway Traffic Safety Administration commissioned a systematic review of published evidence to mitigate fatigue in emergency service personnel [41]. A diverse team of experts in sleep medicine, fatigue science, risk management, and emergency medicine reviewed >38,000 pieces of literature involving EMS personnel or similar shift workers. The systematic review lead to five recommendations: the use of reliable and/or valid fatigue survey tools to diagnose fatigue in the workplace [104]; shifts <24 h in duration [11]; the use of caffeine as a fatigue countermeasure [105]; napping [106]; and sleep health education and training [107].

The EMS systematic review revealed that sleep health education and training improved patient safety, personal safety, and ratings of acute fatigue and reduced stress and burnout. Further, a meta-analysis of the literature showed improvement in sleep quality [107]. Additional research is necessary to further dissemination of successful fatigue risk management programs in public safety personnel.

Different branches of public safety and different departments and organizations within the same branch have different resources and needs. With sleep health education and sleep disorders screening programs, it has been shown that several forms of implementation (i.e., expert-led, train-the-trainer, online) can be successful in improving sleep health in firefighters [108]. The most pertinent information to include in an education program, the length of the program and the durability of the benefits remain to be determined [107].

CONCLUSION

The sleep health of public safety personnel is a major concern. Rapid backward shift rotation, prolonged weekly work hours, and extended duration shifts contribute to sleep deficiency and circadian misalignment. Poor sleep quality and fatigue increase the risk of adverse safety outcomes and contribute to chronic health problems. Furthermore, there is an epidemic of sleep disorders among police, firefighters, and EMS providers. Fatigue risk management programs are effective, but remain underutilized. Future research should seek to develop and test schedules which align with sleep and circadian principles. Advocacy efforts should promote increased adoption of sleep health education, sleep disorder screening, and fatigue risk management programs. As a society, we rely on the public safety net to act swiftly and appropriately in emergencies. Efforts to improve sleep have the potential to vastly improve the safety, health, and performance of this vulnerable occupational group, benefiting not only them, but the public that they serve.

REFERENCES

[1] Martin JS, Laberge L, Sasseville A, et al. Day and night shift schedules are associated with lower sleep quality in evening-types. Chronobiol Int 2015;32:627–36.

[2] Elliot DL, Kuehl KS. Effects of sleep deprivation on fire fighters and EMS responders. The International Association of Fire Chiefs; 2007. Available at: https://www.iafc.org/docs/default-source/uploaded-documents/progssleep-sleepdeprivationreport.pdf?sfvrsn=20e9df0d_0&download=true.

[3] Patterson PD, Weaver MD, Frank RC, et al. Association between poor sleep, fatigue, and safety outcomes in emergency medical services providers. Prehosp Emerg Care 2012;16:86–97.

[4] Weaver MD, Patterson PD, Fabio A, Moore CG, Freiberg MS, Songer TJ. An observational study of shift length, crew familiarity, and occupational injury and illness in emergency medical services workers. Occup Environ Med 2015;72:798–804.

[5] Brown Jr. WE, Dickison PD, Misselbeck WJ, Levine R. Longitudinal emergency medical technician attribute and demographic study (LEADS): an interim report. Prehosp Emerg Care 2002;6:433–9.

[6] Barger LK, Rajaratnam SM, Wang W, et al. Common sleep disorders increase risk of motor vehicle crashes and adverse health outcomes in firefighters. J Clin Sleep Med 2015;11:233–40.

[7] Rajaratnam SM, Barger LK, Lockley SW, et al. Sleep disorders, health, and safety in police officers. JAMA 2011;306:2567–78.

[8] James L. The stability of implicit racial bias in police officers. Police Q 2018;21:30–52.
[9] Blake DM. Factoring fatigue into police deadly force encounters: decision making and reaction time Kaplan University; 2014.
[10] Kong PW, Beauchamp G, Suyama J, Hostler D. Effect of fatigue and hypohydration on gait characteristics during treadmill exercise in the heat while wearing firefighter thermal protective clothing. Gait Posture 2010;31:284–8.
[11] Patterson PD, Runyon MS, Higgins JS, et al. Shorter versus longer shift durations to mitigate fatigue and fatigue-related risks in emergency medical services personnel and related shift workers: a systematic review. Prehosp Emerg Care 2018;22:28–36.
[12] Schafer K, Sutter R, Gibbons S. Characteristics of individuals and employment among first responders. Washington, DC: US Department of Labor, Chief Evaluation Office; 2015.
[13] Patterson PD, Weaver MD, Hostler D, Guyette FX, Callaway CW, Yealy DM. The shift length, fatigue, and safety conundrum in EMS. Prehosp Emerg Care 2012;16:572–6.
[14] Luckhaupt SE, Cohen MA, Li J, Calvert GM. Prevalence of obesity among U.S. workers and associations with occupational factors. Am J Prev Med 2014;46:237–48.
[15] Bentley MA, Shoben A, Levine R. The demographics and education of emergency medical services (EMS) professionals: a National Longitudinal Investigation. Prehosp Disaster Med 2016;31:S18–29.
[16] Beaton RD, Murphy SA. Sources of occupational stress among firefighter/EMTs and firefighter/paramedics and correlations with job-related outcomes. Prehosp Disaster Med 1993;8:140–50.
[17] Hipple SF. Multiple jobholding during the 2000s. Mon Labor Rev 2010;133:21–32.
[18] Spiegel U, Gonen L, Weber M. Duration and optimal number of shifts in the labour market. Appl Econ Lett 2014;21:429–32.
[19] Dembe AE. Ethical issues relating to the health effects of long working hours. Springer J Bus Ethics 2009;84(Suppl 2):195–208.
[20] Rosenbluth G, Landrigan CP. Sleep science, schedules, and safety in hospitals: challenges and solutions for pediatric providers. Pediatr Clin North Am 2012;59:1317–28.
[21] Radstaak M, Geurts SA, Beckers DG, Brosschot JF, Kompier MA. Recovery and well-being among Helicopter Emergency Medical Service (HEMS) pilots. Appl Ergon 2014;45:986–93.
[22] Amendola KL, Weisburd D, Hamilton E, et al. The shift length experiment: what we know about 8-, 10-, and 12-hour shifts in policing. Police Foundation; 2011. Available at: https://www.policefoundation.org/wp-content/uploads/2015/12/Shift-Length-Experiment-Practitioner-Guide.pdf.
[23] Poole TL. The 48/96 work schedule: a viable alternative? Fire Eng 2012;165:85–9.
[24] Vila B. Tired cops: the importance of managing police fatigue. Washington, DC: Police Executive Research Forum; 2000. Available at: https://www.amazon.com/Tired-cops-importance-managing-fatigue/dp/1878734679.
[25] Vila B, Morrison GB, Kenney DJ. Improving shift schedule and work-hour policies and practices to increase police officer performance, health, and safety. Police Q 2002;5:4–24.
[26] Weaver MD, Patterson PD, Fabio A, Moore CG, Freiberg MS, Songer TJ. The association between weekly work hours, crew familiarity, and occupational injury and illness in emergency medical services workers. Am J Ind Med 2015;58:1270–7.
[27] Patterson PD, Buysse DJ, Weaver MD, Callaway CW, Yealy DM. Recovery between work shifts among emergency medical services clinicians. Prehosp Emerg Care 2015;19:365–75.
[28] Fekedulegn D, Burchfiel CM, Charles LE, Hartley TA, Andrew ME, Violanti JM. Shift work and sleep quality among urban police officers: the BCOPS study. J Occup Environ Med 2016;58:e66–71.
[29] Fekedulegn D, Burchfiel CM, Hartley TA, et al. Shiftwork and sickness absence among police officers: the BCOPS study. Chronobiol Int 2013;30:930–41.
[30] Violanti JM, Burchfiel CM, Hartley TA, et al. Atypical work hours and metabolic syndrome among police officers. Arch Environ Occup Health 2009;64:194–201.
[31] Violanti JM, Fekedulegn D, Andrew ME, et al. Shift work and the incidence of injury among police officers. Am J Ind Med 2012;55:217–27.
[32] Riedel M, Berrez S, Pelisse D, et al. 24-hour pattern of work-related injury risk of French firemen: nocturnal peak time. Chronobiol Int 2011;28:697–705.
[33] Haddock CK, Poston WS, Jitnarin N, Jahnke SA. Excessive daytime sleepiness in firefighters in the Central United States. J Occup Environ Med 2013;55:416–23.
[34] Choi B, Schnall P, Dobson M. Twenty-four-hour work shifts, increased job demands, and elevated blood pressure in professional firefighters. Int Arch Occup Environ Health 2016;89:1111–25.
[35] Soteriades ES, Smith DL, Tsismenakis AJ, Baur DM, Kales SN. Cardiovascular disease in US firefighters: a systematic review. Cardiol Rev 2011;19:202–15.
[36] Allen TL, Delbridge TR, Stevens MH, Nicholas D. Intubation success rates by air ambulance personnel during 12-versus 24-hour shifts: does fatigue make a difference? Prehosp Emerg Care 2001;5:340–3.
[37] Thomas F, Hopkins RO, Handrahan DL, Walker J, Carpenter J. Sleep and cognitive performance of flight nurses after 12-hour evening versus 18-hour shifts. Air Med J 2006;25:216–25.
[38] Guyette FX, Morley JL, Weaver MD, Patterson PD, Hostler D. The effect of shift length on fatigue and cognitive performance in air medical providers. Prehosp Emerg Care 2013;17:23–8.
[39] Pirrallo RG, Loomis CC, Levine R, Woodson T. The prevalence of sleep problems in emergency medical technicians. Sleep Breath 2012;16:149–62.
[40] FEMA. Fire department overall run profile as reported to the national fire incident reporting system. 2014. p. 2017.
[41] Patterson PD, Higgins JS, Van Dongen HPA, et al. Evidence-based guidelines for fatigue risk Management in Emergency Medical Services. Prehosp Emerg Care 2018;22:89–101.
[42] Vetter C, Fischer D, Matera JL, Roenneberg T. Aligning work and circadian time in shift workers improves sleep and reduces circadian disruption. Curr Biol 2015;25:907–11.
[43] Dijk DJ, Duffy JF, Czeisler CA. Circadian and sleep/wake dependent aspects of subjective alertness and cognitive performance. J Sleep Res 1992;1:112–7.
[44] Johnson MP, Duffy JF, Dijk DJ, Ronda JM, Dyal CM, Czeisler CA. Short-term memory, alertness and performance: a reappraisal of their relationship to body temperature. J Sleep Res 1992;1:24–9.
[45] Dijk DJ, Shanahan TL, Duffy JF, Ronda JM, Czeisler CA. Variation of electroencephalographic activity during non-rapid eye movement and rapid eye movement sleep with phase of circadian melatonin rhythm in humans. J Physiol (Lond) 1997;505(3):851–8.
[46] Czeisler CA, Dijk DJ, Duffy JF. Entrained phase of the circadian pacemaker serves to stabilize alertness and performance throughout the habitual waking day. In: Ogilvie RD, Harsh JR, editors. Sleep onset: normal and abnormal processes. Washington, DC: American Psychological Association; 1994. p. 89–110.

[47] Dinges DF. The nature of sleepiness: causes, contexts, and consequences. In: Stunkard AJ, Baum A, editors. Perspectives in behavioral medicine: eating, sleeping, and sex. Hillsdale, NJ: Lawrence Erlbaum Associates; 1989. p. 147–79.

[48] Carskadon MA, Dement WC. Multiple sleep latency tests during the constant routine. Sleep 1992;15:396–9.

[49] Pilcher JJ, Huffcutt AI. Effects of sleep deprivation on performance: a meta-analysis. Sleep 1996;19:318–26.

[50] Koslowsky M, Babkoff H. Meta-analysis of the relationship between total sleep deprivation and performance. Chronobiol Int 1992;9:132–6.

[51] Fröberg JE, Karlsson CG, Levi L, Lidberg L. Circadian rhythms of catecholamine excretion, shooting range performance and self-ratings of fatigue during sleep deprivation. Biol Psychol 1975;2:175–88.

[52] Carskadon MA, Dement WC. Cumulative effects of sleep restriction on daytime sleepiness. Psychophysiology 1981;18:107–13.

[53] Carskadon MA, Roth T. Sleep restriction. In: Monk TH, editor. Sleep, sleepiness and performance. John Wiley & Sons Ltd; 1991. p. 155–67.

[54] Gillberg M, Åkerstedt T. Sleep restriction and SWS-suppression: effects on daytime alertness and night-time recovery. J Sleep Res 1994;3:144–51.

[55] Blagrove M, Alexander C, Horne JA. The effects of chronic sleep reduction on the performance of cognitive tasks sensitive to sleep deprivation. Appl Cogn Psychol 1995;9:21–40.

[56] Brunner DP, Dijk DJ, Borbély AA. Repeated partial sleep deprivation progressively changes the EEG during sleep and wakefulness. Sleep 1993;16:100–13.

[57] Van Dongen HPA, Maislin G, Mullington JM, Dinges DF. The cumulative cost of additional wakefulness: dose-response effects on neurobehavioral functions and sleep physiology from chronic sleep restriction and total sleep deprivation. Sleep 2003;26:117–26.

[58] Achermann P, Werth E, Dijk DJ, Borbély AA. Time course of sleep inertia after nighttime and daytime sleep episodes. Arch Ital Biol 1995;134:109–19.

[59] Jewett ME, Kronauer RE. Interactive mathematical models of subjective alertness and cognitive throughput in humans. J Biol Rhythms 1999;14:588–97.

[60] Dinges DF. Sleep inertia. In: Carskadon MA, editor. Encyclopedia of sleep and dreaming. New York: Macmillillan Publishing Company; 1993. p. 553–4.

[61] Jewett ME, Wyatt JK, Ritz-De Cecco A, Khalsa SB, Dijk DJ, Czeisler CA. Time course of sleep inertia dissipation in human performance and alertness. J Sleep Res 1999;8:1–8.

[62] Richardson GS, Carskadon MA, Orav EJ, Dement WC. Circadian variation of sleep tendency in elderly and young adult subjects. Sleep 1982;5(Suppl 2):S82–94.

[63] Wyatt JK, Ritz-De Cecco A, Czeisler CA, Dijk DJ. Circadian temperature and melatonin rhythms, sleep, and neurobehavioral function in humans living on a 20-h day. Am J Physiol 1999;277:R1152–63.

[64] Klein DC, Moore RY, Reppert SM. Suprachiasmatic nucleus: the mind's clock. USA: Oxford University Press; 1991.

[65] Czeisler CA, Weitzman E, Moore-Ede MC, Zimmerman JC, Knauer RS. Human sleep: its duration and organization depend on its circadian phase. Science 1980;210:1264–7.

[66] Strogatz SH, Kronauer RE, Czeisler CA. Circadian regulation dominates homeostatic control of sleep length and prior wake length in humans. Sleep 1986;9:353–64.

[67] Dijk DJ, Czeisler CA. Contribution of the circadian pacemaker and the sleep homeostat to sleep propensity, sleep structure, electroencephalographic slow waves, and sleep spindle activity in humans. J Neurosci 1995;15:3526–38.

[68] Colquhoun WP. Experimental studies of shiftwork. Westdeutscher Verlag; 1975.

[69] Åkerstedt T, Torsvall L, Gillberg M. Sleep-wake disturbances in shift work: implications of sleep loss and circadian rhythms. Sleep Res 1983;12:359.

[70] Vidaček S, Kaliterna L, Radošević-Vidaček B, Folkard S. Productivity on a weekly rotating shift system: circadian adjustment and sleep deprivation effects? Ergonomics 1986;29:1583–90.

[71] Folkard S, Tucker P. Shift work, safety and productivity. Occup Med (Lond) 2003;53:95–101.

[72] Barger LK, Lockley SW, Rajaratnam SM, Landrigan CP. Neurobehavioral, health, and safety consequences associated with shift work in safety-sensitive professions. Curr Neurol Neurosci Rep 2009;9:155–64.

[73] Cohen DA, Wang W, Wyatt JK, et al. Uncovering residual effects of chronic sleep loss on human performance. Sci Transl Med 2010;2:14ra3.

[74] Kecklund G, Axelsson J. Health consequences of shift work and insufficient sleep. BMJ 2016;355:i5210.

[75] Zimmerman FH. Cardiovascular disease and risk factors in law enforcement personnel: a comprehensive review. Cardiol Rev 2012;20:159–66.

[76] Colten HR, Altevogt BM, Institute of Medicine (U.S.). Committee on Sleep Medicine and Research. Sleep disorders and sleep deprivation: an unmet public health problem. Washington, DC: Institute of Medicine: National Academies Press; 2006.

[77] Kapur VK, Redline S, Nieto FJ, Young TB, Newman AB, Henderson JA. The relationship between chronically disrupted sleep and healthcare use. Sleep 2002;25:289–96.

[78] Malhotra A, White DP. Obstructive sleep apnoea. Lancet 2002;360:237–45.

[79] Kapur V, Blough DK, Sandblom RE, et al. The medical cost of undiagnosed sleep apnea. Sleep 1999;22:749–55.

[80] Hossain JL, Shapiro CM. The prevalence, cost implications, and management of sleep disorders: an overview. Sleep Breath 2002;6:85–102.

[81] Kessler RC, Berglund PA, Coulouvrat C, et al. Insomnia and the performance of US workers: results from the America insomnia survey. Sleep 2011;34:1161–71.

[82] Weaver MD, Sullivan JP, O'Brien CS, Qadri S, Czeisler CA, Barger LK. Sleep disorders are common risk factors for occupational injury. Prehosp Emerg Care 2018;21:137.

[83] Kessler RC, Chiu WT, Demler O, Merikangas KR, Walters EE. Prevalence, severity, and comorbidity of 12-month DSM-IV disorders in the National Comorbidity Survey Replication. Arch Gen Psychiatry 2005;62:617–27.

[84] Slaven JE, Mnatsakanova A, Burchfiel CM, et al. Association of sleep quality with depression in police officers. Int J Emerg Ment Health 2011;13:267–77.

[85] American Psychiatric Association. DSM-5 task force. Diagnostic and statistical manual of mental disorders: DSM-5. 5th ed. Washington, DC: American Psychiatric Association; 2013.

[86] Edwards C, Mukherjee S, Simpson L, Palmer LJ, Almeida OP, Hillman DR. Depressive symptoms before and after treatment of obstructive sleep apnea in men and women. J Clin Sleep Med 2015;11:1029–38.

[87] Yon A, Scogin F, DiNapoli EA, et al. Do manualized treatments for depression reduce insomnia symptoms? J Clin Psychol 2014;70:616–30.

[88] Klingaman EA, Brownlow JA, Boland EM, Mosti C, Gehrman PR. Prevalence, predictors and correlates of insomnia in US army soldiers. J Sleep Res 2018;27(3):e12612. https://doi.org/10.1111/jsr.12612. [Epub 11 October 2017].

[89] Studnek JR, Fernandez AR. Characteristics of emergency medical technicians involved in ambulance crashes. Prehosp Disaster Med 2008;23:432–7.

[90] Butt M, Dwivedi G, Shantsila A, Khair OA, Lip GY. Left ventricular systolic and diastolic function in obstructive sleep apnea: impact of continuous positive airway pressure therapy. Circ Heart Fail 2012;5:226–33.

[91] Engleman HM, Asgari-Jirhandeh N, McLeod AL, Ramsay CF, Deary IJ, Douglas NJ. Self-reported use of CPAP and benefits of CPAP therapy: a patient survey. Chest 1996;109:1470–6.

[92] Cassel W, Ploch T, Becker C, Dugnus D, Peter JH, von Wichert P. Risk of traffic accidents in patients with sleep-disordered breathing: reduction with nasal CPAP. Eur Respir J 1996;9:2606–11.

[93] Krieger J, Meslier N, Lebrun T, et al. Accidents in obstructive sleep apnea patients treated with nasal continuous positive airway pressure: a prospective study. The working group ANTADIR, Paris and CRESGE vol. 112. Lille, France: Association Nationale de Traitement a Domicile des Insuffisants Respiratoires Chest; 1997. p. 1561–6.

[94] Albarrak M, Banno K, Sabbagh AA, et al. Utilization of healthcare resources in obstructive sleep apnea syndrome: a 5-year follow-up study in men using CPAP. Sleep 2005;28:1306–11.

[95] Bahammam A, Delaive K, Ronald J, Manfreda J, Roos L, Kryger MH. Health care utilization in males with obstructive sleep apnea syndrome two years after diagnosis and treatment. Sleep 1999;22:740–7.

[96] Schutte-Rodin S, Broch L, Buysse D, Dorsey C, Sateia M. Clinical guideline for the evaluation and management of chronic insomnias in adults. J Clin Sleep Med 2008;4:487–504.

[97] Wu JQ, Appleman ER, Salazar RD, Ong JC. Cognitive behavioral therapy for insomnia comorbid with psychiatric and medical conditions: a meta-analysis. JAMA Intern Med 2015;175:1461–72.

[98] Ritterband LM, Thorndike FP, Ingersoll KS, et al. Effect of a web-based cognitive behavior therapy for insomnia intervention with 1-year follow-up: a randomized clinical trial. JAMA Psychiat 2017;74:68–75.

[99] FAA. Flightcrew member duty and rest requirements: 14 CFR Part 117, 119, and 121. Washington, DC: FAA; 2012. p. 2012.

[100] Quan SF, Barger LK. Brief review: sleep health and safety for transportation workers. Southwest J Pulm Crit Care 2015;10(3):130–9.

[101] Czeisler CA. Role of sleep medicine and chronobiology for optimizing productivity, safety and health in the workplace. 86th annual meeting, Japan society for occupational health; May 16, 2013; Matsuyama, Ehinme, Japan.

[102] James L, Samuels CH, Vincent F. Evaluating the effectiveness of fatigue management training to improve police sleep health and wellness: a pilot study. J Occup Environ Med 2018;60:77–82.

[103] Sullivan JP, O'Brien CS, Barger LK, Rajaratnam SM, Czeisler CA, Lockley SW. Randomized, prospective study of the impact of a sleep health program on firefighter injury and disability. Sleep 2017;40.

[104] Patterson PD, Weaver MD, Fabio A, et al. Reliability and validity of survey instruments to measure work-related fatigue in the emergency medical services setting: a systematic review. Prehosp Emerg Care 2018;22:17–27.

[105] Temple JL, Hostler D, Martin-Gill C, et al. Systematic review and meta-analysis of the effects of caffeine in fatigued shift workers: implications for emergency medical services personnel. Prehosp Emerg Care 2018;22:37–46.

[106] Martin-Gill C, Barger LK, Moore CG, et al. Effects of napping during shift work on sleepiness and performance in emergency medical services personnel and similar shift workers: a systematic review and meta-analysis. Prehosp Emerg Care 2018;22:47–57.

[107] Barger LK, Runyon MS, Renn ML, et al. Effect of fatigue training on safety, fatigue, and sleep in emergency medical services personnel and other shift workers: a systematic review and meta-analysis. Prehosp Emerg Care 2018;22:58–68.

[108] Barger LK, O'Brien CS, Rajaratnam SM, et al. Implementing a sleep health education and sleep disorders screening program in fire departments: a comparison of methodology. J Occup Environ Med 2016;58:601–9.

Index

Note: Page numbers followed by *f* indicate figures and *t* indicate tables.

A

Academic achievement, 449–450
Accelerometer, 149, 154
Acculturation, 71–72
Acetaldehyde, 269
Acetate, 269
Acetylcholine, 8
Acquired immunity, 319–322
Actigraphy (ACT), 8–10, 21, 57, 392–393, 436, 441, 466–467
    commercially-available sleep trackers, 155
    dynamic range, 154
    frequency response, 154
    high and low frequency limit, 154
    identifying sleep stages with, 152–153
    insufficient sleep, 204
    limitations, 150–152
    movement recording, 153
    noise, 154
    off-wrist time, 154
    older adults, 34
    output deviation, 154
    vs. polysomnography, 147, 148*t*
    recording modes, 153–154
    scientific guidelines, 154–155
    scoring algorithms, 147–149
    temperature sensitivity and range, 154
    types of devices, 149–150
Actillume, 149–150
Activator protein 1 (AP-1), 322–323
Actiwatch, 149–150
Adaptive servoventilation, 413
Adenosine, 260
    receptors, 309
    in sleep-wake cycle, 308–309
Adenosine A_{2A} receptor (ADORA2A) gene, 309, 344
Adenosine deaminase (ADA), 309, 344
Adolescents, 447, 450
    mental health and sleep improvements in, 439–442
        clinicians, 441
        families, 440
        policy makers, 441–442
        schools, 440–441
    recommended sleep duration for, 425
    sleep, 22–24
Adrenocorticotropic hormone (ACTH), 401
Adult sleep, 78–79
Aerobic capacity, 364
Aggressive/punitive responses, 353
Aging immune system, 320
Air quality, impact on sleep, 92–93
Alameda 7 study, 424
Alcohol, 138–139, 269
    dehydrogenase, 269
    metabolism, 269, 270*f*
Alcohol dependence, 270–271
    active, 272
    in acute withdrawal, 272
    behavioral treatments for insomnia and, 273
    circadian disruption, 274
    early recovery, 272
    insomnia in, 272
    obstructive sleep apnea, 275–276
    pharmacologic treatments for insomnia in, 273
    sustained recovery, 273
Alcohol use
    breathing related sleep disorders and, 275–276
    circadian rhythms and, 273–274
    insomnia and, 269–273
    napping, 277
    neurobiology of, 269
    parasomnias and, 277
    during pregnancy, 277
    sleep duration abnormalities and, 274–275
    sleep-related movement disorders and, 276–277
Alcohol use disorder (AUD), 381
Aldehyde dehydrogenase, 269
Alertness, 334–335, 492–493
Allostatic Load Model, 458
Alternative explanations, 173
Alzheimer's disease (AD), 33
Ambulatory Monitoring, Inc. (AMI), 149–150
American Academy of Sleep Medicine (AASM), 121, 425, 429, 429*t*, 458, 482–483, 484*t*
American College of Physicians (ACP), 402–403
American Psychiatric Association's (APA) diagnostic nomenclature, 374
American Time Use Survey (ATUS), 13, 107, 108–109*f*, 459–461
Anterior cingulate cortex (ACC), 191, 347
Antigen presenting cells (APCs), 319–320
Anti-retroviral therapy (ART), 293
Anxiety, 435
    disorders, 380
    exercise, 260
Anxious children, 438
Apnea, 275, 409, 413
Apnea-hypopnea index (AHI), 36, 118, 261, 392–393
Apnea Positive Pressure Long-term Efficacy Study (APPLES), 414
2-Arachidonoylglycerol (2-AG), 247
Arrhythmias, 413
Artificial light at night (ALAN), 86–88
ASV, in heart failure, 413
Atrial fibrillation, 413
Attention, 361–362
Attention-deficit/hyperactivity disorder (ADHD), 380–381
Attitudes, 111
Autism spectrum disorder (ASD), 382
Autonomic nervous system (ANS), 8, 233–234, 326

B

Balloon Analog Risk Task (BART), 353
B cells, 319–320
Bedroom environment, 139–140
Behavioral economics, 180–181
Behavioral intention, 175–176
Behavioral model, 375
Behavioral rhythms, 232–233
Behavioral Risk Factor Surveillance System (BRFSS), 12, 12*f*, 58, 286–287
Behavior change
    causation in, 172–174
    developed instrument, 182
    develop items, 182–183
    draft instrument, 183
    foundation of theory for, 171–172
    instrument, purpose of, 181
    limitations, 184
    models and theories, 181–184
    objects of interest
        constitutively define, 181
        identify, 181
        operationalization, 181–182
    panel of experts, 183
    pilot test, 183
    readability test, 183
    reliability and validity, 183–184
    review previously developed instruments, 182
    select appropriate scales, 182
Benzodiazepines, 295
Berlin questionnaire, 121, 122*f*
Bi-directional regulation, 235–236
Binge-watching television, 109
Biobehavioral state, 160–163
Biological rhythms, 233–236

Biomarkers, 343
Bipolar disorder (BPD), 379–380
Blood alcohol level, 269, 270f
Blood pressure (BP), 206–207, 393–398
Body mass index (BMI), 22, 421–423, 483–484
Body temperature, 260
Brain derived neurotrophic factor (BDNF) gene, 348–349
Brain oxygenation level-dependent (BOLD) activity, 362–363
Brainstem, 7
Breathing related sleep disorders, 275–276
Brief Index of Sleep Control (BrISC), 113–114
Bright light therapy, 441
Bronfenbrenner's ecological systems theory, 439–440, 442
Buffalo Cardio-Metabolic Occupational Police Stress (BCOPS), 491
Bupropion, 292
B vitamins, 250

C

Caffeine, 138, 303, 352–353, 352f
  adverse effects, 310
  alcohol use, 311
  beverages and contents, 304t
  chronotype/time of intake, 310
  circadian misalignment, 312
  comorbid burden of disease, 310
  concurrent use of, 313
  epidemiology of sleep in, 304–306
  health implications of, 311–312
  insomnia, 308
  intake of, 303–304, 310
  intoxication, 304–305
  non-specific sleep-disturbances, 311–312
  physiology, in sleep-wake cycle, 308–311, 309f
  recovery sleep, 308
  short sleep duration, 305, 311
  sleep deprivation, 307–308
  sleepiness, 307–308
  sociodemographic pattern, 310–311
  withdrawal, 310
Caffeine-induced anxiety disorder, 304–305
Caffeine-induced sleep disorder, 304–305
Caffeine related disorder, 304–305
Caloric consumption, 247
Caloric intake, 189–191, 194–195
Candidate Gene Association Resource (CARe), 219
CANPAP trial, 413
Carbohydrates, 248–249
Cardiometabolic health
  behavioral rhythms and, 232–233
  biological rhythms and, 233–236
  circadian disruption and, 229–230
  environmental rhythms and, 231–232
Cardiometabolic syndrome, 227–228, 230–231
Cardiovascular functioning, 230
Cardiovascular health, 35–36
Catecholamine, 400

CBT-I. *See* Cognitive behavioral therapy for insomnia (CBT-I)
CDC. *See* Centers for Disease Control and Prevention (CDC)
Cellular senescence, 320
Centers for Disease Control and Prevention (CDC), 12, 421–423, 426, 473
Central nervous system (CNS), 21–22
Central sleep apnea (CSA), 261, 275
Cerebrovascular disease, 412
Cerebrum, 7–8
Chalder Fatigue Questionnaire (CFQ), 491
Cheyne-Stokes respiration (CSR), 36, 208–209
Childhood obesity
  causes, 423
  consequences, 423
  energy homeostasis, 425–426
  epidemic, reduction, 423–424
  sleep duration, 424
  sleep parameters, 424–425
Childhood sleep, 22
Children
  diabetes and, 429–431
  mental health and sleep improvements in, 439–442
    clinicians, 441
    families, 440
    policy makers, 441–442
    schools, 440–441
  obesity in, 421–423, 422f
  overweight in, 421–423, 422f
  recommended sleep duration for, 429t
  sleep and, 429
Chronic disease, mobile technology, 163
Chronic heart failure (CHF), 36
Chronic insufficient sleep, 118
Chronic pain, 108
Chronic partial sleep deprivation, 11
Chronic sleep restriction, 308
Chronotypes, 273, 361, 491–492
Cigarette smoking
  epidemiology of, 283–284
  smokers *vs.* nonsmokers
    daytime sleepiness, 287
    objective and subjective sleep metrics in, 285–286t
    sleep architecture, 284–286
    sleep continuity, 286–287
    sleep fragmentation, 287
Circadian control, 230–231
Circadian disruption, 229–230
  mobile technology, 165–166
  and social jetlag, 229
Circadian dysfunction, 162
Circadian misalignment, 312, 450, 463
  and diabetes, 220
  pathophysiological effects, 221f
Circadian process, 339–340, 340f, 342, 342f
Circadian rhythm, 71, 228–229, 424–425
  alcohol use
    alcohol dependent individuals, 274
    chronopharmacokinetic studies, 274
    clinical findings on shiftwork, 274

  changes in, 31
  and diabetes, 218–220
  disorders, 125–128
  endogenous, 219–220
  exogenous, 220
  sleep and, 6–7
Clockwatching, 140
Clonazepam, 277
Cognitive Activation Theory of Stress, 458
Cognitive behavioral therapy for insomnia (CBT-I), 32, 137, 273, 294, 365, 382–383, 402–403, 441, 495
Cognitive control
  cognitive interference, 348–349
  flexible attentional control, 349–350
  multi-tasking, 348
  task-switching, 348
Cognitive flexibility, 347–348
Cognitive framing, 351
Cognitive function, 33–35, 163
Cognitive interference, 348–349
Cognitive model, 377
Cognitive stability, 347–348
Commercially-available sleep trackers, 155
Commercial motor vehicle (CMV) operators, obstructive sleep apnea
  current regulation, 482, 483f
  diagnosis, 484
  education, 485
  initial evaluation, 484
  PAP therapy, 485
  prevalence, 481–482
  screening, 482–484, 484t
  treatment, 484–485
Community level theories, 179–181
Confounding variables, 173
Consistency, 173
Consumer sleep technologies, 128–130
Continuous positive airway pressure (CPAP), 33, 122, 195, 365, 411
Continuum theory, 174–175
  behavioral economics, 180
  social cognitive theory, 178
  social network theory, 179
Convergent thinking, 350
Coronary artery disease, 412
Coronary Artery Risk Development in Young Adults (CARDIA) sleep study, 206
Coronary heart disease (CHD), 207, 399
Cortisol, 234, 326, 401
C-reactive protein (CRP), 320, 401
Cues to action, 175
CYP1A2 gene polymorphism, 309
Cytokines, 319–320

D

Daytime napping, 34
Daytime sleepiness, 414
  excessive, 217
  in smokers *vs.* nonsmokers, 287
Default mode network, 340–341
Delay discounting, 362
Delaying school start time, 448–449

Depression, 435
  exercise, 260
  insomnia, 378–379
  smoking, 293
Dextroamphetamine, 307, 352–353, 352f
Diabetes
  circadian misalignment and, 220
  circadian rhythm and, 218–220
  daytime sleepiness, 217
  endogenous circadian rhythm and, 219–220
  epidemiological evidence, 218
  exogenous circadian rhythm and, 220
  indirect effects of sleep on, 220–221
  insufficient/short sleep, 213–215
  longer sleep and
    duration, 215
    indirect relationship between, 216
    negative associations, 215–216
    protective effects of, 215
  mechanistic studies, 218
  mortality among, 221–222
  physiological and biological mechanisms, 217
  population level studies, 218
  qualitative sleep parameters, 216–217
  risk reduction, 222
  sleep disorders and, 217–218
  sleep duration and, 213–216
  sleep parameters and, 213–222
  sleep quality, 216
Diabetes mellitus (DM), 426–427
Diabetic retinopathy (DR), 221–222
Diffusion tensor imaging (DTI), 343–344
Diffusion theory, 179–180
Dim-light melatonin onset (DLMO), 273–274, 380–381
Disparities, 464–466
Divergent thinking, 350–351
Dopamine, 8
Dopamine D_2 receptor (DRD2), 349–350
Dopamine transporter (DAT1) gene, 344, 345f
Dose-response relationship, 173
Drowsy driving, 481
Dual-energy-X-ray absorptiometry (DEXA), 425

E
Eating behavior, 359–360
Ecological momentary assessment (EMA), 461
Effort discounting, 362
Effort-Recovery Model, 458
Electroencephalography (EEG), 4, 398
Electronics, removal of, 140
Emotion dysregulation, 293–294
Emotion regulation, 437
Endogenous circadian rhythm, 219–220
Energy balance, 193–196
Energy drink supplements, 306–307
Energy expenditure, 425
Energy homeostasis, 425–426
Energy metabolism, 231
Enteric nervous system, 326
Environmental exposures on sleep
  physical environment
    air quality, 92–93
    light, 85–89
    noise, 90–91
    seasonality and latitude/longitude, 93–94
    temperature, 89–90
    vibrations, 91–92
  social environment
    interpersonal relationships, 96–98
    neighborhood environment, 95
    psychosocial stress, 94
    racism, 94–95
    socioeconomic status, 94–95
    work environment, 96
Environmental rhythms, 231–232
Epidemiologic Catchment Area study (ECA), 270–271
Epidemiology
  in caffeine, 304–306
  in energy drink supplements, 306–307
  in psychostimulants, 307
  work, 458–459
Epworth sleepiness scale (ESS), 122–123, 217
Established Populations for Epidemiologic Studies of the Elderly study (EPESES), 57
Excess adiposity, 423
Excessive daytime sleepiness (EDS), 217
  cognitive function, 34–35
  older adults, 32
Executive functions, 344–351, 345f, 362
  cognitive control, 347–350
  inhibitory control, 347
  problem solving, 350–351
  working memory, 346–347
Exercise, 137–138, 257
  acute, 258–259
  chronic exercise training, 259
  experimental research, 258–259
  observational research, 257–258
  potential mechanisms
    adenosine, 260
    anxiolytic and antidepressant effects, 260
    body temperature effects, 260
    circadian phase-shifting effects, 260
  sedentary behavior, 259–260
  sleep disorders, impacts on
    insomnia, 260–261
    periodic limb movements during sleep, 261–262
    restless legs syndrome, 261–262
    sleep-disordered breathing, 261
Exogenous circadian rhythm, 220
Exosystem, 46
Experimental evidence, 173

F
Facebook®, 159
Fat, 249
Fatigue, 489
  neurobiology, 339–340
  risk management, 495–496
Fear of hypoglycemia (FOH), 216
Federal Motor Carrier Safety Administration (FMCSA), 481–482
Federal Road Administration (FRA), 482
Fitbit, 129–130
Flexible attentional control, 349–350
Food and Drug Administration (FDA), 310
Food intake, 139
  alternative medicine, 251
  caloric consumption, 247
  carbohydrates, 248–249
  fat, 249
  homeostatic mechanisms, 245–246
  protein, 247–248
  relation between sleep and, 245–247
  sleep loss and, 244
  total dietary approaches, 251
  vitamins and supplements, 246–247
Food labeling practices, 312–313
Framing, 351
Fruits, 250
Functional outcomes of sleep questionnaire (FOSQ-30), 123

G
Gabapentin, 276
Genetic polymorphisms, 344
Ghrelin, 217, 235, 425
Global Burden of Disease (GBD), 427
Glucagon-like peptide 1 (GLP-1), 234, 246
Glucose metabolism, 230–231, 344, 346f
Granulocytes, 320
Gross domestic product (GDP), 57
Group differences, 192–193
Growth hormone, 235
Gut microbiome, 233

H
Habitual sleep duration, 11
Health behavior, 46–47, 400–402
  circadian disruption, 359, 361, 363
  defined, 359
  exposure, 361
  influences on sleep and, 359–360
  neurocognitive factors, 361–362
  neuroimaging, 362–363
  short sleep duration, 360
  sleep loss, 359, 361, 363–364
  sleep related changes in neurocognitive function, 362
  utility of theory, 172
Health belief model (HBM), 110–111, 111f, 174–175
Health-damaging behavior, 474
Health differences, 474–476
Health disparities, 464–466, 476
  distal factors, 476
  history and definition, 58
  intermediate factors, 476
  proximal factors, 476
Health equity, 474
Health promotion, 442
Health-related quality of life (HRQoL), 423
Healthy sleep, 222, 427–428
Heart disease, 399–400

Heart failure
    insufficient sleep and, 208–209
    obstructive sleep apnea, 412–413
Heart rate variability (HRV), 398
Hedonic eating, 359–360
Hepatitis B vaccination, 321–322
High heat capacity mattress (HHCM), 90
Histamine, 8
HIV-associated neurocognitive disorder (HAND), 293
Homeostasis, 245–246, 339–340, 340f, 342, 342f, 359
Home sleep apnea testing (HSAT), 118, 484
Home sleep testing (HST), 275
Horne-Ostberg morningness-eveningness questionnaire (MEQ), 125–127
HPA. *See* Hypothalamic-pituitary-adrenal (HPA) axis
HSAT. *See* Home sleep apnea testing (HSAT)
Hypercapnia, 409
Hypersomnolence, 121–124
Hypertension (HTN), 36, 206–207, 209, 393–398, 411
Hypopneas, 275
Hypothalamic-pituitary-adrenal (HPA) axis, 326, 400
Hypothalamus, 7
Hypoxia, 409, 414

I

Immune system, 319, 400–402
    acquired, 319–320
    aging, 320
    innate, 320
Immunological aging, 323–324
Impedance cardiography, 400
Impulsive behavior, 353
Impulsivity, 362
Individual differences, 193
Infant sleep, 21–22
Infectious disease, 320–322
Inflammaging, 320
Inflammation, 217
Inflammatory cytokines, 322
Inflammatory disease, 322–323
Inflammatory response, 322–323
Information technology (IT), 458, 459f
Inhibitory control, 347
Innate immunity, 320, 322–323
Innovative thinking, 350–351
Innovators, 179–180
Insomnia, 11, 16–17, 31, 360, 378f
    acute, 142
    in alcoholics, 272–273
    alcohol use, 269–273
        adult sleep problems and, 270–271
        behavioral treatments, 273
        childhood sleep problems, 270
        clinical findings, 271–272
        epidemiology, 269–270
        genetic studies, 271
        pharmacologic treatments, 273
        spectral PSG studies, 271
    behavioral treatment, 375, 382–383
    caffeine, 308
    and cardiometabolic disease risk
        blood pressure, 393–398
        health behaviors, 400–402
        heart disease, 399–400
        hypertension, 393–398
        immunity, 400–402
        insulin resistance, 398–399
        meta-analyses, 394t
        objective sleep measures, 396–397t
        outcomes, 393t
        public health and clinical implications, 402–403
        stress, 400–402
        stroke, 399–400
        type 2 diabetes, 398–399
    cardiovascular health, 36
    chronic, 141–143
    cognitive behavioral therapy for insomnia, 382–383
    cognitive function, 35
    cognitive model, 377
    definitions, 373–374, 392t
    diabetes, 218
    diagnostic criteria, 391–392
    DSM-5 criteria for, 374
    etiological models, 141–142
    etiology, 374–377
    exercise, 260–261
    identifying risk, 142–143
    incidence, 374
    mobile technology, 167
    neighborhood factors, 79
    neurocognitive model, 375–377
    older adults, 32
    parallel process, 376f, 377
    prevalence, 24–25, 374
    psychiatric illness, 37
    and psychiatric morbidity
        alcohol use disorder, 381
        anxiety disorders, 380
        attention-deficit/hyperactivity disorder, 380–381
        autism spectrum disorder, 382
        bipolar disorder, 379–380
        depressive disorders, 378–379
        post-traumatic stress disorder, 380
        schizophrenia, 382
        suicide, 379
    psychobiological inhibition model, 377
    racial/ethnic groups, 66–69, 67–68t
    sleep hygiene, 140–141
    socioeconomic status, 69
    stimulus control model, 375
    suvorexant for, 295
    symptoms, 391–393
Insomnia severity index (ISI), 124, 402
Insufficient sleep
    by age, 13–14, 13f
    and blood pressure, 206–207
    coldspots, 15, 16f
    and coronary heart disease, 207–208
    defining, 11–12, 204–205
    diabetes, 213–215
    by geography, 15
    and heart failure, 208–209
    hotspots, 15, 16f
    limitations, 15–16
    pathophysiology, 205–210
    in population, 12–13
    prevalence of, 12–15
    by race/ethnicity, 14
    by sex, 14
    by socioeconomic status, 14–15
    and stroke, 209–210
Insulin, 234, 427–428
Insulin resistance (IR), 217, 398–399, 425–428, 430
    insufficient/short sleep duration, 213–215
    long sleep duration, 215
Integrated behavioral model, 111–112, 112f
Interleukin-6 (IL-6), 217, 401
International Classification of Sleep Disorders (ICSD), 141
International Dark Sky Association, 88–89
International restless legs syndrome scale (IRLS), 128
Interpersonal relationships, 96–98
Interpersonal theories, 178–179
Intrapersonal theories, 174–178
Intrinsically photosensitive retinal ganglion cells (ipRGC), 160f
Iowa Gambling Task (IGT), 352–353, 352f
IR. *See* Insulin resistance (IR)
Isoflavones, 250
Item Response Theory (IRT), 125

J

Jawbone UP, 130
Job Demand Control (JD-C) model, 458–460
Job Demand-Resources (JD-R) model, 458–460

K

Kantar World Panel, 306
Karolinska sleepiness scale (KSS), 123–124
Kava, 251

L

Laggards, 179–180
Large neutral amino acids (LNAA), 248f
Leptin, 191, 217, 235, 425
Light, impact on sleep, 85–89
Lipid metabolism, 231
Lipopolysaccharide (LPS), 322–323
Liquid intake, 139
Locomotor Inactivity During Sleep (LIDS), 152, 152f
Logical deduction, 350
Longer sleep, and diabetes
    duration, 215
    indirect relationship between, 216
    negative associations, 215–216
    protective effects of, 215
Longitudinal EMT Attribute Demographic Study (LEADS), 491
Lymphocytes, 319–320

M

Macronutrient intake, 190, 192–193
Macrophages, 320
Macrosystem, 46
Magnesium, 250, 276
Maintenance of wakefulness test (MWT), 121
Major depressive disorder (MDD), 378–379
Major depressive episode (MDE), 378–379
Meal timing, 190–191, 193
Measurability, 173–174
Mediating variable, 173
Mediterranean Diet, 251
Melatonin, 232, 235, 249–250, 273
    nicotine withdrawal, 294–295
    REM sleep behavior disorder, 277
Mental health, 450–451
    and sleep duration, 436–438
    and sleep improvements, in children and adolescents, 439–442
        clinicians, 441
        families, 440
        policy makers, 441–442
        schools, 440–441
    and sleep quality, 438–439
Mental illness, 435
Mesosystem, 46
Meta-analysis, 462–463
Metabolic disease, 426
Metabolic syndrome (MetS). *See* Cardiometabolic syndrome
Metabolic system, 230–231
Microsystem, 46
Middle-aged sleep, 24–25
Mini-Mitter Inc., 149–150
Mobile technology, 159
    artificial light, 165–166
    chronic disease, 163
    circadian disruption, 165–166
    cognitive function, 163
    emergence of, 163–164
    insomnia, 167
    sleep behavior and, 161
    sleep deficiency, 161–162
    sleep loss, 161–162, 164–167
    time displacement, 164–165
    video gaming, 166–167
Modafinil, 307, 352–353, 352*f*
Model of goal-directed behavior (MGDB), 176–177
Moderate to vigorous physical activity (MVPA), 221
Moderating variable, 173
Moral judgment, 354
Motionlogger, 149–150
Motivation, 364–365
Movement behaviors, 262, 263*f*
Multiple sleep latency test (MSLT), 121, 395–398, 482
Multi-tasking, 348
Munich chronotype questionnaire (MCTQ), 127–128
Myocardial infarction (MI), 399, 412

N

Napping, 34, 277
Narcolepsy, 70–71, 195
Nasal mask, 411
National Health and Nutrition Examination Survey (NHANES), 12–13, 206, 227–228, 286–287, 305–307, 320
National Health Interview Survey, 360
National Heart Lung and Blood Institute (NHLBI), 474–475
National Highway Traffic Safety Administration (NHTSA), 481, 491, 496
National Institute of Child Health and Human Development (NICHD) study, 305
National Sleep Foundation, 16, 449, 462
National Survey of Children's Health, 429
Natural killer (NK) cells, 320
Negative emotions, 363
Neighborhood environment, impact on sleep, 95
Neighborhood factors
    accessing physical activity, 78
    adult sleep, 78–79
    evidence from natural experiments, 80
    interventions and policies, 80–81
    limitations, 79–80
    long-term trajectories, studying, 80
    pediatric sleep, 78
    and sleep health, 77
    technological advances, 80
    theoretical justification, 77–78
    violence and safety concerns, 78
Neighborhood socioeconomic status (NSES), 78
Neurobehavioral consequences, 333
Neurobiology, 339–340
Neurocognitive factors, 361–362, 375–377
Neuromodulators, 8
NHTSA. *See* National Highway Traffic Safety Administration (NHTSA)
Nicotine, 139
Nicotine replacement therapy (NRT), 284, 291–292
Nicotine withdrawal, 287
    cognitive-deficits, 293
    melatonin, 294
Nightmares, 438
Nocturnal hypoglossal nerve stimulation, 409–411
Noise, impact on sleep, 90–91
Nonhomeostatic mechanisms, 246–247
Non-rapid eye movement (NREM) sleep, 4–5, 258, 284
    parasomnias, 277
    stage dissection, 5
Non-verbal cues, 457
Norepinephrine, 8, 400
Notice of Proposed Rulemaking (NPRM), 482
Nuclear factor kB (NF-kB), 322–323

O

Obesity
    in children, 421–423, 422*f*
    insufficient/short sleep duration, 213–215
    long sleep duration, 215
    relationship between sleep and diabetes, 221
    sleep in individuals with, 196
Obesogenic behaviors, 189–191
Objectively measured sleep disturbances, 34
Objective reported sleep duration, 61
Obstructive sleep apnea (OSA), 195, 275, 365, 428, 493–494
    apneic events, 414
    arrhythmias, 413
    cerebrovascular disease, 412
    in commercial motor vehicle operators
        current regulation, 482, 483*f*
        diagnosis, 484
        education, 485
        initial evaluation, 484
        PAP therapy, 485
        prevalence, 481–482
        screening, 482–484, 484*t*
        treatment, 484–485
    coronary artery disease, 412
    diabetes, 217
    diagnosis of, 275
    exercise, 261
    heart failure, 412–413
    hypertension, 411
    neighborhood factors, 79
    older adults, 32
    prevalence, 411
    sedentary behavior, 259–260
    treatment, 409–411
    upper airways, 409
Occupational Fatigue Exhaustion Recovery (OFER) scale, 491
Older adults
    cardiovascular health, 35–36
    circadian rhythm, 31
    cognitive function, 33–35
    health outcomes, 39*t*
    insomnia, 32
    pain, 37–38
    poor sleep, 38*t*
    psychiatric illness, 36–37
    sleep and health, 25–26, 33–38
    sleep-disordered breathing, 32–33
    sleep homeostasis, 31
    sleep parameters, 31
On-demand culture, 109
Orbitofrontal cortex (OFC), 348
Orexin, 8, 295
OSA. *See* Obstructive sleep apnea (OSA)
Osteoarthritis (OA), 37–38
Overweight, in children, 421–423, 422*f*

P

Paid employment, 442
Pain, 37–38
Parallel process, insomnia, 376*f*, 377
Parasomnias
    and alcohol use, 277
    defined, 277
Parasympathetic nervous system (PNS), 326
Paroxysmal nocturnal dyspnea, 208–209

Partial sleep deprivation, 11
Patient-reported outcomes measurement information system (PROMIS™), 125
Pediatric sleep, 78
People living with HIV (PLWH), 293
Perceived barriers, 113, 175
Perceived benefits, 113, 175
Perceived discrimination, 72
Perceived susceptibility, 175
PERIOD3 (PER3), 346
Periodic leg movement disorder (PLMD)
 and alcohol use, 276
 magnesium, 276
Periodic limb movements during sleep (PLMS), 69–70, 261–262
Peripheral blood mononuclear cells (PBMCs), 324
Personal agency, 111
Physical activity, 78, 364
Pittsburgh Sleep Quality Index (PSQI), 124, 272, 324, 491
Polysomnography (PSG), 8, 10, 21, 147, 204–205, 275, 284–288, 392–393, 435–436, 481–482, 484
 actigraphy vs., 147, 148t
 older adults, 32, 34
 screening, 118
Poor decision-making, 352
Poor sleep, 473
 Americans, 71–72
 health, 473
 quality, 16–17
 screening, 118
Population density, 78
Positional therapy, 409–411
Positive airway pressure (PAP) therapy
 benefits, 485
 monitoring, 485
Positive airway pressure (PAP) treatments, 35
Positive predictive value (PPV), 118–121
Post-hoc analysis, 496
Post-traumatic stress disorder (PTSD), 5, 380
Potential physiological mechanisms, 191–192
Pramipexole, 276
Predictability, 173–174
Prefrontal cortex (PFC), 344, 348
Prepotent response, 348–349
Present-tense bias, 180
Programme for International Student Assessment (PISA), 441
Progressive muscle relaxation (PMR), 273
Prophylactic vaccination, 321
Proportional integral mode (PIM), 153–154
Protein, 247–248
PSG. See Polysomnography (PSG)
Psychiatric illness, 36–37
Psychobiological inhibition model, 377
Psychoeducation, 441
Psychological stress, 325–326, 325t, 326f
Psychomotor vigilance test (PVT), 340–342, 341f, 343–344f
 sleep loss, 334–335, 334f
 duration, 335–336
 outcome metrics, 336
 software and hardware, 335
 sustained attention, 340
Psychosocial stress, impact on sleep, 94
Psychostimulants, 307
Public health, 473
Public safety, 489
 extended-duration shifts, 492–493
 personnels, 489–490, 490t, 493
 physiological determinants of fatigue in, 492–493, 492f
 sleep deficiency, 493
 sleep disorders, 493–495, 494–495f
PVT. See Psychomotor vigilance test (PVT)

Q

Qualitative sleep parameters, 216–217
Quetiapine, 273

R

Racial/ethnic disparities, 464–466
Racial/ethnic groups
 insomnia, 66–69, 67–68t
 objective reported sleep duration, 61
 self-reported sleep duration, 58–61, 59–60t
 sleep architecture, 62, 63t
 sleep disordered breathing, 64–65
 sleep duration, 61
Racism, impact on sleep, 94–95
Randomized controlled trials (RCTs), 425
Rapid-eye movement (REM) sleep, 89, 258, 284, 450
 description, 5
 parasomnias, 277
Reactive oxygen species (ROS), 414
Readiness, 114
Reasons for Geographic And Racial Differences in Stroke (REGARDS) study, 209
REM sleep behavior disorder (RBD), 277
Response times (RT), 334
Restless legs syndrome (RLS), 69–70, 128
 and alcohol use, 276
 exercise, 261–262
Reversal learning paradigms, 349, 350f
Reward-based learning, 352–353
Rhinovirus, 321
RICCADSA trials, 412, 414
Risk perception, 72
Risky behavior, 450–451
Risky decision-making
 aggressive/punitive responses, 353
 altered expectations of reward, 351–352
 cognitive framing, 351
 impulsive behavior, 353
 moral judgment, 354
 reward-based learning, 352–353
Ropinirole, 276

S

SAVE trials, 412, 414
Schizophrenia, 382
Screening
 Berlin questionnaire, 121
 circadian rhythm disorders, 125–128
 consumer sleep technologies, 128–130
 Epworth sleepiness scale, 122–123
 Fitbit, 129–130
 functional outcomes of sleep questionnaire, 123
 Horne-Ostberg morningness-eveningness questionnaire, 125–127
 hypersomnolence, 121–124
 insomnia severity index, 124
 international restless legs syndrome scale, 128
 Jawbone, 130
 Karolinska sleepiness scale, 123–124
 Munich chronotype questionnaire, 127–128
 patient-reported outcomes measurement information system, 125
 Pittsburgh sleep quality index, 124
 questionnaires, 118, 119–120t
 restless legs syndrome, 128
 sleep-disordered breathing, 118–121
 SleepScore max, 130
 stanford sleepiness scale, 123
 STOP-BANG questionnaire, 121
 using data, 131
SDB. See Sleep-disordered breathing (SDB)
Seasonal changes, impact on sleep, 93–94
Sedentary behavior, 259–260
Self-efficacy, 113–114, 175
Self-rated risk propensity, 351
Self-reported sleep complaints, 34
Self-reported sleep duration, 58–61, 59–60t
Senescence associated secretory phenotype (SASP), 320
Serotonin, 8, 248f
SERVE-HF trial, 413
Sex differences
 adolescent sleep, 22–24
 biologically based, 21
 childhood sleep, 22
 infant sleep, 21–22
 middle-aged sleep, 24–25
 older adult sleep, 25–26
 young adult sleep, 24
Shift duration, 490
Shift work, 460–461, 465
Short sleep, 203–210, 213–215
Sleep
 architecture, 284, 374
 across racial/ethnic groups, 62, 63t
 in smokers vs. nonsmokers, 284–286
 and socioeconomic status, 64
 autonomic nervous system, 8
 as biobehavioral state, 160–163
 brainstem, 7
 cerebrum, 7–8
 and circadian rhythms, 6–7
 definition of, 3
 as domain of health behavior, 46–47
 epidemiology, 15–16
 as health behavior, 3–4
 hypothalamus, 7
 neuromodulators, 8
 objective measures, 9–10

as physiological process, 4–6
at population level, 11
process model, 3f
quantifying, 8–10
questionnaires, 9t
stages, 6
subjective measures, 9
thalamus, 7
Sleep ability, 3
Sleep apnea, 25
Sleep behavior, 161
Sleep changes, 31
Sleep complaints (SC), 17, 57–58
Sleep continuity, 284, 286–287
Sleep deficiency, 161–162, 493
Sleep deprivation (SD), 11–12, 162, 334–335, 427–428
  adolescents, 451
  glucose metabolism, 344, 346f
  psychomotor vigilance test, 340–341, 341f
  self-rated risk propensity, 351
  time displacement, 164–165
  Wisconsin Card Sorting Test, 350
Sleep diaries, 435–436
Sleep disorder, 162, 195–196, 493–495, 494–495f
  and diabetes, 217–218
  prevalence of, 16–17
Sleep-disordered breathing (SDB), 31, 64, 398, 412, 474–475
  cardiovascular health, 36
  cognitive function, 35
  exercise, 261
  first-line therapy, 33
  older adults, 32–33
  psychiatric illness, 37
  racial/ethnic groups, 65
  risk factors, 64–65
  screening, 118–121
  and socioeconomic status, 65–66
  symptoms, 64–65
Sleep disturbances
  caffeine, 311–312
  factors associated with, 33
  in older adults, 32–33
Sleep-dose-response experiments, 333
Sleep duration, 11–12, 14–15, 189–193
  alcohol consumption
    clinical findings in adolescents and young adults, 275
    long sleep duration and, 275
    short sleep duration and, 274
  cardiovascular health, 35–36
  cognitive function, 33–34
  defined, 274
  and diabetes, 213–216
  inadequate, 79
  and mental health, 436–438
  and obesity in pediatric populations, 424
  objective reported, 61
  psychiatric illness, 36–37
  within racial/ethnic groups, 61
  self-reported, 58–61
  social cognitive factors, 360

socioeconomic status, 61–62
virus exposure, 321, 321f
Sleep efficiency (SE), 258
Sleep extension, 375
Sleep fragmentation, 162, 287–288, 414
  in smokers vs. nonsmokers, 287
  time displacement, 165
Sleep gap, 474–476
Sleep health, 292–293, 473–474
  health belief model, 110–111
  integrated behavioral model, 111–112
  perceived barriers, 113
  perceived benefits, 113
  readiness, 114
  real-world barriers
    chronic pain, 108
    distractions, 109
    health conditions, 108
    lack of time, 107
    on-demand culture, 109
    physical environment, 108
    social norms and beliefs, 107
    substance use, 108–109
  self-efficacy and control, 113–114
  social norms, 113
  transtheoretical stages-of-change model, 112
Sleep health equity, 476–477, 478f, 479t
Sleep Heart Health Study (SHHS), 399–400, 412
Sleep homeostasis, 31
Sleep hygiene
  alcohol, 138–139
  bedroom environment, 139–140
  caffeine, 138
  clockwatching, 140
  conflict of interest, 143
  definition, 137
  exercise, 137–138
  food intake, 139
  insomnia, 140–141
  liquid intake, 139
  measuring, 140
  nicotine, 139
  practices, 312
  recommendations, 138t
  removal of electronics, 140
Sleep inertia, 307–308
Sleep loss, 342–343, 437
  differential vulnerability, 333
  and food intake, 244
  individual differences, 343–344
  mobile technology, 161–162
  neurobehavioral consequences of acute and chronic, 333
  psychomotor vigilance test, 334–335, 334f
    duration, 335–336
    outcome metrics, 336
    software and hardware, 335
  on vigilant attention, 333–334
Sleep medicine, 141
Sleep need, 3
Sleep onset latency (SOL), 258–259, 286
Sleep opportunity, 3–4, 72
Sleep parameters

changes in, 31
and diabetes risk, 213–222
objective parameters, 213
subjective parameters, 213
Sleep physiology, 160–161
Sleep quality, 216
  and mental health, 438–439
  psychiatric illness, 37
  screening, 124–125
Sleep reactivity, 325
Sleep-related bruxism, 276
Sleep-related eating disorder, 277
Sleep-related movement disorders, 276–277
Sleep-related workplace, 463
Sleep Research Society (SRS), 458
Sleep restriction, 11, 162, 425
  on inflammation, 322–323
  therapy, 294
  time displacement, 165
SleepScore MAX device, 130, 150, 153
Sleep timing, 79, 193–195
Sleep trackers, 155
Sleep-wake cycle
  adenosine and caffeine in, 308–309
  environmental factors and response to caffeine, 310–311
  genetic factors and response to caffeine, 309
  orexin in, 295
Sleep-wake rhythms, 228–229
Sleep-wake times, 229
Sleepwalking, 277
Slow waves, 4–5
Slow-wave sleep (SWS), 5, 196, 258, 428–431
Smokers vs. nonsmokers
  daytime sleepiness, 287
  objective and subjective sleep metrics in, 285–286t
  sleep architecture, 284–286
  sleep continuity, 286–287
  sleep fragmentation, 287
Smoking abstinence. See Smoking cessation
Smoking cessation
  behavioral treatments, 294–295
  changes in sleep, 287–288
  outcomes, 288–291
    poor sleep, 292–294
    sleep metrics and, 289–290t
  pharmacotherapy, 295
    bupropion, 292
    nicotine replacement therapy, 291–292
    varenicline, 292
  sleep deficits, 287, 288f
Snoring, 32, 275
Social cognitive factors, 360
Social cognitive theory (SCT), 178
Social determinants, 474, 475t
Social ecological model, 45–46, 46f
  applications of model, 51
  conceptualizing sleep, 47
  downstream consequences, 50–51
  individual level, 47–48, 48f
  social level, 48–49, 48f
  societal-level factors, 49–50
  upstream influences, 50–51

Social jetlag, 217, 229
Social network theory (SNT), 179
Social norms
　addressing, 113
　and beliefs, 107, 110f
Society of Behavioral Sleep Medicine (SBSM), 154
Socio-ecological model, 175f
Socioeconomic status (SES)
　impact on sleep, 94–95
　insomnia, 69
　sleep architecture and, 64
　sleep disordered breathing, 65–66
　sleep duration across, 61–62
SOL. See Sleep onset latency (SOL)
Spielman's 3P model. See Behavioral model
Stage theory, 179–180
Stanford sleepiness scale (SSS), 123
Stimulus control model, 375
STOP-BANG questionnaire, 121
Stress, 400–402
　hormones, 430
　influence immunity, 324–325
　psychological, 325–326, 325t, 326f
Stroke, 209–210, 399–400
Stroop task, 348–349
Subjective norms, 111
Subjective socioeconomic status (SES), 327
Substance use, 108–109
Sudden Infant Death Syndrome (SIDS), 21–22
Suicide, 379
Suprachiasmatic nucleus (SCN), 7, 160, 228, 273, 339–340, 424–425
Sustained attention, 340
Suvorexant, 295
Switch cost, 348
SWS. See Slow-wave sleep (SWS)
Sympathetic nervous system (SNS), 326
Syndrome X. See Cardiometabolic syndrome

T

Target, action, context, and time (TACT), 172
Task-switching, 348
T cells, 319–320
Telomeres, 320
Temperature, impact on sleep, 89–90
Temporality, 173
Thalamus, 7
Theory of planned behavior (TPB), 175–177
Theory of reasoned action (TRA), 175–177
Time above threshold (TAT), 153–154
Time in bed (TIB), 465

Time-on-task effect, 342
Total dietary approaches, 251
Total sleep deprivation (TSD), 11, 245
Total sleep time (TST), 11, 31, 57, 258–259
Train-the-trainer approach, 496
Transportation Safety Awareness Task Force (TSATF), 483, 484t
Transtheoretical model (TTM)
　application, 178
　stage theory, 177–178
Transtheoretical stages-of-change model, 112
Trazodone, 273, 403
Tricarboxylic acid (TCA) cycle, 269
Tryptophan, 248–249, 248f
TTM. See Transtheoretical model (TTM)
Tumor necrosis factor alpha (TNFα) gene, 344
Twitter®, 159
Two Process Model of Sleep, 6–7, 6f
Type 1 diabetes (T1D), 426–427
Type 2 diabetes (T2D), 398–399, 427–429
　insufficient/short sleep duration, 213–215
　long sleep duration, 215

U

UK Prospective Diabetes Mellitus Study (UKPDS), 427
Unhealthy behavior, 361
Unintentional injury, 451
Upper respiratory infection (URI), 321
Urbanicity, and population density, 78
Uvulopalatopharyngoplasty (UPPP), 409–411

V

Valerian, 251
Varenicline
　side-effects of, 292
　smoking cessation, 292
Vibrations, impact on sleep, 91–92
Video gaming, 166–167
Vigilance decrement, 334–335
Vigilant attention, sleep loss on, 333–334
Visceral adiposity, 221
Visceral factors, 180
Visual working memory (VWM), 347

W

Wake-after-sleep-onset (WASO), 151f, 258–260, 424, 458
Wakefulness, 4, 340–341
Wake state instability, 342–343
Wanting, 364

WASO. See Wake-after-sleep-onset (WASO)
WatchPAT, 152–153
Weight loss interventions, 196–197
Willis-Ekbom disease. See Restless legs syndrome (RLS)
Wisconsin Card Sorting Test (WCST), 350
Women's Health Initiative Insomnia Rating Scale (WHIIRS), 57–58
Work
　demands, 460–461
　environment, 96
　epidemiology, 458–459
　factors impacts nighttime sleep, 457
　function and productivity, 457
　hours, 490–492
　interpersonal stressors at, 461
　modifiable work factors, 458
　schedules, 489
　　based on sleep and circadian principles, 491–492
　　and health and safety outcomes, 491
　shift duration, 460–461, 490
　stress, 459–460
　stressors, 461
　work-family conflict, 460–461
Workers
　future health risks, 462
　interpersonal stressors, 461
　with multiple jobs, 460–461
　night shift, 492
　poorer sleep health, 462
　self-employed, 460–461
Work-Home Resources (W-HR) model, 460
Working memory, 346–347
Workplace intervention
　business case for sleep, 462
　worksite programs, 463–464
　worksite wellness, 462–463
Work-related injury, 491
Worksite programs, 463–464
Worksite wellness, 462–463
World Health Organization (WHO), 421–423
Worry, 72

Y

Young adult sleep, 24

Z

Zero-crossing mode (ZCM), 153–154
Zolpidem, 403